ONLINE RESOURCES

# IMPORTANT:

## HERE IS YOUR REGISTRATION CODE TO ACCESS
## YOUR PREMIUM McGRAW-HILL ONLINE RESOURCES.

For key premium online resources you need THIS CODE to gain access. Once the code is entered, you will be able to use the Web resources for the length of your course.

If your course is using **WebCT** or **Blackboard**, you'll be able to use this code to access the McGraw-Hill content within your instructor's online course.

Access is provided if you have purchased a new book. If the registration code is missing from this book, the registration screen on our Website, and within your WebCT or Blackboard course, will tell you how to obtain your new code.

## Registering for McGraw-Hill Online Resources

TO gain access to your McGraw-Hill web resources simply follow the steps below:

1. USE YOUR WEB BROWSER TO GO TO:     **www.mhhe.com/ecology**

2. CLICK ON **FIRST TIME USER**.

3. ENTER THE REGISTRATION CODE* PRINTED ON THE TEAR-OFF BOOKMARK ON THE RIGHT.

4. AFTER YOU HAVE ENTERED YOUR REGISTRATION CODE, CLICK **REGISTER**.

5. FOLLOW THE INSTRUCTIONS TO SET-UP YOUR PERSONAL UserID AND PASSWORD.

6. WRITE YOUR UserID AND PASSWORD DOWN FOR FUTURE REFERENCE.
   KEEP IT IN A SAFE PLACE.

**TO GAIN ACCESS** to the McGraw-Hill content in your instructor's **WebCT** or **Blackboard** course simply log in to the course with the UserID and Password provided by your instructor. Enter the registration code exactly as it appears in the box to the right when prompted by the system. You will only need to use the code the first time you click on McGraw-Hill content.

Thank you, and welcome to your McGraw-Hill online Resources!

*YOUR REGISTRATION CODE CAN BE USED ONLY ONCE TO ESTABLISH ACCESS. IT IS NOT TRANSFERABLE.

0-07-283063-8  T/A  MOLLES: ECOLOGY, 3/E

REGISTRATION CODE

7A3U-QE78-9LC5-SG30-XL3R

# ECOLOGY

## Concepts and Applications

### third edition

## Manuel C. Molles Jr.
### University of New Mexico

Boston   Burr Ridge, IL   Dubuque, IA   Madison, WI   New York   San Francisco   St. Louis
Bangkok   Bogotá   Caracas   Kuala Lumpur   Lisbon   London   Madrid   Mexico City
Milan   Montreal   New Delhi   Santiago   Seoul   Singapore   Sydney   Taipei   Toronto

**Higher Education**

ECOLOGY: CONCEPTS AND APPLICATIONS, THIRD EDITION

Published by McGraw-Hill, a business unit of The McGraw-Hill Companies, Inc., 1221 Avenue of the Americas, New York, NY 10020. Copyright © 2005, 2002, 1999 by The McGraw-Hill Companies, Inc. All rights reserved. No part of this publication may be reproduced or distributed in any form or by any means, or stored in a database or retrieval system, without the prior written consent of The McGraw-Hill Companies, Inc., including, but not limited to, in any network or other electronic storage or transmission, or broadcast for distance learning.

Some ancillaries, including electronic and print components, may not be available to customers outside the United States.

 This book is printed on recycled, acid-free paper containing 10% postconsumer waste.

3 4 5 6 7 8 9 0 QPD/QPD 0 9 8 7 6 5

ISBN 0–07–243969–6

Publisher: *Margaret J. Kemp*
Senior developmental editor: *Kathleen R. Loewenberg*
Executive marketing manager: *Lisa L. Gottschalk*
Senior project manager: *Gloria G. Schiesl*
Senior production supervisor: *Laura Fuller*
Lead media project manager: *Judi David*
Senior media technology producer: *Jeffry Schmitt*
Senior coordinator of freelance design: *Michelle D. Whitaker*
Cover/interior designer: *Jamie E. O'Neal*
Cover image: © *Daryl Benson/Masterfile*
Lead photo research coordinator: *Carrie K. Burger*
Photo research: *Toni Michaels/PhotoFind, LLC*
Supplement producer: *Brenda A. Ernzen*
Compositor: *Precision Graphics*
Typeface: *10/12 Times Roman*
Printer: *Quebecor World Dubuque, IA*

The credits section for this book begins on page 607 and is considered an extension of the copyright page.

**Library of Congress Cataloging-in-Publication Data**

Molles, Manuel C. (Manuel Carl), 1948–
  Ecology : concepts and applications / Manuel C. Molles, Jr. — 3rd ed.
    p. cm.
  Includes bibliographical references (p.    ) and index.
  ISBN 0–07–243969–6 (hard : alk. paper).
  1. Ecology. I. Title.

QH541.M553    2005
577—dc22

                                        2003025674
                                        CIP

www.mhhe.com

# About the Author

**Manuel C. Molles Jr.** is Professor of Biology at the University of New Mexico, where he has been a member of the faculty and curator in the Museum of Southwestern Biology since 1975. He received his B.S. from Humboldt State University and his Ph.D. from the Department of Ecology and Evolutionary Biology at the University of Arizona. Seeking to broaden his geographical perspective, he has taught and conducted ecological research in Latin America, the Caribbean, and Europe. He was awarded a Fulbright Research Fellowship to conduct research on river ecology in Portugal and has held visiting professor appointments in the Department of Zoology at the University of Coimbra, Portugal, in the Laboratory of Hydrology at the Polytechnic University of Madrid, Spain, and at the University of Montana's Flathead Lake Biological Station.

Originally trained as a marine ecologist and fisheries biologist, the author has worked mainly on river and riparian ecology at the University of New Mexico. His research has covered a wide range of ecological levels, including behavioral ecology, population biology, community ecology, ecosystem ecology, biogeography of stream insects, and the influence of a large-scale climate system (El Niño) on the dynamics of southwestern river and riparian ecosystems. His current research concerns the effects of flooding and exotic vegetation on the structure and dynamics of the Rio Grande riparian ecosystem. Throughout his career, Dr. Molles has attempted to combine research, teaching, and service, involving undergraduate as well as graduate students in his ongoing projects. At the University of New Mexico, he has taught a broad range of lower division, upper division, and graduate courses, including Principles of Biology, Evolution and Ecology, Stream Ecology, Limnology and Oceanography, Marine Biology, and Community and Ecosystem Ecology. He has taught courses in Global Change and River Ecology at the University of Coimbra, Portugal, and General Ecology, and Groundwater and Riparian Ecology at the Flathead Lake Biological Station. Dr. Manuel Molles was named Teacher of the Year by the University of New Mexico for 1995–96 and Potter Chair in Plant Ecology in 2000.

To Mary Anne and
Misha

# Contents

**Section IV**
Interactions

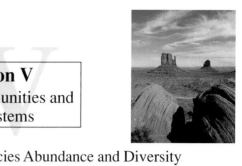

## Section VI
### Large-Scale Ecology

# Preface

The accelerating pace of ecological discovery makes staying current with the field very difficult. The challenge faced by ecology instructors and their students is captured by a tale told by the Greek philosopher Zeno about 2,500 years ago. In the tale of Achilles and the Tortoise, called Zeno's Paradox, Zeno argued that if a fast runner, such as the mythical Achilles, gave a tortoise a head start in a race, he could never overtake it. Zeno proposed that since there are an infinite number of points between him and the tortoise, Achilles would never catch up. Modern mathematics has solved this paradox and we can take comfort that, even in Zeno's theoretical universe, Olympic runners can overtake tortoises. However, there is a place where Zeno's arguments appear to hold, and that is in the world of teaching dynamic scientific disciplines such as ecology.

The challenge to ecology instructors and their students is much greater than that faced by Achilles—greater because instructors and students are matched against a much swifter opponent with a longer head start. As they attempt to cover the space between the beginning and end of this subject, the rapid pace of discovery moves the limits of the discipline ahead, not at the speed of a tortoise but at that of a hare. Zeno would be very happy in this universe because, here, the instructor and students can never catch up. However, with careful organization and modern tools, such as the Internet, they can come close.

## Unique Approach

In an address at the 1991 meeting of the Ecological Society of America in San Antonio, Texas, eminent ecologist Paul Risser challenged ecology instructors to focus their attention on the major concepts of the field. If we subdivide a large and dynamic subject, such as ecology, too finely, we cannot cover it in one or two academic terms. Risser proposed that by focusing on major concepts, however, we may provide students with a robust framework of the discipline upon which they can build.

*"What primarily motivated me to adopt* Ecology: Concepts and Applications *is the way the author emphasizes a few key ecological concepts in each chapter and then uses relevant studies to demonstrate how scientists have "discovered" these concepts. I find this emphasis on concepts and the science behind them to be a refreshing change from the typical textbook that tends to present the science of ecology as a rather dull collection of facts. I feel that the students who read this text will come away with a better understanding not only of ecology, but also of the method of doing science."*

—Tim Maret,
*Shippensburg University*

This book attempts to address Risser's challenge. **Each chapter is organized around two to four major concepts, presenting the student with a manageable and memorable synthesis of the subject.** I have found that while beginning ecology students can absorb a few central concepts well, they can easily get lost in a sea of details. Each concept is supported by concept discussions that provide evidence for the concept and introduce students to the research approaches used in the various areas of ecology. Wherever possible, the original research and the scientists who did the research are presented. Allowing the scientists who created this field to emerge from the background and lead students through the discipline breathes life into the subject and helps students retain information.

*"The Molles text is by far the best ecology text available in terms of the combination of presentation and accessibility of the material as well as breadth of coverage. The material is up-to-date, very well illustrated, and has a clear focus on concepts rather than details that distract a student's understanding of process."*

—Craig Williamson
*Lehigh University*

# Introductory Audience

I have written this book for students taking their first undergraduate course in ecology. I have assumed that students in this one-semester course have some knowledge of basic chemistry and mathematics and that they have had a course in general biology that included introductions to physiology, biological diversity, and evolution.

*"I receive positive feedback about the text from my students. During or after the course, some students majoring in other fields have expressed an interest in switching to ecology as a major, and I believe the text contributes toward that interest."*

—Carolyn Meyer
*University of Wyoming*

# New to This Edition

***All 23 chapters of the book have been revised*** following the suggestions of numerous reviewers. An attempt was made to address reviewers' concerns, to update material where needed, add missing perspectives, correct errors, and generally freshen the treatment. Additional suggested readings, drawn mainly from literature published since the publication of the second edition, have been added to each chapter.

*Chapter 8, "Population Genetics and Natural Selection," has been thoroughly rewritten* based on reviewer feedback. The principle concepts listed in the chapter have been replaced. The historical treatment has been reduced and replaced with more modern examples, particularly molecular approaches to the study of populations. The contribution of environment to phenotypic variation is more clearly stated and supported. The discussion of the process of natural selection has been expanded under a new concept. The major forms of natural selection are included in this expanded treatment. The dependence of evolution by natural selection on the extent of genetic variation is discussed, including the concept of heritability. The discussion of heritability is supported by an explanation of how heritability may be estimated using regression analysis.

*New "Investigating the Evidence" boxed discussions focus on the scientific method,* emphasizing statistics and study design. The boxes are intended to support the concept discussions by presenting step-by-step clarification within a broader treatment of the process of science. The readings begin in chapter 1 with an overview of the scientific method, providing a conceptual context for the boxes in the next 22 chapters:

- Determining the Sample Mean
- Determining the Sample Median
- Laboratory Experiments
- Sample Size
- Variation in Data
- Scatter Plots and the Relationship Between Variables
- Estimating Heritability Using Regression Analysis
- Clumped, Random, and Regular Distributions
- Hypotheses and Statistical Significance
- Frequency of Alternative Phenotypes in a Population
- A Statistical Test for Distribution Pattern
- Field Experiments
- Standard Error of the Mean
- Confidence Intervals
- Estimating the Number of Species in Communities
- Using Confidence Intervals to Compare Populations
- Comparing Two Populations with the *t*-Test
- Assumptions for Statistical Tests
- Variation Around the Median
- Comparison of Two Samples Using a Rank Sum Test
- Sample Size Revisited
- Discovering What's Been Discovered

***Expanded appendixes*** now include commonly used statistical tables and an extended list of abbreviations used in the text.

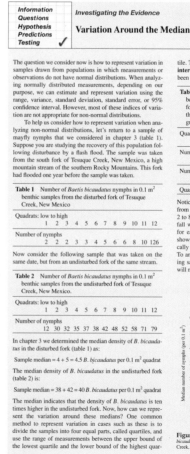

| Information Questions Hypothesis Predictions Testing ✓ | Investigating the Evidence
**Variation Around the Median** |

The question we consider now is how to represent variation in samples drawn from populations in which measurements or observations do not have normal distributions. When analyzing normally distributed measurements, depending on our purpose, we can estimate and represent variation using the range, variance, standard deviation, standard error, or 95% confidence interval. However, most of these indices of variation are not appropriate for non-normal distributions.

To help us consider how to represent variation when analyzing non-normal distributions, let's return to a sample of mayfly nymphs that we considered in chapter 3 (table 1). Suppose you are studying the recovery of this population following disturbance by a flash flood. The sample was taken from the south fork of Tesuque Creek, New Mexico, a high mountain stream of the southern Rocky Mountains. This fork had flooded one year before the sample was taken.

**Table 1** Number of *Baetis bicaudatus* nymphs in 0.1 m² benthic samples from the disturbed fork of Tesuque Creek, New Mexico.

Quadrats: low to high

| 1 | 2 | 3 | 4 | 5 | 6 | 7 | 8 | 9 | 10 | 11 | 12 |

Number of nymphs

| | | 2 | 2 | 3 | 4 | 5 | 6 | 6 | 8 | 10 | 126 |

Now consider the following sample that was taken on the same date, but from an undisturbed fork of the same stream.

**Table 2** Number of *Baetis bicaudatus* nymphs in 0.1 m² benthic samples from the undisturbed fork of Tesuque Creek, New Mexico.

Quadrats: low to high

| 1 | 2 | 3 | 4 | 5 | 6 | 7 | 8 | 9 | 10 | 11 | 12 |

Number of nymphs

| 12 | 30 | 32 | 35 | 37 | 38 | 42 | 48 | 52 | 58 | 71 | 79 |

In chapter 3 we determined the median density of *B. bicaudatus* in the disturbed fork (table 1) as:

Sample median = 4 + 5 = 4.5 *B. bicaudatus* per 0.1 m² quadrat

The median density of *B. bicaudatus* in the undisturbed fork (table 2) is:

Sample median = 38 + 42 = 40 *B. bicaudatus* per 0.1 m² quadrat

The median indicates that the density of *B. bicaudatus* is ten times higher in the undisturbed fork. Now, how can we represent the variation around these medians? One common method to represent variation in cases such as these is to divide the samples into four equal parts, called quartiles, and use the range of measurements between the upper bound of the lowest quartile and the lower bound of the highest quar-

tile. This representation of variation in a sample is called the **interquartile range**. In table 3, the data in tables 1 and 2 have been divided into quartiles with different colors:

**Table 3** Number of *Baetis bicaudatus* nymphs in 0.1 m² benthic samples from the undisturbed and disturbed forks of Tesuque Creek, New Mexico. The first, second, third, and fourth quartiles are shaded orange, yellow, green, and blue, respectively.

Quadrats: low to high

| 1 | 2 | 3 | 4 | 5 | 6 | 7 | 8 | 9 | 10 | 11 | 12 |

Number of nymphs, disturbed fork

| | | 2 | 2 | 3 | 4 | 5 | 6 | 6 | 8 | 10 | 126 |

Number of nymphs, undisturbed fork

| 12 | 30 | 32 | 35 | 37 | 38 | 42 | 48 | 52 | 58 | 71 | 79 |

| Quartiles | 1st | | 2nd | | 3rd | | 4th | | | | |

Notice that the interquartile range for the undisturbed fork is from 32 to 58; for the disturbed fork, the interquartile range is 2 to 8. Notice that 50% of the quadrat counts in each sample fall within this range. The medians and interquartile ranges for each of the populations are plotted in figure 1, which shows that they do not overlap. However, is there a statistically significant difference in density in the two stream forks? To answer that question, we will need a method for comparing samples that does not assume a normal distribution. We will make that comparison in chapter 21.

Notice that unlike standard error bars around a sample mean, interquartile ranges can be asymmetrical around the sample median.

**Figure 1** Medians and interquartile ranges of mayfly nymphs. *Baetis bicaudatus*, in 0.1 m² quadrats in disturbed and undisturbed forks of Tesuque Creek, New Mexico.

# Organized Around Key Concepts

An evolutionary perspective forms the foundation of the entire textbook, as it is needed to support understanding of major concepts. The textbook begins with a brief introduction to the nature and history of the discipline of ecology, followed by section I, which includes two chapters on natural history—life on land and life in water. Sections II through VI build a hierarchical perspective through the traditional subdisciplines of ecology: section II concerns the ecology of individuals; section III focuses on population ecology; section IV presents the ecology of interactions; section V summarizes community and ecosystem ecology; and finally, section VI discusses large-scale ecology and includes chapters on landscape, geographic, and global ecology. These topics were first introduced in section I within a natural history context. In summary, the book begins with the natural history of the planet, considers portions of the whole in the middle chapters, and ends with another perspective of the entire planet in the concluding chapter.

# Features Designed with the Student in Mind

The features of this textbook are unique and were carefully planned to enhance the students' comprehension of ecology. All chapters beyond the introductory chapter 1 are based on a distinctive learning system, featuring the following key components:

**Introduction:** The introduction to each chapter presents the student with the flavor of the subject at hand and important background information. Some introductions include historical events related to the subject; others present an example of an ecological process. All attempt to engage students and draw them into the discussion that follows.

**Concepts:** The goal of this book is to build a foundation of ecological knowledge around key concepts. These key concepts are listed after the chapter introduction to alert the student to the major topics to follow, and to

provide a place where the student can find a list of the important points of each chapter.

**Concept Discussions:** These sections reinforce the aforementioned concepts with a focus on published studies. This case-study approach supports the concepts with evidence, and introduces students to the methods and people that have created the discipline of ecology.

**Applications & Tools:** Many undergraduate students want to know how abstract ideas and general relationships can be applied to the ecological problems facing us all. They are concerned with the practical side of ecology and want to know more about the tools of science. Including a few applications in each chapter motivates students to learn more of the underlying principles of ecology. In addition, it seems that environmental problems are now so numerous and so pressing that they have erased a once easy distinction between general and applied ecology.

**"Investigating the Evidence" Boxes:** These important readings offer "mini-lessons" on the scientific method, emphasizing statistics and study design. They are intended to supplement the chapter's Concept Discussions. They present a broad outline of the process of science, while also providing step-by-step explanation. The series of boxes begins in chapter 1 with an overview of the scientific method, which provides a conceptual context for more specific material in the next 21 chapters. The last reading wraps up the series with a discussion on electronic literature searches.

---

APPLICATIONS & TOOLS

Estimating Abundance—from Whales to Sponges

The abundance of organisms and how abundance changes in time and space are among the most fundamental concerns of ecology. These factors are so basic that some authors define ecology as the study of distribution and abundance of organisms. Because abundance is so important, ecologists should understand how to estimate it for a wide variety of organisms. Keep in mind, however, that ecologists do not measure abundance as an end in itself but as a tool to understand the ecology of populations. Knowing how abundant an organism is can tell us whether its population is growing, declining, or stable. As we saw in the previous section, population size is one of the characteristics that helps ecologists assess a species' vulnerability to extinction. However, to estimate the abundance of species the ecologist must contend with a variety of practical challenges and conceptual subtleties. Some of these are discussed here.

**Estimating Whale Population Size**

In 1989, the journal *Oceanus* published a table that listed the estimated sizes of whale populations. The table included the following note: "All estimates . . . are highly speculative." Why

is it difficult to provide firm estimates of whale population size? Briefly, whales live at low population densities and may be distributed across vast expanses of ocean. They also spend much time submerged and move around a great deal. As large as they are, you cannot count all the whales in the ocean. Instead, marine ecologists rely on population estimation. Each method of estimation has its own limitations and uncertainties.

One method used to estimate population sizes of elusive animals involves marking or tagging some known number of individuals in the population, releasing the marked individuals so they will mix with the remainder of the population, and then sampling the population at some later time. The ratio of marked to unmarked individuals in the sample gives an estimate of population size. The simplest formula expressing this relationship is the Lincoln-Peterson index:

$$M/N = m/n$$

where:

M = the number of individuals marked and released

N = the actual size of the study population

m = the number of marked individuals in a sample of the population

n = the total number of individuals in the sample

---

I n nature, the consumer eventually becomes the consumed. A moose browses intently on the twigs and buds of a willow barely protruding above the deep snow of midwinter (fig. 14.1). With each mouthful it chews and swallows, the moose reduces the mass of the willows and adds to the growing energy store in its own large and complex stomach, energy stores that the moose will need to make it through one more northern winter. Then, a familiar scent catches the moose's attention and startled, it runs off.

Suddenly, the clearing where the moose had been feeding is a blur of bounding forms dashing headlong in the direction the moose has gone—a pack of wolves in pursuit of its own meal. A portion of the pack has already run ahead of the moose and is cutting off its retreat. This time, unlike so many times before, the old moose will not escape. After a fierce struggle, the moose is down and the wolves settle in to feed.

But the wolves are not the only organisms to benefit from this great quantity of food. Within the intestines of the wolves live several species of parasitic worms that will soon claim their share of the wolves' hard-won feast. The worms will turn some of the energy and structural compounds they absorb into the infective stages of their own kind, which after being shed into the environment may attach themselves to other hosts, who will serve as their unwitting providers.

Some of the strongest links between populations are those between herbivore and plant, between predator and prey, and between parasite or pathogen and host. The conceptual thread that links these diverse interactions between species is that the interaction enhances the fitness of one individual—the predator, the pathogen, etc.—while reducing the fitness of the exploited individual—the prey, host, etc. Because of this common thread we can group these interactions under the heading of *exploitation*.

Let's consider some of the most common means of exploitation. Herbivores consume live plant material but do not usually kill plants. **Predators** kill and consume other organisms. Typical predators are animals that feed on other animals—wolves that eat moose, snakes that eat mice, etc. **Parasites** live on the tissues of their host, often reducing the fitness of the host, but not generally killing it. A **parasitoid** is an insect whose larva consumes its host and kills it in the process; parasitoids are functionally equivalent to predators. **Pathogens** induce disease, a debilitating condition, in their hosts.

As clear as all these definitions may seem, they are fraught with semantic problems. Once again, we are faced with capturing the full richness of nature with a few restrictive definitions. For instance, not all predators are animals, a few are plants, some are fungi, and many are protozoans. When an herbivore kills the plant upon which it feeds, should we call it a predator? If an herbivore does not kill its food plants, would it be better to call it a parasite? What do we do with a parasite that kills its host? Is it then a predator or perhaps a pathogen? The point of these questions is not to argue for more terminology but to argue for fewer, less restrictive terms. As is often the case, we are faced with a continuum of interesting and sometimes bewildering interactions involving

**Figure 14.1** This moose exploits the twigs and buds of woody plants for the food it needs to survive the cold northern winter. Eventually, wolves may prey upon the moose to meet their needs for food.

millions of organisms. Let's recognize the diversity and continuous variation facing the ecologist, put the restrictive definitions aside for the moment, and recognize what is common to all these interactions: *exploitation*, that is, one organism makes its living at the expense of another.

CONCEPTS

- Exploitation weaves populations into a web of relationships that defy easy generalization.
- Predators, parasites, and pathogens influence the distribution, abundance, and structure of prey and host populations.
- Predator-prey, host-parasite, and host-pathogen relationships are dynamic.
- To persist in the face of exploitation, hosts and prey need refuges.

CONCEPT DISCUSSION

Complex Interactions

Exploitation weaves populations into a web of relationships that defy easy generalization.

By conservative estimate the number of species in the biosphere is on the order of 10 million. As huge as this number may seem, the number of exploitative interactions between

**Illustrations:** A great deal of effort has been put into the development of illustrations, both photographs and line art. The goal has been to create more effective pedagogical tools through skillful design and use of color, and to rearrange the traditional presentation of information in figures and captions. Much explanatory material has been moved from captions to within the illustrations, providing students with key information where they need it most. The numerous explanatory boxes used in this text are unique and not available in any other ecology textbook.

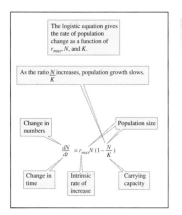

*"I love the boxes in the illustrations! I think they facilitate both the reading and comprehension for the students. I think this style is the most helpful with the graphs and charts."*

—Tatiana Roth
*Coppin State College*

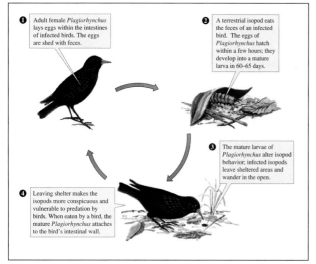

**Summary:** The chapter summary reviews the main points of the content. The concepts around which each chapter is organized are boldfaced and redefined in the summary to emphasize once again the main points of the chapter.

**End-of-Chapter Material:**

- *Review Questions* The review questions are designed to help students think more deeply about each concept and to reflect on alternative views. They also provide a place to fill in any remaining gaps in the information presented and take students beyond the foundation established in the main body of the chapter.

- *Suggested Readings* Each chapter ends with a list of suggested readings. Though all of the readings offer the student coverage beyond the chapter content, they have been chosen to serve a variety of purposes. Some are books that provide a broad overview; others are papers that trace the development of particular topics or controversies in ecology. I have provided a brief description and rationale for each.

- *On the Net* The World Wide Web provides one of the most powerful tools to help keep abreast of changes in ecology. A broad range of applicable Internet sites are just a click away on the Online Learning Center that accompanies this edition. Practice quizzing, articles on current ecological and environmental issues, and a variety of teaching aids are also available online at http://www.mhhe.com/ecology.

**End-of-Book Material:**

- *Appendixes* Three appendixes, "Abbreviations Used in This Text," "List of Chapter Concepts," and "Statistical Tables" are available to the student for reference and as study aids.

- *Glossary*

- *References* References are an important part of any scientific work. However, many undergraduates are distracted by a large number of references within the text. One of the goals of a general ecology course should be to introduce these students to the primary literature without burying them in citations. The number of citations has been reduced to those necessary to support detailed discussions of particular research projects.

- *Index*

# Useful Supplements

**Digital Content Manager (DCM) CD-ROM:** This multimedia collection of visual resources allows instructors to utilize all of the artwork from the text in multiple formats to create customized classroom presentations, visually based tests and/or quizzes, dynamic course website content, or attractive printed support materials.

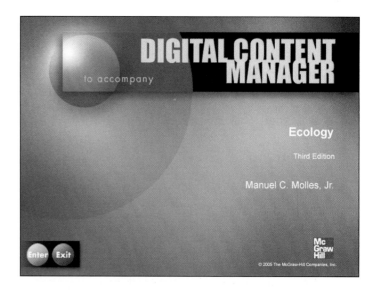

**Instructor's Testing and Resource CD-ROM:** This cross-platform CD-ROM provides a computerized test bank utilizing Brownstone Diploma@ testing software to quickly create customized exams. The user-friendly program allows instructors to search for questions by topic, format, or difficulty level; edit existing questions or add new ones; and scramble questions and answer keys for multiple versions of the same test.

**Transparencies:** A set of 100 transparencies is available to users of the text. These acetates include key figures from the text, including new art from this edition.

**Online Learning Center (OLC)** (http://www.mhhe.com/ecology/): This comprehensive website offers numerous resources for both students and instructors.

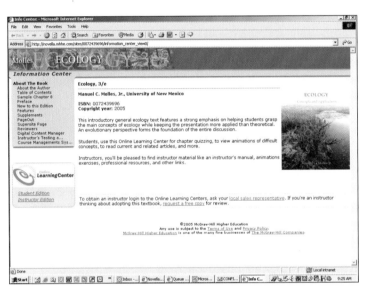

### Student Resources—Everything You Need in One Place:

- Practice quizzing
- Hyperlinks on chapter topics
- Animations
- Guide to electronic research
- Regional perspectives (case studies)
- Ecology/environmental science issues world map
- Key term flashcards
- Career information
- PowerWeb's hundreds of current articles and daily news items have been integrated into each chapter on the OLC.
- Access Science offers the advantage of an online, interactive encyclopedia.

### Instructor Resources—In Addition to All of the Above, You'll Receive:

- Answers to review questions
- Class activities
- PowerPoint lectures
- Interactive world maps
- PageOut (create your own course website)

**Ecology Essential Study Partner CD-ROM:** A complete, interactive student study tool, this CD features animations, videos, and learning activities. From quizzes to interactive diagrams, you'll find that there has never been a better study partner to ensure the mastery of core concepts. Best of all, it's provided free with a new textbook purchase in an optional package.

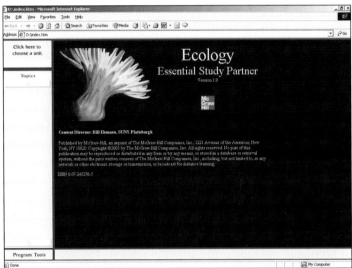

# Packaging Opportunities

McGraw-Hill offers packaging opportunities that not only provide students with valuable ecology-related material, but also a substantial cost savings. Instructors, ask your McGraw-Hill sales representative for information on discounts and special ISBNs for ordering a package that contains one or more of the following:

*Exploring Environmental Science with GIS:* This short book provides a set of exercises aimed at students and instructors who are new to GIS (Geographic Information Systems), but are familiar with the Windows operating system. The exercises focus on improving analytical skills, understanding spatial relationships, and understanding the nature and structure of environmental data. Because the software used is distributed free of charge, this text is appropriate for courses and schools that are not yet ready to commit to the expense and time involved in acquiring other GIS packages.

*General Ecology Lab Manual, 8th Edition,* by Cox: Designed for juniors and seniors, this one-semester laboratory manual is based on mathematical statistics. Author George Cox begins with exercises covering library research, designing an ecological study, and other introductory concepts. He then proceeds to an examination of specific types of measurement and an analysis of various aspects of ecology. Many of these laboratories are tied to current, commercially available computer programs and software packages.

*Field and Lab Methods for General Ecology, 4th Edition*, by Brower et al.: This introductory ecology lab manual focuses on the process of collecting, recording and analyzing data, and equips students with the tools they need to function in more advanced science courses. It reflects the most current techniques for data gathering so that students can obtain the most accurate samples. Balanced coverage of plant, animal, and physical elements offers a diverse range of exercises. The lab manual includes an exercise on writing research reports.

*Conservation Biology Workbook* by Van Dyke: The purpose of this workbook is to teach solutions to actual problems that are presented to conservation biologists. The exercises chosen are practical in that they are selected from the wide variety of problems that working, contemporary conservation biologists are attempting to solve today. The unifying feature of these exercises is that they promote the following skills: critical reading and writing, applying appropriate field techniques, using computer simulations and applications, and synthesizing information to create comprehensive conservation applications and multidimensional conservation plans.

*Annual Editions: Environment 04/05* This 23rd edition is a compilation of current articles from the best of the public press. The selections explore the global environment, the world's population, energy, the biosphere, natural resources, and pollution.

*Taking Sides: Clashing Views on Controversial Environmental Issues, Revised 10th Edition* This represents the arguments of leading environmentalists, scientists, and policymakers. The issues reflect a variety of viewpoints and are staged as "pro" and "con" debates. Topics are organized around four core areas: general philosophical and political issues, the environment and technology, disposing of wastes, and the environment and the future.

*Sources: Notable Selections in Environmental Studies, 2nd Edition* This volume brings together primary source selections of enduring intellectual value—classic articles, book excerpts, and research studies—that have shaped environmental studies and our contemporary understanding of it. The book includes carefully edited selections for the works of the most distinguished environmental observers, past and present. Selections are organized topically around the following major areas of study: energy, environmental degradation, population issues and the environment, human health and the environment, and environment and society.

*Student Atlas of Environmental Issues* by Allen This is an invaluable pedagogical tool for exploring the human impact on the air, waters, biosphere, and land in every major world region. This informative resource provides a unique combination of maps and data helping students understand the dimensions of the world's environmental problems and the geographical basis of these problems.

*You Can Make a Difference: Be Environmentally Responsible, 2nd Edition* by Getis This book is organized around the three parts of the biosphere: land, water, and air. Each section contains descriptions of the environmental problems associated with that part of the biosphere. Immediately following each problem or "challenge" are suggested ways that individuals can help solve or alleviate them. This book has been written to provide the reader with some easy and practical ways to protect the Earth and to help understand why the task is so important.

# Acknowledgments

A complete list of the people who have helped me with this project would be impossibly long. However, during the development of this third edition, several colleagues freely shared their ideas and expertise, reviewed new sections, and offered the encouragement a project like this needs to keep it going: Eric Charnov, Scott Collins, John Craig, Cliff Crawford, Cliff Dahm, Tim Lowrey, Mary Anne Nelson, Randy Thornhill, Eric Toolson, and Roman Zlotin. In addition, I am indebted to the many student readers of earlier editions who have helped by contacting me with questions and suggestions for improvements.

I also wish to acknowledge the skillful guidance throughout the publishing process given by many professionals associated with McGraw-Hill during this project, including Marge Kemp, Lisa Gottschalk, Kathy Loewenberg, Gloria Schiesl, Linda Gomoll, Carrie Burger, Toni Michaels, Michelle Whitaker, Jamie O'Neal, Laura Fuller, Judi David, and Brenda Ernzen.

Finally, I wish to thank all my family for support given throughout the project, especially Mary Ann Esparza, Dan Esparza, Hani Molles, and Anders Molles.

I gratefully acknowledge the many reviewers who, over the course of the last several revisions, have given of their time and expertise to help this textbook evolve to its present third edition. I honestly could not have done it without them.

## Reviewers for the Third Edition

Sina Adl  *Dalhousie University, Canada*
Harvey J. Alexander  *College of Saint Rose*
Peter Alpert  *University of Massachusetts–Amherst*
Julie W. Ambler  *Millersville University*
Robert K. Antibus  *Bluffton College*
Tom L. Arsuffi  *Southwest Texas State University*
Claude D. Baker  *Indiana University*
Ellen H. Baker  *Santa Monica College*
Charles L. Baube  *Oglethorpe University*
Edmund Bedecarrax  *City College of San Francisco*
Jerry Beilby  *Northwestern College*
R.P. Benard  *American International College*
Erica Bergquist  *Holyoke Community College*
Richard A. Boutwell  *Missouri Western State College*
Ward Brady  *Arizona State University East–Mesa*
Fred J. Brenner  *Grove City College*
Robert Brodman  *Saint Joseph's College*
Elaine R. Brooks  *San Diego City College*
Evert Brown  *Casper College*
Stephanie Brown Fabritius  *Southwestern University*
Rebecca S. Burton  *Alverno College*
James E. Byers  *University of New Hampshire*
Guy Cameron  *University of Cincinnati*
Geralyn M. Caplan  *Owensboro Community & Technical College*
Walter P. Carson  *University of Pittsburgh*
Ben Cash III  *Maryville College*
Young D. Choi  *Purdue University Calumet*
Ethan Clotfelter  *Providence College*
Liane Cochran-Stafira  *Saint Xavier University*
Joe Coelho  *Culver-Stockton College*

Jerry L. Cook   *Sam Houston State University*
Tamara J. Cook   *Sam Houston State University*
Erica Corbett   *Southeastern Oklahoma State University*
Tim Craig   *University of Minnesota*
Jack A. Cranford   *Virginia Tech*
Greg Cronin   *University of Colorado–Denver*
Todd Crowl   *Utah State University*
Richard J. Deslippe   *Texas Tech University*
Kenneth M. Duke   *Brevard College*
Andy Dyer   *University of South Carolina*
Ginny L. Eckert   *University of Alaska*
J. Nicholas Ehringer   *Hillsborough Community College*
George F. Estabrook   *University of Michigan*
Richard S. Feldman   *Marist College*
Charles A. Francis   *University of Nebraska–Lincoln*
Carl Freeman   *Wayne State University*
J. Phil Gibson   *Agnes Scott College*
Robert R. Glesener   *Brevard College*
Michael L. Golden   *Grossmont College*
Paul Grecay   *Salisbury University*
Lana Hamilton   *Northeast State Tech Community College*
Brian Helmuth   *University of South Carolina*
James R. Hodgson   *Saint Norbert College*
Jeremiah N. Jarrett   *Central Connecticut State University*
Krish Jayachandran   *Florida International University*
Seema Sanjay Jejurikar   *Bellevue Community College*
Mark Jonasson   *Crafton Hills College*
Thomas W. Jurik   *Iowa State University*
Karen L. Kandl   *University of New Orleans*
Robert Keys   *Cornerstone University*
Mark E. Knauss   *Shorter College*
Jean Knops   *University of Nebraska*
Anthony J. Krzysik   *Embry-Riddle Aeronautical University*
Eddie N. Laboy-Nieves   *InterAmerican University of Puerto Rico*
Vic Landrum   *Washburn University*
Michael T. Lanes   *University of Mary*
Tom Langen   *Clarkson University*
Kenneth A. LaSota   *Robert Morris College*
Hugh Lefcort   *Gonzaga University*
Peter V. Lindeman   *Edinboro University of Pennsylvania*
John F. Logue   *University of South Carolina–Sumter*
John S. Mackiewicz   *State University of New York–Albany*
Tim Maret   *Shippensburg University*
Ken R. Marion   *University of Alabama–Birmingham*
Vicky Meretsky   *Indiana University*
John C. Mertz   *Delaware Valley College*
Carolyn Meyer   *University of Wyoming*
Sheila G. Miracle   *Southeast Community College–Bell City*
Timothy Mousseau   *University of South Carolina*
Virginia Naples   *Northern Illinois University*
Peter Nonacs   *University of California Los Angeles*
Mark H. Olson   *Franklin & Marshall College*
David W. Onstad   *University of Illinois–Champaign*
Fatimata A. Palé   *Thiel College*
Mary Lou Peltier   *Saint Martin's College*
Carolyn Peters   *Spoon River College*
Kenneth L. Petersen   *Dordt College*
Eric R. Pianka   *University of Texas*
Raymond Pierotti   *University of Kansas–Lawrence*
David Pindel   *Corning Community College*
Jon K. Piper   *Bethel College*
Thomas E. Pliske   *Florida International University*
Michael V. Plummer   *Harding University*
Ellen Porter Holtman   *Virginia Western Community College*
Diane Post   *University of Texas Permian Basin*
Kathleen Rath Marr   *Lakeland College*
Brian C. Reeder   *Morehead State University*
Seth R. Reice   *University of North Carolina–Chapel Hill*
Robin Richardson   *Winona State University*
Carol D. Riley   *Gainesville College*
Marianne W. Robertson   *Millikin University*

Tom Robertson   *Portland Community College*
Bernadette M. Roche   *Loyola College in Maryland*
Tatiana Roth   *Coppin State College*
Neil Sabine   *Indiana University East*
Timothy Savisky   *University of Pittsburgh*
Josh Schimel   *University of California Santa Barbara*
Michael G. Scott   *Lincoln University*
Erik R. Scully   *Towson University*
Michael J. Sebetich   *William Paterson University*
Walter M. Shriner   *Mount Hood Community College*
John Skillman   *California State University–San Bernardino*
Jerry M. Skinner   *Keystone College*
Garriet W. Smith   *University of South Carolina–Aiken*
Stacy Smith   *Lexington Community College*
Joseph Stabile   *Iona College*
Alan Stam   *Capital University*
Alan Stiven   *University of North Carolina–Chapel Hill*
Eric D. Storie   *Roanoke-Chowan Community College*
William A. Szelistowski   *Eckerd College*
Robert Tatina   *Dakota Wesleyan University*
Nina N. Thumser   *California University of Pennsylvania*
John A. Tiedemann   *Monmouth University*
Anne H. Todd Bockarie   *Philadelphia University*
Conrad Toepfer   *Millikin University*
Donald E. Trisel   *Fairmont State College*
Dessie L.A. Underwood   *California State University–Long Beach*
Carl Von Ende   *Northern Illinois University*
Fred E. Wasserman   *Boston University*
Phillip L. Watson   *Ferris State University*
Donna Wear   *Augusta State University*
John F. Wegner   *Emory State University*
Matt R. Whiles   *Southern Illinois University*
Howard Whiteman   *Murray State University*
Craig E. Williamson   *Lehigh University*
Gordon Wolfe   *California State University–Chico*
Derek Zelmer   *Emporia State University*
Douglas Zook   *Boston University*

## Reviewers for Previous Editions

Earl Aagaard   *Pacific Union College*
Marc C. Albrecht   *University of Nebraska–Kearney*
Jane Aloi   *Saddleback College*
Clifford Amundsen   *University of Tennessee–Knoxville*
Walt Anderson   *Prescott College*
Nick Ashbaugh   *Lambuth University*
Bob Bailey   *Central Michigan University*
Gerald J. Bakus   *University of Southern California*
Mary Ball   *Carson-Newman College*
Susan C. Barber   *Oklahoma City University*
James W. Bartolome   *University of California–Berkeley*
Andrew M. Barton   *University of Maine–Farmington*
David Bass   *University of Central Oklahoma*
Thomas L. Beitinger   *University of Northern Texas*
Mark C. Belk   *Brigham Young University*
Dan Benjamin   *Central Michigan University*
Arthur C. Benke   *University of Alabama*
C. R. Blem   *Virginia Commonwealth University*
George C. Boone   *Susquehanna University*
April M. Boulton   *Vanderbilt University*
Randy Brooks   *Florida Atlantic University*
Gordon Brown   *St. John's University*
David C. Brubaker   *Seattle University*
Willodean D. S. Burton   *Austin Peay State University*
Michael S. Capp   *Carlow College*
William Patrick Carew   *Dominican College*
Robert J. Caron   *Bristol Community College*
Steven B. Carroll   *Truman State University*
William H. Chrouser   *Warner Southern College*
Gary K. Clambey   *North Dakota State University*
Patricia J. Clark   *Cumberland College*

Phillip D. Clem   *University of Charleston*
George R. Cline   *Jacksonville State University*
Jay P. Clymer III   *Marywood University*
Marty Condon   *Cornell College*
Alan Covich   *Colorado State University*
Mitchell B. Cruzan   *University of Tennessee*
Cliff Dahm   *University of New Mexico*
Karen J. Dalton   *Catonsville Community College*
Randi Darling   *Westfield State College*
Veronique A. Delesalle   *Gettysburg College*
Paul H. Demchick   *Barton College*
Elizabeth A. Desy   *Southwest State University*
Gary Dolph   *Indiana University Kokomo*
Donald Dorfman   *Monmouth University*
William J. Ehmann   *Drake University*
Inge W. Eley   *Hudson Valley Community College*
Paul Ellefson   *University of Minnesota*
Charles Elliott   *Eastern Kentucky University*
Gina Erickson   *Highline Community College*
John Faaborg   *University of Missouri*
Lloyd C. Fitzpatrick   *University of Northern Texas*
Jonathan Frye   *McPherson College*
James C. Gibson   *Chadron State College*
Sandra L. Gilchrist   *New College of University of South Florida*
Frank S. Gilliam   *Marshall University*
Mac F. Given   *Neumann College*
Jim Goetze   *Laredo Community College*
Brent Graves   *Northern Michigan University*
Don Hall   *Michigan State University*
Peter R. Hannah   *University of Vermont*
Christopher Hartleb   *University of Wisconsin–Stevens Point*
Stephen B. Heard   *University of New Brunswick*
Thomas E. Hemmerly   *Middle Tennessee State University*
Linda L. Hensel-Burke   *Merrer University*
Graham C. Hickman   *Texas A & M University–Corpus Christi*
Nelda W. Hinckley   *John A. Logan College*
Carl W. Hoagstrom   *Ohio Northern University*
Ken Hoover   *Jacksonville University*
Henry S. Horn   *Princeton University*
Michael G. Hosking   *Davidson College*
Jodee Hunt   *Grand Valley State University*
L.E. Hurd   *Washington & Lee University*
Laura Gough   *University of Alabama*
Dan F. Ippolito   *Anderson University*
Jeffrey D. Jack   *Western Kentucky University*
Alan R. P. Journet   *Southeast Missouri State University*
Thomas W. Jurik   *Iowa State University*
Gail E. Kantak   *Saginaw Valley State University*
Donald E. Keith   *Tarleton State University*
David H. Kesler   *Rhodes College*
Sekender A. Khan   *Elizabeth City State University*
Robert W. Kingsolver   *Kentucky Wesleyan College*
Tom H. Klubertanz   *Peru State College*
Richard C. Knaub   *Colorado Northwestern Community College*
David Knowles   *East Carolina University*
John Korstad   *Oral Roberts University*
Edward S. Kubersky   *Felician College*
Frank T. Kuserk   *Moravian College*
Allan J. Landwer   *Hardin-Simmons University*
James W. Langdon   *University of South Alabama*
Eric Larsen   *Villanova University*
David C. LeBlanc   *Ball State University*
Susan Lewis   *Carroll College*
Rex L. Lowe   *Bowling Green State University*
Richard Lowell   *Ramapo College of New Jersey*
Jeffrey R. Lucas   *Purdue University*
James O. Luken   *Northern Kentucky University*
Jon R. Maki   *Eastern Kentucky University*
Michael Howard Marcovitz   *Midland Lutheran College*
Bradford D. Martin   *La Sierra University*
Lee Anne Martinez   *University of Southern Colorado*
David L. McNeely   *University of Texas–Brownsville*

C. Neal McReynolds   *Blue Mountain College*
John C. Mertz   *Delaware Valley College*
Harry A. Meyer   *McNeese State University*
George Middendorf   *Howard University*
Paul A. Mills   *Hannibal-La Grange College*
Richard F. Modlin   *University of Alabama–Huntsville*
L. Maynard Moe   *California State University–Bakersfield*
Patricia Mosto   *Rowan University*
Juliana Mulroy   *Denison University*
Allan Nelson   *Tarleton State University*
Howard S. Neufeld   *Appalachian State University*
Robert A. Nicholson   *Fort Hays State University*
Robert K. Noyd   *U.S. Air Force Academy*
Walter C. Oechel   *San Diego State University*
Steve L. O'Kane, Jr.   *University of Northern Iowa*
Laura Pannaman   *Jersey City State College*
Ken Parejko   *University of Wisconsin– Stout*
Craig Plante   *College of Charleston*
Mary Power   *University of California*
Carl Quertermus   *State University of West Georgia*
Frank J. Rahel   *University of Wyoming*
Marcel Rejmanek   *University of California–Davis*
Carlton Lee Rockett   *Bowling Green State University*
Larry L. Rockwood   *George Mason University*
Neil Sabine   *Indiana University–East*
Anna Sala   *University of Montana*
Thomas J. Sarro   *Mount Saint Mary College*
Angela M. Sauro   *Mount Mary College*
Udo M. Savalli   *Fordham University*
Jan Savitz   *Loyola University of Chicago*
Neil B. Schanker   *College of the Siskiyous*
Bruce A. Schulte   *Providence College*
Wendy E. Sera   *Baylor University*
Shaukat M. Siddiqi   *Virginia State University*
Geoffrey R. Smith   *William Jewel College*
Walker Smith   *University of Tennessee*
Rosemary J. Smith   *Nebraska Wesleyan University*
George G. Spomer   *University of Idaho*
Frank G. Stanton   *Leeward Community College*
Nancy L. Stanton   *University of Wyoming*
Benjamin Steele   *Colby-Sawyer College*
Steven J. Stein   *Eastern Washington University*
Margaret M. Stewart   *University at Albany*
Judith Stribling   *Salisbury State University*
Merrill H. Sweet   *Texas A & M University*
Swee May Tang   *Bethel College*
Max R. Terman   *Tabor College*
Kenneth W. Thompson   *Lock Haven University*
Harry M. Tiebout III   *West Chester University*
C. Richard Tracy   *University of Nevada–Reno*
James Traniello   *Boston University*
Nancy Tuchman   *Loyola University of Chicago*
Sarah Twombly   *University of Rhode Island*
Delmar Vander Zee   *Dordt College*
Neal J. Voelz   *St. Cloud State University*
Stephen M. Wagener   *Western Connecticut State University*
Joseph M. Wahome   *Mississippi Valley State University*
John F. Weishampel   *University of Central Florida*
Lynn D. Wike   *University of South Carolina–Aiken*
Stephen W. Wilson   *Central Missouri State University*
Erwin B. Wingfield   *Virginia Military Institute*
Barbara L. Winternitz   *The Colorado College*
Daniel Wivagg   *Baylor University*
David M. Wood   *California State University–Chico*
Wade B. Worthen   *Furman University*
Richard J. Wright   *Valencia Community College*
Jianguo Wu   *Arizona State University*
Bruce A. Wunder   *Colorado State University*
Brenda L. Young   *Daemen College*
A. L. Youngman   *Wichita State University*
Shep Zedaker   *Virginia Polytechnic Institute and State University*
Gregory Zimmerman   *Lake Superior State University*

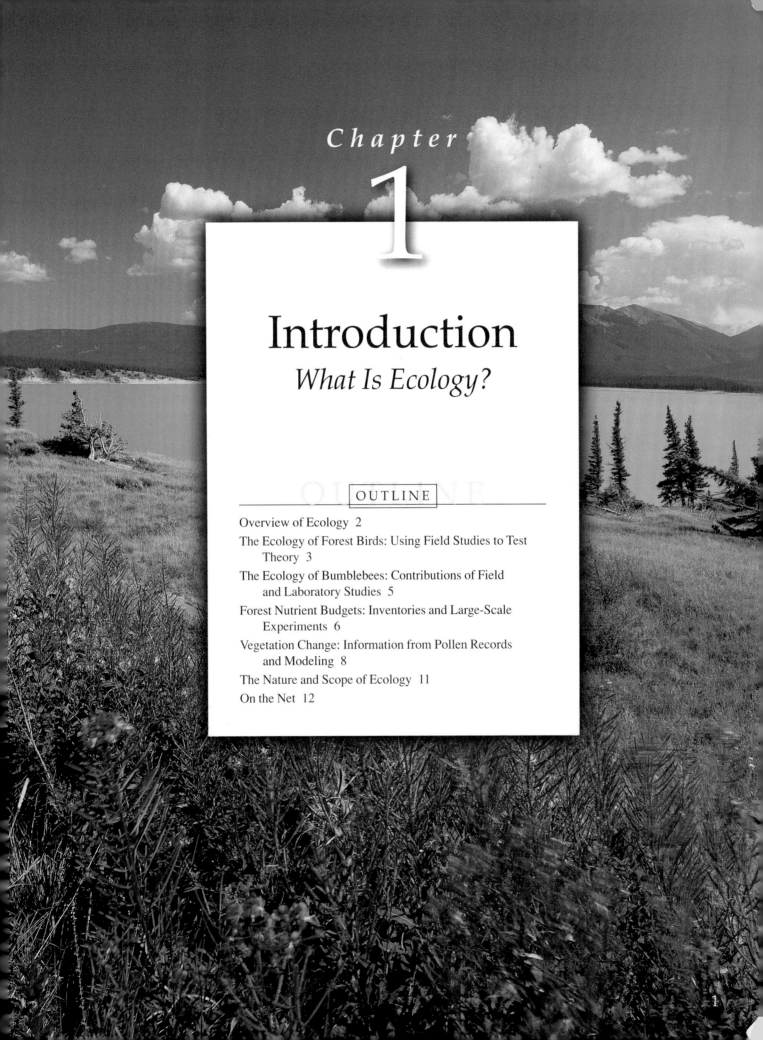

# Chapter 1

# Introduction
## *What Is Ecology?*

OUTLINE

What is ecology? **Ecology** can be defined as the study of relationships between organisms and the environment. Humans have been students of ecology as long as we have existed as a species. Our survival has depended upon how well we could observe variations in the environment and predict the responses of organisms to those variations. The earliest hunters and gatherers had to be familiar with the habits of their animal prey. They also had to know where to find food plants and when they would ripen. Later, farmers and ranchers had to be aware of variations in weather and soils and of how they might affect their crops and livestock.

Today, most of earth's human population lives in cities and most of us have little direct contact with nature. More than ever before, though, the future of our species depends on how well we understand the relationships between organisms and the environment. We must study these relationships because our species is rapidly changing earth's environment, yet we do not fully understand the consequences of these changes. For instance, human activity has greatly increased the quantity of nitrogen cycling through the biosphere, changed land use across the globe, and increased the atmospheric concentration of $CO_2$. Changes such as these threaten the diversity of life on earth and may endanger our life support system. At the dawn of the twenty-first century, it is imperative that we once again become ardent students of ecology.

Behind the simple definition of ecology lies a broad scientific discipline that almost defies definition. Ecologists may study individual organisms, entire forests or lakes, or even the whole earth. The measurements made by ecologists include counts of individual organisms, rates of reproduction, or rates of processes such as photosynthesis and decomposition. Ecologists often spend as much time studying nonbiological components of the environment, such as temperature or soil chemistry, as they spend studying organisms. Meanwhile, the "environment" of organisms in some ecological studies may be other species. While you may think of ecologists as typically studying in the field, some of the most important conceptual advances in ecology have come from ecologists who build theoretical models of ecological systems or do ecological research in the laboratory. Clearly, our simple definition of *ecology* does not communicate the great breadth of the discipline or the diversity of its practitioners. To get a better idea of what ecology is, let's briefly review the scope of the discipline.

## Overview of Ecology

The discipline of ecology addresses environment relationships ranging from those of individual organisms to factors influencing the state of the entire biosphere. This broad range of subjects can be organized and facilitated by arranging them as levels in a hierarchy of ecological organization, such as that imbedded in the brief table of contents and sections II to VI of this book. Figure 1.1 attempts to display such a hierarchy graphically, including the relationships of the various levels to each other and to the types of problems addressed by the various subdisciplines of ecology.

Historically, the ecology of individuals, which is presented at the base of figure 1.1, has been the domain of physiological ecology and behavioral ecology. Physiological ecologists have emphasized the physiological and anatomical mechanisms by which organisms solve problems posed by physical and chemical variation in the environment. Meanwhile, behavioral ecologists have focused principally on the ways that animals (and to a limited extent, plants) use behavior to deal with environmental variation, including variation in the social environment. However, the distinction between physiological ecology and behavioral ecology is often blurred. For instance, both behavioral ecologists and physiological ecologists study the temperature relations and the energy and nutrient relations of organisms.

There is a strong conceptual linkage between studies of the ecology of individuals and population studies, particularly where they concern evolutionary processes. Population ecology is centered on the factors influencing population structure and process, where a population is a group of individuals of a single species inhabiting a defined area. The processes studied by population ecologists include adaptations, extinction, the distribution and abundance of species, population growth and regulation, and variation in the reproductive ecology of species. Population ecologists are particularly interested in how these processes are influenced by the environment, including nonbiological and biological components of the environment.

Bringing biological components of the environment into the picture takes us to the next level of organization studied by ecologists, the ecology of interactions such as predation, parasitism, and competition. Ecologists that study interactions between species have often emphasized the evolutionary effects of the interaction on the species involved. A second approach explores the effect of the interaction on population structure or on the properties of ecological communities.

The definition of an ecological community as an association of interacting species links community ecology with the ecology of interactions. Community and ecosystem ecology have a great deal in common, since both are concerned with the factors controlling multispecies systems. However, the objects of their study are slightly different. While community ecologists concentrate on the organisms inhabiting an area, ecosystem ecologists include the ecological community in an area plus all of the physical and chemical factors influencing the community. The goal of ecosystem ecology is to understand the controls on nutrient cycling and energy flow through ecosystems.

To simplify their studies, ecologists have long attempted to identify and study isolated communities and ecosystems. However, all communities and ecosystems on earth are open systems subject to exchanges of materials, energy, and organisms with other communities and ecosystems. The study of these exchanges, especially among ecosystems, is the intellectual territory of landscape ecology. However, landscapes are not isolated either but part of geographical regions subject to large-scale and long-term regional processes. These regional processes are the subjects of geographic ecology. Geographic

**Biosphere**

What role does concentration of atmospheric $CO_2$ play in the regulation of global temperature?

**Region**

How has geologic history influenced regional diversity within certain groups of organisms?

**Landscape**

How do hedgerows and other vegetated corridors affect the rate of movement by mammals among isolated forest fragments?

**Ecosystem**

What factors control rate of energy fixation by ecosystems?

**Community**

How does disturbance influence the number of species in communities?

**Interactions**

How does soil nutrient availability affect the exchange of materials between plants and mycorrhizal fungi (fungi associated with plant roots?

**Population**

What factors control population growth rates?

**Individuals**

How do plants or animals regulate their internal water balance?

**Figure 1.1** Levels of ecological organization and examples of the kinds of questions asked by ecologists working at each level. These ecological levels correspond broadly to sections II to VI of this book.

ecology in turn leads us to the largest spatial scale and highest level of ecological organization—the **biosphere,** which falls within the realm of global ecology.

While this description of ecology provides a brief preview of the material covered in this book, it is by necessity a rough sketch and highly abstract. To move beyond the abstraction represented by figure 1.1, we need to connect it to the work of the scientists that have created the discipline of ecology. To do so, let's briefly review the research approaches of several ecologists working at a broad range of ecological levels.

# The Ecology of Forest Birds: Using Field Studies to Test Theory

Robert MacArthur gazed intently through his binoculars. He was watching a small bird, called a warbler, searching for insects in the top of a spruce tree. To the casual observer it might have seemed that MacArthur was a weekend bird-watcher. Yes, he was intensely interested in the birds he was watching, but he was just as interested in testing ecological theory.

The year was 1955, and MacArthur was studying the ecology of five species of warblers that live together in the spruce forests of northeastern North America. All five warbler species, Cape May (*Dendroica tigrina*), yellow-rumped (*D. coronata*), black-throated green (*D. virens*), blackburnian (*D. fusca*), and bay-breasted (*D. castanea*), are about the same size and shape and feed on insects. Theory predicted that two species with identical ecological requirements would compete with each other and that, as a consequence, they could not live in the same environment indefinitely. MacArthur wanted to understand how several warbler species with apparently similar ecological requirements could live together in the same forest.

The warblers fed mainly by gleaning insects from the bark and foliage of trees. MacArthur predicted that these warblers might be able to coexist and not compete with each other if they fed on the insects living in different zones within trees. To map where the warblers fed, he subdivided trees into vertical and horizontal zones. He then carefully recorded the amount of time warblers spent feeding in each.

MacArthur's prediction proved to be correct. His quantitative observations demonstrated that the five warbler species in his study area fed in different zones in spruce trees. As figure 1.2 shows, the Cape May warbler fed mainly among new needles and buds at the tops of trees. The feeding zone of the blackburnian warbler overlapped broadly with that of the Cape May warbler but extended farther down the tree. The black-throated green warbler fed toward the trees' interiors. The bay-breasted warbler concentrated its feeding in the interior of trees. Finally, the yellow-rumped warbler fed mostly on the ground and low in the trees. MacArthur's observations showed that though these warblers live in the same forest, they extract food from different parts of that forest. He concluded that feeding in different zones may reduce competition among the warblers of spruce forests.

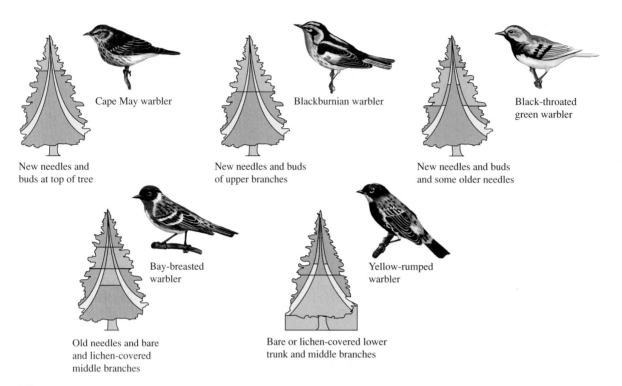

New needles and
buds at top of tree

Cape May warbler

New needles and buds
of upper branches

Blackburnian warbler

New needles and buds
and some older needles

Black-throated
green warbler

Bay-breasted
warbler

Old needles and bare
and lichen-covered
middle branches

Yellow-rumped
warbler

Bare or lichen-covered lower
trunk and middle branches

**Figure 1.2** Warbler feeding zones. The several warbler species that coexist in the forests of northeastern North America feed in distinctive zones within forest trees.

MacArthur's study (1958) of foraging by warblers is a true classic in the history of ecology. However, like most studies it raised as many questions as it answered. Scientific research is important both for what it teaches us directly about nature and for how it stimulates other studies that improve our understanding. MacArthur's work stimulated numerous studies of competition among many groups of organisms, including warblers. Some of these studies produced results that supported his work and others produced different results. All added to our knowledge of competition between species and of warbler ecology.

One ecologist whose studies extended our knowledge of warbler ecology a great deal was Douglass Morse (1980, 1989). His research addressed several questions raised by MacArthur's work, including whether warblers use the same feeding zones in the absence of one or more of the other species. Morse studied this possibility by comparing the feeding zones of warblers living in the presence or absence of other warbler species.

Morse compared the feeding zones of warblers in spruce forests on the mainland of Maine to their feeding zones on small islands. The islands were 0.2 to 1.5 km offshore and were inhabited by one to three species of warblers. Two of the warbler species that lived on the islands, the black-throated green warbler and the yellow-rumped warbler, also lived in MacArthur's study areas. Morse found that the black-throated green warbler maintained approximately the same feeding zone whether it lived on the mainland, with many other warbler species, or on islands, with only two other warbler species. In contrast, the yellow-rumped warbler moved its feeding zone upward on islands where the black-throated

green warbler was absent. This shift in feeding zone by island populations of yellow-rumped warblers is shown in figure 1.3.

Why doesn't the yellow-rumped warbler feed higher in the trees when black-throated green warblers are present? Morse found that the feeding zones of spruce-forest warblers are at least partially maintained by aggressive interactions between species. He also discovered that the black-throated green warbler is socially dominant over yellow-rumped warblers. Consequently, aggression by black-throated green warblers can exclude yellow-rumped warblers from potential feeding areas. Morse proposed that aggressive interactions between warbler species help maintain differences in feeding zones of the kind described by MacArthur.

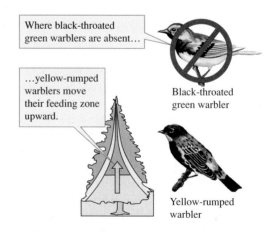

Where black-throated
green warblers are absent…

Black-throated
green warbler

…yellow-rumped
warblers move
their feeding zone
upward.

Yellow-rumped
warbler

**Figure 1.3** The upward shift in the feeding zone of yellow-rumped warblers in the absence of black-throated warblers suggests that the typical yellow-rumped warbler feeding zone is maintained by competition.

The studies of MacArthur and Morse show how field studies can be used to address important ecological questions. Field studies can also be combined with laboratory studies to yield even more detailed information about ecological systems. As we shall see in the following example, this approach has revealed a great deal about the ecology of bumblebees.

## The Ecology of Bumblebees: Contributions of Field and Laboratory Studies

Complex ecological problems may require a combination of field and laboratory studies. Field studies provide information within a natural context. Laboratory studies can provide precise measurements within controlled environments. The two approaches provide complementary information.

Bernd Heinrich has tackled many complex ecological problems using a combination of field and laboratory studies. In one of his research projects he pointed out that bumblebees live in most of the cool regions of the earth. They live in all temperate regions, on cool tropical mountaintops, and even above the Arctic Circle. Two bumblebee species live farther north than any humans. In all these regions, bumblebees keep their thoraxes, the part of the body that houses the flight muscles and to which the wings and legs are attached, warm when they are active. Maintaining a warm body temperature in a cool environment requires energy. Heinrich realized that to understand the ecology of bumblebees he needed to quantify their gains and losses of energy.

Figure 1.4 summarizes how Heinrich used a combination of field and laboratory studies to estimate the energy budgets of bumblebees feeding on different kinds of flowers at different temperatures. First, what is an energy budget? The budget that Heinrich had in mind is similar to a financial budget. He wanted to know the rate at which feeding bumblebees take in energy (income) relative to the rate at which they expend energy (cost). Why should an ecologist be interested in an **energy budget?** An energy budget would give Heinrich an estimate of the amount of energy available for maintaining the bumblebee colony. The difference between energy intake and energy expended while feeding on particular flowers at particular temperatures is the energy gain or loss from foraging in that environment. Energy gains may be invested in the production of honey to feed the bumblebee colony and in reproduction. Foraging that results in a net loss of energy means less food and dwindling reproduction and cannot be sustained for long.

Heinrich followed individual bumblebees as they gathered nectar from various species of flowers to estimate their rate of energy intake. He recorded the number and kinds of flowers visited and also measured the volume and sugar content of nectar produced by each species. He followed one bumblebee as it visited 145 flowers. This particular bumblebee only visited a species of hawkweed that produces orange flowers. Later, Heinrich followed a second bumblebee as it visited 184 flowers.

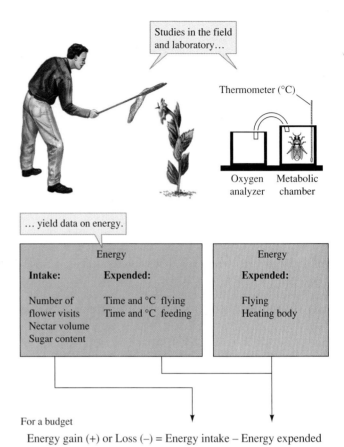

**Figure 1.4** Estimating the energy budgets of foraging bumblebees.

This bumblebee fed almost exclusively on the yellow flowers of a different species. The behavior of these two bumblebees was typical; that is, most of the bumblebees usually visited only a single flower species. This fidelity to a single plant species simplified Heinrich's estimates of the amount of sugar gathered by a bee during a foraging trip.

To estimate the energy expended by a foraging bumblebee, Heinrich needed to know how much energy bumblebees use while flying and feeding. To make these estimates, he needed to know the amount of time bumblebees spend flying and feeding and the temperatures of their thoraxes while doing so. In the field, Heinrich used a stopwatch to measure the time they spent flying and feeding. He also caught flying and feeding bumblebees and measured their thoracic temperatures.

In the laboratory, Heinrich estimated the amount of energy used during flight and the amount of energy bumblebees expend to heat their bodies. He estimated the energetic costs of flight by measuring the rate at which flying bumblebees consume oxygen. How can oxygen consumption be used to estimate energy expenditure? Remember that during respiration all aerobic organisms derive energy from the *oxidation* of organic molecules such as sugars and fats. Consequently, the amount of oxygen consumed can be converted directly into the amount of organic molecules oxidized. The amount of organic molecules can be converted directly into energy equivalents such as calories. Heinrich found that the amount of energy expended during flight was approximately the same regardless of air temperature. This was convenient because he needed only to record the time bumblebees spent flying to know their energy expenditure for flight.

Heinrich found that bumblebees maintain the temperature of their thoraxes at 30° to 37°C, even at air temperatures as low as 0°C. How do bumblebees in cold environments maintain elevated thoracic temperatures? They elevate the temperatures of their thoraxes by contracting their flight muscles. When they are heating, bumblebees simultaneously contract the muscles that raise the wings and those that lower them. Consequently, instead of flying, they shiver.

The amount of energy that bumblebees expend to heat their thoraxes decreases as air temperature increases. In other words, a bumblebee expends less energy to heat itself and can heat itself faster in warm environments than in cold environments (fig. 1.5). Heinrich calculated that a bumblebee perched on a flower in an air temperature of 25°C does not have to produce any heat beyond that produced during flights between flowers to maintain its thoracic temperature at 30°C. In contrast, a bumblebee perched on a flower in an air temperature of 5°C must burn enough sugar to produce half a calorie of heat per minute to maintain a thoracic temperature of 30°C.

Heinrich estimated that a bumblebee feeding on fireweed flowers would have to visit one to two flowers per minute to obtain enough energy for 1 minute of flying. However, he found that bumblebees generally visited fireweed flowers at a rate of 20 to 30 per minute, gathering enough energy to meet their energy needs during flight, to keep themselves warm on cold days, and to take honey back to the colony.

Bernd Heinrich's research (1979, 1993) on bumblebees has focused on the behavior and physiology of individuals in their natural environments. He has combined field measurements with laboratory studies of physiology to estimate the energy budgets of bumblebees. Such studies help us to understand the ecology of individual organisms, single species, and interactions between species. Other ecologists have been concerned with the ecology of entire forests, lakes, or grasslands, which they treat as ecosystems. An **ecosystem** includes all the organisms that live in an area and the physical environment with which those organisms interact. Many ecosystem studies have focused on **nutrients,** the raw materials that an organism must acquire from the environment to live.

# Forest Nutrient Budgets: Inventories and Large-Scale Experiments

For ecologists who study the budgets of nutrients such as nitrogen, phosphorus, or calcium, one of the first steps is to inventory their distribution within an ecosystem. Inventories by Nalini Nadkarni (1981, 1984a, 1984b) changed our ideas of how tropical and temperate rain forests are structured and how they function. With the aid of mountain-climbing equipment, Nadkarni slowly made her first ascent into the canopy of the Costa Rican rain forest, a world explored by few others and where she was to become a pioneer (fig. 1.6). She stood

**Figure 1.6** Exploring the rain forest canopy. What Nalini Nadkarni discovered helped solve a puzzle

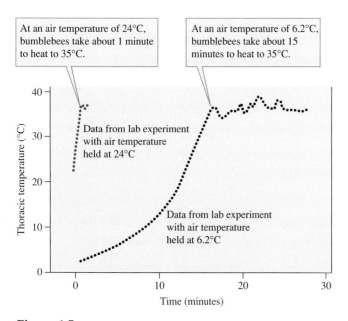

At an air temperature of 24°C, bumblebees take about 1 minute to heat to 35°C.

At an air temperature of 6.2°C, bumblebees take about 15 minutes to heat to 35°C.

Data from lab experiment with air temperature held at 24°C

Data from lab experiment with air temperature held at 6.2°C

**Figure 1.5** Bumblebee heating rates measured in laboratory experiments show the influence of environmental temperatures on rate of thoracic heating (data from Heinrich 1993).

on the rain forest floor and wondered about the diversity of organisms and ecological relationships that might be hidden in the canopy high above. Her wonder soon gave way to determination, and she not only visited the canopy but was among the first to explore the ecology of this unseen world.

Because of leaching by heavy rains, many rain forest soils are poor in nutrients such as nitrogen and phosphorus. The low availability of nutrients in many rain forest soils has produced one of ecology's puzzles. Ecologists have often asked how the prodigious life of rain forests can be maintained on such nutrient-poor soils. Many factors contribute to the maintenance of this intense biological activity. Nadkarni's research in the treetops uncovered one of those factors, a significant store of nutrients in the rain forest canopy.

The nutrient stores in the rain forest canopy are associated with epiphytes. **Epiphytes** are plants, such as many orchids and ferns, that live on the branches and trunks of other plants. Epiphytes are not parasitic: they do not derive their nutrition from the plant they grow on. As they grow on the branches of a tree they begin to trap organic matter, which eventually forms a mat. Epiphyte mats increase in thickness up to 30 cm, providing a complex structure that supports a diverse community of plants and animals.

Epiphyte mats contain significant quantities of nutrients. Nadkarni estimated that these quantities in some tropical rain forests are equal to about half the nutrient content of the foliage of the canopy trees. In the temperate rain forests of the Olympic Peninsula in Washington, the mass of epiphytes is four times the mass of leaves on their host trees.

Nadkarni's research showed that in both temperate and tropical rain forests, trees access these nutrient stores by sending out roots from their trunks and branches high above the ground. These roots grow into the epiphyte mats and extract nutrients from them. As a consequence of this research, we now know that to understand the nutrient economy of rain forests the ecologist must venture into the treetops.

Easier means of working in the rain forest canopy have been developed, and this research is no longer limited to the adventurous and agile. New ways to get into the forest canopy range from hot air balloons and aerial trams to large cranes. The Wind River Canopy Crane offers scientists access to any level within a 70 m tall coniferous forest in a 2.3 ha area near the Columbia River Gorge in Washington (fig. 1.7). Research projects supported—and made far easier—by this crane have included the ecology of migratory birds in the forest canopy, photosynthesis by epiphytes living at different canopy heights,

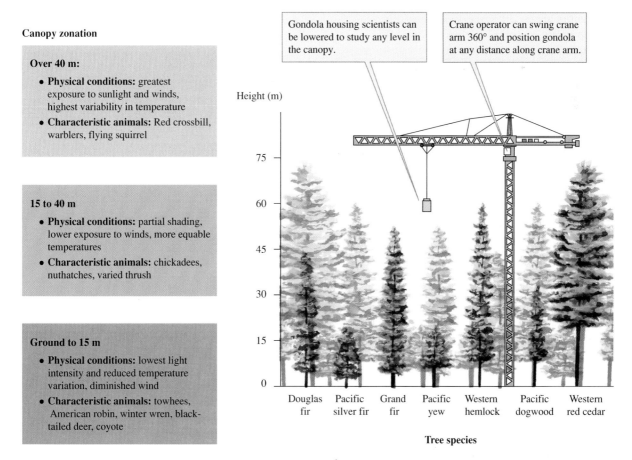

**Figure 1.7**  The Wind River Canopy Crane provides access to the forest canopy for a broad range of ecology and ecological studies.

and vertical stratification of habitat use by bats. Nadkarni points out that the canopy as a physical frontier may be closing, but its exploration as a scientific frontier is just beginning.

The researchers we have considered so far have described a population or an environment by working with small pieces of their ecological system. While Heinrich followed individual bumblebees and Nadkarni climbed individual trees in search of ecological knowledge, other researchers manipulated entire lakes or forests. One team of researchers studied how forests affect the movement of plant nutrients across landscapes.

As Gene Likens and Herbert Bormann watched, work crews felled the trees covering an entire stream basin in the Hubbard Brook Experimental Forest of New Hampshire. The felling of these trees was a key part of an experiment that Likens and Bormann had designed to study how forests affect the loss of nutrients, such as nitrogen, from forested lands (Bormann and Likens 1994, Likens and Bormann 1995). They had studied two small stream valleys for 3 years before cutting the trees in one of the valleys. The undisturbed stream valley would act as a control against which to compare the response of the deforested stream valley. Likens and Bormann combined biology with physical sciences, including geochemistry, hydrology, micrometeorology, and applied disciplines, including forestry.

The central hypothesis guiding their experiment was that organisms, especially plants, regulate the rate of nutrient loss from northern hardwood forests. Their study area, the Hubbard Brook Experimental Forest, covers approximately 3,000 ha and ranges in altitude from 200 to 1,000 m. The Hubbard Brook valley was nearly completely deforested by 1917, and most of the present-day forest has grown up since that time. The forest is fairly representative of second-growth forests across northern New England and is dominated by sugar maple, beech, and yellow birch, along with some red spruce, balsam fir, and white birch.

The researchers organized their studies around small stream basins that included small tributaries of Hubbard Brook. The natural topographic boundaries of these stream basins offered the opportunity for measuring the movement of nutrients. Before they deforested the experimental basin, Likens and Bormann inventoried the distribution of nutrients. Those measurements indicated that over 90% of the nutrients in the ecosystem were tied up in soil organic matter. Most of the rest, 9.5%, was in vegetation. They estimated the rates at which some organisms fix atmospheric nitrogen and the rates at which weathering releases nutrients from the granite bedrock of the stream basins. They also measured the input of nutrients to the forest ecosystem from precipitation and nutrient outputs with stream water. The nutrient outputs in streamflow amounted to less than 1% of the amount contained within the forest ecosystems.

After this preliminary work, they cut the trees on their experimental stream basin. They used herbicides to suppress regrowth of vegetation at their experimental site and continued to apply herbicides for 3 years. As figure 1.8 indicates, cutting the forest dramatically increased rates of nutrient loss from the experimental stream basin. Losses of nitrate ($NO_3^-$) were approximately 40 to 50 times higher. The average con-

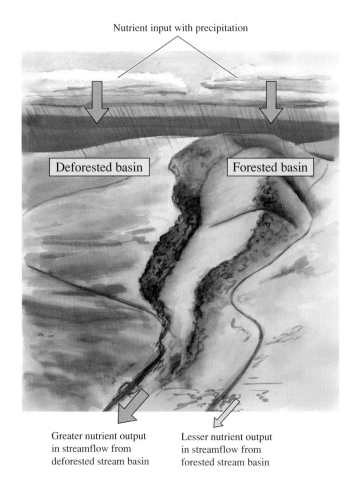

Nutrient input with precipitation

Deforested basin

Forested basin

Greater nutrient output in streamflow from deforested stream basin

Lesser nutrient output in streamflow from forested stream basin

**Figure 1.8** This whole stream basin manipulation demonstrated the influence of forest trees on nutrient budgets of northeastern hardwood forests.

centrations of other major elements in the stream draining the deforested basin increased by 177% to 1,558%. Clearly, this type of temperate forest exerts strong controls on the movement of nutrients across the landscape.

This study by Likens and Bormann gave ecologists new insights into the influences of vegetation on nutrient movements. The study is also notable because it was conducted on a much larger scale than most ecological studies. However, there are ecologists who think and work at even larger scales.

# Vegetation Change: Information from Pollen Records and Modeling

The earth and its life are always changing. However, many of the most important changes occur over such a long period of time or at such large spatial scales that they are difficult to study. Two approaches that provide insights into long-term and large-scale processes are studies of pollen preserved in lake sediments and theoretical modeling.

Margaret Davis (1983, 1989) carefully searched through a sample of lake sediments for pollen. The sediments had come from a lake in the Appalachian Mountains, and the pollen they contained would help her document changes in the plants living near the lake during the past several thousands of years. Davis is a paleoecologist trained to think at very large spatial scales and over very long periods of time. She has spent much of her professional career studying changes in the distributions of plants during the Quaternary period, particularly during the most recent 20,000 years.

Some of the pollen produced by plants that live near a lake falls on the lake surface, sinks, and becomes trapped in lake sediments. As lake sediments build up over the centuries, this pollen is preserved and forms a historical record of the kinds of plants that lived nearby. As the lakeside vegetation changes, the mix of pollen preserved in the lake's sediments also changes. In the example shown in figure 1.9, the earliest appearance of pollen from spruce trees, *Picea* spp., is in the lake sediments from about 12,000 years ago and pollen from beech, *Fagus grandifolia,* first appears in the sediments from about 8,000 years ago. Chestnut pollen does not appear in the sediments until about 2,000 years ago. The pollen from all three tree species continues in the sediment record until about 1920, when chestnut blight killed most of the chestnut trees in the vicinity of the lake. Thus, the pollen preserved in the sediments of individual lakes can be used to reconstruct the history of vegetation in the area.

By studying many different lakes, Davis could study changes in vegetation across an entire continent. Her studies have demonstrated how forests in eastern North America changed with changing climate. Some of her work has been particularly valuable in reconstructing the history of the deciduous forests of eastern North America.

Today, the greatest diversity of deciduous trees in eastern North America is in the Appalachian Mountains. This pattern led ecologists to propose that these mountains have long been a center of diversity for deciduous trees in North America. One hypothesis proposed that deciduous trees survived the last glacial period in the southern Appalachian Mountains. However, by studying the pollen record in lakes across eastern North America, Davis showed that during the height of the last glaciation, about 18,000 years ago, the southern Appalachian Mountains were covered by coniferous trees. At that time, the nearest deciduous forests were in the lower Mississippi Valley, a pattern largely unsuspected until Davis and others studied pollen preserved in lake sediments.

Theoretical models can also provide insights into long-term ecological change. Theoretical analyses by Bruce Milne and his colleagues (1996) provide ecologists with ways to characterize spatial and temporal changes in vegetation. Some of their work has focused on ecotones. **Ecotones** are transitions from one type of ecosystem to another, for instance the transition from a woodland to a grassland. Milne modeled ecotones as a kind of **phase transition.** Typical phase transitions involve changes in the state of matter, such as the change of water from a liquid to a solid state as temperature decreases. The change from liquid water to ice involves fundamental changes in the organization of water, including the average distance between water molecules and their distributions. The change along an ecotone involves analogous changes in the structure of vegetation. In the case of an ecotone, vegetation changes from one place to another rather than over time. Consequently, Milne and his colleagues thought in terms of spatial phase transitions.

Often, a phase transition takes place abruptly under some critical conditions. For instance, water shifts abruptly from liquid to ice as its temperature falls to 0°C. Milne's analyses of the ecotones between woodland and grassland attempted to identify the critical tree densities at which there is an abrupt transition from one vegetation type to another. This transition occurs at the densities where tree cover changes abruptly from a fragmented landscape of small patches of trees to a landscape where the tree canopies are interconnected. While

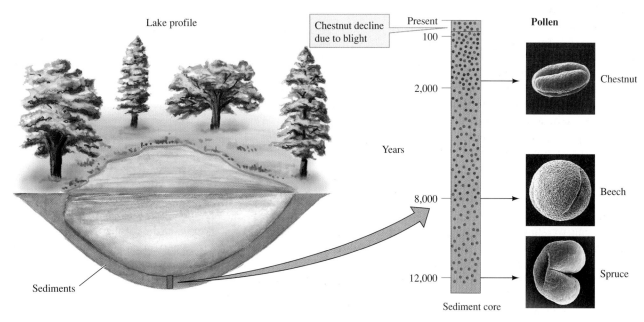

**Figure 1.9** The vegetation history of landscapes can be reconstructed using the pollen contained within the sediments of nearby lakes.

| Information ✓ | *Investigating the Evidence* |
| Questions | |
| Hypothesis | **The Scientific Method—Questions and Hypotheses** |
| Predictions | |
| Testing | |

Ecologists explore the relationships between organisms and environment using the methods of science. The series of boxes called "Investigating the Evidence" that are found throughout the chapters of this book are discussions of various aspects of the scientific method and its application to ecology. While each box describes only a small part of science, taken together, they represent a substantial introduction to the philosophy, techniques, and practice of ecological science.

Let us begin this distributed discussion with the most basic point. What is science? The word *science* comes from a Latin word meaning "to know." Broadly speaking, science is a way of obtaining knowledge about the natural world using certain formal procedures. Those procedures, which make up what we call, "the scientific method," are outlined in figure 1. Despite a great diversity of approaches to doing science, sound scientific studies have many methodological characteristics in common. The most universal and critical aspects of the scientific method are: asking interesting questions and forming testable hypotheses.

## Questions and Hypotheses

What do scientists do? Simply put, scientists ask and attempt to find answers to questions about the natural world. Questions are the guiding lights of the scientific process. Without them, exploration of nature lacks focus and yields little understanding of the world. Let's consider some questions asked by the ecologists discussed in this chapter. The main question asked by Robert MacArthur in his studies of warblers (pp. 3–5) was something like the following: "How can several species of insect-eating warblers live in the same forest without one species eventually excluding the others through competition?" In their studies of northern hardwood forests, Likens and Bormann (p. 8) asked, "What influence does forest vegetation have on the rate of nutrient loss from northern hardwood forests?" While this focus on questions may seem obvious, one of the most common questions asked of scientists at seminars and professional meetings is, "What is your question?"

If scientists are in the business of asking questions about nature, where does a hypothesis enter the process? A hypothesis is a possible answer to a question. MacArthur's main hypothesis (possible answer to his question) was: "Several warbler species are able to coexist because each species feeds on insects living in different zones within trees." The central hypothesis of

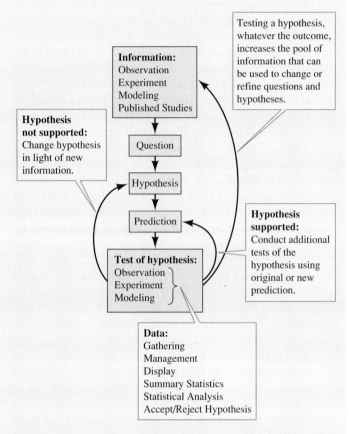

**Figure 1** Graphic summary of the scientific method. The scientific method centers on the use of information to propose and test hypotheses through observation, experiment, and modeling.

the Likens and Bormann study was: "Organisms, especially plants, regulate the rate of nutrient loss from northern hardwood forests." Again, the hypothesis proposed by Likens and Bormann is a tentative answer to their underlying question.

Once a scientist or team of scientists proposes a hypothesis (or multiple alternative hypotheses), the next step in the scientific method is to determine its validity by testing predictions that follow from the hypothesis. Three fundamental ways to test hypotheses are through observation, experiments, and modeling. These approaches, which are all represented in the research presented here in chapter 1 and in figure 1, will be discussed in detail in the *Investigating the Evidence* boxes in later chapters.

**Figure 1.10** Boundaries within landscapes created by human activity (background) and natural environmental gradients (foreground).

this may seem like a purely geometrical exercise, Milne pointed out that the locations of these critical densities may be used to identify the critical environmental conditions that regulate the transition from one vegetation to another.

We may ask where the edge is between one type of vegetation and another. Human intervention makes the edge between a cultivated field and a woodlot obvious. However, the edge between one natural vegetation type and another may be difficult to identify. Figure 1.10 contrasts the sharpness of ecotones in landscapes with and without strong human influences.

Milne and his research group searched for the edge that defines critical densities of vegetation along ecotones. Their analyses led to a number of significant insights about the geometry of ecotones and their sensitivity to environmental change. One of their results concerned the distances between edges defined by different plant densities. For instance, the edge between a woodland and a grassland might be defined by the places on the landscape where there is 41% tree cover or 59% tree cover. The distance between ecotonal edges defined by different densities of vegetation may indicate differences in the environmental gradient along the ecotone.

Milne pointed out that where edges overlap or are very close to each other, the environmental gradient is likely to be steep. Where edges defined by different densities of vegetation are more widely spaced, the environmental gradient is likely to be more gradual. Milne and his colleagues suggested that it is these areas of gradual environmental change within a landscape where we are most likely to see biological responses to environmental change. Davis's studies of pollen trapped in lake sediments document changes in vegetation in response to past climate change, while Milne's studies thrust us into the future. His analyses of the geometry of ecotones suggest where we should concentrate our studies of ecological response to future environmental change.

## The Nature and Scope of Ecology

With this brief review of research approaches and topics, we return to the question asked at the beginning of the chapter: What is ecology? Ecology is indeed the study of relationships between organisms and the environment. However, as you can see from the studies we have reviewed, ecologists study those relationships over a large range of temporal and spatial scales using a wide variety of approaches. Ecology includes Heinrich's studies of bumblebees living around a single bog in New England and Davis's studies of vegetation moving across the North American continent. Ecology also includes the observational studies of MacArthur and Morse. Ecologists may study processes on plots measured in square centimeters or, like Likens and Bormann, study an entire stream valley. Important ecological discoveries have come from Nadkarni's probing of the rain forest canopy and Milne's manipulating landscape images on a computer screen. Ecology includes all these approaches and many more.

In the remainder of this book we will fill in the details of the sketch of ecology presented in this chapter. This brief survey has only hinted at the conceptual basis for the research described. Throughout this book we emphasize the conceptual foundations of ecology. Each chapter focuses on a few ecological concepts. We also explore some of the applications and tools associated with the concepts introduced. Of course the most important tool used by ecologists is the scientific method, which is introduced on page 10.

We continue our exploration of ecology in section I with the natural history of life on land and in water. Natural history was the foundation upon which ecologists built modern ecology. A major premise of this book is that knowledge of natural history improves our understanding of ecological relationships.

# On the Net

Visit this textbook's accompanying website at www.mhhe.com/ecology (click on the book's title) to take advantage of practice quizzing, study/writing tips, timely news articles, and additional URLs for research on the topics in this chapter.

Introductory Materials and Government Sites
Environmental and Ecological Organization Sites
Science as a Process
Introductory Sites
Writing Papers and Study Tips
Glossaries and Dictionaries
Careers in Science
Utility and Organizational Sites
Scientific Method
General Environmental Sites

Individual Contributions to Environmental Issues
Environmental Organizations
Miscellaneous Environmental Resources
General Ecology Sites
Animal Population Ecology
Community Ecology
Field Methods for Studies of Populations
Field Methods for Studies of Ecosystems
The Science of Marine Biology
Inorganic Chemistry

# SECTION I
## Natural History

*"One touch of nature
makes the whole world kin."*

William Shakespeare, *Troilus and Cressida*, 1601–3

# 2

# Life on Land

Detailed knowledge of natural history is proving invaluable to restoration of natural ecosystems across the globe. One of the most dramatic restoration successes that incorporated natural history into its approach comes from Costa Rica. Daniel Janzen's goal was to restore tropical dry forest, a forest nearly as rich in species as tropical rain forest, to Guanacaste National Park, Costa Rica. As he studied the guanacaste tree, *Enterolobium cyclocarpum* (fig. 2.1), however, he realized that something was missing from the present-day forest. The guanacaste tree, a member of the pea family, produces disk-shaped fruit about 10 cm in diameter and 4 to 10 mm thick. Each year, a large tree produces up to 5,000 of these fruits, which fall to the ground when ripe. Janzen asked, Why does the guanacaste tree produce so much fruit? His answer to this question was that the fruit of the tree should promote seed dispersal by animals.

As Janzen looked out on the remaining dry forest, however, he saw no native animals of the size and behavior that would make them dependable dispersers of guanacaste seeds. Dependable dispersers would be necessary to speed restoration of tropical dry forest across Guanacaste National Park. True, some large herbivores fed on guanacaste fruits and dispersed the seeds with their feces. But most of these dispersers were cattle and horses, which were introduced during the Spanish colonial period. Had the guanacaste tree evolved an elaborate fruit and made thousands of them each year in the absence of native dispersers? On the surface, it appeared so.

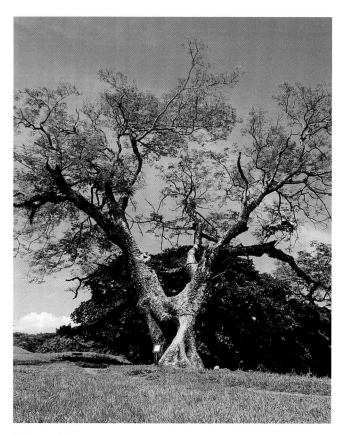

**Figure 2.1** A guanacaste tree, *Enterolobium cyclocarpum,* growing in Costa Rica. Guanacaste trees, which produce large amounts of edible fruit, require large herbivores to disperse their seeds.

Janzen's restoration of tropical dry forest was guided by his knowledge of **natural history,** the study of how organisms in a particular area are influenced by factors such as climate, soils, predators, competitors, and evolutionary history. Natural history eventually led Janzen to an understanding of the fruiting biology of the guanacaste tree. As he considered the long-term natural history of Central American dry forest, he found what he was looking for: a whole host of large herbivorous animals, including ground sloths, camels, and horses. The dry forest had once supported plenty of potential dispersers of guanacaste seeds. However, all these large animals became extinct about 10,000 years ago; overhunting by humans may have been a contributory factor. For thousands of years following these extinctions the guanacaste tree prepared its annual feast of fruits, but there were few large animals to consume them. Then about 500 years ago, Europeans introduced horses and cattle, which ate the fruits of the guanacaste tree and dispersed its seeds around the landscape (fig. 2.2). Janzen recognized the practical value of livestock as seed dispersers and included them in his plan for tropical dry forest restoration.

In the mind of someone else, the natural history of the guanacaste tree, with its odd twists and turns, might have remained a footnote in the history of tropical dry forest. However, Janzen put his knowledge to work. First, he tested and published the hypothesis that contemporary horses can act as effective seed dispersers for the guanacaste tree. After this test, he applied his knowledge by incorporating horses into the management plan for Guanacaste National Park. The guanacaste tree and other trees in a similar predicament would have their dispersers, and restoration of tropical dry forest would be accelerated.

The breadth and depth of perspective offered by natural history, including long-term evolutionary history, gave Janzen the vantage point from which he could look beyond appearances. Through the lens of natural history, Janzen could look beyond the horse as an "alien interloper" and see the species as a key to restoring the tropical dry forest. You might ask, Who, even without studying the natural history of tropical dry forest, would overlook the absence of animals as large as horses? Before Janzen, no one noticed the absence of these animals and their important ecological influence on tropical dry forest species.

Janzen's natural history of tropical dry forest also includes people, unlike most natural histories. He worked closely with people from all parts of Costa Rican society, from the president of the country to local schoolchildren. He realized that long-term support for Guanacaste National Park depended upon its contribution to the economic and cultural well-being of local people. It's the people in Janzen's natural history that stand guard over the Guanacaste project. Janzen calls his approach "biocultural restoration," an approach that seeks to preserve tropical dry forest for its own sake and as a place that provides a host of human benefits, ranging from drinking water to intellectual stimulation. Using natural history as their guide, Janzen and the people of Costa Rica are restoring tropical dry forest in Guanacaste National Park.

Janzen's work (1981a, 1981b) shows how natural history can be used to address a practical problem. Natural history also

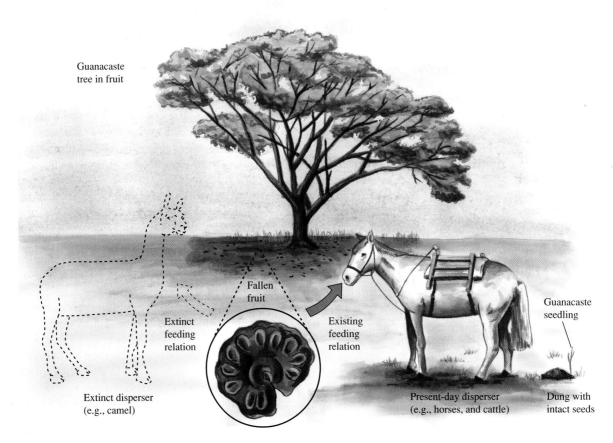

Guanacaste
tree in fruit

Fallen
fruit

Extinct
feeding
relation

Existing
feeding
relation

Guanacaste
seedling

Extinct disperser
(e.g., camel)

Present-day disperser
(e.g., horses, and cattle)

Dung with
intact seeds

**Figure 2.2** Dispersers of guanacaste seeds—past and present. Most of the original dispersers of guanacaste seeds went extinct over 10,000 years ago. Now the tree depends on introduced domestic livestock for its dispersal.

formed the foundation upon which modern ecology developed. Because ecological studies continue to be built upon a solid foundation of natural history, we devote chapters 2 and 3 to the natural history of the biosphere. In chapter 2, we examine the natural history of life on land. Before we begin that discussion, we need to introduce terrestrial biomes, the concept around which this chapter is built. We also discuss the development and structure of soils, the foundation supporting terrestrial biomes.

## Terrestrial Biomes

Chapter 2 focuses on major divisions of the terrestrial environment called **biomes.** Biomes are distinguished primarily by their predominant plants and are associated with particular climates. They consist of distinctive plant formations such as the tropical rain forest biome and the desert biome. Because tropical rain forest and desert are characterized by very different types of plants and animals and occur in regions with very different climates, the natural histories of these biomes differ a great deal. The student of ecology should be aware of the major features of those differences.

The main goal of chapter 2 is to take a large-scale perspective of nature before delving, in later chapters, into finer details of structure and process. We pay particular attention to

the geographic distributions of the major biomes, the climate associated with each, their soils, their salient biological relationships, and the extent of human influences. In other words, we discuss the natural history of life on land using biomes as our organizer. Now let's move on to the central concepts of this chapter, which concern patterns of climatic variation, soil structure and formation, and the distribution of the major biomes in relation to climate.

### CONCEPTS

- **Uneven heating of the earth's spherical surface by the sun and the tilt of the earth on its axis combine to produce predictable latitudinal variation in climate.**
- **Soil structure results from the long-term interaction of climate, organisms, topography, and parent mineral material.**
- **The geographic distribution of terrestrial biomes corresponds closely to variation in climate, especially prevailing temperature and precipitation.**

## CONCEPT DISCUSSION

### Large-Scale Patterns of Climatic Variation

**Uneven heating of the earth's spherical surface by the sun and the tilt of the earth on its axis combine to produce predictable latitudinal variation in climate.**

In chapter 1, ecology was defined as the study of the relationships between organisms and the environment. Consequently, geographic and seasonal variations in temperature and precipitation are fundamental aspects of terrestrial ecology and natural history. Several attributes of climate vary predictably over the earth. For instance, average temperatures are lower and more seasonal at middle and high latitudes. Temperature generally shows little seasonality near the equator, while rainfall may be markedly seasonal. Deserts, which are concentrated in a narrow band of latitudes around the globe, receive little precipitation, which generally falls unpredictably in time and space. What mechanisms produce these and other patterns of climatic variation?

# Temperature, Atmospheric Circulation, and Precipitation

Much of earth's climatic variation is caused by uneven heating of its surface by the sun. This uneven heating results from the spherical shape of the earth and the angle at which the earth rotates on its axis as it orbits the sun. Because the earth is a sphere, the sun's rays are most concentrated where the sun

is directly overhead. However, the latitude at which the sun is directly overhead changes with the seasons. This seasonal change occurs because the earth's axis of rotation is not perpendicular to its plane of orbit about the sun but is tilted approximately 23.5° away from the perpendicular (fig. 2.3).

Because this tilted angle of rotation is maintained throughout earth's orbit about the sun, the amount of solar energy received by the Northern and Southern Hemispheres changes seasonally. During the northern summer the Northern Hemisphere is tilted toward the sun and receives more solar energy than the Southern Hemisphere. During the northern summer solstice on approximately June 21, the sun is directly overhead at the tropic of Cancer, at 23.5° N latitude. During the northern winter solstice, on approximately December 21, the sun is directly overhead at the tropic of Capricorn, at 23.5° S latitude. During the northern winter, the Northern Hemisphere is tilted away from the sun and the Southern Hemisphere receives more solar energy. The sun is directly overhead at the equator during the spring and autumnal equinoxes, on approximately March 21 and September 23. On those dates, the Northern and Southern Hemispheres receive approximately equal amounts of solar radiation.

This seasonal shift in the latitude at which the sun is directly overhead drives the march of the seasons. At high latitudes, in both the Northern and Southern Hemispheres, seasonal shifts in input of solar energy produce winters with low average temperatures and shorter day lengths and summers with high average temperatures and longer day lengths. In many areas at middle to high latitudes there are also significant seasonal changes in precipitation. Meanwhile, between the tropics of Cancer and Capricorn, seasonal variations in temperature and day length are slight, while precipitation may vary a great deal. What produces spatial and temporal variation in precipitation?

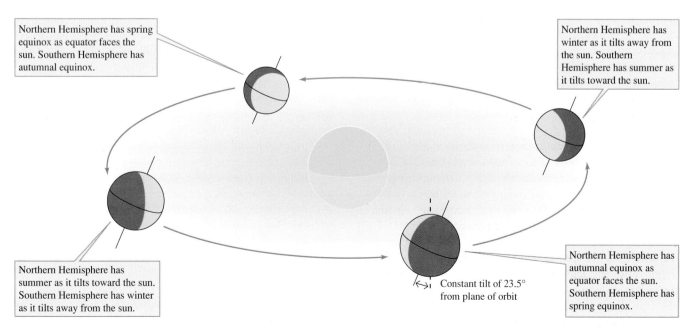

Northern Hemisphere has spring equinox as equator faces the sun. Southern Hemisphere has autumnal equinox.

Northern Hemisphere has winter as it tilts away from the sun. Southern Hemisphere has summer as it tilts toward the sun.

Northern Hemisphere has summer as it tilts toward the sun. Southern Hemisphere has winter as it tilts away from the sun.

Constant tilt of 23.5° from plane of orbit

Northern Hemisphere has autumnal equinox as equator faces the sun. Southern Hemisphere has spring equinox.

**Figure 2.3** The seasons in the Northern and Southern Hemispheres.

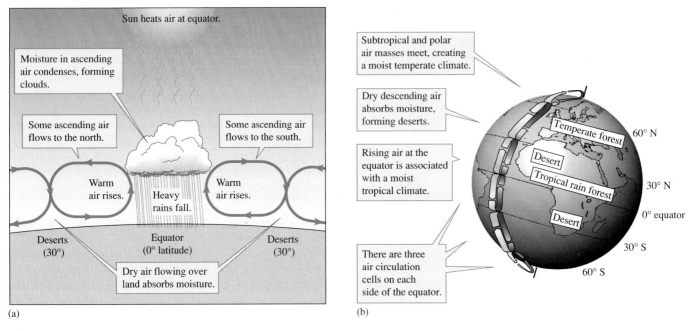

**Figure 2.4**  (*a*) Solar-driven air circulation. (*b*) Latitude and atmospheric circulation.

Heating of the earth's surface and atmosphere drives circulation of the atmosphere and influences patterns of precipitation. As shown in figure 2.4*a*, the sun heats air at the equator, causing it to expand and rise. This warm, moist air cools as it rises. Since cool air holds less water vapor than warm air, the water vapor carried by this rising air mass condenses and forms clouds, which produce the heavy rainfall associated with tropical environments.

Eventually, this equatorial air mass ceases to rise and spreads north and south. This high-altitude air is dry, since the moisture it once held fell as tropical rains. As this air mass flows north and south, it cools, which increases its density. Eventually it sinks back to the earth's surface at about 30° latitude and spreads north and south. This air draws moisture from the lands over which it flows and creates deserts in the process.

Air moving from 30° latitude toward the equator completes an atmospheric circulation cell at low latitudes. As figure 2.4*b* shows, there are three such cells on either side of the equator. Air moving from 30° latitude toward the poles is part of the atmospheric circulation cell at middle latitudes. This warm, moist air flowing from the south rises as it meets cold polar air flowing from the north. As this air mass rises, moisture picked up from desert regions at lower latitudes condenses to form the clouds that produce the abundant precipitation of temperate regions. The air rising over temperate regions spreads northward and southward at a high altitude, completing the middle- and high-latitude cells of general atmospheric circulation.

The patterns of atmospheric circulation shown in figure 2.4*b* suggest that air movement is directly north and south. However, this does not reflect what we observe from the earth's surface as the earth rotates from west to east. An observer at tropical latitudes observes winds that blow from the northeast in the Northern Hemisphere and from the southeast in the Southern Hemisphere (fig. 2.5). These are the

*northeast* and *southeast trades.* Someone studying winds within the temperate belt between 30° and 60° latitude would observe that winds blow mainly from the west. These are the *westerlies* of temperate latitudes. At high latitudes, our observer would find that the predominant wind direction is from the east. These are the *polar easterlies.*

Why don't winds move directly north to south? The prevailing winds do not move in a straight north–south direction because of the **Coriolis effect.** In the Northern Hemisphere, the Coriolis effect causes an apparent deflection of winds to the right of their direction of travel and to the left in the Southern Hemisphere. We say "apparent" deflection because we see this deflection only if we make our observations from the surface of the earth. To an observer in space, it would appear that winds move in approximately a straight line, while the earth rotates beneath them. However, we need to keep in mind that the perspective from the earth's surface is the ecologically relevant perspective. The biomes that we discuss in chapter 2 are as earth-bound as our hypothetical observer. Their distributions across the globe are substantially influenced by global climate, particularly geographic variations in temperature and precipitation.

Geographic variation in temperature and precipitation is very complex. How can we study and represent geographic variation in these climatic variables without being overwhelmed by a mass of numbers? This practical problem is addressed by a visual device called a climate diagram.

## Climate Diagrams

**Climate diagrams** were developed by Heinrich Walter (1985) as a tool to explore the relationship between the distribution of terrestrial vegetation and climate. Climate diagrams

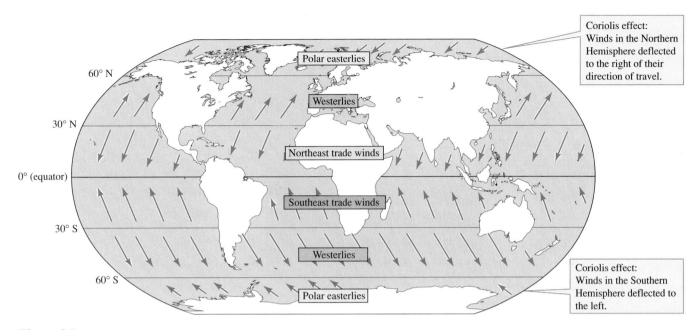

**Figure 2.5** The Coriolis effect and wind direction.

summarize a great deal of useful climatic information, including seasonal variation in temperature and precipitation, the length and intensity of wet and dry seasons, and the portion of the year during which average minimum temperature is above and below 0°C.

As shown in figure 2.6, climate diagrams summarize climatic information using a standardized structure. The months of the year are plotted on the horizontal axis, beginning with January and ending with December for locations in the Northern Hemisphere and beginning with July and ending with June in the Southern Hemisphere. Temperature is plotted on the left vertical axis and precipitation on the right vertical axis. Temperature and precipitation are plotted on different scales so that 10°C is equivalent to 20 mm of precipitation. Climate diagrams for wet areas such as tropical rain forest compress the precipitation scale for precipitation above 100 mm so that 10°C is equivalent to 200 mm of precipitation. With this change in scale, rainfall data from very wet climates can be fit on a graph of convenient size. This change in scale is represented by darker shading in the climate diagram for Kuala Lumpur, Malaysia (fig. 2.7a). Notice that the precipitation at Kuala Lumpur exceeds 100 mm during all months of the year.

Because the temperature and precipitation scales are constructed so that 10°C equals 20 mm of precipitation, the relative positions of the temperature and precipitation lines reflect water availability. Theoretically, adequate moisture for plant growth exists when the precipitation line lies above the temperature line. These moist periods are indicated in the figure by blue shading. When the temperature line lies above the precipitation line, potential evaporation rate exceeds precipitation. These dry periods are indicated by gold shading in the climate diagram. Notice that gold shading of the climate diagram for Yuma, Arizona (fig. 2.7b), indicates year-round drought, while

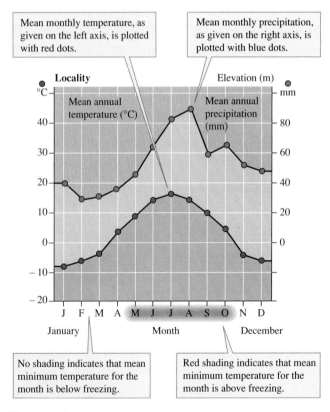

**Figure 2.6** The structure of climate diagrams.

the blue shading of the climate diagram for Kuala Lumpur (fig. 2.7a) indicates moist conditions year-round.

The climate diagram for Dzamiin Uuded, Mongolia (fig. 2.7c), is much more complex than those of either the rain forest or hot desert. This complexity results from the much greater seasonal change in the cold desert climate. Dzamiin Uuded is moist from October to April. These moist periods

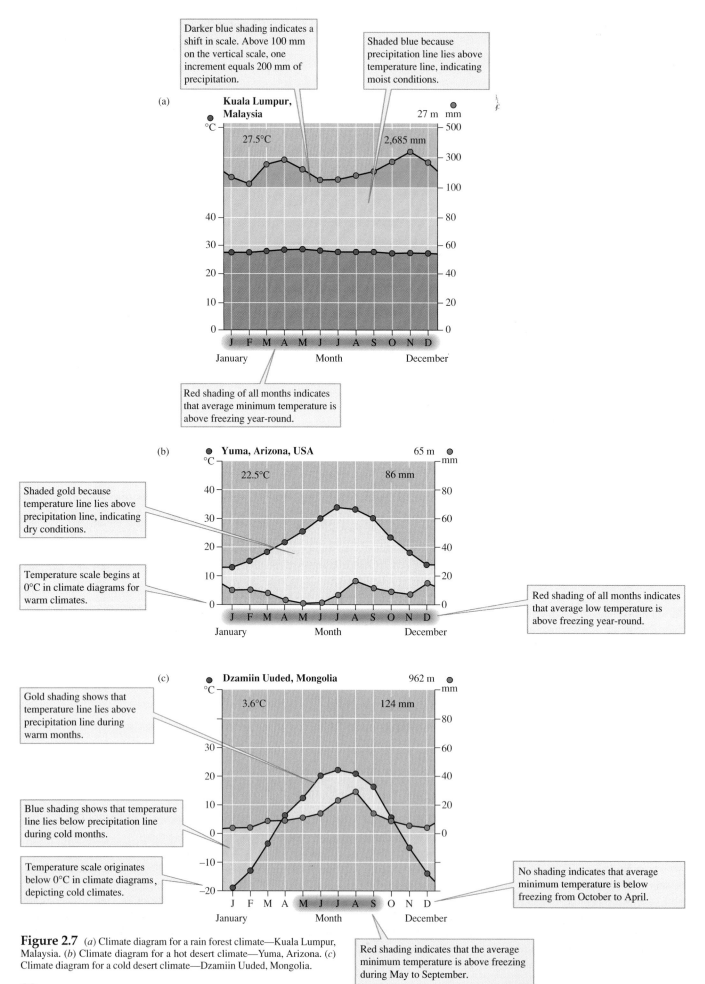

**Figure 2.7** (*a*) Climate diagram for a rain forest climate—Kuala Lumpur, Malaysia. (*b*) Climate diagram for a hot desert climate—Yuma, Arizona. (*c*) Climate diagram for a cold desert climate—Dzamiin Uuded, Mongolia.

The following text labels appear within the figure:

(a) **Kuala Lumpur, Malaysia** — 27 m — 27.5°C — 2,685 mm

Darker blue shading indicates a shift in scale. Above 100 mm on the vertical scale, one increment equals 200 mm of precipitation.

Shaded blue because precipitation line lies above temperature line, indicating moist conditions.

Red shading of all months indicates that average minimum temperature is above freezing year-round.

(b) **Yuma, Arizona, USA** — 65 m — 22.5°C — 86 mm

Shaded gold because temperature line lies above precipitation line, indicating dry conditions.

Temperature scale begins at 0°C in climate diagrams for warm climates.

Red shading of all months indicates that average low temperature is above freezing year-round.

(c) **Dzamiin Uuded, Mongolia** — 962 m — 3.6°C — 124 mm

Gold shading shows that temperature line lies above precipitation line during warm months.

Blue shading shows that temperature line lies below precipitation line during cold months.

Temperature scale originates below 0°C in climate diagrams, depicting cold climates.

No shading indicates that average minimum temperature is below freezing from October to April.

Red shading indicates that the average minimum temperature is above freezing during May to September.

20

Information
Questions
Hypothesis
Predictions
Testing ✓

*Investigating the Evidence*

# Determining the Sample Mean

One of the most common and important steps in the processing of data is the production of summary statistics. First, what is a statistic? A statistic is a number that is used by scientists to estimate a measurable characteristic of an entire population. Population characteristics of interest to an ecologist might include features such as average mass, height, temperature, or growth rate. In order to determine the exact average value of such population characteristics, the ecologist would have to measure every individual in the population. Clearly, the opportunity to measure or test all the individuals in a population for any characteristic is extremely rare. For instance, an ecologist studying reproductive rate in a population of birds would be unlikely to locate and study all the nests in the population. As a consequence, ecologists generally estimate reproductive rates for birds, or other characteristics of any population, using samples drawn from the population. An ecologist working with a population of rare plants, for example, might locate 11 seedlings and calculate the average height of these 11 individual plants. This average calculated from the sample of 11 seedlings would be the **sample mean.** The sample mean is a statistical estimate of the true population mean.

The sample mean is one of the most common and useful of summary statistics. It is a statistic that we use extensively in this chapter as we discuss average temperature or average precipitation for biomes around the world. How is the sample mean calculated? Consider the following sample of the heights of 11 seedlings measured in cm:

| Seedling number | 1 | 2 | 3 | 4 | 5 | 6 | 7 | 8 | 9 | 10 | 11 |
|---|---|---|---|---|---|---|---|---|---|---|---|
| Height in cm | 3 | 6 | 8 | 7 | 2 | 4 | 9 | 4 | 5 | 7 | 8 |

What was the average height of seedlings in the population at the time of the study? Since we did not locate all of the seedlings, we cannot know the true population mean, or parameter. However, our sample of 11 seedlings allows us to calculate a sample mean as follows:

$$\text{Sum of measurements} = \Sigma X$$
$$\Sigma X = 3 + 6 + 8 + 7 + 2 + 4 + 9 + 4 + 5 + 7 + 8$$
$$\Sigma X = 63$$

We calculate the sample mean by dividing the sum of measurements by the number of seedlings measured:

$$\text{Sample mean} = \overline{X}$$
$$n = \text{sample size, or } 11$$

$$\overline{X} = \frac{\Sigma X}{n}$$

$$\overline{X} = \frac{63}{11}$$

$$\overline{X} = 5.7 \text{ cm}$$

Again, 5.7 cm, the sample mean, is the ecologist's estimate of the true mean height of seedlings in the entire population at the time of the study.

are separated by the months of May to September, when the temperature line rises above the precipitation line, indicating drought. During October to April, the mean *minimum* temperature at Dzamiin Uuded is below freezing (0°C). The months when the mean minimum temperature is above freezing are May through September.

Climate diagrams also include the mean annual temperature, which is presented in the upper left corner (e.g., 27.5°C at Kuala Lumpur). The mean annual precipitation (e.g., 86 mm at Yuma, Arizona) is presented in the upper right corner of each climate diagram. The elevation of each site, in meters above sea level (e.g., 962 m at Dzamiin Uuded), is also presented in the upper right corner.

As you can see, climate diagrams efficiently summarize important environmental variables. In the following Concept Discussion section, we use climate diagrams to represent the climates associated with major terrestrial biomes.

## CONCEPT DISCUSSION

### Soil: The Foundation of Terrestrial Biomes

**Soil structure results from the long-term interaction of climate, organisms, topography, and parent mineral material.**

Soil is a complex mixture of living and nonliving material upon which most terrestrial life depends. Here we summarize the general features of soil structure and development. The biome discussions that follow include specific information about the soils associated with each.

**Soil horizons**

) O    Organic horizon. Upper layer contains loose, somewhat fragmented plant litter. Litter in lower layer is highly fragmented.

) A    Mineral soil mixed with some organic matter. Clay, iron, aluminum, silicates, and soluble organic matter are gradually leached from A horizon.

) B    Depositional horizon. Materials leached from A horizon are deposited in B horizon. Deposits may form distinct banding patterns.

) C    Weathered parent material. The C horizon may include many rock fragments. It often lies on bedrock.

**Figure 2.8**  Generalized soil profile, showing O, A, B, and C horizons.

Soil structure can be observed by digging a soil pit, a hole in the ground 1 to 3 m deep. In a soil pit you see one of the most significant aspects of soil structure, its vertical layering. Though soil structure usually changes gradually with depth, soil scientists generally divide soils into several discrete horizons. In the classification system used here the soil profile is divided into O, A, B, and C horizons (fig. 2.8). The **O**, or **organic, horizon** lies at the top of the profile. The most superficial layer of the O horizon is made up of freshly fallen organic matter, including whole leaves, twigs, and other plant parts. The deeper portions of the O horizon consist of highly fragmented and partially decomposed organic matter. Fragmentation and decomposition of the organic matter in this horizon are mainly due to the activities of soil organisms, including bacteria, fungi, and animals ranging from nematodes and mites to burrowing mammals. This horizon is usually absent in agricultural soils and deserts. At its deepest levels, the O horizon merges gradually with the A horizon.

The **A horizon** contains a mixture of mineral materials, such as clay, silt, and sand, and organic material derived from the O horizon. The A and O horizons both support high levels of biological activity. Burrowing animals, such as earthworms, mix organic matter from the O horizon into the A horizon. The A horizon is generally rich in mineral nutrients. It is gradually leached of clays, iron, aluminum, silicates, and humus, which is partially decomposed organic matter. These substances slowly move down through the soil profile until they are deposited in the B horizon.

The **B horizon** contains the clays, humus, and other materials that have been transported by water from the A horizon. The deposition of these materials often gives the B horizon a distinctive color and banding pattern. This horizon is also occupied by the roots of many plants. The B horizon gradually merges with the C horizon.

The **C horizon** is the deepest layer in our soil pit. It consists of weathered parent material, which has been worked by the actions of frost, water, and the deeper penetrating roots of plants. Weathering slowly breaks the parent material into smaller and smaller fragments to produce sand, silt, and clay-sized particles. Because weathering is incomplete and less intense than in the A and B horizons, the C horizon may contain many rock fragments. Under the C horizon we find unweathered parent material, which is often bedrock.

The soil profile gives us a snapshot of soil structure. However, soil structure is in a constant state of flux as a consequence of several influences. Those influences were summarized by Hans Jenny (1980) as climate, organisms, topography, parent material, and time. Climate affects the rate of weathering of parent materials, the rate of leaching of organic and inorganic substances, the rate of erosion and transport of mineral particles, and the rate of decomposition of organic matter. Climate also influences the kinds of vegetation and animals that occupy an area. These organisms, in turn, influence the quantity and quality of organic matter added to soil and the rate of soil mixing by burrowing animals. Topography affects the rates and direction of water flow and patterns of erosion. Meanwhile, parent materials, such as granite, volcanic rock, and wind- or water-transported sand, set the stage for all other influences. Last is the matter of time. Soil age influences soil structure.

In short, soil is a complex and dynamic entity. It forms the medium in which organisms grow, and the activities of those organisms, in turn, affect soil structure. As with many aspects of ecology, it is often difficult to separate organisms from their environment. The biome discussions that follow provide additional information on soils by including aspects of soil structure and chemistry characteristic of each biome.

## CONCEPT DISCUSSION

### Natural History and Geography of Biomes

**The geographic distribution of terrestrial biomes corresponds closely to variation in climate, especially prevailing temperature and precipitation.**

Early in the twentieth century, many plant ecologists studied how climate and soils influence the distribution of vegetation. Later ecologists concentrated on other aspects of plant ecology. Today, as we face the prospect of global warming (see chapter 23), ecologists are once again studying climatic influences on the distribution of vegetation. International teams of ecologists, geographers, and climatologists are exploring the influences of climate on vegetation with renewed interest and with much more powerful analytical tools.

In this section, we discuss the climate, soils, and organisms of the earth's major biomes and how they have been influenced by humans. However, don't be concerned if you know of some places that don't quite fit any of the biomes discussed. For instance, though our focus is on the biomes that occur in the absence of human disturbance, many of these regions now contain only remnants of their natural plant and animal associations. Though biomes are presented here as distinctive entities, they change gradually along environmental gradients. In addition, no two rain forests are exactly alike, nor are any two prairies or deserts. Discovery of the full extent of nature's diversity and of the mechanisms that produce and maintain that diversity remains an uncompleted task for future ecologists—perhaps for you. Now, let's have a look at natural history around the globe.

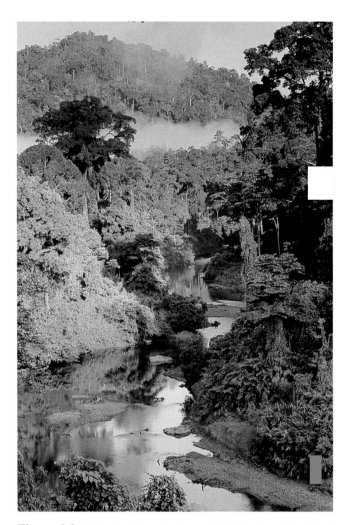

**Figure 2.9** Tropical rain forest in Borneo. Within the three-dimensional framework of tropical rain forests live a higher diversity of organisms than in any other terrestrial biome.

## Tropical Rain Forest

**Tropical rain forest** is nature's most extravagant garden (fig. 2.9). Beyond its tangled edge, a rain forest opens into a surprisingly spacious interior, illuminated by dim greenish light shining through a ceiling of leaves. High above towers the forest canopy, home to many rain forest species and the aerial laboratory of a few intrepid rain forest ecologists (see fig. 1.6). The architecture of rain forests, with their vaulted ceilings and spires, has invited comparisons to cathedrals and mansions. However, this cathedral is alive from ceiling to floor, perhaps more alive than any other biome on the planet. In the rain forest, the sounds of evening and morning, the brilliant flashes of color, and rich scents carried on moist night air speak of abundant life, in seemingly endless variety.

### Geography

Tropical rain forests straddle the equator in three major regions: Southeast Asia, West Africa, and South and Central America (fig. 2.10). Most rain forest occurs within 10° of latitude north or south of the equator. Outside this equatorial band are the rain forests of Central America and Mexico, southeastern Brazil, eastern Madagascar, southern India, and northeastern Australia.

### Climate

The global distribution of rain forests corresponds to areas where conditions are warm and wet year-round (fig. 2.10). Temperatures in tropical rain forests vary little from month to month and often change as much in a day as they do over the entire year. Though rain forests have a reputation for being extremely hot places, they are not. Average temperatures are about 25° to 27°C, lower than average maximum summer temperatures in many deserts and temperate regions. Annual rainfall ranges from about 2,000 to 4,000 mm, and some rain forests receive even more precipitation. To support a tropical rain forest, rainfall must be fairly evenly distributed throughout the year. Notice that the climate diagrams for Belem,

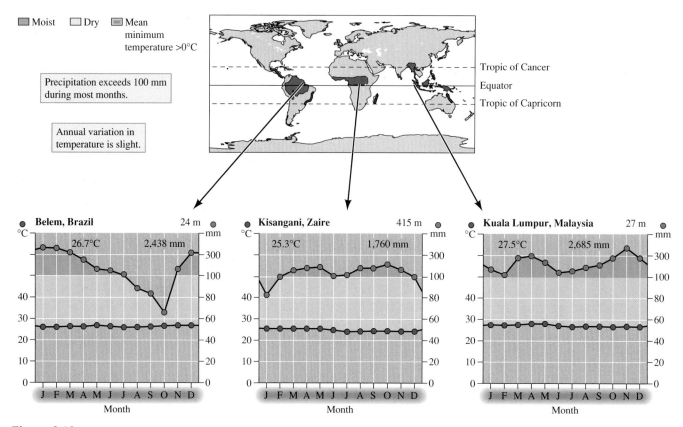

**Figure 2.10** Tropical rain forest distribution and climate.

Kisangani, and Kuala Lumpur indicate moist conditions throughout the year. In a rain forest, a month with less than 100 mm of rain is considered dry. In summary, the tropical rain forest climate is warm, moist, and one of the most equable on earth.

## Soils

Heavy rains gradually leach nutrients from rain forest soils and rapid decomposition in the warm, moist rain forest climate keeps the quantity of soil organic matter low. Consequently, rain forest soils are often nutrient-poor, acidic, thin, and low in organic matter. In many rain forests, more nutrients are tied up in living tissue than in soil. Rain forest plants are adept at conserving nutrients. They get help in gathering nutrients from infertile soils from fungi associated with their roots, through mutually beneficial partnerships called **mycorrhizae.** Free-living fungi, bacteria, and soil animals, such as mites and springtails, rapidly scavenge nutrients from plant litter (leaves, flowers, etc.) and animal wastes, further tightening the nutrient economy of the rain forest.

Some rain forests, however, occur where soils are very fertile. For instance, rain forests grow on young volcanic soils that have not yet been leached of their nutrients by heavy tropical rains. Fertile rain forest soils also occur along rivers, where a fresh nutrient supply is delivered with each flood.

## Biology

Humans live mostly in two dimensions and are most at home on the rain forest floor. In contrast, many organisms in the rain forest have evolved to use the vertical dimension provided by trees. Trees dominate the rain forest landscape and average about 40 m in height. However, some reach 50, 60, or even 80 m tall. The massive weight of these rain forest giants is often supported by well-developed buttresses. The diversity of rain forest trees is also impressive. One hectare (100 m × 100 m) of temperate forest may contain a few dozen tree species; 1 ha of tropical rain forest may contain up to 300 tree species.

The three-dimensional framework formed by rain forest trees is festooned with other plant growth forms. The trees are trellises for climbing vines and growing sites for epiphytes, plants that grow on other plants (fig. 2.11). The great diversity and sheer mass of epiphytes and vines give an impression of great biological richness, of a forest teeming with life. Look closely at rain forest animals and that impression is amplified.

Rain forest animals, from parrots and bats to sloths, snakes, frogs, and monkeys, are also strongly arboreal and the insects of the rain forest canopy are the most diverse of all. A single rain forest tree may support several thousand species of insects, many of which have not been described by scientists. Biologists now estimate that tens of millions of undiscovered insect species may live in tropical rain forests.

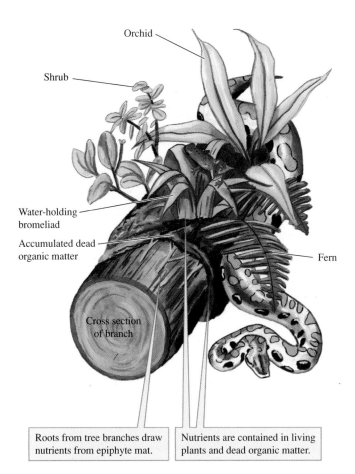

Orchid

Shrub

Water-holding
bromeliad

Accumulated dead
organic matter

Fern

Cross section
of branch

| Roots from tree branches draw nutrients from epiphyte mat. | Nutrients are contained in living plants and dead organic matter. |

**Figure 2.11** An epiphyte mat in the tropical rain forest canopy. Epiphyte mats store a substantial fraction of the nutrients in tropical rain forests and support a high diversity of plant and animal species.

bananas, and sugarcane, and approximately 25% of all prescription drugs, were originally derived from tropical plants. Many more species, directly useful to humans, may await discovery. In addition, the tropics continue to harbor important genetic varieties of domesticated plant species. Unfortunately, tropical rain forests are fast disappearing. Without them, our understanding of the causes and maintenance of biological diversity will remain forever impoverished.

Humans have exploited tropical rain forests for thousands of years through a mixture of hunting and gathering and shifting agriculture. The hunter-gatherer cultures of rain forests, whether in Asia, Africa, or the Americas, have each used hundreds of species for everything from food and building materials to medicines. Today, we are destroying rain forest for timber, minerals, and short-lived agricultural profits. In response to demographic pressures from exploding human populations in tropical countries and economic pressures from the developed countries, traditional systems of exploitation have given way to the bulldozer and chain saw. Can modern tropical societies create a balanced contract with nature that preserves the invaluable biological resources of the rain forest and provides a decent livelihood for local people? This question will be answered by the present generation.

# Tropical Dry Forest

During the dry season, the **tropical dry forest** is all earth tones; in the rainy season, it's an emerald tangle (fig. 2.12). Life in the tropical dry forest responds to the rhythms of the annual solar cycle, which drives the oscillation between wet and dry seasons. During the dry season, most trees in the tropical dry forest are dormant. Then, as the rains approach, trees flower and insects appear to pollinate them. The pace of life quickens. Eventually, as the first storms of the wet season arrive, the trees produce their leaves and transform the landscape.

## Geography

Tropical dry forests occupy a substantial portion of the earth's surface between about 10° and 25° latitude (fig. 2.13). In Africa, tropical dry forests are found both north and south of the central African rain forests. In the Americas, tropical dry forests are the natural vegetation of extensive areas south and north of the Amazon rain forest. Tropical dry forests also extend up the west coast of Central America and into North America along the west coast of Mexico. In Asia, tropical dry forests are the natural vegetation of most of India and the Indochina peninsula. Australian tropical dry forests form a continuous band across the northern and northeastern portions of the continent.

## Climate

The climate of tropical dry forests is more seasonal than that of tropical rain forests. The three climate diagrams shown in

The rain forest is not, however, just a warehouse for a large number of dissociated species. Intricate relationships weave these species into a living green tapestry. Most rain forest plant species depend on animals, from bats and birds to butterflies and bees, to pollinate the flowers for which the rain forest is famous. The plants also produce a striking variety of fruit and rely on animals to disperse their seeds. For the service of seed dispersal, the plants trade a nutritious meal. In the tropical rain forest there are plants that cannot live without particular species of ants, mites that make their homes in the flowers of plants and depend on hummingbirds to get them from flower to flower, and trees and vines that compete continuously for light and space. How did this incredible diversity of species and relationships arise? How is it maintained? Does the health of the rain forest depend on this stunning biological diversity? We do not yet know the answers to these questions. They can only be answered by studying an intact rain forest, where nature is at its most prodigious.

## Human Influences

People from all over the globe owe more to the tropics than is generally realized. Many of the world's staple foods, including maize (called corn in North America and Australia), rice,

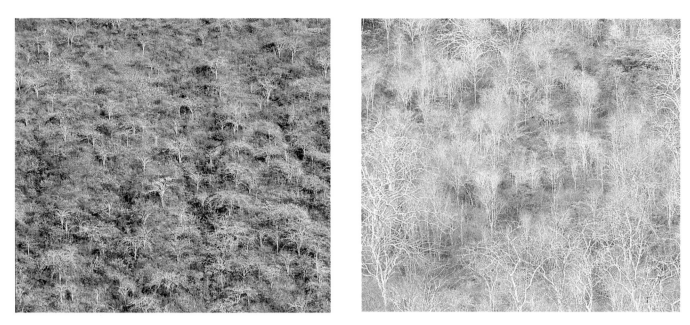

**Figure 2.12** Tropical dry forest in the Galápagos Islands during the wet and dry seasons.

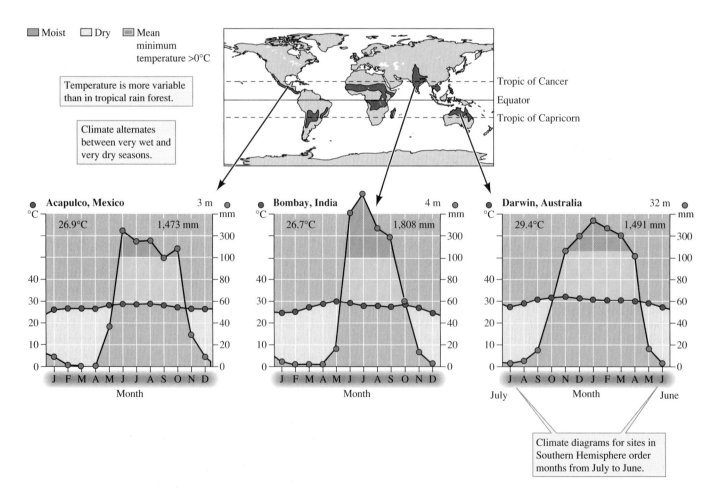

**Figure 2.13** Tropical dry forest distribution and climate.

figure 2.13, for example, show a dry season lasting for 6 to 7 months, followed by a season of abundant rainfall. This wet season lasts for about 5 months in Acapulco, Mexico, and Bombay, India, and about 6 months in Darwin, Australia. Heavy rains occur during the wet season at all three sites. The climate diagrams also indicate more seasonal variation in temperature than we saw in the climate diagrams for the tropical rain forest. Notice that the seasonal rains in the tropical dry forest come during the warmer part of the year.

## Soils

The soils of many tropical dry forests are of great age, particularly those in the parts of Africa, Australia, India, and Brazil that were once part of the ancient southern continent of Gondwana. The soils of tropical dry forests tend to be less acidic than those of rain forests and are generally richer in nutrients. However, the annual pulses of torrential rain make the soils of tropical dry forest highly vulnerable to erosion, particularly when deforested and converted to agriculture.

## Biology

The plants of the tropical dry forest are strongly influenced by physical factors. For example, the height of the dry forest is highly correlated with average precipitation. Trees are tallest in the wettest areas. In the driest places, where the trees are smallest and the landscape more open, the tropical dry forest may appear similar to the tropical savanna or even desert. In addition, in the driest habitats, all trees drop their leaves during the dry season; in wetter areas over 50% may be evergreen. Many plant species in the tropical dry forest produce fruits that are attractive to animals and, as in the tropical rain forest, have animal-dispersed seeds. However, wind-dispersed seeds are also common in more open tropical dry forests.

The tropical dry forest shares many animal species with the rain forest and savanna, including monkeys, parrots, and large cats such as the tiger in Asia and the jaguar in the Americas. The lives of dry forest animals, like those of its plants, are organized around alternating wet and dry seasons. However, how might animal responses to the dry season differ from those of plants? While plants have to "sit and take it" during the dry season, animals can move and many do. Many dry forest birds, mammals, and even insects make seasonal migrations to wetter habitats along rivers or to the nearest rain forest. The discovery of these migrations has led to other questions that await study: Do dry forest species require nearby rain forests for refuge during the dry season? To what extent may some "rain forest" species depend upon intact dry forest?

## Human Influences

Peter Murphy and Ariel Lugo (1986) studied the patterns of human settlement in the tropical forests of Central America. They divided the types of forests into rain forest, wet forest,

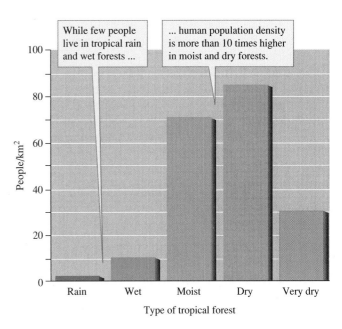

**Figure 2.14**  Human population density in the tropical forests of Central America (data from Murphy and Lugo 1986, after Tosi and Voertman 1964).

moist forest, dry forest, and very dry forest. Their analysis showed a very uneven pattern of human settlement. Murphy and Lugo calculated that tropical rain forest and tropical wet forest include approximately 7% of the human population of Central America. In contrast, tropical dry forest and moist forest include about 79% of the Central American population. As figure 2.14 shows, the population density—the number of people per square kilometer—in tropical dry and moist forests is more than 10 times higher than in tropical wet and rain forests.

Heavy human settlement has devastated the tropical dry forest. While the world's attention has been focused on the plight of rain forests, intact tropical dry forests have nearly disappeared. Why have tropical dry forests been more densely settled? The relatively fertile soil of tropical dry forests has attracted agricultural development. Extensive clearing for agriculture has reduced tropical dry forests in Central America and Mexico to less than 2% of its former area. People have replaced tropical dry forests with cattle ranches, grain farms, and cotton fields. Tropical dry forests are more vulnerable to human exploitation than tropical rain forests because the dry season makes them more accessible and easier to burn. Murphy and Lugo also suggested that there may be less impact from human diseases in tropical dry forests.

Intensive settlement and agricultural development have whittled away at tropical dry forests over a period of centuries. In contrast, rain forests are disappearing in a recent push involving far fewer people but an enormous application of mechanical energy. This energy is directed at the extraction of lumber and minerals and at large-scale conversion of land from rain forest to agriculture. Though human impacts on tropical dry forests and rain forests have differed in tempo, their ecological consequences appear to be the same—massive loss of biological diversity.

The loss of the dry forest is significant because, while rain forests may support a somewhat greater number of species, many dry forest species are found nowhere else. However, out of this devastation has come Guanacaste National Park in Costa Rica, a model attempt to restore a tropical dry forest in a way that also helps serve the cultural and economic needs of local people (see the introduction to chapter 2).

# Tropical Savanna

Stand in the middle of a savanna, a tropical grassland dotted with scattered trees, and your eye will be drawn to the horizon for the approach of thunderstorms or wandering herds of wildlife (fig. 2.15). The **tropical savanna** is the kingdom of the farsighted, the stealthy, and the swift and is the birthplace of humankind. It was from here that humans eventually moved out into every biome on the face of the earth. Now, most humans live away from this first home. The first naturalists, our ancient ancestors, knew the tropical savanna biome best, and the fascination continues.

## *Geography*

Most tropical savannas occur north and south of tropical dry forests within 10° to 20° of the equator. In Africa south of the Sahara Desert, tropical savannas extend from the west to the east coasts, cut a north–south swath across the east African highlands, and reappear in south-central Africa (fig. 2.16). In South America, tropical savannas occur in south-central Brazil and cover a great deal of Venezuela and Columbia. Tropical savannas are also the natural vegetation of much of northern Australia in the region just south of the tropical dry forest. The savanna is the natural vegetation of an area in southern Asia just east of the Indus River in eastern Pakistan and northwestern India.

## *Climate*

As in the tropical dry forest, life on the savanna cycles to the rhythms of alternating dry and wet seasons (fig. 2.16). Here, however, seasonal drought combines with another important physical factor, fire. The rains come in summer and are accompanied by intense lightning. This lightning often starts fires, particularly at the beginning of the wet season when the savanna is tinder dry. These fires kill young trees while the grasses survive

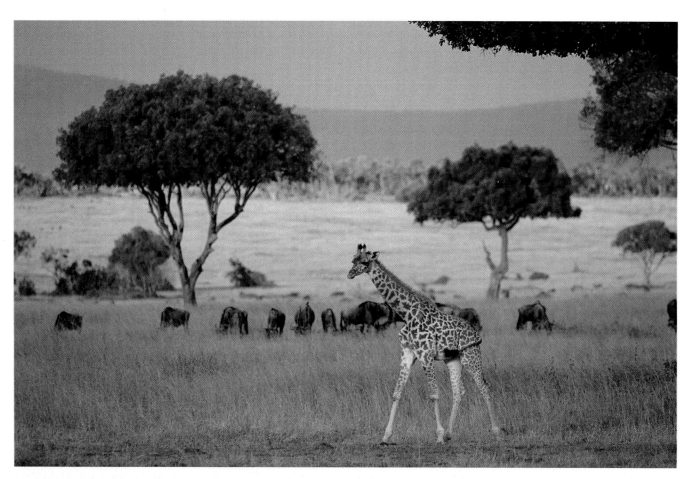

**Figure 2.15** Tropical savanna and grazers in Kenya. The tropical savanna landscape is partially maintained by periodic fires that help control the density of woody vegetation.

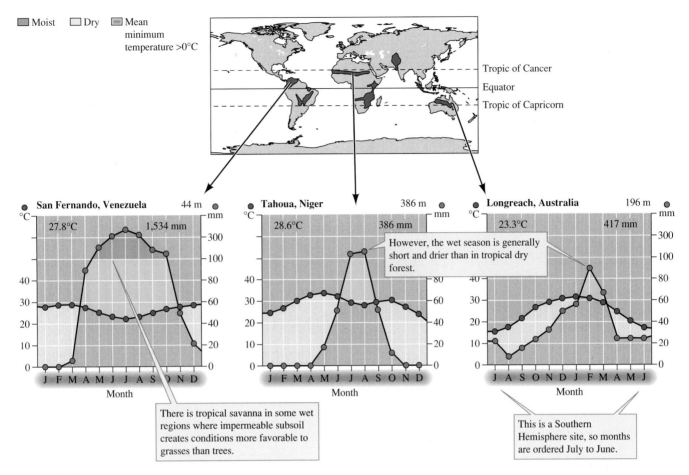

**Figure 2.16** Tropical savanna distribution and climate.

and quickly resprout. Consequently, fires help maintain the tropical savanna as a landscape of grassland and scattered trees.

The savanna climate is generally drier than that of tropical dry forest. The longer dry season and lower annual precipitation of most savannas are clearly shown by the climate diagrams for Tahoua, Niger, and Longreach, Australia (fig. 2.16). The mean rainfall for these two areas is within the range (300–500 mm) occurring on most savannas. However, the climate diagram for San Fernando, Venezuela (fig. 2.16) shows that some savannas receive as much rainfall as a tropical dry forest. Other savannas occur in areas that are as dry as deserts. What keeps the wet savannas near San Fernando from being replaced by forest and how can savannas persist under desert-like conditions? The answer lies deep in the savanna soils.

## Soils

Soil layers with low permeability to water play a key role in maintaining many tropical savannas. For instance, because a dense, impermeable subsoil retains water near the surface, savannas occur in areas of southwest Africa that would otherwise support only desert. Impermeable soils also help savannas persist in wet areas, particularly in South America. Trees do not move onto savannas where an impermeable subsoil keeps sur-

face soils waterlogged during the wet season. In these landscapes, scattered trees occur only where soils are well drained.

## Biology

As you gaze out across the savanna landscape, think back to the tropical rain forest and dry forest. How is the savanna different? One difference is that trees don't completely dominate the landscape. Compared to tropical forests, the savanna landscape is more two-dimensional. Consequently, a greater proportion of the biological activity on the savanna takes place near ground level.

The tropical savanna is populated by wandering animals that move in response to seasonal and year-to-year variations in rainfall and food availability. The wandering consumers of the Australian savannas include kangaroos, large flocks of birds, and, for at least 40,000 years, humans. During droughts, some of these Australian species travel thousands of kilometers in search of suitable conditions. The African savanna is home to a host of well-known mobile consumers, such as elephants, wildebeest, giraffes, zebras, lions, and, again, humans (see fig. 2.15).

The parklike landscape of the savanna is maintained by a dynamic interplay of physical and biological forces. The

**Figure 2.17** Domestic livestock, such as these cattle on an African savanna, have had a major impact on tropical savannas around the world.

diverse mammalian herbivores of the African savanna harvest all parts of the vegetation, from low herbs to the tops of trees. As noted, fire plays a key role in maintaining the savanna landscape. Frequent fires have selected for fire resistance in the savanna flora. The few tree species on the savanna resist fire well enough to be unaffected by low-intensity fires.

## Human Influences

Humans are, in some measure, a product of the savanna and the savanna, in turn, has been influenced by human activity.

One of the factors that forged an indelible link between us and this biome is fire. Long before the appearance of hominids, fire played a role in the ecology of the tropical savanna. Later, the savanna was the classroom where early humans observed and learned to use, control, and make fire. Eventually, humans began to purposely set fire to the savanna, which, in turn, helped to maintain and spread the savanna itself. We had entered the business of large-scale manipulation of nature.

Originally, humans subsisted on the savanna by hunting and gathering. In time, they shifted from hunting to pastoralism, replacing wild game with domestic grazers and browsers. Today, livestock ranching is the main source of livelihood in all the savanna regions. In Africa, livestock raising has coexisted with wildlife for millennia. In modern-day subsaharan Africa, however, the combination of growing human populations, high density of livestock, and drought has devastated much of the region known as the Sahel (fig. 2.17).

## Desert

In the spare **desert** landscape, sculpted by wind and water, the ecologist grows to appreciate geology, hydrology, and climate as much as organisms (fig. 2.18). In the desert, drought and flash floods, and heat and bitter cold, often go hand in hand. Yet, the often repeated description of life in the desert as "life on the edge" betrays an outsider's view. Life in the desert is not luxuriant, but it does not follow that living conditions there are necessarily harsh. For many species, the desert is the center of their world, not the edge. In their own way, many desert organisms flourish on meager rations of water, high temperatures, and saline soils. To understand life in the desert, the ecologist must see it from the perspective of its natural inhabitants. The ecologist who can peer out at the desert environment

**Figure 2.18** Life on the edge. Two dormant acacia trees living on the boundary between gravel plain and sand dunes in the Namib Desert of southwestern Africa.

from under the skin of a cactus or sand viper is on the threshold of understanding.

## Geography

Deserts occupy about 20% of the land surface of the earth. Two bands of deserts ring the globe, one at about 30° N latitude and one at about 30° S (fig. 2.19). These bands correspond to latitudes where dry subtropical air descends (see fig. 2.4), drying the landscape as it spreads north and south. Other deserts are found either deep in the interior of continents, for example, the Gobi of central Asia, or in the rain shadow of mountains, for example, the Great Basin Desert of North America. Still others are found along the cool western coasts of continents, for example, the Atacama of South America and the Namib of southwestern Africa, where air circulating across a cool ocean delivers a great deal of fog to the coast but little rain.

## Climate

Environmental conditions vary considerably from one desert to another. Some, such as the Atacama and central Sahara, receive very little rainfall and fit the stereotype of deserts as extremely dry places. Other deserts, such as some parts of the Sonoran Desert of North America, may receive nearly 300 mm of rainfall annually. Whatever their mean annual rainfall, however, water loss in deserts due to evaporation and transpiration by plants exceeds precipitation during most of the year.

Figure 2.19 includes the climate diagrams of two hot deserts. Notice that drought conditions prevail during all months and that during some months average temperatures exceed 30°C at both Yuma, Arizona, and Faya Largeau, Chad. The maximum shade temperatures in any biome, greater than 56°C, were recorded in the deserts of North Africa and western North America. However, some deserts can be bitterly cold. For example, average winter temperatures at Dzamiin Uuded, Mongolia, in the Gobi Desert of central Asia sometimes fall to –20°C. Notice that the average annual temperature at Dzamiin Uuded is only 3.6°C and the growing season, as indicated by the months shaded red on the climate diagram, lasts only 5 months. The relatively moist period at Dzamiin Uuded occurs during the cold season when average temperatures are below 0°C.

## Soils

Desert plants and animals can turn this landscape into a mosaic of diverse soils. Desert soils are generally so low in organic matter that they are sometimes classified as **lithosols,** which means stone or mineral soil. However, the soils under desert shrubs often contain large amounts of organic matter and form

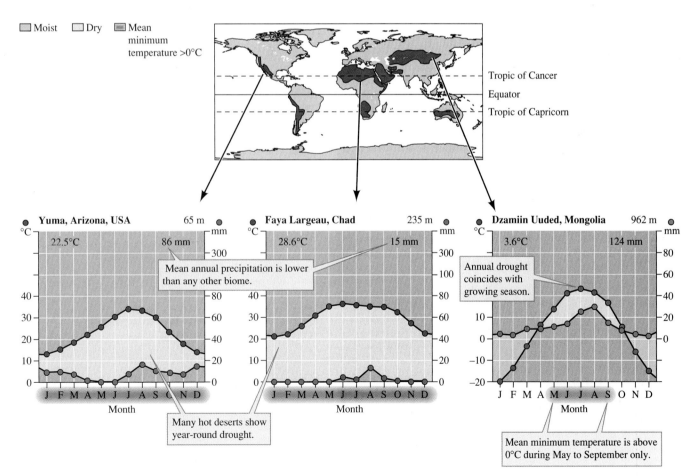

**Figure 2.19**  Desert distribution and climate.

islands of fertility. Desert animals can also affect soil properties. For example, in North America, kangaroo rats change the texture and elevate the nutrient content of surface soils by burrowing and hoarding seeds. In Middle Eastern deserts, porcupines and isopods strongly influence a variety of soil properties.

Desert soils, particularly those in poorly drained valleys and lowlands, may contain high concentrations of salts. Salts accumulate in these soils as water evaporates from the soil surface, leaving behind any salts that were dissolved in the water. Salt accumulation increases the aridity of the desert environment by making it harder for plants to extract water from the soils. As desert soils age they tend to form a calcium carbonate–rich hardpan horizon called **caliche.** The extent of caliche formation has proved a useful tool for aging these soils.

## Biology

The desert landscape presents an unfamiliar face to the visitor from moist climates. Plant cover is absent from many places, exposing soils and other geological features. Where there is plant cover, it is sparse. The plants themselves look unfamiliar. Desert vegetation often cloaks the landscape in a gray-green mantle. This is because many desert plants protect their photosynthetic surfaces from intense sunlight and reduce evaporative water losses with a dense covering of plant hairs. Other plant adaptations to drought include small leaves, producing leaves only in response to rainfall and then dropping them during intervening dry periods, or having no leaves at all (fig. 2.20). Some desert plants avoid drought almost entirely by remaining dormant in the soil as seeds, which germinate and grow only during infrequent wet periods.

In deserts, animal abundance tends to be low but diversity can be high. Most desert animals use behavior to avoid environmental extremes. In summer, many avoid the heat of the day by being active at dusk and dawn or at night. In winter, the same species may be active during the day. Animals (as well as plants) use body orientation to minimize heat gain in the summer. What might the adaptations of desert species teach us about living in environments in which water is scarce?

## Human Influences

Many exquisitely adapted human cultures have arisen independently in the deserts of North America, Australia, Africa, and Asia. Desert peoples have flourished where nature is stingiest. Compared to true desert species, however, humans are profligate water users. Consequently, human populations in desert regions are concentrated around oases and river val-

(a)

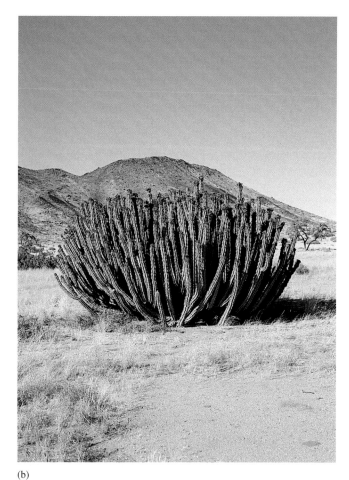

(b)

**Figure 2.20**  Similarity among desert plants: (*a*) cactus in North America, (*b*) *Euphorbia* in Africa.

leys, and wherever they live, desert people dream of the sound of water. That dreaming, gone awry, litters desert landscapes around the world with outrageous fountains, ancient and modern water diversions, and failed agricultural schemes to make the desert bloom. Pushed, the desert blooms. Unfortunately, many desert landscapes have been pushed until they now grow little but salt crystals.

The desert is the one biome that, because of human activity, is increasing in area. We must stop the spread of deserts that comes at the expense of other biomes. We must also establish a balanced use of deserts that safeguards their inhabitants, human and nonhuman alike.

# Mediterranean Woodland and Shrubland

The **Mediterranean woodland and shrubland** climate was the climate of the classical Greeks and the coastal Native American tribes of Old California. The mild temperate climate experienced by these cultures was accompanied by high biological richness (fig. 2.21). The richness of the Mediterranean woodland flora is captured by a folk song from the Mediterranean region that begins: "Spring has already arrived. All the countryside will bloom; a feast of color!" To this visual feast, Mediterranean woodlands and shrublands around the Mediterranean Sea add a chorus of bird song and the smells of aromatic plants, including rosemary, thyme, and laurel.

## *Geography*

Mediterranean woodlands and shrublands occur on all the continents except Antarctica (fig. 2.22). They are most extensive around the Mediterranean Sea and in North America, where

**Figure 2.21** A Mediterranean woodland in California shown during the cool moist season, when the herbaceous vegetation is still green.

they extend from California into northern Mexico. They are also found in central Chile, southern Australia, and southern Africa. Under present climatic conditions Mediterranean woodlands and shrublands grow between about 30° and 40° latitude. This position places the majority of this biome north of the subtropical deserts in the Northern Hemisphere, and south of them in the Southern Hemisphere. The farflung geographic distribution of Mediterranean woodland and shrubland is reflected in the diversity of names for this biome. In western North America, it is called chaparral. In Spain, the most common name for Mediterranean woodland and shrubland is *matoral*. Farther east in the Mediterranean basin the biome is referred to as *garrigue*. Meanwhile in the Southern Hemisphere, South Africans call the biome *fynbos*, while Australians refer to at least one form of it as *mallee*. While the names for this biome vary widely, its climate does not.

## *Climate*

The Mediterranean woodland and shrubland climate is cool and moist during fall, winter, and spring. In most regions the Mediterranean woodland and shrubland summers are hot and dry. The danger of frost varies considerably from one Mediterranean woodland and shrubland region to another. When they do occur, however, frosts are usually not severe. The combination of dry summers and dense vegetation, rich in essential oils, creates ideal conditions for frequent and intense fires.

## *Soils*

The soils of Mediterranean woodlands and shrublands are generally of low to moderate fertility and have a reputation for being fragile. Some soils, such as those of the South African fynbos, have exceptionally low fertility. Fire in the Mediterranean woodlands and shrublands of southern California can cause 40-fold increases in soil losses through erosion. Fire coupled with overgrazing has stripped the soil from some Mediterranean woodland and shrubland landscapes. Elsewhere, these landscapes, under careful stewardship, have maintained their integrity for thousands of years.

## *Biology*

The plants and animals of Mediterranean woodlands and shrublands, like their desert neighbors, show several adaptations to drought. Trees and shrubs are typically evergreen and have small, tough leaves. This vegetation conserves both water and nutrients; many plants of Mediterranean woodlands and shrublands have well-developed mutualistic relationships with microbes that fix atmospheric nitrogen.

The process of decomposition is greatly slowed during the dry summer and then started again with the coming of fall and winter rains. Curiously, this intermittent decomposition may speed the process sufficiently so that average rates of decomposition are comparable to those in temperate forests

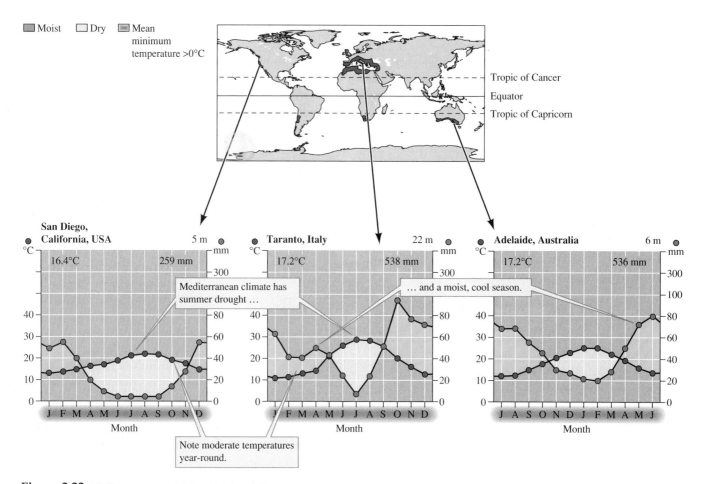

**Figure 2.22** Mediterranean woodland and shrubland distribution and climate.

Fire, a common occurrence in Mediterranean woodlands and shrublands, has selected for fire-resistant plants. Many Mediterranean woodland trees have thick, tough bark that is resistant to fire (fig. 2.23). In contrast, many shrubs in Mediterranean woodlands are rich in oils and burn readily but resprout rapidly. Most herbaceous plants grow during the cool, moist season and then die back in summer, thus avoiding both drought and fire.

The animals of the Mediterranean woodlands and shrublands, both vertebrates and invertebrates, are highly diverse. In addition to resident species, many migratory birds and some migratory insects spend the winter in Mediterranean woodland and shrubland climates, as do some human populations. The plants and animals of the far-flung Mediterranean woodlands have been derived from diverse evolutionary lineages. For example, in the Mediterranean woodlands and shrublands of North America, the native browsers, herbivores that feed on the buds, twigs, and bark of woody plants, are deer. Around the Mediterranean Sea, in addition to deer, there are wild sheep and goats. In southern Africa, the native browsers are small antelope, and in Australia, they are kangaroos.

**Figure 2.23** The thick bark of the cork oak of the Mediterranean region protects the tree from fire.

## Human Influences

Human activity has had a substantial influence on the structure of landscapes in Mediterranean woodlands and shrublands. For example, the open oak woodlands of southern Spain and Portugal are the product of an agricultural management system that is thousands of years old. In this system, cattle graze on grasses, pigs consume acorns produced by the oaks, and cork is harvested from cork oaks as a cash crop. Selected areas are planted in wheat once every 5 to 6 years and allowed to lie fallow the remainder of the time. This system of agriculture, which emphasizes low-intensity cultivation and long-term sustainability, may offer clues for long-term sustainable agriculture in other regions.

Whether in southern France or southern California, people generally find the Mediterranean woodland and shrubland climate agreeable. High population densities coupled with a long history of human occupation have left an indelible mark on Mediterranean woodlands and shrublands. Early human impacts included clearing of forests for agriculture, setting fires to control woody species and encourage grass, harvesting brush for fuel, and grazing and browsing by domestic livestock. Today, Mediterranean woodlands and shrublands around the world are being covered by human habitations. In Mediterranean woodlands and shrublands, as is often the case with fragile things that we hold dear, enjoyment and destruction are close kin.

## Temperate Grassland

In their original state, **temperate grasslands** extended unbroken over vast areas (fig. 2.24). Nothing this side of the open sea feels quite like standing in the middle of unobstructed prairie, under a dome of blue sky. It is no accident that early visitors from forested Europe and eastern North America often referred to the prairie in the American Midwest as a "sea of grass" and to the wagons that crossed them as "prairie schooners." Prairies were the home of the bison and pronghorn and of the nomadic cultures of Eurasia and North America.

## Geography

Temperate grassland is the largest biome in North America and is even more extensive in Eurasia (fig. 2.25). In North America, the prairies of the Great Plains extend from southern Canada to the Gulf of Mexico and from the Rocky Mountains to the deciduous forests of the east. Additional grasslands are found on the Palouse prairies of Idaho and Washington and in the central valley and surrounding foothills of California. In Eurasia, the temperate grassland biome forms a virtually unbroken band from eastern Europe all the way to eastern China. In the Southern Hemisphere, temperate grassland occurs in Argentina, Uruguay, southern Brazil, and New Zealand.

## Climate

Temperate grasslands receive between 300 and 1,000 mm of precipitation annually. Though wetter than deserts, temperate grasslands do experience drought, and droughts may persist for several years. The maximum precipitation usually occurs in summer during the height of the growing season. This summer peak in precipitation is clearly shown in the climate diagrams for Manhattan, Kansas; Magnitogorsk, Russia; and Taiyuan, China (fig. 2.25). Notice that the climate diagrams for these three areas show a significantly shorter growing season compared to most of the areas graphed so far. Winters in temperate grasslands are generally cold and summers hot.

**Figure 2.24**  Pronghorn, native grazers of the temperate grasslands of North America.

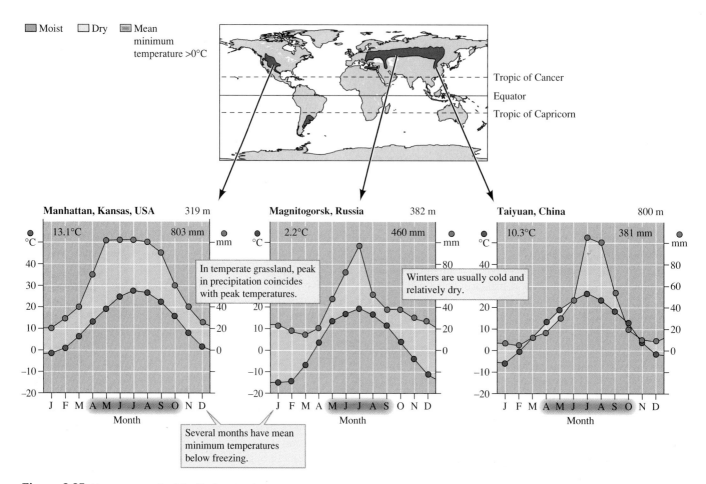

**Figure 2.25** Temperate grassland distribution and climate.

## Soils

Temperate grassland soils are derived from a wide variety of parent materials. The best temperate grassland soils are deep, basic or neutral, and fertile and contain large quantities of organic matter. The black prairie soils of North America and Eurasia, famous for their fertility, contain the greatest amount of organic matter. The brown soils of the more arid grasslands contain less organic matter.

## Biology

Temperate grassland is thoroughly dominated by herbaceous vegetation. Drought and high summer temperatures encourage fire. As in tropical savannas, fire helps exclude woody vegetation from temperate grasslands, where trees and shrubs are often limited to the margins of streams and rivers. In addition to grasses, which make up the bulk of vegetative biomass, there can be a striking diversity of other herbaceous vegetation. Spring graces temperate grasslands with showy anemones, ranunculus, iris, and other wild flowers; up to 70 species can bloom simultaneously on the species-rich North American prairie. The height of grassland vegetation varies from about 5 cm in dry, short-grass prairies to over 200 cm in the wetter tall-grass prairies. The root systems of grasses and forbs form a dense network of sod that resists invasion by both trees and the plow.

Temperate grasslands once supported huge herds of roving herbivores: bison and pronghorns in North America; wild horses and Saiga antelope in Eurasia (see fig. 2.24). As in the open sea, the herbivores of the open grassland banded together in social groups; as did their attendant predators, the steppe and prairie wolves. The smaller animals, such as grasshoppers and mice, inconspicuous among the herbaceous vegetation, were even more numerous than the large herbivores. Grassland animals of intermediate size generally had one of two lifestyles: there were the burrowing, like the badger and prairie dog, and the fleet, like the swift fox and prairie falcon. With domestication of the horse, the human cultures of temperate grasslands joined this second group.

## Human Influences

The first human populations on temperate grasslands were nomadic hunters. Next came the nomadic herders. Later, with their plows, came the farmers, who broke the sod and tapped into fertile soils built up over thousands of years. Under the plow, temperate grasslands have produced some of the most

**Figure 2.26** Once the most extensive biome in North America, temperate grasslands have been largely converted to agriculture.

fertile farm lands on earth and fed much of the world (fig. 2.26). However, much of this productivity depends on substantial additions of inorganic fertilizers, and we are "mining" the fertility of prairie soils. For example, prairie soils have lost as much as 35% to 40% of their organic matter in just 35 to 40 years of cultivation. In addition, the more arid grasslands, with their frequent droughts, do not appear capable of supporting sustainable farming. The future of agriculture in temperate grasslands hinges on several unanswered questions, among them: Can the losses of organic matter and nutrients be reversed? What level of agricultural production can be sustained over the long term?

# Temperate Forest

Old-growth **temperate forest** offers *Homo sapiens* an alternative measure (fig. 2.27). The largest living organisms on earth, perhaps the largest that have ever lived, the sequoias of western North America and the giant *Eucalyptus* trees of southern Australia, live in temperate forests. The temperate forests of eastern North America, Europe, and Asia still harbor ancient trees that are no less impressive. Enter the subdued light of this cool, moist realm, this world of mushrooms and decaying leaves, and feel yourself shrink before the giants of the biosphere. At dusk, in the heart of old-growth forest, it is easy to understand how cultures around the earth came to make the temperate forest the haunt of diverse mythical creatures such as the nymphs and elves of European folk tales. If you can, visit one of the remaining stands of old-growth temperate forest. There, among giant and ancient trees, you may find a fresh perspective on yourself and life.

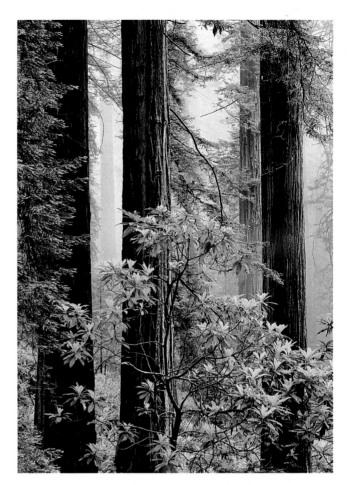

**Figure 2.27** Old-growth redwood forest in western North America. Redwoods are the tallest trees in the world, with some individual trees growing to heights of over 100 m.

## Geography

Temperate forest can be found between 30° and 55° latitude. However, the majority of this biome lies between 40° and 50° (fig. 2.28). In Asia, temperate forest originally covered much of Japan, eastern China, Korea, and eastern Siberia. In western Europe, temperate forests extended from southern Scandinavia to northwestern Iberia and from the British Isles through eastern Europe. North American temperate forests are found from the Atlantic sea coast to the Great Plains and reappear on the West Coast as temperate coniferous forests that extend from northern California through southeastern Alaska. In the Southern Hemisphere, temperate forests are found in southern Chile, New Zealand, and southern Australia.

## Climate

Temperate forests, which may be either coniferous or deciduous, occur where temperatures are not extreme and where annual precipitation averages anywhere from about 650 mm to over 3,000 mm (fig. 2.28). These forests generally receive

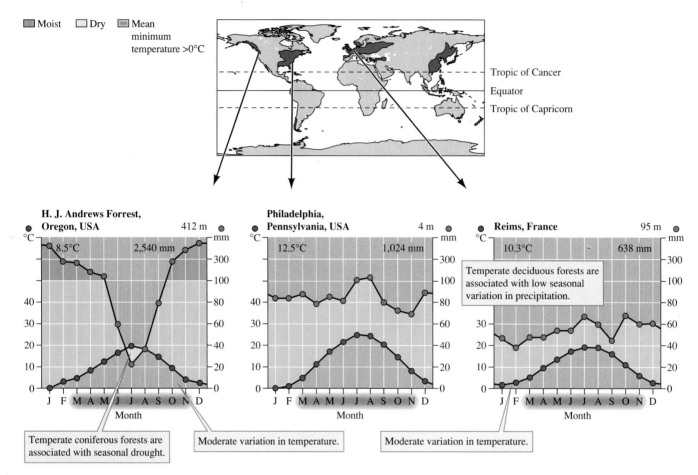

**Figure 2.28**  Temperate forest geography and climate.

more winter precipitation than temperate grasslands. Deciduous trees usually dominate temperate forests, where the growing season is moist and at least 4 months long. In deciduous forests, winters last from 3 to 4 months. Though snowfall may be heavy, winters in deciduous forests are relatively mild. Where winters are more severe or the summers drier, conifers are more abundant than deciduous trees. The temperate coniferous forests of the Pacific Coast of North America receive most of their precipitation during fall, winter, and spring and are subject to summer drought. Summer drought is shown clearly in the climate diagram for the H. J. Andrews Forest of Oregon (fig. 2.28). The few deciduous trees in these coniferous forests are largely restricted to streamside environments, where water remains abundant during the drought-prone growing season.

## Soils

Temperate forest soils are usually fertile. The most fertile soils in this biome develop under deciduous forests, where they are generally neutral or slightly acidic and rich in both organic matter and inorganic nutrients. Rich soils may develop under coniferous forests but conifers are also able to grow on poorer, acidic soils. Nutrient movement between soil and vegetation tends to be slower and more conservative in coniferous forests;

nutrient movement within deciduous forests is generally more dynamic. For example, each year, temperate deciduous forests recycle about twice the amount of nitrogen recycled by temperate coniferous forests of similar age.

## Biology

While the diversity of trees found in temperate forests is lower than that of tropical forests, temperate forest biomass can be as great, or greater. Like tropical rain forests, temperate forests are vertically stratified. The lowest layer of vegetation, the herb layer, is followed by a layer of shrubs, then shade-tolerant understory trees, and finally the canopy, formed by the largest trees. The height of this canopy varies from approximately 40 m to over 100 m. Birds, mammals, and insects make use of all layers of the forest from beneath the forest floor through the canopy. Small arboreal mammals such as deer mice, tree squirrels, and flying squirrels use the tree canopy. Other mammals such as deer, bear, and fox make their livings on the forest floor. Still others burrow into the rich forest soil. But the most important consumers of all are the fungi and bacteria, largely unnoticed members of entirely different kingdoms. They, along with a diversity of microscopic invertebrate animals, consume the large quantities of wood stored on the floor of old-growth temperate forest (fig. 2.29). The activities of these

**Figure 2.29** Massive amounts of wood are stored on the forest floor of the virgin temperate coniferous forests of the Pacific Northwest, such as this one in southeastern Alaska.

organisms recycle nutrients, a process upon which the health of the entire forest depends. Thus, the temperate forest, realm of the giants of the biosphere, emerges as a partnership of the great and the very small.

## *Human Influences*

What, besides being large cities, do Tokyo, Beijing, Moscow, Warsaw, Berlin, Paris, London, New York, Washington, D.C., Boston, Toronto, Chicago, and Seattle have in common? They are all built on lands that once supported a temperate forest. Major population centers have grown up in the temperate forest regions of Europe, eastern China, Japan, and North America. The first human settlements in temperate forests were concentrated along forest margins, usually along streams and rivers. Eventually, agriculture was practiced in these forest clearings, and animals and plant products were harvested from the surrounding forest. This was the circumstance several thousand years ago, in Europe and Asia, and five centuries ago, in North America. Since those times, most of the ancient forests have fallen before ax and saw. For example, the Black Forest of central Europe, where forest still survives, has been largely replaced by tree plantations. The true Black Forest, the archetype of wild and tangled brooding nature, lives on only in legends and fairy tales as part of our collective memory. The ancient forests of North America have not fared much better. Few tracts of the virgin deciduous forest that once covered most of the eastern half of the continent remain, and disparate interests struggle over the fate of the remaining 1% to 2% of old-growth forests in western North America. Fortunately, forests are living, self-renewing systems because now there is nearly as much demand for wild space as for "forestry products." If we are decisive, we can have both.

## Boreal Forest

The **boreal forest,** or **taiga,** is a world of wood and water that covers over 11% of the earth's land area. This biome extends right around the globe in a repeating pattern: forest-water, forest-water, etc. (fig. 2.30). On the surface, the boreal forest is the essence of monotony. However, if you pay attention you are rewarded with plenty of variety. In places, the trees stand so close together you can barely walk through them. Elsewhere, so many trees have been toppled by wind that you

**Figure 2.30** Boreal forests, such as this one in Siberian Russia, are dominated by a few species of conifer trees.

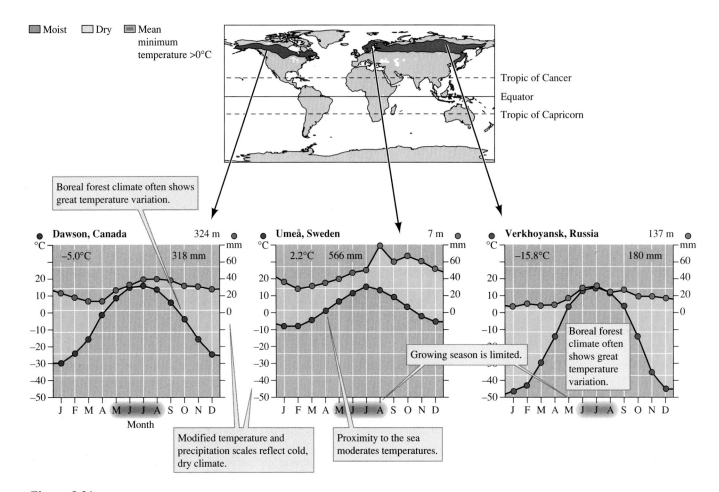

**Figure 2.31** Boreal forest geography and climate.

can walk on their piled trunks, 1 to 2 m above the ground, for many kilometers. In still other places, the forest is open and you can wander wherever you like on its soft floor of needles and duff. Here and there, where light penetrates, are berry bushes of many varieties where wildlife and humans alike pause and snack. A trek through a boreal forest eventually leads to the shore of a lake or river, where shade and cover give way to light and space. Along the lake margins grow willows and other water- and light-loving plants. The summer forest is colored green, gray, and brown; the autumn adds brilliant splashes of yellow and red; and the long northern winter turns the boreal forest into a land of white solitude.

## Geography

Boreal comes from the Greek word for north, reflecting the fact that boreal forests are confined to the Northern Hemisphere. Boreal forests extend from Scandinavia, through European Russia, across Siberia, to central Alaska, and across all of central Canada in a band between 50° and 65° N latitude (fig. 2.31). These forests are bounded in the south either by temperate forests or temperate grasslands and in the north by tundra. Fingers of boreal forest follow the Rocky Mountains south along the spine of North America, and patches of boreal forest reappear on the mountain slopes of south-central Europe and Asia.

## Climate

The boreal forest is found where winters are too long, usually longer than 6 months, and the summers too short to support temperate forest (fig. 2.31). The boreal forest zone includes some fairly moderate climates, such as that at Umeå, Sweden, where the climate is moderated by the nearby Baltic Sea. However, boreal forests are also found in some of the most variable climates on earth. For instance, the temperature at Verkhoyansk, Russia, in central Siberia, ranges from about −70°C in winter to over 30°C in summer, an annual temperature range of about 100°C! Precipitation in the boreal forest is moderate, ranging from about 200 to 600 mm. Yet, because of low temperatures and long winters, evaporation rates are low, and drought is either infrequent or brief. When droughts do occur, however, forest fires can devastate vast areas of boreal forest.

## Soils

Boreal forest soils tend to be of low fertility, thin, and acidic. Low temperatures and low pH impede decomposition of plant litter and slow the rate of soil building. As a consequence, nutrients are largely tied up in a thick layer of plant litter that carpets the forest floor. In turn, most trees in boreal forests have a dense network of shallow roots that, along with associated mycorrhizal fungi, tap directly into the nutrients bound up in this litter layer. The topsoil, which underlies the litter layer, is thin. In the more extreme boreal forest climates, the subsoil is permanently frozen in a layer of "permafrost" that may be several meters thick.

## Biology

The boreal forest is generally dominated by evergreen conifers such as spruce, fir, and, in some places, pines. Larch, a deciduous conifer, dominates in the most extreme Siberian climates. Deciduous aspen and birch trees grow here and there in mature conifer forests and may dominate the boreal forest during the early stages of recovery following forest fires. Willows grow along the shores of rivers and lakes. There is little herbaceous vegetation under the thick forest canopy, but small shrubs such as blueberry and shrubby junipers are common.

The boreal forest is home to many animals. This is the winter home of migratory caribou and reindeer and the year-round home of moose and woodland bison. The wolf is the major predator of the boreal forest. This biome is also inhabited by black bears and grizzly bears in North America and the brown bear in Eurasia. A variety of smaller mammals such as lynx, wolverine, snowshoe hare, porcupines, and red squirrels also live in boreal forests. The boreal forest is the nesting habitat for many birds that migrate from the tropics each spring and the year-round home of other birds such as crossbills and spruce grouse.

Our survey of the biosphere has taken us far from the rain forest, where we started. Let's reflect back on the tropical rain forest and where we've come. What has changed? Well, we're still in forest but a very different one. In the rain forest, a single hectare could contain over 300 species of trees; here, in the boreal forest, you can count the dominant trees on one hand. What about epiphytes and vines? The vines are gone and the epiphytes are limited to lichens and some mistletoe. In addition, most of the intricate relationships between species that we saw in the rain forest are absent. All the trees are wind pollinated, and none produce fleshy fruits like bananas or papayas. Now listen to the two forests at night. The rich tropical rain forest chorus is hushed. The silence of the boreal forest is broken by few animal voices—the howl of a wolf, the hoot of an owl, the cry of the loon, soloists of the northern forest—accompanied by incessant wind through the trees.

## Human Influences

Ancient cave paintings in southern France and northern Spain, made during the last ice age when the climate was much colder, reveal that humans have lived off boreal forest animals, such as the migratory reindeer, for tens of thousands of years. In Eurasia, from Lapland in Scandinavia to Siberia, hunting of reindeer eventually gave way to domestication and herding. In northern Canada and Alaska, where some Native Americans still rely on wild caribou for much of their food, we find a reminder of the earliest human ways of making a living in these northern lands. Northern peoples have also long harvested the berries that grow in abundance in the boreal forest. The many berry dishes of Scandinavia are living testimony to this heritage.

For most of history, human intrusion in the boreal forest was relatively light. More recently, however, harvesting of both animals and plants has become intense. Hunting and trapping have devastated many wildlife species. Boreal forests are being rapidly cut for lumber and pulp (fig. 2.32). Human influences on the boreal forest are now substantial.

**Figure 2.32** Deforestation in the boreal forest.

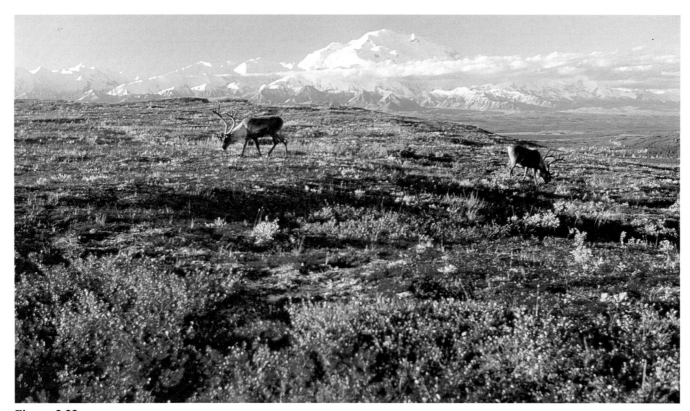

**Figure 2.33** The vegetation of the tundra is made up mostly of low-growing mosses, lichens, perennial herbaceous plants, and dwarf willows and birches.

# Tundra

Follow the caribou north as they leave their winter home in the boreal forest and you eventually reach an open landscape of mosses, lichens, and dwarf willows, dotted with small ponds and laced with clear streams (fig. 2.33). This is the **tundra.** If it is summer and surface soils have thawed, your progress will be cushioned by a spongy mat of lichens and mosses and punctuated by sinking into soggy accumulations of peat. The air will be filled with the cries of nesting birds that have come north to take advantage of the brief summer population explosion of their plant and animal prey. You may find the air surprisingly warm. Just as often, you will feel the bite of a midsummer snowstorm. After the long, deep winter, the midnight sun signals an annual celebration of light and life.

## *Geography*

Like the boreal forest, the arctic tundra rings the top of the globe, covering most of the lands north of the Arctic Circle (fig. 2.34). The tundra extends from northern-most Scandinavia, across northern European Russia, through northern Siberia, and right across northern Alaska and Canada. It reaches far south of the Arctic Circle in the Hudson Bay region of Canada and is also found in patches on the coast of Greenland and in northern Iceland.

## *Climate*

The tundra climate is typically cold and dry. However, temperatures are not quite as extreme as in the boreal forest; the tundra usually doesn't get quite as cold in the winter or quite as warm in the summer. As a consequence, though winter temperatures are less severe, the summers are shorter. Notice that the red months indicating the growing season in figure 2.34 are shorter for tundra sites compared to boreal forest sites in the same region (see fig. 2.31): Point Barrow versus Dawson, Vardo versus Umeå, or Tiksi versus Verkhoyansk. In this comparison, Vardo emerges as the site with the most moderate temperatures, which reflects the moderating influence of the Norwegian Current. Precipitation on the tundra varies from less than 200 mm to a little over 600 mm. Still, because average annual temperatures are so low, precipitation exceeds evaporation. As a consequence, the short summers are soggy and the tundra landscape is alive with ponds and streams.

## *Soils*

Soil building is slow in the cold tundra climate. Because rates of decomposition are low, organic matter accumulates in deposits of peat and humus. Surface soils thaw each summer but are often underlain by a layer of permafrost that may be many meters thick. The annual freezing and thawing of surface soil combines with the actions of water and gravity to

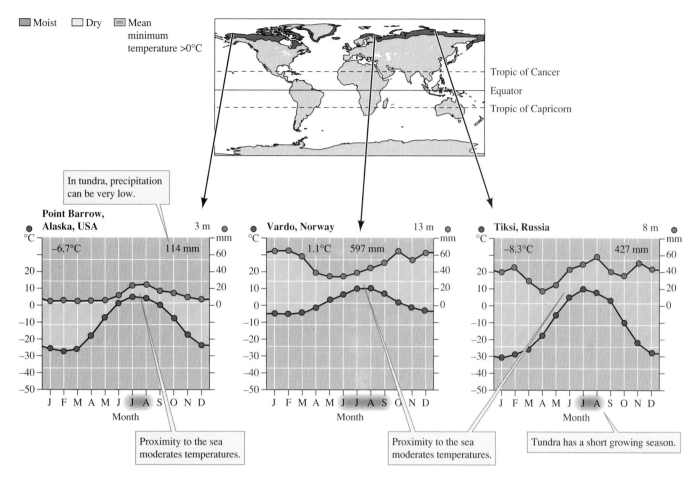

Moist ☐ Dry ☐ Mean minimum temperature >0°C

Tropic of Cancer
Equator
Tropic of Capricorn

In tundra, precipitation can be very low.

**Point Barrow, Alaska, USA**   3 m
−6.7°C   114 mm

**Vardo, Norway**   13 m
1.1°C   597 mm

**Tiksi, Russia**   8 m
−8.3°C   427 mm

Proximity to the sea moderates temperatures.

Proximity to the sea moderates temperatures.

Tundra has a short growing season.

**Figure 2.34** Tundra geography and climate.

produce a variety of surface processes that are largely limited to the tundra. One of these processes, **solifluction,** slowly moves soils down slopes. In addition, freezing and thawing brings stones to the surface of the soil, forming a netlike, or polygonal, pattern on the surface of tundra soils (fig. 2.35).

## *Biology*

The open tundra landscape is dominated by a richly textured patchwork of perennial herbaceous plants, especially grasses, sedges, mosses, and lichens. The lichens, associations of fungi and algae, are eagerly eaten by reindeer and caribou. The woody vegetation of the tundra consists of dwarf willows and birches along with a variety of low-growing shrubs.

The tundra is one of the last biomes on earth that still supports substantial numbers of large native mammals, including caribou, reindeer, musk ox, bear, and wolves. Small mammals such as arctic foxes, weasels, lemmings, and ground squirrels are also abundant. Resident birds such as the ptarmigan and snowy owl are joined each summer by a host of migratory bird species. Insects, though not as diverse as in biomes farther south, are very abundant. Each summer, swarms of mosquitoes and black flies emerge from the many tundra ponds and streams.

**Figure 2.35** Freezing and thawing forms netlike polygons on the surface of the tundra as seen here in an aerial photo of Alaska.

## Human Influences

Until recently, human presence in the tundra was largely limited to small populations of hunters and nomadic herders. As a consequence, the tundra has been viewed as one of the last pristine areas of the planet. Recently, however, human intrusion has increased markedly. This biome has been the focus of intense oil exploration and extraction. Airborne pesticides and radionuclides, which originate in distant human population centers, have been deposited on the tundra, sometimes with devastating results. For example, radioactive cesium-137 from the Chernobyl power plant disaster of 1986 was deposited with rainfall, more than 2,000 km away, on the tundra of Norway. In some areas, cesium-137 became so concentrated as it passed through the food chain from lichens to reindeer that both the milk and meat of reindeer were rendered unfit for human consumption. Such incidents have shattered the illusion of the tundra as an isolated biome and the last earthly refuge from human influence.

# Mountains: Islands in the Sky

We now shift our attention to mountains, though they do not represent a specific biome. Because of the environmental changes that occur with altitude, several biomes may be found on a single mountainside. This environmental and biological diversity is something common to mountains. We include mountains here because they often introduce unique environmental conditions and organisms to regions around the globe. No discussion of the biosphere would be complete without an introduction to their unique characteristics.

Everywhere, mountains capture the imagination as places of geological, biological, and climatic diversity and as places with a view (fig. 2.36). You can stand with eagles and gaze on the plains below, an experience that in the days before air travel was unique to mountains. Mountains have long offered refuge for special flora and fauna and humans alike. Like oceanic islands, they offer unique insights into evolutionary and ecological processes.

## Geography

Mountains are built by geological processes, such as volcanism and movements of the earth's crust that elevate and fold the earth's surface. These processes operate with greater intensity in some places than others, and so mountains are concentrated in belts where these geological forces have been at work (fig. 2.37). In the Western Hemisphere, these forces have been particularly active on the western sides of both North and South America, where a chain of mountain ranges extends from northern Alaska across western North America to Tierra del Fuego at the tip of South America. Ancient low mountain ranges occupy the eastern sides of both continents. In Africa, the major mountain ranges are the Atlas Mountains of northwest Africa and the mountains of East Africa that run from the highlands of Ethiopia to southern Africa like beads on a string. In Australia, the flattest of the continents, mountains extend down the eastern side of the continent. Eurasian mountain ranges include the Pyrenees, the Alps, the Caucasus, and of course, the Himalayas, the highest of them all.

**Figure 2.36** Mount Denali, Alaska. Environmental conditions and organisms vary greatly from low to high elevations on mountains.

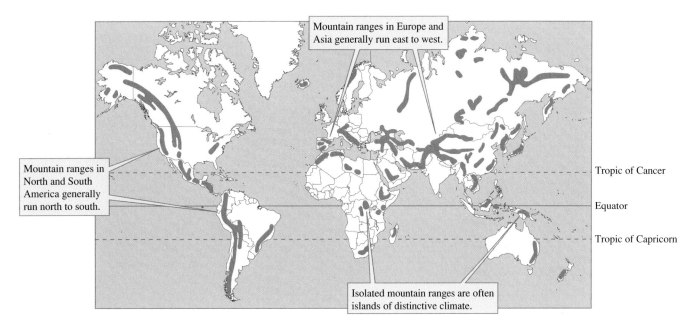

**Figure 2.37**  Mountain geography.

## Climate

On mountains, climates change from low to high elevation, but the specific changes are different at different latitudes. On mountains at middle latitudes, the climate is generally cooler and wetter at higher altitudes (fig. 2.38). In contrast, there is less precipitation at the higher elevations of polar mountains and on some tropical mountains. In other tropical regions, precipitation increases up to some middle elevation and then decreases higher up the mountain. On high tropical mountains, warm days are followed by freezing nights. The organisms on these mountains experience summer temperatures every day and winter temperatures every night. The changes in climate that occur up the sides of mountains have profound influences on the distribution of mountain organisms.

## Soils

Mountain soils change with elevation and have a great deal in common with the various soils we've already discussed. However, some special features are worth noting. First, because of the steeper topography, mountain soils are generally well drained and tend to be thin and vulnerable to erosion. Second, persistent winds blowing from the lowlands deposit soil particles and organic matter on mountains, materials that can make a significant contribution to local soil building. In some locations in the southern Rocky Mountains, coniferous trees draw the bulk of their nutrition from materials carried by winds from the valleys below, not from local bedrock.

## Biology

Climb any mountain that is high enough and you will notice biological and climatic changes. Whatever the vegetation at the base of a mountain, that vegetation will change as you climb and the air becomes cooler. The sequence of vegetation up the side of a mountain may remind you of the biomes we encountered on our journey from the equator to the poles. In the cool highlands of desert mountains in the southwestern United States, you can hike through spruce and fir forests much like those we encountered far to the north. However, what you see on these desert mountains differs substantially from boreal forests. These mountain populations have been isolated from the main body of the boreal forest for over 10,000 years; in the interim, some populations have become extinct, some teeter on the verge of extinction, while others have evolved sufficiently to be recognized as separate species or subspecies. On these mountains, time and isolation have forged distinctive gene pools and mixes of species.

The species on high equatorial mountains are even more isolated. Think for a moment of the geography of high tropical mountains: some in Africa, some in the highlands of Asia, and the Andes of South America. The high-altitude communities of Africa, South America, and Asia share very few species. On the other hand, despite differences in species composition, there are structural similarities among the organisms on these mountains (fig. 2.39). These similarities suggest there may be general rules for associating organisms with environments.

## Human Influences

Because mountains differ in climate, geology, and biota (plants and animals) from the surrounding lowlands, they have been useful as a source of raw materials such as wood, forage for animals, medicinal plants, and minerals. Some of these uses, such as livestock grazing, are highly seasonal. In temperate regions, livestock are taken to mountain pastures during the summer and back down to the lowlands in winter.

■ Moist    □ Dry    ▨ Mean minimum temperature >0°C

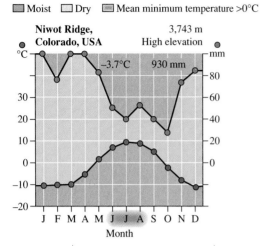

**Niwot Ridge, Colorado, USA**    3,743 m    High elevation

−3.7°C    930 mm

Elevation ↑    Temperature ↓    Precipitation ↑

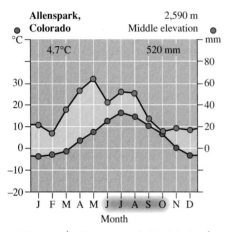

**Allenspark, Colorado**    2,590 m    Middle elevation

4.7°C    520 mm

Elevation ↑    Temperature ↓    Precipitation ↑

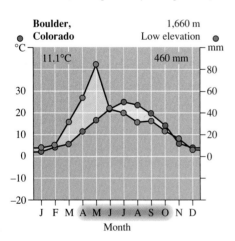

**Boulder, Colorado**    1,660 m    Low elevation

11.1°C    460 mm

**Figure 2.38** Mountain climates along an elevational gradient in the Colorado Rockies. Temperatures decrease and precipitation increases from low to high elevations in these midlatitude mountains.

(a)

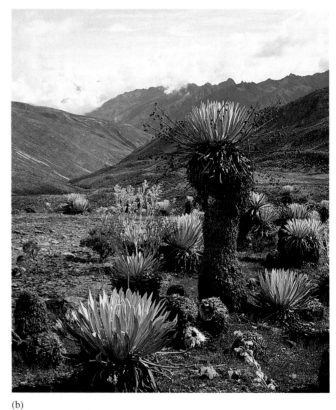

(b)

**Figure 2.39** Convergence among tropical alpine plants: (*a*) *Senecio* trees on Mount Kilimanjaro, Africa; (*b*) *Espeletia* in the Andes of South America.

Human exploitation of mountains has produced ecological degradation in many places and surprising balance in others. Increased human pressure on mountain environments has sometimes created conflict between competing economic interests, between recreation seekers and livestock ranchers, and even between groups of scientists. Because of their compressed climatic gradients and biological diversity, mountains offer living laboratories for the study of ecological responses to climatic variation.

## APPLICATIONS & TOOLS

### Climatic Variation
### and the Palmer Drought Severity Index

In this chapter we've used Walter climate diagrams to represent the climates of earth's biomes. Climate diagrams capture some of the climatically significant differences among the climates experienced by the various biomes. However since they focus on average climatic conditions, they emphasize only one aspect of climate. Recall that climate diagrams plot average (mean) monthly precipitation, which is plotted as a line graph (e.g., fig. 2.38), and average monthly temperature, also plotted as a line graph but connecting red dots. Including mean annual temperature and mean annual precipitation in each climate diagram further reinforces the focus on average conditions. However as we all know, climates everywhere vary substantially from the average conditions presented in climate diagrams.

Here we explore a climatic index, the **Palmer Drought Severity Index,** which can be used to characterize climatic variation. While this index has been historically used to assess drought conditions, it indicates wet periods as well. First, what is a drought? A **drought** can be defined as an extended period of dry weather during which precipitation is reduced sufficiently to damage crops, impair the functioning of natural ecosystems, or cause water shortages for human populations. While such a definition may be sufficient for some needs, climatologists have tried to create quantitative indices of drought. The Palmer Drought Severity Index, or PDSI, is such an index. The PDSI uses temperature and precipitation to calculate moisture conditions relative to long-term averages for a particular region. Negative values of the PDSI reflect drought conditions, while positive values indicate relatively moist periods. What do zero values of the Palmer Drought Severity Index indicate? Values near zero indicate approximately average conditions in a particular region.

Figure 2.40 shows a plot of the Palmer Drought Severity Index for the region around Manhattan, Kansas, from 1895 through 2002. To ease interpretation, negative values of the

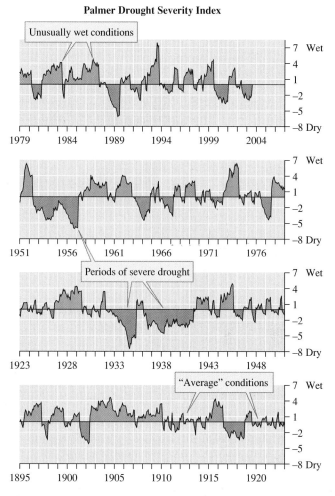

**Figure 2.40** The Palmer Drought Severity Index for Kansas region 3 near Manhattan, Kansas, plotted for the years 1895 to 2002 indicates substantial climatic variation (data from www.cpc.ncep.noaa.gov/oa/climate/onlineprod/drought/main.html).

Palmer Drought Severity Index are shaded red, indicating drought. Periods during which the index was positive are shaded blue, indicating moist conditions. The area of Kansas from which the climate data are plotted in figure 2.40 falls within the temperate grassland biome. What does figure 2.40 suggest about moisture availability in the region around Manhattan, Kansas? One of the most apparent characteristics of this plot is its great variability. Clearly the availability of water in the region is far from constant. Now compare figure 2.40 with the representation of climate for Manhattan, Kansas, shown in figure 2.25. How do the two figures compare? While the climate diagram and the PDSI represent climate from the same geographic location, the climate diagram, because it draws our attention to average climatic conditions, suggests climatic stability. Meanwhile the PDSI shows that the climate around Manhattan, Kansas is in fact highly variable.

Temporal climatic variability is matched or exceeded by climatic variation in space. Spatial variation in climatic

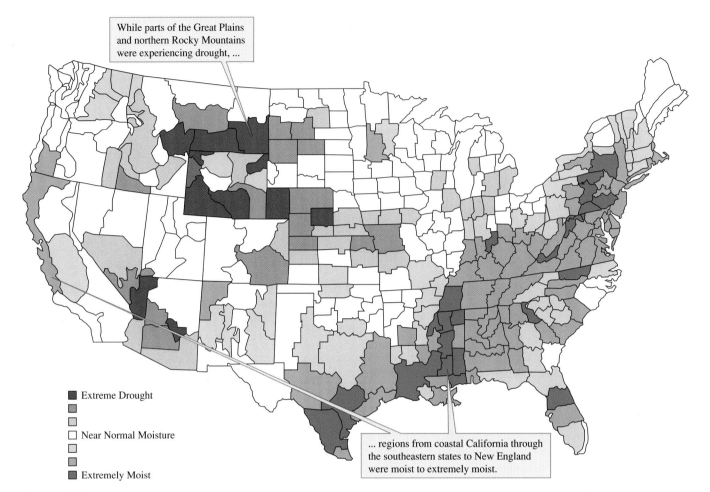

While parts of the Great Plains and northern Rocky Mountains were experiencing drought, ...

■ Extreme Drought
■
□
□ Near Normal Moisture
□
■
■ Extremely Moist

... regions from coastal California through the southeastern states to New England were moist to extremely moist.

**Figure 2.41** Regional variation in moisture conditions for the week of January 4, 2003, as indicated by the Palmer Drought Severity Index (data from www.cpc.ncep.noaa.gov/oa/climate/onlineprod/drought/main.html).

conditions can also be represented using the Palmer Drought Severity Index. For instance, figure 2.41 maps values of the Palmer Drought Severity Index across the United States for the week ending January 4, 2003. Notice that during this period, moisture conditions varied widely across this portion of the North American continent. While some parts of the desert Southwest and northern Rocky Mountain regions experienced severe to extreme drought conditions, other regions of the United States were very moist or extremely moist. In still other regions, the conditions were near normal.

The levels of regional and temporal variability shown on figures 2.40 and 2.41 are not exceptional. Similar levels of spatial variation occur on all continents. However, when considering temporal variation in climate, some regions are climatically more variable than others. For example, those regions under the influence of the El Niño Southern Oscillation (p. 563) are particularly variable. Ecologists study the relationships between organisms and environment. As these examples show, in the study of those relationships both averages and variation in environmental factors need to be considered.

SUMMARY

Natural history is helping with the difficult task of restoring tropical dry forest in Costa Rica. Natural history also formed the foundation upon which modern ecology developed. Because ecological studies continue to be built upon a solid foundation of natural history, this chapter is devoted to the natural history of terrestrial biomes. Biomes are distinguished

primarily by their predominant vegetation and are associated with particular climates.

**Uneven heating of the earth's spherical surface by the sun and the tilt of the earth on its axis combine to produce predictable latitudinal variation in climate.** Because the earth is a sphere, the sun's rays are most concentrated at the latitude where the sun is directly overhead. This latitude changes with the seasons because the earth's axis of rotation is not perpendicular to its plane of orbit about the sun but is tilted approximately 23.5° away from the perpendicular. The sun is directly overhead at the tropic of Cancer, at 23.5° N latitude during the northern summer solstice. During the northern winter solstice the sun is directly overhead at the tropic of Capricorn, at 23.5° S latitude. The sun is directly overhead at the equator during the spring and autumnal equinoxes. During the northern summer the Northern Hemisphere is tilted toward the sun and receives more solar energy than the Southern Hemisphere. During the northern winter, the Northern Hemisphere is tilted away from the sun and the Southern Hemisphere receives more solar energy.

Heating of the earth's surface and atmosphere drives atmospheric circulation and influences global patterns of precipitation. As the sun heats air at the equator, it expands and rises, spreading northward and southward at high altitudes. This high-altitude air cools as it spreads toward the poles, eventually sinking back to the earth's surface. Rotation of the earth on its axis breaks up atmospheric circulation into six major cells, three in the Northern Hemisphere and three in the Southern Hemisphere. These three circulation cells correspond to the trade winds north and south of the equator, the westerlies between 30° and 60° N or S latitude, and the polar easterlies above 60° latitude. These prevailing winds do not blow directly south because of the Coriolis effect.

As air rises at the tropics it cools, and the water vapor it contains condenses and forms clouds. Precipitation from these clouds produces the abundant rains of the tropics. Dry air blowing across the lands at about 30° latitude produces the great deserts that ring the globe. When warm, moist air flowing toward the poles meets cold polar air it rises and cools, forming clouds that produce the precipitation associated with temperate environments. Complicated differences in average climate can be summarized using a climate diagram.

**Soil structure results from the long-term interaction of climate, organisms, topography, and parent mineral material.** Terrestrial biomes are built upon a foundation of soil, a vertically stratified and complex mixture of living and nonliving material. Most terrestrial life depends on soil. Soil structure varies continuously in time and space. Soils are gen-

erally divided into O, A, B, and C horizons. The O horizon is made up of freshly fallen organic matter, including leaves, twigs, and other plant parts. The A horizon contains a mixture of mineral materials and organic matter derived from the O horizon. The B horizon contains clays, humus, and other materials that have been transported from the A horizon. The C horizon consists of weathered parent material.

**The geographic distribution of terrestrial biomes corresponds closely to variation in climate, especially prevailing temperature and precipitation.** The major terrestrial biomes and climatic regimes are: *Tropical rain forest:* Warm; moist; low seasonality; infertile soils; exceptional biological diversity and intricate biological interactions. *Tropical dry forest:* Warm and cool seasons; seasonally dry; biologically rich; as threatened as tropical rain forest. *Tropical savanna:* Warm and cool seasons; pronounced dry and wet seasons; impermeable soil layers; fire important to maintaining dominance by grasses; still supports high numbers and diversity of large animals. *Desert:* Hot or cold; dry; unpredictable precipitation; low productivity but often high diversity; organisms well-adapted to climatic extremes. *Mediterranean woodland and shrubland:* Cool, moist winters; hot, dry summers; low to moderate soil fertility; organisms adapted to seasonal drought and periodic fires. *Temperate grassland:* Hot and cold seasons; peak rainfall coincides with growing season; droughts sometimes lasting several years; fertile soils; fire important to maintaining dominance by grasses; historically inhabited by roving bands of herbivores and predators. *Temperate forest:* Moderate, moist winters; warm, moist growing season; fertile soils; high productivity and biomass; dominated by deciduous trees where growing seasons are moist, winters are mild, and soils fertile; otherwise dominated by conifers. *Boreal forest:* Long, severe winters; climatic extremes; moderate precipitation; infertile soils; permafrost; occasional fire; extensive forest biome, dominated by conifers. *Tundra:* Cold; low precipitation; short, soggy summers; poorly developed soils; permafrost; dominated by low vegetation and a variety of animals adapted to long, cold winters; migratory animals, especially birds, make seasonal use. *Mountains:* Temperature, precipitation, soils, and organisms shift with elevation; mountains are climatic and biological islands.

# Review Questions

1. Daniel Janzen (1981a, 1981b) proposed that the seeds of the guanacaste tree were once dispersed by several species of large mammals that became extinct following the end of the Pleistocene about 10,000 years ago. There may have been other plant species with a similar relationship with large herbivorous mammals. How do you think the distributions of these plant species may have changed from the time of the extinctions of Pleistocene mammals until the introduction of other large herbivores such as horses? How might the introduction of horses about 500 years ago have affected the distribution of these species? How could you test your ideas?

2. Draw a typical soil profile, indicating the principal layers, or horizons. Describe the characteristics of each layer.

3. Describe global patterns of atmospheric heating and circulation. What mechanisms produce high precipitation in the tropics? What mechanisms produce high precipitation at temperate latitudes? What mechanisms produce low precipitation in the tropics?

4. Use what you know about atmospheric circulation and seasonal changes in the sun's orientation to earth to explain the highly seasonal rainfall in the tropical dry forest and tropical savanna

biomes. (Hint: Why does the rainy season in these biomes come during the warmer months?)

5. We focused much of our discussion of biomes on their latitudinal distribution. The reasonably predictable relationship between latitude and temperature and precipitation provides a link between latitude and biomes. What other geographic variable might affect the distribution of temperature and precipitation and, therefore, of biomes?

6. You probably suggested altitude in response to the previous question because of its important influence on climate. Some of the earliest studies of the geographic distribution of vegetation suggested a direct correspondence between latitudinal and altitudinal variation in climate, and our discussion in this chapter stressed the similarities in climatic changes with altitude and latitude. Now, what are some major climatic differences between high altitude at midlatitudes and high altitude at high latitudes?

7. How is the physical environment on mountains at midlatitudes similar to that in tropical alpine zones? How do these environments differ?

8. English and other European languages have terms for four seasons: spring, summer, autumn, and winter. This vocabulary summarizes much of the annual climatic variation at midlatitudes in temperate regions. Are these four seasons useful for summarizing annual climatic changes across the rest of the globe? Look back at the climate diagrams presented in this chapter. How many seasons would you propose for each of these environments? What would you call these seasons?

9. Biologists have observed much more similarity in species composition among boreal forests and among areas of tundra in Eurasia and North America than among tropical rain forests or among Mediterranean woodlands around the globe. Can you offer an explanation of this contrast based on the global distributions of these biomes shown in figures 2.10, 2.22, 2.31, and 2.34?

10. To date, which biomes have been the most heavily affected by humans? Which seem to be the most lightly affected? How would you assess human impact? How might these patterns change during the coming century? (You may need to consult the discussion of human population growth in the Applications & Tools section of chapter 11).

# Suggested Readings

Attenborough, D. 1987. *The First Eden: The Mediterranean World and Man.* New York: Little, Brown.

*A natural history of the Mediterranean region, written as only David Attenborough can. A splendid model for developing a natural history perspective on a particular region.*

Attenborough, D., P. Whitfield, P. D. Moore, and B. Cox. 1989. *The Atlas of the Living World.* Boston: Houghton Mifflin.

*A survey of the biosphere written for the general reader. Richly illustrated and well written.*

Janzen, D. H. 1981. Guanacaste tree seed-swallowing by Costa Rican range horses. *Ecology* 62:587–92.

Janzen, D. H. 1981. *Enterolobium cyclocarpum* seed passage rate and survival in horses, Costa Rican Pleistocene seed dispersal agents. *Ecology* 62:593–601.

*These two papers are excellent case studies involving experimental field ecology, with implications for the management and restoration of tropical dry forest.*

Janzen, D. H. 1983. *Costa Rican Natural History.* Chicago: University of Chicago Press.

*Another regional natural history. This one focuses on one of the most diverse countries in the world—Costa Rica.*

Jenny, H. 1980. *The Soil Resource.* New York: Springer-Verlag.

*This is a classic in the field of soil science. Jenny provides a masterful and lively introduction to the ecology of soils.*

Walter, H. 1985. *Vegetation of the Earth.* 3d ed. New York: Springer-Verlag.

*Heinrich Walter reviews global patterns of climate and their correspondence to major classes of vegetation. He employs the climate diagrams that he developed throughout the presentation.*

Wilson, E. O. 1992. *The Diversity of Life.* New York: W. W. Norton.

*A highly acclaimed synthesis on the patterns and threats to biological diversity—an engaging and provocative discussion.*

# On the Net

Visit this textbook's accompanying website www.mhhe.com/ecology (click on the book's title) to take advantage of practice quizzing, study/writing tips, timely news articles, and additional URLs for research on the topics in this chapter.

Biomes and Environmental Habitats

Tropical Rainforests

Temperate Forests

Savannah

Grasslands

Deserts

Tundra

Taiga

Land Use: Forests and Rangelands

Tropical Rain Forests and Land Use Issues

Nontropical Forests and Land Use Issues

Rangelands and Land Use Issues

Atmosphere, Climate, and Weather

Soil and Its Uses

Restoration Ecology

Geographic Ecology

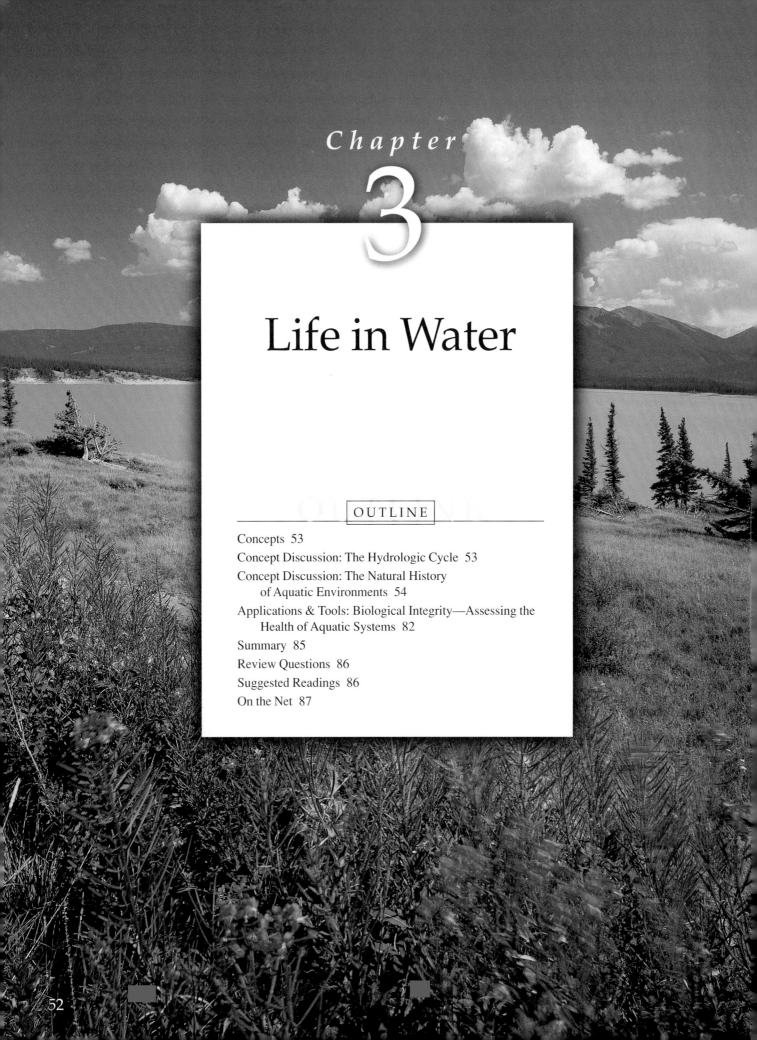

*Chapter*

# 3

# Life in Water

OUTLINE

The names that people around the world have given to our planet reveal a perspective consistent across cultures. Those names, whether in English (earth), Latin (*terra*), Greek (Γεοσ, *geos*), or Chinese ( 地球 , *di qiu*), all refer to land or soil, revealing that cultures everywhere hold a land-centered perspective. The Hawaiians, Polynesian inhabitants of the most isolated specks of land on earth, call the planet *ka honua,* an allusion to a level landing place or dirt embankment. This universal land-centered perspective may partly explain why portraits of earth transmitted from space are so stunning. Those images challenge our sense of place by portraying our planet as a shining blue ball, as a landing place in space covered not by land but mostly by water (fig. 3.1).

Life originated in water but from our perspective as terrestrial organisms, the aquatic realm remains an alien environment governed by unfamiliar rules. In the aquatic environment, life is often most profuse where conditions appear most hostile to us: along cold, wave-swept seacoasts, in torrential mountain streams during the depths of winter, in murky waters where rivers meet the sea. The goal of this chapter is to make this realm more familiar; we'll take a look at the aquatic environment and its inhabitants and gain a general sense of the natural history of life in water that will prepare the way for more detailed and abstract studies of ecology.

**Figure 3.1** From space earth shows itself as a planet covered mostly by water.

## CONCEPTS

- **The hydrologic cycle exchanges water among reservoirs.**
- **The biology of aquatic environments corresponds broadly to variations in physical factors such as light, temperature, and water movements and to chemical factors such as salinity and oxygen.**

## CONCEPT DISCUSSION

### The Hydrologic Cycle

**The hydrologic cycle exchanges water among reservoirs.**

Over 71% of the earth's surface is covered by water. This water is unevenly distributed among aquatic environments such as lakes, rivers, and oceans; most is seawater. The oceans contain over 97% of the water in the biosphere, and the polar ice caps and glaciers contain an additional 2%. Less than 1% is freshwater in rivers, lakes, and actively exchanged groundwater. The situation on earth is indeed as Samuel Coleridge's ancient mariner saw it: "Water, water, everywhere, nor any drop to drink."

The distribution of water across the biosphere is not static, however. Figure 3.2 summarizes the dynamic exchanges called the **hydrologic cycle.** The various aquatic environments such as lakes, rivers, and oceans plus the atmosphere, ice, and even organisms can be considered as "reservoirs" within the hydrologic cycle, places where water is stored for some period of time. The water in these reservoirs is renewed, or turned over.

As a result of the hydrologic cycle, water is constantly entering each reservoir either as precipitation or as surface or subsurface flow and leaving each reservoir either as evaporation or as flow. The hydrologic cycle is powered by solar energy, which drives the winds and evaporates water, primarily from the surface of the oceans. Water vapor cools as it rises from the ocean's surface and condenses, forming clouds. These clouds are then blown by solar-driven winds across the planet, eventually yielding rain or snow, the majority of which falls back on the oceans and some of which falls on land. The water that falls on land has several fates. Some immediately evaporates and reenters the atmosphere; some is consumed by terrestrial organisms; some percolates through the soil to become groundwater; and some ends up in lakes and ponds or in streams and rivers, which eventually find their way back to the sea.

Turnover time is the time required for the entire volume of a particular reservoir to be renewed. Because reservoir size and rates of water exchange differ, water turnover occurs at vastly different rates. The water in the atmosphere turns over about every 9 days. The renewal time for river water, 12 to 20 days, is nearly as rapid. Lake renewal times are longer, ranging anywhere from days to centuries, depending on lake depth, area, and rate of drainage. But the biggest surprise is the renewal time for the largest reservoir of all, the oceans.

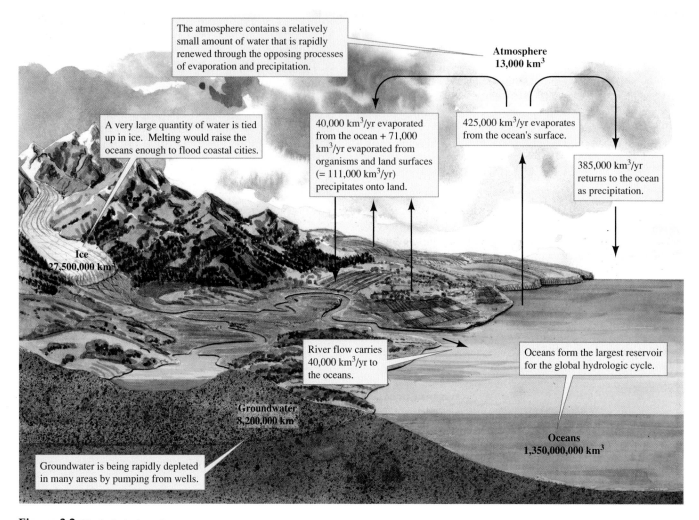

The atmosphere contains a relatively small amount of water that is rapidly renewed through the opposing processes of evaporation and precipitation.

Atmosphere
13,000 km$^3$

A very large quantity of water is tied up in ice. Melting would raise the oceans enough to flood coastal cities.

40,000 km$^3$/yr evaporated from the ocean + 71,000 km$^3$/yr evaporated from organisms and land surfaces (= 111,000 km$^3$/yr) precipitates onto land.

425,000 km$^3$/yr evaporates from the ocean's surface.

385,000 km$^3$/yr returns to the ocean as precipitation.

Ice
27,500,000 km$^3$

River flow carries 40,000 km$^3$/yr to the oceans.

Oceans form the largest reservoir for the global hydrologic cycle.

Groundwater
8,200,000 km$^3$

Oceans
1,350,000,000 km$^3$

Groundwater is being rapidly depleted in many areas by pumping from wells.

**Figure 3.2**  The hydrologic cycle.

With a renewal time of only 3,100 years, the total volume of the oceans, over 1.3 billion km$^3$ of water, has turned over more than 30 times in the last 100,000 years or so, since the first *Homo sapiens* gazed out on the deep blue sea.

## CONCEPT DISCUSSION

### The Natural History of Aquatic Environments

**The biology of aquatic environments corresponds broadly to variations in physical factors such as light, temperature, and water movements and to chemical factors such as salinity and oxygen.**

Our discussion of the natural history of aquatic environments begins with the natural history of the oceans, the largest aquatic environment on the planet. We continue our tour with

environments found along the margins of the oceans, including kelp forests and coral reefs, the intertidal zone, and salt marshes. We then venture up rivers and streams, important avenues for exchange between terrestrial and aquatic environments. Finally, we consider lakes, inland aquatic environments that are similar in many ways to the oceans where we begin.

## The Deep Blue Sea

The blue solitude of open ocean is something palpable, a sensation you can almost taste. As we have seen, the only terrestrial biomes that evoke anything close to the feeling of this place are the open prairies and deserts like the Namib, where the extensive dunes are called the "sand sea." But there is a difference between these terrestrial environments and the sea. On the open ocean, all is blue—blue sea stretching to the horizon, where it meets blue sky (fig. 3.3).

Experience with terrestrial organisms cannot prepare you for what you encounter in samples taken from the deep ocean. We dream of unknown extraterrestrial beings, some friendly and some monstrous, all with strange and shocking anatomy. We

**Figure 3.3** The open ocean is the most extensive biome on earth.

parade them through science fiction literature and films, while, unknown to most of us, creatures as odd and wonderful, some beyond imagining, live in the deep blue world beyond the continental shelves. Figure 3.4 shows one of the species found in the deep sea—a female deep-sea anglerfish with her male partner.

## Geography

The world ocean covers over 360 million km$^2$ of earth's surface and consists of one continuous, interconnected mass of water. This water is spread among three major ocean basins: the Pacific, Atlantic, and Indian, each with several smaller seas along its margins. The largest of the ocean basins, the Pacific,

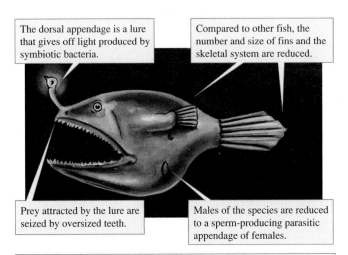

The dorsal appendage is a lure that gives off light produced by symbiotic bacteria.

Compared to other fish, the number and size of fins and the skeletal system are reduced.

Prey attracted by the lure are seized by oversized teeth.

Males of the species are reduced to a sperm-producing parasitic appendage of females.

The darkness, low food availability, and high pressures of the deep-sea environment have selected for organisms quite different from those typical of either shallow seas or the terrestrial environment. Only the females of this deep-sea anglerfish species are active predators.

**Figure 3.4** Deep-sea anglerfish.

has a total area of nearly 180 million km$^2$ and extends from the Antarctic to the Arctic Sea. In the Pacific Ocean, the major seas include the Gulf of California, the Gulf of Alaska, the Bering Sea, the Sea of Okhotsk, the Sea of Japan, the China Sea, the Tasman Sea, and the Coral Sea. The second largest basin, the Atlantic, has a total area of over 106 million km$^2$ and also extends nearly from pole to pole. The major seas of the Atlantic are the Mediterranean, the Black Sea, the North Sea, the Baltic Sea, the Gulf of Mexico, and the Caribbean Sea. The smallest of the three oceans, the Indian, with a total area of just under 75 million km$^2$, is mostly confined to the Southern Hemisphere. Its major seas are the Bay of Bengal, the Arabian Sea, the Persian Gulf, and the Red Sea. Figure 3.5 maps the world's oceans and their adjoining seas.

The Pacific is also the deepest ocean, with an average depth of over 4,000 m. The average depths of the Atlantic and Indian Oceans are approximately equal, at just over 3,900 m. Undersea mountains stud the floor of the deep sea, some isolated and some in long chains that run as ridges for thousands of kilometers. Undersea trenches, some of great depth and volume, rip through the seafloor. One such trench, the Marianas, in the western Pacific Ocean, is over 10,000 m deep—deep enough to engulf Mount Everest with 2 km to spare. The peak of Mauna Loa in Hawaii is a bit over 4,000 m above sea level, a modest height for a mountain. But Mauna Loa hides a secret below its sea apron. The base of Mauna Loa extends 6,000 m below sea level, making it, from base to peak, one of the tallest mountains on earth. What new biological discoveries might await future ecologists along this undersea slope?

## Structure

The oceans can be divided into several vertical and horizontal zones. The shallow shoreline under the influence of the rise and fall of the tides is called the **littoral,** or **intertidal, zone.**

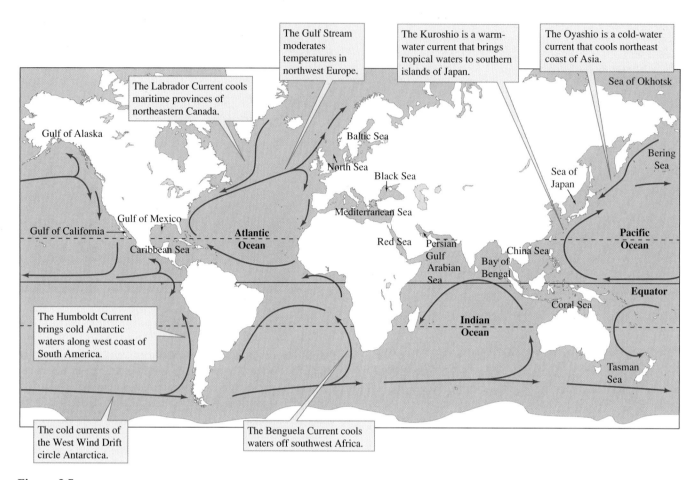

**Figure 3.5** Oceanic circulation, which is driven mainly by the prevailing winds, moderates earth's climate.

The **neritic zone** extends from the coast to the margin of the continental shelf, where the ocean is about 200 m deep. Beyond the continental shelf lies the **oceanic zone.** The ocean is also generally divided vertically into several depth zones. The **epipelagic zone** is the surface layer of the oceans that extends to a depth of 200 m. The **mesopelagic zone** extends from 200 to 1,000 m, and the **bathypelagic zone** extends from 1,000 to 4,000 m. The layer from 4,000 to 6,000 m is called the **abyssal zone,** and finally the deepest parts of the oceans belong to the **hadal zone.** Habitats on the bottom of the ocean, and other aquatic environments, are referred to as **benthic,** while those off the bottom, regardless of depth, are called **pelagic.** Each of these zones supports a distinctive assemblage of marine organisms. Figure 3.6 sketches the general structure of the oceans.

## *Physical Conditions*

### Light

Approximately 80% of the solar energy striking the ocean is absorbed in the first 10 m. Most ultraviolet and infrared light is absorbed in the first few meters. Within the visible range, red, orange, yellow, and green light are absorbed more rapidly than blue light. Consequently, the open ocean appears blue—

the wavelength most likely scattered back to our eyes. In the first 10 m, the marine environment is bright with all the colors of the rainbow; below 50 or 60 m it is a blue twilight. Even in the clearest oceans on the brightest days, the amount of sunlight penetrating to a depth of 600 m is approximately equal to the intensity of starlight on a clear night. That leaves, on average, about 3,400 m of deep black water in which the only light is that produced by bioluminescent fishes and invertebrates. Figure 3.7 compares the colors seen by a scuba diver in deep and shallow water to demonstrate the selective absorption of light by water.

### Temperature

The sunlight absorbed by water increases the *kinetic state,* or velocity of motion, of water molecules. We detect this increased kinetic state as increased temperature. Because more rapid molecular motion decreases water density, warm water floats on cold water. As a consequence, surface water warmed by the sun floats on the colder water below. These warm and cold layers are separated by a **thermocline,** a layer of water through which temperature changes rapidly with depth. This layering of the water column by temperature, which is called *thermal stratification,* is a permanent feature of tropical seas. Temperate oceans are stratified only during the summer, and

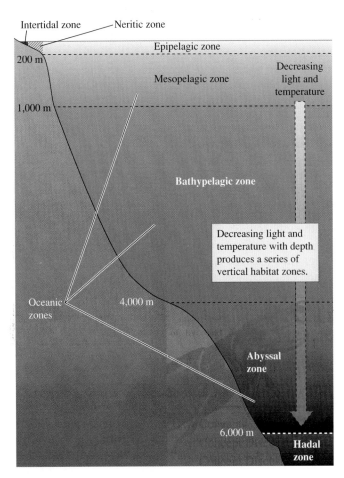

**Figure 3.6** Vertical structuring of the oceans is associated with substantial variation in light and temperature with depth.

the thermocline breaks down as surface waters cool during fall and winter. At high latitudes, thermal stratification is only weakly, if ever, developed. As we shall see, these differences in thermal conditions at different latitudes have far-reaching consequences to the ecological functioning of the oceans.

At the ocean surface, average annual temperature and annual variation in temperature change with latitude but, at all latitudes, oceanic temperatures are much more stable than terrestrial temperatures. The lowest average oceanic temperature, about −1.5°C, is around the Antarctic. The highest average surface temperatures, a bit over 27°C, occur near the equator. Maximum annual variation in surface temperature, approximately 7° to 9°C, occurs in the temperate zone above 40° N latitude. Near the equator, as in the tropical rain forest, the total annual range in temperature, about 1°C, approximately equals the daily range. The greatest stability in oceanic temperatures, however, is below the surface, where, at just 100 m depth, annual variation in temperature is often less than 1°C.

## Water Movements

The oceans are never still. Prevailing winds drive currents that transport nutrients, oxygen, and heat, as well as organisms, across the globe. These currents moderate climates, fertilize the surface waters off the continents, stimulate photosynthesis, and promote gene flow among populations of marine organisms. For example, wind-driven surface currents sweep across vast expanses of open ocean to create great circulation systems called **gyres** that move to the right in the Northern Hemisphere and to the left in the Southern Hemisphere. The great oceanic gyres transport warm water from equatorial regions toward the poles, moderating climates at middle and high latitudes. A segment of one of these gyres, the Gulf Stream, moderates the climate of northwest Europe. One branch of the Gulf Stream, the Norwegian Current, moderates climate all the way to Vardo, Norway, which lies above the Arctic Circle. We saw the climate diagram for Vardo in figure 2.34. Without the moderating influence of the Gulf Stream, the climate of Vardo would be more similar to that of Point Barrow, Alaska, or Tiksi, Russia, the other two sites shown in that figure. Figure 3.5 shows the locations and movements of the oceanic gyres.

In addition to surface currents, there are deepwater currents such as those produced as cooled, high-density water sinks at the Antarctic and Arctic and then moves along the ocean floor. Deep water may also be moved to the surface in a process called **upwelling.** Upwelling occurs along the west coasts of continents and around Antarctica, where winds blow surface water offshore, allowing colder water to rise to the surface. These various water movements are like undersea winds

(a)                              (b)

**Figure 3.7** Changes in light quality with depth: (*a*) the rich colors on a shallow coral reef; (*b*) the blue of the deeper reef.

but with a difference: water is vastly more dense than air. How might this difference in density affect the anatomy, behavior, and distributions of marine organisms?

## Chemical Conditions

### Salinity

The amount of salt dissolved in water is called **salinity.** Salinity varies with latitude and among the seas that fringe the oceans. In the open ocean, it varies from about 34 g of salt per kilogram of water ($^0/_{00}$ or parts per thousand) to about 36.5 $^0/_{00}$. The lowest salinities occur near the equator and above 40° N and S latitudes, where precipitation exceeds evaporation. The excess of precipitation over evaporation at these latitudes is clearly shown by the climate diagrams for temperate forests, boreal forests, and tundra that we examined in chapter 2 (see figs. 2.28, 2.31, and 2.34). Highest salinities occur in the subtropics at about 20° to 30° N and S latitudes, where precipitation is low and evaporation high—precisely those latitudes where we encountered deserts (see fig. 2.19). Salinity varies a great deal more in the small, enclosed basins along the margins of the major oceans. The Baltic Sea, which is surrounded by temperate and boreal forest biomes and receives large inputs of freshwater, has local salinities of 7 $^0/_{00}$ or lower. In contrast, the Red Sea, which is surrounded by deserts, has surface salinities of over 40 $^0/_{00}$.

Despite considerable variation in total salinity, the relative proportions of the major ions (e.g., sodium [$Na^+$], magnesium [$Mg^{+2}$], and chloride [$Cl^-$]) remain approximately constant from one part of the ocean to another. This uniform composition, which is a consequence of continuous and vigorous mixing of the entire world ocean, underscores the connections between different regions of the world's oceans.

### Oxygen

Oxygen is present in far lower concentrations and varies much more in the oceans than in aerial environments. A liter of air contains about 200 ml of oxygen at sea level, while a liter of seawater contains a maximum of about 9 ml of oxygen. Typically, oxygen concentration is highest near the ocean surface and decreases progressively with depth to some intermediate depth. The depth at which oxygen reaches a minimum is usually less than 1,000 m. From this minimum, oxygen concentration increases progressively to the bottom. However, some marine environments such as in the deep waters of the Black Sea and the Norwegian sill fjords are devoid of oxygen.

## Biology

A century and a quarter of research on the open ocean has revealed close correspondence between physical and chemical conditions and the diversity, composition, and abundance of oceanic organisms. For instance, because of the limited penetration of sunlight into seawater, photosynthetic organisms are limited to the brightly lighted upper epipelagic zone of the ocean

(see fig. 3.6). The most significant photosynthetic inhabitants of this zone, also called the *photic zone,* are microscopic organisms called **phytoplankton** that drift with the currents in the open sea. The small animals that drift with these same currents are called **zooplankton.** While there is no ecologically significant photosynthesis below the photic zone, there is no absence of deep-sea organisms. Fishes, ranging from small bioluminescent forms to giant sharks, whales, and invertebrates from tiny crustaceans to giant squid, prowl the entire water column, from the surface of the oceans to the bottom. There is life even in the deepest trenches, below 10,000 m.

Most deep-sea organisms are nourished—whatever their place in the food chain—by organic matter fixed by photosynthesis near the surface. It was long assumed that the rain of organic matter from above was the *only* source of food for deep-sea organisms. Then, about two decades ago the sea surprised everyone. There are entire biological communities on the seafloor that are nourished not by photosynthesis at the surface but by chemosynthesis on the ocean floor (see chapter 6). These oases of life are associated with undersea hot springs and harbor many life-forms entirely new to science. Figure 3.8 shows the great density of organisms found on the ocean floors near an undersea hydrothermal vent.

The deep ocean shines with the blue of pure water and is often called a "biological desert." This description suggests that the open ocean is an area nearly devoid of life—a wasteland, perhaps—that can be dismissed. While it is true that the average rate of photosynthesis per square meter of ocean surface is similar to that of terrestrial deserts, the oceans, because they are so vast, contribute approximately one-fourth of the total photosynthesis in the biosphere. This oceanic production constitutes a substantial contribution to the global carbon and oxygen budget. So why "desert"? Oceanic populations live at

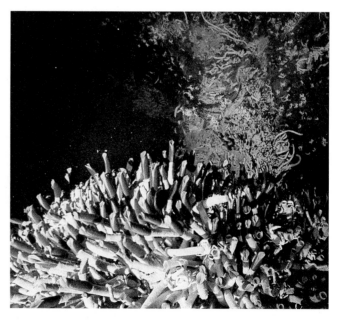

**Figure 3.8** Chemosynthesis-based community on the East Pacific Rise.

such low densities that there is little in the open ocean that can be economically harvested for direct human consumption. J. H. Ryther (1969) estimated that the open ocean contains less than 1% of the harvestable fish stocks. Most fish are found along the coasts. We can, however, appreciate the open ocean from other perspectives.

The open ocean is home, the only home, for thousands of organisms with no counterparts on land. The terrestrial environment supports 11 animal phyla, only 1 of which is endemic to the terrestrial environment—that is, found in no other environments. Fourteen phyla live in freshwater environments but none are endemic. Meanwhile the marine environment supports 28 phyla, 13 of which are endemic to the marine environment. Figure 3.9 compares the number of phyla in terrestrial, freshwater, and marine environments.

Does the greater diversity of phyla in the marine environment shown in figure 3.9 contradict our impression of high biological diversity in biomes such as the tropical rain forest? No, it does not. The terrestrial environment is extraordinarily diverse because there are many species in a few animal and plant phyla, especially arthropods and flowering plants. Still, the number of marine species may also be very high. J. F. Grassle (1991) estimated that the number of bottom-dwelling, or benthic, marine species may exceed 10 million, a level of species diversity that would rival that of the tropical rain forest. We still have not documented the full extent of species diversity in the oceans.

### Human Influences

Human impact on the oceans was once less than on other parts of the biosphere. For most of our history, the vastness of the oceans has been a buffer against human intrusions, but our influence is growing. The decline of large whale populations

around Antarctica and elsewhere sounded a warning of what we can do to the open ocean system. The killing of whales has been curtailed, but there are plans to harvest the great whales' food supply, the small planktonic crustaceans known as *krill.* Although we may find them less engaging than their predators, the large whales, these zooplankton may be more important to the life of the open ocean. Whales are not the only marine populations that have collapsed. Overfishing has led to great declines in commercially important fish stocks, such as the Grand Banks cod population. Many marine fish populations, which once seemed inexhaustible, are now all but gone and fishing fleets sit idle in ports all over the world.

Another threat to marine life is the possibility of dumping wastes of all sorts, including nuclear and chemical wastes, into the deep ocean. In recent years, chemical pollution of the sea has increased substantially, and chemical pollutants are accumulating in deep-sea sediments. Assaults such as these will continue as long as the deep sea is considered by most to be a biological desert. The threats to this blue wilderness could be reduced by changes in human activities based on an appreciation of the great biological richness of the oceans and their critical importance to global carbon and oxygen budgets.

## Life in Shallow Marine Waters: Kelp Forests and Coral Gardens

The shallow waters along continents and around islands support marine communities of very high diversity and biomass. Imagine yourself snorkeling along a marine shore, beyond the intertidal zone. If you are at temperate latitudes and over a solid bottom, you are likely to swim through groves of brown seaweed called *kelp.* Along many coasts, kelp grows so tall, over 40 m in some places, and in such densities that they resemble submarine forests (fig. 3.10).

If you snorkel in the tropics, you may come across a coral reef so diverse in color and texture that it appears to be a well-tended garden. But these are forests and gardens with a difference. Here, you can soar through the canopy with fish so graceful they are called "eagle" rays or float leisurely along with "butterfly" fish as they tend their coral "flowers." The colors on a coral reef rival that of any terrestrial biome (fig. 3.11).

In a kelp forest and coral reef, chance meetings with large carnivorous sea animals seldom fade from the memory and are reminders that here you are not at the top of the food chain. The enchantment runs deep in these environments. The kelp that form the canopy and understory of the temperate submarine forest are not members of the plant kingdom but are, at least in some current classifications, gigantic photosynthetic protists. The corals that form the framework of the coral garden are not plants either but animals that secrete a stony skeleton. Corals are indeed animals but they depend for their survival on photosynthesis by photosynthetic protists called zooxanthellae that live in their tissues. These nearshore marine environments are worlds of surprise and enigma.

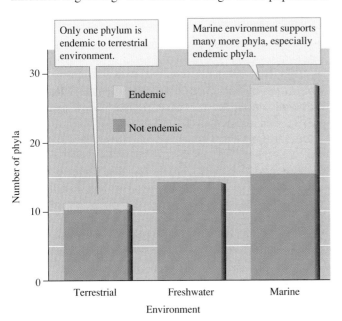

**Figure 3.9** Distribution of animal phyla among terrestrial, freshwater, and marine environments (data from Grassle 1991).

<table>
<tr><td>

*Information*
*Questions*
*Hypothesis*
*Predictions*
*Testing* ✓

</td><td>

*Investigating the Evidence*

# Determining the Sample Median

</td></tr>
</table>

In chapter 2 (p. 21) we determined the sample mean. However, while the sample mean is one of the most common and useful of summary statistics, it is not the most appropriate statistic for some situations. One of the assumptions underlying the use of the sample mean is that the observations used to calculate it are drawn from a population with a normal, or bell-shaped, distribution. However, where the distribution of values within a population deviates substantially from a normal distribution, it may be better to use another estimator of the population "average." An alternative statistic is the **sample median.** The sample median is the middle value in a sample. Let's determine the median for the sample we used to determine a sample mean in chapter 2. Here is the table summarizing that sample:

**Table 1** Heights of 11 seedlings of a rare plant species

| Seedling number | 1 | 2 | 3 | 4 | 5 | 6 | 7 | 8 | 9 | 10 | 11 |
|---|---|---|---|---|---|---|---|---|---|---|---|
| Height in cm | 3 | 6 | 8 | 7 | 2 | 4 | 9 | 4 | 5 | 7 | 8 |

To determine the sample median it's convenient to reorder the observations in a sample from lowest to highest:

**Table 2** Heights of 11 seedlings of a rare plant species with observations ordered from lowest to highest

| Samples: low to high | 1 | 2 | 3 | 4 | 5 | 6 | 7 | 8 | 9 | 10 | 11 |
|---|---|---|---|---|---|---|---|---|---|---|---|
| Height in cm | 2 | 3 | 4 | 4 | 5 | 6 | 7 | 7 | 8 | 8 | 9 |

Because there was an odd number of observations in this sample, 11, there is a middle value in the series of observations, with 5 observations with higher value and 5 with lower value. That middle value occurs at sample rank number 6 and the height of the seedling with this rank happens to be 6 cm. So, the sample median is 6 cm. Notice that this value is very similar to the

sample mean we calculated in chapter 2, which was 5.7 cm. In this case, the sample mean and sample median give similar estimates of the average within the population.

However, where a population contains a few very large or very small values, the sample median may give a better estimate of the average within the population. Consider the following sample of the abundance of mayfly nymphs living within $0.1$ m$^2$ areas of stream bottom. The sample was taken from a high mountain stream of the southern Rocky Mountains and the mayfly species was *Baetis bicaudatus*:

**Table 3** Number of *Baetis bicaudatus* nymphs in $0.1$ m$^2$ benthic samples

| Samples: low to high | 1 | 2 | 3 | 4 | 5 | 6 | 7 | 8 | 9 | 10 | 11 | 12 |
|---|---|---|---|---|---|---|---|---|---|---|---|---|
| Number of nymphs | 2 | 2 | 2 | 3 | 3 | 4 | 5 | 6 | 6 | 8 | 10 | 126 |

In this case, because one of the samples contained 126 nymphs, the sample median and sample mean give very different estimates of the average number of *B. bicaudatus* living within a $0.1$ m$^2$ area of the study stream. *Because this sample has an even number of observations the sample median is determined as the average of the two middle observations.* That is:

Sample median $= \dfrac{4+5}{2} = 4.5$ *B. bicaudatus* per $0.1$ m$^2$.

The sample mean is $\overline{X} = \Sigma X/n = 177/12 = 14.8$ *B. bicaudatus* per $0.1$ m$^2$. This estimate of the population mean is more than three times the estimate provided by the sample median. In this case it is clear that the sample median, which again is the middle value in an ordered sample of observations, more closely estimates the number of *B. bicaudatus* that you are likely to encounter within $0.1$ m$^2$ of benthic habitat in this mountain stream.

At this time, there is no universally accepted mathematical symbol for sample median.

**Figure 3.10** Giant submarine forest off the California coast. Kelp forests have structural features suggestive of terrestrial forests.

## Geography

The nearshore marine environment and its inhabitants vary with latitude. In temperate to subpolar regions, wherever there is a solid bottom and no overgrazing, there are profuse growths of kelp. As you get closer to the equator, these kelp forests are gradually replaced by coral reefs. Coral reefs are confined to middle latitudes between 30° N and S latitudes. Figure 3.12 shows the global distributions of coral reefs and kelp forests.

## Structure

Charles Darwin (1842) was the first to place coral reefs into three categories: fringing reefs, barrier reefs, and atolls. **Fringing reefs** hug the shore of a continent or island. Barrier reefs, such as the Great Barrier Reef, which stretches for nearly 2,000 km off the northeast coast of Australia, stand some distance offshore. A **barrier reef** stands between the open sea and a lagoon. Coral **atolls,** which dot the tropical Pacific and Indian Oceans, consist of coral islets that have built up from a submerged oceanic island and ring a lagoon. Darwin's theory of the long-term development of reefs and the structure of fringing reefs, barrier reefs, and atolls is presented in figure 3.13.

Distinctive habitats associated with coral reefs include the *reef crest,* where corals grow in the surge zone created by waves coming from the open sea. The reef crest extends to a

**Figure 3.11** Coral reefs, such as this one in Manado, Indonesia, support one of the most diverse assemblages of organisms on the planet.

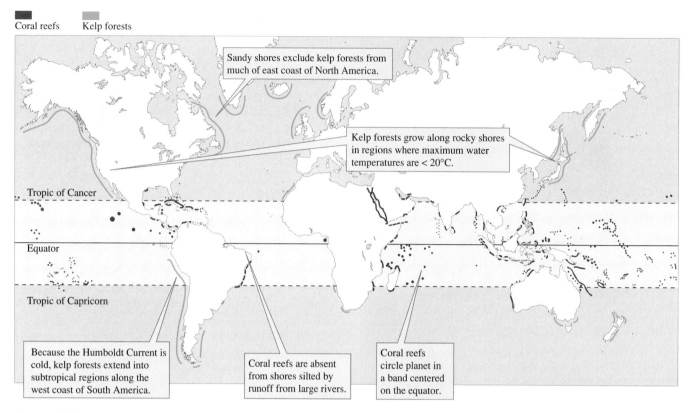

**Figure 3.12**  Distribution of kelp forests and coral reefs (data from Barnes and Hughes 1988, after Schumacher 1976).

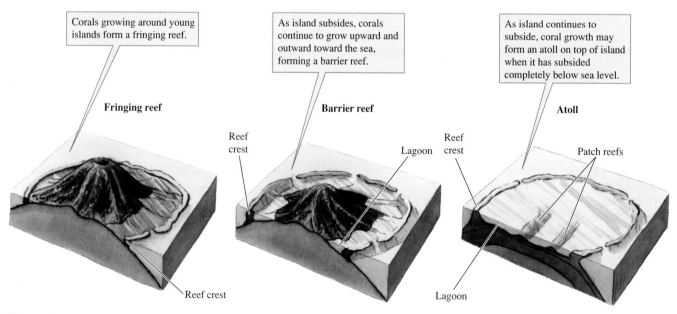

**Figure 3.13**  Types of coral reefs.

depth of about 15 m. Below the reef crest is a *buttress zone,* where coral formations alternate with sand-bottomed canyons. Behind the reef crest lies the *lagoon,* which contains numerous small coral reefs called *patch reefs* and sea grass beds.

Beds of kelp, particularly those of giant kelp, have structural features similar to those of terrestrial forests. At the water's surface is the *canopy,* which may be more than 25 m above the seafloor. The *stems,* or *stipes,* of kelp extend from the canopy to the bottom and are anchored with structures called *holdfasts.* On the stipes and fronds of kelp grow numerous species of epiphytic algae and sessile invertebrates. Other seaweed species of smaller stature usually grow along the bottom, forming an understory to the kelp forest (fig. 3.14).

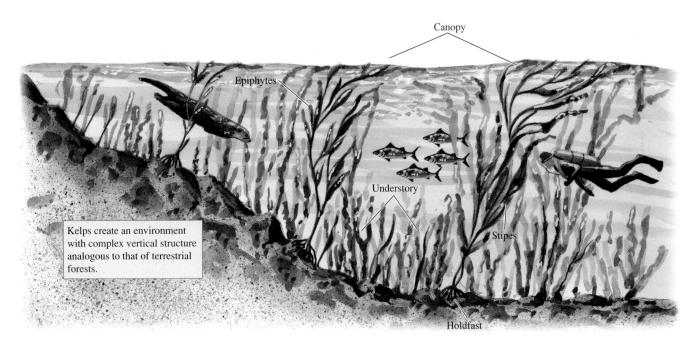

Canopy

Epiphytes

Understory

Stipes

Holdfast

Kelps create an environment with complex vertical structure analogous to that of terrestrial forests.

**Figure 3.14** Kelp forest structure.

## *Physical Conditions*

### Light

Both seaweeds and reef-building corals grow only in surface waters, where there is sufficient light to support photosynthesis. The depth of light penetration sufficient to support kelp and coral varies with local conditions from a few meters to nearly 100 m.

### Temperature

Temperature limits the distribution of both kelp and coral. Most kelp are limited to temperate shores, to those regions where temperatures may fall below 10°C in winter and rise to a bit above 20°C in the summer. Corals are restricted to warm waters, to those regions where the minimum temperature does not fall below about 18° to 20°C and average temperatures usually vary from about 23° to 25°C. Reef-building corals are also sensitive to high temperatures, however, and temperatures above about 29°C are usually lethal.

### Water Movements

Coral reefs and kelp beds are continuously washed by oceanic currents. These currents deliver oxygen and nutrients and remove waste products. The biological productivity of kelp beds and coral reefs may depend upon the flushing action of these currents. However, extremely strong currents and wave action, as during hurricanes, can detach entire kelp forests and flatten entire coral reefs built up over many centuries. Periodic disturbance is a characteristic of both the kelp bed and the coral reef, and both may require some abiotic disturbance for their long-term survival.

## *Chemical Conditions*

### Salinity

Coral reefs grow only in waters with fairly stable salinity. Heavy rainfall or runoff from rivers that reduces salinity below about 27% of seawater can be lethal to corals. Kelp beds appear to be more tolerant of freshwater runoff and grow well along temperate shores, where surface salinities are substantially reduced by runoff from large rivers.

## *Oxygen*

Coral reefs and kelp beds occur where waters are well oxygenated.

## *Biology*

Coral reefs also face intense, and sometimes complex, biological disturbance. Periodic outbreaks of the predatory crown-of-thorns sea star, *Acanthaster planci,* which eats corals, have devastated large areas of coral reef in the Indo-Pacific region. In a Caribbean coral reef community, populations of a sea star relative, the sea urchin *Diadema antillarum,* were infected by a pathogen and crashed to 1% to 5% of previous densities. It turns out that urchins, which eat both algae and corals, may benefit the corals. In the absence of urchins, algal biomass increased greatly, covering previously bare areas needed by young corals to establish themselves. Algal populations, no longer held in check by predation, compete for space with young corals. In the long run, reducing populations of urchins may reduce coral reproductive success. This is a good example

**Figure 3.15** The sea urchin *Diadema* on a coral reef. Feeding by *Diadema* appears to play a key role in the interaction between reef-building corals and benthic algae.

of the complexity and indirect effects that characterize ecological relationships. Figure 3.15 shows one of these sea urchins on a coral reef in the Caribbean Sea.

Corals also compete vigorously among themselves. Reminiscent of rain forest trees and vines, corals engage in a ceaseless struggle for light and space. The corals, however, add a new dimension to the struggle. They actively attack and kill neighboring corals of other clones that differ genetically from themselves.

Coral reefs and kelp beds are among the most productive and diverse of all ecological systems in the biosphere. Robert Whittaker and Gene Likens (1973) estimated that the rate of primary production on coral reefs and algal beds exceeds that of tropical rain forests. The center of diversity for reef-building corals is the western Pacific and eastern Indian Oceans, where there are over 600 coral species and over 2,000 species of fish. By comparison, the western Atlantic Ocean supports about 100 species of corals. Biotic diversity on reefs is also impressive on a small scale. A single coral head may support over 100 species of polychaete worms (Grassle 1973) and over 75 species of fish (Smith and Tyler 1972).

On the coral reef, the ecologist is faced with the same seeming paradox encountered in the tropical rain forest: over-whelming diversity and high primary production in an ecosystem that is nutrient-poor. For the coral reef and for the rain forest, ecologists explain that the answer lies with the organisms themselves and their biotic interactions, including mutualisms, and with rapid recycling and retention of nutrients in the biological parts of the ecosystem.

## Human Influences

Coral reefs and kelp forests are increasingly exploited for a variety of purposes. Tons of kelp are harvested for use as a food additive and for fertilizer. Fortunately, most of this harvest is quickly replaced by kelp growth. Corals, however, which are intensively harvested and bleached for decorations, do not quickly replace themselves. The fish and shellfish of kelp forests and coral reefs have also been heavily exploited. Once again, it appears that coral reefs are more vulnerable. Some coral reefs have been so heavily fished, both for food and for the aquarium trade, that most of the larger fish are rare. Unfortunately, some especially destructive means of fishing are used on coral reefs, including dynamite and poison, with disastrous results. In the Philippines, over 60% of the area once covered by coral has been destroyed by these techniques during recent years. While an appreciation of the threats to rain forests grows, there is less said of the plight of the rain forest's marine cousin, the coral reef, as it is changed from marine garden to wasteland. Again, the question is how can local people and coral reefs thrive together.

# Marine Shores: Life Between High and Low Tides

At its shore, the ocean pulses, gurgles, crashes, hisses, and booms like a living being. The rise and fall of the tides make the shore one of the most dynamic environments in the biosphere. The intertidal zone is a magnet for the curious naturalist and one of the most convenient places to study ecology. Where else in the biosphere does the structure of the landscape change several times each day? Where else does nature expose entire aquatic communities for leisurely exploration? Where else are environmental and biological gradients so compressed? It should be no surprise that here in the intertidal zone, immersed in tide pools, salt spray, and the sweet smell of kelp, ecologists have found the inspiration and circumstance for some of the most elegant experiments and most enduring generalizations of ecology. The intertidal zone, the area covered by waves at high tide and exposed to air at low tides, has proved to be an illuminating window to the world. Figure 3.16 shows the tangle of diverse life that can be observed on a rocky shore during low tide.

## Geography

Countless thousands of kilometers of coastline around the world have intertidal zones. From a local perspective, it is significant to distinguish between exposed and sheltered shores.

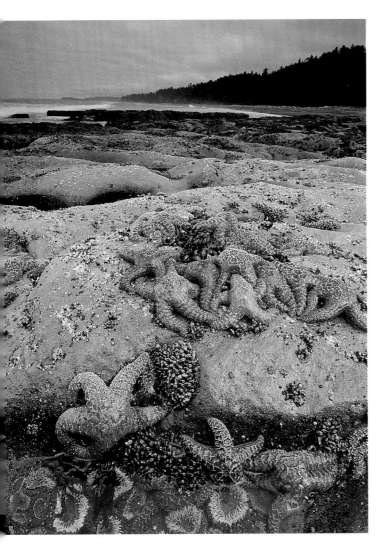

**Figure 3.16** A rocky shore at low tide, showing the great abundance that can be attained by populations of intertidal organisms.

Battered by the full force of ocean waves, exposed shores support very different organisms from those found along sheltered shores on the inside of headlands or in coves and bays. A second important distinction is between rocky and sandy shores.

## Structure

The intertidal zone can be divided into several vertical zones (fig. 3.17). The highest zone is called the *supratidal fringe*, or *splash zone*. The supratidal fringe is seldom covered by high tides but is often wetted by waves. Below this fringe is the intertidal zone proper. The upper intertidal zone is covered only during the highest tides, and the lower intertidal zone is uncovered only during the lowest tides. Between the upper and lower intertidal zones is the middle intertidal zone, which is covered and uncovered during average tides. Below the intertidal zone is the *subtidal zone,* which remains covered by water even during the lowest tides. As we shall see in the next two sections (Physical Conditions and Chemical Conditions),

tidal fluctuation produces steep gradients of physical and chemical conditions within the intertidal zone.

## Physical Conditions

### Light

Intertidal organisms are exposed to wide variations in light intensity. At high tide, water turbulence reduces light intensity. At low tide, intertidal organisms are exposed to the full intensity of the sun. How might this variation in light intensity affect the distribution of photosynthetic organisms in the intertidal zone? How vulnerable are intertidal organisms to damage by sunlight, compared to organisms from other marine environments?

## Temperature

Because the intertidal zone is exposed to the air once or twice each day, intertidal temperatures are always changing. At high latitudes, tide pools, small basins that retain water at low tide, can cool to freezing temperatures during low tides, while tide pools along tropical and subtropical shores can heat to temperatures in excess of 40°C. The dynamic intertidal environment contrasts sharply with the stability of most marine environments and presents substantial environmental challenges to organisms of marine origin.

## Water Movements

The two most important water movements affecting the distribution and abundance of intertidal organisms are the waves that break upon the shore and the tides. The tides vary in magnitude and frequency. Most tides are *semidiurnal,* that is, there are two low tides and two high tides each day. However, in seas, such as the Gulf of Mexico and the South China Sea, there are *diurnal* tides, that is, a single high and low tide each day. The total rise and fall of the tide varies from a few centimeters along some marine shores to 15 m at the Bay of Fundy in northeastern Canada (fig. 3.18).

The sun and moon and local geography determine the magnitude and timing of tides. The main tide-producing forces are the gravitational pulls of the sun and moon on water. Of the two forces, the pull of the moon is greater because, although the sun is far more massive, the moon is much closer. Tidal fluctuations are greatest when the sun and moon are working together, that is, when the sun, moon, and earth are in alignment, which happens at full and new moons. These times of maximum tidal fluctuation are called *spring tides.* Tidal fluctuation is least when the gravitational effects of the sun and moon are working in opposition, that is, when the sun and moon, relative to earth, are at right angles to each other, as they are at the first and third quarters of the moon. These times of minimum tidal fluctuation are called *neap tides.* The size and geographic position of a bay, sea, or section of coastline determine whether the influences of sun and moon are amplified or damped and are responsible for the variations in tides from place to place.

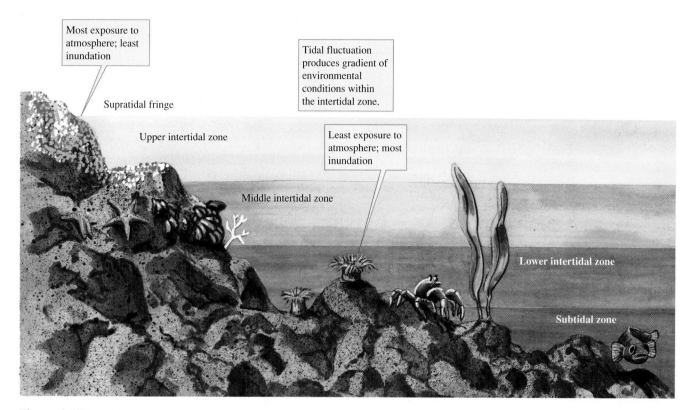

Most exposure to atmosphere; least inundation

Supratidal fringe

Upper intertidal zone

Tidal fluctuation produces gradient of environmental conditions within the intertidal zone.

Middle intertidal zone

Least exposure to atmosphere; most inundation

Lower intertidal zone

Subtidal zone

**Figure 3.17**  Intertidal zonation.

(a)

(b)

**Figure 3.18**  The Bay of Fundy, a site of some of the greatest tidal fluctuations anywhere, at: (*a*) high tide; and (*b*) low tide.

Intertidal organisms have a lot to withstand—not only exposure to air during low tide but also the pounding of waves breaking on the seashore. The amount of wave energy to which intertidal organisms are exposed varies considerably from one section of coast to another; this variation affects the distribution and abundance of intertidal species. Exposed headlands are hit by high waves (fig. 3.19), and they are also subjected to strong currents, which are at times as strong as those of swift rivers. Coves and bays are the least exposed to waves, but even the most sheltered areas may be subjected to intense wave action during storms.

## Chemical Conditions

### Salinity

Salinity in the intertidal zone varies much more than in the open sea, especially within tide pools isolated at low tide. Rapid evaporation during low tide increases the salinity

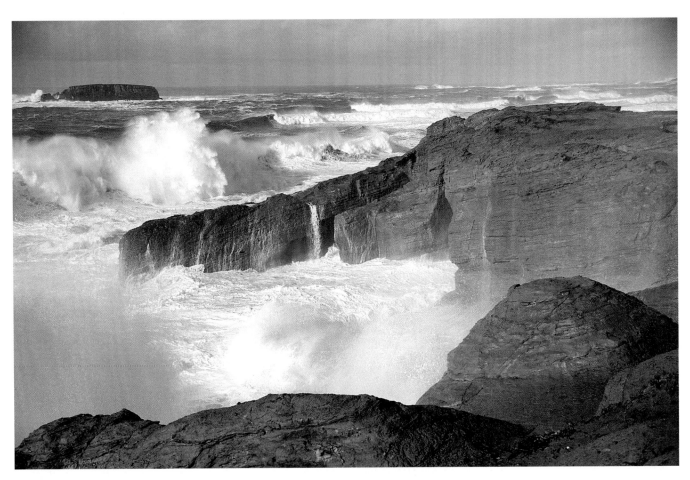

**Figure 3.19**  Storm waves such as these pounding a rocky headland have an important influence on the distribution and abundance of intertidal organisms.

within tide pools along desert shores. Along rainy shores at high latitudes and in the tropics during the wet season, tide pool organisms can experience much reduced salinity.

### Oxygen

Oxygen does not generally limit the distributions of intertidal organisms for two major reasons. First, intertidal species are exposed to air at each low tide. Second, the water of wave-swept shores is thoroughly mixed and therefore well oxygenated. An intertidal environment where oxygen availability may be low is in interstitial water within the sediments along sandy or muddy shores, especially in sheltered bays, where water circulation is weak.

## Biology

The inhabitants of the intertidal zone are adapted to an amphibious existence, partly marine, partly terrestrial. All intertidal organisms are adapted to periodic exposure to air, but some species are better equipped than others to withstand that exposure. This fact produces one of the most noticeable intertidal features, **zonation of species.** Some species inhabit the highest levels of the intertidal zone, are exposed by almost all tides, and remain exposed the longest. Others are exposed during the lowest tides only, perhaps once or twice per month, or even less frequently. On an even finer spatial scale, microtopography influences the distribution of intertidal organisms. Tide pools support very different organisms than sections of the intertidal zone from which the water drains completely. The channels in which seawater runs, like a salty stream, during the ebb and flow of the tides offer yet another habitat.

The substratum also affects the distribution of intertidal organisms. Hard, rocky substrates support a biota different from that on sandy or muddy shores. You can see an obvious profusion of life on rocky shores because most species are attached to the surface of the substratum (see fig. 3.16). The residents of the rocky intertidal zone you will likely see are sea stars, barnacles, mussels, and seaweeds. If there are sea urchins, you will notice their spiky presence. But even here, where low tide seems to freely yield the secrets of the sea, all is not obvious. Most organisms take shelter at low tide, some among the fronds and holdfasts of kelp and others under boulders. There are even animals that burrow into and live inside rocks. As we shall see when we discuss competition in chapter 13 and predation in chapter 14, biological interactions make major contributions to the distributions of intertidal organisms.

On soft bottoms some species wander the surface of the substrate, but most are burrowers and shelter themselves within the sand or mud bottom. Here, nature does not give up its secrets easily. To thoroughly study the life of sandy shores you must separate organisms from sand or mud. Perhaps this is the reason rocky shores have gotten more attention by researchers and why we know far less about the life of sandy shores. Beaches, like the open ocean, have been considered biological deserts. Careful studies, however, have shown that the intensity and diversity of life on sandy shores rivals that of any benthic aquatic community (MacLachlan 1983).

## Human Influences

People have long sought out intertidal areas, first for food and later for recreation, education, and research. Shell middens, places where prehistoric people piled the remains of their seafood dinners, from Scandinavia to South Africa, bear mute testimony to the importance of intertidal species to human populations for over 100,000 years. Today, each low tide still finds people all over the world scouring intertidal areas for mussels, oysters, clams, and other species. But the intertidal zone, which resists, and even thrives, in the face of twice daily exposure to air and pounding surf, is easily devastated by the trampling feet and probing hands of a few human visitors. Relentless exploitation has severely reduced many intertidal populations. Exploitation for food is not the only culprit, however. Collecting for education and research also takes its toll. The intertidal zone is also vulnerable to devastation by oil spills, which have damaged intertidal areas around the world.

Each low tide, the intertidal community offers itself up. Don't abuse this rare openness; walk lightly and look closely.

Take photographs; make sketches and notes; take away only inspiration and renewal and you can come back for more when you need it.

# Estuaries, Salt Marshes, and Mangrove Forests

**Estuaries** are found wherever rivers meet the sea. **Salt marshes** and **mangrove forests** are concentrated along low-lying coasts with sandy shores and may, like estuaries, be associated with the mouths of rivers. All three are at the transition between one environment and another—salt marshes and mangrove forests at the transition between land and sea, and estuaries at the transition between river and sea. Because these areas are transitions between very different environments, they have a great deal in common physically, chemically, and biologically. These are environments that pulse to the rhythm of lunar-driven tides and teem with life. Figure 3.20 shows a rich salt marsh landscape, and figure 3.21 shows the structurally complex environment provided by dense populations of mangroves and their many prop roots.

## Geography

Salt marshes, which are dominated by herbaceous vegetation, are concentrated along sandy shores from temperate to high latitudes. At tropical and subtropical latitudes the herb-dominated salt marsh is replaced by mangrove forests. Mangroves are associated with the terrestrial climates of the tropical rain forest, tropical dry forest, savanna, and desert, due mainly to the sensitivity of mangroves to frost. Figure 3.22 maps the global distributions of salt marshes and mangrove forests.

**Figure 3.20**   The salt marsh landscape is inhabited by few species of plants but their productivity is exceptionally high.

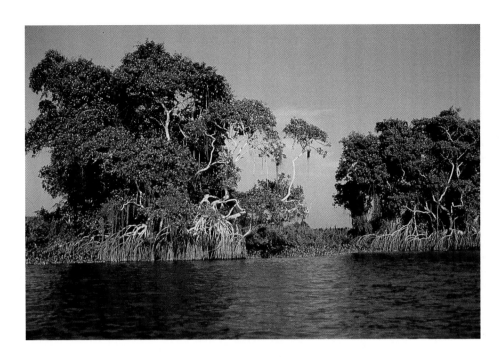

**Figure 3.21** The prop roots of mangroves provide a complex habitat for a high diversity of marine fish and invertebrates.

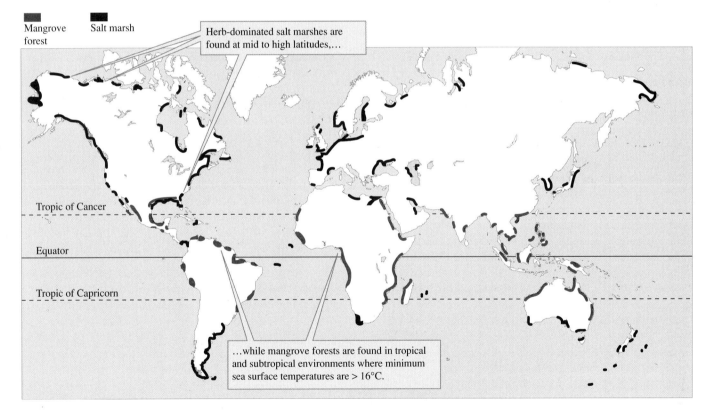

Mangrove forest   Salt marsh

Herb-dominated salt marshes are found at mid to high latitudes,…

Tropic of Cancer

Equator

Tropic of Capricorn

…while mangrove forests are found in tropical and subtropical environments where minimum sea surface temperatures are > 16°C.

**Figure 3.22** Salt marsh and mangrove forests (data from Chapman 1977, Long and Mason 1983).

## Structure

Salt marshes generally include channels, called tidal creeks, that fill and empty with the tides. These meandering creeks can create a complex network of channels across a salt marsh (fig. 3.23). Fluctuating tides move water up and down these channels, or tidal creeks, once or twice each day. These daily movements of water gradually sculpt the salt marsh into a gently undulating landscape. Tidal creeks are generally bordered by natural levees. Beyond the levees are marsh flats, which may include small basins called *salt pans* that periodically collect water that eventually evaporates, leaving a layer

**Figure 3.23** Viewing a salt marsh from the air reveals great structural complexity.

of salt. This entire landscape is flooded during the highest tides and drains during the lowest tides. A typical cross section of a salt marsh is shown in figure 3.24.

The mangrove trees of different species are usually distributed according to height within the intertidal zone. For instance, in mangrove forests near Rio de Janeiro, Brazil, the mangroves growing nearest the water belong to the genus *Rhizophora*. At this level in the intertidal zone, *Rhizophora* is inundated by average high tides. Above *Rhizophora* grow other mangroves such as *Avicennia,* which is flooded by the average spring tides, and *Laguncularia,* which is touched only by the highest tides. Figure 3.25 shows zonation within a mangrove forest.

Estuaries vary vertically and longitudinally, especially in regard to the amount of dissolved salts (see the discussion on salinity in the Chemical Conditions section).

## Physical Conditions

### Light

Estuaries, salt marshes, and mangrove forests experience significant fluctuations in tidal level. Consequently, the organisms in these environments are exposed to highly variable light conditions. They may be exposed to full sunlight at low tide and very little light at high tide. The waters of these areas are usually turbid because shifting currents, either from the tides or rivers, keep fine organic and inorganic materials in suspension.

### Temperature

Several factors make the temperatures of estuaries, salt marshes, and mangrove forests highly variable. First, because their waters are generally shallow, particularly at low tide, water temperature varies with air temperature. Second, the temperatures of seawater and river water may be very different. If so, the temperature of an estuary may change with each high and low tide. Salt marshes at high latitudes may freeze during the winter. In contrast, mangroves grow mainly along desert and tropical coasts, where the minimum annual temperature is about 20°C. The shallows in these environments can heat up to over 40°C.

### Water Movements

Ocean tides and river flow drive the complex currents in estuaries. These currents are at the heart of the ecological processes of the estuary because they transport organisms, renew nutrients and oxygen, and remove wastes. Complex tidal currents also flow in salt marshes and mangrove forests, where they are involved in these processes and also fragment and transport the litter produced by salt marsh and mangrove vegetation. Once or twice a day, high tides create saltwater currents that move up the estuaries of rivers and the channels within salt marshes and mangrove forests. Low tides reverse these currents and saltwater moves seaward. Tidal height may fluctuate far from where an estuary meets the sea. For example, tidal fluctuations occur over 200 km upstream from where the Hudson River flows into

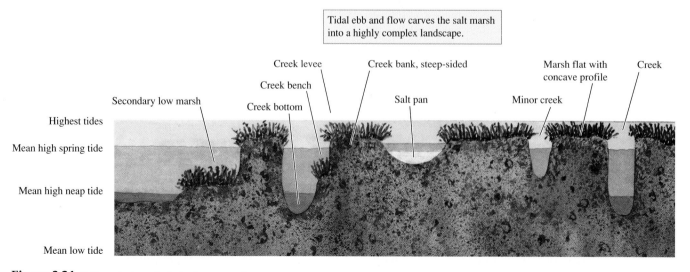

**Figure 3.24** Salt marsh channels shown in cross section.

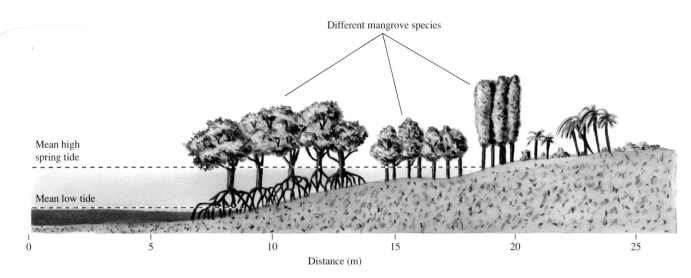

**Figure 3.25** Where mangrove diversity is high, mangrove species show clear patterns of vertical zonation relative to tidal level.

the sea. The vigorous mixing, in more than one direction, makes these transitional environments some of the most physically dynamic in the biosphere. Penetration of light and water movements vary over short distances and in the course of a day. This physical variability is reflected in highly variable chemical conditions.

## Chemical Conditions

### Salinity

The salinity of estuaries, salt marshes, and mangrove forests may fluctuate widely, particularly where river and tidal flow are substantial. In such systems, the salinity of seawater can drop to nearly that of freshwater an hour after the tide turns.

Because estuaries are places where rivers meet the sea, their salinity is generally lower than that of seawater. In hot, dry climates, however, evaporation often exceeds freshwater inputs and the salinity in the upper portions of estuaries may exceed that of the open ocean.

Estuarine waters are also often stratified by salinity, with lower-salinity, low-density water floating on a layer of higher-salinity water, isolating bottom water from the atmosphere. On the incoming tide, seawater coming from the ocean and river water are flowing in opposite directions. As seawater flows up the channel, it mixes progressively with river water flowing in the opposite direction. Due to this mixing, the salinity of the surface water gradually increases down river from less than 1 $^0/_{00}$ to salinities approaching that of seawater at the river mouth (fig. 3.26).

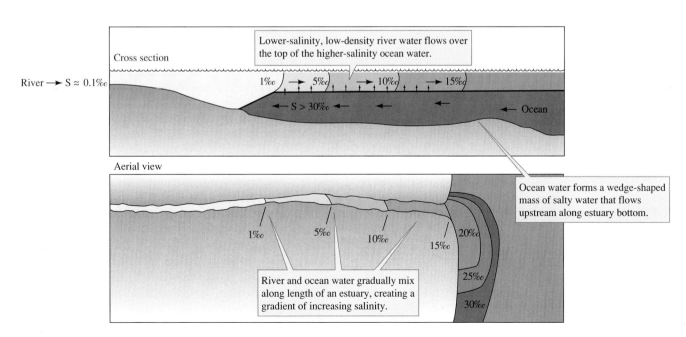

**Figure 3.26** Structure of a salt wedge estuary.

## Oxygen

In estuaries, salt marshes, and mangrove forests, oxygen concentration is highly variable and often reaches extreme levels. Decomposition of the large quantities of organic matter produced in these environments can deplete dissolved oxygen to very low levels, and isolation of saline bottom water from the atmosphere adds to the likelihood that oxygen will be depleted. At the same time, however, high rates of photosynthesis can increase dissolved oxygen concentrations to supersaturated levels. Again, the oxygen concentrations to which an organism is exposed can change with each turn of the tide.

## *Biology*

The salt marshes of the world are dominated by grasses such as *Spartina* spp. and *Distichlis* spp., by pickleweed, *Salicornia* spp., and by rushes, *Juncus* spp. The mangrove forest is dominated by mangrove trees belonging to many genera. The species that make up the forest change from one region to another; however, within a region, there is great uniformity in species composition.

Estuaries and salt marshes don't support a great diversity of species, but their primary production is very high. These are places where some of the most productive fisheries occur and where aquatic and terrestrial species find nursery grounds for their young. Most of the fish and invertebrates living in estuaries evolved from marine ancestors, but estuaries also harbor a variety of insects of freshwater origin. Whatever their origins, however, the species that inhabit estuaries and salt marshes have to be physiologically tough. Estuaries and salt marshes also attract birds, especially water birds. In the mangrove forest, birds are joined by crocodiles, alligators, and, in the Indian subcontinent, by tigers. From both academic and practical perspectives, the ecological importance of these environments cannot be overstated.

## *Human Influences*

Estuaries, salt marshes, and mangrove forests are extremely vulnerable to human interference. People want to live and work at the coast, but building sites are limited. One solution to the problem of high demand for coastal property and low supply has been to fill and dredge salt marshes, replacing wildlife habitat with human habitat (fig. 3.27). Because cities benefit from access to the sea, many, such as Boston, San Francisco, and London, have been built on estuaries. As a consequence, many estuaries have been polluted for centuries. The discharge of wastes depletes oxygen supplies, which physiologically stresses aquatic organisms. The discharge of organic wastes depletes oxygen directly as it decomposes, and the addition of nutrients such as nitrogen can lead to oxygen depletion by stimulating primary production. Heavy metals discharged into estuaries and salt marshes are incorporated into plant and animal tissues and have been, through the process of bioaccumulation, elevated to toxic levels in some food species. The assaults on estuaries and salt

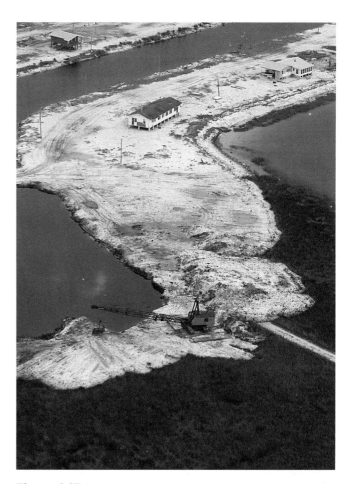

**Figure 3.27** Dredging and filling of salt marshes, such as the operation shown in this photo, has destroyed vast areas of these highly productive biomes.

marshes have been chronic and intense, but the interest and concerns of people grow steadily. In these resilient environments, hope comes with each change of the tide.

# Rivers and Streams: Life Blood and Pulse of the Continents

We become aware of the importance of rivers in human history and economy as we name the major ones: Nile, Danube, Tigris, Euphrates, Yukon, Indus, Tiber, Mekong, Ganges, Rhine, Mississippi, Missouri, Yangtze-Kiang, Amazon, Seine, Zaire, Volga, Thames, Rio Grande. The names of these rivers, and many others great and small, ring with a thousand images of history, geography, and poetry. The importance of rivers to human history, ecology, and economy is inestimable. However, river ecology has lagged behind the ecological study of lakes and oceans and is one of the youngest of the many branches of aquatic ecology. In the past couple of decades, however, river ecology has, as all youthful sciences do, exploded with published research, competing theories, controversies, and international symposia and now claims a well-earned place beside its more mature cousins.

What might rivers offer to the science of ecology? Their most notable feature is their dynamism. In art and literature, this characteristic has made rivers symbols of ceaseless change. For example, Leonardo da Vinci wrote: "In rivers the water you touch is the last of what has passed and the first of that which comes. So with time present." The ancient Greeks said simply: "You never step in the same river twice." In ecology, we call dynamic ecosystems such as rivers "nonequilibrial." **Nonequilibrial theory,** one of the newest branches of theoretical ecology, may find, as has art and literature, precisely the metaphor it needs in the rivers of the world. The meandering pattern of the river shown in figure 3.28 suggests the dynamism of river ecosystems.

## Geography

Rivers drain most of the landscapes of the world. When rain falls on a landscape, a portion of it runs off, either as surface or subsurface flow. Some of this runoff water eventually collects in small channels, which join to form larger and larger water courses until they form a network of channels that drains the landscape. A river basin is that area of a continent or island that is drained by a river drainage network, such as the Mississippi River basin in North America or the Congo River basin in Africa. Rivers eventually flow out to sea or to some interior basin like the Aral Sea or the Great Salt Lake. River basins are separated from each other by watersheds, that is, by topographic high points. For instance, the peaks of the Rocky Mountains divide runoff from snowmelt. Water on the east side of the peaks flows to the Atlantic ocean, while water on the west side flows to the Pacific Ocean. Figure 3.29 shows the distribution of the major rivers of the planet.

## Structure

Rivers and streams can be divided along three dimensions (fig. 3.30). They can be divided along their *lengths* into pools, runs, riffles, and rapids and, because of variation in flow, rivers can also be divided across their *widths* into wetted channels and active channels. A wetted channel contains water even during low flow conditions. An active channel, which extends out from one or both sides of a wetted channel, may be dry during part of the year but is inundated annually during high flows. Outside the active channel is the **riparian zone,** a transition between the aquatic environment of the river and the upland terrestrial environment.

Rivers and streams can be divided *vertically* into the water surface, the water column, and the bottom, or benthic, zone. The benthic zone includes the surface of the bottom substrate and the interior of the substrate through depths at which substantial surface water still flows. Below the benthic zone is the **hyporheic zone,** a zone of transition between areas of surface water flow and groundwater. The area containing groundwater below the hyporheic zone is called the **phreatic zone.** Each part of a river or stream is a physically and chemically distinctive environment and each supports different organisms. Streams and rivers within a drainage network can be classified by where they occur in the drainage network. This classification is based on a system called **stream order.** In this system, headwater streams are first order, while a stream formed by the joining of two first order streams is a second order stream. A third order stream results from the joining of two second order streams and so on. In this system, a lower order stream, say a first order, joining a higher order stream, for instance, a second order stream, does not raise the order of the stream below the junction. In this case, that stream would remain a second order stream.

## Physical Conditions

### Light

There are two principal aspects of light to consider in relation to rivers and streams. First, how far light penetrates into the water column and second, how much light shines on the surface of a river. Streams and rivers vary considerably in water clarity. Generally, however, even the clearest streams are much more turbid than clear lakes or seas. The reduced clarity of rivers results from two main factors. First, rivers are in intimate contact with the surrounding landscape, and inorganic and organic

**Figure 3.28** The meandering Okavango River, Botswana.

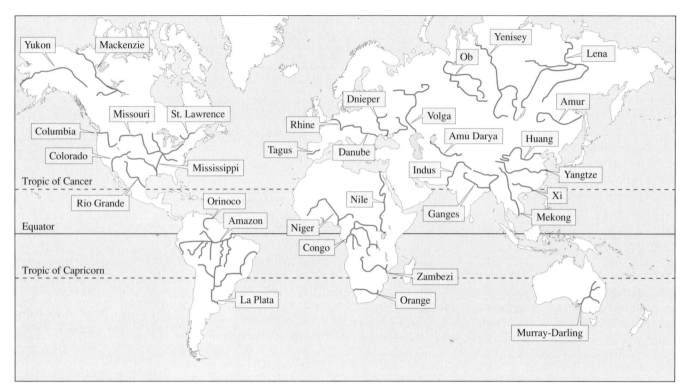

**Figure 3.29**  Major rivers.

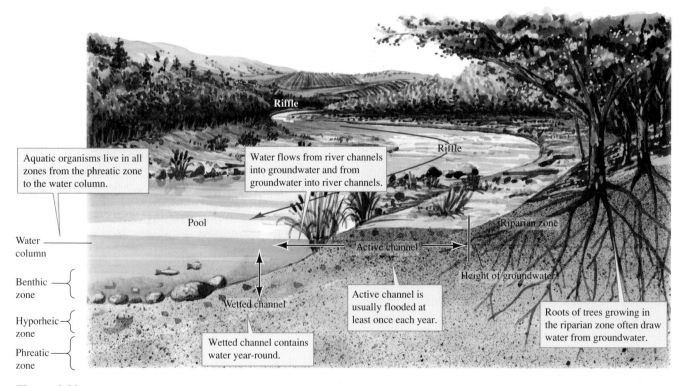

**Figure 3.30**  The three dimensions of stream structure.

materials continuously wash, fall, or blow into rivers. Second, river turbulence erodes bottom sediments and keeps them in suspension, particularly during floods. The headwaters of rivers are generally shaded by riparian vegetation. Shading may be so thorough along some streams that there is very little photosynthesis by aquatic primary producers. The extent of shading decreases progressively downstream as stream width increases. In desert regions, headwater streams usually receive large

(a)

(b)

**Figure 3.31** Headwater streams in: (*a*) forested Great Smoky Mountains; and (*b*) Sonoran Desert. The consumers in headwater streams draining forested lands generally depend on energy from the surrounding forest. Meanwhile desert streams are open to sunlight and support high levels of photosynthesis by stream algae, the source of most energy for desert-stream consumers.

amounts of solar radiation and support high levels of photosynthesis. Figure 3.31 contrasts the environments of headwater streams flowing through a forest and a desert.

## Temperature

The temperature of rivers closely tracks air temperature but does not reach the extremes of terrestrial habitats. The coldest river temperatures, those of high altitudes and high latitudes, may drop to a minimum of 0°C. The warmest rivers are those flowing through deserts, but even desert rivers seldom exceed 30°C. The outflows of hot springs can be boiling in their upper reaches, but populations of thermophilic bacteria live in even the hottest of these.

### Water Movements

What is notable about a river is the continuous movement of water. Viewed environmentally, river currents deliver food, remove wastes, renew oxygen, and strongly affect the size, shape, and behavior of river organisms. Currents in quiet pools may flow at only a few millimeters per second, while water in the rapids of swift rivers in a flood stage may flow at 6 m per second. Contrary to popular belief, the currents of large rivers may be as swift as those in the headwaters.

The amount of water carried by rivers, which is called *river discharge,* differs a lot from one climatic regime to another. River flows are often unpredictable and "flashy" in arid and semiarid regions, where extended droughts may be followed by torrential rains. The flow in tropical rivers varies considerably. Many tropical rivers, which flow very little during the dry season, become torrents during the wet season. Some of the most constant flows are found in forested temperate regions, where, as we saw in chapter 2, precipitation is fairly evenly distributed throughout the year (see fig. 2.28). Forested landscapes can damp out variation in flow by absorbing excessive rain during wet periods and acting as a reservoir for river flow during drier periods. Figure 3.32 compares the annual flows of rivers of moist temperate and semiarid climates.

It appears that the health and ecological integrity of rivers and streams depend upon keeping the natural flow regime for a region intact. Historical patterns of flooding have particularly important influences on river ecosystem processes, especially on the exchange of nutrients and energy between the river channel and the floodplain and associated wetlands. This idea, which was first proposed as the **flood pulse concept,** is supported by a growing body of evidence from research conducted on rivers on virtually every continent.

## *Chemical Conditions*

### Salinity

Water flowing across landscapes or through soils picks up dissolved materials. The amount of salt dissolved in river water reflects the history of leaching that has gone on in its basin. As we saw in chapter 2, annual rainfall is high in tropical regions. Consequently, many tropical soils have been leached of much of their soluble materials and it is in the tropics that the salinity of river water is often very low. Desert rivers generally have the highest salinities. Figure 3.33 shows that the salinity of river water from different regions may show 10- to 100-fold differences.

### Oxygen

The oxygen content of river water is inversely correlated with temperature. Oxygen supplies are generally richest in cold, thoroughly mixed headwater streams and lower in the warm downstream sections of rivers. However, because the waters in

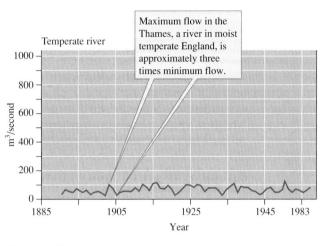

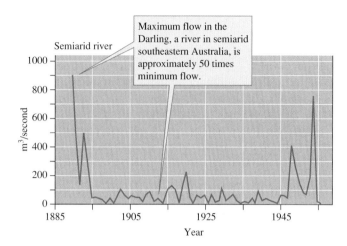

**Figure 3.32** Annual flow of rivers in moist temperate and semiarid climates (data from Calow and Petts 1992).

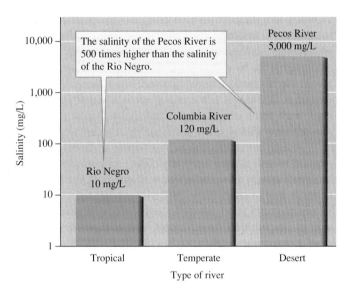

**Figure 3.33** Salinities of tropical, temperate, and arid land rivers (data from Gibbs 1970).

streams and rivers are continuously mixed, oxygen is generally not limiting to the distribution of river organisms. The major exception to this generalization is in sections of streams and rivers receiving organic wastes from cities (wastes with high biochemical oxygen demand or BOD) and industry. Only organisms tolerant of low oxygen concentrations can inhabit these sections.

## Biology

As in the terrestrial biomes, large numbers of species inhabit tropical rivers. The number of fish species in tropical rivers is much higher than in temperate rivers. For example, the Mississippi River basin, which supports one of the most diverse temperate fish faunas, is home to about 300 fish species. By contrast, the tropical Congo River basin contains about 669 species of fish, of which over 558 are found nowhere else. The most impressive array of freshwater fish is that of the Amazon River basin, which contains over 2,000 species, approximately 10% of all the fish species on the planet.

The organisms of river systems change from headwaters to mouth. These patterns of biological variation along the courses of rivers have given rise to a variety of theories that predict downstream change in rivers and their inhabitants. One of these theories is the **river continuum concept** (Vannote et al. 1980). According to this concept, in temperate regions, leaves and other plant parts are often the major source of energy available to the stream ecosystem. Upon entering the stream, this coarse particulate organic matter (CPOM) is attacked by aquatic microbes, especially fungi. Colonization by fungi makes CPOM more nutritious for stream invertebrates. The stream invertebrates of headwater streams are usually dominated by two feeding groups: shredders, which feed on CPOM, and collectors, which feed on fine particulate organic matter (FPOM). The fishes in headwater streams are usually those, such as trout, that require high oxygen concentrations and cool temperatures.

The river continuum concept predicts that the major sources of energy in medium-sized streams will be FPOM washed down from the headwater streams and algae and aquatic plants. Algae and plants generally grow more profusely in medium streams because they are too wide to be entirely shaded by riparian vegetation. Because of the different food base, shredders make up a minor portion of the benthic community, which is dominated by collectors and grazers on the abundant algae and aquatic plants. The fishes of medium streams generally tolerate somewhat higher temperatures and lower oxygen concentrations than headwater fishes.

In large rivers, the major sources of energy are FPOM and, in some rivers, phytoplankton. Consequently, the benthic invertebrates of large rivers are dominated by collectors, which make their living by filtering FPOM from the water column. In large rivers, there are also zooplankton. The fish found in large temperate rivers are those, such as carp and catfish, that are more tolerant of lower oxygen concentrations and higher water temperatures. Because of the development of a plankton community, plankton-feeding fish also live in large rivers. The major changes in temperate river systems predicted by the river continuum concept are summarized in figure 3.34.

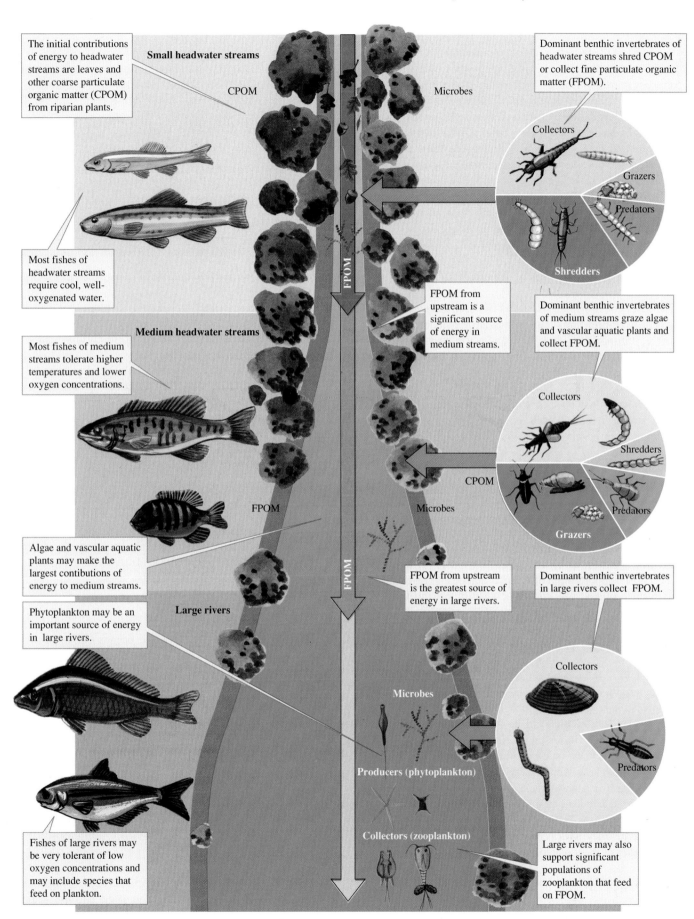

The initial contributions of energy to headwater streams are leaves and other coarse particulate organic matter (CPOM) from riparian plants.

**Small headwater streams**

CPOM

Microbes

Dominant benthic invertebrates of headwater streams shred CPOM or collect fine particulate organic matter (FPOM).

Collectors

Grazers

Predators

**Shredders**

Most fishes of headwater streams require cool, well-oxygenated water.

FPOM

FPOM from upstream is a significant source of energy in medium streams.

Dominant benthic invertebrates of medium streams graze algae and vascular aquatic plants and collect FPOM.

**Medium headwater streams**

Most fishes of medium streams tolerate higher temperatures and lower oxygen concentrations.

Collectors

Shredders

CPOM

Microbes

Predators

FPOM

**Grazers**

Algae and vascular aquatic plants may make the largest contibutions of energy to medium streams.

FPOM from upstream is the greatest source of energy in large rivers.

Dominant benthic invertebrates in large rivers collect FPOM.

Phytoplankton may be an important source of energy in large rivers.

**Large rivers**

FPOM

Microbes

Collectors

Predators

**Microbes**

**Producers (phytoplankton)**

Fishes of large rivers may be very tolerant of low oxygen concentrations and may include species that feed on plankton.

**Collectors (zooplankton)**

Large rivers may also support significant populations of zooplankton that feed on FPOM.

**Figure 3.34** The river continuum.

Most of the invertebrates of streams and rivers live on or in the sediments; that is, most are benthic. These benthic organisms are influenced substantially by the type of bottom sediments. Stony substratum in the riffles and runs of rivers harbor fauna and flora that are different from those in sections with silt or sand bottoms mainly because of differences in the structure and stability of these bottom types. River ecologists have recently discovered that a great number and diversity of invertebrate animals live deep within the sediments of rivers in both the hyporheic and phreatic zones. These species may be pumped up with well water many kilometers from the nearest river. We know very little about the lives of these organisms. Once more, nature has yielded another surprise.

## Human Influences

The influence of humans on rivers has been long and intense. Rivers have been important to human populations for commerce, transportation, irrigation, and waste disposal. Because of their potential to flood, they have also been a constant threat. In the service of human populations, rivers have been channelized, poisoned, filled with sewage, dammed, filled with nonnative fish species, and completely dried. One of the most severe human impacts on river systems has been the building of reservoirs. Reservoirs eliminate the natural flow regime—including flood pulses—alter temperatures, and impede the movements of migratory fish. Because of the rapid turnover of their waters, however, rivers have a great capacity for recovery and renewal. The River Thames in England was severely polluted in the Middle Ages and remained so until recent times. During recent decades, great efforts have been made to reduce the amount of pollution discharged into the Thames, and the river has recovered substantially. The Thames once again supports a run of Atlantic salmon and gives hope to all the beleaguered river conservationists of the world.

# Lakes: Small Seas

In 1892, F. A. Forel defined the scientific study of lakes as the *oceanography of lakes*. On the basis of a lifetime of study, Forel concluded that lakes are much like small seas (fig. 3.35). Differences between lakes and the oceans are due, principally, to the smaller size of lakes and their relative isolation. Perhaps because they are cast on a more human scale, lakes have long captured the imagination of everyone from poets to scientists. For poets such as Henry David Thoreau (1854), they have been sources of inspiration and mirrors of inner truth. For scientists such as Stephen A. Forbes (1887), who wrote, "The lake as a microcosm," they have been mirrors of the outside world and microcosms of the ecological universe.

## Geography

Lakes are simply basins in the landscape that collect water like so many rain puddles. Most lakes are found in regions

**Figure 3.35** Oligotrophic Lake Baikal in Siberian Russia contains 20% of all the surface freshwater on earth.

worked over by the geological forces that produce these basins. These forces include shifting of the earth's crust (tectonics), volcanism, and glacial activity.

Most of the world's freshwater resides in a few large lakes. The Great Lakes of North America together cover an area of over 245,000 km$^2$ and contain 24,620 km$^3$ of water, approximately 20% of all the freshwater on the surface of the planet. An additional 20% of freshwater is contained in Lake Baikal, Siberia, the deepest lake on the planet (1,600 m), with a total volume of 23,000 km$^3$. Much of the remainder is contained within the rift lakes of East Africa. Lake Tanganyika, the second deepest lake (1,470 m), alone has a volume of 23,100 km$^3$, virtually identical to that of Lake Baikal. Still, the world contains tens of thousands of other smaller, shallow lakes, usually concentrated in "lake districts" such as northern Minnesota, much of Scandinavia, and vast regions across north-central Canada and Siberia. Figure 3.36 shows the locations of some of the larger lakes.

## Structure

Lake structure parallels that of the oceans but on a much smaller scale (fig. 3.37). The shallowest waters along the lake shore, where rooted aquatic plants may grow, is called the littoral zone. Beyond the littoral zone in the open lake is the **limnetic zone.** Lakes are generally divided vertically into three main depth zones. The **epilimnion** is the warm surface layer of lakes. Below the epilimnion is the thermocline, or **metalimnion.** The thermocline is a zone through which temperature changes substantially with depth, generally about 1°C per meter of depth. Below the thermocline are the cold dark waters of the **hypolimnion.** Each of these zones supports a distinctive assemblage of lake organisms.

## Physical Conditions

### Light

Lake color ranges from the deep blue of the clearest lakes to yellow, brown, or even red. The color, which depends on light absorption within a lake, is influenced by many factors but

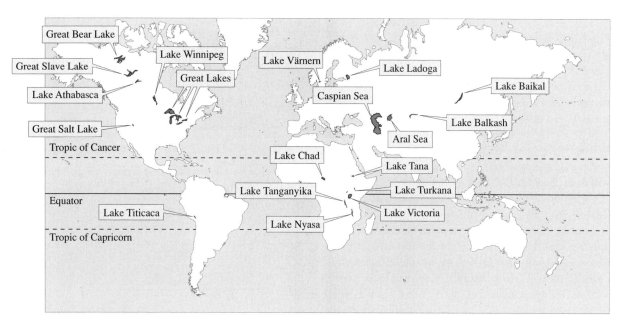

**Figure 3.36** Distributions of some major lakes.

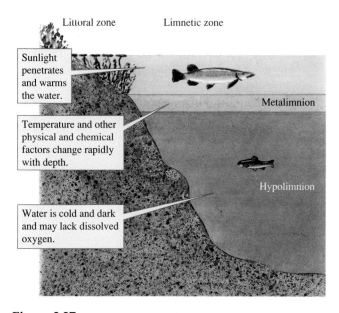

**Figure 3.37** Lake structure.

especially lake chemistry and biological activity. In lakes where the surrounding landscape delivers large quantities of nutrients, primary production is high and phytoplankton populations reduce light penetration. These highly productive lakes are usually a deep green. They are also often shallow and surrounded by cultivated lands or cities. Dissolved organic compounds, such as humic acids leached from forest soils, increase absorption of blue and green light. Absorption in this range shifts lake color to the yellow-brown end of the spectrum. These acid-stained lakes are generally of low productivity. In deep lakes where the landscape delivers low quantities of either nutrients or dissolved organic compounds, phytoplankton production is generally low and light penetrates to

great depths. These lakes, such as Lake Baikal in Siberia, Lake Tahoe in California, and Crater Lake in Oregon, are nearly as blue as the open ocean.

## Temperature

As in the oceans, lakes become thermally stratified as they heat. Consequently, during the warm season, they are substantially warmer at the surface than they are below the thermocline. Temperate lakes are stratified during the summer, while lowland tropical lakes are stratified year-round. As in temperate seas, thermal stratification breaks down in temperate lakes as they cool during the fall. Where lakes freeze over in winter, the water immediately under the ice is approximately 0°C. Meanwhile, bottom water is a comparatively warm 4°C, the temperature at which the density of water is highest. In spring, once the ice has melted, temperate lakes spend a period without thermal stratification. As summer approaches, they gradually become stratified again. In high-elevation tropical lakes, a thermocline may form every day and break down every night! This dynamic situation occurs on the same tropical mountains where, as we saw in chapter 2, terrestrial organisms experience winter temperatures every night and summer temperatures every day. As in the oceans, these patterns of thermal stratification determine the frequency and extent of mixing of the water column. The seasonal dynamics of thermal stratification and mixing in temperate lakes are shown in figure 3.38.

## Water Movements

Wind-driven mixing of the water column is the most ecologically important water movement in lakes. As we have just seen, temperate zone lakes are thermally stratified during the summer, a condition that limits wind-driven mixing to surface waters above the thermocline. During winter on these lakes, ice forms a surface barrier that prevents mixing. In the spring and

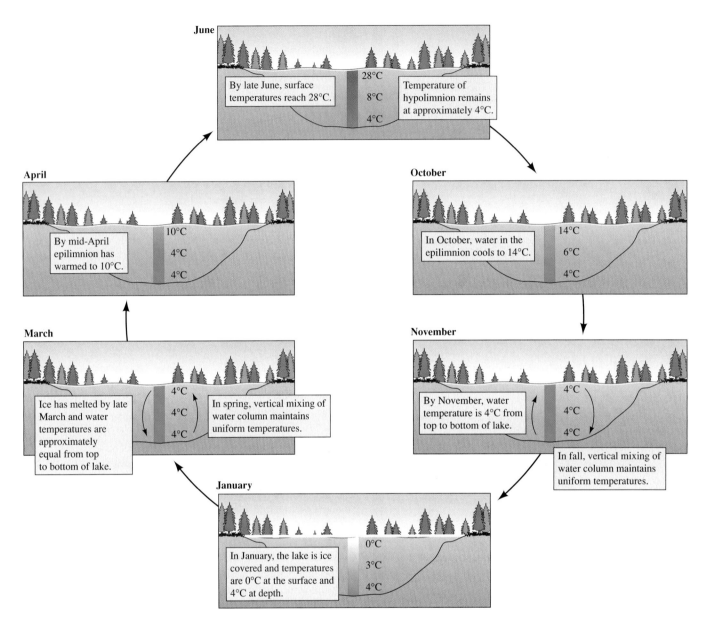

**Figure 3.38** Seasonal changes in temperature in a temperate lake (data from Wetzel 1975).

fall, however, stratification breaks down and winds drive vertical currents that can mix temperate lakes from top to bottom. These are the times when a lake renews oxygen in bottom waters and replenishes nutrients in surface waters. Like tropical seas, tropical lakes at low elevations are permanently stratified. Of the 1,400 m of water in Lake Tanganyika, for example, only about the upper 200 m is circulated each year. Tropical lakes at high elevations heat and stratify every day and cool sufficiently to mix every night. Patterns of mixing have profound consequences to the chemistry and biology of lakes.

## Chemical Conditions

### Salinity

The salinity of lakes is much more variable than that of the open ocean. The world average salinity for freshwater, 120 mg per liter (approximately 0.120 $^0/_{00}$), is a tiny fraction of the salinity of the oceans. Lake salinity ranges from the extremely dilute waters of some alpine lakes to the salt brines of desert lakes. For instance, the Great Salt Lake in Utah sometimes has a salinity of over 200 $^0/_{00}$, which is much higher than oceanic salinity. As we shall see in chapter 23 (see fig. 23.10), the salinity of desert lakes may also change over time, particularly where variations in precipitation, runoff, and evaporation combine to produce wide fluctuations in lake volume.

### Oxygen

Mixing and biological activities have profound effects on lake chemistry. Well-mixed lakes of low biological production, which are called **oligotrophic,** are nearly always well oxygenated. Lakes of high biological production, which are called **eutrophic,** may be depleted of oxygen. Oxygen depletion is particularly likely during periods of thermal stratification, when decomposing organic matter accumulates below the thermocline and consumes oxygen. In eutrophic lakes, oxygen concentrations may be depleted from surface waters at night as respiration continues in the absence of photosynthesis. Oxygen is also often depleted in winter, especially under the ice of productive temperate lakes. In tropical lakes, water below the euphotic zone is often permanently depleted of dissolved oxygen.

## Biology

In addition to their differences in oxygen availability, oligotrophic and eutrophic lakes also differ in factors such as availability of inorganic nutrients and temperature (fig. 3.39). Because aquatic organisms differ widely in their environmental requirements, oligotrophic and eutrophic lakes generally support distinctive biological communities. In temperate regions, oligotrophic lakes generally support the highest diversity of phytoplankton. These lakes are also usually inhabited by fish requiring high oxygen concentrations and relatively low temperatures, such as trout and whitefish. The benthic faunas of these lakes are rich in species and include the larvae of mayflies and caddisflies, small clams, and, along wave-swept shores, the larvae of stoneflies. Eutrophic temperate lakes, which tend to be warmer and, as we have seen, periodically depleted of oxygen, are inhabited by fish tolerant of high temperatures and low oxygen concentrations, such as carp and catfish, or fish that can breathe air in an emergency, such as gars and bowfins. The benthic invertebrate faunas of these lakes

also tend to be tolerant of low oxygen concentrations; for example, midge larvae and tubificid worms, common in such lakes, have hemoglobin that helps them extract oxygen from oxygen-poor waters.

Much less is known about the biology of tropical lakes, however, a few generalizations are possible. Tropical lakes can be very productive. Also, their fish faunas may include a great number of species. Three East African lakes, Lake Victoria, Lake Malawi, and Lake Tanganyika, contain over 700 species of fish, approximately the number of freshwater fish species in all of the United States and Canada; all of western and central Europe and the former Soviet Union together contain only about 400 freshwater fish species. The invertebrates and algae of tropical lakes are much less studied, but it appears that the number of species may be similar to that of temperate zone lakes.

## Human Influences

Human populations have had profound, and usually negative, influences on the ecology of lakes. In addition to examples of ecological degradation, however, are cases of amazing resilience and recovery—resilience in the face of fierce ecological challenge and recovery to substantial ecological integrity. Because lakes offer ready access to water for domestic and industrial uses, many human population centers have grown up around them. In both the United States and Canada, for example, large populations surround the Great Lakes. The human population around Lake Erie, one of the most altered of the Great Lakes, grew from 2.5 million in the 1880s to over

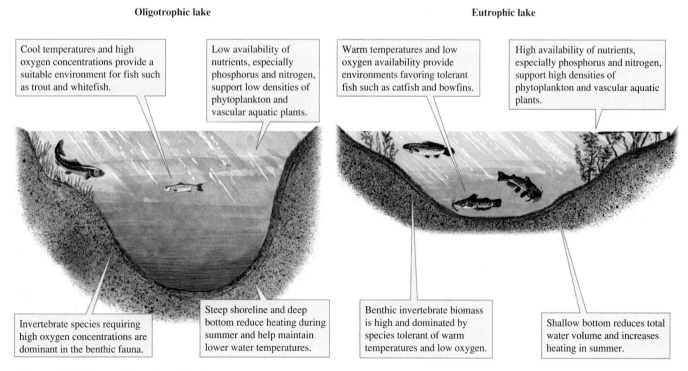

Oligotrophic lake

Cool temperatures and high oxygen concentrations provide a suitable environment for fish such as trout and whitefish.

Low availability of nutrients, especially phosphorus and nitrogen, support low densities of phytoplankton and vascular aquatic plants.

Invertebrate species requiring high oxygen concentrations are dominant in the benthic fauna.

Steep shoreline and deep bottom reduce heating during summer and help maintain lower water temperatures.

Eutrophic lake

Warm temperatures and low oxygen availability provide environments favoring tolerant fish such as catfish and bowfins.

High availability of nutrients, especially phosphorus and nitrogen, support high densities of phytoplankton and vascular aquatic plants.

Benthic invertebrate biomass is high and dominated by species tolerant of warm temperatures and low oxygen.

Shallow bottom reduces total water volume and increases heating in summer.

**Figure 3.39** Oligotrophic and eutrophic lakes.

13 million in the 1980s. The primary ecological impact of these populations has been the dumping of astounding quantities of nutrients and toxic wastes into Lake Erie. By the mid-1960s, the Detroit River alone was dumping 1.5 billion gallons of waste water into Lake Erie each day. The Cuyahoga River, which flows through Cleveland before reaching the lake, was so fouled with oil in the 1960s that it would catch fire. In the face of such ecological challenges, much of Lake Erie, particularly the eastern end, was transformed from a healthy lake with a rich fish fauna to one that was, for a time, essentially an algal soup in which only the most tolerant fish species could live. With greater controls on waste disposal, the process of degradation began to reverse itself, and Lake Erie recovered much of its former health and vitality by the 1980s.

Nutrients aren't the only things that people put into lakes, however. Fish and other species are constantly moved around, either intentionally or unintentionally. For instance, the canals that were dug to connect the Great Lakes with each other and to bypass Niagara Falls inadvertently introduced two species of fish, the sea lamprey and the alewife, that seriously disrupted the biology of the lakes. Once in the Great Lakes, sea lampreys fed mainly on lake trout, lake herring, and chubs. This predation, combined with intense fishing, devastated these commercially important fish populations. As these populations declined, alewife populations exploded. With exploding alewife populations came periodic and massive die-offs that littered beaches with tons of rotting fish. Massive efforts at controlling the sea lamprey by the United States and Canada have been reasonably successful.

These early introductions of fish into the Great Lakes were just a preview of future biological challenges, however. The rogues' gallery of introductions to the Great Lakes, which now includes species such as the zebra mussel, the river ruffe, and the spiny water flea, continues to grow, and there appears to be no end in sight. As figure 3.40 shows, 139 species of fish, invertebrates, plants, and algae had been introduced to the Great Lakes by 1990.

The population growth of many introduced species has been explosive and has had great ecological and economic impacts. One such introduction was that of the zebra mussel, *Dreissena polymorpha,* a bivalve mollusk native to the drainages emptying into the Aral, Caspian, and Black Seas. Zebra mussels disperse by means of pelagic larvae but spend their adult lives attached to the substrate by means of byssal threads. Their pelagic larvae allows them to disperse at a high rate. Though they spread throughout western Europe by the early 1800s, zebra mussels were not recorded in North America until the late 1980s. In 1988, they were collected in Lake Saint Clair, which connects Lake Huron and Lake Erie. In just 3 years, zebra mussels spread to all the Great Lakes and to most of the major rivers of eastern North America.

Locally, zebra mussels have established very dense populations within the Great Lakes. Shells from dead mussels have accumulated to depths of over 30 cm along some shores. Such dense populations threaten the native mussels of the Great Lakes with extinction. Zebra mussels are also fouling water

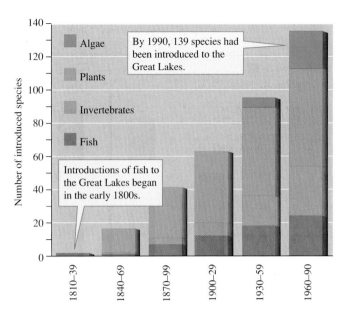

**Figure 3.40** Cumulative number of species introduced to the Great Lakes (data from Mills et al. 1994).

intake structures of power plants and municipal water supplies, which may result in billions of dollars in economic impact. Biologists are working furiously to document and understand the impact of zebra mussels and other species introduced in the Great Lakes. Meanwhile, the governments of Canada and the United States are taking steps to reduce the rate of biological invasion of the Great Lakes. As a consequence of introductions of zebra mussels and other species, the Great Lakes have become a laboratory for the study of human-caused biological invasions (fig. 3.41).

## APPLICATIONS & TOOLS

### Biological Integrity—Assessing the Health of Aquatic Systems

How can we put our knowledge of the natural history of aquatic life to work? A major question that biologists often face is whether a particular influence impairs the health of an aquatic system. Natural history information can play a significant role in making that judgment. Given the complex array of potential human impacts on aquatic systems, what might we use as indicators of health? An answer to this question has been proposed by James Karr and his colleagues, who suggest that we consider what they call "biological integrity," which they define as "a balanced, integrated, adaptive community of organisms having a species composition, diversity, and functional organization comparable to that of the natural habitat of the region" (Karr and Dudley 1981). These

(a)

(b)

**Figure 3.41** Two invaders of the Great Lakes: (*a*) sea lamprey; and (*b*) zebra mussels. Invading species, such as these, have created ecological disasters in freshwater ecosystems around the globe.

researchers proposed that a healthy aquatic community is one that is similar to the community in an undisturbed habitat in the same region. The community should be "balanced" and "integrated." Deciding what constitutes this state requires judgment based on broad knowledge of the habitats in question and their inhabitants—that is, knowledge of natural history. If we could assess the health, as defined by Karr, of a community of aquatic organisms, we would have gone a long way toward assessing the health of the aquatic system in which this community is contained.

Moving beyond general definitions and broad goals, Karr developed an Index of Biological Integrity (IBI) and applied his index to fish communities. Fish communities were chosen

because we know a lot about fish and their habitat requirements and they are relatively easy to sample. Karr's index has three categories for rating a stream or river:

1. number of species and species composition, which includes the number, kinds, and tolerances of fish species;
2. trophic composition, which considers the dietary habits of the fish making up the community;
3. fish abundance and condition.

Under these three categories are 12 attributes of the fish community. The stream is assigned a score of 5, 3, or 1 for each attribute, where 5 equals best and 1 equals worst. The scores on all the attributes are added to give a total score that ranges from 12 (poor biological integrity) to 60 (excellent biological integrity). Notice that Karr has built a safeguard into his index. Judging several attributes of the fish community eliminates the bias that might creep in if assessments were made from only one or a very few attributes. We will examine the three categories of community characteristics in turn.

# Number of Species and Species Composition

In this category the numbers of native and nonnative species are considered. The reason for including the number of native species should be apparent from our discussions, both in this chapter and in chapter 2. Heavy human impact generally reduces the number of native species in a community while increasing the number of nonnative species. The kinds of species that make up the community should also be telling, because some fish, such as trout, are intolerant of poor water quality while others, such as carp, are highly tolerant of poor water quality. The designation of *tolerant* versus *intolerant* species must be tailored for local, or at least regional, circumstance and requires a thorough knowledge of the natural history of the waters under study, as does scoring the number and abundance of species.

# Trophic Composition

The dietary habits of the fish that make up a community reflect kinds of food available in a stream as well as the quality of the environment. The attributes rated in this category are the percentage of fish such as carp that eat a wide range of food and are called **omnivores** by ecologists, the percentage of fish such as trout and bluegill that feed on insects, called **insectivores,** and the percentage of fish such as pike and largemouth bass that feed on other fish, called **piscivores.** Degradation of aquatic systems generally increases the proportion of omnivores and decreases the proportion of insectivores and piscivores in the community. Notice that here, again, scoring the attributes must be based on a thorough understanding of natural history.

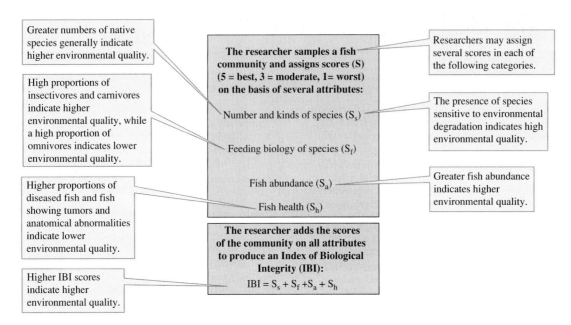

**Figure 3.42** Calculating an Index of Biological Integrity.

# Fish Abundance and Condition

Fish are often less abundant in degraded situations and their condition is often adversely affected. Two aspects of condition are considered for the index. First, what percentage of the individuals are hybrids between different species? Second, what percentage of individuals have noticeable disease, tumors, fin damage, or skeletal deformities—all strong indicators of poor environmental quality. Figure 3.42 summarizes the process of calculating Karr's Index of Biological Integrity.

The Index of Biological Integrity has been successfully applied to a large number of environmental situations. But, as with any tool, full appreciation of the details of the index only comes with use. Let's examine one application of the index.

# An Application

Paul Leonard and Donald Orth (1986) tested Karr's Index of Biological Integrity in seven tributary streams of the New River, which flows through the Appalachian Plateau region of West Virginia. Leonard and Orth had to adapt the index to reflect conditions in their region. In their study streams the number of darter species, small benthic fish in the family Percidae, indicates high environmental quality, while increasing numbers of creek chubs indicate increasing pollution. In addition, high proportions of insectivores indicate excellent environmental conditions, while high proportions of generalist feeders, or omnivores, indicate poor conditions. High densities of fish were taken as a sign of high environmental quality, while the presence of diseased or deformed individuals indicated environmental problems.

Leonard and Orth assigned scores of 1 (worst conditions), 3 (fair conditions), or 5 (best conditions) for each of the variables they studied at each of their sampling sites in the study

streams. They then summed the scores for the seven variables at each site to determine an Index of Biological Integrity. The minimum possible value was 7, poorest conditions, and the maximum possible value was 35, best conditions. They next made independent estimates of levels of pollution at each study site. Their estimates were based upon the daily discharge of municipal sewage and the local densities of septic tanks, roads, and mines. The study streams showed a wide range of environmental pollution due to sewage, mining, and urban development. Leonard and Orth found that the Index of Biological Integrity correlated well with independent estimates of pollution at each study site (fig. 3.43).

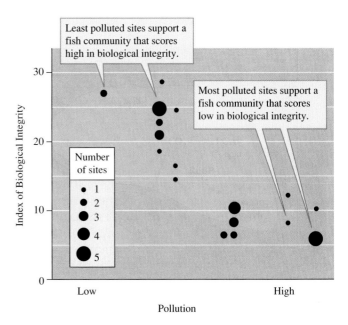

**Figure 3.43** Pollution and the Index of Biological Integrity (data from Leonard and Orth 1986).

Many other investigators have tested the ability of the Index of Biological Integrity to represent the extent of environmental degradation in rivers and lakes. The index is effective in a wide range of regions and aquatic environments. The important point here is that natural history is being put to work to address important environmental problems. The foundation of natural history built in this chapter and in chapter 2 is useful now as we go forward to study ecology at levels of organization ranging from individual species through the entire biosphere.

## SUMMARY

Humans everywhere hold a land-centered perspective of the planet. Consequently, aquatic life is often most profuse where conditions appear most hostile to people, for example, along cold, wave-swept seacoasts, in torrential mountain streams, and in the murky waters where rivers meet the sea.

**The hydrologic cycle exchanges water among reservoirs.** Of the water in the biosphere, the oceans contain 97% and the polar ice caps and glaciers an additional 2%, leaving less than 1% as freshwater. The turnover of water in the various reservoirs of the hydrologic cycle ranges from only 9 days for the atmosphere to 3,100 years for the oceans.

**The biology of aquatic environments corresponds broadly to variations in physical factors such as light, temperature, and water movements and to chemical factors such as salinity and oxygen.**

The *oceans* form the largest continuous environment on earth. An ocean is generally divided vertically into several depth zones, each with a distinctive assemblage of marine organisms. Limited light penetration restricts photosynthetic organisms to the photic, or epipelagic, zone and leads to thermal stratification. Oceanic temperatures are much more stable than terrestrial temperatures. Tropical seas are more stable physically and chemically; temperate and high-latitude seas are more productive. Highest productivity occurs along coastlines. The open ocean supports large numbers of species and is important to global carbon and oxygen budgets.

*Kelp forests* are found mainly at temperate latitudes. *Coral reefs* are limited to the tropics and subtropics to latitudes between 30° N and S latitudes. Coral reefs are generally one of three types: fringing reefs, barrier reefs, and atolls. Kelp beds share several structural features with terrestrial forests. Both seaweeds and reef-building corals grow only in surface waters, where there is sufficient light to support photosynthesis. Kelp forests are generally limited to areas where temperature ranges from about 10° to 20°C, while reef-building corals are limited to areas with temperatures of about 18° to 29°C. The diversity and productivity of coral reefs rival that of tropical rain forests.

The *intertidal* zone lines the coastlines of the world. It can be divided into several vertical zones: the supratidal, high intertidal, middle intertidal, and low intertidal. The magnitude and timing of the tides is determined by the interaction of the gravitational effects of the sun and moon with the configuration of coastlines and basins. Tidal fluctuation produces steep gradients of physical and chemical conditions within the intertidal zone. Exposure to waves, bottom type, height in the intertidal zone, and biological interactions determine the distribution of most organisms within this zone.

*Salt marshes, mangrove forests,* and *estuaries* occur at the transitions between freshwater and marine environments and between marine and terrestrial environments. Salt marshes, which are dominated by herbaceous vegetation, are found mainly at temperate and high latitudes. Mangrove forests grow in the tropics and subtropics. Estuaries are extremely dynamic physically, chemically, and biologically. The diversity of species is not as high in estuaries, salt marshes, and mangrove forests as in some other aquatic environments but productivity is exceptional.

*Rivers* and *streams* drain most of the land area of the earth and reflect the land use in their basins. Rivers and streams are very dynamic systems and can be divided into several distinctive environments: longitudinally, laterally, and vertically. Periodic flooding has important influences on the structure and functioning of river and stream ecosystems. The temperature of rivers follows variation in air temperature but does not reach the extremes occurring in terrestrial habitats. The flow and chemical characteristics of rivers change with climatic regime. Current speed, distance from headwaters, and the nature of bottom sediments are principal determinants of the distributions of stream organisms.

*Lakes* are much like small seas. Most are found in regions worked over by tectonics, volcanism, and glacial activity, the geological forces that produce lake basins. A few lakes contain most of the freshwater in the biosphere. Lake structure parallels that of the oceans but on a much smaller scale. The salinity of lakes, which ranges from very dilute waters to over 200 $^0/_{00}$, is much more variable than that of the oceans. Lake stratification and mixing vary with latitude. Lake flora and fauna largely reflect geographic location and nutrient content.

Potential threats to all these aquatic systems include overexploitation of populations and waste dumping. Reservoir construction and flow regulation have had major negative impacts on river ecosystems and biodiversity. Freshwater environments are particularly vulnerable to the introduction of exotic species. The nature of fish assemblages is being used to assess the "biological integrity" of freshwater communities. The application of this Index of Biological Integrity depends on detailed knowledge of the natural history of regional fish faunas.

# Review Questions

1. Review the distribution of water among the major reservoirs of the hydrologic cycle. What are the major sources of freshwater? Explain why according to some projections availability of freshwater may limit human populations and activity.

2. The oceans cover about 360 million $km^2$ and have an average depth of about 4,000 m. What proportion of this aquatic system receives sufficient light to support photosynthesis? Make the liberal assumption that the photic zone extends to a depth of 200 m.

3. Below about 600 to 1,000 m in the oceans there is no sunlight. However, many of the fish and invertebrates at these depths have eyes. In contrast, fish living in caves are often blind. What selective forces could maintain eyes in populations of deep-sea fish? (Hint: Many species of deep-sea invertebrates are bioluminescent.)

4. Darwin (1842) was the first to propose that fringing reefs, barrier reefs, and atolls are different stages in a developmental sequence that begins with a fringing reef and ends with an atoll. Outline how this process might work. How would you test your ideas?

5. How does feeding by urchins, which prey on young corals, improve establishment by young corals? Use a diagram outlining interactions between urchins, corals, and algae to help in the development of your explanation.

6. How might a history of exposure to wide environmental fluctuation affect the physiological tolerances of intertidal species compared to close relatives in subtidal and oceanic environments? How might salinity tolerance vary among organisms living at different levels within the intertidal?

7. How might oxygen concentration of interstitial water be related to the grain size of the sand or mud sediment? How might the oxygen concentrations of tide pools in sheltered bays compare to those on the shores of exposed headlands?

8. According to the river continuum model, the organisms inhabiting headwater streams in temperate forest regions depend mainly upon organic material coming into the stream from the surrounding forests. According to the model, photosynthesis within the stream is only important in the downstream reaches of these stream systems. Explain. How would you go about testing the predictions of the river continuum model?

9. How could you test the generalization that lake primary production and the composition of the biota living in lakes are strongly influenced by the availability of nutrients such as nitrogen and phosphorus? Assume that you have unlimited resources and that you have access to several experimental lakes.

10. Biological interactions may also affect lake systems. How does the recent history of the Great Lakes suggest that the kinds of species that inhabit a lake influence the nature of the lake environment and the composition of the biological community?

# Suggested Readings

Barnes, R. S. K. and R. N. Hughes. 1988. *An Introduction to Marine Ecology*. Oxford: Blackwell Scientific Publications.

*An excellent and readable introduction to marine ecology.*

Carson, R. 1951. *The Sea Around Us*. London: Oxford University Press.

Carson, R. 1955. *The Edge of the Sea*. Boston: Houghton Mifflin.

*Two evocative masterpieces on the natural history of the sea. Models for natural history writing on any ecosystem.*

Chapman V. J. 1977. *Ecosystems of the World 1. Wet Coastal Ecosystems*. Amsterdam: Elsevier Scientific Publishing.

Cushing, C. E., K. W. Cummins, and G. W. Minshall. 1995. *Ecosystems of the World 22. River and Stream Ecosystems*. Amsterdam: Elsevier Scientific Publishing.

Dubinsky, Z. 1990. *Ecosystems of the World 25. Coral Reefs*. Amsterdam: Elsevier Scientific Publishing.

*This series by Elsevier Scientific Publishing provides detailed information on most of the world's terrestrial and aquatic ecosystems.*

Grassle, J. F. 1991. Deep-sea benthic biodiversity. *BioScience* 41:464–69.

Grassle, J. F., P. Lasserre, A. D. McIntyre, and G. C. Ray. 1991. Marine biodiversity and ecosystem function. *Biology International Special Issue* 23:i–iv, 1–19.

Jackson, J. B. C. 1991. Adaptation and diversity of reef corals. *BioScience* 41:475–82.

Thorne-Miller, B. and J. Catena. 1991. *The Living Ocean: Understanding and Protecting Marine Biodiversity*. Washington, D.C.: Island Press.

*Explorations on the frontiers of marine biodiversity. Much biodiversity is left to be discovered in marine environments.*

Karr, J. R. 1991. Biological integrity: a long-neglected aspect of water resource management. *Ecological Applications* 1:66–84.

Woodley, S., J. Kay, and G. Francis. 1993. *Ecological Integrity and the Management of Ecosystems*. Delray Beach, Fla.: St. Lucie Press.

*Two useful presentations on the concept of ecological integrity.*

Ketchum, B. H. 1983. *Ecosystems of the World 26. Estuaries and Enclosed Seas*. Amsterdam: Elsevier Scientific Publishing.

Mills, E. L., J. H. Leach, J. T. Carlton, and C. L. Secor. 1994. Exotic species and the integrity of the Great Lakes. *BioScience* 44:666–76.

Nalepa, T. F. and D. W. Schloesser. 1992. *Zebra Mussels Biology, Impacts, and Control*. Ann Arbor: Lewis Publishers.

*The causes and consequences of introductions of exotic species are presented in these two references.*

Pomeroy, L. R. and R. G. Wiegert. 1981. *The Ecology of a Salt Marsh, Ecological Studies 38*. New York: Springer-Verlag.

Teal, J. and M. Teal. 1969. *Life and Death of the Salt Marsh*. Boston: Little, Brown.

*Two presentations of salt marsh ecology. The first is a technical case study. The second highlights natural history.*

Ricketts, E. F., J. Calvin, J. W. Hedgpeth, and D. W. Phillips. 1985. *Between Pacific Tides*. 5th ed. Stanford, Conn.: Stanford University Press.

*A classic on intertidal ecology of western North America.*

# On the Net

Visit this textbook's accompanying website at www.mhhe.com/ecology (click on the book's title) to take advantage of practice quizzing, study/writing tips, timely news articles, and additional URLs for research on the topics in this chapter.

Freshwater Habitats

Marine Ecology

The Marine Pelagic Zone

The Marine Deep Sea Zone

Kelp Forests

The Coast and the Continental Shelf

Coral Reefs

Rocky Shore Communities

Beaches and Mud Flats

Shallow Subtidal Communities

Estuaries

Macroscopic Algae and Sea Grasses

Mangroves

Water Use and Management

# SECTION II

## *Individuals*

*"Learn from the birds what food the thickets yield;
Learn from the beasts the physic of the field."*

Alexander Pope, *An Essay on Man*, 1733

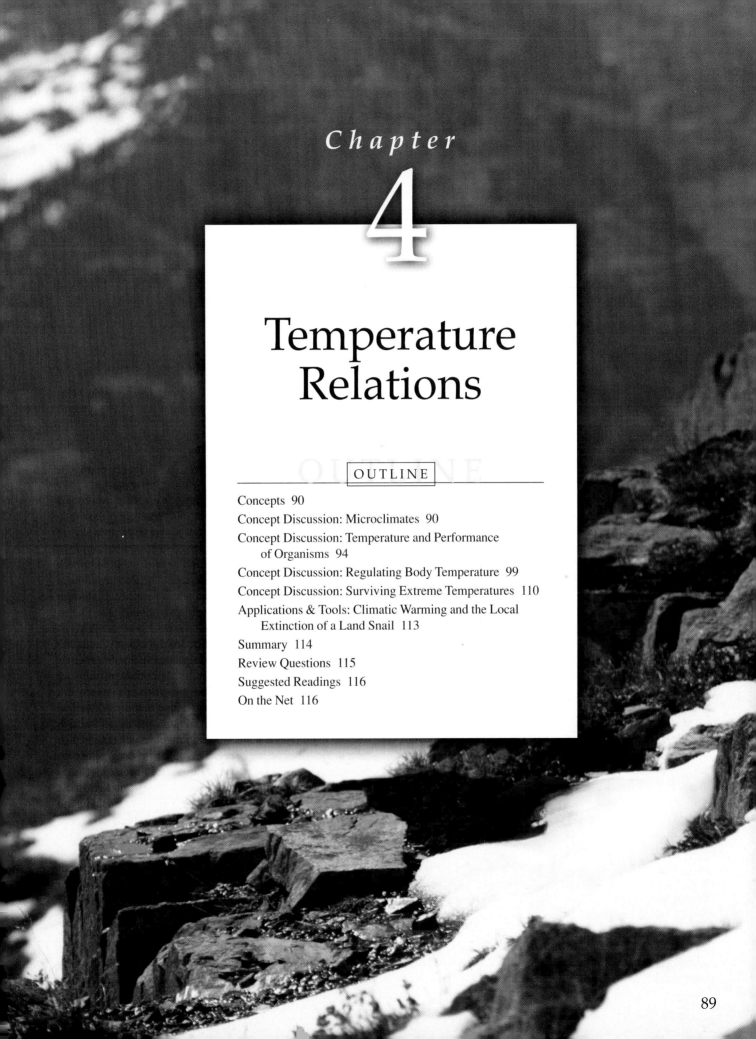

# Chapter

# 4

# Temperature Relations

Many organisms regulate the temperature of their bodies or the temperature of parts of their anatomy. At least one plant of the arctic tundra regulates the temperature of its reproductive structures. Peter Kevan had come to Ellesmere Island, which lies at about 82° N latitude in the Northwest Territories of Canada, to study sun-tracking behavior by arctic flowers. It was summer, there was little wind, and the sun stayed above the horizon 24 hours each day. As the sun's position in the arctic sky changed, one of the common tundra flowers, *Dryas integrifolia* (fig. 4.1), like the sunflowers of lower latitudes, followed.

Kevan found that the sun-tracking behavior of *Dryas* increased the temperature of its flowers. Though the air temperature hovered around 15°C, the temperature of the *Dryas* flowers was nearly 25°C. Kevan discovered that the flowers act like small solar reflectors; their parabolic shape reflects and concentrates solar energy on the reproductive structures. He also observed that many species of small insects, attracted by their warmth, basked in the sun-tracking *Dryas* flowers, elevating their body temperatures as a consequence (fig. 4.1). *Dryas* depends on these insects to pollinate its flowers.

How does *Dryas* and its insect visitors benefit from their basking behaviors? How does cloud cover affect the temperature and sun-tracking behavior of *Dryas* flowers? These are the kinds of questions addressed by Kevan (1975) and other ecologists who study the ecology of temperature relations, one of the most fundamental aspects of ecology. In their quest for answers to questions like these, ecologists learn how the world works.

The thermometer was one of the first quantitative instruments to appear in the scientific tool kit, and we have been measuring and reporting temperatures ever since. Human concern for temperature shows itself everywhere. Local television reviews the high and low temperatures of the preceding day and forecasts temperatures for the coming day. Daily newspapers report temperatures from nearly every corner of the globe. If two people from different regions meet, the first questions they ask concern the weather: Are the summers very hot? Are the winters cold? We wear our endurance of extreme temperatures like badges of heroism; yet today we listen apprehensively to the forecast of a small temperature change—the prospect of global warming.

Why is *Homo sapiens* so concerned with temperature? For us and all other species, the impact of extreme temperatures can range from discomfort, at a minimum, to extinction. Long-term changes in temperature have set entire floras and faunas marching across continents, some species thriving, some holding on in small refuges, and others becoming extinct. Areas now supporting temperate species were at times tropical and at other times the frigid homes of reindeer and woolly mammoths.

We defined ecology as the study of the relationships between organisms and their environments. In chapter 4, we examine the relationship between individual organisms and temperature, one of the most important environmental factors in the lives of organisms.

## CONCEPTS

- **Macroclimate interacts with the local landscape to produce microclimatic variation in temperature.**
- **Most species perform best in a fairly narrow range of temperatures.**
- **Many organisms have evolved ways to compensate for variations in environmental temperature by regulating body temperature.**
- **Many organisms survive extreme temperatures by entering a resting stage.**

## CONCEPT DISCUSSION

### Microclimates

**Macroclimate interacts with the local landscape to produce microclimatic variation in temperature.**

Microclimate is a fundamental aspect of environmental variation. What do we mean by macroclimate and microclimate? **Macroclimate** is what weather stations report and

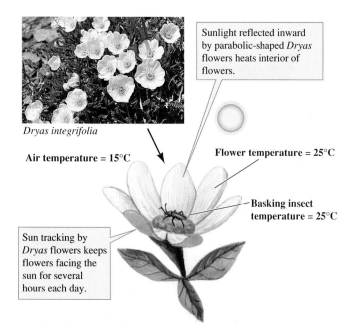

*Dryas integrifolia*

Sunlight reflected inward by parabolic-shaped *Dryas* flowers heats interior of flowers.

Air temperature = 15°C

Flower temperature = 25°C

Basking insect temperature = 25°C

Sun tracking by *Dryas* flowers keeps flowers facing the sun for several hours each day.

**Figure 4.1** Sun-tracking behavior of the arctic plant, *Dryas integrifolia,* heats the reproductive parts of its flowers, making them attractive to pollinating insects.

what we represented with climate diagrams in chapter 2. **Microclimate** is climatic variation on a scale of a few kilometers, meters, or even centimeters, usually measured over short periods of time. You acknowledge microclimate when you choose to stand in the shade on a summer's day or in the sun on a winter's day. Macroclimate and microclimate are usually substantially different. Many organisms live out their lives in very small areas during periods of time ranging from days to a few months. For these organisms macroclimate may be less important than microclimate. Microclimate is influenced by landscape features such as altitude, aspect, vegetation, color of the ground, and presence of boulders and burrows. The physical nature of water reduces temperature variation in aquatic environments.

## Altitude

As we saw in chapter 2 (see fig. 2.38), temperatures are generally lower at high elevations. Along the elevational gradient presented in figure 2.38, average annual temperature is 11.1°C at 1,660 m compared to –3.7°C at 3,743 m. Lower average temperatures at higher elevations are a consequence of several factors. First, because atmospheric pressure decreases with elevation, air rising up the side of a mountain expands. The energy of motion (kinetic energy) required to sustain the greater movement of air molecules in the expanding air mass is drawn from the surroundings, which cool as a result. A second reason that temperatures are generally lower at higher elevations is that there is less atmosphere to trap and radiate heat back to the ground.

## Aspect

Topographic features such as hills, mountains, and valleys create microclimates that would not occur in a flat landscape. Mountains and hillsides create these microclimates by shading parts of the land. In the Northern Hemisphere, the shaded areas are on the north-facing sides, or *northern aspects,* of hills, mountains, and valleys, which face away from the equator. In the Southern Hemisphere, the *southern aspect* faces away from the equator.

You can see the effect of aspect, in miniature, around buildings. If you want to warm yourself on a sunny winter's day in the Northern Hemisphere, you go to the south side of a building, to its southern aspect, which faces the equator. In the Southern Hemisphere, you would generally find the warmest spot on the north side of a building. Similarly, the northern and southern aspects of mountains and valleys offer organisms contrasting microclimates. The microclimates of north- and south-facing aspects of hillsides may support very different types of vegetation (fig. 4.2).

**Figure 4.2**  The north-facing slope at this site supports a Mediterranean woodland, while the vegetation on the south-facing slope is mainly grassland.

The greater density of oaks and shrubs on the north-facing slope shown in figure 4.2 is paralleled in miniature on north- and south-facing dune slopes in the Negev Desert, where north-facing slopes support a higher density of crust-forming mosses. G. Kidron, E. Barzilay, and E. Sachs, earth scientists from Hebrew University Jerusalem, documented a possible physical basis for the differences in moss cover (Kidron, Barzilay, and Sachs 2000). They found that north-facing dune slopes are cooler: 7.8° to 9.2°C cooler at midday in winter and 1.8° to 2.5°C cooler at midday in summer. These earth scientists also found north-facing slopes remain moist approximately 2.5 times longer than south-facing slopes following rainfall. They suggested that lower evaporation rates on north-facing slopes are at least partly responsible. Detailed physical studies such as this one by Kidron, Barzilay, and Sachs provide for a basic understanding of the distribution of organisms, especially vegetation.

## Vegetation

Because they also shade the landscape, plants create microclimates. For instance, trees, shrubs, and plant litter (fallen leaves, twigs, and branches) produce ecologically important microclimates in deserts. The desert landscape, which often consists of a mosaic of vegetation and bare ground, is also a patchwork of sharply contrasting thermal environments. Such a patchwork is apparent near Kemmerer, Wyoming, a cold desert much like the Gobi in Mongolia (see fig. 2.19). Like the Gobi, Kemmerer can be bitterly cold in winter and blistering in summer. One summer's day Robert Parmenter and his colleagues (1989) measured the temperatures in various parts of the Kemmerer landscape. Parmenter found that while the

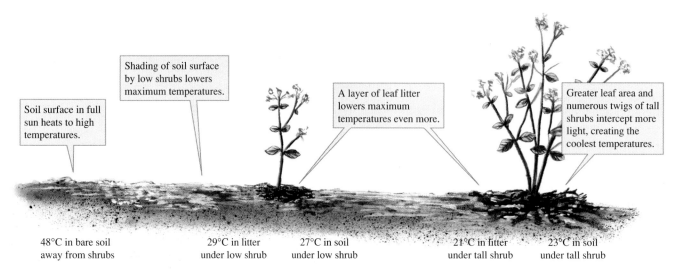

**Figure 4.3** Desert shrubs create distinctive thermal microclimates in the desert landscape (data from Parmenter, Parmenter, and Cheney 1989).

temperature on bare soil soared to 48°C, a few meters away in plant litter under a tall shrub the temperature was a moderate 21°C (fig. 4.3). Meanwhile, temperatures under low shrubs with less leaf area were a bit warmer but still not as hot as soil in the open. A small organism in this landscape could choose microclimates differing in temperature by 27°C.

# Color of the Ground

Another factor that can significantly affect temperatures is the color of the ground. This statement may sound a bit odd if you are from a moist climate, either temperate or tropical, where vegetation usually covers the ground. But, as we have just seen, much of the arid or semiarid landscape is bare ground, which can vary widely in color. Colors have been used to name deserts around the world, such as the central Asian

deserts called Kara Kum, which means black sand in Turkish, and Kyzyl Kum, or red sand, and White Sands, New Mexico (fig. 4.4).

Bare ground is the dominant environment offered by beaches. Neil Hadley and his colleagues (1992) studied the beaches of New Zealand, which range in color from white to black and offer a wide range of microclimates to beach organisms. These beaches heat up under the summer sun, but black beaches heat up faster and to higher temperatures. The black beaches heat up more because they absorb more visible light than do the white beaches (fig. 4.5). When air temperatures at both beaches hovered around 30°C, Hadley and his colleagues found that the temperature of the sand on the white beach averaged around 45°C. In contrast, they measured sand temperatures on the black beach as high as 65°C. Though these white and black beaches are exposed to nearly identical macroclimates, they have radically different microclimates.

(a)

(b)

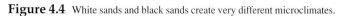

**Figure 4.4** White sands and black sands create very different microclimates.

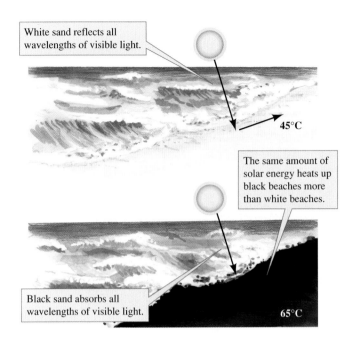

**Figure 4.5** By reflecting most visible light, white sands create a much cooler microclimate than black sands (data from Hadley, Savill, and Schultz 1992).

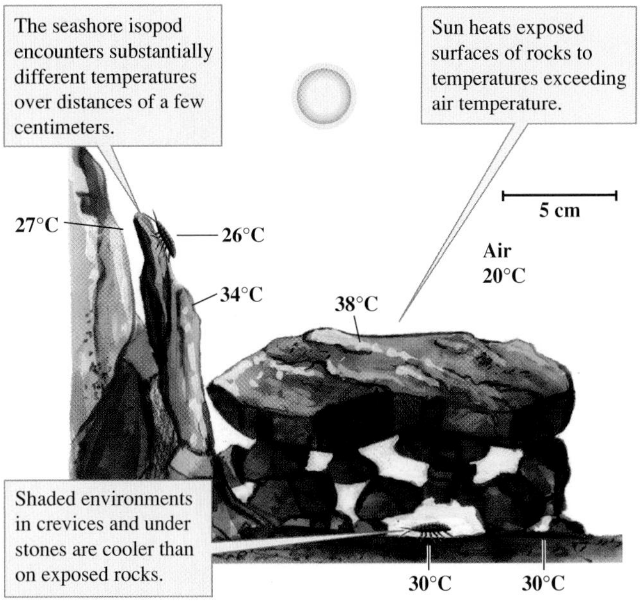

**Figure 4.6** Stones create distinctive microclimates (data from Edney 1953).

# Presence of Boulders and Burrows

Many children soon discover that the undersides of stones harbor a host of organisms seldom seen in the open. This is partly because the stones create distinctive microclimates. E. B. Edney's classic studies (1953) of the seashore isopod *Ligia oceanica* documented the effect of stones on microclimate. Edney found that over the space of a few centimeters, *Ligia* could choose air temperatures ranging from 20°C in the open to 30°C in the air spaces under stones, which heated to between 34° and 38°C. This small-scale variation in temperature is summarized in figure 4.6.

Animal burrows also have their own microclimates, in which temperatures are usually more moderate than at the soil surface. For example, while daily temperatures under a shrub in the Chihuahuan Desert ranged from 17.5° to 32°C, temperatures in a nearby mammal burrow ranged from 26° to 28°C. This burrow was cooler than the surface during the day and warmer at night. What do these data suggest about the microclimates experienced by plant roots, soil bacteria, and burrowing animals?

# Aquatic Temperatures

As we saw in chapter 2, air temperature generally fluctuates more than water temperature. The thermal stability of the aquatic environment derives partly from the high capacity of water to absorb heat energy without changing temperature (a capacity called *specific heat*). This capacity is about 3,000 times higher for water than for an equal volume of air. It takes approximately 1 calorie of energy to heat 1 cm³ of water 1°C. For an equal volume of air, this temperature rise requires only about 0.0003 calories.

A second cause of the thermal stability of aquatic environments is the large amount of heat absorbed by water as it evaporates (which is called the *latent heat of vaporization*). This amounts to about 584 calories per gram of water at 22°C and 580 calories per gram of water at 35°C. So, 1 g of water evaporating from the surface of a desert stream, a lake, or a tide pool at 35°C draws 580 calories of heat from its surroundings. From the definition of a calorie, this is enough energy to cool 580 g of water 1°C. What makes evaporative coolers, whether mechanical or biological, so effective? The answer is that you need to evaporate relatively little water to cool a great deal of either air or living matter.

A third cause of the greater thermal stability of aquatic environments is the heat energy that water gives up to its environment as it freezes (the *latent heat of fusion*). Water gives up approximately 80 calories as 1 g of water freezes because the energy of motion of water molecules decreases as they leave the liquid state and become incorporated into the crystalline latticework of ice. So, as 1 g of pond water freezes, it gives off sufficient energy to heat 80 g of water 1°C, thus retarding further cooling.

The aquatic environments with greatest thermal stability are generally large ones, such as the open sea. These are environments that store large quantities of heat energy and where daily fluctuations are often less than 1°C. Even the temperatures of small streams, however, usually fluctuate less than the temperatures of nearby terrestrial habitats. Figure 4.7 summarizes the daily range in air temperature with that in the Coal

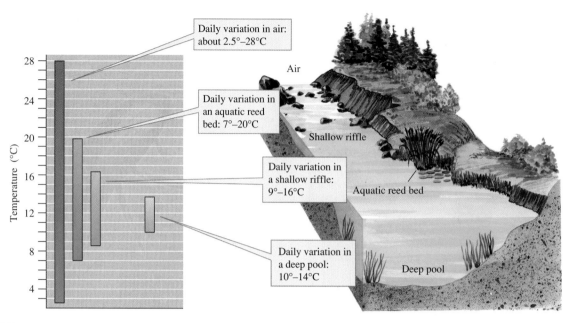

**Figure 4.7** Aquatic microclimates; aquatic environments generally show less temperature variation compared to terrestrial environments (data from Ward 1985).

River in Tasmania. While air temperature ranged from 2.5° to 28°C, daily surface temperatures in the river ranged from 7° to 20°C. In other words, the range in air temperature was nearly twice that of water temperature. Meanwhile, the temperature 60 cm below the surface of the Coal River ranged from 10° to 14°C, a fraction of the daily variation in air temperature.

Other factors besides the physics of water can affect the temperature of aquatic environments. **Riparian vegetation,** that is, vegetation that grows along rivers and streams, influences the temperature in streams in the same way that vegetation modifies the temperature of desert soils—by providing shade. Shading by riparian vegetation reduces temperature fluctuations by insulating the stream environment.

## CONCEPT DISCUSSION

### Temperature and Performance of Organisms

**Most species perform best in a fairly narrow range of temperatures.**

Ecologists concerned with the ecology of individual organisms study how environmental factors, such as temperature, water, and light, affect the physiology and behavior of organisms: how fast they grow; how many offspring they produce; how fast they run, fly, or swim; how well they avoid preda-

tors; and so on. We can group these phenomena and say that ecologists study how environment affects the "performance" of organisms.

Whether in response to variations in temperature, moisture, light, or nutrient availability, most species perform best in a fairly narrow range of environmental conditions. The influence of temperature on performance can tell us how organisms respond to physical and chemical factors generally. In this section, we discuss the influence of temperature on animal performance, on photosynthesis by plants, and on microbial activity.

## Temperature and Animal Performance

Let's begin our discussion of temperature and animal performance by reviewing the influence of temperature on enzyme function. The influence of temperature on the performance of organisms begins at the level of biomolecules, which often perform their functions by balancing opposing tendencies. Consider enzymes. Because they must match the shape of the substrate upon which they act, enzymes must assume a specific shape for proper function. Most enzymes have a rigid, predictable shape at low temperatures, but rates of chemical reactions tend to be low at these low temperatures. Also, rigidity does not help an enzyme perform its function. Their functioning often depends upon flexibility, the ability to assume another shape after binding with the substrate. Enzymes have greater flexibility at higher temperatures, but excessively high temperatures destroy their shape. Temperatures at either extreme thus impair the functioning of enzymes.

Enzymes usually work best in some intermediate range of temperatures, neither too hot nor too cold, where they retain both proper shape and sufficient flexibility. In other words, there is usually some optimal range of temperatures for most enzymes. How might you determine the optimal temperature for an enzyme? One way that molecular biologists assess the optimal conditions for enzyme performance is to determine the concentration of substrate required for an enzyme to work at a particular rate. If this concentration is low, the enzyme is performing well at low concentrations of the substrate; that is, the enzyme has a high affinity for the substrate. The affinity of an enzyme for its substrate is one measure of its performance.

John Baldwin and P. W. Hochachka (1970) studied the influence of temperature on the activity of acetylcholinesterase, an enzyme produced at the synapse between neurons. This enzyme promotes the breakdown of the neurotransmitter acetylcholine to acetic acid and choline and so turns off neurons, a process critical for proper neural function. The researchers found that rainbow trout, *Oncorhynchus mykiss,* produce two forms of acetylcholinesterase. One form has highest affinity for acetylcholine at 2°C, that is, at winter temperatures. However, the affinity of this enzyme for acetylcholine declines rapidly above 10°C. The second form of acetylcholinesterase shows highest affinity for acetylcholine at 17°C, at summer temperatures. However, the affinity of this second form of acetylcholinesterase falls off rapidly at both higher and lower temperatures. In other words, the optimal temperatures for the two forms of acetylcholinesterase are 2° and 17°C (fig. 4.8).

This influence of temperature on the performance of acetylcholinesterase makes sense if you consider the temperatures of the rainbow trout's native environment. Rainbow trout are native to the cool, clear streams and rivers of western North America. During winter, the temperatures of these streams hover between 0° and 4°C, while summer temperatures approach 20°C. These environmental temperatures are similar to the temperatures at which the acetylcholinesterase of rainbow trout performs optimally.

Today, rainbow trout have been introduced around the world but are still largely confined to cold waters that don't get much warmer than about 20°C, even at the height of summer. Now you have a biochemical mechanism to explain these distributional limits. At temperatures above 20°C, what happens to the performance of the acetylcholinesterase produced by rainbow trout? Some signs of thermal stress in fish are loss of equilibrium, swimming on their sides, and swimming in spirals. Can you explain these responses using what you now know about the influence of temperature on the performance of the acetylcholinesterase?

Studies of reptiles, especially lizards and snakes, are offering additional valuable insights into the influence of temperature on animal performance. Widely distributed species often offer the opportunity for studies of local variation in ecological relationships, including the influ-

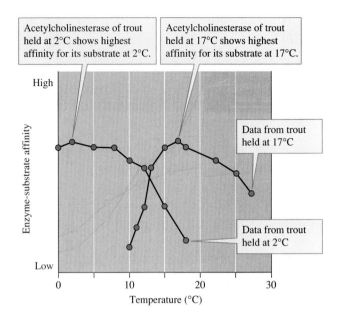

**Figure 4.8**   Enzyme activity is affected substantially by temperature (data from Baldwin and Hochachka 1970).

ence of temperature on performance. For example, the eastern fence lizard, *Sceloporus undulatus,* is found across approximately two-thirds of the United States, living in a broad diversity of climatic zones. Taking advantage of this wide range of environmental conditions, Michael Angilletta (2001) studied the temperature relations of *S. undulatus* over a portion of its range. In one of his studies, Angilletta determined how temperature influences metabolizable energy intake or MEI. He measured MEI as the amount of energy consumed (C) minus energy lost in feces (F) and uric acid (U), which is the nitrogen waste product produced by lizards. We can summarize MEI in equation form as:

$$MEI = C - F - U$$

Angilletta studied two populations from New Jersey and South Carolina, regions with substantially different climates. He collected a sample of lizards from both populations and maintained portions of his samples from both populations at 30°, 33°, and 36°C. Angilletta kept his study lizards in separate enclosures and provided them with crickets that he had weighed to the nearest 0.1 mg as food. Since he had determined the energy content of an average cricket, Angilletta was able to determine the energy intake by each lizard by counting the number of crickets they ate and calculating the energy content of that number. He determined the energy lost as feces (F) and uric acid (U) by collecting all the feces and uric acid produced by each lizard and then drying and weighing this material. He estimated the average energy content of feces and uric acid using a bomb calorimeter.

*Investigating the Evidence*

## Laboratory Experiments

One of the most powerful ways to test a hypotheses is through an experiment. Experiments used by ecologists generally fall into one of two categories—field or laboratory experiments. Field and laboratory experiments generally provide complementary information or evidence, and differ somewhat in their design. Here we discuss the design of laboratory experiments.

In a laboratory experiment, the researcher attempts to control all factors but one. That one factor, which is not controlled, is the one of interest to the experimenter and it is the one that the experimenter varies across experimental conditions. Let's draw an example of a laboratory experiment discussed in chapter 4. Based upon published studies, Michael Angilletta (2001), concluded that geographically separated populations of the eastern fence lizard, *Sceloporus undulatus,* may differ physiologically or behaviorally. Because of his interest in the potential differences among populations *S. undulatus,* he designed a number of field and laboratory studies to test for those differences.

Angilletta designed one laboratory experiment to determine if the rate of metabolizable energy intake by two geographically separated populations is influenced by temperature. The results of that experiment are summarized by figure 4.9. What we want to consider here is the design of the experiment that produced those results. What factors do you think Angilletta may have attempted to control in this experiment? First, he used similar numbers of lizards from the two populations. He tested 20 lizards from both populations at 33°C, 13 from New Jersey at 30° and 36°C, and 14 from South Carolina at 30° and 36°C. A second factor that Angilletta controlled was lizard size. Lizards

from both populations used in the experiments had an average body mass of approximately 5.4 g. Since males and females may differ physiologically, Angilletta included approximately equal numbers of males and females in his experiments. He also was careful to expose all the lizards to the same quality of light and to the same numbers of hours of light and darkness and he maintained them in the same kinds of experimental enclosures. Angilletta also fed all the lizards in his experiment the same type of food: live crickets. The list could go on but these are the major factors controlled in this experiment.

Now, what factors did Angilletta vary in that experiment? For each study population, New Jersey or South Carolina, he varied a single factor: temperature. In the experiment, Angilleta maintained lizards from New Jersey and South Carolina at three temperatures: 30°, 33°, and 36°C and estimated their rates of metabolizable energy intake at these three temperatures. Angilletta's experiment revealed that lizards from both populations have a maximum metabolizable energy intake at 33°C. This result suggests that the optimum temperature for feeding does not differ for the two populations. However, the experiment also showed that at 33°C *S. undulatus* from South Carolina have a higher metabolizable energy intake compared to lizards from New Jersey. This result provides evidence of the geographic differences that Angilletta thought might exist across the range of *S. undulatus.* The power of this experiment to reveal the influence of temperature on lizard performance resulted from the ability of the researcher to control all significant factors but the one of interest. In this case the main factor of interest was temperature.

The results of Angilletta's experiment, which are shown in figure 4.9, show clearly that MEI is highest in both populations of lizards at the intermediate temperature of 33°C. Note that the differences in performance indicated by this experiment were observed over a relatively small range of temperatures: from 30° to 36°C. This result is consistent with the concept that most species perform best in a fairly narrow range of temperatures. Analogous influences of temperature on performance have also been well documented in plants.

## Extreme Temperatures and Photosynthesis

One of the most fundamental characteristics of plants is their ability to photosynthesize. **Photosynthesis,** the conversion of light energy to the chemical energy of organic molecules, is

the basis for the life of plants—their growth, reproduction, and so on—and the ultimate source of energy for most heterotrophic organisms.

Photosynthesis can be summarized by the following equation:

$$6\,CO_2 + 12\,H_2O \xrightarrow{\text{Light} \;\; \text{Chlorophyll}} C_6H_{12}O_6 + 6\,O_2 + 6\,H_2O$$

This equation indicates that as light interacts with chlorophyll, carbon dioxide and water combine to produce sugar and oxygen.

Extreme temperatures generally reduce the rate of photosynthesis by plants. Figure 4.10 shows the influence of temperature on rate of photosynthesis by a moss from the boreal forest, *Pleurozium schreberi,* and a desert shrub, *Atriplex*

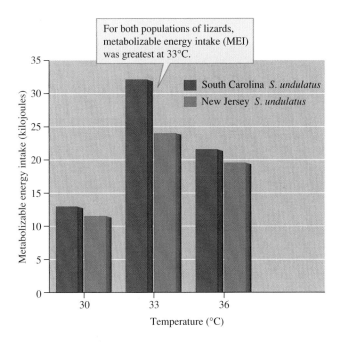

**Figure 4.9** The rate of metabolizable energy intake by two populations of the eastern fence lizard, *Sceloporus undulatus,* peaks at the same temperature (data from Angilletta 2001).

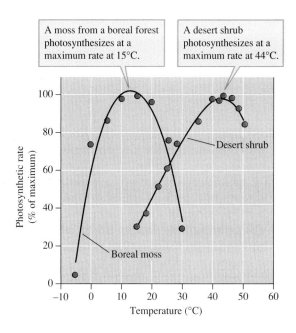

**Figure 4.10** The optimal temperatures for photosynthesis by a boreal forest moss and a desert shrub differ substantially (data from Kallio and Kärenlampi 1975, Pearcy and Harrison 1974).

*lentiformis.* The moss and the desert shrub both photosynthesize at a maximum rate over some narrow range of temperatures. Both plants photosynthesize at lower rates at temperatures above and below this range. How do the responses of the boreal moss and desert shrub to temperature differ? The major difference is that their rates of photosynthesis peak at different temperatures. The moss photosynthesizes at a maximum rate at about 15°C, while the desert shrub photosynthesizes at a maximum rate at 44°C.

The results shown in figure 4.10 demonstrate that the moss and the shrub have substantially different optimal temperatures for photosynthesis. At 15°C, where the moss photosynthesizes at a maximum rate, the desert shrub photosynthesizes at about 25% of its maximum. At 44°C, where the desert shrub is photosynthesizing at its maximum rate, the moss would probably die. These physiological differences clearly reflect differences in the environments where these species live and seem to say something about their evolutionary histories. While the moss lives in the cool boreal forests of Finland, the study population of the desert shrub, *A. lentiformis,* lives near Thermal, California, in one of the hottest deserts on earth.

The pattern of photosynthetic response to temperature by these two species is remarkably similar to the response of acetylcholinesterase to temperature (see fig. 4.8). Why might this be? (Hint: What roles do enzymes play in the process of photosynthesis?)

Plant responses to temperature, as well as those of animals, can also reflect the short-term physiological adjustments called **acclimation.** Acclimation involves physiological, not genetic, changes in response to temperature; acclimation is generally reversible with changes in environmental conditions.

Studies of *A. lentiformis* by Robert Pearcy (1977) clearly demonstrate the effect of acclimation on photosynthesis. Pearcy located a population of this desert shrub in Death Valley and grew plants for his experiments from cuttings. By propagating plants from cuttings, he was able to conduct his experiments on genetically identical clones. The clones from the Death Valley plants were grown under two temperature regimes: one set in "hot" conditions of 43°C during the day and 30°C at night; the other set under cool conditions of 23°C during the day and 18°C at night.

Pearcy then measured the photosynthetic rates of the two sets of plants. The plants grown in a cool environment photosynthesized at a maximum rate at about 32°C. Those grown in a hot environment photosynthesized at a maximum rate at 40°C, a difference in the optimum temperature for photosynthesis of 8°C. Figure 4.11 summarizes the results of Pearcy's experiment. How can we be sure that the different responses to temperature shown by *A. lentiformis* grown under cool and hot conditions were due to physiological adjustments to their growing conditions and not to genetic differences between the plants? Remember that the experimental plants were clones grown from cuttings. Pearcy used clones so he could control for the effects of genes and uncover the effects of physiological adjustment through acclimation.

The physiological adjustments made by *A. lentiformis* correspond to what these plants do during an annual cycle. The plant is evergreen and photosynthesizes throughout the year, in the cool of winter and in the heat of summer. The physiological adjustments suggest that acclimation by *A. lentiformis* may shift its optimal temperature for photosynthesis to match seasonal changes in environmental temperature.

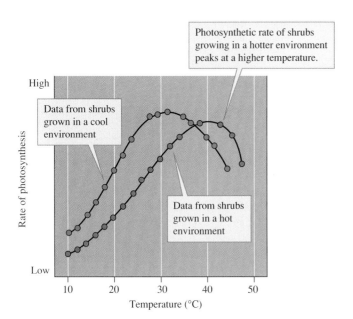

**Figure 4.11** Growing the same species of shrub in cool versus hot environments altered their optimal temperature for photosynthesis. This change was a short-term physiological adjustment due to acclimation (data from Berry and Björkman 1980, after Pearcy 1977).

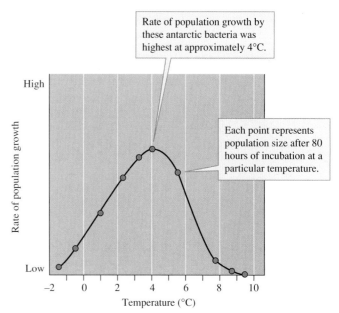

**Figure 4.12** Antarctic bacteria have a very low optimal temperature for population growth (data from Morita 1975).

# Temperature and Microbial Activity

Microbes appear to have adapted to all temperatures at which there is liquid water, from the frigid waters around the Antarctic to boiling hot springs. However, while each of these environments harbors one or more species of microbes, no known species thrives in all these conditions. All microbes that have been studied perform best over a fairly narrow range of temperatures. Let's look at two microbes that live in environments at opposite extremes of the aquatic temperature spectrum.

In chapter 3, we saw that most of the oceanic environment, the largest continuous environment on the earth, lies below the well-lighted surface waters. The organisms that live in the deep oceans live in darkness. Their environment is also cold, generally below 5°C. This cold water environment extends to the surface in the Arctic and Antarctic. A wide variety of organisms live in these cold waters. How do you think the performance of these organisms is affected by temperature?

Richard Morita (1975) studied the effect of temperature on population growth among cold-loving, or **psychrophilic,** marine bacteria that live in the waters around Antarctica. He isolated and cultured one of those bacteria, *Vibrio* sp., in a temperature-gradient incubator for 80 hours. During the experiment, the temperature gradient within the incubator ranged from about –2°C to just over 9°C. The results of the experiment show that this *Vibrio* sp. grows fastest at about

4°C. At temperatures above and below this, its population growth rate decreases. As figure 4.12 shows, Morita recorded some growth in the *Vibrio* population at temperatures approaching –2°C, however, populations did not grow at temperatures above 9°C. Morita has recorded population growth among some cold-loving bacteria at temperatures as low as –5.5°C.

Some microbes can live at very high temperatures. Microbes have been found living in all of the hot springs that have been studied. Some of these heat-loving, or **thermophilic,** microbes grow at temperatures above 40°C in a variety of environments. The most heat-loving microbes are the hyperthermophiles, which have temperature optima above 80°C. Some hyperthermophiles grow best at 110°C! Some of the most intensive studies of thermophilic and hyperthermophilic microbes have been carried out in Yellowstone National Park by Thomas Brock (1978) and his students and colleagues. One of the genera they have studied is *Sulfolobus,* a member of the microbial Domain Archaea, which obtains energy by oxidizing elemental sulfur. Jerry Mosser and colleagues (1974) used the rate at which *Sulfolobus* oxidizes sulfur as an index of its metabolic activity. They studied the microbes from a series of hot springs in Yellowstone National Park that ranged in temperature from 63° to 92°C. The temperature optimum for the *Sulfolobus* populations ranged from 63° to 80°C and was related to the temperature of the particular spring from which the microbes came. For instance, one strain isolated from a 59°C spring oxidized sulfur at a maxi-

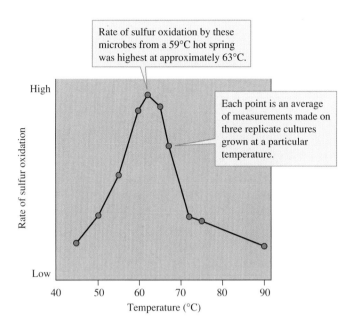

**Figure 4.13** Hot spring microbes have a very high optimal temperature for population growth (data from Mosser, Mosser, and Brock 1974).

mum rate at 63°C. This *Sulfolobus* population oxidizes sulfur at a high rate within a temperature range of about 10°C (fig. 4.13). Outside of this temperature range, its rate of sulfur oxidation is much lower.

New research tools are creating a new frontier in ecology, the ecology of microbes. Tools developed in molecular biology and in phylogenetics, the study of evolutionary relationships among organisms, are helping microbial ecologists explore the diversity of microbes and develop approaches to studying their ecology (e.g., Huber, Huber, and Stetter 2000). For instance, Anna-Louise Reysenbach, Marissa Ehringer, and Karen Hershberger (2000) used some of these modern tools to uncover the existence of previously unknown microbial lineages within the well-studied hot springs of Yellowstone National Park. Meanwhile, other researchers (e.g., Ishii and Marumo 2002) are using these modern tools to probe the microbial diversity of seafloor hydrothermal systems. Wherever these researchers apply their modern approaches, however, it remains correct to say that temperature plays a key role in determining the distribution of individual species and the composition of communities.

We have reviewed how temperature can affect microbial activity, plant photosynthesis, and animal performance. These examples demonstrate that most organisms perform best over a fairly narrow range of temperatures. Consider the effects of temperature on the performance of organisms relative to our discussion of how temperatures can vary greatly over small distances. In addition, the climate diagrams presented in chapter 2 showed us that temporal variation in temperature can also be substantial. In the next Concept Discussion, we review how some organisms respond to variation in environmental temperatures.

## CONCEPT DISCUSSION

### Regulating Body Temperature

**Many organisms have evolved ways to compensate for variations in environmental temperature by regulating body temperature.**

So, how do organisms respond to the juxtaposition of thermal heterogeneity in the environment and their own fairly narrow thermal requirements? Do they sit passively and let environmental temperatures affect them as they will, or do they take a more active approach? Many organisms have evolved ways to regulate body temperatures.

## Balancing Heat Gain Against Heat Loss

Organisms regulate body temperature by manipulating heat gain and loss. An equation, used by K. Schmidt-Nielsen (1983), can help us understand the components of heat that may be manipulated:

$$H_s = H_m \pm H_{cd} \pm H_{cv} \pm H_r - H_e$$

Here, $H_s$, the total heat stored in the body of an organism, is made up of $H_m$, heat gained from metabolism; $H_{cd}$, heat gained or lost through conduction; $H_{cv}$, heat lost or gained by convection; $H_r$, heat gained or lost through electromagnetic radiation; and $H_e$, heat lost through evaporation. These heat components represent ways that heat is transferred between an organism and its environment. **Metabolic heat,** $H_m$, is the energy released within an organism during the process of cellular respiration. **Conduction** is the movement of heat between objects in physical contact, as occurs when you sit on a stone bench on a cold winter's day; **convection** is the process of heat flow between a solid body and a moving fluid, such as wind or flowing water. During the process of conduction or convection, $H_{cd}$ and $H_{cv}$, the direction of heat flow is always from the warmer region to the colder.

Heat may also be transferred through electromagnetic radiation. This transfer of heat, $H_r$, is often called simply **radiation.** All objects above absolute 0, above –273°C, give off electromagnetic radiation, but the most obvious source in our environment is the sun. Curiously, we are blind to most of this heat flux, because at sea level over half of the energy content of sunlight falls outside our visible range. Much of this radiation that we cannot see is in the infrared part of the spectrum. The electromagnetic radiation emitted by most objects in our environment, including our own bodies, is also infrared light. Infrared light is responsible for most of the warmth you feel when standing in front of a fire or that you feel radiating from the sunny side of a building on a winter's day. The chilling effect of standing outdoors under a clear, cold night sky with no wind is also mainly due to radiative heat flux, in this case from your body to the surroundings, including the night sky.

Heat, $H_e$, may be lost by an organism through **evaporation.** In general, we need only consider the heat lost as water evaporates from the surface of an organism. The ability of water to absorb a large amount of heat as it evaporates makes cooling systems based on the evaporation of water very effective. Figure 4.14 summarizes the potential pathways by which heat can be transferred between an organism and the environment.

So how can organisms regulate body temperature? First of all, many organisms don't. The body temperature of these organisms, called **poikilotherms,** varies directly with environmental temperatures. Poikilotherms are commonly called *cold-blooded.* Of the organisms that regulate body temperature, most use external sources of energy and a combination of anatomy and behavior to manipulate $H_c$, $H_r$, and $H_e$. Animals that rely mainly on external sources of energy for regulating body temperature are called **ectotherms.** Organisms that rely heavily on internally derived metabolic heat energy, $H_m$, are called **endotherms.** Among endotherms birds and mammals use metabolic energy to heat most of their bodies. Other endothermic animals, including certain fish and insects, use metabolic energy to selectively heat critical organs. Endotherms that use metabolic energy to maintain a relatively constant body temperature are called **homeotherms.** Homeothermic animals, which include birds and mammals, are often called *warm-blooded.* It is wise to reflect a bit on these terms. They can help guide our discussion of temperature relations of organisms. However, this small number of terms cannot represent the rich variety of biology that occurs in nature. Let's take a closer look at how plants and animals thermoregulate.

Temperature regulation presents both plants and ectothermic animals with a similar problem. Both groups of organisms rely primarily on external sources of energy. Despite the much greater mobility of most ectothermic animals, the ways in which plants and ectothermic animals solve these problems are similar. We begin by discussing temperature regulation by desert plants and then move to the opposite environmental extreme and consider temperature regulation by arctic and alpine plants. We then discuss temperature regulation by lizards and then by tiger beetles living on beaches. Each of the environments we consider presents a different environmental problem; each of the organisms we consider demonstrates a different solution to those problems.

# Temperature Regulation by Plants

What sorts of environments are best for studying temperature regulation by plants? Plant ecologists have typically concentrated their studies in extreme environments, such as the desert and tundra, where the challenges of the physical environment are greater and where ecologists believed they would find the most dramatic adaptations.

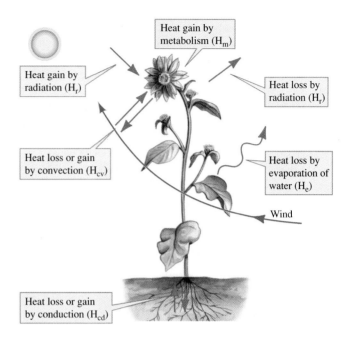

**Figure 4.14** There are multiple pathways for heat exchange between organisms and the environment.

## Desert Plants

The desert environment challenges plants to avoid overheating; that is, plants are challenged to reduce their heat storage, $H_s$. How do desert plants meet this challenge? They, like plants from other environments, use morphology and behavior to alter heat exchange with the environment. Plants engage in a number of behaviors—the sunflower orienting its inflorescence toward the sun is one of the best-known behaviors. Plants just engage in fewer behaviors compared to animals, and they do them more slowly. Evaporative cooling of leaves, which would increase heat loss, $H_e$, is not a workable option because desert plants usually have inadequate supplies of water. Also, for most plants, we can ignore $H_m$. Most produce only a small quantity of heat by metabolism. So, for a plant in a hot desert environment, our equation for heat balance reduces to:

$$H_s = H_{cd} \pm H_{cv} \pm H_r$$

To avoid heating, plants in hot deserts have three main options: decreasing heating by conduction, $H_{cd}$, increasing rates of convective cooling, $H_{cv}$, and reducing rates of radiative heating, $H_r$. Many desert plants place their foliage far enough above the ground to reduce heat gain by conduction. Many desert plants have also evolved very small leaves and an open growth form, adaptations that give high rates of convective cooling because they increase the ratio of leaf surface area to volume and the movement of air around the plant's stems and foliage. Some desert plants have low rates of radiative heat gain, $H_r$, because they have evolved reflective surfaces. As we observed in chapter 2, many desert plants cover their leaves with a dense coating of white plant hairs. These hairs reduce H$_r$ gain by reflecting visible light, which constitutes nearly half the energy content of sunlight.

We can see how natural selection has adapted plants to different temperature regimes by comparing species in the genus *Encelia,* which are distributed along a temperature and moisture gradient from the coast of California to Death Valley. James Ehleringer (1980) showed that the leaves of the coastal species, *Encelia californica,* lack hairs entirely and reflect only about 15% of visible light. He also found that two other species that grow part way between the cool coast regions and Death Valley produce leaves that are somewhat pubescent and reflect about 26% of visible light. The desert species, *Encelia farinosa,* produces two sets of leaves, one set in the summer and another when it's cooler. The summer leaves are highly pubescent (hairy) and reflect more than 40% of solar radiation. What do you think the cool season leaves are like? If you predict that they are much less pubescent than summer leaves you are correct. Why is that? We know the benefits of leaf pubescence. What might be some costs? (Hint: What do plants do with visible light other than heat up?)

Plants can also modify radiative heat gain, H$_r$, by changing the orientation of leaves and stems. Many desert plants reduce heating by orienting their leaves parallel to the rays of the sun or by folding them at midday, when sunlight is most intense. Figure 4.15 portrays the main processes involved in heat balance in desert plants.

## Arctic and Alpine Plants

How might the means of regulating temperature by desert plants compare to those of arctic and alpine plants? As you would probably predict, the plants in cold regions do, in most cases, exactly the opposite of what desert plants do. We can use the same equation we used for heat regulation in desert plants: H$_s$ = H$_{cd}$ ± H$_{cv}$ ± H$_r$.

The problem here, though, is staying warm, and arctic and alpine plants have two main options: increase their rate of radiative heating, H$_r$, and/or decrease their rate of convective cooling, H$_{cv}$. It appears that many have evolved to do both and, as a result, can heat up to temperatures far above air temperature. So, while favoring desert plants that reflect light, natural selection has favored arctic and alpine plants that absorb light with dark pigments. These dark pigments increase radiative heat gain, H$_r$. Arctic and alpine plants, such as the *Dryas integrifolia* (see fig. 4.1), also increase their H$_r$ gain by orienting their leaves and flowers perpendicular to the sun's rays. In addition, many plants increase their H$_r$ gain from the surroundings by assuming a "cushion" growth form that "hugs" the ground. The ground often warms to temperatures exceeding that of the overlying air and radiates infrared light, which can be absorbed by cushion plants. Cushion plants can also gain heat from warm substrate through conduction, H$_{cd}$.

The cushion growth form also reduces convective heat loss, H$_{cv}$, in two main ways. First, growing close to the ground gives them some shelter from the wind. Second, the compact,

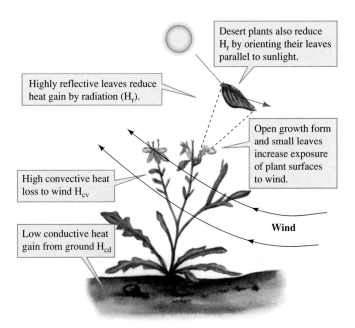

**Figure 4.15** The form and orientation of desert plants reduces heat gain from the environment and facilitates cooling.

hemispherical growth form of cushion plants reduces the ratio of surface area to volume, which slows the movement of air through the interior of the plant. Growing close to the reduced surface area also reduces the rate of radiative heat loss.

Figure 4.16 summarizes the processes involved in thermal regulation by a cushion plant. As a consequence of these processes, cushion plants are often warmer than the surrounding air and than plants with other growth forms. Y. Gauslaa (1984), who studied the heat budgets of a variety of Scandinavian plants, documented the thermal consequences of the cushion growth form. He found that while the temperature of plants with an open growth form closely matches air temperature, the temperature of cushion plants can be over 10°C higher than air temperature. The results of one of Gauslaa's comparisons is shown in figure 4.17.

## Tropical Alpine Plants

Some of the most amazing examples of thermoregulation occur among the plants that inhabit the far-flung world of the tropical alpine zone. As we saw in chapter 2, this zone is a unique environment with little annual variation in temperature but with so much daily fluctuation that freezing temperatures at night are often followed by "summer" temperatures during the day. In this environment, natural selection has produced one of the most remarkable examples of convergence, the giant rosette plants that cloak the sides of tropical mountains throughout the world (see fig. 2.39).

The giant rosette growth form has a number of features that buffer the plant against the extreme daily temperature fluctuations of the tropical alpine zone. Rosette plants generally retain their dead leaves, which insulate the stem and protect it from freezing. A dense pubescence, which may be 2 to 3 mm thick,

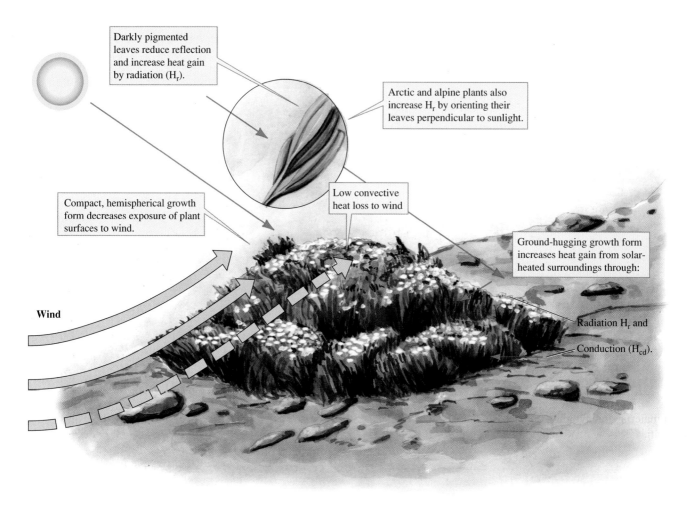

Darkly pigmented leaves reduce reflection and increase heat gain by radiation ($H_r$).

Arctic and alpine plants also increase $H_r$ by orienting their leaves perpendicular to sunlight.

Compact, hemispherical growth form decreases exposure of plant surfaces to wind.

Low convective heat loss to wind

Ground-hugging growth form increases heat gain from solar-heated surroundings through:

Radiation $H_r$ and

Conduction ($H_{cd}$).

Wind

**Figure 4.16** Arctic and alpine cushion plant form and orientation increases heat gain from sunlight and the surrounding landscape and conserves any heat gained.

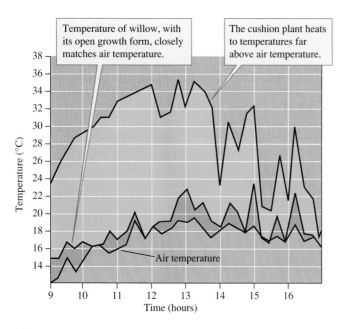

Temperature of willow, with its open growth form, closely matches air temperature.

The cushion plant heats to temperatures far above air temperature.

Air temperature

Temperature (°C)

Time (hours)

**Figure 4.17** An arctic cushion plant maintains significantly higher temperatures compared to plants with a more open growth form, such as willows (data from Fitter and Hay 1987, after Gauslaa 1984).

covers the living leaves of most species. This thick pubescence helps increase leaf temperature in the cool alpine environment by creating a dead air space above the leaf surface, which reduces convective heat losses. Leaf pubescence on these tropical alpine plants acts like a kind of plant fur. The rosettes of many species secrete and retain several liters of fluids within their rosettes or within large hollow inflorescences. Retaining these large volumes of water increases the capacity of the rosette or inflorescence to store heat. Greater heat storage, $H_s$, during the day means a lower probability of freezing at night. Much as the *Dryas* flowers we discussed at the beginning of chapter 4, the leaves of some species act as parabolic mirrors and improve radiative heating, $H_r$, of the apical bud and expanding leaves. The rosettes of some tropical alpine plants even close over the apical bud at night, which protects the bud from freezing.

# Temperature Regulation by Ectothermic Animals

Like plants, the vast majority of animals, including fish, amphibians, reptiles, and invertebrates of all sorts, use external sources of energy to regulate body temperature. These ectothermic animals use means analogous to those used by plants, including variations in body size, shape, and pigmentation. The obvious difference between plants and ectothermic animals is that the animals have more options for using behavior to thermoregulate. Yet, as we shall see, the difference between the behavior of these animals and that of plants is more a matter of degree than of kind.

Can thermoregulation by ectotherms be either effective or precise? Let's allow an ectotherm from a rigorous environment to answer our question.

## *Liolaemus* Lizards

Oliver Pearson (1954) studied *Liolaemus multiformis,* an unusual lizard because it thrives in a cold environment. This lizard lives in the high Andes Mountains of South America at altitudes over 4,800 m. In these mountains, it is cold year-round, with morning temperatures falling as low as −5°C. The lizard spends the night in burrows, where its rate of cooling is lower than it would be in the open. However, Pearson found that during the night, its body temperature may still fall to as low as 2.5°C. Even at these temperatures, the lizard emerges from its burrows early each morning and immediately begins to bask, usually on a mat of plant material. By perching on plant material and avoiding contact with stones, it reduces its rate of heat loss by conduction to the ground.

While basking, *Liolaemus* orients its back toward the sun, which increases radiative heat gain. It also presses itself flat against the substrate, which reduces its exposure to the wind and heat losses by convection. In addition, Pearson observed that cold lizards emerging from their burrows are dark. He proposed that this dark pigmentation increases the rate of radiative heat gain by basking lizards.

Pearson demonstrated that these behaviors produce a rapid rise in body temperature. After an hour of basking, as air temperature rises to about 1.5°C, the body temperature of *Liolaemus* rises to about 33°C, over 30°C above that of the surrounding air! As the day progresses, and air temperature continues to rise, the lizard maintains a more or less constant body temperature of 35°C. Figure 4.18 summarizes this basking behavior. We asked whether thermoregulation by an ectotherm can be either effective or precise. Pearson's studies suggest that thermoregulation by ectotherms can be both.

While the early studies of Oliver Pearson show that thermoregulation by ectotherms can be both effective and precise, we are left with many important biological questions. For instance, what relationships are there between temperature regulation, temperature preference, and optimal performance by a particular species? Insights into these relationships have come from studies of the eastern fence lizard, *Sceloporus undulatus.* Research by Michael Angilletta showed that the rate of metabolizable energy intake is maximized at a temperature of 33°C (see figure 4.9). Now, what relationship does this optimal temperature bear to the preferred temperature of *S. undulatus?* Angilletta (2001) explored this relationship by placing *S. undulatus* from New Jersey and South Carolina in a

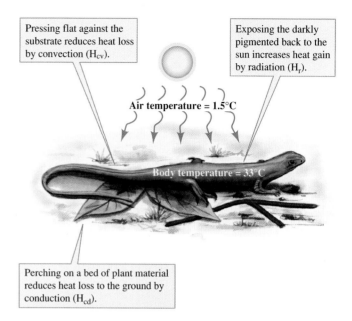

Pressing flat against the substrate reduces heat loss by convection ($H_{cv}$).

Exposing the darkly pigmented back to the sun increases heat gain by radiation ($H_r$).

Air temperature = 1.5°C

Body temperature = 33°C

Perching on a bed of plant material reduces heat loss to the ground by conduction ($H_{cd}$).

**Figure 4.18** The high-elevation lizard, *Liolaemus multiformis,* uses a combination of behavior and dark pigmentation to increase solar heat gain and elevate its body temperature (data from Pearson 1954).

temperature gradient that ranged from 26°C at one end to 38°C at the other end. He determined preferred temperature early each morning by quickly measuring the body temperature of each lizard. Body temperature would indicate where each lizard had been in the temperature gradient, that is, its "preferred" temperature. Angilletta examined thermoregulation by measuring the body temperatures of active individuals in the field.

The results of Angilletta's study provide strong evidence for a correspondence between preferred temperature, thermoregulation, and optimal temperatures in *S. undulatus* (fig. 4.19). Lizards from New Jersey and South Carolina had virtually identical preferred temperatures: 32.8° versus 32.9°C respectively. The body temperatures found by Angilletta in the field were also very similar. The body temperatures of *S. undulatus* measured in the field in New Jersey averaged 34.0°C, while the body temperatures of *S. undulatus* taken in South Carolina averaged 33.1°C. As shown in figure 4.19, both preferred temperatures determined in the laboratory and the body temperatures of *S. undulatus* measured in the field are very close to the temperature that maximizes metabolizable energy intake by these lizards. The following example shows that effective thermoregulation by ectotherms is not limited to lizards.

## *Grasshoppers: Some Like It Hot*

Many grasshoppers also bask in the sun, elevating their body temperature to 40°C or even higher. R. I. Carruthers and his colleagues (1992) described how some species of grasshoppers even adjust their capacity for radiative heating, $H_r$, by varying the intensity of their pigmentation during development. When

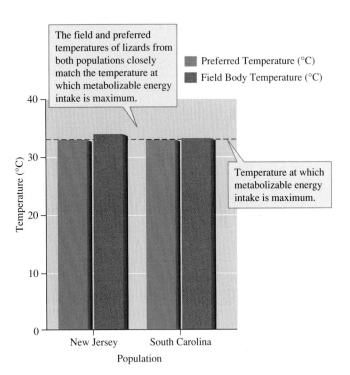

Figure 4.19 Two populations of the eastern fence lizard, *Sceloporus undulatus*, both regulate their body temperatures to match closely the temperature of maximum metabolizable energy intake (data from Angilletta 2001).

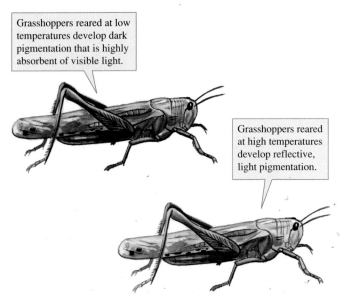

**Figure 4.20** Rearing temperatures influence the pigmentation of the clear-winged grasshopper.

reared at low temperatures, these species appear to compensate by developing dark pigmentation; while at higher developmental temperatures, they produce less pigmentation (fig. 4.20). How would changing pigmentation in response to developmental temperatures affect thermoregulation by these grasshoppers? Because grasshoppers reared at low temperatures develop darker pigmentation, they increase their potential for $H_r$ gain. Because those reared at high temperatures develop lighter pigmentation, they reduce their potential for $H_r$ gain.

The clear-winged grasshopper, *Camnula pellucida*, inhabits subalpine grasslands in the White Mountains of eastern Arizona, where the cool mornings warm up quickly under the mountain sun. During early morning, *Camnula*, like *Liolaemus*, orients its body perpendicular to the sun's rays and quickly heats to 30° to 40°C. Given the opportunity, young *Camnula* will maintain a body temperature around 38° to 40°C, very close to its optimal temperature for development. In the laboratory, *Camnula* is able to elevate its body temperature to 12°C above air temperature and maintain it within a very narrow range (± 2°C) for many hours.

Carruthers and his colleagues divided a sample of *Camnula* into two groups, which were kept at an air temperature of about 18°C. One of the groups also had access to light, while the other was restricted to the shade. The grasshoppers that had access to light basked and elevated their body temperatures about 10°C above air temperature. Meanwhile, the body temperatures of the grasshoppers kept in the shade remained close to air temperature (fig. 4.21).

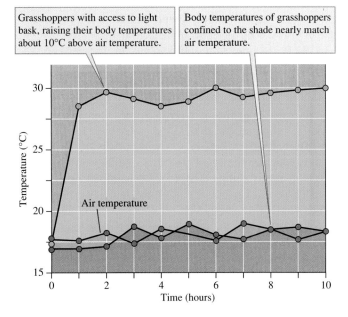

**Figure 4.21** Basking allows the clear-winged grasshopper to elevate its body temperature significantly (data from Carruthers et al. 1992).

Why does *Camnula* bask and maintain a body temperature above air temperature? The researchers estimated that by basking in the sun the grasshopper develops faster than it would if it allowed its body temperature to match air temperature. What other benefits might *Camnula* gain by maintaining a high body temperature? The grasshopper may raise its body temperature to 38° to 40°C to control *Entomophaga grylli*, a fungus that infects and kills grasshoppers.

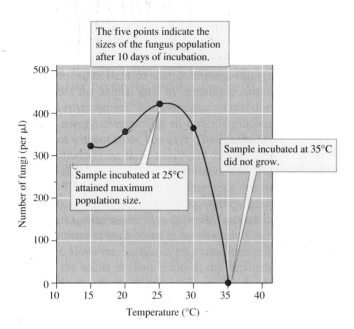

The five points indicate the sizes of the fungus population after 10 days of incubation.

Sample incubated at 25°C attained maximum population size.

Sample incubated at 35°C did not grow.

**Figure 4.22** High temperatures inhibit growth by *Entomophaga grylli* (data from Carruthers et al. 1992).

The idea that high temperatures could control *Entomophaga* was tested by growing the fungus in artificial media at 15°, 20°, 25°, 30°, 35°, and 45°C. The populations grew fastest at 25°C; above and below 25°C the fungus populations grew at a slower rate; they did not grow at 35°C and were killed at 45°C (fig. 4.22).

What are the limitations of the experiment summarized in figure 4.22? This experiment was conducted on fungus populations growing on artificial media. Therefore, we should be cautious about predicting how temperature may affect *Entomophaga* growing inside of living grasshoppers. After studying the growth of the fungus in artificial media, the researchers studied how temperature influences mortality among grasshoppers infected with the fungus. They found that exposure to 40°C temperatures for as few as 4 hours each day significantly reduced the numbers of grasshoppers dying of *Entomophaga* infections. The results of these experiments support the hypothesis that by maintaining body temperatures of 38° to 40°C, clear-winged grasshoppers create an environment unsuitable for one of their most serious pathogens. How does the effect of temperature on population growth by *Entomophaga* compare to the effect of temperature on population growth by bacteria as shown in figures 4.12 and 4.13?

# Temperature Regulation by Endothermic Animals

Do endothermic animals thermoregulate differently than the other organisms we've discussed? Endotherms use all the anatomical and behavioral tricks used by other organisms to manipulate heat exchange with the environment. So, our basic equation for temperature regulation, $H_s = H_m \pm H_{cd} \pm H_{cv} \pm H_r - H_e$, still applies but with some changes in the relative importance of the terms. Most significantly, endotherms rely a great deal more on metabolic heat, $H_m$, to maintain constant body temperature.

## Environmental Temperature and Metabolic Rates

P. F. Scholander and his colleagues (1950) studied thermoregulation in several endothermic species by monitoring metabolic rate while exposing them to a range of temperatures. The range of environmental temperatures over which the metabolic rate of a homeothermic animal does not change is called its **thermal neutral zone.** When environmental temperatures are within the thermal neutral zone of an endothermic animal, its metabolic rate stays steady at resting metabolism. An endotherm's metabolic rate will rapidly increase to two or even three times resting metabolism if the environmental temperature falls below or rises above the thermal neutral zone.

What causes metabolic rates to rise when environmental temperatures are outside the thermal neutral zone? We can use humans as a model for the responses of endotherms generally. At low temperatures, we start shivering, which generates heat by muscle contractions. We also release hormones that increase our metabolic rate, the rate at which we metabolize our energy stores, which are mainly fats. Increasing metabolic rate increases the rate at which we generate metabolic heat, $H_m$. At high temperatures, heart rate and blood flow to the skin increase. This increased blood flow transports heat from the body core to the skin, where an evaporative cooling system based on sweating accelerates unloading of heat to the external environment. Many large endotherms, such as horses and camels, also cool by sweating. Other endotherms do not sweat but evaporatively cool by other means: dogs and birds pant and marsupials and rodents moisten their body surfaces by salivating and licking.

The breadth of the thermal neutral zone varies a great deal among endothermic species. Scholander and his colleagues suggested that differences in the width of the thermal neutral zone defines two groups of organisms: tropical species, with narrow thermal neutral zones, and arctic species, with broad thermal neutral zones. The researchers pointed out that the narrow thermal neutral zone of *Homo sapiens* is similar to that of several species of rain forest mammals and birds. Meanwhile, arctic species, such as the arctic fox, have impressively broad thermal neutral zones.

Since the normal body temperature of most endotherms varies from about 35° to 40°C, it is no surprise that this range of temperatures falls within the thermal neutral zone of both tropical and arctic species. What distinguishes tropical and arctic species is the great tolerance that arctic species have for cold. For instance, the arctic fox can tolerate environmental temperatures down to −30°C without showing any increase in metabolic rate. Meanwhile the metabolic rate of some tropical

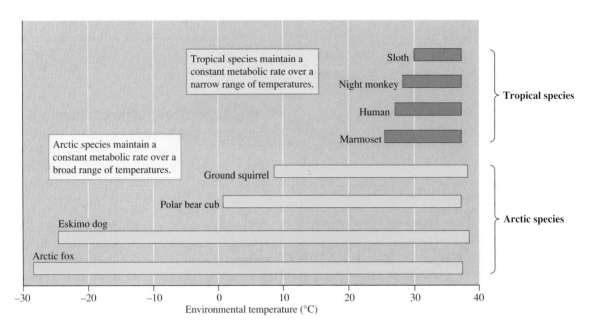

**Figure 4.23** Temperature and the thermal neutral zone of arctic and tropical mammals. Bars indicate range of temperatures over which metabolic rate does not change for each species (data from Scholander et al. 1950).

species begins to increase when air temperature falls below 29°C. Figure 4.23 contrasts the thermal neutral zones of some arctic and tropical species.

Is this classification of thermal responses consistent with what we learned about the relative temperature variation in tropical versus high-latitude environments? Yes, it is. (Compare the climate diagrams for the tropical rain forest and tropical savanna with those for the boreal forest and tundra—figs. 2.10, 2.16, 2.31, and 2.34.) Does *H. sapiens* somehow violate this classification? We live virtually everywhere on the planet yet our thermal neutral zone suggests that we are tropical. How can you explain this?

While humans live virtually everywhere on earth, we do so by virtue of our ability to manipulate the environment. We spend a great deal of time and energy compensating for variation in temperature, using devices as simple as clothing or as complex as houses fitted with mechanical systems for heating and cooling. We put on extra clothing when it's cold and wear less when it's warm. When it's hot, we decrease radiative heating by wearing a straw hat. When it's cold, we decrease radiative cooling by wearing a woolen cap. Our ability to create large, temperature-controlled environments in the interiors of schools, office buildings, sports stadiums, and theaters is unique in the biosphere. Yet these controlled environments, like our narrow thermal neutral zone, suggest a tropical heritage. Can you imagine the temperature controls within a shopping mall designed by arctic foxes?

From an ecological perspective, the important point of this discussion is that thermoregulation outside the thermal neutral zone costs energy that could be otherwise directed toward reproduction. How might such energetic costs affect the distribution and abundance of organisms in nature? This is one of the central questions of ecology.

## Aquatic Birds and Mammals

Now let's turn to thermoregulation by aquatic endotherms, where the aquatic environment limits the possible ways organisms can regulate their body temperatures. Why is that? First, as we have seen, the capacity of water to absorb heat energy without changing temperature is about 3,000 times that of air. Second, conductive and convective heat losses to water are much more rapid than to air, over 20 times faster in still water and up to 100 times faster in moving water. Thus, the aquatic organism is surrounded by a vast heat sink. The potential for heat loss to this heat sink is very great, particularly for gill-breathing species that must expose a large respiratory surface in order to extract sufficient oxygen from water. In the face of these environmental difficulties, only a few aquatic species are truly endothermic.

Aquatic birds and mammals, such as penguins, seals, and whales, can be endothermic in an aquatic environment for two major reasons: First, they are all air breathers and do not expose a large respiratory surface to the surrounding water. Second, many endothermic aquatic animals, including penguins, seals, and whales, are well insulated from the heat-sapping external environment by a thick layer of fat while others, such as the sea otter, are insulated by a layer of fur that traps air. The parts of these animals that are not well insulated, principally appendages, are outfitted with *countercurrent heat exchangers*, vascular structures that reduce the rate of heat loss to the surrounding aquatic environment. Figure 4.24 diagrams the structure and functioning of a countercurrent heat exchanger in the flipper of a dolphin.

Heat exchangers are so efficient at conserving heat that some species of fish are able to maintain a significant thermal gradient between some of their muscles and the external environment. While not capable of regulating the temperature of

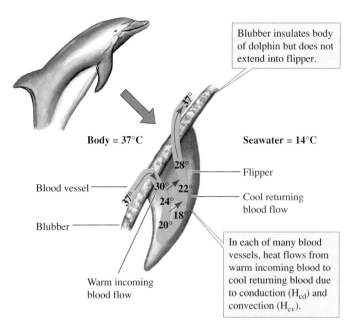

**Figure 4.24** Countercurrent heat exchange in dolphin flippers promotes conservation of body heat.

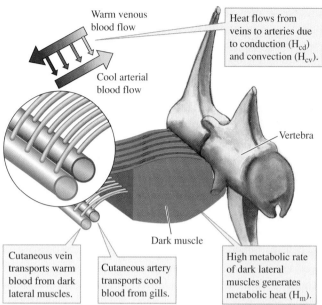

**Figure 4.25** Countercurrent heat exchange in the lateral muscles of bluefin tuna.

their body core, as aquatic birds and mammals do, these fish can selectively heat certain muscle groups and perhaps increase their swimming performance over a larger range of temperatures as a consequence.

## Warming the Swimming Muscles of Large Marine Fish

Francis Carey (1973) was fascinated to learn that some fishes, such as tuna and mackerel sharks, have body temperatures *above* that of the surrounding water, a fact that seemed to contradict the physics of heat exchange. Consequently, Carey and his colleagues at the Woods Hole Oceanographic Institution set out to determine how this could be. As a consequence of their research program, we now know a great deal more about the temperature relations of large endothermic fishes.

The lateral swimming muscles of endothermic fish, such as tunas and mackerel sharks, are well supplied with blood vessels that function as countercurrent heat exchangers. These heat exchangers heat cool arterial blood as it carries oxygen to the lateral swimming muscles, and by the time this blood delivers its supply of oxygen and nutrients it has been heated to the same temperature as the active muscles. On the return trip the heat in this warm blood is used to heat the newly arriving blood and so, when blood exits the swimming muscles, it is again approximately the same temperature as the surrounding water. The countercurrent heat exchangers of tuna are efficient enough at conserving heat that these fish can elevate the temperature of their swimming muscles up to 14°C above the temperature of the surrounding water. The anatomy of the countercurrent heat exchange in bluefin tuna muscle is presented in figure 4.25.

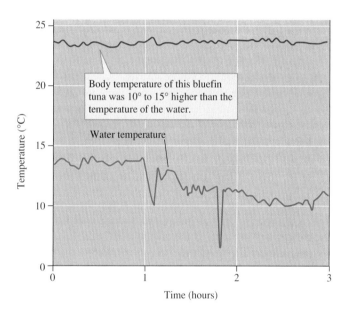

**Figure 4.26** The muscle temperature of an actively swimming bluefin tuna is elevated above that of the surrounding ocean water (data from Carey 1973).

Carey and his colleagues implanted devices that would measure and transmit the temperature of the muscles of bluefin tuna and of the surrounding water. Their tracking boat could usually follow a released fish carrying a temperature-sensing implant for a few hours, which provided enough time to collect data that revealed a great deal about their temperature relations. As one of the monitored fish swam through water varying in temperature from 7° to 14°C, the temperature of its swimming muscles remained a constant 24°C. These results, shown in figure 4.26, demonstrate that a bluefin tuna can maintain a remarkably constant muscle temperature even in the face of substantial variation in water temperature. More recent work has shown that other organs, such as the stomach, of bluefin tuna vary in temperature much more than do the swimming muscles (Stevens, Kanwisher, and Carey 2000).

Now, let's move from the sea and the giant bluefin tuna, which can reach up to 1,000 kg, to land, where we find some of the smallest endotherms. Many terrestrial insects have evolved the capacity to heat their flight muscles.

## Warming Insect Flight Muscles

Have you ever gone outside on a cool fall or spring morning when few insects were active and yet met with bumblebees visiting flowers? Were you surprised? While you may have taken the meeting for granted, these early morning forays by bumblebees require some impressive physiology. Most insects use external sources of energy to heat their bodies, but there are some notable exceptions. As we saw in chapter 1, bumblebees maintain the temperature of their thoraxes, which house the flight muscles, at 30° to 37°C regardless of air temperature. Because they can warm their flight muscles, bumblebees can fly when environmental temperatures are as low as 0°C. A number of other insects use metabolic heat, $H_m$, to warm their flight muscles, including large nocturnal moths, which were the subject of some of the earliest studies of endothermic insects.

Bernd Heinrich (1993) has spent a great deal of his professional life studying thermoregulation by insects. Some of the inspiration that launched this work came to him when he was a graduate student recording the body temperatures of moths in the highlands of New Guinea. Heinrich relates how as he captured moths flying to a sheet illuminated by a lantern, air temperatures were about 9°C. Despite these low temperatures, some of the larger moths captured had thoracic temperatures of 46°C, 9°C higher than Heinrich's own body temperature. It was at this point that he became convinced that some insects can thermoregulate by endothermic means. However, you don't have to travel to the highlands of New Guinea to meet endothermic insects. Some of Heinrich's most elegant studies of thermoregulation have been done on moths from temperate latitudes.

Studies of temperature regulation by moths began in the early 1800s. Many of these studies were focused on moths of the family Sphingidae, the sphinx moths. Sphinx moths are convenient insects for study because many reach impressive sizes, large enough to be mistaken for hummingbirds. Heinrich's dissertation focused on thermoregulation by the sphinx moth *Manduca sexta,* whose large green caterpillars feed on a wide variety of plants including tobacco and tomato plants. *M. sexta* is among the larger sphinx moths and weighs 2 to 3 g—which is heavier than some hummingbirds and shrews, the smallest of the birds and mammals.

Since the nineteenth century, researchers have been aware that active sphinx moths have elevated thoracic temperatures. These early researchers also knew that temperature increases within the thorax were due to activity of the flight

muscles contained within the thorax that vibrated the wings. Later researchers discovered that during flight, the muscles responsible for the upstroke of the wings and those responsible for the downstroke contracted sequentially. However, during preflight warm-up, the upstroke and downstroke muscles contracted nearly simultaneously. Consequently, the wings of a moth warming its flight muscles only vibrated. Once warmed up and actively flying, sphinx moths maintained a relatively constant thoracic temperature over a broad range of environmental temperatures. It was clear. Sphinx moths thermoregulate.

You can see that a lot was known before Heinrich began his dissertation research. However, a significant problem remained. No one knew how sphinx moths accomplished thermoregulation. Phillip Adams and James Heath (1964) proposed that the moths thermoregulate by changing their metabolic rate in response to changing environmental temperatures. In terms of our equation for thermoregulation, Adams and Heath proposed that the moths increased $H_m$ when environmental temperatures fell and decreased $H_m$ when environmental temperatures rose.

Several observations led Heinrich to propose an alternative hypothesis, however. He proposed that active sphinx moths have a fairly constant metabolic rate and so generate metabolic heat, $H_m$, at a constant rate. Heinrich also proposed that sphinx moths thermoregulate by changing their rates of heat loss to the environment. In terms of our equation for thermoregulation, the moths *decrease* their rate of cooling by convection and conduction when environmental temperatures fall and when temperatures rise, sphinx moths *increase* their cooling rates.

Heinrich tested his hypothesis with a series of pioneering experiments that demonstrated *M. sexta* cools its thorax by using its circulatory system to transport heat to the abdomen. In other words, the blood of these moths acts as a coolant. In his first experiment, he immobilized a moth and heated its thorax with a narrow beam of light while monitoring the temperature of the thorax and abdomen. Because it was narrow, the light beam increased radiative heat gain, $H_r$, of the thorax only. Heinrich used the beam to simulate metabolic heat production by the flight muscles. He observed that the thoracic temperature of these heated moths stabilized at about 44°C. Meanwhile, their abdominal temperatures gradually increased.

These results indicated that heat within the thorax was transferred to the abdomen. Heinrich proposed that blood flowing from the thorax to the abdomen was the means of heat transfer. To confirm this, he conducted a second experiment. He tied off blood flow to the thorax using a fine human hair. With this blood flow stopped, flying moths overheated and stopped flying. Instead of stabilizing at 44°C, the thoracic temperatures approached the lethal limit of 46°C. An interesting debate between two groups of researchers with competing hypotheses was decided by two decisive experiments, which are summarized by figure 4.27.

Endothermic insects were a surprise to many biologists. The existence of endothermic plants was even more surprising.

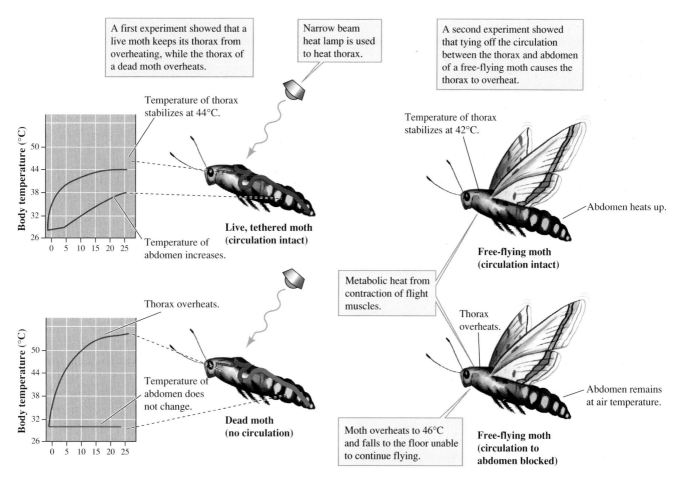

**Figure 4.27** The circulatory system plays a central role in thermoregulation by the moth, *Manduca sexta* (data from Heinrich 1993).

# Temperature Regulation by Thermogenic Plants

Roger Knutson (1974, 1979) visited a marsh on a cold February day in northeast Iowa. There he saw eastern skunk cabbage, *Symplocarpus foetidus,* emerging from the frozen landscape. Each plant was surrounded by a melted circle in the snow. It appeared as though the skunk cabbage had generated enough heat to melt its way through the snow. Knutson returned the next day with a thermometer and so began a research project that produced some surprising observations of thermoregulation by plants.

Almost all plants are poikilothermic ectotherms. However, plants in the family Araceae have the unusual habit of using metabolic energy to heat their flowers. Some of the temperate species in this mostly tropical family use this ability to protect their inflorescences from freezing and to attract pollinators. One of the most studied of these temperate species is the eastern skunk cabbage, which lives in the deciduous forests of eastern North America. This skunk cabbage blooms from February to March, when air temperatures vary between –15° to 15°C. During this period, the inflorescence of the plant, which weighs from 2 to 9 g, maintains a tempera-

ture 15° to 35°C above air temperature. As Knutson observed, this temperature is warm enough so that *S. foetidus* can melt its way through snow. The plant's inflorescences can maintain these elevated temperatures for up to 14 days. During this period, it functions as an endothermic organism.

How does the skunk cabbage fuel the heating of its inflorescence? It has a large root in which it stores large quantities of starch. Some of this starch is translocated to the inflorescence, where it is metabolized at a high rate, generating large quantities of heat in the process. This heat, besides keeping the inflorescence from freezing, may help attract pollinators. Various pollinators are attracted to both the warmth and the sweetish scent given off by the plant. Some of the biology of this interesting plant is summarized in figure 4.28.

The inflorescence of the skunk cabbage maintains a high respiratory rate, equivalent to that of a small mammal of similar size. However, its metabolic rate is not constant. The plant adjusts its metabolic rate to changes in environmental temperatures. The metabolic rate increases with decreasing temperature, which increases the rate of metabolic heat production. By adjusting its metabolic rate, the plant can maintain its inflorescence at a similar temperature despite substantial variation in environmental temperature (fig. 4.29).

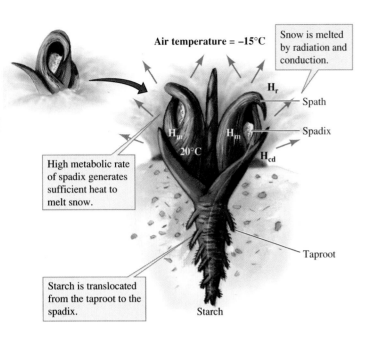

**Air temperature = –15°C**

Snow is melted by radiation and conduction.

$H_r$

Spath

$H_m$    $H_m$

20°C

$H_{cd}$

Spadix

High metabolic rate of spadix generates sufficient heat to melt snow.

Taproot

Starch is translocated from the taproot to the spadix.

Starch

**Figure 4.28** Eastern skunk cabbage, an endothermic plant, can melt its way up through spring snow cover (data from Knutson 1974).

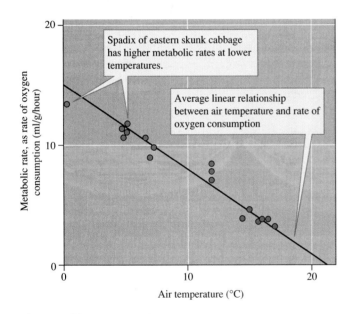

Spadix of eastern skunk cabbage has higher metabolic rates at lower temperatures.

Average linear relationship between air temperature and rate of oxygen consumption

Metabolic rate, as rate of oxygen consumption (ml/g/hour)

Air temperature (°C)

**Figure 4.29** Air temperature has a clear influence on the metabolic rate of eastern skunk cabbage (data from Knutson 1974).

In this section, we have considered how various organisms regulate their body temperatures by using external sources of energy, internal sources of energy, or both. Thermoregulation is possible where organisms face temperatures within their range of tolerance. However, organisms do not always respond to variation in environmental temperatures by thermoregulating. In many circumstances, they use various means to survive extreme environmental temperatures, as we shall discuss next.

## CONCEPT DISCUSSION

### Surviving Extreme Temperatures

**Many organisms survive extreme temperatures by entering a resting stage.**

Think of an environment that is either very cold or very hot, perhaps a temperate forest in winter or a desert in the middle of a hot summer's day. If you have been in such an environment you may have noticed less obvious biological activity than at other times of the year. Many plants are dormant, few birds are active, and insects may not be evident. Many organisms have evolved ways to avoid extreme environmental temperatures by entering a resting stage. This stage may be as simple as resting in a sheltered spot during the heat of the day or may involve elaborate physiological and behavioral adjustments. Let's examine some of the ways organisms avoid extreme temperatures.

## Inactivity

A simple way to avoid extreme environmental temperatures is to seek shelter during the hottest or coldest times of the day. We have already seen one example of this behavior. During the cold nights of the tropical alpine zone, *Liolaemus* lizards take shelter in burrows, where temperatures are several degrees warmer than on the surface. Now, let's consider some beetles that must take shelter during the middle of the day, when environmental temperatures are too high.

Many organisms live on the beaches of New Zealand, including small predatory tiger beetles. One species of black tiger beetle, *Neocicindela perhispida campbelli,* lives on black sand beaches. As we saw earlier in the chapter, these black sand beaches heat up rapidly in the morning sun and reach higher temperatures than nearby white sand beaches (see fig. 4.5). The black beetles that live on these beaches also heat up quickly. By basking in the morning sun they warm enough to become active early in the day. However, later in the day, they must work very hard to avoid overheating.

The beetles maintain their body temperature at about 36.4°C by shuttling between sun and shade and by facing into the sun to orient themselves parallel to the sun's rays. The beetles also reduce heating by increasing their rate of convective cooling. They do this by "stilting," standing on the tips of their feet and extending their legs, to get themselves a bit higher into the air. With this combination of behaviors, the beetle maintains its body temperature substantially below the temperature of the sand.

Thermoregulation becomes difficult by midday, however, when sand temperatures may reach 70°C. Most of the beetles simply avoid these high temperatures by leaving. As shown in figure 4.30, those beetles that remain active are mostly in the shade.

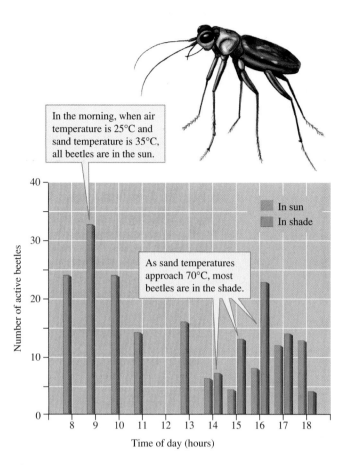

In the morning, when air temperature is 25°C and sand temperature is 35°C, all beetles are in the sun.

As sand temperatures approach 70°C, most beetles are in the shade.

In sun
In shade

Number of active beetles

Time of day (hours)

**Figure 4.30** Tiger beetles' avoidance of high environmental temperatures helps them avoid elevating their body temperatures above acceptable levels (data from Hadley, Savill, and Schultz 1992).

Now let's look at hummingbirds that live in environments that get cold at night. Maintaining an elevated body temperature during cold nights requires a great deal of energy. In some environments, hummingbirds may save energy by reducing their metabolic rate during the night and allowing their body temperature to decrease.

## Reducing Metabolic Rate

Hummingbirds are small birds that depend upon a diet of nectar and insects to maintain a high metabolic rate and a body temperature of about 39°C. When food is abundant, they maintain these high rates throughout the day and night. However, when food is scarce and night temperatures are cold, hummingbirds may enter a state of torpor. **Torpor** is a state of low metabolic rate and lowered body temperature. During torpor, a hummingbird's body temperature is about 12° to 17°C, quite a reduction from 39°C. Because this lower body temperature is a direct consequence of a lower metabolic rate, a hummingbird in torpor saves a great deal of energy.

How much energy is saved? F. L. Carpenter and colleagues (1993) estimated that rufus hummingbirds that maintain full body temperature all night lost (metabolized) about

0.24 g of fat. They estimated that birds in a state of torpor lost only 0.02 g of fat, resulting in an energy savings of over 90%.

This much was known when William Calder (1994) began a study to discover the circumstances under which hummingbirds use torpor. When he began his work there were two major hypotheses. The "routine" hypothesis proposed that hummingbirds go into torpor regularly, perhaps every night. The "emergency-only" hypothesis proposed that hummingbirds go into torpor only when food supplies are inadequate.

Since wild hummingbirds are difficult to follow, how could Calder determine whether they go into torpor at night? By weighing hummingbirds just before they went to their night roosts and then again first thing in the morning, he could estimate the amount of fat metabolized during the night. A hummingbird in torpor would lose much less weight. To weigh hummingbirds, Calder either captured them in mist nets or rigged the perch on a hummingbird feeder with an electronic balance.

Calder's observations did not support the routine hypothesis. His measurements indicated that hummingbirds generally lost 15 times the weight that would be required to meet the energy demands while in torpor. Clearly, hummingbirds usually have elevated metabolic rates during the night and go into torpor only occasionally. Calder found that hummingbirds use torpor under two main circumstances: (1) when they arrive at breeding or wintering sites before flowers are abundant, and (2) when their food intake is reduced by decreased nectar production by the flowers they visit or by storms that interfere with their feeding. As with Heinrich's work on moth thermoregulation, Calder's observations nicely resolved the debate between two competing hypotheses (fig. 4.31).

While hummingbirds may go into torpor for several hours each night, other animals can go into a state of reduced metabolism that may last several months. If this state occurs mainly in winter, it is called **hibernation.** If it occurs in summer, it is called **estivation.** During hibernation, the body temperature of arctic ground squirrels may drop to 2°C. The metabolic rates of hibernating marmots may fall to 3% of levels during active periods. During estivation, the metabolic rate of long-neck turtles may fall to 28% of their normal metabolic rate. Such reductions in metabolic rate allow these animals to survive long arctic and alpine winters or hot, dry periods in the desert, during which they must rely entirely on stored energy reserves. Under some conditions, even tropical species may hibernate.

## Hibernation by a Tropical Species

While most studies of hibernation have focused on temperate and arctic species, there are tropical animals that hibernate. One of those species is a primate called the fat-tailed dwarf lemur, *Cheirogaleus medius* (fig. 4.32). *C. medius* lives in the tropical dry forests of western Madagascar, where it is active during 5 months and hibernates for 7 months. As its common

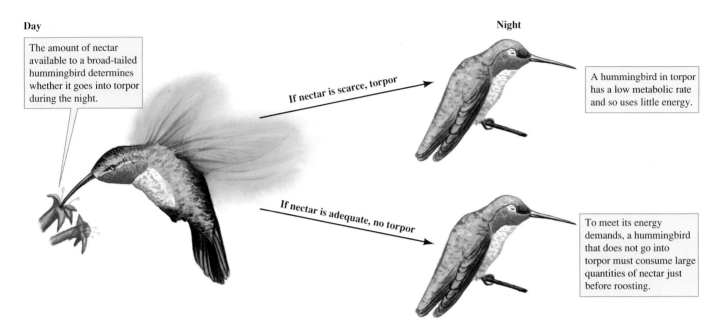

**Day**

The amount of nectar available to a broad-tailed hummingbird determines whether it goes into torpor during the night.

**Night**

If nectar is scarce, torpor

A hummingbird in torpor has a low metabolic rate and so uses little energy.

If nectar is adequate, no torpor

To meet its energy demands, a hummingbird that does not go into torpor must consume large quantities of nectar just before roosting.

**Figure 4.31** The availability of nectar affects whether broad-tailed hummingbirds enter torpor at night.

**Figure 4.32** The fat-tailed dwarf lemur, *Cheirogaleus medius,* inhabits tropical dry forests in western Madagascar where it hibernates through most of the dry season.

name implies, *C. medius,* is a small primate, with a body length of about 20 cm and a tail of about the same length. As adults they weigh about 140 g. The tail of *C. medius* is a primary site for storing the fat that the species uses for energy during its long hibernation. *C. medius* lives in trees, sleeping in tree cavities during the day in groups of up to five individuals and foraging in the canopy at night. In contrast to its day-time roosting behavior, *C. medius* is entirely solitary while foraging at night. The main foods of *C. medius* are fruits and flowers but it also eats some insects, and small vertebrate animals such as chameleons.

Why do these tropical primates hibernate? Photos of tropical dry forest landscape during wet and dry seasons and climate diagrams of tropical dry forest provide a suggestion (see figs. 2.12 and 2.13). During the wet season, tropical dry forests are very productive of the fruits and flowers eaten by *C. medius*. However during the dry months, these foods are scarce. The forests inhabited by *C. medius* in Madagascar have a dry season that lasts for 8 months, which for *C. medius* represents a long lean time.

The physiology of hibernation by free-ranging *C. medius* was studied by Joanna Fietz, Kathrin Dausmann, Frieda Tataruch, and Jörg Ganzhorn (2003) in the Kirindy forest of western Madagascar. This team of researchers from two German and one Austrian university found that during hibernation, the body temperature of *C. medius* varies from about 18° to 31°C. By lowering its body temperature, *C. medius* saves energy during the food-scarce dry season. As a result of its lower energy requirement, *C. medius* is able to live off the fat it stores in its tail and in other parts of the body. These fat stores are gradually depleted as *C. medius* passes its annual seven months of hibernation. Fietz and her colleagues found that during hibernation the body mass of *C. medius* decreases by approximately 34%, while the volume of the tail is reduced by nearly 58%. Studies of hibernation among tropical species, such as this one, broaden our understanding of hibernation generally and underscore the need of homeothermic animals for access to adequate supplies of energy for maintaining their relatively constant body temperatures. Hibernation, and its associated energy savings, has selective advantage when energy supplies are inadequate for supplying these metabolic needs. This appears to be the case in cold as well as tropical regions.

The temperature relations of organisms, a fundamental aspect of ecology, is attracting increased attention. This interest is fueled by concerns about the ecological consequences of global warming. Though we discuss this issue in detail in chapter 23, in the Applications & Tools section we look at how studies of temperature relations and climatic warming are helping explain the local extinction of a species.

# APPLICATIONS & TOOLS

## Climatic Warming and the Local Extinction of a Land Snail

Between 1906 and 1908, a Ph.D. candidate named G. Bollinger (1909) studied land snails in the vicinity of Basel, Switzerland. Eighty-five years later, Bruno Baur and Anette Baur (1993) carefully resurveyed Bollinger's study sites near Basel for the presence of land snails. In the process, they found that at least one snail species, *Arianta arbustorum,* had disappeared from several of the sites. This discovery led the Baurs to explore the mechanisms that may have produced extinction of these local populations.

*A. arbustorum* is a common land snail in meadows, forests, and other moist, vegetated habitats in northwestern and central Europe. The species lives at altitudes up to 2,700 m in the Alps. The Baurs report that the snail is sexually mature at 2 to 4 years and may live up to 14 years. Adult snails have shell diameters of 16 to 20 mm. The species is hermaphroditic. Though individuals generally mate with other *A. arbustorum,* they can fertilize their own eggs. Adults produce one to three batches of 20 to 80 eggs each year. They deposit their eggs in moss, under plant litter, or in the soil. Eggs generally hatch in 2 to 4 weeks, depending upon temperature. The egg is an especially sensitive stage in the life cycle of land snails. *A. arbustorum* often lives alongside *Cepea nemoralis,* a land snail with a broader geographic distribution that extends from southern Scandinavia to the Iberian peninsula.

How did the Baurs document local extinctions of *A. arbustorum?* If you think about it a bit, you will probably realize that it is usually easier to determine the presence of a species than its absence. If you do not encounter a species during a survey, it may be that you just didn't look hard enough. Fortunately, the Baurs had over 13 years of experience doing field work on *A. arbustorum* and knew its natural history well. For instance, they knew that it is best to search for the snails after rainstorms, when up to 70% of the adult population is active. Consequently, the Baurs searched Bollinger's study sites after heavy rains. They concluded that the snail was absent at a site only after two 2-hour surveys failed to turn up either a living individual or an empty shell of the species.

The Baurs found *A. arbustorum* still living at 13 of the 29 sites surveyed by Bollinger near Basel. Eleven of these remaining populations lived in deciduous forests and the other two lived on grassy riverbanks. However, the Baurs could not find the snail at 16 sites. Eight of these sites had been urbanized, which made the habitat unsuitable for any land snails because natural vegetation had been removed. Between 1900 and 1990 the urbanized area of Basel had increased by 500%. However, the eight other sites where *A. arbustorum* had disappeared were still covered by vegetation that appeared suitable. Four of these sites were covered by deciduous forest, three were on riverbanks, and one was on a railway embankment. These vegetated sites also supported populations of five other land snail species, including *C. nemoralis.*

What caused the extinction of *A. arbustorum* at sites that still supported other snails? The Baurs compared the characteristics of these sites with those of the sites where *A. arbustorum* had persisted. They found no difference between these two groups of sites in regard to slope, percent plant cover, height of vegetation, distance from water, or number of other land snail species present. The first major difference the Baurs uncovered was in altitude. The sites where *A. arbustorum* was extinct have an average altitude of 274 m. The places where it survived have an average altitude of 420 m. The places where the snail had survived were also cooler.

A thermal image of the landscape taken from a satellite showed that surface temperatures in summer around Basel ranged from about 17° to 32.5°C. Surface temperatures where *A. arbustorum* had survived averaged approximately 22°C, while the sites where the species had gone extinct had surface temperatures that averaged approximately 25°C. The sites where the snail was extinct were also much closer to very hot areas with temperatures greater than 29°C. Figure 4.33 is based on the Baurs' thermal image of the area around Basel and shows where the snail was extinct and where it persisted.

The Baurs attributed the higher temperatures at the eight sites where the snail is extinct to heating by thermal radiation from the urbanized areas of the city. Buildings and pavement store more heat than vegetation. In addition, the cooling effect of evaporation from vegetation is lost when an area is built over. Increased heat storage and reduced cooling make urbanized landscapes thermal islands. Heat energy stored in urban centers is transferred to the surrounding landscape through thermal radiation, $H_r$.

The Baurs documented higher temperatures at the sites near Basel where *A. arbustorum* is extinct and identified a well-studied mechanism that could produce the higher temperatures of these sites. However, are the temperature differences they observed sufficient to exclude *A. arbustorum* from the warmer sites? The researchers compared the temperature relations of *A. arbustorum* and *C. nemoralis* to find some clues. They concentrated their studies on the influence of temperature on reproduction by these two snail species.

The eggs of each species were incubated at four temperatures—19°, 22°, 25°, and 29°C. Notice that these temperatures fall within the range measured by the satellite

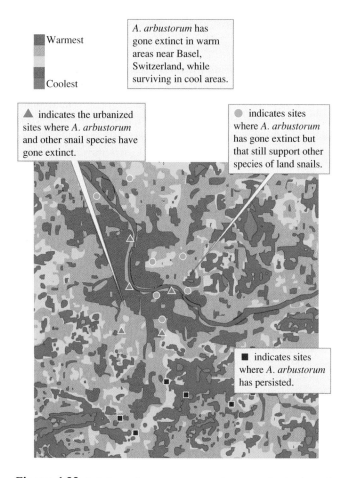

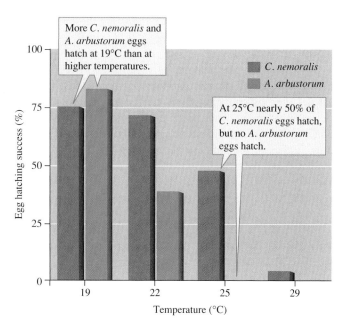

Figure 4.33 Relative surface temperatures and patterns of extinction and persistence by the snail *Arianta arbustorum* around Basel, Switzerland (data from Baur and Baur 1993).

Figure 4.34 Temperature and hatching success of two snail species; the eggs of *Arianta arbustorum* are sensitive to high temperatures (data from Baur and Baur 1993).

image (fig. 4.33). The eggs of both species hatched at a high rate at 19°C. However, at higher temperatures, their eggs hatched at significantly different rates. At 22°C, less than 50% of *A. arbustorum* eggs hatched, while the eggs of *C. nemoralis* continued to hatch at a high rate. At 25°C, no *A.*

*arbustorum* eggs hatched, while approximately 50% of the *C. nemoralis* eggs hatched. At 29°C, the hatching of *C. nemoralis* eggs was also greatly reduced. Figure 4.34 summarizes the results of this hatching experiment.

The results of this study show that the eggs of *A. arbustorum* are more sensitive to higher temperatures than are the eggs of *C. nemoralis*. This greater thermal sensitivity can explain why *A. arbustorum* is extinct at some sites, while *C. nemoralis* survived. These results also suggest that climatic warming can lead to the local extinction of species. As we face the prospect of warming on a global scale, studies of temperature relations will assume greater importance. In chapter 5, we look at a related topic, water relations.

## SUMMARY

**Macroclimate interacts with the local landscape to produce microclimatic variation in temperature.** The sun's uneven heating of the earth's surface and earth's permanent tilt on its axis produce macroclimate. Macroclimate interacts with the local landscape—mainly altitude, aspect, vegetation, color of the ground, and small-scale structural features such as boulders and burrows—to produce microclimates. For the individual organism, macroclimate may be less significant than microclimate. The physical nature of water limits temperature variation in aquatic environments.

**Most species perform best in a fairly narrow range of temperatures.** The influence of temperature on the performance of organisms begins at the molecular level, where extreme temperatures impair the functioning of enzymes. Rates of photosynthesis and microbial activity generally peak in a narrow range of temperatures and are much lower outside of this optimal temperature range. How temperature affects the performance of organisms often corresponds to the current distributions of species and their evolutionary histories.

**Many organisms have evolved ways to compensate for variations in environmental temperature by regulating body temperature.** Temperature regulation balances heat gain against heat loss. Plants and ectothermic animals use morphology and behavior to modify rates of heat exchange with the environment. Birds and mammals rely heavily on metabolic energy to regulate body temperature. The physical nature of the aquatic environment reduces the possibilities for temperature regulation by aquatic organisms. Most endothermic aquatic species are air breathers. Some organisms, mainly flying insects and some large marine fish, improve performance by selectively heating parts of their anatomy. The energetic requirements of thermoregulation may influence the geographic distribution of species.

**Many organisms survive extreme temperatures by entering a resting stage.** This stage may be as simple as resting in a sheltered spot during the heat of the day or may involve elaborate physiological adjustments. Hummingbirds may enter a state of torpor, a state of low metabolic rate and lowered body temperature, when food is scarce and night temperatures cold. Other animals can go into a state of reduced metabolism that may last several months. If this state occurs mainly in winter, it is called hibernation. If it occurs in summer, it is called estivation. Such reductions in metabolic rate allow these animals to survive extreme environmental conditions during which they must rely entirely on stored energy reserves.

Long-term studies of populations of land snails around Basel, Switzerland, have documented local extinctions of these land snails. These extinctions are attributable to habitat destruction and climatic warming. The results of these studies suggest that climatic warming can lead to the local extinction of species. As we face the prospect of climatic warming at a global scale, studies of temperature relations will assume greater importance.

# Review Questions

1. Many species of plants and animals that are associated with boreal forests also occur on mountains far to the south of the boreal forests. Using what you have learned about microclimates, predict how aspect and elevation would influence their distributions on these southern mountains.

2. Imagine a desert beetle that uses behavior to regulate its body temperature above 35°C. How might this beetle's use of microclimates created by shrubs, burrows, and bare ground change with the season?

3. J. L. Mosser and colleagues (1974) found that populations of the bacterium *Sulfolobus* living at different temperatures had different optimal temperatures for sulfur oxidation. Use natural selection to explain these patterns. Design an experiment to test your explanation. Assume you can create artificial springs and regulate their temperature as you like.

4. Figure 4.8 shows how temperature influences the activity of acetylcholinesterase in rainbow trout. Assuming that the other enzymes of rainbow trout show similar responses to temperature, how would trout swimming speed change as environmental temperature increases above 20°C?

5. The Applications & Tools section reviews how the studies of Bruno Baur and Anette Baur (1993) have documented the local extinction of the land snail *Arianta arbustorum*. Their research also shows that these extinctions may be due to reduced egg hatching at higher temperatures. Do these results show conclusively that the direct effect of higher temperatures on hatching success is responsible for the local extinctions of *A. arbustorum*? Propose and justify alternative hypotheses. Be sure you take into account all of the observations of the Baurs.

6. Butterflies, which are ectothermic and diurnal, are found from the tropical rain forest to the Arctic. They can elevate their body temperatures by basking in sunlight. How would the percentage of time butterflies spend basking versus flying change with latitude? Would the amount of time butterflies spend basking change with daily changes in temperature?

7. When we reviewed how some organisms use torpor, hibernation, and estivation to avoid extreme temperatures, we discussed the idea of energy savings. However, organisms do not always behave in a way that saves energy. For instance, when food is abundant hummingbirds do not go into torpor at night. This suggests that there may be some disadvantages associated with torpor. What are some of those potential disadvantages?

8. The section on avoiding temperature extremes focused mainly on animals. What are some of the ways in which plants avoid temperature extremes? Bring cold and hot environments into your discussion. Some of the natural history included in chapter 2 might be useful as you formulate an answer.

9. Some plants and grasshoppers in hot environments have reflective body surfaces, which make their radiative heat gain, $H_r$, less than it would be otherwise. If you were to design a tiger beetle that could best cope with thermal conditions on black beaches (see fig. 4.5), what color would it be? The beetles on the black beaches of New Zealand are black, and the beetles on the white beaches are white. What do the matches between the color of these beetles and their beaches tell us about the relative roles of thermoregulation and predation pressure in determining beetle color? What does this example imply about the ability of natural selection to "optimize" the characteristics of organisms?

10. In most of the examples discussed in chapter 4, we saw a close match between the characteristics of organisms and their environment. However, natural selection does not always produce an optimal, or even a good, fit of organisms to their environments. To verify this you need only reflect on the fact that most of the species that have existed are now extinct. What are some of the reasons for a mismatch between organisms and environments? Develop your explanation using the environment, the characteristics of organisms, and the nature of natural selection.

# Suggested Readings

Angilletta, M. J., Jr. 2001. Thermal and physiological constraints on energy assimilation in a widespread lizard (*Sceloporus undulatus*). *Ecology* 82:3044–56.

Angilleta, M. J., Jr., T. Hill, and M. A. Robson. 2002. Is physiological performance optimized by thermoregulatory behavior?: a case study of the eastern fence lizard, *Sceloporus undulatus*. *Journal of Thermal Biology* 27:199–204.

*Exemplary studies of the temperature relations of an ectotherm, with clear explanations of study design.*

Heinrich, B. 1993. *The Hot Blooded Insects.* Cambridge, Mass.: Harvard University Press.

Heinrich, B. 1996. *The Thermal Warriors: Strategies of Insect Survival.* Cambridge, Mass.: Harvard University Press.

*Heinrich combines the knowledge of a scientist with the skills of artist and poet to produce two books that make the technical details of insect thermal ecology accessible and interesting to general readers. Valuable for readers at any level of biological knowledge.*

Huber, R., H. Huber, and K. O. Stetter. 2000. Towards the ecology of hyperthermophiles: biotopes, new isolation strategies and novel metabolic properties. *FEMS Microbiology Reviews* 24:615–623.

Ishii, K. and K. Marumo. 2002. Microbial diversity in hydrothermal systems and their influence on geological environments. *Resource Geology* 52:135–46.

Reysenbach, A.-L., M. Ehringer, and K. Hershberger. 2000. Microbial diversity at 83°C in Calcite Springs, Yellowstone National Park: another environment where the Aquificales and "Korarchaeota" coexist. *Extremophiles* 4:61–67.

Staley, J. T. and A.-L. Reysenbach. 2002. *Biodiversity of Microbial Life: Foundation of Earth's Biosphere.* New York: Wiley.

*References that provide a window to the world of the prokaryotes found in extreme thermal environments.*

Calder, W. A. 1994. When do hummingbirds use torpor in nature? *Physiological Zoology* 67:1051–76.

Carpenter, F. L., M. A. Hixon, C. A. Beuchat, R. W. Russell, and D. C. Patton. 1993. Biphasic mass gain in migrant hummingbirds: body composition changes, torpor, and ecological significance. *Ecology* 74:1173–82.

*Two excellent papers concerning the ecology and physiology of torpor in hummingbirds. These researchers show some of the challenges that ecologists must overcome when working on small, highly mobile organisms.*

Carruthers, R. I., T. S. Larkin, H. Firstencel, and Z. Feng. 1992. Influence of thermal ecology on the mycosis of a rangeland grasshopper. *Ecology* 73:190–204.

*Carruthers and colleagues explore the details of how grasshoppers use behavioral thermoregulation to combat a potentially deadly pathogen.*

Fietz, J., F. Tataruch, K. H. Dausmann, and J. U. Ganzhorn. 2003. White adipose tissue composition in the free-ranging fat-tailed dwarf lemur (*Cheirogaleus medius*; Primates), a tropical hibernator. *Journal of Comparative Physiology B: Biochemical, Systemic, and Environmental Physiology* 173:1–10.

*Fascinating details of hibernation by a tropical primate.*

Knutson, R. M. 1974. Heat production and temperature regulation in eastern skunk cabbage. *Science* 186:746–47.

Knutson, R. M. 1979. Plants in heat. *Natural History* 88:42–47.

Uemura, S., K. Ohkawara, G. Kudo, N. Wada, and S. Higashi. 1993. Heat-production and cross-pollination of the Asian skunk cabbage *Symplocarpus renifolius* (Araceae). *American Journal of Botany* 80:635–40.

*Fascinating accounts of the ecology of the physiologically unique thermogenic skunk cabbages. Knutson's pioneering research sets the stage for the exploration; Uemura and colleagues extend the research to the Asian skunk cabbage.*

# On the Net

Visit this textbook's accompanying website at www.mhhe.com/ecology (click on the book's title) to take advantage of practice quizzing, study/writing tips, timely news articles, and additional URLs for research on the topics in this chapter.

Biological Structure and Function

Temperature Regulation

Thermal Relations

# Water Relations

Water plays a central role in the lives of all organisms. However, water acquisition and conservation is particularly critical for desert organisms. As a consequence, many ecologists studying the water relations of organisms have focused their attention on desert species. The steady buzzing of the Sonoran Desert cicada, *Diceroprocta apache,* seemed to amplify the withering heat. Air temperature in the shade hovered around 46°C, and the ground surface temperature was over 70°C. All other animals had taken refuge from the desert heat. Nothing else called, and nothing moved, except a lone biologist with an insect net who stalked in the direction of the calling cicada.

The biologist was Eric Toolson. Toolson was well acquainted with the calls of all the cicadas in the region and he knew their natural history. He knew when they were active, where they fed, and their natural enemies. Toolson associated the call of *Diceroprocta* with the hottest hours of the desert day, when air temperatures often exceeded the lethal limit for the species. His goal was to understand the ecology of this extraordinary cicada. If Toolson could capture the calling cicada, he would put it in an environmental chamber in his laboratory and measure its body temperature and water loss rates under a variety of conditions. Later, he would release it to resume its midday serenade.

Questions raced through Toolson's mind as he made his way through the shimmering desert air toward the cicada. Above all, how could this species be active in apparently lethal air temperatures? We might ask the same question of Toolson himself. How did he maintain a body temperature of approximately 37°C in this desert heat? Humans evaporatively cool by sweating. If we placed humidity sensors just above Toolson's skin, they would show that he sweated profusely as he picked his way through the cactus. To keep from becoming dehydrated, he took frequent drinks from the water bottle at his side. This enabled him to maintain sufficient internal water and continue to evaporatively cool by sweating.

During his pauses to drink, more questions came. Did the cicada keep cool by using small, shady microclimates in the mesquite tree from which it called? Did the cicada somehow manage to evaporatively cool? This seemed unlikely, since biologists had long assumed that insects were too small and vulnerable to water loss to do so. If *Diceroprocta* did evaporatively cool, how did it avoid desiccating in the desert heat? It did not, like Toolson, have a water bottle strapped to its waist.

As Toolson stalked the cicada, he pursued an even greater prize: an understanding of how *Diceroprocta* can regulate the temperature and water content of its body while living in such an extreme environment. This second pursuit would lead Toolson to discover an unsuspected physiological process in these desert insects. Like Bernd Heinrich, who had discovered the mechanisms by which sphinx moths thermoregulate (see chapter 4), Toolson and his students, Stacy Kaser and Jon Hastings, would be the first to comprehend a bit of nature that had escaped the notice of all researchers before them. Few scientists make such a fundamental discovery. Those that do

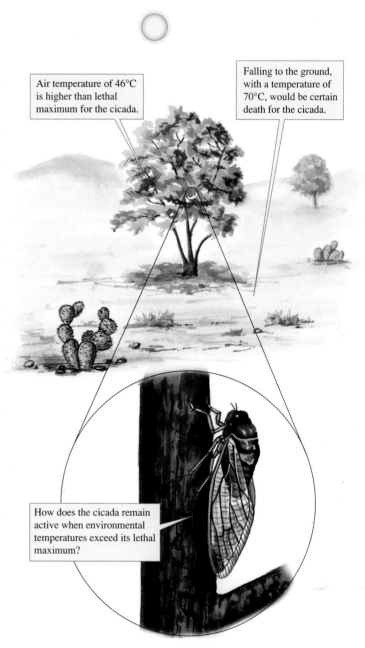

**Figure 5.1** An ecological puzzle: the cicada, *Diceroprocta apache,* is active when air temperatures would appear to be lethal for the species.

never forget the thrill. Figure 5.1 summarizes the extreme physical conditions under which *Diceroprocta* lives that inspired Toolson and his colleagues to study its ecology (Toolson 1987, Toolson and Hadley 1987).

Before we discuss the ecology of these desert cicadas, we need to introduce some background information. Water and life on earth are closely linked. The high water content of most organisms, which ranges from about 50% to 90%, reflects life's aquatic origins. Life on earth originated in salty aquatic environments and is built around biochemistry within an aquatic medium. To survive and reproduce, organisms must maintain appropriate internal concentrations of water and dissolved substances. To maintain these internal concentrations organisms must balance water losses to the environment with water intake. How organisms maintain this water balance is called their water relations, which is the subject of chapter 5.

In some environments, organisms face the problem of water loss. Elsewhere, water streams in from the environment. The problem of maintaining proper water balance is especially strong for those organisms, such as *Diceroprocta,* that live in arid terrestrial environments. A parallel challenge faces organisms that live in aquatic environments with a high salinity. In these extreme environments, the water relations of organisms stand out in bold relief. However, most organisms must expend energy to maintain their internal pool of water. In the study of relationships between organisms and the environment, which we call ecology, the study of water relations is fundamental.

## CONCEPTS

- **The movement of water down concentration gradients in terrestrial and aquatic environments determines the availability of water to organisms.**
- **Terrestrial plants and animals regulate their internal water by balancing water acquisition against water loss.**
- **Marine and freshwater organisms use complementary mechanisms for water and salt regulation.**

## CONCEPT DISCUSSION

### Water Availability

**The movement of water down concentration gradients in terrestrial and aquatic environments determines the availability of water to organisms.**

The tendency of water to move down concentration gradients and the magnitude of those gradients from an organism to its environment determine whether an organism tends to lose or gain water from the environment. In chapter 1 we pointed out that ecologists may study purely physical aspects of the environment to understand the ecology of organisms. To understand the water relations of organisms, we must understand at least the basic physical behavior of water in terrestrial and aquatic environments.

In chapter 2, we saw that water availability on land varies tremendously, from the tropical rain forest with abundant moisture throughout the year (see fig. 2.10), to hot deserts with

year-round drought (see fig. 2.19). In chapter 3, we reviewed the considerable variation in salinity among aquatic environments, ranging from the dilute waters of tropical rivers draining highly weathered landscapes to hypersaline lakes. The majority of aquatic environments, including the sea, fall somewhere between these extremes. Salinity, as we shall see, reflects the relative "aridity" of aquatic environments.

These preliminary descriptions in chapters 2 and 3 do not include the situations faced by individual organisms within their microclimates—microclimates such as those experienced by a desert animal that lives at an oasis, where it has access to abundant moisture, or a rain forest plant that lives in the forest canopy, where it is exposed to full tropical sun and drying winds. As with temperature, to understand the water relations of an organism we must consider its microclimate. For instance, consider the microclimates available to the seashore isopod, *Ligia oceanica* (see fig. 4.6). In addition to temperature, those microclimates vary considerably in relative humidity, a measure of the relative water content of air. The relative humidity of air within crevices is usually close to 100%, while in the open it can be as low as 70%. Such differences in relative humidity can significantly affect rates of water loss by terrestrial organisms. To fully understand a comparison of microclimates such as this one, however, you need to understand the measure, such as relative humidity, being used. This is why we turn now to the measurement of water in the environment.

## Water Content of Air

As we saw when we reviewed the hydrologic cycle in chapter 3, water vapor is continuously added to air as water evaporates from the surfaces of oceans, lakes, and rivers. On land, evaporation also accounts for much of the water lost by organisms. The potential for such evaporative water loss depends upon the temperature and water content of the air around the organisms. As the amount of water vapor in the surrounding air increases, the water concentration gradient from organisms to the air is reduced and the rate at which organisms lose water to the atmosphere decreases. This is the reason that evaporative air coolers work poorly in humid climates, where the water content of air is high. These coolers work best in arid climates, where there is a steep gradient of water concentration from the cooler to the air. A steep gradient of water concentration produces a high rate of evaporation.

We know how temperature is measured, but how is the water content of air measured? The quantity of water vapor in the air can be expressed conveniently in relative terms. Since air rarely contains all the water vapor it can hold, we can use its degree of saturation with water vapor as a relative measure of water content. The most familiar measure of the water content of air relative to its content at saturation is **relative humidity,** defined as:

$$\text{Relative humidity} = \frac{\text{Water vapor density}}{\text{Saturation water vapor density}} \times 100$$

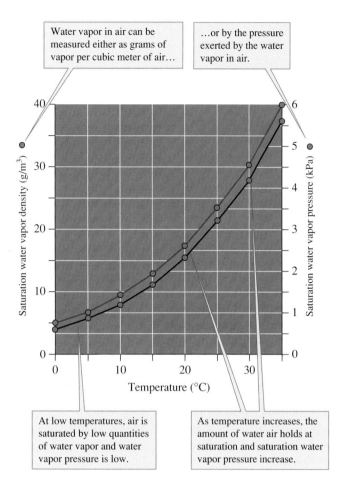

Water vapor in air can be measured either as grams of vapor per cubic meter of air...

...or by the pressure exerted by the water vapor in air.

At low temperatures, air is saturated by low quantities of water vapor and water vapor pressure is low.

As temperature increases, the amount of water air holds at saturation and saturation water vapor pressure increase.

**Figure 5.2** The relationship between air temperature and two measures of water vapor saturation of air.

The actual amount of water in air is measured directly as the mass of water vapor per unit volume of air. This quantity, the *water vapor density,* is the numerator in the relative humidity equation and is given either as milligrams of water per liter of air (mg $H_2O$/L) or as grams of water per cubic meter of air (g $H_2O$/m$^3$). The quantity of water vapor that air potentially can hold is its *saturation water vapor density,* the denominator in the relative humidity equation. Saturation water vapor density changes with temperature and, as you can see from the red curve in figure 5.2, warm air can hold more water vapor than cold air, something that no one who has experienced a really hot and humid climate will doubt.

One of the most useful ways of expressing the quantity of water in air is in terms of the pressure it exerts. If we express the water content of air in terms of pressure, we can use similar units to consider the water relations of organisms in air, soil, and water. Using pressure as a common currency to represent water relations in very different environments helps us unify our understanding of this very important area of ecology. We usually think in terms of *total atmospheric pressure,* the pressure exerted by all the gases in air, but you can also calculate the partial pressures due to individual atmospheric gases such as oxygen, nitrogen, or water vapor.

We call this last quantity **water vapor pressure.** You probably learned long ago that at sea level, atmospheric pressure averages approximately 760 mm of mercury, the height of a column of mercury supported by the combined force (pressure) of all the gas molecules in the atmosphere. The international convention for representing water vapor pressure, however, is in terms of the pascal (Pa), where 1 Pa is 1 newton of force per square meter. Using this convention, 760 mm of mercury, or one atmosphere of pressure, equals approximately 101,300 Pa, 101.3 kilopascals (kPa), or .101 megapascals (MPa = $10^6$ Pa).

The pressure exerted by the water vapor in air that is saturated with water is called **saturation water vapor pressure.** As the black curve in figure 5.2 shows, this pressure increases with temperature and closely parallels the increase in saturation water vapor density shown by the red curve. Why are the two curves so similar? They are similar because the amount of pressure exerted by a gas, in this case water vapor, is directly proportional to its density.

We can also use water vapor pressure to represent the *relative* saturation of air with water. You calculate this measure, called the **vapor pressure deficit,** as the difference between the actual water vapor pressure and the saturation water vapor pressure at a particular temperature. In terrestrial environments, water flows from organisms to the atmosphere at a rate influenced by the vapor pressure deficit of the air surrounding the organism. Figure 5.3 shows the relative rates of water loss by an organism exposed to air with a low versus high vapor pressure deficit. Again, one of the most useful features of vapor pressure deficit is that it is expressed in units of pressure, generally kilopascals.

# Water Movement in Aquatic Environments

In aquatic environments, water moves down its concentration gradient. It may sound silly to speak of the amount of water in an aquatic environment but, as we saw in chapter 2, all aquatic environments contain dissolved substances. These dissolved substances, however slightly, dilute the water. While oceanographers and limnologists (those who study bodies of freshwater) generally focus on salt content, or salinity, we take the opposite point of view in order to build a consistent perspective for considering water relations in air, water, and soil. From this perspective, water is more concentrated in freshwater environments than in the oceans. The oceans, in turn, contain more water per liter than do saline lakes such as the Dead Sea or the Great Salt Lake. The relative concentration of water in each of these environments strongly influences the biology of the organisms that live in them.

The body fluids of all organisms contain water and solutes, including inorganic ions and amino acids. We can think of aquatic organisms and the environment that surrounds them as two aqueous solutions separated by a selec-

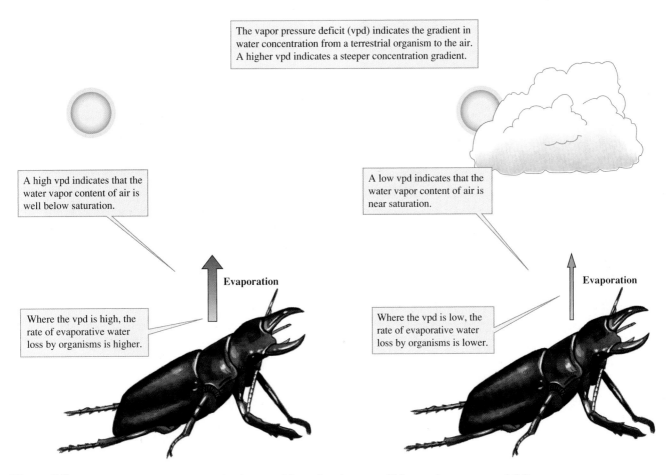

The vapor pressure deficit (vpd) indicates the gradient in water concentration from a terrestrial organism to the air. A higher vpd indicates a steeper concentration gradient.

A high vpd indicates that the water vapor content of air is well below saturation.

A low vpd indicates that the water vapor content of air is near saturation.

Evaporation

Evaporation

Where the vpd is high, the rate of evaporative water loss by organisms is higher.

Where the vpd is low, the rate of evaporative water loss by organisms is lower.

**Figure 5.3** The potential for evaporative water loss by terrestrial organisms increases with increased vapor pressure deficit.

tively permeable membrane. If the internal environment of the organism and the external environment differ in concentrations of water and salts, these substances will tend to move down their concentration gradients. This movement is the process of **diffusion.** We give the diffusion of water across a semipermeable membrane a special name, however: **osmosis.**

In the aquatic environment, water moving down its concentration gradient produces osmotic pressure. Osmotic pressure, like vapor pressure, can be expressed in pascals. The strength of the osmotic pressure across a semipermeable membrane, such as the gills of a fish, depends upon the difference in water concentration across the membrane. Larger differences, between organism and environment, generate higher osmotic pressures.

Aquatic organisms generally live in one of three environmental circumstances. Organisms with body fluids containing the same concentration of water as the external environment are **isosmotic.** Organisms with body fluids with a higher concentration of water (lower solute concentration) than the external medium are **hypoosmotic** and tend to lose water to the environment. Those with body fluids with a lower concentration of water (higher solute concentration) than the external

medium are **hyperosmotic** and are subject to water flooding inward from the environment. In the face of these osmotic pressures, aquatic organisms must expend energy to maintain a proper internal environment. How much energy the organism must expend depends upon the magnitude of the osmotic pressure between them and the environment and the permeability of their body surfaces. Figure 5.4 summarizes the movement of water and salts into and out of isosmotic, hyperosmotic, and hypoosmotic organisms.

## Water Movement Between Soils and Plants

On land, water flows from the organism to the atmosphere at a rate influenced by the vapor pressure deficit of the air surrounding the organism. In the aquatic environment, water may flow either to or from the organism, depending on the relative concentrations of water and solutes in body fluids and the surrounding medium. But here too, water flows down its concentration gradient. As shown in figure 5.5, water moving from the soil through a plant and into the atmosphere flows down a gradient of **water potential.** Water in soils and plants

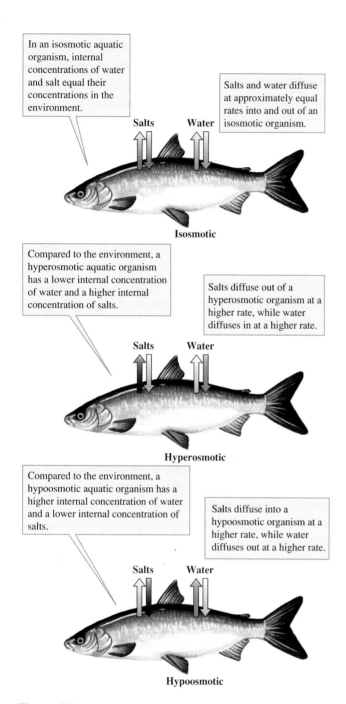

In an isosmotic aquatic organism, internal concentrations of water and salt equal their concentrations in the environment.

Salts and water diffuse at approximately equal rates into and out of an isosmotic organism.

**Salts**    **Water**

**Isosmotic**

Compared to the environment, a hyperosmotic aquatic organism has a lower internal concentration of water and a higher internal concentration of salts.

Salts diffuse out of a hyperosmotic organism at a higher rate, while water diffuses in at a higher rate.

**Salts**    **Water**

**Hyperosmotic**

Compared to the environment, a hypoosmotic aquatic organism has a higher internal concentration of water and a lower internal concentration of salts.

Salts diffuse into a hypoosmotic organism at a higher rate, while water diffuses out at a higher rate.

**Salts**    **Water**

**Hypoosmotic**

**Figure 5.4** Water and salt regulation by isosmotic, hyperosmotic, and hypoosmotic aquatic organisms.

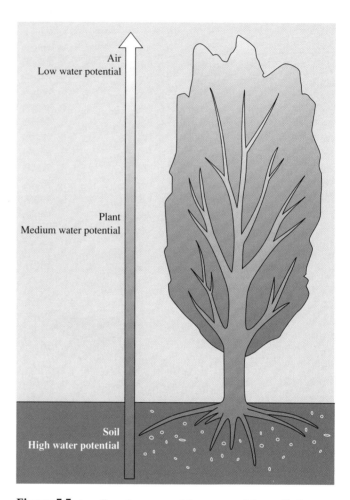

Air
Low water potential

Plant
Medium water potential

**Soil**
**High water potential**

**Figure 5.5** A gradient of water potential: water potential generally decreases from soil to plant to air.

moves through the small pore spaces within soils and within the small water-conducting cells of plants. Therefore, water potential in soils and plants is determined by the concentration gradient of water plus other factors related to the movement of water through these small spaces. Understanding water potential takes some patience, but that patience will be paid off by a significant improvement in understanding the water relations of terrestrial plants.

We can define water potential as the capacity of water to do work. Flowing water has the capacity to do work such as turning the water wheel of an old-fashioned water mill or the turbines of a hydroelectric plant. The capacity of water to do work depends upon its free energy content. Water flows from positions of higher to lower free energy. Under the influence of gravity, water flows downhill from a position of higher free energy, at the top of the hill, to a position of lower free energy, at the bottom of the hill.

In the section "Water Movement in Aquatic Environments," we saw that water flows down its concentration gradient, from locations of higher water concentration (hypoosmotic) to locations of lower water concentration (hyperosmotic). The measurable "osmotic pressure" generated by water flowing down these concentration gradients shows that water flowing in response to osmotic gradients has the capacity to do work. Which has a higher free energy content, pure water or seawater? Since osmotic flow would be from pure water to seawater and water flows from high to low free energy, pure water must have a higher free energy content than seawater. We measure water

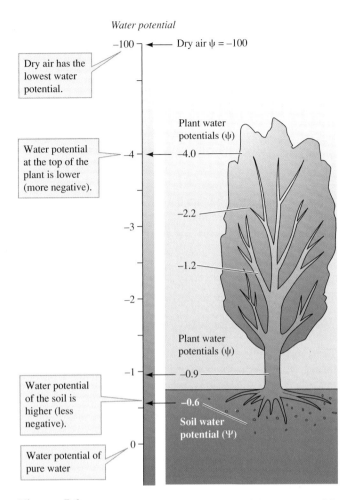

*Water potential*

Dry air $\psi = -100$

Dry air has the lowest water potential.

Plant water potentials ($\psi$)

Water potential at the top of the plant is lower (more negative).

−4.0

−2.2

−1.2

Plant water potentials ($\psi$)

−0.9

Water potential of the soil is higher (less negative).

−0.6

**Soil water potential ($\Psi$)**

Water potential of pure water

**Figure 5.6** A quantified gradient of water potential: water potentials become more negative from soil to plant to air (data from Wiebe et al. 1970).

potential, like vapor pressure deficit and osmotic pressure, in pascals, usually megapascals (MPa = Pa $\times 10^6$). By convention, water potential is represented by the symbol $\psi$ and the water potential of pure water is set at 0. If the water potential of pure water is 0, then the water potential of a solution, such as seawater, must be negative (i.e., < 0).

In nature, water potentials are generally negative. Why is that? This must be so since all water in nature, even rainwater, contains some solute or occupies spaces where matric forces (see following) are significant. So, gradients of water potential in nature are generally from less negative to more negative water potential. This convention takes a bit of getting used to! Once familiar, however, the convention is useful for representing and thinking about the water relations of plants. Figure 5.6, a quantified version of the gradient of water potential we saw in figure 5.5, shows that water is flowing down a gradient of water potential that goes from a slightly negative water potential in the soil through the moderately negative water potentials of the plant to the highly negative water potential of dry air.

Now let's look at some of the mechanisms involved in producing a gradient of water potential such as that shown in figure 5.6. We can express the water potential of a solution as:

$$\psi = \psi_{solutes}$$

$\psi_{solutes}$ is the *reduction* in water potential due to dissolved substances, which is a negative number.

Within small spaces, such as the interior of a plant cell or the pore spaces within soil, other forces, called **matric forces,** are also at work. Matric forces are a consequence of water's tendency to adhere to the walls of containers such as cell walls or the soil particles lining a soil pore. Matric forces lower water potential. The water potential for fluids within plant cells is approximately:

$$\psi_{plant} = \psi_{solutes} + \psi_{matric}$$

In this expression, $\psi_{matric}$ is the *reduction* in water potential due to matric forces within plant cells. At the level of the whole plant, another force is generated as water evaporates from the surfaces of leaves into the atmosphere. Evaporation of water from the surfaces of leaves generates a negative pressure, or tension, on the column of water that extends from the leaf surface through the plant all the way down to its roots. This negative pressure reduces the water potential of plant fluids still further.

So, the water potential of plant fluids is affected by solutes, matric forces, and the negative pressures exerted by evaporation. Consequently, we can represent the water potential of plant fluids as:

$$\psi_{plant} = \psi_{solutes} + \psi_{matric} + \psi_{pressure}$$

Here again, $\psi_{pressure}$ is the *reduction* in water potential due to negative pressure created by water evaporating from leaves.

Meanwhile, the solute content of soil water is often so low that soil matric forces account for most of soil water potential:

$$\psi_{soil} \cong \psi_{matric}$$

Matric forces vary considerably from one soil to another, depending primarily upon soil texture and pore size. Coarser soils, such as sands and loams, with larger pore sizes exert lower matric forces, while fine clay soils, with smaller pore sizes, exert higher matric forces. So, while clay soils can hold a higher quantity of water compared to sandy soils, the higher matric forces within clay soils bind that water more tightly. As long as the water potential of plant tissues is less than the water potential of the soil, $\psi_{plant} < \psi_{soil}$ water flows from the soil to the plant.

The higher water potential of soil water compared to the water potential of roots induces water to flow from the soil into plant roots. As water enters roots from the surrounding soil, it joins a column of water that extends from the roots through the water-conducting cells, or xylem, of the stem to the leaves. Hydrogen bonds between adjacent water molecules bind the

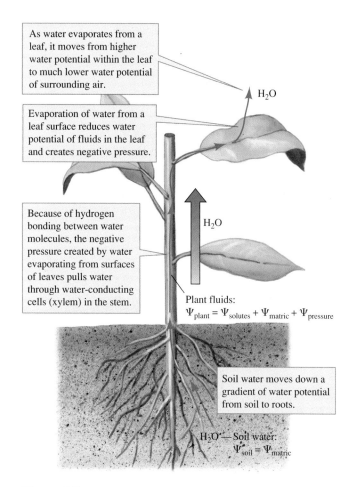

As water evaporates from a leaf, it moves from higher water potential within the leaf to much lower water potential of surrounding air.

Evaporation of water from a leaf surface reduces water potential of fluids in the leaf and creates negative pressure.

Because of hydrogen bonding between water molecules, the negative pressure created by water evaporating from surfaces of leaves pulls water through water-conducting cells (xylem) in the stem.

$H_2O$

$H_2O$

Plant fluids:
$\Psi_{plant} = \Psi_{solutes} + \Psi_{matric} + \Psi_{pressure}$

Soil water moves down a gradient of water potential from soil to roots.

$H_2O$ — Soil water:
$\Psi_{soil} = \Psi_{matric}$

**Figure 5.7** Mechanisms of water movement from soil through plants to the atmosphere.

water molecules in this water column together. Consequently, as water molecules at the upper end of this column evaporate into the air at the surfaces of leaves, they exert a tension, or negative pressure, on the entire water column. This negative pressure further reduces the water potential of plant fluids and helps power uptake of water by terrestrial plants. Figure 5.7 summarizes the mechanisms underlying the flow of water from soil to plants.

As plants draw water from the soil, they soon deplete the water held in the larger soil pore spaces, leaving only water held in the smaller pores. Within these smaller soil pores matric forces are greater than in the larger pores. Consequently, as soil dries, soil water potential becomes more and more negative and the remaining water becomes harder and harder to extract.

This section has given us a basis for considering the availability of water to organisms living in terrestrial and aquatic environments. Let's use the foundation we have built here to explore the water relations of organisms on land and in water. To survive and reproduce, organisms must maintain appropriate internal concentrations of water and dissolved substances. As a consequence, in the face of variation in water availability, organisms have been selected to regulate their internal water.

## CONCEPT DISCUSSION

### Water Regulation on Land

**Terrestrial plants and animals regulate their internal water by balancing water acquisition against water loss.**

When organisms moved into the terrestrial environment, they faced two major environmental challenges: potentially massive losses of water to the environment through evaporation and reduced access to replacement water. Many adaptations helped terrestrial organisms meet these challenges and acquire the capacity to regulate their internal water content. We can summarize water regulation by terrestrial animals as:

$$W_{ia} = W_d + W_f + W_a - W_e - W_s$$

This says simply that the internal water of an animal ($W_{ia}$) results from a balance between water acquisition and water loss. The major sources of water are:

$$W_d = \text{water taken by drinking}$$

$$W_f = \text{water taken in with food}$$

$$W_a = \text{water absorbed from the air}$$

The avenues of water loss are:

$$W_e = \text{water lost by evaporation}$$

$$W_s = \text{water lost with various secretions and excretions}$$
including urine, mucus, and feces

We can summarize water regulation by terrestrial plants in a similar way:

$$W_{ip} = W_r + W_a - W_t - W_s$$

The internal water concentration of a plant ($W_{ip}$) results from a balance between gains and losses, where the major sources of water for plants are:

$$W_r = \text{water taken from soil by roots}$$

$$W_a = \text{water absorbed from the air}$$

The major ways that plants lose water are:

$$W_t = \text{water lost by transpiration}$$

$$W_s = \text{water lost with various secretions and reproductive}$$
structures including nectar, fruit, and seeds

The main avenues of water gain and loss by terrestrial plants and animals are summarized in figure 5.8. The figure presents a generalized picture of the water relations of terres-

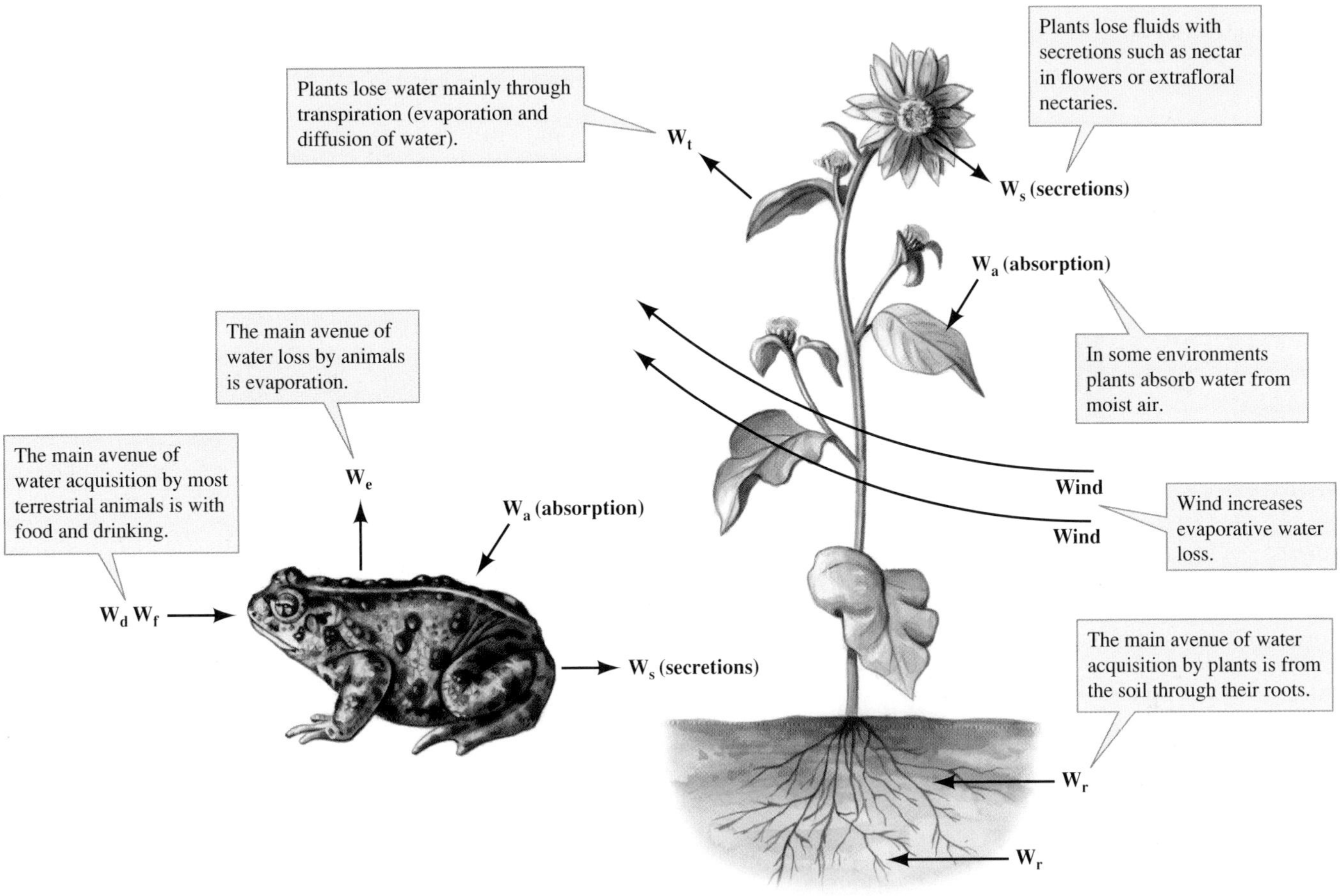

**Figure 5.8** Terrestrial plants and animals can be characterized by analogous pathways for water gain and loss.

trial organisms. However, organisms in different environments face different environmental challenges and they have evolved a wide variety of solutions to those problems. Let's now look at the diverse ways in which terrestrial plants and animals regulate their internal water.

## Water Acquisition by Animals

Many small terrestrial animals can absorb water from the air. Most terrestrial animals, however, satisfy their need for water either by drinking or by taking in water with food. In moist climates, there is generally plenty of water, and, if water becomes scarce, the mobility of most animals allows them to go to sources of water to drink. In deserts, animals that need abundant water must live near oases. Those that live out in the desert itself, away from oases, have evolved adaptations for living in arid environments.

Some desert animals acquire water in unusual ways. Coastal deserts such as the Namib Desert of southwest Africa receive very little rain but are bathed in fog. This aerial moisture is the water source for some animals in the Namib. One of these, a beetle in the genus *Lepidochora* of the family Tenebrionidae, takes an engineering approach to water acqui-

sition. These beetles dig trenches on the face of sand dunes to condense and concentrate fog. The moisture collected by these trenches runs down to the lower end, where the beetle waits for a drink. Another tenebrionid beetle, *Onymacris unguicularis,* collects moisture by orienting its abdomen upward. Fog condensing on this beetle's body flows to its mouth. Figure 5.9 shows this beetle's unique means of obtaining drinking water. *Onymacris* also takes in water with its food. Some of this water is absorbed within the tissues of the food. The remaining water is produced when the beetle metabolizes the carbohydrates, proteins, and fats contained in its food. We can see the source of this "metabolic water" if we look at an equation for oxidation of glucose:

$$C_6H_{12}O_6 + 6 O_2 \rightarrow 6 CO_2 + \textbf{6 H}_2\textbf{O}$$

As you can see, cellular respiration liberates the water that combined with carbon dioxide during the process of photosynthesis (see chapter 4). The water released during cellular respiration is called **metabolic water.**

Paul Cooper (1982) estimated the water budget for free-ranging *Onymacris* from the Namib Desert near Gobabeb. He estimated the rate of water intake by this beetle at 49.9 mg of $H_2O$ per gram of body weight per day. Of this total, 39.8 mg

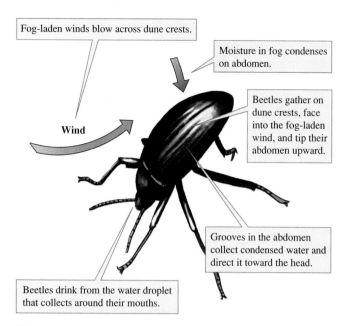

Fog-laden winds blow across dune crests.

Moisture in fog condenses on abdomen.

Beetles gather on dune crests, face into the fog-laden wind, and tip their abdomen upward.

Wind

Grooves in the abdomen collect condensed water and direct it toward the head.

Beetles drink from the water droplet that collects around their mouths.

**Figure 5.9** Some beetles of the Namib Desert can harvest sufficient moisture from fog to meet their needs for water.

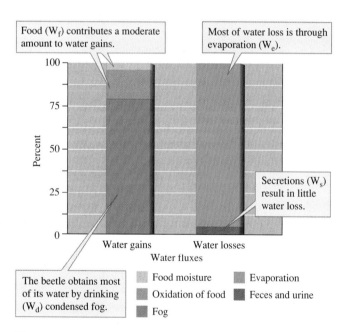

Food ($W_f$) contributes a moderate amount to water gains.

Most of water loss is through evaporation ($W_e$).

Secretions ($W_s$) result in little water loss.

Water gains    Water losses
Water fluxes

The beetle obtains most of its water by drinking ($W_d$) condensed fog.

Food moisture    Evaporation
Oxidation of food    Feces and urine
Fog

**Figure 5.10** Water budget of the desert beetle, *Onymacris unguicularis* (data from Cooper 1982).

came from fog, 1.7 mg came from moisture contained within food, and 8.4 mg came from metabolic water. The rate of water loss by these beetles, 41.3 mg of $H_2O$ per gram per day, was slightly less than water intake. Of this total, 2.3 mg were lost with feces and urine, and 39 mg by evaporation. The water budget of the beetle studied by Cooper is shown in figure 5.10.

While *Onymacris* gets most of its water from fog, other small desert animals get most of their water from their food. Kangaroo rats of the genus *Dipodomys* in the family Hetero-

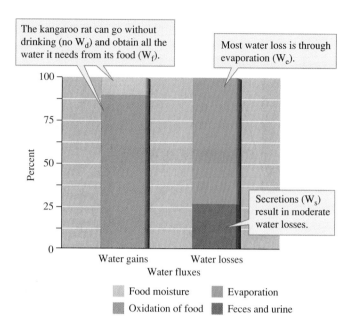

The kangaroo rat can go without drinking (no $W_d$) and obtain all the water it needs from its food ($W_f$).

Most water loss is through evaporation ($W_e$).

Secretions ($W_s$) result in moderate water losses.

Water gains    Water losses
Water fluxes

Food moisture    Evaporation
Oxidation of food    Feces and urine

**Figure 5.11** Water budget of Merriam's kangaroo rat *Dipodomys merriami* (data from Schmidt-Nielsen 1964).

myidae (see fig. 13.23) don't have to drink at all and can survive entirely on metabolic water. Knut Schmidt-Nielsen (1964) showed that the approximately 60 ml of water gained from 100 g of barley makes up for the water a Merriam's kangaroo rat, *D. merriami,* loses in feces, urine, and evaporation while metabolizing the 100 g of grain. The 100 g of barley contains only 6 ml of absorbed water, that is, water that can be driven off by drying. The remaining 54 ml of water is released as the animal metabolizes the carbohydrates, fats, and proteins in the grain. The importance of metabolic water in the water budget of Merriam's kangaroo rat is pictured in figure 5.11.

While animals generally obtain most of their water by drinking or with their food, these options are not available to plants. Though many plants can absorb some water from the air, most get the bulk of their water from the soil through their roots.

## Water Acquisition by Plants

The extent of root development by plants often reflects differences in water availability. Studies of root systems in different climates show that plants in dry climates grow more roots than do plants in moist climates. In dry climates, plant roots tend to grow deeper in the soil and to constitute a greater proportion of plant biomass. The tap roots of some desert shrubs can extend 9 or even 30 m down into the soil, giving them access to deep groundwater. Roots may account for up to 90% of total plant biomass in deserts and semiarid grasslands. In coniferous forests, roots constitute only about 25% of total plant biomass.

You don't have to compare forests and deserts, however, to observe differences in root development. R. Coupland and R. Johnson (1965) compared the rooting characteristics of

On dry sites, the forb grows a dense network of deeply penetrating roots.

On moist sites, the forb grows a sparse network of shallow roots.

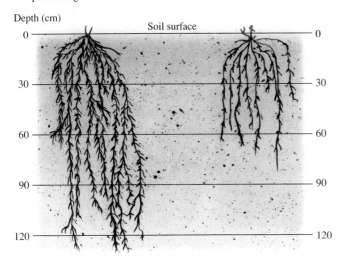

**Figure 5.12** Soil moisture influences the extent of root development by this desert forb (data from Coupland and Johnson 1965).

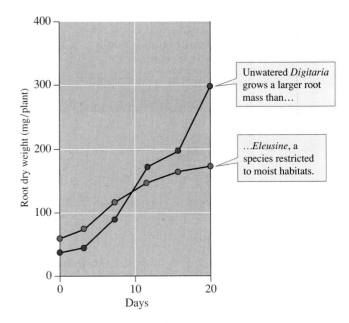

**Figure 5.13** A grass from a dry habitat responded to a simulated drought by greater root growth compared to a grass from a moist habitat (data from Park 1990).

plants growing in the temperate grasslands of western Canada. During their study, they carefully excavated the roots of over 850 individual plants, digging over 3 m deep to trace some roots. They found that many species have lower root biomass and higher aboveground biomass in moist microclimates within the temperate grassland biome. For instance, the roots of *Artemesia frigida* penetrate over 120 cm into the soil on dry sites; on moist sites, its roots grow only to a depth of about 60 cm (fig. 5.12). This contrast in root development is only one of many uncovered by the work of Coupland and Johnson.

Deeper roots often help plants from dry environments extract water from deep within the soil profile. This generalization is supported by studies of two common grasses that grow in Japan, *Digitaria adscendens* and *Eleusine indica*. The grasses overlap broadly in their distributions in Japan, however, only *Digitaria* grows on coastal sand dunes, which are among the most drought-prone habitats in Japan.

Y.-M. Park (1990) was interested in understanding the mechanisms allowing *Digitaria* to grow on coastal dunes where *Eleusine* could not. Because of the potential for drought in coastal dunes, Park studied the responses of the two grasses to water stress. He grew both species from seeds collected at the Botanical Gardens at the University of Tokyo. Seeds were germinated in moist sand and the seedlings were later transplanted into 10 cm by 90 cm PVC tubes filled with sand from a coastal dune. Park planted two seedlings of *Digitaria* in each of 36 tubes and two of *Eleusine* in 36 other tubes. He watered all 72 tubes with a nutrient solution every 10 days for 40 days. At the end of the 40 days, Park divided the 36 tubes of each species into two groups of 18. One group of each species was kept well watered for the next 19 days, while the other group remained unwatered.

Unwatered *Digitaria* and *Eleusine* responded differently. The root mass of *Digitaria* increased almost sevenfold over the 19 days of no watering, while the root mass of *Eleusine* increased about threefold. In addition, the roots of *Digitaria* were still growing at the end of the experiment, while those of *Eleusine* stopped growing about 4 days before the end of the experiment. Figure 5.13 summarizes these results.

Park found that the differences in root growth were greatest in the deeper soil layers. Below 60 cm in the growing tubes, the unwatered group of *Eleusine* showed suppressed root growth, while *Digitaria* did not. With its greater mass of more deeply penetrating roots, *Digitaria* maintained high leaf water potential throughout the 19 days of no watering. During this same period, *Eleusine* showed a substantial decline in leaf water potential. The leaf water potentials of *Digitaria* and *Eleusine* over the 19 days are shown in figure 5.14.

Park's results suggest that *Digitaria* can be successful in the drier dune habitat because it grows longer roots, which exploit deeper soil moisture. With these deeper roots, *Digitaria* can keep the water potential of its tissues high even in relatively dry soils, where *Eleusine* suffers lowered water potential. In other words, *Digitaria* maintains higher leaf water potentials because its greater root development maintains a higher rate of water intake—higher $W_r$.

The examples we've just reviewed concern rooting by individual plant species either in the field or under experimental conditions. An important question that we might ask is whether there have been enough root studies to make tentative generalizations about the rooting biology of plants. Jochen Schenk and Robert Jackson (2002) conducted an analysis of 475 root profile (see fig. 5.12) studies from 209 geographic localities from around the world. Based in this

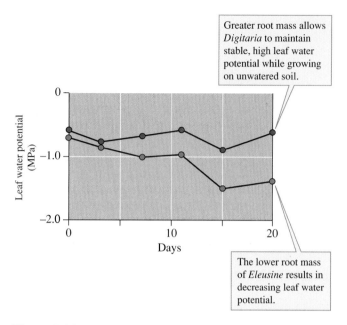

**Figure 5.14** A grass from a dry habitat maintained a higher water potential during a simulated drought compared to a grass from a moist habitat (data from Park 1990).

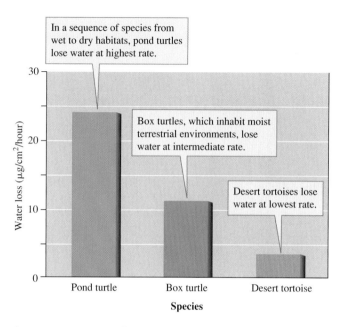

**Figure 5.15** Rates of water loss by two turtles and a tortoise indicate an inverse relationship between the dryness of the habitat and water loss rates (data from Schmidt-Nielsen 1969).

massive analysis, Schenk and Jackson reached the following conclusions. In over 90% of the 475 root profiles at least 50% of roots were in the top 0.3 m of the soil and at least 95% of roots were in the upper 2 m. However, they also found pronounced geographic differences in rooting depth. Schenk and Jackson found that rooting depth increases from 80° to 30° latitude, that is from arctic tundra to Mediterranean woodlands and shrublands and deserts. However, there were no clear trends in rooting depth in the tropics. Consistent with our present discussion, Schenk and Jackson found that deeper rooting depths occur mainly in water-limited ecosystems.

# Water Conservation by Plants and Animals

Another way to balance a water budget is by reducing water losses. One of the most obvious ways to cut down on water losses is by waterproofing to reduce evaporation. Many terrestrial plants and animals cover themselves with a fairly waterproof "hide" impregnated with a variety of waterproofing waxes. However, some organisms are more waterproof than others, and rates of evaporative water loss vary greatly from one animal or plant species to another.

Why do the water loss rates of organisms differ? One reason is that species have evolved in environments that differ greatly in water availability. As a consequence, selection for water conservation has been more intense in some environments than others. Species that evolved in warm deserts are generally much more resistant to desiccation than relatives

that evolved in moist tropical or temperate habitats. In general, populations that evolved in drier environments lose water at a slower rate. For instance, turtles from wet and moist habitats lose water at a much higher rate than do desert tortoises (fig. 5.15). As the following example shows, however, the water loss rates of even closely related species can differ substantially.

Neil Hadley and Thomas Schultz (1987) studied two species of tiger beetles in Arizona that occupy different microclimates. *Cicindela oregona* lives along the moist shoreline of streams and is active in fall and spring. In contrast, *Cicindela obsoleta* lives in the semiarid grasslands of central and southeastern Arizona and is active in summer. The researchers suspected that these differences in microclimate select for differences in waterproofing of the two tiger beetles.

Hadley and Schultz studied the waterproofing of the tiger beetles by comparing the amount of water each species lost while held in an experimental chamber. They pumped dry air through the chamber at a constant rate and maintained its temperature at 30°C. They weighed each beetle at the beginning of an experiment and then again after 3 hours in the chamber. The difference between initial and final weights gave them an estimate of the water loss rate of each beetle. By determining water loss for several individuals of each species, they estimated the average water loss rates for *C. oregona* and *C. obsoleta*. Hadley and Schultz found that *C. oregona* loses water two times as fast as *C. obsoleta* (fig. 5.16). In other words, the species from the drier microclimate, *C. obsoleta* appears to be more waterproofed.

Information
Questions
Hypothesis
Predictions
Testing ✓

*Investigating the Evidence*

## Sample Size

The number of observations included in a sample, that is, sample size, has an important influence on the level of confidence we place on conclusions based on that sample. Let's examine a localized example of how sample size affects our estimate of some ecological feature. Consider an ecologist interested in how disturbance by flash flooding may affect the number of benthic insect species living in a stream. The stream is Tesuque Creek at about 3,000 m elevation in the mountains above Santa Fe, New Mexico. A flash flood, which completely disrupted one fork of Tesuque Creek, left a second similar-sized fork undisturbed. Nine months after the flood, samples were taken to determine if there was a difference in the number of species of mayflies (Order Ephemeroptera), stoneflies (O. Plecoptera), caddisflies (O. Trichoptera), and beetles (O. Coleoptera) living in similar-sized reaches of the two forks. Samples of the benthic community were taken at 5-m intervals with a Surber sampler, which has a 0.1 m² metal frame, or quadrat, and an attached net. As a stream ecologist disturbs the bottom material within the quadrat of a Surber sampler, the net trailing in the current catches benthic organisms that are dislodged. In the study of Tesuque Creek, the number of benthic insect species captured in each 0.1 m² sample ranged from 1 to 6 in the disturbed fork and from 2 to 8 in the undisturbed fork. However, our question concerns the total number of species in each fork and the number of benthic samples required to make a good estimate of that number of species.

Figure 1 plots the data in a way that provides an answer to both questions. The Surber samples are plotted in the exact order they were taken, beginning with the first that was taken at the downstream end of each study reach and ending with the twelfth sample taken 55 m upstream from the first. As shown in figure 1, each of the first few samples adds to the cumulative number of species collected at each site, which rises steeply at first and then levels off at a maximum number of species in each study reach. The cumulative number of species stopped increasing at a sample size of 7 quadrats in the undisturbed study reach and at 5 quadrats in the disturbed study reach.

How many samples should a researcher take? In the case of the benthic community just examined, 7 replicate counts from 0.1 m² quadrats appears to be sufficient to estimate the number of benthic mayfly, stonefly, caddisfly, and beetle species living in a short reach of a small, high-elevation stream in the Rocky Mountains. The twelve samples actually taken in the study certainly is sufficient for a reasonable estimate of the number of species in these orders. In contrast, to make generalizations about global patterns of rooting among plants, Schenk and Jackson (2002) reported on 475 root profiles at 209 locations (see p. 127). The number of samples necessary depends on the amount of variability in the system under study and the spatial and temporal scope of the study. However whether the scope of a project is large or small, sample size is one of the most important components of study design.

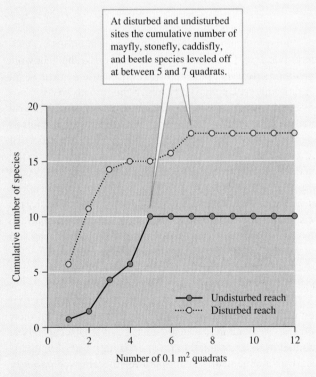

At disturbed and undisturbed sites the cumulative number of mayfly, stonefly, caddisfly, and beetle species leveled off at between 5 and 7 quadrats.

**Figure 1** The cumulative number of species increased with the number of quadrats studied in both disturbed and undisturbed streams, eventually leveling off at a sample size of 5 to 7 quadrats.

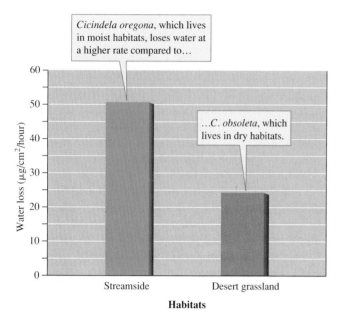

**Figure 5.16** A tiger beetle species from a moist habitat lost water at a higher rate than one from a dry habitat (data from Hadley and Schultz 1987).

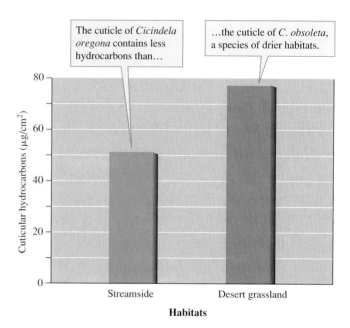

**Figure 5.17** The cuticles of tiger beetles from dry habitats tend to contain a higher concentration of waterproofing hydrocarbons compared to tiger beetles from moist habitats (data from Hadley and Schultz 1987).

Waterproofing of the cuticles of terrestrial insects is usually provided by hydrocarbons. Hydrocarbons include organic compounds such as lipids and waxes. Because of their influences on waterproofing, Hadley and Schultz analyzed the cuticles of the two species of tiger beetles for their hydrocarbon content. They found that the concentration of hydrocarbons in the cuticle of *C. obsoleta* is 50% higher than in the cuticle of *C. oregona* (fig. 5.17). In addition, the two species differ in the percentages of cuticular hydrocarbons that are saturated with hydrogen. Fully saturated hydrocarbons are much more effective at waterproofing. One hundred percent of the hydrocarbons in the cuticle of *C. obsoleta* are saturated. In contrast, only 50% of the cuticular hydrocarbons of *C. oregona* are saturated. These results support the hypothesis that *C. obsoleta* loses water at a lower rate because its cuticle contains a higher concentration of waterproofing hydrocarbons.

Merriam's kangaroo rats conserve water sufficiently that they can live entirely on the moisture contained within their food and on "metabolic water" (see fig. 5.11). This capacity is assumed to be an adaptation to desert living. Over long periods of time as the American Southwest became increasingly arid, the ancestors of today's Merriam's kangaroo rats were subject to natural selection that favored a range of adaptations to dry environments, including water conservation. However, Merriam's kangaroo rat is a widespread species that lives from 21° N latitude in Mexico to 42° N latitude in northern Nevada. Over this large geographic range, Merriam's kangaroo rat populations are exposed to a very broad range of environmental conditions.

Intrigued by their large geographic range and exceptional adaptation to desert living, Richard Tracy and Glenn Walsberg studied three populations of Merriam's kangaroo rats across a climatic gradient. Their main objective was to determine if different populations of Merriam's kangaroo rat vary in their degree of adaptation to living in dry environments (Tracy and Walsberg 2000, 2001, 2002). The three populations studied by Tracy and Walsberg live in southwest Arizona near Yuma, central Arizona, and north-central Arizona, at elevations of 150 m, 400 m, and 1200 m respectively. Mean annual maximum temperatures at the study sites are 31.5°, 29.1°, and 23.5°C and mean annual precipitation at the three sites is 10.6 cm, 33.6 cm, and 43.6 cm. The differences in climates at the three study sites are reflected in the vegetation. The habitat at the driest site consists of sand dunes with scattered shrubs; the intermediate site is a desert shrubland; and the vegetation at the moist site consists of a pinyon-juniper woodland.

One of the main questions asked by Tracy and Walsberg was whether rates of evaporative water loss would differ among the Merriam's kangaroo rats at dry, intermediate, and moist sites. The results of this study showed clear differences among the study populations. The mean rate of evaporative water loss at the dry site was 0.69 mg of water per g per hour, compared to 1 mg $H_2O$/g/h and 1.08 mg $H_2O$/g/h at the intermediate and moist sites respectively (fig. 5.18). Tracy and Walsberg expressed the rate of water loss by the kangaroo rats on a per gram basis because the kangaroo rats from the three sites differ significantly in size. The average mass of individuals from the moist site was approximately 33% greater than the mass of rats from the dry site. In additional studies, Tracy and Walsberg found that acclimating animals to laboratory conditions did not eliminate the differences in water conservation among populations. In other words, even after being kept in the laboratory under controlled conditions, Merriam's kangaroo rats from the driest study site continued to lose water at a

Kangaroo rats from the driest site showed much lower rates of water loss.

**Figure 5.18** Water loss rates by Merriam's kangaroo rats from across a moisture gradient suggest adaptation to local climate by each of the populations (data from Tracy and Walsberg 2001).

lower rate. The evidence from these studies supports the conclusion that these three populations differ in their degree of adaptation to desert living.

Animals adapted to dry conditions have many other water conservation mechanisms besides waterproofing. These mechanisms include producing concentrated urine or feces with low water content, condensing and reclaiming the water vapor in breath, and restricting activity to times and places that decrease water loss.

Plants have also evolved a wide variety of means for conserving water. How much water a plant can conserve depends in part on its leaf area relative to its root area or length. Plants with more leaf surface per length of root lose more water. Compared to plants from moist climates, arid land plants generally have less leaf area per unit area of root. Many plants reduce leaf area over the short term by dropping leaves in response to drought. Some desert plants produce leaves only in response to soaking rains and then shed them when the desert dries out again. These plants reduce leaf area to zero in times of drought. Figure 5.19 shows one of these plants, the ocotillo of the Sonoran Desert of North America.

Other plant adaptations that conserve water include thick leaves, which have less transpiring leaf surface area per unit volume of photosynthesizing tissue than thin leaves do; few stomata on leaves rather than many; structures on the stomata that impede the movement of water; dormancy during times when moisture is unavailable; and alternative, water-conserving pathways for photosynthesis. (We discuss these alternative pathways for photosynthesis in chapter 6.)

We should remember that plants and animals in terrestrial environments other than deserts also show evidence of selection for water conservation. For instance, Nona Chiariello and

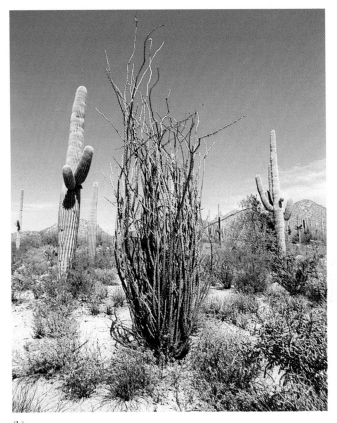

(a)

(b)

**Figure 5.19** Changing leaf area: (*a*) following rainfall ocotillo plants of the Sonoran Desert develop leaves and flower; (*b*) during dry periods they lose their leaves and blossoms.

her colleagues (1987) discovered an intriguing example of adjusting leaf area in the moist tropics. *Piper auritum*, a large-leafed, umbrella-shaped plant, grows in clearings of the rain forest. Because it grows in clearings, the plant often faces drying conditions during midday. However, it reduces the leaf area it exposes to the midday sun by wilting. Wilting at midday reduces leaf area exposed to direct solar radiation by about 55% and leaf temperature by up to 4° to 5°C. These reductions decrease the rate of transpiration by 30% to 50%,

In a shaded portion of a greenhouse, the leaves of the rain forest plant are unwilted and fully exposed to incoming light.

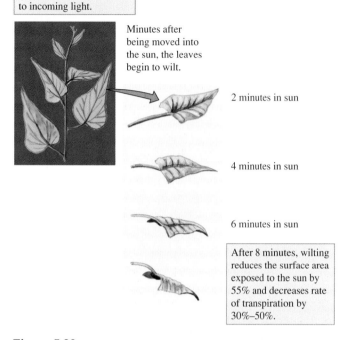

Minutes after being moved into the sun, the leaves begin to wilt.

2 minutes in sun

4 minutes in sun

6 minutes in sun

After 8 minutes, wilting reduces the surface area exposed to the sun by 55% and decreases rate of transpiration by 30%–50%.

**Figure 5.20** Temporary wilting by this rain forest plant decreases rates of water loss (data from Chiariello, Field, and Mooney 1987).

which is a substantial water savings. The behavior of this tropical rain forest plant reminds us that even the rain forest has its relatively dry microclimates, such as the forest clearings where *P. auritum* grows. The rapidity of *P. auritum*'s wilting response is shown in figure 5.20.

Organisms balance their water budgets in numerous ways. Some rely mainly on water conservation. Others depend upon water acquisition. However, every biologist who studies organisms in their natural environment knows that nature is marked by diversity and contrast. To sample nature's variety, let's review the variety of approaches to desert living.

# Dissimilar Organisms with Similar Approaches to Desert Life

On the surface, camels and saguaro cactus appear entirely different (fig. 5.21). If you look deeper into their biology, however, you find that they take very similar approaches to balancing their water budgets. Both the camel and the saguaro cactus acquire massive amounts of water when water is available, store water, and conserve water.

The camel can go for long periods in intense desert heat without drinking, up to 6 to 8 days in conditions that would kill a person within a day. During this time, the animal survives on

(a)

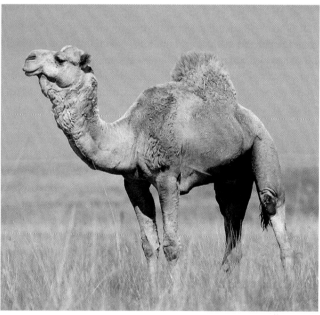

(b)

**Figure 5.21** Two desert dwellers: (*a*) saguaro cactus; and (*b*) camel as different as they are seem to show parallel adaptations to desert environments.

the water stored in its tissues and can withstand water losses of up to 20% of its body weight without harm. For humans, a loss of about 10% to 12% is near the fatal limit. When the camel has the opportunity, it can drink and store prodigious quantities of water, up to one-third of its body weight at a time.

Between opportunities to drink, the camel is a master of water conservation. One way it conserves body water is by reducing its rate of heat gain. Like overheating tiger beetles

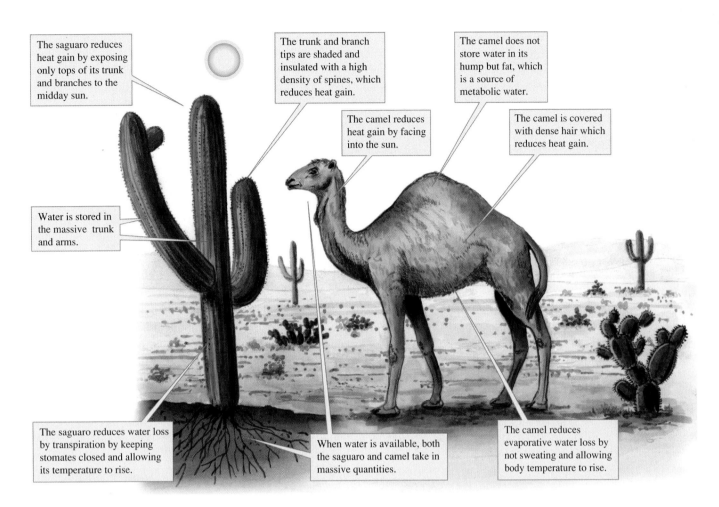

The saguaro reduces heat gain by exposing only tops of its trunk and branches to the midday sun.

The trunk and branch tips are shaded and insulated with a high density of spines, which reduces heat gain.

The camel does not store water in its hump but fat, which is a source of metabolic water.

The camel reduces heat gain by facing into the sun.

The camel is covered with dense hair which reduces heat gain.

Water is stored in the massive trunk and arms.

The saguaro reduces water loss by transpiration by keeping stomates closed and allowing its temperature to rise.

When water is available, both the saguaro and camel take in massive quantities.

The camel reduces evaporative water loss by not sweating and allowing body temperature to rise.

**Figure 5.22**   Dissimilar organisms with similar approaches to desert living.

(see chapter 4), the camel faces into the sun, reducing the body surface it exposes to direct sunlight. In addition, its thick hair insulates it from the intense desert sun, and rather than sweating sufficiently to keep its body temperature down, the camel allows its body temperature to rise by up to 7°C. This reduces the temperature difference between the camel and the environment and so decreases the rate of additional heating. Reduced heating translates into reduced water loss by evaporation.

The saguaro cactus takes a similar approach. The trunk and arms of the plant act as organs in which the cactus can store large quantities of water. During droughts, the saguaro draws on these stored reserves and so can endure long periods without water. When it rains, the saguaro, like a camel at an oasis, can ingest great quantities of water but instead of drinking, the saguaro gets its water through its dense network of shallow roots. These roots extend out in a roughly circular pattern to a distance approximately equal to the height of the cactus. For a 15 m tall saguaro this means a root coverage of over 700 m$^2$ of soil.

The saguaro also reduces its rate of evaporative water loss in several ways. First, like other cactus, it keeps its stomata closed during the day when transpiration losses would be highest. In the absence of transpiration, in full sun, the inter-

nal temperature of the saguaro rises to over 50°C, which is among the highest temperatures recorded in plants. However, as we noted for the camel, higher body temperature can be an advantage because it reduces the rate of additional heating. The saguaro's rate of heating is also reduced by the shape and orientation of its trunk and arms. At midday, when the potential for heating is greatest, the saguaro exposes mainly the tips of its arms and trunk to direct sunlight. However, the tips of the saguaro's arms and trunk are insulated by a layer of plant hairs and a thick tangle of spines, which reflect sunlight and shade the growing tips of the cactus.

The parallel approaches to desert living seen in saguaro cactus and camels are outlined in figure 5.22. Now let's examine two organisms that live in the same desert but have very different water relations.

## Two Arthropods with Opposite Approaches to Desert Life

Though both are arthropods and may live within a few meters of each other, cicadas and scorpions take sharply contrasting approaches to living in the desert. The scorpion's approach is to slow down, conserve, and stay out of the sun. Scorpions are relatively large and long-lived arthropods with very low metabolic rates. A low rate of metabolism means that they can subsist on low rations of food and lose little water during respiration. In addition, scorpions conserve water by spending most of their time in their burrows, where the humidity is higher than at the surface. They come out to feed and find mates only at night, when it's cooler. In addition, desert scorpions are well waterproofed; hydrocarbons in their cuticles seal in moisture. With this combination of water-conserving characteristics, scorpions can easily satisfy their need for water by consuming the moisture contained in the bodies of their arthropod prey. Figure 5.23 summarizes the habits of desert scorpions.

In comparison to desert scorpions, the cicada's approach to desert living may seem out of place. As we saw in the introduction to chapter 5, the Sonoran Desert cicada, *Diceroprocta apache,* is active on the hottest days, when air temperature is near its lethal limit. How can *Diceroprocta* do this and not die? The solution to this puzzle begins with a study by James Heath and Peter Wilkin (1970). These researchers observed that just before sunrise *Diceroprocta* perches on large branches of mesquite trees, *Prosopis juliflora,* where it feeds on the fluids in the tree's xylem. As the sun rises, the cicadas move from their feeding sites to leaves and twigs exposed to full sun. They remain motionless on these perches until midmorning, when the temperature rises above 35°C. At this high temperature, birds and wasps, the main predators of the cicadas, seek shelter from the heat.

As their predators retire to the shade, the cicadas reach their peak of activity. When body temperature reaches 39°C, they shift their position to the shade of large mesquite branches, where the males sing. If the temperature does not rise above 40°C, they sing throughout midday. However, when air temperature reaches 48°C, *Diceroprocta* will sit quietly from about 1:00 to 3:00 P.M. and then resume singing.

Heath and Wilkin found that singing *Diceroprocta* have a body temperature considerably below the temperature of the surrounding air. They also found that the temperature close to the surface of the branches where the cicadas perch is cooler than other potential perches. Heath and Wilkin concluded that the cicadas keep cool by remaining in these small patches of cool air on the shady sides of large branches. They suggested that these cool microclimates are too small to be exploited by birds, the chief predators of *Diceroprocta.*

Are cool microclimates the only means of thermoregulation used by *Diceroprocta?* A study of another cicada by two graduate students suggested that *Diceroprocta* might evaporatively cool. Stacy Kaser and Jon Hastings (1981) found evidence of evaporative cooling by *Tibicen duryi,* which lives in the same region as *Diceroprocta* but at higher elevations, where it feeds on the xylem fluids of pinyon pine trees, *Pinus edulis.* Kaser and Hastings found that the abdominal temperature of *Tibicen* remained lower than air temperature when air temperature rose above 36°C. The cicadas were able to maintain these reduced abdominal temperatures as long as they had

**Figure 5.23** These two desert arthropods, a scorpion and a cicada, have evolved very different approaches to living in the desert.

access to xylem fluids. However, when Kaser and Hastings denied their access to water, their abdominal temperatures rose. Though never observed before, these results suggested that *Tibicen* is capable of evaporative cooling.

These observations led to a series of investigations and papers by Eric Toolson and Neil Hadley which showed conclusively that cicadas, including *Diceroprocta,* are capable of evaporative cooling. In one of these studies, Eric Toolson (1987) collected *Diceroprocta* from a mesquite tree and placed them in an environmental chamber. The chamber temperature was kept at 45.5°C; however, *Diceroprocta* was able to main-

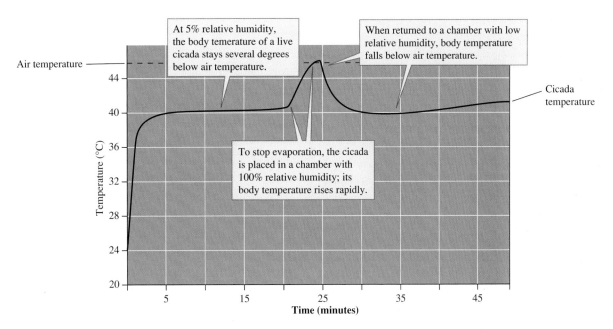

**Figure 5.24** A laboratory experiment verified evaporative cooling by the cicada, *Diceroprocta apache* (data from Toolson 1987).

tain its body temperature at least 2.9°C lower. Since the cicadas within the chamber did not have access to any cool microclimates, Toolson concluded that they must be evaporatively cooling. To verify this hypothesis, he placed cicadas in the environmental chamber and then raised the relative humidity to 100%. At 100% relative humidity, the body temperatures of the cicadas quickly increased to the temperature of the environmental chamber. When Toolson reduced relative humidity to 0%, the cicadas cooled approximately 4°C within minutes. The results of this experiment are outlined in figure 5.24.

How do the results of Toolson's experiment support the hypothesis of evaporative cooling? Remember that air with a relative humidity of 100% contains all the water vapor it can hold (see p. 119). Consequently, by raising the humidity of the air surrounding the cicadas to 100%, Toolson shut off any evaporative cooling that might be taking place. When he reintroduced dry air, he created a gradient of water concentration from the cicada to the air and evaporative cooling resumed. This experiment by Toolson was analogous to Heinrich's tying off the circulatory system of a sphinx moth to determine the role of the circulatory system in thermoregulation (see chapter 4).

Toolson's results are consistent with the hypothesis that *Diceroprocta* evaporatively cools but does not demonstrate that capacity directly. Consequently, Toolson and Hadley (1987) conducted observations to make a direct demonstration. First, they placed a live *Diceroprocta* in an environmental chamber with a humidity sensor just above its cuticle. If *Diceroprocta* evaporatively cools, then this sensor would detect higher humidity as the temperature of the environment was increased. This is exactly what occurred. As the temperature was increased from 30° to 43°C, the rate of water movement across the cicada's cuticle increased in three steps. When

Toolson and Hadley increased the temperature from 37° to 39°C, water loss increased from 5.7 to 9.4 mg $H_2O$ per square centimeter per hour. At 41°C, water loss increased from 9.4 to 36.1 mg $H_2O$ per square centimeter per hour and at 43°C water loss increased from 36.1 to 61.4 mg $H_2O$ per square centimeter per hour. These results are graphed in figure 5.25.

The rate of water loss by *Diceroprocta* is among the highest ever reported for a terrestrial insect. How does water cross the cuticle of this cicada at such a high rate? Toolson and Hadley searched the cuticle of *Diceroprocta* for avenues of water movement. They found three areas on the dorsal surface with large pores that might be involved in evaporative cooling (fig. 5.26). When they plugged these pores *Diceroprocta* could no longer cool itself. In summary, Toolson and Hadley verified a previously unknown phenomenon, evaporative cooling by cicadas, and carefully demonstrated the underlying mechanisms.

So, it turns out that these cicadas can sing in the hottest hours of the desert day because they sweat! *Diceroprocta* is able to maintain this seemingly impossible lifestyle because it has tapped into a rich supply of water. Cicadas are members of the order Homoptera and distant relatives of the aphids. Like aphids, cicadas feed on plant fluids. So, though the cicada lives in the same macroclimate as the scorpion, it has tapped into a totally different microclimate. The cicada's scope for water acquisition is extended up to 30 m deep into the soil by the tap roots of its mesquite host plant, *P. juliflora*. *Diceroprocta* can sustain high rates of water loss through evaporation, high $W_e$, because it is able to balance these losses with a high rate of water acquisition, high $W_d$. Figure 5.27 illustrates how *Diceroprocta* uses mesquite trees to get access to deep soil moisture.

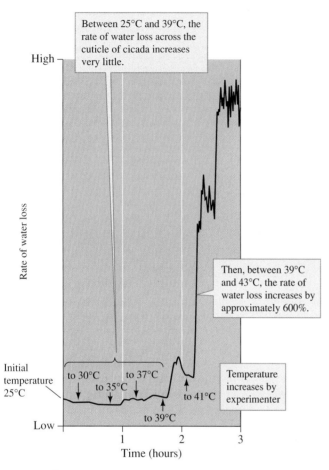

Between 25°C and 39°C, the rate of water loss across the cuticle of cicada increases very little.

Then, between 39°C and 43°C, the rate of water loss increases by approximately 600%.

Temperature increases by experimenter

High

Rate of water loss

Initial temperature 25°C

Low

to 30°C    to 37°C
to 35°C
to 39°C
to 41°C

Time (hours)
1    2    3

**Figure 5.25** High temperatures induce massive rates of water loss by the cicada, *Diceroprocta apache* (data from Toolson and Hadley 1987).

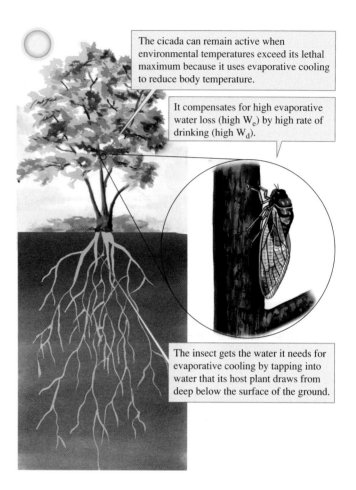

The cicada can remain active when environmental temperatures exceed its lethal maximum because it uses evaporative cooling to reduce body temperature.

It compensates for high evaporative water loss (high $W_e$) by high rate of drinking (high $W_d$).

The insect gets the water it needs for evaporative cooling by tapping into water that its host plant draws from deep below the surface of the ground.

**Figure 5.27** An ecological puzzle solved.

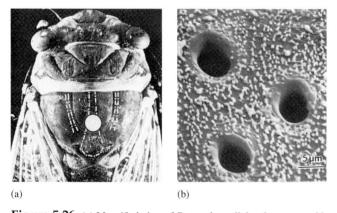

(a)    (b)

5 μm

**Figure 5.26** (*a*) Magnified view of *D. apache* outlining three areas with high densities of small pores; (*b*) dorsal pores under high magnification.

Sometimes, similar organisms employ radically different approaches to balancing their water budgets. Sometimes, organisms of very different evolutionary lineages employ functionally similar approaches. In short, the means by which terrestrial organisms balance water acquisition against water loss are almost as varied as the organisms themselves. Similar variation occurs among aquatic organisms.

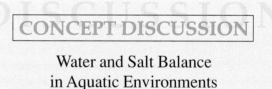

## CONCEPT DISCUSSION

### Water and Salt Balance in Aquatic Environments

**Marine and freshwater organisms use complementary mechanisms for water and salt regulation.**

Aquatic organisms, like their terrestrial kin, regulate internal water, $W_i$, by balancing water gain against water loss. We can represent water regulation in aquatic environments by modifying our equation for terrestrial water balance to:

$$W_i = W_d - W_s \pm W_o$$

Drinking, $W_d$, is a ready source of water for aquatic organisms. Secretion of water with urine, $W_s$, is an avenue of water loss. By osmosis, $W_o$, an aquatic organism may either gain or lose water, depending on the organism and the environment.

# Marine Fish and Invertebrates

Most marine invertebrates maintain an internal concentration of solutes equivalent to that in the seawater around them. What does the animal gain by remaining isosmotic with the external environment? The isosmotic animal does not have to expend energy, overcoming an osmotic gradient. This strategy is not without costs, however. Though the total concentration of solutes is the same inside and outside the animal, there are still differences in the concentrations of some individual solutes. These concentration differentials can only be maintained by active transport, which consumes some energy.

Sharks, skates, and rays generally elevate the concentration of solutes in their blood to levels slightly hyperosmotic to seawater. However, inorganic ions constitute only about one-third of the solute in shark's blood; the remainder consists of the organic molecules urea and trimethylamine oxide, or TMAO. As a consequence of being slightly hyperosmotic, sharks slowly gain water through osmosis, that is, $W_o$ is slightly positive. The water that diffuses into the shark, mainly across the gills, is pumped out by the kidneys and exits as urine. Sodium, because it is maintained at approximately two-thirds its concentration in seawater, diffuses into sharks from seawater across the gill membranes and some sodium enters with food. Sharks excrete excess sodium mainly through a specialized gland associated with the rectum called the salt gland. The main point here is that sharks and their relatives reduce the costs of osmoregulation by decreasing the osmotic gradient between themselves and the external environment (fig. 5.28).

In contrast to most marine invertebrates and sharks, marine bony fish have body fluids that are strongly hypoosmotic to the surrounding medium. As a consequence, they lose water to the surrounding seawater, mostly across their gills. Marine bony fish make up these water losses by drinking seawater. However,

drinking adds to salt influxes through their gills. The fish rid themselves of excess salts in two ways. Specialized "chloride" cells at the base of their gills secrete sodium and chloride directly to the surrounding seawater, while the kidneys excrete magnesium and sulfate. These ions exit with the urine. The urine, because it is hypoosmotic to the body fluids of the fish, represents a loss of water. However, the loss of water through the kidneys is low because the quantity of urine is low.

The larvae of some mosquitoes in the genus *Aedes* live in saltwater. These larvae meet the challenge of a high-salinity environment in ways analogous to those used by marine bony fish. Like marine bony fish, saltwater mosquitoes are hypoosmotic to the surrounding environment, to which they lose water. Saltwater mosquitoes also make up this water loss by drinking large amounts of seawater, up to 130% to 240% of body volume per day. This would even impress a camel! While this prodigious drinking solves the problem of water loss, it imports another: large quantities of salts that must be eliminated. Saltwater mosquitoes secrete these salts into the urine using specialized cells that line the posterior rectum. Here, saltwater mosquitoes do something that marine bony fish cannot. They excrete a urine that is hyperosmotic to their body fluids, which reduces water loss through the urine. The parallels in water and salt regulation by marine bony fish and saltwater mosquitoes are outlined in figure 5.29.

# Freshwater Fish and Invertebrates

Freshwater bony fish face an environmental challenge opposite to that faced by marine bony fish. Freshwater fish are hyperosmotic; they have body fluids that contain more salt and less water than the surrounding medium. As a consequence,

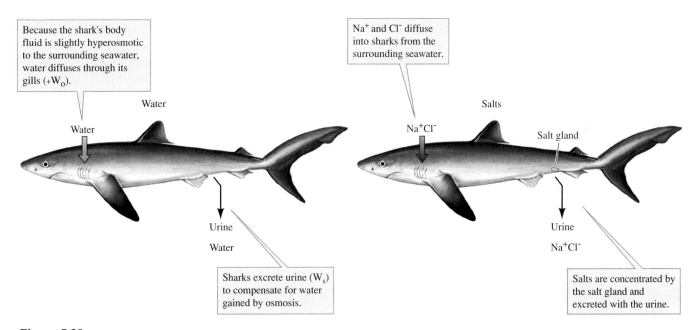

**Figure 5.28** Osmoregulation by sharks.

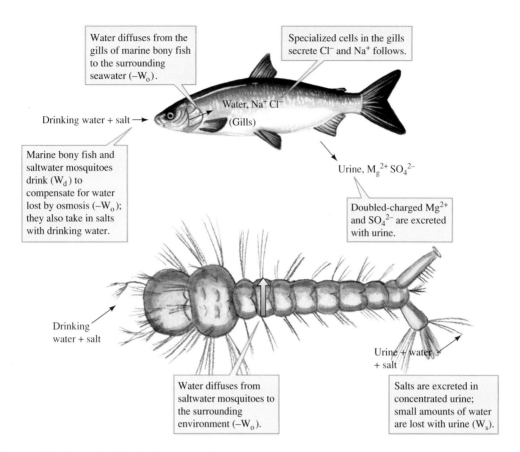

**Figure 5.29** Osmoregulation by marine fish and saltwater mosquitoes.

water floods inward and salts diffuse outward across their gills. Freshwater fish excrete excess internal water as large quantities of dilute urine. They replace the salts they lose to the external environment in two ways. Chloride cells at the base of the gill filaments absorb sodium and chloride from the water, while other salts are ingested with food.

Like freshwater fish, freshwater invertebrates are hyperosmotic to the surrounding environment. Freshwater invertebrates must expend energy to pump out the water that floods their tissues. They also expend energy by actively absorbing salts from the external environment. However, the concentration of solutes in the body fluids of freshwater invertebrates ranges from between about one-half and one-tenth that of their marine relatives. This lower internal concentration of solutes reduces the osmotic gradient between freshwater and the outside environment and so reduces the energy freshwater invertebrates must expend to osmoregulate.

Freshwater mosquito larvae are a good model for osmoregulation by freshwater invertebrates. The larvae of approximately 95% of mosquito species live in freshwater, where they face osmotic challenges very similar to those faced by freshwater fish. Like freshwater fish, mosquito larvae must solve the twin problems of water gain and ion loss. In response, they drink very little water. They conserve ions taken with the diet by absorbing them with cells that line the midgut and rectum, and they secrete a dilute urine. Freshwater mosquito larvae replace the ions lost with urine by actively absorbing $Na^+$

and $Cl^-$ from the water with cells in their anal papillae. Freshwater mosquitoes and fish use totally different structures to meet nearly identical environmental challenges. Figure 5.30 compares water and salt regulation by freshwater fish and mosquitoes.

In chapter 5, we have reviewed the water relations of individual organisms. The relationship between individual organisms and the environment is a fundamental aspect of ecology. However, ecologists are also concerned with levels of organization above the individual, such as the ecology of populations or of entire biomes. The following example in the Applications & Tools section shows how an understanding of the water relations of individual organisms is helping ecologists study the distribution of biomes across a continent.

## APPLICATIONS & TOOLS

### Using Stable Isotopes to Study Water Uptake by Plants

In order to fully understand the ecology of an individual plant or the dynamics of an entire landscape, ecologists need information about what happens below the earth's surface as

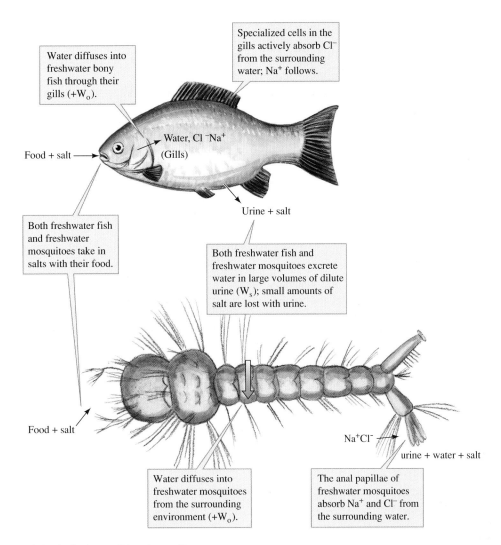

Water diffuses into freshwater bony fish through their gills ($+W_o$).

Specialized cells in the gills actively absorb $Cl^-$ from the surrounding water; $Na^+$ follows.

Food + salt

Water, $Cl^-$ $Na^+$ (Gills)

Urine + salt

Both freshwater fish and freshwater mosquitoes take in salts with their food.

Both freshwater fish and freshwater mosquitoes excrete water in large volumes of dilute urine ($W_s$); small amounts of salt are lost with urine.

Food + salt

$Na^+Cl^-$

urine + water + salt

Water diffuses into freshwater mosquitoes from the surrounding environment ($+W_o$).

The anal papillae of freshwater mosquitoes absorb $Na^+$ and $Cl^-$ from the surrounding water.

**Figure 5.30**  Osmoregulation by freshwater fish and mosquitoes.

well as about surface structure and processes. However, ecologists have produced much more information about the surface realm than about the subsurface, the domain of soil microbes, burrowing animals, and of roots. While many ecologists have worked very hard to fill this gap in our knowledge, their work on belowground ecology has been historically slow. Fortunately, progress has accelerated in recent years. A major contributor to recent progress in belowground ecology has been the development of new tools. One of the most important of those tools is **stable isotope analysis,** which involves the analysis of the relative concentrations of stable isotopes, such as the stable isotopes of carbon $^{13}C$ and $^{12}C$, in materials. Stable isotope analysis is increasingly used in ecology to study the flow of energy and materials through ecosystems (Dawson et al. 2002). For instance, stable isotope analysis has proved a very powerful tool in studies of water uptake by plants. To understand the applications of this analytical tool, we need to know a little about the isotopes themselves and about their behavior in ecosystems.

# Stable Isotope Analysis

Most chemical elements include several stable isotopes, which occur in different concentrations in different environments or differ in concentration from one organism to another. Stable isotopes of hydrogen include $^1H$ and $^2H$, which is generally designated as D, an abbreviation of deuterium. Stable isotopes of carbon, for example, include $^{13}C$ and $^{12}C$; stable isotopes of nitrogen include $^{15}N$ and $^{14}N$; and stable isotopes of sulfur include $^{34}S$ and $^{32}S$. The relative concentrations of these stable isotopes can be used to study the flow of energy and materials through ecosystems because different parts of the ecosystem often contain different concentrations of the light and heavy isotopes of these elements.

Different organisms contain different ratios of light and heavy stable isotopes because they use different sources of these elements, because they preferentially use (fractionate) different stable isotopes, or because they use different sources and fractionate. For instance, the lighter isotope of nitrogen,

[14]N, is preferentially excreted by organisms during protein synthesis. As a consequence of this preferential excretion of [14]N, an organism becomes relatively enriched in [15]N compared to its food. Therefore, as materials pass from one trophic level to the next, tissues become richer in [15]N. The highest trophic levels within an ecosystem contain the highest relative concentrations of [15]N, while the lowest trophic levels contain the lowest concentrations. Stable isotope analysis can also measure the relative contribution of $C_3$ and $C_4$ plants to a species' diet. This is possible because $C_4$ plants are relatively richer in [13]C. Other processes affect the relative concentrations of stable isotopes of sulfur. Because different sources of water often have different ratios of D to [1]H, for example, shallow soil moisture versus deep soil moisture, hydrogen isotope analyses have been valuable aids to identifying where plants acquire their water.

The concentrations of stable isotopes are generally expressed as differences in the concentration of the heavier isotope relative to some standard. The units of measurement are differences ($\pm$) in parts per thousand ($\pm\,^0/_{00}$). These differences are calculated as:

$$\delta X = \left[\left(\frac{R_{sample}}{R_{standard}}\right) - 1\right] \times 10^3$$

where:

$$\delta = \pm$$

$X$ = the relative concentration of the heavier isotope, for example, D, [13]C, [15]N, or [34]S in $^0/_{00}$

$R_{sample}$ = the isotopic ratio in the sample, for example, D:[1]H, [13]C:[12]C, or [15]N:[14]N

$R_{standard}$ = the isotopic ratio in the standard, for example, D:[1]H, [13]C:[12]C, or [15]N:[14]N

The reference materials used as standards in the isotopic analyses of hydrogen, nitrogen, carbon, and sulfur are the D:[1]H ratio in Standard Mean Ocean Water, the [15]N:[14]N ratio in atmospheric nitrogen, the [13]C:[12]C ratio in PeeDee limestone, and the [34]S:[32]S in the Canyon Diablo meteorite.

The ecologist measures the ratio of stable isotopes in a sample and then expresses that ratio as a difference relative to some standard. If $\delta X = 0$, then the ratios of the isotopes in the sample and the standard are the same; if $\delta X = -X\,^0/_{00}$, the concentration of the heavier isotope is lower (e.g., [15]N) in the sample compared to the standard, and if $\delta X = +X\,^0/_{00}$, the concentration of the heavier isotope is higher in the sample compared to the standard. The important point here is that these isotopic ratios are generally different in different parts of ecosystems. Therefore, ecologists can use isotopic ratios to study the structure and processes in ecosystems. Here is an example of how hydrogen isotope ratios have been used to study the uptake of water by plants in a natural ecosystem.

# Using Stable Isotopes to Identify Plant Water Sources

The laboratory of James Ehleringer has taken a leadership role in the development of stable isotope analysis as a tool for assessing water relations among plants and within ecosystems (e.g., Ehleringer, Roden, and Dawson 2000). In an early study Ehleringer and several colleagues (Ehleringer et al. 1991) used deuterium:hydrogen (D:[1]H) ratios, or $\delta$D, to explore the use of summer versus winter rainfall by various plant growth forms in the deserts of southern Utah. They could use $\delta$D to determine the relative utilization of these two water sources since summer rains are relatively enriched with D and winter rains are relatively depleted of D. The $\delta$D of summer and winter rains in southern Utah at the time of Ehleringer's study were $-25\,^0/_{00}$ and $-90\,^0/_{00}$ respectively (fig. 5.31).

Ehleringer measured $\delta$D in the xylem fluid of several plant growth forms during spring, when soil moisture at all rooting depths would be predominantly from winter precipitation and summer, when summer precipitation would be present as moisture in surface soils and winter precipitation would predominate at deeper soil layers. Ehleringer and his

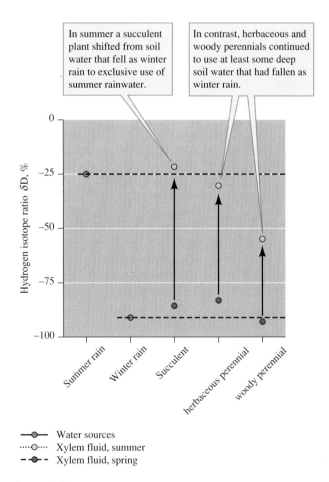

**Figure 5.31** Stable isotope analysis identified the water sources used by three groups of desert plants during the spring and summer (data from Ehleringer et al. 1991).

research team found that a succulent, several herbaceous perennials, and several woody perennials used winter moisture in the spring (fig. 5.31). However, when summer rains fell the succulent species shifted entirely to using soil moisture from summer rains that were stored mainly at shallow soil depths. Meanwhile herbaceous and woody perennials continued to use significant amounts of deeper soil moisture that fell the previous winter. So, stable isotope analysis opens a window to the water relations of plants that would not be accessible without this innovative tool. We will explore the use of stable isotope analysis further in chapter 18, where we discuss its use in studies of energy flow in ecosystems.

## SUMMARY

**The movement of water down concentration gradients in terrestrial and aquatic environments determines the availability of water to organisms.** The most familiar relative measure of the water content of air is relative humidity, defined as water vapor density divided by saturation water vapor density multiplied by 100. On land, the tendency of water to move from organisms to the atmosphere can be approximated by the vapor pressure deficit of the air. Vapor pressure deficit is calculated as the difference between the actual water vapor pressure and the saturation water vapor pressure.

In the aquatic environment, water moves down its concentration gradient, from solutions of higher water concentration and lower salt content (hypoosmotic) to solutions of lower water concentration and higher salt content (hyperosmotic). This movement of water creates osmotic pressure. Larger osmotic differences, between organism and environment, generate higher osmotic pressures.

In the soil-plant system water flows from areas of higher water potential to areas of lower water potential. The water potential of pure water, which by convention is set at zero, is reduced by adding solute and by matric forces, the tendency of water to cling to soil particles and to plant cell walls. Typically, the water potential of plant fluids is determined by a combination of solute concentrations and matric forces, while the water potential of soils is determined mainly by matric forces. In saline soils, solutes may also influence soil water potential. Water potential, osmotic pressure, and vapor pressure deficit can all be measured in pascals (newtons/m$^2$), a common currency for considering the water relations of diverse organisms in very different environments.

**Terrestrial plants and animals regulate their internal water by balancing water acquisition against water loss.** Water regulation by terrestrial animals is summarized by $W_{ia} = W_d + W_f + W_a - W_e - W_s$, where $W_d$ = drinking, $W_f$ = taken in with food, $W_a$ = absorption from the air, $W_e$ = evaporation, and $W_s$ = secretions and excretions. Water regulation by terrestrial plants is summarized by $W_{ip} = W_r + W_a - W_t - W_s$, where $W_r$ = uptake by roots, $W_a$ = absorption from the air, $W_t$ = transpiration, and $W_s$ = secretions and reproductive structures. Some very different terrestrial plants and animals, such as the camel and saguaro cactus, use similar mechanisms to survive in arid climates. Some organisms, such as scorpions and cicadas, use radically different mechanisms. Comparisons such as these suggest that natural selection is opportunistic.

**Marine and freshwater organisms use complementary mechanisms for water and salt regulation.** Marine and freshwater organisms face exactly opposite osmotic challenges. Water regulation in aquatic environments is summarized by: $W_i = W_d - W_s \pm W_o$, where $W_d$ = drinking, $W_s$ = secretions and excretions, $W_o$ = osmosis. An aquatic organism may either gain or lose water through osmosis, depending on the organism and the environment. Many marine invertebrates reduce their water regulation problems by being isosmotic with seawater. Some freshwater invertebrates also reduce the osmotic gradient between themselves and their environment. Sharks, skates, and rays elevate the urea and TMAO content of their body fluids to the point where they are slightly hyperosmotic to seawater. Marine bony fish and saltwater mosquito larvae are hypoosmotic relative to their environments, while freshwater bony fish and freshwater mosquito larvae are hyperosmotic.

While the strength of environmental challenge varies from one environment to another, and the details of water regulation vary from one organism to another, all organisms in all environments expend energy to maintain their internal pool of water and dissolved substances.

Stable isotope analysis, an important new tool in ecology, involves the analysis of the relative concentrations of stable isotopes in materials. Examples of stable isotopes include the stable isotopes of hydrogen $^2H$ (which is usually symbolized by D, referring to deuterium) and $^1H$, and the stable isotopes of carbon, $^{13}C$ and $^{12}C$. Stable isotope analysis has proved a very powerful tool in studies of water uptake by plants. For example deuterium:hydrogen (D:$^1H$) ratios, or $\delta D$, has been used to quantify the relative use of summer versus winter rainfall by various plant growth forms in the deserts of southern Utah.

# Review Questions

1. Turn back to chapter 4 and examine figure 4.6. Notice that the body temperature of the isopod *Ligia oceanica* is 30°C under stones but 26°C on the surface, where it is exposed to full sun. Edney (1953) proposed that the isopods in the open had lower body temperatures because they evaporatively cooled in the open air. Explain why evaporative cooling would be effective in the open air but nearly impossible under stones.

2. Distinguish between vapor pressure deficit, osmotic pressure, and water potential. How can all three phenomena be expressed in the same units of measure: pascals?

3. Leaf water potential is typically highest just before dawn and then decreases progressively through midday. Should lower leaf water potentials at midday increase or decrease the rate of water movement from soil to a plant? Assume soil water potential is approximately the same in early morning and midday. Are the water needs of the plant greater in early morning or at midday?

4. Compare the water budgets of the tenebrionid beetle, *Onymacris,* and the kangaroo rat, *Dipodomys,* shown in figures 5.10 and 5.11. Which of these two species obtains most of its water from metabolic water? Which relies most on condensation of fog as a water source? In which species do you see greater losses of water through the urine?

5. In chapter 5, we discussed water relations of tenebrionid beetles from the Namib Desert. However, members of this family also occur in moist temperate environments. How should water loss rates vary among species of tenebrionids from different environments? On what assumptions do you base your prediction? How would you test your prediction?

6. In the Sonoran Desert, the only insects known to evaporatively cool are cicadas. Explain how cicadas can employ evaporative cooling while hundreds of other insect species in the same environment cannot.

7. Many desert species are well waterproofed. Evolution cannot, however, eliminate all evaporative water loss. Why not? (Hint: Think of the kinds of exchanges that an organism must maintain with its environment.)

8. While we have concentrated in chapter 5 on regulation of water and salts, most marine invertebrates are isosmotic with their external environment. What is a potential benefit of being isosmotic?

9. Review water and salt regulation by marine and freshwater bony fish. Which of the two is hypoosmotic relative to its environment? Which of the two is hyperosmotic relative to its environment? Some sharks live in freshwater. How should the kidneys of marine and freshwater sharks function?

10. The model built by Ronald Neilson and his colleagues (1992, 1995) uses the physiological requirements of plant growth forms to predict the responses of vegetation to climate change. In chapter 1, we briefly discussed the studies of Margaret Davis (1983, 1989) that reconstructed the movement of vegetation across eastern North America. She made this reconstruction using the pollen preserved in lake sediments. How might the results of paleoecological studies such as this be used to refine Neilson's model? (Assume that you can also reasonably reconstruct the climate when historic changes in vegetation occurred.)

# Suggested Readings

Dawson, T. E., S. Mambelli, A. H. Plamboeck, P. H. Templer, and K. P. Tu. 2002. Stable isotopes in plant ecology. *Annual Review of Ecology and Systematics* 33:507–99.

Ehleringer, J. R., J. Roden, and T. E. Dawson. 2000. Assessing ecosystem-level water relations through stable isotope ratio analyses. In O. E. Sala, R. B. Jackson, H. A. Mooney, and R. W. Howarth. eds. *Methods in Ecosystem Science.* New York: Springer.

*Overview and introduction to stable isotope analysis in ecology.*

Dong, X. J. and X. S. Zhang. 2001. Some observations of the adaptations of sandy shrubs to the arid environment of the Mu Us Sandland: leaf water relations and anatomic features. *Journal of Arid Environments* 48:41–48.

Ohte, N., K. Koba, K. Yoshikawa, A. Sugimoto, N. Matsuo, N. Kabeya, and L. H. Wang. 2003. Water utilization of natural and planted trees in the semiarid desert of Inner Mongolia, China. *Ecological Applications* 13:337–51.

*Introduction to some of the plants and application of stable isotope analysis to understanding one of the great deserts of the world.*

Schenk, H. J. and R. B. Jackson. 2002. The global biogeography of roots. *Ecological Monographs* 72:311–28.

*Intriguing analysis of a massive set of 475 root profiles from 209 localities from around the earth.*

Heath, J. E. and P. J. Wilkin. 1970. Temperature responses of the desert cicada, *Diceroprocta apache* (Homoptera, Cicadidae). *Physiological Zoology* 43:145–54.

Neilson, R. P. 1995. A model for predicting continental-scale vegetation distribution and water balance. *Ecological Applications* 5:362–85.

*This paper showcases Neilson's continental-scale modeling efforts. Tough going in places but worth the effort. The colored maps capture the essence of model results.*

Toolson, E. C. 1987. Water profligacy as an adaptation to hot deserts: water loss rates and evaporative cooling in the Sonoran Desert cicada, *Diceroprocta apache* (Homoptera, Cicadidae). *Physiological Zoology* 60:379–85.

Toolson, E. C. and N. F. Hadley. 1987. Energy-dependent facilitation of transcuticular water flux contributes to evaporative cooling in the Sonoran Desert cicada, *Diceroprocta apache* (Homoptera, Cicadidae). *Journal of Experimental Biology* 131:439–44.

*This series of papers traces how biologists determined the means by which the Sonoran Desert cicada,* Diceroprocta apache, *manages to remain active in extremely high environmental temperatures.*

Tracy, R. L. and G. E. Walsberg. 2000. Prevalence of cutaneous evaporation in Merriam's kangaroo rat and its adaptive variation at the subspecific level. *Journal of Experimental Biology* 203:773–81.

Tracy, R. L. and G. E. Walsberg. 2001. Intraspecific variation in water loss in a desert rodent, *Dipodomys merriami. Ecology* 82:1130–37.

Tracy, R. L. and G. E. Walsberg. 2002. Kangaroo rats revisited: re-evaluating a classic case of desert survival. *Oecologia* 133:449–57.

*Studies revealing adaptation to local climates by a desert rodent long studied by physiological ecologists.*

# On the Net

Visit this textbook's accompanying website at www.mhhe.com/ecology (click on the book's title) to take advantage of practice quizzing, study/writing tips, timely news articles, and additional URLs for research on the topics in this chapter.

Homeostasis

Water and Osmotic Regulation in Aquatic Organisms

Water and Osmotic Regulation in Terrestrial Organisms

Vertebrates: Macroscopic Anatomy of Excretory Organs

Water Relations in Plants

Leaves and the Movement of Water

Water Use and Management

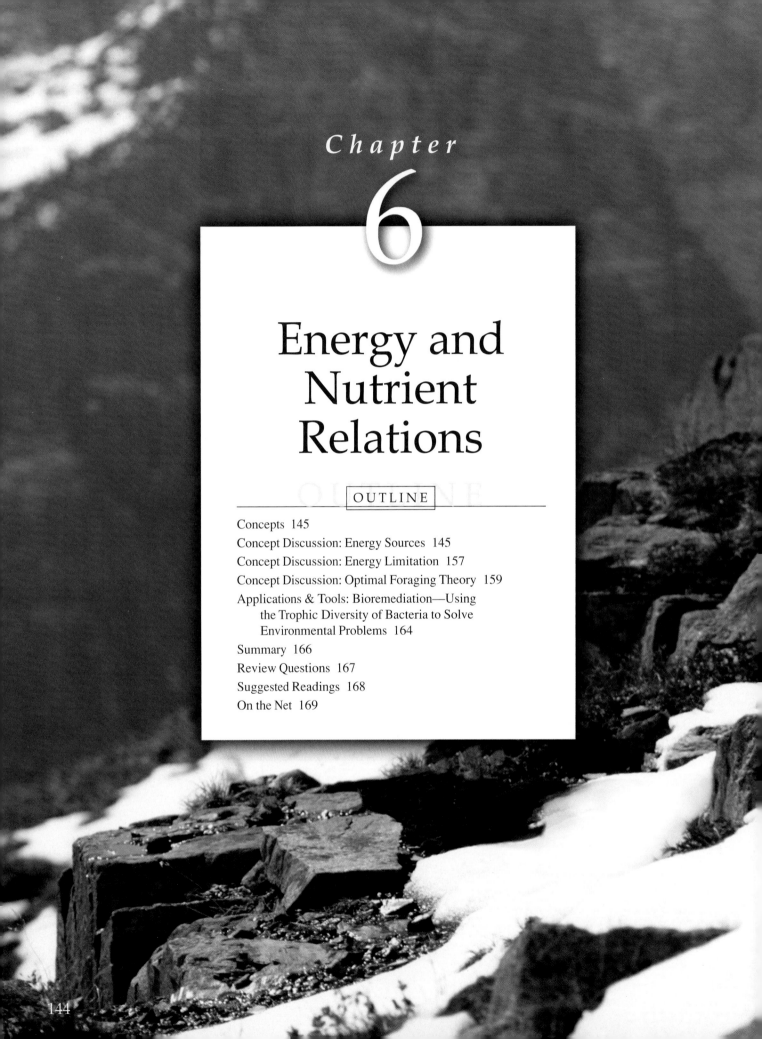

# *Chapter*

# 6

# Energy and Nutrient Relations

Evidence of nutrient and energy acquisition by animals and plants can be seen everywhere in nature. A scorpion fish lies half buried in the sand near the edge of a coral reef; the only clues to its presence are the telltale movements of its gill covers. Its head looks so much like an algae-covered stone that several tiny shrimp gather over it and swim lazily in the current. A small fish on the nearby reef sees the shrimp and darts over to feed on them. The scorpion fish opens its mouth and swallows the small fish in a lightning quick movement. However, before the scorpion fish can settle back into the sand, a green moray eel, nearly 2 m long, darts from the reef, grabs the scorpion fish with its razor-sharp teeth, and swallows it (fig. 6.1).

An herbaceous plant with broad leaves and slender stems grows in the half light of the rain forest floor. It is difficult to understand how it can live in such dim light. However, as you watch, a small shaft of intense sunlight pierces an unseen hole in the rain forest canopy and shines on one of the plant's leaves. The photosynthetic machinery of the plant takes advantage of the situation, and for a few minutes the plant uses the energy of the tiny sun fleck. Nearby is the buttressed trunk of a gigantic tree that has grown tall enough to emerge from the forest canopy and count itself among the rain forest giants, seemingly a more secure position than that of the understory herb. However, a small vine has begun to grow up the side of the tree. It will grow quickly upward, winding its way toward the sun and exploiting the woody support of the tree. Soon the vine will overwhelm and kill the tree, which will be reduced to a trellis for the vine.

Whether on coral reef, rain forest, or abandoned urban lot, organisms engage in an active search for energy and nutrients. For most organisms, life boils down to converting energy and nutrients into descendants. The energy used by different organisms comes in the form of light, organic molecules, or inorganic molecules. Nutrients are the raw materials an organism must acquire from the environment to live. Because organisms acquire energy and nutrients in diverse ways, we need to organize our discussion under the umbrella of major concepts. In chapter 6, we focus on three.

**Figure 6.1** The moray eel meets its energy and nutrient needs by being an effective predator.

## CONCEPTS

- **Organisms use one of three main sources of energy: light, organic molecules, or inorganic molecules.**
- **The rate at which organisms can take in energy is limited.**
- **Optimal foraging theory attempts to model how organisms feed as an optimizing process.**

## CONCEPT DISCUSSION

### Energy Sources

**Organisms use one of three main sources of energy: light, organic molecules, or inorganic molecules.**

How do we group organisms? We generally group organisms on the basis of shared evolutionary histories, creating taxa such as vertebrate animals, insects, coniferous trees, and orchids. However, we can also classify organisms by how they obtain energy—that is, by their **trophic (feeding) biology.** Organisms that use inorganic sources of both carbon and energy are called **autotrophs** ("self-feeders") and are of two types, **photosynthetic** and **chemosynthetic.** Photosynthetic autotrophs use carbon dioxide ($CO_2$) as a source of carbon and light as a source of energy. This group

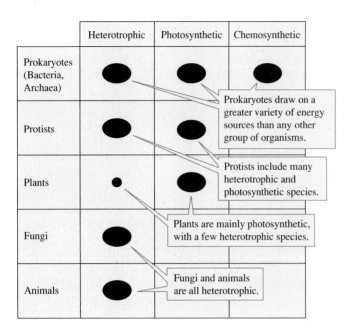

**Figure 6.2** A plot of trophic diversity across the major groups of organisms shows highest trophic diversity among the prokaryotic bacteria and archaea.

includes the plants, photosynthetic protists, and photosynthetic bacteria. Chemosynthetic autotrophs use inorganic molecules as a source of carbon and energy. These are made up of a highly diverse group of chemosynthetic bacteria. **Heterotrophs** ("other-feeders") are organisms that use organic molecules both as a source of carbon and as a source of energy. The heterotrophs include bacteria, fungi, protists, animals, and parasitic plants.

**Prokaryotes** show more trophic diversity than the other biological kingdoms (fig. 6.2). Prokaryotes, which have cells with no membrane bound nucleus or organelles, include the bacteria and the **archaea.** The archaea are prokaryotes distinguished from bacteria on the basis of structural, physiological, and other biological features. Though first discovered in association with extreme environments, the archaea are now known to be widely spread in the biosphere, particularly in the oceans. The protists are either photosynthetic or heterotrophic, most plants are photosynthetic, and all fungi and animals are heterotrophic. In contrast, the prokaryotes include photosynthetic, chemosynthetic, and heterotrophic species, making them, as a group, the most trophically diverse organisms in the biosphere.

Some of the most ecologically significant discoveries of prokaryotic trophic diversity have come from studies of marine prokaryotes. For instance, Oded Béjà and Edward Delong and their research team from the Monterey Bay Aquarium Research Institute and the University of Texas Medical School discovered a new type of energy production from light, involving bacterial **rhodopsin** (Béjà et al. 2000). Rhodopsins are light-absorbing pigments found in animal eyes and in the bacteria and archaea. The rhodopsin in bacteria and archaea performs a variety of functions, including that of a proton pump involved

in ATP synthesis, that is, in the production of energy. Further study by Béjà, Delong, and their research team (Béjà et al. 2001) has shown that bacterial rhodopsin is widespread through the oceans. In one particularly intriguing aspect of their study, they found that the light sensitivity of bacterial rhodopsin appears adapted to local variations in light quality. For instance, bacterial rhodopsin from deep clear waters absorbs light most strongly within the blue range of the visible spectrum, while that from shallow coastal waters absorbs most strongly in the green range. Discoveries such as these and others (e.g., Kolber et al. 2000, Béjà et al. 2002) are rapidly revolutionizing our understanding of how the biosphere works.

# Using Light and $CO_2$

Because photosynthetic organisms use light as a source of energy, we need to learn about light. We also need to understand how photosynthetic organisms use $CO_2$. These are topics we investigate next.

## *The Solar-Powered Biosphere*

As we saw in chapters 2 and 3, solar energy powers the winds and ocean currents, and annual variation in sunlight intensity drives the seasons. In chapter 4, we also discussed how organisms use sunlight to regulate body temperature. Here, building on those discussions, we look at light as a source of energy for photosynthesis.

Light propagates through space as a wave, with all the properties of waves such as frequency and wavelength. When light interacts with matter, however, it acts not as a wave but as a particle. Particles of light, called *photons,* bear a finite quantity of energy. Longer wavelengths, such as *infrared light,* carry less energy than shorter wavelengths, such as *visible* and *ultraviolet light.*

Infrared light, as we saw in chapter 4 (see fig. 4.14), is very important for temperature regulation by organisms. This is because the main effect of infrared light on matter is to increase the motion of whole molecules, which we measure as increased temperature. However, infrared light does not carry enough energy to drive photosynthesis. At the other end of the solar spectrum, ultraviolet light carries so much energy that it breaks the covalent bonds of many organic molecules. Because it can break down organic molecules, ultraviolet light can destroy the complex biochemical machinery of photosynthesis. Between these extremes is the light we can see, so-called visible light, which is also called **photosynthetically active radiation,** or **PAR.** PAR, with wavelengths between about 400 and 700 nm, carries sufficient energy to drive the light-dependent reactions of photosynthesis but not so much as to destroy organic molecules. PAR makes up about 45% of the total energy content of the solar spectrum at sea level, while infrared light accounts for about 53% and ultraviolet light for the remainder.

## Measuring PAR

Ecologists quantify PAR as photon flux density. **Photon flux density** is the number of photons striking a square meter surface each second. The number of photons is expressed as micromoles ($\mu$mol), where 1 mole is Avagadro's number of photons, $6.023 \times 10^{23}$. To give you a point of reference, a photon flux density of about 4.6 $\mu$mol per square meter per second equals a light intensity of about 1 watt per square meter. Measuring light as photosynthetic photon flux density makes sense ecologically because chlorophyll absorbs light as photons.

Light changes in quantity and quality with latitude, with the seasons, with the weather, and with the time of day. In addition, landscapes, water, and even organisms themselves change the amount and quality of light. For example, in aquatic environments (see chapter 3), only the superficial euphotic zone receives sufficient light to support photosynthetic organisms. In addition, light changes in quality, as well as quantity, within the euphotic zone, which ranges in depth from a few meters to about 100 m (see fig. 3.7).

As in the sea, sunlight changes as it shines through the canopy of a forest. A mature temperate or tropical forest can reduce the total quantity of light reaching the forest floor to about 1% to 2% of the amount shining on the forest canopy (fig. 6.3). However, forests also change the quality of sunlight. Within the range of photosynthetically active radiation, leaves absorb mainly blue and red light and transmit mostly green light with a wavelength of about 550 nm. As in the deep sea, the organisms on the forest floor live in a kind of twilight. Only here, the twilight is green (see fig. 2.9).

## Alternative Photosynthetic Pathways

During photosynthesis, the photosynthetic pigments of plants, algae, or bacteria absorb light and transfer their energy to electrons. Subsequently, the energy carried by these electrons is used to synthesize ATP and NADPH. These molecules, in turn, serve as donors of electrons and energy for the synthesis of sugars. In this way, photosynthetic organisms convert the electromagnetic energy of sunlight into energy-rich organic molecules, the fuel that feeds most of the biosphere. Within photosynthetic organisms, specific biochemical pathways carry out this energy conversion; three different biochemical pathways are known: $C_3$ photosynthesis, $C_4$ photosynthesis, and CAM photosynthesis. These are found in ecologically different organisms.

Biologists often speak of photosynthesis as "carbon fixation," which refers to the reactions in which $CO_2$ becomes incorporated into a carbon-containing acid. In the photosynthetic pathway used by most plants and all algae, the $CO_2$ first combines with a five-carbon compound called *ribulose bisphosphate,* or *RuBP.* The product of this initial reaction, which is catalyzed by the enzyme RuBP carboxylase, is *phosphoglyceric acid,* or *PGA,* a three-carbon acid. Therefore, this photosynthetic pathway is usually called **$C_3$ photosynthesis** and the plants that employ it are called $C_3$ plants (fig. 6.4).

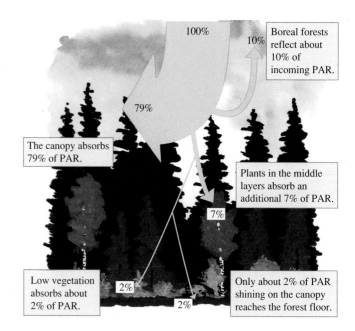

100%

10%

Boreal forests reflect about 10% of incoming PAR.

79%

The canopy absorbs 79% of PAR.

Plants in the middle layers absorb an additional 7% of PAR.

7%

Low vegetation absorbs about 2% of PAR.

2%

2%

Only about 2% of PAR shining on the canopy reaches the forest floor.

**Figure 6.3** Photosynthetically active radiation (PAR) diminishes substantially with passage through the canopy of a boreal forest (data from Larcher 1995, after Kairiukstis 1967).

To fix carbon, plants must open their stomata to let $CO_2$ into their leaves, but as $CO_2$ enters, water exits. Water vapor flows out faster than $CO_2$ flows in. The movement of water is more rapid because the gradient in water concentration from the leaf to the atmosphere is much steeper than the gradient in $CO_2$ concentration from the atmosphere to the leaf. In $CO_3$ plants, there is another factor that contributes to a low rate of $CO_2$ uptake: RuBP carboxylase has a low affinity for $CO_2$. Relatively high rates of water loss are not a problem for plants that live in cool, moist conditions but in hot, dry climates, high rates of water loss can close the stomata and shut down photosynthesis.

In arid environments, two alternative photosynthetic pathways have evolved. Both pathways fix and store $CO_2$ in acids containing four carbon atoms. Light plays no part in carbon fixation, but the reactions that follow depend on light. Both alternative pathways separate the initial fixation of carbon from the light-dependent reactions.

One of these alternative pathways, **$C_4$ photosynthesis,** separates carbon fixation and the light-dependent reactions of photosynthesis into separate cells (fig. 6.5). $C_4$ plants fix $CO_2$ in mesophyll cells by combining it with *phosphoenol pyruvate,* or *PEP,* to produce an acid. This initial reaction, which is catalyzed by PEP carboxylase, concentrates $CO_2$. Because PEP carboxylase has a high affinity for $CO_2$, $C_4$ plants can reduce their internal $CO_2$ concentrations to very low levels. Low internal concentration of $CO_2$ increases the gradient of $CO_2$ from atmosphere to leaf, which in turn increases the rate of diffusion of $CO_2$ inward. Consequently, compared to $C_3$ plants, $C_4$ plants need to open fewer stomata to deliver sufficient $CO_2$ to photosynthesizing cells. By having fewer stomata open, $C_4$ plants conserve water.

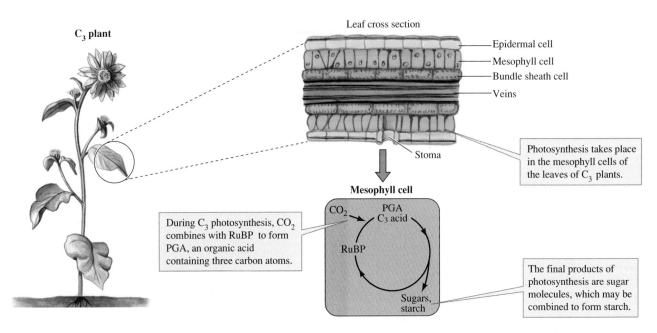

**C₃ plant**

Leaf cross section

Epidermal cell
Mesophyll cell
Bundle sheath cell
Veins

Stoma

Photosynthesis takes place in the mesophyll cells of the leaves of C₃ plants.

**Mesophyll cell**

$CO_2$    PGA
C₃ acid

RuBP

During C₃ photosynthesis, $CO_2$ combines with RuBP to form PGA, an organic acid containing three carbon atoms.

The final products of photosynthesis are sugar molecules, which may be combined to form starch.

Sugars, starch

**Figure 6.4**  C₃ photosynthesis.

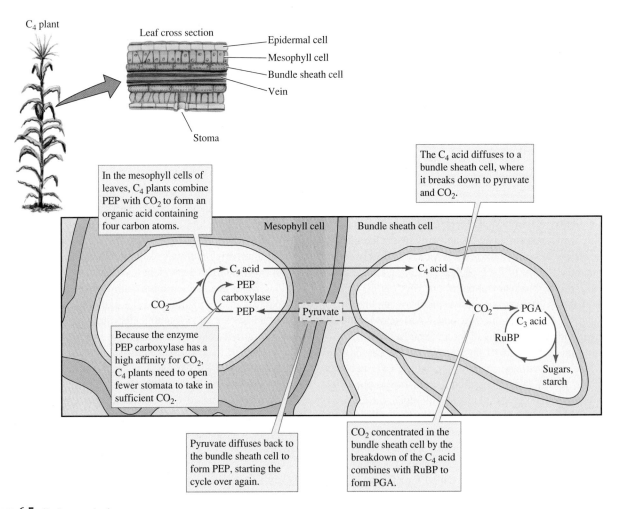

C₄ plant

Leaf cross section

Epidermal cell
Mesophyll cell
Bundle sheath cell
Vein

Stoma

In the mesophyll cells of leaves, C₄ plants combine PEP with $CO_2$ to form an organic acid containing four carbon atoms.

The C₄ acid diffuses to a bundle sheath cell, where it breaks down to pyruvate and $CO_2$.

Mesophyll cell    Bundle sheath cell

C₄ acid    C₄ acid

PEP carboxylase

$CO_2$    PEP    Pyruvate    $CO_2$    PGA
C₃ acid

RuBP

Because the enzyme PEP carboxylase has a high affinity for $CO_2$, C₄ plants need to open fewer stomata to take in sufficient $CO_2$.

Sugars, starch

Pyruvate diffuses back to the bundle sheath cell to form PEP, starting the cycle over again.

$CO_2$ concentrated in the bundle sheath cell by the breakdown of the C₄ acid combines with RuBP to form PGA.

**Figure 6.5**  C₄ photosynthesis.

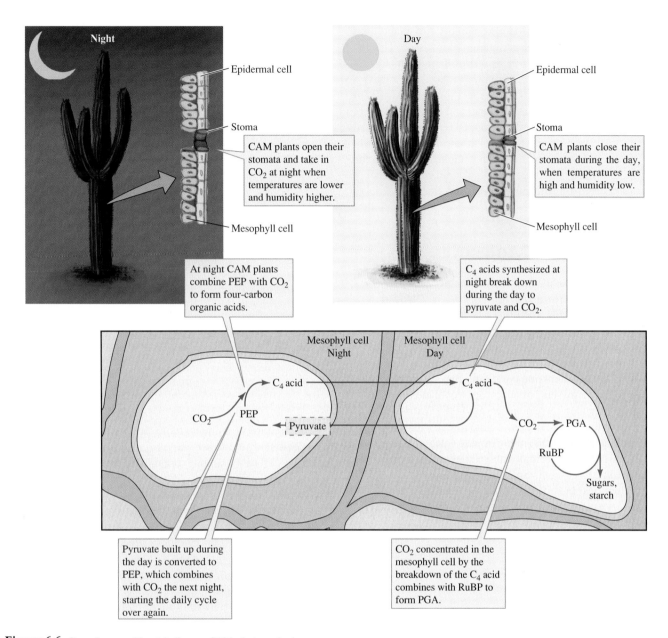

**Figure 6.6** Crassulacean acid metabolism, or CAM, photosynthesis.

In $C_4$ plants, the acids produced during carbon fixation diffuse to specialized cells surrounding a structure called the **bundle sheath.** There, deeper in the leaf, the four-carbon acids are broken down to a three-carbon acid and $CO_2$. In this way, $C_4$ plants can build up the $CO_2$ concentration in bundle sheath cells to high levels, increasing the efficiency with which RuBP carboxylase combines $CO_2$ with RuBP to produce PGA. $C_4$ plants do better than $C_3$ plants under conditions of high temperature, high light intensity, and limited water. Review $C_3$ and $C_4$ photosynthesis (figs. 6.4 and 6.5) before considering the third major photosynthetic pathway.

**CAM** (crassulacean acid metabolism) **photosynthesis** is largely limited to succulent plants in arid and semiarid environments. In this pathway, carbon fixation takes place at night, when lower temperatures reduce the rate of water loss during $CO_2$ uptake. CAM plants fix carbon by combining $CO_2$ with PEP to form four-carbon acids. These acids are stored until daylight, when they are broken down into pyruvate and $CO_2$, which then enters the $C_3$ photosynthetic pathway (fig. 6.6). In CAM plants, all these reactions take place in the same cells. While CAM plants do not normally show very high rates of photosynthesis, their water use efficiency, as estimated by the mass of $CO_2$ fixed per kilogram of water used, is higher than that of either $C_3$ or $C_4$ plants.

Separating initial carbon fixation from the other reactions reduces water losses during photosynthesis: $C_3$ plants lose from about 380 to 900 g of water for every gram (dry weight) of tissue produced. $C_4$ plants lose from about 250 to 350 g of water per gram of tissue produced, while CAM plants lose approximately 50 g of water per gram of new tissue. The differences in these numbers give us one of the reasons $C_4$ and CAM plants do well in hot, dry environments.

Whether the pathway of carbon fixation is CAM, $C_3$, or $C_4$, plants and photosynthetic algae and bacteria capture energy from sunlight and carbon from $CO_2$. These photosynthesizers package this energy and carbon in organic molecules. The photosynthesizers and other autotrophs opened the way for the evolution of organisms that could get their energy and carbon from organic molecules. And this new trophic level did indeed evolve.

# Using Organic Molecules

Heterotrophic organisms use organic molecules both as a source of carbon and as an energy source. They depend, ultimately, on the carbon and energy fixed by autotrophs. Heterotrophs have evolved numerous ways of feeding. This trophic variety has stimulated ecologists to invent numerous terms to describe the ways heterotrophs feed; a few of these terms have already crept into our discussions. In chapter 2, we referred to the "browsers" of temperate woodlands and shrublands and in chapter 3 when discussing James Karr's Index of Biological Integrity, we defined "omnivores," "insectivores," and "piscivores." A full list of the trophic categories proposed by ecologists would be impossibly long and not especially useful to this discussion. So, we will concentrate on three major categories: **herbivores,** organisms that eat plants; **carnivores,** organisms that mainly eat animals; and **detritivores,** organisms that feed on nonliving organic matter, usually the remains of plants. While these categories do not capture all the trophic diversity in nature, they are not arbitrary. Herbivores, carnivores, and detritivores must solve fundamentally different problems in order to obtain adequate supplies of energy and nutrients.

## *Chemical Composition and Nutrient Requirements*

We can get some idea of the nutrient requirements of organisms by examining their chemical composition. Biologists have found that the chemical composition of organisms is very similar. Just five elements (carbon [C], oxygen [O], hydrogen [H], nitrogen [N], and phosphorus [P]) make up 93% to 97% of the biomass of plants, animals, fungi, and bacteria. Of these four groups, plants are the most distinctive chemically. Plant tissues generally contain lower concentrations of phosphorus and nitrogen. The nitrogen content of plant tissues averages about 2%, while in fungi, animals, and bacteria it averages about 5% to 10%. Ecologists often express the relative nitrogen content of whole organisms or tissues as the ratio of carbon to nitrogen (C:N ratios). A high C:N ratio indicates low nitrogen content. The C:N ratio of plants averages about 25:1, which is substantially higher than

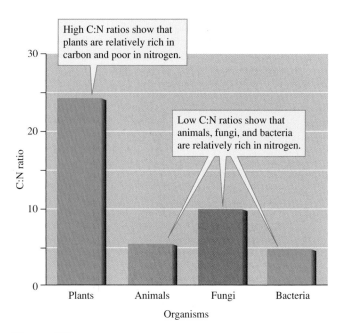

**Figure 6.7** On average, the ratio of carbon to nitrogen is much higher in terrestrial plants than in other major groups of organisms (data from Spector 1956).

the C:N ratios of animals, fungi, and bacteria, which average approximately 5:1 to 10:1 (fig. 6.7). Differences in C:N ratios among tissues or among organisms significantly influence what organisms eat, how rapidly consumers reproduce, and how rapidly organisms decompose.

If carbon, oxygen, hydrogen, nitrogen, and phosphorus make up 93% to 97% living biomass, then what accounts for the remainder? Dozens of other elements occur in the tissues of organisms. Essential plant nutrients include potassium (K), calcium (Ca), magnesium (Mg), sulfur (S), chlorine (Cl), iron (Fe), manganese (Mn), boron (B), zinc (Zn), copper (Cu), and molybdenum (Mo). Most of these nutrients are also essential for other organisms. Some organisms require additional nutrients. For instance, animals also require sodium (Na) and iodine (I).

Plants obtain carbon from the air through their stomata. They obtain other essential nutrients from the soil through their roots. For the most part, animals obtain both the energy they require and essential nutrients with their food. Let's now turn to the energy and nutrient relations of herbivorous, detritivorous, and carnivorous animals.

## *Herbivores*

While a band of zebra grazing on the plains of Africa or a sea turtle munching on sea grass in a tropical lagoon may suggest a life of ease, this image does not accurately represent the life of an herbivore. Herbivores face substantial problems that begin at the level of nutritional chemistry. Most plant tissues contain a great deal of carbon but low concentrations of nitrogen (fig. 6.7).

**Figure 6.8** Herbivores must overcome the wide variety of physical and chemical defenses evolved by plants.

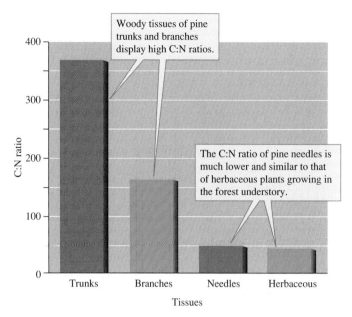

Woody tissues of pine trunks and branches display high C:N ratios.

The C:N ratio of pine needles is much lower and similar to that of herbaceous plants growing in the forest understory.

**Figure 6.9** C:N ratios differ a great deal among the tissues of pines and between the woody tissues of pines and those of herbaceous plants on the forest floor (data from Klemmedson 1975).

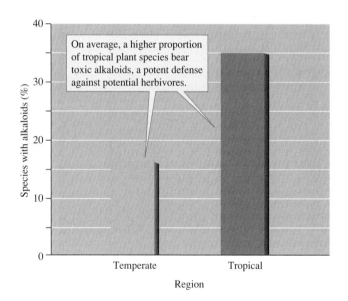

On average, a higher proportion of tropical plant species bear toxic alkaloids, a potent defense against potential herbivores.

**Figure 6.10** Proportion of temperate and tropical plants bearing toxic alkaloids (data from Coley and Aide 1991).

Herbivores must also overcome the physical and chemical defenses of plants. Some physical defenses are obvious, such as thorns that deter some herbivores entirely and slow the rate of feeding of others (fig. 6.8). However, plants also often deploy a variety of more subtle physical defenses. Grasses incorporate large amounts of abrasive silica into their tissues, which makes feeding on them difficult and which has apparently selected for specialized dentition among grazing mammals. Many plants toughen their tissues with large quantities of cellulose and lignin, producing leaves that are fibrous and difficult to chew.

The use of cellulose and lignin to strengthen tissues may also provide plants with a kind of chemical defense. Increasing the cellulose and lignin content of tissues increases their C:N ratios. An increased C:N ratio decreases the nutritional value of plant tissues. Some plant tissues have C:N ratios that are far higher than the average values we saw in figure 6.7. For instance, the tree trunks that make up most of the plant biomass in a pine forest have a C:N ratio of over 300:1 (fig. 6.9) . This ratio is much higher than that of either branches or needles. The living needles of pine trees have C:N ratios very similar to those of understory herbs living on the forest floor.

In addition, most animals cannot digest either cellulose or lignin. Those that can generally do so with the help of bacteria, fungi, or protists that live in their digestive tracts. This suggests that the cellulose and lignin in plants may be a first line of chemical defense against herbivores, a defense that most herbivores overcome with the help of other organisms.

When ecologists talk about plant chemical defenses, however, they are generally referring to two other classes of chemicals, toxins and digestion-reducing substances. Toxins are chemicals that kill, impair, or repel most would-be consumers. Digestion-reducing substances are generally phenolic compounds such as tannins that bind to plant proteins, inhibiting their breakdown by enzymes and further reducing the already low availability of nitrogen in plant tissues.

Chemists have isolated thousands of toxins from plant tissues, and the list continues to grow. The great variety of plant toxins defies easy description and generalization. However, one interesting pattern is that more tropical plants contain toxic alkaloids than do temperate species (fig. 6.10). In addition, on average, the alkaloids produced by tropical plants are more toxic than their temperate counterparts. Despite these higher levels of chemical defense, herbivores appear to remove approximately 11% to 48% of leaf biomass in tropical forests, while in temperate forests they remove about 7%. These higher levels of herbivore attack on tropical plants suggest that natural selection for chemical defense is more intense in tropical plant populations.

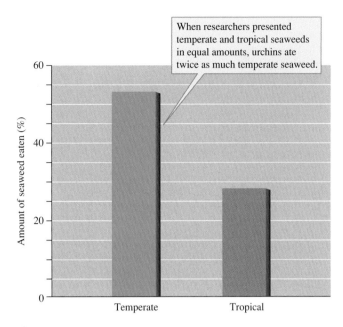

When researchers presented temperate and tropical seaweeds in equal amounts, urchins ate twice as much temperate seaweed.

**Figure 6.11** Sea urchin preference for temperate versus tropical seaweeds suggests that tropical seaweeds are better defended against attack by herbivores (data from Bolser and Hay 1996).

The generalization about higher levels of chemical defense also appears to apply to marine algae. Robin Bolser and Mark Hay (1996) tested the hypothesis that tropical seaweeds have more chemical defenses than temperate seaweeds. They gathered several species of seaweeds from the coast of temperate North Carolina and from the tropical Bahama Islands. Bolser and Hay were careful to pick the same species of seaweed in the two study sites or at least to pick species that belonged to the same genus. They tested the relative palatability of temperate and tropical seaweeds using temperate and tropical sea urchins.

The researchers were very careful to preserve any chemical defenses their study algae might contain. They cleaned and then froze the seaweeds in a freezer (–20°C) on the research ship. On shore, the seaweeds were transferred to a colder freezer (–70°C) to minimize chemical changes.

To remove the potential confounding effect of various physical factors, Bolser and Hay created artificial algae to test their hypothesis. They did this by freeze-drying samples of each seaweed and grinding them up in a coffee mill. The powdered algae was then mixed with agar at a concentration of 0.1 g alga per milliliter of agar. The warm alga and agar mixture was poured into a mold set on screening. As the mixture gelled it attached to the screening. The result was strips of artificial seaweed that could be cut up into equal-sized squares and presented to sea urchins in equal numbers. This method of presentation also provided an easy means of quantifying the actual amount of seaweed eaten.

The results of this study showed a clear preference for temperate species of seaweed (fig. 6.11). When given a choice the urchins removed approximately twice as much of the available temperate seaweed. In addition, both temperate and tropical urchins showed a similar preference for temperate

seaweeds. What caused the lower palatability of the tropical seaweeds? In additional tests, Bolser and Hay showed that the tropical seaweeds have more potent chemical defenses. So we see that this study produced a pattern in the sea that parallels the better-studied pattern known for tropical and temperate forests. Tropical plants and algae appear to possess stronger chemical defenses.

No defense is perfect; the defenses of most plants work against some herbivores, but not all. The tobacco plant uses nicotine, a toxic alkaloid, to repel herbivorous insects, most of which die suddenly after ingesting nicotine. However, several insects specialize in eating tobacco plants and manage to avoid the toxic effects of nicotine. Some simply excrete nicotine, while others convert it to nontoxic molecules. Similarly, toxins and repellents produced by plants in the cucumber family repel most herbivorous insects but attract the spotted cucumber beetle. This beetle is a specialist that feeds mainly on members of the cucumber family. Some specialized herbivores go even further by using plant toxins as a source of nutrition!

The effectiveness of phenolic compounds as digestion-reducing substances is also uneven. For example, while the tannins in oak leaves deter feeding by some insects, they only reduce the growth and development of the winter moth, *Operophtera brumata,* which specializes on oak leaves. Meanwhile, other insects appear to be unaffected by moderate concentrations of tannins or may even be stimulated to feed by the presence of tannins.

The world may appear green to us, but to herbivores only some shades of green are edible. Plant defenses and the adaptations of herbivores that overcome those defenses are complex.

## Detritivores

The problems faced by herbivores in their search for energy and nutrients are related to those faced by detritivores, which feed on dead plant material. These organisms consume food that is rich in carbon and energy but very poor in nitrogen. In fact, plant tissues, already relatively low in nitrogen when living (see figs. 6.7 and 6.9), are even lower in nitrogen content when cast off by plants as detritus. Keith Killingbeck and Walt Whitford (1996) averaged the nitrogen contents of living and dead leaves of many plant species of environments from tropical rain forests through deserts and temperate forests. Their results show that in all these environments, living leaves contain about twice the nitrogen as dead leaves (fig. 6.12).

In addition, fresh detritus may retain levels of chemical defenses high enough to reduce its use by detritivores. I. Middleton (1984) suggested that plant chemical defenses may evolve because reducing the rate of decomposition increases a plant's fitness. Why? Middleton proposed that toxic chemicals may delay the rate of decomposition just enough to give the plant more control over limited nutrients. Delayed decomposition may also promote the buildup of organic matter in soils. Like herbivores, detritivores may have important influences on the evolution of plant chemical defenses.

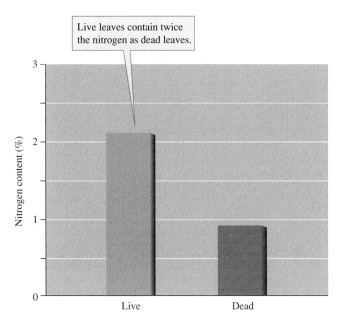

Figure 6.12 Nitrogen content of live and dead leaves (data from Killingbeck and Whitford 1996).

## Carnivores

Carnivores consume prey that are nutritionally rich. However, carnivores cannot go out into their environment and choose prey at will. Most prey species are masters of defense. One of the most basic prey defenses is camouflage. Predators cannot eat prey they cannot find. Other prey defenses include anatomical defenses such as spines, shells, repellents, and poisons and behavioral defenses such as flight, taking refuge in burrows, banding together in groups, playing dead, fighting, flashing bright colors, spitting, hissing, and screaming at predators. It's enough to spoil your appetite!

Prey that carry a threat to predators often advertise that fact, usually by being brightly colored or conspicuous in some other way. The conspicuous, or *aposematic,* colors of many distasteful or toxic butterflies, snakes, and nudibranchs warn predators that "feeding on me may be hazardous to your health." Many noxious organisms, such as stinging bees and wasps, poisonous snakes, and butterflies, seem to mimic each other. This form of comimicry among several species of noxious organisms is called **Müllerian mimicry** (fig. 6.13a). In addition, many harmless species appear to mimic noxious ones. For instance, king snakes mimic coral snakes, and syrphid flies mimic bees. This form of mimicry is called **Batesian mimicry.** In Batesian mimicry, the noxious species serves as the model and the harmless species is the mimic (fig. 6.13b).

How have prey populations evolved their defenses? The predators themselves are usually the agents of selection for refined prey defense. In one of the most thoroughly studied cases of natural selection for prey defense, H. Kettlewell (1959) found that predation by birds favors camouflage among peppered moths, *Biston betularia.* Birds eliminate the more conspicuous members of the peppered moth population,

(a)

(b)

Figure 6.13 Poisonous Müllerian (a) and nonpoisonous Batesian (b) mimics.

leaving the better camouflaged (fig. 6.14). In general, predators eliminate poorly defended individuals and leave the well defended. Consequently, the average level of defense in the prey population increases with time.

As a consequence of prey defenses, the rate of prey capture by predators is often low. For instance, wolves on Isle Royale in Lake Superior capture moose only about 8% of the times they try. Bernd Heinrich (1984) found that predatory bald-faced hornets have an even lower success rate. The hornets hunt other insects by flying rapidly among the plants in their environment and pouncing on objects that may be prey. Because their prey are well camouflaged, the hornets often pounce on inanimate objects. Heinrich was able to observe 260 pounces by hornets. About 72% of these pounces were directed at inanimate objects, such as bird droppings and brown spots on leaves. Another 21% of the pounces were directed at insects, such as bumblebees and other wasps, that are too well defended to be prey. Only 7% of the pounces were

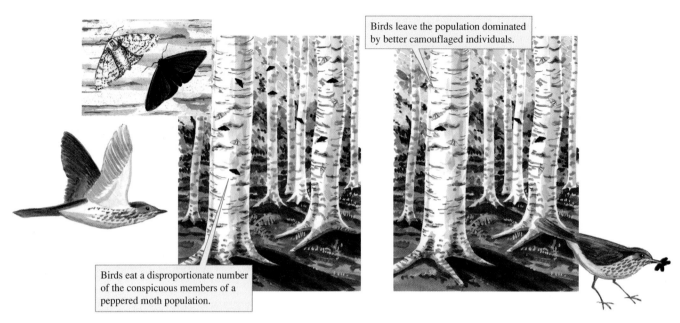

Birds leave the population dominated by better camouflaged individuals.

Birds eat a disproportionate number of the conspicuous members of a peppered moth population.

**Figure 6.14** Birds and other predators act as agents of natural selection for improved prey defense.

on potential insect prey. The hornets managed to capture two of these, a moth and a fly. Heinrich's observations indicate bald-faced hornets have a prey capture rate of less than 1%.

Though elusive, the prey of carnivores are generally similar in nutrient content. Consequently, carnivores, which are often widely distributed geographically, can vary their diets from one region to another. The Eurasian otter, *Lutra lutra,* which is distributed from Europe and North Africa through northern and central Asia, changes its diet based on the local availability of prey. Manuel Graça and F. X. Ferrand de Almeida (1983) compared otter diets along a gradient from northern to southern Europe (fig. 6.15). On the Shetland Islands, otter diets are over 91% fish, with the remainder consisting almost entirely of crabs. To this staple diet of fish the otters of England add frogs, mammals, and birds. Meanwhile, the diets of otters in central Portugal are less than one-third fish, with the remainder consisting of frogs, water snakes, birds, and aquatic insects. While some of the items that Graça and Ferrand de Almeida found on the otters' menus may be esthetically unacceptable to us, they are all fairly similar in terms of their carbon, nitrogen, and phosphorus content; they are just packaged differently.

Because predators must catch and subdue their prey, they often select prey by size, a behavior that ecologists call **size-selective predation.** Because of this behavior, prey size is often significantly correlated with predator size, especially among solitary predators. One such solitary predator, the puma, or mountain lion, *Felis concolor,* ranges from the Canadian Yukon to the tip of South America. Puma size changes substantially along this latitudinal gradient. Mammals make up over 90% of the puma's diet and large mammals, especially deer, are its main prey in the northern part of its range in North America. However, Augustin Iriarte and his colleagues (1990) found that

as pumas decrease in size southward, the average size of their prey also decreases (fig. 6.16). In the tropics, the puma feeds mainly on medium and small prey, especially rodents. Then, as pumas again increase in size south of the equator, large mammals form an increasing portion of their diet. Why should different-sized pumas feed on different-sized prey? One reason is that large prey may be difficult to subdue and may even injure the predator, while small prey may be difficult to find or catch. As we shall see later in chapter 6, size-selective predation may also have an energetic basis.

In summary, predators consume nutritionally rich but elusive and often well-defended prey. As a consequence, predators and their prey appear engaged in a coevolutionary race. In this race, predators eliminate poorly defended individuals in the population and average prey defenses improve. As average prey defenses improve, the poorer hunters go hungry and leave fewer offspring. Consequently, improved hunting skills evolve in the predator population, which exerts further selection on the prey population.

Now let's turn from typical heterotrophs such as the puma to organisms that obtain their energy from inorganic molecules. These are the chemosynthetic autotrophs. Though less familiar to most of us, chemosynthesis may be the oldest way of making a living.

## Using Inorganic Molecules

In 1977, a routine dive by a small submersible carried scientists exploring the Galápagos rift to a grand discovery. Their discovery changed our view of how a biosphere can be structured. Ecologists had long assumed that photosynthesis provides the energy for nearly all life in the sea. However, these unsuspect-

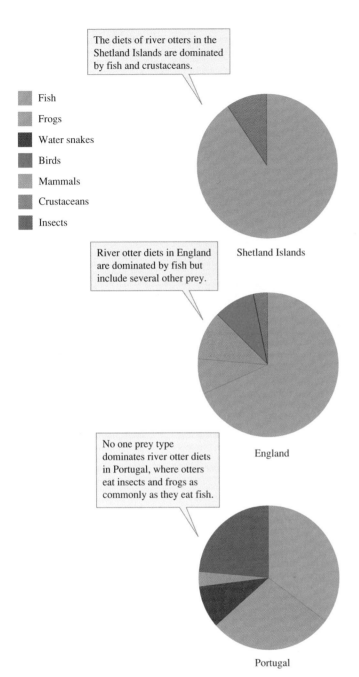

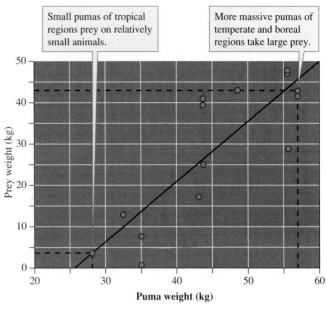

**Figure 6.16** The size of pumas and their prey change with latitude (data from Iriarte et al. 1990).

**Figure 6.15** Like many other widespread carnivores, the diets of river otters show great geographic variation (data from Graça and Ferrand de Almeida 1983).

ing scientists came across a world based upon an entirely different energy source, energy captured by chemosynthesis. The world they discovered was inhabited by giant worms up to 4 m long with no digestive tracts, by filter-feeding clams, and by carnivorous crabs tumbling over each other in tangled abundance (see fig. 3.8). These organisms lived on nutrients discharged by deep-sea volcanic activity through an oceanic rift, a crack in the seafloor. Interconnected systems of rifts extend tens of thousands of kilometers along the seafloor. Subsequent explorations have confirmed that chemosynthetic communities exist at many points of volcanic discharge along the seafloor.

The autotrophs upon which these submarine oases depend are chemosynthetic bacteria. Some of the most common are sulfur oxidizers, bacteria that use $CO_2$ as a source of carbon and get their energy by oxidizing elemental sulfur, hydrogen sulfide, or thiosulfite. The submarine volcanic vents with which these organisms are associated discharge large quantities of sulfide-rich warm water. The sulfur-oxidizing bacteria that exploit this resource around the vents are of two types: free-living forms and those that live within the tissues of a variety of invertebrate animals, including the giant tube worms (fig. 6.17). Other communities dependent upon sulfur-oxidizing bacteria have been discovered in thermal vents in deep freshwater lakes, in surface hot springs, and in caves.

Other chemosynthetic bacteria oxidize ammonium ($NH_4^+$), nitrite ($NO_2^-$), iron ($Fe^{2+}$), hydrogen ($H_2$), or carbon monoxide (CO). Of these, the nitrifying bacteria, which oxidize ammonium to nitrite and nitrite to nitrate, are undoubtedly among the most ecologically important organisms in the biosphere. Figure 6.18 summarizes one of the energy-yielding reactions exploited by nitrifying bacteria. The importance of these bacteria is due to their role in cycling nitrogen. As we saw earlier in chapter 6, nitrogen is a key element in the chemical makeup of individual organisms. It also plays a central role in the economy of the entire biosphere. Nitrogen will frequently enter our discussions in later chapters. In the Applications & Tools section of chapter 6, we will see how nitrifying bacteria have contributed to a pollution problem associated with an old gold mine.

As you can see, the trophic diversity among organisms is great. However, at least one ecological characteristic is shared by all organisms, regardless of the trophic group to which they belong—all organisms take in energy at a limited rate.

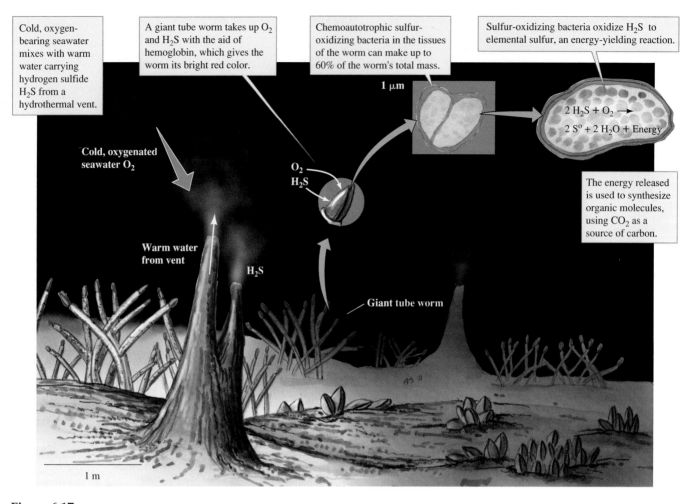

Cold, oxygen-bearing seawater mixes with warm water carrying hydrogen sulfide H₂S from a hydrothermal vent.

A giant tube worm takes up O₂ and H₂S with the aid of hemoglobin, which gives the worm its bright red color.

Chemoautotrophic sulfur-oxidizing bacteria in the tissues of the worm can make up to 60% of the worm's total mass.

Sulfur-oxidizing bacteria oxidize H₂S to elemental sulfur, an energy-yielding reaction.

Cold, oxygenated seawater O₂

$$2\,H_2S + O_2 \rightarrow 2\,S^o + 2\,H_2O + Energy$$

1 μm

O₂
H₂S

The energy released is used to synthesize organic molecules, using CO₂ as a source of carbon.

Warm water from vent

H₂S

Giant tube worm

1 m

**Figure 6.17** Hydrogen sulfide as an energy source for chemoautotrophic bacteria in the deep sea.

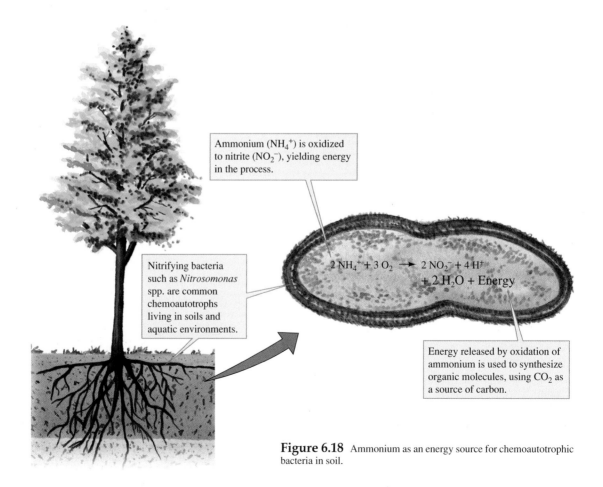

Ammonium (NH₄⁺) is oxidized to nitrite (NO₂⁻), yielding energy in the process.

Nitrifying bacteria such as *Nitrosomonas* spp. are common chemoautotrophs living in soils and aquatic environments.

$$2\,NH_4^+ + 3\,O_2 \rightarrow 2\,NO_2^- + 4\,H^+ + 2\,H_2O + Energy$$

Energy released by oxidation of ammonium is used to synthesize organic molecules, using CO₂ as a source of carbon.

**Figure 6.18** Ammonium as an energy source for chemoautotrophic bacteria in soil.

## CONCEPT DISCUSSION

### Energy Limitation

**The rate at which organisms can take in energy is limited.**

As children, many of us imagined that if we had free access to a candy or ice cream shop we would consume an infinite quantity of goodies. But even if we were given a chance to do this, our rate of intake would be limited, not by supply but by the rate at which we could process what we ate. In reality, the rate of intake of candy or ice cream for most children is limited, at least in the short term, by how much is available. The same is true in nature. If organisms are not limited by the availability of energy in the environment, their energy intake is limited by internal constraints. Limits on the potential rate of energy intake by animals have been demonstrated by studying how feeding rate increases as the availability of food increases. Limits on rates of energy intake by plants have been demonstrated by studying how photosynthetic rate responds to photon flux density.

## Photon Flux and Photosynthetic Response Curves

Plant physiologists generally test the photosynthetic potential of plants in environments that are ideal for the particular species being studied. These environments have abundant nutrients and water, normal concentrations of oxygen and carbon dioxide, ideal temperatures, and high humidity. If you gradually increase the intensity of light shining on plants growing under these conditions, that is, if you increase the photon flux density, the plants' rates of photosynthesis gradually increase and then level off. At low light intensities, photosynthesis increases linearly with photon flux density. At intermediate light intensities, photosynthetic rate rises more slowly. Finally, at high light intensity, but well below that of full sunlight, photosynthesis levels off. Data that show this type of photosynthetic response curve have been collected in studies of terrestrial plants, lichens, planktonic algae, and benthic algae.

Let's examine the structure of a theoretical photosynthetic response curve before comparing the photosynthetic responses of representative plant species. The response curves for different plant species generally level off at different maximum rates of photosynthesis. This rate in figure 6.19 is indicated as $P_{max}$. A second difference among photosynthetic response curves is the photon flux density, or **irradiance,** required to produce the maximum rate of photosynthesis. The irradiance required to saturate photosynthesis is shown in figure 6.19 as $I_{sat}$.

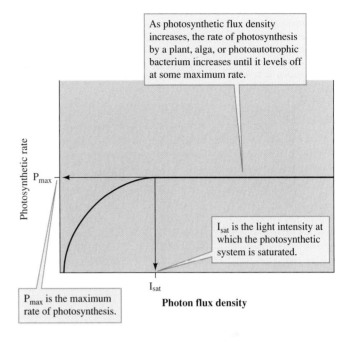

**Figure 6.19**   A theoretical photosynthetic response curve.

Differences in photosynthetic response curves have been used to divide plants into "sun" and "shade" species. The response curves of plants from shady habitats suggest selection for efficiency at low light intensities. The photosynthetic rate of shade plants levels off at lower light intensities, and they are often damaged by intense light. However, at very low light intensities, shade plants usually have higher photosynthetic rates than sun plants. Park Nobel (1977) determined the photosynthetic response curve for the maidenhair fern, *Adiantum decorum.* This plant generally grows at low light intensities in forests. In one of Park's trials, the maximum rate of photosynthesis by *A. decorum*, $P_{max}$, was approximately 9 μmol of $CO_2$ per square meter per second. The amount of light required to achieve this maximum rate of photosynthesis, $I_{sat}$, was a PAR photon flux density of about 300 μmol per square meter per second (fig. 6.20). The values of $P_{max}$ and $I_{sat}$ shown by *A. decorum* are much lower than those observed in plants that have evolved in sunny environments.

Herbs and short-lived perennial shrubs that have evolved in sunny environments show high maximum rates of photosynthesis, $P_{max}$, at relatively high light intensities, $I_{sat}$. One such plant, *Encelia farinosa,* grows in the hot deserts of North America. James Ehleringer and his colleagues (1976) found that *E. farinosa* has a high $P_{max}$, more than four times that of *A. decorum.* In addition, *E. farinosa* reaches these maximum rates of photosynthesis at a photon flux density of about 2,000 μmol per square meter per second (fig. 6.20). This combination of high $P_{max}$ and $I_{sat}$ allows *E. farinosa* to fix energy at a high rate during the infrequent times when water is plentiful in its desert environment.

Whether of shade or sun plant, photosynthetic response curves eventually level off. In other words, the rate at which photosynthetic organisms can take in energy is limited. As we shall now see, animals also take in energy at a limited rate.

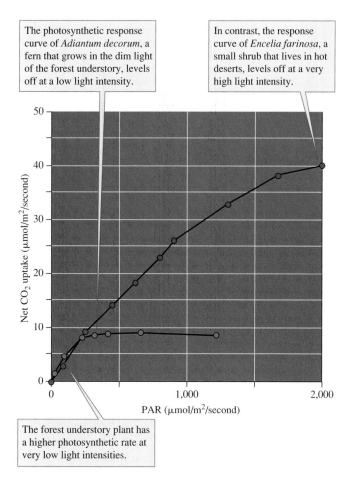

The photosynthetic response curve of *Adiantum decorum*, a fern that grows in the dim light of the forest understory, levels off at a low light intensity.

In contrast, the response curve of *Encelia farinosa*, a small shrub that lives in hot deserts, levels off at a very high light intensity.

The forest understory plant has a higher photosynthetic rate at very low light intensities.

**Figure 6.20** Contrasting photosynthetic response curves indicate adaptation to very different environmental light conditions (data from Ehleringer, Björkman, and Mooney 1976, after Nobel 1977).

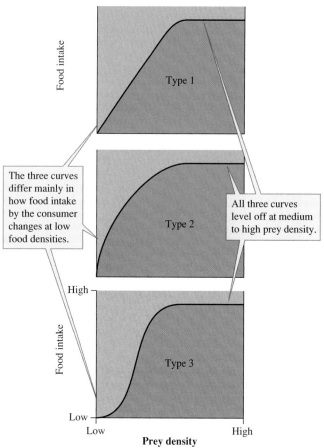

The three curves differ mainly in how food intake by the consumer changes at low food densities.

All three curves level off at medium to high prey density.

**Figure 6.21** Three theoretical functional response curves.

# Food Density and Animal Functional Response

If you gradually increase the amount of food available to a hungry animal, its rate of feeding increases and then levels off. This relationship is called the **functional response.** Ecologists use graphs to describe functional responses. C. S. Holling (1959) described three types of functional responses, all of which level off at a maximum feeding rate (fig. 6.21).

Type 1 functional responses are those in which feeding rate increases linearly (as a straight line) as food density increases and then levels off abruptly at some maximum feeding rate. The only animals that have type 1 functional responses are consumers that require little or no time to process their food; for example, some filter-feeding aquatic animals that feed on small prey.

In a type 2 functional response, feeding rate at first rises linearly at low food density, rises more slowly at intermediate food density, and then levels off at high densities. At low food densities, feeding rate appears limited by how long it takes the animal to find food. At intermediate food densities, the animal's feeding rate is partly limited by the time spent searching for food and partly by the time spent handling food. "Handling" refers to

such activities as cracking the shells of nuts or snails, removing distasteful scent glands from prey, and chasing down elusive prey. At high food density an animal does not have to search for food at all and feeding rate is determined almost entirely by how fast the animal can handle its food. At these very high densities, the animal, in effect, has "all the food it can handle."

The type 3 functional response is S-shaped. What mechanisms may be responsible for the more complicated shape of the type 3 functional response? Why does feeding rate increase slowly at low densities? At low density, food organisms may be better protected from predators because they occupy relatively protected habitats, or "safe sites." In addition, animals often ignore uncommon foods. Many animals seem to focus most of their attention on more abundant foods, switching to less common food only when it exceeds some threshold density. Animals may also require some learning to exploit food at a maximum rate. At low food densities they do not have sufficient exposure to a particular food item to fully develop their searching and handling skills. Holling's research provided a theoretical basis for later empirical studies of animal functional response.

Of the hundreds, perhaps thousands, of functional response curves described by ecologists, the most common is the type 2 functional response. Here are some examples. John Gross and several colleagues (1993) conducted a well-controlled study of

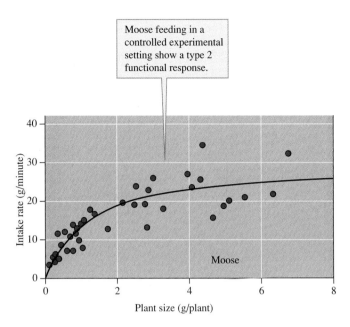

Moose feeding in a controlled experimental setting show a type 2 functional response.

**Figure 6.22** A functional response by moose (data from Gross et al. 1993).

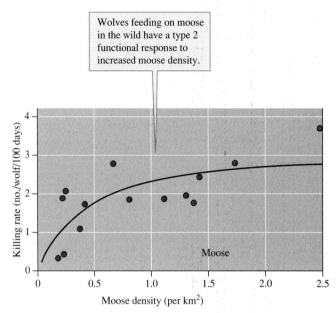

Wolves feeding on moose in the wild have a type 2 functional response to increased moose density.

**Figure 6.23** Wolf functional response (data from Messier 1994).

the functional responses of 13 mammalian herbivore species. The researchers manipulated food density by offering each herbivore various densities of fresh alfalfa, *Medicago sativa.* The rate of food intake was measured as the difference between the amount of alfalfa offered to an animal at the beginning of a trial and how much was left over at the end. Gross and his colleagues ran 36 to 125 feeding trials for each herbivore species for a total of over 900 trials. Every species of herbivore examined, from moose to lemmings to prairie dogs, showed a type 2 functional response. Figure 6.22 shows the type 2 functional response shown by moose, *Alces alces.*

Gross and his colleagues worked in a controlled experimental environment. Do consumers in natural environments also show a type 2 functional response? To answer this question, let's examine the functional response of wolves, *Canis lupus,* feeding on moose. François Messier (1994) examined the interactions between moose and wolves in North America. He focused on areas where moose are the dominant large prey species eaten by wolves. When moose density in various regions was plotted against the rate at which they are killed by wolves, the result was a clear type 2 functional response (fig. 6.23). While these examples of functional response emphasize visual feeders, functional response should occur regardless of the sense used by consumers to detect their food. Evidence for such a functional response among nonvisual feeders was found by a team of marine ecologists, working at the Hatfield Marine Science Center in Newport, Oregon (Ryer et al. 2002). The team, headed by Clifford Ryer, studied the feeding behavior of juveniles of two fish species, walleye pollock, *Theragra chalcogramma,* and sablefish, *Anoplopoma fimbria.* The research team studied the functional response in the light and in the dark. In darkness, these fish detect their prey using their lateral line system, which has sensory cells that respond to water movements. Though both fish species

consumed plankton at a lower rate in the dark than in the light, their rates of consumption increased in the dark with increases in prey density. The increase in feeding rate in walleye pollock followed a type 1 functional response, while that of sablefish followed a type 2 functional response. Studies such as this one provide evidence for the generality of functional responses among a diversity of consumer organisms.

Type 2 functional responses are remarkably similar to the photosynthetic response curves shown by plants (see fig. 6.20) and have the same implications. Even if you provide an animal with unlimited food, its energy intake eventually levels off at some maximum rate. This is the rate at which energy intake is limited by internal rather than external constraints. What conclusions can we draw from this parallel between plants and animals? We can conclude that even under ideal conditions, organisms as different as wolves, moose, and the plants eaten by moose take in energy at a limited rate. As we shall now see, limited energy intake is a fundamental assumption of optimal foraging theory.

## CONCEPT DISCUSSION

### Optimal Foraging Theory

**Optimal foraging theory attempts to model how organisms feed as an optimizing process.**

Evolutionary ecologists predict that if organisms have limited access to energy, then natural selection is likely to favor individuals within a population that are more effective at acquiring energy. This prediction spawned an area of ecological inquiry called **optimal foraging theory.** Optimal foraging theory assumes that if energy supplies are limited, organisms cannot simultaneously maximize all of life's functions; for example, allocation of energy to one function, such as growth or reproduction, reduces the amount of energy available to other functions, such as defense. As a consequence, there must be compromises between competing demands. This seemingly inevitable conflict between energy allocations has been called the **principle of allocation.**

Optimal foraging theory attempts to model how organisms feed as an optimizing process, a process that maximizes or minimizes some quantity. In some situations, the environment may favor individuals that assimilate energy or nutrients at a high rate (e.g., some filter-feeding zooplankton and short-lived weedy annual plants growing in disturbed habitats). In other situations, selection for minimum water loss appears much stronger (e.g., cactus and scorpions in the desert). Optimal foraging theory attempts to predict what consumers will eat, and when and where they will feed. Early work in this area concentrated on animal behavior. More recently the acquisition of energy and nutrients by plants has been modeled, using ideas borrowed from economic theory.

# Testing Optimal Foraging Theory

How can you test optimal foraging theory? Unfortunately, you cannot test this theory, or any other complex theory, directly in one grand experiment. Consequently, researchers chip away at the problem by testing specific predictions of the theory. One of the most productive avenues of research has been to use optimal foraging theory to predict the composition of animal diets.

When ecologists consider potential prey for a consumer, they try to identify the prey attributes that may affect the rate of energy intake by the predator. One of the most important factors is the abundance of a potential food item. All things being equal, a more abundant prey item yields a larger energy return than an uncommon prey. In optimal foraging studies, prey abundance is generally expressed as the number of the prey encountered by the predator per unit of time, $N_e$. Another prey attribute is the amount of energy, or costs, expended by the predator while searching for prey, $C_s$. A third characteristic of potential prey that could affect the energy return to the predator is the time spent processing prey in activities such as cracking shells, fighting, removing noxious scent glands, and so forth. Time spent in activities such as these are summarized as handling time, $H$. Ecologists ask, given the searching and handling capabilities of an animal and a certain array of available prey, Do animals select their diet in a way that yields the maximum rate of energy intake? We can rephrase this question mathematically by incorporating the terms for prey encounter rate, $N_e$, searching costs, $C_s$, and handling time, $H$, into a model.

# A Model for Diet Breadth

One of the most basic questions that we might ask about feeding by a predator concerns the number of prey items that should be included in its diet. Put another way, what mix of prey will maximize energy intake by a predator feeding under a particular set of circumstances? Early theoretical work on this question was published by MacArthur and Pianka (1966) and Charnov (1973) and several others. We can represent the rate of energy intake of a predator as $E/T$, where $E$ is energy and $T$ is time. Earl Werner and Gary Mittelbach (1981) modeled the rate of energy intake for a predator feeding on a single prey species as follows:

$$\frac{E}{T} = \frac{N_{e1}E_1 - C_s}{1 + N_{e1}H_1}$$

In this equation, $N_{e1}$ is the number of prey 1 encountered per unit of time. $E_1$ is the energy gained by feeding on an individual of prey 1 minus the costs of handling. $C_s$ is the cost of searching for the prey. $H_1$ is the time required for "handling" an individual of prey 1. Once again, this equation expresses the net rate at which a predator takes in energy when it feeds on a particular prey species.

What would be the rate of energy intake if the predator fed on two types of prey? The rate is calculated as follows:

$$\frac{E}{T} = \frac{(N_{e1}E_1 - C_s) + (N_{e2}E_2 - C_s)}{1 + N_{e1}H_1 + N_{e2}H_2}$$

This is an extension of the first equation. Here, we've added encounter rates for prey 2, $N_{e2}$, the energetic return from feeding on prey 2, $E_2$, and the handling time for prey 2, $H_2$. The searching costs, $C_s$, are assumed to be the same for prey 1 and prey 2.

The rate of energy intake by a predator feeding on several prey can be represented as:

$$\frac{E}{T} = \frac{\sum_{i=1}^{n} N_{ei}E_i - C_s}{1 + \sum_{i=1}^{n} N_{ei}H_i}$$

Here, $\Sigma$ means "the sum of" and $i$ equals 1, 2, 3, etc., to $n$, where $n$ is the total number of prey. Remember that this equation gives an estimate of the rate of energy intake. The question that optimal foraging theory asks is whether organisms feed in a way that maximizes the rate of energy intake, $E/T$.

Optimal foraging theory predicts that a predator will feed exclusively on prey 1, ignoring other available prey, when:

$$\frac{N_{e1}E_1 - C_s}{1 + N_{e1}H_1} > \frac{(N_{e1}E_1 - C_s) + (N_{e2}E_2 - C_s)}{1 + N_{e1}H_1 + N_{e2}H_2}$$

This expression says that the rate of energy intake is greater if the predator feeds only on prey 1. If the predator feeds on both prey species, the rate will be lower.

Optimal foraging theory predicts that predators will include a second prey species in their diet when:

$$\frac{(N_{e1}E_1 - C_s) + (N_{e2}E_2 - C_s)}{1 + N_{e1}H_1 + N_{e2}H_2} > \frac{N_{e1}E_1 - C_s}{1 + N_{e1}H_1}$$

In this case, feeding on two prey species gives the predator a higher rate of energy intake than if it feeds on one. The general prediction is that predators will continue to add different types of prey to their diet until the rate of energy intake reaches a maximum. This is called **optimization.**

Now let's get back to our basic question: Do animals select food in a way that maximizes their rate of energy intake? Testing such a prediction requires a great deal of information. Fortunately, mathematical models such as this one help focus experiments and observations on a few key variables.

## *Foraging by Bluegill Sunfish*

Some of the most thorough tests of optimal foraging theory have been conducted on the bluegill sunfish, *Lepomis macrochirus.* The bluegill is a medium-sized fish native to eastern and central North America, where it inhabits a wide range of freshwater habitats, from small streams to the shorelines of small and large lakes. Bluegills feed mainly on benthic and planktonic crustaceans and aquatic insects, prey that differ in size and habitat and in ease of capture and handling. Bluegills often choose prey by size, feeding on organisms of certain sizes and ignoring others. This behavior is convenient because it gives the ecologist a relatively simple measure to describe the composition of the available prey and the composition of the theoretically optimal diet.

Werner and Mittelbach used published studies to estimate the amount of energy expended by bluegills while they search for ($C_s$) and handle prey. They used laboratory experiments to estimate handling times ($H$) and encounter rates ($N_e$) for various prey. For these laboratory experiments, they constructed approximations of the places where bluegills forage in nature—open water, sediments, and vegetation. These model habitats were constructed in large aquaria and stocked with some of the important prey of bluegills: damselfly larvae, midge larvae, and *Daphnia.* These experiments showed that encounter rates increase as fish size, prey size, and prey density increase and that handling time depends on the relative sizes of predator and prey. Small bluegills require a relatively long time to handle large prey, while large bluegills expend little time handling small prey.

The energy content of prey was calculated by measuring the lengths of prey available in lakes and ponds; prey length was converted to mass, and then mass was converted to energy content using published values. With this information, Werner and Mittelbach characterized the prey available in Lawrence Lake, Michigan, and then estimated the diet that would maximize the rate of energy intake. They then sampled the bluegills of Lawrence Lake and examined their stomach contents to see how closely their diet approximated the diet predicted by optimal foraging theory.

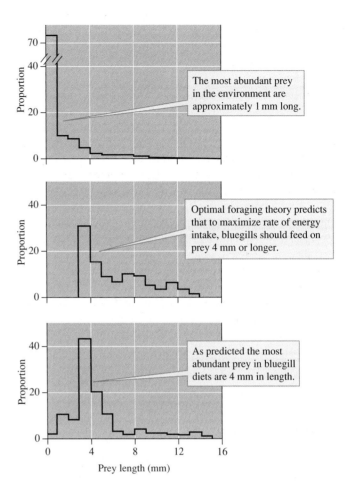

The most abundant prey in the environment are approximately 1 mm long.

Optimal foraging theory predicts that to maximize rate of energy intake, bluegills should feed on prey 4 mm or longer.

As predicted the most abundant prey in bluegill diets are 4 mm in length.

**Figure 6.24** Optimal foraging theory predicts composition of bluegill sunfish diets (data from Werner and Mittelbach 1981).

The upper graph in figure 6.24 shows the size distribution of potential prey in vegetation in Lawrence Lake. The middle graph shows the composition of the optimal diet as predicted by the optimal foraging model just presented. Finally, the bottom graph shows the actual composition of the diets of bluegills from Lawrence Lake. Bluegills feeding in vegetation selected prey that were uncommon and larger than average. The match between the optimal diet and the prey that bluegills in Lawrence Lake actually ate seems uncanny. A similar match was obtained for bluegills feeding on zooplankton in open water.

Werner and Mittelbach found that optimal foraging theory provides reasonable predictions of prey selection by natural populations of bluegills. Ecologists studying plants have developed an analogous predictive framework for foraging by plants.

## Optimal Foraging by Plants

How do plants "forage"? What animals do with behavior, plants do with growth. Plants forage by growing and orienting structures that capture either energy or nutrients. They grow leaves or other green surfaces to capture light and roots to capture nutrients. Terrestrial plants harvest energy from sunlight aboveground and nutrients and water from soil.

| Information |
| Questions |
| Hypothesis |
| Predictions |
| Testing ✓ |

*Investigating the Evidence*

# Variation in Data

We reviewed how to determine the "average" within a sample in two ways. In chapter 2 we calculated the sample mean and in chapter 3 we determined the sample median. The mean and median are different ways of representing the middle, or typical, within a sample of measurements or other types of observations. Another important question we can ask is how much *variation* is there around the average. This is important for several reasons. For example, two or more samples may have the same mean but have quite different amounts of variation among the observations within each sample. Knowing the variation within a sample of observations, as well as the mean or median, is critical to statistical comparisons of samples.

Suppose you are studying the feeding ecology of a species of fish and you need to report the variation in size of the fish used in your study. Consider the following measurements of body length for a sample of loach minnows, *Tiaroga cobitis*, from a tributary of the San Francisco River in southwestern New Mexico:

| Specimen | 1 | 2 | 3 | 4 | 5 | 6 | 7 | 8 | 9 | 10 |
|---|---|---|---|---|---|---|---|---|---|---|
| Total length (mm) | 60 | 62 | 56 | 53 | 53 | 59 | 62 | 41 | 58 | 58 |

The simplest index of variation would be the **range,** which is the difference between the largest and the smallest observation. In this case the range would be:

$$\text{Range} = 62 - 41 = 21 \text{ mm}$$

The range does not represent variation in samples well since very different sets of observations can have the same range. A better representation of the variation in a sample is one that factors in all the observations relative to the sample mean. One such index that is most commonly used is called the sample **variance.** The variance is calculated as follows:

First we calculate the sample mean as we did in chapter 2.

$\overline{X} = 56.2$ total length (For practice you could calculate this sample mean using the data given above.)

The variance is calculated by squaring the differences between the sample mean and each of the observations, adding them up to produce the "sum of squares," and dividing by the sample size minus one. Let's do this in steps. First let's calculate the sum of squares.

$$\text{Sum of squares} = \sum(X - \overline{X})^2$$

Using the fish length measurements given in the table above:

$$\sum(X - \overline{X})^2 = (60 - 56.2)^2 + (62 - 56.2)^2 + (56 - 56.2)^2 + \\ (53 - 56.2)^2 + (53 - 56.2)^2 + (59 - 56.2)^2 + \\ (62 - 56.2)^2 + (41 - 56.2)^2 + (58 - 56.2)^2 + \\ (58 - 56.2)^2$$

Taking the differences gives:

$$\sum(X - \overline{X})^2 = (3.8)^2 + (5.8)^2 + (-0.2)^2 + (-3.2)^2 + \\ (-3.2)^2 + (2.8)^2 + (5.8)^2 + (-15.2)^2 + \\ (1.8)^2 + (1.8)^2$$

Squaring the differences yields:

$$\sum(X - \overline{X})^2 = 14.44 + 33.64 + .04 + 10.24 + 10.24 + \\ 7.84 + 33.64 + 231.04 + 3.24 + 3.24$$

Adding the squared differences gives the sum of squares:

$$\text{Sum of squares} = \sum(X - \overline{X})^2 = 347.6 \text{ mm}^2$$

The sample variance is calculated by dividing the sum of squares by the sample size minus 1. The sample size in this case is 10 measurements.

$$\text{Sample variance} = s^2 = \frac{\sum(X - \overline{X})^2}{n - 1}$$

Now putting in the values,

$$s^2 = \frac{347.6 \text{ mm}^2}{9}$$

and dividing,

$$s^2 = 38.6 \text{ mm}^2$$

This then is the sample variance. However, notice that the units of the sample variance is square mm, not mm. Because the sample variance is expressed in squares of the original units, generally we take the square root of the variance to calculate a measure of variation called the sample **standard deviation.**

$$\text{Standard deviation} = s = \sqrt{s^2}$$

Calculating the standard deviation for our data:

$$s = \sqrt{38.6 \text{ mm}^2} = 6.2 \text{ mm}$$

While it took a little effort to calculate it, the standard deviation, 6.2 mm, provides us with a standardized index of the variation in length of the fish used in our feeding study. Fortunately most electronic calculators make these calculations automatically, once the data from a sample are entered. The sample standard deviation along with the sample mean enable us to make statistical comparisons of samples.

Because of the structure of their environment and the distribution of their resources, plants forage in two directions at once. Like animals, however, plants face limited supplies of energy and nutrients and so face the prospect of compromises between competing demands for energy. Allocation of energy to leaves and stems reduces the amount of energy available for root growth. Increased allocation to root growth means less energy available for leaves and stems.

In some environments, such as the desert, plants have access to an abundance of light but face shortages of water. In other environments, such as in the understory of temperate forests, there is little light but the soil may be rich in moisture and nutrients. In the face of such environmental heterogeneity, how do plants invest their energy? Using economic theory, Arnold Bloom and his colleagues (1985) suggested that plants adjust their allocation of energy to growth in such a way that all resources are equally limited. They predicted that plants in environments with abundant nutrients but little light would invest more energy in the growth of stems and leaves and less in roots to match their supply of energy to the supply of nutrients. They predicted that in environments rich in light but poor in nutrients, plants would invest more in roots.

The predictions of this economically based model have been supported by numerous studies showing that plants in light-poor environments invest more aboveground, while plants from nutrient-poor environments invest more belowground. So, it appears that plants allocate energy for growth to those structures that gather the resources that most limit growth in a particular environment. Some of the most revealing tests of these predictions come from studies of plants growing along gradients of nutrient availability.

Experimental studies show that the same plant species grown in nutrient-poor soils often develop a higher ratio of root biomass to shoot biomass, the so-called root:shoot ratio, than when grown on nutrient-rich soils. For example, when H. Setälä and V. Huhta (1991) grew birch tree seedlings in boreal forest soils of low and high nitrogen content, those in the nitrogen-poor soils developed higher root:shoot ratios (fig. 6.25).

David Tilman and M. Cowan (1989) obtained similar results when they grew four species of grass and four species of forbs on soils of different nitrogen content. They created a nitrogen gradient by mixing three different soils in different proportions. The soils were a subsoil (B horizon, see fig. 2.8) containing approximately 25 mg of nitrogen per kilogram of soil, a topsoil (A horizon) with 350 mg of nitrogen per kilogram of soil, and a black loam topsoil (A horizon) from a nearby site containing 5,000 mg of nitrogen per kilogram of soil. These soils were mixed to produce seven levels of soil nitrogen ranging from about 125 to 1,800 mg N per kilogram of soil. Several other nutrients were added to the experimental soils so that other nutrients would not limit growth of the experimental plants.

Tilman and Cowan conducted their growth experiments in 504 flowerpots that were 30 cm wide by 30 cm deep. They filled several pots with each of their seven soil mixtures and grew each of the eight study species from seed at high densities, 100 plants per pot, and low densities, 7 plants per pot. In each of the

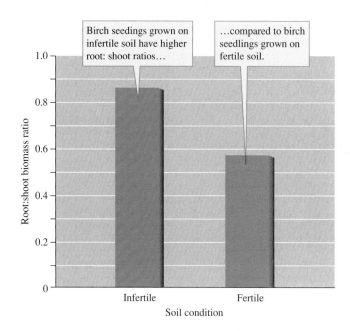

**Figure 6.25** Soil fertility and ratio of root biomass to shoot biomass (data from Setälä and Huhta 1991).

soil types, six pots of each species were planted at low density and three pots of each species at high density. Why did Tilman and Cowan plant several pots of each species in each of their growing conditions? Replicating their experimental conditions allowed them to test the statistical significance of their results.

The plants were grown outdoors in full sun and watered twice a week throughout the growing season except when there was adequate rain. The pots used for determining how nitrogen availability affects root:shoot ratios were maintained through two growing seasons. At the end of the second growing season, the researchers harvested the plants in these 280 pots, separating aboveground shoots from roots and carefully washing the soil from roots. They then dried roots and shoots for 1 week in an oven at 50°C. Finally, they weighed their dried samples and calculated root:shoot ratios of each study species grown at each nitrogen level.

Virtually all the species at both high and low densities had lower root:shoot ratios when grown with more nitrogen. Figure 6.26 shows the pattern for the grass *Sorghastrum nutans* when grown at high density. Like the other species, *S. nutans* reduced its root biomass and increased its shoot biomass in the presence of higher nitrogen availability. In a study of plants of the Great Plains of North America, Nichole Levang-Brilz and Mario Biondini found that 62% of the 55 plant species in their study increased root:shoot ratios in the face of reduced nitrogen availability (Levang-Brilz and Biondini 2002). These results indicate that many plants modify their anatomy to match environmental circumstance in the direction predicted by economic theory.

In summary, ecologists, using simple models of feeding behavior, have successfully predicted the feeding behavior of organisms as different as predatory animals and plants. Many animals forage in a way that tends to maximize their rate of energy intake. Meanwhile, plants, from birch trees to temperate

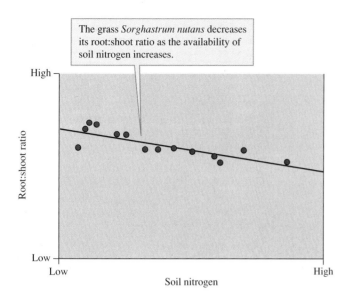

The grass *Sorghastrum nutans* decreases its root:shoot ratio as the availability of soil nitrogen increases.

**Figure 6.26** The root:shoot ratios of a grass species changes along a gradient of nitrogen availability (data from Tilman and Cowan 1989).

grasses and forbs, appear to allocate energy to growth in a way that increases the rate at which they acquire those resources in shortest supply. While the optimal foraging approach to trophic ecology is far from complete, it is a substantial improvement over a body of knowledge consisting of a long list of species-specific descriptions of diet and feeding habits.

In the following example in the Applications & Tools section, we see how biologists are using some of the patterns and concepts we have discussed in chapter 6 to address important environmental problems.

## APPLICATIONS & TOOLS:

### Bioremediation—Using the Trophic Diversity of Bacteria to Solve Environmental Problems

Imagine yourself in the center of a densely populated region with a mountain of sewage to dispose of or with thousands of leaky gasoline tanks contaminating the groundwater. How would you solve these environmental problems? Where would you turn for help? Increasingly, we are turning to nature's own cleanup crew, the bacteria. Environmental managers are taking advantage of the exceptional trophic diversity of bacteria to perform a host of environmental chores.

## Removing a Mountain of Sewage

Human populations produce prodigious amounts of sewage. Europe alone produces approximately 6 million tons per year. This sewage presents a tremendous disposal problem.

Historically, much of this sewage was dumped into the sea, but European countries have agreed to stop this practice. Consequently, alternative disposal strategies must be developed, some of which are spreading sewage sludge on agricultural lands or burning it. Both these alternatives are expensive and both pose some environmental hazards because sewage is generally contaminated by heavy metals. One of the most promising disposal techniques involves bacteria.

John Pirt of Kings College London discovered bacteria living in horse manure that can break down sewage at temperatures of 80°C (Coghlan 1993). He used these bacteria to develop a system for sewage disposal that alternates these high-temperature bacteria with bacteria that grow at 37°C. Pirt's system nearly eliminates the organic material in sewage sludge, leaving water and some mineral waste. The mineral waste contains heavy metal contaminants that can be isolated and either disposed of or recycled.

In the first stage of the process, the high-temperature bacteria consume about 55% of the organic material in sewage sludge (fig. 6.27). Some of this organic matter is used as an energy source and converted into $CO_2$ during respiration. The remainder is used as a source of carbon to construct more bacteria and is converted into bacterial biomass. The mixture of high-temperature bacteria and residual sludge from stage 1 is cooled to 37°C and moved to another chamber. In stage 2, bacteria consume the biomass created by high-temperature bacteria in the previous stage and reduce the organic content of the material by another 5%. In the third stage high-temperature bacteria remove another 25% of the original organic matter. The remaining 15% of the organic matter is removed in the fourth stage of the process, which is carried out at 37°C.

This process takes advantage of the biology of heterotrophic bacteria that use organic molecules as a source of energy and structural carbon. Some of the organic molecules are broken down into $CO_2$ and $H_2O$ during respiration. Some are incorporated into the organic molecules that make up the bacteria. By using two different groups of bacteria in alternating high- and medium-temperature environments the system eventually converts almost all of the organic matter in sewage to carbon dioxide and water.

The process is projected to cost much less than burning sludge or spreading it on agricultural lands. It also prevents contamination with heavy metals. Pirt points out that he did not have to go to the far-off hot springs of Yellowstone National Park to find his thermophilic bacteria but just to the nearest pile of horse dung. As we shall see in the next two examples, other ecologists are using local bacteria to meet daunting environmental challenges.

## Leaking Underground Storage Tanks

Gasoline and other petroleum derivatives are stored in underground storage tanks all over the planet. Those that leak are a serious source of pollution. Maribeth Watwood and Cliff

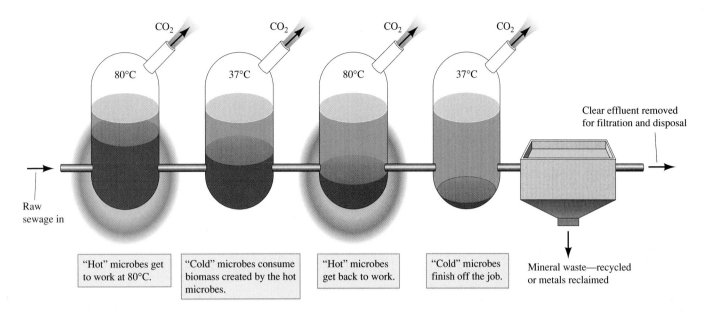

**Figure 6.27** Sewage digestion using bacteria with different temperature requirements may offer a solution to the global problem of sewage disposal (data from Coghlan 1993).

Dahm (1992) have been exploring the possibility of using bacteria to clean up soils and aquifers contaminated by leaking storage tanks. The first step in their work was to determine if there are naturally occurring populations of bacteria that can break down complex petroleum derivatives such as benzene.

Watwood and Dahm collected sediments from a shallow aquifer that contained approximately $8.5 \times 10^8$ bacterial cells per gram of wet sediment. Of these, $6.55 \times 10^4$ bacterial cells per milliliter were capable of living on benzene as their only source of carbon and energy. By exposing sediments from the aquifer to benzene for 6 months, the researchers increased the populations of benzene-degrading bacteria approximately 100 times.

How rapidly can these bacteria break down benzene? Watwood and Dahm found that with no prior exposure, bacterial populations could break down 90% of the benzene in their test flasks within 40 days (fig. 6.28). Exposing sediments to benzene prior to their tests increased the rate of breakdown.

Briefly, this study demonstrated that naturally occurring populations of bacteria can rapidly break down benzene leaking from underground storage tanks. This study suggests that these bacteria will eventually clean up the organic contaminants from leaking gasoline storage tanks without manipulation of the environment. However, in the next example, environmental managers found that they had to manipulate the environment to stimulate the desired bacterial cleanup of a contaminant.

## Cyanide and Nitrates in Mine Spoils

Many gold mines were abandoned when they could not be mined profitably with the mining technology of the nineteenth and early twentieth centuries. Then, in the 1970s, techniques

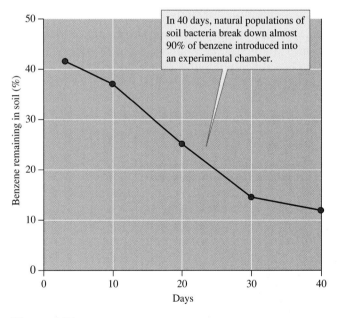

In 40 days, natural populations of soil bacteria break down almost 90% of benzene introduced into an experimental chamber.

**Figure 6.28** Benzene breakdown by soil bacteria (data from Watwood and Dahm 1992).

were developed to economically extract gold from low-grade ores. One of the main extraction techniques was to leach ore with cyanide (CN). Dissolved CN forms chemical complexes with gold and other metals. The solution containing gold-bearing CN can be collected and the gold and CN removed by filtering the solution with activated charcoal.

This new method of mining solved a technical problem but contaminated soils and groundwater. When the leaching process is finished, the leached ore is stored in piles; however, much CN remains. Several kinds of bacteria can break down

CN and produce $NH_3$. This $NH_3$ can, in turn, be used by nitrifying bacteria as an energy source, producing $NO_3$. Thus, leaching gold-bearing ores and subsequent microbial activity can contaminate soil and groundwater with CN, a deadly poison, and with nitrate, another contaminant.

Carleton White and James Markwiese (1994) studied a gold mine that had been worked with the CN leaching process. The leached ores from the mine were gradually releasing CN and $NO_3$ into the environment. The researchers looked to bacteria to solve this environmental problem. They first documented the presence of CN degraders by looking for bacterial growth in a diagnostic medium. This medium contained CN as the only source of carbon and nitrogen. Using this growth medium, White and Markwiese estimated that each gram of ore contained approximately $10^3$ to $10^5$ cells of organisms capable of growing on CN.

The leached ores presented bacteria with a rich source of nitrogen in the form of CN and $NO_3$ but the ores contained little organic carbon. White and Markwiese predicted that adding a source of carbon to the residual ores would increase the rate at which bacteria break down CN and reduce the concentration of $NO_3$ in the environment. Why should adding organic molecules rich in carbon increase bacterial use of nitrogen in the environment? Look back at figure 6.7, which shows that bacteria have a carbon:nitrogen ratio of about 5:1. In other words, growth and reproduction by bacteria require about five carbon atoms for each nitrogen atom.

White and Markwiese tested their ideas in the laboratory. In one experiment, they added enough sucrose to produce a C:N ratio of 10:1 within leached ores. This experiment included two controls, both of which contained leached ores without sucrose. One of the controls was sterilized to kill any bacteria. The other control was left unsterilized.

Bacteria in the treatments containing sucrose broke down all the CN within the leached ore in 13 days. Meanwhile, only a small amount of CN was broken down in the unsterilized control and no CN was broken down in the sterilized control (fig. 6.29). Why did the researchers include a sterilized control? The sterilized control demonstrated that nonbiological processes were not responsible for the observed breakdown of CN.

Figure 6.29 shows that adding sucrose to the residual ore stimulates the breakdown of CN. However, remember that this process ultimately leads to the production of $NO_3$. Does

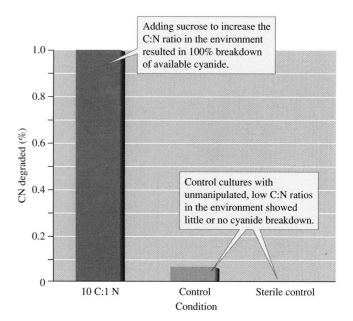

**Figure 6.29** Manipulating C:N ratios to stimulate breakdown of cyanide (CN) (data from White and Markwiese 1994).

adding sucrose to eliminate CN lead to the buildup of $NO_3$, trading one pollution problem for another? No, it does not. In another experiment White and Markwiese showed that adding sucrose also stimulates uptake of $NO_3$ by heterotrophic bacteria and fungi. These organisms use organic molecules, in this case sucrose, as a source of energy and carbon and $NO_3$ as a source of nitrogen. The nitrogen taken up by bacteria and fungi becomes incorporated in biomass as complex organic molecules. Nitrogen in this form is recycled within the microbial community and is not a source of environmental pollution.

White and Markwiese recommended that sucrose be added to leached gold-mining ores to stimulate breakdown of CN and uptake of $NO_3$ by bacteria. This environmental cleanup project was successful because the researchers were thoroughly familiar with the energy and nutrient relations of bacteria and fungi. Another key to the project's success was the great trophic diversity of bacteria. Bacteria will likely continue to play a great role as we address some of our most vexing environmental problems.

## SUMMARY

**Organisms use one of three main sources of energy: light, organic molecules, or inorganic molecules.** Photosynthetic plants and algae use $CO_2$ as a source of carbon and light, of wavelengths between 400 and 700 nm, as a source of energy. Light within this band, which is called photosynthetically active radiation, or PAR, accounts for about 45% of the total energy content of the solar spectrum at sea level. PAR can be quantified as photosynthetic photon flux density, generally reported as µmol per square meter per second. Among plants, there are three major alternative photosynthetic pathways, $C_3$, $C_4$, and CAM. $C_4$ and CAM plants are more efficient in their use of water than are $C_3$ plants. Heterotrophs use organic molecules both as a source of carbon and as a source of energy. Herbivores, carnivores, and detritivores face fundamentally different trophic problems. Herbivores feed on plant tissues, which often contain a great deal of carbon but little nitrogen. Herbivores must also overcome the physical and chemical defenses of plants. Detritivores feed on dead plant material, which is even lower in nitrogen than living plant tissues. Carnivores consume prey that are nutritionally rich but very well defended. Chemosynthetic autotrophs, which consist of a highly diverse group of chemosynthetic bacteria, use inorganic molecules as a source of energy. Bacteria are the most trophically diverse organisms in the biosphere.

**The rate at which organisms can take in energy is limited,** either by external or internal constraints. The relationship between photon flux density and plant photosynthetic rate is called photosynthetic response. Herbs and short-lived perennial shrubs from sunny habitats have high maximum photosynthetic rates that level off at high light intensities. The lowest maximum rates of photosynthesis occur among plants from shady environments. The relationship between food density and animal feeding rate is called the functional response. The shape of the functional response is generally one of three types. The forms of photosynthetic response curves and type 2 animal functional responses are remarkably similar. Energy limitation is a fundamental assumption of optimal foraging theory.

**Optimal foraging theory attempts to model how organisms feed as an optimizing process.** Evolutionary ecologists predict that if organisms have limited access to energy, natural selection is likely to favor individuals that are more effective at acquiring energy and nutrients. Many animals select food in a way that appears to maximize the rate at which they capture energy. Plants appear to allocate energy to roots versus shoots in a way that increases their rate of intake of the resources that limit their growth. Plants in environments with abundant nutrients but little light tend to invest more energy in the growth of stems and leaves and less in roots. In environments rich in light but poor in nutrients, plants tend to invest more energy in the growth of roots.

The trophic diversity of bacteria, which is critical to the health of the biosphere, can also be used as a tool to address some of our most challenging waste disposal problems. Bacteria can be used to eliminate the huge quantities of sewage produced by human populations, clean up soils and aquifers polluted by petroleum products such as benzene, and eliminate the pollution caused by some kinds of mine waste. The success of these projects requires that ecologists understand the energy and nutrient relations of bacteria. Bacteria will likely continue to play a great role as we address some of our most vexing environmental problems.

# Review Questions

1. Why don't plants use highly energetic ultraviolet light for photosynthesis? Would it be impossible to evolve a photosynthetic system that uses ultraviolet light? Does the fact that many insects see ultraviolet light change your mind? Would it be possible to use infrared light for photosynthesis? (Photosynthetic bacteria tap into the near infrared range.)

2. In what kinds of environments would you expect to find the greatest predominance of $C_3$, $C_4$, or CAM plants? How can you explain the co-occurrence of two, or even all three, of these types of plants in one area? (Think about the variations in microclimate that we considered in chapters 4 and 5.)

3. In chapter 6, we emphasized how the $C_4$ photosynthetic pathway saves water, but some researchers suggest that the greatest advantage of $C_4$ over $C_3$ plants occurs when $CO_2$ concentrations are low. What is the advantage of the $C_4$ pathway when $CO_2$ concentrations are low? As we shall see in chapter 23, the atmospheric concentration of $CO_2$ has been increasing for the last century or so. If this trend continues, and if the interactions between $C_4$ and $C_3$ plants are influenced by atmospheric $CO_2$ concentrations, how might the geographic distributions of $C_3$ and $C_4$ plants change?

4. What are the relative advantages and disadvantages of being an herbivore, detritivore, or carnivore? What kinds of organisms were left out of our discussions of herbivores, detritivores, and carnivores? Where do parasites fit? Where does *Homo sapiens* fit?

5. What advantage does advertising give to noxious prey? How would convergence in aposematic coloration among several species of Müllerian mimics contribute to the fitness of *individuals* in each species? In the case of Batesian mimicry, what are the costs and benefits of mimicry to the model and to the mimic?

6. Design a planetary ecosystem based entirely on chemosynthesis. You might choose an undiscovered planet of some distant star or one of the planets in our own solar system, either today or at some distant time in the past or future.

7. What kinds of animals would you expect to have type 1, 2, or 3 functional responses? How should natural selection for better prey defense affect the height of functional response curves? How should natural selection for more effective predators affect the height of the curves? What net effect should natural selection on predator and prey populations have on the height of the curves?

8. The rivers of central Portugal have been invaded, and densely populated, by the Louisiana crayfish, *Procambarus clarki*, which looks like a freshwater lobster about 12 to 14 cm long. The otters of these rivers, which were studied by Graça and Ferrand de Almeida (1983), can easily catch and subdue these crayfish. Using the model for prey choice:

$$\frac{E}{T} = \frac{\sum_{i=1}^{n} N_{ei}E_i - C_s}{1 + \sum_{i=1}^{n} N_{ei}H_i}$$

explain why the diets of the otters of central Portugal would shift from the highly diverse menu shown in figure 6.15, which included fish, frogs, water snakes, birds, and insects, to a diet dominated by crayfish. For the crayfish, assume low handling time, very high encounter rates, and high energy content.

9. The data of Iriarte and colleagues (1990) suggest that prey size may favor a particular body size among pumas (see fig. 6.16). However, this variation in body size also correlates well with latitude; the larger pumas live at high latitudes. Consequently, this variation in body size has been interpreted as the result of selection for efficient temperature regulation. Homeothermic animals are often larger at high latitudes, a pattern called Bergmann's rule. Larger animals, with lower surface area relative to their mass, would be theoretically better at conserving heat. Smaller animals, with higher surface area relative to their mass, would be theoretically better at keeping cool. So what determines predator size? Is predator size determined by climate, predator-prey interactions, or both? Design a study of the influence of the environment on the size of homeothermic predators.

10. How is plant allocation to roots versus shoots similar to plant regulation of temperature and water? (We discussed these topics in chapters 4 and 5.) Consider discussing these processes under the more general heading of homeostasis. (Homeostasis is the maintenance of a relatively constant internal environment in the face of variation in the external environment.)

# Suggested Readings

Béjà, O., L. Aravind, E. V. Koonin, M. T. Suzuki, A. Hadd, L. P. Nguyen, S. B. Jovanovich, C. M. Gates, R. A. Feldman, J. L. Spudich, E. N. Spudich, and E. F. Delong. 2000. Bacterial rhodopsin: evidence for a new type of phototrophy in the sea. *Science* 289:1902–6.

Béjà, O., E. N. Spudich, J. L. Spudich, M. Leclerc, and E. F. Delong. 2001. Proteorhodopsin phototrophy in the ocean. *Nature* 411:786–89.

Béjà, O., M. T. Suzuki, J. F. Heidelberg, W. C. Nelson, C. M. Preston, T. Hamada, J. A. Eisen, C. M. Fraser, and E. F. Delong. 2002. Unsuspected diversity among marine aerobic anoxygenic phototrophs. *Nature* 415:630–33.

Kolber, Z. S., C. L. Van Dover, R. A. Niederman, and P. G. Falkowski. 2000. Bacterial photosynthesis in surface waters of the open ocean. *Nature* 407:177–79.

*Four papers that give insights into one of the exciting frontiers of modern ecology, microbial ecology.*

Bolser, R. C. and M. E. Hay. 1996. Are tropical plants better defended? Palatability and defenses of temperate vs. tropical seaweeds. *Ecology* 77:2269–86.

Coley, P. D. and J. A. Barone. 1996. Herbivory and plant defenses in tropical forests. *Annual Review of Ecology and Systematics* 27:305–35.

*These two papers address the ecology and evolution of plant and algal defenses in tropical versus temperate marine and terrestrial environments.*

Ehleringer, J. R. and R. K. Monson. 1993. Evolutionary and ecological aspects of photosynthetic pathway variation. *Annual Review of Ecology and Systematics* 24:411–39.

*This paper gives an excellent review of the literature concerning the ecology of alternative photosynthetic pathways.*

Forde, B. and H. Lorenzo. 2001. The nutritional control of root development. *Plant and Nutrition* 232:51–68.

Levang-Brilz, N. and M. E. Biondini. 2002. Growth rate, root development and nutrient uptake of 55 plant species from the Great Plains Grasslands, USA. *Plant Ecology* 165:117–44.

Wijesinghe, D. K., E. A. John, S. Beurskens, and M. J. Hutchings. 2001. Root system size and precision in nutient foraging: responses to spatial pattern of nutrient supply in six herbaceous species. *Journal of Ecology* 13:337–51.

*Studies that provide an introduction to another frontier in ecology, the ecology of the subsurface; these emphasize the ecology of roots.*

Fortin, D., J. M. Fryxell, L. O'Brodovich, and D. Frandsen. 2003. Foraging ecology of bison at the landscape and plant community levels: the applicability of energy maximization principles. *Oecologia* 134:219–27.

Iason, G. R., T. Manso, D. A. Sim, and F. G. Hartley. 2002. The functional response does not predict the local distribution of European Rabbits (*Oryctolagus cuniculus*) on grass swards: experimental evidence. *Functional Ecology* 16:394–02.

Morris, D. W. and D. L. Davidson. 2000. Optimally foraging mice match patch use with habitat differences in fitness. *Ecology* 81:2061–66.

Plath, K. and M. Boersma. 2001. Mineral limitation of zooplankton: stoichiometric constraints and optimal foraging. *Ecology* 82:1260–69.

*Four studies of animal foraging that show the complexity of foraging decisions made by animals feeding in complex environments.*

Gross, J. E., L. A. Shipley, N. T. Hobbs, D. E. Spalinger, and B. A. Wunder. 1993. Functional response of herbivores in food-concentrated patches: tests of a mechanistic model. *Ecology* 74:778–91.

Holling, C. S. 1959. The components of predation as revealed by a study of small mammal predation of the European pine sawfly. *The Canadian Entomologist* 91:293–320.

Messier, F. 1994. Ungulate population models with predation: a case study with the North American moose. *Ecology* 75:478–88.

*This series of papers gives very good insights into theoretical, experimental, and observational approaches to studying animal functional response.*

Ritland, D. B. 1991. Unpalatability of viceroy butterflies (*Limenitis archippus*) and their purported mimicry model, Florida queens (*Danaus gilippus*). *Oecologia* 88:102–8.

Ritland, D. B. and L. P. Brower. 1991. The viceroy butterfly is not a Batesian mimic. *Nature* 350:497–8.

*These two papers show how experiments have been used to distinguish between Batesian and Müllerian mimicry among butterflies.*

Werner, E. E. and D. J. Hall. 1974. Optimal foraging and the size selection of prey by the bluegill sunfish (*Lepomis macrochirus*). *Ecology* 55:1042–52.

Werner, E. E. and D. J. Hall. 1979. Foraging efficiency and habitat switching in competing sunfishes. *Ecology* 60:256–64.

Werner, E. E. and G. G. Mittelbach. 1981. Optimal foraging: field tests of diet choice and habitat switching. *American Zoologist* 21:813–29.

*A series of classic papers showing the complementary use of theory and experiments to study optimal foraging by animals.*

# On the Net

Visit this textbook's accompanying website at www.mhhe.com/ecology (click on the book's title) to take advantage of practice quizzing, study/writing tips, timely news articles, and additional URLs for research on the topics in this chapter.

Photosynthesis

Food Webs

Nutrient Cycling

Restoration Ecology

Animal Population Ecology

Optimal Foraging and Energetics

Community Ecology

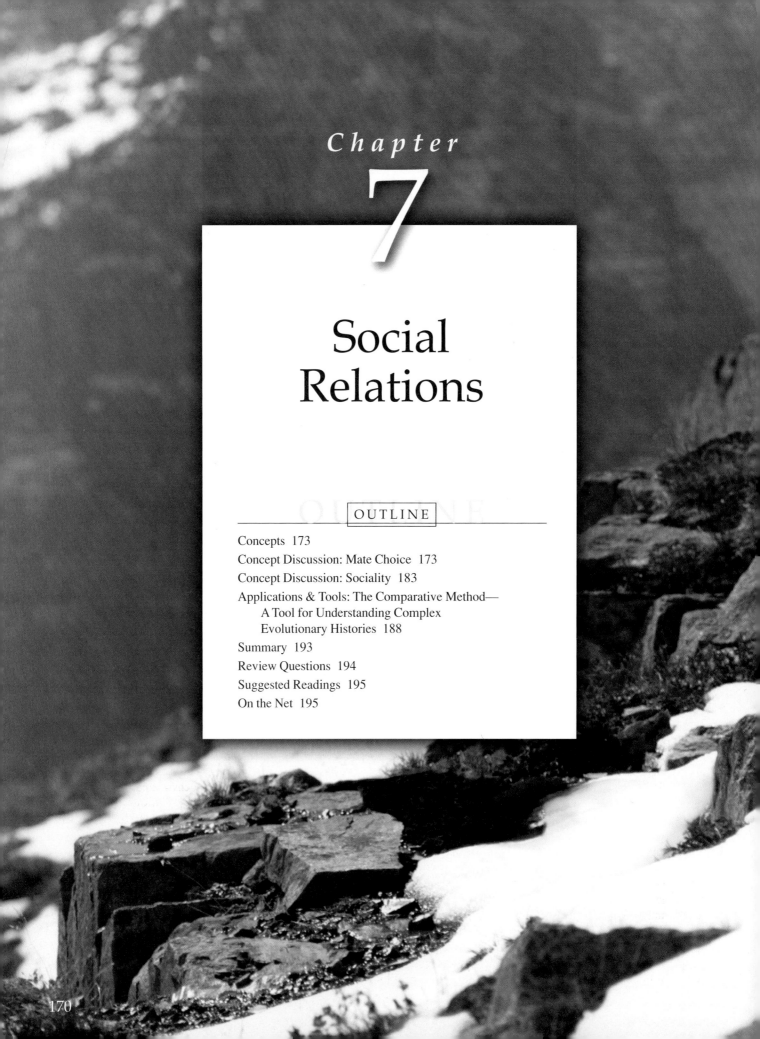

# Chapter
# 7

# Social Relations

Nowhere are the social interactions and other behaviors of vertebrate animals easier to observe than on tropical reefs. In early evening as the sun's rays shine obliquely through the clear waters over a coral reef, the activity of some of its inhabitants quickens. As if activated by some remote switch, a vast school of fish that had remained in the lagoon all day begins to move steadily toward an opening in the reef. The school is leaving the lagoon's protection and going out to the open sea for a night of feeding. Living in a school appears to have favored uniformity among its members. Approached underwater, the edge of the school looks like a giant translucent curtain stamped with the silhouettes of thousands of identical fish. Their coloration, countershaded dark above and silvery below, their similar size, highly coordinated movements, and great numbers give the fish within the school some protection from predators. Though seabirds and predaceous fish ambush the school as it makes its way, the schooling fish are so numerous and their individual movements so difficult to follow that only a small proportion of them are eaten. Gradually the school, moving like a gigantic, shape-shifting organism, passes through the channel connecting the lagoon to the open sea. The school of fish will be back by daybreak only to repeat its seaward journey next evening in a cycle of comings and goings that helps mark the rhythm of life on the reef.

Meanwhile along the reef, damselfish are distributed singly on territories. The damselfish retain exclusive possession of their territorial patches of coral rubble, living coral, and sand by patrolling the boundaries and driving off any fish attempting to intrude, especially other damselfish that would take their territory or other fish that would prey on eggs or consume food within the territory. Each day at this time, however, some territory-holding males are joined by females. For the space of time that they court and deposit eggs and sperm on the nest site prepared by the male, the territory contains two fish. Once mating is complete, however, the male is again alone on the territory, guarding the food and shelter contained within its boundaries as well as the newly deposited eggs that he fertilized minutes before.

Higher along the reef face a male bluehead wrasse mates with a member of the harem of females that live within his territory (fig. 7.1). In contrast to the male with his blue head, black bars, and green body, the female is mostly yellow with a large black spot on her dorsal fin. As the male bluehead extrudes sperm to fertilize the eggs laid by the female, small males, similar in color to the female, streak by the mating pair, discharging a cloud of sperm as they do. Some of the female's eggs will be fertilized by the large territorial bluehead male while others will be fertilized by the sperm discharged by the smaller yellow streakers. In addition to differences in color and courtship behavior, bluehead and yellow males have distinctive histories. While the yellow males began their lives as males, the bluehead male began life as a female and only transformed to a male when the local bluehead male was eaten by a predator or met some other end. At that point, because she was the largest yellow phase among the local females and males, she was in line to become the dominant local male and so changed from the yellow to the

**Figure 7.1** Bluehead wrasse males with yellow females of the species. If the bluehead male is removed from a territory, the largest female in the territory can change to a fully functional bluehead male within days.

bluehead form of the species. Within a week the former female was producing sperm and fertilizing the eggs produced by the females in the territory.

While male bluehead wrasses patrol their individual mating territories and male damselfish fight with each other at the boundaries of theirs, elsewhere on the reef groups of snapping shrimp live cooperatively in colonies that may contain over 300 individuals. Most of the individuals in the colonies are juveniles or males and each contains a single reproductive female. The female snapping shrimp, which plays a role much like the queen ant in an ant colony, breeds continuously and so is easily identified by her ripe ovaries or by the eggs she carries. Meanwhile the males of the colony, most of which will probably never mate, vigorously defend the nest site, with its "queen" shrimp and numerous juveniles against intruders. In this shrimp society most males serve the colony and its queen by protecting her offspring and the sponge where they live. While the queen reproduces profusely, the chance to reproduce is probably rare for an individual male. The colony thrives but reproduction is restricted to a few individuals in the population.

During a short swim over a coral reef you can observe great variation in social interactions among individuals belonging to the same species. Analogous variation can be found in terrestrial environments. In chapters 4, 5, and 6 of section II, we considered the relations of organisms to physical and chemical aspects of the environment, including temperature, water, energy, and nutrients. However, to an individual organism, other members of its own species are a part of the environment as significant to it as temperature, food, or the quantity and quality of available water. In chapter 7 we will consider some of the interrelations among individuals under the heading of social relations.

The study of social relations is the territory of **behavioral ecology,** which concentrates on relationships between organisms and environment that are mediated by behavior. In the case of social relations, other individuals of a species are the part of the environment of particular interest. A branch of biology concerned with the study of social relations is **sociobiology.** Social relations, from dominance relationships and reproductive interactions to cooperative behaviors, are important since they often directly and obviously impact the reproductive contribution of individuals to future generations, a key component of Darwinian or evolutionary fitness, usually referred to simply as **fitness.** Fitness can be defined as the number of offspring, or genes, contributed by an individual to future generations, which can be substantially influenced by social relations within a population.

One of the most fundamental social interactions between individuals takes place during sexual reproduction. The timing of those interactions and their nature is strongly influenced by the reproductive system of a species. The behavioral ecologist considers several factors. Does the population engage in sexual reproduction? Are the sexes separate? How are the sexes distributed among individuals? Are there several forms of one sex or the other? Questions such as these have drawn the attention of biologists since Darwin (1862) who wrote, "We do not even in the least know the final cause of sexuality; why new beings should be produced by the union of the two sexual elements, instead of by a process of parthenogenesis [production of offspring from unfertilized eggs] . . . The whole subject is as yet hidden in darkness." As you will see, behavioral and evolutionary ecologists have learned a great deal about the evolution and ecology of reproduction in the nearly one and a half centuries since Darwin published this statement. However, much remains to be discovered.

Since mammals and birds reproduce sexually, from a human perspective sexual reproduction may appear the norm. However, asexual reproduction is common among many groups of organisms such as bacteria, protozoans, plants, and some vertebrates. However, most described species of plants and animals include male and female functions, sometimes in separate individuals or within the same individual. This brings us to a fundamental question in biology. What is female and what is male? From a biological perspective, the answer is simple. **Females** produce larger, more energetically costly gametes (eggs or ova), while **males** produce smaller, less costly gametes (sperm or pollen). Because of the greater energetic cost of producing their gametes, female reproduction is thought to be generally limited by access to the necessary resources. In contrast, male reproduction is generally limited by access to female mates. Biologists long ago proposed that this difference in investment in gametes has usually led to a fundamental dichotomy between actively courting males and highly selective females.

Despite the basic differences between males and females, distinguishing the two sexes in nature is sometimes difficult. While it is easy to distinguish between males and females in species where males and females differ substantially in external morphology, the males and females of other species appear very similar and are very difficult to distinguish using only external anatomy. Still other species are **hermaphrodites,** organisms that combine male and female function in the same individual (fig. 7.2). The most familiar examples of hermaphrodites are plants, among which the vast majority of species produce flowers that have both male and female parts. Among animals, fish provide many interesting examples of hermaph-

(a)

(b)

**Figure 7.2**  Male and female function: (*a*) male and female Canada geese, a species in which males and females have very similar external anatomy (i.e., are monomorphic); (*b*) a "perfect" flower, which includes both male (stamens) and female (pistil) parts and function.

roditism. For instance, many species of small seabasses are hermaphrodites. As pairs within these species court, one member of the pair performs male-specific courtship behaviors while the other member of the pair produces eggs. As the eggs are laid, the first member of the pair fertilizes them. Later, the two fish may switch roles, with the second individual behaving as a male while the first assumes the female role and lays eggs.

Many aspects of sexual function that we may take for granted represent complex biological problems that have puzzled biologists for generations. For example, what factors have favored separate sexes in some species and hermaphrodites in others? Eric Charnov, J. Maynard Smith, and James Bull (1976) addressed this question in a classic paper titled, "Why Be a Hermaphrodite?" These authors identified three conditions that should favor a hermaphroditic population over one with separate sexes: (1) low mobility, which limits the opportunities for male to male competition, which often depends on structures designed to find and compete aggressively for females, (2) low overlap in resource demands by female and male structures and functions, such as in plants, where pollen production often occurs earlier in the season than seed maturation, and (3) sharing of costs for male and female function, for instance in insect-pollinated plants where attractive flowers promote both male and female reproductive success.

Clearly, the way populations are divided between the sexes will influence social relations, which will in turn affect the fitness of individuals, particularly through influences on their reproductive rates. Here are two concepts that have emerged from studies of social relations that provide examples of the complex relationships between social interactions and fitness. These concepts form the framework of chapter 7.

## CONCEPTS

- **Mate choice by one sex and/or competition for mates among individuals of the same sex can result in selection for particular traits in individuals, a process called sexual selection.**
- **The evolution of sociality is generally accompanied by cooperative feeding, defense of the social group, and restricted reproductive opportunities.**

## CONCEPT DISCUSSION

### Mate Choice

**Mate choice by one sex and/or competition for mates among individuals of the same sex can result in selection for particular traits in individuals, a process called sexual selection.**

Darwin (1871) proposed that the social environment, particularly the mating environment, could exert significant influence on the characteristics of organisms. He was particularly intrigued by the existence of what he called "secondary sexual characteristics," the origins of which he could not explain except by the advantages they gave to individuals during competition for mates. Darwin used the term *secondary sexual characteristics* to mean characteristics of males or females not directly involved in the process of reproduction. Some of the traits that Darwin had in mind were "gaudy colors and various ornaments . . . the power of song and other such characters." How do we explain the existence of characteristics such as the antlers of male deer, the bright peacock's tail, or the gigantic size

and large nose of the male elephant seal? In order to explain the existence of such secondary sexual characteristics, Darwin proposed a process that he called **sexual selection.** Sexual selection results from differences in reproductive rates among individuals as a result of differences in their mating success.

Sexual selection is thought to be important under two circumstances. The first is where individuals of one sex compete among themselves for mates, which results in a process called **intrasexual selection.** For instance, when male mountain sheep or elephant seals fight among themselves for dominance or mating territories, the largest and strongest generally win such contests. In such situations the result is often selection for larger body size and more effective weapons such as horns or teeth. Since this selection is the result of contests within one sex, it is called intrasexual selection.

Sexual selection can also occur when members of one sex consistently choose mates from among members of the opposite sex on the basis of some particular trait. Because two sexes are involved, this form is called **intersexual selection.** Examples of traits used for mate selection include female birds choosing among potential male mates based on the brightness of their feather colors or on the quality of their songs. Darwin proposed that once individuals of one sex begin to choose mates on the basis of some anatomical or behavioral trait, sexual selection would favor elaboration of the trait. For instance, the plumage of male birds' color might become brighter over time or their songs more elaborate or both.

However, how much can sexual selection elaborate a trait before males in the population begin to suffer higher mortality due to other sources of natural selection, such as that exerted by predators? Darwin proposed that sexual selection will continue to elaborate a trait until balanced by other sources of natural selection, such as predation. Since Darwin's early work on the subject, research has revealed a great deal about how organisms choose mates and the basis of sexual selection. An excellent model for such studies is the guppy, *Poecilia reticulata.*

## Mate Choice and Sexual Selection in Guppies

It would be difficult for experimental ecologists interested in mate choice and sexual selection to design a better experimental animal than the guppy (fig. 7.3). Guppies are native to the streams and rivers of Trinidad and Tobago, islands in the southeastern Caribbean, and in the rivers draining nearby parts of the South American mainland. The waters inhabited by guppies range from small clear mountain streams to murky lowland rivers. Along this gradient of physical conditions, guppies also encounter a broad range of biological situations. In the headwaters of streams above waterfalls, guppies live in the absence of predaceous fish or with the killifish *Rivulus hartii,* which preys mainly on juveniles and is not a very effective predator on adult guppies. In contrast, guppies in lowland rivers live with a wide variety of predaceous fish, including the pike cichlid, *Crenicichla alta,* a very effective visual predator of adult guppies.

**Figure 7.3** A colorful male guppy courting a female guppy: What are the influences of mate selection by female guppies and natural selection by predators?

Male guppies show a broad range of coloration both within and among populations. What factors may produce this range of variation? It turns out that female guppies, if given a choice, will mate with more brightly colored males. However, brightly colored males are attacked more frequently by visual predators. This trade-off between higher mating success by bright males but

greater vulnerability to predators provides a mechanistic explanation for variation in male coloration among different habitats. The most brightly colored male guppies are found in populations exposed to few predators, while those exposed to predators, such as the pike cichlid, are much less brightly colored (Endler 1995). Thus the coloration of male guppies in local populations may be determined by a dynamic interplay between natural selection exerted by predators and by female mate choice.

While field observations are consistent with a trade-off between sexual selection due to mate choice and natural selection due to predation, the evidence would be more convincing with an experimental test. John Endler (1980) performed such a test in an exemplary study of natural selection for color pattern in guppies.

## Experimental Tests

Endler performed two experiments, one in artificial ponds in a greenhouse at Princeton University (fig. 7.4) and one at field sites (fig. 7.5). For the greenhouse experiments, Endler constructed 10 ponds designed to approximate pools in the streams of the Northern Range in Trinidad. Four of the ponds were of a size (2.4 m × 1.2 m × 40 cm) typical of the pools inhabited by a

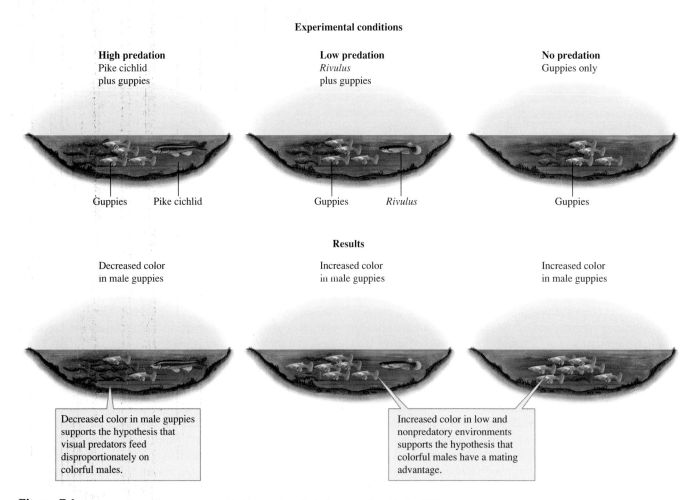

**Experimental conditions**

| **High predation** Pike cichlid plus guppies | **Low predation** *Rivulus* plus guppies | **No predation** Guppies only |

Guppies    Pike cichlid         Guppies    *Rivulus*         Guppies

**Results**

Decreased color in male guppies         Increased color in male guppies         Increased color in male guppies

Decreased color in male guppies supports the hypothesis that visual predators feed disproportionately on colorful males.

Increased color in low and nonpredatory environments supports the hypothesis that colorful males have a mating advantage.

**Figure 7.4** Summary of greenhouse experimental design and results (information from Endler 1980).

**Experimental design**

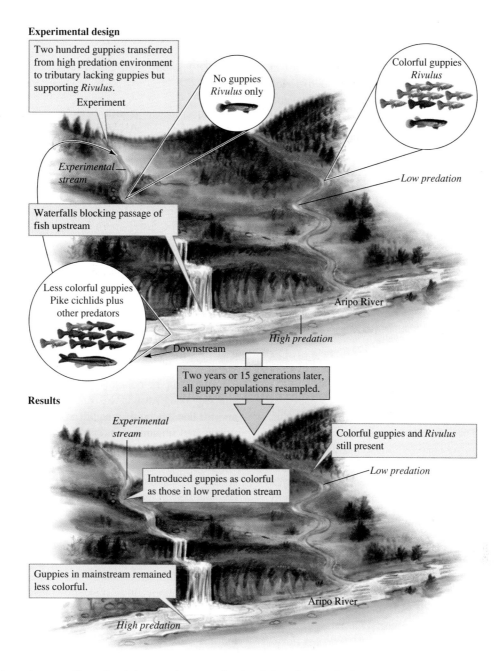

Two hundred guppies transferred from high predation environment to tributary lacking guppies but supporting *Rivulus*.
Experiment

No guppies *Rivulus* only

Colorful guppies *Rivulus*

*Experimental stream*

*Low predation*

Waterfalls blocking passage of fish upstream

Less colorful guppies Pike cichlids plus other predators

Aripo River

*High predation*

Downstream

Two years or 15 generations later, all guppy populations resampled.

**Results**

*Experimental stream*

Colorful guppies and *Rivulus* still present

*Low predation*

Introduced guppies as colorful as those in low predation stream

Guppies in mainstream remained less colorful.

Aripo River

*High predation*

**Figure 7.5** Field experiment on effects of predation on male guppy coloration (information from Endler 1980).

single pike cichlid in smaller streams. During the final phase of the experiment, Endler placed a single pike cichlid in each of these ponds. The six other ponds were similar in size (2.4 m × 1.2 m × 15 cm) to stream pools in the headwaters which contain approximately 6 *Rivulus*. Endler eventually placed 6 *Rivulus* in 4 of these ponds and maintained the other two ponds with no predators as controls. What did Endler create with this series of pools and predator combinations? These three groups of ponds represented three levels of predation: high predation (pike cichlid), low predation (*Rivulus*), and no predation.

However, before introducing predators, Endler established similar physical environments in the pools and stocked them with carefully chosen guppies. He lined all ponds with commercially available dyed gravel, taking care to put the same proportions of gravel colors in each of the ponds. The gravel he used in all ponds was 31.4% black, 34.2 % white, 25.7 % green, plus 2.9% each of blue, red, and yellow. Why did Endler take great care to put the same colors of gravel in the same proportions into all of his ponds? One of the most critical elements of the experiment was to standardize the background colors across all of the ponds. The influence of prey color on vulnerability to predators depends on the background against which the prey is viewed by visual predators. As a consequence, controlling background color was of critical importance to Endler's experiments.

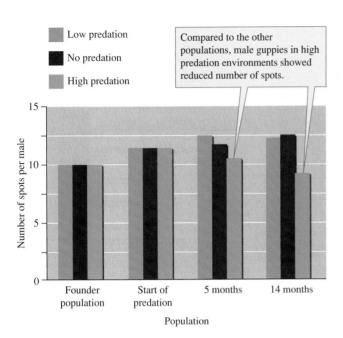

**Figure 7.6** Results of greenhouse experiment, which exposed populations of guppies to no predation, low predation (killifish), and high predation (pike cichlid) environments (data from Endler 1980).

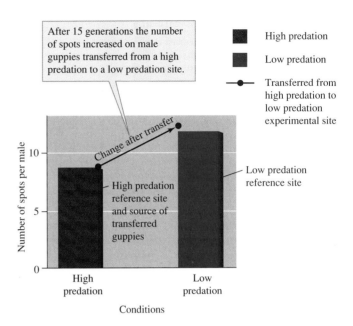

**Figure 7.7** Results of field experiment involving transfer of guppies from high predation site to site with killifish, a fairly ineffective predator (data from Endler 1980).

Endler stocked the experimental pond with 200 guppies, which were descended from 18 different populations in Trinidad and Venezuela. By drawing guppies from so many populations, Endler ensured that the experimental populations would include a substantial amount of color variation. As we will see in chapter 8, genetic variation is an essential requirement for evolutionary change in populations.

Endler's second experiment was conducted in the field within the drainage network of the Aripo River (fig 7.5), where he encountered three distinctive situations within a few kilometers. Within the mainstream of the Aripo River, guppies coexisted with a wide variety of predators, including pike cichlids, which provided a "high predation" site. Upstream from the high predation site, Endler discovered a small tributary which flowed over a series of waterfalls near its junction with the mainstream. Because the waterfalls prevented most fish from swimming upstream, this tributary was entirely free of guppies but supported a population of the ineffective predator *Rivulus*. This potential "low predation" site provided an ideal situation for following the evolution of male color. The third site, which was a bit farther upstream, was a small tributary that supported guppies along with *Rivulus*. This third site gave Endler a low predation reference site for his study. Endler captured 200 guppies in the high predation environment, measured the coloration of these guppies, and then introduced them to the site lacking guppies. Six months later the introduced guppies and their offspring had spread throughout the previously guppy-free tributary. Finally 2 years or about 15 guppy generations after the introduction, Endler returned and sampled the guppies at all three study sites.

The results of the greenhouse and field experiments supported each other. As shown in figure 7.6, the number of colored spots on male guppies increased in the greenhouse ponds with no predators and with *Rivulus* but decreased in the high predation ponds containing pike cichlids. Figure 7.7, which summarizes the results of Endler's field experiment, compares the number of spots on males in high predation and low predation stream environments with guppies transferred from the high predation environment to a low predation environment. Notice that the transplanted population converged with the males at the low predation reference site during the experiment. In other words, when freed from predation, the average number of spots on male guppies increased substantially. This result, along with the results of the greenhouse experiment, supports the hypothesis that predation reduces male showiness in guppy populations.

Research by many other researchers supports the impact of predators on male ornamentation. However, the observation that male colorfulness increased in the absence of predators or in the presence of weak predation both in the field and the laboratory invites explanation. Why did male color increase rather that just remain static? The observed changes imply that colorful males enjoy some selective advantage. That advantage appears to result from how female guppies choose their mates.

## Mate Choice by Female Guppies

What cues do female guppies use to choose their mates? Anne Houde (1997), who summarized the findings of numerous studies, found that several male traits were associated with greater mating success. The weight of the evidence supports the conclusion that male coloration contributes significantly to male mating success. Color characteristics that have been shown to confer a male mating advantage include "brightness," number of red spots, number of blue spots, iridescent area, total pigmented area, and carotenoid or orange area.

These results appear to account for the increase in male colorfulness observed by Endler in the absence of predation or in the presence of low predation pressure. That is, female preference for more colorful males gave them greater fitness in the absence of strong predation. As a consequence, male colorfulness increased in the study populations in low predation or no predation environments. Male behavior, especially their rate of making courtship displays, has also been found associated with increased male mating success.

There have been fewer studies of how competition among males, that is, intrasexual selection, may influence male mating success. Let's look at one of the few studies that does. Astrid Kodric-Brown (1993) studied whether competitive interactions among males contribute to variation in male mating success. She obtained guppies for her behavioral experiments from stock John Endler had originally collected from the Aripo and Paria Rivers in Trinidad. Males and females used in the behavioral experiments were reared separately. Males were kept in 95-liter aquaria in populations consisting of 10 males and 20 females. Meanwhile virgin females were reared in all-female groups of sisters until they were 6 months old. During this period they had no visual contact with males. Both males and females were fed a standardized diet and maintained at the same temperatures and exposed to the same numbers of hours of dark and light.

From her stock populations, Kodric-Brown chose 59 pairs of males with contrasting colors and 59 females. To test female preference she placed a single female into the central chamber of a test tank and each member of a male pair in the side chambers flanking the central chamber. Screens covering glass partitions prevented visual contact between males and females initially. After 10 minutes of acclimation by the guppies, Kodric-Brown removed the screens. Once the screens were removed males would usually begin courtship displays and the female would inspect the males through the glass partition. Kodric-Brown recorded the behavior of males and females in the display tank for 10 minutes, recording the time and the rates at which males displayed, and the amount of time the female spent within 5 cm of the glass partition of each male. She designated the male that the female spent the most time with as the preferred or attractive male and the male with which the female spent less time as the nonpreferred or unattractive male.

After this initial 10-minute period during which females indicated their preferences, Kodric-Brown removed the glass partitions separating the guppies, allowing interactions among the males and the female. Kodric-Brown observed that males engaged in agonistic interactions, such as chasing and nipping, in over 94% of mating trials, which gave her a basis for determining which males were behaviorally dominant and which were subordinate. Kodric-Brown recorded the interactions between the two males and between the males and the female until 5 minutes after a copulation. After a mating trial the female was moved to a rearing tank where she eventually gave birth.

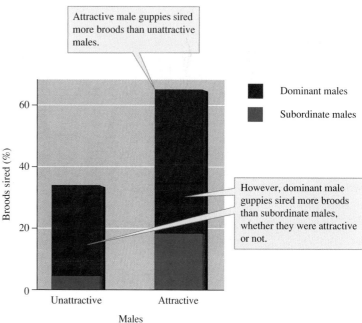

**Figure 7.8**  Relative reproductive success by attractive versus unattractive and dominant versus subordinate male guppies (data from Kodric-Brown 1993).

The offspring from each female were raised separately. In order to establish paternity, male offspring were raised to maturity when they expressed their full coloration, which is inherited from their fathers.

The results of Kodric-Brown's experiments indicate that reproductive success was determined by a combination of male attractiveness and male dominance status. Female mate preference, which was determined when the guppies had visual contact only, was highly correlated with subsequent male mating success (fig. 7.8). The males that attracted females when viewed through the glass partition subsequently sired a greater percentage of broods than did unattractive males. Approximately 67% of the broods were sired by attractive males compared to 33% that were sired by unattractive males. However, it appears that male dominance status also contributes to male reproductive success. Among unattractive males that sired broods, 87.5% were dominant. The conclusion that reproductive success is determined by a combination of competition between males and female choice is reinforced by the low reproductive success by males that were neither attractive nor dominant. These males, which lacked the apparent advantages associated with either dominance or attractiveness, sired only 4% of the broods. The result indicates that reproductive opportunities are highly restricted for these males.

The characteristics associated with male mating success among guppies are often correlated. Kodric-Brown observed that attractive males tended to be dominant, court more, and have more and brighter orange and iridescent spots. These characteristics are closely associated with a male's anatomy and physiology. Let's look at a mating system where male attractiveness is dependent upon complex behaviors that in effect extend the male phenotype.

**Figure 7.9** Male scorpionflies such as this one compete vigorously with each other for female scorpionflies.

# Mate Choice Among Scorpionflies

Scorpionflies (fig. 7.9) belong to the order Mecoptera, a group of insects most closely related to the caddisflies (order Trichoptera) and the moths and butterflies (order Lepidoptera). The common name "scorpionfly" is related to the way that males hold their genitalia over the back of their abdomens in a position that suggests a scorpion's sting. Despite their appearance, male scorpionflies are entirely harmless to people. Compared to insects such as moths or beetles, there are relatively few scorpionfly species alive today. However, they have been a rich source of information on behavioral ecology, particularly on the evolution and ecology of mating systems. Randy Thornhill has been a central figure in research on mate choice and sexual selection and his studies of scorpionfly mating systems are regarded as classic studies of the evolution and ecology of mating systems (e.g., Thornhill 1981, Thornhill and Alcock 1983).

Adult scorpionflies in the genus *Panorpa* feed on dead arthropods in the shrub and herb understory of forests. Several lines of evidence suggest that the supply of dead arthropods available to scorpionflies is limited and that the intensity of competition for dead arthropods is intense, especially among males. Thornhill observed that male scorpionflies fight over dead arthropods and even steal them from spider webs, a behavior which leads to significant scorpionfly mortality. Why do scorpionflies compete so vigorously and risk death over dead arthropods? One reason they fight is that male *Panorpa* use dead arthropods to attract females. If a male finds a dead arthropod and can successfully defend it from other males, he will stand next to the arthropod and secrete a pheromone, which can attract females from several meters away. A female attracted by the pheromone will usually feed on it while the male mates with her. However, if an arthropod is not available

as a nuptial offering, males will secrete a mass of saliva from their enlarged salivary glands and use that to attract females. Finally, males without gifts may attempt forced copulations.

In a series of experimental studies, Thornhill explored the details of alternative male mating strategies and the ecological conditions associated with each. In one of his most basic studies, he asked whether there is a difference in mating success among males using different mating strategies. Thornhill created an enclosed environment where he could control the availability of dead arthropods and the number of male *Panorpa* competing for them. He set up 12 replicate environments in 10-gallon terrariums. He included 6 dead crickets, 2 large, 2 medium, and 2 small, in each terrarium and added 12 male *P. latipennis* to each. Male aggression over crickets, which began soon after they were introduced, was finished after about 3 hours. At that time each of the crickets had been won by a single male, which stood near its respective prize and secreted pheromone. The majority of the remaining 6 males secreted a mass of saliva, which they guarded, while secreting pheromones. Finally, some males had no nuptial offerings.

Once the competition among male *Panorpa* for possession of the dead crickets had been decided, Thornhill introduced 12 females and recorded mating activity once per hour for 3 hours. Across the 12 terrariums, there were 144 male *Panorpa* and 144 females. Of the males, 72 males took possession of crickets, 45 had secreted salivary masses, and 27 had no nuptial offerings. How did mating success differ among these groups of males? Figure 7.10 shows that males with a medium or large

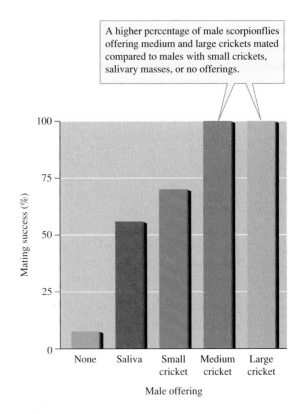

**Figure 7.10** Influence of alternative nuptial offerings on mating success by male scorpionflies, *Panorpa latipennis* (data from Thornhill 1981).

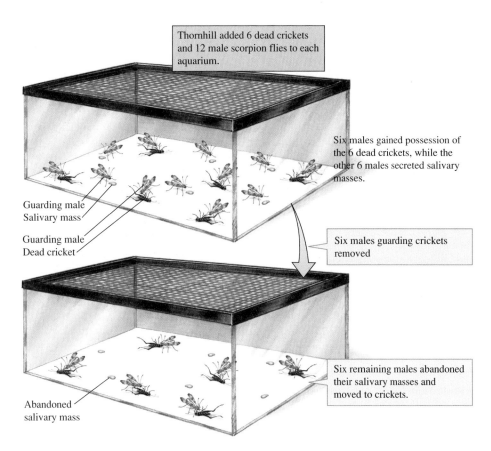

**Figure 7.11**  Experimental test of the influence of nuptial offerings on mating success by male scorpionflies (information from Thornhill 1981).

cricket as a nuptial offering had a clear advantage over those that offered females a small cricket, a salivary mass, or no nuptial offering.

What benefit do females gain by mating with males that offer larger arthropods? One of the clearest benefits is that females feeding on the arthropods offered by males do not have to forage for their own and avoid the risk of being eaten by a spider or other predator as they fly through the forest understory. In addition, feeding on these larger nuptial offerings gives females a reproductive advantage. Thornhill documented that rate of egg laying is higher among females mating with males that provide arthropod prey compared to females that mate with males offering saliva only. Meanwhile females mated to males with no offering lay very few eggs. What produces this contrast among females mated to males with different nuptial gifts? Thornhill's results likely reflect the greater nutritional benefit of arthropod prey versus saliva and the lack of a nutritional contribution by males without gifts.

Next, Thornhill asked what factors determine whether males compete successfully for arthropod offerings or resort to the alternatives of salivary masses or no nuptial gifts? One of the most basic questions that one could ask is whether males are fixed in particular behaviors. That is, if males that have not competed successfully for possession of a dead arthropod are given access to one, will they take possession of it and advertise their possession by secreting pheromone? Thornhill addressed

this question with a series of controlled experiments with enclosures. Again, he placed 6 crickets, all medium, in each of 12 terrariums and added 12 male *Panorpa* to each. As in previous experiments, 6 of the males took possession of the dead crickets in each aquarium, leaving 6 males without arthropods. Again, the males without arthropods secreted a salivary mass which they stood beside as they secreted pheromone. At this point, Thornhill removed all the males possessing crickets in all the terrariums. Within half an hour, almost all the remaining males moved from their salivary masses to the available crickets and secreted pheromone (fig. 7.11). It therefore appears that given the opportunity, male *Panorpa* will take possession of and guard dead arthropods.

What factors determine whether male *Panorpa* will be able to successfully claim a dead arthropod in a competitive environment? Males contesting over a dead arthropod will usually first display to each other. However, visual displays often quickly escalate to head butting and lashing at each other with the scorpion-like genital bulb with its pair of sharp claspers. The claspers of male scorpionflies are capable of tearing wings or other body parts of an opponent. As a consequence, these battles over bugs can be dangerous to both opponents. Because male body size varies widely within populations and male aggression over dead arthropods often involves direct combat, Thornhill predicted that larger males would be most successful as competitors over arthropods.

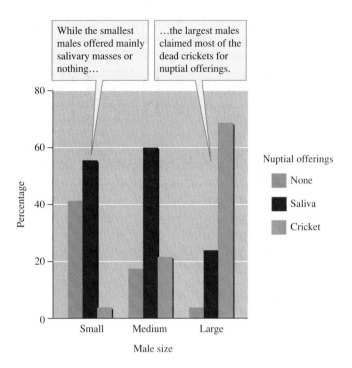

**Figure 7.12** Type of nuptial offering has a significant influence on mating success among male scorpionflies (data from Thornhill 1981).

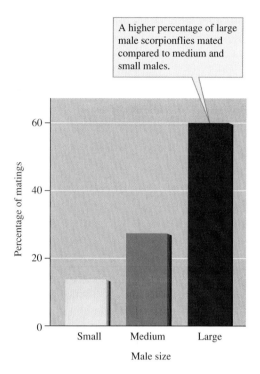

**Figure 7.13** Male size has a significant influence on mating success among male scorpionflies (data from Thornhill 1981).

Thornhill tested the relationship of male size on their ability to compete for and retain possession of arthropod prey in another experiment. This time he conducted his experiment in 14 larger, 3' × 3' × 3' screen enclosures set out on the forest floor of his study area. Because the enclosures had no bottom panel they just enclosed a 9-square-foot area of the herbaceous vegetation growing on the forest floor. In effect they enclosed a bit of scorpionfly habitat. Thornhill placed 4 crickets in seven of the enclosures and 2 in the other seven. He then added 10 female and 10 male scorpionflies to each of the enclosures, which were similar to natural population densities. The males in each enclosure consisted of 3 large males (55–64 mg), 4 medium males (42–53 mg), and 3 small males (33–41 mg). Because scorpionflies are nocturnal, Thornhill monitored the scorpionflies from sunset to sunrise with night vision equipment. Observations continued every night for a week during which Thornhill periodically added fresh dead crickets and replaced any female or male scorpionflies that died with new individuals to keep population densities constant.

The results of the field experiment clearly support the hypothesis that during competition for dead arthropods larger males have an advantage over small males. Figure 7.12 compares the nuptial offerings of small, medium, and large males in the enclosures with 2 crickets. While most small males either had no offerings or had a salivary mass, medium males generally offered salivary masses and occasionally competed successfully for a cricket. In contrast, large males generally offered females a cricket and only occasionally offered saliva or had no nuptial offering.

Thornhill's study revealed the mechanism underlying variation among males in their ability to compete for nuptial offerings. Larger males are more likely to successfully defend the available arthropod offerings due to their advantages in aggressive encounters. Now, does this difference in offerings translate into different mating success among males? The answer is given in figure 7.13, which shows the percentage of matings observed by Thornhill in cages with 2 crickets. Large males were involved in 60% of the matings observed, compared to 27% for medium males and 13% for small males. Clearly, the ability of large males to defend higher quality nuptial offerings translates directly into higher mating success.

Let us leave animals now and consider plants. Though we know much less about the mating behavior of plants, it appears that their reproductive ecology also includes the potential for mate choice and sexual selection. One of the best studied mating systems in plants is that of the wild radish, *Raphanus sativus*.

# Nonrandom Mating Among Wild Radish

Wild radish grows as an annual weed in California where it can be commonly seen along roadways and in abandoned fields (fig. 7.14). The seeds of wild radish germinate in response to the first winter rains of California's Mediterranean climate (see fig. 2.22) and the plants flower by January. Flowering may continue to late spring or early summer, depending on the length of the wet sea-

**Figure 7.14** The wild radish, *Raphanus sativus,* has become a model for studying the mating behavior of plants.

son. During their flowering season, wild radishes are pollinated by a wide variety of insects, including honeybees, syrphid flies (see fig. 6.13*b*), and butterflies. Wild radish flowers have both male (**stamens**) and female (**pistils**) parts and produce both pollen and ovules. However, a wild radish plant cannot pollinate itself, a condition called **self-incompatibility.** Because they must mate with other plants, a researcher working on wild radish can more easily control matings between plants.

Diane Marshall has used the many advantages offered by wild radish, such as its rapid growth to maturity and self-incompatibility, to explore the topic of mate selection in plants. Marshall and Michael Folsom (1992) listed a number of other characteristics of wild radish that make it convenient for study. For instance, its fruits contain several seeds which allows the possibility of multiple paternity of offspring. However, the seeds are not so numerous that the researcher is overwhelmed by a vast number of seeds. In addition, each plant produces numerous flowers allowing the possibility of several kinds of matings per plant and several replications of each mating experiment on the same plant. The seeds, which weigh about 10 mg, are also a convenient size for handling and weighing. Finally, there is sufficient genetic variation among individual radish plants to identify the male parent of each seed using electrophoresis of isozymes (see Applications & Tools in chapter 8).

The insects that pollinate wild radish generally arrive at flowers carrying pollen from several different plants, and as a consequence a wild radish plant typically has about seven mates. Under these circumstances of multiple mates, Marshall

asked whether siring of offspring is a random process. In other words, do the seven mates of a typical wild radish plant have an equal probability of fertilizing the available ovules? The alternative, nonrandom mating, would suggest the potential for mate choice and sexual selection. What mechanisms might produce nonrandom mating among wild radish? Nonrandom mating could result from maternal control over the fertilization process, competition among pollen, or a combination of the two processes. If it does occur in plants, nonrandom mating establishes the conditions necessary for sexual selection in plants. However, as Marshall and Folsom (1991) pointed out, though sexual selection is well documented in animals, its occurrence among plants remains a controversial and open question.

While the existence of sexual selection in plants remains controversial, nonrandom mating is well documented. Marshall and her colleagues have repeatedly demonstrated nonrandom mating in wild radish. For instance, Marshall (1990) carried out greenhouse experiments that showed nonrandom mating among 3 maternal plants and 6 pollen donors. In this experiment Marshall mated 3 seed parents or maternal plants with 6 pollen donors, the plants that would act as sources of pollen to pollinate the flowers of the seed plants.

Marshall used the 6 pollen donors to make 63 kinds of crosses, 6 single donor crosses plus 57 mixed donor crosses, on each maternal plant. Her crosses included all possible mixtures of pollen from 1 to 6 donors. Plants were pollinated in the greenhouse by hand. All pollinations were performed on freshly opened flowers in the morning when the temperature was cool enough for researchers to work comfortably. Pollen was collected by tapping flowers lightly on the bottom of small petri dishes from an equal number of flowers of each pollen donor. Pollen was then mixed and applied to the stigmas of flowers on the maternal plant using forceps wrapped in tissue. Sufficient pollen was applied to cover each stigma. Because each cross was replicated from 2 to 20 times depending on the type of cross, the total number of pollinations performed on each plant was 300. This is a good example of the unique opportunities for experimental work offered by plants.

One of the ways that Marshall assessed the possibility of nonrandom mating was through performance of pollen donors. She estimated pollen donor performance in three ways: (1) number of seeds sired in mixed pollinations, (2) positions of seeds sired, and (3) weight of seeds sired. The results of this analysis are shown in figure 7.15. What would you expect to see in figure 7.15 if performance was equal across pollen donors? If performance was equal, the heights of the bars would be approximately equal for all pollen donors. However they are not, and figure 7.15 indicates clearly that pollen donors vary widely in their performance. In other words, mating in this experiment was nonrandom.

Because Marshall conducted her 1990 study under greenhouse conditions, we might ask whether nonrandom mating also occurs under field conditions. In other words, could the nonrandom mating she documented have been an artifact of greenhouse conditions? Marshall and Ollar Fuller (Marshall

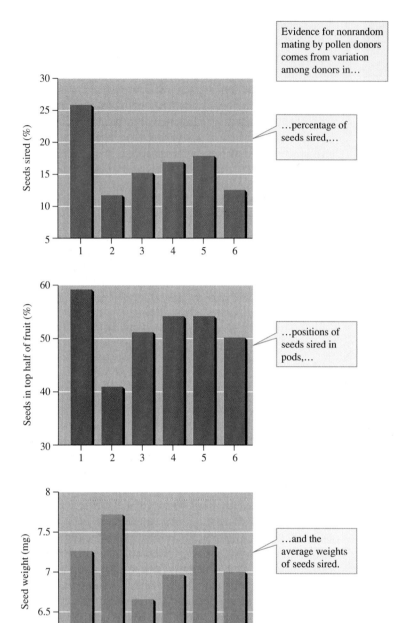

Evidence for nonrandom mating by pollen donors comes from variation among donors in…

…percentage of seeds sired,…

…positions of seeds sired in pods,…

…and the average weights of seeds sired.

**Figure 7.15** Evidence for unequal mating success among wild radish pollen donors in a greenhouse environment (data from Marshall 1990).

Marshall and Fuller chose four maternal plants and grew their offspring in a field setting. Three other maternal lineages, (A, B, and C) were chosen to act as pollen donors. In the field, the maternal plants were covered with fine mesh nylon bags until the experimental pollinations were completed. Using the forceps and tissue method described earlier, Marshall and Fuller performed several kinds of hand-pollinations, including mixed pollinations using pollen from all three pollen donors. Once the hand pollinations were completed the nylon mesh bags were removed from the flowers.

The result of this experiment provided clear support for nonrandom mating in the field population. Figure 7.16 shows that during the mixed pollen donor pollinations, pollen donor C1 (56.5%) sired a much greater proportion of seeds compared to pollen donors A1 (24.8%) and B1 (18.7%). This finding suggests that the nonrandom matings observed in prior greenhouse pollination studies were not an artifact of greenhouse conditions.

Additional work by Marshall and her colleagues (Marshall et al. 1996) suggests that competition between pollen grains may contribute to nonrandom mating in wild radish populations. They used three maternal plants in these crosses, which they crossed with seven pollen donors (A, B, C, D, E, F, Z). The maternal plants were pollinated with pollen from single donors and from pairs of donors. The paired pollinations (A+B, C+D, etc.) were done in two ways. In one set of experiments, the pollen from the two donors was mixed as in the previous experiments described earlier. Because the two pollen types were in physical contact with each other in these "mixed" pollinations, this method of pollinating increased the opportunity for interaction between pollen types. In the second set of experiments, the pollen of the two donors used was not mixed. Each was applied to adjacent halves of the stigma, the tip of the pistil that acts as a pollen-receptive area. Since the two pollen types did not contact each other in these "adjacent" pollinations, there was a reduced chance that they would interact. Pollen response to these conditions was measured as the percentage of pollen that germinated within 90 minutes of pollination. Reduced percentage of germination would indicate lower pollen responsiveness and the possibility of inhibition of pollen response, either through pollen to pollen interactions or through maternal tissue effects expressed through the stigma.

Some of the results of this experiment are shown in figure 7.17. The percentage of pollen that germinated after 90 minutes was essentially the same in the single donor and adjacent pollinations. Meanwhile, the rate of germination when pollen from the two donors was mixed was much reduced. This reduced germination, where pollen grains of different pollen donors were in contact with each other, indicates that interactions between pollen inhibited pollen germination. These results sug-

and Fuller 1994) designed a study to address this question. Why might nonrandom mating be limited to the greenhouse environment? Marshall and Fuller point out that the harsh and variable environments to which plants are exposed in nature might mean that the condition of the maternal plant may be of overwhelming importance in determining the amount of seed produced, the weight of seeds, and so forth. Under such conditions nonrandom pollination which produces differences in seed weight in the greenhouse, might be undetectable and biologically insignificant.

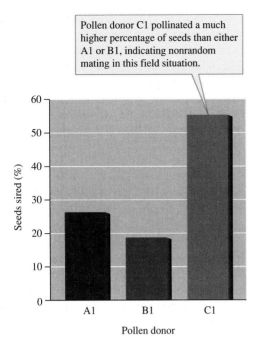

**Figure 7.16** Variation in wild radish pollen donor mating success in a field environment (data from Marshall and Fuller 1994).

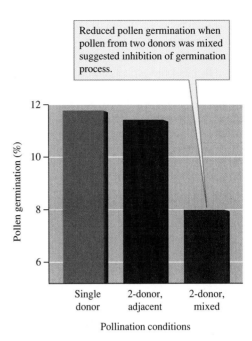

**Figure 7.17** Competition between pollen from different donors (data from Marshall et al. 1996).

gest **interference competition** among pollen grains, which usually involves some form of aggressive or inhibitory interactions between individuals.

Experiments such as this one are revealing the details of plant ecology. While ecological interactions between plants are often much less obvious than those of animals, careful and ingenious experiments such as those of Marshall and her colleagues are proving that they are every bit as rich and fascinating.

In this section we have seen how organisms as different as fish, insects, and plants compete for and select mates. While competition for mates may be intense, the vast majority of mature females in most populations mate and a large proportion of males may also mate. In populations that have evolved a high degree of sociality, however, the opportunities for mating are often restricted to relatively few individuals.

## CONCEPT DISCUSSION

### Sociality

**The evolution of sociality is generally accompanied by cooperative feeding, defense of the social group, and restricted reproductive opportunities.**

Chapters 4 through 6 focused on the ecology of individual organisms, mainly on how individuals solve environmental problems. Some of the problems we considered were how animals maintain a particular range of body temperatures in the face of much greater variation in environmental temperatures or how

plants sustain high rates of photosynthesis while avoiding excessive water loss. In the preceding parts of chapter 7 we've also concentrated on the ecology of individuals, examining how individuals choose mates. However, a fundamental change in relationships among individuals within a population takes place when individuals begin living in groups, such as colonies, herds, or schools and begin to cooperate with each other. Cooperation generally involves exchanges of resources between individuals or various forms of assistance, such as defense of the group against predators. Group living and cooperation signal the beginnings of **sociality.** The degree of sociality in a social species ranges from acts as simple as mutual grooming or group protection of young to highly complex, stratified societies such as those found in colonies of ants or termites. This more complex level of social behavior, which is considered to be the pinnacle of social evolution, is called **eusociality.** Eusociality is generally thought to include three major characteristics: (1) Individuals of more than one generation living together, (2) cooperative care of young, and (3) division of individuals into sterile, or nonreproductive, and reproductive castes.

Because individuals in social species often appear to have fewer opportunities to reproduce compared to individuals in nonsocial species, the evolution of sociality has drawn a great deal of attention from behavioral ecologists. The apparent restriction of reproductive opportunities that comes with sociality appears to challenge the idea that the fitness of an individual is determined by the number of offspring it produces. How does sociality challenge this concept of fitness? The challenge emerges from the observation that in many situations, individuals in social species do not reproduce themselves, while helping others in the population to reproduce. How can we explain such behavior that on first glance appears to be self-sacrificing? It can be argued that such behavior should be quickly eliminated from populations. However, since eusocial species such as bees and ants have survived for millions of years, behavioral ecologists have assumed that in some circumstances, the benefits of sociality must outweigh the costs.

Behavioral ecologists have assumed that the key to understanding the evolution of sociality will result from careful assessment of its costs and benefits. The ultimate goal of sociobiology has been a comprehensive theory capable of explaining the evolution of the various forms of sociality, particularly its most specialized form, eusociality. However, in our quest for such a theory, where should we begin the accounting of costs and benefits? David Ligon (1999) pointed the way when he wrote, "Most, if not all, of the important issues relevant to cooperative breeding systems are . . . related to the costs and benefits of sociality." Following Ligon's suggestion, the case histories begin with cooperative breeders.

# Cooperative Breeders

Species that live in groups often cooperate or help during the process of producing offspring. Help may include defending the territory or the young, preparing and maintaining a nest or den, or feeding young. Since the young which receive the care are not the offspring of the helpers, one of the most basic questions that we can ask about these breeding systems is why do helpers help? In other words, what benefits do helpers gain from their cooperation?

Sociobiologists have offered two main reasons. First, helpers may increase their own evolutionary, or genetic, fitness by improving the rates of survival and reproduction of relatives. Sociobiologists have suggested that investing resources, such as time or energy, in genetically related individuals that are not offspring (for instance, siblings, cousins, nieces, and so forth) may add to an individual's **inclusive fitness.** The concept of inclusive fitness, which was developed by William D. Hamilton (1964), proposes that an individual's inclusive, or overall, fitness is determined by its own survival and reproduction plus the survival and reproduction of individuals with whom the individual shares genes. Under some conditions individuals can increase their inclusive fitness by helping increase the survival and reproduction of genetic relatives that are not offspring. Because this help is given to relatives, or kin, the evolutionary force favoring such helping behavior is called **kin selection.** Hamilton proposed that selection will favor diverting resources to kin under conditions where its benefit to the helper, measured as improved survival and reproduction of kin, exceeds its cost to the helper. Benefit is scaled by the genetic relatedness of individuals.

The second reason offered to explain the evolution of cooperative breeding is that helping may improve the helper's own probability of successful reproduction. Because helping gives the helper experience in raising young, helping may increase the helper's chances of successfully raising young of its own and recruiting helpers of its own. In addition, where suitable breeding habitat is limited, helpers may have a better chance of inheriting the breeding territory from the reproductive individuals they help. Again, they are improving their chances of eventually raising their own young.

What sorts of species engage in cooperative breeding? Approximately 100 species of birds are cooperative breeders. In addition, several species of mammals such as wolves, wild dogs, and African lions engage in cooperative breeding. Let's review two intensely studied species where several benefits of cooperative breeding have been demonstrated.

## Green Woodhoopoes

We know a great deal about the cooperative breeding and general ecology of green woodhoopoes due to the pioneering, long-term studies of J. David Ligon and Sandra Ligon (Ligon and Ligon 1978, 1982, 1989, 1991). Adult green woodhoopoes, *Phoeniculus purpureus,* have reddish-orange bills and feet and black feathers with a metallic green and blue-purple sheen (fig. 7.18). Meanwhile, juvenile green woodhoopoes have black bill and feet, which allowed the Ligons to distinguish between mature and immature individuals in the field. Of the eight species of woodhoopoes, all of which are restricted to sub-Saharan Africa, the green woodhoopoe is the most common and widespread. Green woodhoopoes live in a wide variety of habitats at elevations from sea level to over 2,000 m. However, their most common habitat is open woodlands with trees large enough to provide cavities for nesting and roosting. For instance, the Ligon's long-term study site was located near Lake Naivasha in the central rift valley of Kenya in a woodland dominated by yellow-barked acacia.

Tree cavities keep the birds warm at night and provide some protection from predators. The habit of cavity roosting also makes green woodhoopoes ideal for field studies. To place unique color bands on green woodhoopoes in their study area, all the Ligons had to do was plug the opening to a roosting cavity after dark and then place a clear plastic bag over the opening in the morning to catch the woodhoopoes as they left the roost. Using this technique, they placed unique color bands on 386 green woodhoopoes. By closely studying the movements and interactions of banded individuals over a long period of time, the Ligons learned a great deal of the social relations of green woodhoopoes. For instance, they eventually knew the parentage of over 93% of the birds in their study area, the number and fates of offspring produced by each flock, and the identity of all breeders and nonbreeders in each flock. The results of this long-term study provide clues to the costs and benefits of cooperative breeding.

The Ligons found that green woodhoopoes live in territories that are occupied and defended by flocks of 2 to 16 individuals. Average flock size varied from approximately 4 to 6 over the course of their studies. Within a group only one pair breeds, while the remainder act as helpers. Males, which are approximately 20% larger than females, are particularly vigorous in their defense of breeding territories. The Ligons (1989) suggested that the larger body size of males is related to their intense competition with other males over territories and females. Territory defense is very important because territories appear to vary widely in quality.

**Figure 7.18** Studies of African green woodhoopoes have made major contributions to our understanding of the evolution of cooperative breeding among vertebrate animals.

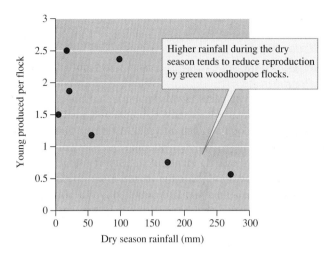

Higher rainfall during the dry season tends to reduce reproduction by green woodhoopoe flocks.

**Figure 7.19** Rainfall substantially influences the rate of reproduction by flocks of green woodhoopoes (data from Ligon and Ligon 1989).

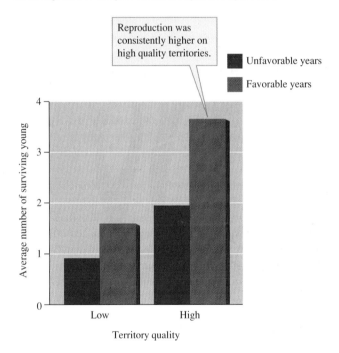

Reproduction was consistently higher on high quality territories.

Unfavorable years

Favorable years

**Figure 7.20** Relationship between territory quality and reproduction by green woodhoopoe flocks (data from Ligon and Ligon 1989).

One of the most obvious differences among territories is the quality of the cavities they contain. Cavity characteristics are very important because one of the major sources of mortality is predation while the birds are in their roosting or nesting cavities. The Ligons documented annual mortality rates of 30% for females and 40% for males, most of which was due to predation. Predators on nestlings include driver ants, hawks, and owls. Roosting adults are attacked at night by driver ants and large-spotted genets, small slender predators related to the mongoose. The vulnerability of nestlings and roosting adults to these predators depends on the characteristics of the cavity, especially its depth, the size of the opening, and the soundness of the wood.

Green woodhoopoes stay very close to their **natal territories,** that is the territories where they were raised. Out of 38 females that the Ligons banded as nestlings or fledglings and later observed breeding, 18 bred on their natal territory, 14 bred on an adjacent territory, and 6 only two or three territories away from their natal territory. Male dispersal is also limited. In other words, this population of green woodhoopoes shows a great deal of **philopatry.** Philopatry, which means literally "love of place," is a term that behavioral ecologists use to describe the tendency of some organisms to remain in the same area throughout their lives.

Why do green woodhoopoes stay at home and help raise young, which are close relatives, rather than disperse to produce their own offspring? The Ligons suggested that the major factor producing this high degree of philopatry is that roost cavities on which green woodhoopoes depend in the highlands of Kenya are scarce. By staying home a young green woodhoopoe gets a warm and relatively safe place to roost at night and may eventually inherit the territory and its cavities.

Over the course of their study, the Ligons found that 91% of females and 89% of males died without leaving any descendants. However, they also documented very high reproductive success among some woodhoopoes. Variation in reproductive success within the study area seemed to have two major sources, spatial and temporal variation. Year-to-year variation in breeding success was largely a result of variation in rainfall and its influence on the woodhoopoes food supply. The main food that

woodhoopoes give to nestlings are moth larvae that pupate in the soil and are sensitive to soil moisture. In general, rainfall during the savanna dry season kills these larvae and produces a food shortage. The result of rain during the dry season is usually reproductive failure among the woodhoopoes (fig. 7.19).

The second source of variation in reproductive success appears to be differences in territory quality. The Ligons found that territories fell into two clearly distinctive groups which they called high quality and low quality territories. Territory quality appeared to be mainly determined by the availability of roosting cavities capable of protecting the birds from predators. Figure 7.20 compares the average number of young produced per year on low quality and high quality territories. As you can see, the number of offspring produced on

| Information | |
|---|---|
| Questions | |
| Hypothesis | |
| Predictions | |
| Testing | ✓ |

*Investigating the Evidence*

# Scatter Plots and the Relationship Between Variables

Ecologists are often interested in the relationship between two variables, which we might call X and Y. For example, in chapter 7 we have reviewed studies of how the size of male scorpionflies (X) affects their mating success (Y). We also examined the influence of rainfall (X) on reproductive rate by green woodhoopoes (Y). One way of visualizing such relationships between two variables is with an X-Y scatter plot (e.g., see fig. 7.19). The scatter plot shown in figure 7.19 is one of an infinite number of possible relationships between two variables.

Let's consider just three of the possible relationships, which are shown in figure 1. The most basic scatter plot is one in which there is no relationship between X and Y. This situation is represented by figure 1*a*, where there is no correlation between values of X and values of Y. As a result, the scatter plot forms a more or less circular pattern. In contrast the pattern shown in figure 1*b*, which represents the situation where as X increases, Y decreases, follows an approximately linear pattern that slopes downward to the right. This pattern is similar to the plot of the relationship between rainfall and green woodhoopoe reproduction (see fig. 7.19). This type of relationship between X and Y is called a negative correlation. The opposite pattern, shown by figure 1*c*, is called a positive correlation. When two variables are positively correlated, increases in X are associated with increases in Y. For instance, increased body size in populations of pumas are correlated with increased size of the prey they eat (see fig. 6.16).

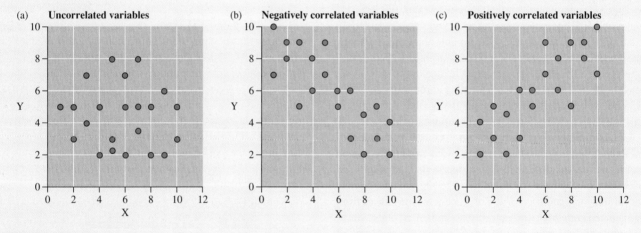

**Figure 1**  A scatter plot is a useful tool for exploring relationships between any two variables X and Y.

high quality territories is approximately twice as high as on low quality territories during both favorable and unfavorable years.

While the birds can do nothing about the chances of rainfall during the dry season, they can and do compete for territories. Those flocks that successfully compete for the best territories, have a clear reproductive advantage. So, returning to our original question, why do green woodhoopoes stay home and help? The first reason seems to be that by helping to raise and protect close relatives, the helpers may increase their inclusive fitness. The Ligons found that the bulk of the young tended by helpers ranged from half siblings to full siblings. We should keep in mind that a full sibling, on average, would share as many genes (50%) with the helper as its own son or daughter. The second and perhaps clearest potential benefit to a helper is that since

high quality territories are limited in number, the chance of inheriting the natal territory and advancing to breeding status may be greater than finding another suitable territory elsewhere.

What might we learn about the evolution of cooperative breeding from other species? Several cooperative species live in sub-Saharan Africa. For example, the African lion, a species that shares the same landscape with green woodhoopoes, also seems to be forced by a variety of environmental circumstances into a cooperative social system.

## African Lions

At about the same time that the Ligons were studying cooperative breeding among green woodhoopoes, Craig Packer and Anne E. Pusey were studying cooperation among African

**Figure 7.21** African lions are highly social predators.

lions in the Serengeti (Packer and Pusey 1982, 1983, 1997, Packer et al. 1991). Their studies have revealed a great deal of complexity in lion societies. Female lions live in groups of related individuals called prides (fig. 7.21). Prides of female lions generally include 3 to 6 adults but may contain as many as 18 or as few as 1. In addition to adult females, prides also include their dependent offspring and a coalition of adult males. Male coalitions may be made up of closely related individuals or of unrelated individuals.

Within lion society one can observe many forms of cooperation. Female lions nurse each other's cubs. They also cooperate when hunting large, difficult-to-kill game such as zebra and buffalo. In addition, females cooperatively defend their territory against encroaching females. However, the most critical form of cooperation among females is their group defense of the young against infanticidal males. These attacks on the young generally take place as a male coalition is displaced by another invading coalition. While a single female lion has little chance in a fight against a male lion, which are nearly 50% larger, cooperating females are often successful at repelling attacking males. Males, in turn, cooperate in defending the territory against invading males, which threaten the young they have sired, and against threats from other predators such as hyenas. The challenge for the behavioral ecologist has been to determine whether these various forms of cooperation can be reconciled with evolutionary theory.

Since the females in lion prides are always close relatives, their cooperative behavior can be readily explained within the conceptual framework of kin selection. As females cooperate

in nursing or defending young against males, they contribute to the growth and survival of their own offspring or to those of close kin. Cooperative hunting and sharing the kill also contribute to the welfare of offspring and close relatives. All these contributions add to the inclusive fitness of individual females.

In contrast, because male coalitions are sometimes made up of close relatives and sometimes not, cooperation within coalitions has represented a greater challenge to evolutionary theory. However, on close consideration Packer and colleagues (Packer et al. 1991) discovered that the rules associated with the formation and behavior of coalitions are consistent with predictions of evolutionary theory. Single males have virtually no chance of claiming and defending a pride of female lions. Therefore they must form coalitions with other males. This represents a type of ecological constraint on viable choices open to males. If males form a coalition with brothers and cousins, cooperative behavior that increases the production and survival of offspring of the coalition will increase an individual male's inclusive fitness. However, theoretically, a male within a coalition with unrelated males must produce some offspring of his own or he is merely increasing the fitness of others at the expense of his own fitness.

The first question we should ask is, Do all males within a coalition have an equal opportunity to reproduce? If all males within a coalition have an equal probability of reproducing, then forming coalitions with unrelated males is easier to reconcile with evolutionary theory. However, if there is significant variation in reproductive opportunities within coalitions, then cooperating with unrelated males is more difficult to reconcile with

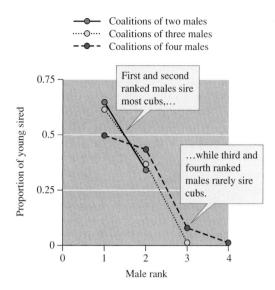

**Figure 7.22** Male lion rank and proportion of cubs sired in male coalitions of different sizes (data from Packer et al. 1991).

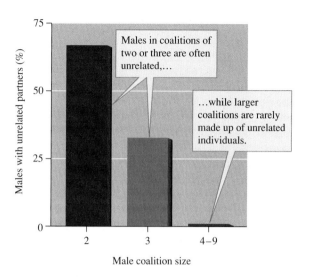

**Figure 7.23** Relatedness and size of male coalitions among African lions (data from Packer et al. 1991).

theories predicting that individuals will attempt to maximize their inclusive fitness. It turns out that the probability of a male siring young depends on his rank within a coalition and on coalition size. As shown in figure 7.22, males in coalitions of two sire a relatively similar proportion of the young produced by the pride. In addition, these proportions are close to the proportions sired by the two top ranked males in coalitions of three and four. However, the third ranked males in coalitions of three and the third and fourth ranked in coalitions of four sire almost no young lions. Packer and his team concluded from these data that variation in reproductive success is much higher in coalitions of three and four than in coalitions of two. In other words, the chance of reproducing is less evenly distributed among males in coalitions of three and four than in coalitions of two.

What implications do the results of Packer's studies have to the formation of coalitions containing unrelated individuals? One of the implications is that an unrelated male in a coalition of three or more runs the risk of investing time and energy in helping maintain a pride without an opportunity to reproduce himself and without improving his inclusive fitness since the other coalition members are not relatives. This result suggests that males should avoid joining larger coalitions of unrelated males, and this is just what Packer and his colleagues found (fig. 7.23). Figure 7.23 shows the percentage of males with unrelated partners in coalitions of different sizes. These patterns show clearly that males that team up with unrelated individuals mostly do so in coalitions of two or three. Larger coalitions of four to nine individuals are almost entirely made up of relatives. What are the implications of these data? They suggest that males avoid joining larger coalitions unless the coalition consists of relatives. Such a strategy avoids the risk of helping without gaining in inclusive fitness.

In summary, cooperation among green woodhoopoes and African lions appears to be a response to environmental conditions that require cooperation for success. In the case of green woodhoopoes, the scarcity of high quality territories and intense competition between flocks for those territories create conditions that favor staying in the natal territory and helping raise related young and perhaps inheriting the territory at a later date. Packer and Pusey (1997) captured the situation facing African lions in a fascinating article titled, "Divided We Fall: Cooperation Among Lions." To survive, reproduce, and successfully raise offspring to maturity, African lions must work in cooperative groups. The lone lion has no chance of meeting the ecological challenges presented by living on the Serengeti in lion society with its aggressive prides and invasive and infanticidal male coalitions. However, as we have seen, within the constraints set by their environments both green woodhoopoes and African lions appear to behave in a way that contributes positively to their overall fitness.

While the complexities of African lion and green woodhoopoe societies have taken decades to uncover, they pale beside the intricacies of life among eusocial species such as bees, termites, and ants. Let's explore eusociality in some animal populations to get some insights into the evolution of these complex social systems and to introduce the comparative method, one of the most valuable tools in evolutionary ecology.

## APPLICATIONS & TOOLS

### The Comparative Method—A Tool for Understanding Complex Evolutionary Histories

Behavioral ecologists are concerned with both how particular social systems work and with determining the mechanisms responsible for their evolution and maintenance. In most cases, however, the evolutionary origins of biological traits lie deep in the past and biologists cannot observe their evolution directly. So, how do scientists construct evolutionary hypotheses, test them, and eventually construct evolutionary theories? Many tools are used in such a process. We have already employed one of those tools, in a rudimentary way, without giving it a name. As we explored mate choice and sexual selection using guppies, scorpionflies, and wild radish as case histories to provide insights into the evolution of sociality, we were employing one of the most valuable tools available to evolutionary biologists. That tool is the **comparative method.**

The comparative method involves comparisons of the characteristics of different species or populations of organisms in a way that attempts to isolate a particular variable or characteristic of interest, such as sociality. Randy Thornhill (1984) suggested that in the ideal application of the comparative method, the influence of confounding, or confusing, variables on the variable of interest are randomized across the species or populations in the study. In his discussion of the comparative method, Thornhill reviewed its use to test whether or not **polygynous species,** species in which some males have several mates, show greater sexual dimorphism (males larger than females) than weakly polygynous or **monogamous species,** species in which males and females have a single mate. He pointed out that since the time of Darwin, who used the comparative method extensively, biologists have assumed that there is a connection between degree of sexual dimorphism and degree of polygyny. However, it was not until approximately a century after Darwin's work that evolutionary biologists carefully tested the idea using the comparative method. Their approach was to choose a wide variety of mammals that differed in their mating systems and degree of polygyny and then statistically test the relationship between the two variables.

When they did so, they found a significant positive relationship between the degree of sexual dimorphism and degree of polygyny among several groups of very different mammals, including hoofed mammals, primates, and seals and sea lions. It turned out that the most polygynous species in all groups showed the greatest degree of sexual dimorphism. Closely related species in these groups that differed in degree of polygyny also differed markedly in their degree of sexual dimorphism. In addition, distantly related species that have similar degree of polygyny, converged in their degree of sexual dimorphism. This convergence in dimorphism among distantly related taxa is analogous to the findings of Ligon and Ligon (1991) and Packer and Pusey (1997) that demonstrated that cooperative breeding by both green woodhoopoes and African lions is associated with significant environmental pressures that favor remaining in a group.

Let's review a remarkable case of convergence in social organization between a eusocial insect and a eusocial mammal. The main purpose of this comparison is to see the extent to which unrelated organisms can converge in biology and to suggest that such comparisons, the foundation of the comparative method, if quantified and replicated across many species, can help disentangle the evolution of complex characteristics, including the evolution of social systems.

**Figure 7.24** Leafcutter ants carrying leaf fragments back to the nest where they will be processed to create a substrate for growing the fungi that the ants eat. Smaller ants riding on leaf fragments offer protection from aerial attack by parasitoid flies.

# Eusocial Species

Probably the most thoroughly studied of eusocial species are the ants. Ants and their complex behaviors have attracted the attention of people from the earliest times and appear in the oldest writings such as the Bible and the classical writings of ancient Greece. Such written records were likely predated by older folktales. Bert Hölldobler and Edward O. Wilson (1990) pointed out that many of the earliest accounts of ants focused on ant species that make their living by harvesting seeds. These seed-harvesting species were serious agricultural pests around the Mediterranean Sea and their dependence on grains paralleled remarkably the economy of the human populations of the region.

In the intervening centuries since these earliest writings, we have learned much more about ants. Taxonomists have described nearly 9,000 species of ants, all belonging to the family Formicidae, which along with their relatives the wasps and bees, are members of the insect order Hymenoptera. Hölldobler and Wilson (1990) wrote a monumental summary of what was known about ants near the end of the twentieth century in a book titled simply *The Ants.* However, despite that book and the hundreds of studies done on ants since its publication, much is left to learn about this group of insects that Hölldobler and Wilson referred to as the "culmination of insect evolution."

One of the most socially complex groups of ants are the leafcutters (fig. 7.24). The 39 described species of leafcutter ants, which belong to two genera, are found only in the Americas, from the southern United States to Argentina. Leafcutter ants make their living by cutting and transporting leaf fragments to their nest, where the leaf material is fragmented and used as a substrate upon which to grow fungi. The fungi provide the primary food source for leafcutter ants.

Among the various species of leafcutter ants, some of the most thoroughly studied are species belonging to the genus *Atta. Atta* species live mainly in tropical Central and South America. However, at least two species reach as far north as Arizona and Louisiana in the United States. Leafcutter ants

**Figure 7.25** Naked mole rats live in colonies of closely related individuals ruled by a single dominant female, or queen, shown here resting on top of members of her colony.

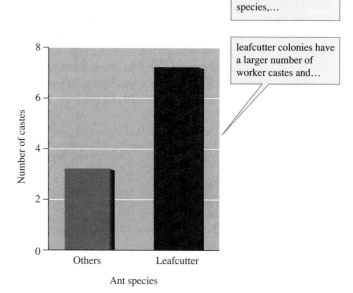

Compared to other ant species,...

leafcutter colonies have a larger number of worker castes and...

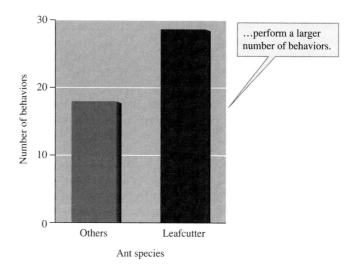

...perform a larger number of behaviors.

**Figure 7.26** Comparison of the number of castes and number of behaviors in a colony of leafcutter ants, *Atta sexdens,* and in colonies of three other ant species (data from Wilson 1980).

are important consumers in the tropical ecosystems, where they move large amounts of soil and process large quantities of leaf material in their nests. The nests of leafcutter ants can attain great size. For instance, the nests of *A. sexdens* can include over 1,000 entrance holes and nearly 2000 occupied and abandoned chambers. In one excavation of an *A. sexdens* nest (cited in Hölldobler and Wilson 1990), researchers estimated that the ants had moved more than 22 m$^3$ of soil, which weighed over 40,000 kg. Within this nest, the occupants had stored nearly 6,000 kg of leaves. Mature nests of *A. sexdens* contain a queen, various numbers of winged males and females which disperse to mate and found colonies elsewhere, and up to 5 to 8 million workers.

Though involving far fewer individuals, there are striking analogies between the organization of ant colonies and colonies of naked mole rats, *Heterocephalus glaber,* one of the few species of eusocial mammals (fig. 7.25). Despite their common name, naked mole rats are not completely naked and they are neither moles nor rats. Like moles, naked mole rats live underground but they are rodents not moles. However, the family of rodents to which they belong is more closely related to porcupines and chinchillas than to rats.

Naked mole rats live in underground colonies in the arid regions of Kenya, Somalia, and Ethiopia. Colonies often include 70 to 80 individuals but can sometimes contain as many as 250 individuals. The burrow system of a single colony of naked mole rats is extensive and can cover up to approximately 100,000 m$^2$, or about 20 football fields. Most of the digging required to maintain their large burrow systems is done with the naked mole rats' teeth and massive jaws. It turns out that the jaw muscles of naked mole rats make up about 25% of their entire muscle mass. This would be approximately equivalent to having muscles the size of those in your legs powering your jaws!

Both naked mole rats and leafcutter ants live in social groups in which individuals are divided among **castes** that engage in very different activities. We can define a caste as a group of physically distinctive individuals that engage in specialized behavior within the colony. E. O. Wilson (1980) studied

how labor is divided among castes of ants in a laboratory colony of *A. sexdens* that he established and studied over a period of 8 years. During this period, Wilson carefully cataloged the behaviors of individual colony members. Because the colony lived in a closed series of clear plastic containers, their behavior could be studied easily. In addition to recording behaviors, Wilson also estimated the sizes of individuals engaging in each behavior by measuring their head widths to the nearest 0.2 mm. He made his estimates visually by comparing an ant to a standard array of preserved *A. sexdens* specimens of known size.

When Wilson compared the leafcutter ant *A. sexdens* with three non-leafcutter ant species, he found that the leafcutter ants included a larger number of castes and engaged in a wider variety of behaviors (fig. 7.26). Wilson identified a total of 29 distinctive tasks performed by the leafcutter ants compared to an average of 17.7 tasks performed by the three other species. He found that the division of labor within the *A. sexdens* colony was mainly based on size. Possibly because of the large number of specialized tasks that need to be performed by leafcutter ants, they have one of the most complex social structures and one of the greatest size ranges found among the ants. Within *A. sexdens* colonies, the head width of the largest individuals (5.2 mm) is nearly nine times the head width of the smallest individuals (0.6 mm). On the basis of size, Wilson identified four castes within his leafcutter colony. However, because the tasks performed by some of the size classes change as they age, Wilson discovered three additional temporal or developmental castes for a total of seven castes within the colony, compared to an average of three castes in the non-leafcutter ant species he studied.

As a consequence of this great variation, someone watching a trail of leafcutter ants bring freshly cut leaf fragments back to their nest is treated to a rich display of size and behavioral diversity. While medium-sized ants carry the leaf fragments above their heads, the largest ants line the trail like sentries, guarding against ground attacks on the column of ants carrying leaf fragments. Very small ants ride on many of the leaf fragments, protecting the ant carrying a leaf fragment from aerial attacks by parasitic flies. Meanwhile, other size classes of leafcutters performing behaviors associated with processing leaves, tending larvae, and maintaining fungal gardens remain hidden in the nest. It was the activity of these smaller individuals that Wilson's laboratory colony was able to reveal so clearly.

Careful study has revealed some remarkable parallels in the structures of naked mole rat and leafcutter ant societies. The social behavior of naked mole rats was first reported by Jennifer Jarvis, professor at the University of Cape Town, South Africa, in a paper in the journal *Science* (Jarvis 1981). Her published study was based on more than 6 years of observation and experimentation with colonies of naked mole rats that she had established in the laboratory. Jarvis dug up a number of colonies and relocated them to a laboratory habitat analogous to that used by Wilson in his study of leafcutter ants. She waited approximately a year after bringing the naked mole rats to the laboratory before attempting to quantify their behaviors. Once this period of acclimation was over, Jarvis spent approximately 100 hours detailing how the members of her laboratory population of naked mole rats spent their time.

The picture of naked mole rat society that emerged from Jarvis' study was immediately intriguing to behavioral ecologists. The social organization of the colony appeared more similar to an ant colony than to any other mammal population known. Jarvis' paper in *Science* stimulated dozens of studies of naked mole rats and of related species. The results provide interesting insights into the evolution of social behavior. Within a colony of naked mole rats, one female and only a few males breed. This group of reproductive individuals functions basically as a queen and her mates, while all of the rest of the colony is nonreproductive. Behavioral ecologists have found that life in a naked mole rat colony centers on the queen and her offspring, and the queen's behavior appears to maintain this

focus. She is the most active member of the colony and literally pushes her way around the colony. By physically pushing individuals she appears to call them to action when there is work to be done or when the colony is threatened and needs defending. The aggressiveness of the queen also appears to maintain her dominance over other females in the colony and prevent them from coming into breeding condition. If the queen dies or is removed from the colony, one of the other females in the colony will assume the role of queen. If two or more females compete for the position of queen, they may fight to the death during the process of establishing the new social hierarchy.

In contrast to leafcutter ant colonies, where all workers are females, both males and females work in naked mole rat colonies. Jarvis found that work is divided among colony members, as in leafcutter ant colonies, according to size. However, in contrast to leafcutter ants, colonies of naked mole rats include only two worker size classes, small and large. Small workers are the most active. Small workers excavate tunnels, build the nest, which is deeper than most of the passage ways, and line the nest with plant materials for bedding. In addition, small workers also harvest food, mainly roots and tubers, and deliver it to other colony members, including the queen for feeding. Since they spend most of their time sleeping, the role of large nonbreeders was unclear for some time. However, eventually researchers working in the field were able to observe these large nonbreeders in action. It turns out that the large workers, as in ant colonies, are a caste specializing in defense. If the tunnel system is breached by members of another colony, the large nonbreeders move out quickly from their resting places to defend the colony from the invaders, literally throwing themselves into the breach. Eventually the large nonbreeders push up enough soil to wall off the intruders. However, they may be most important in defending against snakes, the most dangerous predators of naked mole rats. When confronted with a snake, the large nonbreeders will try to kill the snake or spray it with soil until it is driven off or buried.

## Evolution of Eusociality

Despite their distinctive evolutionary histories and other biological differences, the studies of Wilson and Jarvis suggest interesting parallels in the organizations of leafcutter ant and naked mole rat colonies (fig. 7.27). Similarities include division of labor within colonies based on size, with smaller workers specializing in foraging, nest maintenance, and excavation of extensive burrow systems. Meanwhile, larger workers in both species specialize in defense. In addition, reproduction in both species is limited to a single queen and her mates. These areas of convergence in social organization between such different organisms may help shed light on the forces responsible for the evolution of eusociality. Such comparisons form the basis of the comparative method.

What factors may have been important in the evolution and maintenance of naked mole rat and leafcutter ant sociality? Kin selection may play a role. Leafcutter ants along with

Division of labor in both
leafcutter ant colonies and naked
mole rat colonies is based on size.

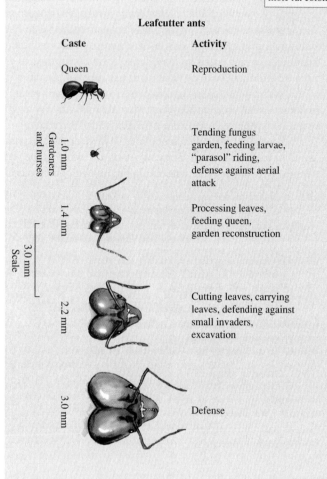

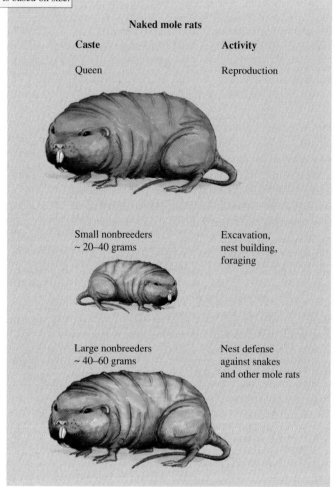

**Figure 7.27** Division of labor among castes of leafcutter ants, *Atta sexdens,* and naked mole rats, *Heterocephalus glaber.* Ant sizes are head widths of workers typically engaged in each activity (data from Wilson 1980, Jarvis 1981, and Sherman, Jarvis, and Braude 1992).

other Hymenoptera, such as bees and wasps, have an inheritance system called **haplodiploidy.** The term haplodiploid refers to the number of chromosome sets possessed by males and females. In haplodiploid systems males develop from unfertilized eggs and are haploid, while females develop from fertilized eggs and so are diploid. One of the consequences of haplodiploidy is that worker ants within a colony can be very similar genetically. In an ant colony where there is a single queen that mated with a single male, the workers will be more related to each other than they would be to their own offspring. W. D. Hamilton (1964) was the first to point out that under these conditions the average genetic similarity among workers would be 75%, while their relationship with any offspring they might produce would be 50% (fig. 7.28).

What is the source of this high degree of relatedness? The queen mates only during her mating flight and stores the sperm she receives to fertilize all the eggs she lays to produce daughters. If she mates with a single male, since he is haploid, all her daughters will receive the same genetic information from their male parent. As a consequence, the 50% of the genetic makeup that workers receive from their male parent will be identical. In addition, workers will share an average of 25% of their genes through those that they receive from the queen, yielding an average genetic relatedness of 50% + 25% = 75%. The important point here is that the activity of workers promotes the production of closely related individuals, their sisters, an activity that should be favored by kin selection.

Because naked mole rat colonies are relatively closed to outsiders, the individuals within each colony, like the workers within leafcutter ant colonies, are also very similar genetically. Paul Sherman, Jennifer Jarvis, and Stanton Braude (1992) reported that approximately 85% of matings within a colony of naked mole rats are between parents and offspring or between siblings. As a consequence of these matings between close relatives, the relatedness between individuals within a colony is about 81%, suggesting that kin selection may be involved in the maintenance of nonreproductive helpers in colonies of naked mole rats.

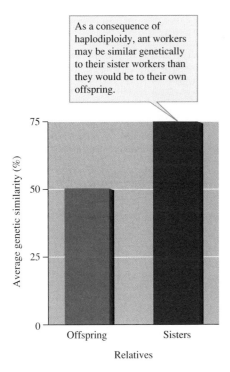

**Figure 7.28**  Average genetic similarity between ant workers and sisters (other workers) versus their offspring if they reproduced.

What factors other than kin selection may have contributed to the evolution of eusociality? Many factors have been implicated. While researchers working on ants and other social Hymenoptera have emphasized the potential importance of kin selection, studies of cooperative-breeding vertebrate species have emphasized ecological constraints. What sorts of ecological common constraints are faced by leafcutter ants and naked mole rats? One of the most obvious is the work associated with the creation, maintenance, and defense of extensive burrow systems. The more social organisms are studied, the less likely it has become that one or a few simple mechanisms will be adequate to explain their evolution. However, the results of studies such as those of Wilson and Jarvis should encourage continued careful comparative studies as a means for eventual understanding of the evolution of sociality. In the application of the comparative method, species such as the leafcutter ant *A. sexdens* and naked mole rats function as invaluable tools.

## SUMMARY

Social relations are important since they often directly and obviously impact the reproductive contribution of individuals to future generations, a key component of evolutionary fitness, the number of offspring, or genes, contributed by an individual to future generations. One of the most fundamental social interactions between individuals takes place during sexual reproduction.

**Mate choice by one sex and/or competition for mates among individuals of the same sex can result in selection for particular traits in individuals, a process called sexual selection.** Sexual selection results from differences in reproductive rates among individuals as a result of differences in their mating success. Sexual selection is thought to work either through intrasexual selection, where individuals of one sex compete with each other for mates, or intersexual selection, when members of one sex consistently choose mates from among members of the opposite sex on the basis of some particular trait.

Experimental evidence supports the hypothesis that the coloration of male guppies in local populations is determined by a dynamic interplay between natural selection exerted by predators, under which less-colorful males have higher survival, and by female mate choice, which results in higher mating success by more-colorful males. Among scorpionflies,

larger males are more likely to successfully defend available arthropod offerings due to their advantages in aggressive encounters and consequently mate more frequently than smaller males without arthropod offerings. Studies of mating in the wild radish, *Raphanus sativus,* in greenhouse and field experiments indicate nonrandom mating and suggest interference competition among pollen from different pollen donors.

**The evolution of sociality is generally accompanied by cooperative feeding, defense of the social group, and restricted reproductive opportunities.** The degree of sociality in a social species ranges from acts as simple as mutual grooming or group protection of young to highly complex, stratified societies such as those found in colonies of ants or termites. This more complex level of social behavior, which is considered to be the pinnacle of social evolution, is called eusociality. Eusociality is generally thought to include three major characteristics: (1) individuals of more than one generation living together, (2) cooperative care of young, and (3) division of individuals into sterile, or nonreproductive, and reproductive castes.

Cooperation among green woodhoopoes and African lions appears to be a response to environmental conditions that require cooperation for success. For green woodhoopoes, the scarcity of high quality territories and intense competition between flocks for those territories create conditions that favor staying in the natal territory and helping raise related young and perhaps inheriting the territory at a later date. To survive, reproduce, and successfully raise offspring to maturity, African lions must work in cooperative groups of females, which are called prides, and of males, which are called coalitions.

One of the most valuable tools available to evolutionary biologists is the comparative method. The comparative method examines the characteristics of different species or populations of organisms in a way that attempts to isolate a particular variable or characteristic of interest, such as sociality, while randomizing the influence of confounding, or confusing, variables on the variable of interest. The comparative method has been used to study the evolution of eusociality among a wide variety of animal species including leafcutter ants and naked mole rats, both of which live in social groups in which individuals are divided among castes that engage in very different activities. Compared to other ant species, leafcutter ant colonies have a larger number of castes that engage in a wider variety of behaviors. In contrast to leafcutter ant colonies, where all workers are females, both males and females work in naked mole rat colonies. However, as in leafcutter ant colonies, work in naked mole rat colonies is divided among members according to their size. Many factors have likely contributed to the evolution of eusociality in leafcutter ants and naked mole rats, including kin selection and ecological constraints.

# Review Questions

1. The introduction to chapter 7 included sketches of the behavior and social systems of several fish species. Using the concepts that you have learned in the chapter, revisit those examples and predict the forms of sexual selection occurring in each species.

2. One of the basic assumptions of the material presented in chapter 7 is that the form of reproduction will exert substantial influence on social interactions within a species. How might interactions differ in populations that reproduce asexually versus ones that engage in sexual reproduction? How might having separate sexes versus hermaphrodites affect the types of social interactions within a population? How should having several forms of one sex, for example, large and small males, influence the diversity of behavioral interactions within the population?

3. Endler (1980) pointed out that though field observations are consistent with the hypothesis that predators may exert natural selection on guppy coloration, some other factors in the environment could be affecting variation in male color patterns among guppy populations. What other factors, especially physical and chemical factors, might affect male colors and should each influence male color?

4. Endler set up two experiments, one in the greenhouse and one in the field. What were the advantages of the greenhouse experiments? What were the shortcomings of the greenhouse experiments? Endler also set up field experiments along the Aripo River. What were the advantages of the field experiments and what were their shortcomings?

5. Examine figure 7.8. While most of the male guppies that successfully mated were dominant, a substantial proportion of attractive males that sired broods were subordinate. How might we interpret this reproductive success by attractive but subordinate males? What might these results indicate about the potential influence of female choice on mating success among male guppies?

6. Using the studies of Kodric-Brown and Thornhill, compare guppy and scorpionfly mating systems. Pay particular attention to the potential roles of intersexual and intrasexual selection in each species. What are the similarities between the two species? What are some apparent differences?

7. The results of numerous studies indicate nonrandom mating among plants at least under some conditions. These results lead to questions concerning the biological mechanisms that produce these nonrandom matings. How might the maternal plant control or at least influence the paternity of her seeds? What role might competition between pollen determine in the nonrandom patterns observed?

8. The details of experimental design are critical for determining the success or failure of both field and laboratory experiments. Results often depend on some small details. For instance, why did Jennifer Jarvis wait one year after establishing her laboratory colony of naked mole rats before attempting to quantify the behavior of the laboratory population? What might have been the consequence of beginning to quantify the behavior of the colony soon after it was established?

9. Behavioral ecologists have argued that naked mole rats are eusocial. What are the major characteristics of eusociality and which of those characteristics are shared by naked mole rats?

10. Choose a problem in the ecology of social relations, formulate a hypothesis, and design a study to test your hypothesis. Take two approaches. In one approach use field and laboratory experiments to test your ideas. In the second design a study that will employ the comparative method.

# Suggested Readings

Bonduriansky, R. 2001. The evolution of male mate choice in insects: a synthesis of ideas and evidence. *Biological Reviews* 76:305–39.

*An interesting review of mate choice among insects from the perspective of males.*

Brooks, R. 2002. Variation in female mate choice within guppy populations: population divergence, multiple ornaments and the maintenance of polymorphism. *Genetica* 116:343–58.

Brooks, R. and J. A. Endler. 2001. Female guppies agree to differ: phenotypic and genetic variation in mate-choice behavior and the consequences for sexual selection. *Evolution* 55:1644–55.

Gamble, S., A. K. Lindholm, J. A. Endler, and R. Brooks. 2003. Environmental variation and the maintenance of polymorphism: the effect of ambient light spectrum on mating behaviour and sexual selection in guppies. *Ecology Letters* 6:463–72.

Godin, J. G. J. and H. E. McDonough. 2003. Predator preference for brightly colored males in the guppy: a viability cost for a sexually selected trait. *Behavioral Ecology* 14:194–200.

Kodric-Brown, A. and P. F. Nicoletto. 2001. Age and experience affect female choice in the guppy (*Poecilia reticulata*). *American Naturalist* 157:316–23.

*This set of five papers gives a thorough update of research on mate selection in guppies.*

Choe, J. C. and B. J. Crespi. 1997. *The Evolution of Social Behavior in Insects and Arachnids.* Cambridge: Cambridge University Press.

*Excellent compendium of social behavior among a wide variety of species.*

Clutton-Brock, T. 2002. Breeding together: kin selection and mutualism in cooperative vertebrates. *Science* 296:69–72.

*A broad overview of cooperative behavior among vertebrate animals.*

Courchamp, F. and D. W. Macdonald. 2001. Crucial importance of pack size in the African wild dog *Lycaon pictus. Animal Conservation* 4:169–74.

*A detailed study of the social dynamics of African wild dogs from a conservation standpoint.*

Duffy J. E., C. L. Morrison., and R. Rios. 2000. Multiple origins of eusociality among sponge-dwelling shrimps (*Synalpheus*). *Evolution* 54:503–16.

*Fascinating analysis of the remarkable discovery of eusociality among marine crustaceans.*

Hölldobler, B. and E. O. Wilson. 1990. *The Ants.* Cambridge, Mass.: Harvard University Press.

*A landmark book which presents the biology of ants in great detail. The many high quality original illustrations make this as much a work of art as a biological monograph.*

Houde, A. E. 1997. *Sex, Color, and Mate Choice in Guppies.* Princeton, N. J.: Princeton University Press.

*A detailed and readable account of one of the most important vertebrate "models" for the study of animal behavior.*

Marshall, D. L. and P. K. Diggle. 2001. Mechanisms of differential pollen donor performance in wild radish, *Raphanus sativus* (Brassicaceae). *American Journal of Botany* 88:242–57.

Marshall, D. L. and D. M. Oliveras. 2001. Does differential seed string success change over time or with pollination history in wild radish, *Raphanus sativus* (Brassicaceae)?

Skogsmyr, I. and A. Lankinen. 2000. Potential selection for female choice in *Viola tricolor. Evolutionary Ecology Research* 2:965–79.

*Detailed recent studies of mating ecology of two species of plants.*

Packer, C. and A. E. Pusey. 1997. Divided we fall: cooperation among lions. *Scientific American* 276(5):52–59.

*Very interesting summary of Packer and Pusey's long-term studies of African lions.*

Sherman, P.W., J. U. M. Jarvis, and S. H. Braude. 1992. Naked mole rats. *Scientific American* 257(8):72–78.

*Introduction to naked mole rat behavior and ecology, with its notable similarity to the eusocial insects.*

# On the Net

Visit this textbook's accompanying website at www.mhhe.com/ecology (click on the book's title) to take advantage of practice quizzing, study/writing tips, timely news articles, and additional URLs for research on the topics in this chapter.

Animal Behavior
Sociobiology
Natural Selection

Flowers and Fruits
Flowers

# SECTION III
## Population Ecology

*"All the flowers of all the tomorrows are in the seeds of today."*

Chinese proverb

# Population Genetics and Natural Selection

arwin's theory of evolution by natural selection, the unifying concept of modern biology, was crystallized by his observations in the Galápagos Islands. In mid October of 1835 under a bright equatorial sun, a small boat moved slowly from the shore of a volcanic island to a waiting ship. The boat carried a young naturalist who had just completed a month of exploring the group of islands known as the Galápagos, which lie on the equator approximately 1,000 km west of the South American mainland (fig. 8.1). As the seamen rowed into the oncoming waves, the naturalist, Charles Darwin, mused over what he had found on the island. His observations had confirmed expectations built on information gathered earlier on the other islands he had visited in the archipelago. Later Darwin recorded his thoughts in his journal which he later published (Darwin 1839), "The distribution of the tenants of this archipelago would not be nearly so wonderful, if, for instance, one island had a mocking-thrush, and a second island some other quite distinct genus— if one island had its genus of lizard and a second island another distinct genus, or none whatever. . . . But it is the circumstance, that several of the islands possess their own species of the tortoise, mocking-thrush, finches, and numerous plants, these species having the same general habits, occupying analogous situations, and obviously filling the same place in the natural economy of this archipelago, *that strikes me with wonder* [emphasis added].

Darwin wondered at the sources of the differences among clearly related populations and attempted to explain the origin of these differences. He would later conclude that these populations were descended from common ancestors whose descendants had changed after reaching each of the islands. The ship to which the seamen rowed was the H.M.S. *Beagle,* halfway through a voyage around the world. The main objective of the *Beagle*'s mission, charting the coasts of southern South America would be largely forgotten, while the thoughts of the young Charles Darwin would eventually develop into one of the most significant theories in the history of science. Darwin's wondering, carefully organized and supported by a lifetime of observation, would become the theory of evolution by natural selection, a theory that would transform the prevailing scientific view of life on earth and rebuild the foundations of biology.

Darwin left the Galápagos Islands convinced that the various populations on the islands were gradually modified from their ancestral forms. In other words, Darwin concluded that the island populations had undergone a process of **evolution,** a process that changes populations of organisms over time. Though Darwin left the Galápagos convinced that the island populations had evolved, he had no mechanism to explain the evolutionary changes that he was convinced they had undergone. However, a plausible mechanism to produce evolutionary change in populations came to Darwin almost exactly 3 years after his taking leave of the Galápagos Islands. In October of 1838 while reading the essay on populations by Thomas Malthus, Darwin was convinced that during competition for limited resources, such as food or space, among individuals

**Figure 8.1** On the Galápagos Islands Charles Darwin encountered many examples of readily observed plants and animal species that differed physically from one island to another island. Here a Galápagos hawk lands on a giant tortoise for which the islands are named.

within populations, some individuals would have a competitive advantage. He proposed that the characteristics producing that advantage would be "preserved" and the unfavorable characteristics of other individuals would be "destroyed." As a consequence of this process of selection by the environment, populations would change over time. With this mechanism for change in hand, Darwin sketched out the first draft of his theory of natural selection in 1842. However, it would take him many years and many drafts before he honed the theory to its final form and amassed sufficient supporting information. Darwin's theory of **natural selection** can be summarized as follows:

1. Organisms beget like organisms. (Offspring appear, behave, function, and so forth like their parents.)

2. There are chance variations between individuals in a species. Some variations (differences among parents) are heritable (are passed on to offspring).

3. More offspring are produced each generation than can be supported by the environment.

4. Some individuals, because of their physical or behavioral traits, have a higher chance of surviving and reproducing than other individuals in the same population.

Darwin (1859) proposed that differential survival and reproduction of individuals would produce changes in species populations over time. That is, the environment acting on variation among individuals in populations would result in **adaptation** of the population to the environment. He now had a mechanism to explain the differences among populations that he had observed on the Galápagos Islands. Still, Darwin was keenly aware of a major insufficiency in his theory. The theory of natural selection depended upon the passage of "advantageous" characteristics from one generation to the next. The problem was that the mechanisms of inheritance were unknown in Darwin's time. In addition, the prevailing idea at the time, blending inheritance, suggested that rare traits, no matter how favorable, would be blended out of a population, preventing change as a consequence.

Darwin worked for nearly half a century to uncover the laws of inheritance. However, he did not. To do so required a facility with mathematics that Darwin had not developed. In a short autobiography, Darwin himself (1859) remarked, "I attempted mathematics, and even went during the summer of 1828 with a private tutor . . . but I got on very slowly. The work was repugnant to me, chiefly from my not being able to see any meaning in the early steps in algebra. This impatience was very foolish, and in after years I have deeply regretted that I did not proceed far enough at least to understand something of the great leading principles of mathematics, *for men thus endowed seem to have an extra sense*" [emphasis added].

As Darwin explored the Galápagos Islands, halfway around the world in central Europe a schoolboy named Johann Mendel was studying under difficult conditions and developing the facility with mathematics necessary to complete Darwin's theory of natural selection. At thirteen, Johann was half Darwin's age, yet he had already set a course for a life of study which he followed as resolutely as the crew of the *Beagle* on their voyage around the world. At the end of his scientific voyage, Mendel would uncover the basic mechanisms of inheritance.

Mendel was the oldest child of a family that farmed a small landholding near Brno, a town in what is now the Czech Republic. He would have had little schooling if it were not for the philanthropy of the countess Walpurga Truchsess-Zeil who ruled the district in which Mendel's family lived. The countess had a standing order to her advisors that they should identify all of the promising boys and girls living within her domain and send them to school, where she paid their room and board. Mendel had been one of those children. The countess was more than a philanthropist, however. She also paid attention to details, including the curriculum of her school, which she specified should include the natural sciences. Thus from the outset, Mendel's studies included a firm grounding in the sciences. A countess with foresight, intelligence, and heart and her perceptive advisors had discovered an intellectual treasure and provided for the blossoming of one of biology's great geniuses.

Johann would be renamed Gregor Mendel when he joined the Augustinian order of monks that maintained a monastery near his birthplace. In a garden within the walls of the abbey, Mendel would discover what Darwin's around-the-world voyage would not reveal. The two keys to Mendel's discoveries would be excellent training in mathematics and physics from which he derived a sense of quantitative relationships and the power of experimental approaches to the study of the natural world.

What did Mendel discover? Briefly he discovered what we now call "Mendelian genetics," including the very fundamental concept of particulate inheritance. That is the concept that characteristics pass from parent to offspring in the form of discrete packets of information that we now call genes. Mendel also determined that genes come in alternative forms, which we term **alleles.** For instance, Mendel worked with alleles such as round versus wrinkled seeds and tall versus short plants. In addition, he found that some alleles prevent

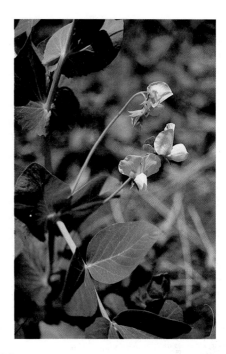

**Figure 8.2** Garden pea plant in flower. Because the garden pea normally self-pollinates, Mendel could keep track of and control mating in his study plants.

the expression of other alleles. We call such alleles "dominant" and the alleles that they suppress "recessive." Mendel's work also revealed the distinction between genotype and phenotype and the difference between homozygous and heterozygous genotypes. Mendel's work, which revealed still other aspects of the laws of inheritance, laid a solid foundation for the science of genetics.

How did Mendel succeed, while so many others had failed? The sources of his success can be traced to his education and his own special genius. Mendel's education at the University of Vienna exposed him to some of the best minds working in the physical sciences and to an approach to science that emphasized experimentation. His introduction to the physical sciences included a solid foundation in mathematics, including probability and statistics. As a consequence, Mendel could quantify the results of his experimental research.

Mendel chose to work with plants which could be maintained in the abbey garden. His most famous and influential work was done on the garden pea, *Pisum sativum,* that has many desirable traits (fig. 8.2). Many domestic varieties of peas, which showed a great deal of physical variation, with its attendant underlying genetic variation, were available to Mendel. However, he subjected the phenotypes of his study organisms to careful analysis. Rather than treat the phenotype as a whole, Mendel subdivided the organism into a set of manageable characteristics such as seed form, stem length, and so forth, which it turned out were controlled by individual genes. This analytical perspective of his study organisms was probably another legacy of his training in the physical sciences. Finally, to his excellent education and genius, Mendel added a lot of hard work and perseverance. For a full discussion of Mendel, his work and its ongoing analysis, including controversial aspects of the work, see the excellent biography by Orel (1996).

Darwin and Mendel complemented each other perfectly and their twin visions of the natural world revolutionized biology. The synthesis of the theory of natural selection and genetics gave rise to modern evolutionary ecology, a very broad field of study. Here we examine five major concepts within that broad discipline.

## CONCEPTS

- Phenotypic variation among individuals in a population results from the combined effects of genes and environment.
- The Hardy-Weinberg equilibrium model helps identify evolutionary forces that can change gene frequencies in populations.
- Natural selection is the result of differences in survival and reproduction among phenotypes.
- The extent to which phenotypic variation is due to genetic variation determines the potential for evolution by natural selection.
- Random processes, such as genetic drift, can change gene frequencies in populations, especially in small populations.

## CONCEPT DISCUSSION

### Variation Within Populations

**Phenotypic variation among individuals in a population results from the combined effects of genes and environment.**

Because phenotypic variation is the substrate upon which the environment acts during the process of natural selection, determining the extent and sources of variation within populations is one of the most fundamental considerations in evolutionary studies. The following examples review variation in representative plant and animal populations and some of the early methods used to uncover that variation.

## Variation in Plant Populations

Darwin's theory of natural selection sparked a revolution in thinking among biologists, who responded almost immediately by studying variation among organisms in all sorts of

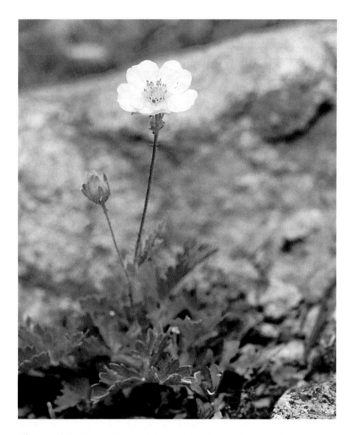

**Figure 8.3** *Potentilla glandulosa,* sticky cinquefoil, grows from sea level to over 3,000 m elevation and shows remarkable morphological variation along this elevational gradient.

environments. The first of these biologists to conduct truly thorough studies of variation and to incorporate experimentation in their studies, focused on plants.

## Phenotypic and Genetic Variation in *Potentilla glandulosa*

Jens Clausen, David Keck, and William Hiesey, who worked at Stanford University in California, conducted some of the most widely cited studies of plant variations. Their studies provided deep insights into the extent and sources of morphological variation in plant populations, including both the influence of environment and genetics. Though this research group and its successors studied nearly 200 species, it is best known for its work on *Potentilla glandulosa* or sticky cinquefoil (fig. 8.3) (Clausen, Keck, and Hiesey 1940).

Clausen and his research team worked with clones of several populations of *P. glandulosa,* which they grew in three main experimental gardens—one at Stanford near the coast at an elevation of 30 m, another in a montane environment at Mather at an elevation of 1,400 m in the Sierra Nevada, and a third garden in an alpine environment at Timberline at 3,050 m. By cloning lowland, mid-elevation, and alpine plants and growing them in experimental gardens, Clausen, Keck, and Hiesey established experimental condi-

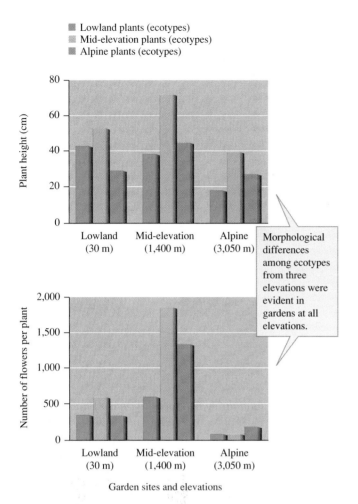

**Figure 8.4** Growth responses by *Potentilla glandulosa* grown at three elevations (data from Clausen, Keck, and Hiesey 1940).

Other information indicates that genetic differences among the plants are associated with adaptation to the environments of the native elevation. It is clear that lowland ecotypes of *P. glandulosa* are excluded from the alpine. Most died during their first winter in the alpine garden and those that survived flowered during the following summer but their fruits did not mature. Mid-elevation *P. glandulosa* also survived poorly in the alpine and their fruits often failed to mature. Alpine plants showed the opposite trends. They had poor survival in the lowland garden and went dormant in winter, while the lowland plants remained active. In summary, the experiments of Clausen, Keck, and Hiesey demonstrated genetic differences among populations and adaptation to their natural environments. Ecologists call such locally adapted and genetically distinctive populations within a species **ecotypes.** Applying this term then, we can conclude that the lowland, mid-elevation, and alpine populations studied by Clausen, Keck, and Hiesey were ecotypes. Using transplant and common garden approaches ecologists have learned a great deal about genetic variation among and within plant populations. These classical approaches combined with modern molecular techniques are rapidly increasing our knowledge of genetic variation in natural populations.

## Combining Molecular and Morphological Information

In the previous classic example, Clausen, Keck, and Hiesey used differences in growth form of *P. glandulosa* plants grown in common gardens to infer genetic differences among populations of this species. More recently, Kjell Hansen, Reidar Elven, and Christian Brochmann combined molecular and morphological techniques to explore genetic variation in populations of *Potentilla* species living on Spitsbergen Island in the high Arctic. Spitsbergen is the largest island in the Svalbard archipelago, a far northern part of the country of Norway (fig. 8.5).

Hansen, Elven, and Brochmann (2000) were interested in whether a combination of genetic and morphological information could help them understand the complex variation of the type seen in *Potentilla* species (see fig. 8.4). One of the questions addressed by this team from the Botanical Garden and Museum of the University of Oslo, Norway concerned a group of forms known as the *Potentilla nivea* complex. Based on morphological evidence, the complex had been divided into three species: *P. chamissonis, P. insularis,* and *P. nivea.* A second problem addressed by the researchers was whether three distinctive forms of *P. pulchella* should be recognized as different taxa, perhaps varieties or subspecies, within *P. pulchella.* The typical form of *P. pulchella* is large and hairy and grows in a variety of habitats, including cliffs, in cliff meadows where seabirds deposit significant quantities of feces, and on ridges. A second form of *P. pulchella,* which is small and lacks abundant hairs, grows on gravel terraces along shorelines. The third form is small and hairy and grows on silty shoreline terraces.

tions that could reveal potential genetic differences among populations. In addition, because they studied the responses of plants from all populations to environmental conditions in lowland, mid-elevation, and alpine gardens, their experiment could demonstrate adaptation by *P. glandulosa* populations to local environmental conditions.

Two responses of *P. glandulosa* to environmental conditions at the three common garden sites are summarized in figure 8.4. Plant height differed significantly among the study sites, which shows an environmental effect on plant morphology, but the lowland, mid-elevation and alpine plants responded differently to the three environments. For instance, while the mid-elevation and alpine plants attained their greatest height in the mid-elevation garden, the lowland plants grew the tallest in the lowland garden. In the gardens corresponding to their natural elevation, the mid-elevation and alpine plants produced more flowers than the other two ecotypes. The lowland ecotype, in contrast, did not produce the most flowers in any of the experimental gardens. These differences in response by different ecotypes indicate genetic differences among populations of *P. glandulosa.*

**Figure 8.5** High above the Arctic Circle, the island of Spitzbergen presents an extreme environment for terrestrial plants.

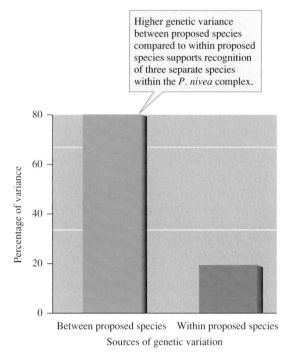

Higher genetic variance between proposed species compared to within proposed species supports recognition of three separate species within the *P. nivea* complex.

**Figure 8.6** Sources of genetic variance between and among proposed species within the *Potentilla nivea* complex.

Hansen, Elven, and Brochmann sampled 17 populations of *Potentilla*. The area on Spitsbergen where Hansen, Elven, and Brochmann collected *Potentilla* extended from about 78° to over 79° N latitude. Ten of these populations were of the *P. nivea* complex and seven were populations of *P. pulchella*. Using these collections, Hansen, Elven, and Brochmann studied 64 morphological characters of 146 plants and they did genetic analyses of 136 plants. Genetic analyses were done using the randomly amplified polymorphic DNA, or RAPD, method (see Applications & Tools). Again, the question addressed by the researchers was whether genetic information combined with morphology would support the earlier recognition of three species within the *P. nivea* complex and the subdivision of *P. pulchella* into three different taxa.

The results of this study demonstrate the utility of joining morphological information with genetic information. The RAPD method identified three genetically distinct groups of plants within the *P. nivea* complex, which we can call "RAPD phenotypes." Significantly, most of the genetic variance within the *P. nivea* complex was due to variation between the proposed species, while much less was due to variation within each of the proposed species (fig. 8.6). The three RAPD phenotypes were also separated clearly on the basis of several morphological characters. It turned out that the separation of plants achieved by Hansen, Elven, and Brochmann, which was based on combined genetic and morphological data, corresponded precisely to the three previously proposed species: *P. chamissonis, P. insularis,* and *P. nivea.* These results support the continued recognition of these taxa.

In contrast, the results of the study did not support recognizing the three morphologically distinctive forms of *P. pulchella* as separate taxa. Despite their substantial morphologi-

cal differences, the most common RAPD phenotype was observed in all three forms of *P. pulchella*. From this result, Hansen, Elven, and Brochmann concluded that the morphologically distinctive forms in *P. pulchella* result from plastic growth responses to local environments or perhaps are due to the effects of a small number of genes. As a consequence, the researchers concluded that the three forms of *P. pulchella* should not be recognized as separate taxa.

The ability of researchers to study the genes of organisms directly has revolutionized evolutionary and ecological studies. However, the older experimental garden approaches remain essential for answering some types of scientific questions, particularly in studies of plants. As the following example shows, however, these approaches have also been used successfully by ecologists studying animal populations.

## Variation in Animal Populations

Studies of phenotypic and genetic variation among animal populations are usually more difficult than similar studies of plant populations. However, the chuckwalla, *Sauromalus obesus,* a large herbivorous lizard of the southwestern United States and northwestern Mexico (fig. 8.7) has been studied almost as thoroughly as some of the plant species just discussed. *Sauromalus* prefers to feed on annual forbs and grasses but will feed on the leaves of shrubs if its preferred and more nutritious foods are not available. Though it grows most rapidly when young, the species continues growing throughout life, reaching a body length of over 220 mm (excluding the tail) and a mass of about 400 g.

**Figure 8.7** A chuckwalla, *Sauromalus obesus*. Chuckwallas are large herbivorous lizards living in the southwestern United States and northwestern Mexico.

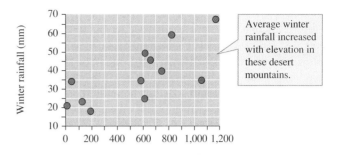

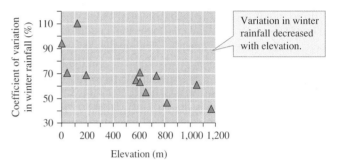

**Figure 8.8** Average winter rainfall and variation in rainfall among sites inhabited by *Suromalus obesus* (data from Case 1976).

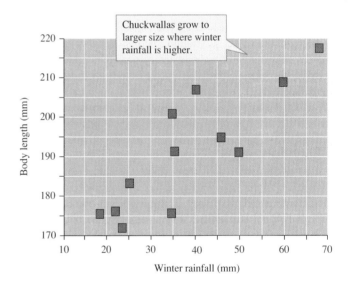

**Figure 8.9** Relationship between winter rainfall and chuckwalla, *Sauromalus*, size (data from Case 1976).

Ted Case (1976) explored variation in body size among *Sauromalus* populations at twelve sites distributed across its geographic range. Because the environments in which *Sauromalus* lives vary greatly across its range, we might expect that environmental selection has favored different characteristics in different parts of the species range. Case found that average summer temperatures at his desert study sites ranged from 23.8° to 35°C, while average annual rainfall varied from approximately 35 to 194 mm.

Clearly, *Sauromalus* lives in hot, dry places. Just how hot and dry some of these places are is shown by the climate graph for one of Case's study sites, Yuma, Arizona (see fig. 2.19). However, Case found considerable variation in climate over the elevational range of 4 to 1,166 m where *Sauromalus* lives. Elevation was especially well correlated with winter weather (fig. 8.8). As you can see in figure 8.8, average winter rainfall increases with elevation, from less than 20 mm at the lowest elevations to over 60 mm at the highest elevations. Winter rain is critical for growing the annual herbaceous plants which *Sauromalus* prefers to eat and the amount of winter rainfall largely determines the amount of plant growth in these desert environments.

Higher average rainfall at higher elevations translates into more food available for *Sauromalus*. However, the higher elevations inhabited by *Sauromalus* not only receive higher average rainfall, they also show less year-to-year variation in amount of rainfall. At the other end of the environmental spectrum, the *Sauromalus* at lower elevations lives in environments where much less rain falls and where there is more year-to-year variation in rainfall. What does variation in rainfall mean to *Sauromalus*? Variation in rainfall translates into variation in food availability. The lizards at lower elevations, on average, have access to less food and the amount available on any given year is unpredictable. Meanwhile the lizards at higher elevations live in a relatively food-rich environment where food availability is much more constant.

Case found that the lizards from the food-rich higher elevations are approximately 25% longer than those from lower elevations. This difference in body length translates into a twofold difference in body weight! What is the source of these size differences among populations? Of the many environmental variables that he measured, Case determined that the best predictor of *Sauromalus* body length across his study sites is average winter rainfall (fig. 8.9).

Case uncovered substantial variation in size among *Sauromalus* populations. This variation is analogous to the variation in plant sizes observed by plant ecologists along elevational gradients. How might we determine whether the differences in body size among *Sauromalus* populations Case observed are due to differences in food availability or due to genetic differences among populations? Like the plant ecologists Clausen, Keck, and Hiesey, we could rear individuals from low- and high-elevation populations in a common environment. That is, we could construct a kind of common garden for lizards. This is precisely what was done by Christopher R. Tracy (1999).

Tracy collected 12 to 15 juvenile *Sauromalus* from six populations in Arizona, California, and Nevada, living at elevations ranging from 200 to 890 m. He then raised these juvenile lizards under identical environmental conditions in a laboratory. By growing juvenile *Sauromalus* under identical environmental conditions, Tracy could determine the contributions of environmental versus genetic factors to size differences among *Sauromalus* populations.

Tracy set up the laboratory environment in a way that simulated late spring conditions, including 14 hours of light and 10 hours of darkness daily. These conditions provided the lizards with long periods for daily activity. He provided rocks for shelter and a heat lamp for basking. The laboratory environment maintained a temperature gradient from room temperature to 42°C under a heat lamp, which allowed the lizards the opportunity to use behavior to maintain their body temperatures at a preferred 36°C. Tracy also made an abundance of high quality food and vitamins available at all times so that food would not limit rates of lizard growth. In addition, he took *Sauromalus* social life into account. Observations by other ecologists had shown that *Sauromalus* eats more and grows faster when living in small groups than when isolated from other *Sauromalus*. Therefore, Tracy kept his lizards in groups of 3 to 5 while he followed their growth under laboratory conditions for 462 days.

How did *Sauromalus* from different elevations respond to Tracy's laboratory conditions? Lizards from all populations grew well in the laboratory. However, they showed marked different patterns of growth. First, females grew slower than males but individuals of both sexes grew faster before reaching sexual maturity. However, before sexual maturity, the fastest growth was shown by lizards from low elevations. After maturity, however, the lizards from higher elevations grew faster. Despite these complications the outcome of the experiment was clear. Lizard size at the end of the laboratory experiment was highly and positively correlated with the elevation at which they had been collected as juveniles (Fig. 8.10). In the end lizards from the higher elevations grew to a larger size, approximating in a laboratory common garden for lizards the pattern of variation in body size found in the field.

What do the results of Tracy's experiment indicate about variation in body size among *Sauromalus* populations? One important conclusion is that the differences in body size observed in the field are at least partly determined by genetic differences among populations. It appears that natural selection has favored different sized individuals at different elevations. Tracy's study of *Sauromalus* demonstrates how traditional morphological and laboratory studies continue to make

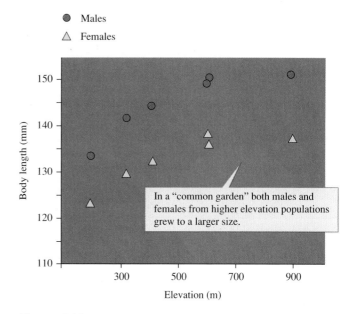

**Figure 8.10** Chuckwalla body lengths at the end of a laboratory rearing experiment (data from Tracy 1999).

significant contributions to our understanding of variation in animal populations. However, modern molecular approaches dominate contemporary studies of genetic variation in animal populations. The following study shows how molecular studies of genetic variation may be combined with morphological studies to explore the distribution and extent of genetic variation in animal populations, even where the historical patterns have been obscured by human interference.

## Genetic Variation in Alpine Fish Populations

The Alps rise out of the landscape of south central Europe, forming a moist and cool high-elevation environment. The Alps' deep winter snows and glaciers make them the origin of four important rivers: the Danube and Rhine Rivers, which flow out of the northern Alps, and the Po and Rhone Rivers, which flow out of the southern Alps. Because the headwater streams of these rivers are cool, they became refuges for cold-water aquatic organisms following the last Ice Age. As temperatures of the surrounding lowlands began to warm at the end of the Pleistocene, approximately 12,000 years ago, aquatic species requiring cold water migrated to the headwaters of these rivers. The movement of cold-adapted aquatic species into the headwater streams and lakes of the glacial valleys that lace the Alps created clusters of geographically isolated populations. This isolation reduced movements of individuals between populations. With reduced gene flow, populations could diverge genetically. Such genetic divergence would increase the genetic variation among populations.

Morphological differences among populations of headwater fish species in the Alps have long suggested genetic differences among them. Nowhere has morphological variation among populations been better studied and documented than among the whitefishes. Whitefish are relatives of the trout and salmon and are classified in the genus *Coregonus* (fig. 8.11). Marlis Douglas and Patrick Brunner (2002) explored the genetic and phenotypic variation among populations of *Coregonus* in the central Alps. Douglas and Brunner pointed out that ichthyologists have described 19 indigenous *Coregonus* populations from the central Alps. However, there has been significant disagreement over the taxonomic status of these 19 populations. The classification of these populations ranges from that of a single variable species with 19 distinctive populations to dividing the 19 populations into more than a dozen separate species.

The taxonomic status of *Coregonus* populations in the central Alps is made more difficult by a one-hundred-year history of intensive fisheries management. Douglas and Brunner review this history, which included raising *Coregonus* in hatcheries and moving fish between lakes. One of the main purposes of the study by Douglas and Brunner was to describe the genetic variation among the present-day populations of *Coregonus* in order to determine if there is evidence for significant genetic differences among historically recognized populations. A second purpose was to examine the genetic similarity between introduced *Coregonus* populations and the populations from which they were drawn. Using this information, Douglas and Brunner intended to offer suggestions for the management and conservation of *Coregonus* in the central Alps.

Douglas and Brunner collected 907 *Coregonus* specimens from 33 populations in 17 lakes in the Central Alpine Region of Europe. They used a mixture of anatomical and genetic features to characterize the fish collected from the study populations. The anatomical features were the number of rays in the dorsal, anal, pelvic, and pectoral fins, the extent of pigmentation in these fins, and the number of gill rakers on the first gill arch. The study populations were characterized genetically by using specific primers to amplify six different loci on **microsatellite DNA,** tandemly repetitive nuclear DNA, 10–100 base pairs long.

Genetic analyses by Douglas and Brunner demonstrated a moderate to high level of genetic variation within all 33 study populations. They also found that genetic and morphological analyses distinguished the 19 historically recognized *Coregonus* populations of the central Alps. Genotypic differences among populations were sufficient to correctly assign individual fish to the indigenous population from which they were sampled with approximately a 71% probability. Fin ray counts correctly assigned fish to the 19 indigenous populations with a 69% probability, while pigmentation could identify them with a 43% probability. Combining genetic and phenotypic data increased the correct assignment of specimens to the populations from which they were drawn to 79%. Genetic analyses of the introduced *Coregonus* populations revealed their genetic similarity to the populations from which they were stocked. However, these analyses also showed that the introduced populations have become genetically distinctive from their source populations.

The conclusion that Douglas and Brunner drew from these results was that the *Coregonus* of the central Alps is made up of a highly diverse set of populations that show a high level of genetic

**Figure 8.11**  Whitefish, *Coregonus* sp., are adapted to cold, highly oxygenated waters like their relatives the trout and salmon. Because they are valued food fishes, whitefish have been intensively managed particularly in the central Alps.

differentiation. They suggest that these populations should be considered as an "evolutionarily significant unit." They further conclude that the distinctiveness of local *Coregonus* populations is sufficient so that they should be managed as separate units. Douglas and Brunner recommend that *Coregonus* should not be moved from one lake basin to another.

The studies of plants and animals that we have reviewed have repeatedly demonstrated genetic variation in populations. The ecological literature contains thousands of such demonstrations. What can we conclude from this? One of the major conclusions that we can reach is that the potential for evolution by natural selection, which requires genetic variation in populations, is great. However, in order to better understand how such evolutionary change may come about, we need to first understand some aspects of the genetics of populations, or **population genetics.** The theoretical foundations of population genetics were established early in the twentieth century by two investigators named Hardy and Weinberg.

## CONCEPT DISCUSSION

### Hardy-Weinberg

**The Hardy-Weinberg equilibrium model helps identify evolutionary forces that can change gene frequencies in populations.**

We defined evolution as a change in a population over time. Since evolution ultimately involves changes in the frequency of heritable traits in a population, we can define evolution more precisely as a change in gene frequencies in a population.

**Figure 8.12** Two color forms of *Harmonia axyridis,* the Asian lady beetle. The genetic basis of the color forms of *H. axyridis* is well studied, making it a useful species for studies of population genetics and natural selection.

Therefore a thorough understanding of evolution must include some knowledge of population genetics. Though Mendel is not generally credited with studying the genetics of populations, he included a population level analysis in his paper on inheritance in garden peas (Mendel 1866). In a section of this paper titled, "The Subsequent Generations from the Hybrids," Mendel demonstrated mathematically that if self-fertilization was the only form of fertilization in a population consisting of three genotypes, *AA* (homozygous dominant), *Aa* (heterozygous), and *aa* (homozygous recessive) present in a ratio of one *AA* individual : two *Aa* individuals : one *aa* individual, the frequency of homozygous recessive (*aa*), and homozygous dominant (*AA*) individuals would increase in the population. Mendel did not consider what would happen to gene frequencies in his theoretical population if breeding occurred through something other than self-fertilization. Still, his analysis anticipated the field of population genetics, the foundations of which would be laid 42 years later.

## Calculating Gene Frequencies

Consider a population of Asian lady beetles of the species *Harmonia axyridis* (fig. 8.12). *Harmonia* populations generally include a great deal of variation in color pattern on the wing covers, or elytra, and over 200 color variants are known. Many color forms are so distinctive that early taxonomists described them as different species or even different genera. Genetists in the first half of the twentieth century, especially Chia-Chen Tan and Ju-Chi Li (1934, 1946) and Theodosius Dobzhansky (1937), determined that the variation in color patterns shown by *Harmonia* is due to the effects of more than a dozen alternative alleles for color pattern. The phenotypic expressions of two of those alleles are shown in figure 8.13. The homozygous "19-signata" genotype of *Harmonia*, which we can represent as *SS,* has yellow elytra with several black spots, while the homozygous "aulica" genotype, represented here as *AA,* has elytra with prominent black borders and a large

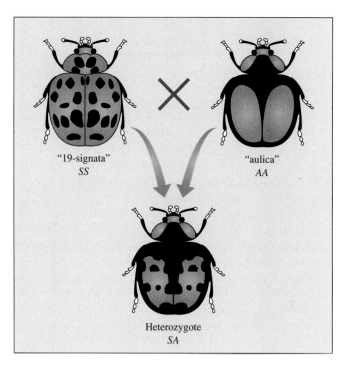

**Figure 8.13** Color patterns in the Asian lady beetle, *Harmonia axyridis* (after Dobzhansky 1937 and Tan 1946).

oval area of yellow or orange. Tan and Li, who did extensive breeding experiments using *Harmonia* that they collected in southwestern China, found that crosses between 19-signata and aulica genotypes produce heterozygous offspring, indicated here as *SA,* with a color pattern that includes elements of both the 19-signata and the aulica parental forms (fig. 8.13). One of the convenient features of knowing so much about color pattern inheritance in *Harmonia* is that color pattern can be used to determine the genotypes of many individuals.

   Now suppose that you sampled the genotypes of *Harmonia* in a tract of forest in Asia and found that the frequency of beetles with genotype *SS* is 0.81 (81%), the frequency of the *SA* genotype is 0.18 (18%), and the frequency of the *AA* genotype is 0.01 (1%). What is the frequency of the *S* and *A* alleles in this population? The frequency in the *S* allele is:

$$\text{Frequency of } SS + 1/2(\text{Frequency of } SA)$$
$$= 0.81 + 1/2(0.18) = 0.81 + 0.09 = 0.90$$

The frequency of the *A* allele is:

$$\text{Frequency of } AA + 1/2(\text{Frequency of } SA)$$
$$= 0.01 + 1/2(0.18) = 0.01 + 0.09 = 0.10$$

These calculations show that the frequency of the *S* allele in this lady beetle population is 0.90, while the frequency of the *A* allele is 0.10.

   Evolutionary ecologists are interested in knowing what factors may change allele frequencies in a population such as that of our hypothetical population of *Harmonia*. Those factors, which we can consider as evolutionary forces, are revealed indirectly by the **Hardy-Weinberg principle.** The Hardy-Weinberg principle states that in a population mating at random in the absence of evolutionary forces, allele frequencies will remain constant.

George H. Hardy, a British mathematician, and Wilhelm Weinberg, a German physician, established their principle, one of the most fundamental of population genetics, in 1908. They did so to address a growing controversy surrounding the applicability of Mendelian genetics to human populations. Hardy was addressing the assertion by a contemporary biologist that a genetically dominant gene introduced to a randomly breeding population would increase in frequency until it reached a frequency of 0.5, producing a ratio of genotypes of one homozygous dominant individual: two heterozygous individuals: one homozygous recessive individual. Because some genetically dominant human traits, such as brachydactyly which produces short fingers, remain rare and do not occur in such simple "Mendelian" ratios, some biologists of the early 1900s claimed that Mendelian genetics does not apply to human populations. Hardy and Weinberg independently revealed the flaws in this line of reasoning and established the Hardy-Weinberg principle.

Let us review how random mating will influence gene frequencies in the *Harmonia* beetle population we just reviewed. Assuming equal fertility of the *SS, SA,* and *AA* genotypes, the proportion of *S* and *A* alleles in the population, 0.9 and 0.1, are also the proportions of eggs and sperm carrying the two alleles. With random mating, the probability that any two alleles will be paired in a zygote is determined by the frequency of the alleles in our hypothetical population as follows:

Proportion of matings that will pair an *S* sperm with an *S* egg = $0.9 \times 0.9 = 0.81$,

Proportion of matings that will pair an *S* sperm with an *A* egg = $0.9 \times 0.1 = 0.09$,

Proportion of matings that will pair an *A* sperm with an *S* egg = $0.1 \times 0.9 = 0.09$

and

Proportion of matings that will pair an *A* sperm with an *A* egg = $0.1 \times 0.1 = 0.01$

The proportion of the three genotypes produced by this random mating will be: *SS* = 0.81, *SA* = 0.09 + 0.09 = 0.18, and *AA* = 0.01. Notice that the proportions of these genotypes in the parents and offspring in the population are the same. If you calculate the **allele frequencies** from the genotype frequencies in the offspring you will find that they remain at *S* = 0.90 and *A* = 0.10, which is what the Hardy-Weinberg principle predicts when mating in a population is random.

We can represent these relationships in a more general way using some basic algebra, if we let *p* equal the frequency of one allele and *q* the frequency of the second allele. In the case of the *Harmonia* example just discussed, let *p* = the frequency of the *S*

allele and *q* = the frequency of the *A* allele. Expressing these frequencies in numbers, *p* = 0.90 and *q* = 0.10. For a population in Hardy-Weinberg equilibrium in a situation where there are only two alleles at a particular locus, *p* + *q* = 1.0. Again referring to the *Harmonia* example, *p* + *q* = 0.90 + 0.10 = 1.0. Using this relationship we can calculate the frequency of genotypes in a population in Hardy-Weinberg equilibrium as:

$$(p + q)^2 = (p + q) \times (p + q) = p^2 + 2pq + q^2 = 1.0$$

The result of this calculation is:

$$(0.90)^2 + 2(0.90 \times 0.10) + (0.10)^2 = 0.81 + 0.18 + 0.01 = 1.0$$

According to this equation, the frequencies of the genotypes in our hypothetical *Harmonia* population are:

$$p^2 = (0.90)^2 = 0.81 = \text{frequency of the } SS \text{ genotype,}$$

$$2pq = 2(0.90 \times 0.10) = 0.18 = \text{frequency of the } SA \text{ genotype,}$$

and

$$q^2 = (0.10)^2 = 0.01 = \text{frequency of the } AA \text{ genotype.}$$

These calculations are equivalent to the combining of alleles that would occur if individuals in the *Harmonia* population mated at random. The mathematics of the Hardy-Weinberg model are further dissected in figure 8.14.

In the equations we just explored, random mating is sufficient to maintain constant genotype and allele frequencies. However, Hardy pointed out in his 1908 paper that in natural populations, other conditions are also required to maintain constant allele frequencies. For instance, Hardy recognized that nonrandom mating or differences in fertility among genotypes can change allele frequencies in a population. The conditions necessary to maintain constant allele frequencies in a population, what is called Hardy-Weinberg equilibrium, are as follows:

1. *Random mating.* Nonrandom or preferential mating, in which the probability of pairing alleles is either greater or lower than would be expected based on their frequency in the population, can change the frequency of genotypes.

2. *No mutations.* Mutations which add new alleles to the population or change an allele from one form to another have the potential to change allele frequencies in a population and therefore disrupt Hardy-Weinberg equilibrium.

3. *Large population size.* Small population size increases the probability that allele frequencies will change from one generation to the next due to chance alone. Change in allele frequencies due to chance or random events is called **genetic drift.** Genetic drift reduces genetic variation in populations over time by increasing the frequency of some alleles and reducing the frequency of some alleles or eliminating others.

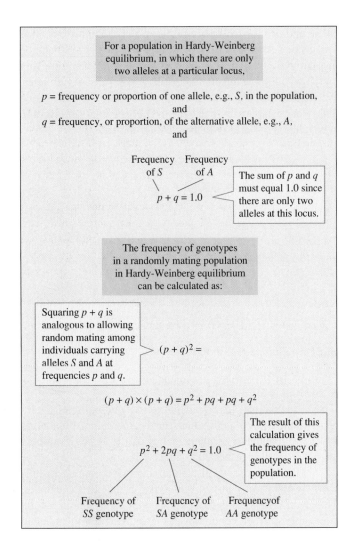

For a population in Hardy-Weinberg equilibrium, in which there are only two alleles at a particular locus,

$p$ = frequency or proportion of one allele, e.g., $S$, in the population, and

$q$ = frequency, or proportion, of the alternative allele, e.g., $A$, and

Frequency of $S$    Frequency of $A$

$p + q = 1.0$

The sum of $p$ and $q$ must equal 1.0 since there are only two alleles at this locus.

The frequency of genotypes in a randomly mating population in Hardy-Weinberg equilibrium can be calculated as:

Squaring $p + q$ is analogous to allowing random mating among individuals carrying alleles $S$ and $A$ at frequencies $p$ and $q$.

$(p + q)^2 =$

$(p + q) \times (p + q) = p^2 + pq + pq + q^2$

$p^2 + 2pq + q^2 = 1.0$

The result of this calculation gives the frequency of genotypes in the population.

Frequency of $SS$ genotype     Frequency of $SA$ genotype     Frequency of $AA$ genotype

**Figure 8.14**  Anatomy of a Hardy-Weinberg equilibrium equation.

4. *No immigration.* Immigration can introduce new alleles into a population or, because allele frequencies are different among immigrants, alter the frequency of existing alleles. In either case immigration will disrupt Hardy-Weinberg equilibrium.

5. *All genotypes have equal fitness, where fitness is the probability of surviving and reproducing.* If different genotypes survive and reproduce at different rates, then gene and genotype frequencies will change in populations.

Hardy-Weinberg equilibrium requires that all five of these conditions be met. How likely is it that all the conditions required for Hardy-Weinberg equilibrium will be present in a natural population? In places and at times the conditions appear to be present. However, it is very likely that one or more of these conditions will not be met and allele frequencies will change. While at first thought it may not appear that the Hardy-Weinberg principle is an important contribution to biology, it is in fact very important. By carefully defining the

highly restrictive conditions under which evolution is not expected, the analysis by Hardy and Weinberg leads us to conclude that the potential for evolutionary change in natural populations is often very great.

When a population is not in Hardy-Weinberg equilibrium, the principle helps us to identify the evolutionary forces that may be in play. Observations of natural populations of *Harmonia* indicate that they are often not in Hardy-Weinberg equilibrium. For instance, Dobzhansky (1937) did extensive surveys of *Harmonia* across Asia and found the aulica form in many sites along with the 19-signata form (see fig. 8.13). However, he did not report the intermediate form, 19-signata crossed with aulica (see fig. 8.13). The absence of this intermediate phenotype from Dobzhansky's surveys suggests that the populations he studied were not in Hardy-Weinberg equilibrium.

Why would these intermediate types not be present in sufficient numbers for Dobzhansky to report them? One possible reason is nonrandom mating within the populations. Is there evidence of nonrandom mating by *Harmonia?* Substantial work on associations of color variants of *Harmonia* has been done in Japan. Taku Komai and Yasushi Hosino (1951) found that *Harmonia* with different color patterns had different habitat associations in a village landscape near Nagoya, Japan. Differences in habitat preferences among variants within a population can contribute to nonrandom mating. In addition, other Japanese researchers have more recently made direct observations of nonrandom mating in *Harmonia.* Naoya Osawa and Takayoshi Nishida (1992) observed preferential mating based on color pattern in a population of *Harmonia* near Kyoto, Japan. In 1998, H. Ueno, Y. Sato, and K. Tsuchida observed preferential mating in another *Harmonia* population in Japan based on size not on color pattern.

Meanwhile, other researchers have documented changes in gene frequencies in *Harmonia* populations near Vladisvostok, Russia, that have taken place since the 1920s, when they were studied by Dobzhansky. L. Bogdanov and N. Gagal'chii (1986) collected *Harmonia* near Vladisvostok and compared the frequencies of color variants within their collections to those found by Dobzhansky (1937) approximately one-half century earlier. What they found was a great departure from Hardy-Weinberg equilibrium. While most color variants decreased in frequency, 19-signata increased by 30%. Meanwhile, the aulica color variant had disappeared entirely. The work by Bogdanov and Gagal'chii clearly documents changes in genotype frequencies within these populations. In other words, though they did not document the mechanisms involved, they found evidence for evolutionary change.

In the remaining sections of chapter 8 we will discuss examples in which one or more of the conditions for Hardy-Weinberg equilibrium have not been met and where evolutionary change has occurred in populations as a consequence. We begin this discussion with a general overview of the process of natural selection.

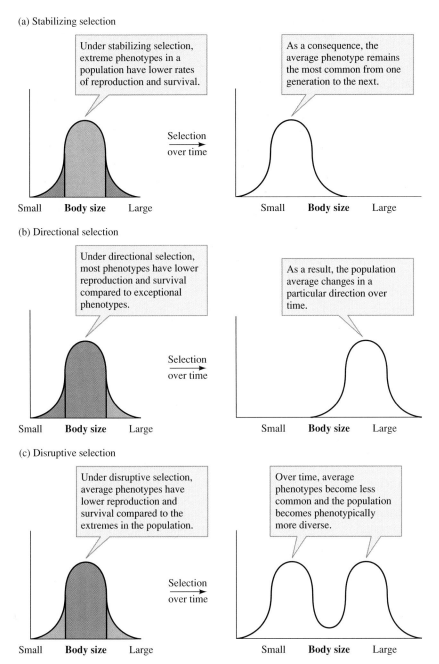

(a) Stabilizing selection

Under stabilizing selection, extreme phenotypes in a population have lower rates of reproduction and survival.

Selection over time

Small   **Body size**   Large

As a consequence, the average phenotype remains the most common from one generation to the next.

Small   **Body size**   Large

(b) Directional selection

Under directional selection, most phenotypes have lower reproduction and survival compared to exceptional phenotypes.

Selection over time

Small   **Body size**   Large

As a result, the population average changes in a particular direction over time.

Small   **Body size**   Large

(c) Disruptive selection

Under disruptive selection, average phenotypes have lower reproduction and survival compared to the extremes in the population.

Selection over time

Small   **Body size**   Large

Over time, average phenotypes become less common and the population becomes phenotypically more diverse.

Small   **Body size**   Large

**Figure 8.15**   Three principle forms of natural selection: (*a*) stabilizing selection, (*b*) directional selection, and (*c*) disruptive selection.

## CONCEPT DISCUSSION

### The Process of Natural Selection

**Natural selection is the result of differences in survival and reproduction among phenotypes.**

As we saw in the introduction to chapter 8, Darwin was one of the first people to recognize the biological significance of varia-

tion among individuals in a population. The biological significance of the variation that Darwin recognized stemmed from an inference that he drew. His inference was that some phenotypes in a population would have an advantage over others under particular environmental circumstances. That is, the phenotypic characteristics of some individuals, for instance, larger or smaller size, different body proportions, lighter or darker pigmentation, or higher or lower metabolic rate, would result in higher rates of reproduction and survival compared to other individuals with other phenotypic characteristics. In other words, some individuals in a population, because of their phenotypic characteristics, produce more offspring that themselves live to reproduce.

While the basic concept of natural selection is easy enough to grasp, natural selection does not a take the same form everywhere and at all times. Rather, natural selection can act against different segments of the population under different circumstances and can produce quite different results. Natural selection can lead to change in populations but it can also serve as a conservative force, impeding change in a population. Natural selection can increase diversity within a population or decrease diversity. Let's begin our discussion of natural selection with a process that conserves population characteristics.

## Stabilizing Selection

One of the conclusions that we might draw from the discussion of the Hardy-Weinberg equilibrium model is that most populations have a high potential for evolutionary change. However, our observations of the natural world suggest that species can remain little changed for generation after generation. If the potential for evolutionary change is high in populations, why does it not always lead to obvious evolutionary change at least on the short term? There are many reasons for apparent absence of change in populations. For example, one form of natural selection, called **stabilizing selection,** can act to impede changes in populations.

Stabilizing selection acts against extreme phenotypes and as a consequence favors the average phenotype. Figure 8.15*a* pictures stabilizing selection, using a normal distribution of body size. Under the influence of stabilizing selection, individuals of average size have higher survival and reproductive rates, while the largest and smallest individuals in the population have lower rates of survival and reproduction. As a consequence of stabilizing selection, a population tends to sustain the same phenotype over time. Stabilizing selection occurs where average individuals in a population are best adapted to a given set of environmental conditions. If a population is well adapted to a given set of environmental circumstances, stabilizing selection may sustain the match between prevailing environmental conditions and the average phenotype within a population. However, stabilizing selection for a particular trait can be challenged by environmental change. In the face of environmental change the dominant form of selection may be directional.

## Directional Selection

If we examine the fossil record or trace the history of well-studied populations over time, we can find many examples of how populations have changed in many characteristics over time. For instance, there have been remarkable changes in body size or body proportions in many evolutionary lineages. Such changes may be the result of **directional selection.**

Directional selection favors an extreme phenotype over other phenotypes in the population. Figure 8.15b presents an example of directional selection, again, using a normal distribution of body size. In this hypothetical situation, larger individuals in the population realize higher rates of survival and reproduction, while average and small individuals have lower rates of survival and reproduction. As a consequence of these differences in survival and reproduction, the average phenotype under directional selection changes over time. In the example shown in figure 8.15b, average body size increases with time. Directional selection occurs where one extreme phenotype has an advantage over all other phenotypes. However, there are circumstances in which more than one extreme phenotype may have an advantage over the average phenotype. Such a circumstance can lead to diversification within a population.

## Disruptive Selection

There are populations that do not show a normal distribution of characteristics such as body size. In a normal distribution such as those depicted in figures 8.15a and 8.15b, there is a single peak, which coincides with the population mean. That is, the average phenotype in the population is the most common and all other phenotypes are less common. However, in some populations there may be two or more common phenotypes. In many animal species, for example, males may be of two or more discrete sizes. For example, it appears that in some animal populations small and large males have higher reproductive success than males of intermediate body size. In such populations, natural selection seems to have produced a diversity of male sizes. One way to produce such diversity is through **disruptive selection.**

Disruptive selection favors two or more extreme phenotypes over the average phenotype in a population. In figure

8.15c, individuals of average body size have lower rates of survival and reproduction than individuals of either larger or smaller body. As a consequence, both smaller and larger individuals increase in frequency in the population over time. The result is a distribution of body sizes among males in the population with two peaks. That is, the population has many large males and many small males but few of intermediate body size.

Figure 8.15b and 8.15c indicate change in the frequencies of phenotypes in the two hypothetical populations after a period of natural selection. This change depends on the extent to which genes determine the phenotype upon which natural selection acts. This dependence is the focus of the following concept discussion.

---

### CONCEPT DISCUSSION

#### Evolution by Natural Selection

**The extent to which phenotypic variation is due to genetic variation determines the potential for evolution by natural selection.**

---

The most general postulate of the theory of natural selection is that the environment determines the evolution of the anatomy, physiology, and behavior of organisms. This is what Darwin surmised as he studied variation among populations and species in different environments. Coincidentally, one of the clearest demonstrations of natural selection has resulted from studies of populations of Galápagos finches, which are reviewed in chapter 11 (pp. 285–289) and chapter 13 (pp. 328–330). Those studies showed that the quantity and quality of available food exerts strong selection on beak size in finch populations. Here we review additional studies that also provide evidence for Darwin's bold hypothesis that natural selection by the environment can result in evolutionary change in populations.

## Evolution by Natural Selection and Genetic Variation

Darwin was keenly aware that the only way natural selection can produce evolutionary change in a population is if the phenotypic traits upon which natural selection acts can be passed from generation to generation. In other words, evolution by natural selection depends upon the heritability of traits. We can define **heritability** of a trait—usually symbolized as $h^2$—in a broad sense as the proportion of total phenotypic variation in a trait, such as body size or pigmentation, that is attributable to genetic variance. In equation form, heritability can be expressed as:

$$h^2 = V_G/V_P$$

Here $V_G$ represents genetic variance and $V_P$ represents phenotypic variance. (We reviewed how to calculate variance in chapter 6, p. 162.) Many different factors contribute to the amount of phenotypic variance in a population. We will subdivide phenotypic variance into only two components: variance in phenotype due to genetic effects, $V_G$, and variance in phenotype due to environmental effects on the phenotype, $V_E$. Subdividing $V_P$ in the heritability equation given above produces the following:

$$h^2 = V_G/(V_G + V_E)$$

This highly simplified expression for heritability has important implications so let's take a little space to examine it. First, let's consider environmental variance, $V_E$. Environment has substantial effects on many aspects of the phenotype of organisms. For instance, the quality of food eaten by an animal can contribute significantly to the growth rate of the animal and to its eventual size. Similarly, the amount of light, nutrients, temperature, and so forth, affect the growth form and size of plants. So, when we consider a population of plants or animals, some of the phenotype that we might measure will be the result of environmental effects, that is, $V_E$. However, we are just as familiar with the influence of genes on phenotype. For example, some of the variation in stature that we see in a population of animals or plants will generally result from genetic variation among individuals in the population, that is, $V_G$.

What our equation says is that the heritability of a particular trait depends on the relative sizes of genetic versus environmental variance. Heritability increases with increased $V_G$ and decreases with increased $V_E$. Imagine a situation in which all phenotypic variation is the result of genetic differences between individuals and none results from environmental effects. In such a situation, $V_E$ is zero and $h^2 = V_G/(V_G + V_E)$ is equal to $h^2 = V_G/V_G$ (since $V_E = 0$), which equals 1.0. In this case since all phenotypic variation is due to genetic effects, the trait is perfectly heritable. We can also imagine the opposite circumstance in which none of the phenotypic variation that we observe is due to genetic effects. In this case, $V_G$ is zero and so the expression $h^2 = V_G/(V_G + V_E)$ also equals zero. Because all of the phenotypic variation we observe in this population is due to environmental effects, natural selection cannot produce evolutionary change in the population. Generally, heritability of traits falls somewhere in between these extremes in the very broad region where both environment and genes contribute to the phenotypic variance shown by a population. For instance, Peter Boag and Peter Grant (1978) estimated bill width in the Galápagos finch *Geospiza fortis* to have a heritability of 0.95. By comparison they estimated that bill length in the species has a heritability of 0.62. In a study of morphological variation in the water lily leaf beetle, a team of Dutch scientists (Pappers et al. 2002) found that body length and mandible width had heritabilities of between 0.53 and 0.83. Now that we have established the requirement of heritable variation in a trait for evolution in that trait, let's review studies that have explored evolution by natural selection in nature.

**Figure 8.16** A brown anole, *Anolis sagrei*, jumping. Limb length is known to be highly correlated with the types of perches used by *Anolis* species.

# Adaptive Change in Colonizing Lizards

As we reviewed cases of physiological, anatomical, or behavior features of organisms, especially in section II of the text, we assumed that they were the result of adaptation of populations through the process of natural selection. However, we have reviewed few studies that have documented the process of natural selection. Why is it so important to make this distinction? In science we must always guard against mixing pattern and process or evidence and interpretation. In this section we address this omission by reviewing elegant studies that have documented natural selection in progress.

One of those studies was conducted by Jonathan Losos, Kenneth Warheit, and Thomas Schoener on lizards of the genus *Anolis* (Losos, Warheit, and Schoener 1997). Approximately 150 species of *Anolis* inhabit the islands of the Caribbean Sea and another 250 are found in Central and South America (fig. 8.16). This great diversity of lizards in a single genus includes a great amount of variation in size and body proportions. The anatomy of *Anolis* lizards, especially the length of their hind limbs, appears to reflect selection for effective use of vegetation. The attribute of vegetation that appears to be most significant in selection for hind limb length is the diameter of surfaces available for perching. Hind limb length in *Anolis* populations appears to be the result of a trade-off between selection for maximum speed (lizards with longer hind limbs run faster) and selection for moving efficiently on narrow branch surfaces (lizards with shorter hind limbs move more efficiently on narrow surfaces).

Losos, Warheit, and Schoener used replicated field experiments to study natural selection for changes in morphology in *Anolis* lizard populations. They designed their experiments in such a way that they could make very specific predictions concerning expected morphological changes among lizard populations. Losos and his colleagues captured adult *Anolis sagrei* on Staniel Cay in the Bahama Islands and then introduced them in groups of 5 to 10 lizards, at a ratio of 2 males:3 females, to 11 small islands in 1977 and to three more in 1981. None of these small islands had their own lizard populations, probably because hurricanes periodically eliminate lizards from them. The islands also differed greatly in their vegetative cover, which ranged in maximum height from 1 to 3 m on the different islands, but all had substantially lower vegetation than Staniel Cay, which supports some trees over 10 m tall.

Let us reflect on the conditions of the experiment. All the introduced lizards were drawn from the same source population on Staniel Cay, which could serve as a reference population. The islands onto which the lizards were introduced each supported somewhat different vegetation. Therefore, if vegetation is a primary agent selecting for differences in hind limb size, the morphology of the introduced populations should change from that of the source populations, but they should also differ from each other, depending on the vegetation on each small island. Losos, Warheit, and Schoener made two specific predictions: (1) the extent to which the colonizing populations change morphologically from the source population will correlate with the amount of difference in vegetative structure on the experimental islands and Staniel Cay, and (2) the *Anolis* populations on the experimental islands and Staniel Cay should show a significant correlation between relative hind limb length within populations and average perch diameter used on the islands.

After the lizards had occupied the experimental islands for 10 to 14 years, Losos and his colleagues returned to the islands and measured lizard morphology and their distributions on the local vegetation. Both predictions of the researchers were well supported by the results of their study. First, they found a positive correlation between the difference in vegetative height on experimental islands compared to Staniel Cay and the degree to which introduced lizards diverged from the ancestral population (fig. 8.17). Second, the hind limb length in the lizard populations was positively correlated with the average perch diameter the lizards used on each island (fig. 8.18). That is, on islands where lizards use perches of larger diameter, they have longer hind limbs.

The researchers point out that their results indicate that colonizing populations can adapt rapidly to new environmental conditions. However, they also caution that while their results are consistent with the effects of natural selection for changed morphology, they do not demonstrate unequivocally that the colonizing populations have evolved. What would we have to know to demonstrate an evolutionary response? We would have to know that the composition of the founding populations had changed genetically and that some of those genetic changes were responsible for the changes in morphology observed by Losos and his research partners. If the founding populations have not changed genetically, what is another possible source of their changed morphology? The environ-

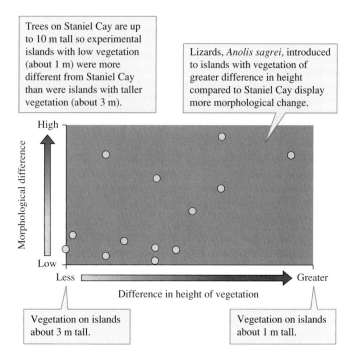

**Figure 8.17** Relationship between the difference in height of vegetation between the home island, Staniel Cay, and island of introduction and change in lizard morphology after their introduction (data from Losos et al. 1997).

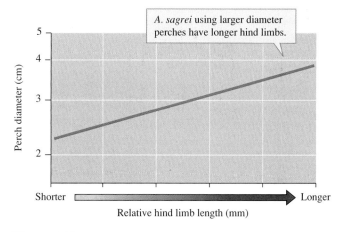

**Figure 8.18** Relationship between hind limb length in *Anolis sagrei* and perch diameters on experiment islands (data from Losos et al. 1997).

mental differences on the different islands, especially perch diameter, may have induced different developmental patterns that resulted in different hind limb lengths in the different lizard populations. At this point we cannot rule out the possibility that lizards on experimental islands underwent a developmental change and not an evolutionary change. To eliminate this possibility requires genetic studies. In the next study on rapid adaptation by soapberry bugs, the researchers collected extensive genetic information to document the operation of natural selection.

| Information |
| Questions |
| Hypothesis |
| Predictions |
| Testing ✓ |

*Investigating the Evidence*

# Estimating Heritability Using Regression Analysis

As we have seen, the extent to which phenotypic variation in a trait is determined by genetic variation affects its potential for that trait to evolve by natural selection. In other words, the potential for a trait to evolve is affected by the trait's heritability. How can we estimate the heritability of a particular trait? One common method is through regression analysis. Regression analysis is a statistical technique used to explore the extent to which one factor, called the **independent variable** (usually symbolized as X) determines the value of another variable, which we call the **dependent variable** (usually represented by the symbol Y). In regression analysis, we construct X-Y plots as we did when we explored scatter plots and correlation (Investigating the Evidence, Chapter 7, p. 186). However, regression analysis is used to determine the equation for a line, called a **regression line,** that best fits the relationship between X and Y. When the relationship between X and Y follows a straight line, the regression equation takes the following form:

$$Y = bX + a$$

In this equation, $a$ is the point at which the line crosses the Y axis, which is called the Y intercept, and $b$, which is the slope of the line, is the **regression coefficient.**

Let's use a natural system to learn more about regression analysis and its use in heritability studies. In heritability studies, we are interested in the extent to which the characteristics of parents determine the characteristics of offspring. For instance, the team of Dutch scientists studying water lily leaf beetles (Pappers et al. 2002) explored the heritability of body length in different populations of the beetle. To determine the heritability of body length, they conducted regression analyses using the body length of parents as the independent variable, and body length of the offspring as the dependent variable. Because each of the parents contributes to the genotype of the offspring, the value used for parental body length is the "mid-parent body length," which is the average of the two parents' body lengths. Let's consider the relationships between length of parents and offspring, and use regression analysis to estimate heritability of body length in some hypothetical populations of water lily leaf beetles.

Consider the three scatter plots shown in figure 1 and the lines drawn through the scatter of points. Again, these are much like the scatter plots we examined in chapter 7 but with regression lines drawn through each. The regression coefficient in each of the graphs indicates the level of heritability in the three hypothetical populations. In population a, the regression coefficient of 0.00 indicates that there is no relationship between parental body length and the body length of offspring. This result is apparent from just the scatter plot, which shows that parents of any length, large or small, can have small or large offspring. In this population it appears that the variation in body length among the offspring is determined entirely by environmental effects. In contrast, body length has a heritability of 0.52 in population b and 1.00 in population c. What do these values indicate? With a heritability of 0.52, we can conclude that about half of the variation in body length in population b results from genetic effects, and about half from environmental effects, such as food quality, temperature, and so forth. The regression coefficient of 1.00 in population c indicates that all the variation in body length in the offspring in that population is the result of genetic effects.

What are the evolutionary implications of the patterns shown in figure 1? The main evolutionary consequence is that natural selection on body size could lead to evolutionary change in body size in populations b and c but not in population a.

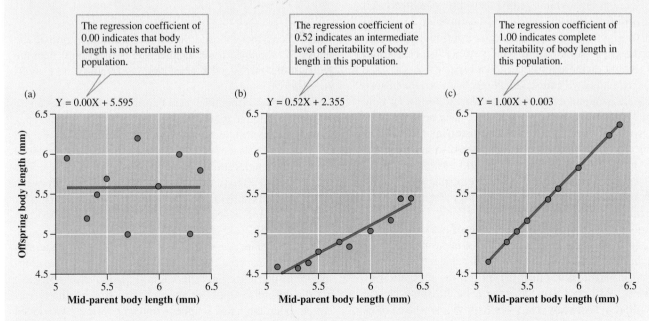

Figure 1 Regression analyses indicating degree of heritability of body length in three hypothetical populations of water lily leaf beetles.

# Rapid Adaptation by Soapberry Bugs to New Host Plants

As discussed in chapter 6, herbivores must overcome a wide variety of physical and chemical defenses evolved by plants. As a consequence, plants theoretically exert strong selection on herbivore physiology, behavior, and anatomy. While herbivore adaptation to plant defenses are generally inferred from the juxtaposition of plant defenses and herbivore characteristics, few studies have documented the process of herbivore adaptation. A notable exception is provided by studies of the soapberry bug and its evolution on new host plants.

The soapberry bug, *Jadera haematoloma,* feeds on seeds produced by plants of the family Sapindaceae. Soapberry bugs use their slender beaks to pierce the walls of the fruits of their host plants. To allow the bug to feed on the seeds within the fruit, the beak must be long enough to reach from the exterior of the fruit to the seeds. The distance from the outside of the fruit wall to the seeds varies widely among potential host species. Thus beak length should be under strong selection for appropriate length.

Scott Carroll and Christin Boyd (1992) reviewed the history and biogeography of the colonization of new host plants by soapberry bugs. Historically, soapberry bugs fed on three main host plants in the family Sapindaceae: the soapberry tree, *Sapindus saponaria* v. *drummondii,* in the southcentral region of the United States, the serjania vine, *Serjania brachycarpa,* in southern Texas, and the balloon vine, *Cardiospermum corindum,* in southern Florida. During the second half of the twentieth century three additional species of the plant family Sapindaceae were introduced to the southern United States. The round-podded golden rain tree, *Koelreuteria paniculata,* from east Asia and the flat-podded golden rain tree, *K. elegans,* from southeast Asia are both planted as ornamentals, while the subtropical heartseed vine, *Cardiospermum halicacabum,* has invaded Louisiana and Mississippi. At some point after their introduction, some soapberry bugs shifted from their native host plants and began feeding on these introduced plant species.

Carroll and Boyd painstakingly reconstructed the history of the colonization of the southern United States by new species of host plants and colonization of these new plants by soapberry bugs. Fortunately, extensive historical museum collections of plants and insects allowed them to assemble the history of a fascinating host shift by an herbivorous insect. They were particularly interested in determining whether the beak length had changed in soapberry bugs that shifted from native to introduced host plants.

Figure 8.19 contrasts the fruit radius of native and introduced host plants in Florida and the south central United States. In Florida the fruit of the native host plant *C. corindum* has a much larger radius than the fruit of the introduced *K. elegans* (11.92 mm versus 2.82 mm). In the south central United States soapberry bugs shifting to introduced host

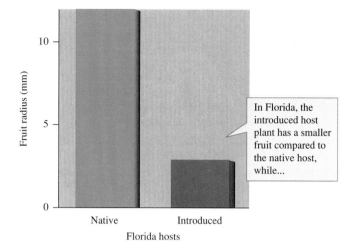

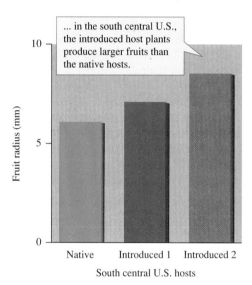

**Figure 8.19** Comparison of the radius of fruits produced by native and introduced species of Sapindaceae (data from Carroll and Boyd 1992).

plants faced the opposite situation. There, the fruit of the native *S. saponaria* has a smaller radius (6.05 mm) than the fruits of the introduced *K. paniculata* (7.09 mm) and *C. hali-cacabum* (8.54 mm).

Carroll and Boyd reasoned that if beak length was under natural selection to match the radius of host plant fruits, bugs shifting to the introduced plants in Florida should be selected for reduced beak length, while those shifting to introduced hosts in the south central United States should be selected for longer beaks. Figure 8.20 shows the relationship between soapberry beak length and the radius of fruits of their host plants. As you can see, there is a close correlation between fruit radius and beak length.

At this point we should ask whether the differences in beak length observed by Carroll and Boyd might be developmental responses to the different host plants. In other words, are the differences in beak length due to genetic differences among populations of soapberry bugs or were they induced

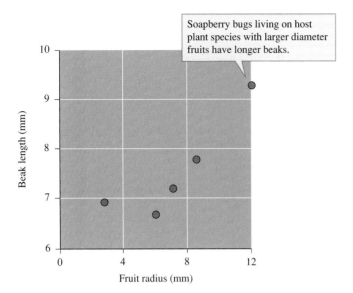

**Figure 8.20** Relationship between fruit radius and beak length in populations of native and introduced species of soapberry bugs (data from Carroll and Boyd 1992).

by the different host plants? Fortunately, Carroll reared juvenile bugs from the various populations on alternative host plants so we can answer this question. As it turns out, the differences in beak length observed in the field among bugs feeding on the various native and introduced host plants were retained in bugs that developed on alternative hosts. Thus, we have more information than is available for the *Anolis* lizard study reviewed earlier. Here we have evidence for a genetic basis for interpopulational differences among soapberry bugs. Consequently, we can conclude that the differences in beak length documented by Carroll and Boyd were likely the result of natural selection for increased or decreased beak length.

Scott Carroll, Stephen Klassen, and Hugh Dingle (1997, 1998) have done extensive additional studies of soapberry bugs that document substantial genetic differences between populations living on native versus introduced plants in the family Sapindaceae. Significantly, from the perspective of natural selection, the differences between these populations of soapberry bugs are great enough that both show reduced reproduction and survival when forced to live on the alternative host plants. That is, when soapberry bugs that normally live on native host plants are moved to introduced plants, their survival and reproductive rates decrease. However, when soapberry bugs that now live on introduced plants are moved to native plants, which their ancestors fed on only 30 to 100 years ago, their reproductive and survival rates also decrease. These additional studies of the genetic differences between soapberry bug populations provide additional evidence that populations of these bugs living on different host plants have undergone natural selection for traits that favor their survival and reproduction on their plant hosts.

## Change Due to Chance

**Random processes, such as genetic drift, can change gene frequencies in populations, especially in small populations.**

While we may often think of evolutionary change as a consequence of predictable forces such as natural selection which favors, or disfavors, particular genotypes over others, allele frequencies can change as a consequence of random processes such as genetic drift. Genetic drift is theoretically most effective at changing gene frequencies in small populations such as those that inhabit islands. In the following examples, we consider the effects of genetic drift on populations on isolated mountaintops and on islands.

# Evidence of Genetic Drift in Chihuahua Spruce

One of the greatest concerns associated with fragmentation of natural ecosystems due to human land use is that reducing habitat availability will decrease the size of animal and plant populations to the point where genetic drift will reduce the genetic diversity within natural populations. Are these concerns well-founded? The Hardy-Weinberg principle predicts that reducing small population sizes will lead to reduced genetic variation. However, we do not have to rely solely on theory to learn of the effects of habitat fragmentation on genetic diversity.

Many natural populations have undergone fragmentation as a consequence of changing climates and natural habitat fragmentation. One of those is the Chihuahua spruce, *Picea chihuahuana,* which is now restricted to the peaks of the Sierra Madre Occidental in northern Mexico. During the Pleistocene glacial period when the global climate was much cooler, spruce were found much farther south in Mexico and in more extensive populations. However, following the end of the Pleistocene and the onset of the warmer recent, or Holocene period, spruce populations moved northward and to higher elevations. Today, all spruce populations in Mexico are restricted to small, highly fragmented areas of subalpine environment in the mountains of states of Chihuahua and Durango. On these high mountains, Chihuahua spruce lives in an 800 km long band along the crest of the Sierra Madre Occidental at elevations between 2,200 and 2,700 m. On a local scale, the species is mainly found on cooler north-facing slopes along well-watered stream corridors, which are the microclimates where you would expect to find the descendants of an ice age relictual population. In these mountain refuges, Chihuahua spruce persists as far south as 23°30′ N latitude, just south of the Tropic of Cancer.

While the spruce of Durango have not been censused yet, all the Chihuahua spruce in the State of Chihuahua have been located and counted. Local populations of the species range in size from 15 to 2,441 individuals. This situation presents itself as a natural experiment on the effects of population size and habitat fragmentation on genetic diversity in populations. The opportunity for such studies was pursued by a joint team of U.S. and Mexican scientists (Ledig et al. 1997). F. Thomas Ledig and Paul D. Hodgskiss from the USDA Forest Service and Virginia Jacob-Cervantes and Teobaldo Eguiluz-Piedra of the Universidad Autonoma of Chapingo, Mexico, combined efforts to determine whether Chihuahua spruce has lost genetic diversity as a consequence of reduced population size following climatic warming after the end of the last ice age. They were also interested in whether reduced genetic diversity may be contributing to continuing decline of the species and its potential for extinction.

Ledig and his colleagues were particularly interested in the relationship between genetic diversity and population size. They used a technique called starch gel electrophoresis to determine the number of alleles present for 16 enzyme systems. Enzymes are of course gene products, and greater numbers of the various forms of an enzyme, which are called **allozymes,** indicate higher levels of genetic diversity in a population. The team assayed allozyme diversity for 24 genes, or **loci,** in seven populations ranging in size from 17 to 2,441 individuals.

As you might predict from the Hardy-Weinberg principle, Ledig and his colleagues found a significant positive correlation between population size and genetic diversity of their study populations. Figure 8.21 indicates that the smallest populations of Chihuahua spruce have much lower levels of genetic diversity than the largest populations. These results are consistent with the Hardy-Weinberg principle, which predicts that genetic drift will be most important in small populations.

How might drift occur in populations of spruce living on isolated mountain peaks in western Mexico and how might genetic drift reduce genetic variation in spruce populations? Imagine a population of 15 Chihuahua spruce on a mountain peak in the Sierra Madre Occidental at the beginning of July when the summer rains begin. The forest is dry after a long spring drought and as the lightning produced by a thunderstorm begins to strike the mountain, one bolt hits one of the spruce trees. The tree explodes as its interior water is turned into superheated steam, sending showers of splintered wood 50 m in all directions. The spruce tree then catches fire and the flames engulf two neighboring spruce trees before the ensuing torrential rains put out the fire. The result is a small spot fire that has killed three trees. The deaths of three trees would make very little difference in a population of several thousand. However, in a population of just 15, three trees represent 20% of the individuals. When individuals are removed from very small populations their removal often reduces the frequency of some alleles; such events will eventually eliminate some alleles entirely from a small population.

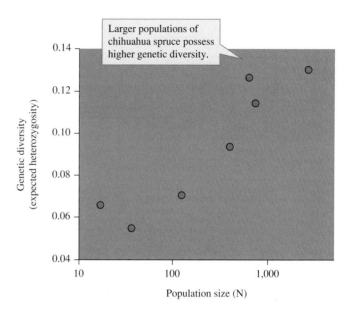

**Figure 8.21** Relationship between population size and genetic diversity of chihuahua spruce, *Picea chihuahuana,* populations (data from Ledig et al. 1997).

It seems likely that genetic drift is changing allele frequencies and reducing overall genetic diversity in populations of Chihuahua spruce. However, this is one species occupying relictual environments in one corner of North America. Would we see consistent reductions in genetic diversity if we examined a larger number of populations inhabiting insular or fragmented environments? The next study addresses this question for both plants and animals.

# Genetic Variation in Island Populations

Richard Frankham (1997) of the Centre for Biodiversity and Bioresources at Macquarie University in Sydney, Australia, compared the genetic diversity of island and mainland populations of both animals and plants. His study was motivated by the fact that rates of extinction in historic times have been much higher for island populations compared to mainland populations. Frankham developed the idea that because lower genetic variation within a population indicates lower potential for evolutionary responses to environmental challenge, lower genetic variation within island populations may be partly responsible for their greater vulnerability to extinction compared to mainland populations. However, when he reviewed what was known about the relative genetic variation in island and mainland populations, he encountered a great deal of uncertainty. Frankham undertook his study to fill this information gap. He posed two main questions. Do island populations of sexually reproducing species have lower genetic variation than comparable mainland populations? Do **endemic** island populations, which have lived in isolation on islands long enough to diverge substantially from mainland populations, have lower genetic variation than nonendemic mainland populations?

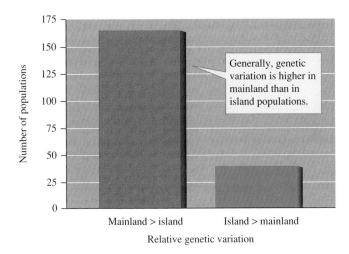

**Figure 8.22** Comparison of genetic variation in mainland versus island populations (data from Frankham 1997).

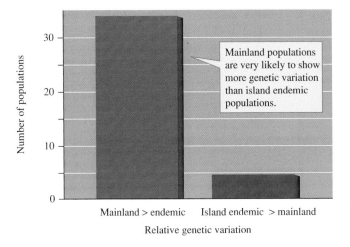

**Figure 8.23** Comparison of genetic variation in mainland versus island populations (data from Frankham 1997).

Frankham addressed these questions by thoroughly searching the extensive literature on genetic variation in animal and plant populations. His research uncovered 202 comparisons of genetic diversity in island versus mainland populations and 38 comparisons of genetic diversity in endemic species on islands versus related mainland species populations. The organisms in the analysis ranged from moose and wolves to toads, insects, and trees. The results of Frankham's analyses clearly support the hypothesis that genetic diversity is lower in island populations (fig. 8.22). Out of 202 mainland-island comparisons, 165 showed higher genetic variation in mainland populations compared to 37 which indicated higher genetic variation in island populations. Frankham found that the trend toward higher genetic variation in mainland populations was even stronger when he compared island endemic populations versus mainland populations of closely related species (fig. 8.23). Out of 38 endemic island-mainland comparisons, 34 showed higher genetic variation in mainland pop-

ulations compared to 4 which indicated higher genetic variation in endemic island populations.

Frankham's analysis takes us well beyond the study of how population size is related to genetic variation in populations of Chihuahua spruce (Ledig et al. 1997). It appears that in general, genetic variation is lower in isolated and generally smaller, island populations. What is the ecological significance of this result? One very fundamental point of interest is that genetic variation is the substrate upon which the environment can act to produce evolutionary change by natural selection. Reduced genetic variation indicates a lower potential for a population to evolve. One of Frankham's motivations for his study was to explore the possibility that lower genetic variation in island populations may contribute to the higher rates of extinction of island populations. By demonstrating that island populations have lower genetic variation than mainland populations, he shows that genetic factors cannot be eliminated as a contributor to the higher extinction rates observed on islands. However, while this study keeps genetic diversity alive as a viable hypothesis, it does not in itself demonstrate a connection between extinction rates and genetic diversity. That connection was made in a study published a year after Frankham's results appeared in print.

# Genetic Diversity and Butterfly Extinctions

The landscape of Åland in southwestern Finland is a patchwork of lakes, wetlands, cultivated fields, pastures, meadows, and forest (see fig. 21.12). Here and there in this well-watered landscape you can find dry meadows that support populations of plants, *Plantago lanceolata* and *Veronica spicata,* that act as hosts for the Glanville fritillary butterfly, *Melitaea cinxia* (fig. 8.24). As discussed in chapter 21, the meadows where *Melitaea* lives vary greatly in size, and *Melitaea* population size increases directly with the size of meadows (see fig. 21.13).

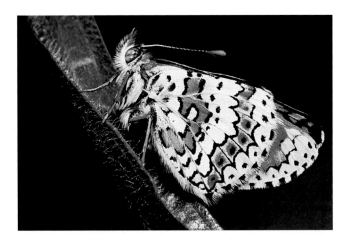

**Figure 8.24** Long-term studies of the Glanville fritillary butterfly, *Melitaea cinxia,* have provided exceptional insights into the relationship between population size and genetic diversity.

Careful studies of these populations by Ilkka Hanski, Mikko Kuussaari, and Marko Nieminen (1994) showed that small populations of *Melitaea* living in small meadows were most likely to go extinct.

Several factors likely influence the greater vulnerability of small populations to extinction. However, what role might genetic factors, especially reduced genetic variation, play in the vulnerability of small populations to extinction? Richard Frankham and Katherine Ralls (1998) point out that one of the contributors to higher extinction rates in small populations may be **inbreeding.** Inbreeding, which is mating between close relatives, is more likely in small populations. Combining already low genetic variation in small populations with a high rate of inbreeding has several negative impacts on populations, including reduced fecundity, lower juvenile survival, and shortened life span.

Ilik Saccheri and five coauthors (1998) reported one of the first studies giving direct evidence that inbreeding contributes to extinctions in wild populations. Saccheri and his colleagues studied 1,600 dry meadows and found *Melitaea* in 524, 401, 384, and 320 of the meadows in 1993, 1994, 1995, and 1996, respectively. Over this period they documented an average of 200 extinctions and 114 colonizations of meadows annually. As you can see, these populations are highly dynamic. In order to determine the extent that genetic factors, especially inbreeding, may contribute to extinctions, Saccheri and his colleagues conducted genetic studies on populations of *Melitaea* in 42 of the meadows. They estimated heterozygosity, an indicator of genetic variability, with respect to seven enzyme systems and one locus of nuclear microsatellite DNA. The researchers used the level of heterozygosity within each meadow population as an indicator of inbreeding, with low heterozygosity indicating high levels of inbreeding.

The results of the study indicated that influence of inbreeding on the probability of extinction was very significant. It turned out that the populations with the highest levels of inbreeding (lowest heterozygosity) had the highest probabilities of extinction. Saccheri and his colleagues found a connection between heterozygosity and extinction through effects on larval survival, adult longevity, and egg hatching. Females with low levels of heterozygosity produced smaller larvae fewer of which survived to the winter dormancy period. Pupae of mothers with low heterozygosity also spent more time in the pupal stage, exposing them to greater attack by parasites. In addition adult females with low heterozygosity had lower survival and laid eggs with a 24 to 46% lower rate of hatching. These effects have the potential to reduce the viability of local populations of *Melitaea,* which are made up of individuals of low heterozygosity (low genetic variation), and increase their risk of local extinction.

We have seen how the small population size and isolation can influence the genetic structure of populations of many kinds of organisms, including the Chihuahua spruce isolated in cool moist microenvironments in the mountains of Mexico and the Glanville fritillary, *Melitaea* in the dry meadow environments of southwestern Finland. In situations like these, chance plays a significant role in determining the genetic structure of populations.

## APPLICATIONS & TOOLS
### Estimating Genetic Variation in Populations

In chapter 8 we have focused considerable attention on genetic variation in populations. Here we return to genetic variation to review some historical and recent methods used to measure this significant aspect of population structure. How did the scientists whose work we discussed in this chapter study genetic variation in populations? The earlier research by Clausen, Keck, and Hiesey (1940) used transplant experiments to detect genetic differences among populations. Later research on genetic variation in Chihuahua spruce (Ledig et al. 1997) and in the Glanville fritillary butterfly (Saccheri et al. 1998) used techniques developed in molecular biology. Though the number of molecular-based studies of genetic variation is growing at a tremendous rate, transplant experiments remain a useful tool for studying genetic differences among populations.

## Transplant Experiments

The classical studies of variation among *Potentilla glandulosa* by Clausen, Keck, and Hiesey (1940) provide a model for the design and interpretation of transplant, or common garden, studies. Figure 8.25 shows photos of the transplant gardens used by Clausen and his colleagues at Stanford (lowland elevation 30 m), Mather (mid-elevation 1,400 m), and Timberline (alpine 3,050 m). Because these photos show the local natural vegetation in the background, they give a visual impression of the biomes in which the gardens were established. The natural vegetation at the sites were temperate woodland at the lowland elevation site, temperate coniferous forest at the mid-elevation site, and subalpine forest grading into alpine meadow at the alpine site.

As you would predict from our earlier review of the influence of elevation on climate (see fig. 2.38), the climates at the three study sites differed substantially. The growing seasons were 12 months at the lowland elevation site, 5 1/2 months at the mid-elevation site, and approximately 2 months at the alpine site. Minimum monthly temperatures ranged from $-2°C$ at the lowland elevation site and $-10°C$ at the mid-elevation site down to $-22°C$ at the alpine site. Maximum monthly temperatures ranged from $35°C$ at the lowland and mid-elevation sites to $25°C$ at the alpine site. While there was no snow at the lowland site, snow cover at the mid-elevation site generally persisted from October to April. Meanwhile, at the alpine site snows began in September and continued to approximately the first of July. This range of conditions certainly offers the potential for local adaptation and genetic variation among

**Figure 8.25** Photos of gardens used by Clausen, Keck, and Hiesey (1940) in transplant experiments with *Potentilla glandulosa*. The photos show (a) the Timberline (3,050 m), (b) Mather (1,400 m), and (c) Stanford (30 m) sites.

local populations of *P. glandulosa.* Clausen and his colleagues designed their transplant experiments to reveal those differences if they existed.

Figure 8.26 summarizes the details of the *P. glandulosa* transplant experiments. The upper panel of figure 8.26 sketches how plants from each study area were transplanted to the other garden sites where they were grown beside the local plants. How did Clausen, Keck, and Hiesey's transplant experiment indicate genetic differences among local populations of *P. glandulosa?* To understand how their results showed genetic differences we need to consider how the results would have looked if there were no genetic differences among local populations. This hypothetical situation is presented as a null hypothesis in the middle panel of figure 8.26. If there were no genetic differences among populations, all plants would have shown the same characteristics at each site. Contrast these uniform responses, expected if the null hypothesis were true, with the representation of the actual results in the lower panel. Each population showed unique growth responses at each of the transplant gardens. On the basis of differences in growth, flower number (see fig. 8.4), survival, and several other characteristics, Clausen and his coauthors concluded that the study populations of *P. glandulosa* differed genetically.

The continued utility of transplant experiments is shown by the results of Tracy's (1999) study of variation among chuckwalla lizard, *Sauromalus,* populations (see fig. 8.10). In that study Tracy transplanted lizards from different regions into a controlled laboratory environment. Since modern molecular methods allow us to look directly at genetic differences among populations, why would some biologists continue to use transplant experiments? One advantage of transplant experiments is that they are simple and require little investment in technology. What are some of the disadvantages of transplant experiments? They often require more time and labor to carry out and they can be applied to a limited number of organisms. While transplant experiments continue to be useful, modern molecular techniques are allowing evolutionary ecologists to explore details of genetic variation within and among populations that would be impossible without these modern techniques.

## Molecular Approaches to Genetic Variation

The tools of molecular biology can be used to determine the genotypes of individuals either by looking at products of genes, such as enzymes, or by analyzing DNA directly. In chapter 8, Ledig and his coauthors (1997) estimated genetic variation in populations of Chihuahua pine by measuring variation in the allozymes of 16 different enzyme systems (see fig. 8.21). Because allozymes of the same enzyme are the products of different alleles of the same gene locus, the number of allozymes produced by a population can be used as an indicator of genetic variation within the population. Many studies of enzymes examine all **isozymes,** which are all enzymes with the same biochemical function. Different isozymes may be produced by the same or

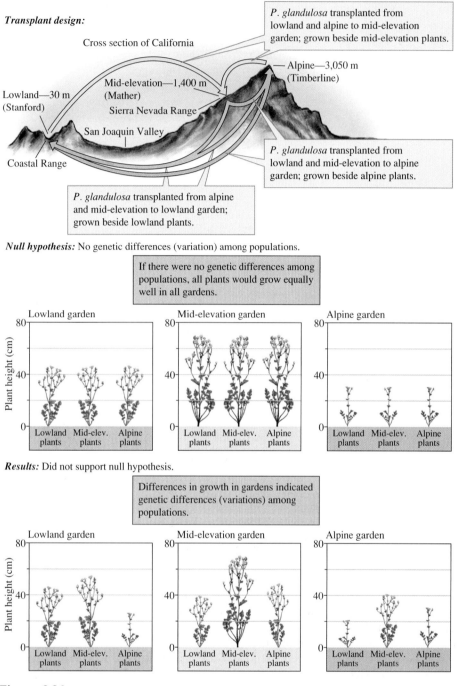

**Figure 8.26** A common garden approach to studying genetic variation among populations of *Potentilla glandulosa* (data from Clausen, Keck, and Hiesey 1940).

different loci. Though enzyme studies remain a useful and powerful tool in evolutionary studies, genetic variation is increasingly assessed by looking directly at DNA. For instance, Saccheri and his colleagues (1998) used a combination of enzyme and direct DNA studies to characterize the genetic structure of populations of the Glanville fritillary butterfly. A detailed review of molecular methods used to study genetic variation is well beyond the scope of this discussion. However, reviewing at least the basics of some of the common molecular methods used

to study genetic variation will offer an entry to this powerful set of modern tools.

In enzyme studies the tissues of organisms are generally mechanically homogenized and the resulting homogenate analyzed for the presence and kinds of enzymes. Generally, larger tissue samples are required for enzyme studies than for studies of DNA. Since DNA studies may be performed on very small samples, biologists may sample populations without damaging them. Nondestructive sampling is especially important in the study of endangered species or in any study following known individuals over long periods of time. For instance, the grizzly bears of Glacier National Park are being counted and mapped using the DNA in hair that the bears leave on scratching trees and on baited hair traps (USGS 2000). To obtain sufficient quantities of DNA for analysis, such as that contained within a hair follicle, biologists generally use one of two techniques to amplify the quantity of DNA present in a sample. DNA is usually cloned either by using bacteria and recombinant DNA technology or by a procedure called polymerase chain reaction or PCR (Hillis et al. 1996). During the PCR process, short, single-stranded DNA is used as primers for DNA synthesis. Each primer is highly specific for a given nucleotide sequence and can be used to amplify a specific locus or gene. However in one approach, designed simply to identify genetic differences or similarities, somewhat randomly chosen primers are used to amplify unspecified DNA sequences. This is the so-called random amplified polymorphic DNA, or RAPD, method. These techniques are well presented in many introductory biology texts.

Once a sufficient quantity of DNA has been obtained, the sample may be analyzed in several ways. One commonly applied method uses **restriction enzymes,** enzymes produced naturally by bacteria to cut up foreign DNA. Restriction enzymes cut DNA molecules at particular places called **restriction sites.** The locations of restriction sites along a DNA molecule are determined by the locations of specific **nucleotide** sequences. Nucleotides are the basic building blocks of nucleic acids and are made up of a five-carbon sugar (deoxyribose or ribose), a phosphate group, and a nitrogenous base (guanine, cytosine, adenine, or thymine). The nucleotide sequences determining restriction sites along the length of a DNA molecule are different for different restriction enzymes. Because restriction sites are determined by a specific sequence of base pairs on the DNA molecule, differences in number and location of restriction sites reflect differences in DNA structure. When exposed to a particular restriction enzyme, a given DNA molecule will be broken up into a series of DNA fragments of precise number and lengths. The number and lengths of DNA fragments, called **restriction fragments,** are determined by the number and location of restriction sites for a particular restriction enzyme. Therefore, if DNA samples from different organisms exposed to the same restriction enzyme yield different numbers and lengths of DNA fragments, we can conclude that those organisms differ genetically.

The number and sizes of restriction fragments resulting from treating a DNA molecule with restriction enzymes or the number of isozymes present in the homogenized tissues of an organism may be analyzed using a technique called **electrophoresis.** Electrophoresis uses the rate at which enzymes, DNA fragments, or other macromolecules move in an electrical field as a means of identifying the molecules (fig. 8.27). When placed in an electrical field, a molecule will move either toward the positive or negative end of the field. Negatively charged molecules will move toward the positive end, while positively charged molecules will move toward the negative pole. Smaller molecules move more rapidly than larger molecules. Due to the influences of molecule size and charge on rates of movement, isozymes or DNA restriction fragments of different structure will migrate at different rates during electrophoresis. Consequently during a given time interval, molecules of different sizes will migrate different distances from the point where they are initially placed in the electrical field.

Electrophoresis is generally referred to as gel electrophoresis because migration of molecules generally takes place in one of several possible types of gels. Various stains and other techniques have been developed to detect the locations of DNA fragments or of specific enzymes within the gel after an electrophoresis run. The result is a pattern of banding in a gel that generally allows the biologist to identify genetic differences among individuals. By sampling many individuals from a population researchers can characterize the genetic structure of the population and determine if populations differ genetically.

What do the banding patterns, such as those shown in figure 8.27, reveal about the genetics of individuals and populations? We can say that the sample of four hypothetical individuals depicted in figure 8.27 includes three different genotypes. Individuals I and II have the same genotype, while individuals III and IV are of two other genotypes. By sampling many individuals and many enzyme systems or genetic loci in a population, the biologist will be able to estimate the genetic variation and genetic composition of a population. After characterizing several populations, we can test questions such as whether population size influences genetic variation in species such as the Chihuahua spruce (see fig. 8.21) or whether island populations of a species have less genetic variation than mainland populations of the same species (see fig. 8.23).

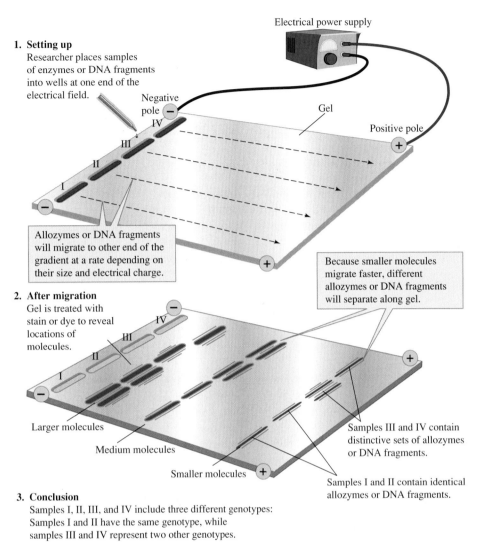

**1. Setting up**
Researcher places samples of enzymes or DNA fragments into wells at one end of the electrical field.

Electrical power supply

Negative pole

Positive pole

Gel

Allozymes or DNA fragments will migrate to other end of the gradient at a rate depending on their size and electrical charge.

**2. After migration**
Gel is treated with stain or dye to reveal locations of molecules.

Because smaller molecules migrate faster, different allozymes or DNA fragments will separate along gel.

Larger molecules

Medium molecules

Smaller molecules

Samples III and IV contain distinctive sets of allozymes or DNA fragments.

Samples I and II contain identical allozymes or DNA fragments.

**3. Conclusion**
Samples I, II, III, and IV include three different genotypes: Samples I and II have the same genotype, while samples III and IV represent two other genotypes.

**Figure 8.27**  Gel electrophoresis can be used to study genetic variation.

An approach that gives a very high resolution picture of the genetic makeup of individuals and populations and that is receiving increasing attention is **DNA sequencing.** Because sequencing reveals the sequence of nucleic acids along DNA molecules, this tool gives the ultimate genetic information. The number of DNA sequences described is increasing rapidly and our ability to interpret and compare DNA sequence data is also increasing at an impressive rate (Hillis et al. 1996). While the human genome project has assumed center stage (DOE 2000), the genomes of many other species are completely described or will be soon.

David Hillis and his coauthors (1996) suggest that DNA sequencing can be used as a powerful tool for studying genetic variation within and among populations. Some of the areas where sequencing might be applied include geographic variation among populations and gene flow among popula-

tions. However, Hillis and his team point out that there are trade-offs. Obtaining and interpreting the highly detailed information provided by sequencing for one or two loci necessarily limits the number of loci that the biologist can study. Where the emphasis is on studying larger numbers of loci, isozyme studies or restriction fragment analyses allow the researcher to study larger numbers of loci. At this point in time, the biologist's choice of methods is governed by these trade-offs.

Future advances in DNA sequencing and analysis will very likely improve the potential for comparing large numbers of loci using sequence data. Regardless of future development, ecologists now have many powerful tools for assessing the extent of genetic variation in populations. These tools will be invaluable as this generation of ecologists works to conserve the earth's biodiversity.

## SUMMARY

Darwin and Mendel complemented each other well and their twin visions of the natural world revolutionized biology. The synthesis of the theory of natural selection and genetics gave rise to modern evolutionary ecology. Here we examine five major concepts within the area of population genetics and natural selection.

**Phenotypic variation among individuals in a population results from the combined effects of genes and environment.** The first biologists to conduct thorough studies of phenotypic and genotypic variation and to incorporate experiments in their studies, focused on plants. Clausen, Keck, and Hiesey explored the extent and sources of morphological variation in plant populations, including both the influences of environment and genetics. Case determined that the best predictor of chuckwalla, *Sauromalus,* body length was average winter rainfall. Tracy's laboratory growth experiments indicated that variation in body size among chuckwalla populations is at least partly determined by genetic differences among populations.

**The Hardy-Weinberg equilibrium model helps identify evolutionary forces that can change gene frequencies in populations.** Because evolution involves changes in gene frequencies in a population, a thorough understanding of evolution must include the area of genetics known as population genetics. One of the most fundamental concepts in population genetics, the Hardy-Weinberg principle, states that in a population mating at random in the absence of evolutionary forces, allele frequencies will remain constant. For a population in Hardy-Weinberg equilibrium in a situation where there are only two alleles at a particular locus, $p + q = 1.0$. The frequency of genotypes in a population in Hardy-Weinberg equilibrium can be calculated as $(p + q)^2 = (p + q) \times (p + q) = p^2 +$ $2pq + q^2 = 1.0$. The conditions necessary to maintain constant allele frequencies in a population are: (1) random mating, (2) no mutations, (3) large population size, (4) no immigration, and (5) equal survival and reproductive rates for all genotypes. When a population is not in Hardy-Weinberg equilibrium, the Hardy-Weinberg principle helps us to identify the evolutionary forces that may be in play.

**Natural selection is the result of differences in survival and reproduction among phenotypes.** Natural selection can lead to change in populations but it can also serve as a conservative force, impeding change in a population. Stabilizing selection acts against extreme phenotypes and as a consequence, favors the average phenotype. By favoring the average phenotype, stabilizing selection decreases phenotypic diversity in populations. Directional selection favors an extreme phenotype over other phenotypes in the population. Under directional selection, the average of the trait under selection can change over time. Disruptive selection favors two or more extreme phenotypes over the average phenotype in a population, leading to a increase in phenotypic diversity in the population.

**The extent to which phenotypic variation is due to genetic variation determines the potential for evolution by natural selection.** The most general postulate of the theory of natural selection is that the environment determines the evolution of the anatomy, physiology, and behavior of organisms. Some of the clearest demonstrations of natural selection have resulted from studies of populations of Galápagos finches. Losos, Warheit, and Schoener used replicated field experiments to study natural selection for changes in morphology in *Anolis* lizard populations. Their results indicate that colonizing populations can adapt rapidly to new environmental conditions. Studies by Carroll and several colleagues show that soapberry bug populations living on native and introduced host plants have undergone natural selection for traits that favor their survival and reproduction on particular host plant species. Hundreds of other examples of natural selection have been brought to light during the nearly one and a half century since Darwin published his theory. Still, evolutionary ecology remains a vigorous field of inquiry with plenty of debate, self-criticism, and significant work yet to be done.

The earlier research on adaptation of populations to local environmental conditions used transplant experiments to detect genetic differences among populations. More recent research on genetic variation within and among populations has applied techniques developed in molecular biology. Ecologists now have many powerful tools, ranging from classical techniques to modern technologically sophisticated approaches, for assessing the extent of genetic variation within and among populations and meeting the challenge of documenting and conserving biodiversity.

**Random processes, such as genetic drift, can change gene frequencies in populations, especially in small populations.** Genetic drift is theoretically most effective at changing gene frequencies in small populations such as those that inhabit islands. One of the greatest concerns associated with fragmentation of natural ecosystems due to human land use is that reducing habitat availability will decrease the size of animal and plant populations to the point where genetic drift will reduce the genetic diversity within natural populations. Ledig and his colleagues found a significant positive correlation between population size and genetic diversity in populations of Chihuahua spruce, a naturally fragmented population of trees living on mountain islands. Frankham showed that compared to mainland populations, island populations generally include less genetic variation. Saccheri and his colleagues found that higher heterozygosity (genetic diversity) was associated with lower rates of population extinction through the effects of heterozygosity on larval survival, adult longevity, and egg hatching in populations of the Glanville fritillary butterfly, *Melitaea cinxia*.

# Review Questions

1. Contrast the approaches of Charles Darwin and Gregor Mendel to the study of populations. What were Darwin's main discoveries? What were Mendel's main discoveries? How did the studies of Darwin and Mendel prepare the way for the later studies reviewed in chapter 8?

2. Review the historical studies of genetic and phenotypic variation among populations of plants using transplant experiments. How did the studies of Clausen, Keck, and Hiesey complement these earlier studies?

3. What environmental variable did Ted Case determine to be the best predictor of variation in body size among populations of chuckwallas? Did Case's studies of chuckwallas demonstrate genetic differences among his study populations? What did the more recent studies by Christopher Tracy add to our understanding of variation among chuckwalla populations?

4. What is the Hardy-Weinberg principle? What is Hardy-Weinberg equilibrium? What conditions are required for Hardy-Weinberg equilibrium?

5. Review the Hardy-Weinberg equilibrium equation. What parts of the equation represent gene frequencies? What elements represent genotype frequencies and phenotype frequencies? Are genotype and phenotype frequencies always the same? Use a hypothetical population to specify alleles and allelic frequencies as you develop your presentation.

6. What is genetic drift? Under what circumstances do you expect genetic drift to occur? Under what circumstances is genetic drift unlikely to be important? Does genetic drift increase or decrease genetic variation in populations.

7. Suppose you are a director of a captive breeding program for a rare species of animal, such as Siberian tigers, that are found in many zoos around the world but are increasingly rare in the wild. Design a breeding program that will reduce the possibility of genetic drift in captive populations.

8. Jonathan Losos, Kenneth Warheit, and Thomas Schoener's studies of *Anolis* populations demonstrated significant morphological change following introduction of the lizards to various islands differing in vegetative structure. Design an experiment to determine whether the morphological changes in the study populations were based on genetic changes. Can you adapt the methods of Christopher Tracy to this project?

9. How did the studies of Scott Carroll and his colleagues demonstrate rapid evolutionary adaptation to introduced soapberry plants? What advantages do a group of organisms, such as soapberry bugs, offer to researchers studying natural selection compared to larger organisms such as Chihuahua pines and chuckwalla lizards?

10. How do classical approaches to genetic studies, such as common garden experiments, and modern molecular techniques, such as DNA sequencing, complement each other? What are the advantages and disadvantages of each?

# Suggested Readings

Carroll, S. P. and C. Boyd. 1992. Host race radiation in the soapberry bug: natural history with the history. *Evolution* 46:1052–69.

Carroll, S. P., S. P. Klassen, and H. Dingle. 1998. Rapidly evolving adaptations to host ecology and nutrition in the soapberry bug. *Evolutionary Ecology* 12:955–68.

*This pair of papers traces the fascinating story of some of the research that has revealed one of the best documented cases of natural selection of herbivorous insect populations for living as specialists on particular plant species. These papers showcase well designed and carefully executed studies of evolutionary ecology.*

Carroll, S. P., H. Dingle, T. R. Famula, and C. W. Fox. 2001. Genetic architecture of adaptive differentiation in evolving host races of the soapberry bug, *Jadera haematoloma. Genetica* 112/113:257–72.

*Further studies of the genetic basis for adaptation of soapberry bug populations to new host plants.*

Case, T. J. 1976. Body size differences between populations of the chuckwalla, *Sauromalus obesus. Ecology* 57:313–23.

Tracy, C. R. 1999. Differences in body size among chuckwalla (*Sauromalus obesus*) populations. *Ecology* 80:259–271.

*Separated by over 20 years, these companion papers explore the relationship between climate, variation in morphology, and local adaptation by chuckwallas—careful work on an interesting animal.*

Clausen, J., D. D. Keck, and W. M. Hiesey. 1940. *Experimental Studies on the Nature of Species. I. The Effect of Varied Environments on Western North American Plants.* Washington, DC: Carnegie Institution of Washington Publication no. 520.

*This classic in evolutionary plant ecology shows the great care and thoroughness of some of the early work on plant evolution in North America.*

Darwin, C. 1839. *Journal of Researches into the Geology and Natural History of the Various Countries Visited During the Voyage of H.M.S. 'Beagle' Under the Command of Captain FitzRoy, R.N., From 1832–1836.* London: Henry Colborn. [also available on the Internet—see "On the Net" section.]

*A fascinating account of one of the most important intellectual voyages in the history of science told by the young Charles Darwin. Recommended for anyone interested in the history of evolutionary ideas.*

Darwin, C. 1859. *The Origin of Species by Means of Natural Selection, or the Preservation of Favored Races in the Struggle for Life.* New York: Modern Library [also available on the Internet—see "On the Net" section.]

*A landmark publication in the history of science.*

Douglas, M. R. and P. C. Brunner. 2002. Biodiversity of central alpine *Coregonus* (Salmoniformes): impact of one-hundred years of management. *Ecological Applications* 12:154–72.

*A thorough and modern exploration of the genetic morphologic structure of a complex evolutionary lineage, made even more complex by intensive management.*

Frankham, R. and K. Ralls. 1998. Inbreeding leads to extinction. *Nature* 392:441–42.

Saccheri, I., M. Kuussaari, M. Kankare, P. Vikman, W. Fortelius, I. Hanski. 1998. Inbreeding and extinction in a butterfly metapopulation. *Nature* 392:491–94.

*Articles summarizing a landmark research project that demonstrates a long-suspected link between inbreeding and extinction. The Frankham and Ralls paper provides a broad interpretive framework for the study by Saccheri and colleagues.*

Hansen, K. T., R. Elven, and C. Brochmann. 2000. Molecules and morphology in concert: tests of some hypotheses in arctic *Potentilla* (Rosaceae). *American Journal of Botany* 87:1466–79.

*Application of genetic techniques to the study of speciation and morphological variation among arctic populations of* Potentilla.

Hillis, D. M., C. Moritz, and B. K. Mable. 1996. *Molecular Systematics.* Sunderland, Mass.: Sinauer Associates, Inc.

*This is a highly detailed treatment of molecular approaches to studying the genetics of populations; suggested here as a reference.*

Losos, J. B., I. Warheit, and T. W. Schoener. 1997. Adaptive differentiation following experimental island colonization in *Anolis* lizards. *Nature* 387:70–73.

*Short article on an experiment destined to become a classic in the literature on evolutionary ecology.*

Losos, J. B., T. W. Schoener, K. I. Warheit, and D. Creer. 2001. Experimental studies of adaptive differentiation in Bahamian *Anolis* lizards. *Genetica* 112/113:399–415.

*Study of the extent of morphological plasticity in populations of* Anolis *lizards.*

Orel, V. 1996. *Gregor Mendel: The First Geneticist.* Oxford: Oxford University Press.

*A fascinating account of the life of one of the most important figures in the history of biology. Orel includes many details absent from other biographies of Mendel.*

Pappers, S. M., G. van der Velde, N. J. Ouborg, and J. M. van Groenendael. 2002. Genetically based polymorphisms in morphology and life history associated with putative host races of the water lily leaf beetle *Galerucella nymphaeae. Evolution* 56:1610–21.

*Carefully controlled studies of heritability of polymorphic traits in populations of the water lily leaf beetle.*

# On the Net

Visit this textbook's accompanying website at www.mhhe.com/ecology (click on the book's title) to take advantage of practice quizzing, study/writing tips, timely news articles, and additional URLs for research on the topics in this chapter.

Evolution

Evidences for Evolution

Natural Selection

Speciation

Macroevolution

Microevolution

Recombinant Technology

# *Chapter*

# 9

# Population
# Distribution
# and Abundance

The distributions and dynamics of populations vary widely among species. While some populations are small and have highly restricted distributions, other populations number in the millions and may range over vast areas of the planet. Standing on a headland in central California overlooking the Pacific Ocean, a small group of students spots a group of gray whales, *Eschrichtilus robustus*, rising to the surface and spouting water as they swim northward (fig. 9.1*a*). The whales are rounding the point of land on their way to feeding grounds off the coasts of Alaska and Siberia. This particular group is made up of females and calves. The calves were born during the previous winter along the coast of Baja California, the gray whale's wintering grounds. Over the course of the spring, the entire population of over 20,000 gray whales will round this same headland on their way to the Bering and Chukchi Seas. Gray whales travel from one end of their range to the other twice each year, a distance of about 18,000 km. Home to the gray whale encompasses a swath of seacoast extending from southern Baja California to the coast of northeast Asia.

The grove of pine trees on the headland where the students stand gazing at the whales is winter home to another long distance traveler: monarch butterflies, *Danaus plexippus* (fig. 9.1*b*). The lazy flying of the bright orange and black monarch butterflies gives no hint of their capacity to migrate. Some of the butterflies flew to the grove of pines the previous autumn from as far away as the Rocky Mountains of southern Canada. As the students watch the whales, the male monarch butterflies pursue and mate with the female monarch butterflies. After mating, the males die, while the females begin a migration that leads inland and north. The females stop to lay eggs on any milkweeds they encounter along the way and eventually die; however, their offspring continue the migration. The monarch caterpillars grow quickly on their diet of milkweed and then transform to pupae contained within cocoons. The monarch butterflies that emerge from the cocoons mate and, like the previous generation, fly northward and inland. By moving farther north and inland each generation, some of the monarch butterflies eventually reach the Rocky Mountains of southern Canada, far from where their ancestors fluttered around the group of students on the pine-covered coastal headland.

Then as the autumn days grow shorter, the monarch butterflies begin their long flight back to the coastal grove of pines. This autumn generation, which numbers in the millions, flies southwest to their wintering grounds on the coast of central and southern California. Some of them might fly over 3,000 km. The monarch butterflies that survive the trip to the pine grove overwinter, hanging from particular roost trees in the thousands. They mate in the following spring and start the cycle all over again.

Gray whales and monarch butterflies, as different as they may appear, lead parallel lives. The Monterey pines, *Pinus radiata,* covering the headland where the monarch butterflies overwinter and by which the gray whales pass twice each year are quite different. The Monterey pine population does not migrate each generation and has a highly restricted distribution. The current natural range of the Monterey pine is limited to a few sites on the coast of central and northern California and to two islands off the coast of western Mexico. These scattered populations are

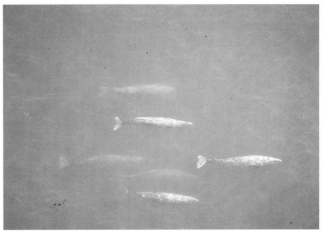

(a)

(b)

**Figure 9.1** (*a*) During their annual migration, the entire population of gray whales migrates from subtropical waters off Baja California to the Arctic and back again. (*b*) Some of the monarch butterflies roosting on these trees flew thousands of kilometers from the Rocky Mountains to reach their winter roost. In contrast, the entire natural population of the Monterey pine, *Pinus radiata,* is restricted to five small areas along the coast of California.

the remnants of a large continuous population that extended for over 800 km along the California coast during the cooler climate of the last glacial period.

With these three examples, we begin to consider the ecology of populations. Ecologists usually define a **population** as a group of individuals of a single species inhabiting a specific area. A population of plants or animals might occupy a mountaintop, a river basin, a coastal marsh, or an island, all areas defined by natural boundaries. Just as often, the populations studied by biologists occupy artificially defined areas such as a particular country, county, or national park. The areas inhabited by populations range in size from the few cubic centimeters occupied by the bacteria in a rotting apple to the millions of square kilometers occupied by a population of migratory whales. A population studied by ecologists may consist of a highly localized group of individuals representing a fraction of the total population of a species, or it may consist of all the individuals of a species across its entire range.

Ecologists study populations for many reasons. Population studies hold the key to saving endangered species, controlling pest populations, and managing fish and game populations. They also offer clues to understanding and controlling disease epidemics. Finally, the greatest environmental challenge to biological diversity and the integrity of the entire biosphere is at its heart a population problem—the growth of the human population.

All populations share several characteristics. The first is its distribution. The distribution of a population includes the size, shape, and location of the area it occupies. A population also has a characteristic pattern of spacing of the individuals within it. It is also characterized by the number of individuals within it and their **density,** which is the number of individuals per unit area. Additional characteristics of populations—their age distributions, birth and death rates, immigration and emigration rates, and rates of growth—are the subject of chapters 10 and 11. In chapter 9 we focus on two population characteristics: **distribution** and **abundance**.

## CONCEPTS

- **The physical environment limits the geographic distribution of species.**
- **On small scales, individuals within populations are distributed in patterns that may be random, regular, or clumped; on larger scales, individuals within a population are clumped.**
- **Many populations are subdivided into subpopulations called metapopulations.**
- **Population density declines with increasing organism size.**
- **Commonness and rarity of species are influenced by population size, geographic range, and habitat tolerance.**

## CONCEPT DISCUSSION

### Distribution Limits

**The physical environment limits the geographic distribution of species.**

A major theme in chapters 4, 5, and 6 is that individual organisms have evolved physiological, anatomical, and behavioral characteristics that compensate for environmental variation. Organisms compensate for temporal and spatial variation in the environment by regulating body temperature and water content and by foraging in a way that maintains energy intake at relatively high levels. However, there are limits on how much organisms can compensate for environmental variation.

While there are few environments on earth without life, no single species can tolerate the full range of earth's environments. For each species some environments are too warm, too cold, too saline, or unsuitable in other ways. As we saw in chapter 6, organisms take in energy at a limited rate. It appears that at some point, the metabolic costs of compensating for environmental variation may take up too much of an organism's energy budget. Partly because of these energy constraints, the physical environment places limits on the distributions of populations. Let's now turn to some actual species and explore the factors that limit their distributions.

## Kangaroo Distributions and Climate

The Macropodidea includes the kangaroos and wallabies, which are some of the best known of the Australian animals. However, this group of large-footed mammals includes many less familiar species, including rat kangaroos and tree kangaroos. While some species of macropods can be found in nearly every part of Australia, no single species ranges across the entire continent. All are confined to a limited number of climatic zones and biomes.

G. Caughley and his colleagues (1987) found a close relationship between climate and the distributions of the three largest kangaroos in Australia (fig. 9.2). The eastern grey kangaroo, *Macropus giganteus,* is confined to the eastern third of the continent. This portion of Australia includes several biomes (see chapter 2). Temperate forest grows in the southeast and tropical forests in the north. Mountains, with their varied climates, occupy the central part of the eastern grey kangaroo's range (see figs. 2.13, 2.28, and 2.37). The climatic factor that distinguishes these varied biomes is little seasonal variation in precipitation or dominance by summer precipitation. The western grey kangaroo, *M. fuliginosus,* lives mainly in the southern and western regions of Australia. Most of the western grey kangaroo's range coincides with the distribution

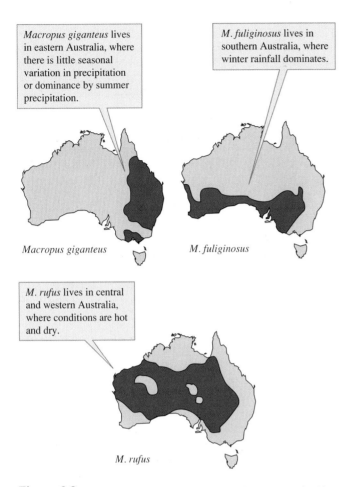

*Macropus giganteus* lives in eastern Australia, where there is little seasonal variation in precipitation or dominance by summer precipitation.

*M. fuliginosus* lives in southern Australia, where winter rainfall dominates.

*Macropus giganteus*

*M. fuliginosus*

*M. rufus* lives in central and western Australia, where conditions are hot and dry.

*M. rufus*

**Figure 9.2** Climate and the distributions of three kangaroo species (data from Caughley et al. 1987).

of the temperate woodland and shrubland biome in Australia. The climatically distinctive feature of this biome is a predominance of winter rainfall (see fig. 2.22). Meanwhile, the red kangaroo, *M. rufus,* wanders the arid and semiarid interior of Australia. The biomes that cover most of the red kangaroo's range are savanna and desert (see figs. 2.16 and 2.19). Of the three species of large kangaroos, the red kangaroo occupies the hottest and driest areas.

The distributions of these three large kangaroo species cover most of Australia. However, as you can see in figure 9.2, none of these species lives in the northernmost region of Australia. Caughley and his colleagues explain that these northern areas are probably too hot for the eastern grey kangaroo, too wet for the red kangaroo, and too hot in summer and too dry in winter for the western grey kangaroo. However, they are also careful to point out that these limited distributions may not be determined by climate directly. Instead, they suggest that climate often influences species distributions through factors such as food production, water supply, and habitat. Climate also affects the incidence of parasites, pathogens, and competitors.

Regardless of how the influences of climate are played out, the relationship between climate and the distributions of species can be stable over long periods of time. The distribu-

tions of the eastern grey, western grey, and red kangaroos have been stable for at least a century. In the next example, we discuss a species of beetle that appears to have maintained a stable association with climate for 10,000 to 100,000 years.

## A Tiger Beetle of Cold Climates

Tiger beetles have entered our discussions several times. In chapter 4, we saw how one species regulates body temperatures on hot black beaches in New Zealand. In chapter 5, we compared the water loss rates of tiger beetles from desert grasslands and riparian habitats in Arizona. Here we consider the distribution of a tiger beetle that inhabits the cold end of the range of environments occupied by tiger beetles.

The tiger beetle *Cicindela longilabris* lives at higher latitudes and higher elevations than just about any other species of tiger beetle in North America. In the north, *C. longilabris* is distributed from the Yukon Territory in northwestern Canada to the maritime provinces of eastern Canada (fig. 9.3). This northern band of beetle populations coincides with the distribution of northern temperate forest and boreal forest in North America (see figs. 2.28 and 2.31). *C. longilabris* also lives as far south as Arizona and New Mexico. However, these southern populations are confined to high mountains, where *C. longilabris* is associated with montane coniferous forests. As we saw in chapter 2, these high mountains have a climate similar to that of boreal forest (see fig. 2.38).

Ecologists suggest that during the last glacial period *C. longilabris* lived far south of its present range limits. Then with climatic warming and the retreat of the glaciers, the tiger beetles followed their preferred climate northward and up in elevation into the mountains of western North America (fig. 9.3). As a consequence, the beetles in the southern part of this species range live in isolated mountaintop populations. This hypothesis is supported by the fossil records of many beetle species.

Intrigued by the distribution and history of *C. longilabris,* Thomas Schultz, Michael Quinlan, and Neil Hadley (1992) set out to study the environmental physiology of widely separated populations of the species. Populations separated for many thousands of years may have been exposed to significantly different environmental regimes. If so, natural selection could have produced significant physiological differences among populations. The researchers compared the physiological characteristics of beetles from populations of *C. longilabris,* from Maine, Wisconsin, Colorado, and northern Arizona. Their measurements included water loss rates, metabolic rates, and body temperature preferences.

Schultz and his colleagues found that the metabolic rates of *C. longilabris* are higher and its preferred temperatures lower than those of most other tiger beetle species that have been studied. These differences support the hypothesis that *C. longilabris* is adapted to the cool climates of boreal and montane forests. In addition, the researchers found that none of their measurements differed significantly among populations

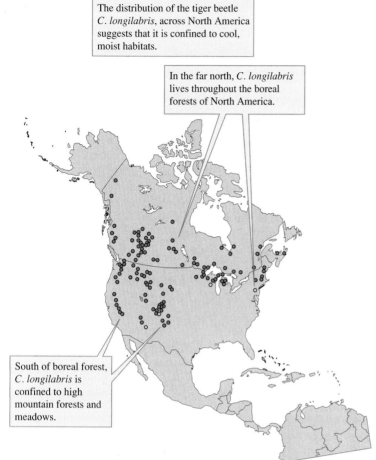

The distribution of the tiger beetle *C. longilabris*, across North America suggests that it is confined to cool, moist habitats.

In the far north, *C. longilabris* lives throughout the boreal forests of North America.

South of boreal forest, *C. longilabris* is confined to high mountain forests and meadows.

**Figure 9.3** A tiger beetle, *Cicindela longilabris,* confined to cool environments. Physiological studies conducted on populations indicated by yellow dots (data from Schultz, Quinlan, and Hadley 1992).

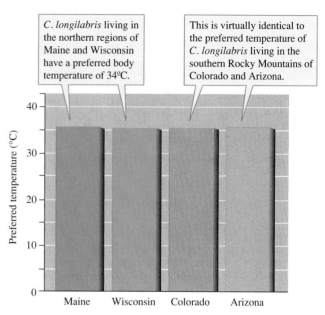

*C. longilabris* living in the northern regions of Maine and Wisconsin have a preferred body temperature of 34ºC.

This is virtually identical to the preferred temperature of *C. longilabris* living in the southern Rocky Mountains of Colorado and Arizona.

**Figure 9.4** Uniform temperature preference across an extensive geographic range (data from Schultz, Quinlan, and Hadley 1992).

of *C. longilabris,* Figure 9.4 illustrates the remarkable similarity in preferred body temperature shown by foraging *C. longilabris* from populations separated by as much as 3,000 km and, perhaps, by 10,000 years of history. These results support the generalization that the physical environment limits the distributions of species. It also suggests that those limits may be stable for long periods of time.

Now, let's consider how the physical environment may limit the distribution of plants. Our example is drawn from the arid and semiarid regions of the American Southwest.

## Distributions of Plants along a Moisture-Temperature Gradient

In chapter 4, we discussed the influence of pubescence on leaf temperature in plants of the genus *Encelia*. Variation in leaf pubescence among *Encelia* species appears to correspond directly to the distributions of these species along a moisture-temperature gradient from the California coast eastward (Ehleringer and Clark 1988). *Encelia californica,* the species

with the least pubescent leaves, occupies a narrow coastal zone that extends from southern California to northern Baja California (fig. 9.5). Inland, *E. californica* is replaced by *E. actoni,* which has leaves that are slightly more pubescent. Still farther to the east, *E. actoni* is in turn replaced by *E. frutescens* and *E. farinosa.*

These geographic limits to these species' distributions correspond to variations in temperature and precipitation. The coastal environments where *E. californica* lives are all relatively cool. However, average annual precipitation differs a great deal across the distribution of this species. Annual precipitation ranges from about 100 mm in the southern part of its distribution to well over 400 mm in the northern part. By comparison, *E. actoni* occupies environments that are only slightly warmer but considerably drier. The rainfall in areas occupied by *E. frutescens* and *E. farinosa* is similar to the amount that falls in the areas occupied by *E. actoni* and *E. californica.* However, the environments of *E. frutescens* and *E. farinosa* are much hotter.

Variation in leaf pubescence does not correspond entirely to the macroclimates inhabited by *Encelia* species. The leaves of *E. frutescens* are nearly as free of pubescence as the coastal species *E. californica.* However, *E. frutescens* grows side by side with *E. farinosa* in some of the hottest deserts in the world. Because they are sparsely pubescent, the leaves of *E. frutescens* absorb a great deal more radiant energy than the leaves of *E. farinosa* (fig. 9.6). Under similar conditions, however, leaf temperatures of the two species are nearly identical. How does *E. frutescens* avoid overheating? The leaves do not overheat because they transpire at a high rate and are evaporatively cooled as a consequence.

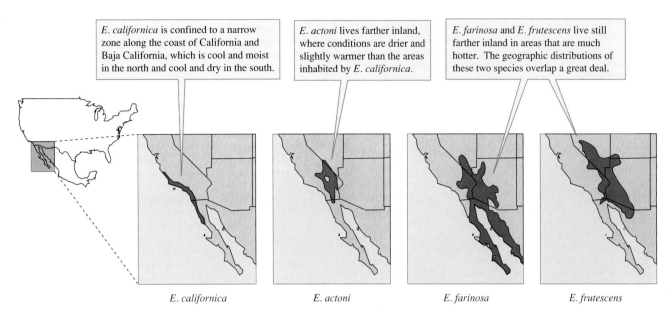

E. californica is confined to a narrow zone along the coast of California and Baja California, which is cool and moist in the north and cool and dry in the south.

E. actoni lives farther inland, where conditions are drier and slightly warmer than the areas inhabited by E. californica.

E. farinosa and E. frutescens live still farther inland in areas that are much hotter. The geographic distributions of these two species overlap a great deal.

E. californica    E. actoni    E. farinosa    E. frutescens

**Figure 9.5** The distributions of four *Encelia* species in southwestern North America (data from Ehleringer and Clark 1988).

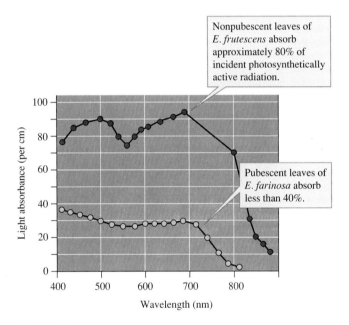

Nonpubescent leaves of *E. frutescens* absorb approximately 80% of incident photosynthetically active radiation.

Pubescent leaves of *E. farinosa* absorb less than 40%.

**Figure 9.6** Light absorption by leaves of *Encelia frutescens* and *E. farinosa* (data from Ehleringer and Clark 1988).

Evaporative cooling solves one ecological puzzle but appears to create another. Remember that these two shrubs live in some of the hottest and driest deserts in the world. Where does *E. frutescens* get enough water to evaporatively cool its leaves? Though the distributions of *E. frutescens* and *E. farinosa* overlap a great deal on a geographic scale, these two species occupy distinctive microenvironments. As shown in figure 9.7, *E. farinosa* grows mainly on upland slopes, while *E. frutescens* is largely confined to ephemeral stream channels, or desert washes. Along washes, runoff combined with deep soils increases the availability of soil moisture. This example

reminds us of a principle that we first considered in chapter 4: organisms living in the same macroclimate may, because of slight differences in local distribution, experience substantially different microclimates. This is certainly true of the two barnacle species we consider in the following example.

## Distributions of Barnacles along an Intertidal Exposure Gradient

The marine intertidal zone presents a steep gradient of physical conditions from the shore seaward. As we saw in chapter 3, the organisms high in the intertidal zone are exposed by virtually every tide while the organisms that live at lower levels in the intertidal zone are exposed by the lowest tides only. Exposure to air differs at different levels within the intertidal zone. Organisms that live in the intertidal zone have evolved different degrees of resistance to drying, a major factor contributing to zonation among intertidal organisms (see fig. 3.17).

Barnacles, one of the most common intertidal organisms, show distinctive patterns of zonation within the intertidal zone. For example, Joseph Connell (1961) described how along the coast of Scotland, adult *Chthamalus stellatus* are restricted to the upper levels of the intertidal zone, while adult *Balanus balanoides* are limited to the middle and lower levels (fig. 9.8). What role does resistance to drying play in the intertidal zonation of these two species? Unusually calm and warm weather combined with very low tides gave Connell some insights into this question. In the spring of 1955, warm weather coincided with calm seas and very low tides. As a consequence, no water reached the upper intertidal zone occupied by both species of barnacles. During this period, *Balanus* in the upper intertidal zone suffered much higher mortality than *Chthamalus* (fig. 9.9). Meanwhile, *Balanus* in

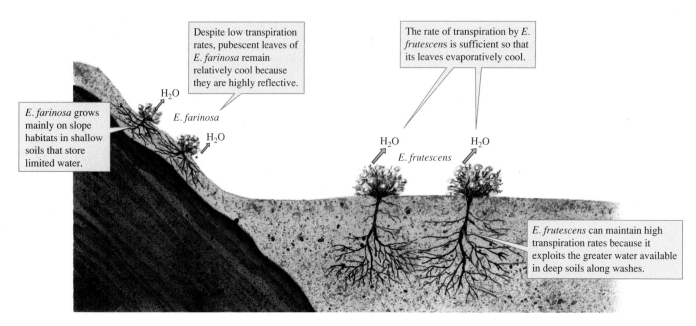

**Figure 9.7** Temperature regulation and distributions of *Encelia farinosa* and *E. frutescens* across microenvironments.

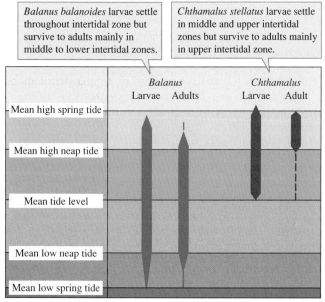

**Figure 9.8** Distributions of two barnacle species within the intertidal zone (data from Connell 1961).

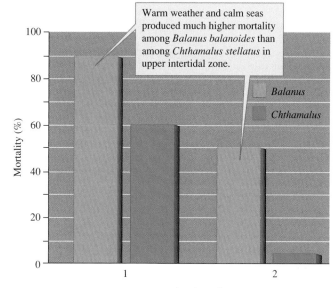

**Figure 9.9** Barnacle mortality in the upper intertidal zone (data from Connell 1961).

the lower intertidal zone showed normal rates of mortality. Of the two species, *Balanus* appears to be more vulnerable to desiccation. Higher rates of desiccation may exclude this species of barnacle from the upper intertidal zone.

Vulnerability to desiccation, however, does not completely explain the pattern of intertidal zonation shown by *Balanus* and *Chthamalus*. What excludes *Chthamalus* from the lower intertidal zone? Though the larvae of this barnacle settle in the lower intertidal zone, the adults rarely survive there. Connell explored this question by transplanting adult *Chthamalus* to the lower intertidal zone and found

that transplanted adults survive in the lower intertidal zone very well. If the physical environment does not exclude *Chthamalus* from the lower intertidal zone, what does? It turns out that this species is excluded from the lower intertidal zone by competitive interactions with *Balanus*. We discuss the mechanisms by which this competitive exclusion is accomplished in chapter 13, which covers interspecific competition.

These barnacles remind us that the environment consists of more than just physical and chemical factors. An organism's environment also includes biological factors. In many situations, biological factors may be as important or even more important than physical factors in determining the distribution and abundance of species. Often the influences of biological factors remain hidden, however, because of the difficulty of demonstrating them. In ecology, we must usually probe deeper to see beyond outward appearances, as Connell did when he transplanted *Chthamalus* from the upper to the lower intertidal zone. The influence of biological factors, such as competition, predation, and disease, on the distribution and abundance of organisms is a theme that enters our discussions frequently in the remainder of this book, especially in chapters 13, 14, and 15.

Now that we have considered factors limiting the distributions of individuals, let's consider the patterns of distribution of individuals within their habitat. Let's begin by considering three basic patterns of distribution.

## CONCEPT DISCUSSION

### Distribution Patterns

**On small scales, individuals within populations are distributed in patterns that may be random, regular, or clumped; on larger scales, individuals within a population are clumped.**

We have just considered how the environment limits the distributions of species. When you map the distribution of a species such as the red kangaroo in Australia (see fig. 9.2), or the zoned distribution of *Chthamalus* and *Balanus* in the intertidal zone (see fig. 9.8), the boundaries on your map indicate the range of the species. In other words, your map shows where at least some individuals of the species live and where they are absent. Knowing a species' range, as defined by presence and absence, is useful, but it says nothing about how the individuals that make up the population are distributed in the areas where they are present. Are individuals randomly distributed across the range? Are they regularly distributed? As we shall see, the distribution pattern observed by an ecologist is strongly influenced by the scale at which a population is studied.

Ecologists refer frequently to large-scale and small-scale phenomena. What is "large" or "small" depends on the size of organism or other ecological phenomenon under study. For this discussion, **small scale** refers to distances of no more than a few hundred meters, over which there is little environmental change significant to the organism under study. **Large scale** refers to areas over which there is substantial environmental change. In this sense, large scale may refer to patterns over an entire continent or patterns along a mountain slope, where environmental gradients are steep. Let's begin our discussion with patterns of distribution observed at small scales.

## Distributions of Individuals on Small Scales

Three basic patterns of distribution are observed on small scales: random, regular, or clumped. A **random distribution** is one in which individuals within a population have an equal chance of living anywhere within an area. A **regular distribution** is one in which individuals are uniformly spaced. In a **clumped distribution,** individuals have a much higher probability of being found in some areas than in others (fig. 9.10).

These three basic patterns of distribution are produced by the kinds of interactions that take place between individuals within a population, by the structure of the physical environment, or by a combination of interactions and environmental structure. Individuals within a population may *attract* each other, *repel* each other, or *ignore* each other. Mutual attraction creates clumped, or aggregated, patterns of distribution. Regular patterns of distribution are produced when individuals avoid each other or claim exclusive use of a patch of landscape. Neutral responses contribute to random distributions.

The patterns created by social interactions may be reinforced or damped by the structure of the environment. An environment with patchy distributions of nutrients, nesting sites, water, and so forth fosters clumped distribution patterns. An environment with a fairly uniform distribution of resources and frequent, random patterns of disturbance (or mixing) tends to reinforce random or regular distributions. Let's now consider factors that influence the distributions of some species in nature.

### Distributions of Tropical Bee Colonies

Stephen Hubbell and Leslie Johnson (1977) recorded a dramatic example of how social interactions can produce and enforce regular spacing in a population. They studied competition and nest spacing in populations of stingless bees in the family Trigonidae. The bees they studied live in tropical dry forest in Costa Rica. Though these bees do not sting, rival colonies of some species fight fiercely over potential nesting sites.

Stingless bees are abundant in tropical and subtropical environments, where they gather nectar and pollen from a wide variety of flowers. They generally nest in trees and live in colonies made up of hundreds to thousands of workers. Hubbell and Johnson observed that some species of stingless bees are highly aggressive to other members of their species from other colonies, while others are not. Aggressive species usually forage in groups and feed mainly on flowers that occur in high-density clumps. The nonaggressive species feed singly or in small groups and on more widely distributed flowers.

Hubbell and Johnson studied several species of stingless bees to determine whether there is a relationship between aggressiveness and patterns of colony distribution. They predicted that the colonies of aggressive species would show regular distributions while those of nonaggressive species would

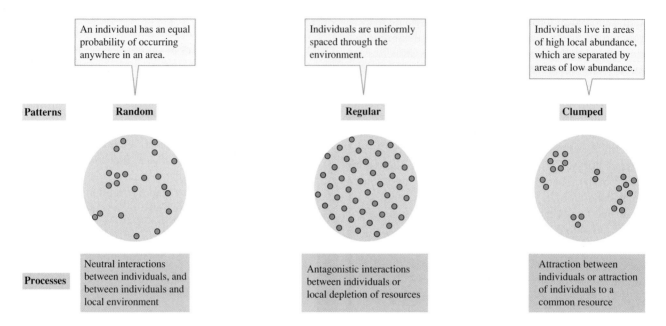

**Patterns**    **Random**

An individual has an equal probability of occurring anywhere in an area.

**Regular**

Individuals are uniformly spaced through the environment.

**Clumped**

Individuals live in areas of high local abundance, which are separated by areas of low abundance.

**Processes**

Neutral interactions between individuals, and between individuals and local environment

Antagonistic interactions between individuals or local depletion of resources

Attraction between individuals or attraction of individuals to a common resource

**Figure 9.10** Random, regular, and clumped distributions.

show random or clumped distributions. They concentrated their studies on a 13 ha tract of tropical dry forest that contained numerous nests of nine species of stingless bees.

Though Hubbell and Johnson were interested in how bee behavior might affect colony distributions, they recognized that the availability of potential nest sites for colonies could also affect distributions. So, in one of the first steps in their study, they mapped the distributions of trees suitable for nesting. They found that potential nest trees were distributed randomly through the study area and that the number of potential nest sites was much greater than the number of bee colonies. What did these measurements tell the researchers? They indicated that the number of colonies in the study area was not limited by availability of suitable trees and that clumped and regular distribution of colonies would not be due to an underlying clumped or regular distribution of potential nest sites.

Hubbell and Johnson were able to map the nests of five of the nine species of stingless bees accurately. The nests of four of these species were distributed regularly. As they had predicted, all four species with regular nest distributions were highly aggressive to bees from other colonies of their own species. The fifth species, *Trigona dorsalis,* was not aggressive and its nests were randomly distributed over the study area. Figure 9.11 contrasts the random distribution of *T. dorsalis* with the regular distribution of one of the aggressive species, *T. fulviventris.*

The researchers also studied the process by which the aggressive species establish new colonies. In the process, they made observations that provide insights into the mechanisms that establish and maintain the regular nest distributions of species such as *T. fulviventris.* This species and the other aggressive species apparently mark prospective nest sites with a pheromone. **Pheromones** are chemical substances secreted by some animals for communication with other

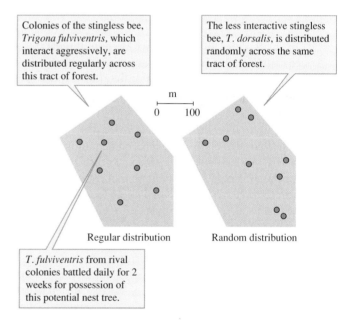

Colonies of the stingless bee, *Trigona fulviventris,* which interact aggressively, are distributed regularly across this tract of forest.

The less interactive stingless bee, *T. dorsalis,* is distributed randomly across the same tract of forest.

m
0    100

Regular distribution

Random distribution

*T. fulviventris* from rival colonies battled daily for 2 weeks for possession of this potential nest tree.

**Figure 9.11** Regular and random distributions of stingless bee colonies in the tropical dry forest (data from Hubbell and Johnson 1977).

members of their species. The pheromone secreted by these stingless bees attracts and aggregates members of their colony to the prospective nest site; however, it also attracts workers from other nests.

If workers from two different colonies arrive at the prospective nest, they may fight for possession. Fights may be escalated into protracted battles. Hubbell and Johnson observed battles over a nest tree that lasted for 2 weeks. Each dawn, 15 to 30 workers from two rival colonies arrived at the contested nest site. The workers from the two rival colonies faced off in two swarms and displayed and fought with each other. In the displays, pairs of bees faced each other, slowly flew vertically to a height of about 3 m, and then grappled each other to the ground. When the two bees hit the ground, they separated, faced off, and performed another aerial display. Bees did not appear to be injured in these fights, which were apparently ritualized. The two swarms abandoned the battle at about 8 or 9 A. M. each morning, only to re-form and begin again the next day just after dawn. While this contest over an unoccupied nest site produced no obvious mortality, fights over occupied nests sometimes killed over 1,000 bees in a single battle. These tropical bees space their colonies by engaging in pitched battles, but as we see next, plants space themselves by more subtle means.

## Distributions of Desert Shrubs

Half a century ago desert ecologists suggested that desert shrubs tend to be regularly spaced due to competition between the shrubs. You can see the patterns that inspired these early ecologists by traveling across the seemingly endless expanses of the Mojave Desert in western North America. One of the most common plants that you will see is the creosote bush, *Larrea tridentata,* which dominates thousands of square kilometers of this area. As you look out across landscapes dominated by creosote bushes it may appear that the spacing of these shrubs is regular (fig. 9.12). In places, their spacing is so uniform that they appear to have been planted by some very careful gardener. As we shall see, however, visual impressions can be deceiving.

Quantitative sampling and statistical analysis of the distributions of creosote bushes and other desert shrubs led to a controversy that took the better part of two decades to settle. In short, when different teams of researchers quantified the distributions of desert shrubs, some found the regular distributions reported by earlier ecologists. Others found random or clumped distributions. Still others reported all three types of distributions.

Though we are generally accustomed to having one answer to our questions, the answers to ecological questions are often more complex. Research by Donald Phillips and James MacMahon (1981) showed that the distribution of creosote bushes changes as they grow. They mapped and analyzed the distributions of creosote bushes and several other shrubs at nine sites in the Sonoran and Mojave Deserts. Because earlier researchers had suggested that creosote bush spacing changed with available moisture, they chose sites with different average precipitations. Precipitation at the study sites ranged from 80 to 220 mm, and average July temperature varied from 27° to 35°C. Phillips and MacMahon took care to pick sites with similar soils and with similar topography. They studied populations growing on sandy to sandy loam soils with less than 2% slope with no obvious surface runoff channels.

The results of this study indicate that the distribution of desert shrubs changes from clumped to random to regular distribution patterns as they grow. The young shrubs tend to be clumped for three reasons: because seeds germinate at a lim-

**Figure 9.12** Are local populations of the creosote bush, *Larrea tridentata,* distributed regularly?

ited number of "safe sites," because seeds are not dispersed far from the parent plant, or because asexually produced offspring are necessarily close to the parent plant. Phillips and MacMahon proposed that as the plants grow, some individuals in the clumps die, which reduces the degree of clumping. Gradually, the distribution of shrubs becomes more and more random. However, competition among the remaining plants produces higher mortality among plants with nearby neighbors, which thins the stand of shrubs still further and eventually creates a regular distribution of shrubs. This hypothetical process is summarized in figure 9.13.

Phillips and MacMahon and other ecologists proposed that desert shrubs compete for water and nutrients, a competition that takes place belowground. How can we study these belowground interactions? Work by Jacques Brisson and James Reynolds (1994) provides a quantitative picture of the belowground side of creosote bush distributions. These researchers carefully excavated and mapped the distributions of 32 creosote bushes in the Chihuahuan Desert. They proposed that if creosote bushes compete, their roots should grow in a way that reduces overlap with the roots of nearby individuals.

The 32 excavated creosote bushes occupied a 4 by 5 m area on the Jornada Long Term Ecological Research site near Las Cruces, New Mexico. The creosote bush was the only shrub within the study plot. Their roots penetrated to only 30 to 50 cm, the depth of a hardpan calcium carbonate deposition layer. Because they did not have to excavate to great depths, Brisson and Reynolds were able to map more root systems than previous researchers. Still, their excavation and mapping of roots required 2 months of intense labor.

The complex pattern of root distributions uncovered confirmed the researchers proposal: Creosote bush roots grow in a pattern that reduces overlap between the roots of adjacent plants (fig. 9.14*a*). We can make the root distributions of individual plants clearer by plotting their perimeters only. Figure 9.14*b* shows the hypothetical distributions of creosote bushes with circular root systems, while figure 9.14*c* shows their

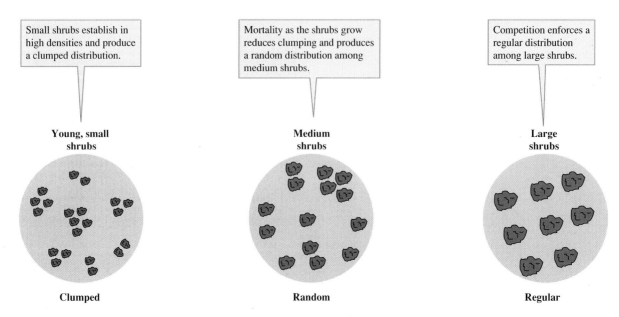

**Figure 9.13**  Hypothetical change in shrub distributions with increasing shrub size.

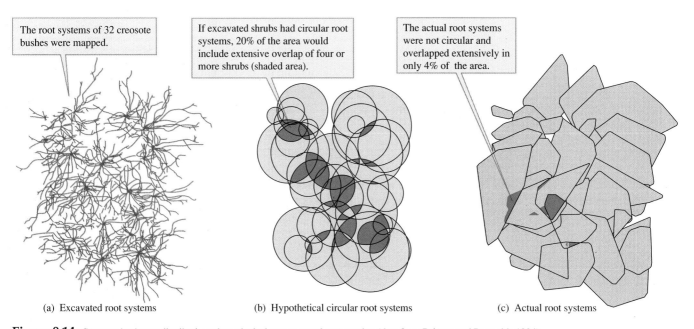

**Figure 9.14**  Creosote bush root distributions: hypothetical versus actual root overlap (data from Brisson and Reynolds 1994).

actual root distributions. Notice that the root systems of creosote bushes overlap much less than they would if they had circular distributions. Brisson and Reynolds conclude that competitive interactions with neighboring shrubs influence the distribution of creosote bush roots. Their work suggests that creosote bushes compete for belowground resources.

After more than two decades of work on this single species of plant, desert ecologists have a much clearer understanding of the factors that influence the distribution of individuals on a small scale. On small scales, the creosote bush may have clumped, random, or regular distributions. Hubbell and Johnson (1977) showed that stingless bee colonies may

also show different patterns of distribution, depending on the level of aggression between colonies. As we shall see in the following section, however, on distributions on larger scales, individuals have clumped distributions.

# Distributions of Individuals on Large Scales

We have considered how individuals within a population are distributed on a small scale: how bee colonies are distributed within a few acres of forest and how shrubs are distributed

*Investigating the Evidence*

# Clumped, Random, and Regular Distributions

Imagine sampling a population of plants or animals to determine the distribution of individuals across the habitat. One of the most basic questions that you could ask is, "How are individuals in the population distributed across the study area?" How might they be distributed? The three basic patterns that we've discussed in this section are clumped, random, and regular distributions. The first step toward testing statistically between these three types of distributions is to sample the population to estimate the mean (p. 21) and variance (p. 162) in density of the population across the study area. The theoretical relationships between variance in density and mean density in clumped, random, and regular distributions are as follows:

| Distributions | Relation of variance to mean |
|---|---|
| Clumped | Variance > Mean, or Variance/Mean > 1 |
| Random | Variance = Mean, or Variance/Mean = 1 |
| Regular | Variance < Mean, or Variance/Mean < 1 |

How do we connect these relationships between variance and mean density with what we see on the ground? In a clumped distribution, many sample plots will contain few or no individuals while some will contain a large number. As a consequence, the variance among sample plots will be high and the variance in density will be greater than the mean. In contrast, sample plots of a population with a regular distribution will all include a similar number of individuals. As a result, the variance in density across samples will be low when taken from a population with a regular distribution; therefore, the variance will be less than the mean. Meanwhile, in randomly distributed population, the variance in density across the habitat will be approximately equal to the mean density.

Consider the following samples of three different populations of herbaceous plants growing on a desert landscape. Each sample is the number counted in a randomly located 1 $m^2$ area at the study site.

| Sample | a | b | c | d | e | f | g | h | i | j | k |
|---|---|---|---|---|---|---|---|---|---|---|---|
| Number of Species A | 10 | 2 | 8 | 1 | 2 | 5 | 9 | 4 | 5 | 16 | 8 |
| Number of Species B | 5 | 6 | 6 | 5 | 5 | 5 | 5 | 6 | 4 | 5 | 5 |
| Number of Species C | 20 | 1 | 2 | 1 | 15 | 3 | 1 | 1 | 10 | 2 | 2 |

The distribution of individuals among the samples of species A, B, and C are quite different. For instance, each of the samples contained approximately the same number of individuals of species B. In contrast, the numbers of species C varied widely among samples. Meanwhile, counts of species A showed a level of variation somewhere in between variations in species B and C. The samples of species A, B, and C may give the impression of random, regular, and clumped distributions. We can quantify our visual impressions by calculating the sample means (p. 21) and sample variances (p. 162) for the densities of species A, B, and C:

| Statistic | Species A | Species B | Species C |
|---|---|---|---|
| Sample mean, $\overline{X}$ | 5.27 | 5.18 | 5.27 |
| Sample variance, $s^2$ | 5.22 | 0.36 | 44.42 |
| $\dfrac{s^2}{\overline{X}}$ | 0.99 | 0.07 | 8.43 |

While the mean density calculated from the samples was very similar for all three species, their variance in density among samples was quite different. As a consequence, the ratios of sample variance to sample means, $\dfrac{s^2}{\overline{X}}$, were also different. While $\dfrac{s^2}{\overline{X}}$ for species A was nearly 1, this ratio was much less than 1 for species B, and much greater than 1 for species C. These results show how the $\dfrac{s^2}{\overline{X}}$ ratio can capture the visual impression of random, regular, and clumped distributions that we formed when we inspected the samples of the three species. Can we conclude from the $\dfrac{s^2}{\overline{X}}$ ratios that we calculated that species A has a random distribution, that species B has a regular distribution, and that C has a clumped distribution? While it is likely that they do, in science we need to attach probabilities to such conclusions. To do that, we need to consider these samples of species A, B, and C from a statistical perspective. We will look at the statistics of these samples in chapter 10 (see p. 267).

within a small stand. Now let's step back and ask how individuals within a population are distributed on a larger scale over which there is significant environmental variation. For instance, how does the density of individuals vary across the entire range of a species? Is population density fairly regularly distributed across the entire area occupied by a species, or are there a few centers of high density surrounded by areas in which the species is present but only in low densities?

# Bird Populations Across North America

Terry Root (1988) mapped patterns of bird abundance across North America using the "Christmas Bird Counts." These bird counts provide one of the few data sets extensive enough to study distribution patterns across an entire continent. Christmas Bird Counts, which began in 1900, involve annual counts of birds during the Christmas season. The first Christmas Bird Count was attended by 27 observers, who counted birds in 26 localities—2 in Canada and the remainder in 13 states of the United States. In the 1985–86 season, 38,346 people participated in the Christmas Bird Count. The observers counted birds in 1,504 localities throughout the United States and most of Canada. The Christmas Bird Count marked its centennial anniversary in the year 2000. It continues to produce a unique record of the distribution and population densities of wintering birds across most of a continent.

Root's analysis centers around a series of maps that show patterns of distribution and population density for 346 species of birds that winter in the United States and Canada. Although species as different as swans and sparrows are included, the maps show a consistent pattern. At the continental scale, bird populations show clumped distributions. Clumped patterns occur in species with widespread distributions, such as the American crow, *Corvus brachyrhynchos*, as well as in species with restricted distributions, such as the fish crow, *C. ossifragus*. Though the winter distribution of the American crow includes most of the continent, the bulk of individuals in this population are concentrated in a few areas. These areas of high density, or "hot spots," appear as red dots in figure 9.15*a*. For the American crow population, hot spots are concentrated along river valleys, especially the Cumberland, Mississippi, Arkansas, Snake, and Rio Grande. Away from these hot spots the winter abundance of American crows diminishes rapidly.

The fish crow population, though much more restricted than that of the American crow, is also concentrated in a few areas (fig. 9.15*b*). Fish crows are restricted to areas of open water near the coast of the Gulf of Mexico and along the southern half of the Atlantic coast of the United States. Within this restricted range, however, most fish crows are concentrated in a few hot spots—one on the Mississippi Delta, another on Lake Seminole west of Tallahassee, Florida, and a third in the everglades in southern Florida. Like the more widely distributed

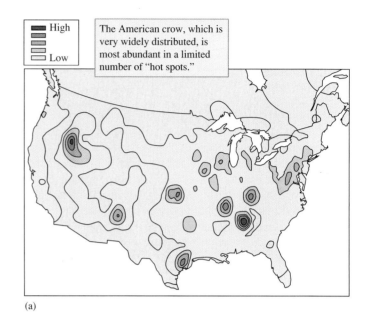

(a)

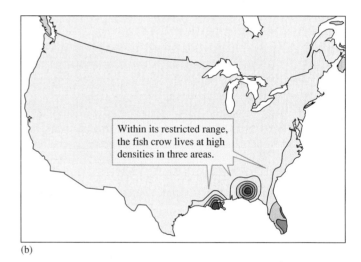

(b)

**Figure 9.15** (*a*) Winter distribution of the American crow, *Corvus brachyrynchos*. (*b*) Winter distribution of the fish crow, *C. ossifragus* (data from Root 1988).

American crow, the abundance of fish crows diminishes rapidly away from these centers of high density.

Might bird populations have clumped distributions only on the wintering grounds? James H. Brown, David Mehlman, and George Stevens (1995) analyzed large-scale patterns of abundance among birds across North America during the breeding season, the opposite season from that studied by Root. In their study these researchers used data from the Breeding Bird Survey, which consists of standardized counts by amateur ornithologists conducted each June at approximately 2,000 sites across the United States and Canada under the supervision of the Fish and Wildlife Services of the United States and Canada. For their analyses, they chose species of birds whose geographic ranges fall mainly or completely within the eastern

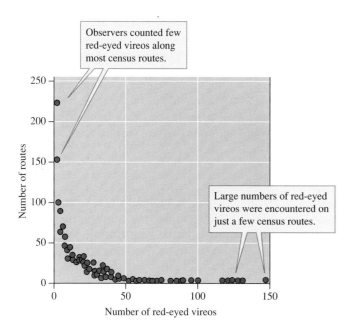

**Figure 9.16** Red-eyed vireos, *Vireo olivaceus,* counted along census routes of the Breeding Bird Survey (data from Brown, Mehlman, and Stevens 1995).

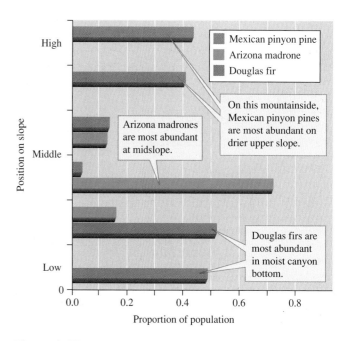

**Figure 9.17** Abundances of three tree species on a moisture gradient in the Santa Catalina Mountains, Arizona (data from Whittaker and Niering 1965).

and central regions of the United States, which are well covered by study sites of the Breeding Bird Survey.

Like Root, Brown and his colleagues found that a relatively small proportion of study sites yielded most of the records of each bird species. That is, most individuals were concentrated in a fairly small number of hot spots. For instance, the densities of red-eyed vireos are low in most places (fig. 9.16). Clumped distributions were documented repeatedly. When the numbers of birds across their ranges were totaled, generally about 25% of the locations sampled supported over half of each population. By combining the results of Root and Brown and his colleagues we can say confidently that at larger scales, bird populations in North America show clumped patterns of distribution. In other words, most individuals within a bird species live in a few hot spots, areas of unusually high population density.

Brown and his colleagues propose that these distributions are clumped because the environment varies and individuals aggregate in areas where the environment is favorable. What might be the patterns of distribution for populations distributed along a known environmental gradient? Studies of plant populations provide interesting insights.

## Plant Abundance Along Moisture Gradients

Several decades ago, Robert Whittaker gathered information on the distributions of woody plants along moisture gradients in several mountain ranges across North America. As we saw in chapter 2 (see fig. 2.38) environmental conditions on

mountainsides change substantially with elevation. These steep environmental gradients provide a compressed analog of the continental-scale gradients to which the birds studied by Root and Brown and his colleagues were presumably responding.

Let's look at the distributions of some tree species along moisture gradients in two of the mountain ranges studied by Whittaker. Robert Whittaker and William Niering (1965) studied the distribution of plants along moisture and elevation gradients in the Santa Catalina Mountains of southern Arizona. These mountains rise out of the Sonoran Desert near Tucson, Arizona, like a green island in a tan desert sea. Vegetation typical of the Sonoran Desert, including the saguaro cactus and creosote bush, grow in the surrounding desert and on the lower slopes of the mountains. However, the summit of the mountains is topped by a mixed conifer forest. Forests also extend down the flanks of the Santa Catalinas in moist, shady canyons.

There is a moisture gradient from the moist canyon bottoms up the dry southwest-facing slopes. Whittaker and Niering found that along this gradient the Mexican pinyon pine, *Pinus cembroides,* is at its peak abundance on the uppermost and driest part of the southwest-facing slope (fig. 9.17). Along the same slope, Arizona madrone, *Arbutus arizonica,* reaches its peak abundance at middle elevations. Finally, Douglas firs, *Pseudotsuga menziesii,* are restricted to the moist canyon bottom. Mexican pinyon pines, Arizona madrone, and Douglas fir are all clumped along this moisture gradient, but each reaches peak abundance at different positions on the slope. These positions appear to reflect the different environmental requirements of each species.

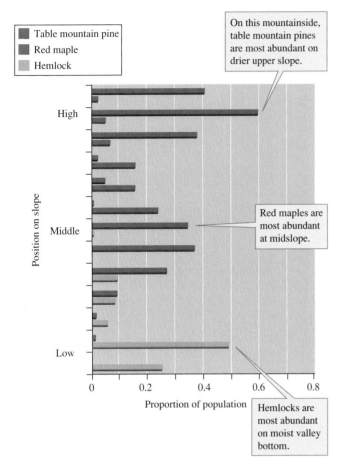

**Figure 9.18** Abundance of three tree species on a moisture gradient in the Great Smoky Mountains, Tennessee (data from Whittaker 1956).

Whittaker (1956) recorded analogous tree distributions along moisture gradients in the Great Smoky Mountains of eastern North America. Again, the gradient was from a moist valley bottom to a drier southwest-facing slope. Along this moisture gradient, hemlock, *Tsuga canadensis,* was concentrated in the moist valley bottom and its density decreased rapidly upslope (fig. 9.18). Meanwhile red maple, *Acer rubrum,* grew at highest densities in the middle section of the slope, while table mountain pine, *Pinus pungens,* was concentrated on the driest upper sections. As in the Santa Catalina Mountains of Arizona, these tree distributions in the Great Smoky Mountains reflect the moisture requirements of each tree species.

The distribution of trees along moisture gradients seems to resemble the clumped distributional patterns of birds across the North American continent but on a smaller scale. All species of trees discussed here showed a highly clumped distribution along moisture gradients, and their densities decreased substantially toward the edges of their distributions. In other words, like birds, tree populations are concentrated in hot spots. As we shall see in the next concept discussion, some populations are subdivided into separate subpopulations that exchange individuals over time.

## CONCEPT DISCUSSION:

### Metapopulations

**Many populations are subdivided into subpopulations called metapopulations.**

Populations of many species occur not as a single continuously distributed population but in spatially isolated patches, with significant exchange of individuals among patches. A group of subpopulations living on such patches connected by exchange of individuals among patches make up a **metapopulation.** The population of Glanville fritillary butterflies, *Melitaea cinxia,* which lives in dry meadows scattered through the landscape of southern Finland (see chapter 8, p. 217), is a metapopulation. In our earlier discussion of this butterfly metapopulation, we reviewed how the exchange of individuals among subpopulations has been well documented (Saccheri et al. 1998). However, the metapopulation of *Melitaea* in southern Finland is only one of many that are well known. Here is another example of a butterfly metapopulation.

## A Metapopulation of an Alpine Butterfly

Once population biologists began to include the concept of metapopulations in their thinking, they found them everywhere. Butterflies have been well represented in studies of metapopulations. One of these butterflies is the Rocky Mountain Parnassian butterfly, *Parnassius smintheus* (figure 9.19). The range of *P. smintheus* extends from northern New Mexico along the Rocky Mountains to southwest Alaska. Along this range *P. smintheus* caterpillars feed mainly on the leaves and flowers of stonecrop, *Sedum* sp., in areas of open forest and meadows. Because of their tie to a narrow range of host plants, *P. smintheus* populations are often distributed among the habitat patches occupied by their host plant, appearing to form metapopulations.

One such metapopulation was studied by Jens Roland, Nusha Keyghobadi and Sherri Fownes of the University of Alberta in Edmonton, Canada (Roland, Keyghobadi, and Fownes 2000). Roland, Keyghobadi, and Fownes focused their attention on a series of 20 alpine meadows on ridges in the Kananaskis region of the Canadian Rocky Mountains. The study meadows ranged in area from about 0.8 ha to 20 ha. While some meadows were adjacent to each other, others were separated by up to 200 m of coniferous forest. The host plant of *P. smintheus* in the study meadows was the lanceleaf stonecrop, *Sedum lanceolatum.*

A combination of fire suppression and global warming appears to be decreasing the size of alpine meadows and increasing their isolation from each other by intervening forest.

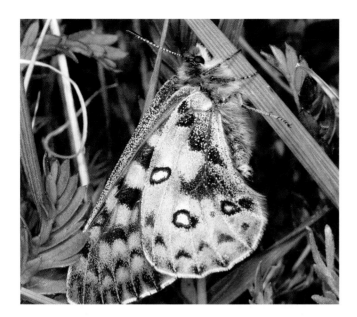

**Figure 9.19** The Rocky Mountain Parnassian butterfly, *Parnassius smintheus*, is found in alpine meadows in the Rocky Mountains from northern New Mexico to southwestern Alaska. Because it is tied to meadows and open forest where its larval host plants grow, *P. smintheus* lives in scattered subpopulations connected by dispersal and forming metapopulations.

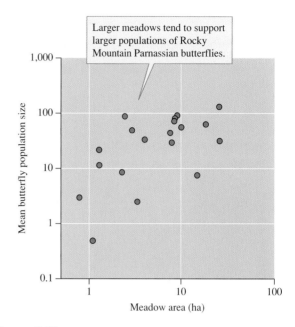

Larger meadows tend to support larger populations of Rocky Mountain Parnassian butterflies.

**Figure 9.20** The relationship between meadow area and the size of Rocky Mountain Parnassian butterfly, *Parnassius smintheus*, populations. With forest encroachment into alpine meadows in the Rocky Mountains, populations of *P. smintheus* will likely decline.

In 1952, the study meadows averaged approximately 36 ha in area. By 1993, the average area of these meadows had declined to approximately 8 ha, a decrease in area of approximately 77%. These changes motivated the research team of Roland, Keyghobadi, and Fownes to study the influences of meadow size and isolation on movements of *P. smintheus* during the summers of 1995 and 1996.

The research team used mark and recapture techniques (see Applications & Tools) to estimate population size in each meadow and to follow their movements. Butterflies were hand netted and marked on the hind wing with a three-letter identification code, using a fine-tipped permanent marker. The team recorded the sex of *P. smintheus* captured and its location within 20 m. Upon recapture, dispersal distance of an individual was estimated as the straight line distance from its last point of capture.

Roland, Keyghobadi, and Fownes marked 1,574 *P. smintheus* in 1995 and 1,200 in 1996. Of these marked individuals, they recaptured 726 in 1995 and 445 in 1996. Over the course of the study, the size of *P. smintheus* populations in the 20 study meadows ranged from 0 to 230. The average movement distance by males and females in 1995 was approximately 131 m. In 1996, the average movement distances of males and females was 162 m and 118 m respectively. The maximum dispersal distance for a butterfly in 1995 was 1,729 m and in 1996 the maximum dispersal distance was 1,636 m. Most of the movements determined by recaptures were the result of dispersal within meadows. In 1995 only 5.8% of documented dispersal movements were from one meadow to another and in 1996 dispersal between meadows accounted for 15.2% of total recaptures.

One of the questions posed by Roland, Keyghobadi, and Fownes was how meadow size and population size might affect dispersal by *P. smintheus*. As shown in figure 9.20, average butterfly population size increased with meadow area. It turned out that butterflies are more likely to leave small populations than large populations. Butterflies leaving small populations generally immigrate to larger populations. The results of this study by Roland, Keyghobadi, and Fownes indicate that as alpine meadows in the Rocky Mountains decline in area, populations of *P. smintheus* will become progressively more compressed into fewer and fewer small meadows, perhaps disappearing entirely in parts of their range. Some of the patterns of dispersal within this metapopulation of alpine butterflies have been observed in a recent study of dispersal in a metapopulation of a small falcon.

## Dispersal Within a Metapopulation of Lesser Kestrels

The lesser kestrel, *Falco naumanni*, is a small migratory falcon that breeds in colonies of monogamous pairs in Eurasia and spends its winters in sub-Saharan Africa (fig. 9.21). Lesser kestrels have suffered a high rate of decline across their range and are listed as a globally threatened species. In contrast to its global circumstance, the lesser kestrel population of the Ebro River valley of northeast Spain has grown dramatically in recent years. David Serrano and Jose Tella, two ecologists who have conducted long-term studies of this population (Serrano and Tella 2003), documented growth in this population from 224 pairs distributed among 4 subpopulations in 1993 to 787

**Figure 9.21** The lesser kestrel, *Falco naumanni,* breeds in scattered colonies that collectively form metapopulations. Most populations of this species have declined dramatically with modernization of agriculture.

pairs living in 14 subpopulations in 2000. Serrano and Tella attribute this regional growth to sustained traditional farming practices in the Ebro River valley. However, they warn that plans to modernize farming practices in the Ebro River valley may lead to the population declines seen elsewhere.

Serrano and Tella used numbered, color leg bands to mark and keep track of individual kestrels in their study population. From 1993 to 1999, they banded 4,901 fledgling kestrels and 640 adults. Because lesser kestrels breed within colonies, they are easier to track during the breeding season. Once locating a colony, Serrano and Tella would observe the colony members from a blind, record the numbers of pairs within the colony and, using a telescope, read the numbers on the leg band of any banded adult birds in the colony. Serrano and Tella were able to obtain accurate counts of the entire breeding population of lesser kestrels within the Ebro River valley each year. They could also use their observations to plot the movements of any banded birds they saw. Within colonies, the percentage of banded adults of known age ranged from 60% to 90%.

The data gathered by Serrano and Tella indicate that a substantial percentage of birds leave the breeding colony where they hatched to join other subpopulations in their first year of breeding. However, females in this species are more likely to move than males. The rate of emigration by first-breeding females is approximately 30% versus 22% for first-breeding males. Meanwhile, less than 4% of adults emigrate from a colony on any given year. Though some lesser kestrels in the study population dispersed more than 100 km, Serrano and Tella found a negative correlation between distance between colonies and the frequency of dispersal between them.

The lesser kestrels of the Ebro River valley and the Rocky Mountain Parnassian butterflies of southern Alberta, Canada, interact with their environments on greatly different scales. In addition, the butterfly population appears to be contracting spatially; the lesser kestrel population is expanding. However, these populations also have a number of features in common. First, they both are spatially organized into metapopulations. Another feature that the two populations share is the influence of local population size on the tendency to disperse and the direction of dispersal. Like *P. smintheus,* lesser kestrels in smaller subpopulations are more likely to emigrate than are individuals in larger subpopulations. Second, lesser kestrels are more likely to disperse from small colonies to larger colonies.

In the last two concept discussions, we have reviewed patterns of distribution within populations. We have seen that those patterns vary from one population to another and may depend upon the scale at which ecologists make their observations. Now we turn from patterns of spatial variation within populations to compare the average densities of different populations. Is there any way to predict the average population density of populations? While it is not possible to make precise predictions, the following examples show that population densities are very much influenced by organism size.

## CONCEPT DISCUSSION

### Organism Size and Population Density

**Population density declines
with increasing organism size.**

If you estimate the densities of organisms in their natural environments, you will find great ranges. While bacterial populations in soils or water can exceed $10^9$ per cubic centimeter and phytoplankton densities often exceed $10^6$ per cubic meter, populations of large mammals and birds can average considerably less than one individual per square kilometer. What factors produce this variation in population density? The densities of a wide variety of organisms are highly correlated with body size. In general, densities of animal and plant populations decrease with increasing size.

While it makes common sense that small animals and plants generally live at higher population densities than larger ones, quantifying the relationship between body size and population density provides valuable information. First, quantification translates a general qualitative notion into a more precise quantitative relationship. For example, you might want to know how much population density declines with increased body size. Second, measuring the relationship between body size and population density for a wide variety of species reveals different relationships for different groups of organisms. Differences in the relationship between size and population density can be seen among major groups of animals.

# Animal Size and Population Density

John Damuth (1981) produced one of the first clear demonstrations of the relationship between body size and population density. He focused his analysis on herbivorous mammals. The size of herbivorous mammals included in the analysis ranged from small rodents, with a mass of about 10 g, to large herbivores such as rhinoceros, with a mass well over $10^6$ g.

Meanwhile, average population density ranged from about 1 individual ($10^{-1}$) per 10 km$^2$ to about 10,000 ($10^4$) per 1 km$^2$, which spans approximately five "orders of magnitude," or powers of 10, in population density. As figure 9.22 shows, Damuth found that the population density of 307 species of herbivorous mammals decreases, from species to species, with increased body size. The line in the graph, which is called the *regression line,* shows the average decrease in population density with increased body size.

Building on Damuth's analysis, Robert Peters and Karen Wassenberg (1983) explored the relationship between body size and average population density for a wider variety of animals. Their analysis included terrestrial invertebrates, aquatic invertebrates, mammals, birds, and poikilothermic vertebrates. They included animals representing a great range in size and population density. Animal mass ranged from $10^{-11}$ to about $10^{2.3}$ kg, while population density ranged from less than 1 per square kilometer to nearly $10^{12}$ per square kilometer. When Peters and Wassenberg plotted animal mass against average density, they, like Damuth, found that population density decreased with increased body size.

If you look closely at the data in figure 9.23, however, it is clear that there are differences among the animal groups. First, aquatic invertebrates of a given body size tend to have higher population densities, usually one or two orders of magnitude higher, than terrestrial invertebrates of similar size. Second, mammals tend to have higher population densities than birds of similar size. Peters and Wassenberg suggest that it may be appropriate to analyze aquatic invertebrates and birds separately from the other groups of animals.

The general relationship between animal size and population density has held up under careful scrutiny and reanalysis. Plant ecologists have found a qualitatively similar relationship in plant populations, as we see next.

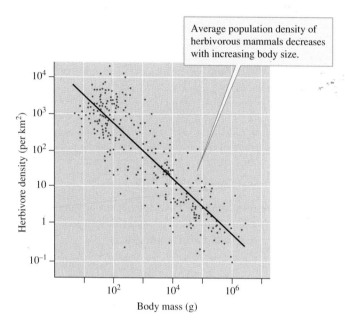

Average population density of herbivorous mammals decreases with increasing body size.

**Figure 9.22** Body size and population density of herbivorous mammals (data from Damuth 1981).

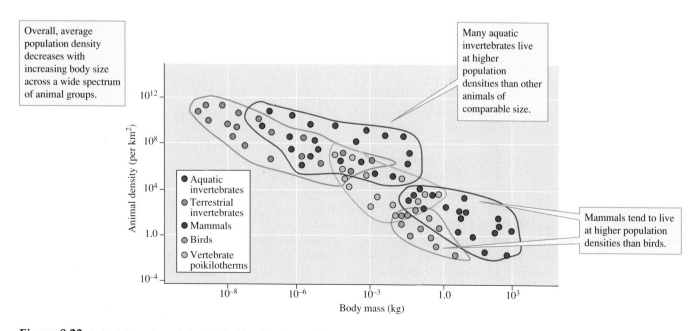

Overall, average population density decreases with increasing body size across a wide spectrum of animal groups.

Many aquatic invertebrates live at higher population densities than other animals of comparable size.

Mammals tend to live at higher population densities than birds.

**Figure 9.23** Animal size and population density (data from Peters and Wassenberg 1983).

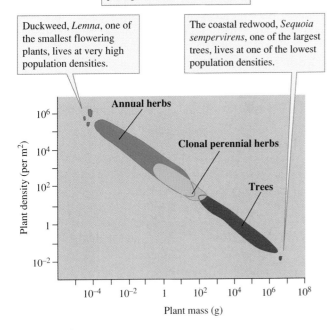

As in animals, plant population density decreases with increasing plant size across a wide range of plant growth forms.

Duckweed, *Lemna*, one of the smallest flowering plants, lives at very high population densities.

The coastal redwood, *Sequoia sempervirens*, one of the largest trees, lives at one of the lowest population densities.

**Figure 9.24** Plant size and population density (data from White 1985).

# Plant Size and Population Density

James White (1985) pointed out that plant ecologists have been studying the relationship between plant size and population density since early in the twentieth century. He suggests that the relationship between size and density is one of the most fundamental aspects of population biology. White summarized the relationship between size and density for a large number of plant species spanning a wide range of plant growth forms (fig. 9.24).

The pattern in figure 9.24 illustrates that as in animals, plant population density decreases with increasing plant size. However, the biological details underlying the size–density relationship shown by plants are quite different from those underlying the size–density patterns shown by animals. The different points in figures 9.22 and 9.23 represent different species of animals. A single species of tree, however, can span a very large range of sizes and densities during its life cycle. Even the largest trees, such as the giant sequoia, *Sequoia gigantea,* start life as small seedlings. These tiny seedlings can live at very high densities. As the trees grow, density declines progressively until the mature trees live at low densities. We discuss this process, which is called *self-thinning,* in chapter 13. Thus, the size–density relationship changes dynamically within plant populations and also differs significantly between populations of plants that reach different sizes at maturity. Despite differ-

ences in the underlying processes, the data summarized in figure 9.24 indicate a predictable relationship between plant size and population density.

The value of such an empirical relationship, whether for plants or animals, is that it provides a standard against which we can compare measured densities and gives an idea of expected population densities in nature. For example, suppose you go out into the field and measure the population density of some species of animal. How would you know if the densities you encounter are unusually high, low, or about average for an animal of the particular size and taxon? Without an empirical relationship such as that shown in figures 9.23 and 9.24 or a list of species densities, it would be impossible to make such an assessment. One question that we might attempt to answer with a population study is whether a species is rare. As we shall see in the next Concept Discussion, "Commonness and Rarity," rarity is a more complex consideration than it might seem at face value.

DISCUSSION

## CONCEPT DISCUSSION

### Commonness and Rarity

**Commonness and rarity of species are influenced by population size, geographic range, and habitat tolerance.**

Viewed on a long-term, geological timescale, populations come and go and extinction seems to be the inevitable punctuation mark at the end of a species' history. However, some populations seem to be more vulnerable to extinction than others. What makes some populations likely to disappear, while others persist through geological ages? At the heart of the matter are patterns of distribution and abundance. Species that are rare seem to be more vulnerable to extinction. In order to understand and, perhaps, prevent extinction, we need to understand the seven forms of rarity.

## Seven Forms of Rarity and One of Abundance

Deborah Rabinowitz (1981) devised a classification of *commonness* and *rarity,* based on combinations of three factors: (1) the geographic range of a species (*extensive* versus *restricted*), (2) habitat tolerance (*broad* versus *narrow*), and (3) local population size (*large* versus *small*). Habitat tolerance is related to the range of conditions in which a species can live. For instance, some plant species can tolerate a broad range of soil texture, pH, and organic matter content, while other plant species are confined to a single soil type. As we shall see, tigers have broad habitat tolerance; however, within the tiger's historical range in Asia lives the snow leopard, which

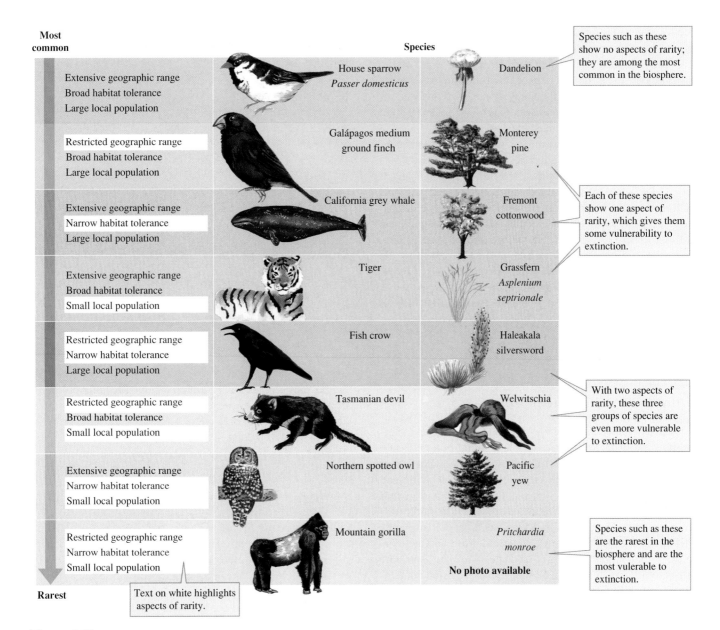

**Most common**

| | |
|---|---|
| Extensive geographic range<br>Broad habitat tolerance<br>Large local population | House sparrow<br>*Passer domesticus* |
| Restricted geographic range<br>Broad habitat tolerance<br>Large local population | Galápagos medium<br>ground finch |
| Extensive geographic range<br>Narrow habitat tolerance<br>Large local population | California grey whale |
| Extensive geographic range<br>Broad habitat tolerance<br>Small local population | Tiger |
| Restricted geographic range<br>Narrow habitat tolerance<br>Large local population | Fish crow |
| Restricted geographic range<br>Broad habitat tolerance<br>Small local population | Tasmanian devil |
| Extensive geographic range<br>Narrow habitat tolerance<br>Small local population | Northern spotted owl |
| Restricted geographic range<br>Narrow habitat tolerance<br>Small local population | Mountain gorilla |

Species: Dandelion, Monterey pine, Fremont cottonwood, Grassfern *Asplenium septrionale*, Haleakala silversword, Welwitschia, Pacific yew, *Pritchardia monroe* (No photo available)

Species such as these show no aspects of rarity; they are among the most common in the biosphere.

Each of these species show one aspect of rarity, which gives them some vulnerability to extinction.

With two aspects of rarity, these three groups of species are even more vulnerable to extinction.

Species such as these are the rarest in the biosphere and are the most vulerable to extinction.

Text on white highlights aspects of rarity.

**Rarest**

**Figure 9.25** Commonness, rarity, and vulnerability to extinction.

is confined to a narrow range of conditions in the high mountains of the Tibetan Plateau. Small geographic range, narrow habitat tolerance, and low population density are attributes of rarity.

As shown in figure 9.25, there are eight possible combinations of these factors, seven of which include at least one attribute of rarity. The most abundant species and those least threatened by extinction have extensive geographic ranges, broad habitat tolerances, and large local populations at least somewhere within their range. Some of these species, such as starlings, Norway rats, and house sparrows, are associated with humans and are considered pests. However, many species of small mammals, birds, and invertebrates not associated with humans, such as the deer mouse, *Peromyscus maniculatus,* or the marine zooplankton, *Calanus finmarchicus,* also fall into this most common category.

Ecologists exploring the relationship between size of geographic range and population size have found that they are not independent. Instead, there is a strong positive correlation between the two variables for most groups of organisms. In other words, species abundant in the places where they occur are generally widely distributed within a region, continent, or ocean, while species living at low population densities generally have small, restricted distributions. The positive relationship between range and population density was first brought to the attention of ecologists by Ilka Hanski (Hanski 1982) and James H. Brown (Brown 1984). Kevin Gaston (Gaston 1996, Gaston et al. 2000) points out that in the two decades since the early work by Hanski and Brown, ecologists have found a positive relationship between range and population density for many groups, including plants, grasshoppers, scale insects, hoverflies, bumblebees, moths, beetles, butterflies, birds, frogs, and mammals. Several mechanisms have been proposed to explain the positive relationship between local abundance and range size. Many of the explanations focus on breadth of environmental tolerances and differences in metapopulation dynamics. However, as Gaston and his colleagues point out (Gaston et al. 2000) there is still no concensus on the most likely explanations.

Most species, are uncommon; seven combinations of range, tolerance, and population size each create a kind of rarity. As a consequence, Rabinowitz referred to "seven forms of rarity." Let's look at species that represent the two extremes of Rabinowitz's seven forms of rarity. The first two discussions concern species that are rare according to only one attribute. These are species that, before they become extinct, may seem fairly secure. The final discussion concerns the very rarest species, which show all three attributes of rarity. Though these rarest species are the most vulnerable to extinction, rarity in any form appears to increase vulnerability to extinction.

**Figure 9.26**  The peregrine falcon, *Falco peregrinus,* is found throughout the Northern Hemisphere but lives at low population densities throughout its range.

## Rarity I: Extensive Range, Broad Habitat Tolerance, Small Local Populations

It is easy to understand how people were drawn to the original practice of falconry. The sight and sound of a peregrine falcon, *Falco peregrinus,* in full dive at over 200 km per hour must have been one of the great experiences of a lifetime (fig. 9.26). To have held such a bird on your arm and launched it at its avian prey must have seemed liked controlling the wind. The peregrine, which has a geographic range that circles the Northern Hemisphere and broad habitat tolerance, is uncommon throughout its range. Apparently, this one attribute of rarity was enough to make the peregrine vulnerable to extinction. The falcon's feeding on prey containing high concentrations of DDT, which produced thin eggshells and nesting failure, was enough to drive the peregrine to the brink of extinction. Peregrine falcons were saved from extinction by control of the use of DDT, strict regulation of the capture of the birds, captive breeding, and reintroduction of the birds to areas where local populations had become extinct.

The range of the tiger, *Panthera tigris,* once extended from Turkey to eastern Siberia, Java, and Bali and included environments ranging from boreal forest to tropical rain forest. The tigers in this far-flung population varied enough from place to place in size and coloration that many local populations were described as separate subspecies, including the Siberian, Bengal, and Javanese tigers. Like peregrine falcons, tigers had an extensive geographic range and broad habitat tolerance but low population density. Over the centuries, relentless pursuit by hunters reduced the tiger's range from nearly half of the largest continent on earth to a series of tiny fragmented populations. Many local populations have become extinct and others, such as the magnificent Siberian

tiger, teeter on the verge of extinction in the wild. These populations may survive only through captive breeding programs in zoos. The next example shows that narrow habitat tolerance can also lead to extinction.

## Rarity II: Extensive Range, Large Populations, Narrow Habitat Tolerance

When Europeans arrived in North America, they encountered one of the most numerous birds on earth, the passenger pigeon. The range of the passenger pigeon extended from the eastern shores of the present-day United States to the Midwest, and its population size numbered in the billions. However, the bird had one attribute of rarity: it had a narrow requirement for its nesting sites. The passenger pigeon nested in huge aggregations in virgin forests. As virgin forests were cut, its range diminished and market hunters easily located and exploited its remaining nesting sites, finishing off the remainder of the population. By 1914, when the last passenger pigeon died in captivity, one of the formerly most numerous bird species on earth was extinct. Extensive range and high population density alone do not guarantee immunity from extinction.

The rivers in the same region inhabited by the passenger pigeon harbored an abundant, widely distributed but narrowly tolerant fish, the harelip sucker, *Lagochila lacera.* This fish was found in streams across most of the east-central United States and was abundant enough that early ichthyologists cited it as one of the commonest and most valuable food fishes in the region. However, the harelip sucker, like the passenger pigeon, had narrow habitat requirements. It was restricted to large pools with rocky bottoms in clear, medium-sized streams about 15 to 30 m wide. This habitat was eliminated by the silting of rivers that followed defor-estation and by the erosion of poorly managed agricultural lands. The last individuals of this species collected by ichthy-ologists came from the Maumee River in northwestern Ohio in 1893.

## Extreme Rarity: Restricted Range, Narrow Habitat Tolerance, and Small Populations

Species that combine small geographic ranges with narrow habitat tolerances and low population densities are the rarest of the rare. This group includes species such as the mountain gorilla, the giant panda, and the California condor. Species showing this extreme form of rarity are clearly the most vul-nerable to extinction. Many island species have these attrib-utes, so it is not surprising that island species are especially vulnerable. Of the 171 bird species and subspecies known to have become extinct since 1600, 155 species have been restricted to islands. Of the 70 species and subspecies of birds known to have lived on the Hawaiian Islands, 24 are now extinct and 30 are considered in danger of extinction.

Organisms on continents that are restricted to small areas, have narrow habitat tolerance, and small population size are also vulnerable to extinction. Examples of popula-tions in such circumstances are common. More than 20 species of plants and animals are confined to about 200 km$^2$ of mixed wetlands and upland desert in California called Ash Meadows. The Ash Meadows stick-leaf, *Mentzelia leu-cophylla,* inhabits an area of about 2.5 km$^2$ and has a total population size of fewer than 100 individuals. Another plant, the Ash Meadows milk vetch, *Astragalus phoenix,* has a total population of fewer than 600 individuals. Human alteration of Ash Meadows appears to have caused the extinction of at least one native species, the Ash Meadows killifish, *Empetrichthys merriami.*

Amazingly, there are species with ranges even more restricted than those of Ash Meadows, California. In 1980, the total population of the Virginia round-leaf birch, *Betula uber,* was limited to 20 individuals in Smyth County, Virginia. Until recently the total habitat of the Socorro iso-pod, *Thermosphaeroma thermophilum,* of Socorro, New Mexico, was limited to a spring pool and outflow with a sur-face area of a few square meters. Meanwhile, a palm species, *Pritchardia monroi,* which is found only on the island of Maui in the Hawaiian Islands, has a total population in nature of exactly one individual!

Examples such as these fill books listing endangered species. In nearly all cases, the key to a species' survival is increased distribution and abundance. One of the most funda-mental needs for managing species, endangered or not, is making accurate estimates of population size. Some of the conceptual and practical issues that population ecologists must consider when censusing a population are the subject of the Applications & Tools section.

## APPLICATIONS & TOOLS

### Estimating Abundance—From Whales to Sponges

The abundance of organisms and how abundance changes in time and space are among the most fundamental concerns of ecology. These factors are so basic that some authors define ecology as the study of distribution and abundance of organ-isms. Because abundance is so important, ecologists should understand how to estimate it for a wide variety of organisms. Keep in mind, however, that ecologists do not measure abun-dance as an end in itself but as a tool to understand the ecology of populations. Knowing how abundant an organism is can tell us whether its population is growing, declining, or stable. As we saw in the previous section, population size is one of the charac-teristics that helps ecologists assess a species' vulnerability to extinction. However, to estimate the abundance of species the ecologist must contend with a variety of practical challenges and conceptual subtleties. Some of these are discussed here.

## Estimating Whale Population Size

In 1989, the journal *Oceanus* published a table that listed the estimated sizes of whale populations. The table included the following note: "All estimates . . . are highly speculative." Why is it difficult to provide firm estimates of whale population size? Briefly, whales live at low population densities and may be dis-tributed across vast expanses of ocean. They also spend much time submerged and move around a great deal. As large as they are, you cannot count all the whales in the ocean. Instead, marine ecologists rely on population estimation. Each method of estimation has its own limitations and uncertainties.

One method used to estimate population sizes of elusive animals involves marking or tagging some known number of individuals in the population, releasing the marked individu-als so they will mix with the remainder of the population, and then sampling the population at some later time. The ratio of marked to unmarked individuals in the sample gives an esti-mate of population size. The simplest formula expressing this relationship is the Lincoln-Peterson index:

$$M/N = m/n$$

where:

> M = the number of individuals marked and released
>
> N = the actual size of the study population
>
> m = the number of marked individuals in a sample of the population
>
> n = the total number of individuals in the sample

**Figure 9.27** A humpback whale, *Megaptera novaeangliae.*

The major assumption of the Lincoln-Peterson index is that the ratio of marked to unmarked individuals in the population as a whole equals the ratio of marked to unmarked individuals in a sample of the population. If this is approximately so, then the population size is estimated as:

$$N = Mn/m$$

However, on average, the Lincoln-Peterson index overestimates population size. To reduce this tendency to overestimate, N. Bailey (1951,1952) proposed a corrected formula:

$$N = M (n + 1)/m + 1$$

Some of the assumptions of mark and recapture studies are:

- All individuals in the population have an equal probability of being captured.

- The population is not increased by births or immigration between marking and recapture.

- Marked and unmarked individuals die and emigrate at the same rates.

- No marks are lost.

Although real populations rarely meet all these assumptions, mark and recapture estimates of population size are often the best estimates available.

Whale populations have been studied using mark and recapture techniques for some time. In the early days of whale population studies, population biologists marked whales by shooting a numbered metal dart into their blubber. The idea was that the dart would be recovered when marked whales were caught and processed during whaling operations. However, the accuracy of this method was limited by several factors. First, it is difficult to mark a free-swimming whale from a moving boat on the open sea, and so biologists never knew exactly how many whales shot at were actually marked. Second, the recovery of darts during processing was poor. It is easy to overlook a relatively small dart on a huge whale. Experiments showed that Japanese whalers recovered 60% to 70% of marks from whales known to be marked, while other whaling fleets recovered an

(a)

(b)

**Figure 9.28** Unique markings identify individual humpback whales. A humpback whale called "Siphon," #700, photographed in Frenchman Bay, Maine: (*a*) in 1995 and (*b*) in 1993.

even lower proportion. Still, the greatest limitation of this early mark and recapture technique is that it requires killing whales during the recapture phase, which is unacceptable when studying protected or endangered species.

Refined mark and recapture methods do not require artificially marking or capturing whales. In the "marking" phase of newer procedures, a whale is photographed and its distinguishing marks are identified. These photographs, along with information such as where the photograph was taken and whether the whale was accompanied by an offspring, are catalogued for future reference. In the "recapture" phase the whale is photographed at a later date and identified from previous photos. This method is called *photoidentification.*

For more than two decades, Steven Katona (1989) has used photoidentification to study the humpback whales, *Megaptera novaeangliae,* of the North Atlantic (fig. 9.27). Humpback whales are particularly rich in individual marks, especially on the tail or flukes. This is convenient for photographic studies because humpback whales generally raise their flukes above the water before they dive. This behavior, called "fluking," exposes the flukes to the photographer and reveals potentially unique markings (fig. 9.28).

Katona points out that the unique markings on a humpback whale's flukes result from a combination of genetics and accidents. In humpback whales, fluke pigmentation ranges from completely black to white. This variation in pigmentation probably reflects genetic differences between individuals. Injuries often produce scars that superimpose other marks on the basic pigmentation of the flukes. Injuries may be the result of fighting between humpback whales, attacks by sharks, or attachment by parasites. The scars produced when these injuries heal create black marks on white flukes and white marks on black flukes. Using photographs of these marks, Katona and his colleagues have produced the North Atlantic Humpback Catalog, which includes photographs of more than 4,000 individual whales. The photographs included in the catalog, along with information on where each photograph was taken, whether the whale was accompanied by an offspring, and other available observations, are curated for future reference. This photographic record is an invaluable source of information for determining the migration routes, feeding grounds, breeding grounds, and size of the North Atlantic humpback whale population (fig. 9.29).

From 1979 to 1986, Scott Baker, Janice Straley, and Anjanette Perry (1992) photographed and identified 257 humpback whales along the coast of southeastern Alaska. In one part of their study the researchers used photoidentification to estimate the number of humpback whales in Frederick Sound, Alaska. In their first sampling period, from July 31 to August 3, 1986, the team photographed and identified 72 humpback whales. In a second sampling period, from August 29 to September 1, 1986, they photographed and identified 78 humpback whales. Of the 78 whales photographed in the second sampling period, 56 were photographed for the first time, while 22 had also been photographed during the first sampling period. These 22 whales were the "recaptures." We can use these data and the corrected Lincoln-Peterson index to estimate the total number of whales in Frederick Sound from August 29 to September 1, 1986:

$$N = M\,(n + 1)/m + 1 = 72\,(78 + 1)/22 + 1 = 247$$

In other words, Baker and his colleagues estimated that there were 247 humpback whales in Frederick Sound during the time of their study. Population estimates such as this are very important for monitoring the state of populations. In addition, photographic studies provide information on movements, calving intervals, and survival because photoidentified whales can go on yielding information throughout their lives.

Population ecologists have now applied the photoidentification techniques to several whale species including gray, right, blue, fin, sei, and killer whales. For instance, by 1986, scientists had photographed about 200 of the 300 to 400 North Atlantic right whales remaining in the Atlantic. From the growing photographic record of the North Atlantic right whale population, ecologists have estimated a 4- to 7-year calving interval for the population. This long period between offspring alone gives a clue to why this population is so vulnerable to extinction in the face of whaling pressure. In

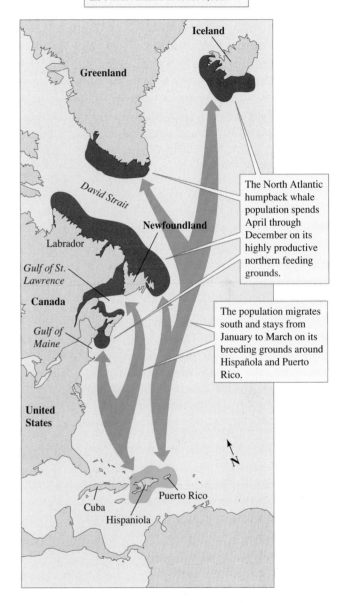

**Figure 9.29** Photoidentification and the North Atlantic humpback whale population.

another study, population ecologists photographed and identified the entire population of 325 killer whales in Puget Sound and around Vancouver Island, revealing information about births, deaths, and the size of the population. An advance in this research approach is the use of computerized image analysis to help identify and match photos of whales. While not providing all the information necessary for conserving and managing whale populations, photographic studies are clearly making an important contribution.

Though it may be more challenging physically, the process of counting whales is much like counting many other kinds of animals such as humans, lynxes, trout, or lady beetles.

However, ecologists must use different methods to estimate the abundance of organisms that have a more variable growth form or differ greatly in size. As we shall see in the next example, this is particularly true when the relative abundances of very different organisms are compared.

# The Relative Abundance of Corals, Algae, and Sponges

The reefs along the north coast of Jamaica were once dominated by corals. Thickets of staghorn coral rose from the seafloor like marine bramble bushes, and elkhorn coral grew in abundance in the surging waves. Then dramatic change came in 1980 with Hurricane Allen, which generated waves large enough to flatten the staghorn coral thickets and dislodge elkhorn corals. Most of the branching corals in shallow water were reduced to rubble. However, Hurricane Allen and its devastation was not the only problem. The reefs seemed to be changing even at depths below 25 m, where there was little hurricane damage. Assessing the extent and nature of changes on the reef would require detailed population studies.

Fortunately, Terence Hughes (1996) had started long-term studies of coral populations near Discovery Bay, Jamaica, in 1977, 3 years before Hurricane Allen. One of his goals was to document and understand apparent shifts in dominance from corals to algae. He was also concerned with documenting possible changes in sponge populations. But how can the relative abundance of organisms so different in size and growth form be estimated? A coral colony may cover several square meters or just a few square centimeters. Sponges also differ greatly in size. Algae of several species may grow together in a tangled mat covering several square meters or as a few isolated individuals. Ecologists studying terrestrial plants encounter similar size differences among plants of different ages and species. In the face of this variation, ecologists resort to measures of abundance that take into account differences in size. For instance, ecologists studying shrubs, herbaceous vegetation, or marine organisms such as corals, algae, and sponges often measure coverage, the area of landscape or reef covered by a species.

Hughes estimated percent cover by corals, algae, and sponges on his study reef nearly every year from 1977 to 1993. He made his estimates from photographs of 12 study plots 1 m$^2$ in size taken from a standard distance using slide film. He then mapped the positions and sizes of all coral and

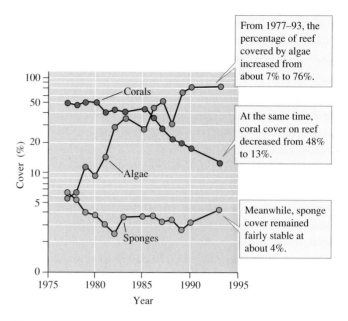

**Figure 9.30** Estimating abundance as percent cover: corals, sponges, and algae (data from Hughes 1996).

sponge colonies within the quadrats. He used a computer program to measure the areas on his maps that were covered by corals and sponges and then converted these measurements of absolute area to percent cover. Hughes estimated percent cover of algae by projecting the slide images of his study plots on a screen and superimposing a uniform grid of points on the image. The grid contained 100 points per square meter. The percentage of these points contacting algae in the image gave him an estimate of percent algal cover.

Hughes' 16-year study shows clear changes in the abundances of corals and algae (fig. 9.30). During the study, algal cover increased tenfold from 7% to 76%. At the same time, coral cover decreased from about 48% to 13%, while sponge populations remained fairly constant. Verifying the status of populations such as these is one of the most basic aspects of ecological research. Such measurements are the first step toward identifying the factors that determine the distribution and abundance of organisms. With his photographs Hughes was able to show that this shift from a coral-dominated reef to one dominated by algae was due to increased mortality of coral colonies and reduced recruitment of new corals to the population. These are aspects of population dynamics that we cover in chapter 10.

Ecologists define a population as a group of individuals of a single species inhabiting an area delimited by natural or human-imposed boundaries. Population studies hold the key to solving practical problems such as saving endangered species, controlling pest populations, or managing fish and game populations. All populations share a number of characteristics. Chapter 9 focused on two population characteristics: distribution and abundance.

While there are few environments on earth without life, no single species can tolerate the full range of earth's environments. Because all species find some environments too warm, too cold, too saline, and so forth, **the physical environment limits the geographic distribution of species.** For instance, there is a close relationship between climate and the distributions of the three largest kangaroos in Australia. The tiger beetle *Cicindela longilabris* is limited to cool boreal and mountain environments. Large- and small-scale variation in temperature and moisture limits the distributions of certain desert plants, such as shrubs in the genus *Encelia*. However, differences in the physical environment only partially explain the distributions of barnacles within the marine intertidal zone, a reminder that biological factors constitute an important part of an organism's environment.

**On small scales, individuals within populations are distributed in patterns that may be random, regular, or clumped.** Patterns of distribution can be produced by the social interactions within populations, by the structure of the physical environment, or by a combination of the two. Social organisms tend to be clumped; territorial organisms tend to be regularly spaced. An environment in which resources are patchy also fosters clumped distributions. Aggressive species of stingless bees live in regularly distributed colonies, while the colonies of nonaggressive species are randomly distributed. The distribution of creosote bushes changes as they grow. **On larger scales, individuals within a population are clumped.** In North America, populations of both wintering and breeding birds are concentrated in a few hot spots of high population density. Clumped distributions are also shown by plant populations living along steep environmental gradients on mountainsides.

**Many populations are subdivided into subpopulations called metapopulations.** Populations of many species occur not as a single continuously distributed population but in spatially isolated patches, with significant exchange of individuals among patches. A group of subpopulations living on such patches connected by exchange of individuals among patches make up a metapopulation. Populations of the Rocky Mountain Parnassian butterfly, *Parnassius smintheus,* in Alberta, Canada, and of the lesser kestrel, *Falco naumanni,* consist of metapopulations. In both metapopulations movement of individuals is predominantly from smaller subpopulations to larger subpopulations.

**Population density declines with increasing organism size.** In general, animal population density declines with increasing body size. This negative relationship holds for animals as varied as terrestrial invertebrates, aquatic invertebrates, birds, poikilothermic vertebrates, and herbivorous mammals. Plant population density also decreases with increasing plant size. However, the biological details underlying the size–density relationship shown by plants are quite different from those underlying the size–density patterns shown by animals. A single species of tree can span a very large range of sizes and densities during its life cycle. The largest trees start life as small seedlings that can live at very high population densities. As trees grow, their population density declines progressively until the mature trees live at low densities.

**Commonness and rarity of species are influenced by population size, geographic range, and habitat tolerance.** Rarity of species can be expressed as a combination of extensive versus restricted geographic range, broad versus narrow habitat tolerance, and large versus small population size. The most abundant species and those least threatened by extinction combine large geographic ranges, wide habitat tolerance, and high local population density. All other combinations of geographic range, habitat tolerance, and population size include one or more attributes of rarity. Rare species are vulnerable to extinction. Populations that combine restricted geographic range with narrow habitat tolerance and small population size are the rarest of the rare and are usually the organisms most vulnerable to extinction.

The abundance of organisms and how abundance changes in time and space are among the most fundamental concerns of ecology. To estimate the abundance of species the ecologist must contend with a variety of practical challenges and conceptual subtleties. Mark and recapture methods are useful in the study of populations of active, elusive, or secretive animals. Mark and recapture techniques, which use natural distinguishing marks, are making an important contribution to the study of populations of endangered whales. Ecologists studying organisms, such as corals, algae, and sponges or many types of terrestrial plants, that differ a great deal in size and form often estimate abundance as coverage, the area covered by a species. Patterns of distribution and abundance are ultimately determined by underlying population dynamics.

# Review Questions

1. What confines *Encelia farinosa* to upland slopes in the Mojave Desert? Why is it uncommon along desert washes, where it would have access to much more water? What may allow *E. frutescens* to persist along desert washes while *E. farinosa* cannot?

2. Spruce trees, members of the genus *Picea,* occur throughout the boreal forest and on mountains farther south. For example, spruce grow in the Rocky Mountains south from the heart of boreal forest all the way to the deserts of the southern United States and Mexico. How do you think they would be distributed in the mountains that rise from the southern deserts? In particular, how do altitude and aspect (see chapter 4) affect their distributions in the southern part of their range? Would spruce populations be broken up into small local populations in the southern or the northern part of the range? Why?

3. What kinds of interactions within an animal population lead to clumped distributions? What kinds of interactions foster a regular distribution? What kinds of interactions would you expect to find within an animal population distributed in a random pattern?

4. How might the structure of the environment, for example, the distributions of different soil types and soil moisture, affect the patterns of distribution in plant populations? How should interactions among plants affect their distributions?

5. Suppose one plant reproduces almost entirely from seeds, and that its seeds are dispersed by wind, and a second plant reproduces asexually, mainly by budding from runners. How should these two different reproductive modes affect local patterns of distribution seen in populations of the two species?

6. Suppose that in the near future, the fish crow population in North America declines because of habitat destruction. Now that you have reviewed the large-scale distribution and abundance of the fish crow (see fig. 9.15*b*), devise a conservation plan for the species that includes establishing protected refuges for the species. Where would you locate the refuges? How many refuges would you recommend?

7. Use the empirical relationship between size and population density observed in the studies by Damuth (1981) (see fig. 9.22) and Peters and Wassenberg (1983) (see fig. 9.23) to answer the following: For a given body size, which generally has the higher population density, birds or mammals? On average, which lives at lower population densities, terrestrial or aquatic invertebrates? Does an herbivorous mammal twice the size of another have on average one-half the population density of the smaller species? Less than half? More than half?

8. Outline Rabinowitz's classification (1981) of rarity, which she based on size of geographic range, breadth of habitat tolerance, and population size. In her scheme, which combination of attributes makes a species least vulnerable to extinction? Which combination makes a species the most vulnerable?

9. Can the analyses by Damuth (1981) and by Peters and Wassenberg (1983) be combined with that of Rabinowitz (1981) to make predictions about the relationship of animal size to its relative rarity? What two attributes of rarity, as defined by Rabinowitz, are not included in the analyses by Damuth and by Peters and Wassenberg?

10. Suppose you have photoidentified 30 humpback whales around the island of Oahu in one cruise around the island. Two weeks later you return to the same area and photograph all the whales you encounter. On the second trip you photograph a total of 50 whales, of which 10 were photographed previously. Use the Lincoln-Peterson index with the Bailey correction to estimate the number of humpback whales around Oahu during your study.

# Suggested Readings

Baker, C. S., J. M. Straley, and A. Perry. 1992. Population characteristics of individually identified humpback whales in southeastern Alaska: summer and fall 1986. *Fishery Bulletin* 90:429–37.

Katona, S.K. 1989. Getting to know you. *Oceanus* 32:37–44.

*These papers provide detailed accounts of using photoidentification to study humpback whale populations.*

Brisson, J. and J. F. Reynolds. 1994. The effects of neighbors on root distribution in a creosote bush (*Larrea tridentata*) population. *Ecology* 75:1693–702.

*The paper by Brisson and Reynolds demonstrates extended studies of creosote bush distributions belowground.*

Brown, J. H., D. W. Mehlman, and G. C. Stevens. 1995. Spatial variation in abundance. *Ecology* 76:2028–43.

Root, T. 1988. *Atlas of Wintering North American Birds.* Chicago: University of Chicago Press.

*These two references provide excellent entries into the area of large-scale distribution patterns in bird populations. These are pioneering efforts.*

Caughley, G., J. Short, G. C. Grigg, and H. Nix. 1987. Kangaroos and climate: an analysis of distribution. *Journal of Animal Ecology* 56:751–61.

Ehleringer, J. R. and C. Clark. 1988. Evolution and adaptation in *Encelia* (Asteraceae). In L. D. Gottlieb and S. K. Jain., eds. *Plant Evolutionary Biology.* London: Chapman and Hall.

*These two papers provide good introductions to climate and the distributions of species within an animal genus and within a plant genus.*

Damuth, J. 1981. Population density and body size in mammals. *Nature* 290:699–700.

Peters, R. H. and K. Wassenberg. 1983. The effect of body size on animal abundance. *Oecologia* 60:89–96.

*These are benchmark papers on the relationship between animal size and population density.*

Freckleton, R. P. and A. R. Watkinson. 2002. Large-scale dynamics of plants: metapopulations, regional ensembles and patchy populations. *Journal of Ecology* 90:419–34.

*Thought-provoking review of the application of the metapopulation concept to plant populations.*

Gaston, K. J. 1996. The multiple forms of the interspecific abundance-distribution relationship. *Oikos* 76:211–20.

Gaston, K. J., T. M. Blackburn, J. J. D. Greenwood, R. D. Gregory, R. M. Quinn, and J. H. Lawton. 2000. Abundance-occupancy relationships. *Journal of Applied Ecology* 37:39–59.

*Two thorough reviews of the relationship between population size and geographic range.*

Phillips, D. L. and J. A. MacMahon. 1981. Competition and spacing patterns in desert shrubs. *Journal of Ecology* 69:97–115.

*The paper by Phillips and MacMahon gives an excellent summary of the history and apparent resolution of a controversy surrounding distributions of creosote bush.*

Rabinowitz, D., S. Cairns, and T. Dillon. 1986. Seven forms of rarity and their frequency in the flora of the British Isles. In M. E. Soule, ed. *Conservation Biology: The Science of Scarcity and Diversity.* Sunderland, Mass.: Sinauer Associates.

*This paper provides an introduction and application of the concept of rarity developed by Deborah Rabinowitz.*

Roland, J., N. Keyghobadi, and S. Fownes. 2000. Alpine Parnassius butterfly dispersal: effects of landscape and population size. *Ecology* 81:1642–53.

*A detailed study of the dynamics of a butterfly metapopulation.*

Serrano, D. and J. L. Tella. 2003. Dispersal within a spatially structured population of lesser kestrels: the role of spatial isolation and conspecific attraction. *Journal of Animal Ecology* 72:400–410.

*A careful long-term study of a growing metapopulation of a species that has been declining over most of its range.*

# On the Net

Visit this textbook's accompanying website at www.mhhe.com/ecology (click on the book's title) to take advantage of practice quizzing, study/writing tips, timely news articles, and additional URLs for research on the topics in this chapter.

Animal Population Ecology
Population Density of Animals
Biodiversity

Endangered Species
Legislation Regarding Endangered Species

# Population Dynamics

Uncovering patterns of survival within natural populations of animals or plants often requires extended field studies. As Adolph Murie watched, the gray wolf ran downhill toward a herd of 20 Dall sheep, *Ovis dalli.* As the wolf approached, the herd of white sheep split into two bands. One band circled the wolf and ran up the slope, while the other ran downhill. In response, the wolf stopped. The two bands of sheep also stopped, only 30 to 40 m away from the wolf. Suddenly the wolf sprinted after the lower band, but they easily outran him on the steep terrain. Again, the sheep and the wolf stopped and rested. After an hour, the wolf broke the stalemate and again charged the lower band. The sheep avoided him, circling the wolf and rejoining the other half of their herd. A few minutes later the wolf abandoned the hunt, trotting away as the herd of Dall sheep watched from the ridge above (fig. 10.1).

Despite this particular wolf's failure, wolves kill enough Dall sheep to cause some people to suggest that wolf populations should be reduced to protect the sheep. Murie (1944) had been hired by the U.S. National Park Service to study the interactions between wolves and Dall sheep in Mount Danali National Park, Alaska. The main purpose of his study was to determine whether wolves kill enough sheep to justify the call for reducing the wolf population.

Murie pursued several lines of research. As in this example, he directly observed wolves and sheep. He also tracked wolves through winter snow to find their kills. The tracks left a record of wolf interactions with their prey. Where wolves had killed Dall sheep, they often left the skulls, which provided a record as rich as the telltale tracks. Murie could age the skulls by the size of the horns. The horns also indicated the sex of the individual. The teeth provided an indication of the sheep's general condition; worn teeth were a sign of poor nutrition and weakness. A careful search of Mount Danali National Park yielded a sample of 608 sheep skulls, which Murie used to explore the causes and age of death. The skulls showed Murie that death within the Dall sheep population fell mainly on the very young and the very old. Most sheep in the population could, as his direct observations had shown, easily avoid attack by wolves.

Fifty years later in the rocky desert terrain of eastern Egypt, Ahmad Hegazy (1990) used similar care to study an endangered plant species. The plant, *Cleome droserifolia,* is heavily exploited by desert dwellers and herbalists as a medicinal plant. Hegazy's study was prompted by concern that harvesting was leading to the extinction of this valuable plant species. Like Murie, Hegazy collected information that would give him insights into patterns of life and death within his study population. He collected information on the number of seeds produced by the population, on seed dispersal, and on the number of seeds in the soil. Hegazy also studied the establishment of seedlings and juvenile plants, as well as the survival of adults.

Though Hegazy did not have to pursue his study organisms through deep winter snows, the *Cleome* population presented him with challenges no less daunting than those faced

**Figure 10.1** Dall sheep, *Ovis dalli,* a mountain sheep of far northern North America, was the subject of one of the classic studies of suvivorship.

by Murie. First, *Cleome* has a much higher number and diversity of life stages, including flowers, fruits, seeds, seedlings, juvenile plants, and adult plants. In addition, where Murie's study population numbered in the hundreds, Hegazy's seed population numbered in the millions. While Murie had to contend with estimating survival and reproduction of an organism with a life span of about a dozen years, Hegazy had to estimate patterns of survival and reproduction of an organism that lives nearly 80 years. Hegazy's careful analysis of the *Cleome* population provided a means for managing the species that promotes its survival and allows its use in traditional medicine.

Adolph Murie's classical studies of wolves and Dall sheep and Hegazy's more recent study of a desert plant introduce us to another area of population biology, population dynamics. In chapter 9, we explored population distribution and abundance. However, to do so, we had to freeze populations at a particular instant in time. In fact, patterns of distribution and abundance result from a dynamic balance between rates of birth, death, immigration, and emigration. These dynamic processes are the subject of chapter 10.

Population dynamics involve what we might call the "behavior" of populations. However, population behavior differs from the behavior of individual organisms. Population dynamics occur at a different level of biological organization, at the level of groups of individuals. Another major difference is that except for highly localized populations of microorganisms, population processes are played out at larger spatial, and longer temporal, scales. This difference in scale largely hides all but the most obvious population phenomena from the unaided human observer. To see and probe the dynamics of populations we need the mathematical tools of population ecology. These quantitative tools provide us with a window to otherwise largely invisible phenomena.

It is difficult to keep track of everything going on in populations. Distributions may expand and contract. Numbers may increase for some time and then fall precipitously. A new population of a previously unrecorded species may suddenly appear in an area, persist for a season or a decade, and then disappear. Estimating characteristics such as survival and birthrates requires a great deal of information. Numbers of individuals alone, which may range from dozens to millions, can overwhelm the population ecologist. In addition, individuals of different ages and sexes may make different contributions to population dynamics and so must be followed separately. To organize our exploration of population dynamics, we first consider patterns of survival in populations and then age distributions. Populations sometimes increase or decrease in size, and we next acquire the quantitative tools for perceiving such changes. Finally, we examine the effects of individuals' moving into and out of populations.

## CONCEPTS

- **A survivorship curve summarizes the pattern of survival in a population.**
- **The age distribution of a population reflects its history of survival, reproduction, and potential for future growth.**
- **A life table combined with a fecundity schedule can be used to estimate net reproductive rate ($R_0$), geometric rate of increase, ($\lambda$), generation time (T), and per capita rate of increase ($r$).**
- **Dispersal can increase or decrease local population densities.**

## CONCEPT DISCUSSSION

### Patterns of Survival

**A survivorship curve summarizes the pattern of survival in a population.**

Patterns of survival vary a great deal from one species to another and, depending on environmental circumstances, can vary substantially even within a single species. Some species produce young by the millions, which, in turn, die at a high rate. Other species produce a few young and invest heavily in their care. The young of species that have evolved this pattern survive at a high rate. Still other species show intermediate patterns of reproductive rate, parental care, and juvenile survival. In response to practical challenges of discerning patterns of survival, population biologists have invented bookkeeping devices called **life tables** that list both the survivorship and the deaths, or *mortality,* in populations.

# Estimating Patterns of Survival

There are three main ways of estimating patterns of survival within a population. The first and most reliable way is to identify a large number of individuals that are born at about the same time and keep records on them from birth to death. A group born at the same time is called a **cohort.** A life table made from data collected in this way is called a **cohort life table.** The cohort studied might be a group of plant seedlings that germinated at the same time or all the lambs born into a population of mountain sheep in a particular year.

While understanding and interpreting a cohort life table may be relatively easy, obtaining the data upon which a cohort life table is based is not. Imagine yourself lying face down in a meadow painstakingly counting thousands of tiny seedlings of an annual plant. You must mark their locations and then come back every week for 6 months until the last member of the population dies. Or, if you are studying a moderately long-lived species, such as a barnacle or a perennial herb like a buttercup, imagine checking the cohort repeatedly over a period of several years. If your study organism is a mobile animal such as a whale or falcon, the problems multiply. If your species is very long-lived, such as a giant sequoia, such an approach is impossible within a single human lifetime. In such circumstances population biologists usually resort to other techniques.

A second way to estimate patterns of survival in wild populations is to record the age at death of a large number of individuals. This method differs from the cohort approach because the individuals in your sample are born at different times. This method produces a **static life table.** The table is called *static* because the method involves a snapshot of survival within a population during a short interval of time. To produce a static life table the biologist often needs to estimate the age at which individuals die. This can be done by tagging individuals when they are born and then recovering the tags after death and recording the age at death. An alternative procedure is to somehow estimate the age of dead individuals. The age of many species can be determined reasonably well. For instance, mountain sheep can be aged by counting the growth rings on their horns. There are also growth rings on the carapaces of turtles, in the trunks of trees, and in the "stems" of soft or hard corals.

A third way of determining patterns of survival is from the **age distribution.** An age distribution consists of the proportion of individuals of different ages within a population. You can use an age distribution to estimate survival by calculating the difference in proportion of individuals in succeeding age classes. This method, which also produces a static life table, assumes that the difference in numbers of individuals in one age class and the next is the result of mortality. What are some other major assumptions underlying the use of age distributions to estimate patterns of survival? This method requires that a population is neither growing nor declining and that it is not receiving new members from the outside or losing members because they migrate away. Since most of

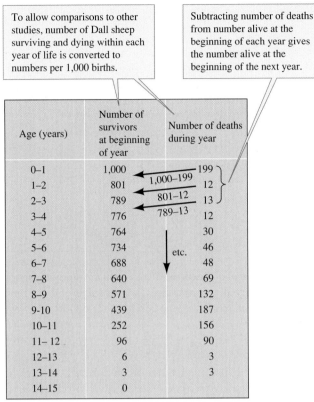

To allow comparisons to other studies, number of Dall sheep surviving and dying within each year of life is converted to numbers per 1,000 births.

Subtracting number of deaths from number alive at the beginning of each year gives the number alive at the beginning of the next year.

| Age (years) | Number of survivors at beginning of year | Number of deaths during year |
|---|---|---|
| 0–1 | 1,000 | 199 |
| 1–2 | 801 | 12 |
| 2–3 | 789 | 13 |
| 3–4 | 776 | 12 |
| 4–5 | 764 | 30 |
| 5–6 | 734 | 46 |
| 6–7 | 688 | 48 |
| 7–8 | 640 | 69 |
| 8–9 | 571 | 132 |
| 9-10 | 439 | 187 |
| 10–11 | 252 | 156 |
| 11– 12 | 96 | 90 |
| 12–13 | 6 | 3 |
| 13–14 | 3 | 3 |
| 14–15 | 0 | |

1,000–199
801–12
789–13
etc.

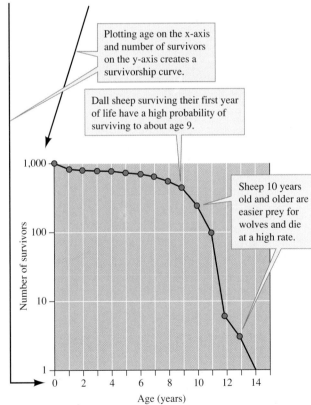

Plotting age on the x-axis and number of survivors on the y-axis creates a survivorship curve.

Dall sheep surviving their first year of life have a high probability of surviving to about age 9.

Sheep 10 years old and older are easier prey for wolves and die at a high rate.

**Figure 10.2** Dall sheep: from life table to survivorship curve (data from Murie 1944).

these assumptions are often violated in natural populations, a life table constructed from this type of data tends to be less accurate than a cohort life table. Static life tables are often useful, however, since they may be the only information available.

# High Survival Among the Young

As we saw in the introduction, Adolph Murie studied patterns of survival among Dall sheep in what is now Mount Danali National Park, Alaska. Murie estimated survival patterns by collecting the skulls of 608 sheep that had died from various causes. He determined the age at which each sheep in his sample died by counting the growth rings on their horns and by studying tooth wear. The major assumption of this study is that the proportion of skulls in each age class represented the typical proportion of individuals dying at that age. For example, the proportion of sheep in the sample that died before the age of 1 year represents the proportion that generally dies during the first year of life. While this assumption is not likely to be strictly true, the pattern of survival that emerges probably gives a reasonable picture of survival in the population, particularly when the sample is as large as Murie's.

Figure 10.2 summarizes the survival patterns for Dall sheep based on Murie's sample of skulls. The upper portion of the figure shows the static life table that Murie constructed. The first column lists the ages of the sheep, the second column lists the number surviving in each age class, and the third column lists the numbers dying in each age class. Notice that although Murie studied only 608 skulls, the numbers in the table are expressed as numbers per 1,000 individuals. This adjustment is made to ease comparisons with other populations.

The upper portion of figure 10.2 also shows how to translate numbers of deaths into numbers of survivors. Plotting number of survivors per 1,000 births against age produces the **survivorship curve** shown in the lower portion of the figure. A survivorship curve shows patterns of life and death within a population. Notice that in this population of Dall sheep, there are two periods when mortality rates are higher: during the first year and during the period between 9 and 13 years. Juvenile mortality and mortality of the aged are higher in this population, while mortality in the middle years is lower. The overall pattern of survival and mortality among Dall sheep is much like that for a variety of other large vertebrates, including red deer, *Cervus elaphus,* Columbian black-tailed deer, *Odocoileus hemionus columbianus,* East African buffalo, *Syncerus caffer,* and humans. The key characteristics of survival among these populations are relatively high rates of survival among the young and middle-aged and high rates of mortality among the older members.

This pattern of survival has also been observed in populations of annual plants and small invertebrate animals. Notice in figure 10.3 that patterns of survival in a population of a plant, *Phlox drummondii,* and a rotifer, *Floscularia conifera,*

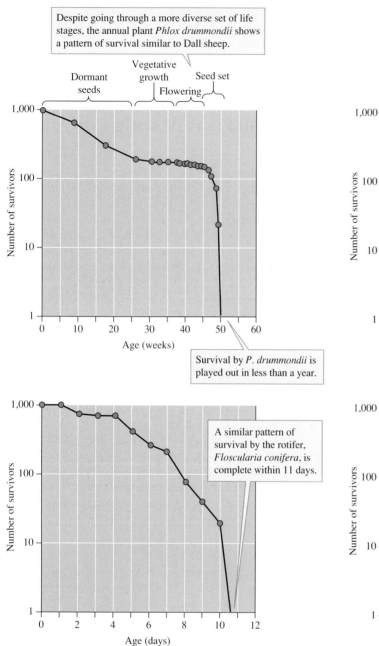

**Figure 10.3** High rates of survival among the young and middle-aged in plant and rotifer populations (data from Deevey 1947, *bottom*, Leverich and Levin 1979, *top*).

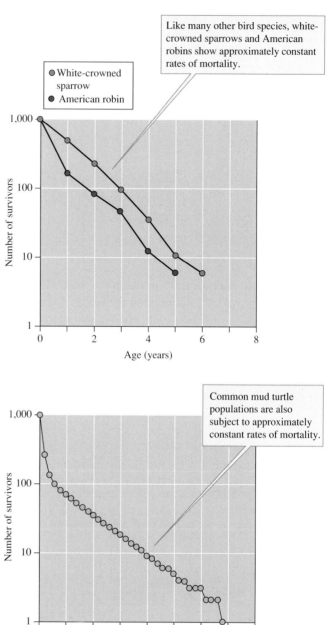

**Figure 10.4** Constant rates of survival (data from Deevey 1947, Baker, Mewaldt, and Stewart 1981, Frazer, Gibbons, and Greene 1991).

are remarkably similar to that of Dall mountain sheep. Following an initial period of higher juvenile mortality, mortality is relatively low for a period, and then mortality is high among older individuals. In the *Phlox* population, however, this pattern of survival is played out in less than 1 year and in the rotifer population in less than 11 days. These survivorship curves are based on cohort life tables.

Survival patterns can be quite different in other species. In the next example, mortality is not delayed until old age but occurs at approximately equal rates throughout life.

## Constant Rates of Survival

The survivorship curves of many species are nearly straight lines. In these populations, individuals die at approximately the same rate throughout life. This pattern of survival has been commonly observed in birds, such as the American robin, *Turdus migratorius,* and the white-crowned sparrow, *Zonotrichia leucophrys nuttalli* (fig. 10.4). Life expectancy remains relatively constant over the whole period a cohort is in existence. While birds are the most well known for showing a linear pattern of survival, many other species do as well. For instance, figure 10.4 also shows the same pattern of survival for a population of the common mud turtle, *Kinosternon subrubrum.* Though the mud turtle has a high rate of mortality during the first year of life, thereafter, survival follows a straight line.

As we shall see next, some organisms die at a much higher rate as juveniles than we have seen in any of the populations we have considered to this point.

## High Mortality Among the Young

Some organisms produce large numbers of young with very high rates of mortality. The eggs produced by marine fish such as the mackerel, *Scomber scombrus,* may number in the millions. Out of 1 million eggs laid by a mackerel, more than 999,990 die during the first 70 days of life either as eggs, larvae, or juveniles. Survival rates are similar in populations of the prawn *Leander squilla* off the coast of Sweden. For each 1 million eggs laid by *Leander,* only about 2,000 individuals survive the first year of life. This period of high mortality among young prawns is followed by a fairly constant mortality over the remainder of the life span.

Similar patterns of survival are shown by other marine invertebrates and fish and by plants that produce immense numbers of seeds. One of these plants is *Cleome droserifolia,* the desert shrub studied by Ahmad Hegazy (1990) that we discussed briefly in the introduction. Hegazy estimated that a local population of approximately 2,000 plants produce almost 20 million seeds each year. Of these, approximately 12,500 seeds germinate and produce seedlings. Only 800 seedlings survive to become juvenile plants. Figure 10.5 traces this pattern of survival by *Cleome* expressed as sur-

vivors per million seeds. Hegazy estimated that for each 1 million seeds produced about 39 survive to the age of 1 year, a survival rate of only 0.0039%. Survival in this desert plant population contrasts sharply with that seen in Dall sheep. The striking difference in patterns of survival between populations such as *Cleome,* birds such as the American robin, and large mammals such as Dall sheep led early population biologists to propose a classification of survivorship curves.

## Three Types of Survivorship Curves

Based on studies of survival by a wide variety of organisms, population ecologists have proposed that most survivorship curves fall into three major categories (fig. 10.6). A relatively high rate of survival among young and middle-aged individuals followed by a high rate of mortality among the aged is known as a **type I survivorship curve.** This is the pattern of survival we saw in populations of Dall sheep, *P. drummondii,* and rotifers (see figs. 10.2 and 10.3). Constant rates of survival throughout life produce the straight-line pattern of survival known as a **type II survivorship curve.** American robins, white-crowned sparrows, and common mud turtles show this pattern of survival (see fig. 10.4). A **type III survivorship curve** is one in which a period of extremely high rates of mortality among the young is followed by a relatively high rate of survival. The desert plant *Cleome* provides an excellent example of a type III survivorship curve (see fig. 10.5).

How well does this classification of survivorship represent natural populations? Most populations do not conform perfectly to any one of the three basic types of survivorship but

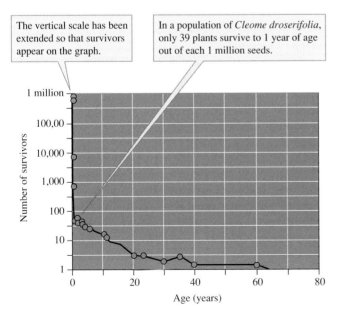

Figure 10.5 A high rate of mortality among the young of a perennial plant population (data from Hegazy 1990).

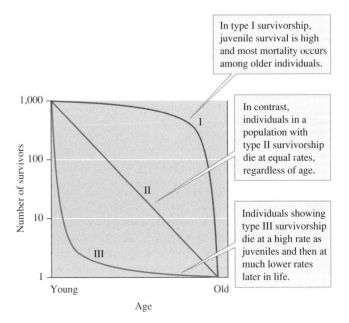

Figure 10.6 Three types of survivorship curves.

show virtually every sort of intermediate form of survivorship between the curves. Even single species can show considerable variation in survivorship from one environment to another. For example, while human survivorship generally follows a type I survivorship curve, in difficult environments, human survivorship approaches a type II curve. This variation in patterns of human survival prompted G. Evelyn Hutchinson (1978) to muse: "One can only conclude that sometimes man is constrained to die randomly like a bird, but in other circumstances he may aspire to as ripe an old age as that of a wild sheep or an African buffalo." If survivorship can be so variable within species, what good are these idealized, theoretical survivorship curves? Their most important value, like most theoretical constructs, is that they set boundaries that mark what is possible within populations. Regardless of how closely actual survivorship curves approximate the theoretical curves, they serve excellent summaries of survival patterns within populations.

We now turn to the age distributions of populations, a topic closely related to survivorship. As we have seen, the age distribution of a population can be used to construct a static life table from which a survivorship curve can be drawn. However, as we shall see next, a population's age distribution offers other insights into population dynamics.

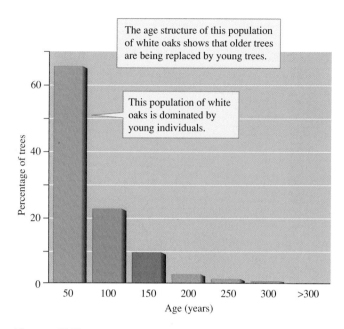

**Figure 10.7** The age distribution of a white oak, *Quercus alba,* population in Illinois (data from Miller 1923).

## CONCEPT DISCUSSION

### Age Distribution

**The age distribution of a population reflects its history of survival, reproduction, and potential for future growth.**

Population ecologists can tell a great deal about a population just by studying its age distribution. Age distributions indicate periods of successful reproduction, periods of high and low survival, and whether the older individuals in a population are replacing themselves or if the population is declining. By studying the history of a population, population ecologists can make predictions about its future.

## Stable and Declining Tree Populations

In 1923, R. B. Miller published data on the age distribution of a population of white oak, *Quercus alba,* in a mature oak-hickory forest in Illinois. In his study, Miller first determined the relationship between the age of a white oak and the diameter of its trunk. To do this, he measured the diameters of 56 trees of various sizes and then took a core of wood from their trunks. By counting the annual growth rings from each of the cores he could determine the ages of the trees in

his sample. With the relationship between oak age and diameter in hand, Miller used diameter to estimate the ages of hundreds of trees.

Most white oaks in Miller's study forest were concentrated in the youngest age class of 1 to 50 years, with progressively fewer individuals in the older age classes (fig. 10.7). The oldest white oaks in the forest were over 300 years old. In other words, the age distribution of white oak in this forest was biased toward the young trees. What might we infer from this age distribution? The age distribution indicates that reproduction is sufficient to replace the oldest individuals in the population as they die. That is, this population of white oaks appeared to be stable, neither growing nor declining, at the time it was studied.

The age distribution of this white oak population contrasts sharply with the age distributions of populations of Rio Grande cottonwoods, *Populus deltoides* spp. *wislizenii.* The most extensive cottonwood forests remaining in the southwestern United States grows along the Middle Rio Grande in central New Mexico. However, studies of age distributions indicate that these populations are declining. Older trees, which can live to a maximum age of about 130 years, are not being replaced by younger trees (fig. 10.8). In contrast to the white oak population in Illinois, the Rio Grande cottonwood population is dominated by older individuals. At the study site represented by figure 10.8, there has been no reproduction for over a decade. At other sites along the Rio Grande there has been little reproduction for over three decades.

Why have Rio Grande cottonwoods failed to reproduce? Reproduction by Rio Grande cottonwoods depends upon seasonal floods, which play two key roles. First, floods create areas of bare soil without a surface layer of organic matter and

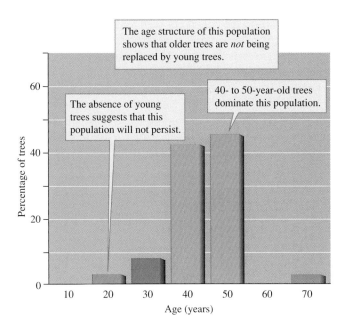

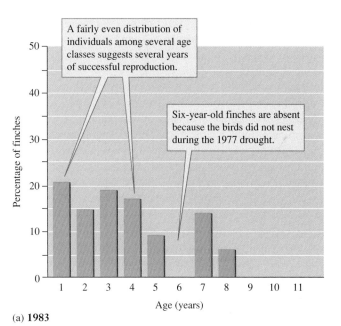

(a) **1983**

**Figure 10.8** The age distribution of a population of Rio Grande cottonwoods, *Populus deltoides* subsp. *wislizenii,* near Belen, New Mexico (data from Howe and Knopf 1991).

without competing vegetation, ideal conditions for germination and establishment of cottonwood seedlings. Floods also keep these nursery areas of bare soil moist until cottonwood seedlings can grow their roots deep enough to tap into the shallow water table. Historically, these conditions were created by spring floods, the timing of which coincided with dispersal of cottonwood seeds by wind. The annual rhythm of seed bed preparation and seeding has been interrupted by the construction of dams on the Rio Grande for flood control and irrigation. The tamed Rio Grande no longer floods, and though Rio Grande cottonwoods produce seeds each year, their age distribution indicates that these seeds find few suitable places to germinate.

The age distributions of tree populations change over the course of many decades or centuries. Meanwhile, other populations can change significantly on much shorter timescales. One of these dynamic populations has been thoroughly studied on the Galápagos Islands.

# A Dynamic Population in a Variable Climate

Rosemary Grant and Peter Grant (1989) have spent decades studying Galápagos finch populations. One of their most thorough studies has concerned the large cactus finch, *Geospiza conirostris,* on the island of Genovesa, which lies in the northeastern portion of the Galápagos archipelago, approximately 1,000 km off the west coast of South America. The Galápagos Islands have a highly variable climate, which is reflected in the highly dynamic populations of the organisms living on the islands, including populations of the large cactus finch.

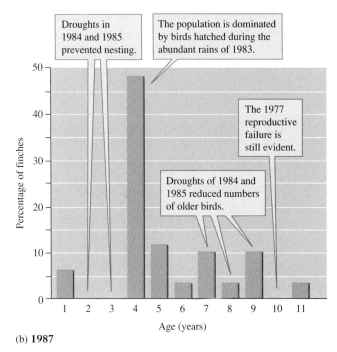

(b) **1987**

**Figure 10.9** The age distribution of a population of large cactus finches, *Geospiza conirostris,* on the island of Genovesa in the Galápagos Islands during 1983 (*a*) and 1987 (*b*) (data from Grant and Grant 1989).

The age distributions of the large cactus finch during 1983 and 1987 show that the population can be very dynamic (fig. 10.9). The 1983 age distribution shows a fairly regular distribution of individuals among age classes. However, there were no 6-year-old individuals in the population. This gap is due to a drought in 1977, during which no finches reproduced. Now, compare the 1983 and 1987 age distributions. The distributions contrast markedly, though they are for the same population separated by only 4 years!

The 1977 gap is still present in the 1987 age distribution and another has been added for 2- and 3-year-old finches. This second gap is the result of 2 years of reproductive failure during a drought that persisted from 1984 to 1985. Another difference is that the 1987 age distribution is dominated by 4-year-old birds that were fledged during 1983. The 1983 class dominates because wet weather that year resulted in very high production of food that the finches depend upon for reproduction. The 1987 age distribution also shows evidence of high mortality among older finches. Might this decline be due to high adult mortality during the 1984–85 drought? Whatever the cause of these declines, the reproductive output of this population of large cactus finches is dominated by birds hatched in one exceptionally favorable year, 1983. This long-term study of the large cactus finch population of Genovesa Island demonstrates the responsiveness of population age structure to environmental variation.

In this section, we have seen that an age distribution tells population ecologists a great deal about the dynamics of a population, including whether a population is growing, declining, or approximately stable. The next section goes beyond these qualitative assessments. By combining information on survival and age structure with reproductive rates, population ecologists can quantify rates of growth or decline.

## CONCEPT DISCUSSION

### Rates of Population Change

**A life table combined with a fecundity schedule can be used to estimate net reproductive rate ($R_0$), geometric rate of increase ($\lambda$), generation time ($T$), and per capita rate of increase ($r$).**

In addition to survival rates, population ecologists are concerned about another major influence on local population density—birthrates. In mammals and other live-bearing organisms, from sharks to humans, the term **birthrate** means the number of young born per female in a period of time. Population biologists also use the term *birth* more generally to refer to any other processes that produce new individuals in the population. In populations of birds, fish, and reptiles, births are usually counted as the number of eggs laid. In plants, the number of births may be the number of seeds produced or the number of shoots produced during asexual reproduction. In bacteria, the birth, or reproductive, rate is taken as the rate of cell division.

Tracking birthrates in a population is similar to tracking survival rates. In a sexually reproducing population, the population biologist needs to know the average number of births per female for each age class and the number of females in each age class. In practice, the ecologist counts the number of eggs produced by birds or reptiles, the number of fawns pro-

duced by deer, or the number of seeds or sprouts produced by plants. The numbers of offspring produced by parents of different ages are then tabulated. The tabulation of birthrates for females of different ages in a population is called a **fecundity schedule.** If we combine the information in a fecundity schedule with that in a life table we can estimate several important characteristics of populations. To a population ecologist, one of the most important things to know is whether a population is growing or declining.

## Estimating Rates for an Annual Plant

Table 10.1 combines survivorship with seed production by the annual plant *P. drummondii*. The first column, *x,* lists age intervals in days. The second column, $n_x$, lists the number of individuals in the population surviving to each age interval. The third column, $l_x$, lists survivorship, the proportion of the population surviving to each age *x*. The fourth column, $m_x$, lists the average number of seeds produced by each individual in each age interval. Finally, the fifth column, $l_x m_x$, is the product of columns 3 and 4.

We've already used the data in column 3, $l_x$, to construct the survivorship curve for this species (see fig. 10.3). Now, let's combine those survivorship data with the seed production for *P. drummondii*, $m_x$, to calculate the **net reproductive rate, $R_0$.** The calculations of reproductive rates in this section assume that birthrates and death rates for each age class in a population are constant and that the population under study has a **stable age distribution.** In a population with a stable age distribution, the proportion of individuals in each of the age classes is constant. In general, the net reproductive rate is the average number of offspring produced by an individual in a population during its lifetime or per generation. In the case of the annual plant *P. drummondii*, the net reproductive rate is the average number of seeds left by an individual. You can calculate the net reproductive rate from table 10.1 by adding the values in the final column. The result is:

$$R_0 = \sum l_x m_x = 2.4177$$

To calculate the total number of seeds produced by this population during the year of study, multiply 2.4177 by 996, which was the initial number of plants in this population. The result, 2,408, is the number of seeds that this population of *P. drummondii* will begin with the next year.

Since *P. drummondii* has pulsed reproduction, we can estimate the rate at which its population is growing with a quantity known as the **geometric rate of increase, $\lambda$.** The geometric rate of increase is the ratio of the population size at two points in time:

$$\lambda = \frac{N_{t+1}}{N_t}$$

## Table 10.1

Combining survivorship with seed production by *P. drummondii* to estimate net reproductive rate, $R_0$

| Age (days) | Number surviving to day $x$ | Proportion surviving to day $x$ | Average number of seeds per individual during time interval | Multiplication of $l_x$ and $m_x$ |
|---|---|---|---|---|
| $x$ | $n_x$ | $l_x$ | $m_x$ | $l_x m_x$ |
| 0–299 | 996 | 1.0000 | 0.0000 | 0.0000 |
| 299–306 | 158 | 0.1586 | 0.3394 | 0.0532 |
| 306–13 | 154 | 0.1546 | 0.7963 | 0.1231 |
| 313–20 | 151 | 0.1516 | 2.3995 | 0.3638 |
| 320–27 | 147 | 0.1476 | 3.1904 | 0.4589 |
| 327–34 | 136 | 0.1365 | 2.5411 | 0.3470 |
| 334–41 | 105 | 0.1054 | 3.1589 | 0.3330 |
| 341–48 | 74 | 0.0743 | 8.6625 | 0.6436 |
| 348–55 | 22 | 0.0221 | 4.3072 | 0.0951 |
| 355–62 | 0 | 0.0000 | 0.0000 | 0.0000 |

Each individual leaves an average of 2.4177 offspring.

Data from Leverich and Levin 1979.

$$R_0 = \sum l_x m_x = 2.4177$$

The value of $R_0$, which is greater than 1.0, indicates that this population of *P. drummondii* is growing.

Summing the final column yields $R_0$, the net reproductive rate per individual.

In this equation, $N_{t+1}$ is the size of the population at some future time and $N_t$ is the size of the population at some earlier time (fig. 10.10). The time interval $t$ may be years, days, or hours; which time interval you use to calculate the geometric rate of increase for a population depends on the organism and the rate at which its population grows.

Let's calculate $\lambda$ for the population of *P. drummondii*. What time interval should we use for our calculation? Since *P. drummondii* is an annual plant, the most meaningful time interval would be 1 year. The initial number, $N_t$, of *P. drummondii* in the population was 996. The number of individuals (seeds) in the population at the end of a year of study was 2,408. This is the number in the next generation, which is $N_{t+1}$. Therefore, the geometric rate of increase for the population over the period of this study was:

$$\lambda = \frac{2,408}{996} = 2.4177$$

This is the same value we got for $R_0$. But, before you jump to conclusions, you should know that $R_0$, which is the number of offspring per female per generation, does not

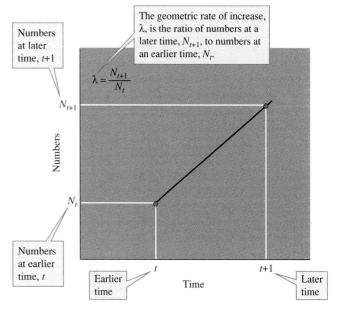

**Figure 10.10** The geometric rate of increase.

always equal λ. In this case, λ equaled $R_0$ because *P. drummondii* is an annual plant with pulsed reproduction. If a species has overlapping generations and continuous reproduction, $R_0$ will usually not equal λ.

How long do you think this plant can continue to reproduce at the rate of λ, or $R_0 = 2.4177$? Not long, but we'll get back to this point in chapter 11. Before we do that, let's do some calculations for organisms with overlapping generations.

# Estimating Rates When Generations Overlap

The population of the common mud turtle, *K. subrubrum,* whose mortality we examined in figure 10.4, contrasts with the *P. drummondii* population in various ways. Let's examine some of the details of this turtle's reproductive patterns in order to calculate the net reproductive rate of this population. About half (0.507) of the turtles nest each year. Of those females that do nest, most nest once during the year. However, some nest twice and a few even nest three times during the year. The average number of nests per year for the nesting turtles is 1.2, which means that 0.2, or one-fifth, of the turtles produce a second nest each year. The average **clutch size,** which is the number of eggs produced by a nesting female, is 3.17. So, the average number of eggs produced by nesting females each year is 3.17 eggs per nest × 1.2 nests per year = 3.8 eggs per year. However, remember that only half the females in the population nest each year. Therefore, the number of eggs per female per year is $0.507 \times 3.8 = 1.927$ eggs per female per year. This is the average total number of eggs per female. On average, half of these eggs will develop into males and half into females. However, population biologists generally keep track of the females and are concerned with the production of daughters. We didn't have to consider the sex of individuals in the *Phlox* population because all individuals have both male and female reproductive organs. Since the sex ratio in this turtle population is 1 male:1 female, we multiply 1.927 by 0.50 to calculate the number of female eggs per adult female in the population, which equals 0.96 female eggs. This is the value listed in column 3 of table 10.2.

Table 10.2 includes the life table information used to construct figure 10.4 plus the fecundity information we just calculated. As in the *Phlox* population, the sum of $l_x m_x$, $\Sigma l_x m_x$ provides an estimate of $R_0$, the net reproductive rate of females in this population. In this case, $R_0 = 0.601$. We can interpret this number as the average number of daughters produced by each female in this population over the course of her lifetime. If this number is correct, the mothers in this population are not producing enough daughters to replace themselves. It

## Table 10.2

Calculating net reproductive rate, $R_0$, and generation time, $T$, for a population of the common mud turtle, *K. subrubrum*.

Age in years, $x$, times $l_x m_x$

| x (years) | $l_x$ | $m_x$ | $l_x m_x$ | $x l_x m_x$ |
|---|---|---|---|---|
| 0 | 1.0000 | 0 | 0 | 0 |
| 1 | 0.2610 | 0 | 0 | 0 |
| 2 | 0.1360 | 0 | 0 | 0 |
| 3 | 0.0981 | 0 | 0 | 0 |
| 4 | 0.0786 | 0.96 | 0.07546 | 0.30184 |
| 5 | 0.0689 | 0.96 | 0.06614 | 0.33070 |
| 6 | 0.0603 | 0.96 | 0.05789 | 0.34734 |
| 7 | 0.0528 | 0.96 | 0.05069 | 0.40523 |
| 8 | 0.0463 | 0.96 | 0.04445 | 0.35560 |
| 9 | 0.0405 | 0.96 | 0.03888 | 0.34992 |
| 10 | 0.0355 | 0.96 | 0.03408 | 0.34080 |
| 11 | 0.0311 | 0.96 | 0.02986 | 0.32846 |
| 12 | 0.0273 | 0.96 | 0.02621 | 0.31452 |
| 13 | 0.0239 | 0.96 | 0.02294 | 0.29822 |
| 14 | 0.0209 | 0.96 | 0.02006 | 0.28084 |
| 15 | 0.0183 | 0.96 | 0.01757 | 0.26355 |
| 16 | 0.0160 | 0.96 | 0.01536 | 0.24576 |
| 17 | 0.0141 | 0.96 | 0.01354 | 0.23018 |
| 18 | 0.0123 | 0.96 | 0.01181 | 0.21258 |
| 19 | 0.0108 | 0.96 | 0.01037 | 0.19703 |
| 20 | 0.00945 | 0.96 | 0.00907 | 0.18140 |
| 21 | 0.00829 | 0.96 | 0.00796 | 0.16716 |
| 22 | 0.00725 | 0.96 | 0.00696 | 0.15312 |
| 23 | 0.00635 | 0.96 | 0.00610 | 0.14030 |
| 24 | 0.00557 | 0.96 | 0.00535 | 0.12840 |
| 25 | 0.00487 | 0.96 | 0.00468 | 0.11700 |
| 26 | 0.00427 | 0.96 | 0.00410 | 0.10660 |
| 27 | 0.00374 | 0.96 | 0.00359 | 0.09693 |
| 28 | 0.00328 | 0.96 | 0.00315 | 0.08820 |
| 29 | 0.00287 | 0.96 | 0.00276 | 0.08004 |
| 30 | 0.00251 | 0.96 | 0.00241 | 0.07230 |
| 31 | 0.00220 | 0.96 | 0.00211 | 0.06541 |
| 32 | 0.00193 | 0.96 | 0.00185 | 0.05920 |
| 33 | 0.00169 | 0.96 | 0.00162 | 0.05346 |
| 34 | 0.00148 | 0.96 | 0.00142 | 0.04828 |
| 35 | 0.00130 | 0.96 | 0.00125 | 0.04375 |
| 36 | 0.00114 | 0.96 | 0.00109 | 0.03924 |
| 37 | <0.00100 | 0 | 0 | 0 |

Data from Frazer, Gibbons, and Greene 1991.

$$R_0 = \sum l_x m_x = 0.601 \qquad \sum x l_x m_x = 6.4$$

The value of $R_0$, which is less than 1.0, indicates that this population is declining.

$$T = \frac{\sum x l_x m_x}{R_0} = \frac{6.4}{0.601} = 10.6$$

Dividing $\sum x l_x m_x$ by $R_0$ gives an estimate of generation time.

The generation time for this population is 10.6 years.

appears that this population is declining. This result makes sense because during the time this study was done, the region of South Carolina in which the turtle population lives was experiencing severe drought. During the drought Ellenton Bay dried from a maximum of 10 ha of open water to about 0.05 ha.

The trend in this mud turtle population appears to reflect the declining quality of the environment. What value of $R_0$ would produce a stable turtle population? In a stable population, $R_0$ would be 1.0, which means that each female would replace just herself during her lifetime. In a growing population, such as the population of *Phlox*, $R_0$ would be greater than 1.0.

Population ecologists are also interested in several other characteristics of populations. One of those is the generation time, *T*, which is the average time from egg to egg, seed to seed, and so forth. We can use the information in table 10.2 to calculate the average generation time for the common mud turtles of Ellenton Bay:

$$T = \frac{\sum x l_x m_x}{R_0}$$

In this equation *x* is age in years. To calculate *T*, sum the last column and divide the result by $R_0$. The result shows that the common mud turtles of Ellenton Bay have an average generation time of 10.6 years.

How could you tell if 10 years is an unusually long, or short, generation time? Figure 10.11 plots the generation time for a broad range of organisms against body size. As we saw for population density in chapter 9 (see figs. 9.22, 9.23, and 9.24), there is a significant positive correlation between body size and generation time. The largest organisms have the longest generation times and the smallest have the shortest. While this relationship might not be particularly surprising, its consistency across such a wide range of organisms is impressive. In addition, the relationship isn't restricted to a narrow taxon such as herbivorous mammals (see fig. 9.22). John Bonner (1965) found the trend shown in figure 10.11, which is rooted in the bacteria and extends all the way to the largest organisms in the biosphere, the giant sequoia, *Sequoia gigantea*. Humans and common mud turtles lie somewhere in the middle range of the distribution. So, what about the common mud turtles of Ellenton Bay? Remember we calculated a mean generation time of about 10 years. Females in the population mature at about 7.5 cm. If we draw a line across from 7.5 cm and up from 10 years, the point where they cross is the position of the common mud turtle population on the graph. The point where these lines cross is not that far off from the remainder of the points on the graph. In other words, the estimated generation time for the common mud turtle of Ellenton Bay is similar to that of organisms of similar size.

Knowing $R_0$ and *T* allows us to estimate *r*, the **per capita rate of increase** for a population:

$$r = \frac{\ln R_0}{T}$$

(ln is the base of the natural logarithms). We can interpret *r* as birthrate minus death rate: $r = b - d$. Using this method, the estimated per capita rate of increase for the common mud turtle population of Ellenton Bay is:

$$r = \frac{\ln 0.601}{10.6} = -0.05$$

The negative value of *r* in this case indicates that birthrates are lower than death rates and the population is declining. A value of *r* greater than 0 would indicate a growing population, and a value equal to 0 would indicate a stable population. While there are ways to make more accurate estimates of *r*, this method is accurate enough for our discussion. We will return to *r* in chapter 11 as we discuss population growth.

In this section we have seen how a life table combined with a fecundity schedule can be used to estimate net reproductive rate, $R_0$, geometric rate of increase, $\lambda$, generation time, *T*, and per capita rate of increase, *r*. Population dynamics are clearly influenced by patterns of survival and reproduction. However, births and deaths are not the only processes that make populations dynamic. As we shall see in the next concept discussion, population dynamics are also influenced by the movements of organisms.

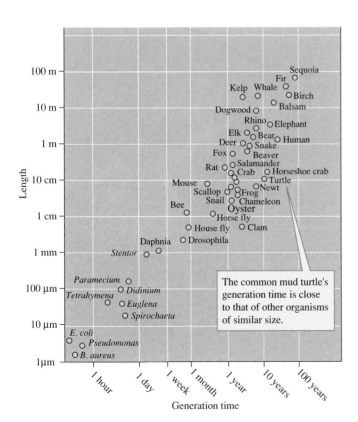

The common mud turtle's generation time is close to that of other organisms of similar size.

**Figure 10.11**  Size and generation time (data from Bonner 1965).

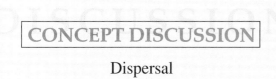

## CONCEPT DISCUSSION

### Dispersal

**Dispersal can increase or decrease local population densities.**

As we saw in chapter 9 where we considered metapopulations (p. 240), dispersal is an important aspect of population dynamics. The seeds of plants disperse with wind or water or may be transported by a variety of mammals, insects, or birds. Adult barnacles may spend their lives attached to rocks, but their larvae travel the high seas on far-ranging ocean currents. A host of other sessile marine invertebrates, algae, and many highly sedentary reef fishes also disperse widely as larvae. Some young spiders spin a small net that catches winds and carries them for distances up to hundreds of kilometers. Young mammals and birds often disperse from the area where they were born and may join other local populations. As a consequence of movements such as these (fig. 10.12), the population ecologist trying to understand local population density must consider dispersal *into* (immigration) and *out of* (emigration) the local population.

Despite its importance, dispersal is one of the least-studied aspects of population dynamics. Its study is clearly a difficult undertaking. But dispersal is worth studying; the health and survival of many local populations may depend upon this underappreciated aspect of population dynamics. One of the richest sources of information on dispersal and some of the clearest examples come from studies of expanding populations.

# Dispersal of Expanding Populations

Expanding populations are in the process of increasing their geographic range. Why should this type of population provide us with some of the best records of species dispersal? The appearance of a new species in an area is quickly noted and recorded, especially if the species impacts the local economy or human health or safety. For instance, the expansion of Africanized bees through South and North America is well documented (fig. 10.13). The legendary aggressiveness of these bees ensures that their dispersal into an area does not escape notice for long.

## *Africanized Honeybees*

Honeybees, *Apis melifera,* evolved in Africa and Europe, where their native range extends from tropical to cold, temperate environments. Across this extensive environmental range, this species has differentiated into a number of locally adapted subspecies. In an attempt to improve the adaptability of managed honeybees to their tropical climate, Brazilian scientists imported queens of the African subspecies *Apis melifera scutellata* in

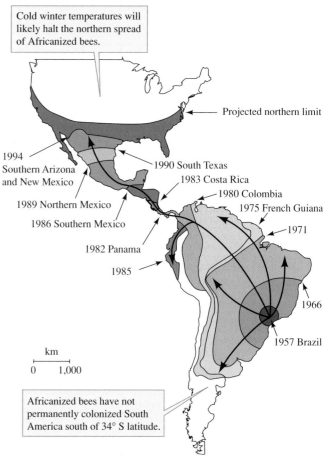

**Figure 10.13** The expansion of Africanized bees from South America through Central and North America, 1956 to 1994 (data from Winston 1992).

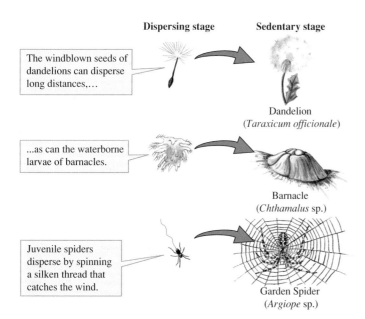

**Figure 10.12** Dispersing and sedentary stages of organisms.

| Information | |
| Questions | |
| Hypothesis | ✓ |
| Predictions | |
| Testing | |

*Investigating the Evidence*

# Hypotheses and Statistical Significance

In chapter 1, we reviewed the roles of questions and hypotheses in the process of science. Briefly, we considered how scientists use information to formulate questions about the natural world and convert their questions to hypotheses. A hypothesis, we said, is a possible answer to a question.

Let's use the distributions we considered in the Investigating the Evidence box in chapter 9 (p. 237) to examine the nature of scientific hypotheses in more detail. In that discussion we examined samples from three populations of plants, from which we calculated the following statistics:

| Statistic | Species A | Species B | Species C |
|---|---|---|---|
| Mean density, $\overline{X}$ | 5.27 | 5.18 | 5.27 |
| Variance in density, $s^2$ | 5.22 | 0.36 | 44.42 |
| Ratio of variance to mean, $s^2/\overline{X}$ | 0.99 | 0.07 | 8.43 |

Recall that in a random distribution the ratio of the variance to the mean equals one, that is, $s^2/\overline{X} = 1.0$.

As we have stated repeatedly, the center of scientific investigation is the hypothesis. In the case of these three populations, an appropriate hypothesis would be that in each case, $s^2/\overline{X}$ does not differ *significantly* from one. In other words, our hypothesis is that each of the populations has a random distribution. How well does this hypothesis match the results of our population counts? While $s^2/\overline{X}$ for species A comes close to 1.0 ($s^2/\overline{X} = 0.99$), the calculated value of $s^2/\overline{X}$ does not exactly equal 1 for any of the populations. However, we need to remember that the numbers listed in the table are statistical *estimates* of the true, or actual, variance to mean ratio for each of the populations. It is unlikely that any of the $s^2/\overline{X}$ values, which were calculated from a sample of observations, would exactly match the true variance to mean ratio in any of the three study populations. Because of our limited sample size, we expect to see some difference between the statistical estimate and the theoretical expectation of $s^2/X = 1.0$, even in a population known to have a random distribution.

The critical point here is to have a basis for judging whether an observed $s^2/\overline{X}$ differs *significantly* from the theoretical expectation of $s^2/\overline{X} = 1.0$. That judgment is made on the basis of the probability of being incorrect if the hypothesis is rejected. By tradition, the level of significance used in most scientific investigations is $P < 0.05$, or less than 1 chance in 20. Again in our example, this is the probability that we will be wrong if we reject the hypothesis of a random distribution.

Let's go back to our populations of plant species A, B, and C. How can we tell if any of the three $s^2/\overline{X}$ values in the table differs significantly from 1? That is, how can we determine if the probability of obtaining each value by chance in a population that actually has a random distribution, is less than 0.05? This is generally done by comparing an observed value with tables of theoretically derived values. For now, we can use our judgment to make some predictions. Consider species A. The probability that we could observe a mean value of $s^2/\overline{X} = 0.99$ by chance in a population with a random distribution is likely to be much greater than 0.05. As a consequence, we are likely to accept the hypothesis that species A has a random distribution. In contrast, the values of $s^2/\overline{X}$ observed for species B and C (0.07 and 8.43) differ so much from 1.0 that the probability of obtaining these results by chance from a population that is actually randomly distributed is likely to be low. If it is less than 0.05, we will reject the hypothesis that these populations are randomly distributed, and tentatively accept the alternative hypotheses of clumped or regular distributions. We will gradually work up to evaluating whether the $s^2/\overline{X}$ ratios listed in the table differ significantly from $s^2/\overline{X} = 1.0$ over the course of chapters 11 and 12.

---

1956. These queens mated with the European honeybees used by Brazilian beekeepers, producing what we now call Africanized bees.

Africanized honeybees differ in several ways from European honeybees. Temperate and tropical environments have apparently selected for markedly different behavior and population dynamics. Natural selection by a high diversity and abundance of nest predators has probably produced the greater aggressiveness shown by Africanized bees. The warmer climate and greater stability of nectar sources eliminates the advantages of storing large quantities of honey and maintaining large colonies for survival through the winter. Most important to this discussion of dispersal, Africanized honeybees produce swarms that disperse to form new colonies at a much higher rate than do European honeybees.

High rates of colony formation and dispersal have caused a rapid expansion of Africanized honeybees through South and North America. Their rate of dispersal has ranged from 300 to 500 km per year. Within 30 years, Africanized honeybees occupied most of South America, all of Central America, and most of Mexico. The estimated number of wild colonies of these bees in South America alone is 50 to 100 million.

Africanized bees reached southern Texas in 1990 and southern Arizona and New Mexico in 1994. The honeybees stopped spreading southward through South America by about 1983, stopping at about 34° S latitude. However, they continue to spread northward through North America and will continue to do so until stopped by cold climates. Population ecologists predict that Africanized honeybees will reach the northern limit of their distribution within North America sometime early in the twenty-first century.

## Collared Doves

Birds provide some of the best examples of rapid population expansion. European starlings and house sparrows, which were purposely introduced into North America, spread across the continent in less than a century. Collared doves, *Streptopelia decaocto,* began to spread from Turkey into Europe after 1900. The expansion of collared doves into Europe was notable in a number of ways. First, the spread began suddenly and, once begun, was relentless. By the 1980s, the doves were found in every country of western and eastern Europe (fig. 10.14). In addition, in contrast to many other recent cases of range expansion, such as that of Africanized honeybees and European star-

lings, the expansion of collared doves did not appear to be influenced by humans.

Another notable feature of the collared dove expansion across Europe is that we know a great deal about the underlying population dynamics. The expansion took place in small jumps. Adult collared doves are highly sedentary, and dispersal is limited to young doves. Most dispersing young stay within a few kilometers of their parent's nest, but some disperse hundreds of kilometers (fig. 10.15). Once they have chosen a mate, the young birds nest and become sedentary like their parents. These pulses of dispersal by young birds spread the collared dove population across Europe at a rate of about 45 km per year.

How does this rate of expansion by collared doves compare to rates of expansion by other populations? Compared with the dispersal rate of Africanized honeybees across the Americas, 45 km per year is a modest rate. However, compared with dispersal rates for most other animals that have been studied, 45 km per year is rapid. Figure 10.16, which summarizes rates of dispersal for a variety of mammals and birds, shows that rates of dispersal differ by three orders of magnitude. While some species such as Africanized bees and collared doves spread at rates of tens or hundreds of kilometers per year, others disperse only a few hundred meters per year. This is about the same rate at which North American trees expanded their distributions following the retreat of the glaciers.

# Range Changes in Response to Climate Change

In response to climate change following retreat of the glaciers northward in North America beginning about 16,000 years ago, organisms of all sorts began to move northward from their

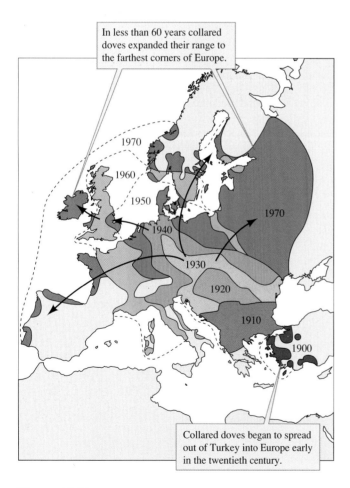

**Figure 10.14** The expansion of collared doves, *Streptopelia decaocto,* across Europe (data from Hengeveld 1988).

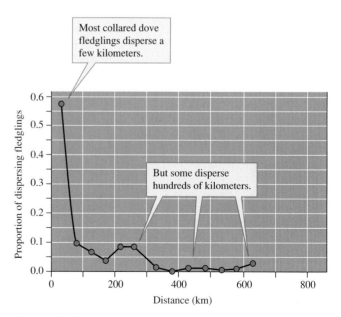

**Figure 10.15** Dispersal distances by collared dove fledglings (data from Hengeveld 1989).

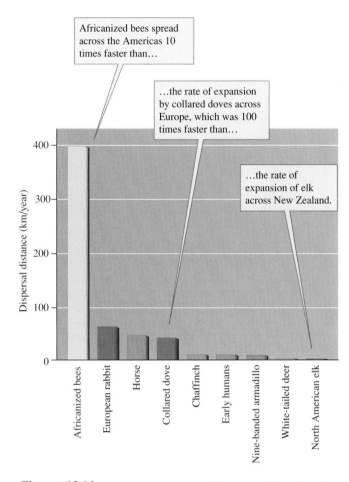

**Figure 10.16** Rates of expansion by animal populations (data from Caughley 1977, Hengeveld 1988, Winston 1992).

ice age refuges. Temperate forest trees have left one of the best preserved records of this northward dispersal. As we saw in chapter 1, the record of tree movements is well preserved in lake sediments (see fig. 1.9). The northward advance of maple and hemlock is shown in figure 10.17.

Figure 10.17 illustrates a number of ecologically significant messages. Though the distributions of maple and hemlock overlap today, they did not during the height of the last ice age. In addition, maple colonized the northern part of its present range from the lower Mississippi Valley region, while hemlock colonized its present range from a refuge along the Atlantic coast. The two trees dispersed at very different rates. Of the two species, maple dispersed faster, arriving at the northern limits of its present-day range about 6,000 years ago. In contrast, hemlock didn't reach the northwestern limit of its present distribution until 2,000 years ago.

The pollen preserved in lake sediments indicates that forest trees in eastern North America spread northward following the retreat of the glaciers at the rate of 100 to 400 m (0.1–0.4 km) per year. This rate of dispersal is similar to that of some large mammals such as the North American elk. However, it is 1/100 the rate of dispersal shown by collared doves in Europe and 1/1,000 the dispersal rate of Africanized bees across South, Central, and North America.

The previous examples concern dispersal by populations in the process of expanding their ranges. Significant dispersal also takes place within established populations whose ranges are not changing. Movements, within established ranges, can be an important aspect of local population dynamics. We will consider two examples.

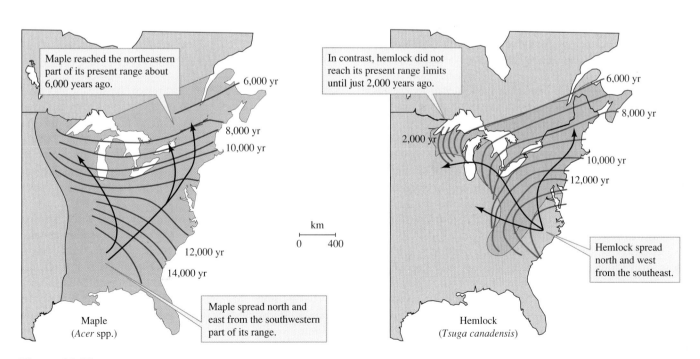

**Figure 10.17** The northward expansion of two tree species in North America following glacial retreat (data from Davis 1981).

# Dispersal in Response to Changing Food Supply

Predators show several kinds of responses to variation in prey density. In addition to the functional response we discussed in chapter 6, C. S. Holling (1959) also observed **numerical responses** to increased prey availability. Numerical responses are changes in the density of predator populations in response to increased prey density. Holling studied populations of mice and shrews preying on insect cocoons and attributed the numerical responses he observed to increased reproductive rates. He commented that "because the reproductive rate of small mammals is so high, there was an almost immediate increase in density with increase in food." However, some other predators, with much lower reproductive rates, also show strong numerical responses. These numerical responses to prey density are almost entirely due to dispersal.

In some years, northern landscapes are alive with small rodents called voles, *Microtus* spp. Go to the same place during other years and it may be difficult to find any voles. In northern latitudes, vole populations usually reach high densities every 3 to 4 years. Between these peak times, population densities crash. Population cycles in different areas are not synchronized, however. In other words, while vole population density is very low in one area, it is high elsewhere.

Erkki Korpimäki and Kai Norrdahl (1991) conducted a 10-year study of voles and their predators. The study began in 1977 during a peak in vole densities of about 1,800 per square kilometer and continued through two more peaks in 1982 (960/km$^2$) and 1985–86 (1,980 and 1,710/km$^2$). The researchers estimated that between these population peaks vole densities per square kilometer fell to as low as 70 in 1980 and 40 in 1984. During this period, the densities of the European kestrel, *Falco tinnunculus,* short-eared owls, *Asio flammeus,* and long-eared owls, *Asio otus,*closely tracked vole densities (fig. 10.18). How do kestrel and owl populations track these variations in vole densities?

What mechanisms produce the numerical responses by kestrels and owls to changing vole densities? Look at figure 10.18 for a clue. The peaks in raptor densities in 1977, 1982, and 1986 match the peaks in vole densities almost perfectly. If reproduction was the source of numerical response by kestrels and owls, there would have been more of a delay, or time lag, in kestrel and owl numerical response. From this close match in numbers, Korpimäki and Norrdahl proposed that kestrels and owls must move from place to place in response to local increases in vole populations.

Is there any supporting evidence for high rates of movement by kestrels and owls? Korpimäki (1988) marked and recaptured 217 kestrels, a large proportion of their study population. Because European kestrel populations have an annual survival rate of 48% to 66%, he predicted a high rate of recapture of the marked birds. However, only 3% of the female and 13% of the male kestrels were recaptured. These very low rates of recapture indicated that kestrels were moving out of the

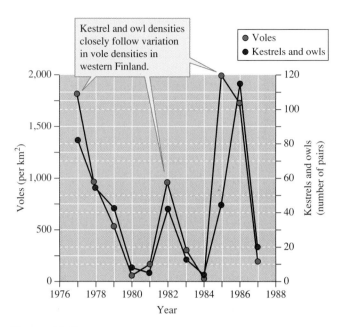

**Figure 10.18** Dispersal and numerical response by predators (data from Korpimäki and Norrdahl 1991).

study area. From their data, Korpimäki and Norrdahl concluded that the hawks and owls in western Finland are nomadic, moving from place to place in response to changing vole densities.

These studies documented the contribution of dispersal to local populations of kestrels and owls. Earlier in this section, we saw how studies of expanding populations have shed light on the contribution of dispersal to local population density and dynamics. Many other local populations are strongly influenced by dispersal. One of the environments in which dispersal has a major influence on local populations is in streams and rivers.

# Dispersal in Rivers and Streams

One of the most distinctive features of the stream and river environment is *current,* the downstream flow of water. What effect does current have on the lives of stream organisms? As you may recall from chapter 3, the effects of current are substantial and influence everything from the amount of oxygen in the water to the size, shape, and behavior of stream organisms. In this section, we stop and consider how stream populations are affected by current.

Let's begin with a question. Why doesn't the flowing water of streams eventually wash all stream organisms, including fish, insects, snails, bacteria, algae, and fungi, out to sea? All stream dwellers have a variety of characteristics that help them maintain their position in streams. Some fish such as trout are streamlined and can easily swim against swift currents, while other fish like sculpins and loaches are well designed for avoiding the full strength of currents by living on the bottom and seeking shelter among or under stones. Microorganisms resist being washed away by adhering to the surfaces of stones, wood, and other substrates. Many stream insects are flattened and so stay out of the main force of the current, while others are streamlined and fast-swimming.

Despite these means of staying in place, stream organisms do get washed downstream in large numbers, particularly during flash floods, or **spates.** To observe this downstream movement of organisms, put a fine or medium mesh net in a stream or river and you will soon capture large numbers of stream insects and algae along with fragments of leaves and wood. If you place some of the organic matter washed into your net under a microscope, you will find it laden with all sorts of microorganisms. Stream ecologists refer to this downstream movement of stream organisms as **drift.** Some drift is due to displacement of organisms during flash floods. However, some is due to the active movement of organisms downstream.

Whatever its cause, stream organisms drift downstream in large numbers. Why doesn't drift eventually eliminate organisms from the upstream sections of streams? Karl Müller (1954, 1974) hypothesized that drift would eventually wash entire populations out of streams unless organisms actively moved upstream to compensate for drift. He proposed that stream populations are maintained through a dynamic interplay between downstream and upstream dispersal that he called the **colonization cycle.** The colonization cycle is a dynamic view of stream populations in which upstream and downstream dispersal, as well as reproduction, have major influences on stream populations (fig. 10.19).

Many studies support Müller's hypothesized colonization cycle, especially among aquatic insects. As larvae, aquatic insects disperse upstream as well as downstream by swimming, crawling, and drifting. Because of continuous dispersal, which reshuffles stream populations, new substrates put into streams are quickly colonized by a wide variety of stream invertebrates, algae, and bacteria. Most of these dynamics are difficult to observe because they occur too quickly, within the substratum, or at night, or they involve microorganisms impossible to observe directly without the aid of a microscope. However, a snail that lives in a tropical stream in Costa Rica provides a well-documented example of the colonization cycle.

The Rio Claro flows approximately 30 km through tropical forest on the Osa Peninsula of Costa Rica before flowing into the Pacific Ocean. One of the most easily observed inhabitants of the Rio Claro is the snail *Neritina latissima,* which occupies the lower 5 km of the river. The eggs of *Neritina* hatch to produce free-living planktonic larvae that drift down to the Pacific Ocean. After the larvae metamorphose into small snails they reenter the Rio Claro and begin moving upstream in huge migratory aggregations of up to 500,000 individuals (fig. 10.20). These aggregations move slowly and may take up to 1 year to reach the upstream limit of the population.

The population of *Neritina* in the Rio Claro consists of a mixture of migrating and stationary subpopulations, with exchange between them. Individual snails migrate upstream for some distance and then leave the migrating wave and

(a)

(b)

**Figure 10.20** The colonization cycle in action. (*a*) A wave of migrating snails, *Neritina latissima,* in the Rio Claro, Costa Rica; (*b*) a close-up of the migrating snails.

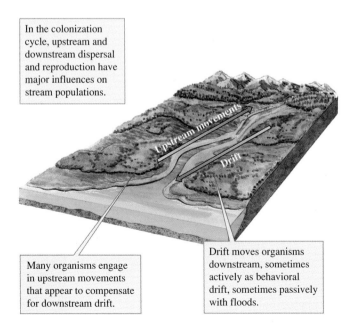

In the colonization cycle, upstream and downstream dispersal and reproduction have major influences on stream populations.

Many organisms engage in upstream movements that appear to compensate for downstream drift.

Drift moves organisms downstream, sometimes actively as behavioral drift, sometimes passively with floods.

**Figure 10.19** The colonization cycle of stream invertebrates.

enter a local subpopulation. At the same time individuals from the local subpopulation enter the migratory wave and move upstream. Thus, individuals move upstream in steps and immigration continuously adds to local subpopulations, while emigration removes individuals. Because an organism that is visible to the naked eye does all this in a clear stream, and does it at a snail's pace, we are provided with a unique opportunity to observe that dispersal can strongly influence local population density. Dispersal dynamics, though difficult to study, deserve greater attention.

How can information on population dynamics be used to address environmental problems? For instance, how would you go about evaluating the possible effects of a potential pollutant on natural populations? Making such a judgment is usually not as simple as it might sound. If a toxin kills everything in its path, at virtually all detectable concentrations, the situation is clear enough but often the effects of pollutants are not so obvious. One of the most promising approaches to assessing the impact of pollutants is to study their effects on population dynamics.

## APPLICATIONS & TOOLS

### Using Population Dynamics to Assess the Impact of Pollutants

Many scientists are studying the effects of sublethal concentrations (concentrations too low to kill within a short period of time) of pollutants on the population biology of aquatic organisms. Some promising research by Donald Baird, Ian Barber, and Peter Calow (1990) of the United Kingdom and Amadeu Soares (Soares, Baird, and Calow 1992) of Portugal has focused on a group of small planktonic crustaceans called cladocera, or water fleas. Their main study species, *Daphnia magna*, is distributed across the Northern Hemisphere. These water fleas have a number of characteristics that make them good candidates for laboratory studies of pollution. They are small, have a short generation time, and filter-feed on algae, which can be easily grown in the laboratory. In addition, *D. magna* reproduce asexually most of the time. Asexual reproduction ensures genetically uniform study populations. Consequently, genetic differences between test organisms can be controlled or accounted for in toxicity studies.

One of the central themes running through the work of these scientists is their attempt to connect the effects of pollutants on physiology with their effects on populations. In terms of the organization of this book, this research bridges the gap between section II, on the ecology of individuals, and the material we are now considering in section III on population ecology. An energy balance equation provides the key to bridging physiological and population ecology:

Energy assimilated = Respiration + Excretion + Production

In this equation, the amount of energy assimilated by an animal equals the sum of that expended in respiration, the amount of energy excreted, and the amount of energy available for production. This production energy is the amount of energy that an organism has at its disposal for growth and reproduction.

How does this equation connect the physiological effects of pollutants with their effects on populations? The connection derives from the principle of allocation, which we first considered in chapter 6. The principle of allocation assumes that energy supplies available to organisms are limited and predicts that any increase in the allocation of energy to any one of life's functions decreases the amount of energy available to other functions. In terms of our energy balance equation, if an organism is exposed to a toxin that induces physiological stress, energy expended in respiration generally increases. This increased respiration includes energy expended to excrete the toxin, to convert the toxin into a nontoxic chemical form, and to repair cellular damage caused by the toxin. The important point is that the processes that *increase* the energy spent for respiration *decrease* the energy available for growth and reproduction. This trade-off between reproduction and respiration provides the bridge between physiological and population ecology.

In their search for a reliable indicator of pollutant effects, Soares, Baird, and Calow examined a number of population-level responses. It turns out that one of the most responsive population characteristics is per capita rate of increase, *r*, which they calculated in the same way we discussed previously:

$$r = \frac{\ln R_0}{T}$$

How did the researchers choose *r* as an indicator of population response to potential of pollutants? To understand their choice, we need to consider *variability* in the responses of organisms to environmental challenges.

First, what do we mean by "variability" in response? Variability means differences in response; for example, the differences in per capita rate of increase, *r*, among several populations exposed to a toxin. What determines such differences? Some are due to genetic differences among populations. Other differences in response may be due to differences in toxin concentrations, that is, variation in the environment. In addition, different genotypes may respond differently to different environments. This is what population ecologists call gene-by-environment interactions. Variation not explained by genetic differences, environmental differences, and gene-by-environment interactions can be due to measurement error and unmeasured environmental variation. This type of variation is usually referred to as residual variation. We can summarize this dividing up, or "partitioning," of variation with an equation:

$$V_t = V_g + V_e + V_{ge} + V_r$$

Where total variation, $V_t$, equals the sum of variation due to genetic differences, $V_g$; variation due to environmental differences, $V_e$; variation due to gene-by-environment interactions, $V_{ge}$; and unexplained residual variation, $V_r$.

So, what does all this partitioning of variation have to do with choosing an indicator of the effects of potential pollutants? The best sort of indicator should vary substantially with environmental differences, in this case with concentration of a toxin. Such an indicator will show consistent responses to pollutants by a variety of genotypes. In general, variation due to the particular environmental effect, $V_e$, should exceed variation in response due to differences in genotypes, $V_g$.

Soares, Baird, and Calow have explored the partitioning of variation in *D. magna* populations by exposing several genotypes of this water flea to variation in a number of environmental factors. In one experiment, they studied the effect of a residue of an organic pesticide, dichloroaniline (DCA), on *r*. The results of their experiments show that the *r* of nine genotypes of *D. magna* changes substantially with concentration of DCA. Approximately 46% of the variation observed in *r* was due to variation in concentration of DCA (fig. 10.21). Of the remaining variation in *r*, 22% was accounted for by genetic differences between *D. magna* populations, and 24% was accounted for by interactions between genotypes and DCA concentration. Approximately 8% of the variation in *r* was unrelated to DCA concentration, genotypes, or interaction between genotype and DCA concentration. In summary, environment accounted for more variation in *r* than any of the other factors considered by the researchers. These results give a biological basis for using per capita rate of increase, *r*, as an indicator of the effects of a potential pollutant.

The results of other studies also indicate that *r* responds consistently to several different toxins. In one of these studies, three different clones of *D. magna* were exposed to several concentrations of DCA. The effects of increasing concentrations of the pesticide residue on the per capita rate of increase, *r*, of three different clones of *D. magna* are shown in figure 10.22. Notice that concentrations of the pesticide residue vary from 0 to 100 micrograms per liter (µg/L), or parts per billion, while the scale for *r* ranges from –0.2 to 0.4. To interpret these results you need to remember the significance of different values of *r*: positive values indicate a growing population, while negative values indicate a declining population. An *r* value of zero indicates a stable population.

Notice that the responses of the three genotypes were fairly similar. Though clone 3 appeared the most tolerant of DCA, all three clones had negative *r* values at DCA concentrations of 50 µg/L and above. We might have predicted this broad consistency in response across genotypes from the

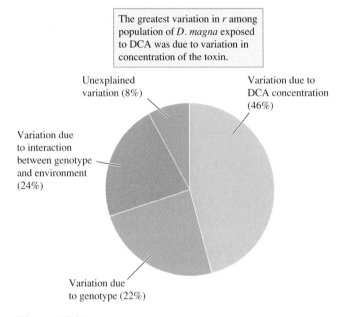

The greatest variation in *r* among population of *D. magna* exposed to DCA was due to variation in concentration of the toxin.

Unexplained variation (8%)

Variation due to DCA concentration (46%)

Variation due to interaction between genotype and environment (24%)

Variation due to genotype (22%)

**Figure 10.21** Partitioning the variation in per capita rate of increase, *r*, of *Daphnia magna* (data from Soares, Baird, and Calow 1992).

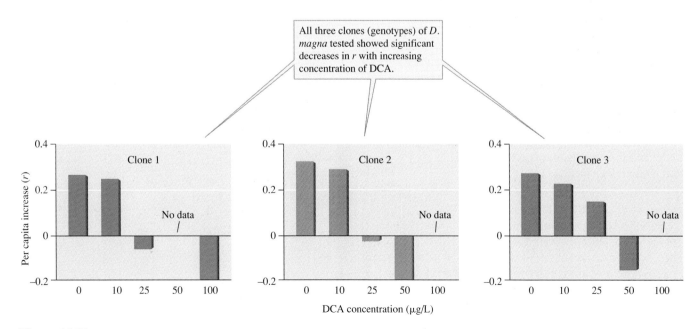

All three clones (genotypes) of *D. magna* tested showed significant decreases in *r* with increasing concentration of DCA.

**Figure 10.22** Effect of dichloroaniline (DCA) concentration on per capita rate of increase, *r*, of *Daphnia magna* clones (data from Baird, Barber, and Calow 1990).

results shown in figure 10.21, which showed that variation in DCA concentration accounted for the greatest amount of variation in the response of nine clones of *D. magna*.

The implications of the work of these researchers goes well beyond the effects of particular environmental variables on the population biology of *D. magna*. Their results suggest that population processes and the mechanisms underlying variation in those processes can be used as sensitive predic-

tors of the ecological effects of environmental change. It is also significant that this research work at the population level is rooted in phenomena at the level of individual organisms. This successful bridging between physiological and population ecology suggests that similar connections may exist between the population level and the higher organizational levels that we examine in the later sections of this book. Keep your eyes open for those possible connections.

## SUMMARY

**A survivorship curve summarizes the pattern of survival in a population.** Patterns of survival can be determined either by following a cohort of individuals of similar age to produce a cohort life table or by determining the age at death of a large number of individuals or the age distribution of a population to produce a static life table. Life tables can be used to draw survivorship curves, which generally fall into one of three categories: (1) type I survivorship, in which there is low mortality among the young but high mortality among older individuals; (2) type II survivorship, in which there is a fairly constant probability of mortality throughout life; and (3) type III survivorship, in which there is high mortality among the young and low mortality among older individuals.

**The age distribution of a population reflects its history of survival, reproduction, and potential for future growth.** Age distributions indicate periods of successful reproduction, high and low survival, and whether the older individuals in a population are replacing themselves or if the population is declining. Population age structure may be highly complicated in variable environments, such as that of the Galápagos Islands. Populations in highly variable environments may reproduce episodically.

**A life table combined with a fecundity schedule can be used to estimate net reproductive rate ($R_0$), geometric rate of increase ($\lambda$), generation time ($T$), and per capita rate of increase ($r$).** Because these population parameters form the core of population dynamics, it is important to understand their derivation as well as their biological meaning. Net reproductive rate, $R_0$, the average number of offspring left by an individual in a population, is calculated by multiplying age-specific survivorship rates, $l_x$, times age-specific birthrates, $m_x$, and summing the results:

$$\sum l_x m_x$$

The geometric rate of increase, $\lambda$, is calculated as the ratio of population sizes at two successive points in time. Generation time is calculated as:

$$T = \frac{\sum x l_x m_x}{R_0}$$

The per capita rate of increase, $r$, is related to generation time and net reproductive rate as:

$$r = \frac{\ln R_0}{T}$$

The per capita rate of increase may be positive, zero, or negative depending on whether a population is growing, stable, or declining.

**Dispersal can increase or decrease local population densities.** The contribution of dispersal to local population density and dynamics is demonstrated by studies of expanding populations of species such as Africanized bees in the Americas and collared doves in Europe. Climate changes can induce massive changes in the ranges of species. As availability of prey changes, predators may disperse, which increases and decreases their local population densities. Stream organisms actively migrating upstream or drifting downstream increase densities of stationary and migrating populations by immigrating and decrease them by emigrating.

Ecologists are using the effects of pollutants on population dynamics to predict the potential ecological impact of these pollutants on populations. Good candidates for indicators of pollution are those aspects of population dynamics that are sensitive to environmental variation. Based on its environmental sensitivity, per capita rate of increase, $r$, appears to be an excellent predictor of the impact of a wide range of potential pollutants. The results of this research suggest that population processes and the mechanisms underlying variation in those processes can be used as sensitive predictors of the ecological effects of environmental change.

# Review Questions

1. Compare cohort and static life tables. What are the main assumptions of each? In what situations or for what organisms would it be practical to use either?

2. Of the three survivorship curves, type III has been the least documented by empirical data. Why is that? What makes this pattern of survivorship difficult to study?

3. Population ecologists have assumed that populations of species with very high reproductive rates, those with offspring sometimes numbering in the millions per female, must have a type III survivorship curve even though very few survivorship data exist for such species. Why is this a reasonable assumption? In general, what is the expected relationship between reproductive rate and patterns of survival?

4. Draw hypothetical age structures for growing, declining, and stable populations. Explain how the age structure of a population with highly episodic reproduction might be misinterpreted as indicating population decline. How might population ecologists avoid such misinterpretations?

5. The third concept in chapter 10 says that we can use the information in life tables and fecundity schedules to estimate some characteristics of populations ($R_0$, $T$, $r$). Why does the second concept use the word "estimate" rather than "calculate"? In putting together your answer, think about the population of *P. drummondii* studied by Leverich and Levin (1979). We calculated $R_0$ for this population by summing the $l_x m_x$ column. When we did, the number we got was $R_0 = 2.4177$. Assuming that Leverich and Levin accurately counted seeds and surviving plants, is 2.4177 an estimate of the average reproductive rate of the 996 individual *P. drummondii* in their study? Right, 2.4177 is not an estimate; it's the actual average number of seeds produced by these 996 individuals. So, what's this estimate business about? If Leverich and Levin had studied a second (or third, fourth, etc.) group of 996 individuals in their *P. drummondii* population, do you think it's likely that they would have gotten an $R_0$ exactly equal to 2.4177?

6. What values of $R_0$ indicate that a population is growing, stable, or declining? What values of $r$ indicate a growing, stable, or declining population?

7. From a life table and a fecundity schedule, you can estimate the geometric rate of increase, $\lambda$, the average reproductive rate, $R_0$, the generation time, $T$, and the per capita rate of increase, $r$. That is a lot of information about a population. What minimum information do you need to construct a life table and fecundity schedule?

8. C. S. Holling (1959) observed predator numerical responses to changes in prey density. He attributed the numerical responses to changes in the reproductive rates of the predators. Discuss a hypothetical example of reproductive-rate numerical response by a population of predators in terms of changes in fecundity schedules and life tables. Include the terms $R_0$, $T$, and $r$ in your discussion.

9. Outline Müller's (1954, 1974) colonization cycle. If you were studying the colonization cycle of the freshwater snail *N. latissima*, how would you follow colonization waves upstream? How would you verify that these colonization waves gain individuals from local populations and also contribute individuals to those same local populations?

10. In our discussions of the research of Baird, Soares, and their colleagues (Baird, Barber, Calow 1990, Soares, Baird, Calow 1992), we focused on the effects of a residue of an organic pesticide, dichloroaniline (DCA), on the per capita rate of increase, $r$, of *D. magna*. These researchers also found that the $r$ of *D. magna* responds significantly to a variety of inorganic pollutants. What do these results indicate about the usefulness of $r$ as an indicator of the ecological impact of potential pollutants?

# Suggested Readings

Baird, D., I. Barber, and P. Calow. 1990. Clonal variation in general responses of *Daphnia magna* Straus to toxic stress. I. Chronic life-history effects. *Functional Ecology* 4:399–407.

Soares, A. M. V. M., D. J. Baird, P. Calow. 1992. Interclonal variation in the performance of *Daphnia magna* Straus in chronic bioassays. *Environmental Toxicology and Chemistry* 11:1477–83.

*These papers demonstrate the application of population dynamics to the study of pollution.*

Carey, J. R. 2001. Insect biodemography. *Annual Review of Entomology* 46:79–110.

*A comprehensive review of life tables and survivorship in insect populations, including a complete cohort life table for 1.2 million Mediterranean fruit flies.*

Grant, R. B. and P. R. Grant. 1989. *Evolutionary Dynamics of a Natural Population.* Chicago: University of Chicago Press.

*Exceptional long-term study of the large cactus finch on Genovesa Island—destined to be a classic study in ecology.*

Hellgren, E. C., R. T. Kazmaier, D. C. Ruthven, and D. R. Synatzske. 2000. Variation in tortoise life history: demography of *Gopherus berlandieri*. *Ecology* 81:1297–1310.

*Excellent study providing a life table and estimates of demographic parameters along with comparisons to other tortoise populations.*

Hengeveld, R. 1989. *Dynamics of Biological Invasions.* New York: Chapman and Hall.

*Excellent and readable introduction to biological invasions. Includes a quantitative approach to studying population expansion.*

Hyrenbach, K. D. and R. C. Dotson. 2003. Assessing the susceptibility of female black footed albatross (*Phoebastria nigripes*) to longline fisheries during the post-breeding dispersal: an integrated approach. *Biological Conservation* 112:391–404.

*A modern approach to following dispersal of a far-ranging pelagic seabird, using satellite tracking.*

Leverich, W. J. and D. A. Levin. 1979. Age-specific survivorship and reproduction in *Phlox drummondii*. *American Naturalist* 113:881–903.

Sarukhán, J. and J. L. Harper. 1973. Studies on plant demography: *Ranunculus repens* L. and *R. acris* L. I. population flux and survivorship. *Journal of Ecology* 61:675–716.

*These two classic papers provide an excellent introduction to detailed studies of plant population dynamics.*

Müller, K. 1974. Stream drift as a chronobiological phenomenon in running water ecosystems. *Annual Review of Ecology and Systematics* 5:309–23.

*This review article lays the foundation for Müller's colonization cycle hypothesis.*

Schneider, D. W. and J. Lyons. 1993. Dynamics of upstream migration in two species of tropical freshwater snails. *Journal of the North American Benthological Society* 12:3–16.

*Fascinating account and photos of migratory dynamics of a snail in a clear tropical stream.*

Winston, M. L. 1992. Biology and management of Africanized bees. *Annual Review of Entomology* 37:173–93.

*Detailed review of the spread of Africanized honeybees from their point of introduction in Brazil. Management implications and potential economic impact integrated with basic biology.*

# On the Net

Visit this textbook's accompanying website at www.mhhe.com/ecology (click on the book's title) to take advantage of practice quizzing, study/writing tips, timely news articles, and additional URLs for research on the topics in this chapter.

Animal Population Ecology
Population Density of Animals
Population Growth
Extinction Issues

Movement of Populations
Field Methods for Studies of Populations
Field Methods for Studies of Ecosystems

# *Chapter* 11

# Population Growth

Given suitable environmental conditions, aquatic and terrestrial populations will manifest their great capacity for growth. Each spring, in temperate seas and lakes around the globe, planktonic populations of diatoms take advantage of the increasing availability of sunlight and an abundance of nutrients. The numbers of diatoms explode as these single-celled protists survive, mature, and reproduce. Populations of zooplankton respond to the "spring blooms" of diatoms, on which they feed, by increasing their own numbers in turn (fig. 11.1). Later in the annual cycle, the numbers of individuals in the diatom and zooplankton populations decrease, responding to decreases in sunlight and nutrients and increases in competition and predation. Populations are dynamic—increasing, decreasing, and responding to changes in the biotic and abiotic environments.

Sizes of populations fluctuate in terrestrial, as well as aquatic, environments. Some of the most variable terrestrial populations are found on the Galápagos Islands. The sizes of populations on these islands vary a great deal because they are subject to exceptional environmental fluctuations. Much of this fluctuation is produced by a large-scale climatic system commonly called El Niño (see chapter 23). El Niño warms the waters around the Galápagos Islands and brings higher than average rainfall once or twice each decade. This increased rainfall stimulates the germination and growth of plants. These plants produce an abundance of seeds upon which Galápagos finches depend for food. In response to increased seed production, the size of the finch populations can increase several fold in 1 year. However, these same populations are also exposed to periodic droughts. During droughts, which can be severe, both plant and finch populations decline dramatically. Again whether in the sea or on land, populations are dynamic.

In chapter 11 we examine the factors that determine rates and patterns of population growth. We also review the environmental forces that limit population size. At this juncture in the history of our own species, there is no more important ecological topic.

We look at population growth in the presence of abundant resources, growth where resources are limiting, how the environment can act to change birth and death rates, and, finally, how rates of population growth are affected by the size of organisms. The concept discussions we review reflect the historical development of population ecology. That history has involved two complementary approaches. One approach uses mathematics to model population growth. The second approach focuses on studies of laboratory and natural populations. Our knowledge of population growth has progressed through an interplay between modeling and observations of actual populations. Let's turn now to studies of population growth in the presence of abundant resources.

(a)

(b)

**Figure 11.1** Lake plankton populations undergo explosive population growth each spring in mid- and high-latitude lakes as a result of favorable environmental conditions. Shown here are (*a*) diatoms and (*b*) a copepod.

## CONCEPTS

- **In the presence of abundant resources, populations can grow at geometric or exponential rates.**
- **As resources are depleted, population growth rate slows and eventually stops; this is known as logistic population growth.**
- **The environment limits population growth by changing birth and death rates.**
- **On average, small organisms have higher rates of per capita increase, $r_{max}$, and more variable populations, while large organisms have lower rates of per capita increase and less variable populations.**

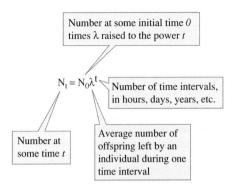

## CONCEPT DISCUSSION

### Geometric and Exponential Population Growth

**In the presence of abundant resources, populations can grow at geometric or exponential rates.**

Suppose a population had access to abundant resources, such as food, space, nutrients, and so forth. How fast could it grow? Imagine a plant, animal, or bacterial population reproducing at its maximum reproductive rate. What would the resulting pattern of population growth be? Regardless of the species you choose, the pattern will be the same. A population growing at its maximum rate grows slowly at first and then faster and faster. In other words, population growth accelerates.

When growing at their maximum rates, some populations are said to grow *geometrically* and others *exponentially.* We examine what causes these two ways of modeling population growth in this section, and we once again make use of the per capita rate of increase, $r$.

## Geometric Growth

Because it is an annual plant, populations of *Phlox drummondii* grow in discrete annual pulses. Populations of insects that produce a single generation a year also grow in pulses. Growth by any population with pulsed reproduction can be modeled as **geometric population growth,** in which successive generations differ in size by a constant ratio.

We can use the population of *Phlox* studied by Leverich and Levin (1979) to build a model of geometric population growth. In chapter 10, we calculated a geometric rate of increase, $l = N_{t+1}/N_t$, for this population of 2.4177. At the end of that discussion, we asked how long the *Phlox* population could continue growing at this rate. Let's address that question here.

As we saw in chapter 10, we can compute the growth of a population of organisms whose generations do not overlap by simply multiplying $\lambda$ times the size of the population at the beginning of each generation. The initial size of the population studied by Leverich and Levin was 996 (see table 10.1) and the number of offspring produced by this population during their year of study was $N_1 = N_0 \times \lambda$, or $996 \times 2.4177 = 2,408$. Now let's repeat this calculation for a few generations. The population size at the beginning of the next generation, $N_2$, would be $N_1 \times \lambda$. However, because $N_1 = N_0 \times \lambda$, $N_2 = N_0 \times \lambda \times \lambda$, or $N_0 \times \lambda^2$, which is $996 \times 2.4177 \times 2.4177 = 5,822$. At the third generation, $N_3$, $N_0 \times \lambda^3 = 14,076$ and in general, the size of a population growing geometrically at any time, $t$, can be modeled as:

$$N_t = N_0\lambda^t$$

**Figure 11.2** Anatomy of the equation for geometric population growth.

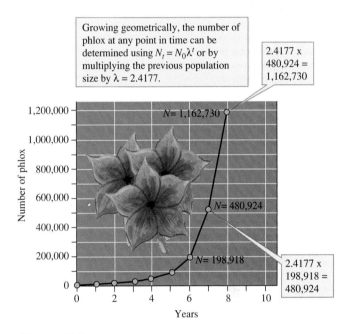

**Figure 11.3** Geometric growth by a hypothetical population of *Phlox drummondii.*

In this model, $N_t$ is the number of individuals at any time $t$, $N_0$ is the initial number of individuals, $\lambda$ is the geometric rate of increase, and $t$ is the number of time intervals or generations. The interpretation of this model and the definitions of each of its terms are summarized in figure 11.2. We can use this model to project the future size of our hypothetical *Phlox* population. Notice in figure 11.3 that in only 8 years the population has grown from 996 to $1.16 \times 10^6$, to over 1 million individuals. By 16 years, the population would be over a billion, by 24 years the population would top 1 trillion individuals, and by year 40 it would increase to over $10^{18}$, or 1 billion billion individuals.

We can get a feeling for how large this hypothetical *Phlox* population would be by calculating how much space the growing population would occupy. Since the *Phlox* population studied by Leverich and Levin was from Texas, let's confine our hypothetical population to North America and scale population growth against the area of the North American continent, which is about 24 million km². Assuming a uniform density across the continent, by 32 years our population would reach a density of nearly 80 million individuals per square kilometer or about 80 individuals per square meter across the entire continent, from southern Mexico to northern Canada and Alaska. Eight years later, the density would be nearly 90,000 individuals per square meter!

There are many reasons why this exercise is unrealistic. The population would soon be so dense that plants would die because they lacked sufficient nutrients, light, and water; and the population would soon spread beyond the physical climates to which *P. drummondii* is adapted. However, out of this unrealistic exercise comes an important fact about the natural world. Clearly, natural populations have a tremendous capacity for increase, and geometric population growth cannot be maintained in any population for very many generations.

Now let's consider population growth by organisms such as bacteria, forest trees, and humans, which have overlapping generations. Because growth by these populations can be continuous, the geometric model is usually not appropriate.

## Exponential Growth

Continuous population growth in an unlimited environment can be modeled as **exponential population growth:**

$$\frac{dN}{dt} = r_{max}N$$

The exponential growth equation (fig. 11.4) expresses the rate of population growth, $dN/dt$, which is the change in numbers with change in time, as the per capita rate of increase, $r_{max}$, times population size, $N$. The exponential model is appropriate for populations with nonpulsed reproduction because it represents population growth as a continuous process. Notice that the per capita rate of increase, $r_{max}$, has a subscript $_{max}$. The subscript here indicates that this is the *maximum* per capita rate of increase, achieved by a species under ideal environmental conditions, where birthrates, death rates, and age structure are constant. The per capita rate of increase attained under such circumstances, $r_{max}$, is called the **intrinsic rate of increase.** When we calculated the rate of increase from a life table in chapter 10, we determined $r$, the *realized* or *actual* per capita rate of increase. As we saw, realized $r$ may be positive, zero, or negative, depending on environmental conditions. Because natural populations are usually subject to factors such as disease, competition, and so forth, the actual per capita rate of increase, realized $r$, is generally less than $r_{max}$. In the exponential model, $r_{max}$ is a constant, while $N$ is a variable. Therefore, as population size, $N$, increases the rate of population increase, $dN/dt$, gets larger and larger. The rate of increase gets larger because the constant $r$ is multiplied by a larger and larger population size, $N$. Consequently, during exponential growth, the rate of population growth increases over time.

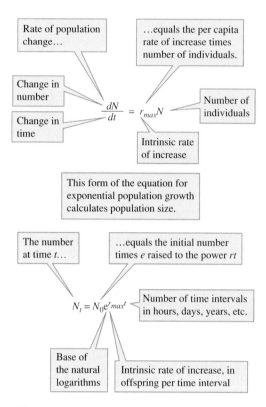

**Figure 11.4** Anatomy of equations for exponential population growth.

For a population growing at an exponential rate, the population size at any time $t$ can be calculated as:

$$N_t = N_0 e^{r_{max}t}$$

In this form of the exponential growth model, $N_t$ is the number of individuals at time $t$, $N_0$ is the initial number of individuals, $e$ is the base of the natural logarithms, $r_{max}$ is the intrinsic rate of increase, and $t$ is the number of time intervals. Notice that this form of the exponential model of population growth is virtually the same as our equation for geometric growth but with $e^{r_{max}}$ taking the place of $\lambda$. The two forms of the exponential growth equation are presented and explained in figure 11.4.

## Exponential Growth in Nature

Some of the assumptions of the exponential growth model, such as a constant rate of per capita increase, may seem a bit unrealistic, so it is reasonable to ask whether populations in nature ever grow at an exponential rate. The answer is a qualified yes. Natural populations may grow at exponential rates for relatively short periods of time in the presence of abundant resources.

## Exponential Growth by Tree Populations

As we saw in chapters 1 and 10 (see figs. 1.9 and 10.17), as the last ice age was ending, tree populations in the Northern Hemisphere followed the retreating glaciers northward. Ecologists have documented these movements by studying the sediments of lakes, where the pollen of wind-pollinated tree species is especially abundant. The appearance of pollen of a tree species in a lake sediment is a record of its establishment near the lake. The date of each establishment can be determined using carbon-14 concentration to determine the age of organic matter along a sediment profile.

Pollen records have also been used to estimate the growth of several postglacial tree populations in Britain. K. Bennett (1983) estimated population sizes and growth by counting the number of pollen grains of each tree species deposited within lake sediments. By counting the number of pollen grains per square centimeter deposited each year, Bennett was able to reconstruct changes in tree population densities in the surrounding landscape. This approach is a bit different from going out in a forest and estimating population density directly by counting trees. What is the main assumption of this method? Bennett's assumption was that the rate of pollen deposition is proportional to the size of tree populations around a lake. This assumption, which seems reasonable, leads to an interesting picture of growth by postglacial tree populations in the British Isles. Populations of the tree species studied grew at exponential rates for 400 to 500 years following their initial appearance in the pollen record. Figure 11.5 shows the exponential increase in abundance of Scots pine, *Pinus sylvestris,* which first appeared in the pollen record of the study lake about 9,500 years ago.

## Conditions for Exponential Growth

Natural populations of organisms as different as diatoms, whales, and trees can grow at exponential rates. However, as different as these organisms are, the circumstances in which their populations grow at exponential rates have a great deal in common. All begin their exponential growth in favorable environments at low population densities. The trees studied by Bennett began at low densities because they were invading new territory previously unoccupied by the species. Spring blooms of planktonic diatoms are the result of exponential population growth in response to seasonal increases in nutrients and light. As we will see in the section of chapter 11 titled "Growth of a Whale Population," the California gray whale population, which had been reduced to low densities by whaling, grew exponentially once whaling was stopped.

The whooping crane provides another example of exponential growth following protection and careful management. Hunting and habitat destruction reduced the population of whooping cranes to 15 individuals by 1941–42. At that time it was known that this remnant population of whooping cranes winters on the Texas Gulf Coast but its northern breeding grounds were unknown. It was later discovered that they breed in Wood Buffalo National Park in Canada. Under the full protection and careful management in both Canada and the United States, the migratory whooping crane population has grown exponentially from 15 in 1942 to over 180 individuals in 2003 (fig. 11.6).

These examples suggest that exponential population growth may be very important to populations during the process of establishment in new environments, during exploitation of transient, favorable conditions, and during the process

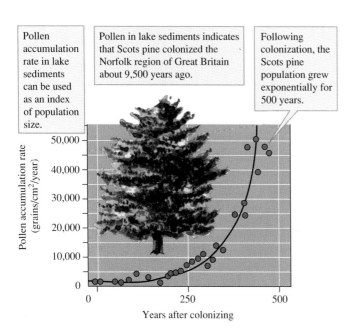

**Figure 11.5** Exponential growth of a colonizing population of Scots pine, *Pinus sylvestris* (data from Bennett 1983).

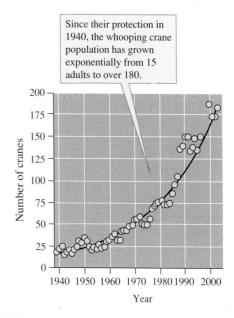

**Figure 11.6** Hunting and habitat destruction reduced the whooping crane, which is endemic to North America, to a single natural population. Protection and intensive management of this population has led to its dramatic recovery (data from USGS 1995, USFWS Whooping Crane Recovery Plans).

of recovery from some form of exploitation. However, as we saw with *P. drummondii,* geometric or exponential growth cannot continue indefinitely. In nature, population growth must eventually slow and population size level off.

## Slowing of Exponential Growth

As we saw in chapter 10, the collared dove, *Streptopelia decaocto,* expanded beyond its historical range into western Europe during the latter half of the twentieth century. As the bird spread into new territory, its populations grew at exponential rates for a decade or more. For instance, from 1955 to 1972, the expanding population in the British Isles followed a typical exponential curve (fig. 11.7). However, if you examine figure 11.7 closely, you will see evidence that the rate of growth by the collared dove population began to slow by 1970.

The pattern for the collared dove indicates that its population grew at a higher rate, from 1955 to 1964, and then, between 1965 and 1970, its rate of population growth began to slow. This slowdown suggests that between 1965 and 1970, this invading population was approaching some environmental limits. Environmental limitation is incorporated into another model of population growth called **logistic population growth.**

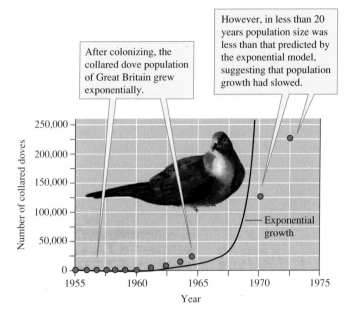

**Figure 11.7** Exponential growth of the collared dove population of Great Britain (data from Hengeveld 1988).

## CONCEPT DISCUSSION

### Logistic Population Growth

**As resources are depleted, population growth rate slows and eventually stops; this is known as logistic population growth.**

Obviously, exponential growth cannot continue indefinitely. Eventually, populations run up against environmental limits to further increase. The effect of the environment on population growth is reflected in the shapes of population growth curves. As population size increases, growth rate eventually slows and then ceases as population size levels off. This pattern of growth produces a **sigmoidal,** or S-shaped, **population growth curve** (fig. 11.8). The population size at which growth stops is generally called **carrying capacity,** or **K,** which is the number of individuals of a particular population that the environment can support. At carrying capacity, because population size is approximately constant, birthrates must equal death rates and population growth is zero.

Sigmoidal growth curves have been observed in a wide variety of populations. In the course of his laboratory experiments, G. F. Gause (1934) obtained sigmoidal growth curves for populations of several species of yeast (fig. 11.9) and protozoa (fig. 11.10). Similar patterns of population growth have been recorded for other populations, including barnacles (fig. 11.11) and African buffalo (fig. 11.12).

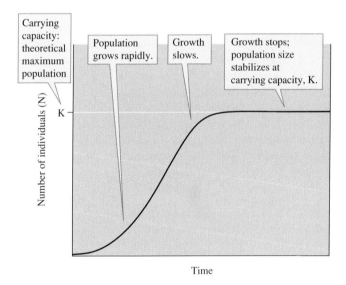

**Figure 11.8** Sigmoidal, or logistic, population growth results from environmental limitation on population size.

What causes these populations to slow their rates of growth and eventually stop growing at carrying capacity? The idea behind the concept of carrying capacity is that a given environment can only support so many individuals of a particular species. For the barnacles studied by J. H. Connell (1961), carrying capacity is largely determined by the amount of space available on rocks for attachment by new barnacles. For African buffalo, carrying capacity appears largely determined by the amount of grass available as food (Sinclair 1977). Yeast feed on sugars and produce alcohol. As the density of a population of yeast increases, their environment contains less and less sugar and more and more alcohol, which is

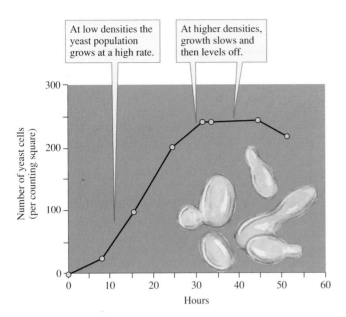

**Figure 11.9** Sigmoidal growth by a population of the yeast *Saccharomyces cerevisiae* (data from Gause 1934).

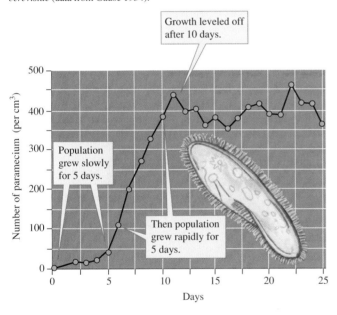

**Figure 11.10** Sigmoidal growth by a population of *Paramecium caudatum* (data from Gause 1934).

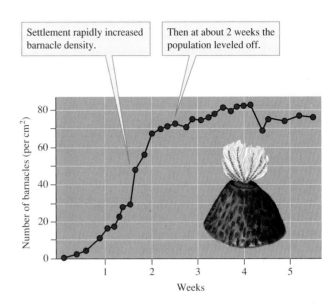

**Figure 11.11** Settlement by the barnacle *Balanus balanoides* in the intertidal zone (data from Connell 1961).

**Figure 11.12** Sigmoidal population growth by African buffalo, *Syncerus caffer,* on the Serengeti Plain (data from Sinclair 1977).

toxic to them. So, yeast populations are eventually limited by their own waste products. For most species, carrying capacity is likely determined by a complex interplay among factors such as food, parasitism, disease, and space. While we can discuss these factors in a general way, the mathematical models of population biology help us to discuss population processes in a more precise way.

The logistic model was proposed to account for the patterns of growth shown by populations as they begin to deplete environmental resources. Population ecologists built the logistic growth model by modifying the exponential growth model. The exponential model of population growth, $dN/dt = r_{max}N$,

can be modified to produce a model in which population growth is sigmoidal. The simplest way to do this is to add an element that slows growth as population size approaches carrying capacity, *K:*

$$\frac{dN}{dt} = r_{max}N\left(\frac{K-N}{K}\right)$$

The inventor of this equation for sigmoidal population growth, P. F. Verhulst, called it the **logistic equation** (Verhulst and Quetelet 1838). Rearranging the logistic equation shows more clearly the influence of population size, *N*, on rate of population growth:

$$\frac{dN}{dt} = r_{max}N\left(\frac{K}{K} - \frac{N}{K}\right) = r_{max}N\left(1 - \frac{N}{K}\right)$$

In the logistic equation, the rate of population growth, $dN/dt$, slows as population size increases because the difference, $(1 - N/K)$, becomes a smaller and smaller decimal fraction until $N$ equals $K$. When $N$ equals $K$, the right side of the equation becomes zero. Therefore, as population size increases, the logistic growth rate becomes a smaller and smaller fraction of the exponential growth rate and when $N = K$, population growth ceases (fig. 11.13). Logistic population growth is highest when $N = K/2$.

The ratio $N/K$ has been called the "environmental resistance" to population growth. As the size of a population, $N$, gets closer and closer to carrying capacity, environmental factors increasingly impede further population growth.

In the logistic growth model, the *realized* per capita rate of increase, which is $r = r_{max}(1 - N/K) = r_{max} - r_{max}(N/K)$, depends upon population size. Therefore, when population size, $N$, is very small, the per capita rate of increase is approximately $r_{max}$. As $N$ increases, however, realized $r$ decreases until $N$ equals $K$. At that point, realized $r$ is zero. The relationship between realized $r$ and population size in the logistic model, which follows a straight line, is shown in figure 11.14.

When working with mathematical models, it is always useful for the ecologist to keep the biology behind the model firmly in mind. In the case of models of population growth, we should remember that $r$ is the difference between birth and death rates in a population. Let's think about figure 11.14 from this perspective. At very low population size, the per capita birthrate, $b$, greatly exceeds the per capita death rate, $d$. As population size increases, the logistic model assumes that per capita birthrates will decrease and per capita death rates will increase. Then, when population size reaches carrying capacity, or $K$, $b = d$ and since $b - d = 0$, population growth stops.

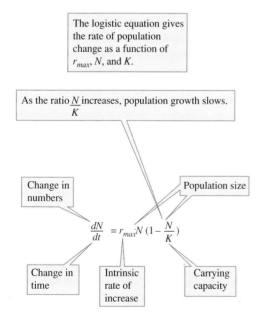

**Figure 11.13** Anatomy of the logistic equation for population growth.

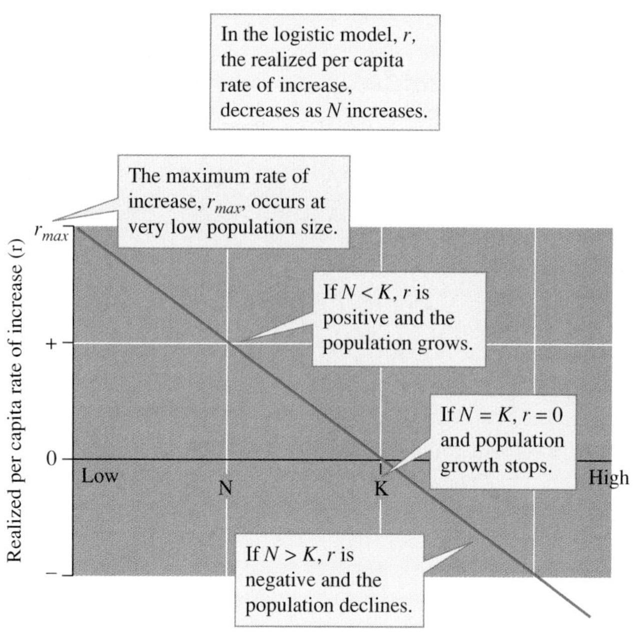

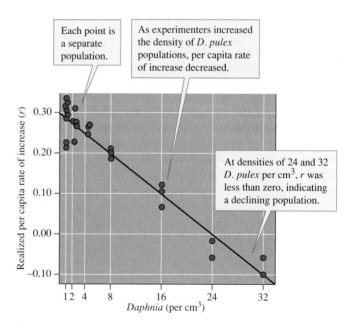

**Figure 11.14** The relationship between population size, $N$, and realized per capita rate of increase, $r$, in the logistic model of population growth.

**Figure 11.15** Relationship of density to per capita rate of increase in populations of *Daphnia pulex* (data from Frank, Boll, and Kelly 1957).

The response of per capita rate of increase by *Daphnia pulex*, a water flea in the same genus as *D. magna*, which we discussed in chapter 10, to population density closely matches the assumptions of the logistic growth model. When *D. pulex* are grown at densities ranging from 1 to 32 individuals per cubic centimeter, $r$ decreases with increasing population size (fig. 11.15). As assumed by the logistic growth model, per capita rate of increase was highest at the lowest population densities. Per capita rate of increase was positive in *D. pulex* populations with densities of 16 individuals per cubic centimeter or lower. However, at densities of 24 and 32 individuals per cubic centimeter, per capita rate of increase was negative.

Ultimately the environment limits the growth of populations by modifying birth and death rates. In the following concept discussion section, we examine in detail a few examples of environmental effects on population growth.

(a)

(b)

**Figure 11.16** The abundant rains of 1983 (*a*) greatly increased plant growth on the Galápagos Islands compared to (*b*) periods of lower rainfall.

## CONCEPT DISCUSSION

### Limits to Population Growth

**The environment limits population growth by changing birth and death rates.**

Most of us could recite an impressive list of factors affecting the size of populations. Such lists generally include food, shelter, rainfall, disease, floods, and predators—a mixture of abiotic and biotic factors. Ecologists have long been concerned with the effects of environmental factors such as these on populations. Out of this concern came a long period of debate between the champions of the importance of abiotic factors and those who argued for the importance of biotic factors. Because the effects of biotic factors, such as disease and predation, are often influenced by population density, biotic factors are often referred to as **density-dependent factors.** Meanwhile, abiotic factors, such as floods and extreme temperature, can exert their influences independently of population density and so are often called **density-independent factors.** However, many ecologists were (and are) quick to point out that abiotic factors can influence populations in a density-dependent fashion. For instance, think of the effect on mortality of an unusually cold period. At high population densities a larger proportion of the population may inhabit less sheltered sites and so mortality rate in the population is greater at high population density than at low population density. Similarly, biotic factors such as disease can affect populations in a density-independent way—for example, a particularly virulent pathogen, such as Dutch elm disease, which causes total mortality in infected populations regardless of their local density. The major point of this section is that biotic and abiotic factors act on populations by modifying birth and death rates. The significance of biotic and abiotic factors on populations has been well demonstrated by studies of Galápagos finches and their major food sources.

## Environment and Birth and Death Among Galápagos Finches

Since Charles Darwin's visit in the 1830s, the Galápagos Islands have continued to provide scientists with a rich source of information concerning ecological and evolutionary processes. More than two decades ago Peter Grant and Rosemary Grant and their students and colleagues began a long-term study of the evolution and ecology of Galápagos finches (see chapter 10). This long-term project has yielded information extending well beyond the finch populations and the Galápagos Islands. Knowledge of the influences of the environment on birth and death rates in natural populations could not have been gained from a short-term study.

Highly variable rainfall and responsive plant populations provided the environmental setting for these finch studies (fig. 11.16). In 1976, P. T. Boag and Peter Grant began a study of the populations of Darwin's finches inhabiting Daphne Major, an island of only 0.4 km$^2$ situated in the middle of the Galápagos Archipelago (Boag and Grant 1984). The numerically dominant finch on Daphne Major at the beginning of the study was the medium ground finch, *Geospiza fortis,* with about 1,200 individuals. In 1977, a drought struck the Galápagos Islands and by the end of the year the population of *G. fortis* had fallen to about 180 individuals. This decrease represents a decline in population size of about 85% in just 1 year.

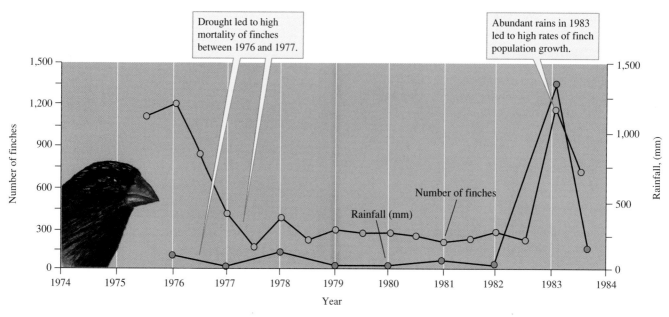

**Figure 11.17**  Rainfall and the medium ground finch, *Geospiza fortis,* population of Daphne Major Island (data from Gibbs and Grant 1987).

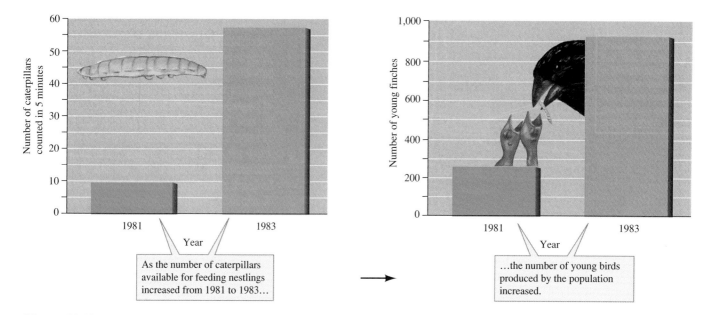

**Figure 11.18**  Availability of caterpillars and fledgling of young medium ground finches on Daphne Major (data from Gibbs and Grant 1987).

Though a few birds may have emigrated to other nearby islands, most of this population decline was due to starvation. During the drought, the plants that normally produce an annual crop of seeds, upon which the finches depend for food, failed to do so. From 1977 to 1982, the population of *G. fortis* on Daphne Major averaged about 300 individuals. Then in 1983, about 10 times the average amount of rainfall fell and the population grew to about 1,100 individuals (fig. 11.17).

This population growth was due to an increased birthrate as a consequence of an abundance of seeds that the adult finches eat and an abundance of caterpillars that the finches feed to their young (fig. 11.18). As you can see, *G. fortis* populations declined in 1977 because death rates due to starvation far exceeded birthrates. However, the situation was reversed in 1983, when, in the presence of abundant food, birthrates greatly exceeded death rates.

Information
Questions
Hypothesis
Predictions
Testing ✓

*Investigating the Evidence*

# Frequency of Alternative Phenotypes in a Population

Ecologists often ask questions about observed frequencies of individuals in a population relative to some theoretical or expected frequencies. For example, an ecologist studying the nesting habits of Galápagos finches may be interested in the frequency that finches nest in alternative nest sites relative to the availability of the alternative nest sites in the environment. Another ecologist studying the habitat association of a plant may be interested in its relative frequencies in sandy, loamy, or clay soils. An ecologist studying the mating behavior of some species may want to determine whether alternative male phenotypes (males with different physical and/or behavioral characteristics) occur at different frequencies in the population.

A common method to test hypotheses concerning the relationship between observed and hypothesized frequencies is the **chi-square** $(\chi^2)$ "goodness of fit" test. This test is used to judge how well an observed distribution of frequencies matches one "expected" from a particular hypothesis. Let's explore this test using the frequency of alternative male phenotypes in a population of side-blotched lizards, *Uta stansburiana*. Barry Sinervo (Sinervo and Lively 1996) found that males in populations of side-blotched lizards of the coastal range of central California include three male phenotypes: very aggressive orange-throat males, moderately aggressive blue-throat males, and sneaker yellow-throat males. Sinervo and his colleagues also found that these three male phenotypes vary in their frequencies over time. Let's consider the following hypothetical table of data from a population and test the hypothesis that the three male phenotypes are present in equal frequencies in the population.

| Male phenotype | Observed frequency (O) | Expected frequency (E) |
|---|---|---|
| Orange-throat | 12 | 17 |
| Blue-throat | 30 | 17 |
| Yellow-throat | 9 | 17 |

Before proceeding with our test, let's reflect on what we are doing. We would like to know if there are differences in the frequencies of male phenotypes in this population of side-blotched lizards. So, we went out and obtained a sample of 51 males from the population. This sample included 12 orange-throat males, 30 blue-throat males, and 9 yellow-throat males. Our sample is an estimate of the actual frequencies of the three phenotypes in the larger population that we are studying. Because our hypothesis is that there are no differences in frequencies among the male phenotypes, the "expected" frequencies, in the third column of our table, are equal $\left(\frac{51}{3} = 17\right)$.

Let's use the chi-square $(\chi^2)$ "goodness of fit" test to determine how well our observed frequency distribution matches the expected frequency distribution. The value of $\chi^2$ is calculated as follows:

$$\chi^2 = \sum \frac{(O-E)^2}{E}$$

In this equation, $O$ is the observed frequency of a particular phenotype and $E$ is the expected frequency. If we enter the values from the table, we get the following:

$$\chi^2 = \frac{(12-17)^2}{17} + \frac{(30-17)^2}{17} + \frac{(9-17)^2}{17}$$

$$\chi^2 = \frac{(-5)^2}{17} + \frac{(13)^2}{17} + \frac{(-8)^2}{17}$$

$$\chi^2 = \frac{25}{17} + \frac{169}{17} + \frac{64}{17}$$

$$\chi^2 = 1.47 + 9.94 + 3.76$$

$$\chi^2 = 15.17$$

The next step in the chi-square test is to determine whether the difference between observed and expected values, as indexed by our calculated $\chi^2$, is "significant." We determine significance by comparing our calculated value of $\chi^2$ with a table of $\chi^2$ values, which are included in most statistics textbooks. In order to find the appropriate, or critical, value of $\chi^2$ from such a table, we need to know two things. First, we need to choose a level of significance, which, as we saw in chapter 10 (p. 269), is generally P < 0.05. Second, we need to know the "degrees of freedom," which, in this case, is the number of male phenotypes (3), or *n*, minus 1.

$$\begin{aligned} \text{degrees of freedom} &= n - 1 \\ &= 3 - 1 \\ &= 2 \end{aligned}$$

What does "degrees of freedom" mean? It is the number of values we can pick freely without being constrained by other values within a set. In the case of the frequency of three male phenotypes, given a particular sample size, once we know the frequencies of two of the phenotypes, we automatically know the third. For instance, with a sample of 51 lizards, once we determine that there are 12 orange-throat males and 30 blue-throat males in our sample, the frequency of yellow-throat males is constrained to be 9. So, in this sample of three male phenotypes, there are two degrees of freedom.

In a table of critical values of chi-square, you will find that for a significance level of P = 0.05 and 2 degrees of freedom, the critical $\chi^2$ value is 5.991. Since our calculated value of $\chi^2$ (15.17) is greater than 5.991, we reject the hypothesis that the three male phenotypes are present in equal frequencies in our study population. In other words, we have evidence of significant differences in frequency among the three phenotypes. And, we can attach a probability statement to this conclusion, which is P < 0.05.

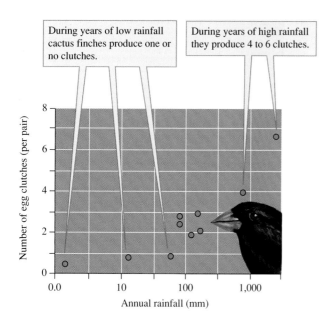

**Figure 11.19** Relationship between annual rainfall and the number of egg clutches produced by large cactus finches, *Geospiza conirostris,* on Genovesa Island (data from Grant and Grant 1989).

**Figure 11.20** High rainfall during the El Niño of 1983 caused increased mortality of the cactus *Opuntia helleri* on Genovesa Island.

Over this same period, Rosemary Grant and Peter Grant (1989) were studying a population of the large cactus finch, *Geospiza conirostris,* on Genovesa, a small, highly isolated island in the extreme northeastern portion of the Galápagos Archipelago. The study continued from 1978 to 1988, long enough for the researchers to observe the effects of two droughts and two wet periods on reproductive biology. In this population of cactus finches, there was a positive correlation between the number of clutches of eggs laid by birds and the total annual rainfall (fig. 11.19). This study also showed how wet and drought cycles and cactus finches affect populations of prickly pear cactus.

## Rainfall, Cactus Finches, and Cactus Reproduction

The Galápagos finches harvest a variety of foods from several species of prickly pear cactus. Two species of finches, *Geospiza scandens* and *G. conirostris,* are well-known specialists on cacti. The Grants documented several ways in which these finches make use of cacti, including (1) opening flower buds in the dry season to eat pollen, (2) consuming nectar and pollen from mature flowers, (3) eating a seed coating called the aril, (4) eating seeds, and (5) eating insects from rotting cactus pads and from underneath bark. In return, the finches disperse some cactus seeds and pollinate cactus flowers.

Finches also damage many cactus flowers, however. When they open flower buds or partially opened flowers, they snip the style and destroy the stigmas. As a consequence, the ovules of these flowers cannot be fertilized and they do not produce seeds. The Grants found that up to 78% of a popula-

tion of flowers can be damaged in this way. These activities, which take place during the wet season, may reduce the seeds available to finches during the dry season.

*Opuntia helleri,* one of the main sources of food for cactus finches on Genovesa Island, was negatively impacted by the El Niño of 1983. This El Niño damaged the cacti in three ways: (1) Many *O. helleri* simply absorbed so much water that their roots could no longer support them and they were blown over by wind; (2) *O. helleri* on sea cliffs were bathed in salt spray during the many storms that hit the island during 1983, which may have produced osmotic stress (see chapter 5); and (3) increased rainfall stimulated growth by a fast-growing vine that smothered many *O. helleri* (fig. 11.20). Though outright mortality of the cactus was not common, flower and fruit production was severely reduced for several years.

Reduced reproductive output by *O. helleri* was at least partly due to the activities of the cactus finches on Genovesa. The stigma snipping behavior of cactus finches was especially damaging during the drought years of 1984 and 1985. During normal years, stigma damage is mainly confined to the early part of the wet season from January to March. During the extremely wet 1983 season, there was very little damage to the stigmas of cactus flowers. However, during the drought years of 1984 and 1985, up to 95% of stigmas were snipped (fig. 11.21). This extensive damage to flowers helped delay recovery of flower and fruit production until 1986, when another El Niño brought heavy rains to the Galápagos Islands.

Populations of Galápagos finches and their food plants are an instructive model of how the environment can affect birth and death rates. Sometimes, as when the cactus fell because they were engorged with water during the El Niño of 1983, the effect of the physical environment is clear and direct. Sometimes, as when *G. fortis* starved in response to reduced seed supplies during the drought of 1977, the effect of the physical environment on a population is clearly mediated through a biological resource (in this case, seeds). In other cases, such as reduced fruit production by *O. helleri* on

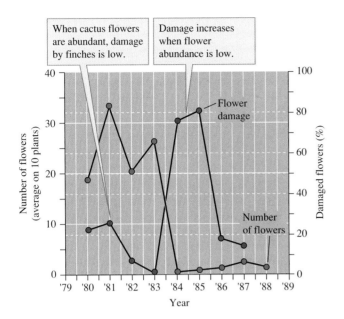

**Figure 11.21** Cactus flower abundance on Genovesa Island and extent of flower damage by large cactus finches (data from Grant and Grant 1989).

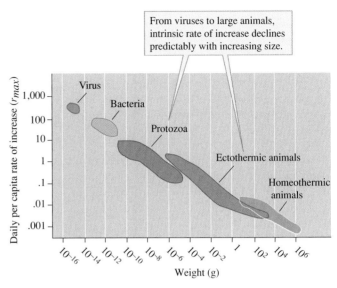

**Figure 11.22** Body size and intrinsic rate of increase (data from Fenchel 1974).

Genovesa, populations respond to a complex mixture of abiotic (drought) and biotic (damage by finches) factors that are themselves interrelated. The message to remember from these detailed studies is that both biotic and abiotic factors have important influences on birth and death rates in populations and that their effects are often tightly interconnected. In the examples just presented, environmental variation essentially changed the carrying capacity (K) of the environment for Galápagos finch populations. In section II, we have focused on how various aspects of the physical environment affect the performance of organisms, including their reproductive performance. In chapters 13, 14, and 15 of section IV we will consider at length how biological interactions affect populations. Before we do that, however, let's consider how the potential rate of population growth is related to organism size.

## CONCEPT DISCUSSION

### Small and Fast Versus Large and Slow—the Intrinsic Rate of Increase

**On average, small organisms have higher rates of per capita increase, $r_{max}$, and more variable populations, while large organisms have lower rates of per capita increase and more stable populations.**

You can think of the intrinsic rate of increase as the potential for growth by a population. As you might expect, $r_{max}$ varies a great deal among organisms. Tom Fenchel (1974) found, however, that

much of the variation can be predicted on the basis of body size. In general, $r_{max}$ decreases with increasing body size (fig. 11.22). Because it is well known that small species, such as mice, reproduce faster than larger species, such as elephants, this negative relationship between body size and $r_{max}$ may not be particularly surprising to you. However, the degree of difference in reproductive rates by large and small organisms remains impressive. The values of $r_{max}$ in figure 11.22 range from $10^{-3}$ per day for a large homeothermic animal such as the domestic cow to 300 for the virus T phage, a range of $10^5$, or 100,000, times!

It seems difficult for any human to imagine the reproductive tempo of very small organisms such as bacteria and viruses that have intrinsic rates of increase 10,000 to 100,000 times our own. However, we can use the tools of population biology to compare the tempo of population growth by small and large organisms. Let's compare population growth rate of a small marine invertebrate animal with a total mass of about 1 g to that of a whale with a mass of about 25 million g. This comparison provides an example of how intrinsic rate of increase changes with body size.

## Population Growth by Small Marine Invertebrates

Populations of the marine pelagic tunicate called *Thalia democratica* grow at exponential rates in response to phytoplankton blooms. The tunicates are members of the phylum Chordata, the same phylum to which we belong. At least as larvae, they share a number of anatomical features with other chordates, including pharyngeal gill slits, a notochord, and a dorsal hollow nerve chord. However, beyond these morphological features we have little in common with *T. democratica*.

The life cycle of tunicates combines a complex mixture of processes, including sex change from female to male, alternating sexual and asexual stages, cloning of asexually produced offspring, direct maternal nutrition of embryos, and live birth. The sexual stage in the life cycle gives birth to a single asexual offspring that is nourished as an embryo through a structure analogous to the mammalian placenta. Following its birth, the asexual stage produces 20 to 80 genetically identical sexual embryos. These embryos enter the juvenile stage at birth, when they are almost immediately fertilized internally by sperm produced by mature sexual adults that developed testes immediately after they gave birth to their own embryos. These juveniles of the sexual stage grow to adults as they nourish their own offspring and complete the life cycle.

Somehow out of this tangled life cycle comes the capacity to respond to increased food availability with explosive population growth. *T. democratica* lives in the epipelagic zone of tropical and subtropical oceans, where these tunicates make their living by filter-feeding on phytoplankton and where they are often the most abundant planktonic animal. One reason for their great abundance is that, compared to other filter-feeding animals in the same waters, they show a much more rapid numerical response (see chapter 9) to phytoplankton blooms. Following spring blooms of algae off the coast of Sydney, Australia, *T. democratica* occurs in dense swarms that feed on the abundant algae well before any other filter feeders can respond. This numerical response can increase the population size 1,000 times in just a few weeks.

The rapid numerical response cannot be due to dispersal, such as occurs in boreal forest populations of hawks and owls (see chapter 10), because *T. democratica,* like other plankton, mainly drifts with oceanic currents and has very limited swimming capacity. A. Heron (1972a, 1972b) proposed that the very rapid numerical response of *T. democratica* is due to exceptionally high reproductive rates. His laboratory studies eventually confirmed this hypothesis. Heron found that these tunicates have a generation time of only 2 days and can double their population size in less than a day. When he grew the organism in the laboratory, he observed per capita rates of increase of 0.47 to 0.91 per day and estimated that the intrinsic rate of increase, $r_{max}$, may be as high as 1.0.

Let's see if these rates of growth can account for the rapid numerical response of *T. democratica.* We'll begin with an initial population density of 1 individual per cubic meter of water and see how long it would take for the population to reach 1,000 individuals per cubic meter. With an *r* of 0.47 it would take less than 15 days for the population to grow from 1 to over 1,000 individuals per cubic meter ($N_{15} = e^{0.47(15)} = 1,153$; remember $N_0 = 1$). With an *r* of 0.91, the same increase in density would take less than 8 days ($N_8 = e^{0.91(8)} = 1,451$). Clearly, reproductive rates alone can account for the exceptionally high rate of numerical response shown by *T. democratica.* Now, because exponential growth is often associated with small organisms, let's examine a case of exponential growth by a population of whales, which are among the largest of marine organisms.

# Growth of a Whale Population

Over the past few decades, no environmental issue has attracted more public attention or caused more international tension than the exploitation and conservation of whales. The whaling controversy still rages and will likely continue for a variety of reasons, some cultural, some economic, and some technical. However, amid the controversy, the population of at least one species of whale has grown at an exponential rate.

Pacific gray whales, *Eschrichtius robustus,* are divided into two subpopulations: the western Pacific (or Korean) population and the eastern Pacific (or California) population. While the Korean population is seriously depleted, the California population has recovered from its earlier decimation by whaling. Whaling for gray whales along the coasts of Baja California and California began around 1845 and resulted in the killing of about 8,000 whales by 1874. After 1874, the population density was so low that it was not profitable to hunt the gray whales and whaling for them was largely abandoned. Still, an additional 1,000 gray whales were killed during the period from 1914 to 1946. In 1946, an international agreement protected the California gray whale against all but aboriginal subsistence whaling. This protection has given this species the opportunity to demonstrate the growth potential of a depleted whale population.

Each year, California gray whales migrate between their feeding grounds in the Bering and Chukchi Seas and their breeding grounds along the coast of Baja California. During this journey the animals swim up to 18,000 km. The bulk of the population remains on the feeding grounds from about May through October, with a few remaining scattered along the west coast of North America. After October, the whales begin migrating southward to their wintering grounds, reaching the warm waters off western Baja California and the southern Gulf of California. The females give birth to their calves in lagoons along the west coast of Baja California by January or February. On average, females give birth every other year.

D. Rice and A. Wolman (1971) estimated trends in the California gray whale population by constructing a life table and examining whales killed during whaling operations. They estimated an average annual mortality rate among gray whales of .089. By examining the reproductive condition of females killed during whaling, they calculated an annual birthrate in the gray whale population of 0.13. The difference between annual birth and death rates gives an estimate of annual per capita population growth, which in this case is $0.13 - 0.089 = .041$. This positive value of *r* indicates that the California gray whale population was growing at a rate of about 4.1% per year during the study.

Reilly, Rice, and Wolman (1983) used a different approach to estimate population trends in the same gray whale population. Their approach relied on annual counts of gray whales as they migrated southward to their wintering grounds. These yearly counts, which were made along the coast near Monterey, California, showed an annual population growth of about 2.5% from 1967 to 1980, which translates into an annual

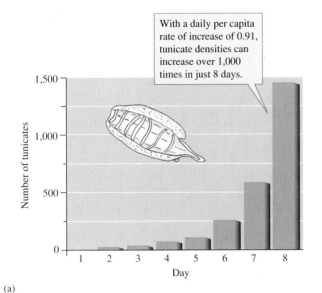

(a)

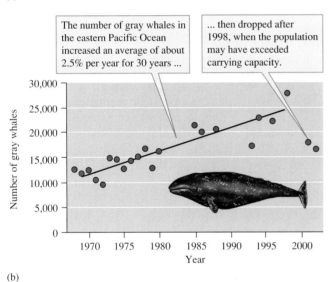

(b)

**Figure 11.23** Small and fast versus large and slow: population growth by *Thalia democratica,* and the eastern North Pacific population of the gray whale, *Eschrichtius robustus* (data from Heron 1972b, Rugh et al. 2003).

per capita rate of increase of $r = 0.025$. During the time when the researchers were studying this population, subsistence whaling by Siberian natives on the gray whale's summer feeding grounds harvested about 1.2% of the population. Summing 2.5% and 1.2% gives an estimate of potential population growth of 3.7%. It is interesting and reassuring that this estimate is very close to the earlier estimate by Rice and Wolman, which was based on survivorship and fecundity schedules.

From 1967 to 1980, the pattern of growth in the California gray whale population fit the exponential model:

$$N_t = N_0 e^{0.025t}$$

In other words, the California gray whale population grew at an exponential rate from 1967 to 1980 and continued to do so for some time (fig. 11.23). By 1993, the California gray whale

population reached its prewhaling level of about 21,000 individuals, a recovery so dramatic that the U.S. National Marine Fisheries Service recommended that the species be removed from the U.S. endangered species list.

More recent analyses by a team of researchers led by David Rugh of the U.S. National Marine Fisheries Service (Rugh et al. 2003) indicate that the gray whale population sustained an annual growth rate of about 2.5% until 1998. Rugh and his research team estimated that the gray whale population in 1998 numbered nearly 28,000. However, the population declined substantially in 2001 and 2002 (fig. 11.23). Several population ecologists have suggested that the population may have exceeded its long-term carrying capacity in 1998 and that the declines in population size of 2001 and 2002 were normal population fluctuations. Increases in the numbers of gray whale calves born in 2002 and 2003 suggest that the gray whale population is growing again.

Gray whales and *T. democratica* differ a great deal in their per capita rates of increase. Heron had estimated the daily intrinsic rate of increase of *T. democratica* as $r_m = 1.0$. For comparison, we can convert the annual per capita rate of increase for gray whales made by Reilly and colleagues to a daily rate: $r = .037/365 = 0.0001$. While this is not the intrinsic rate of increase for the gray whale, it may not be far off. The contrast between these two organisms demonstrates the wide variation in rates of increase by different species. In addition, the contrast reinforces the major point of this section: Small organisms have much higher per capita rates of increase than do large organisms. In addition, the history of the gray whale population shows that even populations of large organisms can grow exponentially. The potential for increase by large organisms is probably most dramatically demonstrated by the global human population.

## APPLICATIONS & TOOLS

### The Human Population

Most of the significant environmental problems on earth today trace their origins to the effects of the human population on the environment. Therefore, it is very important that students of ecology understand the history, current state, and projected growth of human populations. Let's use some of the conceptual tools we discussed in chapters 9 and 10 and in this chapter to review patterns of human distribution and abundance, population dynamics, and growth.

## Distribution and Abundance

One of the most distinctive features of the human population is its distribution. Our species is virtually everywhere. We occupy all the continents—even the Antarctic includes a population of scientists and support staff—and most oceanic islands. What other species, except those dependent upon humans, is so ubiquitous? Except for the Antarctic population, the current distribution of humans did not require modern technological advances. People with stone-age technology nearly reached the present limits of our distribution over 10,000 years ago. Colonization of only the most isolated oceanic islands had to await the development of sophisticated navigational techniques by the Polynesians and Europeans.

Like other populations, human populations are highly clumped at large scales (see chapter 9). In 2003, 60.6% of the global population, or about 3.8 billion people, were concentrated in Asia (fig. 11.24). In turn, most Asians live in two countries, China and India, the most populous countries on the planet. The remainder of the human population is spread across Africa (13.6%), Europe (11.6%), North America (7.9%), and South America (5.8%). The remainder (0.5%) live in Oceania (Australia and scattered oceanic islands).

Within continents, human populations attain their highest densities in eastern, southeastern, and southern Asia. Other areas of high population density include western and central Europe, northern and western Africa, and eastern and western North America. The patterns shown in figure 11.25 suggest that the highest human population densities are in coastal areas and along major river valleys.

There is even more variation in human population density if viewed on a smaller scale. Within Asia, Bangladesh has a population density of nearly 1,000 persons per square kilometer, while Mongolia has a population density of only

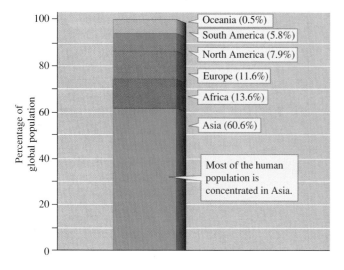

**Figure 11.24** Distribution of the human population by continent in 2003 (data from the U.S. Bureau of the Census, International Data Base).

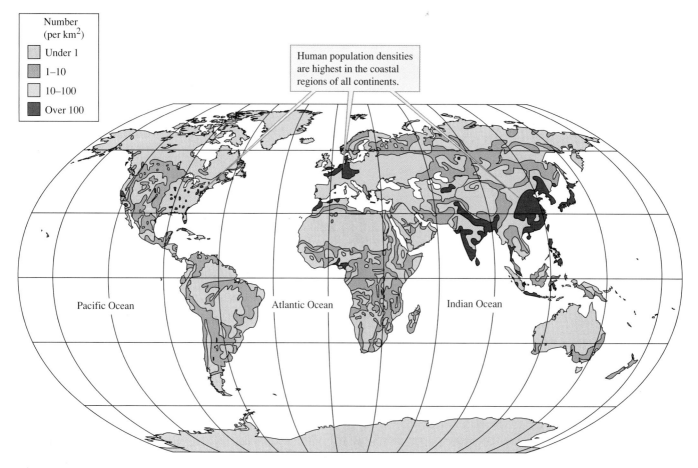

**Figure 11.25** Variation in human population density (data from the United Nations Population Information Network).

1.5 persons per square kilometer. This is slightly less than the density on the continent of Australia, which is 2.5 persons per square kilometer. Within Europe, The Netherlands harbors about 450 persons per square kilometer, while Spain and Greece have population densities of about 80 per square kilometer. In North America, the United States has an average population density of 30 per square kilometer. However, most of this population is concentrated east of the Mississippi and on the West Coast. On a state-by-state basis, population density within the United States varies from that of New Jersey, with approximately 400 people per square kilometer, to Alaska, with a density of less than 1 person per square kilometer. Meanwhile Canada has an average population density of 2.5 per square kilometer. Again, on a large scale, human populations are highly clumped and as a consequence, population density is highly variable. Population dynamics also vary a great deal.

## Population Dynamics

Population dynamics vary widely from region to region and from country to country. Let's examine the age distributions, birthrates, and death rates of three countries that have stable, declining, and rapidly growing populations. As we saw in chapter 10, population ecologists can surmise a great deal about a population by examining its age distribution. Sweden has an age distribution that is approximately the same width near its base as it is higher up (fig. 11.26). This indicates that the individuals in this population are producing just enough offspring to approximately replace losses due to death. Compare this distribution with that of Hungary. The age distribution of Hungary's population is much narrower at its base, which indicates a declining population. In contrast, the very broad base of Rwanda's age distribution indicates a rapidly growing population.

The impressions we get by examining the age distributions of these three countries are confirmed if we calculate their birth and death rates. In 2000, the annual per capita birthrate, $b$, of Sweden's population was 0.010. This exactly matched the death rate, $d$, in Sweden's population, which was 0.010. If we subtract Sweden's death rate from its birthrate (0.010 − 0.010), the result is a zero per capita rate of increase, $r$, of 0.000. In contrast, Hungary's birthrate (0.010) was lower than its death rate (0.013), which results in a per capita rate of increase, $r$, of −0.003. This negative value for $r$ confirms our impression that Hungary's population is declining. At the other end of the population dynamics spectrum, Rwanda's population has a birthrate that is nearly two times its death rate. As a consequence, this country's annual per capita rate of increase is 0.019, which is strongly positive growth. Let's move from these estimates of the present rates of change to examine the longer-term population trends in these countries.

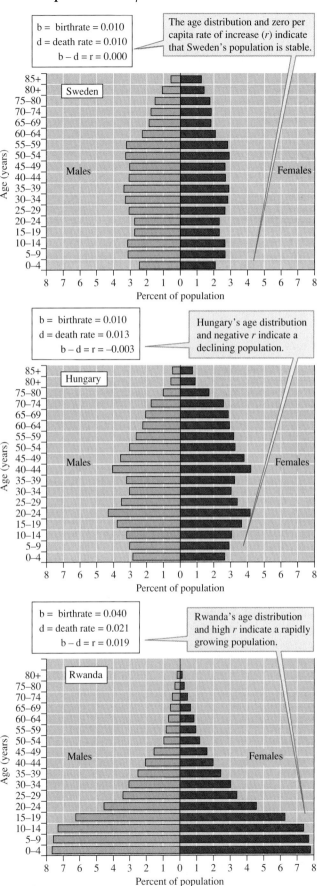

**Figure 11.26** Age distributions for human populations in countries with stable, declining, and rapidly growing populations (data from the U.S. Bureau of the Census, International Data Base 2003).

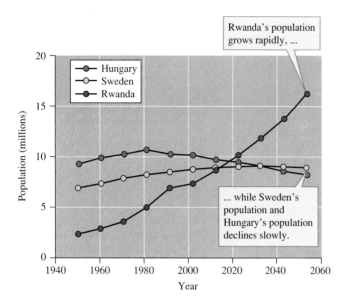

**Figure 11.27** Historical and projected human populations of countries with growing, declining, and stable populations (data from the U.S. Bureau of the Census, International Data Base).

# Population Growth

Figure 11.27 presents the historical and projected populations of Sweden, Hungary, and Rwanda. In 1950, the population of Rwanda was much smaller than either Hungary's or Sweden's population. Rwanda's population is projected to continue growing into the next century and to exceed that of Hungary and Sweden in the year 2020. Meanwhile, Sweden's population is expected to stabilize, while the population of Hungary declines.

How is the global human population changing? While the populations of many developed countries are either stable or declining, those of most developing countries are growing, and the trend for the entire global population is continued growth. While the rate of growth has begun to slow, the global population is expected to exceed 9 billion by the middle of the twenty-first century (fig. 11.28*a*).

There are signs that global population growth is slowing. While the global population continues to grow, it is not now growing exponentially. The *rate* of global population growth has declined substantially over the past 40 years, as shown in figure 11.28*b*. The size of the global population is not rising as steeply as it once was and is projected to level off sometime after the middle of the twenty-first century. Figure 11.28*b* also displays the proximate cause of this leveling off in population size, a decline in annual growth of the global population. The rate of annual growth by the global population rose steadily from 1950 to 1957 and then took a sharp dip during a major famine in China that lasted from 1958 to 1961, resulting in the deaths of an estimated 16 to 33 million Chinese. Annual growth rate, which peaked from 1962 to 1963 at 2.19%, has been decreasing in the four decades since, reaching 1.16% in 2003. The global growth rate is projected to decline to less than 0.5% by 2050. However, this is a projection based on current conditions and recent dynamics of the global and regional popula-

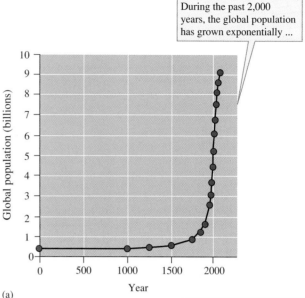

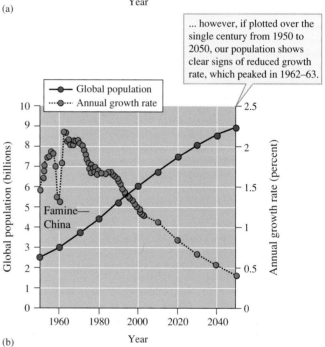

**Figure 11.28** Temporal perspectives on global population growth: (*a*) Exponential growth during the past 2000 years is evident but (*b*) the past 40 years have been a period of slowing growth by the global human population; growth is projected to continue to slow over the next half century (data from the United Nations Population Information Network and the U.S. Bureau of the Census, International Data Base 2003).

tions. Since rates of growth in human populations are currently very dynamic, projections of future global population sizes are being adjusted frequently. During the past 5 years, most of these adjustments have produced lowered estimates of future global population size. However, the cost that the present human population exacts upon the global environment is already substantial (see chapter 23). One of the greatest environmental challenges of the twenty-first century will be to establish a sustainable global population.

**In the presence of abundant resources, populations can grow at geometric or exponential rates.** Population growth by organisms with pulsed reproduction can be described by the geometric model of population growth. Population growth that occurs as a continuous process, as in human or bacterial populations, can be described by the exponential model of population growth. Examples of exponential growth from natural populations suggest that this type of growth may be very important to populations during establishment in new environments, during recovery from some form of exploitation, or during exploitation of transient, favorable conditions.

**As resources are depleted, population growth rate slows and eventually stops; this is known as logistic population growth.** As population size increases, population growth eventually slows and then ceases, producing a sigmoidal, or S-shaped, population growth curve. Population growth stops when populations reach a maximum size called the carrying capacity, the number of individuals of a particular population that the environment can support. Sigmoidal population growth can be modeled by the logistic growth equation, a modification of the exponential growth equation that includes a term for environmental resistance. In the logistic model, the rate of population growth decreases as population density increases. Research on laboratory populations indicates that zero population growth at carrying capacity may be attained by many combinations of reduced birthrates and increased death rates.

**The environment limits population growth by changing birth and death rates.** The factors affecting population size and growth include biotic factors such as food, disease, and predators and abiotic factors such as rainfall, floods, and temperature. Because the effects of biotic factors, such as disease and predation, are often influenced by population density, biotic factors are often referred to as density-dependent factors. Meanwhile, abiotic factors such as floods and extreme temperature can exert their influences independently of population density and so are often called density-

independent factors. As we have already seen, both abiotic and biotic forces have important influences on populations. The significant effects of biotic and abiotic factors on populations have been well-demonstrated by studies of Galápagos finches and their major food sources.

**On average, small organisms have higher rates of per capita increase, $r_{max}$, and more variable populations, while large organisms have lower rates of per capita increase and more stable populations.** The intrinsic rate of increase, $r_{max}$, is the maximum rate of increase for a given species. This rate of increase would occur in the absence of negative environmental influences. The per capita rate of increase, $r$, which is the realized rate of increase, is generally less than $r_{max}$. One of the best predictors of $r_{max}$ is body size. In general, $r_{max}$ decreases with increasing body size, from over 100,000 times from viruses to large vertebrate animals. Though the biological implications of this variation in intrinsic rate of increase are difficult to grasp, we can use the tools of population biology to compare the tempo of population growth by small and large organisms.

The present state of the human population can be examined using the conceptual tools of population biology discussed in chapters 9 and 10 and in chapter 11. Though humans live on every continent, their population density differs by several orders of magnitude in different regions. In 2003, 60.6% of the global population, or about 3.8 billion people, were concentrated in Asia. The remainder of the human population was spread across Africa (13.6%), Europe (11.6%), North America (7.9%), South America (5.8%), and Oceania (0.5%). Population densities in different regions vary from less than 1 person per square kilometer to nearly 1,000 persons per square kilometer. While the populations of some countries are stable, and some are declining, the global population is expected to continue growing past the year 2050. One of the greatest environmental challenges of the twenty-first century will be to establish a sustainable global human population.

# Review Questions

1. For what types of organisms is the geometric model of population growth appropriate? For what types of organisms is the exponential model of population growth appropriate? In what circumstances would a population grow exponentially? In what circumstances would a population not grow exponentially?

2. While populations of gray and blue whales have grown rapidly, the North Atlantic right whale population remains dangerously small despite many decades of complete protection. Assuming that differences in population growth rates are determined by

whale life histories and not by external factors such as pollution, what information would you need to explain the slower growth by the right whale populations? (Hint: While female gray and blue whales give birth every 2 years, right whales give birth only every 4 to 7 years.)

3. How do you build the logistic model for population growth from the exponential model? What part of the logistic growth equation produces the sigmoidal growth curve?

4. In question 3, you thought about how the logistic growth equation produces a sigmoidal growth curve. Now, let's think about nature. What is it about the natural environment that produces sigmoidal growth? Pick a real organism living in an environment with which you are familiar and list the things that might limit the growth of its population.

5. What is the relationship between per capita rate of increase, $r$, and the intrinsic rate of increase, $r_{max}$? In chapter 10, we estimated $r$ from the life tables and fecundity schedules of two species. How would you estimate $r_{max}$?

6. Both abiotic and biotic factors influence birth and death rates in populations. Make a list of abiotic and biotic factors that are potentially important regulators of natural populations.

7. Population biologists may refer to abiotic factors, such as temperature and moisture, as density-independent because such factors can affect population processes independently of local population density. At the same time, biotic factors, such as disease and competition, are called density-dependent factors because their effects may be related to local population density. Explain how abiotic factors can influence populations in a way that is independent of local population density. Explain why the influence of a biotic factor is often affected by local population density. Now, explain how the impact of an abiotic factor may also be affected by the local population density, that is, may behave at least partly as a density-dependent factor.

8. How does intrinsic rate of increase vary with body size? What did we learn in chapter 9 about the relation of body size to population density? How might the variation in intrinsic rate of increase and in population density with body size be related to the vulnerability of species to extinction?

9. Where on earth is human population density highest? Where is it lowest? Where on earth do no people live? Where are human populations growing the fastest? Where are they approximately stable?

10. What factors will determine the earth's carrying capacity for *Homo sapiens*? Explain why the earth's long-term (thousands of years) carrying capacity for the human population may be much lower than the projected population size for the year 2050. Now argue the other side. Explain how the numbers projected for 2050 might be sustained over the long term.

# Suggested Readings

Bennett, K.D. 1983. Postglacial population expansion of forest trees in Norfolk, UK. *Nature* 303:164–67.

*This short paper offers a reconstruction of population growth by trees following the end of the last glacial period.*

Dobson, F.S. and M.K. Oli. 2001. The demographic basis of population regulation in Columbian ground squirrels. *The American Naturalist* 158:236–47.

*Clear demonstration of the influence of food availability on populations of Columbian ground squirrels. This study explores the details of population regulation in the study populations.*

Gibbs, H.L. and P.R. Grant. 1987. Ecological consequences of an exceptionally strong El Niño event on Darwin's finches. *Ecology* 68:1735–46.

*The influence of El Niño on a population of Galápagos finches is clearly presented in this excellent study.*

Heron, A.C. 1972. Population ecology of a colonizing species: the pelagic tunicate *Thalia democratica* I. individual growth rate and generation time. *Oecologia* 10:269–93.

Heron, A.C. 1972. Population ecology of a colonizing species: the pelagic tunicate *Thalia democratica* II. population growth rate. *Oecologia* 10:293–312.

*Excellent studies of the population dynamics of an unusual organism.*

Hixon, M.A., S.W. Pacala, and S.A. Sandin. 2002. Population regulation: historical context and contemporary challenges of open vs. closed systems. *Ecology* 83:1490–1508.

*Careful overview of the concept of population regulation which brings the historical ideas into a contemporary focus.*

Jones, M.L., S.L. Swartz, and S. Leatherwood. 1984. *The Gray Whale Eschrichtius robustus*. New York: Academic Press.

*A collection of papers on a well-studied population. Worth reading by anyone interested in the history and conservation of whale populations.*

Perryman, W.L., M.A. Donahue, P.C. Perkins, and S.B. Reilly. 2002. Gray whale calf production 1994–2000: are observed fluctuation related to changes in seasonal ice cover? *Marine Mammal Science* 18:121–44.

*Detailed study of calf production that reports a possible environmental regulation of this much studied whale population.*

Reilly, S.B., D.W. Rice, and A.A. Wolman. 1983. Population assessment of the gray whale, *Eschrichtius robustus*, from California shore censuses, 1967–80. *Fishery Bulletin* 81:267–81.

*A detailed description of a shore-based study of the California gray whale population.*

Roman, J. and S.R. Palumbi. 2003. Whales before whaling in the North Atlantic. *Science* 301:508–10.

*A fascinating study that uses genetic variation in whale populations to estimate the population sizes of humpback, fin, and minke whales prior to intensive whaling. The controversial conclusion reached is that whale populations may have been up to 20 times larger than previously estimated.*

Snell, T.W., B.J. Dingmann, and M. Serra. 2001. Density-dependent regulation of natural and laboratory rotifer populations. *Hydrobiologia* 446/447:39–44.

Yoshinaga, T., A. Hagiwara, and K. Tsukamoto. 2001. Why do rotifer populations present a typical sigmoid growth curve? *Hydrobiologia* 446/447:99–105.

*A pair of papers that thoroughly document density-dependent regulation of r and the resulting pattern of sigmoidal or logistic population growth in natural and laboratory populations of rotifers.*

# On the Net

Visit this textbook's accompanying website at www.mhhe.com/ecology (click on the book's title) to take advantage of practice quizzing, study/writing tips, timely news articles, and additional URLs for research on the topics in this chapter.

Animal Population Ecology
Population Growth
Human Population Growth
Ecological Economics

Field Methods for Studies of Populations
Field Methods for Studies of Ecosystems
Species Abundance, Diversity, and Complexity

# Chapter

# 12

# Life Histories

ifferent species, often living side-by-side, reproduce at vastly different rates over lifetimes that may differ by several orders of magnitude. On a rare sunny day in temperate rain forest, a redwood tree, *Sequoia sempervirens*, shades a nearby stream. Bathed in fog all summer, soaked by rain during fall, winter, and spring, the redwood has lived through 2,000 annual cycles (see fig. 2.27). The tree was well established when Rome invaded Britain and had produced seeds for 500 years when William the Conqueror invaded the island from across the English Channel. It was 1,800 years old when rag tag colonials wrenched an American territorial prize from William's heirs, claiming it as their own country. Within a mere century the descendants of the colonial rebels had expanded their territory 2000 miles westward and were chopping down redwood trees for lumber. Other human populations had long lived near the base of the tree and had cut some trees. However, no population had been so relentless as the newcomers who stripped vast areas of all trees. Somehow, before all the redwoods were gone the cutting stopped and the grove of the giant redwood was protected. With luck the tree would live through several more centuries of summer fog and winter rain, during which the human order of the world would surely change many times more.

On this summer morning other life was stirring in the nearby stream. A female mayfly along with thousands of others of her species were shedding their larval exoskeletons as they transformed from their robust crawling aquatic stage to graceful flying adults (fig. 12.1). As a larva the mayfly had lived in the stream for a year, but her adult stage would last just this one day, during which she would mate, deposit her eggs in the stream, and then die. She had just this one chance to successfully complete her life cycle. For an adult mayfly, there is no tomorrow—one chance and no more. As the mayflies swarmed, some would be eaten by birds nesting in alder trees that grew along the stream, and some would be caught by bats that found roosting sites on the giant redwood. Some of the mayflies would be eaten by fish that they had successfully eluded for a year of larval life, and still others would be snared by spiders that spun their webs in azalea shrubs that grew between the alders and the redwood. However, this particular mayfly escaped all predators, mated, and laid her eggs.

Spent by the effort of depositing her eggs, the mayfly was caught by the current and washed downstream. Fifty meters from where she emerged that morning, the mayfly was taken from the surface by a trout as she floated past an old redwood log where the trout sheltered. The small splash of the feeding trout caught the attention of a man and a woman who had been studying the stream. They knew the stream well and knew the pool where the big trout lived and they knew the trout, which they had tried to catch many times. Their grandparents' generation had cut the redwood forests. Later, their parents had worked to protect this remnant grove. The man and woman had played in the grove as children and courted there as young adults. Now that their own children were grown, they fished the stream frequently, sharing the place once again.

Redwood, mayfly, fish, and humans, lives intertwined in a web of ecological relationships but vastly different in scale and timing. All four are players in an ecological and evolutionary

**Figure 12.1** Adult mayflies generally live one day only.

drama stretching into a vast past and into an unknown future. Made of the same elements and with their genetic inheritance encoded by DNA of the same basic structure, the four species have inherited vastly different lives. While the redwood has produced seeds numbering in the millions over a lifetime that has stretched for millennia, the mayfly spends a year in the stream and then emerges to lay eggs that will number in the hundreds. The trout's spawn has numbered in the thousands, deposited during the several years of her life. Meanwhile the man and woman have produced two children during their lifetime, investing time and energy into them over a period of decades.

What are the selective forces that created and maintain this vast range of biology? Under what conditions will organisms mature at an early age and small size instead of later at a larger size? What are the costs and benefits of producing millions of tiny offspring, such as the seeds of the redwood tree, versus a few that are large and well cared for? These are the sorts of questions pondered by ecologists who study **life history.** Life history consists of the adaptations of an organism that influence aspects of its biology such as the number of offspring it produces, its survival, and its size and age at reproductive maturity. Chapter 12 presents concept discussions bearing on some of the central concepts of life history ecology.

## CONCEPTS

- **Because all organisms have access to limited energy and other resources, there is a trade-off between the number and size of offspring; those that produce larger offspring are constrained to produce fewer, while those that produce smaller offspring may produce larger numbers.**

- **Where adult survival is lower, organisms begin reproducing at an earlier age and invest a greater proportion of their energy budget into reproduction; where adult survival is higher, organisms defer reproduction to a later age and allocate a smaller proportion of their resources to reproduction.**

- **The great diversity of life histories may be classified on the basis of a few population characteristics. Examples include fecundity or number of offspring, $m_x$, survival, $l_x$, and age at reproductive maturity, $\alpha$.**

## CONCEPT DISCUSSION

### Offspring Number Versus Size

**Because all organisms have access to limited energy
and other resources, there is a trade-off between
the number and size of offspring; those that
produce larger offspring are constrained
to produce fewer, while those that
produce smaller offspring may
produce larger numbers.**

The discussions of photosynthetic response by plants (see figs. 6.19, 6.20) and functional response by foraging animals (see figs. 6.21, 6.22, 6.23) led us to conclude that all organisms take in energy at a limited rate. As we saw, rate of energy of intake is limited either by conditions in the external environment, such as food availability, or by internal constraints such as the rate at which the organism can process food. These constraints led in turn to the principle of allocation. The principle of allocation underscores the fact that if an organism uses energy for one function such as growth, it reduces the amount of energy available for other functions such as reproduction. This tension between competing demands for resources leads inevitably to trade-offs between functions. One of those is the trade-off between number and size of offspring. Organisms that produce many offspring are constrained, because of energy limitation, to produce smaller offspring (seeds, eggs, or live young). Viewed from the opposite perspective, organisms that produce large, well cared for offspring are constrained to produce fewer. Let's begin our review of concept discussions bearing on this generalization with a survey of patterns among fish, a vertebrate group with especially large variation in life history characteristics.

## Egg Size and Number in Fish

Because of their great diversity (more than 20,000 existing species) and the wide variety of environments in which they live, fish offer many opportunities for studies of life history. Kirk Winemiller (1995) pointed out that fish show more variation in many life history traits than any other group of animals. For instance, the number of offspring they produce per brood, that is, their clutch size, ranges from the one or two large live young produced by mako sharks to the 600,000,000 eggs per clutch laid by the ocean sunfish. However, many variables other than offspring number and size change from sharks to sunfish. Therefore, more robust patterns of variation can be obtained by analyzing relationships within closely related species, such as within families or genera.

In a study of gene flow among populations of darters, small freshwater fish in the perch family, or Percidae, Tom

**Figure 12.2** Darters such as this male orangethroat darter form a diverse and distinctive subfamily of fishes within the perch family. They live only in North America.

Turner and Joel Trexler tried to determine the extent to which life history differences among species might influence gene flow between populations. Turner and Trexler (1998) pointed out that in such a study, it is best to focus on a group of relatively closely related organisms with a shared evolutionary history. They were particularly interested in determining the relationship between egg size and egg number, or **fecundity,** and the extent of gene flow among populations. Fecundity is simply the number of eggs or seeds produced by an organism. Turner and Trexler proposed that gene flow would be higher among populations producing more numerous smaller eggs, that is, among populations with higher fecundity.

Turner and Trexler chose the darters for their studies because they are an ideal study group. Darters are small, streamlined benthic fishes that live in rivers and streams throughout eastern and central North America. Male darters are usually strikingly colored during the breeding season (fig. 12.2). The darters consist of 174 species in three genera within the family Percidae, which makes them one of the most species-rich groups of vertebrates in North America. The most diverse genus, *Etheostoma,* alone includes approximately 135 species. However, despite the fact that the darters as a whole live in similar habitats and have similar anatomy, they vary widely in their life histories. The genera most similar to the ancestors of the darters, *Crystallaria* (one species) and *Percina* (38 species), are larger and produce more eggs than species in the genus *Etheostoma.* However, *Etheostoma* species also vary substantially in their life histories.

Turner and Trexler sampled 64 locations on streams and rivers in the Ohio, Ozark, and Ouachita Highlands regions of Ohio, Arkansas, and Missouri, the heart of freshwater fish diversity in North America, which supports one of the most diverse temperate freshwater fish faunas on earth. Of the darters they collected at these locations, they chose 15 species, 5 in the genus *Percina* and 10 *Etheostoma* species, for detailed study. Turner and Trexler chose darter species that included a wide range of variation in life history traits, especially variation in body size, number of eggs laid, and egg size.

The species in the study ranged in length from 44 to 127 mm and the number of mature eggs that they produced ranged from 49 to 397. Meanwhile, the size of eggs produced by the

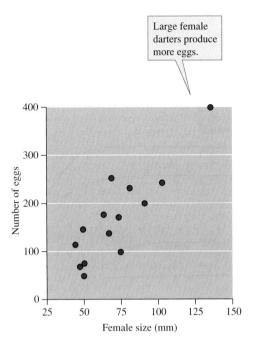

**Figure 12.3** Relationship between female darter size and number of eggs (data from Turner and Trexler 1998).

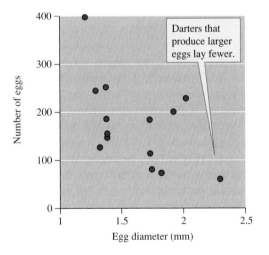

**Figure 12.4** Relationship between the size of eggs laid by darters and the number of eggs laid (data from Turner and Trexler 1998).

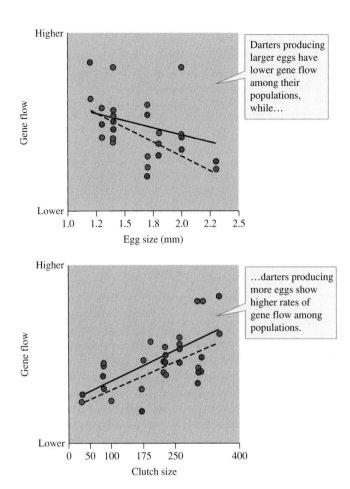

**Figure 12.5** Egg size, egg number, and gene flow among darter populations (data from Turner and Trexler 1998).

study species varied from 0.9 to 2.3 mm in diameter. As they expected, Turner and Trexler found that larger darter species produce larger numbers of eggs (fig. 12.3). Their results also support the generalization that there is a trade-off between offspring size and number. On average, darters that produce larger eggs produce fewer eggs (fig. 12.4).

Turner and Trexler characterized the genetic structure of darter populations using electrophoresis of allozymes produced by 21 different loci (see chapter 8). They chose 21 loci out of 40 that they examined because they were polymorphic. A **polymorphic locus** is one that occurs as more than one allele. In this case each allele synthesizes a different allozyme. Turner

and Trexler assessed genetic structure using allelic frequencies. Allelic frequencies were measured as the frequencies of allozymes across the 21 different study loci. Populations with similar allelic frequencies were taken as genetically similar, while those that differed in allelic frequencies were concluded to be different genetically. Gene flow was estimated by the degree of similarity in allelic frequencies between populations.

How can the number and kinds of allozymes synthesized by a series of populations be used to determine the extent of gene flow among populations? Turner and Trexler assumed that the populations differing in allelic frequencies have lower gene flow between them than populations that have similar allelic frequencies. In other words, they assumed that genetic similarity between populations is maintained by gene flow, while genetic differences arise in the absence or restriction of gene flow.

What relationship is there between egg size and number and gene flow between populations? Turner and Trexler found a negative relationship between egg size and gene flow but a strong positive relationship of gene flow with the number of eggs produced by females (fig. 12.5). That is, populations of darter species that produce many small eggs showed less difference in allelic frequencies across the study region than did populations that produce fewer larger eggs.

How do differences in egg size and number translate into differences in gene flow among populations? It turns out that the larvae of darters that hatch from larger eggs are larger when they hatch. These larger larvae begin feeding on prey that live on the streambed at an earlier age, and spend less time drifting with the water current. Consequently, larvae hatching from larger eggs disperse shorter distances and therefore carry their genes shorter distances. As a result, populations of species producing fewer larger eggs will be more isolated genetically from other populations. Because of their greater isolation, such populations will differentiate genetically more rapidly compared to populations of species that produce many smaller larvae that disperse longer distances.

Turner and Trexler's study not only provides a case history consistent with the generalization that there is a trade-off between offspring size and number, it also reveals some of the evolutionary consequences of that trade-off.

Trade-offs between offspring number and size have been found in populations of many kinds of organisms. For instance, ecologists have found parallel relationships among terrestrial plants, involving seed number and size.

## Seed Size and Number in Plants

Like fish, plants vary widely in the number of offspring they produce, ranging from those that produce many small seeds to those that produce a few large seeds (fig. 12.6). The sizes of seeds produced by plants range over 10 orders of magnitude, from the tiny seeds of orchids that weigh 0.000002 g to the giant double coconut palm with seeds that weigh up to 27,000 g. While some orchids are known to produce billions of seeds, coconut palms produce small numbers of huge seeds. At this scale it is clear that there is a trade-off between seed size and seed number and while there are complexities that must be accounted for (Harper, Lovell, and Moore 1970), botanists long ago described a negative relationship between seed size and seed number (Stevens 1932). Figure 12.7 shows the relationship between average seed mass and the number of seeds per plant among species in four families of plants, daisies (Asteraceae), grasses (Poaceae), mustards (Brassicaceae), and beans (Fabaceae). In all four families, species producing larger numbers of seeds on average produce fewer seeds.

Having documented a trade-off between seed size and number, plant ecologists searched for the mechanisms favoring many small seeds in some environments and few larger seeds in others. However, when venturing into the world of plants, the ecologist should be aware of the subtleties of plant biology, much of which can be inferred from their morphology. For instance, many characteristics of plants correlate with their **growth form** or life-form, which itself constitutes an aspect of the plant life history. Therefore, comparing seed production of orchids and coconut palms, which mixes data from a species having the growth form of an

epiphyte (the orchid) and another with the growth form of a tree (the palm), may not be a valid comparison. Such a comparison may not be valid since growth form may itself influence the number and size of seeds produced by plants.

What other aspects of plant biology might influence seed size? As we saw in chapter 10, dispersal is an important facet of the population biology of all organisms, including plants. For instance, figure 10.17 shows the history of maple and hemlock dispersal northward following glacial retreat beginning approximately 14,000 years ago. One of the notable differences shown in figure 10.17 is that maple dispersed northward much faster than hemlock. What is the source of this

**Figure 12.6** A small sample of the great diversity of seed sizes and shapes.

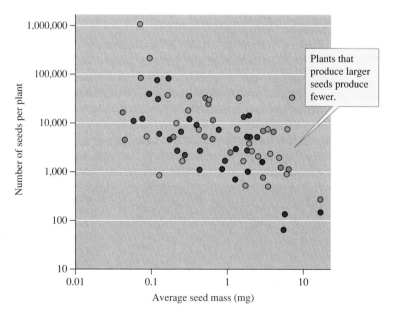

**Figure 12.7** Relationship between seed mass and seed number (data from Stevens 1932).

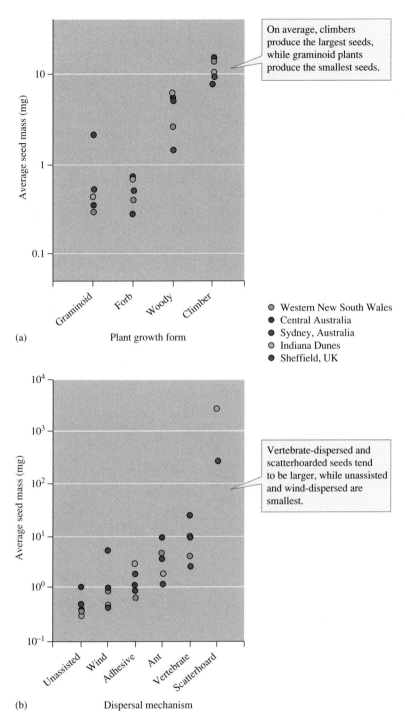

On average, climbers produce the largest seeds, while graminoid plants produce the smallest seeds.

○ Western New South Wales
● Central Australia
● Sydney, Australia
○ Indiana Dunes
● Sheffield, UK

(a)   Plant growth form

Vertebrate-dispersed and scatterhoarded seeds tend to be larger, while unassisted and wind-dispersed are smallest.

(b)   Dispersal mechanism

**Figure 12.8** Plant growth form and dispersal mechanism and seed mass (data from Westoby, Leishman, and Lord 1996).

difference in dispersal rate? Since long-distance dispersal by plants is mainly by means of seeds, we might ask whether there is a relationship between seed characteristics and mode of dispersal.

Aware of the potential influence of growth form and dispersal mode on seed characteristics, Mark Westoby, Michelle Leishman, and Janice Lord (1996), studied the relationship between plant growth form and seed size. Their study included the seeds of 196 to 641 species of plants from five different

regions. Three of their study regions were in Australia: New South Wales, central Australia, and Sydney; one was in Europe: Sheffield, United Kingdom; and one was in North America: Indiana Dunes National Lakeshore. Why did Westoby, Leishman, and Lord include five floras on three continents in their study? By including the plants on three different continents, Westoby, Leishman, and Lord increased their chances of discovering patterns of general importance. If they had worked within a single region, they could not be sure that the patterns they uncovered would hold in other regions.

Westoby, Leishman, and Lord recognized four plant growth forms. Grasses and grasslike plants, such as sedges and rushes, were classified as **graminoids.** Herbaceous plants other than graminoids were assigned to a **forb** category. Species with woody thickening of their tissues were considered as woody plants. Finally, climbing plants and vines were classified as climbers. The results showed a clear association between seed size and plant growth form (fig. 12.8*a*). In most of the floras analyzed by Westoby and his colleagues, the smallest seeds were produced by graminoid plants, followed by the seeds produced by forbs. In all five study regions, woody plants produce seeds that are far larger than those produced by either graminoids or forbs. However, the largest seeds in all regions are produced by vines. The researchers found that the seeds produced by woody plants and vines in the five floras were on average approximately 10 times the mass of seeds produced by either graminoid plants or forbs.

Westoby and his coauthors recognized six dispersal strategies. They classified seeds with no specialized structures for dispersal as unassisted dispersers. If seeds had hooks, spines, or barbs, they were classified as **adhesion-adapted.** Meanwhile, seeds with wings, hairs, or other structures that provide air resistance were assigned to a wind-dispersed category. Animal-dispersed seeds in the study included ant-dispersed, vertebrate-dispersed, and scatterhoarded. Westoby, Leishman, and Lord classified seeds with an **elaiosome,** a structure on the surface of some seeds generally containing oils attractive to ants, as ant-dispersed. Seeds with an **aril,** a fleshy covering of some seeds that attracts birds and other vertebrates, or with flesh were classified as vertebrate-dispersed. Finally they classified as **scatterhoarded** those seeds known to be gathered by mammals and stored in scattered caches or hoards.

Westoby, Leishman, and Lord also found that plants that disperse their seeds in different ways tend to produce seeds of different sizes (fig. 12.8*b*). Plants that they had classified as unassisted dispersers produced the smallest seeds, while wind-dispersed seeds were slightly larger. Adhesion-adapted seeds were of intermediate size, while animal-dispersed seeds were largest. Ant-dispersed seeds were the next largest, vertebrate-dispersed seeds were somewhat larger, and scatterhoarded were the largest by far. Westoby and his team point out that between 21% and 47% of the variation in seed size in the five floras included in their study is accounted for by a combination of growth form and mode of dispersal.

The analyses by Westoby and his colleagues show that both plant growth form and dispersal mode are associated with differences in seed size among plants. Impressively, the relationships between seed size and both growth form and dispersal mode were consistent across widely separated geographic regions. However, Westoby, Leishman, and Lord pointed out that their analysis uncovered wide variation in seed size among plants in all regions. What are the factors that maintain variation in seed size? To maintain such variation, there must be advantages and disadvantages of producing either large or small seeds. What are those advantages and disadvantages? Plants that produce small seeds can produce greater numbers of seeds. Such plants seem to have an advantage where disturbance rates are high and where plants with the capacity to colonize newly opened space appear to thrive. Though plants that produce large seeds are constrained to produce fewer, large seeds produce seedlings that survive at a higher rate in the face of environmental hazards. Those hazards include competition from established plants, shade, defoliation, nutrient shortage, deep burial in soil or litter, and drought.

Anna Jakobsson and Ove Eriksson (2000) of Stockholm University studied the relationships between seed size, seedling size, and seedling recruitment among herbs and grasses living in seminatural grasslands in southeastern Sweden. To estimate the influence of seed size and seedling size, Jakobsson and Eriksson germinated seeds in pots containing a standardized soil mix. The pots were maintained in a greenhouse under standardized conditions and seedlings were harvested and weighed 3 weeks after **germination.** Germination is the process by which seeds begin to grow or develop, producing the small plant called a seedling in the process. Why did Jakobsson and Eriksson conduct this experiment in a greenhouse? The main reason was that their ability to control environmental conditions such as soil type, moisture availability, and temperature in the greenhouse ensured that differences in seedling size would be due mainly to differences in seed size and not due to differences in the environments in which the seeds germinated. The results of this portion of the study showed clearly that larger seeds produced larger seedlings (fig. 12.9).

Jakobsson and Eriksson also investigated the relationship between seed size and recruitment among 50 plant species living in the meadows of their study region, using a field experiment. At their field sites, Jakobsson and Eriksson planted the seeds of each species in 14 small 10 × 10 cm plots. Each plot

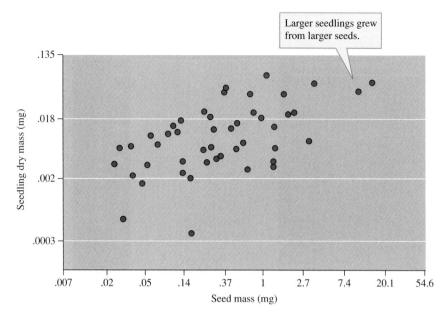

**Figure 12.9** Seed mass and seedling mass among grassland plants in Sweden (data from Jakobsson and Eriksson 2000).

was sown with 50 to 100 seeds of the study species. They left half of the study plots undisturbed, while the other plots were disturbed before planting by scratching the soil surface and removing any accumulated litter. In addition to the 14 plots where seeds were sown, Jakobsson and Eriksson established control plots where they did not plant seeds. Again, half of these were disturbed and half left undisturbed. Why did Jakobsson and Eriksson need to establish these control plots? The control plots allowed them to estimate how much germination of each species would occur in the absence of their sowing new seeds. The seeds of many species can lie dormant in soils for long periods of time and additional seeds of their study species might have dispersed into the study plots during the experiment. Therefore, without the control plots, Jakobsson and Eriksson would have no way of knowing if the seedlings they observed had grown from the seeds they had sown or from other seeds.

Of the 50 species of seeds planted, the seeds of 48 species germinated and those of 45 species established recruits. Jakobsson and Eriksson observed no recruitment of any of the study species on the control plots. Therefore they could be confident that new plants recruited into their experimental plots came from seeds that they had planted. Though plants recruited to both undisturbed and disturbed plots, the number of recruits was generally higher in disturbed plots. Further, eight species of plants recruited only on disturbed plots.

What role did differences in seed size play in the rate of recruitment by different species? Jakobsson and Eriksson calculated recruitment success in various ways. One of the most basic ways that they calculated recruitment was by dividing the total number of recruits by the total number of seeds of a species that they planted, giving the proportion of seeds sown that produced recruits. While 45 of 50 species established new recruits in the experimental plots, the rate at which they established varied widely among species from approximately

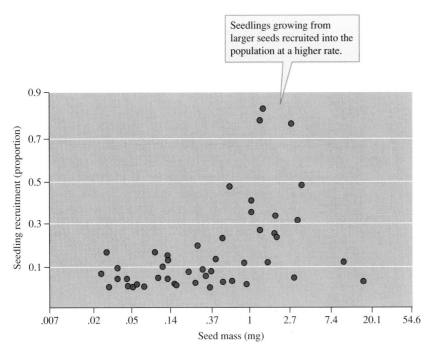

Seedlings growing from larger seeds recruited into the population at a higher rate.

**Figure 12.10** Seed mass and recruitment rates in grassland plants (data from Jakobsson and Eriksson 2000).

research team removed any fruit pulp from the seeds, washed them, and then allowed them to air dry for 24 hours. Seiwa and Kikuzawa then estimated average seed mass by weighing one to five groups of 100 to 1,000 randomly chosen seeds. A week after they collected the fruits, Seiwa and Kikuzawa planted the seeds they contained in the arboretum nursery at the Hokkaido Forest Experimental Station. They planted seeds at depths of 1 to 2 cm in a clay loam soil and watered, until the soil was saturated, three times a week.

Seiwa and Kikuzawa's results showed clearly that larger seeds produced taller seedlings (fig. 12.11). They explained this pattern as the result of the larger seeds providing greater energy reserves to boost initial seedling growth. Seiwa and Kikuzawa observed that seedlings from large-seeded species unfolded all of their leaves rapidly in the spring and shed all of their leaves synchronously in the autumn. They concluded that this timing allows the seedlings from large-seeded species to emerge early in the spring before the trees forming the canopy of the forest have expanded their leaves and have shaded the forest floor. Seiwa and Kikuzawa also pointed out that rapid growth would help seedlings penetrate the thick litter layer on the floor of deciduous forests and help them establish themselves as part of the forest understory.

5% to nearly 90%. Jakobsson and Eriksson found that differences in seed size explained much of the observed differences in recruitment success among species (fig. 12.10). On average, larger seeds, which produce larger seedlings, were associated with a higher rate of recruitment. Therefore it appears that by investing more energy into a seed, the maternal plant increases the probability that the seed will successfully establish itself as a new plant. This advantage associated with large seed size is probably very important in environments such as the grasslands studied by Jakobsson and Eriksson, where competition with established plants is likely to be high.

Jakobsson and Eriksson focused their work on grasslands where the principal growth forms were, using the classification presented in figure 12.8a, graminoid or forbs. However, as shown in figure 12.8a, woody plants and vines produce substantially larger seeds than herbaceous graminoids and forbs. How might patterns in seed and seedling size vary among woody plants? Kenji Seiwa and Kihachiro Kikuzawa (1991) studied the relationship between seed size and seedling size among tree species native to Hokkaido, the northernmost large island of Japan. The results of their work and their interpretation of the results provide clear insights into how seed size may improve the ability of seedlings to survive environmental hazards. Seiwa and Kikuzawa were especially focused on the influences of shade on seedling establishment.

The trees studied by Seiwa and Kikuzawa were all broad-leaved deciduous trees that grow in the temperate deciduous forests of Hokkaido either on mountain slopes between 100 and 200 m in altitude or in riparian forests. The fruits of all the study species were collected from trees growing in the arboretum of the Hokkaido Forest Experimental Station. In the laboratory the

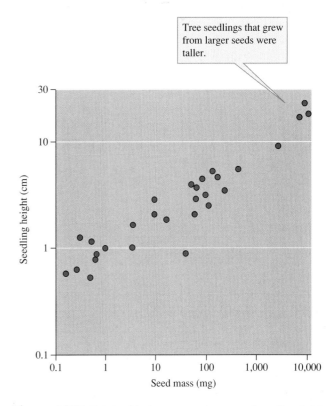

Tree seedlings that grew from larger seeds were taller.

**Figure 12.11** Relationship between seed mass and seedling height among trees (Seiwa and Kikuzawa 1991).

In addition to showing variation in the number and sizes of offspring produced, organisms also show a great deal of variation in the age at which they begin reproducing. They also differ greatly in the relative amount of energy they allocate to reproduction versus growth and maintenance. Over the years life history ecologists have observed patterns in age of reproductive maturity and relative investment in reproduction among species that support some broad generalizations.

## CONCEPT DISCUSSION

### Adult Survival and Reproductive Allocation

**Where adult survival is lower, organisms begin reproducing at an earlier age and invest a greater proportion of their energy budget into reproduction; where adult survival is higher, organisms defer reproduction to a later age and allocate a smaller proportion of their resources to reproduction.**

Is there a relationship between the probability of an organism living from one year to the next and the age at which the organism begins reproducing? What environmental factors are responsible for variation in age at maturity and the amount of energy allocated to reproduction, which has been called **reproductive effort?** (Reproductive effort is the allocation of energy, time, and other resources to the production and care of offspring.) These are two questions central to life history ecology.

Reproductive effort generally involves trade-offs with other needs of the organism, including allocation to growth and maintenance. Because of these trade-offs, allocation to reproduction may reduce the probability that an organism will survive. However, delaying reproduction also involves risk. An individual that delays reproduction runs the risk of dying before it can reproduce. Consequently, evolutionary ecologists have predicted that variation in mortality rates among adults will be in association with variation in the age of first reproduction, or age of reproductive maturity. Specifically, they have predicted that where adult mortality is higher, natural selection will favor early reproductive maturity; and where adult mortality is low, natural selection has been expected to favor delaying reproductive maturity.

## Life History Variation Among Species

The relationship between mortality, growth, and age at first reproduction or reproductive maturity has been examined in a large number of organisms. Early work, which concentrated on fish, shrimp, and sea urchins, suggested linkages between mortality or survival, growth, and reproduction. Richard Shine and Eric Charnov (1992) explored life history variation among snakes and lizards to determine whether generalizations developed through studies of fish and marine invertebrates could be extended to another group of animals living in very different environments.

Shine and Charnov began their presentation with a reminder that, in contrast to most terrestrial arthropods, birds, and mammals, including humans, many animals continue growing after they reach sexual maturity. In addition, most vertebrate species begin reproducing before they reach their maximum body size. Shine and Charnov pointed out that the energy budgets of these other vertebrate species, such as fish and reptiles, are different before and after sexual maturity. Before these organisms reach sexual maturity, energy acquired by an individual is allocated to one of two competing demands: maintenance and growth. However, after reaching sexual maturity, limited energy supplies are allocated to three functions: maintenance, growth, and reproduction. Because they have fewer demands on their limited energy supplies, individuals delaying reproduction until they are older will grow faster and reach a larger size. Because of the increase in reproductive rate associated with larger body size (see fig. 12.3), deferring reproduction would lead to a higher reproductive rate. However again, where mortality rates are high, deferring reproduction increases the probability that an individual will die before reproducing. These relationships suggest that mortality rates will play a pivotal role in determining the age at first reproduction.

Shine and Charnov gathered information from published summaries on annual adult survival and age at which females mature for several species of snakes and lizards. The annual rate of adult survival among snakes in their data set ranged from approximately 35% to 85% of the population, while age at reproductive maturity ranged from 2 to 7 years. Meanwhile, the annual rate of lizard survival ranged from approximately 8% to 67% of the population and their age at first reproduction ranged from a little less than 8 months to 6.5 years. Because most of the species they examined were North American and were members of either one family of snakes or one family of lizards, Shine and Charnov urged that their results not be generalized to snakes and lizards generally until other groups from other regions had been analyzed. Regardless of these cautions, the results of Shine and Charnov's study showed clearly that as survival of adult lizards and snakes increases, their age at maturity also increases (fig. 12.12a).

More recent analyses of the relationship between adult mortality rate and age at maturity among fish species provide additional support for the prediction that high adult survival leads to delayed maturity. Donald Gunderson (1997) explored patterns in adult survival and reproductive effort among several populations of fish. Gunderson suggested that there should be a strong relationship between adult mortality in populations and reproductive effort because some combinations of mortality and reproductive effort have a higher probability of persisting than others. For instance, a population showing a combination of high mortality and high reproductive effort would have a higher chance of persisting than one experiencing high mortality but allocating low

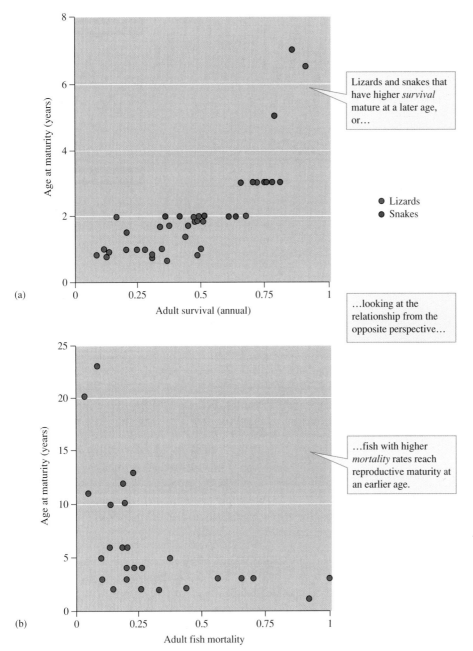

(a)

(b)

**Figure 12.12**  Relationship between (*a*) adult survival among lizards and snakes and (*b*) adult fish mortality and age of reproductive maturity (data from Shine and Charnov 1992 and Gunderson 1997).

sharks, which reproduce only every other year, was divided by 2. However, since most of the species included in the analysis spawn once per year, their ovary weights required no adjustment for GSI calculations.

The fish included in Gunderson's analysis ranged in size from the Puget Sound rockfish, which reaches a maximum size of approximately 15 cm, to northeast Arctic cod that reaches a length of 130 cm. The age at maturation among these fish species ranges from 1 year in northern anchovy populations to 23 years in dogfish shark populations. Like Shine and Charnov, Gunderson gathered information about the life histories of the fish in his analysis from previously published papers and several experts on particular fish species. In his table summarizing life history information for the 28 species included in his analysis, Gunderson lists 72 references. In contrast to Shine and Charnov, Gunderson provides estimates of mortality rates rather than survival rates. In addition his estimates are of "instantaneous" mortality rates instead of annual rates. However, like Shine and Charnov, his results show a clear relationship between adult mortality and age of reproductive maturity (fig. 12.12*b*). These results support the idea that natural selection has acted to adjust age at reproductive maturity to rates of mortality experienced by populations.

Gunderson's analysis also gives information on variation in reproductive effort among species. His calculations of a gonadosomatic index, or GSI, for each of the 28 species included in the analysis spanned more than a 30-fold difference from a value of 0.02 for the rougheye rockfish to 0.65 for the northern anchovy. What do these numbers mean? Remember that the formula for GSI is ovary weight (multiplied by 3 in the case of the northern anchovy because it spawns three times per year) divided by body weight. In other words, reproductive effort is expressed as a proportion of body weight. Converting these proportions to percentages, we can say that the yearly allocation to reproduction by the rougheye rockfish is approximately 2% of its body weight, while the northern anchovy allocates approximately 65% annually! When Gunderson plotted GSI against mortality rates (fig. 12.13), the results supported the prediction from life history theory that species with higher mortality would show higher relative reproductive effort.

reproductive effort. The population with this second combination would likely go extinct in a short period of time.

The life history information Gunderson summarized in his analysis included mortality rate, estimated maximum length, age at reproductive maturity, and reproductive effort. Gunderson estimated reproductive effort as each population's **gonadosomatic index,** or **GSI.** GSI was taken as the ovary weight of each species divided by the species body weight and adjusted for the number of batches of offspring produced by each species per year. For example, because the northern anchovy spawns three times per year, the weight of its ovary was multiplied by 3 for calculating its GSI. Meanwhile, the ovary weight for dogfish

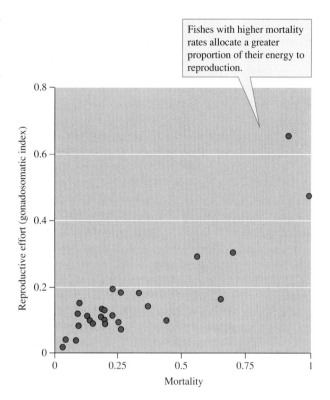

Fishes with higher mortality rates allocate a greater proportion of their energy to reproduction.

**Figure 12.13** Relationship between adult fish mortality and reproductive effort as measured by the gonadosomatic index or GSI (data from Gunderson 1997).

**Figure 12.14** Male pumpkinseed sunfish, *Lepomis gibbosus,* build their nests in the shallows of lakes and ponds. They guard their nests against intrusions by other males and attempt to attract females of their species to deposit eggs within them.

# Life History Variation Within Species

To this point in our discussion we have emphasized life history differences between species, such as the lizard and snake species compared by Shine and Charnov (fig. 12.12*a*) or the fish species compared by Gunderson (fig. 12.12*b*). Is there evidence that life history differences will evolve within species, where different populations experience different rates of adult mortality? The data set analyzed by Shine and Charnov included 9 populations of the eastern fence lizard, *Sceloporus undulatus.* Variation among those populations indicates that age at maturity within lizard populations increases with increased adult survival. Additional evidence for the evolution of such intraspecific differences comes from a comparative study of several populations of the pumpkinseed sunfish, *Lepomis gibbosus* (fig. 12.14).

Kirk Bertschy and Michael Fox (1999) studied the influence of adult survival on pumpkinseed sunfish life histories. One of the major objectives of their study was to test the prediction by life history theory that increased adult survival, relative to juvenile mortality, favors delayed maturity and reduced reproductive effort. What distinguishes this study from those of Shine and Charnov and Gunderson (fig. 12.12)? Again, Bertschy and Fox focused their attention entirely on variation in life histories among populations of *one species.* In other words, their study goal was to explain the evolution of life history variation within a species.

Bertschy and Fox selected five populations of pumpkinseed sunfish living in 5 lakes from a group of 27 lakes in southern Ontario, Canada. Fox had previously studied the pumpkinseed sunfish living in these lakes and so they had a considerable basis for choosing study populations. Bertschy and Fox chose lakes that were similar in area and depth and small enough that they had a reasonable chance of estimating mortality rates and variation in other life history characteristics. Their study lakes varied in area from 7.2 to 39.6 ha and in depth from 2.6 to 11 m. Bertschy and Fox also chose lakes that had no major inflows or outflows. Why did they restrict the study to lakes without major inflows or outflows? One reason is that they wanted to avoid as much movement of individuals in and out of their populations as possible. Such movement could obscure the results of natural selection within the lakes for particular life history characteristics.

Bertschy and Fox estimated life history characteristics from annual samples of approximately 100 pumpkinseed sunfish taken from each of the 5 study lakes. They caught the fish in their shallow (0.5–2 m depth) littoral habitat using funnel traps and beach seine nets. Bertschy and Fox took their annual population sample in late May or early June just before the beginning or right at the beginning of the spawning season. The individuals caught were sacrificed by placing them in an ice slurry and then freezing them for later analysis. They made several measurements on each individual in their samples, including their age (by counting annual rings in scales), weight (to the nearest 0.1 g), length (in mm), sex, and reproductive status. Because female reproductive effort is largely restricted to egg production while male reproductive effort includes activities such as territory guarding and nest building, Bertschy and Fox studied reproductive traits in females only. A female was considered mature if her ovaries contained eggs with yolk. The ovaries of mature females were dissected out and weighed to the nearest 0.01 g. Bertschy and Fox represented female reproductive effort using the gonadosomatic index, GSI, which they calculated as $100 \times$ (ovary mass) $\div$

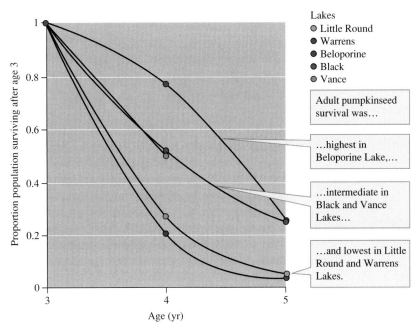

**Figure 12.15** Pumpkinseed sunfish survival after age 3 years in five small lakes (data from Bertschy and Fox 1999).

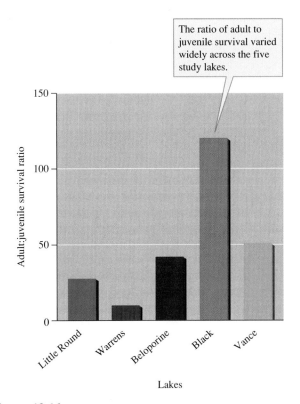

**Figure 12.16** Ratio of adult to juvenile survival in pumpkinseed sunfish populations in five small lakes (data from Bertschy and Fox 1999).

(body mass), which yields GSI values expressed as percentages rather than as proportions.

Bertschy and Fox used mark and recapture surveys (see Applications & Tools, chapter 9) to estimate the number of adult pumpkinseed sunfish and the age structure (e.g., see figure 10.9)

of pumpkinseed populations in each of the study lakes. Ages of fish were estimated from their length using the relationship between length and age of individuals of known age from each population. These surveys, which were conducted each year from 1992 to 1994, gave a basis for estimating rates of adult survival for each age class in each lake's population. The lowest rate, or probability, of adult survival was 0.19, while the highest was 0.65. In other words, the proportion of adults surviving from one year to the next ranged from approximately 1 adult out of 5 (0.19) to about 2 adults out of 3 (0.65). This variation among lakes produced striking differences in the form of survivorship curves (fig. 12.15).

Juvenile survival was estimated by counting the number of pumpkinseed nests and then collecting all the larval fish in a sample of nests. The number of nests in the study lakes varied from 60 to over 1,000 and the number of larval fish produced ranged from approximately 100,000 to over a million. Using their estimate of the number of larvae produced and the number of 3-year-old fish in the same lake, Bertschy and Fox estimated juvenile survival. Juvenile survival to adulthood in the study lakes ranged from 0.004, or about 4 out of 1,000 larvae, to 0.016, or about 16 out of 1,000 larvae. Because they were interested in the relative rates of adult and juvenile survival, Bertschy and Fox represented survival in their study lakes as the ratio of adult to juvenile survival probabilities. Figure 12.16 shows that this ratio ranged widely among study lakes from a low of 10.6 to 116.8, a tenfold difference among lakes.

Bertschy and Fox found significant variation in most life history characteristics across their study lakes. Pumpkinseed sunfish matured at ages ranging from 2.4 to 3.4 years in the different study lakes and they showed reproductive investments (gonadosomatic indexes or GSI) ranging from 6.9% to 9.3%. The relationship between survival rate and age at maturity found by Bertschy and Fox suggests that populations with higher adult survival mature at a greater age (fig. 12.17). The correlation between survival rate and age at maturity was not high enough to be statistically significant; however, the relationship between adult survival and reproductive effort was very clear and highly significant (fig. 12.18). The patterns of life history variation across the pumpkinseed populations studied by Bertschy and Fox support the theory that where adult survival is lower relative to juvenile survival, natural selection will favor allocating greater resources to reproduction.

As we explored the relationship between offspring size and number and the influence of mortality on the timing of maturation and reproductive effort, we've accumulated a large body of information on life histories. Let's step back now and try to organize that information to make it easier to think about life history variation in nature. Several researchers have proposed classification systems for life histories.

*Investigating the Evidence*

# A Statistical Test for Distribution Pattern

Suppose you are studying the life history of three species of herbaceous plants in a desert landscape. As part of that study, you are interested in determining the pattern of distribution of individuals in each population. Your hypothesis states that the individuals in each population are randomly distributed across the landscape. Your alternative hypotheses propose that individuals are either clumped or uniformly distributed. In chapter 9, we reviewed a study that suggested very different patterns of distribution in three plant populations (see p. 237). In that example, the sample means and sample variances of the three hypothetical populations of plant species A, B, and C were as follows:

| Statistic | Species A | Species B | Species C |
|---|---|---|---|
| Sample mean, $\overline{X}$ | 5.27 | 5.18 | 5.27 |
| Sample variance, $s^2$ | 5.22 | 0.36 | 44.42 |
| Ratio of sample variance to sample mean, $s^2/\overline{X}$ | 0.99 | 0.07 | 8.43 |

Also in chapter 9, we reviewed how in a randomly distributed population, variance/mean = 1, while in a regularly distributed population, variance/mean < 1, and in a clumped population, variance/mean > 1. Given these relationships, the $s^2/\overline{X}$ ratios in the above table suggest that species A has a random distribution, species B has a regular distribution, and species C has a clumped distribution. However, since the values in the table are statistical *estimates* of the true variance/mean ratios in the study populations, we need to do a statistical test to determine the significance of our results.

The first step in our test is to establish a hypothesis. In each case, our hypothesis states that the variance/mean ratio of the population equals 1. We next need to determine a significance level, which we have seen (p. 267) is generally P < 0.05. We can use a chi-square test to determine whether a sample variance/sample mean ratio is significantly different from 1 as follows:

$$\chi^2 = \frac{s^2(n-1)}{\overline{X}}$$

Here, n − 1 is the degrees of freedom, which is the sample size minus 1. In the case of our plant study, the sample size, n, is the number of sample quadrats studied for each population, which was 11 (see p. 237).

For our sample of the species A population, the calculation is:

$$\chi^2 = \frac{5.22 \times 10}{5.27}$$
$$\chi^2 = 9.905$$

For our sample of the species B population, the calculation is:

$$\chi^2 = \frac{0.36 \times 10}{5.18}$$
$$\chi^2 = 0.695$$

For our sample of the species C population, the calculation is:

$$\chi^2 = \frac{44.42 \times 10}{5.27}$$
$$\chi^2 = 84.288$$

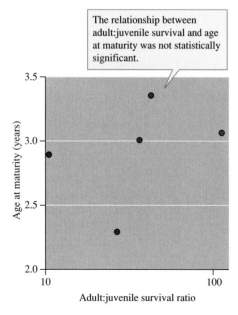

The relationship between adult:juvenile survival and age at maturity was not statistically significant.

**Figure 12.17** Adult:juvenile survival ratios and age at reproductive maturity in populations of pumpkinseed sunfish (data from Bertschy and Fox 1999).

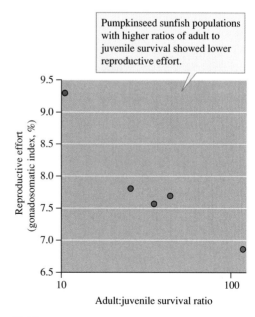

Pumpkinseed sunfish populations with higher ratios of adult to juvenile survival showed lower reproductive effort.

**Figure 12.18** Adult:juvenile survival ratio and reproductive effort as measured by the gonadosomatic index or GSI (data from Bertschy and Fox 1999).

How do we determine if these values of chi-square are statistically significant at P < 0.05? Here, we need to consider whether the $s^2/\overline{X}$ ratios are significantly *greater* than 1, *or* significantly *less* than 1. So, in contrast to the situation that we analyzed in chapter 11, we will compare our chi-square values to two critical values, one small and one large.

The situation we are considering is pictured in figure 1. With degrees of freedom of 10 (11 − 1 = 10), the critical values of chi-square are $\chi^2 < 3.247$ and $\chi^2 > 20.483$. As shown in figure 1, these values of chi-square fall on the lines that form the boundaries of the area shaded green. For values of chi-square within the green area, we accept the hypothesis that the variance/mean ratio in the population equals 1, and that the population has a random distribution. Values of chi-square in the blue zone indicate a clumped distribution, while values in the red zone indicate a regular distribution. Returning to the populations of species A, B, and C, we accept the hypothesis that the variance/mean ratio of species A ($\chi^2 = 9.905$) does not differ significantly from 1, and therefore, we conclude that it has a random distribution. Meanwhile, because the value of chi-square for species B ($\chi^2 = 0.695$) is less than the critical value of 3.247, we reject the hypothesis that the variance/mean ratio for species B equals 1, and conclude that it has a regular distribution. Using similar logic, we conclude that species C ($\chi^2 = 84.288$) has a clumped distribution.

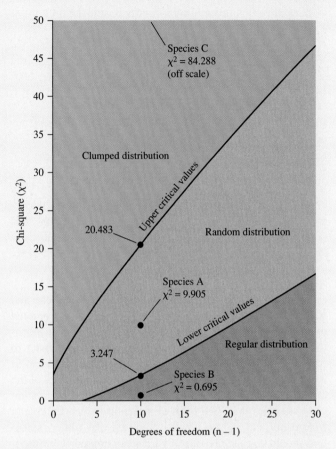

**Figure 1** Figure for determining critical values of chi-square for variance/mean ratio test of random, regular, or clumped distributions.

### Life History Classification

**The great diversity of life histories may be classified on the basis of a few population characteristics. Examples include fecundity or number of offspring, $m_x$, survival, $l_x$, relative offspring size, and age at reproductive maturity, $\alpha$.**

While classification systems never capture the full diversity of nature, they make working with the often bewildering variety of nature much easier. It is important to bear in mind when using classification systems, however, that they are an abstraction from nature and that most species fall somewhere in between the extreme types.

## r and K Selection

One of the earliest attempts to organize information on the great variety of life histories that occur among species was under the heading of *r* selection and K selection (MacArthur and Wilson 1967). The term *r* **selection,** which refers to the per capita rate of increase, *r*, which we calculated in chapter 10, was defined by Robert MacArthur and E. O. Wilson as selection favoring a higher population growth rate. MacArthur and Wilson suggested that *r* selection would be strongest in species often colonizing new or disturbed habitats. Therefore, high levels of disturbance would lead to ongoing *r* selection. MacArthur and Wilson contrasted *r* selected species with those subject mainly to K selection. The term **K selection** refers to the carrying capacity of the logistic growth equation summarized in figure 11.13. MacArthur and Wilson proposed that K selection favors more efficient utilization of resources such as food and nutrients. They envisioned that K selection would be most prominent in those situations where species populations are near carrying capacity much of the time.

## Table 12.1

**Characteristics favored by *r* versus K selection**

| Population attribute | *r* selection | K selection |
|---|---|---|
| Intrinsic rate of increase, $r_m$ | High | Low |
| Competitive ability | Not strongly favored | Highly favored |
| Development | Rapid | Slow |
| Reproduction | Early | Late |
| Body size | Small | Large |
| Reproduction | Single, semelparity | Repeated, iteroparity |
| Offspring | Many, small | Few, large |

Source: After Pianka 1970.

Eric Pianka (1970, 1972) developed the concept of *r* and K selection further in two important papers. Pianka pointed out that *r* selection and K selection are the endpoints on a continuous distribution and that most organisms are subject to forms of selection somewhere in between these extremes. In addition, he correlated *r* and K selection with attributes of the environment and of populations. He also listed the population characteristics that each form of selection favors. Following MacArthur and Wilson, Pianka predicted that while *r* selection should be characteristic of variable or unpredictable environments, fairly constant or predictable environments should create conditions for K selection. In such conditions survivorship among *r* selected species will approximate type III, while K selected species should show type I or II survivorship (see fig. 10.6). Table 12.1 summarizes Pianka's proposed contrast in population characteristics favored by *r* versus K selection.

Pianka's detailed analysis clarified the sharp contrast between the two selective extremes represented by *r* and K selection by revealing biological details. The most fundamental contrasts are of course between intrinsic rate of increase, $r_{max}$, which should be highest in *r* selected species, and competitive ability, which should be highest among K selected species. In addition, according to Pianka, development should be rapid under *r* selection and relatively slow under K selection. Meanwhile, early reproduction and smaller body size will be favored by *r* selection, while K selection favors later reproduction and larger body size. Pianka predicted that reproduction under *r* selection will tend toward a single reproductive event in which many small offspring are produced. This type of reproduction, which is called **semelparity,** occurs in organisms such as annual weeds and salmon. In contrast, K selection should favor repeated reproduction, or **iteroparity,** of fewer larger offspring. Iteroparity, which spaces out reproduction over several reproductive periods during an organism's lifetime, is the type of reproduction seen

**Figure 12.19** The deer mouse and the African elephant represent extremes among mammals of r versus K selection.

in most perennial plants and most vertebrate animals. Pianka's contrast puts a name on and fleshes out the comparison we developed in chapter 11, where we contrasted organisms that are "small and fast," analogous to *r* selected species, with ones that are "large and slow," analogous to K selected species (fig. 12.19).

The ideas of *r* and K selection helped greatly as ecologists and evolutionary biologists attempted to think more systematically about life history variation and its evolution. However, ecologists who found that the dichotomy of *r* versus K did not include a great deal of known variation in life histories have proposed alternative classifications.

## Plant Life Histories

J. P. Grime (1977, 1979) proposed that variation in environmental conditions has led to the development of distinctive strategies or life histories among plants. The two variables that he selected as most important in exerting selective pressure on plants were the intensity of disturbance and the intensity of stress. Grime contrasted four extreme environmental types which he characterized by combinations of disturbance intensity and stress intensity. Four environmental extremes

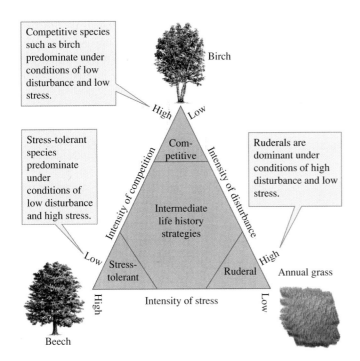

**Figure 12.20** Grime's classification of plant life history strategies (after Grime 1979).

envisioned by Grime were: (1) low disturbance–low stress, (2) low disturbance–high stress, (3) high disturbance–low stress, and (4) high disturbance–high stress. Drawing on his extensive knowledge of plant biology, Grime suggested that plants occupy three of his theoretical environments but that there is no viable strategy among plants for the fourth environmental combination, high disturbance–high stress.

Grime next described plant strategies, or life histories, that match the requirements of the remaining three environments. His strategies were ruderal, stress-tolerant, and competitive (fig. 12.20). **Ruderals** are plants that live in highly disturbed habitats and that may depend on disturbance to persist in the face of potential competition from other plants. Grime summarized several characteristics of ruderals that allow them to persist in habitats experiencing frequent and intense **disturbance,** which he defined as any mechanisms or processes that limit plants by destroying plant biomass. One of the characteristics of ruderals is their capacity to grow rapidly and produce seeds during relatively short periods between successive disturbances. This capacity alone would favor persistence of ruderals in the face of frequent disturbance. In addition, however, ruderals also invest a large proportion of their biomass in reproduction, producing large numbers of seeds that are capable of dispersing to new habitats made available by disturbance. The term ruderal is sometimes used synonymously with the term "weed." Animals that are associated with disturbance, have high reproductive rates, and are good colonists, are also sometimes referred to as ruderals.

Grime (1977) began his discussion of the second type of plant life history, stress-tolerant, with a definition of **stress** as ". . . external constraints which limit the rate of dry matter production of all or part of the vegetation." In other words, stress is induced

by environmental conditions that limit the growth of all or part of the vegetation. What environmental conditions might create such constraints? Our discussions in chapters 4, 5, and 6, where we considered temperature, water, and energy and nutrient relations, provide several suggestions. Stress is the result of extreme temperatures, high or low, extreme hydrologic conditions, too little or too much water, or too much or too little light or nutrients. Because different species are adapted to different environmental conditions, the absolute levels of light, water, temperature, and so forth that constitute stress will vary from species to species. In addition, conditions that induce stress will vary from biome to biome. For instance, the amount of precipitation leading to drought stress is different in rain forest and desert, or the minimum temperatures inducing thermal stress are different in tropical forest compared to boreal forest.

The important point that Grime made, however, was that in every biome, some species are more tolerant to the environmental extremes that occur. These are the species that he referred to as "stress-tolerant." **Stress-tolerant plants** are those that live under conditions of high stress but low disturbance. Grime proposed that, in general, stress-tolerant plants grow slowly, are evergreen, conserve fixed carbon, nutrients, and water, and are adept at exploiting temporary favorable conditions. In addition, stress-tolerant plants are often unpalatable to most herbivores. Because stress-tolerant species endure some of the most difficult conditions a particular environment has to offer, they are there to take advantage of infrequent favorable periods for growth and reproduction.

The third plant strategy proposed by Grime, the competitive strategy, is in many respects intermediate between the ruderal strategy and the stress-tolerant strategy. In Grime's classification **competitive plants** occupy environments where disturbance intensity is low and the intensity of stress is also low. Under conditions of low stress and low disturbance, plants have the potential to grow well. As they do so, however, they eventually compete with each other for resources, such as light, water, nutrients, and space. Grime's model predicts that the plants living under such circumstances will be selected for strong competitive abilities.

How does Grime's system of classification compare with the $r$ and $K$ selection contrast proposed by MacArthur and Wilson and Pianka? Grime proposed that $r$ selection corresponds to his ruderal strategy or life history, while $K$ selection corresponds to the stress-tolerant end of his classification. Meanwhile, he placed the competitive life history category in a position intermediate between the extremes represented by $r$ selection and $K$ selection. However, while attempting this reconciliation of the two classifications, Grime suggested that a linear arrangement of life histories with $r$ selection and $K$ selection occupying the extremes fails to capture the full variation shown by organisms. He suggested that more dimensions are needed and, of course, Grime's triangular arrangement (fig. 12.20) adds another dimension. The factors varying along the edges of Grime's triangle are intensity of disturbance, stress, and competition. Other ecologists have also recognized the need for more dimensions in representing life history diversity.

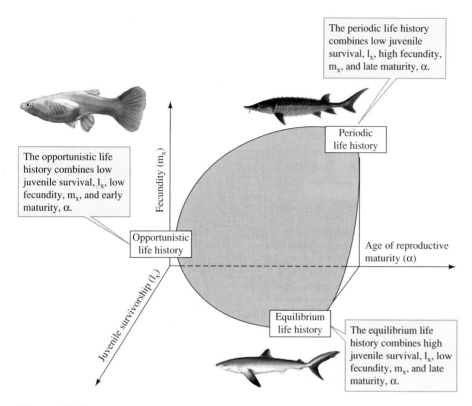

The periodic life history combines low juvenile survival, $l_x$, high fecundity, $m_x$, and late maturity, $\alpha$.

The opportunistic life history combines low juvenile survival, $l_x$, low fecundity, $m_x$, and early maturity, $\alpha$.

Opportunistic life history

Fecundity ($m_x$)

Periodic life history

Age of reproductive maturity ($\alpha$)

Juvenile survivorship ($l_x$)

Equilibrium life history

The equilibrium life history combines high juvenile survival, $l_x$, low fecundity, $m_x$, and late maturity, $\alpha$.

**Figure 12.21** Classification of life histories based on juvenile survival, $l_x$, fecundity, $m_x$, and age at reproductive maturity, $\alpha$ (after Winemiller and Rose 1992).

# Opportunistic, Equilibrium, and Periodic Life Histories

In a review of life history patterns among fish, Kirk Winemiller and Kenneth Rose (1992) proposed a classification of life histories based on some of the aspects of population dynamics that we reviewed in chapter 10. They drew particular attention to survivorship especially among juveniles, $l_x$, fecundity or number of offspring produced, $m_x$, and generation time or age at maturity, $\alpha$. Table 10.2 summarized the relationship between these variables. While the analysis by Winemiller and Rose overlaps those of Pianka and Grime, their system adds coherence to life history classification by its linkage to fundamental elements of population ecology, $l_x$, $m_x$, and $\alpha$.

Winemiller and Rose start, as we began chapter 12, with the concept of trade-offs. Their trade-offs are among fecundity, survivorship, and age at reproductive maturity. Using variation in fish life histories as a model, Winemiller and Rose proposed that life histories should lie on a semitriangular surface as shown in figure 12.21. They called the three endpoints on their surface "opportunistic," "equilibrium," and "periodic" life histories. The opportunistic strategy, by combining low juvenile survival, low numbers of offspring, and early reproductive maturity, maximizes colonizing ability across environments that vary unpredictably in time or space. It is important to keep in mind, however, that while the absolute reproductive output of opportunistic species may be low, the percentage of their energy budget allocated to reproduction is high. Winemiller and Rose's equilibrium

strategy combines high juvenile survival, low numbers of offspring, and late reproductive maturity. Finally, the periodic strategy combines low juvenile survival, high numbers of offspring, and late maturity. Among fish, periodic species tend to be large and produce numerous small offspring. By producing large numbers of offspring over a long life span, periodic species can take advantage of infrequent periods when conditions are favorable for reproduction.

It is difficult to map the exact correspondence of Winemiller and Rose's classification of life history strategies to either the $r$-K continuum of MacArthur and Wilson and Pianka or the triangular classification of plant life histories developed by Grime. For instance, opportunistic species share characteristics with $r$ selected and ruderal species. However, opportunistic species differ from the typical $r$ selected species because they tend to produce small clutches of offspring. The equilibrium strategy, which combines production of high juvenile survival, low numbers of offspring, and late reproductive maturity, approaches the characteristics of typical K selected species. Winemiller and Rose point out, however, that many fish classified as "equilibrium" are small, while typically K selected species tend toward large body size (see table 12.1). Periodic species are not captured by the linear $r$ to K selection gradient. Meanwhile the periodic and equilibrium species in Winemiller and Rose's classification share some characteristics with Grime's stress-tolerant and competitive species but differ in other characteristics.

Thus far in this review of systems for life history classification, we have focused on just three of the many that have been proposed. Even with just these three, however, translation from one classification to another is difficult. What are the sources of these differences in perspective? One of the sources is that different ecologists have worked with different groups of organisms. While MacArthur and Wilson's system was built after years of work on birds and insects, respectively, Pianka had worked mainly with lizards. Grime's classification was built on and intended for plants. Finally, the perspective of Winemiller and Rose was influenced substantially by their work with fish. Because these ecologists worked with such different groups of organisms, it is not surprising that their classifications of life histories do not overlay precisely.

However, it may be that the analysis by Winemiller and Rose has laid the foundation for a more general theory of life histories. By basing their classification system on some of the most basic aspects of population ecology, $l_x$, $m_x$, and $\alpha$, Winemiller and Rose (1992) established a common currency for representing and analyzing life history information for any

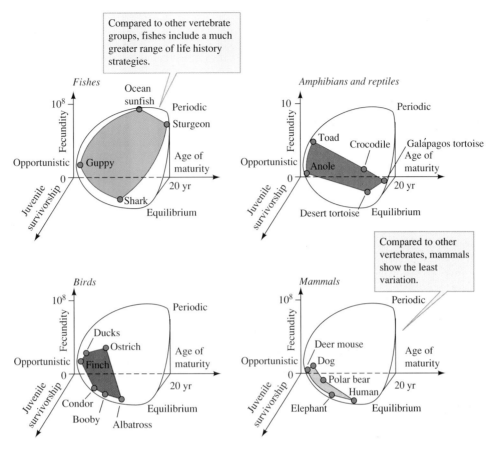

**Figure 12.22**   Variation in life histories within vertebrate animals (after Winemiller 1992).

organism. As a model for how such a translation might be done, Winemiller (1992) plotted the distributions of life history parameters of representative animal groups on their life history classification axes (fig. 12.22). By plotting life history variation among vertebrate groups on the same axes using the same variables, figure 12.22 demonstrates differences in the amount of life history variation between the groups. Notice that fish show the greatest variation and mammals the least, while birds and reptiles and amphibians include intermediate levels of variation.

# Reproductive Effort, Offspring Size, and Benefit-Cost Ratios

In response to the various attempts to classify life histories, Eric Charnov (2002) developed a new approach to life history classification. His goal was to develop a classification free of the influences of size and time that would facilitate the exploration of life history variation within and among groups of closely related taxa. Why remove the influences of size and time? Our discussion of *r* and K selection underscored the relationship between size of organisms and timing of life history features (see table 12.1). The influences of size and timing are responsible for many of the obvious life history differences among species of closely related taxa, for instance, the differences

among large and small mammal species, such as between a deer mouse and an African elephant (see fig. 12.19). By removing size and time effects, we may be able to more clearly detect life history differences among evolutionary lineages.

Charnov's approach was to take a few key life history features and convert them to dimensionless numbers. One of his variables was relative size of offspring. He created this dimensionless variable by dividing the mass of offspring at independence from the parent, I, by the average adult mass, m. The result, I/m, is the size of offspring expressed as a proportion of body mass. While it is clear that an elephant is larger at independence than is a mouse at the same life stage, Charnov's approach allows us to determine whether one is relatively larger than the other. A young, newly independent mouse may represent as large a proportion of its parent's mass as a young elephant. The second variable used by Charnov was a measure of the amount of a lifetime allocated to reproduction. He constructed this variable by dividing the average length (e.g., years) of a species' reproductive life, E, by the average length of time required to reach reproductive age, $\alpha$. Again because Charnov's index is a ratio of numbers, $E/\alpha$, with the same units, the units cancel and the ratio is dimensionless. The third measure was proportion of body adult mass allocated to reproduction per unit time, C, divided by the average adult mortality rate $E^{-1}$. This index $C/E^{-1}$, which is equivalent to $C \cdot E$, scales reproductive effort, C, by mortality

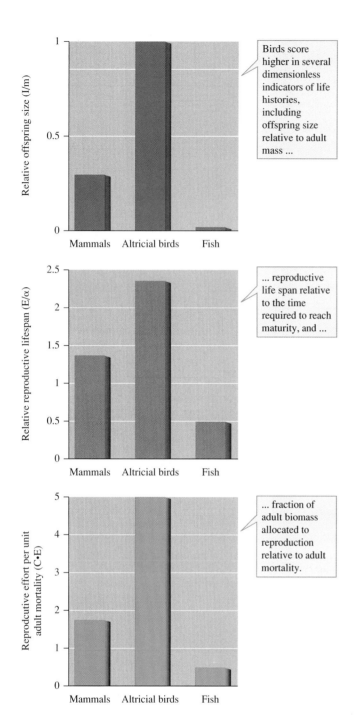

**Figure 12.23** Comparison of life history features mammals, altricial birds, and fish (data from Charnov 2002).

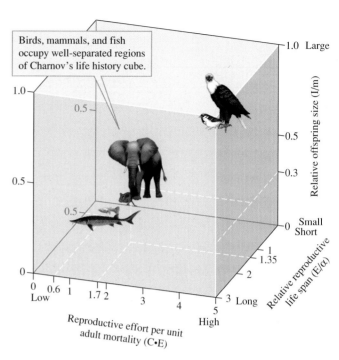

**Figure 12.24** Life history cube, a classification of fish, mammals, and altricial birds based on three dimensionless indices, indicates little variation within taxa but a great deal of difference among taxa (data from Charnov 2002).

cost. As we have seen, higher reproductive effort, a benefit, is associated with higher rates of mortality, a cost (see fig. 12.13). Charnov's C•E index is a benefit-cost ratio without dimensions.

For his initial classification of life histories, Charnov chose three groups of well-studied organisms, mammals, fish, and altricial birds. Altricial birds are those birds, ranging from sparrows to eagles, that are born helpless and depend entirely on parental care to mature to independence. One of the striking results of using Charnov's dimensionless analysis is that while there is little variation within mammals, fish, or birds,

there are substantial differences between these groups of animals. Figure 12.23 shows that I/m, E/α, and C•E are all higher among birds, intermediate among mammals, and lowest among fish. For instance, while most fish produce very tiny offspring, the average value of I/m for mammals is approximately 0.3 and for altricial birds, which raise their young to adult size, I/m = 1.

Previous classifications of life histories have revealed substantial variation within taxa, such as mammals and fish (see fig. 12.22). In contrast, Charnov's classification, by removing the influences of time and size, allows us to see the great similarities within these groups and reveals the substantial differences among them. Charnov placed values of I/m, E/α, and C•E along the edges of a cube to form what he called a "life history cube." Figure 12.24 shows the results of plotting the average values of I/m, E/α, and C•E for mammals, fish, and altricial birds with a life history cube. The striking separation of these taxa within the cube suggests that mammals, fish, and birds have life histories that are fundamentally different.

This analysis is only the beginning, however, since it raises many unanswered questions. Charnov wonders where bats will appear in his life history cube since they raise their offspring to nearly adult size. He also raises a question about precocial birds, such as pheasants and quail, whose offspring are independent at a very small size. In terms of life history, will bats be more like altricial birds, while precocial birds are more like mammals? Then there are the hundreds of thousands of vascular plants to consider.

The knowledge of species life histories revealed by the studies of life history ecologists has produced a subdiscipline

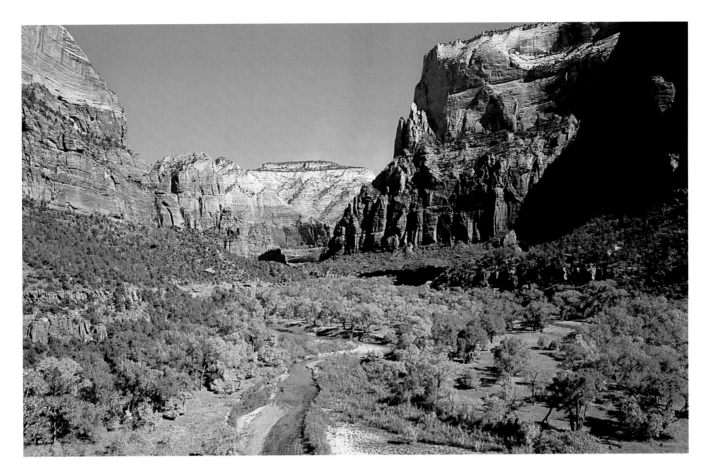

**Figure 12.25** Cottonwood riparian forest: an island of green and diversity in the semiarid landscape.

of ecology rich in both theory and biological detail. In the challenges that lie ahead as we work to conserve endangered species, both theory and detailed knowledge of the life histories of individual species will be important. For instance, life history information is playing a key role in the conservation of riparian forests across western North America.

## APPLICATIONS & TOOLS

### Using Life History Information to Restore Riparian Forests

Riparian ecosystems all over the planet are threatened by human modification of rivers. Impacts on these ecosystems come from a variety of sources, including channelizing of rivers, damming to control flooding, and diverting water for agricultural and urban uses. These modifications have greatly reduced the natural complexity of riverine landscapes and have eliminated the natural flow regime on most rivers. The effects of these modifications on landscape structure and processes within riparian zones are discussed in the Applications & Tools section of chap-

ter 21. Here we consider how these modifications impact key tree species in riparian forests and how ecologists are using their understanding of life histories to restore those forests.

As we saw in chapter 3, riparian zones are transitions between the aquatic environment of the river and the upland terrestrial environment. Because they inhabit this transition zone, riparian organisms are adapted to periodic flooding. However, many riparian species not only tolerate flooding but require it to remain healthy and complete their life cycles. Some of the organisms most dependent on the natural cycle of flooding and drying in riparian zones are the trees that form the dominant structure of riparian ecosystems.

Riparian forests support large numbers of species and high population densities of many species, particularly in arid and semiarid lands (fig. 12.25). Many species of trees inhabit riparian zones at middle latitudes and while the number and kinds of species changes from one region to another, two of the most common riparian trees are willows, *Salix* spp., and cottonwoods, *Populus* spp., both of which depend on flooding to maintain their populations. In western North America, cottonwood-willow riparian forests support a very large proportion of the diversity of the region, particularly among birds, reptiles, amphibians, and invertebrates such as butterflies and ground beetles. Riparian forests also provide critical wintering grounds for populations of large vertebrates such as elk, or wapiti, in the northern regions of the west. During the past century over 90% of riparian forests have been lost across western North America and much of the rest is threatened.

Jeff Braatne, Stewart Rood, and P. E. Heilman (1996) listed 10 major impacts of human activity on cottonwood-willow riparian forests in western North America. One of the chief threats to the riparian forests of the region results from the building of dams and subsequent flow control and diversion of water for irrigation. The negative impact of dams on cottonwoods in western North America has been well documented. Braatne, Rood, and Heilman's list of impacts of dams on cottonwoods include reduced growth by established cottonwood trees, lower cottonwood abundance, increased mortality, altered growth form, and reduced germination and seedling establishment. These impacts are mainly the result of four environmental changes induced by dams and river management: reduced water availability, reduced flooding, stabilized flows, and simplified river channel structure.

Of the many potential effects of dams and their management on riparian cottonwoods, one of the most critical is their negative impact on seed germination and seedling establishment. Without the establishment of young cottonwoods, the entire riparian forest will eventually die and the diversity of organisms that the forest supports will be lost. Can flows from dams be managed to prevent flood damage to property and loss of human life and still maintain cottonwood riparian forests? This is a question addressed by the work of John Mahoney of Alberta Environment Protection, Alberta, Canada, and Stewart Rood of the University of Lethbridge, Alberta, Canada. A key to the success of the work by Mahoney and Rood (1998) is their intimate knowledge of the life history of cottonwood trees.

As we saw in chapter 10 (see fig. 10.8), flood control on the Rio Grande in New Mexico has largely eliminated reproduction by Rio Grande cottonwood trees, *Populus deltoides* subsp. *wislizenii*. Like other cottonwood species, the Rio Grande cottonwood requires flooding to prepare a seedbed of moist bare soil in which its seeds can germinate and its young trees become established. In addition to preparing the seed bed, floods are also critical for keeping soils moist long enough for the roots of young cottonwood trees to grow into the shallow groundwater of riparian zones.

With this background in mind, Mahoney and Rood built a model for flow management by dams that would foster, rather than inhibit, cottonwood germination and establishment (fig. 12.26). The first step in their analysis was to describe the **phenology** of cottonwood trees. Phenology is the study of the relationship between climate and the timing of ecological events such as the date of arrival of migratory birds on their wintering

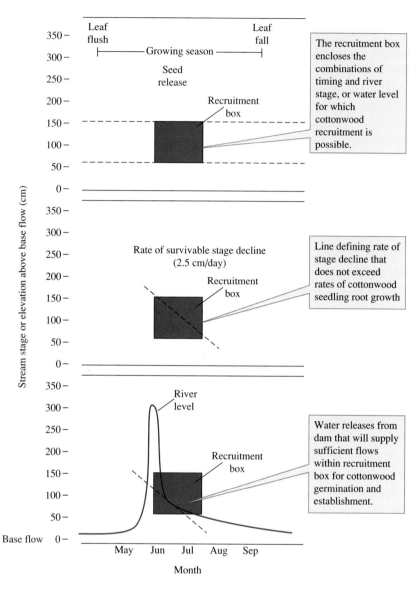

**Figure 12.26** River management to sustain cottonwood riparian forests (after Mahoney and Rood 1998).

grounds or the timing of spring plankton blooms. In the case of cottonwoods, Mahoney and Rood bounded the growing season with cottonwood leaf flush to mark the beginning of the growing season and with leaf fall to mark its end. Because their goal was to build a model that would predict the potential timing of cottonwood germination and seedling establishment, the most critical phenological event within the growing season was the timing of seed release by female cottonwood trees. The timing of river flows to foster cottonwood germination must coincide with the period when cottonwood trees are releasing their seed.

A second step in Mahoney and Rood's model building was to determine how much elevation of river level would be required to flood potential seedbeds for new cottonwood trees. Releases of water from the dam that did not flood these areas would produce no recruitment of young cottonwoods. The results of these first steps produced the "recruitment box" shown in figure 12.26.

Having established when flood flows should be released and what level or stage they should attain, Mahoney and Rood next turned to the maximum rate at which flows could be reduced by river managers and still promote cottonwood seedling establishment. They realized that germination is only one step on the way to successful seedling establishment. Because cottonwood seedlings in semiarid landscapes draw critical water from the shallow groundwater, flows must be drawn down at a rate that does not exceed the rate of seedling root growth. For young cottonwoods to successfully establish, flood waters must recede slowly enough so that the growth of the young cottonwood roots can keep pace. If flood waters fall too quickly, the roots of the young trees will not grow down to the groundwater level and they will die. Using results from their own experiments and the published results of other researchers, Mahoney and Rood proposed a survivable rate of river level, or stage, decline of 2.5 cm per day. This rate of stage decline is given by the dashed line in middle and lower panels in figure 12.26 The most important feature of this line is its slope. If the slope is much steeper than that shown in figure 12.26, the rate of groundwater-level fall will probably exceed the rate of root growth by cottonwood seedlings.

The next stage in the analysis was to specify flow releases that would be workable within existing programs of river management and produce needed cottonwood reproduction. Throughout the intermountain region of western North America where Mahoney and Rood worked on their model, peak flows occur in late spring and then gradually decline into late summer when they approach base flows. Cottonwood seed release across much of western North America extends from May into July, coinciding with peak flows across the region. Two factors were critical in their recommendations to river managers. First, ensuring that peak flows overlap with the "recruitment box" in the model and second, making sure that high spring flows recede at a rate that does not exceed the rate of seedling root growth which is represented by the slope of the dashed line in figure 12.26. Mahoney and Rood point out that to maintain this slope during drawdown, the flow peak which precedes the recruitment box must exceed the height of the recruitment box. Lower peak flows will be followed by a drop in groundwater level that is too rapid to allow cottonwood establishment.

Mahoney and Rood's recruitment box model has the potential to help conserve and sustain cottonwood riparian forest in western North America and with modification could be applied to very different riparian ecosystems. Preliminary indications are that following the recommendations of the model can lead to successful cottonwood recruitment. Managers following recommendations of the model have stimulated successful recruitment of cottonwood seedlings along the Oldman River in Alberta, Canada, and along the Truckee River near Reno, Nevada. Experiments and experiences such as these provide models for collaboration between ecologists, conservationists, and natural resource managers. However, as Mahoney and Rood point out, these exercises also improve our understanding of basic ecological processes such as basic cottonwood reproductive ecology. Improved understanding of basic ecology will in turn help any management programs directed at the species.

## SUMMARY

Life history consists of the adaptations of an organism that influence aspects of their biology such as the number of offspring it produces, its survival, and its size and age at reproductive maturity. This chapter presents concept discussions bearing on some of the central concepts of life history ecology.

**Because all organisms have access to limited energy and other resources, there is a trade-off between the number and size of offspring; those that produce larger offspring are constrained to produce fewer, while those that produce smaller offspring may produce larger numbers.** Turner and Trexler found that larger darter species produce larger numbers of eggs. Their results also support the generalization that there is a trade-off between offspring size and number. On average, darters that produce larger eggs produce few eggs. They found a strong positive relationship between gene flow among darter populations and the number of eggs produced by females and a negative relationship between egg size and gene flow. Plant ecologists have also found a negative relationship between sizes of seeds produced by plants and the number of seeds they produce. Westoby, Leishman, and Lord found that plants of different growth form and different seed dispersal mechanisms tend to produce seeds of different sizes. Larger seeds, on average, produce larger seedlings that have a higher probability of successfully recruiting, particularly in the face of environmental challenges such as shade and competition.

**Where adult survival is lower, organisms begin reproducing at an earlier age and invest a greater proportion of their energy budget into reproduction; where adult survival is higher, organisms defer reproduction to a later age and allocate a smaller proportion of their resources to reproduction.** Shine and Charnov found that as survival of adult lizards and snakes increases, their age at maturity also increases. Gunderson found analogous patterns among fish. In addition, fish with higher rates of mortality allocate a greater proportion of their biomass to reproduction. In other words, they show higher reproductive effort. These generalizations are supported by comparisons both between and within species. For instance, pumpkinseed sunfish allocate greater energy, or biomass, to reproductive effort where adult pumpkinseed survival is lower.

**The great diversity of life histories may be classified on the basis of a few population characteristics. Examples include fecundity or number of offspring, $m_x$, survival, $l_x$, relative offspring size, and age at reproductive maturity, $\alpha$.** One of the earliest attempts to organize information on the great variety of life histories that occur among species was under the heading of $r$ selection and K selection. $r$ selection refers to the per capita rate of increase, $r$, and is thought to favor higher population growth rate. $r$ selection is predicted to be strongest in disturbed habitats. K selection refers to the carrying capacity in the logistic growth equation and is envisioned as a form of natural selection favoring more efficient utilization of resources such as food and nutrients. Grime described plant strategies, or life histories, that match the requirements of three environments: (1) low disturbance–low stress, (2) low disturbance–high stress, (3) high disturbance–low stress. His plant strategies matching these environmental conditions were competitive, stress-tolerant, and ruderal. Based on life history patterns among fish, Kirk

Winemiller and Kenneth Rose proposed a classification of life histories based on survivorship especially among juveniles, $l_x$, fecundity or number of offspring produced, $m_x$, and generation time or age at maturity, $\alpha$. By basing their classification system on some of the most basic aspects of population ecology, $l_x$, $m_x$, and $\alpha$, Winemiller and Rose established a common currency for representing and analyzing life history information for any organism.

Eric Charnov developed a new approach to life history classification free of the influences of size and time that facilitates the exploration of life history variation within and among groups of closely related taxa. Charnov's classification, based on relative offspring size, I/m, relative reproductive life span, E/$\alpha$, and reproductive effort per unit adult mortality, C•E, suggests that mammals, fish, and altricial birds have life histories that are substantially different.

Life history information is playing a key role in the conservation of riparian forests across western North America. As ecologists contribute to the management of endangered populations, they also increase our understanding of basic population ecology.

# Review Questions

1. The discussion of seed size and number focused mainly on the advantages associated with large seeds. However, research by Westoby, Leishman, and Lord has revealed that the plants from widely separated geographic regions produce a wide variety of seed sizes. If this variation is to be maintained, what are some of the advantages associated with producing small seeds?

2. Under what conditions should natural selection favor production of many small offspring versus the production of a few well provisioned offspring?

3. Plant ecologists using experimental studies have verified that seedlings growing from larger seeds have a better chance of surviving environmental challenges such as deep shade, drought, physical injury, and competition from other plants. Explain how growing from larger seeds could give an advantage to seedlings facing strong environmental challenge to their establishing.

4. The studies by Shine and Charnov (1992) and Gunderson (1997) addressed important questions of concern to life history ecologists and their work provided robust answers to those questions. However, the methods they employed differed substantially from those used in most of the studies discussed in this and other chapters. The chief difference is that both relied heavily on data on life histories published previously by other authors. What was it about the nature of the problems addressed by these authors that constrained them to use this approach? In what types of studies would it be most appropriate to perform a synthesis of previously published information?

5. Much of our discussion of life history variation involved variation among species within groups as broadly defined as "fish," "plants," or "reptiles." However, the work of Bertschy and Fox

revealed significant variation in life history within species. In general, what should be the relative amount of variation within a species compared to that among many species? Develop your discussion using relative amounts of genetic variation upon which natural selection might act. You might review the sections discussing the evolutionary significance of genetic variation in chapter 8.

6. Grime's proposed classification of environments based on intensity of disturbance and stress resulted in four environments, three of which he proposed were inhabitable by plants and one of which was not. That fourth environment shows high intensity of disturbance and high stress. What sorts of life histories would an organism have to possess to live in such an environment? What kinds of real organisms can you think of that could live and perhaps thrive in such an environment? (Hint: Look for some ideas in figs. 10.11 and 11.22.)

7. Once established, Rio Grande cottonwoods can live to be well over 100 years old. However, they experience very high rates of mortality as seeds, which only germinate in conditions that occur very unpredictably in time and space. Female cottonwood trees produce about 25 million seeds annually and could produce up to two and a half billion seeds during a lifetime. Which of the life history categories that we've discussed most closely match the life history of the Rio Grande cottonwood.

8. Using what you know about the trade-off between seed number and seed size (e.g., fig. 12.7) and patterns of variation among plants, predict the relative number of seeds produced by the various plant growth forms and dispersal strategies listed on figure 12.8.

9. Apply Winemiller's model to plants. If you were to construct a strictly quantitative classification of plant life histories using Winemiller and Rose's approach, what information would you need about the plants included in your analysis? How many plant species would you need to have an idea of how variation in their life histories compares with those of animals (e.g., as in fig. 12.22)? Try to reconcile Grime's plant classification with the scheme offered by Winemiller and Rose. Where are they similar? How are they different?

10. Suppose you are a river manager and you need to operate the dams on your river in a way that fosters riparian tree growth. However, assume that the dominant trees along the river that you manage are not cottonwoods. In order to apply Mahoney and Rood's recruitment box model, what information do you need to know about your river and the riparian tree species you intend to manage? How would you go about gathering this information?

# Suggested Readings

Bertschy, K. A. and M. G. Fox. 1999. The influence of age-specific survivorship on pumpkinseed sunfish life histories. *Ecology* 80:2299–2313.

*Interesting and well-designed study of variation in life histories between populations of the same species.*

Charnov, E. L. 1993. *Life History Invariants.* Oxford: Oxford University Press.

Charnov, E. L. 2002. Reproductive effort, offspring size and benefit-cost ratios in the classification of life histories. *Evolutionary Ecology Research* 4:1–10.

Charnov, E. L., T. F. Turner, and K. O. Winemiller. 2001. Reproductive constraints and the evolution of life histories with indeterminate growth. *Proceedings of the National Academy of Sciences of the United States of America* 98:9460–64.

*This series of works by Eric Charnov and his colleagues provide thought-provoking, cutting-edge insights into the research of one of the leaders in the exploration of life history evolution.*

Grime, J. P. 1977. Evidence for the existence of three primary strategies in plants and its relevance to ecological and evolutionary theory. *American Naturalist* 111:1169–94.

Grime, J. P. 1979. *Plant Strategies and Vegetation Processes.* New York: John Wiley & Sons.

*Classical publications on plant life histories. These two publications provide a foundation for understanding plant life histories.*

Jakobsson, A. and O. Eriksson. 2000. A comparative study of seed number, seed size, seedling size and recruitment in grassland plants. *Oikos* 88:494–502.

*This paper provides an entry into modern experimental research on plant life histories.*

Mahoney, J. M. and S. B. Rood. 1998. Streamflow requirements for cottonwood seedling recruitment—an integrative model. *Wetlands* 18:634–45.

*Excellent example of how life history ecology can be applied to a significant environmental problem.*

Roff, D. A. 1993. *The Evolution of Life Histories: Theory and Analysis.* New York: Chapman & Hall.

Stearns, S. C. 1992. *The Evolution of Life Histories.* Oxford: Oxford University Press.

*Two texts that give broad introductions to the evolution of life histories.*

Tracy, C. R. 1999. Differences in body size among chuckwalla (*Sauromalus obesus*) populations. *Ecology* 80:259–71.

*A complement to the paper by Jakobsson and Eriksson that provides insights into approaches used by animal ecologists.*

# On the Net

Visit this textbook's accompanying website at www.mhhe.com/ecology (click on the book's title) to take advantage of practice quizzing, study/writing tips, timely news articles, and additional URLs for research on the topics in this chapter.

Life Histories
Animal Population Ecology

# SECTION IV

## Interactions

*"Wild bees, wasps, butterflies sip gently from yellow-white blossoms and assure your future seasons."*

Bruce Noll, *"Elaeagnus, the Russian Olive,"* 1997

# Chapter

# 13

# Competition

Careful observation and experimentation can reveal competition between species in nature. Along a coral reef off the north coast of Jamaica, threespot damselfish guard small territories of less than 1 m² (fig. 13.1). These small territories are regularly dispersed across the reef and contain most of the resources upon which the damselfish depend: nooks and crannies for shelter against predators, a carefully tended patch of fast-growing algae for food, and in the territories of males, an area of coral rubble kept clean for spawning. The damselfish constantly patrol and survey the borders of their territories, vigorously attacking any intruder that presents a threat to their eggs and developing larvae, or to their food supply. If you look carefully, however, you may find that not all members of the population have a territory. Damselfish without territories live in marginal areas around the territorial members, wandering from one part of the reef to another.

If you create a vacancy on the reef by removing one of the damselfish holding a territory, other damselfish appear within minutes to claim the vacant territory. Some of the new arrivals are threespot damselfish like the original resident, and some are cocoa damselfish, which generally live a bit higher on the reef face. These new arrivals fight fiercely for the vacated territory. The damselfish chase each other, nip each other's flanks, and slap each other with their tails. The melee ends within minutes, and life among the damselfish settles back into a kind of tense tranquility. The new resident, which may have driven off a half dozen rivals, is usually another threespot damselfish.

This example demonstrates several things. First, individual damselfish maintain possession of their territories through ongoing competition with other damselfish, and this competition takes the form of *interference competition,* which involves direct aggressive interaction between individuals. Second, though it may not appear so to the casual observer, there is a limited supply of suitable space for damselfish territories, a condition that ecologists call **resource limitation.** Third, the threespot damselfish are subject to **intraspecific competition,** competition with members of their own species, as well as **interspecific competition,** competition between individuals of two species that reduces the fitness of both. The effects of competition on the two competitors may not be equal, however. The individuals of one species may suffer greatly reduced fitness while those of the second are affected very little. The observation that threespots generally win in aggressive encounters with cocoa damselfish suggests this sort of competitive asymmetry.

Competition is not always as dramatic as fighting damselfish nor is it always resolved so quickly. In a mature white pine forest in New Hampshire, tree roots grow throughout the soil taking up nutrients and water as they provide support. In 1931, J. Toumey and R. Kienholz designed an experiment to determine whether the activities of these tree roots suppress the activities of other plants. The researchers cut a trench, 0.92 m deep, around a plot 2.74 m by 2.74 m in the middle of the forest. In so doing, they cut 825 roots, which removed potential competition by these roots for soil resources. They also established control plots on either side of the trenched plot and then watched as the results of their experiment unfolded. The experiment continued for 8

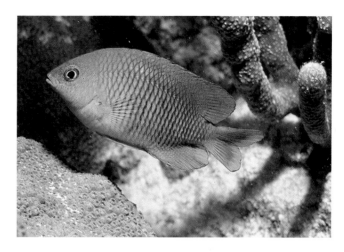

**Figure 13.1** Territorial reef fish, such as this threespot damselfish, *Eupomacentrus planifrons,* compete intensely for space.

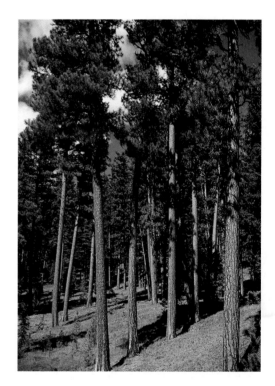

**Figure 13.2** Competition in a forest can be as intense as competition on a coral reef. However, much of the competition in a forest takes place underground, where the roots of plants compete for water and nutrients.

years, with retrenching every 2 years and over 100 roots cut each time. By retrenching, the researchers maintained their experimental treatment, suppression of potential root competition.

In the end, this 8-year experiment yielded results as dramatic as those with the damselfish. Vegetative cover on the section of forest floor that had been released from root competition was 10 times that present on the control plots. Apparently the roots of white pines exert interspecific competition for limited supplies of nutrients and water that is strong enough to suppress the growth of forest floor vegetation (fig. 13.2). Competition involving the use of such limited resources is called **resource**

**competition.** In addition, the growth of young white pines was also much greater within the trenched plots than in the control plots. Therefore, considerable intraspecific competition also occurred on the forest floor.

Ecologists have long thought that both interspecific and intraspecific competition are pervasive in nature. For instance, Darwin thought that interspecific competition was an important source of natural selection. While ecologists have shown that interspecific competition substantially influences the distribution and abundance of many species, they have also questioned the assumption that competition is an all-important organizer of nature. Such questioning has stimulated more careful research and more rigorous testing of the influence of competition on populations, and while this testing continues, sufficient evidence has accumulated to make some tentative generalizations.

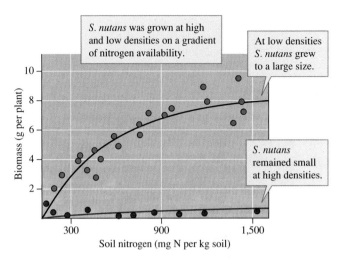

**Figure 13.3** Population density, soil nitrogen, and the size attained by the grass *Sorghastrum nutans* (data from Tilman and Cowan 1989).

- **Studies of intraspecific competition provide evidence for resource limitation.**
- **The niche reflects the environmental requirements of species.**
- **Mathematical and laboratory models provide a theoretical foundation for studying competitive interactions in nature.**
- **Competition can have significant ecological and evolutionary influences on the niches of species.**

## CONCEPT DISCUSSION

### Resource Competition

**Studies of intraspecific competition provide evidence for resource limitation.**

In chapter 11, we saw that slowing population growth at high densities produces a sigmoidal, or S-shaped, pattern in which population size levels off at carrying capacity. Our assumption in that discussion was that intraspecific competition for limited resources plays a key role in slowing population growth at higher densities. The effect of intraspecific competition is included in the model of logistic population growth. If competition is an important and common phenomenon in nature, then we should be able to observe it among individuals of the same species, individuals with identical or very similar resource requirements. Thus we begin our discussion of competition with intraspecific competition.

## Intraspecific Competition Among Herbaceous Plants

In chapter 6, we reviewed experiments by David Tilman and M. Cowan (1989) that showed how plants alter root:shoot ratios in response to availability of soil nitrogen. The plants in these experiments reduced their allocation to roots as soil nitrogen concentration increased. The experiments also included evidence for intraspecific competition. Tilman and Cowan grew the grass *Sorghastrum nutans* at low density (7 plants per pot) and high density (100 plants per pot). The results showed that the root:shoot ratios are higher when the plants are grown at high density, suggesting that competition for nutrients was more intense under these conditions.

The results of Tilman and Cowan's experiments also show that soil nitrogen concentration and population density substantially influence growth rates and individual plant weight. For example, the weight of *S. nutans* increased with increased soil nitrogen (fig. 13.3). Therefore, we can conclude that both these responses were limited by nitrogen availability at the lower concentrations in the experiment. Now compare the growth rates and plant weights shown by plants grown at low and high densities. How are they different? Both growth rate and plant weight are higher in the low-density populations, and we can conclude that competition for nutrients (resources) is more intense at the higher plant population density. Such competition for limited resources in natural populations usually leads to mortality among the competing plants.

### Self-Thinning in Plant Populations

The development of a stand of plants from the seedling stage to mature individuals suggests competition for limited resources. Each spring as the seeds of annual plants germinate, their population density often numbers in the thousands per square meter. However, as the season progresses and individual plants grow, population density declines. This same pattern occurs in the

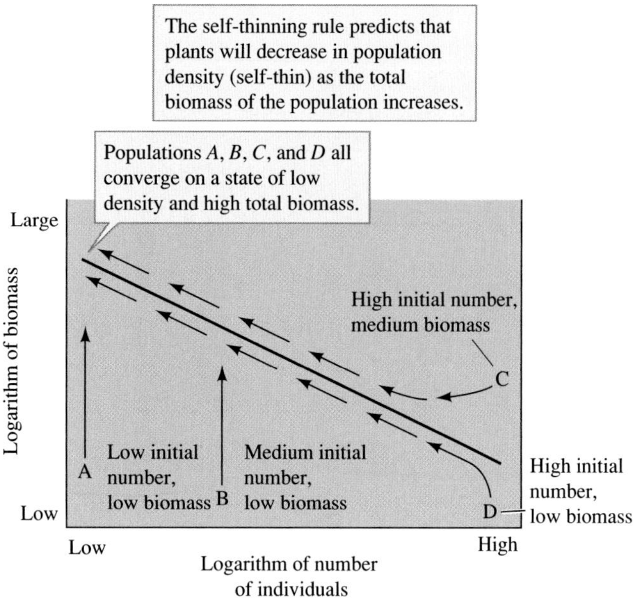

> The self-thinning rule predicts that plants will decrease in population density (self-thin) as the total biomass of the population increases.

> Populations *A*, *B*, *C*, and *D* all converge on a state of low density and high total biomass.

Figure 13.4 shows: Logarithm of biomass (Large to Low) on the y-axis and Logarithm of number of individuals (Low to High) on the x-axis.

- High initial number, medium biomass — C
- Low initial number, low biomass — A
- Medium initial number, low biomass — B
- High initial number, low biomass — D

**Figure 13.4** Self-thinning in plant populations (data from Westoby 1984).

> As plantings of alfalfa, *Medicago sativa*, grew, mortality thinned the stands as surviving plants reached larger size.

> *M. sativa* population at end of the experiment consisted of larger plants growing at lower density.

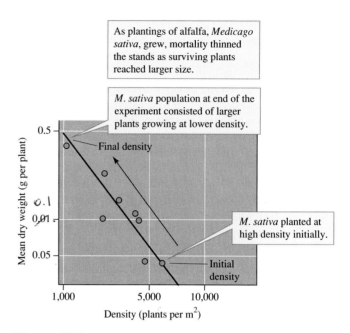

Figure 13.5 shows: Mean dry weight (g per plant) on the y-axis (0.5, 0.1, 0.01, 0.05) and Density (plants per m²) on the x-axis (1,000, 5,000, 10,000).

- Final density
- *M. sativa* planted at high density initially.
- Initial density

**Figure 13.5** Self-thinning in populations of alfalfa, *Medicago sativa* (data from White and Harper 1970).

development of a stand of trees. As the stand of trees develops, more and more biomass is composed of fewer and fewer individuals. This process is called **self-thinning.**

Self-thinning appears to result from intraspecific competition for limited resources. As a local population of plants develops, individual plants take up increasing quantities of nutrients, water, and space for which some individuals compete more successfully. The losers in this competition for resources die, and population density decreases, or "thins," as a consequence. Over time the population is composed of fewer and fewer large individuals.

One way to represent the self-thinning process is to plot total plant biomass against population density. If we plot the logarithm of plant biomass against the logarithm of plant density, the slope of the resulting line averages around $-\frac{1}{2}$. In other words, there is an approximately one-unit increase in total plant biomass with each two-unit decrease in population density; plant population density declines more rapidly than biomass increases (fig. 13.4).

Another way to represent the self-thinning process is to plot the average weight of individual plants in a stand against density (fig. 13.5). The slope of the line in such plots averages around $-\frac{3}{2}$. Because self-thinning by many species of plants comes close to a $-\frac{3}{2}$ relationship, this relationship has come to be called the $-\frac{3}{2}$ **self-thinning rule.** The $-\frac{3}{2}$ self-thinning rule was first proposed by K. Yoda and colleagues (1963) and amplified by White and Harper (1970), who provided many additional examples (e.g., fig. 13.5). Subsequently, the self-thinning rule became widely accepted among ecologists.

Recent analyses have shown that self-thinning in some plant populations deviates significantly from the $-\frac{3}{2}$ (or $-\frac{1}{2}$ for biomass-numbers) slope. However, regardless of the precise trajectory followed by different plant populations, self-thinning of plant populations has been demonstrated repeat-

edly. The important point, from the perspective of our present discussion, is that self-thinning occurs and appears to be the consequence of intraspecific competition for limited resources. Resource limitation has also been demonstrated in experiments on intraspecific competition within animal populations.

## Intraspecific Competition Among Planthoppers

Ecologists have often failed to demonstrate that insects, particularly herbivorous insects, compete. However, one group of insects in which competition has been repeatedly demonstrated are the Homoptera, including the leafhoppers, planthoppers, and aphids. Robert Denno and George Roderick (1992), who studied interactions among planthoppers (Homoptera, Delphacidae), attribute the prevalence of competition among the Homoptera to their habit of aggregating, to rapid population growth, and to the mobile nature of their food supply, plant fluids.

Denno and Roderick demonstrated intraspecific competition within populations of the planthopper *Prokelesisia marginata,* which lives on the salt marsh grass *Spartina alterniflora* along the Atlantic and Gulf coasts of the United States. The population density of *P. marginata* was controlled by enclosing the insects with *Spartina* seedlings at densities of 3, 11, and 40 leafhoppers per cage, densities that are within the range at which they live in nature. At the highest density, *P. marginata* showed reduced survivorship, decreased body length, and increased developmental time (fig. 13.6). These signs of intraspecific competition were probably the result of reduced food quality at high leafhopper densities. Plants heavily populated by planthoppers show reduced concentrations of protein, chlorophyll, and moisture. Therefore, competition between these leafhoppers was probably the result of

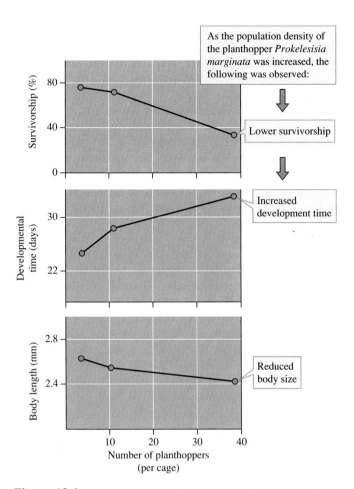

**Figure 13.6** Population density and planthopper performance (data from Denno and Roderick 1992).

limited resource supplies. However, as demonstrated in the following example, intraspecific interference competition may occur in the absence of obvious resource limitation.

# Interference Competition Among Terrestrial Isopods

Edwin Grosholz (1992) used a field experiment to study the effects of a wide range of biotic interactions on the population biology of the terrestrial isopod *Porcellio scaber.* This organism, which is associated with human activities such as farming and gardening and is found throughout the world, sometimes lives at densities in excess of 2,000 individuals per square meter. Such high densities suggest a strong potential for intraspecific competition.

Grosholz conducted his experiments on an outdoor grid of 48, 0.36 m² plots enclosed by aluminum flashing. To control isopod movements, he buried the flashing 12.5 cm into the soil and extended it 12.5 cm above the soil surface. Two experimental treatments were used: (1) to test for food limitation, the food within the enclosures was supplemented by adding sliced carrots and potatoes, and (2) to test for density effects, study plots were stocked with either 100 or 50 *P. scaber.* Supplementing food had no effect on survival by *P. scaber.* However, survival

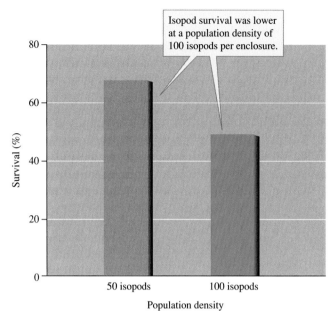

**Figure 13.7** Population density and survival in populations of a terrestrial isopod, *Porcellio scaber* (data from Grosholz 1992).

was lower at the higher population density (fig. 13.7). Grosholz attributed lower survival at the higher density to cannibalism, a common occurrence in terrestrial isopods.

Do you think increasing the population densities in the experiment might have changed the results? Since densities in nature sometimes exceed 2,000 per square meter, food limitation might not be observed until such densities are approached. How might the interpretation of the experiment have been altered if other indicators of competition, such as growth rate, size, and reproductive rate, were measured? While food supplements did not affect survival, these other unmeasured attributes may have been affected. Despite these limitations, the study offers interesting insights into the role that interference may play in intraspecific competition, even in the absence of obvious resource limitation.

As we move from discussions of intraspecific to interspecific competition, we need to back up a bit and consider how we might portray the environmental requirements of species. We do this because interspecific competition usually occurs among species with similar environmental requirements, that is, among species with similar niches.

# Niches

### The niche reflects the environmental requirements of species.

The word *niche* has been in use a long time. Its earliest and most basic meaning was that of a recessed place in a wall where one could set or display items. For about a century, however, ecologists have given a broader meaning to the word. To the ecologist, the **niche** summarizes the environmental factors that influence the growth, survival, and reproduction of a species. In other words, a species' niche consists of all the factors necessary for its existence—approximately when, where, and how a species makes its living.

The niche concept was developed independently by Joseph Grinnell (1917, 1924) and Charles Elton (1927), who used the term *niche* in slightly different ways. In his early writings, Grinnell's ideas of the niche centered around the influences of the physical environment, while Elton's earliest concept included biological interactions as well as abiotic factors. However their thinking and emphasis may have differed, it is clear that the views of these two researchers had much in common and that our present concept of the niche rests squarely on their pioneering work.

The niche concept was developed over a period of several decades; however, it was within the context of interspecific competition that the importance of the niche concept was fully realized. It was the work of G. F. Gause (1934), whose principal interest was interspecific competition, that ensured a prominent place for the niche concept in modern ecology. Particularly important was Gause's **competitive exclusion principle,** which states that two species with identical niches cannot coexist indefinitely. Gause experimented with competition in the laboratory and obtained results indicating that when two species compete, one will be a more effective competitor for limited resources, that is, will be more effective at converting resources into offspring. As a consequence, the more effective competitor will have higher fitness (higher reproductive success) and eventually excludes all individuals of the second species. The competitive exclusion principle set the niche concept in a broader context. After Gause, describing the niches of species was no longer an end in itself but a stepping-stone to understanding interactions between species—a potential key to understanding the organization of nature.

Though the work of Gause played a central role in the development of the niche concept, a rigorous definition of the niche awaited later ecologists. We can now point to a single paper authored by G. Evelyn Hutchinson (1957) as the agent that crystallized the niche concept and stimulated the work of an entire generation of ecologists. In this seminal paper titled simply, "Concluding Remarks," Hutchinson defined the niche as an *n-dimensional hypervolume,* where *n* equals the number of environmental factors important to the survival and reproduction by a species. Hutchinson called this hypervolume, which specifies the values of the *n* environmental factors permitting a species to survive and reproduce, as the **fundamental niche** of the species. The fundamental niche defines the physical conditions under which a species might live, in the absence of interactions with other species. However, Hutchinson recognized that interactions such as competition may restrict the environments in which a species may live and referred to these more restricted conditions as the **realized niche.** While Hutchinson was particularly concerned with the influence of competition on the realized niche, later authors have pointed out that other interactions such as pre-

dation, disease, and parasitism may also be important in restricting the distribution of species.

In a single word, *niche* captures most of what we discussed in sections II and III, where we considered how environment affects the growth, survival, reproduction, distribution, and abundance of species. So, why introduce the niche concept here? The reason is that we, like the first ecologists to use the term, need a concept that represents all the environmental requirements of a species. The niche concept carries us beyond the details of individual species' requirements to a position where we can more easily consider the ecology of interactions between species, interactions such as competition, predation, and mutualism.

Do you think it's possible to completely describe Hutchinson's n-dimensional hypervolume niche for any species? Probably not, since there are so many environmental factors that potentially influence survival and reproduction. Fortunately, it appears that niches are often determined mostly by a few environmental factors and so ecologists are able to apply a simplified version of Hutchinson's comprehensive niche concept. In studies of animals, ecologists have frequently described niches in terms of their feeding biology.

# The Feeding Niches of Galápagos Finches

As we saw in chapter 10, availability of suitable food significantly affects the survival and reproduction of Galápagos finches. In other words, food has a major influence on the niches of Galápagos finches. Because the kinds of food used by birds is largely reflected by the form of their beaks, Peter Grant (1986) and his colleagues were able to represent the feeding niches of Galápagos finches by measuring their beak morphology. For instance, differences in beak size among small, medium, and large ground finches translate directly into differences in diet. The large ground finch, *Geospiza magnirostris,* eats larger seeds; the medium ground finch, *G. fortis,* eats medium-sized seeds; while the small ground finch, *G. fuliginosa,* eats small seeds (fig. 13.8).

The size of seeds that can be eaten by Galápagos finches can be estimated by simply measuring the depths of their beaks. Studies of seed use by *G. fortis* on Daphne Major showed clearly that even within species, beak size affects the composition of the diet. Within this population, individuals with the deepest beaks fed on the hardest seeds, while individuals with the smallest beaks fed on the softest seeds (fig. 13.9).

The importance of beak size to seed use was also demonstrated by the effects of the 1977 drought on the *G. fortis* population of Daphne Major. In chapter 11, we saw how this drought caused substantial mortality in this population (see fig. 11.17). However, this mortality did not fall equally on all segments of the population. As seeds were depleted, the birds ate the smallest and softest seeds first, leaving the largest and toughest seeds (fig. 13.10). In other words, following the drought not only

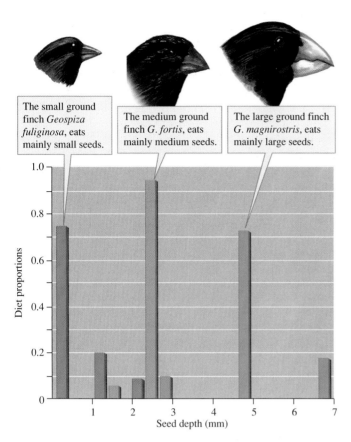

The small ground finch *Geospiza fuliginosa*, eats mainly small seeds.

The medium ground finch *G. fortis*, eats mainly medium seeds.

The large ground finch *G. magnirostris*, eats mainly large seeds.

**Figure 13.8** Relationship between body size and seed size in Galápagos finch species (data from Grant 1986).

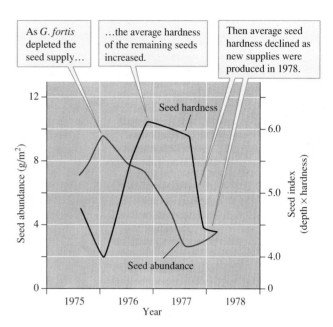

As *G. fortis* depleted the seed supply...

...the average hardness of the remaining seeds increased.

Then average seed hardness declined as new supplies were produced in 1978.

Seed hardness

Seed abundance

**Figure 13.10** Seed depletion by the medium ground finch, *Geospiza fortis,* and average seed hardness (data from Grant 1986).

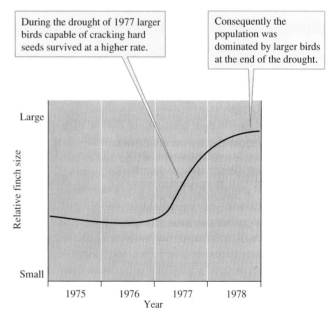

During the drought of 1977 larger birds capable of cracking hard seeds survived at a higher rate.

Consequently the population was dominated by larger birds at the end of the drought.

**Figure 13.11** Selection for larger size among medium ground finches, *Geospiza fortis,* during a drought on the island of Daphne Major (data from Grant 1986).

were seeds in short supply, the remaining seeds were also tougher to crack. Because they could not crack the remaining seeds, mortality fell most heavily on smaller birds with smaller beaks. Consequently, at the end of the drought, the *G. fortis* population on Daphne Major was dominated by larger individuals that had survived by feeding on hard seeds (fig. 13.11).

These studies show that beak size provides significant insights into the feeding biology of Galápagos ground finches. Since food is the major determinant of survival and reproduction among these birds, beak morphology gives us a very good

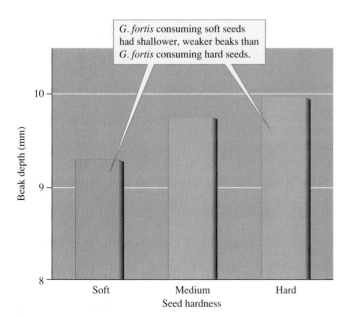

*G. fortis* consuming soft seeds had shallower, weaker beaks than *G. fortis* consuming hard seeds.

**Figure 13.9** Relationship between the hardness of seeds eaten by medium ground finches, *Geospiza fortis,* and beak depth (data from Boag and Grant 1984b).

**Figure 13.12** The salt marsh grass *Spartina anglica,* originated on the coast of England as a hybrid of European and North American species of salt marsh grasses and has since spread to salt marshes in many parts of the world.

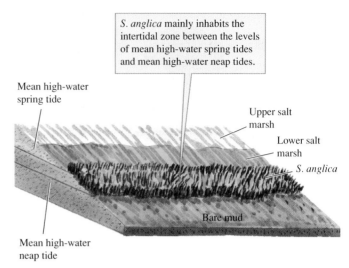

*S. anglica* mainly inhabits the intertidal zone between the levels of mean high-water spring tides and mean high-water neap tides.

Mean high-water spring tide

Upper salt marsh

Lower salt marsh

*S. anglica*

Bare mud

Mean high-water neap tide

**Figure 13.13** The niche of *Spartina anglica* is related to tidal fluctuations.

picture of their niches. However, the niches of other kinds of organisms are determined by entirely different environmental factors. Let's consider the niche of a dominant species in salt marshes.

# The Habitat Niche of a Salt Marsh Grass

Biologists discovered *Spartina anglica* approximately one century ago, as a new species recently produced by **allopolyploidy** (fig. 13.12). Allopolyploidy is a process of speciation initiated by hybridization of two different species. *S. anglica* arose initially as a cross between *S. maritima,* a European species, and *S. alterniflora,* a North American species. At least one of these hybrid plants later doubled its chromosome number, making it capable of sexual reproduction, and produced a new species: *S. anglica.* From its center of origin in Lymington, Hampshire, England, *S. anglica* spread northward along the coasts of the British Isles. During this same period, it colonized the coast of France and was widely planted elsewhere in northwest Europe as well as along the coasts of New Zealand, Australia, and China. The Chinese population of this salt marsh grass, established from only 21 plants in 1963, grew to cover 36,000 ha by 1980. *S. anglica* is extensively planted for stabilizing mudflats because it is more tolerant of periodic inundation and water-saturated soils than most other salt marsh plants. This environmental tolerance is reflected in the distribution of the plant in northwestern Europe, where it generally inhabits the most seaward zone of any of the salt marsh plants.

The local distribution of *S. anglica* in the British Isles is well predicted by a few physical variables related to the duration and frequency of inundation by tides and waves. The lower and upper intertidal limits of the grass are mainly determined by the magnitude of tidal fluctuations during spring tides. Where tidal fluctuations are greater, both the lower and upper limits are higher on the shore. However, throughout its British range, the grass generally occupies the intertidal between mean high-water

spring tides and mean high-water neap tides (fig. 13.13). A second factor that determines the local distribution of *S. anglica* is the **fetch** of the estuary. The fetch of a body of water is the longest distance over which wind can blow and is directly related to the maximum size of waves that can be generated by wind. All other factors being equal, larger waves occur on estuaries with greater fetch. The larger the fetch the higher *S. anglica* must live in an estuary to avoid disturbance by waves.

The upper limit of *S. anglica*'s distribution within the intertidal zone is also negatively correlated with latitude. In northerly locations within the British Isles, the grass does not occur quite as high in the intertidal zone as it does in the south. What factors might restrict the distribution at northern sites? One factor we should consider is that *S. anglica* is a $C_4$ plant. Remember from chapter 6 that $C_4$ grasses generally do better in warm environments. In northerly locations, *S. anglica* is replaced in the upper intertidal zone by $C_3$ plants. Could it be that competition with these $C_3$ plants at northern sites excludes *S. anglica* from the upper intertidal zone? We'll take up this question later in chapter 13 when we discuss experimental approaches to the study of competition.

## CONCEPT DISCUSSION

### Mathematical and Laboratory Models

**Mathematical and laboratory models provide a theoretical foundation for studying competitive interactions in nature.**

As ecologists have used models to explore the ecology of competition, mathematical and laboratory models have played complementary roles. Both mathematical and laboratory models are generally much simpler than the natural circumstances the ecologist wishes to understand. However, while sacrificing accuracy, this simplicity offers a degree of control that ecologists would not have in most natural settings.

D. B. Mertz (1972) began a review of four decades of research on *Tribolium* beetle populations with an astute summary of the characteristics of models in general and of the "*Tribolium* model" in particular: (1) It is an abstraction and simplification, not a facsimile, of nature; (2) except for the beetles themselves, it is a man-made construct, partly empirical and partly deductive; and (3) it is used to provide insights into natural phenomena. The predictions of these simplified models can be tested in natural systems and either supported or falsified. If falsified, a theory can be modified to accommodate the new information. Ideally, scientific understanding proceeds as a consequence of this dialog between theory and observation, between theoretician and empiricist.

# Modeling Interspecific Competition

As we saw in chapter 11, the model of logistic population growth includes a term for intraspecific competition but can be expanded to include the influence of interspecific competition on population growth. The first to do so was Vito Volterra (1926), who was interested in developing a theoretical basis for explaining changes in the composition of a marine fish community in response to reduced fishing during World War I. Alfred Lotka (1932b) independently repeated Volterra's analysis and extended it using graphics to represent changes in the population densities of competing species during competition.

Let's retrace the steps of Lotka's and Volterra's modeling exercise, beginning with the logistic model for population growth discussed in chapter 11:

$$\frac{dN}{dt} = r_{max} N \left( \frac{K - N}{K} \right)$$

We can express the population growth of two species of potential competitors with the logistic equation:

$$\frac{dN_1}{dt} = r_{max1} N_1 \left( \frac{K_1 - N_1}{K_1} \right) \text{ and } \frac{dN_2}{dt} = r_{max2} N_2 \left( \frac{K_2 - N_2}{K_2} \right)$$

Where $N_1$ and $N_2$ are the population sizes of species 1 and 2, $K_1$ and $K_2$ are their carrying capacities, and $r_{max1}$ and $r_{max2}$ are the intrinsic rates of increase for species 1 and 2. In these models, population growth slows as $N$ increases and the relative level of intraspecific competition is expressed as the ratio of numbers to carrying capacity, either $N_1/K_1$ or $N_2/K_2$. The assumption here is that resource supplies will diminish as population size increases due to intraspecific competition for resources. Resource levels can also be reduced by interspecific competition.

Lotka and Volterra included the effect of interspecific competition on the population growth of each species as:

$$\frac{dN_1}{dt} = r_{max1} N_1 \left( \frac{K_1 - N_1 - \alpha_{12} N_2}{K_1} \right) = 0$$

and

$$\frac{dN_2}{dt} = r_{max2} N_2 \left( \frac{K_2 - N_2 - \alpha_{21} N_1}{K_2} \right) = 0$$

In these models, the rate of population growth of a species is reduced both by conspecifics (individuals of the same species) and by individuals of the competing species, that is, interspecific competition. The effects of intraspecific competition ($-N_1$ and $-N_2$) are already included in the logistic models for population growth. The effect of interspecific competition is incorporated into the Lotka-Volterra model by $-\alpha_{12} N_2$ and $-\alpha_{21} N_1$. The terms $\alpha_{12}$ and $\alpha_{21}$ are called **competition coefficients** and express the competitive effects of the competing species. Specifically, $\alpha_{12}$ is the effect of an individual of species 2 on the rate of population growth of species 1, while $\alpha_{21}$ is the effect of an individual of species 1 on the rate of population growth of species 2. In this model, interspecific competitive effects are expressed in terms of intraspecific equivalents. If, for example, $\alpha_{12} > 1$, then the competitive effect of an individual of species 2 on the population growth of species 1 is greater than that of an individual of species 1. If, on the other hand, $\alpha_{12} < 1$, then the competitive effect of an individual of species 2 on the population growth of species 1 is less than that of an individual of species 1.

In general, the Lotka-Volterra model predicts coexistence of two species when, for both species, interspecific competition is weaker than intraspecific competition. Otherwise, one species is predicted to eventually exclude the other. These conclusions come from the following analysis.

Populations of species 1 and 2 stop growing when:

$$\frac{dN_1}{dt} = r_{max1} N_1 \left( \frac{K_1 - N_1 - \alpha_{12} N_2}{K_1} \right) = 0$$

and

$$\frac{dN_2}{dt} = r_{max2} N_2 \left( \frac{K_2 - N_2 - \alpha_{21} N_1}{K_2} \right) = 0$$

That is, when:

$$(K_1 - N_1 - \alpha_{12} N_2) = 0 \text{ and } (K_2 - N_2 - \alpha_{21} N_1) = 0$$

Or, rearranging these equations, we predict that population growth for the two species will stop when:

$$N_1 = K_1 - \alpha_{12} N_2 \text{ and } N_2 = K_2 - \alpha_{21} N_1$$

These are equations for straight lines, called **isoclines of zero population growth,** where everywhere along the lines population growth is stopped:

$$\frac{dN_1}{dt} = 0 \text{ and } \frac{dN_2}{dt} = 0$$

Above an isocline of zero growth, the population of a species is decreasing; below it the population is increasing (fig. 13.14).

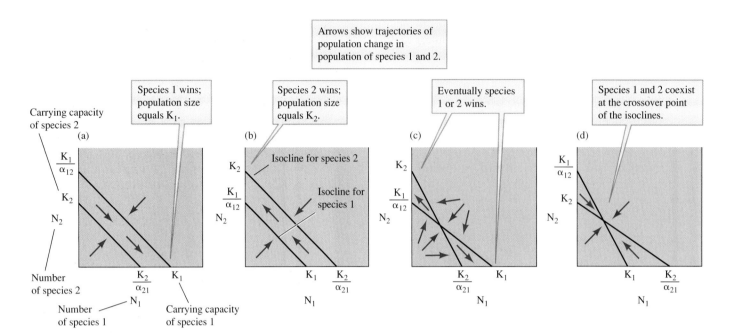

**Figure 13.14** The orientation of isoclines for zero population growth and the outcome of competition according to the Lotka-Volterra competition model.

The isoclines of zero growth show how the environment can be filled up or, in other words, the relative population sizes of species 1 and species 2 that will deplete the critical resources. At one extreme, for example, for species 1, the environment is completely filled by species 1 and species 2 is absent. This occurs where $N_1 = K_1$. At the other extreme, again for species 1, the environment can be saturated entirely by species 2, while species 1 is absent. This occurs where $N_2 = K_1/\alpha_{12}$. In between these extremes, the environment is saturated with a mixture of species 1 and 2. The graph of the isocline for zero growth for species 2 can be interpreted in a similar way.

Putting the isoclines of zero growth for the two species on the same axis allows us to predict if one species will exclude the other or whether the two species will coexist. The precise prediction depends upon the relative orientation of the two isoclines. As shown in figure 13.14, there are four possibilities.

The Lotka-Volterra model predicts that one species will exclude the other when the isoclines do not cross. If the isocline for species 1 lies above that of species 2, species 1 will eventually exclude species 2. This exclusion occurs because all growth trajectories lead to the point where $N_1 = K_1$ and $N_2 = 0$ (fig. 13.14a). Figure 13.14b portrays the opposite situation in which the isocline for species 2 lies completely above that of species 1 and species 2 excludes species 1. In this case, all trajectories of population growth lead to the point where $N_2 = K_2$ and $N_1 = 0$.

Coexistence is possible only in the situations in which the isoclines cross. However, only one of these situations leads to stable coexistence. Figure 13.14c shows the situation in which coexistence is possible at the point where the isoclines of zero population growth cross but coexistence is unstable. In this situation, $K_1 > K_2/\alpha_{21}$ and $K_2 > K_1/\alpha_{12}$ and most population growth trajectories lead either to the points where $N_1 = K_1$ and $N_2 = 0$, or to where $N_2 = K_2$ and $N_1 = 0$. The populations of species 1 and 2 may arrive at the point where the

lines cross, but any environmental variation that moves the populations off this point eventually leads to exclusion of one species by the other. Figure 13.14d represents the only situation that predicts stable coexistence of the two species. In this situation, $K_2/\alpha_{21} > K_1$ and $K_1/\alpha_{21} > K_2$ and all growth trajectories lead to the point where the isoclines of zero growth cross.

What is the biological meaning of saying that all growth trajectories lead to the point where the isoclines of zero growth cross? What this means is that the relative abundances of species 1 and 2 will eventually arrive at the point where the isoclines cross, a point where the abundances of both species are greater than zero. In this situation, each species is limited more by members of their own species than they are by members of the other species. In other words, the Lotka-Volterra model predicts that species coexist when intraspecific competition is stronger than interspecific competition. This prediction is supported by the results of laboratory experiments on interspecific competition.

# Laboratory Models of Competition

## *Experiments with* **Paramecia**

G. F. Gause (1934) used laboratory experiments to test the major predictions of the Lotka-Volterra competition model. During the course of his work Gause experimented with many organisms, but the most well known of his experimental subjects were paramecia. Paramecia are freshwater, ciliated protozoans that offer several advantages for laboratory work. First, since they are small, they can be kept in large numbers in a small space and some of their natural habitats are fairly well simulated by laboratory aquaria. In addition, paramecia feed on microorganisms, which can be cultured in the laboratory and provided in whatever concentration desired by the experimenter.

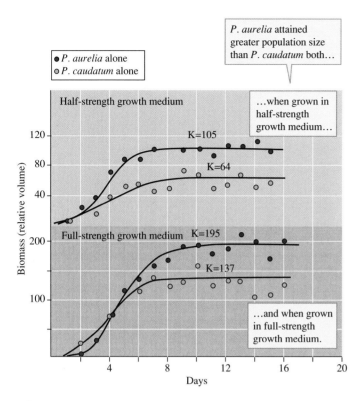

**Figure 13.15** Population growth and population sizes attained by *Paramecium aurelia* and *P. caudatum* grown separately (data from Gause 1934).

In one of his most famous experiments, Gause studied competition between *Paramecium caudatum* and *P. aurelia*. The question he posed was: Would one of these two species drive the other to extinction if grown together in microcosms where they were forced to compete with each other for a limited food supply?

Gause demonstrated resource limitation by growing pure populations of *P. caudatum* and *P. aurelia* in the presence of two different concentrations of their food, the bacterium *Bacillus pyocyaneus*. If food supplies limit the growth of laboratory populations of these paramecia, what kind of population growth would you expect them to show? As you probably expect, Gause observed sigmoidal growth with a clear carrying capacity at both full- and half-strength concentrations of the food supply (fig. 13.15). When grown in the presence of a full-strength concentration of food, the carrying capacity of *P. aurelia* was 195. When food availability was halved, the carrying capacity of this species was reduced to 105. *P. caudatum* showed a similar response to food concentration. In the presence of a full-strength concentration of food, *P. caudatum* had a carrying capacity of 137. At a half-strength concentration, the carrying capacity was 64. The nearly one-to-one correspondence between food level and the carrying capacities of these two species provides evidence that when grown alone, the carrying capacity was determined by intraspecific competition for food. These results set the stage for Gause's experiment to determine whether interspecific competition for food, the limiting resource in this system, would lead to the exclusion of one of the competing species.

When grown together, *P. aurelia* survived, while the population of *P. caudatum* quickly declined. The difference in results obtained at the two food concentrations support the conclusion that competitive exclusion results from competition for food. At a full-strength food concentration, the decline in the *P. caudatum* population was approaching exclusion by 16 days but exclusion was not complete. In contrast, at a half-strength food concentration, *P. caudatum* had been entirely eliminated by day 16. What does this contrast in the time to exclusion suggest about the influence of food supply on competition? It suggests that reduced resource supplies increase the intensity of competition.

## Experiments with Flour Beetles

*Tribolium,* beetles of the family Tenebrionidae, infest stored grains and grain products. The discovery of an infestation of these beetles in an urn of milled grain in the tomb of an Egyptian pharaoh buried about 4,500 years ago suggests that these beetles have been engaged in this occupation for some time. Their habit of attacking stored grains makes them a convenient laboratory model. Since all life stages of *Tribolium* live in finely milled flour, small containers of flour provide all the environmental requirements to sustain a population. R. N. Chapman (1928) began working with laboratory populations of *Tribolium* at the University of Chicago in the 1920s, where ever since, work has focused on two species: *T. confusum* and *T. castaneum*.

Thomas Park (1954) worked extensively on interspecific competition between these two species under six environmental conditions: hot-wet (34°C, 70% RH, relative humidity), hot-dry (34°C, 30% RH), temperate-wet (29°C, 70% RH), temperate-dry (29°C, 30% RH), cool-wet (24°C, 70% RH), and cool-dry (24°C, 30% RH). In environments held at 34°C and 70% relative humidity, both species established healthy populations that persisted over the entire duration of the experiment (fig. 13.16*a*). However, when grown together under these conditions, *T. castaneum* usually excludes *T. confusum* (fig. 13.16*b*). Cool-dry conditions appear to favor *T. confusum*. Even when grown by itself at 24°C and 30% relative humidity, *T. castaneum* does not persist for long (fig. 13.17*a*). This species quickly disappears from mixed cultures held at 24°C and 30% relative humidity (fig. 13.17*b*).

Under intermediate environmental conditions each species did well when grown alone but the outcome of interspecific competition was not completely predictable. *T. castaneum* won 86% of the trials under temperate-wet conditions, while *T. confusum* won 71% to 90% of the trials under hot-dry, temperate-dry, and cool-wet conditions. Notice that under a particular set of conditions either *T. castaneum* or *T. confusum* was usually favored, but not always.

How can we interpret the results of these laboratory experiments in terms of the effects of competition on these species' niches? Growing the two species separately showed that the fundamental niche of *T. castaneum* includes five of the six environmental conditions in the experiment, while the fundamental niche of *T. confusum* includes all six environmental conditions. Growing the two species together suggests that interspecific competition restricts the realized niches of both species to fewer

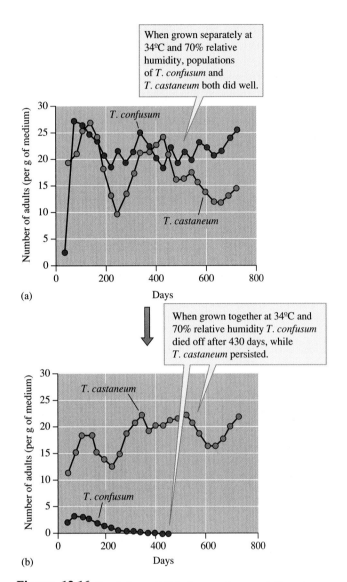

**Figure 13.16** Populations of *Tribolium confusum* and *T. castaneum* grown separately (*a*) and together (*b*) at 34°C and 70% relative humidity (data from Park 1954).

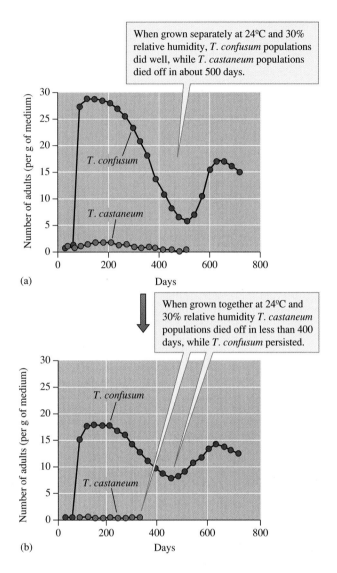

**Figure 13.17** Populations of *Tribolium confusum* and *T. castaneum* grown separately (*a*) and together (*b*) at 24°C and 30% relative humidity (data from Park 1954).

environmental conditions. Based upon these results, can we conclude that interspecific competition restricts the realized niches of species in nature? No, this would be going beyond the proper role for a model system, which is best used to generate hypotheses and guide experimentation on natural populations.

---

**CONCEPT DISCUSSION**

### Competition and Niches

**Competition can have significant ecological and evolutionary influences on the niches of species.**

Competition can have short-term ecological effects on the niches of species by restricting them to realized niches. These

species may retain their capacity to inhabit the fuller range of environments we call the fundamental niche. However, if competitive interactions are strong and pervasive enough, they may produce an evolutionary response in the competitor population—an evolutionary response that changes the fundamental niche. In this section, we explore the evidence for both ecological and evolutionary influences on the niches of natural populations. Field experiments show that interspecific competition may restrict the niches of populations in nature.

## Niches and Competition Among Plants

A. Tansley (1917) conducted one of the first experiments to test whether competition was responsible for the separation of two species of plants on different soil types. In the introduction to his paper, Tansley pointed out that while the separation of closely

**Figure 13.18**  These two species of bedstraw grow predominately on different soil types: *Galium saxatile* (shown here) grows mainly on acidic soils, while *G. sylvestre (G. pumilum)* grows mainly on basic limestone soils.

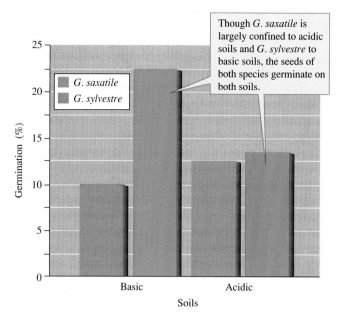

Though *G. saxatile* is largely confined to acidic soils and *G. sylvestre* to basic soils, the seeds of both species germinate on both soils.

**Figure 13.19**  Percentage seed germination by *Galium saxatile* and *G. sylvestre* in basic calcareous soils and acidic peat soil (data from Tansley 1917).

related plants had long been attributed to mutual competitive exclusion, it was necessary to perform manipulative experiments to demonstrate that this interpretation is correct. That is exactly what Tansley did to account for the mutually exclusive distributions of *Galium saxatile* and *G. sylvestre* (now *G. pumilum*), two species of small perennial plants commonly called bedstraw (fig. 13.18). In the British Isles, *G. saxatile* is largely confined to acidic soils and *G. sylvestre* to basic limestone soils.

Tansley conducted his experiment at the Cambridge Botanical Garden from 1911 to 1917, where seeds of the two species of plants were sown in planting boxes of acidic and basic soils. The seeds were sowed in single-species plantings and in mixtures of the two species. Both species germinated on both soil types, in both single- and mixed-species plantings (fig. 13.19). Like the paramecia studied by Gause, both *Galium* species established healthy populations on both soil types when grown by themselves and these single-species plantings persisted to the end of the 6-year study. However, as the two species grew in mixed plantings, Tansley observed clear competitive dominance by each species on its normal soil type.

On limestone soils, *G. sylvestre,* the species naturally found on limestone soils, overgrew and eliminated *G. saxatile,* the acidic soil species, by the end of the first growing season. On acidic soils, the relationship was reversed and *G. saxatile* was competitively dominant but competitive exclusion was not completed. Growth by both species was so slow on the acidic soils that it took until the end of the 6-year experiment for *G. saxatile* to completely cover the planting boxes containing acidic soils, a density attained by *G. sylvestre* on limestone soils in just 1 year. However, among the abundant *G. saxatile* Tansley found a few "quite healthy" plants of *G. sylvestre.* What do you think would have happened to the *G. sylvestre* on acidic soils if the experiment had been continued for a few more years? Of course it's impossible to say with certainty, but it is likely that *G. saxatile* would eventually exclude *G. sylvestre.* The delayed exclusion was probably due to the extremely slow growth of both species on acidic soils.

Tansley was one of the first ecologists to use experiments to demonstrate the influence of interspecific competition on the niches of species. The fundamental niche of both species of *Galium* included a wider variety of soil types than they inhabit in nature. The results of this experiment suggest that interspecific competition restricts the realized niche of each species to a narrower range of soil types.

# Niche Overlap and Competition Between Barnacles

Like salt marsh plants, the barnacles *Balanus balanoides* and *Chthamalus stellatus* are restricted to predictable bands in the intertidal zone. We saw in chapter 9 (see fig. 9.8) that adult *Chthamalus* along the coast of Scotland are restricted to the upper intertidal zone, while adult *Balanus* are concentrated in the middle and lower intertidal zones. Joseph Connell's observations (1961a, 1961b) indicate that *Balanus* is limited to the middle and lower intertidal zones because it cannot withstand the longer exposure to air in the upper intertidal zone. However, physical factors only partially explain the distribution of *Chthamalus.* Connell noted that larval *Chthamalus* readily settle in the intertidal zone below where the species persists as adults but that these colonists die out within a relatively short period. In the course of field experiments, Connell discovered that interspecific competition with *Balanus* plays a key role in determining the lower limit of *Chthamalus* within the intertidal zone.

Because barnacles are sessile, small, and grow in high densities, they are ideal for field studies of survivorship. Their exposure at low tide is an additional convenience for the researcher. Connell established several study sites from the

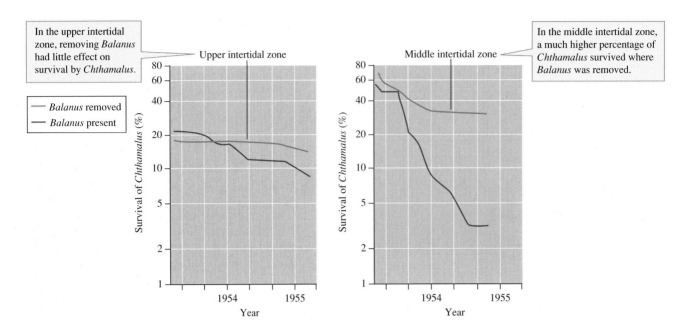

In the upper intertidal zone, removing *Balanus* had little effect on survival by *Chthamalus*.

In the middle intertidal zone, a much higher percentage of *Chthamalus* survived where *Balanus* was removed.

— *Balanus* removed
— *Balanus* present

**Figure 13.20**   A competition experiment with barnacles: removal of *Balanus* and survival by *Chthamalus* in the upper and middle intertidal zones (data from Connell 1961a, 1961b).

upper to the lower intertidal zones where he kept track of barnacle populations by periodically mapping the locations of every individual barnacle on glass plates. He established his study areas and made his initial maps in March and April of 1954, before the main settlement by *Balanus* in late April. He divided each of the study areas in half and kept one of the halves free of *Balanus* by scraping them off with a knife. Connell determined which half of each study site to keep *Balanus*-free by flipping a coin.

By periodically remapping the study sites, Connell was able to monitor interactions between the two species and the fates of individual barnacles. The results showed that in the middle intertidal zone *Chthamalus* survived at higher rates in the absence of *Balanus* (fig. 13.20). *Balanus* settled in densities up to 49 individuals per square centimeter in the middle intertidal zone and grew quickly, crowding out the second species in the process. In the upper intertidal zone, removing *Balanus* had no effect on survivorship by the second species because the population density of *Balanus* was too low to compete seriously. Connell's results provide direct evidence that *Chthamalus* is excluded from the middle intertidal zone by interspecific competition with *Balanus*.

How does interspecific competition affect the niche of *Chthamalus*? In the absence of *Balanus*, it can live over a broad zone from the upper to the middle intertidal zones. Using the terminology of Hutchinson (1957), we can call this broad range of physical conditions the fundamental niche of *Chthamalus*. However, competition largely restricts *Chthamalus* to the upper intertidal zone, a more restricted range of physical conditions constituting the species' realized niche (fig. 13.21).

Does variation in interspecific competition completely explain the patterns seen by Connell? At the lowest levels in the lower intertidal zone, *Chthamalus* suffered high mortality

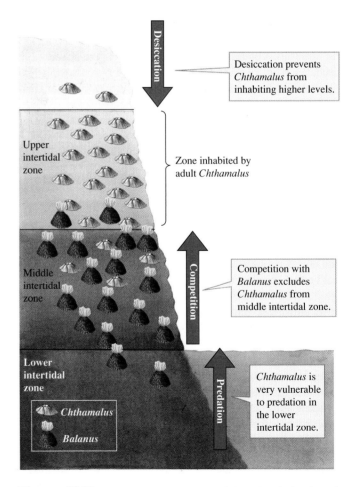

Desiccation

Desiccation prevents *Chthamalus* from inhabiting higher levels.

Upper intertidal zone

Zone inhabited by adult *Chthamalus*

Competition

Competition with *Balanus* excludes *Chthamalus* from middle intertidal zone.

Middle intertidal zone

Lower intertidal zone

Predation

*Chthamalus* is very vulnerable to predation in the lower intertidal zone.

*Chthamalus*

*Balanus*

**Figure 13.21**   Environmental factors restricting the distribution of *Chthamalus* to the upper intertidal zone.

even in the absence of *Balanus* (see fig. 13.20). What other factors might contribute to high rates of mortality by *Chthamalus* in the lower intertidal zone? Experiments have shown that this species can withstand periods of submergence of nearly 2 years, so it seems that it is not excluded by physical factors. It turns out that the presence of predators in the lower intertidal zone introduces complications that we will discuss in chapter 14 when we examine the influences of predators on prey populations.

## Competition and the Habitat of a Salt Marsh Grass

How do you think competition might affect populations of the salt marsh grass *Spartina anglica*, whose niche we discussed earlier in chapter 13? Field experiments have demonstrated that *S. anglica*, like *Chthamalus*, is restricted to its typical intertidal zone partly by interspecific competition with other salt marsh plants. In contrast to *Chthamalus*, however, *S. anglica* receives competitive pressure from the landward side of its intertidal distribution (Scholten and Rozema 1990, Scholten et al. 1987).

Does this reversal in the direction of competitive pressure make sense? It should, since in the case of barnacles we have marine organisms for which greater physical challenge occurs as they inhabit areas higher in the intertidal zone. In the case of the salt marsh plants, we are dealing with organisms descended from terrestrial ancestors that are met with increasing physical challenge as they inhabit areas lower in the intertidal zone. Similar experiments have been conducted on competition among desert rodents.

## Competition and the Niches of Small Rodents

One of the most ambitious and complete of the many field experiments ecologists have conducted on competition among rodents focused on desert rodents in the Chihuahuan Desert near Portal, Arizona. This experiment, conducted by James H. Brown and his students and colleagues (Munger and Brown 1981, Brown and Munger 1985), is exceptional in many ways. First, it was conducted at a large scale; the 20 ha study site includes 24 study plots each 50 m by 50 m (fig. 13.22). Second, the experimental trials have been well replicated, both in space and in time. Third, the project has been long term; it began in 1977 and is ongoing. These three characteristics combine to demonstrate subtle ecological relationships and phenomena that would not otherwise be apparent.

The rodent species living on the Chihuahuan Desert study site can be divided into groups based upon size and feeding habits. Most members of the species are **granivores,** rodents that feed chiefly on seeds. The large granivores consist of three species of kangaroo rats (fig. 13.23*a*) in the genus *Dipodomys*— *D. spectabilis*, 120 g; *D. ordi*, 52 g; and *D. merriami*, 45 g. In

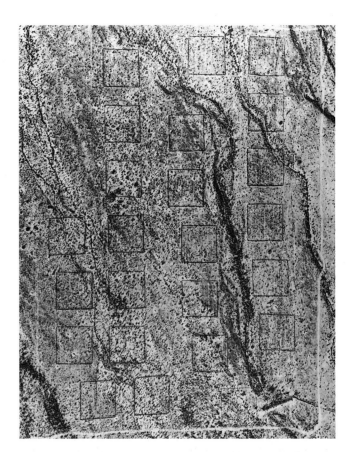

**Figure 13.22** Aerial photo showing the placement of 24 study plots, each 50 m by 50 m, in the Chihuahuan Desert near Portal, Arizona (courtesy of J. H. Brown).

addition, the study site is home to four species of small granivores (fig. 13.23*b*)—*Perognathus penicillatus*, 17 g; *P. flavus*, 7 g; *Peromyscus maniculatus*, 24 g; *Reithrodontomys megalotis*, 11 g—and two species of small insectivorous rodents— *Onychomys leucogaster*, 39 g; and *O. torridus*, 29 g.

In one experiment, Brown and his colleagues set out to determine whether large granivorous rodents (*Dipodomys* spp.) limit the abundance of small rodents on their Chihuahuan Desert study site. They also wanted to know whether the rodents might be competing for food. The researchers addressed their questions with a field experiment in which they enclosed 50 m by 50 m study plots with mouse-proof fences. The fences were constructed with a wire mesh with 0.64 cm openings, which were too small for any of the rodent species to crawl through. They also buried the fencing 0.2 m deep so the mice couldn't dig under it, and they topped the fences with aluminum flashing so the mice couldn't climb over it. This may sound like a lot of work, but to answer their questions, the researchers had to control the presence of rodents on the study plots.

The researchers next cut holes 6.5 cm in diameter in the sides of all the fences to allow all rodent species to move freely in and out of the study plots. With this arrangement in place, the rodents in the study plots were trapped live and marked once a month for 3 months. Following this initial monitoring period, the holes on four of eight study plots were reduced to 1.9 cm, small enough to exclude *Dipodomys* but

(a)

(b)

**Figure 13.23** Two species of granivorous rodents living in the Chihuahuan Desert: (*a*) the kangaroo rat, *Dipodomys* spp., a large granivore; (*b*) a pocket mouse, *Perognathus* sp., a small granivore.

large enough to allow free movement of small rodents. Brown and his colleagues refer to these fences with small holes as semipermeable membranes, since they allow the movement of small rodents but exclude *Dipodomys,* the large granivores in this system.

If *Dipodomys* competes with small rodents, how would you expect populations of small rodents to respond to its removal? The density of small rodent populations should increase, right? If food is the limiting resource, would you expect granivorous and insectivorous rodents to respond differently to *Dipodomys* removal? The researchers predicted that if competition among rodents is mainly for food, then small granivorous rodent populations would increase in response to *Dipodomys* removal, while insectivorous rodents would show little or no response.

The results of the experiment were consistent with the predictions. During the first 3 years of the experiment, small granivores were approximately 3.5 times more abundant on the *Dipodomys* removal plots compared to the control plots, while populations of small insectivorous rodents did not increase significantly (fig. 13.24).

The results presented in figure 13.24 support the hypothesis that *Dipodomys* spp. competitively suppress populations of small granivores. But would they do so again in response to another experimental manipulation? We cannot be certain unless we repeat the experiment. That's just what Edward Heske, James H. Brown, and Shahroukh Mistry (1994) did. In 1988, they selected eight other fenced study plots that they had been monitoring since 1977, installed their semipermeable barriers on four of the plots, and removed *Dipodomys* from them. The result was an almost immediate increase in small granivore populations on the removal plots (fig. 13.25). By reproducing the major results of the first experiment, this second experiment greatly strengthens the case for competition between large and small granivores at this Chihuahuan Desert site.

## Character Displacement

Because interspecific competition reduces the fitness of competing individuals, those individuals that compete less should have higher fitness than individuals that compete more. Because the degree of competition is assumed to depend upon the degree of niche overlap, interspecific competition has been predicted to lead to directional selection for reduced niche overlap. This process of evolution toward niche divergence in the face of competition is called **character displacement.**

The Galápagos finches *Geospiza fortis,* the medium ground finch, and *G. fuliginosa,* the small ground finch, provide one of the most convincing cases of character displacement. These two species occur apart from each other, that is, they are **allopatric,** on Daphne Major and Los Hermanos Island and occur together, that is, they are **sympatric,** on the island of Santa Cruz (fig. 13.26). Where the two species are allopatric, they have very similar beak sizes. However, where they are sympatric, the sizes of their beaks do not overlap. The allopatric *G. fortis* on Daphne Major have smaller beaks than those sympatric with *G. fuliginosa* on Santa Cruz, while the *G. fuliginosa* on Los Hermanos Island have beaks that are significantly larger than those sympatric with *G. fortis* on Santa Cruz. Since beak size correlates with diet in Galápagos finches, we can say that the sympatric populations of the two species on Santa Cruz have different feeding niches. Natural selection has apparently favored divergence in the feeding niches of these sympatric populations (Lack 1947, Schluter, Price, and Grant 1985, and Grant 1986).

A few other studies have demonstrated similar patterns of character displacement among a variety of animal species, including *Cnemidophorus* lizards on islands off Baja California, *Anolis* lizards on Caribbean islands, and sticklebacks inhabiting small lakes around Vancouver Island, Canada. Character displacement has also been observed in laboratory populations of bean weevils. Many studies have provided preliminary data suggesting character displacement among populations but not establishing definitive evidence. Why is that? The main reason is that a definitive demonstration requires a great deal of evidence that is difficult to provide.

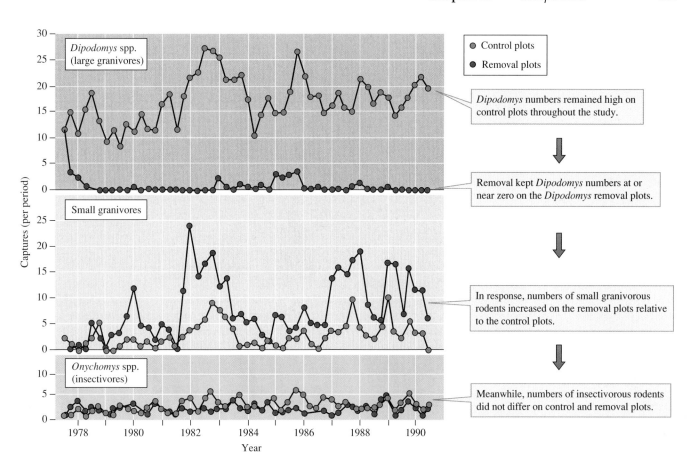

**Figure 13.24** Responses by small granivorous and insectivorous rodents to removal of large granivorous *Dipodomys* species (data from Heske, Brown, and Mistry 1994).

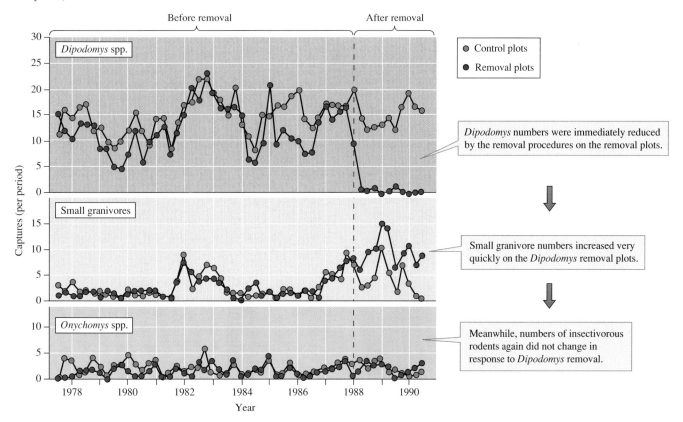

**Figure 13.25** Responses of small granivorous and insectivorous rodents to a second removal experiment, which was preceded by several years of study before initiating *Dipodomys* removal (data from Heske, Brown, and Mistry 1994).

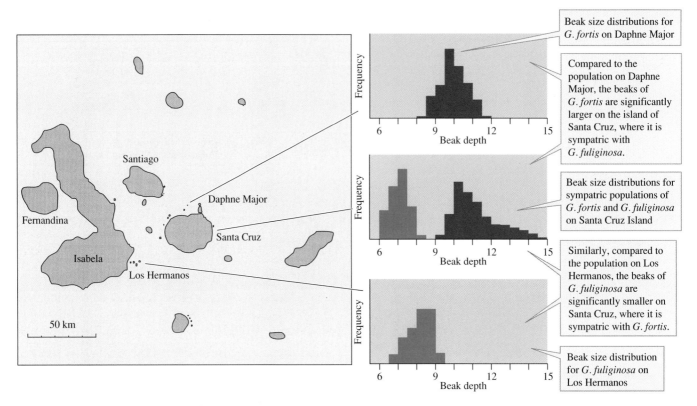

**Figure 13.26** Evidence for character displacement in beak size in populations of the Galápagos finches *Geospiza fortis* and *G. fuliginosa* (data from Grant 1986).

Mark Taper and Ted Case (1992) list six criteria that must be met to build a definitive case for character displacement:

1. Morphological differences between a pair of sympatric species (e.g., *G. fortis* and *G. fuliginosa* on Santa Cruz Island) are statistically greater than the differences between allopatric populations of the same species (*G. fortis* on Daphne Major and *G. fuliginosa* on Los Hermanos Island).

2. The observed differences between sympatric and allopatric populations have a genetic basis.

3. Differences between sympatric and allopatric populations must have evolved in place and they must not be due to the sympatric and allopatric populations having been derived from different founder populations already differing in the character under study (e.g., beak size).

4. Variation in the character (e.g., beak size) must have a known effect on use of resources (e.g., seed sizes).

5. There must be demonstrated competition for the resource under question (e.g., food) and competition must be directly correlated with similarity in the character (e.g., overlap in beak size).

6. Differences in the character cannot be explained by differences in the resources available to sympatric and allopatric populations (e.g., differences in the availability of seeds on one island versus another).

You can see how difficult it would be to satisfy all six of these criteria. It is fitting that one of the few studies that addresses all six criteria reasonably well is that of the Galápagos finches (Grant 1986, Taper and Case 1992), in the place where Darwin started the whole discussion.

# Evidence for Competition in Nature

What have we learned since Darwin initiated our enduring discussion of the role of competition as an organizing force in nature? The study of competition has gone through several phases. There was an early theoretical phase, followed by work with laboratory models, which was in turn followed by intensive observation and experimentation in the field. These phases were followed by a period of vigorous questioning of the assumption that competition is an important force in nature. This questioning forced renewed attention to careful experimental design and stimulated a reanalysis of past research in order to weigh the existing evidence bearing on competition in nature.

Two of the first analyses were those of Thomas Schoener (1983) and Joseph Connell (1983), both of whom reviewed the evidence provided by field experiments. Schoener, who reviewed over 150 field experiments on interspecific competition, reported that competition was found in 90% of the studies and among 76% of the species. Connell, who reviewed 527 experiments on 215 species, found evidence of interspecific competition in about 40% of the experiments and about 50% of the species. Why is there such a difference in these results?

*Investigating the Evidence*

# The Design of Field Experiments

Field experiments have played a key role in the assessment of the importance of competitive interactions in nature. Joseph Connell (1974) and Nelson Hairston, Sr. (1989), two of the pioneers in the use of field experiments in ecology, outlined their proper design and execution. Connell points out that one of the most substantial differences between laboratory and field experiments is that in the laboratory setting, the investigator controls all important factors but one, the factor of interest. In contrast, in field experiments, all factors are allowed to vary naturally (the investigator generally has no choice) while the factor of interest is controlled, or manipulated, by the investigator.

Both laboratory and field experiments have played an important role in ecology, but it is the field experiment that provides the key to unlock the secrets of complex interactions in nature. Why is it that field experiments are more useful in this regard? Connell pointed out that compared to laboratory experiments, the results of field experiments can be more directly applied to understanding relationships in nature because "interactions with other organisms, and the natural variation in the abiotic environment, are included in the experiment." The best field experiments are those that are executed with the least disturbance to the natural community. The utility of field experiments, however, depends upon several design features.

## Knowledge of Initial Conditions

To test for change in response to experimental manipulation, you have to know what conditions were like before the manipulation. Departures from initial conditions indicate a response to the experimental treatments. For instance, in his experiments on competition between barnacles, Connell first estimated the initial population density of one of the species in all his study plots (see fig. 13.20). Brown and his colleagues were also careful to measure the population densities of all rodent species in their study plots several times before excluding large granivorous rodents from their experimental plots.

## Controls

As in laboratory experiments, field experiments must include controls. Without controls it would be impossible to determine whether or not an experimental treatment has had an effect. Tansley created controls for his experiments on competition by growing each of his potential competitors by themselves in acidic and basic soils. What was the control for the experiment on competition among desert rodents? Brown's research team created controls by surrounding study plots with their mouse-proof fence but then cutting holes 6.5 cm in diameter in the fences to allow large granivorous rodents to move freely into and out of the plots.

The field ecologist must take special care in the design of controls. For example, why couldn't Brown's research team have just compared the population densities of small granivores on their large granivore removal plots to the population density of small granivores on the surrounding open desert? Why did they have to go through the trouble of completely fencing their control plots only to cut holes in the sides? The reason is that the fence itself and the activity associated with its construction may have had an effect on rodent densities. Without their carefully constructed controls, they could not have concluded that the increase in small granivore densities on their experimental plots was the result of removing large granivores.

## Replication

Why must experiments be done more than once at one site? The reason is that the same experiment will yield somewhat different results if done at different times or in different plots at the same time. The bottom line is that ecological systems and environmental conditions are variable, both in time and space. Replication captures this variation in response. The question posed by the experimenter is whether an experimental effect is apparent *despite* variation. Ecologists use statistics to make such a judgment. Without replication, you would never know if the results could be repeated either in time or space.

What is considered acceptable study design has changed over the decades, reflecting increasing familiarity and concern for statistical analysis. In Tansley's experiments on how competition may restrict the distribution of *Galium* species to particular soils, replication was totally lacking. In Tansley's experiment each condition (soil type) was represented just one time. Connell's later experiments with barnacles included some replication, but it was still limited at each tidal level. However, since there was a great deal of consistency in response across tidal levels, we can accept that competition acts as a significant force limiting barnacle distributions within the intertidal zone. In contrast to these earlier experiments, the more recent experiments by Brown on competition among desert rodents were replicated sufficiently for statistical analysis and repeated a second time.

The reviews by Connell and Hairston provide a guide to field experimentation as it has been conducted in the past few decades. However, as we shall see in section VI, the need for experimentation at large scales is forcing ecologists to further expand their concept of experimental design.

One reason is that the researchers analyzed different groups of studies and used different criteria for including studies in their analyses. Despite these differences, the analyses indicate that competition is an important force in the lives of many species. However, we may still have a biased estimate of the frequency of competition in nature, since ecologists may focus their studies of competition on species that they expect are competing.

About a decade after the analyses by Schoener and Connell, Jessica Gurevitch and her colleagues (1992) analyzed competition studies using a different statistical approach. Their analysis of studies conducted from 1980 to 1989 indicated a large overall effect of competition but also considerable differences among organisms and experimental approaches. They detected small to medium effects of competition on primary producers and carnivores and large effects on some herbivores and stream arthropods. Their analysis also indicated medium effects on herbivorous marine mollusks and echinoderms but no significant effects on herbivorous terrestrial insects. These researchers also found that larger experiments of longer duration yield less variable results than experiments of smaller scale and shorter duration.

After almost two decades of criticism, reflection, and reanalysis, what can we say about the prevalence and importance of interspecific competition in nature? The evidence supports the conclusion that competition is a common and important force that contributes to the organization of nature. However, the evidence also indicates that competition is neither omnipresent nor omnipotent. What other forces besides competition may be responsible for the patterns of distribution and abundance that we observe in natural populations? We've already reviewed the influences of the physical environment (sections I, II, and III) and in chapters 14 and 15 we consider two other forms of biotic interaction: exploitation (chapter 14) and mutualism (chapter 15). But first, let's consider in detail one of the most important tools in the ecologist's tool kit, the field experiment.

## APPLICATIONS & TOOLS

### Competition Between Indigenous and Introduced Species

Some of the most significant contemporary environmental problems concerns introduced species (see pp. 81–83). Because of the ecological disruption caused by introduced species, it is important to understand the mechanisms allowing them to invade communities of indigenous species. Such studies can also help us improve our understanding of ecological relationships generally.

One of the principal mechanisms thought to allow introduced species to invade communities of indigenous species is superior competitive ability. James Byers (2000) used field experiments to explore the ecological relationships of an indigenous and an introduced species of mud snail on the

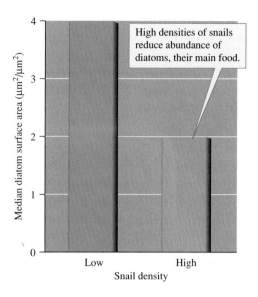

**Figure 13.27** The effects of large *Batillaria attramentaria,* an introduced mud snail, and *Cerithidea californica,* an indigenous mud snail, on the abundance of diatoms, expressed as diatom surface area per unit sediment area, within experimental enclosures after 39 days of residence (data from Byers 2000).

Pacific Coast of North America. The indigenous mud snail, *Cerithidea californica,* occurs from San Ignacio Bay, Mexico, to Tomales Bay, California. Once abundant throughout this range, *Cerithidea* has declined in abundance in some bays while an introduced mud snail, *Batillaria attramentaria,* has increased in abundance. In places along the California coast where *Batillaria* has reached very high densities, for example up to 10,000 individuals per m² in Elkhorn Slough, *Cerithidea* has disappeared entirely. *Batillaria* was introduced accidentally to the California coast with the purposeful introduction of Japanese oysters early in the twentieth century. Because it does not have a planktonic larval stage, *Batillaria* has remained largely restricted to the bays where it was introduced.

Byers conducted his field studies of *Cerithidea* and *Batillaria* in one of the bays where they co-occur: Bolinas Lagoon, approximately 20 km north of San Francisco, California. In a first phase of his study, Byers determined the influence of *Cerithidea* and *Batillaria* on their main food supply, benthic diatoms. He placed each of the snails single-species groups in 35 cm diameter cages, imbedded in the sediments along a tidal channel within Bolinas Lagoon. The densities of *Cerithidea* and *Batillaria* in these cages were 0, 12, 23, 35, 46, 69, and 92 individuals. Byers set up several replicates of each of these snail densities.

When Byers compared the influences of *Cerithidea* and *Batillaria* on diatoms, he found no difference in their effect on diatom densities. Figure 13.27 shows the average density of diatoms after approximately 39 days in cages containing a low density (12 per cage) versus a higher density (92 per cage) of large snails. These results indicate that both snail species reduce diatom cover when they are present at high population densities and that both *Cerithidea* and *Batillaria* have the potential to reduce the availability of their food supply. This

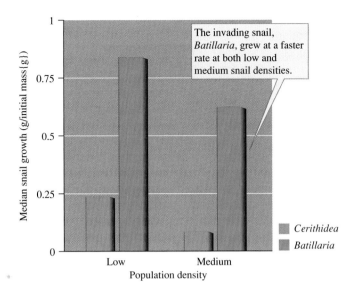

Figure 13.28 Growth, expressed as a proportion of initial mass, of large *Batillaria attramentaria*, an introduced mud snail, and *Cerithidea californica*, an indigenous mud snail, over a period of 60 days within experimental enclosures (data from Byers 2000).

part of Byers' experiment documents a potential for resource competition between the two species.

Having shown the influence of the snails on their food supply, Byers next explored the effects of snail density on their growth rates. He found that the growth rates of both *Cerithidea* and *Batillaria* decline as their population densities increase. However, at all population densities *Batillaria* grows faster than *Cerithidea*. The relative growth rates of large individuals of these two species at densities of 12 and 46 individuals per cage are shown in figure 13.28. These results indicate that *Batillaria* is much more efficient at converting available food into its own biomass. At high densities of 92 per cage, *Batillaria* continued to grow at a relatively high rate, while *Cerithidea* lost weight.

Byers used field experiments to show the potential for competition between an indigenous and an introduced species. His work also identified the mechanism involved, resource competition. In subsequent research, Byers and Goldwasser (2001) used the data generated by these experiments to build a computer simulation of the interaction between *Cerithidea* and *Batillaria*. That experimentally derived model predicts a time to competitive exclusion of *Cerithidea* by *Batillaria* of 55 to 70 years. This result matches the actual times to exclusion that have been recorded in the bays where *Batillaria* has been introduced. In conclusion, detailed experimental work on competitive interactions between indigenous and introduced species have the potential to generate information that will predict the potential pathways of interaction between species and the time to competitive exclusion.

# SUMMARY

Competition, interactions between individuals that reduce their fitness, is generally divided into *intraspecific competition,* competition between individuals of the same species, and *interspecific competition,* competition between individuals of different species. Competition can take the form of interference competition, direct aggressive interactions, or resource competition in which individuals compete through their dependence on the same limiting resources.

**Studies of intraspecific competition provide evidence for resource limitation.** Experiments with herbaceous plants show that soil nutrients may limit plant growth rates and plant weight and that competition for nutrients increases with plant population density. Plants reflect their competition for resources, including water, light, and nutrients, through the process of self-thinning. Resource competition among leafhoppers also varies with population density and is reflected in reduced survivorship, smaller size, and increased developmental time at higher population densities. Experiments with terrestrial isopods show that even where there is no food shortage, intraspecific competition through interference may be substantial.

**The niche reflects the environmental requirements of species.** The niche concept was developed early in the history of ecology and has had a prominent place in the study of interspecific competition because of the competitive exclusion principle: two species with identical niches cannot coexist indefinitely. Hutchinson added the concepts of the *fundamental niche,* the physical conditions under which a species might live in the absence of other species, and the *realized niche,* the more restricted conditions under which a species actually lives as the result of interactions with other species. While a species' niche is theoretically defined by a very large number of biotic and abiotic factors, Hutchinson's n-dimensional hypervolume, the most important attributes of the niche of most species, can often be summarized by a few variables. For instance, the niches of Galápagos finches are largely determined by their feeding requirements, while the niche of a salt marsh grass can be defined by tidal levels.

**Mathematical and laboratory models provide a theoretical foundation for studying competitive interactions in nature.** Lotka and Volterra independently expanded the logistic model of population growth to represent interspecific competition. In the Lotka-Volterra competition model, the growth rate of a species depends both upon numbers of conspecifics and numbers of the competing species. In this model, the effect of one species upon another is summarized by competition coefficients. In general, the Lotka-Volterra competition model predicts coexistence of species when interspecific competition is less intense than intraspecific competition. Competitive exclusion in laboratory experiments suggests the potential for competitive exclusion in nature. Even in the laboratory, however, organisms yield results that are much less predictable than the predictions of the Lotka-Volterra competition equations.

**Competition can have significant ecological and evolutionary influences on the niches of species.** Field experiments involving organisms from herbaceous plants to desert rodents have demonstrated that competition can restrict the niches of species to a narrower set of conditions than they would otherwise occupy in the absence of competition. Theoretically, natural selection may lead to divergence in the niches of competing species, a situation called character displacement. However, stringent requirements for a definitive

demonstration have limited the number of documented cases of character displacement in nature. After many decades of theoretical and experimental work on competition, we can conclude that competition is a common and strong force operating in nature, but not always and not everywhere.

The field experiment is one of the most powerful and important tools at the disposal of the field ecologist. However, the validity of field experiments depends upon several design features, including (1) knowledge of initial conditions, (2) controls, and (3) replication. Detailed experimental work on competitive interactions between indigenous and introduced species have the potential to generate information that will predict the potential pathways of interaction between species and the time to competitive exclusion.

# Review Questions

1. Design a greenhouse (glasshouse) experiment to test for intraspecific competition within a population of herbivorous plants. Specify the species of plant, the volume (or size of pot) and source of soil, the potentially limiting resource you will focus on (e.g., Tilman and Cowan [1989] studied competition for nitrogen) and how you will manipulate it, and the measures of plant performance you will make.

2. How can the results of greenhouse experiments on competition help us understand the importance of competition among natural populations? How can a researcher enhance the correspondence of results between greenhouse experiments and the field situation?

3. Explain how self-thinning in field populations of plants can be used to support the hypothesis that intraspecific competition is a common occurrence among natural plant populations.

4. Researchers have characterized the niches of Galápagos finches by beak size (which correlates with diet) and the niches of salt marsh grasses by position in the intertidal zone. How would you characterize the niches of sympatric canid species such as red fox, coyote, and wolf in North America? Or felids such as ocelots, pumas, and jaguars in South America? What characteristics or environmental features do you think would be useful for representing the niches of desert plants? Or the plants in temperate forest or prairie?

5. Explain why species that overlap a great deal in their fundamental niches have a high probability of competing. Now explain why species that overlap a great deal in their realized niches and live in the same area probably do not compete significantly.

6. Draw the four possible ways in which Lotka's (1932a) isoclines of zero growth (see fig. 13.14) can be oriented with respect to each

other. Label the axes and the points where the isoclines intersect the horizontal and vertical axes. Explain how each situation represented by the graphs leads to either competitive exclusion of one species or the other or to stable or unstable coexistence.

7. How was the amount of food that Gause (1934) provided in his experiment on competition among paramecia related to carrying capacity? In Gause's experiments on competition, *P. aurelia* excluded *P. caudatum* faster when he provided half the amount of food than when he doubled the amount of food. Explain.

8. In his experiments on competition between *T. confusum* and *T. castaneum,* Park (1954) found that one species usually excluded the other species but that the outcome depended upon physical conditions. In which circumstances did *T. confusum* have the competitive advantage? In which circumstances did *T. castaneum* have the competitive advantage? Could Park predict the outcomes of these experiments with complete certainty? What does this suggest about competition in nature?

9. Discuss how mathematical theory, laboratory models, and field experiments have contributed to our understanding of the ecology of competition. List the advantages and disadvantages of each approach.

10. One of the conclusions that seems justified in light of several decades of studies of interspecific competition is that competition is a common and strong force operating in nature, but not always and not everywhere. List the environmental circumstances in which you think intraspecific and interspecific competition would be most likely to occur in nature. In what circumstances do you think competition is least likely to occur? How would you go about testing your ideas?

# Suggested Readings

Aarssen, L. W. and T. Koegh. 2002. Conundrums of competitive ability in plants: what to measure? *Oikos* 96:531–42.
  *A careful analysis of plant competition studies that calls for greater caution when estimating competitive ability. The authors emphasize the need for measuring the allocation of plants to survival and fecundity as indicators of competitive ability.*

Byers, J. E. 2000. Competition between two estuarine snails: implications for invasions of exotic species. *Ecology* 81:1225–39.
  *A modern experimental study that reveals the rich details of interspecific competition between an indigenous and a nonindigenous snail species.*

Byers, J. E. 2002. Impact of non-indigenous species on natives enhanced by anthropogenic alteration of selection regimes. *Oikos* 97:449–58.

*The author reviews how human-caused environmental change can reverse the outcome of competition among species.*

Chuine, E. and E. G. Beaubien. 2001. Phenology is a major determinant of tree species range. *Ecology Letters* 4:500–510.

*An application of the concepts of realized and fundamental niches at a large geographic scale, with applications to studies of plant responses to global warming and the spread of invasive plant species.*

Connell, J. H. 1961. The influence of interspecific competition and other factors on the distribution of the barnacle *Chthamalus stellatus*. *Ecology* 42:710–23.

*This is one of the most enduring classics in the area of competition ecology and one of the early modern field experiments concerning interspecific competition.*

Connell, J. H. 1983. On the prevalence and relative importance of interspecific competition: evidence from field experiments. *American Naturalist* 122:661–96.

Goldberg, D. E. and A. M. Barton. 1992. Patterns and consequences of interspecific competition in natural communities: a review of field experiments with plants. *American Naturalist* 139:771–801.

Gurevitch, J., L. L. Morrow, A. Wallace, and J. S. Walsh. 1992. A meta-analysis of competition in field experiments. *American Naturalist* 140:539–72.

Schoener, T. W. 1983. Field experiments on interspecific competition. *American Naturalist* 122:240–85.

*These four papers trace the history of a controversy concerning the importance of competition in nature. The analyses and the thoughtful presentations by all authors provide insights into some of the uncertainties and means of resolving controversy in ecology.*

Grant, P. R. 1994. Ecological character displacement. *Science* 266:746–47.

Schluter, D., T. D. Price, and P. R. Grant. 1985. Ecological character displacement in Darwin's finches. *Science* 227:1056–59.

Schluter, D. 1994. Experimental evidence that competition promotes divergence in adaptive radiation. *Science* 266:798–801.

*These papers provide some of the best documented examples and concise reviews of the topic of character displacement.*

Heske, E. J., J. H. Brown, and S. Mistry. 1994. Long-term experimental study of a Chihuahuan Desert rodent community: 13 years of competition. *Ecology* 75:438–45.

*This paper reports on the first 13 years of one of the most ambitious experiments on interspecific competition among terrestrial animals. Destined to become a classic.*

Hutchinson, G. E. 1957. Concluding remarks. *Cold Spring Symposia on Quantitative Biology* 22:415–27.

*This is one of the most influential papers ever written on the subject of the niche by one of the most influential ecologists of the twentieth century.*

Lonsdale, W. M. 1990. The self-thinning rule: dead or alive? *Ecology* 71:1373–88.

Weller, D. E. 1987. A reevaluation of the –3/2 power rule of plant self-thinning. *Ecological Monographs* 57:23–43.

Weller, D. E. 1989. The interspecific size–density relationship among crowded plant stands and its implications for the –3/2 power rule of self-thinning. *American Naturalist* 133:20–41.

Westoby, M. 1984. The self-thinning rule. *Advances in Ecological Research* 14:167–255.

*This series of papers reviews the development and refinement of the self-thinning rule—an important principle of plant ecology with implications to the ecology of competition generally.*

Pulliam, H. R. 2000. On the relationship between niche and distribution. *Ecology Letters* 3:349–61.

*In this paper Pulliam revisits and expands the niche concept, particularly as developed by Hutchinson. He suggests further research on habitat suitability for species by studying how demography of species respond to variation in habitat characteristics.*

Schoener, T. W. 1982. The controversy over interspecific competition. *American Scientist* 70:586–95.

Wiens, J. A. 1977. On competition and variable environments. *American Scientist* 65:590–97.

*These two papers represent two sides of a historical debate regarding the importance of competition in nature. This debate had a major influence on the course of ecology during the 1970s and 1980s.*

# On the Net

Visit this textbook's accompanying website at www.mhhe.com/ecology (click on the book's title) to take advantage of practice quizzing, study/writing tips, timely news articles, and additional URLs for research on the topics in this chapter.

Animal Population Ecology

Competition

Field Methods for Studies of Populations

Field Methods for Studies of Ecosystems

Coevolution

# Exploitation
*Predation, Herbivory,
Parasitism, and Disease*

I n nature, the consumer eventually becomes the consumed. A moose browses intently on the twigs and buds of a willow barely protruding above the deep snow of midwinter (fig. 14.1). With each mouthful it chews and swallows, the moose reduces the mass of the willows and adds to the growing energy store in its own large and complex stomach, energy stores that the moose will need to make it through one more northern winter. Then, a familiar scent catches the moose's attention and startled, it runs off.

Suddenly, the clearing where the moose had been feeding is a blur of bounding forms dashing headlong in the direction the moose has gone—a pack of wolves in pursuit of its own meal. A portion of the pack has already run ahead of the moose and is cutting off its retreat. This time, unlike so many times before, the old moose will not escape. After a fierce struggle, the moose is down and the wolves settle in to feed.

But the wolves are not the only organisms to benefit from this great quantity of food. Within the intestines of the wolves live several species of parasitic worms that will soon claim their share of the wolves' hard-won feast. The worms will turn some of the energy and structural compounds they absorb into the infective stages of their own kind, which after being shed into the environment may attach themselves to other hosts, who will serve as their unwitting providers.

Some of the strongest links between populations are those between herbivore and plant, between predator and prey, and between parasite or pathogen and host. The conceptual thread that links these diverse interactions between species is that the interaction enhances the fitness of one individual—the predator, the pathogen, etc.—while reducing the fitness of the exploited individual—the prey, host, etc. Because of this common thread we can group these interactions under the heading of *exploitation.*

Let's consider some of the most common means of exploitation. Herbivores consume live plant material but do not usually kill plants. **Predators** kill and consume other organisms. Typical predators are animals that feed on other animals—wolves that eat moose, snakes that eat mice, etc. **Parasites** live on the tissues of their host, often reducing the fitness of the host, but not generally killing it. A **parasitoid** is an insect whose larva consumes its host and kills it in the process; parasitoids are functionally equivalent to predators. **Pathogens** induce disease, a debilitating condition, in their hosts.

As clear as all these definitions may seem, they are fraught with semantic problems. Once again, we are faced with capturing the full richness of nature with a few restrictive definitions. For instance, not all predators are animals, a few are plants, some are fungi, and many are protozoans. When an herbivore kills the plant upon which it feeds, should we call it a predator? If an herbivore does not kill its food plants, would it be better to call it a parasite? What do we do with a parasite that kills its host? Is it then a predator or perhaps a pathogen? The point of these questions is not to argue for more terminology but to argue for fewer, less restrictive terms. As is often the case, we are faced with a continuum of interesting and sometimes bewildering interactions involving

**Figure 14.1** This moose exploits the twigs and buds of woody plants for the food it needs to survive the cold northern winter. Eventually, wolves may prey upon the moose to meet their needs for food.

millions of organisms. Let's recognize the diversity and continuous variation facing the ecologist, put the restrictive definitions aside for the moment, and recognize what is common to all these interactions: **exploitation,** that is, one organism makes its living at the expense of another.

## CONCEPTS

- **Exploitation weaves populations into a web of relationships that defy easy generalization.**
- **Predators, parasites, and pathogens influence the distribution, abundance, and structure of prey and host populations.**
- **Predator-prey, host-parasite, and host-pathogen relationships are dynamic.**
- **To persist in the face of exploitation, hosts and prey need refuges.**

## CONCEPT DISCUSSION

### Complex Interactions

**Exploitation weaves populations into a web of relationships that defy easy generalization.**

By conservative estimate the number of species in the biosphere is on the order of 10 million. As huge as this number may seem, the number of exploitative interactions between

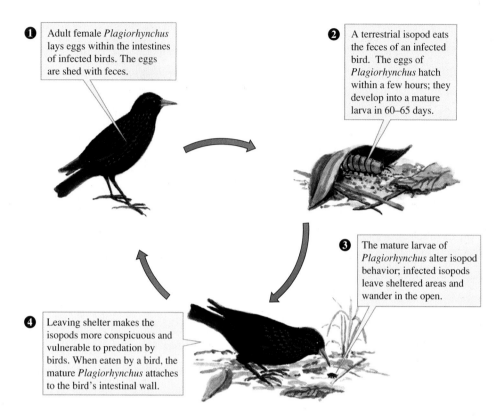

**❶** Adult female *Plagiorhynchus* lays eggs within the intestines of infected birds. The eggs are shed with feces.

**❷** A terrestrial isopod eats the feces of an infected bird. The eggs of *Plagiorhynchus* hatch within a few hours; they develop into a mature larva in 60–65 days.

**❸** The mature larvae of *Plagiorhynchus* alter isopod behavior; infected isopods leave sheltered areas and wander in the open.

**❹** Leaving shelter makes the isopods more conspicuous and vulnerable to predation by birds. When eaten by a bird, the mature *Plagiorhynchus* attaches to the bird's intestinal wall.

**Figure 14.2**   The life cycle of *Plagiorhynchus cylindraceus,* an intestinal parasite of birds.

species is far greater. Why is that? Because every one of those 10 million species is food for a number of other species and is host to a variety of parasites and pathogens. In addition, most feed on other species. Exploitative interactions weave species into a tangled web of relationships. For instance, K. E. Havens (1994) estimated that the approximately 500 known species occupying Lake Okeechobee, Florida, are linked by about 25,000 exploitative interactions—50 times the number of species! Exploitation provides much of the detail in the tapestry we call nature. In this section, we try to capture some of the richness of that tapestry by discussing the natural history of a number of interactions.

## Parasites and Pathogens That Manipulate Host Behavior

The most obvious form of exploitation is when one organism consumes part or all of another. Exploitation, however, can assume far more subtle forms. Some species alter the behavior of the species they exploit.

### Parasites That Alter the Behavior of Their Hosts

A number of parasites alter the behavior of their hosts in ways that benefit transmission and reproduction of the parasite. Acanthocephalans, or spiny-headed worms, change the behav-

ior of amphipods, small aquatic crustaceans, in ways that make it more likely that infected amphipods will be eaten by a suitable vertebrate host, especially ducks, beavers, and muskrats. Uninfected amphipods avoid the light, that is, show **negative phototaxis.** They spend most of their time near the bottoms of ponds and lakes, away from well-lighted surface waters, where the surface-feeding vertebrate hosts of acanthocephalans spend the majority of their time. In contrast, infected amphipods swim toward light, that is, show **positive phototaxis,** a behavior that places them near the pond surface in the path of feeding ducks, beavers, and muskrats (Bethel and Holmes, 1977). Interestingly, amphipod behavior remains unaltered until the acanthocephalan has reached a life stage, called a *cystacanth,* that is capable of infecting the vertebrate host. If eaten earlier, the acanthocephalan would die without completing its life cycle.

Janice Moore (1983, 1984a, and 1984b) studied a similar parasite-host interaction involving an acanthocephalan, *Plagiorhynchus cylindraceus,* a terrestrial isopod or pill bug, *Armadillidium vulgare,* and the European starling, *Sturnus vulgaris.* In this interaction, the pill bug serves as an intermediate host for *Plagiorhynchus,* which completes its life cycle in the starling (fig. 14.2).

At the outset of her research, Moore predicted that *Plagiorhynchus* would alter the behavior of *Armadillidium.* She based this prediction on several observations. One was the relative frequency of infection of *Armadillidium* and starlings by *Plagiorhynchus.* Field studies had demonstrated that even where *Plagiorhynchus* infects only 1% of the *Armadillidium* population, over 40% of the starlings in the area were infected.

Some factor was enhancing rates of transmission to the starlings, and Moore predicted that it was altered host behavior. Moore thought that the size of *Plagiorhynchus* might also be a factor. At maturity, the cystacanth stage of *Plagiorhynchus* grows to about 3 mm, a substantial fraction of the internal environment of an 8 mm pill bug!

Moore brought *Armadillidium* into the laboratory and established two laboratory populations: an uninfected control group and an infected experimental group. She infected half the laboratory populations of *Armadillidium* by feeding them pieces of carrot coated with *Plagiorhynchus* eggs, while keeping the remaining laboratory populations free of *Plagiorhynchus*. After 3 months, the *Plagiorhynchus* in the infected populations matured to the cystacanth stage. At this point Moore mixed the infected and uninfected populations.

Because *Plagiorhynchus* does not alter the outward appearance of *Armadillidium*, Moore could not determine whether an *Armadillidium* was infected or not until she dissected it at the completion of an experiment. Consequently, all the behavioral experiments were conducted "blind"; that is, without the possibility of observer bias due to prior knowledge of the identity of experimental and control animals.

Moore found that *Plagiorhynchus* alters the behavior of *Armadillidium* in several ways. Infected *Armadillidium* spend less time in sheltered areas and more time in low-humidity environments and on light-colored substrates. These changes in behavior would increase the time an *Armadillidium* spends in the open, where it could be easily seen by a bird. In summary, infected *Armadillidium* behave in a way that is likely to increase the probability that they will be discovered by foraging birds.

In laboratory experiments Moore demonstrated that captive starlings consistently captured *Armadillidium* from light rather than dark substrates. She provided caged starlings access to a mixture of 10 infected and 10 uninfected *Armadillidium*, which wandered freely across the bottom of the cage, half of which was covered by black sand and half by white sand. Under these conditions, starlings ate 72% of the infected *Armadillidium* but only 44% of the uninfected individuals (fig. 14.3). The starlings took isopods mainly from the surface of white sand, so it seems that the tendency of *Armadillidium* to seek out light substrates does make them more vulnerable to predation by birds.

A critical step in this research was to determine whether the changed behavior of infected *Armadillidium* translates into their being eaten more frequently by wild birds. Moore collected the arthropods that starlings feed to their nestlings and from these collections estimated the rate at which they delivered *Armadillidium*—about one every 10 hours. Using this delivery rate and the proportion of the *Armadillidium* population infected by *Plagiorhynchus* (about 0.4%), she was able to predict the expected rate of infection among starling nestlings if the adults capture *Armadillidium* at random from the natural population. The proportion of infected nestlings was 32%, about twice the rate of infection predicted if starlings fed randomly on the *Armadillidium* population. These

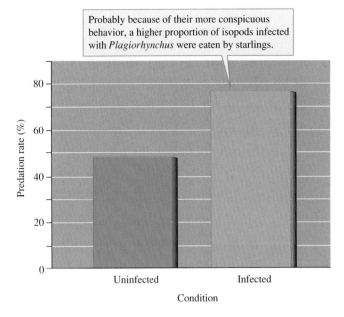

**Figure 14.3** Starling predation on uninfected and infected *Armadillidium vulgare* (data from Moore 1984b).

results support Moore's hypothesis that the altered behavior of infected *Armadillidium* increases their probability of being eaten by starlings.

Moore emphasized that *Plagiorhynchus* does not just alter *Armadillidium*'s behavior but alters its behavior in a particular way—in a way that increases the rate at which the final host of the parasite, starlings, is infected.

## A Plant Pathogen That Mimics Flowers

Every year the slopes of the southern Rocky Mountains are decorated with the colorful blossoms of wildflowers. Some of these wildflowers, however, are not quite what they seem. One bright yellow and sweet-smelling "blossom" is actually produced by a pathogenic fungus that manipulates the growth of its host plant. This pathogen belongs to a group of fungi called rusts because of their rust-colored spores that appear on the surface of the infected host plant. This particular rust is *Puccinia monoica*, and its hosts are mustard plants in the genus *Arabis*. *Arabis* spp. are herbaceous plants that spend anywhere from a few months to several years as a rosette, a low-growing growth form with a high density of leaves. During the rosette stage *Arabis* invests heavily in root development and storage of energy in the roots. At the end of the rosette stage, *Arabis* grows tall quickly, a process called *bolting,* and flowers (fig. 14.4*a*). Once pollinated, the flowers form seeds that mature, completing the life cycle of *Arabis*.

However, *Puccinia* completely alters the life history of *Arabis*. It attacks the rosette stage, manipulating its development to produce a growth form that promotes reproduction by the fungus and usually kills the plant. *Puccinia* infects the rosettes of *Arabis* in late summer and then invades the **meristematic tissue,** the actively dividing tissue responsible for plant growth, during the following winter. As it invades

(a)

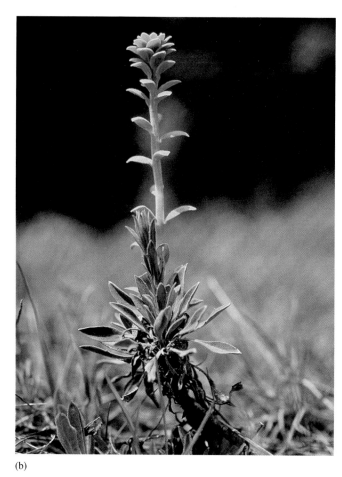

(b)

**Figure 14.4** The effects of the fungus *Puccinia monoica* on morphology: (*a*) a normally developed mustard plant, *Arabis hoelbollii;* (*b*) a pseudoflower formed by *A. hoelbollii* infected by *Puccinia.*

the meristematic tissue, *Puccinia* manipulates future development by the rosette. Infected rosettes elongate rapidly the following spring, maintain a high density of leaves along their entire length, and are topped by a cluster of bright yellow leaves. This cluster of yellow leaves looks very much like the flowers of the buttercup, *Ranunculus* spp. (fig. 14.4*b*).

The bright yellow pseudoflowers of infected rosettes are produced by various fungal structures, including spermogonia containing spermatia (fungal reproductive cells), sexually receptive fungal hyphae, and secretions of sticky, sugar-containing spermatial fluid. Most rusts require outcrossing for sexual reproduction, which is accomplished by insects transferring spermatia from one fungus (thallus) to the receptive hyphae of another thallus. Barbara Roy (1993) found that the combination of yellow color and sugary fluid attracts a wide variety of flower-visiting insects, including butterflies, bees, and flies (fig. 14.5). Flies, the most common visitor to pseudoflowers at her Colorado study site, have been demonstrated to be effective carriers of rust spermatia.

Roy's studies demonstrated that *Puccinia* truncates the life cycle of *Arabis* and in the process generally kills the host plant. The *Arabis* that survive attack by *Puccinia* may go on to flower but none form seeds. Thus, destruction by *Puccinia* is total.

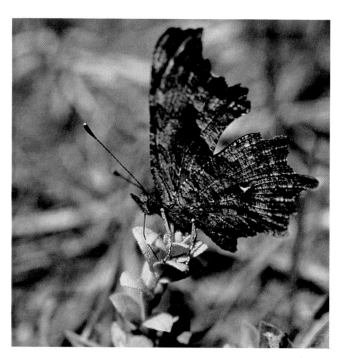

**Figure 14.5** The pseudoflowers formed by *Arabis hoelbollii* infected by the fungus *Puccinia monoica,* such as the one shown here, are attractive to a wide variety of pollinating insects, such as this *Polygonia* butterfly.

# The Entangling of Exploitation with Competition

We often arrange our thoughts about nature in neat categories like the chapters of this book. In chapter 13 we discussed competition, in chapter 14 we discuss exploitation, and in chapter 15 we examine mutualism. Nature itself is not so neatly arranged, nor are natural phenomena so easily isolated. One process is usually connected to several others. The distinction between exploitation and competition is blurred when competitors eat each other.

## *Predation, Parasitism, and Competition in Populations of* Tribolium

Thomas Park and his colleagues (Park 1948, Park et al. 1965) uncovered one of the very first examples of competitors eating each other during their work on competition among flour beetles. As we saw in chapter 13, the outcome of competition between *Tribolium castaneum* and *T. confusum* depended upon temperature and moisture. It turns out that the presence or absence of a protozoan parasite of *Tribolium, Adelina tribolii,* also influences the competitive balance between flour beetle species. The effects of this parasite are also entangled with predation among the flour beetles and cannibalism, which we might think of as a form of intraspecific exploitation.

Park showed that various strains of *T. castaneum* and *T. confusum* differ in their rates of cannibalism. Of the two species, *T. castaneum* is the most cannibalistic but it preys on the eggs of *T. confusum* at an even higher rate than it cannibalizes its own eggs. In the light of its predatory behavior, it's not surprising that *T. castaneum* eliminated *T. confusum* in 84% of 76 competition experiments spanning a period of about 10 years. This predatory strategy works best, however, in the absence of *Adelina.*

*Adelina* invades the cells of its host and lives its life as an intracellular parasite. Beetles become infected when they ingest the oocysts of this parasite, either as they feed on flour or as they consume infected larvae, pupae, or adult beetles. Once in the gut of the beetle, the oocyst eggs rupture, liberating a life stage of *Adelina* called a *sporozoite.* The sporozoites penetrate the beetle's gut and enter the body cavity, or haemocoel. Once in the haemocoel, the sporozoites invade various cells, where they reproduce asexually and produce a second life stage called the *merozoite.* The motile merozoites invade yet other host cells eventually, producing male and female sex cells that combine to form zygotes. The zygotes eventually give rise to new sporozoites, which are encased in oocysts. Ingestion of these oocysts by another beetle completes the life cycle of *Adelina.*

Several biologists before Park had noted that *Adelina* caused "sickness" and death among *Tribolium* populations. It was Park, however, who demonstrated that *Adelina* reduces the density of *Tribolium* populations and can alter

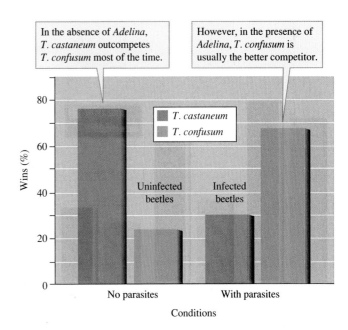

In the absence of *Adelina*, *T. castaneum* outcompetes *T. confusum* most of the time.

However, in the presence of *Adelina*, *T. confusum* is usually the better competitor.

**Figure 14.6** The influence of the protozoan parasite *Adelina tribolii* on competition between the flour beetles *Tribolium castaneum* and *T. confusum* (data from Park 1948).

the outcome of competition between *T. confusum* and *T. castaneum. Adelina* strongly reduces the population density of *T. castaneum* populations but has little effect on *T. confusum* populations. In the absence of the parasite, *T. castaneum* won 12 of 18 competitive contests against *T. confusum.* When the parasite was included, however, *T. confusum* won 11 of 15 contests (fig. 14.6). In other words, parasitism completely reverses the outcome of competitive interactions between the two species. From insects to African lions, interference escalated to the point of predation appears to be a common occurrence among competitors. However, Park's experiments with *Tribolium* indicate that parasites may make the outcome of a predaceous competitive strategy difficult to predict.

## CONCEPT DISCUSSION

### Exploitation and Abundance

**Predators, parasites, and pathogens influence the distribution, abundance, and structure of prey and host populations.**

One of the main reasons ecologists are interested in exploitative interactions between species is that these interactions have the potential to influence prey and host populations. A rapidly growing pool of studies suggest that predators, parasites, and pathogens substantially affect the populations they exploit.

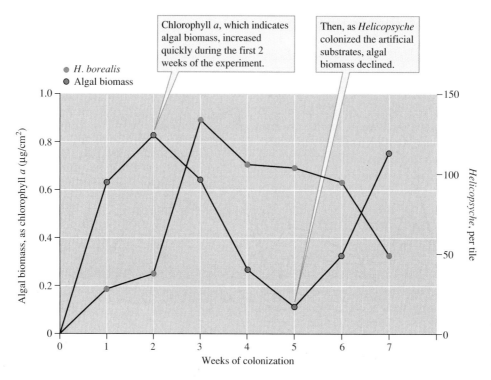

**Figure 14.7** Biomass of algae and numbers of the grazing caddisfly *Helicopsyche borealis* (data from Lamberti and Resh 1983).

# An Herbivorous Stream Insect and Its Algal Food

Gary Lamberti and Vincent Resh (1983) studied the influence of an herbivorous stream insect on the algal and bacterial populations upon which it feeds. The herbivorous insect was the larval stage of the caddisfly (order Trichoptera) *Helicopsyche borealis*. This insect inhabits streams across most of North America and is most notable for the type of portable shelter it builds as a larva. The larvae cement sand grains together to form a helical portable home that looks just like a small snail shell. In fact, the species was originally described as a freshwater snail. Larval *Helicopsyche* graze on the algae and bacteria growing on the exposed surfaces of submerged stones. This feeding habit requires that *Helicopsyche* spend considerable time out in the open, where it would be far more vulnerable to predators were it not for its case.

Lamberti and Resh found that larval *Helicopsyche* grow and develop through the summer and fall, attaining densities of over 4,000 individuals per square meter in Big Sulphur Creek, California. At this density, they make up about 25% of the total biomass of benthic animals. A consumer that reaches such high population densities clearly has the potential to reduce the density of its food supply. Lamberti and Resh got an indication of the potential of *Helicopsyche* to influence its food supply in a preliminary experiment. In this first experiment they placed unglazed ceramic tiles (15.2 cm × 7.6 cm)

on the bottom of the creek and followed colonization of these artificial substrates by algae and *Helicopsyche* over a period of 7 weeks.

Algae rapidly colonized the tiles, reaching peak density 2 weeks after the tiles were placed in Big Sulphur Creek. The *Helicopsyche* population reached its highest density 1 week later. Algal biomass decreased from week 2 to week 5 of the study and then rose again during the last 2 weeks, as *Helicopsyche* numbers declined. These results (fig. 14.7) suggest that the caddisfly larvae depleted their food supply. However, Lamberti and Resh could not be certain. Why is that? First, there are many other benthic invertebrates living in Big Sulphur Creek, some of which might be depleting the algal populations. Second, physical factors could have changed during the 7 weeks of the study, and these changes could have produced the fluctuations in both algal and *Helicopsyche* populations. This initial experiment provided valuable indications but was not a definitive test.

In a follow-up study, the researchers used an exclusion experiment to test for the effect of *Helicopsyche* on its food supply. They placed unglazed ceramic tiles in two 3-by-6 grids of 18 tiles each. One grid was placed directly on the stream bottom, while the other was placed on a metal plate supported by an upside-down J-shaped metal bar. This arrangement, which raised the tiles 15 cm above the bottom but still 35 cm below the stream surface, allowed colonization of tiles by algae and most invertebrates while preventing colonization by *Helicopsyche*. *Helicopsyche*

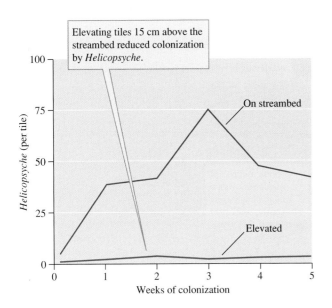

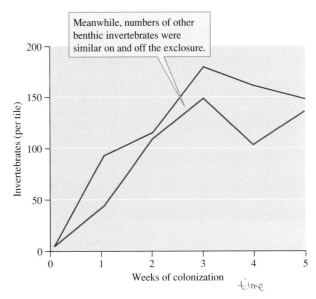

**Figure 14.8** The influence of elevating tiles on colonization by *Helicopsyche borealis* and other benthic invertebrates (data from Lamberti and Resh 1983).

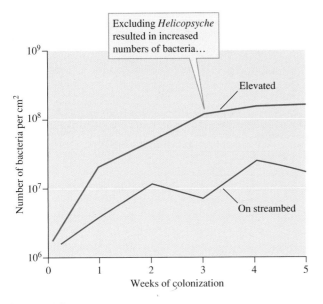

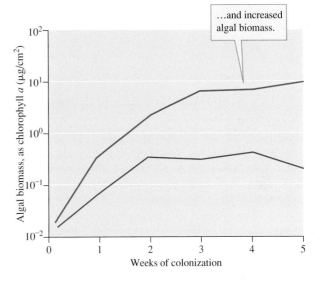

**Figure 14.9** Influence of excluding *Helicopsyche borealis* on abundance of bacteria and algae (data from Lamberti and Resh 1983).

could not colonize the tiles because their heavy snail-shaped case confines them to the stream bottom. To reach the tiles, *Helicopsyche* would have to crawl up the J-shaped support bar, out of the water, and then back down, while most other invertebrates could colonize by either drifting downstream with the current or by swimming to the raised tiles. Lamberti and Resh coated the above-water parts of the bar with an adhesive to prevent adult *Helicopsyche* from crawling down to the tiles to deposit their eggs. As figure 14.8 shows, the experimental arrangement excluded *Helicopsyche* while allowing large numbers of other invertebrates to colonize the raised tiles. Such selective manipulations of natural populations are not easy to attain.

The results of this experiment clearly show that *Helicopsyche* reduces the abundance of its food supply. Figure 14.9 shows that the tiles without *Helicopsyche* supported higher abundances of both algae and bacteria. The large effect of *Helicopsyche* on its food supply is apparent from paired photos of the experimental and control tiles at the beginning and the end of the experiment (fig. 14.10).

The influence of exploitation on populations is often best seen when populations are released from exploitation. This is what we saw when *Helicopsyche* were excluded from experimental habitats. Similar responses occur when exploited populations are introduced into new environments free of significant predators, herbivores, or pathogens.

(a)

(b)

**Figure 14.10** Effects of excluding *Helicopsyche borealis* on benthic algal biomass: (*a*) two sets of tiles at the beginning of experiment; exclusion tiles in foreground; (*b*) same tiles 5 weeks into the experiment.

# An Introduced Cactus and an Herbivorous Moth

In the mid-1800s a prickly pear cactus, *Opuntia stricta,* was introduced to Australia as an ornamental plant and was maintained in gardens for some time. Then, near the turn of the century, *Opuntia* escaped cultivation and became established in the wild, where it found ideal physical conditions for its growth and reproduction. The plant spread quickly, covering over 20 million ha by the late 1920s. By 1930, it appeared that the spread of *Opuntia* had begun to slow, but by that time it covered over 24 million ha.

Nowhere in its natural range in North America does the cactus attain the densities reached in Australia. How did this cactus reach such high densities outside its native range? Biologists trying to control *Opuntia* suggested that its rapid growth in Australia resulted from the absence of "natural enemies"—the herbivores, parasites, and pathogens that attack the cactus in its native environment. The Australian government sent biologists to the native environment to search for biological allies in their fight to control the spread of the cactus.

Biologists eventually identified several insect species that attack the cactus and that might be useful in its control. Of these species, the most effective proved to be a moth eventually described and given the apt name of *Cactoblastis cactorum.* The appropriateness of the name of this species is apparent if you compare the photos showing *Opuntia* densities before and after introduction of *Cactoblastis* (fig. 14.11).

Female *Cactoblastis* deposit eggs on the pads of *Opuntia* in groups of 70 to 90. When the caterpillars hatch, they burrow into the cactus pad and then spend their larval lives dining on the inside of the pad. As the caterpillars burrow, they introduce a diversity of fungi and bacteria that also attack the internal tissues. As a consequence of the combined attack by caterpillars and microbes, cactus tissues are quickly reduced to the consistency of mush and whole thickets of succulent *Opuntia* collapse. The herbivore *Cactoblastis* was so effective at reducing the population of *Opuntia* because it was also a dispersal agent for a variety of pathogens.

**Figure 14.11** Collapse of an *Opuntia stricta* population after introduction of the herbivorous moth *Cactoblastis cactorum.*

Following the release of *Cactoblastis,* the population of cactus collapsed. It took the moth only 2 years to reduce the population of *Opuntia* by three orders of magnitude, from about 12,000 individuals per hectare to 27 per hectare. During this same time, the area covered by *Opuntia* fell from 24 million ha to a few thousand.

Eventually, these two populations reached a kind of cyclic equilibrium (Dodd 1940, 1959). The density of the cactus population is low and highly dispersed and therefore not easily found by the moth. Under these conditions local populations of *Opuntia* can grow. Eventually, however, *Cactoblastis* finds these local populations and destroys them. Meanwhile there will be an outbreak in another area where *Opuntia* has temporarily escaped from *Cactoblastis.* Eventually these are found and also destroyed. Under these conditions, which are more like the conditions under which most species live in their natural environment, it is not obvious what controls the population density of *Opuntia.* As we shall see, this kind of effective and complex exploitation occurs in other populations.

# A Pathogenic Parasite, a Predator, and Its Prey

One of the most impressive aspects of the control of the cactus by the moth *Cactoblastis* is that the events took place over such a large area. One of the great challenges of ecology is to work at large scales. Ecologists rarely have the opportunity to conduct large-scale experiments, however, nature sometimes provides such opportunities. One such opportunity arose in Sweden when a pathogen severely reduced the population of red foxes, *Vulpes vulpes.*

Erik Lindström and his colleagues (1994) at the Grimsö Wildlife Research Station in Örebro County reported that mange mites, *Sarcoptes scabiei,* were first found infesting red foxes in north-central Sweden in 1975. The researchers studied the spread of mange because mange mites are a serious external parasite of foxes that causes hair loss, skin deterioration, and death. Within a decade of its arrival in Sweden, mange had spread over the entire country (fig. 14.12). As it spread, mange reduced the population of red foxes in Sweden by over 70%.

As wildlife ecologists, Lindström's research team was keenly interested in how the prey of red foxes would respond. Would they find evidence of population control by this predator? From 1972 to 1993, the research team studied several prey populations as well as red fox populations. They used many sources of information and conducted their studies at local, regional, and national spatial scales.

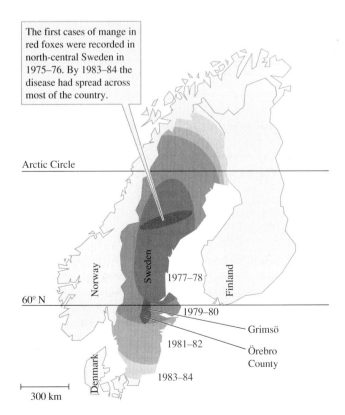

The first cases of mange in red foxes were recorded in north-central Sweden in 1975–76. By 1983–84 the disease had spread across most of the country.

**Figure 14.12** The spread of mange in red foxes across Sweden from 1975 to 1984 (data from Lindström et al. 1994).

Information
Questions
Hypothesis
Predictions
Testing    ✓

*Investigating the Evidence*

# Standard Error of the Mean

When we introduced the sample mean, we pointed out how it is an estimate of the actual, or true, population mean. A second sample from a population would probably have a different sample mean and a third sample would have yet another. How close is a given sample mean to the true population mean? The answer to this question will depend on two factors: the variation within the population and the number of observations or measurements in our sample from the population. Here, we will begin to build a way of representing the precision of a given estimate of a population mean. Our first step will be to calculate a statistic called the standard error of the mean, usually called the **standard error,** $s_{\bar{x}}$. The standard error is calculated from the sample variance and the sample size as follows:

$$s_{\bar{X}} = \sqrt{\frac{s^2}{n}}$$

$$s_{\bar{X}} = \frac{s}{\sqrt{n}}$$

where

$s^2$ = sample variance
$s$ = sample standard deviation
$n$ = sample size

For a concrete example, let's use the body length measurements for a sample of loach minnows, *Tiaroga cobitis,* from a tributary of the San Francisco River in southwestern New Mexico, which we first encountered in chapter 6 (see p. 162). Suppose you are comparing the body sizes of loach minnows in populations exposed to predation by flathead catfish, an introduced species, to populations not exposed to predation by this introduced fish. To do so you need to estimate body sizes in several populations.

Your sample from the San Francisco River was:

| Specimen | 1 | 2 | 3 | 4 | 5 | 6 | 7 | 8 | 9 | 10 |
|---|---|---|---|---|---|---|---|---|---|---|
| Total length (mm) | 60 | 62 | 56 | 53 | 53 | 59 | 62 | 41 | 58 | 58 |

In chapter 6, we calculated mean of this sample as:

$$\bar{X} = 56.2 \text{ mm}$$

And, the standard deviation of this sample was:

$$s = 6.2 \text{ mm}$$

Since the number of fish in our sample is 10, the standard error for this sample of loach minnows is:

$$s_{\bar{X}} = \frac{6.2 \text{ mm}}{\sqrt{10}}$$

$$s_{\bar{X}} = \frac{6.2 \text{ mm}}{3.16}$$

$$s_{\bar{X}} = 1.96 \text{ mm}$$

Now, let's consider the hypothetical a situation in which we obtained a sample of loach minnows from another study site on the Gila River. This second sample happened to yield the same sample mean and the same standard deviation. However, instead of 10 loach minnows, this second sample included 50 loach minnows. The standard error calculated from this sample is:

$$s_{\bar{X}} = \frac{6.2 \text{ mm}}{\sqrt{50}}$$

$$s_{\bar{X}} = \frac{6.2 \text{ mm}}{7.07}$$

$$s_{\bar{X}} = 0.88 \text{ mm}$$

The results of the study were clear. Red foxes in Sweden reduce the populations of their prey, including hares, grouse, and roe deer fawns. Figure 14.13 shows the relationship between numbers of red foxes and mountain hares, *Lepus timidus,* in Sweden. Following the reduction in the red fox population, the number of mountain hares increased two to four times. This is an especially thorough and convincing demonstration of the influence of a terrestrial vertebrate predator on its prey populations. The study also suggests that red foxes may have a significant influence on the cyclic abundance of some of their prey species. The dynamics of prey populations has been the subject of research by ecologists for some time. Studies of predation have been central to this research.

DISCUSSION

## CONCEPT DISCUSSION

### Dynamics

**Predator-prey, host-parasite, and host-pathogen relationships are dynamic.**

In the last section we saw how some predators, parasites, and pathogens affect the populations they exploit. The picture that emerges from these studies is that the biology of exploitation

Notice that because there were more minnows in this second sample, the size of the standard error is considerably smaller. In other words, our second sample mean is a more precise estimate of the true population mean. This is shown in the form of a graph in figure 1. In figure 1*a* the points indicate the sample means for our two samples and the vertical bars, above and below the points, are plus and minus one standard error. The same statistics are plotted in figure 1*b* as a bar graph and only the upper standard error bar is shown, which is

a common way to plot such data. In either case, the smaller standard error around the sample mean for the Gila River indicates that our estimate of the mean length of loach minnows in the population is more precise there than for the population in the San Francisco River. In Investigating the Evidence box of chapter 15 (see p. 387) we will use the standard error to derive a more quantitative expression of precision called the confidence interval.

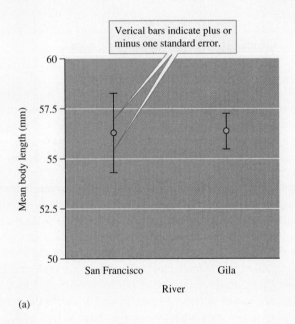

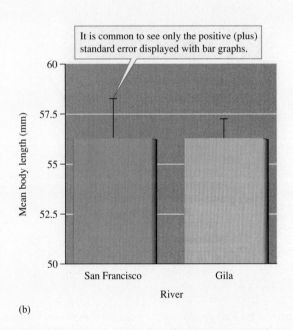

(a)   (b)

**Figure 1**  Average body length of loach minnows and standard errors calculated from samples collected in the San Francisco River (*n* = 10) and the Gila River (*n* = 50). Smaller standard error for the sample from the Gila River is the result of a larger sample size.

is complex. As complex as this emerging picture of exploitation may be, it belies an even deeper underlying complexity. In this section, we add another level of complexity as we take up the topic of *temporal dynamics*. Populations of a wide variety of predators and prey are not static but cycle in abundance over periods of days to decades.

## Cycles of Abundance in Snowshoe Hares and Their Predators

Population cycles are well documented for a wide variety of animals living at high latitudes, including lemmings, voles,

muskrats, red fox, arctic fox, ruffed grouse, and porcupines. We have already seen in chapter 10 how periodic outbreaks of voles lead to local increases in the abundance of avian predators due to numerical responses by owls and hawks (Korpimäki and Norrdahl 1991).

One of the best-studied cases of animal population cycles is that of the snowshoe hare, *Lepus americanus,* and the lynx, *Lynx canadensis,* one of the snowshoe hare's chief predators. The population cycles of these two species are especially well documented because the Hudson Bay Company kept trapping records during most of the eighteenth, nineteenth, and twentieth centuries. Drawing on this unique historical record ecologists were able to estimate the relative abundances of Canadian lynx and snowshoe hare over a period of about 200 years. That record, shown in figure 14.14, demonstrates a remarkable match in the cycles of the two populations.

By the 1950s several hypotheses had been proposed to explain these and other cycles among northern populations. Charles Elton (1924) proposed that cycles of abundance in snowshoe hare and lynx populations are driven by variation in amount of solar radiation as a consequence of sunspot cycles. He proposed that variation in intensity of solar radiation may directly affect snowshoe hares and their food supply and that lynx populations, in turn, respond to the changing abundance of the snowshoe hare, their main prey.

The sunspot hypothesis was rejected by D. MacLulich (1937) and P. Moran (1949), who showed that sunspot cycles do not match snowshoe hare population cycles. The second

group of hypotheses, which Lloyd Keith (1963) referred to as "overpopulation theories," suggested that periods of high population growth are followed by (1) decimation by disease and parasitism, (2) physiological stress at high densities leading to increased mortality as a consequence of nervous disorders, and (3) starvation due to reduced quantity and quality of food at high population densities. An alternative to the overpopulation hypothesis was that cycles like that of the snowshoe hare are driven by predators. According to this hypothesis, predators increase in number in response to increasing prey availability and then eventually reduce prey populations.

Keith observed that none of these hypotheses completely accounts for population cycles in snowshoe hare and other northern populations. He went on to say that "the 10-year cycle is not likely to become better understood by further theorizing. Clearly the present need is for comprehensive long-term investigations by a diversified team of specialists." Heeding his own advice, Keith organized such studies. After three decades of research by his team and several other groups in North America and Europe, we now have a reasonable picture of the roles played by predators and food supply in producing population cycles in the far north.

## The Role of Food Supply

Snowshoe hares live in the boreal forests of North America. As we saw, the boreal forest is dominated by a variety of conifers such as spruce, *Picea* spp., jackpines, *Pinus banksiana,* and tamarack, *Larix laricina,* and deciduous trees such as balsam poplar, *Populus balsamifera,* aspen, *Populus tremuloides,* and paper birch, *Betula papyrifera.* Within the boreal forest, snowshoe hares associate with dense growths of understory shrubs, which provide both cover and winter food, the most critical portion of the snowshoe hare's food supply.

Snowshoe hares have the potential to reduce the quantity and quality of their food supply. The hares live up to the legendary reproductive capacity of rabbits and hares. Estimated geometric rate of increase, $\lambda$ (see chapter 10), during the growth phase of a hare population cycle can average as high as 2.0. In other words, snowshoe hare populations can double in size each

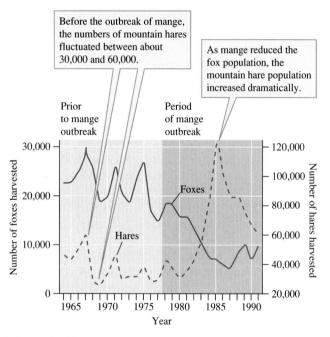

**Figure 14.13** The numbers of foxes and mountain hares in five counties in Sweden estimated from hunters' harvest records (data from Lindström et al. 1994).

**Figure 14.14** Historical fluctuations in lynx and snowshoe hare populations based on the number of pelts purchased by the Hudson Bay Company (data from MacLulich 1937).

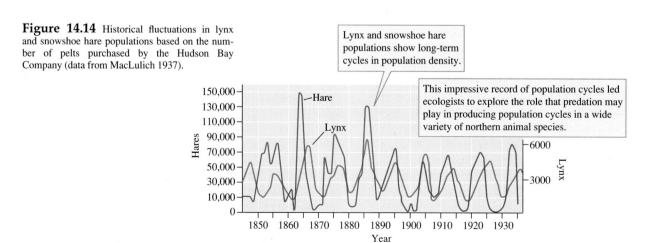

generation. Keith and his colleagues (1984) have observed snowshoe hare population densities of up to 1,100 to 2,300 per square kilometer. However, local densities are highly dynamic. Keith cites 100-fold fluctuations in snowshoe hare densities in some areas and states that 10- to 30-fold fluctuations are common. Similar densities are sometimes observed in populations of the mountain hare, *Lepus timidus,* which shows pronounced population cycles across the Eurasian taiga (Keith 1983) and which destroys considerable vegetation at high densities.

Snowshoe hares spend the long northern winter (6–8 months) browsing on the buds and small stems of shrubs such as rose, *Rosa* spp., and willow, *Salix* spp. Where deep snow provides access, snowshoe hares browse on the saplings of trees such as spruce and aspen. The most nutritious portions of these shrubs and trees are the small stems (< 4–5 mm diameter). Over the winter, each hare requires about 300 g of these stems each day. In some areas, however, snowshoe hares have been observed to remove over 1,500 g of food biomass per day, possibly wasting a great deal of potential food in the process. Feeding at these rates, one population of snowshoe hares reduced food biomass from 530 kg per hectare in late November to 160 kg per hectare by late March. Many ecologists have demonstrated food shortage during winters of peak snowshoe hare density.

Snowshoe hares also influence the quality of their food supply. Feeding by snowshoe hares induces chemical defenses in their food plants, defenses like those we discussed in chapter 6. Shoots produced after substantial browsing contain elevated concentrations of terpene and phenolic resins, defensive chemicals that repel hungry hares. Elevated concentrations of plant defensive chemicals can persist for up to 2 years after browsing by hares. The effect of these induced chemical defenses reduces *usable* food supplies during the population decline. Some ecologists suggest that plant defensive responses may be the "timer" that produces 10-year population cycles in snowshoe hares.

## The Role of Predators

The long historical record of lynx population cycles may have distracted ecologists from the fact that lynxes are only one of several predators that feed on snowshoe hares. Other major predators of snowshoe hares include goshawks, *Accipiter gentilis,* great horned owls, *Bubo virginianus,* mink, *Mustela vison,* long-tailed weasels, *Mustela frenata,* red foxes, *Vulpes vulpes,* and coyotes, *Canis latrans.* Populations of these predators are known to cycle synchronously with snowshoe hare populations. Though the lynx is considered to be a specialist on snowshoe hares, the diet of a generalist predator such as the coyote may also be dominated by snowshoe hares. This is particularly true when snowshoe hare populations are at peak density. A. Todd and L. Keith (1983) report that snowshoe hares made up 67% of the coyote diets in central Alberta, Canada. Ecologists have estimated that predation can account for 60% to 90% of snowshoe hare mortality during peak densities.

Research by Mark O'Donoghue and several colleagues from Canada, Argentina, and Alaska (O'Donoghue et al. 1997, 1998) provides clear evidence of predator *functional response*

(see p. 158) and *numerical response* (see p. 270) to increased hare densities. O'Donoghue and his colleagues focused their research on two of the most important predators of adult snowshoe hares: lynx and coyotes. Their study shows that coyote and lynx numbers increase 6- to 7-fold, a numerical response, following increases in snowshoe hare populations. The two predators also showed functional responses to increased hare densities. However, coyote and lynx functional responses differed in their timing and form. The lynx killed more hares when hare numbers were declining, while coyotes showed higher predation rates when the hare population was increasing. O'Donoghue and colleagues discovered that lynx show a clear type 2 functional response (see fig. 6.21) to increasing hare densities, reaching a maximum number of 1.2 hares per day at medium hare densities. In contrast, coyotes preyed on up to 2.3 hares per day at the highest hare densities and their functional response showed no signs of leveling off. At high hare densities, coyote and lynx predation rates exceeded their daily energetic needs. Coyotes killed more hares early in the winter, caching many and retrieving them later in the season. In some instances individual coyotes returned to eat hares over 4 months after they were cached. The combination of numerical and functional responses by lynx and coyotes indicate great potential for these predators to reduce snowshoe hare populations.

In summary, several decades of research provided evidence that both predation and food can make substantial contributions to snowshoe hare population cycles (Haukioja et al. 1983, Keith 1983, Keith et al. 1984). The food availability and predation hypotheses are not mutually exclusive alternatives but rather are complementary. As hare populations increase, they reduce the quantity and quality of their food supply. Reduced food availability, which leads to starvation and weight loss, would itself likely produce population decline. This potential decline is ensured and accelerated by high rates of mortality due to predation. As hare population density is reduced, predator populations decline in turn, plant populations recover, and the stage is set for another increase in the hare population. This scenario was tested through a series of long-term experiments.

# Experimental Test of Food and Predation Impacts

Charles J. Krebs and several colleagues (Krebs et al. 1995) conducted a large-scale, long-term experiment designed to sort out the tangle of conflicting evidence regarding the impacts of food and predation on snowshoe hare population cycles. Over a period of 8 years, Krebs and colleagues conducted one of the most ambitious field experiments to date. Their experimental plots consisted of nine 1 km$^2$ blocks of undisturbed boreal forest, each separated from other experimental blocks by a minimum of 1 km. Three blocks served as controls for comparison to the six other blocks where experimental treatments were applied. To test the impact of food, hares were given unlimited supplemental food on two experimental blocks during the entire period of the study. To test for the possible influences of plant tissue quality on hare numbers, the researchers applied a nitrogen-potassium-phosphorus fertilizer from the air to two of the experimental blocks. Finally, they built electric fences around two of the 1 km² blocks, which excluded mammalian predators but not hawks and owls. One of these predator reduction blocks received supplemental food. Krebs and his colleagues report that, due to maintenance requirements, they could not replicate the predator reduction and predator reduction + food experimental manipulations. They could not, since the fences on both predator reduction areas (8 km of fence) had to be checked every day through the winter, when temperatures would sometimes dip as low as –45°C. Krebs' research team maintained these experimental conditions through one cycle in snowshoe hare numbers.

During the 8 years of the experiment, the researchers observed an increase in hare numbers to a peak, followed by a decline on all the study plots. The application of fertilizer increased plant growth within the fertilizer treatment blocks but did not increase numbers of snowshoe hares. Meanwhile compared to control plots, hare numbers increased substantially on food addition, predator reduction, and predator reduction + food study plots. Averaged over the peak and decline phases during the study, reducing predators doubled hare density, adding food tripled hare density, and excluding predators and adding food increased hare density to 11 times that of the controls (fig. 14.15). What factors contributed to these increased densities within treatment blocks? Krebs and his colleagues found that higher densities on experimental plots were the result of both higher survival and higher reproduction.

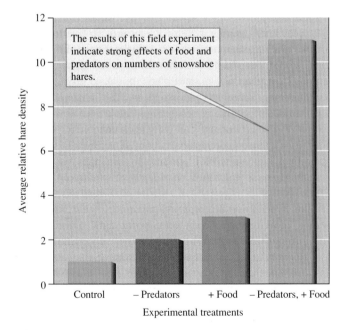

**Figure 14.15** The densities of snowshoe hares averaged from the peak in hare density through the period of declining density observed during the study. Hare densities are expressed relative to the densities on the control plots where no experimental manipulation was applied (data from Krebs et al. 1995, 2001).

After approximately 70 years of research, we can conclude that the population cycle in snowshoe hares is the result of an interaction between three trophic levels: the hares, their food supply, and their predators. Krebs and colleagues (2001) point out, however, that in order to understand the controls on hare numbers, researchers have had to work with all three trophic levels simultaneously. In addition, the critical experiment had to be done on a large scale and in the field. Still, a great deal of insight into this large-scale predator-prey system has come from laboratory and mathematical studies.

## Population Cycles in Mathematical and Laboratory Models

Now let's shift our focus from population cycles in the vast world of the boreal forest to population cycles in mathematical models and controlled laboratory conditions. Mathematical and laboratory models offer population ecologists the opportunity to manipulate variables that they cannot control in the field. Our question here is whether predator-prey or parasite-host cycles can be produced in mathematical and laboratory models without the complications introduced by factors such as the effects of the prey on its food supply and uncontrolled weather cycles. In other words, can the interactions among exploited populations themselves generate population cycles of the type observed in snowshoe hares? The answer to this question is a qualified yes.

### Mathematical Models

The first ecologists to model predator-prey interactions mathematically were Alfred Lotka (1925) and Vito Volterra (1926). Both researchers built their models based on observations of interactions among natural populations. Lotka was impressed by the reciprocal oscillations of populations of moth and butterfly larvae and the parasitoids that attack them. Volterra was inspired by the response of marine fish populations to cessation of fishing during World War I. Volterra observed that the response of fish populations was uneven. Predaceous fish, particularly sharks, increased in abundance, while the populations upon which they fed decreased. This reciprocal change in numbers suggested that predators have the potential to reduce the abundance of their prey. In this single observation, Volterra somehow saw the potential for predator-prey population cycles and suggested that similar cycles should occur in parasite-host and pathogen-host systems, including those in which humans are involved. With these observations in mind, Lotka and Volterra then set out to build mathematical models that would produce the cycles that they thought occurred in nature.

The Lotka-Volterra predator-prey equations demonstrated that very simple models will produce cycling of predator and prey populations. The basic Lotka-Volterra model assumes that the host population grows at an exponential rate and that host population size is limited by its parasites, pathogens, or predators:

$$\frac{dN_h}{dt} = r_h N_h - p N_h N_p$$

represents the host population size, and $r_h N_h$ represents exponential growth by the host population. In the Lotka-Volterra model, exponential growth by the host population is opposed by deaths due to parasitism or predation, which is represented by $-p N_h N_p$, where $p$ is the rate of parasitism or predation, $N_h$ is again the number of hosts, and $N_p$ is the number of parasites or predators.

On the other side of the parasitoid-host system, the Lotka-Volterra model assumes that the rate of growth by the predator or parasite population is determined by the rate at which it converts the hosts it consumes into offspring (new predators or parasitoids) minus the mortality rate of the parasitoid population:

$$\frac{dN_p}{dt} = cp N_h N_p - d_p N_p$$

Here again $N_h$ and $N_p$ are the numbers of hosts and predators or parasites, respectively. The rate at which the predators or parasites convert hosts into offspring is $cp N_h N_p$, which is the rate at which the exploiters destroy hosts, $p N_h N_p$, times a conversion factor, $c$, the rate at which hosts are converted to parasite or predator offspring. In the Lotka-Volterra equation, the growth rate of the predator population is opposed by predator deaths, $d_p N_p$. Notice that in these equations the only variables are $N_h$ and $N_p$. All the other terms in the Lotka-Volterra model, $p$, $c$, $d_p$, and $r_h$, are constants. The Lotka-Volterra predator-prey model is summarized in figure 14.16.

Now let's reflect on the behavior of this model. Because the host population grows at an exponential rate, its population growth accelerates with increasing population size. However, this tendency to grow faster and faster with increasing $N_h$ is opposed by exploitation. As $Nh$ increases the rate of exploitation, $p N_h N_p$, also increases. Consequently, in the Lotka-Volterra model, reproduction by the host is translated immediately into destruction of hosts by the predator. In addition, increased parasitism or predation, $p N_h N_p$, is translated directly and immediately into more parasites or predators by $cp N_h N_p$. Increased numbers of predators and parasites increase the rate of exploitation since increasing $N_p$ increases $p N_h N_p$. Growth of the parasite or predator population eventually reduces the host population, which in turn leads to declines in the predator or parasite population. So, like the host, exploiter success carries the seeds of its own destruction.

These reciprocal effects of host and exploiter produce oscillations in the two populations, which we can represent in two ways. In figure 14.17a, population oscillations are presented as we looked at them in snowshoe hare and lynx populations (see fig. 14.13), while figure 14.17b gives an alternative representation. The time axis has been eliminated and the two remaining axes represent the numbers of predators or parasites and hosts. When we plot population data in this way we see that the Lotka-Volterra model produces oscillations in exploiter and host populations that follow an elliptical path whose size

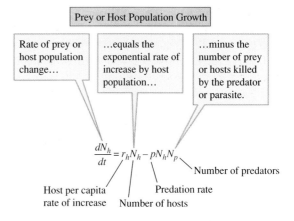

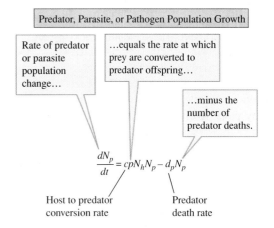

**Figure 14.16** Anatomy of the Lotka-Volterra equations for predator-prey or parasite-host population growth.

depends upon the initial sizes of host and exploiter populations. Whatever the ellipse size, however, the host and exploiter populations just go round and round on the same path forever.

The prediction of eternal oscillations on a very narrowly defined path is obviously unrealistic. Another unrealistic assumption is that neither the host nor the exploiter populations are subject to carrying capacities. Another is that changes in either population are instantaneously translated into responses in the other population. Despite these unrealistic assumptions, Lotka and Volterra made valuable contributions to our understanding of host-parasite and predator-prey systems. They showed that simple models with a minimum of assumptions produce reciprocal cycles in populations of host and parasite and predator and prey analogous to those that biologists had observed in natural populations. They demonstrated that exploitative interactions themselves can, in theory, produce population cycles without any influences from an outside force such as climatic variation.

## Laboratory Models

One of the most successful attempts to produce Lotka-Volterra-type population cycles in the laboratory was that of Syunro Utida (1957) of Kyoto University, Japan. Utida studied interactions between the adzuki bean weevil, *Callosobruchus chinensis,* and

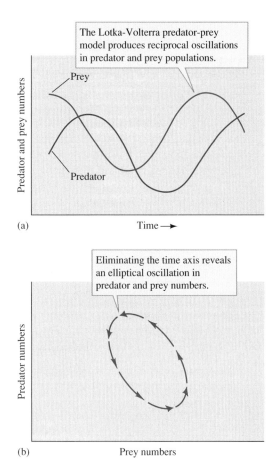

The Lotka-Volterra predator-prey model produces reciprocal oscillations in predator and prey populations.

(a)

Eliminating the time axis reveals an elliptical oscillation in predator and prey numbers.

(b)

**Figure 14.17** A graphical view of the Lotka-Volterra predator-prey model (data from Gause 1934).

a hymenopteran parasitoid wasp, *Heterospilus prosopidis,* which attacks the bean weevil. Adult weevils lay their eggs on adzuki beans, *Paseolus angularis,* and upon hatching the larvae feed on the beans until they metamorphose into pupae. When they emerge from the pupal stage, the adult weevils mate and seek out new beans on which to lay their eggs. The entire life cycle, from egg to egg, takes approximately 20 days. While the weevil works at completing its life cycle, the parasitoid wasp searches for weevil larvae and pupae, where they lay their eggs. The larvae of the wasps feed on the larvae and pupae of the weevils and in the process, kill them. Though the details of their behavior differ, the wasps are predators of the weevils, no less than are lynx predators of snowshoe hares.

Utida's experimental populations lived in petri dishes 1.8 cm tall by 8.5 cm in diameter where temperature was maintained at a constant 30°C and relative humidity at 75%. Within the petri dishes Utida placed 10 g of adzuki beans with a water content of 15% and added a mixture of adult adzuki bean weevils and parasitoid wasps: either 64 weevils and 8 wasps (population A), 8 weevils and 8 wasps (population C), or 512 weevils and 128 wasps (population E). Every 10 days 10 g of fresh beans were added, and the leavings of the old beans were placed in another dish. Any beetles moved with the spent food were recorded over a period of 20 days.

Utida followed population C for 47 beetle generations, approximately 940 days, after which a mistake in handling killed the population. He followed population E for 82 generations, approximately 1,640 days, after which the weevils died out. Population A was followed the longest, 112 generations, over 6 years, after which the population was accidentally destroyed. It was only by following the beetle and wasp populations for so many generations that Utida was able to see the pattern we look at now.

All three of the experimental populations showed the same cyclic behavior (fig. 14.18). For several generations Utida observed reciprocal fluctuations in his beetle and parasitoid populations that look very similar to those we saw for lynx and hare populations (see fig. 14.13). After an initial phase of high-magnitude oscillations the population cycles were decreased in amplitude, remained in a situation of low-amplitude fluctuations for some time, and then increased in amplitude once again. In population A, high-amplitude cycles continued for the first 20 generations, dampened out until about generation 30, and then resumed high-amplitude cycling until about generation 54, when the oscillations dampened out once again.

Utida's results are analogous to the patterns of reciprocal fluctuation seen in the Lotka-Volterra model along with some

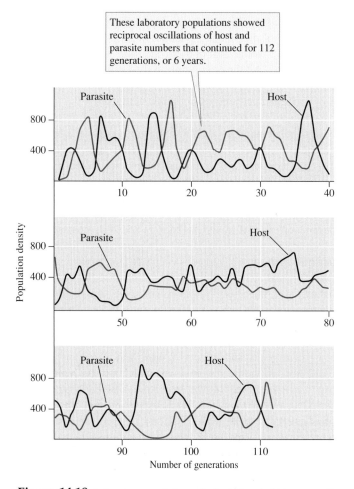

These laboratory populations showed reciprocal oscillations of host and parasite numbers that continued for 112 generations, or 6 years.

**Figure 14.18** Laboratory populations of a host, the adzuki bean weevil, and a parasitoid wasp (data from Utida 1957).

behavior not predicted by the mathematical model. However, despite these differences, like the Lotka-Volterra model, Utida's laboratory model shows that parasitoid-host populations can show reciprocal oscillations without significant temporal variation in the physical environment.

G. Gause (1935) produced similar results when he studied a laboratory population of *Paramecium aurelia* preying upon yeast. He followed the populations through three cycles, which took only 20 days. Though Gause's experiments were much shorter than Utida's, they also produced oscillations like those predicted by the Lotka-Volterra model.

Utida's and Gause's successes make work with laboratory models look far easier than it is. Most attempts to produce Lotka-Volterra-type oscillations in laboratory populations have failed. Most laboratory experiments have led to extinction of the predator or prey population in a fairly short period of time. To sustain oscillations even for a short period researchers have generally had to provide the prey with refuges of some sort, which indicates another generalization about natural predator-prey systems. It appears that to persist in the face of exploitation by predators, parasites, and pathogens, hosts and prey need refuges.

## CONCEPT DISCUSSION

### Refuges

**To persist in the face of exploitation, hosts and prey need refuges.**

This section is about *refuges,* situations in which members of an exploited population have some protection from predators and parasites. When we think of refuges, we generally think of an inaccessible place. There are, however, many other kinds of refuges. Many have nothing to do with places and most do not provide complete security—just enough.

## Refuges and Host Persistence in Laboratory and Mathematical Models

Gause's success at producing cycles in populations of *Paramecium aurelia* and its prey, *Saccharomyces exiguus,* gives no hint of the difficulties he experienced in his earlier attempts. Gause's first attempts to produce Lotka-Volterra population cycles involved *Paramecium caudatum* and one of its predators, another aquatic protozoan called *Didinium nasutum.* If Gause grew these organisms in a simple laboratory microcosm, *Didinium* quickly consumed all the *Paramecium* (fig. 14.19). The absence of a refuge for the prey led eventually to extinction of the predator and prey populations. Gause responded by putting some sediment on the bottom of his microcosm to provide a

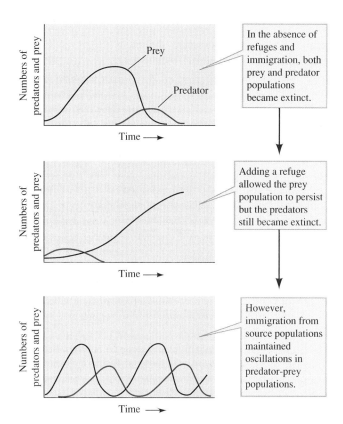

**Figure 14.19** Refuges and the persistence of predator-prey oscillation in laboratory populations of prey (*Paramecium aurelia*) and predators (*Didinium nasutum*) (data from Gause 1934).

refuge for *Paramecium.* In this case once *Didinium* had eaten all of the *Paramecium* not hiding in bottom sediments, it starved and became extinct. Following the disappearance of *Didinium* and the removal of predation pressure, the population of *Paramecium* quickly increased. Here, a simple refuge for the prey population led to extinction of the predator.

Gause was only able to maintain oscillations in predator-prey populations if he periodically restocked the populations from his laboratory cultures. In this experiment, the microcosm contained no refuges for *Paramecium,* but every 3 days Gause would take one of each organism from his pure laboratory cultures and add them to the experimental microcosm. Using these periodic immigrations he was able to produce Lotka-Volterra-type predator-prey oscillations (see fig. 14.19). To do so, however, the experimental system had to include a refuge for the prey and a reservoir for the predator (the laboratory cultures) and Gause had to create periodic immigrations from those populations to the experimental microcosm.

Are these experimental requirements entirely artificial or do they correspond with anything we already know about natural populations? Actually, Gause's experimental results match many of our observations in natural populations. In chapter 9, we saw that on larger scales populations show clumped distributions. Most species are much more common in some parts of their range than in others. Then in chapter 10, we saw how dispersal is an important contributor to population dynamics and that some local populations are maintained entirely by dispersal from other areas. Some biologists have combined observations such as these to hypothesize the existence of population sources and population sinks—local populations maintained by immigration from source populations. In Gause's experiment, the laboratory cultures were population hot spots, or sources, while the microcosms where predator and prey interacted were population sinks. The requirements of Gause's experiment are consistent with the results of later experiments.

C. Huffaker (1958) set out to test whether Gause's results could be reproduced in a situation in which the predator and prey are responsible for their own immigration and emigration among patches of suitable habitat. Huffaker chose the six-spotted mite, *Eotetranychus sexmaculatus,* a mite that feeds on oranges, as the prey and the predatory mite *Typhlodromus occidentalis,* which attacks *E. sexmaculatus,* as the predator. Huffaker's experimental setups, or "universes" as he called them, consisted of various arrangements of oranges, or combinations of oranges and rubber balls, separated by partial barriers to mite dispersal consisting of discontinuous strips of petroleum jelly.

An important point of natural history is that the predatory mite had to crawl in order to disperse from one orange to another, while the herbivorous mite can disperse either by crawling or by "ballooning," a means of aerial dispersal. A mite balloons by spinning a strand of silk that can catch wind currents. Huffaker gave the herbivorous mite the chance to balloon by providing small wooden posts that could serve as launching pads and by having a fan circulate air across his experimental setup.

While Huffaker's simpler experimental universes did not produce predator-prey oscillations, his most elaborate setup of 120 oranges did. These oscillations spanned several months (fig. 14.20). Huffaker observed three oscillations that spanned about 6 months. They were maintained by the dispersal of predator and prey among oranges in a deadly game of hide-and-seek, in which the prey managed to keep ahead of the predator for three full oscillations. These results are similar to those obtained by Gause, but we need to remember that Huffaker did not directly manipulate dispersal. In Huffaker's experiment both predator and prey moved from patch to patch under their own power.

The importance of refuges was recognized by Lotka (1932a) and incorporated into his mathematical theory of predator-prey relations. The starting point for his discussion were the Lotka-Volterra predator-prey equations that we discussed previously:

$$\frac{dN_h}{dt} = r_h N_h - p N_h N_p \text{ and } \frac{dN_p}{dt} = cp N_h N_p - d_p N_p$$

The part of this equation that provided the starting point for Lotka's discussion was $p$, the capture or consumption rate of the predator. Lotka pointed out that while it may be reasonable to assume that $p$ is a constant for a particular environment, its value should change from one environment to another if the environments differ structurally, particularly if there is a difference in the availability of refuges in the two environments.

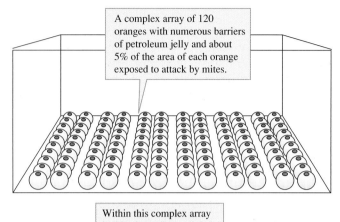

A complex array of 120 oranges with numerous barriers of petroleum jelly and about 5% of the area of each orange exposed to attack by mites.

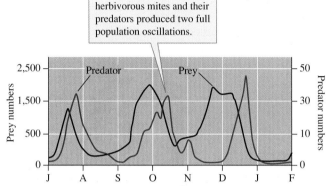

Within this complex array herbivorous mites and their predators produced two full population oscillations.

**Figure 14.20** Environmental complexity and oscillations in laboratory populations of an herbivorous mite and a predatory mite (data from Huffaker 1958).

Specifically, $p$ should be lower where the prey or hosts have access to more refuges. This refinement of the Lotka-Volterra predator-prey model anticipated recent theoretical analysis of the role that refuges and spatial diversity in general play in the persistence of predator-prey and parasite-host systems. While Lotka's analysis concentrated on physical refuges that could shelter terrestrial prey, he recognized the wide variety of forms that refuges could take. He pointed out, for instance, that flight is a refuge for birds from terrestrial predators.

# Exploited Organisms and Their Wide Variety of "Refuges"

## Space

Most of our discussion has focused on what we might call "spatial" refuges, places where members of the exploited population have some protection from predators and parasitoids. Many forms of spatial refuge are familiar: burrows, trees, air, water (if faced with terrestrial predators), and land (if faced with aquatic predators). However, some spatial refuges differ in subtle ways from other areas.

As we saw in a previous section of chapter 14, the cactus *Opuntia stricta,* which had completely covered vast areas of Australia, was controlled by a combination of an herbivorous insect, *Cactoblastis cactorum,* and pathogenic microbes. The introduction of the insects did not drive the cactus to extinction, however. One reason for the persistence of the cactus is that it has a number of spatial refuges. As we saw, small, isolated cactus populations are difficult for *Cactoblastis* to find. This is a spatial refuge much like that designed into Huffaker's experimental arrangement of oranges. In addition, the insects do not vigorously attack the cactus where it grows on nutrient-poor soils and/or above 600 to 900 m elevation, due to low quality of the cactus tissues or low temperatures.

St. John's Wort, *Hypericum perforatum,* persists in similar refuges in the face of attacks by the beetle *Chrysolina quadrigemina,* one of the chief enemies of *Hypericum* in the Pacific Northwest region of the United States. *Hypericum* was introduced into areas along the Klamath River around 1900, and its population quickly grew to cover about 800,000 ha by 1944. Following the release of the beetles, the area covered by St. John's Wort was reduced to less than 1% of its maximum coverage. This remnant population of the plant was concentrated in shady habitats, where, though it grows more poorly than in sunny areas, it is protected from the beetles, which avoid shade.

## Protection in Numbers

Living in a large group provides a type of refuge. Aside from the potential of social groups to intimidate would-be predators, numbers alone can reduce the probability of an individual prey or host being eaten. We can make this prediction based solely on the work of C. S. Holling (1959) on the responses of predators to prey density. In chapter 6 we looked at the functional responses of several predators and herbivores. Briefly, predator functional response results in increasing rate of food intake as prey density increases. Eventually, however, the predator's feeding rate levels off at some maximum rate. In chapter 10, we looked at numerical response, a second component of predator response to prey density that results in increased predator density as prey density increases. As with functional response, the numerical response eventually levels off at the point where further increases in prey density no longer produce increased predator density.

Now let's put functional response and numerical response together to predict the predator's **combined response** to increased prey density. We can combine the two responses by multiplying the number of prey eaten per predator times the number of predators per unit area:

$$\frac{\text{Prey consumed}}{\text{Predator}} \times \frac{\text{Predators}}{\text{Area}} = \frac{\text{Prey consumed}}{\text{Area}}$$

By dividing the prey consumed per unit area by the population density of the prey (prey consumed/area), we can determine the percentage of the prey population consumed by the predator. If we plot percentage of the prey consumed against prey density over a broad range of prey densities, the prediction is that the percentage of the prey population consumed will be lower at high prey densities (fig. 14.21).

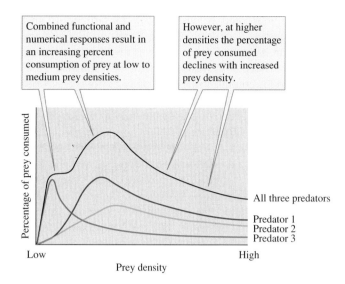

**Figure 14.21** Prey density and the percentage of prey consumed due to combined functional and numerical responses (data from Holling 1959).

Why should the percentage of the prey consumed by the predator decline at high prey densities? The answer to this question, which may not be obvious at first, lies in the predator functional and numerical responses. We see this effect because both numerical and functional responses level off at intermediate prey densities; that is, beyond a certain threshold, further increases in prey density do not lead to either higher predator densities or increased feeding rates. Meanwhile, the density of the prey population continues to increase and the proportion of the prey eaten by predators declines. This work by Holling suggests that prey can reduce their individual probability of being eaten by occurring at very high densities. It appears that this defensive tactic, which is called **predator satiation,** is employed by a wide variety of organisms from insects and plants to marine invertebrates and African antelope.

## Predator Satiation by an Australian Tree

A great number of plants flower and produce seeds synchronously over large areas, including bamboo, pines, beech, and oak. This phenomenon of synchronous widespread seed and fruit production is called **masting.** Daniel Janzen (1978) proposed that a major selective force favoring the production of mast crops is seed predation. He suggested that mast crops may lead to satiation of seed predators, allowing some seeds to escape predation, germinate, and successfully establish.

Many Australian trees in the genus *Eucalyptus* disperse their seeds in large numbers following forest fires. Seeds are produced each year but mostly remain stored in closed seed capsules that are retained on the tree. Following a fire, a massive, synchronized release floods the forest floor with seeds.

Synchronous seed release by *Eucalyptus* may be a defense against seed predators, but which ones? The chief seed predators in Australian forests appear to be ants. We usually think of Australia as the continent of kangaroos and koalas, but the region could be just as well known for its ants. Australia harbors a tremendous diversity of ants. For instance, while North American deserts support about 160 species of ants, the deserts of Australia contain an estimated 2,000 species and the continent as a whole may harbor nearly 4,000 species. The ants in Australian forests have been reported to prevent forest regeneration by removing up to 80% of the seeds broadcast by foresters.

D. O'Dowd and A. Gill (1984) used field experiments to determine whether synchronous seed dispersal by *Eucalyptus* might be a means to reduce seed losses to ants. They set up study plots in two forests of Australian alpine ash, *Eucalyptus delgatensis,* a gigantic tree found in the Australian Alps and Tasmania, where it commonly grows to 50 to 60 m in height. One of the study sites was to serve as a reference site and the second as an experimental site. *E. delgatensis* constituted nearly 90% of the tree biomass at both study sites. The researchers monitored a number of physical and biological variables at the control and experimental sites and then set a controlled high-intensity fire at the experimental site. The area burned was approximately 98 ha; 93% of the trees were killed.

The results of O'Dowd and Gill's field experiments support their hypothesis that synchronous seed dispersal by *Eucalyptus* reduces losses of seeds to ants. As expected, the fire stimulated the release of massive numbers of seeds. During the 3 weeks following the fire, seed fall was approximately 405 fertile seeds per square meter, compared with a peak seed fall of 10 seeds per square meter per week at the control site, which was not burned. The fire also seemed to stimulate ant activity. Prior to the fire, researchers trapped an average of 176 ants belonging to 14 species each week. After the fire, the number of ants trapped each week rose to an average of 680 individuals belonging to 23 species. Despite this strong numerical response, the rate at which ants removed seeds dropped following the fire. The rate of seed removal from seed trays at the experimental site dropped from an average of about 65% per week during the 5 weeks prior to the fire to an average of about 14% per week during the 5 weeks following the fire. This result is consistent with the predator satiation hypothesis proposed by O'Dowd and Gill and is consistent with Holling's prediction of reduced predator combined response at high prey densities.

So, what does O'Dowd and Gill's experiment tell us? We know that *E. delgatensis* stores seeds in closed seed capsules, that these seeds are released synchronously following intense fires, and that a substantial proportion of these seeds escape predation even in the face of strong numerical response by seed-eating ants. Is this information sufficient to conclude that the apparent predator satiation strategy of *E. delgatensis* provides an effective "refuge" from predation? It seems that we should also know whether greater seed survival is translated into greater seedling establishment, an evolutionary bottom line. O'Dowd and Gill also showed that seedling establishment was greater on the burned experimental plot than on the control plot (fig. 14.22). About 1.5 years after the experimental fire, seedling survival at the experimental site was approximately 2 individuals per square meter, or 20,000 individuals per hectare.

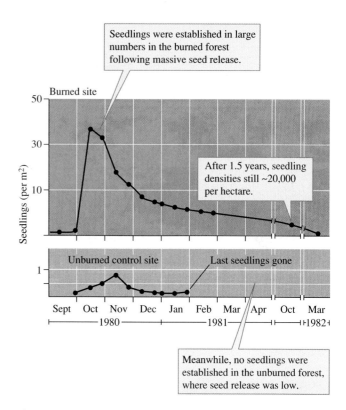

**Figure 14.22** Seedling establishment by *Eucalyptus delgatensis* at burned and unburned sites (data from O'Dowd and Gill 1984).

This seems a respectable level of reproductive success for trees that could eventually reach 50 to 60 m in height. Predator satiation also provides protection to many kinds of insects, but nowhere is this more apparent than among the cicadas.

## Predator Satiation by Periodical Cicadas

Periodical cicadas, *Magicicada* spp., emerge as adults once every 13 years in the southern part of their range in North America and once every 17 years in the northern part of their range. Though these insects emerge only once every 13 or 17 years in any particular area, virtually every year sees a brood emerging somewhere in eastern North America. An emergence of periodical cicadas produces a sudden flush of singing insects whose density can approach $4 \times 10^6$ individuals per hectare, which translates into a biomass of 1,900 to 3,700 kg of cicadas per hectare, the highest biomass of a natural population of terrestrial animals ever recorded.

Periodical cicadas are insects of the order Homoptera, which includes the leafhoppers and aphids. Like their relatives, cicadas make their living by sucking the fluids of plants and spend either 13 or 17 years of their life as nymphs underground, where they feed on the xylem fluids in roots. When mature, nymphs dig their way to the soil surface, where they shed their nymphal skin and emerge as winged adults. Among periodical cicadas this emergence is so synchronized that millions of adults emerge over a period of only a few days. Following emergence males fly to the treetops, where they sing the mating songs to which females are attracted. After they mate, females lay their eggs in living twigs of shrubs and trees. When the nymphs hatch in about 6 weeks, they immediately drop to the ground and burrow down to a root, where they begin to feed, moving around very little for the next 13 or 17 years. A mass emergence of periodical cicadas, one of the most memorable biological phenomena nature has to offer, appears aimed at predator satiation.

Kathy Williams and her colleagues (1993) tested the effectiveness of predator satiation in a population of 13-year periodical cicadas in northwest Arkansas. They monitored emergence of cicadas using conical emergence traps constructed of plastic mesh and inverted their traps to measure predation rates (fig. 14.23). Nymphs emerging from the ground below the traps could be counted to estimate the numbers of emerging nymphs. Then, as adult cicadas died from a variety of factors, including physical factors, senescence, and pathogens, they fell from the trees to the ground, where some were caught in the inverted

traps. Because the major predators were birds, predation rates could be estimated because birds discard the wings of cicadas as they feed upon them. The wings falling into the inverted traps gave an estimate of predation rates.

Patterns of mortality and predation rates relative to population size support the predator satiation hypothesis. Williams and her colleagues estimated that 1,063,000 cicadas emerged from their 16 ha study site and that 50% of these emerged during four consecutive nights. Cicada abundance peaked in late May and then declined rapidly during the first 2 weeks of June. Part of this decline was due to mortality from severe thunderstorms during the first week of June. Figure 14.24 shows that losses due to birds were low throughout the period of peak cicada abundance and then climbed to 100% as cicada populations declined during June. These results indicate that the predator satiation tactic was sufficiently effective to reduce cicada losses to birds to only 15% of the total population.

## Size as a Refuge

We first encountered size-selective predation in chapter 6 among bluegills, *Lepomis macrochirus,* and pumas, *Felis concolor.* However, many other organisms select their prey by size. In fact, average prey size shows a significant correlation with predator size across taxa ranging from lizards to small mammals. The reason for size selective predation among such a diverse array of organisms is that prey capture and consumption are mechanical problems, and as we saw in chapter 6, size can influence the time required to handle prey and therefore the rate of energy intake. The bottom line is that for a given predator some prey are simply too large to be profitable and so are not attacked.

Now let's look at size from the perspective of the prey. If large individuals are ignored by predators, then large size may offer a form of refuge. An obvious example is on the African savanna. While a variety of predators may attack the calves of elephants or rhinoceros, the same predators avoid the adults, which have been observed to kill adult lions (fig. 14.25). On a

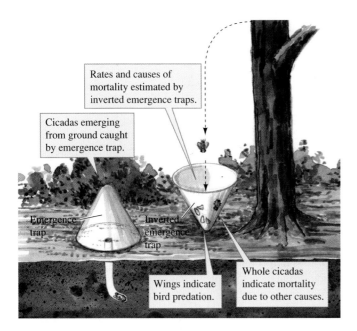

**Figure 14.23** Estimating cicada population size and predation rates by birds.

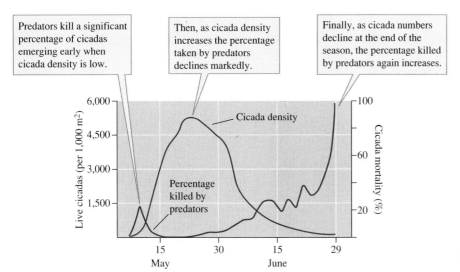

**Figure 14.24** Cicada population density and their percent mortality due to predation (data from Williams, Smith, and Stephan 1993).

**Figure 14.25** Large size can provide a refuge from predators. While young African elephants may be vulnerable to predation by African lions, mature elephants are not.

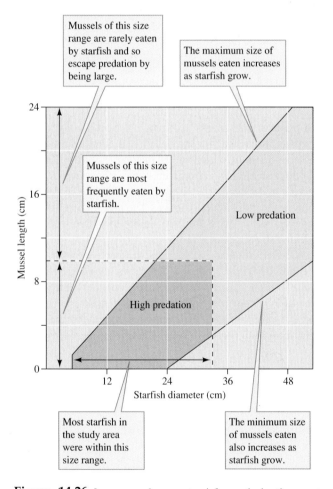

Mussels of this size range are rarely eaten by starfish and so escape predation by being large.

The maximum size of mussels eaten increases as starfish grow.

Mussels of this size range are most frequently eaten by starfish.

Low predation

High predation

Most starfish in the study area were within this size range.

The minimum size of mussels eaten also increases as starfish grow.

**Figure 14.26** Large mussels are eaten infrequently by the sea star *Pisaster ochraceus* (data from Paine 1976).

smaller scale, Robert Paine (1976) found that the sea star *Pisaster ochraceus* does not consume the largest individuals in populations of one of its chief prey species, the mussel *Mytilus californianus.* Figure 14.26 shows that the maximum size of mussels eaten by sea stars is a function of sea star size. Notice that most of the successful predation observed by Paine involved small- to medium-sized sea stars attacking mussels

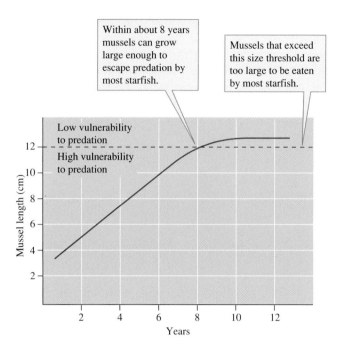

Within about 8 years mussels can grow large enough to escape predation by most starfish.

Mussels that exceed this size threshold are too large to be eaten by most starfish.

Low vulnerability to predation

High vulnerability to predation

**Figure 14.27** Growth by mussels in an intertidal area from which the sea star, *Pisaster ochraceus,* was excluded (data from Paine 1976).

less than 11 cm long. Most sea stars cannot eat the largest mussels, and the largest sea stars that can were limited to a few areas of coastline in the study area. What this means is that if a mussel can manage to escape predation long enough to reach 10 to 12 cm in length, it will be immune from attack by most sea stars. When Paine removed the sea stars from an area of the intertidal zone, resident mussels survived at higher rates and therefore grew to a larger average size (fig. 14.27). When Paine allowed sea stars to recolonize the area, many of the mussels were large enough that they effectively escaped predation by sea stars. This result has implications that reach far beyond higher survival within a single prey population.

If predators pass up prey above a particular size threshold, might natural selection favor organisms that project a "large" body size to some would-be predators? It appears that some aquatic insects have been selected to do just that. Barbara Peckarsky (1980, 1982) observed that mayflies in the family Ephemerellidae would "stand their ground" in the face of a foraging predatory stonefly. In fact, they would not only stand their ground, they would curve their abdomens over their backs and point the tips of their abdominal cerci into the face and antennae of a stonefly, a behavior Peckarsky called a "scorpion" posture (fig. 14.28). Usually a stonefly greeted in this way does not attack. While many other stream ecologists had seen this behavior in ephemerellid mayflies, Peckarsky was the first to suggest that the scorpion posture was a defensive tactic in which the mayfly projected a larger image to a tactile, size-selective predator.

Why should a large stonefly avoid large ephemerellid mayflies? Large ephemerellids have been observed attacking stoneflies trying to prey on them, and so like lions that avoid rhinoceros, stoneflies that avoid ephemerellids may be protect-

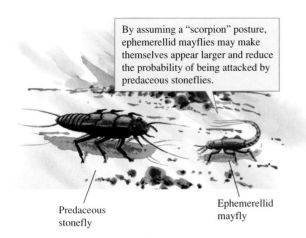

By assuming a "scorpion" posture, ephemerellid mayflies may make themselves appear larger and reduce the probability of being attacked by predaceous stoneflies.

Predaceous stonefly

Ephemerellid mayfly

**Figure 14.28** Posturing by an ephemerellid mayfly confronted by a predaceous stonefly.

ing themselves from injury. Most ephemerellids, however, present no danger to large predaceous stoneflies, so self-protection only partially answers our question. For the bulk of encounters between stoneflies and ephemerellid mayflies, large apparent size would probably indicate low profitability, low E/T in terms of optimal foraging theory (see chapter 6), and send the predator looking for a prey that would yield a higher energy return. It may be that the display by ephemerellids is not a bluff, however, since they require an exceptionally long handling time for a prey of their size. The scorpion posture of ephemerellids may be a case of "truth in advertising."

While we may think of predators as threats to ourselves or to livestock or crops, many predators and parasites have been used to control populations of insects that attack crops or to control invasive weeds. As we shall see in the Applications & Tools section, predators are increasingly used to control parasites that infect humans.

## APPLICATIONS & TOOLS

### Using Predators to Control a Parasite

Parasitic diseases afflict approximately 600 million people across the planet, particularly in tropical and subtropical countries. Despite intensive efforts at control, many parasitic diseases are spreading and the number of cases appears to be increasing. A key factor in this increase appears to be human population growth, which puts additional pressure on sanitation and health care systems and increases the number of hosts for human parasites. The leading parasitic disease in humans is malaria, which is transmitted to humans by mosquitoes and infects an estimated 250 million people. The second most prevalent parasitic disease is schistosomiasis, which infects approximately 200 million people. Schistosomiasis is a debilitating infection caused by blood flukes of the genus *Schistosoma*. Infections by this parasite are particularly debilitating to children. The scope and intensity of infections by *Schistosoma* and other parasites challenge the world health community to develop systems for their control. However, the problem of control is essentially an ecological one and therefore complex.

Much of this complexity is due to the life cycle of the parasite (fig. 14.29). *Schistosoma* spends its larval phase as a parasite in aquatic snails and its adult phase in humans. The *cercariae,* the stage of *Schistosoma* that infects humans, are released by snails into the water. Cercariae penetrate the skin of humans in streams, lakes, or ponds containing infected snails. Some *Schistosoma* infect the human digestive tract while others infect the urinary tract. Humans that either urinate or defecate in water complete the parasite's life cycle by facilitating the infection of snails.

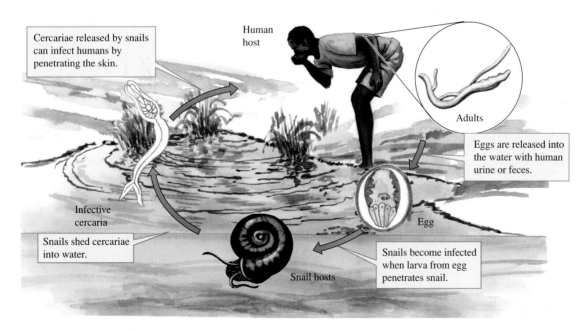

Cercariae released by snails can infect humans by penetrating the skin.

Human host

Adults

Eggs are released into the water with human urine or feces.

Infective cercaria

Snails shed cercariae into water.

Egg

Snails become infected when larva from egg penetrates snail.

Snail hosts

**Figure 14.29** The life cycle of *Schistosoma.*

Schistosomiasis can now be treated with a variety of drugs. However, treated patients are often reinfected. Consequently, schistosomiasis control programs that rely solely on treatment of infected individuals cannot control the spread of infection. Effective control must also include the populations of snails that serve as intermediate hosts. Methods of snail control have included applications of chemicals that kill snails, introduction of other snails that competitively displace the snail species that serve as intermediate hosts for *Schistosoma*, or, increasingly, using predators to control snail populations. This work draws upon one of the main concepts of chapter 14: predators, parasites, and pathogens influence the distribution, abundance, and structure of prey and host populations.

Scientists in several East African countries are researching the potential of a variety of predators to control the host snails infected by *Schistosoma*. One of those countries is Kenya. Kenya's population includes 1 to 2 million people infected with *Schistosoma*. Health officials are concerned that the number of cases may increase due to increased pollution of freshwater environments by untreated sewage and increased construction of dams and irrigation systems, which favor the growth of snail populations. In the face of these threats, Kenyan health officials are developing a comprehensive, multifaceted plan for control of schistosomiasis. One of the elements in their plan is to use predators to control the parasite's host snails.

One of the predators being tested for its effectiveness at snail control is the crayfish *Procambarus clarkii*, which is native to North America. *Procambarus* was introduced into Kenya during the 1970s. Health officials became interested in the potential of the crayfish to control snail populations when snail surveys showed that habitats with *Procambarus* lacked host snails. However, because *Procambarus* is not native to Kenya, ecologists and health officials are proceeding cautiously. They point out that *Procambarus*, which has been introduced to 24 countries worldwide, is a highly invasive species with the potential to disrupt native populations and ecosystems. It has already spread rapidly across Kenya. In addition, the crayfish may threaten rice cultivation in Kenya. Its burrows have damaged rice fields in other regions and it eats rice seedlings. The environments where *Procambarus* may be most useful and where it will likely cause little damage are the thousands of artificial ponds that dot the rural Kenyan landscape. These ponds are used to water livestock and as domestic water supplies.

Ecologists have used a combination of laboratory and field studies to test the potential of *Procambarus* to control host snail populations. In one of the early studies Bruce Hofkin and Sam Loker of the University of New Mexico joined Gerald Mkoji and Davy Koech of the Kenya Medical Research Institute in a survey of the distributions of *Procambarus* and host snails in Kenya (Hofkin et al. 1991b). The research team restricted their survey to areas where both the crayfish and host snails were known to occur and to habitats within those areas that could support both organisms. The survey, which included 53 sites, revealed a highly significant negative association between *Procambarus* and host snails. Nine sites had neither snails nor crayfish. In the 44 other sites, 19 had snails only, 21 had crayfish

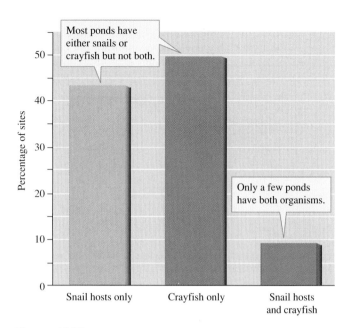

**Figure 14.30** Distributions of snail hosts of *Schistosoma* and crayfish in Kenyan ponds (data from Hofkin et al. 1991b).

but no snails, and only 4 sites had both crayfish and snails (fig. 14.30). The snails at these sites were present in small numbers and consisted of species incapable of transmitting human schistosomiasis. This survey indicates clearly that *Procambarus* has the potential to eliminate host snails from a variety of aquatic habitats in Kenya.

Based on the results of their survey, Hofkin and his colleagues (1991a) conducted a series of laboratory and field trials to test directly for the capacity of crayfish to eliminate host snail populations. The first step in these studies was to determine if *Procambarus* eats the snail species infected by *Schistosoma*. Laboratory feeding studies in 10 L aquaria showed that the crayfish readily eat the main species of host snails. Because the aquaria were bare and no alternative food was provided for crayfish, the conclusions that we can draw from these experiments are limited. Hofkin and his colleagues concluded only that *Procambarus* has the potential to control host snails.

To more closely simulate field conditions, Hofkin and his colleagues next increased the size and complexity of their laboratory environments and provided crayfish with alternative food. In one of these experiments, they used large plastic tanks to which they added 10 cm of soil and then filled with water to 30 cm depth. Each tank contained 50 host snails and four water lilies, which provided both food and shelter for the snails. They added two or four crayfish to each tank and counted the number of remaining snails after 5, 15, or 30 days. *Procambarus* significantly reduced the number of snails after just 5 days and eliminated them entirely after 30 days in tanks containing four crayfish. Crayfish also ate significant quantities of water lily leaves, reducing the habitat and food available to snails. Therefore, *Procambarus* is both a predator and competitor of host snails.

With these results, Kenyan health officials were encouraged to begin testing the effectiveness of crayfish to control host snail populations in the field. However, the potential of

*Procambarus* to damage native ecosystems suggested caution. Consequently, Gerald Mkoji of the Kenya Medical Research Institute and a research team from several other institutions suggested that the crayfish be used only in small artificial impoundments, where they would pose no threat to Kenyan wildlife. These ponds are a major source of infection by *Schistosoma*. It appears that *Procambarus* has eliminated host snail populations from the ponds where it has been introduced. Used judiciously, the crayfish appear to have the potential to reduce the impact of a serious human pathogen without causing significant environmental damage. This result comes directly from the growing understanding of the ecology of exploitation, the central theme of chapter 14.

## SUMMARY

The diversity of interactions between herbivores and plants, between predators and prey, and between parasites, parasitoids, pathogens, and hosts can be grouped under the heading of *exploitation*—interactions between species that enhance the fitness of one individual at the expense of another.

**Exploitation weaves populations into a web of relationships that defy easy generalization.** The number of exploitative interactions between species far exceeds the number of species in the biosphere, and the nature of exploitation goes far beyond the typical consumption of one organism by another. For instance, many parasites and pathogens manipulate host behavior to enhance their own fitness at the expense of the host. Spiny-headed worms alter the behavior of a variety of crustacean hosts in a way that increases the probability that the one host species will be eaten by another. A pathogenic fungus manipulates the growth program of its host plant in a way to produce "pseudoflowers," structures aimed at promoting the reproduction of the pathogen. In the process the pathogen usually kills the host plant and always renders it sterile. Predation by one flour beetle species on another can be used as a potent means of interference competition except in the presence of a protozoan parasite, which seems to give a competitive advantage to less predaceous species.

**Predators, parasites, and pathogens influence the distribution, abundance, and structure of prey and host populations.** Herbivorous stream insects have been shown to control the density of their agal and bacterial food. The herbivorous moth larva *Cactoblastis cactorum* combined with pathogenic microbes reduced the coverage of prickly pear cactus in Australia from millions of hectares to a few thousand. A parasitic infestation reduced the red fox population in Sweden by 70%, which in turn led to increases in the abundance of several prey species eaten by foxes. This parasitic disease revealed the influence of a predator on its prey populations.

**Predator-prey, parasite-host, and host-pathogen relationships are dynamic.** Populations of a wide variety of predators and prey show highly dynamic fluctuations in abundance ranging from days to decades. A particularly well-studied example of predator-prey cycles is that of snowshoe hares and their predators, which have been shown to result from the combined effects of the snowshoe hares on their food and of the predators on the snowshoe hare population. Mathematical models of predator-prey interactions by Lotka and Volterra suggest that exploitative interactions themselves can produce population cycles without any influences from outside forces such as weather. Predator-prey cycles have also been observed in a few laboratory populations under restricted circumstances.

**To persist in the face of exploitation, hosts and prey need refuges.** The refuges that promote the persistence of hosts and prey include secure places to which the exploiter has limited access. However, living in large groups can be considered as a kind of refuge since it reduces the probability that an individual host or prey will be attacked. It appears that predator satiation is a defensive tactic used by a wide variety of organisms from rain forest trees to temperate insects. Growing to large size can also represent a kind of refuge when the prey species is faced by size-selective predators. Size is used as a refuge by prey species ranging from stream insects and intertidal invertebrates to rhinoceros.

Predators and parasites have been used to control populations of insects that attack crops or to control invasive weeds. Recent research in Kenya has shown that a crayfish, *Procambarus clarkii*, controls the snails that act as intermediate hosts for *Schistosoma*, a highly pathogenic human parasite. Preliminary results indicate that crayfish successfully control host snails in the artificial impoundments used for livestock watering and domestic water, important sources of infection by *Schistosoma*.

## Review Questions

1. Predation is one of the processes by which one organism exploits another. Others are herbivory, parasitism, and disease. What distinguishes each of these processes, including predation, from the others? We can justify discussing these varied processes under the heading of exploitation because each involves one organism making its living at the expense of another. By what "currency" would you measure that expense (e.g., energy, fitness)?

2. How are manipulation of host behavior by spiny-headed worms and manipulation of plant growth by the rust *Puccinia monoica* the same? How are they different? The details of these parasitic interactions are very different in many ways from the predatory behavior of lions on the savannas of Africa. How are the activities of spiny-headed worms, rusts, and lions similar?

3. Predation by one flour beetle species on another can be used as a potent means of interference competition. However, the predatory strategy seems to fail consistently in the presence of the protozoan parasite *Adelina tribolii*. Explain how the predatory strategy works in one environmental circumstance and fails in another.

4. In chapter 14 we have seen how an herbivorous stream insect controls the density of its food organisms, how an herbivorous moth larva and pathogenic microbes combine to control an introduced cactus population, and how decimation of a red fox population led to increases in the populations of the foxes' prey. We do not know the specific environmental factors controlling most populations. Explain why such factors must exist. (Hint: Think back to our discussions of geometric and exponential growth in chapter 11).

5. Early work on exploitation focused a great deal of attention on predator-prey relations. However, parasites and pathogens represent a substantial part of the Concept Discussions in chapter 14. Is this representation by parasites and pathogens just the result of biased choices by the author or do you think that parasites and pathogens have the potential to exert significant controls on natural populations? Justify your answer.

6. Researchers have suggested that predators could actually increase the population density of a prey species heavily infected by a pathogenic parasite (Hudson, Dobson, and Newborn 1992). Explain how predation could lead to population increases in the prey population.

7. Explain the roles of food and predators in producing cycles of abundance in populations of snowshoe hare. Populations of many of the predators that feed on snowshoe hares also cycle substantially. Explain population cycles among these predator populations.

8. What contributions have laboratory and mathematical models made to our understanding of predator-prey population cycles? What are the shortcomings of these modeling approaches? What are their advantages?

9. We included spatial refuges, predator satiation, and size in our discussions of the role played by refuges in the persistence of exploited species. How could time act as a refuge? Explain how natural selection could lead to the evolution of temporal "refuges."

Joseph Culp and Gary Scrimgeour (1993) studied the timing of feeding by mayfly larvae in streams with and without fish. These mayflies feed by grazing on the exposed surfaces of stones, where they are vulnerable to predation by fish, which in the streams studied are size-selective feeders and feed predominantly during the day. In the study streams without fish, both small and large mayflies have a slight tendency to feed during the day but feed at all hours of the day and night. In the streams with abundant fish populations, small mayflies fed around the clock, while large mayflies fed mainly at night. Explain these patterns in terms of time as a refuge and size-selective predation.

10. When applying ecological concepts to practical problems, there are often trade-offs between the benefits and costs of a management decision. Review the costs and benefits of using the crayfish *Procambarus clarkii* for control of schistosomiasis in Kenya.

## Suggested Readings

Haukioja, E., K. Kapiainen, P. Niemelä, and J. Tuomi. 1983. Plant availability hypothesis and other explanations of herbivore cycles: complementary or exclusive alternatives? *Oikos* 40:419–32.

Keith, L. B. 1983. Role of food in hare population cycles. *Oikos* 40:385–95.

*These two papers provide comprehensive reviews of the hypotheses proposed to explain hare population cycles.*

Huffaker, C. B. 1958. Experimental studies on predation: dispersion factors and predator-prey oscillations. *Hilgardia* 27:343–83.

Huffaker, C. B., K. P. Shea, and S. G. Herman. 1963. Experimental studies on predation: complex dispersion and levels of food in an acarine predator-prey interaction. *Hilgardia* 34:305–30.

*Two classic papers on predator-prey oscillations, involving complex laboratory studies. Difficult in some places but worth the effort. These two papers have had a major impact on the study of exploitative interactions.*

Krebs, C. J., R. Boonstra, S. Boutin, and A. R. E. Sinclair. 2001. What drives the 10-year cycle of snowshoe hares? *BioScience* 51:25–36.

*In this paper Charles J. Krebs provides an excellent and accessible summary of what is known about the snowshoe hare population cycle, a topic that he has researched for over 40 years.*

Krebs, C. J., S. Boutin, R. Boonstra, A. R. E. Sinclair, J. N. M. Smith, M. R. T. Dale, K. Martin, R. Turkington. 1995. Impact of food and predation on the snowshoe hare cycle. *Science* 269:1112–15.

*A concise report of a critical, large-scale experiment designed to test the influences of food and predation on the snowshoe hare cycle.*

Lamberti, G. A. and V. H. Resh. 1983. Stream periphyton and insect herbivores: an experimental study of grazing by a caddisfly population. *Ecology* 64:1124–35.

*A classic field experiment that reveals the controlling influence of a benthic invertebrate grazer on a stream community. A model for the design of field experiments.*

Lindström, E. R., H. Andrén, P. Angelstam, G. Cederlund, B. Hörnfeldt, L. Jäderberg, P.-A. Lemnell, B. Martinsson, K. Sköld, and J. E. Swenson. 1994. Disease reveals the predator: sarcoptic mange, red fox predation, and prey populations. *Ecology* 75:1042–49.

*Large-scale, long-term study of interactions between a pathogenic parasite, a predator, and its prey. An example of how to use natural experiments to extract information from nature.*

Moore, J. 1984. Parasites that change the behavior of their host. *Scientific American* 250:108–15.

*An excellent review of the influences of parasites on prey behavior. Exceptionally well written and illustrated.*

O'Donoghue, M., S. Boutin, C. J. Krebs, and E. J. Hofer. 1997. Numerical responses of coyotes and lynx to the snowshoe hare cycle. *Oikos* 80:150–62.

O'Donoghue, M., S. Boutin, C. J. Krebs, G. Zuleta, D. L. Murray, and E. J. Hofer. 1998. Functional responses of coyotes and lynx to the snowshoe hare cycle. *Ecology* 79:1193–1208.

*A series of papers in which the authors provide detailed information about the interactions between snowshoe hares and their principal predators, coyotes and lynx.*

Roy, B. A. 1993. Floral mimicry by a plant pathogen. *Nature* 362:56–58.

*Fascinating example of manipulation of a plant and its pollinators by a pathogenic fungus.*

Utida, S. 1957. Cyclic fluctuations of population density intrinsic to the host-parasite system. *Ecology* 38:442–49.

*A concise, readable report of one of the longest-running experiments on parasitoid-host population cycles—a landmark study in the field.*

Williams, K. S., K. G. Smith, and F. M. Stephen. 1993. Emergence of 13-yr periodical cicadas (Cicadidae: *Magicicada*): phenology, mortality, and predator satiation. *Ecology* 74:1143–52.

*Beautifully designed field study of a complex problem. Demonstrates predator satiation by periodical cicadas.*

# On the Net

Visit this textbook's accompanying website at www.mhhe.com/ecology (click on the book's title) to take advantage of practice quizzing, study/writing tips, timely news articles, and additional URLs for research on the topics in this chapter.

Animal Population Ecology
Parasitism, Predation, Herbivory
Field Methods for Studies of Populations

Parasitic Protists
Human Diseases Caused by Nematode
Coevolution

# Chapter

# 15

# Mutualism

Positive interactions between species are found throughout the biosphere. A hummingbird darts among the red blossoms of a plant growing at the edge of a forest glade. As it inserts its bill into a flower, hovering to sip nectar, the hummingbird head brushes up against the anthers of the flower and picks up pollen (fig. 15.1). This pollen will be deposited on the stigmas of other flowers as the hummingbird goes about gathering its meal of nectar. The hummingbird disperses the plant's pollen in trade for a meal of nectar.

Belowground we encounter another partnership. The roots of the hummingbird-pollinated plant are intimately connected with fungi in an association called *mycorrhizae*. The hyphae of the mycorrhizal fungi extend out from the roots, increasing the capacity of the plant to harvest nutrients from the environment. In exchange for the nutrients, the plant delivers sugars and other products of photosynthesis to its fungal partner.

Meanwhile, back aboveground a deer enters the forest glade and wanders over to the plant recently visited by the hummingbird. The deer systematically grazes it to the ground, lightly chews the plant material, and then swallows it. As the plant material enters the deer's stomach, it is attacked by a variety of protozoans and bacteria. These microorganisms break down and release energy from compounds such as cellulose, which the deer's own enzymatic machinery cannot handle. In return, the protozoans and bacteria receive a steady food supply from the feeding activities of the deer as well as a warm, moist place in which to live.

These are examples of **mutualism,** that is, interactions between individuals of different species that benefit both partners. Some species can live without their mutualistic partners and so the relationship is called **facultative mutualism.** Other species are so dependent upon the mutualistic relationship that they cannot live in its absence. Such a relationship is an **obligate mutualism.** It is a curious fact that though observers of nature as early as Aristotle recognized such mutualisms, mutualistic interactions have received much less attention from ecologists than have either competition or exploitation. Does this lack of attention reflect the rarity of mutualism in nature? As you will see in chapter 15, mutualism is virtually everywhere.

Mutualism may be common, but is it important? Does it contribute substantially to the ecological integrity of the biosphere? The answer to both these questions is yes. Without mutualism the biosphere would be entirely different. Let's remove some of the more prominent mutualisms from the biosphere and consider the consequences. An earth without mutualism would lack reef-building corals as we know them. So we can erase the Great Barrier Reef, the largest biological structure on earth, from our hypothetical world. We can also eliminate all the coral atolls that dot the tropical oceans as well as all the fringing reefs. The deep sea would have no bioluminescent fishes or invertebrates. In addition, the deep-sea oases of life associated with ocean floor hydrothermal vents, discovered just two decades ago (see chapter 6), would be reduced to nonmutualistic microbial species.

On land, there would be no animal-pollinated plants: no orchids, no sunflowers, and no apples. The pollinators them-

**Figure 15.1** Hummingbirds feeding on nectar transfer pollen from flower to flower.

selves would also be gone: no bumblebees, no hummingbirds, and no monarch butterflies. Gone too would be all the herbivores that depend on animal-pollinated plants. Without plant-animal mutualisms, tropical rain forests, the most diverse terrestrial biome on the planet, would be all but gone. Many wind-pollinated plants would remain. However, many of these species would also be significantly affected since approximately 90% of all plants form mycorrhizae. Those plants capable of surviving without mycorrhizal fungi would likely be restricted to the most fertile soils.

Even if wind-pollinated, nonmycorrhizal plants remained on our hypothetical world there would be no vast herds of African hoofed mammals, no horses, and no elephants, camels, or even rabbits or caterpillars. There would be few herbivores to feed on the remaining plants since herbivores and detritivores depend upon microorganisms to gain access to the energy and nutrients contained in plant tissues. The carnivores would disappear along with the herbivores. And so it would go. A biosphere without mutualism would be biologically impoverished.

The impoverishment that would follow the elimination of mutualism, however, would go deeper than we might expect. Lynn Margulis and René Fester (1991) have amassed convincing evidence that all eukaryotes, both heterotrophic and autotrophic, originated as mutualistic associations between different organisms. Eukaryotes are apparently the product of mutualistic relationships so ancient that the mutualistic partners have become cellular organelles (e.g., mitochondria and chloroplasts) whose mutualistic origins long went unrecognized. Consequently, without mutualism all the eukaryotes, from *Homo sapiens* to the protozoans, would be gone and the history of life on earth and biological richness would be set back about 1.4 billion years.

But back here in the present, let's accept that mutualism is an integral part of nature and review what is known of the ecology of mutualism. The first part of this brief review emphasizes experimental studies. Then, in the last part of chapter 15, we examine some theoretical approaches to the study of mutualism.

## CONCEPTS

- **Plants benefit from mutualistic partnerships with a wide variety of bacteria, fungi, and animals.**
- **Reef-building corals depend upon mutualistic relationships with algae and animals.**
- **Theory predicts that mutualism will evolve where the benefits of mutualism exceed the costs.**

## CONCEPT DISCUSSION

### Plant Mutualisms

**Plants benefit from mutualistic partnerships with a wide variety of bacteria, fungi, and animals.**

Plants are the center of mutualistic relationships that provide benefits ranging from nitrogen fixation and nutrient absorption to pollination and seed dispersal. It is no exaggeration to say that the integrity of the terrestrial portion of the biosphere depends upon plant-centered mutualism. However, to understand the extent to which ecological integrity may depend upon these relationships we need careful observational studies and experiments. Here are some drawn from studies of mycorrhizae.

# Plant Performance and Mycorrhizal Fungi

The fossil record shows that mycorrhizae arose early in the evolution of land plants, perhaps as long as 400 million years ago. Over evolutionary time, a mutualistic relationship between plants and fungi evolved in which mycorrhizal fungi provide plants with greater access to inorganic nutrients while feeding off the root exudates of plants. The two most common types of mycorrhizae are (1) **arbuscular mycorrhizal fungi (AMF),** in which the mycorrhizal fungus produces **arbuscules,** sites of exchange between plant and fungus, **hyphae,** fungal filaments, and **vesicles,** fungal energy storage organs within root cortex cells, and (2) **ectomycorrhizae (ECM),** in which the fungus forms a mantle around roots and a netlike structure around root cells (fig. 15.2). Mycorrhizae are especially important in increasing plant access to

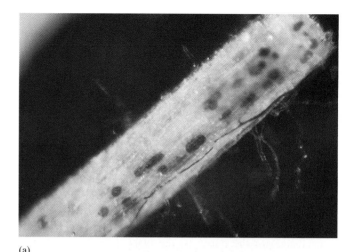

(a)

(b)

**Figure 15.2** Mutualistic associations between fungi and plant roots: (*a*) arbuscular mycorrhizal fungus stained so that fungal structures appear blue; and (*b*) ectomycorrhizae, which give a white fuzzy appearance to these roots.

phosphorus and other immobile nutrients (nutrients that do not move freely through soil) such as copper and zinc, as well as to nitrogen and water.

## *Mycorrhizae and the Water Balance of Plants*

Mycorrhizal fungi appear to improve the ability of many plants to extract soil water. Edie Allen and Michael Allen (1986) studied how mycorrhizae affect the water relations of the grass *Agropyron smithii* by comparing the leaf water potentials of plants with and without mycorrhizae. Figure 15.3 shows that *Agropyron* with mycorrhizae maintained higher leaf water potentials than those without mycorrhizae. This means that when growing under similar conditions of soil moisture, the presence of mycorrhizae helped the grass maintain a higher water potential. Does this comparison show that mycorrhizae are directly responsible for the higher leaf water potential observed in the mycorrhizal grass? No, they do not. These higher water potentials may be an indirect

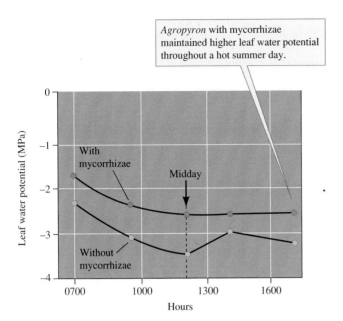

Figure 15.3 caption boxes:

*Agropyron* with mycorrhizae maintained higher leaf water potential throughout a hot summer day.

With mycorrhizae

Midday

Without mycorrhizae

**Figure 15.3** Influence of mycorrhizae on leaf water potential of the grass *Agropyron smithii* (data from Allen and Allen 1986).

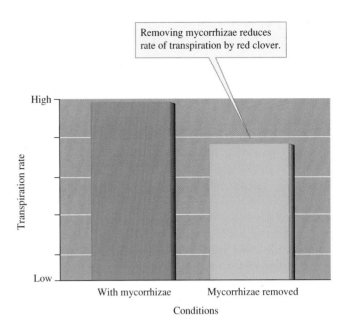

Removing mycorrhizae reduces rate of transpiration by red clover.

With mycorrhizae    Mycorrhizae removed

Conditions

**Figure 15.4** Effect of removing mycorrhizal hyphae on rate of transpiration by red clover (data from Hardie 1985).

effect of greater root growth resulting from the greater access to phosphorus provided by mycorrhizae.

Plants with greater access to phosphorus may develop roots that are more efficient at extracting and conducting water; mycorrhizal fungi may not be directly involved in the extraction of water from soils. Kay Hardie (1985) tested this hypothesis directly with an ingenious experimental manipulation of plant growth form and mycorrhizae. First, she grew mycorrhizal and nonmycorrhizal red clover, *Trifolium pratense,* in conditions in which their growth was not limited by nutrient availability. These conditions produced plants with similar leaf areas and root:shoot ratios. Under these carefully controlled conditions, mycorrhizal red clover showed higher rates of transpiration than nonmycorrhizal plants.

Hardie took her study one step further by removing the hyphae of mycorrhizal fungi from half of the red clover with mycorrhizae. She controlled for possible side effects of this manipulation by using a tracer dye to check for root damage and by handling and transplanting all study plants, including those in her control group. Removing hyphae significantly reduced rates of transpiration (fig. 15.4), indicating a direct role of mycorrhizal fungi in the water relations of plants. Hardy suggests that mycorrhizal fungi improve water relations of plants by giving more extensive contact with moisture in the rooting zone and provide extra surface area for absorption of water.

So far, it seems that plants always benefit from mycorrhizae. That may not always be the case. Environmental conditions may change the flow of benefits between plants and mycorrhizal fungi.

## Nutrient Availability and the Mutualistic Balance Sheet

Mycorrhizae supply inorganic nutrients to plants in exchange for carbohydrates, but not all mycorrhizal fungi deliver nutrients

to their host plants at equal rates. The relationship between fungus and plant ranges from mutualism to parasitism, depending on the environmental circumstance and mycorrhizal species or even strains within species.

Nancy Johnson (1993) performed experiments designed to determine whether fertilization can select for less mutualistic mycorrhizal fungi. Before discussing her experiments, we have to ask what would constitute a "less mutualistic" association. In general, a less mutualistic relationship would be one in which there was a greater imbalance in the benefits to the mutualistic partners. In the case of mycorrhizae, a less mutualistic mycorrhizal fungus would be one in which the fungal partner received an equal or greater quantity of photosynthetic product in trade for a lower quantity of nutrients.

Johnson pointed out that there are several reasons to predict that fertilization would favor less mutualistic mycorrhizal fungi. The first is that plants vary the amount of soluble carbohydrates in root exudates as a function of nutrient availability. Plants release more soluble carbohydrates in root exudates when they grow in nutrient-poor soils and decrease the amount of carbohydrates in root exudates as soil fertility increases. Consequently, fertilization of soils should favor strains, or species, of mycorrhizal fungi capable of living in a low-carbohydrate environment. Johnson suggested that the mycorrhizal fungi capable of colonizing plants releasing low quantities of carbohydrates will probably be those that are aggressive in their acquisition of carbohydrates from their host plants, perhaps at the expense of host plant performance. She addressed this possibility using a mixture of field observations and greenhouse experiments.

In the first phase of her project, Johnson examined the influence of inorganic fertilizers on the kinds of mycorrhizal fungi found in soils. She collected soils from 12 experimental plots in a field on the Cedar Creek Natural History Area in central

Minnesota that had been abandoned from agriculture for 22 years. Six of the study plots had been fertilized with inorganic fertilizers for 8 years prior to Johnson's experiment, while the other six had received no fertilizer over the same period.

Johnson sampled the populations of mycorrhizal fungi from fertilized and unfertilized soils and showed that the composition of mycorrhizal fungi differed substantially. Of the 12 mycorrhizal species occurring in the samples, unfertilized soil supported higher densities of three mycorrhizal fungi, *Gigaspora gigantea*, *G. margarita*, and *Scutellispora calospora*, while fertilized soil supported higher densities of one species, *Glomus intraradix*. Spores of *G. intraradix* accounted for over 46% of the spores recovered from fertilized soils but only 27% of the spores from unfertilized soils.

Johnson used greenhouse experiments to assess how these differences in the composition of mycorrhizal fungi might affect plant performance. She chose big bluestem grass, *Andropogon gerardii*, as a study plant for these experiments because it is native to the Cedar Creek Natural History Area and is well adapted to the nutrient-poor soils of the area. Seedlings of *Andropogon* were planted in pots containing 980 g of a 1:1 mixture of sterilized subsurface sand from the Cedar Creek Natural History Area and river-washed sand. Johnson added a composite sample of other soil microbes living in the soils of fertilized and unfertilized study plots. She prepared the composite by washing a composite soil sample from all fertilized and unfertilized study plots with deionized water and passing this water through a 25 μm screen.

To each pot Johnson added a mycorrhizal "inoculum" of 30 g of soil of one of three types: (1) a fertilized inoculum consisting of 15 g of soil from fertilized study plots mixed with 15 g of sterilized unfertilized soil, (2) an unfertilized inoculum consisting of 15 g of soil from unfertilized study plots mixed with 15 g of sterilized fertilized soil, or (3) a nonmycorrhizal inoculum consisting of 30 g of a sterilized composite from the soils of fertilized and unfertilized study plots. The first two inocula acted as a source of mycorrhizal fungi for colonization of *Andropogon*. The design of Johnson's experiment is summarized in figure 15.5.

Why did Johnson create her inocula by mixing sterilized and unsterilized soils from the fertilized and unfertilized study areas? She did so to control for the possibility that some nonbiological factor such as trace nutrients in one of the two soil types might have a measurable effect on plant performance. The completely sterilized inoculum acted as a control to assess the performance of plants in the absence of mycorrhizae. Why did Johnson's control consist of sterilized composite soil from all the study areas? Again, she had to guard against the possibility that the soils themselves without mycorrhizal fungi might affect plant performance.

Pots were next assigned to one of four nutrient treatments in which Johnson (1) added no supplemental nutrients (None), (2) added phosphorus only (+P), (3) added nitrogen only (+N), or (4) added both nitrogen and phosphorus (+N+P). The subsurface sand from the Cedar Creek Natural History Area contained a fairly low concentration of nitrogen

Question: Does fertilizing soil select for less mutualistic mycorrhizal fungi?

**Experimental Design**

Two sources of mycorrhizal fungi

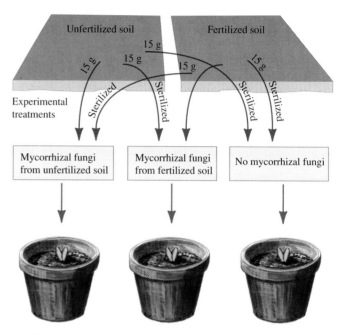

Compare: Growth, root:shoot ratios, and number of inflorescences produced by three treatments.

**Figure 15.5** Testing the effects of long-term fertilizing on interactions between mycorrhizal fungi and plants on agricultural lands.

but considerably higher concentrations of phosphorus. Nutrient additions were adjusted so that the supplemented treatments offered nitrogen and phosphorus concentrations comparable to those of the topsoil in the fertilized study plots.

Johnson harvested five replicates of each of the treatments at two points in time: at 4 weeks, when *Andropogon* was actively growing, and at 12.5 weeks, when the grass was fully grown. At each harvest she measured several aspects of plant performance: plant height, shoot mass, and root mass; and at 12.5 weeks she also recorded the number of inflorescences per plant.

At 12.5 weeks shoot mass was significantly influenced by nutrient supplements and by whether or not plants were mycorrhizal but not by the source of the mycorrhizal inoculum (fig. 15.6). Shoot mass was greatest in the double nutrient supplement treatment (+N+P), somewhat lower in the nitrogen supplement (+N), and very low in the other two treatments (None and +P). Figure 15.6*a* also indicates a definite influence of mycorrhizae on performance. Shoot mass was significantly greater for mycorrhizal plants across all nutrient treatments.

Nutrient supplements and mycorrhizae also significantly influenced root:shoot ratios (fig. 15.6*b*). As we saw in chapter 6, plants invest differentially in roots and shoots depending on nutrient and light availability. It also appears that variation in

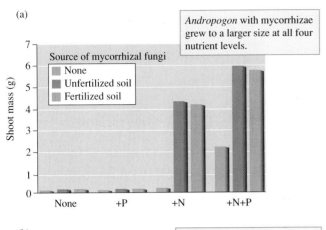

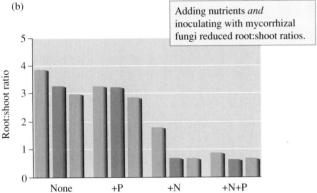

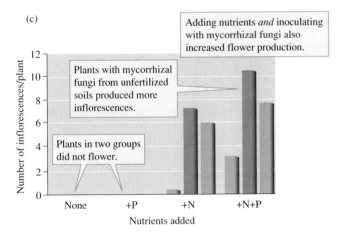

**Figure 15.6** Effect of nutrient additions and mycorrhizae on the grass *Andropogon gerardii* (data from Johnson 1993).

investment is aimed at increasing supplies of resources in short supply. For instance, in nutrient-poor environments many plants invest disproportionately in roots and consequently have high root:shoot ratios, which decline with increasing nutrient availability. The results of Johnson's experiments are consistent with this generalization. Root:shoot ratios were highest in the treatments without nitrogen supplements (None and +P) and lowest in the treatments with nitrogen supplements (+N and +N+P). In other words, higher plant investment in roots in the low-nitrogen treatments suggests greater nutrient limitation than in the high-nitrogen treatments.

Now let's look a bit deeper into Johnson's results, where we find evidence for increased nutrient availability to mycorrhizal plants. In both the +N and "None" treatments, root:shoot ratios were significantly lower in plants with mycorrhizae (fig. 15.6b). Mycorrhizal plants in these treatments invest less in roots, suggesting that they have greater access to nutrients. Here we also see a hint that the source of the inoculum significantly influenced plant performance. Plants in the +N+P treatment that were inoculated with soils from the unfertilized plots had slightly lower root:shoot ratios than those inoculated with soil from fertilized plots. These lower root:shoot ratios suggest that the mycorrhizal fungi from unfertilized soils were supplying their plant partners with more nutrients, freeing the plants to invest more of their energy budget in aboveground photosynthetic tissue.

Inflorescence production provides the strongest evidence for an effect of inoculum source on plant performance (fig. 15.6c). *Andropogon* produced inflorescences only in treatments with nitrogen supplements (+N and +N+P). Within these treatments, the mycorrhizal plants produced the greatest number of inflorescences. In addition, *Andropogon* inoculated with mycorrhizal fungi from the unfertilized plots and grown in the +N+P treatment produced the greatest number of inflorescences of all.

In summary, Johnson's study produced two pieces of evidence that bear on the question posed at the outset of her study: Can fertilization of soil select for less mutualistic mycorrhizal fungi? First, in the early stages of her experiment, *Andropogon* inoculated with fertilized soil had lower shoot mass than those inoculated with unfertilized soil. Second, *Andropogon* inoculated with unfertilized soils produced more inflorescences than did *Andropogon* inoculated with fertilized soils. In other words, *Andropogon* inoculated with mycorrhizal fungi from unfertilized soils showed faster shoot growth as young plants and reproduced at a higher rate when mature. These results suggest that plants receive more benefit from association with the mycorrhizal fungi from unfertilized soils. Johnson's simultaneous studies of the mycorrhizal fungi indicate the mechanisms producing these patterns. It appears that altering the nutrient environment does alter the mutualistic balance sheet, an influence of potential importance to agricultural practice.

Plants engage in a wide variety of mutualisms with many other organisms. One of those involves associations that provide protection from exploiters and competitors. Writing about the natural history of mutualism, Daniel Janzen (1985) included "plant-ant protection mutualisms" as one of his general categories of mutualism. Janzen (1966, 1967a, 1967b) himself is responsible for studying one of the best known of these mutualisms, the obligate mutualism between ants and swollen thorn acacias in Central America.

## Ants and Bullshorn Acacia

The ants mutualistic with swollen thorn acacias are members of the genus *Pseudomyrmex* in the subfamily Pseudomyrmecinae. This subfamily of ants is dominated by genera and species that have evolved close relationships with living plants. *Pseudomyrmex* spp. are generally associated with trees and show several characteristics that Janzen suggested are associated with arboreal living. They are generally fast and agile runners, have good vision, and forage independently. To this list, the *Pseudomyrmex* spp. associated with swollen thorn acacias, or "acacia-ants," add aggressive behavior toward vegetation and animals contacting their home tree, larger colony size, and 24-hour activity outside of the nest. This combination of characteristics means that any herbivore attempting to forage on an acacia occupied by acacia-ants is met by a large number of fast, agile, and highly aggressive defenders and is given this reception no matter what time of the day it attempts to feed. Janzen listed six species of *Pseudomyrmex* with an obligate mutualistic relationship with swollen thorn acacias and refers to three additional undescribed species. His experimental work focused principally on one species, *Pseudomyrmex ferruginea*.

Worldwide, the genus *Acacia* includes over 700 species. Distributed throughout the tropical and subtropical regions around the world, acacias are particularly common in drier tropical and subtropical environments. The swollen thorn acacias, which form obligate mutualisms with *Pseudomyrmex* spp., are restricted to the New World, where they are distributed from southern Mexico, through Central America, and into Venezuela and Columbia in northern South America. Across this region, swollen thorn acacias occur mainly in the lowlands up to 1,500 m elevation in areas with a dry season of 1 to 6 months. Swollen thorn acacias show several characteristics related to their obligate association with ants, including enlarged thorns with a soft, easily excavated pith; year-round leaf production; enlarged foliar nectaries; and leaflet tips modified into concentrated food sources called Beltian bodies. The thorns provide living space, while the foliar nectaries provide a source of sugar and liquid. Beltian bodies are a source of oils and protein. Resident ants vigorously guard these resources against encroachment by nearly all comers, including other plants.

Janzen's detailed natural history of the interaction between bullshorn acacia and ants paints a rich picture of mutual benefits to both partners (fig. 15.7). Newly mated *Pseudomyrmex* queens fly and run through the vegetation searching for unoccupied seedlings or shoots of bullshorn acacia. When a queen finds an unoccupied acacia, she excavates an entrance in one of the green thorns or uses one carved previously by another ant. The queen then lays her first eggs in the thorn and begins to forage on her newly acquired home plant. She gets nectar for herself and her developing larvae from the foliar nectaries and gets additional solid food from the Beltian bodies. As time passes, the number of workers in the new colony increases, and while they take up all the chores of the colony, the queen shifts to a mainly reproductive function; her abdomen enlarges and she becomes increasingly sedentary.

In exchange for food and shelter, ants protect acacias from attack by herbivores and competition from other plants. Workers have several duties, including foraging for themselves, the larvae, and the queen. One of their most important activities is protecting the home plant. Workers will attack, bite, and sting nearly all insects they encounter on their home plant

**Figure 15.7** Split thorn of a bullshorn acacia, revealing a nest of its ant mutualists.

or any large herbivores such as deer and cattle that attempt to feed on the plant. They will also attack and kill any vegetation encroaching on the home tree. Workers sting and bite the branches of other plants that come in contact with their home tree or that grow near its base. These activities keep other plants from growing near the base of the home tree and prevent other trees, shrubs, and vines from shading it. Consequently the home plant's access to light and soil nutrients is increased.

Once a colony has at least 50 to 150 workers, which takes about 9 months, they patrol the home plant day and night. About one-fourth of the total colony is active at all times. Eventually colonies grow so large that they occupy all the thorns on the home tree and may even spread to neighboring acacias. The queen, however, generally remains on the shoot that she colonized originally. When the colony reaches a size of about 1,200 workers, it begins producing a more or less steady stream of winged reproductive males and females, which fly off to mate. The queens among them may eventually establish new colonies on other bullshorn acacias or one of the other Central American swollen thorn acacias. Colonies may eventually reach a total population of 30,000 workers.

## Experimental Evidence for Mutualism

While much of the natural history of this mutualism was known at the time Janzen conducted his studies, no one had experimentally tested the strength of its widely supposed benefits. Janzen took his work beyond natural history to experimentally test for the importance of ants to bullshorn acacias. It was clear that the ant needs swollen thorn acacias, but do the acacias need the ants? Janzen's experiments concentrated on the influence of ants on acacia performance. He also tested the effectiveness of the ants at keeping acacias free of herbivorous insects. Janzen removed ants from acacias by clipping occupied thorns or by cutting out entire shoots with their ants. He then measured the growth rate, leaf production, mortality, and insect population density on acacias with and without ants.

Janzen's experiments demonstrated that ants significantly improve plant performance. Differences in plant performance

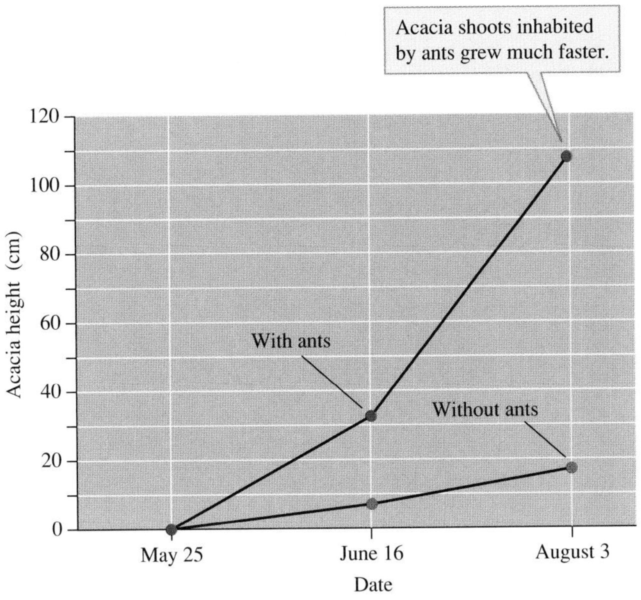

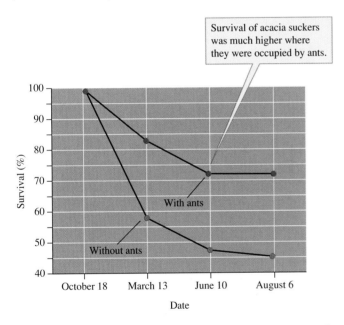

**Figure 15.8** Growth by bullshorn acacia with and without resident ants (data from Janzen 1966).

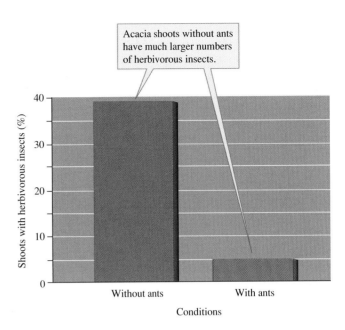

**Figure 15.10** Ants and the abundance of herbivorous insects on bullshorn acacia (data from Janzen 1966).

had more herbivorous insects on them than did acacias with ants (fig. 15.10). Janzen's experiments provide strong evidence that bullshorn acacias need ants as much as the ants need the acacia. It appears that this is a truly mutualistic situation and that it is obligate for both partners.

## Potential Conflict Between Mutualists

Most research on swollen thorn acacias has focused on their ant-protection mutualisms. However, these trees depend on many other mutualistic relationships. Belowground, their roots shelter nitrogen-fixing bacteria in nodules and also harbor mycorrhizal fungi. Aboveground, besides sheltering swarms of *Pseudomyrmex* ants that drive away herbivorous insects, the acacia's flowers depend on other insects, mainly bees, for pollination. The acacia's ant guards could come into conflict with pollinators in two ways. First, the ants could remove nectar from flowers and reduce their attractiveness to potential pollinators. Second, the ants could guard flowers, driving pollinators away.

This potential conflict between mutualists of swollen thorn acacias attracted the attention of Nigel Raine, Pat Willmer, and Graham Stone (Raine, Willmer, and Stone 2002), researchers from three different British universities. They conducted their research at the Chamela Biological Station of the Universidad Nacional Autónoma de México, where they studied *Acacia hindsii* and its ant protector, *Pseudomyrmex veneficus*. Raine, Willmer, and Stone first examined the distribution of ants and pollinators to see if they overlapped in space or time. They found that ant and pollinator activity overlaps in time. However while ants and pollinators are active on *A. hindsii* at the same time, they rarely overlap spatially. The ants rarely visit acacia inflorescences. Why is that? Raine, Willmer, and Stone observed that the foliar nectaries and Beltian bodies used by

**Figure 15.9** Survival of bullshorn acacia shoots with and without resident ants (data from Janzen 1966).

were likely the result of increased competition with other plants and increased attack by insects faced by acacias without their tending ants. Suckers growing from stumps of acacias occupied by ants lengthened at seven times the rate of suckers without ants (fig. 15.8). Suckers with ants were also more than 13 times heavier than suckers without ants and had more than twice the number of leaves and almost three times the number of thorns. Suckers with ants also survived at twice the rate of suckers without ants (fig. 15.9).

What produces the improved performance of acacias with ants? One factor appears to be reduced populations of herbivorous insects. Janzen found that acacias without ants the ants occur on new growth, while flowers are restricted to older shoots. In addition, in contrast to acacia species that do not support protective ants, the inflorescences of *A. hindsii* produce no nectar. Lack of nectar would make the flowers less attractive to patrolling ants. Still because new and older shoots can grow in close proximity, the researchers wondered whether some other factor might keep the ants from patrolling the inflorescences on older shoots.

Since Willmer and Stone (1997) had discovered previously that the flowers of some African acacias contained an ant repellent, they tested for the presence of a repellent in the flowers of *A. hindsii*. They tested for the presence of a repellent by rubbing several acacia tissues on the bark of branches actively patrolled by ants. The tissues tested were new inflorescences, old inflorescences, leaves, and buds. Each of these tissues were rubbed within 3 cm squares marked on the bark of patrolled branches using water-based markers. As a control, Raine, Willmer, and Stone marked one set of squares but did not rub any plant tissues on them. Once experimental and control squares had been established, the researchers watched patrolling ants, noting whether they entered experimental and control squares or avoided crossing them.

Figure 15.11 shows the results of the repellent experiment. Raine, Willmer, and Stone found that new inflorescences were strongly repellent to patrolling ants and that older inflorescences, though repellent, were rejected less often. Meanwhile, leaf and bud rubbings were rejected no more frequently than were control squares. In summary, protection and pollination mutualisms do not come into conflict on *A. hindsii* because of spatial separation of inflorescences and resources used by guarding ants, and because *A. hindsii* inflorescences lack a potential ant attractant (nectar) and instead contain a chemical repellent.

While tropical plant protection mutualisms are most often cited, there are many examples of mutualism between plants and ants in the temperate zone. A particularly well-studied interaction is that between ants and the aspen sunflower, *Helianthella quinquenervis*.

# A Temperate Plant Protection Mutualism

Aspen sunflowers live in wet mountain meadows of the Rocky Mountains from Chihuahua, Mexico, to southern Idaho, at elevations as low as 1,600 m in the northern part of its range to 4,000 m in the south. Ants are attracted to aspen sunflowers because they produce nectar at **extrafloral nectaries,** nectar-producing structures outside of the flowers. In the case of aspen sunflowers the extrafloral nectaries are associated with structures called *involucral bracts,* modified leaves that first enclose the flower head prior to flowering and then surround the base of the flower after it opens. Some early researchers hypothesized that extrafloral nectaries function to attract ants, while others suggested that they are primarily excretory organs.

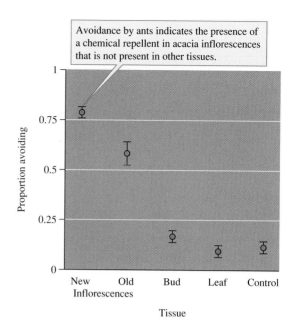

**Figure 15.11** Proportion of ants avoiding control areas and areas rubbed with tissues of new and old inflorescences, buds, and leaves of the swollen thorn acacia, *Acacia hindsii;* values are means ± one standard error (data from Raine, Willmer, and Stone 2002).

The extrafloral nectar produced by aspen sunflowers is rich in sucrose and contains high concentrations of a wide variety of amino acids. So, like the swollen thorn acacias studied by Janzen the aspen sunflower provides food to ants. In contrast to swollen thorn acacias, however, this sunflower does not provide living places. This contrast is general across temperate ant-plant mutualisms, which involve food as an attractant but no living quarters.

David Inouye and Orley Taylor (1979) recorded five species of ants on aspen sunflowers, including *Formica obscuripes, F. fusca, F. integroides planipilis, Tapinoma sessile,* and *Myrmica* sp. These ants are not obligately associated with aspen sunflowers and can be found tending aphids on other species of plants or even collecting flower nectar on some plants. However, Inouye and Taylor never observed them collecting nectar from aspen sunflower blossoms nor tending aphids on this plant. Apparently, the extrafloral nectar produced by the aspen sunflower is a sufficient attractant. Ants find the plant so attractive that Inouye and Taylor observed up to 40 ants on a single flower stalk.

While the ants visiting the extrafloral nectaries of *Helianthella* clearly derive benefit, it is not obvious that the plant receives any benefits from the association. What benefits might this sunflower gain by having ants roaming around its flowers and flower buds? Inouye and Taylor proposed that the ants may protect the sunflower's developing seeds from seed predators. In the central Rocky Mountains, the seeds of aspen sunflowers are attacked by a variety of seed predators, including the larvae of two species of flies in the family Tephritidae, a fly in the family Agromyzidae, and a phycitid moth. These seed predators damaged over 90% of the seeds produced by some of the flowers at one of Inouye and Taylor's study sites.

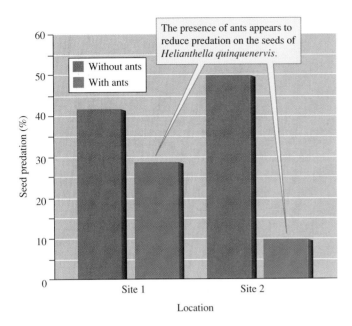

**Figure 15.12** Predation on the seeds of aspen sunflower with and without ants (data from Inouye and Taylor 1979).

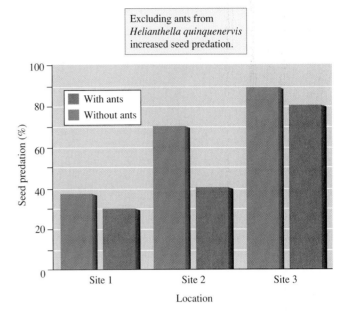

**Figure 15.13** Effect of excluding ants on rates of seed predation on aspen sunflowers (data from Inouye and Taylor 1979).

The high densities of ants that can occur on a single aspen sunflower certainly have the potential to deter seed predators, but these same ants might also interfere with pollination. This potential for interference is not realized, however, because seed predators generally attack aspen sunflowers before they are fully mature and before they have formed ray florets, the "petals" of sunflowers, daisies, etc. Prior to opening of the flower bud, when tephridid and agromyzid flies try to oviposit on the bud, ants visiting the extrafloral nectaries patrol the whole surface of the flower bud in large numbers. Later, the fully formed ray florets, which act as attractants for pollinators (mainly bumblebees), form a screen between the involucral bracts and the flower head, reducing the potential for ants to interfere with pollinators.

The question asked by Inouye and Taylor was whether or not the presence of ants on aspen sunflowers reduces the rate of attack by seed predators. They addressed this question in several ways. First, they compared rates of attack by seed predators on flowers tended by ants with rates of attack on flowers where ants were naturally absent. This comparison showed that flowers without ants suffered two to four times as much seed predation (fig. 15.12). The researchers also found that the average number of ants per flower stalk decreased with distance from an ant nest and that the plants with fewer ants suffered higher rates of seed damage by seed predators.

Next, Inouye and Taylor performed an experiment in which they prevented ants from moving onto some plants by applying a sticky barrier to the base of flower stalks. They used adjacent plants as controls. How did this experimental manipulation strengthen Inouye and Taylor's study? Why wasn't the comparison of plants naturally with and without ants sufficient to assess the influence of ants on rates of seed predation? It might have been that the flowers frequented by

ants are visited less often by seed predators for some other reason. If so, then demonstrating that flowers without ants experience greater seed predation may have simply reflected a low overlap in the distributions of ants and seed predators. The results of Inouye and Taylor's experiment rule out this possibility (fig. 15.13). At two of their study sites, exclusion of ants from flowers resulted in significantly higher rates of seed predation.

As in the tropical swollen thorn acacia-ant mutualism, ants associated with aspen sunflowers provide protection while receiving substantial benefits in the form of food. Unlike the tropical system, the association between aspen sunflowers and ants incorporates a significant degree of flexibility. This flexibility may be a hallmark of many temperate mutualisms.

Why does the relationship between ants and aspen sunflowers remain facultative? In other words, why hasn't there been strong selection for the kind of obligate relationship, such as that between bullshorn acacia and the ant *P. ferruginea?* Continuing studies by David Inouye provide clues. He estimated the abundance of aspen sunflowers on two study plots for more than two decades. This long-term study shows that every few years the flower heads of aspen sunflowers are killed by late frosts. From 1974 to 1995, aspen sunflowers produced few or no flower heads in 1976, 1981, 1985, 1989, and 1992 (fig. 15.14). An ant species with an obligate mutualistic relationship with the aspen sunflower and that relied entirely on it as a source of nectar would not survive long. Inouye points out that paradoxically the frosts are beneficial to aspen sunflowers in the long run because they reduce populations of seed predators such as tephritid flies, which have no place to lay their eggs when hard frost kills the flower heads. In the coevolutionary relationships between the aspen sunflower and its predators, the physical environment plays a

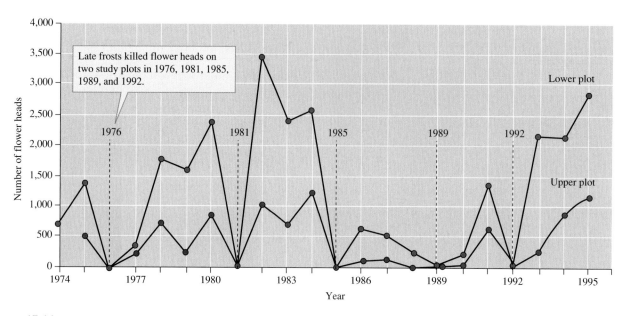

**Figure 15.14** Annual variation in numbers of flower heads produced by aspen sunflowers on two plots at the Rocky Mountain Biological Station (data courtesy of David W. Inouye).

significant role. In temperate climates generally, the physical environment seems to play as large a role as biological relationships in determining ecological patterns and processes.

If we venture into tropical seas and probe their inhabitants, we soon uncover a wide variety of mutualistic relationships at least as rich as those we examined between terrestrial plants and their partners. The most striking marine counterparts to the mutualisms of terrestrial plants are those centered around reef-building corals.

## CONCEPT DISCUSSION

### Coral Mutualisms

**Reef-building corals depend upon mutualistic relationships with algae and animals.**

Because of the importance of mutualism in the lives of reef-building corals, it appears that the ecological integrity of coral reefs depends upon mutualism. Coral reefs show exceptional productivity and diversity. Recent estimates put the number of species occurring on coral reefs at approximately 0.5 million, and coral reef productivity is among the highest of any natural ecosystem. As we saw in chapter 3, the paradox is that this overwhelming diversity and exceptional productivity occurs in an ecosystem surrounded by nutrient-poor tropical seas. The key to explaining this paradox lies with mutualism; in this case, between reef-building corals and unicellular algae called zooxanthellae, members of the phylum Dinoflagellata. Most of these organisms are free-living unicellular marine and freshwater photoautotrophs.

## Zooxanthellae and Corals

The association between corals and zooxanthellae is functionally similar to the relationship between plants and mycorrhizal fungi. Zooxanthellae live within coral tissues at densities averaging approximately 1 million cells per square centimeter of coral surface. Like plants, zooxanthellae receive nutrients from their animal partner. In return, like mycorrhizal fungi, the coral receives organic compounds synthesized by zooxanthellae during photosynthesis.

One of the most fundamental discoveries concerning the relationship between corals and zooxanthellae is that the release of organic compounds by zooxanthellae is controlled by the coral partner. Corals induce zooxanthellae to release organic compounds with "signal" compounds, which alter the permeability of the zooxanthellae cell membrane. Zooxanthellae grown in isolation from corals release very little organic material into their environment. However, when exposed to extracts of coral tissue, zooxanthellae immediately increase the rate at which they release organic compounds. This response appears to be a specific chemically mediated communication between corals and zooxanthellae. Zooxanthellae do not respond to extracts of other animal tissues, and coral extracts do not induce leaking of organic molecules by any other algae that have been studied.

Corals not only control the secretion of organic compounds by zooxanthellae, they also control the rate of zooxanthellae population growth and population density. In corals, zooxanthellae populations grow at rates 1/10 to 1/100 the rates observed when they are cultured separately from corals. Corals exert control over zooxanthellae population density through their influence on organic matter secretion. Normally, unicellular algae show **balanced growth,** growth in which all cell constituents, such as nitrogen, carbon, and DNA, increase at the

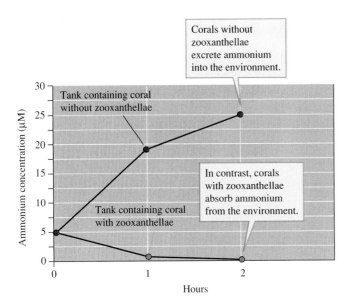

Figure **15.15** Zooxanthellae, corals, and ammonium flux (data from Muscatine and D'Elia 1978).

Figure **15.16** Pistol shrimp will defend their home coral from attacking predators.

same rate. However, zooxanthellae living in coral tissues show unbalanced growth, producing fixed carbon at a much higher rate than other cell constituents. Moreover, the coral stimulates the zooxanthellae to secrete 90% to 99% of this carbon, which the coral uses for its own respiration. Carbon secreted and diverted for use by the coral could otherwise be used to produce new zooxanthellae, which would increase population growth.

What benefits do the zooxanthellae get out of their relationship with corals? The main benefit appears to be access to higher levels of nutrients, especially nitrogen. Corals feed on zooplankton, which gives them a means of capturing nutrients, especially nitrogen and phosphorus. When corals metabolize the protein in their zooplankton prey, they excrete ammonium as a waste product. L. Muscatine and C. D'Elia (1978) showed that coral species such as *Tubastrea aurea* that do not harbor zooxanthellae continuously excrete ammonium into their environment, while corals such as *Pocillopora damicornis* do not excrete measurable amounts of ammonia (fig. 15.15). What happens to the ammonium produced by *Pocillopora* during metabolism of the protein in their zooplankton prey? Muscatine and D'Elia suggested that this ammonium is immediately taken up by zooxanthellae as the coral excretes it. In addition to internal recycling of the ammonium produced by their coral partner, zooxanthellae also actively absorb ammonium from seawater. By absorbing nutrients from the surrounding medium and leaking very little back into the environment, corals and their zooxanthellae gradually accumulate substantial quantities of nitrogen. So, as in tropical rain forest, large quantities of nutrients on coral reefs accumulate and are retained in living biomass.

# A Coral Protection Mutualism

The ant-acacia mutualism that we reviewed previously has a striking parallel on coral reefs. Corals in the genera *Pocillopora*

and *Acropora* host a variety of crabs in the family Xanthidae, mainly *Trapezia* spp. and *Tretralia* spp. as well as a species of pistol shrimp, *Alpheus lottini*. In this mutualistic relationship (fig. 15.16), the crustaceans protect the coral from a wide variety of predators while the coral provides its crustacean partners with shelter and food.

Peter Glynn (1983) surveyed the coral-crustacean mutualism and found that the eastern, central, and western areas of the Pacific Ocean contain 13 species of corals that are protected by crustacean mutualists, including 17 species of crabs and 1 species of shrimp, all of which are found only on corals in what is apparently an obligate mutualism. These crustaceans protect the corals from a variety of sea stars that prey on corals but especially from attacks by the crown-of-thorns sea star, *Acanthaster planci*. At the approach of the sea star, the crabs become highly disturbed and then attack by pinching and clipping the sea star's spines and tube feet, grasping it and jerking it up and down and resisting its retreat. The mutualistic shrimp also attacks the sea star by snipping spines and tube feet and making loud snapping sounds with an enlarged pincer specialized for the purpose. The loud popping sounds, which have given shrimp in the genus *Alpheus* the name "pistol shrimp," are so intense they stun small fish.

Glynn used field and laboratory experiments to test whether this aggression by crustaceans is effective at repelling attacks by predatory sea stars. He conducted a field experiment at 8 to 12 m depth on a reef in Guam, where he removed the crustaceans from an experimental group of corals and gave sea stars a choice between these and an equal number of corals that retained their crustacean partners. Over a period of 2 days the sea stars attacked the unprotected corals at a much higher frequency (fig. 15.17). Glynn obtained similar results in a laboratory study of the corals and crustacean mutualists of Panama, in which sea stars attacked 85% of the unprotected colonies. These results show that the crustacean mutualists of corals substantially improve the chances that a coral will avoid attack by sea stars.

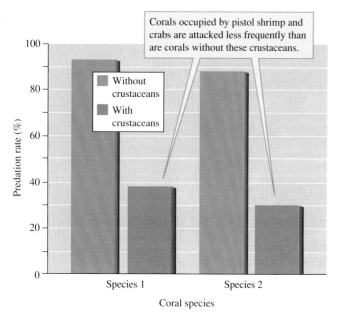

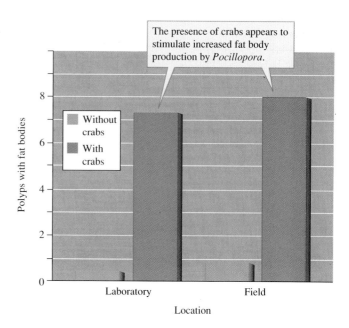

**Figure 15.17** Attacks on corals with and without pistol shrimp and crabs (data from Glynn 1983).

**Figure 15.18** Fat body production by the coral *Pocillopora damicornis* in the presence and absence of crabs (data from Stimson 1990).

Observations by Glynn and also John Stimson (1990) suggest that mutualistic crabs also protect corals from other less conspicuous attackers. Glynn observed that the presence of crabs seems to enhance the condition of coral tissues. Stimson found that when he removed crabs, corals showed tissue death in the deep axils of their branches and that these areas were soon invaded by algae, sponges, and tunicates. It appears that, in addition to protecting corals from the attacks of large predators, the activities of crabs promote the health and integrity of coral tissues. If this is a mutualistic relationship, what do the crabs receive in return for their investment?

Like swollen thorn acacias, corals provide their crustacean mutualists with shelter and food. The corals harboring crabs and pistol shrimp have a tightly branched growth form that offers shelter, and the crustaceans feed on the mucus produced by the corals. *Trapezia* spp., the most common crabs guarding pocilloporid corals, stimulate mucus flow from corals by inserting their legs into coral polyps, a behavior not reported for any other crabs. Corals contain large quantities of lipids that constitute 30% to 40% of the dry weight of their tissues, much of which they release with mucus. This release may constitute up to 40% of the daily photosynthetic production by zooxanthellae.

The pocilloporid corals that host crustaceans concentrate some of this lipid into fat bodies that are 300 to 500 μm in length. Glynn suggested that the fat bodies produced by pocilloporid corals hosting protective crabs may be a part of their mutualistic relationship. Stimson tested this hypothesis by determining whether commensal crabs influence the production of fat bodies by coral polyps. He conducted his

experiments at the Hawaii Institute of Marine Biology on Coconut Island in Kaneohe Bay, Oahu, Hawaii. He collected colonies of *Pocillopora* 8 to 10 cm in diameter from the mid-bay region of Kaneohe Bay, placed them in buckets of seawater, and took them back to the marine laboratory on Coconut Island. There, he divided the corals into experimental and control groups. He then removed crabs and pistol shrimp from the experimental coral colonies by "teasing" them out with a small wire. Corals with and without crabs were then kept separately in outdoor tanks supplied with flowing seawater.

After 24 days, Stimson compared the number of fat bodies on corals with and without crabs. He also compared these experimental results with the density of fat bodies on *Pocillopora* in Kaneohe Bay that naturally hosted or lacked mutualistic crabs. The results of these experiments and field observations show clearly that *Pocillopora* increases its production of fat bodies in the presence of crabs both in the laboratory and in the field (fig. 15.18). Stimson also examined the digestive tract of crabs inhabiting corals and found that they contained large quantities of lipids. At the same time, no significant reductions in either the reproductive rate or growth rates of corals supporting crabs were found. Stimson concluded that the relationship between corals and crabs is a true mutualism, with both partners receiving substantial benefit.

The *extent* of benefit may be the essential factor driving the evolution of mutualisms. In the following Concept Discussion, we review theoretical analyses of how the relative benefits and costs of an association influence the evolution of mutualistic relationships.

*Investigating the Evidence*

# Confidence Intervals

In chapter 14 we reviewed how to calculate the standard error, $s_{\bar{X}}$, which is an estimate of variation among means of samples drawn from a population. Here, we will use the standard error to calculate a **confidence interval.** A confidence interval is a range of values within which the true population mean occurs with a particular probability. That probability, which is called the **level of confidence,** is calculated as one minus the significance level, $\alpha$, which is generally 0.05:

$$\text{Level of confidence} = 1 - \alpha$$
$$\text{Level of confidence} = 1 - 0.05 = 0.95$$

Using this level of confidence produces what is called a 95% confidence interval that is calculated as follows:

$$\text{Confidence interval for } \mu = \bar{X} \pm s_{\bar{X}} \times t$$

where

$\mu$ = true population mean

$\bar{X}$ = sample mean

$s_{\bar{X}}$ = standard error

$t$ = value from a Student's $t$ table

A Student's $t$ table, available in most statistics textbooks, summarizes the values of a statistical distribution known as the Student's $t$ distribution. The value of $t$ we use for calculating a confidence interval is determined by the degrees of freedom ($n - 1$) and the significance level, which in this case is $\alpha = 0.05$.

Let's calculate a 95% confidence interval using the body length measurements for the sample of loach minnows, *Tiaroga cobitis,* that we used to calculate a standard error in chapter 14 (see p. 356).

In chapter 6 (p. 162), we calculated mean of this sample as:

$$\bar{X} = 56.2 \text{ mm}$$

And, in chapter 14 we calculated the standard error of the sample as:

$$s_{\bar{X}} = 1.96 \text{ mm}$$

This sample of body lengths included measurements of 10 fish ($n = 10$) and so the degrees of freedom for this sample ($n - 1$) is 9. Using a significance level of 0.05 and degrees of freedom of 9, we find that the critical value of $t$ from a

Student's $t$ table is 2.26. Therefore, the 95% confidence interval calculated from this sample is:

$$\text{Confidence interval for } \mu = \bar{X} \pm s_{\bar{X}} \times t$$
$$= 56.2 \text{ mm} \pm 1.96 \text{ mm} \times 2.26$$
$$= 56.2 \text{ mm} \pm 4.43 \text{ mm}$$

With this confidence interval, we can say that there is a 95% probability that the true mean body length in this population of loach minnows is somewhere between 60.63 mm (56.2 mm + 4.43 mm) and 51.77 mm (56.2 mm – 4.43 mm).

This is shown graphically in figure 1, along with the mean and 95% confidence interval for the sample of loach minnows from the Gila River that we first considered in chapter 14 (p. 356). Notice that the 95% confidence interval for the Gila River sample is much smaller (56.2 mm ± 1.77 mm). This smaller confidence interval is the result of the larger sample size from the Gila River ($n = 50$), which produced a smaller standard error ($s_{\bar{X}} = 0.88$) and a smaller critical $t$ value (2.01), since the degrees of freedom is 49. As a consequence of having a larger sample, our estimate of the true population mean has been narrowed to a much smaller range for the Gila River population of loach minnows.

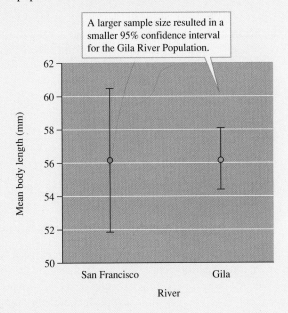

**Figure 1** Average body length of loach minnows and 95% confidence intervals calculated from samples collected in the San Francisco River ($n = 10$) and Gila River ($n = 50$).

**Figure 15.19** A diversity of mutualisms: (*a*) cleaner wrasse and seabass; (*b*) lichens are an association between a fungus and cyanobacteria; (*c*) soybeans fix molecular nitrogen through their association with bacteria within nodules on their roots; (*d*) deer access the energy stored in plant tissues through the activities of a community of mutualistic microorganisms living in their gut.

## CONCEPT DISCUSSION

### Evolution of Mutualism

**Theory predicts that mutualism will evolve where the benefits of mutualism exceed the costs.**

We have reviewed several complex mutualisms both on land and in marine environments. There are many others (fig. 15.19), every one a fascinating example of the intricacies of nature. Ecologists not only study the present biology of those mutualisms but also seek to understand the conditions leading to their evolution and persistence. Theoretical analyses point to the relative costs and benefits of a possible relationship as a key factor in the evolution of mutualism.

Modeling of mutualism has generally taken one of two approaches. The earliest attempts involved modifications of

the Lotka-Volterra equations to represent the population dynamics of mutualism. The alternative approach has been to model mutualistic interactions using cost-benefit analysis to explore the conditions under which mutualisms can evolve and persist. In chapters 13 and 14, where we discussed models of competition and predation, we focused on the population dynamic approach to modeling species interactions. Here, we concentrate on cost-benefit analyses of mutualism.

Kathleen Keeler (1981, 1985) developed models to represent the relative costs and benefits of several types of mutualistic interactions. Among them are two of the mutualistic interactions we discussed in chapter 15: ant-plant protection mutualisms and mycorrhizae. Keeler's approach requires that we consider a population polymorphic for mutualism containing three kinds of individuals: (1) *successful mutualists,* which give and receive measurable benefits to another organism; (2) *unsuccessful mutualists,* which give benefits to another organism but, for some reason, do not receive any benefit in return; and (3) *nonmutualists,* neither giving nor receiving benefit from a mutualistic partner. The bottom line in Keeler's approach is that for a population to be mutualistic, the fitness of successful mutualists must be greater than the fitness of either unsuccessful mutualists or non-mutualists. In addition, the combined fitness of successful and unsuccessful mutualists must exceed that of the fitness of non-mutualists. If these conditions are not met, Keeler proposed that natural selection will eventually eliminate the mutualistic inter-action from the population.

In general, we can expect mutualism to evolve and per-sist in a population when and where mutualistic individuals have higher fitness than nonmutualistic individuals.

Keeler represented the fitness of nonmutualists as:

$$w_{nm} = \text{fitness of nonmutualists}$$

(Fitness has been traditionally represented by the symbol *w* and though it might be clearer to use another symbol, such as *f*, the traditional symbol is used here.) Keeler represents the fitness of mutualists as:

$$w_m = pw_{ms} + qw_{mu} \tag{1}$$

where:

p = the proportion of the population consisting of successful mutualists

$w_{ms}$ = the fitness of successful mutualists

q = the proportion of the population consisting of unsuccessful mutualists

$w_{mu}$ = the fitness of unsuccessful mutualists.

We can represent Keeler's conditions for the evolution and persistence of mutualism as:

$$w_m > w_{nm} \tag{2}$$

or

$$pw_{ms} + qw_{mu} > w_{nm} \tag{3}$$

Keeler predicts that mutualism will persist when the com-bined fitness of successful and unsuccessful mutualists exceeds the fitness of nonmutualists. Why do we have to com-bine the fitness of successful and unsuccessful mutualists? Remember that both confer benefit to their partner, but only the successful mutualists receive benefit in return.

The analysis is more convenient if we think of these rela-tionships in terms of **selection coefficients (s),** the relative selective costs associated with being either a successful mutu-alist, an unsuccessful mutualist, or a nonmutualist:

$$s = 1 - w \text{ and } w = (1 - s).$$

Using selective coefficients, Keeler expressed the selective cost of being a successful mutualist, an unsuccessful mutual-ist, or a nonmutualist as:

$$s_{ms} = (H)(1 - A)(1 - D) + I_A + I_D \tag{4}$$

$$s_{mu} = (H)(1 - D) + I_A + I_D \tag{5}$$

$$s_{nm} = H(1 - D) + I_D \tag{6}$$

where:

H = the proportion of the plant tissue damaged in the absence of any defenses

D = the amount of protection given to the plant tissues by defenses other than ants (e.g., chemical defenses); so, 1 − D is the amount of tissue damage that would occur in spite of these alternative defenses

A = the amount of herbivory prevented by ants (so, again, 1 − A is the amount of herbivory that occurs in spite of ants)

$I_a$ = the investment by the plant in benefits extended to the ants

$I_D$ = investment in defenses other than ants

Using these selective coefficients we can express Keeler's conditions for evolution and persistence of the ant-plant mutu-alism as:

$$p(1 - s_{ms}) - q(1 - s_{mu}) > 1 - s_{nm}$$

into which Keeler substituted the relationships given in equa-tions (4), (5), and (6). By simplifying the resulting equation, she produced the following expression of benefits relative to costs:

$$p[H (1 - D) A] > I_A$$

# Facultative Ant-Plant Protection Mutualisms

Keeler applied her cost-benefit model to facultative mutu-alisms involving plants with extrafloral nectaries and ants that feed at the nectaries and provide protection to the plant in return. These are mutualisms like that between *Helianthella quinquenervis* and ants, which we discussed earlier in chapter 15 in the Concept Discussion section on plant mutualism. Her model is not appropriate for obligate mutualisms like that between swollen thorn acacias and their mutualistic ants. In addition, Keeler wrote her model from the perspective of the plant side of the mutualism. Let's step through the general model and connect each of the terms with the ecology of facultative plant-ant protec-tion mutualisms.

In this model, $w_{ms}$ is the fitness of a plant that produces extrafloral nectaries and that successfully attracts ants effective at guarding it, while $w_{mu}$ is the fitness of a plant that produces extrafloral nectaries but that has not attracted enough ants to mount a successful defense. You may remember that Inouye and Taylor found that *Helianthella* far away from ant nests attracted few ants. These plants would correspond to Keeler's unsuccessful mutualists. In addition, Keeler includes the fitness of nonmutualistic plants, $w_{nm}$, which would be the fitness of individuals of a plant such as *Helianthella* that does not produce extrafloral nectaries. Are there such individuals in the population of *Helianthella* studied by Inouye and Taylor? We don't know, but that is not the point. The reason Keeler includes nonmutualists in her model is to provide an assessment of the potential costs and benefits of such a strategy against which she can weigh the mutualistic strategy.

Keeler's model represents potential benefits to the host plant as:

$$p\,[H\,(1-D)\,A]$$

where:

> $p$ = the proportion of the plant population attracting sufficient ants to mount a defense

Keeler's model represents the plant's costs of mutualism as:

$$I_A = n[m + d\,(a + c + h)]$$

where:

> $n$ = the number of extrafloral nectaries per plant
>
> $m$ = the energy content of nectary structures
>
> $d$ = the period of time during which the nectaries are active
>
> $a$ = costs of producing amino acids in nectar
>
> $c$ = costs of producing the carbohydrates in nectar
>
> $h$ = costs of providing water for nectar

Again, Keeler's hypothesis is that for mutualism to persist, benefits must exceed costs. In terms of her model:

$$p[H(1-D)A] > I_A$$

This model proposes that for a facultative ant-plant mutualism to evolve and persist, the proportion of the plant's energy budget that ants save from destruction by herbivores must exceed the proportion of the plant's energy budget that is invested in extrafloral nectaries and nectar.

The details of Keeler's model offer insights into what conditions may produce higher benefits than costs. First, and most obviously, $I_A$, the proportion of the plant's energy budget that is invested in extrafloral nectaries and nectar should be low. This means that plants living on a tight energy budget, for example, plants living in a shady forest understory, should be less likely to invest in attracting ants than those living in full sun. Higher benefits result from (1) a high probability of attracting ants, that is, high $p$; (2) a high potential for herbivory, $H$; (3) low effectiveness of alternative defenses, low $D$, and (4) highly effective ant defense, high $A$.

The task for ecologists is to determine how well these requirements of the model match values of these variables in nature.

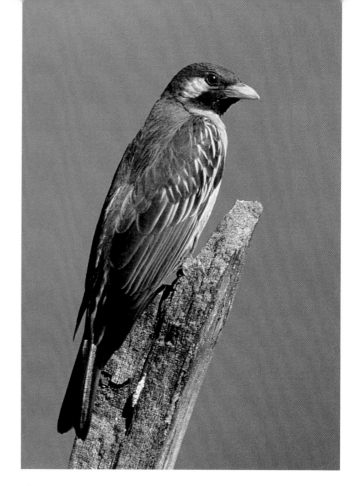

**Figure 15.20**  The greater honeyguide, *Indicator indicator.*

## APPLICATIONS & TOOLS

### Mutualism and Humans

Mutualism has been important in the lives and livelihood of humans for a long time. Historically, much of agriculture has depended upon mutualistic associations between species and much of agricultural management has been aimed at enhancing mutualisms, such as nitrogen fixation, mycorrhizae, and pollination to improve crop production. Agriculture itself has been viewed as a mutualistic relationship between humans and crop and livestock species. However, there may be some qualitative differences between agriculture as it has been generally practiced and mutualisms among other species. How much of agriculture is pure exploitation and how much is truly mutualistic remains an open question.

There is, however, at least one human mutualism that fits comfortably among the Concept Discussions of chapter 15, a mutualism involving communication between humans and a wild species with clear benefit to both. This mutualism joins the traditional honey gatherers of Africa with the greater honeyguide, *Indicator indicator* (fig. 15.20). Honey gathering has long been an important aspect of African cultures, important enough that there are scenes of honey gathering in rock art painted over 20,000 years ago (Isack and Reyer 1989). No one knows how long humans have gathered honey in Africa, but it is difficult to imagine the earliest hominids resisting such sweet temptation. Whenever honey gathering began, humans have apparently had a capable and energetic partner in their searches.

# The Honeyguide

Honeyguides belong to the family Indicatoridae in the order Piciformes, an order that also includes the woodpeckers. The family Indicatoridae includes a total of 17 species, 15 of which are native to Africa. Honeyguides have the unusual habit of feeding on waxes of various sorts—most feed on beeswax and insects. Of the 17 species of honeyguides, only the greater honeyguide, *I. indicator,* is known to guide humans and a few other mammals to bees' nests.

The greater honeyguide is found throughout much of sub-Saharan Africa. It avoids only dense forests and very open grasslands and desert, and its distribution corresponds broadly with the distributions of tropical savanna and tropical dry forest. Like all of the honeyguides, the greater honeyguide is a brood parasite that, like cuckoos, lays its eggs in the nests of other birds. This way of life is reflected in the early morphology of nestling honeyguides, which retain "bill hooks" on their upper and lower bills for the first 14 days of life that they use to lacerate and kill their nest mates. However, nests sometimes contain two honeyguide nestlings, so apparently there is some mechanism by which nestlings of the same species can coexist. After the deaths of their nest mates, honeyguide nestlings receive all the food brought by their foster parents, which continue to feed young honeyguides until they are completely independent, approximately 7 to 10 days after leaving the nest.

Greater honeyguides are capable of completely independent life without mutualistic interactions with humans, so we would classify their mutualism as facultative. Living independently, honeyguides feed on beeswax, and on the adults, larvae, pupae, and eggs of bees. They also feed on a wide variety of other insects. Greater honeyguides show highly opportunistic feeding behavior and sometimes join flocks of other bird species foraging on the insects stirred up by large mammals. The most distinguishing feature of the greater honeyguide, however, is its habit of guiding humans and ratels, or honey badgers, to bees' nests.

# Guiding Behavior

The first written report of the guiding behavior of *I. indicator* was authored in 1569 by João Dos Santos, a missionary in the part of East Africa that is now Mozambique. Dos Santos first noticed honeyguides because they would enter the mission church to feed upon the bits of beeswax on candlesticks. He went on to describe their guiding behavior by saying that when the birds find a beehive, they search for people and attempt to lead them to the hive. He noted that the local people eagerly followed the birds because of their fondness for honey, and he observed that the honeyguide profits by gaining access to the

wax and dead bees left after humans raid the hive. Dos Santos's report of this behavior was confirmed by other European visitors to almost all parts of Africa for the next four centuries. However, it wasn't until the middle of the twentieth century that the mutualism of honeyguides with humans was examined scientifically. The foundation work of these studies was that of H. Friedmann (1955), who reviewed and organized the observations of others, including those of Dos Santos, and who conducted his own extensive research on the honeyguides of Africa.

Friedmann's report of some of the African legends surrounding the greater honeyguide suggests that a wide variety of African cultures prescribed rewarding the bird for its guiding behavior and that native Africans recognized the need for reciprocity in their interactions with honeyguides. One proverb reported by Friedmann was, "If you do not leave anything for the guide [*I. Indicator*], it will not lead you at all in the future." Another proverb stated more ominously, "If you do not leave anything for the guide, it will lead you to a dangerous animal the next time." Friedmann also observed that many African cultures forbid killing a honeyguide and once "inflicted severe penalties" for doing so. These observations suggest long association between humans and honeyguides and that the association has been consciously mutualistic on the human side of the balance sheet.

The mutualistic association between humans and honeyguides may have developed from an earlier association between the bird and the ratel, or honey badger, *Mellivora capensis*. The honey badger is a powerful animal, well equipped with strong claws and powerful muscles to rip open bees' nests, that readily follows honeyguides to bees' nests. The honey badger, though secretive, has been observed often following honeyguides while vocalizing. African honey gatherers also vocalize to attract honeyguides, and Friedmann reported that some of their vocalizations imitate the calls of honey badgers.

The most detailed and quantitative study of this mutualism to date is that of H. Isack of the National Museum of Kenya and H.-U. Reyer of the University of Zurich (Isack and Reyer 1989), who studied the details of the interaction of the greater honeyguide with the Boran people of northern Kenya. The Boran regularly follow honeyguides and have developed a penetrating whistle that they use to attract them. The whistle can be heard over 1 km away, and Isack and Reyer found that it doubles the rate at which Boran honey gatherers encounter honeyguides. If they are successful in attracting a honeyguide, the average amount of time it takes to find a bees' nest is 3.2 hours. Without the aid of a honeyguide the average search time per bees' nest is about 8.9 hours. This is an underestimate of the true time, however, since Isack and Reyer did not include days in which no bees' nests were found in their analysis. The benefit of the association to the bird seems apparent from Isack and Reyer's analysis, since they report that 96% of the nests to which the Boran were guided would have been inaccessible to the birds without human help.

The greater honeyguide attracts the attention of a human by flying close and calling as it does so. Following this initial attention-getting behavior the bird will fly off in a particular

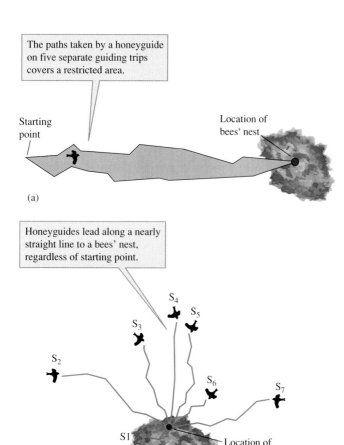

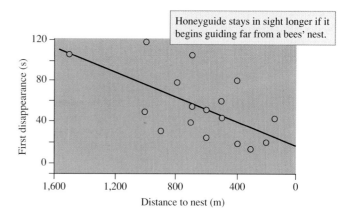

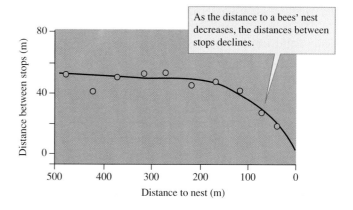

**Figure 15.21** Paths taken by honeyguides leading people to bees' nests (data from Isack and Reyer 1989).

direction and disappears for up to 1 minute. After reappearing, the bird again perches in a conspicuous spot and calls to the following humans. As the honey gatherers follow, they whistle, bang on wood, and talk loudly in order to "keep the bird interested." When the honey gatherers approach the perch from which the honeyguide is calling, the bird again flies off, calling and displaying its white tail feathers as it does so, only to reappear at another conspicuous perch a short time later. This sequence of leading, following, and leading is repeated until the bird and the following honey gatherers arrive at the bees' nest.

Isack, who is a Boran, interviewed Boran honey gatherers to determine what information they obtained from honeyguides. The main purpose of the study was to test assertions by the honey gatherers that the bird informs them of (1) the direction to the bees' nest, (2) the distance to the nest, and (3) when they arrive at the location of the nest. The data gathered by Isack and Reyer support all three assertions.

Honey gatherers reported that the bird indicated direction to the bees' nest on the basis of the direction of its guiding flights. One method used by Isack and Reyer to test how well flight direction indicated direction was to induce honeyguides to guide them from the same starting point to the same known bees' nest on five different occasions. Figure 15.21*a* shows

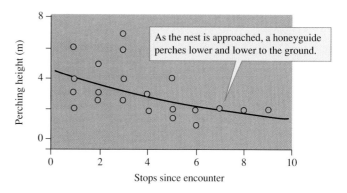

**Figure 15.22** Changes in behavior of the honeyguide as it nears a bees' nest (data from Isack and Reyer 1989).

the highly restricted area covered by these five different guiding trips. Another approach was to induce the bird to guide them to a bees' nest from seven different starting points (fig. 15.21*b*). The result was a consistent tendency by the bird to lead directly to the site of the bees' nest.

The Boran honey gatherers said that three variables decrease as distance to the nest decreases: (1) the time the bird stays out of sight during its first disappearance following the initial encounter, (2) the distance between stops made by the bird on the way to the bees' nest, and (3) the height of the perch on the way to the nest. Data gathered by Isack and Reyer support all three statements (fig. 15.22).

On the way to a bees' nest a honeyguide uses a particular call and responds to a human voice by increasing call frequency.

After arriving at a bees' nest, the honeyguide gives a few distinctive indication calls and then perches silently near the nest.

**Figure 15.23** Vocal communication between honeyguides and humans.

The honey gatherers also report that they can determine when they arrive in the vicinity of a bees' nest by changes in the honeyguide's behavior and vocalizations (fig. 15.23). Isack and Reyer observed several of these changes. While on the path to a bees' nest a honeyguide emits a distinctive guiding call and will answer human calls by increasing the frequency of the guiding call. On arriving at a nest, the honeyguide perches close to the nest and gives off a special "indication" call. After a few indication calls, it remains silent and does not answer to human sounds. If approached by a honey gatherer, a honeyguide flies in a circle around the nest location before perching again nearby.

Isack and Reyer observe that their data do not allow them to test other statements by the Boran honey gatherers, including that when bees' nests are very far away (over 2 km) the honeyguide will "deceive" the gatherers about the real distance to the nest by stopping at shorter intervals. Isack and Reyer add, however, that they have no reason to doubt these other statements, since all others have been supported by the data they were able to collect. What these data reveal is a rich mutualistic interaction between wild birds and humans. The results of Isack and Reyer's study caused Robert May (1989) to wonder how much important ecological knowledge may reside with the dwindling groups of native people living in the tropical regions of the world, regions about which the field of ecology has so little information.

## SUMMARY

*Mutualism,* interactions between individuals that benefit both partners, is a common phenomenon in nature that has apparently made important contributions to the evolutionary history of life and continues to make key contributions to the ecological integrity of the biosphere. Mutualisms can be divided into those that are *facultative,* where species can live without their mutualistic partners, and *obligate,* where species are so dependent on the mutualistic relationship that they cannot live without their mutualistic partners.

**Plants benefit from mutualistic partnerships with a wide variety of bacteria, fungi, and animals.** Mutualism provides benefits to plants ranging from nitrogen fixation and enhanced nutrient and water uptake to pollination and seed dispersal. Ninety percent of terrestrial plants form mutualistic relationships with mycorrhizal fungi, which make substantial contributions to plant performance. Mycorrhizae, which are mostly either vesicular-arbuscular mycorrhizae or ectomycorrhizae, are important in increasing plant access to water, nitrogen, phosphorus, and other nutrients. In return for these nutrients, mycorrhizae receive energy-rich root exudates. Experiments have shown that the mutualistic balance sheet between plants and mycorrhizal fungi can be altered by the availability of nutrients. Plant-ant protection mutualisms are found in both tropical and temperate environments. In tropical environments, many plants provide ants with food and shelter in exchange for protection from a variety of natural enemies. In temperate environments, mutualistic plants provide ants with food but not shelter in trade for protection.

**Reef-building corals depend upon mutualistic relationships with algae and animals.** The coral-centered mutualisms of tropical seas show striking parallels with terrestrial plant-centered mutualisms. Mutualistic algae called zooxanthellae provide reef-building corals with their principal energy source; in exchange for this energy, corals provide zooxanthellae with nutrients, especially nitrogen, a scarce resource in tropical seas. The mutualism between corals and zooxanthellae appears to be largely under the control of the coral partner, which chemically solicits the release of organic compounds from zooxanthellae and controls zooxanthellae population growth. Crabs and shrimp protect some coral species from coral predators in exchange for food and shelter.

**Theory predicts that mutualism will evolve where the benefits of mutualism exceed the costs.** Keeler built a cost-benefit model for the evolution and persistence of facultative plant-ant protection mutualisms in which the benefits of the mutualism to the plant are represented in terms of the proportion of the plant's energy budget that ants protect from damage by herbivores. The model assesses the costs of the mutualism to the plant in terms of the proportion of the plant's energy budget invested in extrafloral nectaries and the water, carbohydrates, and amino acids contained in the nectar. The model predicts that the mutualism will be favored where there are high densities of ants and potential herbivores and where the effectiveness of alternative defenses are low.

Humans have developed a variety of mutualistic relationships with other species, but one of the most spectacular is that between the greater honeyguide and the traditional honey gatherers of Africa. In this apparently ancient mutualism, humans and honeyguides engage in elaborate communication and cooperation with clear benefit to both partners. The mutualism offers the human side a higher rate of discovery of bees' nests, while the honeyguide gains access to nests that it could not raid without human help. Careful observations have documented that the honeyguide informs the honey gatherers of the direction and distance to bees' nests as well as of their arrival at the nest.

# Review Questions

1. List and briefly describe mutualistic relationships that seem to contribute to the ecological integrity of the biosphere.

2. What contributions do mycorrhizal fungi make to their plant partners? What do plants contribute in return for the services of mycorrhizal fungi? How did Hardie (1985) demonstrate that mycorrhizae improve the water balance of red clover? How do mycorrhizae improve the capacity of plants to take up water from their environment?

3. Outline the experiments of Johnson (1993), which she designed to test the possibility that artificial fertilizers may select for less mutualistic mycorrhizal fungi. What evidence does Johnson present in support of her hypothesis?

4. Explain how mycorrhizal fungi may have evolved from ancestors that were originally parasites of plant roots. Do any of Johnson's results (1993) indicate that present-day mycorrhizal fungi may act like parasites? Be specific.

5. Janzen (1985) encouraged ecologists to take a more experimental approach to the study of mutualistic relationships. Outline the details of Janzen's own experiments on the mutualistic relationship between swollen thorn acacias and ants.

6. Inouye and Taylor's study (1979) of the relationship between ants and the aspen sunflower, *Helianthella quinquenervis*, provides a reasonable representative of temperate ant-plant protection mutualisms. Compare this mutualism with that of the tropical mutualism between swollen thorn acacias and ants.

7. How are the coral-centered mutualisms similar to the plant-centered mutualisms we discussed in chapter 15? How are they different? The exchanges between mutualistic partners in both systems revolve around energy, nutrients, and protection. Is this an accident of the cases discussed or are these key factors in the lives of organisms?

8. Outline the benefits and costs identified by Keeler's (1981, 1985) cost-benefit model for facultative ant-plant mutualism. From what perspective does Keeler's model view this mutualism? From the perspective of plant or ant? What would be some of the costs and benefits to consider if the model was built from the perspective of the other partner?

9. How could you change the Lotka-Volterra model of competition we discussed in chapter 13 into a model of mutualism? Would the resulting model be a cost-benefit model or a population dynamic model?

10. Outline how the honeyguide-human mutualism could have evolved from an earlier mutualism between honeyguides and honey badgers. In many parts of Africa today, people have begun to abandon traditional honey gathering in favor of keeping domestic bees and have also begun to substitute refined sugars bought at the market for the honey of wild bees. Explain how, under these circumstances, natural selection might eliminate guiding behavior in populations of the greater honeyguide. (In areas where honey gathering is no longer practiced, the greater honeyguide no longer guides people to bees' nests.)

# Suggested Readings

Allen, M. F. 1991. *The Ecology of Mycorrhizae.* Cambridge, England: Cambridge University Press.

*A thorough, concise, and readable overview of the ecology of mycorrhizae.*

Bever, J. D., P. A. Schultz, A. Pringle, and J. B. Morton. 2001. Arbuscular mycorrhizal fungi: more diverse than meets the eye and the ecological tale of why. *BioScience* 51:923–31.

*Interesting account of diversity among arbuscular mycorrhizal fungi.*

Bronstein, J. L. 1994. Our current understanding of mutualism. *The Quarterly Review of Biology* 69:31–51.

*Bronstein reviews studies of mutualism set within the perspective of approaches used to study competition and predation. She identifies several key research questions to guide future studies of mutualism.*

Buscot, F., J. C. Munch, J. Y. Charcosset, M. Gardes, U. Nehls, and R. Hampp. 2000. Recent advances in exploring physiology and biodiversity of ectomycorrhizas highlight the functioning of these symbioses in ecosystems. *FEMS Microbiology Reviews* 24:601–24.

*This paper provides a microbiological perspective on the connection between the physiology and biodiversity of ectomycorrhizae to ecosystem function.*

Hoeksema, J. D. and E. M. Bruna. 2000. Pursuing the big questions about interspecific mutualism: a review of theoretical approaches. *Oecologia* 125:321–30.

*An overview of the fundamental questions regarding the evolution of interspecific mutualisms.*

Hölldobler B. and E. O. Wilson. 1990. *The Ants.* Cambridge, Mass.: The Belknap Press of Harvard University Press.

*This award-winning book includes descriptions of many fascinating mutualisms involving ants and opens the door to the seldom-observed world of the ants.*

Isack, H. A. and H.-U. Reyer. 1989. Honeyguides and honey gatherers: interspecific communication in a symbiotic relationship. *Science* 243:1343–46.

May, R. M. 1989. Honeyguides and humans. *Nature* 338:707–8.

*Isack and Reyer give a concise account of their scientific exploration of a mutualism involving humans. May provides commentary that sets the work in broader context.*

Janzen, D. H. 1966. Coevolution of mutualism between ants and acacias in Central America. *Evolution* 20:249–75.

*This paper summarizes one of Janzen's classic studies of ant-plant mutualism—a model for field studies of ecological relationships.*

Johnson, N. C. 1993. Can fertilization of soil select less mutualistic mycorrhizae. *Ecological Applications* 3:749–57.

*Johnson's study provides an exemplary application of the scientific method to a complex ecological question.*

Keeler, K. H. 1981. A model of selection for facultative nonsymbiotic mutualism. *American Naturalist* 118:488–98.

*Keeler presents an accessible treatment of a model for the evolution of plant protection mutualisms.*

Lilleskov, E. A., T. J. Fahey, T. R. Horton, and G. M. Lovett. 2002. Belowground ectomycorrhizal fungal community change over a nitrogen deposition gradient in Alaska. *Ecology* 83:104–15.

*Detailed ecological study of changes in diversity and composition of ectomycorrhizae along a gradient in soil nitrogen content.*

Mueller, U. G., T. R. Schultz, C. R. Currie, R. M. M. Adams, and D. Malloch. 2001. The origin of the attine ant-fungus mutualism. *Quarterly Review of Biology* 76:169–97.

*Fascinating consideration of the origin of the mutualism between leafcutter ants and their dietary fungi.*

Raine, N. E., P. Willmer, and G. N. Stone. 2002. Spatial structuring and floral avoidance behavior prevent ant-pollinator conflict in a Mexican ant-acacia. *Ecology* 83:3086–96.

*A detailed study of mediation of potential conflict between two plant mutualists.*

Stimson, J. 1990. Stimulation of fat-body production in the polyps of the coral *Pocillopora damicornis* by the presence of mutualistic crabs of the genus *Trapezia. Marine Biology* 106:211–18.

*Careful work on the ecology of a marine mutualism with discussions of its evolution.*

# On the Net

Visit this textbook's accompanying website at www.mhhe.com/ ecology (click on the book's title) to take advantage of practice quizzing, study/writing tips, timely news articles, and additional URLs for research on the topics in this chapter.

Animal Population Ecology

Mutualism

Specialized Roots

Field Methods for Studies of Populations

Coral Reefs

Coevolution

# Communities and Ecosystems

*"The Greatest beauty is organic wholeness,
the wholeness of life and things,
the divine beauty of the universe."*

Robinson Jeffers, "The Answer," 1933

# Species Abundance and Diversity

Different areas within the same region may differ substantially in the number of species they support. Vast areas of flat or gently sloping land in the hot deserts of North America are dominated by a single species of shrub, the creosote bush, *Larrea tridentata*. While grasses and forbs grow in the spaces between these shrubs, creosote bushes make up most of the plant biomass. In these areas, you can travel many kilometers and see only subtle changes in a landscape dominated by a single species of plant (fig. 16.1).

The uniformity of the creosote flats contrasts sharply with the biological diversity of other places in these hot deserts (fig. 16.2). For instance, a rich variety of plant life-forms cover Organ Pipe National Monument in southern Arizona. Here grow ocotillo, consisting of several slender branches 2 to 3 m tall springing from a common base, palo verde trees with green bark and tiny leaves, and mesquite, which reach the size of medium-sized trees. In addition, there are cactus such as the low-growing prickly pears and the shrublike teddy bear chollas. The most striking are the column-shaped squat barrel cactus, the organ pipe cactus, with its densely packed slender columns, and the saguaro, a massive cactus that towers over all the other plant species. Among these larger plants also grow a wide variety of small shrubs, grasses, and forbs.

The creosote flats, dominated by one species of shrub, convey an impression of great uniformity. The vegetation of Organ Pipe National Monument, consisting of a large number of species of many different growth forms, gives the impression of high diversity. The ecologist is prompted to ask what factors control this difference in diversity? Joseph McAuliffe (1994) has been able to explain much of the variation in woody plant diversity and dominance by *L. tridentata* across Sonoran Desert landscapes by differences in soil age, frequency of disturbance by erosion, and soil depth (see chapter 21).

In chapters 13 to 15 we focused on competition, predation, and mutualism between pairs of species. As you can see, we are now considering patterns and processes that involve a larger number of species. With this shift in focus, we enter the realm of community ecology. A **community** is an association of interacting species inhabiting some defined area. Ecologists may study the plant community on a mountainside, the insect community associated with a particular species of tree, or the fish community on a coral reef. The key point here is that communities generally consist of many species that potentially interact in all of the ways discussed in chapters 13 to 15.

Community ecologists seek to understand how various abiotic and biotic aspects of the environment influence the structure of communities. **Community structure** includes attributes such as the number of species, the relative abundance of species, and the kinds of species comprising a community.

Because it is difficult to study large numbers of species, most community ecologists work with restricted groups of organisms, focusing, for example, on communities of plants, mammals, or insects. Some community ecologists restrict their focus even more by studying guilds of species. A **guild** is a group of organisms that all make their living in a similar way. Examples of guilds include the seed-eating animals in an

**Figure 16.1** Desert landscape dominated by the creosote bush, *Larrea tridentata.*

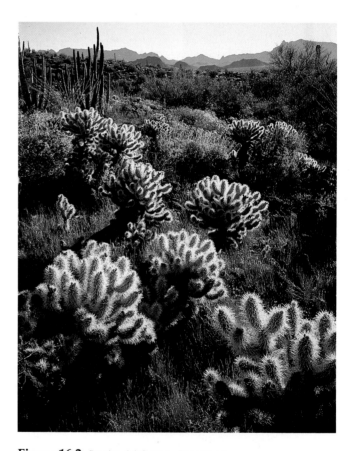

**Figure 16.2** Species-rich Sonoran Desert landscape.

area of desert, the fruit-eating birds in a tropical rain forest, or the filter-feeding invertebrates in a stream. Some guilds consist of closely related species, while others are taxonomically heterogeneous. For instance, the fruit-eating birds on many South Pacific islands consist mainly of pigeons, while the seed-eating guild in the Sonoran Desert includes mammals, birds, and ants.

The main users of the guild concept have been animal ecologists. Similar terms used by botanists are **life-form** or growth form. The life-form of a plant is a combination of its structure and its growth dynamics. Plant life-forms have been classified in various ways and we have used an informal classification since chapter 1, where we discussed life-forms such as trees, vines, annual plants, sclerophylous vegetation, grasses, and forbs.

Like the members of an animal guild, plants of similar life-form exploit the environment in similar ways. As a consequence, plant community ecologists have often concentrated their attention on plants of similar life-form by studying the ecology of tree, herb, or shrub communities. By studying animal guilds or plant life-forms, ecologists focus their energies on a manageable and coherent portion of the community, manageable in terms of number of species and coherent in terms of ecological requirements.

In 1959, G. Evelyn Hutchinson wrote a paper with the captivating title, "Homage to Santa Rosalia or Why Are There So Many Kinds of Animals?" This paper stimulated generations of ecologists to explore biological diversity. One of the most fundamental questions that we still address today is what controls the number and relative abundance of species in communities. These two properties are the focus of chapter 16.

## CONCEPTS

- **Most species are moderately abundant; few are very abundant or extremely rare.**
- **A combination of the number of species and their relative abundance defines species diversity.**
- **Species diversity is higher in complex environments.**
- **Intermediate levels of disturbance promote higher diversity.**

## CONCEPT DISCUSSION

### Species Abundance

**Most species are moderately abundant; few are very abundant or extremely rare.**

The relative abundance of species is one of the most fundamental aspects of community structure. This property is so fundamental that George Sugihara (1980) referred to it as "minimal community structure." We began our discussion of the abundance of species in chapter 9, where we explored the relationship between body size and abundance and considered the various forms of rarity. In this section, we expand our perspective by addressing the following question: What will you find if you go out into a community and quantify the abundance of species within a group of taxonomically or ecologically related organisms such as beetles, birds, shrubs, or diatoms?

It turns out that there are regularities in the relative abundance of species in communities that hold whether you examine plants in a forest, moths in that forest, or algae inhabiting a nearby stream. If you thoroughly sample groups of organisms such as these, you will come across a few abundant species and a few that are very rare. Most species will be moderately abundant. This pattern was first quantified by Frank Preston (1948, 1962a, 1962b), who carefully studied the relative abundance of species in collections and communities. The "distribution of commonness and rarity" among species described by Preston is one of the best documented patterns in natural communities.

## The Lognormal Distribution

How do we think about the abundance of organisms? Preston suggested that we think of abundance in relative terms and say, for example, that one species is twice as abundant as another. This common way of expressing relative abundance led Preston to graph the abundance of species in collections as frequency distributions, where the classes of species abundance were intervals of 1–2, 2–4, 4–8, 8–16, etc., individuals. Preston made each interval twice the preceding one and plotted them on a $\log_2$ scale (e.g., $\log_2$ of 1 = 0, $\log_2$ of 2 = 1, $\log_2$ of 4 = 2, etc.). Preston's graphs plot $\log_2$ of species abundance against the number of species in each abundance interval. When the relative abundance of species were plotted in this way, he consistently obtained results like those shown in figure 16.3.

Figure 16.3*a* shows the relative abundance of desert plants. Robert Whittaker (1965) plotted these abundances using coverage rather than numbers of individuals, which accords well with our discussions in chapter 9 of how to represent the relative abundance of plants. Notice that few species were represented by more than 8% cover or less than 0.15% cover. Most species had intermediate coverage. Whittaker's plot shows the most distinctive feature of Preston's distributions, that is, they are approximately "bell-shaped," or "normal." Since abundance is plotted on a log scale, Preston's curves are called "lognormal" distributions.

Figure 16.3*b* shows the relative abundance of 86 species of birds breeding near Westerville, Ohio, over a 10-year period (Preston 1962a). Notice that few species were represented by over 64 individuals or by a single individual. Like Whittaker's plants, most species showed intermediate levels of abundance, producing another lognormal distribution.

In most lognormal distributions, only a portion of a bell-shaped curve is apparent. For instance, neither figure 16.4*a* nor *b* presents a complete normal curve. However, figure 16.4*b,* a sample of moths from Lethbridge, Alberta, comes closer to a complete curve. Preston suggested that much of the difference between the two curves results from a difference in sample size. While the sample of moths from Saskatoon contained approximately 87,000 individuals belonging to 277 species,

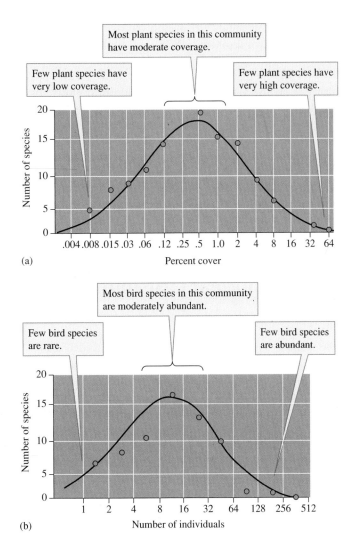

(a)

(b)

**Figure 16.3** Lognormal distributions of (*a*) desert plants, and (*b*) forest birds (data from Whittaker 1965, Preston 1962a).

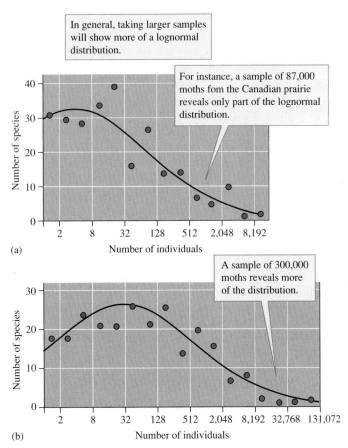

(a)

(b)

**Figure 16.4** Sample size and the lognormal distribution (data from Preston 1948).

the sample from Lethbridge contained an incredible 303,251 individuals belonging to 291 species. If the sample from Saskatoon had contained 300,000 individuals, it would have contained more species, producing a more complete lognormal curve. Ecologists have found that the more you sample a community, the more species you will find. The common species show up in even small samples, but a great deal of sampling effort is needed to capture the rare species.

So, how do we explain the lognormal distribution of commonness and rarity? Robert May (1975) proposed that the lognormal distribution is the product of many random environmental variables acting upon the populations of many species. In other words, the lognormal distribution is a statistical expectation.

Is the lognormal distribution just a mathematical artifact or does it reflect important biological processes? George Sugihara (1980) suggested that the lognormal distribution is a consequence of the species within a community subdividing niche space. However, regardless of its origins, the lognormal distribution is important because it allows us to predict the

distribution of abundance among species. As we shall see in chapter 22, the lognormal distribution has led to other valuable insights into the organization of nature.

DISCUSSION

**CONCEPT DISCUSSION**

## Species Diversity

**A combination of the number of species and their relative abundance defines species diversity.**

Ecologists define **species diversity** on the basis of two factors: (1) the number of species in the community, which ecologists usually call **species richness,** and (2) the relative abundance of species, or **species evenness.** The influence of species richness on community diversity is clear. A community with 20 species is obviously less diverse than one with 80 species. The effects of species evenness on diversity are more subtle but easily illustrated.

Figure 16.5 contrasts two hypothetical forest communities. Both forests contain five tree species, so they have equal levels of species richness. However, community *b* is more diverse than community *a* because its species evenness is

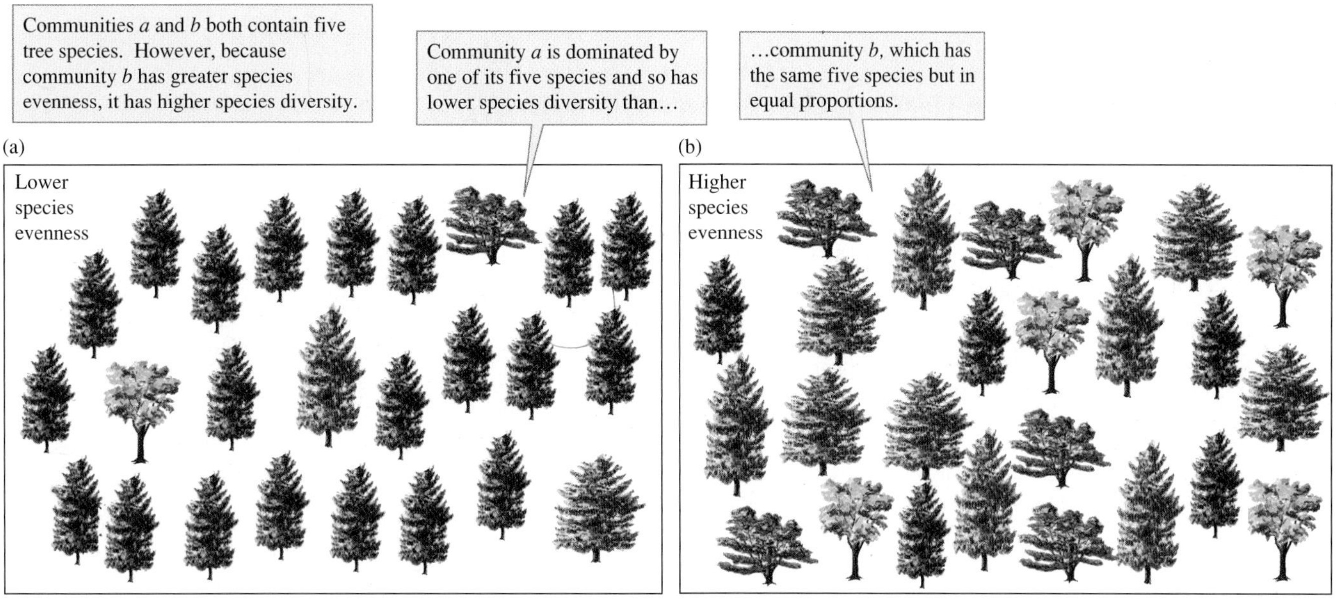


**Figure 16.5** Species evenness and species diversity.

higher. In community *b*, all five species are equally abundant, each comprising 20% of the tree community. In contrast, 84% of the individuals in community *a* belong to one species, while each of the remaining species constitutes only 4% of the community. On a walk through the two forests, you would almost certainly form an impression of higher species diversity in community *b*, despite equal levels of species richness in the two forests. That impression can be quantified.

# A Quantitative Index of Species Diversity

Ecologists have developed many indices of species diversity, the values of which depend upon levels of species richness and evenness. Let's apply one of the commonly used indices of species diversity to our hypothetical forest communities.

A commonly applied measure of species diversity is the Shannon-Wiener index:

$$H' = -\sum_{i=1}^{s} p_i \log_e p_i$$

where:

$H'$ = the value of the Shannon-Wiener diversity index
$p_i$ = the proportion of the ith species
$\log_e$ = the natural logarithm of $p_i$
s = the number of species in the community

To calculate H′, determine the proportions of each species in the study community, $p_i$, and the $\log_e$ of each $p_i$. Next, multiply each $p_i$ times $\log_e p_i$ and sum the results for all species from species 1 to species s, where s = the number of species in the community, that is:

$$\sum_{i=1}^{s}$$

Since this sum will be a negative number, the Shannon-Wiener index calls for taking its opposite, that is:

$$-\sum_{i=1}^{s}$$

The minimum value of H′ is 0, which is the value of H′ for a community with a single species, and increases as species richness and species evenness increase.

Table 16.1 shows how to calculate H′ for our two hypothetical forest communities. The different values of H′ for the two communities reflect the difference in species evenness that we see when we compare the two forests depicted in figure 16.5. H′ for community *b*, the community with higher species evenness, is 1.610, while H′ for community *a* is 0.662. We can also use a graph to contrast communities *a* and *b*.

# Rank-Abundance Curves

We can also portray the relative abundance and diversity of species within a community by plotting the relative abundance of species against their rank in abundance. The resulting **rank-abundance curve** provides us with important information about a community, information accessible at a glance. Figure 16.6 plots the abundance rank of each tree species in communities *a* and *b* (see fig. 16.5) against its proportional abundance. The rank-abundance curve for community *b* shows that all five species are equally abundant, while the rank-abundance curve for community *a* shows its dominance by the most abundant tree species.

Now let's examine the more realistic differences shown by the rank-abundance curves for two actual animal communities. Figure 16.7 shows rank-abundance curves for the Trichoptera (insects called caddisflies that have an aquatic larval stage) emerging from two kinds of aquatic habitat in northern Portugal. The trichopteran community in coastal ponds at Mira contains about 18 species, while the mountain stream community at Relva

## Table 16.1

Calculating species diversity (H′) for two hypothetical communities of forest trees

**Community a**

| Species | Number | Proportion ($p_i$) | $\log_e p_i$ | $p_i \log_e p_i$ |
| --- | --- | --- | --- | --- |
| 1 | 21 | 0.84 | −0.174 | −0.146 |
| 2 | 1 | 0.04 | −3.219 | −0.129 |
| 3 | 1 | 0.04 | −3.219 | −0.129 |
| 4 | 1 | 0.04 | −3.219 | −0.129 |
| 5 | 1 | 0.04 | −3.219 | −0.129 |
| Total | 25 | 1.00 | | −0.662 |

$$H' = -\sum_{i=1}^{s} p_i \log_e p_i = 0.662$$

**Community b**

| Species | Number | Proportion ($p_i$) | $\log_e p_i$ | $p_i \log_e p_i$ |
| --- | --- | --- | --- | --- |
| 1 | 5 | 0.20 | −1.609 | −0.322 |
| 2 | 5 | 0.20 | −1.609 | −0.322 |
| 3 | 5 | 0.20 | −1.609 | −0.322 |
| 4 | 5 | 0.20 | −1.609 | −0.322 |
| 5 | 5 | 0.20 | −1.609 | −0.322 |
| Total | 25 | 1.00 | | −1.610 |

$$H' = -\sum_{i=1}^{s} p_i \log_e p_i = 1.610$$

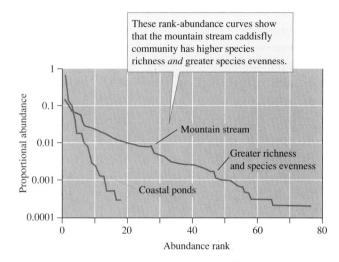

**Figure 16.7** Rank-abundance curves for caddisflies, order *Trichoptera*, of two aquatic habitats in northern Portugal (data courtesy of L. S. W. Terra).

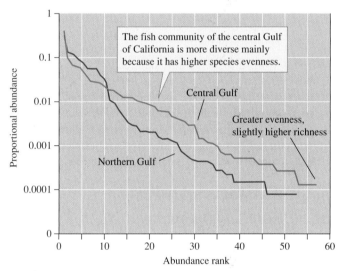

**Figure 16.8** Rank-abundance curves for two reef fish communities in the Gulf of California (data from Molles 1978, Thomson and Lehner 1976, and courtesy of D. A. Thomson and C. E. Lehner 1976).

includes 79 species. In addition, a few abundant species dominate the coastal pond community, while the mountain stream community shows a more even distribution of individuals among species. This difference in species evenness is shown by the much steeper rank-abundance curve for the coastal pond community.

Two reef fish communities from the Gulf of California provide a more subtle contrast in rank-abundance patterns. The reef fish communities yielded approximately similar numbers of species (52 versus 57) but differed substantially in species evenness. The community of the central Gulf of California showed a more even distribution of individuals among species. This greater evenness is depicted in figure 16.8, which shows that after about the tenth most abundant species, the rank-abundance curve for the central Gulf of California lies above the curve for the northern Gulf. Rank-abundance curves will provide a useful representation of community structure in later discussions.

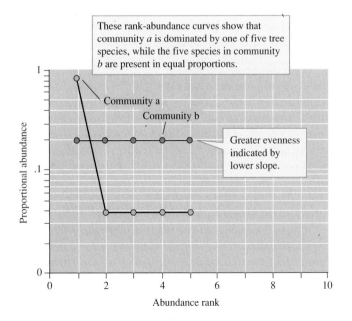

**Figure 16.6** Rank-abundance curves for two hypothetical forests.

| Information | ✓ |
| --- | --- |
| Questions | |
| Hypothesis | |
| Predictions | |
| Testing | |

*Investigating the Evidence*

# Estimating the Number of Species in Communities

How many species are there? This is one of the most fundamental questions that an ecologist can ask about a community. With increasing threats to biological diversity, species richness is also one of the most important community attributes we might measure. For instance, estimates of species richness are critical for determining areas suitable for conservation, for diagnosing the impact of environmental change on a community, or for identifying critical habitat for rare or threatened species. However, determining the number of species in a community is not a simple undertaking. Sound estimates of species richness for most taxa require a carefully designed, standardized sampling program. Here we will review some of the basic factors that an ecologist needs to consider when designing such a sampling program to gather information about species richness within and among communities.

## Sampling Effort

The number of species recorded in a sample of a community increases with higher sampling effort. We reviewed a highly simplified example of this in chapter 5 (see p. 129), where we considered how numbers of quadrats influenced estimates of species richness in the benthic community of a small Rocky Mountain stream. In that example a relatively small sample size was required. However, often far more effort is required. For example, Petri Martikainen and Jari Kouki (2003) estimated that to verify the presence or absence of threatened beetle species in the boreal forests of Finland required a sample of over 400 beetle species. They also suggested that a sample of over 100,000 individual beetles may be required to assess just 10 forest areas in Finland for their suitability to serve as conservation areas for threatened beetle species. To reduce the sampling effort required to estimate species richness, community ecologists and conservationists often focus on groups of organisms that are reliable indicators of species richness.

## Indicator Taxa

Because of the great cost and time of making thorough inventories of species diversity, ecologists have proposed a wide variety of taxa as indicators of overall biological diversity. Indicator taxa have generally been well-known and conspicuous groups of organisms such as birds and butterflies.

However, it appears that indicator taxa need to be chosen with caution. For example, John Lawton of Imperial College in the United Kingdom and twelve colleagues (Lawton et al. 1998) attempted to characterize biological diversity along a disturbance gradient in the tropical forest in Cameroon, Africa, using indicator taxa. In addition to birds and butterflies, Lawton and his colleagues sampled flying beetles, beetles living in the forest canopy, canopy ants, leaf-litter ants, termites, and soil nematodes. They sampled these taxa from 1992 to 1994 and spent several more years sorting and cataloging the approximately 2,000 species collected. This work required approximately 10,000 scientist hours. Unfortunately, the conclusion at the end of this massive study was that no one group serves as a reliable indicator of species richness for other taxonomic groups. Lawton and his colleagues estimated that their survey included from one-tenth to one-hundredth the total number of species in their study site. Citing their own experience, they concluded that characterizing the full biological diversity of just one hectare of tropical forest would require from 100,000 to 1,000,000 scientist hours. As a consequence of these time constraints, ecologists will likely continue to focus their studies of diversity on smaller groups of taxa. However, even with a restricted taxonomic focus, it is important to standardize sampling across study communities.

## Standardized Sampling

Standardizing sampling effort and technique are generally necessary to provide a valid basis for comparing species richness across communities. A standardized sampling program involves several different facets of a sampling regime. The most basic factor to consider is collecting the same number of samples from each community, or, in the case of observational studies, spending the same amount of time searching the community. It is also critical to use the same sampling methods in each study area. Sampling methods include the sampling devices used, the way the devices are employed, the time when sampling is done, and the number of habitats sampled. It is particularly important to sample the environment in a way that takes into account the environmental requirements and preferences of the study taxa. Standardization of procedures is no substitute for ecological understanding.

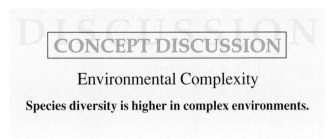

## CONCEPT DISCUSSION

### Environmental Complexity

**Species diversity is higher in complex environments.**

How does environmental structure affect species diversity? This is one of the most fundamental questions we can ask about communities. In general, species diversity increases with environmental complexity or heterogeneity. However, an aspect of environmental structure important to one group of organisms may not have a positive influence on another group. Consequently, you must know something about the ecological requirements of species to predict how environmental structure affects their diversity. In other words, you must know something about their niches.

## Forest Complexity and Bird Species Diversity

In chapter 13, we saw that competition can significantly influence the niches of species. If competition acts to produce divergence in the niches of species, what would you expect to find if you characterized the niches of closely related, coexisting species? The competitive exclusion principle (see chapter 13) leads us to predict that coexisting species will have significantly different niches. As we saw in chapter 1, that is precisely what Robert MacArthur (1958) found when he examined the ecology of five species of warblers that live together in the forests of northeastern North America (see fig. 1.2).

What does MacArthur's study of warbler niches have to do with the influence of environmental complexity on species diversity? MacArthur's results suggest that since these species forage in different vegetative strata, their distributions may be influenced by variation in the vertical structure of vegetation. He explored this possibility on Mount Desert Island, Maine, where he measured the relationship between volume of vegetation above 6 m and the abundance of warblers (fig. 16.9). The number of warbler species at the study sites increased with forest stature. The study sites with greater volume of vegetation above 6 m supported more warbler species. In other words, MacArthur found that warbler diversity increased as the stature of the vegetation increased. These results formed the foundation of later studies of how foliage height diversity influences bird species diversity.

MacArthur was one of the first ecologists to quantify the relationship between species diversity and environmental heterogeneity. He quantified the diversity of species and the complexity of the environment using the Shannon-Wiener index, H'. He measured environmental complexity as foliage height diversity, which increased with the number of vegetative layers and with an even distribution of vegetative biomass among three vertical layers, 0 to 0.6 m, 0.6 to 7.6 m, and > 7.6 m. MacArthur's foliage

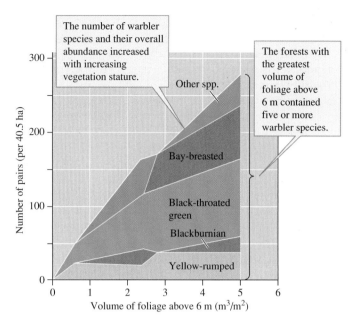

**Figure 16.9** Stature of vegetation and number of warbler species (data from MacArthur 1958).

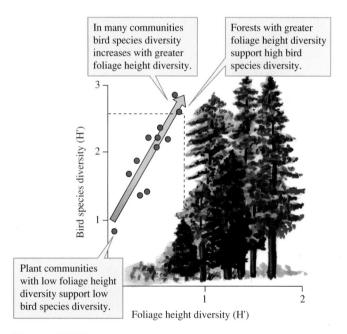

**Figure 16.10** Foliage height diversity and bird species diversity (data from MacArthur and MacArthur 1961).

height diversity, like species diversity, increases with richness (the number of vegetative layers) and evenness (how evenly vegetative biomass is distributed among layers).

Robert MacArthur and John MacArthur (1961) measured foliage height diversity and bird species diversity in 13 plant communities in northeastern North America, Florida, and Panama. The vegetative communities included in their study ranged from grassland to mature deciduous forest, with foliage height diversity that ranged from 0.043 to 1.093. Plant communities with greater foliage height diversity supported more diverse bird communities (fig. 16.10). MacArthur and his colleagues went on to study the relationship between foliage height diversity and bird species diversity in a wide variety of temperate, tropical, and island settings, from North America to Australia. They again found a positive correlation between foliage height diversity and bird species diversity. The combined weight of the evidence from North and Central America and Australia suggests that the relationship is not one of chance but reflects something about the way that birds in these environments subdivide space.

How is environmental complexity related to the diversity of other organisms besides birds? Ecological studies have shown positive relationships between environmental complexity and species diversity for many groups of organisms, including mammals, lizards, plankton, marine gastropods, and reef fish. Notice, however, that this list of organisms is dominated by animals. How does environmental complexity affect diversity of plants?

## Niches, Heterogeneity, and the Diversity of Algae and Plants

The existence of approximately 300,000 species of terrestrial plants presents a multitude of opportunities for specialization by animals. Consequently, high plant diversity can explain much of animal diversity. However, how do we explain the diversity of primary producers? G. Evelyn Hutchinson (1961) described what he called "the paradox of the plankton." He suggested that communities of phytoplankton present a paradox because they live in relatively simple environments (the open waters of lakes and oceans) and compete for the same nutrients (nitrogen, phosphorus, silica, etc.), yet many species can coexist without competitive exclusion. This situation seemed paradoxical because it appears to violate the competitive exclusion principle. The diversity of terrestrial plants presents a similar paradox. This paradox is sufficiently vexing that Joseph Connell (1978) proposed that environmental heterogeneity is not sufficient to account for terrestrial plant diversity, especially in tropical rain forests.

After some decades of theoretical and empirical work, however, it appears that environmental complexity can account for a significant portion of the diversity among both planktonic algae and terrestrial plants. As with animals, in order to study the influence of environment on diversity of plants and algae, we need to understand the nature of their niches.

## The Niches of Algae and Terrestrial Plants

The niches of algae appear to be defined by their nutrient requirements. The importance of nutrient requirements to the niches of phytoplankton was demonstrated by David Tilman (1977). Tilman conducted experiments on competition between freshwater diatoms. His experiments were similar to those conducted by G. F. Gause (1934) (see chapter 13) on

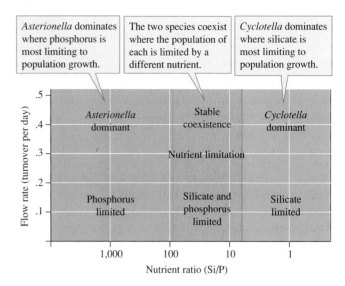

**Figure 16.11**   The ratio of silicate ($SiO_2^{-2}$) to phosphate ($PO_4^{-3}$) and competition between the diatoms *Asterionella formosa* and *Cyclotella meneghiniana* (data from Tilman 1977).

competition between paramecium. However, in addition to demonstrating competitive exclusion, Tilman's experiments also showed the conditions that allowed coexistence of diatom species. Exclusion or coexistence depended upon the ratio of two essential nutrients, silicate, $SiO_2^{-2}$, and phosphate, $PO_4^{-3}$.

When Tilman grew the diatoms *Asterionella formosa* and *Cyclotella meneghiniana* by themselves, they established and maintained stable populations. However, when he grew them together, *Asterionella* sometimes excluded *Cyclotella,* and sometimes the two species coexisted. The outcome of Tilman's experiments depended upon the ratio of silicate to phosphate (fig. 16.11). At high ratios *Asterionella* eventually excluded *Cyclotella*. However, at lower ratios the two species coexisted. At the lowest ratio, *Cyclotella* was numerically dominant over *Asterionella*.

How are Tilman's results different from those obtained by Thomas Park (1954) when he studied competition between *Tribolium* beetles (see chapter 13)? In Park's experiments, the outcome of competition experiments also depended upon physical conditions (temperature and moisture) but in all cases, Park eventually observed competitive exclusion of one *Tribolium* species or the other. In contrast, Tilman found experimental conditions allowing coexistence. Do Tilman's results violate the competitive exclusion principle? They do not because the diatoms he studied had different nutrient requirements. In other words, *Cyclotella* and *Asterionella* have different trophic niches.

How can we explain Tilman's results? It turns out that *Asterionella* takes up phosphorus at a much higher rate than does *Cyclotella*. Tilman reasons that at high ratios of silicate to phosphate *Asterionella* is able to deplete the environment of phosphorus and consequently eliminate *Cyclotella*. However, when ratios are low, silicate limits the growth rate of *Asterionella* and it cannot deplete phosphate. Consequently, when ratios are low, *Asterionella* cannot exclude *Cyclotella*. At these low ratios, silicate limits the growth rate of *Asterionella*, while phosphate limits the growth rate of *Cyclotella*. Consequently, in the presence of low ratios of silicate to phosphate, the two diatoms coexist.

What do the results of Tilman's experiments have to do with the relationship of environmental complexity to species diversity? The implication is that if the ratio of silicate to phosphate varies across a lake, then *Asterionella* will dominate some areas, while elsewhere *Cyclotella* will dominate.

Now, how might we characterize the niches of terrestrial plants? A. Tansley's experiments (1917) on competition between *Galium* species, which we discussed in chapter 13, provide insights into the niches of terrestrial plants. You may recall that Tansley studied two species: *G. saxatile,* which grows mainly on acidic soils, and *G. sylvestre,* which grows mainly on basic soils. When these two plants competed against each other in an experimental garden, each did best on the soil type that it occupies in nature. Like the diatoms *Asterionella* and *Cyclotella,* the niches of *G. saxatile* and *G. sylvestre* are significantly influenced by the chemical characteristics of the environment, in this case of the soil.

So what does this say about environmental complexity from the viewpoint of algae and plants? Because of the feeding niches of the warblers he studied, MacArthur could quantify environmental complexity from his birds' perspectives as foliage height diversity. Forests with higher foliage height diversity provided more distinctive environments for foraging birds. We can define the niches of algae and plants on the basis of their nutrient requirements and responses to constraining physical or chemical conditions, such as moisture and pH. Therefore, from the perspective of plants and algae, variation in the availability of limiting nutrients, such as silicate and phosphate, and variation in physical and chemical conditions, such as temperature, moisture, and pH, contribute to environmental complexity.

## Complexity in Plant Environments

How much do plant nutrients vary across the environments inhabited by plants and algae? Let's look first at environmental heterogeneity in an aquatic environment. Martin Lebo and his colleagues (1993) studied spatial variation in nutrient and particulate concentrations in Pyramid Lake, Nevada, which has a surface area of approximately 450 km$^2$ and a maximum depth of 102 m.

Pyramid Lake, like other lakes, is not a uniform chemical solution. All of the nutrients studied by the researchers showed substantial variation across the lake. Figure 16.12 shows that nitrate (NO$_3$) ranged from > 20 µg per liter (L$^{-1}$) near the inflow of the Truckee River to < 5 µgL$^{-1}$ along the western and northeastern shores. Silicate (SiO$_2$) reached maximum concentrations of > 300 µgL$^{-1}$ at the inflow of the Truckee River and then decreased progressively northward, reaching a minimum of < 200 µgL$^{-1}$ in the north-central portion of the lake. Other nutrients also showed substantial variation across Pyramid Lake, but their pattern of variation differed from that shown by nitrate and silicate. In other words, different parts of the lake offer distinctive growing conditions

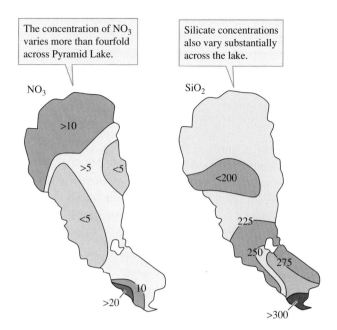

**Figure 16.12** Concentrations (µg/L) of nitrate (NO$_3$) and silicate (SiO$_2$) in the surface waters of Pyramid Lake, Nevada (data from Lebo et al. 1993).

for phytoplankton. This environmental complexity should allow for phytoplankton diversity.

Now, let's look at variation in nutrient concentrations in a terrestrial environment. Our example concerns an abandoned agricultural field, a situation where we might expect low environmental heterogeneity. We can expect reduced heterogeneity in an abandoned field because agricultural practices such as plowing, land leveling, and fertilizer applications would reduce spatial variation across fields.

G. Robertson and a team of researchers (1988) quantified variation in nitrogen and moisture across an abandoned agricultural field. Their study site was located in southeast Michigan, on the E. S. George Reserve, a 490 ha natural area maintained by the University of Michigan. Farmers cleared the field of its original oak-hickory forest and plowed the land sometime before 1870. Crop raising continued on the field until the early 1900s, when most of the land was converted to pasture. Then in 1928, the cattle were removed and the nature reserve was established. Though grazing by cattle has ceased, a dense population of white-tail deer, *Odocoileus virginianus,* continue to graze the site.

Robertson and his colleagues focused their measurements on a 0.5 ha (69 m × 69 m) subplot within the old field in which they measured several soil variables, including nitrate concentration and soil moisture, at 301 sampling points. This large number of sampling points over a small area provided sufficient data to construct a detailed map of soil properties. Figure 16.13 shows considerable patchiness in both nitrate and moisture. Both variables show at least tenfold differences across the study plot. In addition, nitrate concentration and moisture don't appear to correlate well with each other; hot spots for nitrates were not necessarily hot spots for moisture. The researchers concluded that soil conditions show sufficient spatial variability to affect the structure of plant communities.

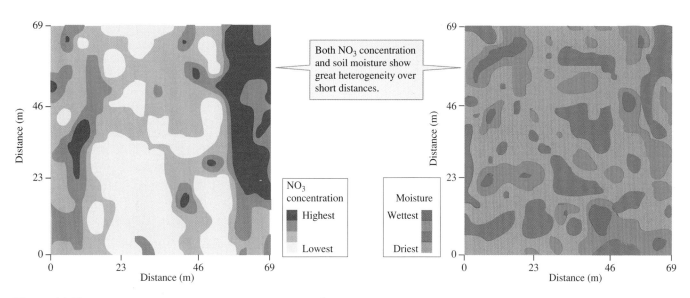

**Figure 16.13**  Variation in nitrate (NO₃) and soil moisture in a 4,761 m² area in an old agricultural field (data from Robertson et al. 1988).

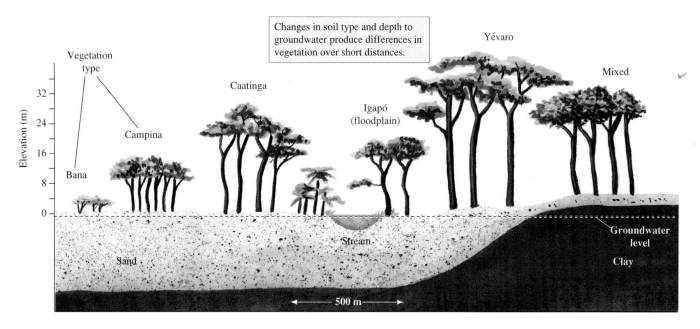

**Figure 16.14**  Variation in vegetation along a gradient of soil and moisture conditions (data from Jordan 1985).

We can see from these studies that algal and plant resources change substantially across aquatic and terrestrial environments. Now let's examine how spatial heterogeneity in these resources may affect the distribution and diversity of plants.

## Soil and Topographic Heterogeneity and the Diversity of Tropical Forest Trees

Carl Jordan (1985) studied the relationship between vegetation and soils in the Amazon forest. His studies led him to conclude that tropical forest diversity is organized in two ways: (1) a large number of *species* live within most tropical forest communities; (2) there are a large number of plant *communities* in a given area, with each community being distinctive in regard to species composition.

Jordan's studies showed that variation in soil characteristics influences the number of plant communities in an area. He found that slight differences in soil properties foster entirely different plant communities. Figure 16.14 shows six different plant communities that Jordan observed in a distance of just 500 m and an elevation range of less than 8 m. In the study area, the subsoil was clay weathered from a granitic bedrock. Sand, deposited on top of the clay, varied in thickness depending upon local topography. Topography also determined the depth of the soil above groundwater.

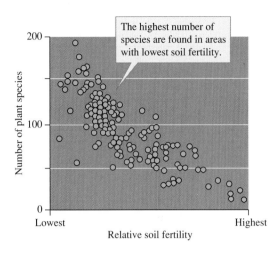

**Figure 16.15** Soil fertility and the number of plant species in 0.1 ha plots of rain forest in Ghana, Africa (data from Hall and Swaine 1976).

Diverse mixed forests occurred on the tops of hills, where clay was close to the surface. Where the topography dipped toward streambeds and the thickness of the sand layer increased, Jordan found a forest dominated by a tree in the legume family called yévaro, *Eperua purpurea*. Though the yévaro community was less diverse than the mixed forest, it was taller and supported higher plant growth rates. In addition, a distinctive forest called igapó grew along the edges of the streams in areas that flooded seasonally.

Away from streams on sandy soil, small changes in elevation produced substantial changes in water availability and plant communities. Near streams but above the level of seasonal flooding, there was another plant community known as caatinga. Still on sand but 1 to 2 m above the caatinga there was a low stature forest known as campina. Finally, at elevations greater than 2 m above stream level, where water drains through the coarse sandy soil fast enough to induce water stress, Jordan found a shrub community known as bana. He observed that while their local names vary, similar plant communities, associated with local variations in soil quality and topography, occur throughout the Amazon basin. Analogous variation in plant communities in response to differences in soil properties have also been observed in temperate regions.

## Algal and Plant Species Diversity and Increased Nutrient Availability

Ecologists have repeatedly observed a negative relationship between nutrient availability and algal and plant species diversity. In other words, as nutrient supplies increase, diversity of plants and algae declines. Michael Huston (1980, 1994a) reported a negative correlation between nutrient availability and plant species diversity in Costa Rican forests, a relationship also reported for African and Asian forests. Figure 16.15 shows this relationship for a series of study plots in rain forests in Ghana. A similar negative correlation has been found between nutrient availability and diversity of diatoms.

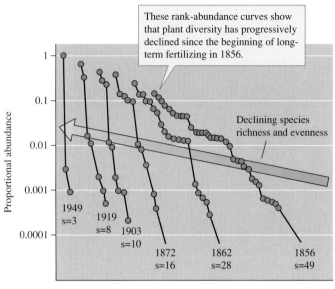

**Figure 16.16** Fertilization and plant diversity at Rothamsted, England (data from Kempton 1979, after Brenchley 1958).

Adding nutrients to water or soils generally reduces the diversity of plants and algae. The results of such experiments suggest a causal linkage between nutrient availability and diversity. For instance, in the Park Grass experiment, researchers have fertilized a grassland at the Rothamsted Experimental Station in Great Britain since 1856 (Kempton 1979). One result of that experiment has been a steady decline in plant diversity on the fertilized plots (fig. 16.16). While control plots have retained their diversity for over 100 years, the number of species on the fertilized plots has declined from 49 to 3. Also notice that figure 16.16 shows that rank-abundance curves have gotten steeper over time, indicating declining species evenness.

What does increasing nutrient availability have to do with environmental complexity? Increasing nutrient availability, whether due to natural variation or fertilization, reduces the number of limiting nutrients. Eventually, when or where all nutrients are abundant, light remains as the single limiting resource. Under these conditions the algal or plant species most effective at competing for light will dominate the community and species diversity will decline. Increased nutrient availability has also been associated with reduced fungal diversity.

## Nitrogen Enrichment and Ectomycorrhizal Fungus Diversity

Ecologists working in areas of high deposition of atmospheric nitrogen have recorded apparent declines in fungal diversity. However, most of these observations of reduced diversity have been based on declines in diversity of aboveground fruiting bodies, such as mushrooms. Such evidence may not reflect declines in fungal diversity but shifts from aboveground to belowground growth. To address this possibility, Erik Lilleskov, Timothy Fahey, Thomas Horton, and Gary Lovett (2002) studied the

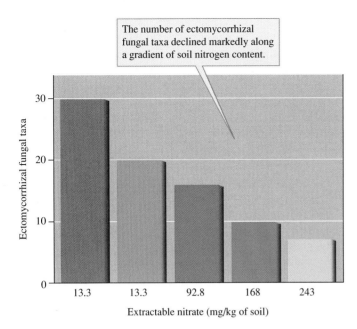

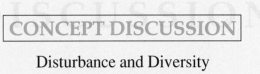

The number of ectomycorrhizal fungal taxa declined markedly along a gradient of soil nitrogen content.

**Figure 16.17** Relationship between soil nitrogen (KCl-extractable) and ectomycorrhizal community diversity near Kenai, Alaska (data from Lilleskov et al. 2002).

diversity and composition of ectomycorrhizal fungi along a gradient of nitrogen deposition on the Kenai Peninsula, Alaska. The study was focused on areas downwind from a fertilizer plant that emits gaseous ammonia. In 1992, nitrogen was deposited on the forested landscape at rates ranging from 20 kg per hectare per year in areas near the fertilizer plant to 1 kg per hectare per year several kilometers away.

Lilleskov and his colleagues sampled soil nutrients, especially nitrogen, and ectomycorrhizal fungi at five sites. The sites all supported mature stands of white spruce, *Picea glauca*, and Alaska paper birch, *Betula kenaica*. The research team sampled the ectomycorrhizal fungi associated with the roots of white spruce trees and identified them using a mix of morphological and molecular techniques. Lilleskov and his colleagues documented a strong gradient in soil nitrogen at their study sites, particularly in the organic horizon. The gradient in soil nitrogen corresponded to a decline in pH. Associated with this gradient in soil nitrogen was a clear decline in the number of ectomycorrhizal fungi taxa across the sites (fig. 16.17). The research team hypothesized that the shift in diversity and composition was the result of a change from species specialized for efficient uptake of nitrogen under conditions of low availability to dominance by acid-tolerant ectomycorrhizal fungi specialized for conditions of high soil fertility. The research team recommends tying future studies of ectomycorrhizal fungal communities with studies of ecosystem processes. It would also be interesting to determine whether nitrogen deposition creates less spatially complex environments for ectomycorrhizal fungi as it apparently does for primary producers.

So, it appears that environmental complexity can account for a portion of plant species diversity. Can complexity account for all of plant diversity? Environmental diversity across Jordan's Amazonian study sites accounts for a great deal of plant

diversity but it does not tell us how, for instance, over 300 tree species can coexist on a single hectare of Amazonian rain forest. To explain such high diversity within relatively homogeneous areas, ecologists have turned to the influences of disturbance.

## CONCEPT DISCUSSION

### Disturbance and Diversity

**Intermediate levels of disturbance promote higher diversity.**

# The Nature of Equilibrium

For several chapters we have assumed that environmental conditions remain more or less stable. Ecologists refer to this state as one of **equilibrium.** In an equilibrial system, stability is maintained by opposing forces. The Lotka-Volterra competition models (see chapter 13) and predator-prey models (see chapter 14) assumed a constant physical environment. In laboratory studies of competition, researchers have generally maintained constant environmental conditions. Even in chapter 16 when we discussed the influences of environmental complexity on species diversity, there was an underlying assumption of a stable environmental equilibrium. However, most natural environments are subject to various forms of disturbance.

# The Nature and Sources of Disturbance

What is disturbance? The answer to this question is not as simple as it may seem. What constitutes disturbance varies from one organism to another and from one environment to another. A disturbance for one organism may have little or no impact on another, and the nature of disturbance may be quite different in different environments. It is difficult to define disturbance because it involves a departure from average conditions. Because the organisms of a particular environment have been selected to cope with average conditions, disturbance must be defined in terms of average conditions. Average conditions for a particular environment may involve substantial variation. Normal daily variation in the salinity of an estuary would cause massive disruption on a coral reef. Similarly, the normal seasonal variation in temperature experienced in a temperate deciduous forest would devastate populations in a tropical rain forest. Conversely, stabilizing either an estuary or the seasonal variation in a deciduous forest, which would be departures from average conditions, would disturb the biota of those systems.

Wayne Sousa (1984), who examined the role of disturbance in structuring natural communities, defined disturbance as "a discrete, punctuated killing, displacement, or damaging

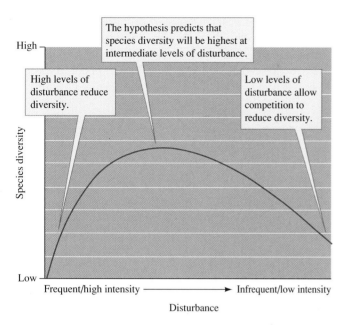

**Figure 16.18** The intermediate disturbance hypothesis (data from Connell 1978).

of one or more individuals (or colonies) that directly or indirectly creates an opportunity for new individuals (or colonies) to become established." P. S. White and S. Pickett (1985) defined disturbance as "any relatively discrete event in time that disrupts ecosystem, community or population structure and changes resources, substrate availability, or the physical environment." They also caution, however, that we must be mindful of spatial and temporal scale. For instance, disturbance to bryophyte (mosses and liverworts) communities growing on boulders along the margin of a stream can occur at spatial scales of fractions of meters and annual temporal scales that are irrelevant to the surrounding forest community.

There are innumerable potential sources of disturbance to communities. White and Pickett listed 26 major sources of disturbance roughly divided into abiotic forces such as fire, hurricanes, ice storms, and flash floods; biotic factors such as disease and predation; and human-caused disturbance. Regardless of the source, we can classify disturbances by a smaller set of characteristics. We will focus our discussion of disturbance on two characteristics: frequency and intensity.

## The Intermediate Disturbance Hypothesis

Joseph Connell (1975, 1978) proposed that disturbance is a prevalent feature of nature that significantly influences the diversity of communities. He questioned the assumption of equilibrial conditions made by most competition-based models of diversity. Instead, he proposed that high diversity is a consequence of continually changing conditions, not of competitive accommodation at equilibrium, and predicted that intermediate levels of disturbance promote higher levels of diversity (fig. 16.18).

Connell suggested that both high and low levels of disturbance would lead to reduced diversity. He reasoned that if disturbance is frequent and intense, the community will consist of those few species able to colonize and complete their life cycles between the frequent disturbances. He also predicted that diversity will decline if disturbances are infrequent and of low intensity. In the absence of significant disturbance, the community is eventually limited to the species that are the most effective competitors, effective either because they are the most efficient at using limited resources or the most effective at interference competition.

How can intermediate levels of disturbance promote higher diversity? Connell suggested that at intermediate levels of disturbance there is sufficient time between disturbances for a wide variety of species to colonize but not enough time to allow competitive exclusion.

## Disturbance and Diversity in the Intertidal Zone

Wayne Sousa (1979a) studied the effects of disturbance on the diversity of marine algae and invertebrates growing on boulders in the intertidal zone. Disturbance to this community comes mainly from ocean waves generated by winter storms. These waves, which can exceed 2.5 m in height, are large enough to overturn intertidal boulders, killing the algae and barnacles growing on their upper surfaces. Meanwhile, the newly exposed underside of the boulder is available for colonization by algae and marine invertebrates.

Because boulders of different sizes turn over at different frequencies and in response to waves of different heights, Sousa predicted that the level of disturbance experienced by the community living on boulder surfaces depends upon boulder size. Smaller boulders are turned over more frequently and therefore experience a high frequency of disturbance, middle-sized boulders experience an intermediate level of disturbance, and large boulders experience the lowest frequency of disturbance.

Sousa quantified the relationship between boulder size and probability of being moved by waves by measuring the force required to dislodge boulders of different sizes. He measured the exposed surface area of a series of boulders and then measured the force required to dislodge each. To make a measurement he wrapped a chain around a boulder, attached a spring scale to the chain, and pulled in the direction of incoming waves until the boulder moved. He recorded the number of kilograms registered on the scale when the boulder moved and then converted his measurements to force expressed in newtons (newtons [N] = kg × 9.80665). As you might expect, there was a positive relationship between size and the force required to move boulders. What was Sousa assuming as he made these measurements? He assumed that the force required to move a boulder with his apparatus was proportional to the force required for waves to move it.

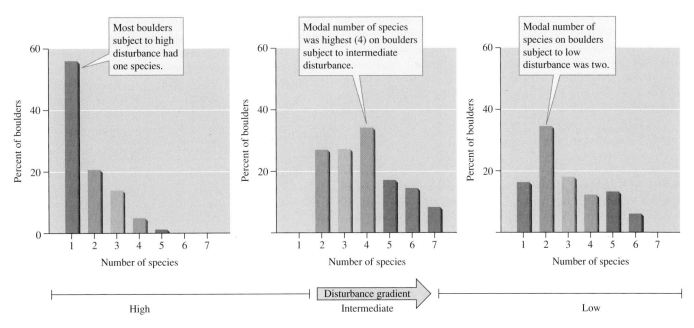

**Figure 16.19** Levels of disturbance and diversity of marine algae and invertebrates on intertidal boulders (data from Sousa 1979a).

Sousa verified this assumption by documenting the relationship between his force measurements and movement by waves. He established six permanent study sites and measured the force required to move the boulders in each. He next mapped the locations of boulders by photographing the study plots and then checked for boulder movements by taking additional photographs monthly for 2 years. Sousa divided the boulders in the study sites into three classes based on the force required for movement: (1) ≤ 49 N, (2) 50 to 294 N, and (3) > 294 N. These classes translated into frequent movement (42% per month = frequent disturbance), intermediate movement (9% per month = intermediate disturbance), and infrequent movement (1% per month = infrequent disturbance).

The number of species living on boulders varied with frequency of disturbance (fig. 16.19). Most of the frequently disturbed boulders supported a single species, few supported five species, and none supported six or seven species. Most of the boulders experiencing a low frequency of disturbance supported one to three species, few supported six species, and none supported seven. The boulders supporting the greatest diversity of species were those subject to intermediate levels of disturbance. Most of these supported three to five species, many supported six species, and some supported seven species.

## Disturbance and Diversity in Temperate Grasslands

Can several species coexist where there is a single limiting resource? The Lotka-Volterra competition equations (see chapter 13) predict that under such circumstances, one species will eventually exclude all others. However, as Sousa's work shows, even where species compete for a single resource, such as space in the intertidal zone, several species may coexist if disturbance prevents competitive exclusion. David Tilman (1994) reached a similar conclusion in regard to plant diversity within North American prairies.

What sorts of disturbance have been important in grasslands? Historically, the magnitude of disturbance on the North American prairie ranged from trampling by bison herds and fire to the death of an individual plant. One of the most important and ubiquitous sources of disturbance to grasslands is burrowing by mammals.

April Whicker and James Detling (1988) proposed that prairie dogs (*Cynomys* spp.), which occupied about 40 million ha of North American grasslands as late as 1919, were an important source of disturbance on the North American prairies. Prairie dogs are herbivorous rodents that weigh approximately 1 kg as adults and live in colonies containing 10 to 55 individuals per hectare. Prairie dogs build extensive burrow systems that are 1 to 3 m deep and about 15 m long, with tunnel diameters of 10 to 13 cm and two entrances. To build a burrow with these dimensions a prairie dog must excavate 200 to 225 kg of soil, which it deposits in mounds 1 to 2 m in diameter around burrow entrances.

Burrowing and grazing by prairie dogs have substantial effects on the structure of plant communities at several spatial scales. Figure 16.20 shows the areas of Wind Cave National Park occupied by prairie dogs. Because of the activities of these rodents, each of these areas supports vegetative communities distinctive from the surrounding landscape. Within a colony, prairie dog activities create patchiness on a smaller scale, with areas of forbs and shrubs, grass and forbs, and grass within the surrounding matrix of prairie grassland. Whicker and Detling estimate that plant species diversity is greatest in areas experiencing intermediate levels of disturbance by prairie dogs (fig. 16.21).

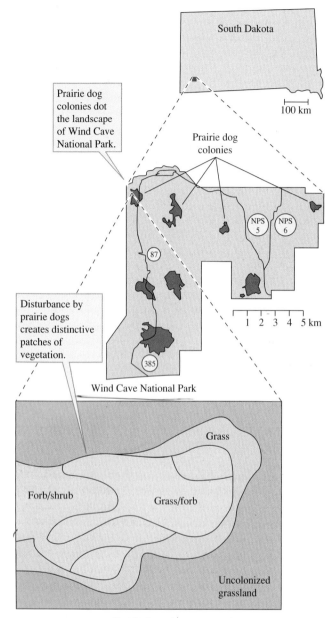

**Figure 16.20** Disturbance by prairie dogs and patchiness of vegetation (data from Coppock et al. 1983, Whicker and Detling 1988).

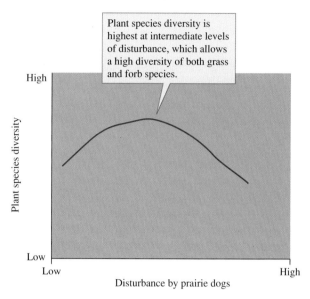

**Figure 16.21** Disturbance by prairie dogs and plant species diversity (data from Whicker and Detling 1988).

How does disturbance by prairie dogs foster higher diversity? The mechanisms underlying this effect are essentially the same as those operating in the intertidal boulder field studied by Sousa. By burrowing and piling earth and by grazing and clipping vegetation, prairie dogs remove vegetation from areas around their burrows. These bare patches are then open for colonization by plants. However, some plant species are more likely to colonize these open patches. Those species investing most heavily in dispersal are usually the first to arrive. However, these early colonists can be displaced by better competitors that arrive later. The persistence of both good colonizers and good competitors in a plant community depends upon intermediate levels of disturbance. Too much disturbance and the community is dominated by the good colonizers; too little disturbance and the better competitors dominate.

Because they have been considered an agricultural pest, various control programs have reduced prairie dog populations by about 98% during the last century. The extermination also eliminated their dynamic influences on plant communities. However, other burrowing mammals remain in large numbers. One of the most important of these are the pocket gophers of the family Geomyidae. Though pocket gophers are much smaller than prairie dogs, weighing from 60 to 900 g, their effects on grassland and arid land communities are considerable. The mounds that gophers create during their burrowing may cover as much as 25% to 30% of the ground surface, which increases heterogeneity in light availability and soil nitrogen, which in turn fosters increased plant species richness.

The influences of prairie dogs and pocket gophers on plant communities are a consequence of a combination of physical disturbances due to burrowing and to feeding, since both are herbivores. How might disturbance by humans affect the diversity of plant and animal communities? We are all well acquainted with examples of how severe disturbance by humans reduces biological diversity. In the following section, we consider the effects of moderate disturbance by humans on diversity.

## APPLICATIONS & TOOLS

### Disturbance by Humans

The effects of disturbance by humans are all around us. Housing developments cover the countryside as human populations continue their rapid growth. Deforestation continues at an alarming rate in both temperate and tropical regions. Industries

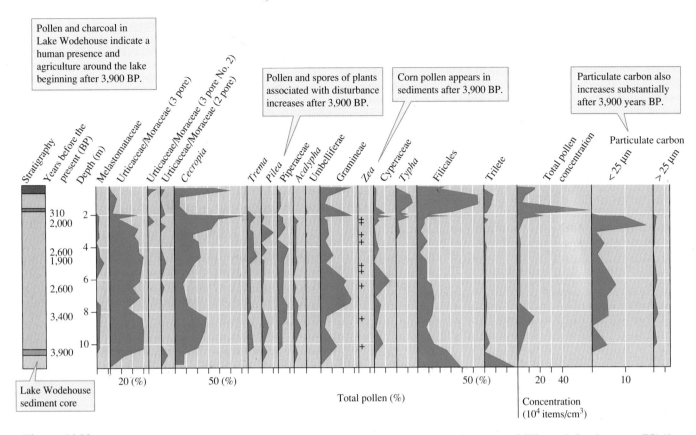

**Figure 16.22** Pollen and particulate carbon in the sediments of Lake Wodehouse, Panama, dating from the present to 3,900 years before the present (BP) (data from Bush and Colinvaux 1994).

pollute air and water. While the destructive effects of disturbance by humans may seem outside the realm of what we've discussed so far, they are not. This section emphasizes two points: (1) disturbance by humans has been around for a long time, even in many apparently pristine habitats, and (2) disturbance by humans does not always reduce species diversity.

## Human Disturbance: An Ancient Feature of the Biosphere

Near the border between Panama and Columbia there is a region so laced with rivers and swamps and so thick with tropical rain forest that the Pan American Highway cannot pass. This region, called the Darien, has become a symbol of raw tropical nature, impenetrable and pristine. Though the Darien has resisted the passage of the Pan American Highway, it is not as pristine as it has seemed. Native Americans occupied and farmed the Darien continuously for about 4,000 years. They abandoned the region only after the Spanish conquest about three centuries ago. This challenge to the generally held view of the Darien comes from work by Mark Bush and Paul Colinvaux (1994). As a consequence of their studies, we now know that these rain forests are not "virgin" but have grown up since the abandonment of agriculture, covering most traces of previous human occupation.

How did Bush and Colinvaux penetrate the secrets of the Darien? They used the same paleoecological methods used by

Margaret Davis to reconstruct the floral history of North American forests that we discussed in chapter 1. Bush and Colinvaux looked for lakes and swamps whose sediments might preserve a record of the Darien's past. They found such a record in a swamp and a small lake, Lake Wodehouse, approximately 15 km apart, near a small mining settlement called Cana near Panama's border with Columbia.

The researchers collected 10 m cores of sediment from Lake Wodehouse and Cana Swamp. At the base (bottom) of the cores from Lake Wodehouse were an abundance of spores from ferns (e.g., *Triletes*) and pollen from other forest plants and little pollen from plants associated with disturbed landscapes (fig. 16.22). This portion of the pollen record indicates a swamp community dominated by ferns surrounded with mature tropical forest. Then, about 3,900 years ago, the sediment record changed abruptly with the appearance of charcoal and corn pollen, which are present throughout the subsequent record until nearly 310 years ago. Similar patterns occurred in Cana Swamp except that corn pollen constituted a greater proportion of the total pollen (about 2%), indicating that corn was being grown right on the swamp during the dry season. The sediments containing corn pollen and charcoal also contained pollen from wild plants that live in disturbed areas (e.g., Melastomataceae, *Cecropia, Trema,* and *Pilea*). Then, about 310 years ago, about 1.6 to 1.7 m deep in the Lake Wodehouse sediment core, charcoal and corn pollen disappear, while the pollen of forest species increases in abundance.

Are the sedimentary records of Lake Wodehouse and Cana Swamp atypical of tropical Central and South America? No, they are not. Other studies reveal even longer histories of human occupation and disturbance. In another study, Mark Bush, Dolores Piperno, and Paul Colinvaux (1989) provide evidence of 6,000 years of corn cultivation in the Amazon River basin. They recovered this record from the sediments of Lake Ayauch, Ecuador, a small lake within the eastern Amazon River basin at an elevation of 500 m.

The longest sediment core retrieved from Lake Ayauch was 3.26 m long and contained sediments deposited during the last 7,100 years. The earliest portion of the sediment record indicates that the lake was surrounded by mature forest. Then about 6,000 years ago, corn pollen appears in lake sediments. The appearance of corn pollen in the sediments of Lake Ayauch coincided with an increase in the pollen of *Cecropia* and herbaceous vegetation, a pattern that Bush and Colinvaux also observed in Lake Wodehouse, Panama. These changes indicate forest clearing and soil disturbance associated with agriculture. However, the occurrence of corn pollen within the sediments of Lake Ayauch is sporadic, a pattern suggestive of shifting agriculture. It appears that the farmers around the lake cleared forest for their fields, farmed the land for a period, and then moved on. After some time, the same land could be cleared once again. This kind of shifting agriculture continued for about 5,000 years and then corn pollen disappears from the record about 800 years ago. The researchers suggest that people may have stopped farming in the lake basin due to climatic change. Other studies suggest that cultivation of corn in Panama began about 7,000 years ago.

As impressive as this 7,000-year history of agriculture may be, human disturbance in the New World tropics is more ancient still. Substantial human disturbance of the forests of Panama began about 11,000 years ago. Evidence comes from the sediments of Lake La Yeguada in central Panama. Lake La Yeguada yielded a 17.5 m sediment core containing a 14,300-year record of the surrounding landscape. This sediment core contains no record of human disturbance for about 3,300 years. Then, about 11,000 years ago, the charcoal content of sediment rises abruptly.

The increase in charcoal in the sediment record of La Yeguada also coincided with the presence of Paleo-Indian artifacts across Panama. We can conclude that humans have played a substantial role in shaping the communities and ecosystems of Panamanian rain forests for at least 11,000 years. At the very least we can say that the exceptionally high diversity of the New World tropics has coexisted with moderate levels of human disturbance for at least 11,000 years. However, the present rate of clear-cutting of Amazonian rain forest has no historical precedent in the basin and the predictions of the intermediate disturbance model are clear. Continued severe disturbance of these tropical rain forests will lead to reduced diversity. There is, however, at least one plant community where human disturbance has fostered higher species diversity.

# Disturbance by Humans and the Diversity of Chalk Grasslands

As we have seen, the North American prairie supports over 100 species of herbaceous plants. On a local scale, this long species list translates into about 18 species per square meter. As impressive as this herbaceous diversity may be, the chalk grasslands of Europe are even richer in species. Dutch ecologists studying the chalk grasslands of the Netherlands have recorded up to 50 species of herbaceous plants in a single square meter, and it appears that this exceptional diversity is maintained by moderate levels of human disturbance.

Chalk grasslands, which grow on thin, infertile soils that overlie calcareous bedrock, became established in western Europe about 10,000 years ago as humans cleared forests. Since then, chalk grasslands have persisted as a result of human activity. Farmers used these grasslands mainly for livestock grazing and for harvesting hay for winter fodder. Both grazing and mowing, which prevented takeover by woody plants or competitively dominant herbaceous plants, maintained elevated plant diversity. However, the importance of disturbance for maintaining the diversity of chalk grasslands became apparent only after they, and their traditional uses, were abandoned.

Bobbink and Willems (1987, 1991) studied the ecology of chalk grasslands in the southern province of Limburg in the Netherlands, where farmers abandoned traditional use of chalk grasslands after World War II. Many of these areas were declared nature reserves. However, reserve managers soon found that they had to actively manage the grasslands in order to maintain their diversity. Without disturbance, plant diversity quickly declined (fig. 16.23), and they became a near monoculture of the grass *Brachypodium pinnatum*.

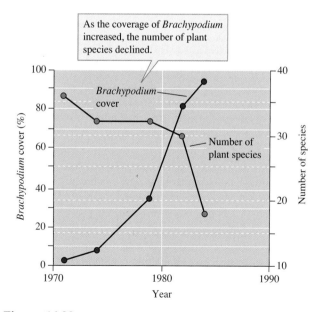

**Figure 16.23** Changes in number of plant species and coverage by the grass *Brachypodium pinnatum* following abandonment of a chalk grassland (data from Bobbink and Willems 1987).

Bobbink and Willems asked whether declines in grassland diversity could be reversed using traditional mowing. Farmers traditionally mowed the grasslands in autumn to harvest hay for winter livestock feed. The researchers compared the influence of the normal autumn mowing with mowing during summer on several study plots from 1982 to 1986. During this period, the biomass of *Brachypodium* on the plots mowed in summer decreased from 77% to 34%, while on the plots mowed in autumn, it remained relatively constant at 75% to 80%. During this period, the number of species per 0.25 m$^2$ increased from 15.6 to 21.2 on the plots mowed in summer, while remaining low on the plots mowed in autumn. These results show that the traditional mowing regime does not restore grassland diversity.

How do summer and autumn mowing differ in their effects on *Brachypodium?* Bobbink and Willems addressed this question by studying the response of the grass to four different mowing regimes from 1984 to 1987: (1) autumn mowing, (2) late summer mowing, (3) early summer mowing, or (4) two mowings, early summer and autumn. It turns out that two mowings and summer mowing greatly reduce the nonstructural carbohydrates in the rhizomes of the grass. Nonstructural carbohydrates are important for shoot formation, and their reduction would reduce shoot formation during the growing season following cutting, an effect that would reduce the competitive ability of *Brachypodium.*

The studies of Bobbink and Willems have improved our understanding of the factors controlling plant species diversity in the chalk grasslands of the Netherlands. However, we are left with a question. Why is one of the traditional management practices on chalk grasslands, autumn mowing for hay harvest, no longer sufficient to maintain the historical diversity of this community? It appears that human activity has produced another disturbance to these communities: increased nitrogen in soils. Increased soil nitrogen appears to strongly favor dominance by *Brachypodium,* and traditional autumn mowing cannot reverse this dominance. As we saw earlier, increased nutrient availability reduces plant diversity by favoring those species that are the most effective competitors for the remaining resources. We will discuss the effects of nitrogen enrichment of ecosystems in more detail in chapter 19, which focuses on nutrient cycling.

## SUMMARY

A *community* is an association of interacting species inhabiting some defined area. Examples of communities include the plant community on a mountainside, the insect community associated with a particular species of tree, or the fish community on a coral reef. Community ecologists often restrict their studies to groups of species that all make their living in a similar way. Animal ecologists call such groups *guilds,* while plant ecologists use the term *life-form.* The field of community ecology concerns how the environment influences *community structure,* including the relative abundance and diversity of species, the subjects of this chapter.

**Most species are moderately abundant; few are very abundant or extremely rare.** Frank Preston (1948) graphed the abundance of species in collections as distributions of species abundance, with each abundance interval twice the preceding one. Preston's graphs were approximately "bell-shaped" curves and are called "lognormal" distributions. Lognormal distributions, which describe the relative abundance of organisms ranging from algae and terrestrial plants to birds, may result from many random environmental variables acting upon the populations of a large number of species or may be a consequence of how species subdivide resources. Regardless of the underlying mechanisms, the lognormal distribution is one of the best described patterns in community ecology.

**A combination of the number of species and their relative abundance defines species diversity.** Two major factors define the diversity of a community: (1) the number of species in the community, which ecologists usually call *species richness,* and (2) the relative abundance of species, or *species evenness.* One of the most commonly applied indices of species diversity is the Shannon-Wiener index:

$$H' = -\sum_{i=1}^{s} p_i \log_e p_i$$

The relative abundance and diversity of species can also be portrayed using *rank-abundance curves.* Accurate estimates of species richness require carefully designed sampling programs.

**Species diversity is higher in complex environments.** Robert MacArthur (1958) discovered that five coexisting warbler species feed in different layers of forest vegetation and that the number of warbler species in North American forests increases with increasing forest stature. Various investigators have found that the diversity of forest birds increases with increased foliage height diversity. The niches of algae can be defined by their nutrient requirements. Heterogeneity in physical and chemical conditions across aquatic and terrestrial environments can account for a significant portion of the diversity among planktonic algae and terrestrial plants. Soil characteristics and depth to groundwater strongly influence the nature of local plant communities in the Amazon River basin. Increased nutrient availability correlates with reduced algal and plant diversity.

**Intermediate levels of disturbance promote higher diversity.** Joseph Connell (1975, 1978) proposed that high diversity is a consequence of continually changing conditions, not of competitive accommodation at equilibrium. He predicted that intermediate levels of disturbance would foster higher levels of diversity. At intermediate levels of disturbance, a wide array of species can colonize open habitats, but there is not enough time for the most effective competitors to exclude the other species. Wayne Sousa (1979a), who studied the effects of disturbance on the diversity of sessile marine algae and invertebrates growing on intertidal boulders, found support for the intermediate disturbance hypothesis. Diversity in prairie vegetation also appears to be higher in areas receiving intermediate levels of disturbance. The effect of disturbance on diversity appears to depend upon a trade-off between dispersal and competitive abilities.

Human disturbance is an ancient feature of the biosphere. Human influences touch every portion of the biosphere and have done so for thousands of years. For instance, humans disturbed tropical rain forests in Central America beginning about 11,000 years ago. The effects of human disturbance fall within the framework of the intermediate disturbance hypothesis. Though intense human disturbance reduces species diversity, moderate levels of disturbance may increase the diversity of some communities such as the European chalk grasslands.

# Review Questions

1. What is the difference between a community and a population? What are some distinguishing properties of communities? What is a guild? Give examples. What is a plant life-form? Give examples.

2. Draw a "typical" lognormal distribution. Include properly labeled horizontal ($x$) and vertical ($y$) axes. You can use the lognormal distributions included in chapter 16 as models.

3. Suppose you are a biologist working for an international conservation organization concerned with studying and conserving biological diversity. On one of your assignments you are sent out to explore the local biotas of several regions. As part of your survey work you are to take large quantitative samples of the copepods of the North Atlantic, the butterflies of central New Guinea, and the ground-dwelling beetles of southwest Africa. Using the lognormal distribution, predict the patterns of relative abundance of species you expect to see within each of these groups of organisms.

4. What are species richness and species evenness? How does each of these components of species diversity contribute to the value of the Shannon-Wiener diversity index (H')? How do species evenness and richness influence the form of rank-abundance curves?

5. Compare the "trophic" niches of warblers and diatoms as described by MacArthur (1958) and Tilman (1977). Why is it important that the ecologist be familiar with the niches of study organisms before exploring relationships between environmental complexity and species diversity?

6. Communities in different areas may be organized in different ways. For instance, C. Ralph (1985) found that in Patagonia in Argentina, as foliage height diversity increases, bird species diversity decreases. This result is exactly the opposite of the pattern observed by MacArthur (1958) and others reviewed in chapter 16. Design a study aimed at determining the environmental factors determining variation in bird species diversity across Ralph's Patagonian study sites.

7. According to the intermediate disturbance hypothesis, both low and high levels of disturbance can reduce species diversity. Explain possible mechanisms producing this relationship. Include trade-offs between competitive and dispersal abilities in your discussion.

8. The dams that have been built on many rivers often stabilize river flow by increasing flows below the dam during droughts and decreasing the amount of flooding during periods of high rainfall. Explain how these stabilized flows can be considered as a "disturbance." Using the intermediate disturbance hypothesis, predict how stabilized flows would affect the diversity of river organisms below reservoirs.

9. Humans have been living in the tropical rain forests of the New World for at least 11,000 years. During this period, disturbance by humans has been a part of these tropical rain forests. Use the intermediate disturbance hypothesis to explain how recent disturbances threaten the biological diversity of these forests, while earlier disturbances apparently did not.

10. Why do introduced predators possibly threaten the species diversity of a community such as Lake Victoria, while indigenous predators do not? Think in evolutionary timescales as you develop your answer to this question.

# Suggested Readings

Bush, M. B. and P. A. Colinvaux. 1994. Tropical forest disturbance: paleoecological records from Darien, Panama. *Ecology* 75:1761–68.

Bush, M. B., D. R. Piperno, and P. A. Colinvaux. 1989. A 6,000 year history of Amazonian maize cultivation. *Nature* 340:303–5.

Bush, M. B., D. R. Piperno, P. A. Colinvaux, P. E. De Oliveira, L. A. Krissek, M. C. Miller, and W. E. Rowe. 1992. A 14,300-yr paleoecological profile of lowland tropical lake in Panama. *Ecological Monographs* 62:251–75.

Colinvaux, P. A. 1989. The past and future Amazon. *Scientific American* 260:102–8.

*This series of papers reviews the long history of human disturbance in tropical communities of Central and South America.*

Cao, Y., D. D. Williams, and D. P. Larsen. 2002. Comparison of ecological communities: the problem of sample representativeness. *Ecological Monographs* 72:41–56.

*In this study, Cao, Williams, and Larsen point out that equalizing sampling efforts across communities may not result in representative estimates of species richness.*

Connell, J. H. 1978. Diversity in tropical rain forests and coral reefs. *Science* 199:1302–10.

Weins, J. A. 1977. On competition and variable environments. *American Scientist* 65:590–97.

*Two papers that helped change the way ecologists view natural communities and controls over their species diversity. These papers present a case for the influence of environmental variability on communities.*

Gotelli, N. J. and R. K. Colwell. 2001. Quantifying biodiversity: procedures and pitfalls in the measurement and comparison of species richness. *Ecology Letters* 4:379–91.

*This study examines some of the subtle issues involved in making quantitative comparisons of species richness.*

Huston, M. 1994. *Biological Diversity.* New York: Cambridge University Press.

Ricklefs, R. E. and D. Schluter. 1993. *Species Diversity in Ecological Communities.* Chicago: The University of Chicago Press.

*These two books provide broad overviews of biological diversity—excellent references and thought-provoking discussions.*

Kitching, R. L., D. Li, and N. E. Stork. 2001. Assessing biodiversity 'sample packages': how similar are arthropod assemblages in different tropical rainforests? *Biodiversity and Conservation* 10:793–813.

*A report on the effectiveness and biases of different methods for sampling the arthropods of tropical forests.*

Lilleskov, E. A., T. J. Fahey, T. R. Horton, and G. M. Lovett. 2002. Belowground ectomycorrhizal fungal community change over a nitrogen deposition gradient in Alaska. *Ecology 83:104–15.*

*Fascinating study of the effects of nitrogen enrichment on diversity of mycorrhizal fungi.*

MacArthur, R. H. and J. W. MacArthur. 1961. On bird species diversity. *Ecology* 42:594–98.

Ralph, C. J. 1985. Habitat association patterns of forest and steppe birds of northern Patagonia, Argentina. *The Condor* 87:471–83.

Recher, H. F. 1969. Bird species diversity and habitat diversity in Australia and North America. *American Naturalist* 103:75–85.

*A series of classical studies on the relationship between bird species diversity and foliage height diversity.*

Martikainen, P. and J. Kouki. 2003. Sampling the rarest: threatened beetles in boreal forest biodiversity inventories. *Biodiversity and Conservation* 12:1815–31.

*A detailed analysis of the sampling efforts required to sample rare beetle species in boreal forests. This study shows that the amount of earth's biodiversity and the challenges of characterizing that diversity are substantial even far from the tropics.*

May, R. M. 1988. How many species are there on Earth? *Science* 341:1441–49.

May, R. M. 1992. How many species inhabit the earth? *Scientific American* 267:42–48.

*These two classic papers by one of the world's most prominent ecologists examine one of the most significant ecological questions.*

Sousa, W. P. 1979a. Disturbance in marine intertidal boulder fields: the nonequilibrium maintenance of species diversity. *Ecology* 60:1225–39.

Whicker, A. D. and J. K. Detling. 1988. Ecological consequences of prairie dog disturbances. *BioScience* 38:778–85.

*Detailed studies of the effects of disturbance on the structure of intertidal and grassland communities.*

Tilman, D. 1977. Resource competition between planktonic algae: an experimental and theoretical approach. *Ecology* 58:338–48.

Tilman, D. 1994. Competition and biodiversity in spatially structured habitats. *Ecology* 75:2–16.

*Pioneering studies on diversity among organisms that contrast sharply with the birds studied by MacArthur and his colleagues.*

# On the Net
INTERNET

Visit this textbook's accompanying website at www.mhhe.com/ecology (click on the book's title) to take advantage of practice quizzing, study/writing tips, timely news articles, and additional URLs for research on the topics in this chapter.

Community Ecology
Field Methods for Studies of Ecosystems
Species Abundance, Diversity, and Complexity
Biodiversity

Species Preservation
Conservation and Management of Habitats and Species
Trees of Life

# Species Interactions and Community Structure

Feeding relationships provide some of the most easily documented examples of interactions within communities. The ocean around Antarctica is one of the most productive marine environments on earth. Phytoplankton, especially diatoms, thrive in these frigid, windswept seas, where they are food for grazing zooplankton. One of the most important of these zooplankton are krill, shrimplike crustaceans named *Euphausia superba*. Krill are prey for a wide variety of larger plankton-feeding species, including crabeater seals, penguins, flying seabirds, and many species of fish and squid. The best known of the krill feeders are the baleen whales that once gathered in huge numbers to feed in Antarctic waters (fig. 17.1).

The krill-feeding fishes and squid are eaten by predaceous species, including emperor penguins, larger fish, and Weddell and Ross seals. Leopard seals, a highly carnivorous species, feed on penguins and the smaller seals. Finally, the ultimate predators in this community are the killer whales, which eat seals, including leopard seals, and even attack and consume baleen whales. Huge populations of organisms live in the oceans surrounding Antarctica, all bound together in a tangle of feeding relationships.

How can we go beyond a confusing verbal description to a useful and easily understood summary of the feeding relationships within communities? One of the earliest approaches to the study of communities was to describe who eats whom. Since the beginning of the twentieth century, ecologists have meticulously described the feeding relationships in hundreds of communities. The resulting tangles of relationships came to be called food webs. If we define a community as an association of interacting species, a moment's reflection will show that a **food web,** a summary of the feeding interactions within a community, is one of the most basic and revealing descriptions of community structure. A food web is, essentially, a community portrait (fig. 17.2). Other community portraits could be produced using competitive or mutualistic relationships.

### CONCEPTS

- **A food web summarizes the feeding relations in a community.**
- **The feeding activities of a few keystone species may control the structure of communities.**
- **Exotic predators can collapse and simplify the structure of food webs.**
- **Mutualists can act as keystone species.**

### CONCEPT DISCUSSION

## Community Webs

**A food web summarizes the feeding relations in a community.**

The earliest work on food webs concentrated on simplified communities. In 1927, Charles Elton pointed out that the number of well-described food webs, which he called "food cycles," could be counted on the fingers of one hand. One of the first of those food webs described the feeding relations on Bear Island in the high Arctic (fig. 17.3). Summerhayes and Elton (1923) studied the feeding relations there because they believed that the high Arctic, with few species, would be the best place to begin the study of food webs.

Summerhayes and Elton used a food web to present the feeding relations on Bear Island in a single picture. The primary producers in the Bear Island food web are terrestrial plants and aquatic algae. These primary producers are fed upon by several kinds of terrestrial and aquatic invertebrates, which are in turn consumed by birds. The birds on Bear Island are attacked by arctic foxes. Arctic foxes also feed on marine mammals that have washed up onto beaches and on the dung of polar bears. The polar bears of Bear Island subsist on a diet of seals and beached marine animals. Seabirds harvest food from the sea around Bear Island but enter the Bear Island food web because they are attacked by foxes, feed on beached marine animals and on the freshwater invertebrates of Bear Island, and contribute dung that fertilizes the primary producers of the island.

The work of Summerhayes and Elton revealed that even in these "impoverished" feeding relations are complex and difficult to document. For instance, they failed to document several probable feeding relations in their Bear Island food web, which they indicated with dotted lines. However, the level of food web complexity increased dramatically as ecologists studied more diverse communities.

**Figure 17.1** A marine food web in action: feeding baleen whales and birds.

**Figure 17.2** The Antarctic pelagic food web.

Killer whale

Leopard seal

Skua

Ross seal

Weddell seal

Larger fish

Emperor penguin

Blue whale

Crabeater seal

Flying birds

Adele penguin

Small fish and squid

Krill

Diatoms

# Detailed Food Webs Reveal Great Complexity

Now let's go from the high arctic to the tropics, where Kirk Winemiller (1990) describe the feeding relations among freshwater fishes. Winemiller studied the aquatic food webs at two locations in the savannas, "llanos," of Venezuela and at two other sites in the lowlands of Costa Rica. His study sites supported from 20 to 88 fish species. One of Winemiller's least species-rich study sites was a medium-sized stream called Caño Volcán. This stream flows through the piedmont of the Andes and supports 20 fish species.

Winemiller represented the food web of his study sites in various ways. In some he only included the "common" fish species whose aggregate abundance exceeded 95% of

the individuals in his collections. These common-fish webs excluded many rare species. He also drew "top-predator sink webs," food webs consisting of all prey consumed by the top predator in a community, all items consumed by the prey of the top predator, and so on down to the base of the food web. Third, Winemiller constructed food webs that excluded the weakest trophic links, those comprising less than 1% of the diet.

Let's look at the results from Caño Volcán, the simplest fish community. Figure 17.4 shows that even when only the 10 most common fish are included in the food web, it remains remarkably complex. The most comprehensible of Winemiller's food webs were those that focused on the strongest trophic links.

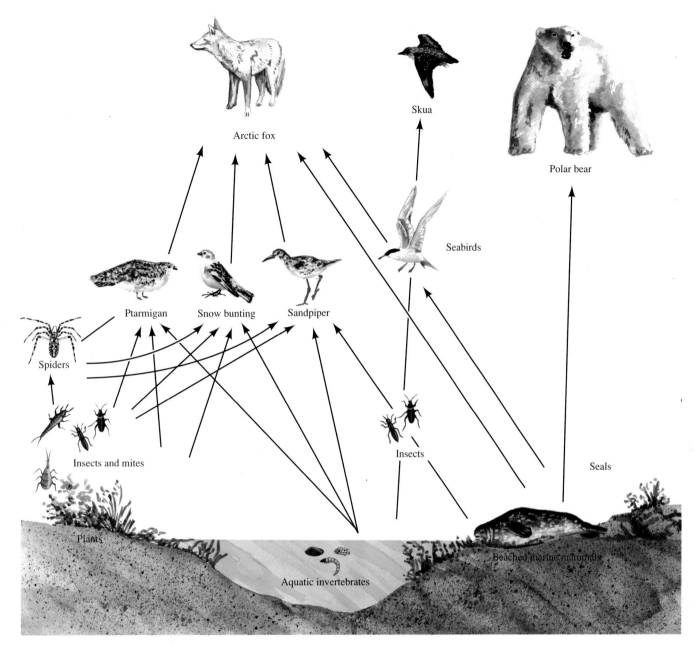

**Figure 17.3** Simple food web of an Arctic island.

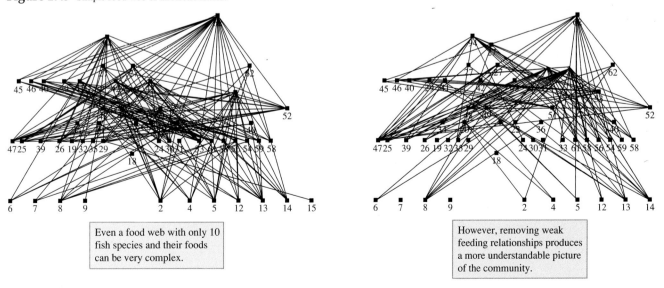

Even a food web with only 10 fish species and their foods can be very complex.

However, removing weak feeding relationships produces a more understandable picture of the community.

**Figure 17.4** Food web representing the feeding relations of the 10 most common fish species at Caño Volcán, Venezuela (data from Winemiller 1990).

# Strong Interactions and Food Web Structure

Robert Paine (1980) suggested that, in many cases, the feeding activities of a few species have a dominant influence on community structure. He called these influential trophic relations **strong interactions.** Paine also suggested that the defining criterion for a strong interaction is not necessarily quantity of energy flow but rather degree of influence on community structure. We will revisit this topic later in the following section on keystone species, but for now, let's look at how recognizing interaction strength can simplify depictions of food webs.

Paine's distinction between strong and weak interactions within food webs has been used to model the interactions within at least one terrestrial food web. Teja Tscharntke (1992) has worked intensively on a food web associated with the wetland reed *Phragmites australis.* This reed grows in large stands along the shores of rivers and other wetlands. Tscharntke's study site was along the Elbe River near Hamburg in northwest Germany. Along the river, *Phragmites* is attacked by *Giraudiella inclusa,* a fly in the family Cecidomyidae, whose larvae develop within galls called "ricegrain" galls. At the study sites *Phragmites* is also attacked by *Archanara geminipuncta,* a moth in the family Noctuidae, whose larvae bore into the stems of *Phragmites.* Stem-boring by *A. geminipuncta* induces *Phragmites* to form side shoots, a response that provides additional sites for oviposition by the gall maker *G. inclusa.*

Tscharntke discovered that at least 14 species of parasitoid wasps attack *G. inclusa.* How can so many species attack a single host species and continue to coexist? Does this seem to violate the competitive exclusion principle (see chapter 13)? Tscharntke explains this apparent paradox by pointing out that each parasitoid species appears to specialize on attacking *G. inclusa* at different times and on different parts of *Phragmites.* In winter, blue tits, *Parus caeruleus,* move into stands of *Phragmites,* where they peck open the galls formed by *G. inclusa* and eat the larvae, causing mortality in this population as well as in its parasitoids.

Tscharntke represented these trophic interactions with a food web that captures the essential interactions among species in this community (fig. 17.5). Even though there are fewer interactions than in Winemiller's tropical fish webs (see fig. 17.4), Tscharntke's web still contains plenty of complexity. However, figure 17.5 focuses the reader on the most important interactions in the community by distinguishing between strong, weak, and weakest interactions by representing this gradient in interaction strength by red (strong), blue (weak), or green (weakest) lines.

Figure 17.5 suggests that feeding by blue tits strongly influences the parasitoids *Aprostocetus calamarius* and *Torymus arundinis* and their host, *G. inclusa,* in large gall clusters on main shoots. The other series of strong interactions involves the parasitoids *Aprostocetus gratus* and *Platygaster quadrifarius,* which attack the *G. inclusa* that inhabit small gall clusters in side shoots of *Phragmites.* These side shoots are in turn stimulated by the stem-boring larvae of the moth *A. geminipuncta.* Notice that blue tits only weakly influence populations on this side of the web.

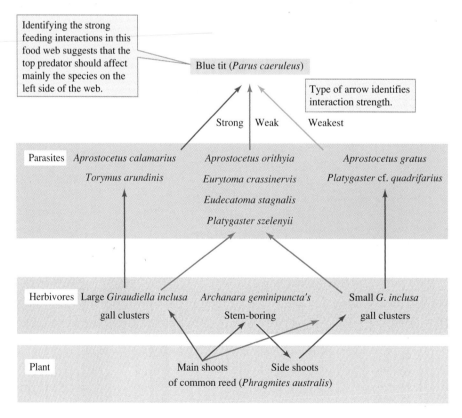

**Figure   17.5** Food web associated with *Phragmites australis* (data from Tscharntke 1992).

By distinguishing between weak and strong interactions, Tscharntke produced an easily understood food web to represent the study community. Identifying strong interactions allows us to determine which species may have the most significant influences on community structure. Those with substantial influence we now call **keystone species.**

## CONCEPT DISCUSSION

### Keystone Species

**The feeding activities of a few keystone species may control the structure of communities.**

Robert Paine (1966, 1969) proposed that the feeding activities of a few species have inordinate influences on community structure. He called these keystone species. Paine's keystone species hypothesis emerged from a chain of reasoning. First, he proposed that predators might keep prey populations below their carrying capacity. Next, he reasoned the potential for competitive exclusion would be low in populations kept below carrying capacity. Finally, he concluded that if keystone species reduce the likelihood of competitive exclusion, their activities would increase the number of species that could coexist in communities. In other words, Paine predicted that some predators may increase species diversity.

## Food Web Structure and Species Diversity

Paine began his studies by examining the relationship between overall species diversity within food webs and the proportion of the community represented by predators. He cited studies that demonstrated that as the number of species in marine zooplankton communities increases, the proportion that are predators also increases. For instance, the zooplankton community in the Atlantic Ocean over continental shelves includes 81 species, 16% of which are predators. In contrast, the zooplankton community of the Sargasso Sea contains 268 species, 39% of which are predators. Paine set out to determine if similar patterns occur in marine intertidal communities.

Paine described a food web from the intertidal zone at Mukkaw Bay, Washington, which lies in the north temperate zone at 49° N. This food web is typical of the rocky shore community along the west coast of North America (fig. 17.6). The base of this food web consists of nine dominant intertidal invertebrates: two species of chitons, two species of limpets, a mussel, three species of acorn barnacles, and one species of gooseneck barnacle. Paine pointed out that *Pisaster* commonly consumes two other prey species in other areas, a snail and another bivalve, bringing the total food web diversity to 13 species. Ninety percent of the energy consumed by the middle level predator, *Thais,* consists of barnacles. Meanwhile the top predator, *Pisaster,* obtains 90% of its energy from a mixture of chitons (41%), mussels (37%), and barnacles (12%).

Paine also described a subtropical food web (31° N) from the northern Gulf of California, a much richer web that included 45 species. However, like the food web at Mukkaw Bay, Washington, the subtropical web was topped by a single predator, the starfish *Heliaster kubinijii* (fig. 17.6). However, six predators occupy middle levels in the subtropical web, compared to one middle level predator at Mukkaw Bay. Because four of the five species in the snail family Columbellidae are also predaceous, the total number of predators in the subtropical web is 11. These predators feed on the 34 species that form the base of the food web. Despite the presence of many more species in this subtropical web, the top predator, *Heliaster,* obtains most of its energy from sources similar to those used by *Pisaster* at Mukkaw Bay. *Heliaster* obtains 74% of its energy directly from a mixture of bivalves, herbivorous gastropods, and barnacles.

Paine found that as the number of species in his intertidal food webs increased, the proportion of the web represented by predators also increased, a pattern similar to that described by G. Grice and A. Hart (1962) when they compared zooplankton communities. As Paine went from Mukkaw Bay to the northern Gulf of California, overall web diversity increased from 13 species to 45 species, a 3.5-fold increase. However, at the same time, the number of predators in the two webs increased from 2 to 11, a 5.5-fold increase. According to Paine's predation hypothesis, this higher proportion of predators produces higher predation pressure on prey populations, which in turn promotes the higher diversity in the Gulf of California intertidal zone.

Does this pattern confirm Paine's predation hypothesis? No, it does not. First, Paine studied a small number of webs—not enough to make broad generalizations. Second, while the patterns described by Paine are consistent with his hypothesis, they may be consistent with a number of other hypotheses. To evaluate the keystone species hypotheses, Paine needed a direct experimental test.

## Experimental Removal of Starfish

For his first experiment, Paine removed the top predator from the intertidal food web at Mukkaw Bay and monitored the response of the community. He chose two study sites in the middle intertidal zone that extended 8 m along the shore and 2 m vertically. One site was designated as a control and the other as an experimental site. He removed *Pisaster* from the experimental site and relocated them in another portion of the intertidal zone. Each week Paine checked the experimental site for the presence of *Pisaster* and removed any that might have colonized since his last visit.

Paine followed the response of the intertidal community for 2 years. Over this interval, the diversity of intertidal invertebrates in the control plot remained constant at 15, while the diversity within the experimental plot declined from 15 to 8, a loss of 7 species. This reduction in species diversity supported

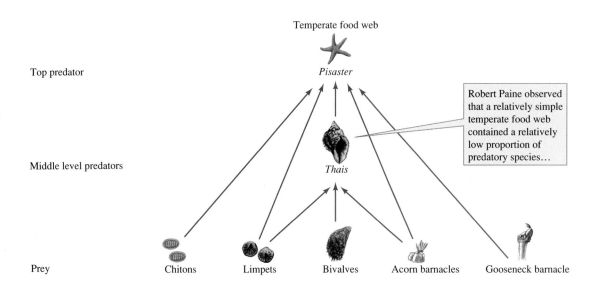

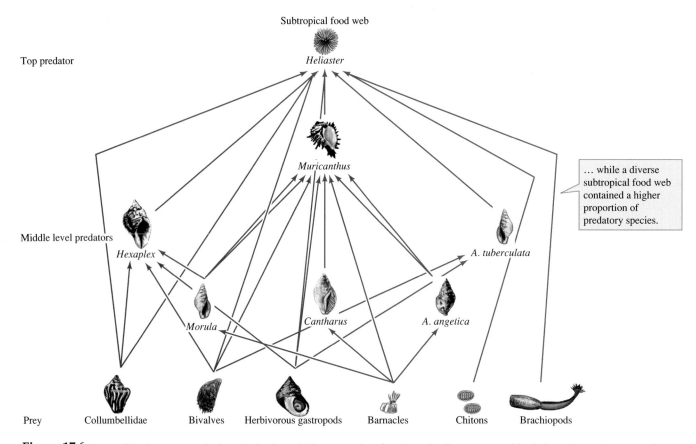

**Figure 17.6** Roots of the keystone species hypothesis: does a higher proportion of predators in diverse communities indicate that predators contribute to higher species diversity?

Paine's keystone species hypothesis. However, if this reduction was due to competitive exclusion, what was the resource over which species competed?

As we saw in chapter 11, the most common limiting resource in the rocky intertidal zone is space. Within 3 months of removing *Pisaster* from the experimental plot, the barnacle *Balanus glandula* occupied 60% to 80% of the available space. One year after Paine removed *Pisaster, B. glandula* was

crowded out by mussels, *Mytilus californianus,* and gooseneck barnacles, *Pollicipes polymerus.* Benthic algal populations also declined because of a lack of space for attachment. The herbivorous chitons and limpets also left, due to a lack of space and a shortage of food. Sponges were also crowded out and a nudibranch that feeds on sponges also left. After 5 years, the *Pisaster* removal plot was dominated by two species: the mussel, *M. californianus,* and the gooseneck barnacle, *P. polymerus.*

This experiment showed that *Pisaster* is a keystone species. When Paine removed it from his study plot, the community collapsed. However, did this one experiment demonstrate the general importance of keystone species in nature? To demonstrate this we need more experiments and observations across a wide variety of communities. Paine followed his work at Mukkaw Bay with a similar experiment in New Zealand.

The intertidal community along the west coast of New Zealand is similar to the intertidal community along the Pacific coast of North America. The top predator is a starfish, *Stichaster australis,* that feeds on a wide variety of invertebrates, including barnacles, chitons, limpets, and a mussel, *Perna canaliculus.* During 9 months following Paine's removal of the starfish, the number of species in the removal plot decreased from 20 to 14 and the coverage of the area by the mussel increased from 24% to 68%. As in Mukkaw Bay, the removal of a predaceous starfish produced a decrease in species richness and a significant increase in the density of a major prey species. Again, the mechanism underlying disappearance of species from the experimental plot was competitive exclusion due to competition for space.

These results show that intertidal communities thousands of kilometers apart that do not share any species of algae or genera of invertebrates are influenced by similar biological processes (fig. 17.7). This is reassurance to ecologists seeking general ecological principles. However, the two communities are not identical. The New Zealand intertidal community includes a large brown alga, *Durvillea antarctica,* that vigorously competes for space with the mussel *Perna.* The mussel *M. californianus* does not face such a competitive challenge in the North American intertidal zone.

In a second removal experiment, Paine removed both the starfish *Stichaster* and the large brown alga *Durvillea* from two different study plots. The result was far more dramatic than when Paine had removed the starfish only. After only 15 months, *Perna* dominated the study area and excluded nearly all other flora and fauna, covering 68% to 78% of the space in the two removal sites. Paine's studies in North America and New Zealand provide substantial support for the keystone species hypothesis. Many other studies quickly followed the lead taken by Paine's pioneering work.

# Consumers' Effects on Local Diversity

Jane Lubchenko (1978) observed that previous studies had indicated that herbivores sometimes increase plant diversity, sometimes decrease plant diversity, and sometimes seem to do both. She proposed that to resolve these apparently conflicting results it would be necessary to understand (1) the food preferences of herbivores, (2) the competitive relationships among plant species in the local community, and (3) how competitive relationships and feeding preferences vary across environments. Lubchenko used these criteria to guide her study of the influences of an intertidal snail, *Littorina littorea,* on the structure of an algal community.

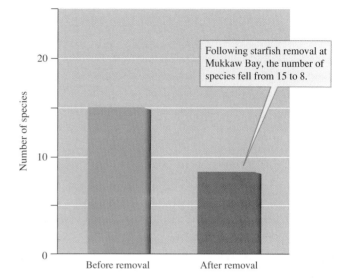

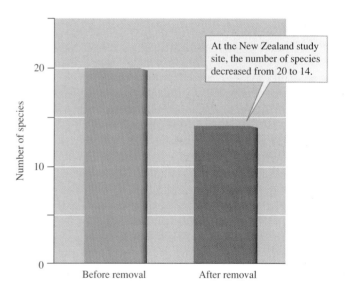

**Figure 17.7**   The effect of removing a top predator from two intertidal food webs (data from Paine 1966, 1971).

Lubchenko studied the feeding preferences of *Littorina* in the laboratory. Her experiments indicated that algae fell into low, medium, or high preference categories. Generally, highly preferred algae were small, ephemeral, and tender like the green algae, *Enteromorpha* spp., while most tough, perennial species like the red alga *Chondrus crispus* were never eaten or eaten only if the snail was given no other choice.

Lubchenko also studied variation in the abundance of algae and *Littorina* in tide pools. She found that tide pools with high densities of *Enteromorpha,* one of the snail's favorite foods, contained low densities (4/m$^2$) of snails. In contrast, pools with high densities of *Littorina* (233–267/m$^2$)

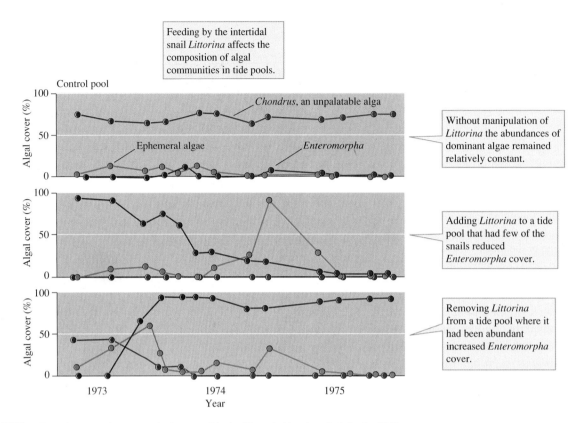

**Figure 17.8** Effect of *Littorina littorea* on algal communities in tide pools (data from Lubchenko 1978).

were dominated by *Chondrus,* a species for which the snail shows low preference. Lubchenko reasoned that in the absence of *Littorina, Enteromorpha* competitively displaces *Chondrus.* She tested this idea by removing the *Littorina* from one of the pools in which they were present in high density and introducing them to a pool in which *Enteromorpha* was dominant. Lubchenko monitored a third pool with a high density of the snails as a control.

The results of Lubchenko's removal experiment were clear (fig. 17.8). While the relative densities of *Chondrus, Enteromorpha,* and other ephemeral algae remained relatively constant in the control pool, the density of *Enteromorpha* declined with the introduction of *Littorina.* Meanwhile, *Enteromorpha* quickly increased in density and came to dominate the pool from which Lubchenko had removed the snails. In addition, as the *Enteromorpha* population in this pool increased, the population of *Chondrus* declined. Lubchenko began another addition and removal experiment in two other pools in the fall to check for seasonal effects on feeding and competitive relations. This second removal experiment produced results almost identical to the first. Where *Littorina* were added, the *Enteromorpha* population declined, while the *Chondrus* population increased. Where the snails were removed, the *Chondrus* population declined, while the *Enteromorpha* population increased.

How can we explain Lubchenko's results? *Littorina* prefers to feed on *Enteromorpha,* a species that can outcompete *Chondrus* in tide pools. So, in the absence of the snails, *Chondrus* is competitively displaced by *Enteromorpha.* However, where it is present in high densities, *Littorina* grazes down the

*Enteromorpha* population, releasing *Chondrus* from competition with *Enteromorpha.*

What controls the local population density of *Littorina?* Apparently, the green crab, *Carcinus maenus,* which lives in the canopy of *Enteromorpha,* preys upon young snails and can prevent the juveniles from colonizing tide pools. Adult *Littorina* are much less vulnerable to *Carcinus* but rarely move to new tide pools. Populations of *Carcinus* are in turn controlled by seagulls. Here again, we begin to see the complexity of a local food web and the influences that trophic interactions within webs can have on community structure.

So, within tide pools *Enteromorpha* can outcompete the other tide pool algae for space and *Enteromorpha* is the preferred food of *Littorina.* How might feeding by the snails affect the diversity of algae within tide pools? The relationship between the snails and the algal species they exploit is similar to the situation studied by Paine, where mussels were the competitively dominant species and one of the major foods of the starfish *Pisaster.*

Lubchenko examined the influence of *Littorina* on algal diversity by observing the number of algal species living in tide pools occupied by various densities of snails (fig. 17.9). As the density increased from low to medium, the number of algal species increased. Then, as the density increased further, from medium to high, the number of algal species declined.

How would you explain these results? At low density, the feeding activity by *Littorina* is not sufficient to prevent *Enteromorpha* from dominating a tide pool and crowding out some other species. At medium densities, the snail's feeding,

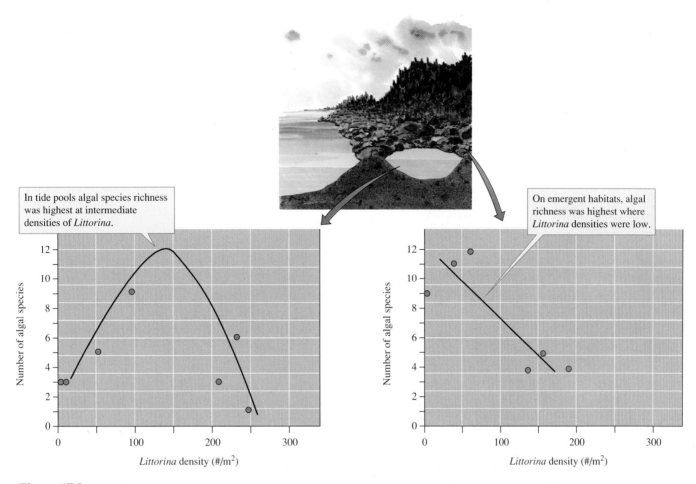

In tide pools algal species richness was highest at intermediate densities of *Littorina*.

On emergent habitats, algal richness was highest where *Littorina* densities were low.

**Figure 17.9** Effect of *Littorina littorea* on algal species richness in tide pools and emergent habitats (data from Lubchenko 1978).

which concentrates on the competitively dominant species, prevents competitive exclusion and so increases algal diversity. However, at high densities the feeding requirements of the population are so high that the snails eat their preferred algae as well as less preferred species. Consequently, intense grazing by snails at high density reduces algal diversity.

What would happen if *Littorina* preferred to eat competitively inferior species of algae? This is precisely the circumstance that occurs on emergent substrata, rock surfaces that are not submerged in tide pools during low tide. On these emergent habitats the competitively dominant algae are species in the genera *Fucus* and *Ascophyllum,* algae for which the snails show low preference. On emergent substrata, the snails continue to feed on ephemeral, tender algae such as *Enteromorpha,* largely ignoring *Fucus* and *Ascophyllum.* In this circumstance, Lubchenko found that algal diversity was highest when *Littorina* densities were low (fig. 17.9).

What produces this reduction in diversity? Let's think first about the effect of competition by *Fucus* and *Ascophyllum.* In the absence of disturbance, these two algae will gradually cover all emergent substrata, crowding out other algal species in the process. Feeding by snails accelerates elimination of the competitively subordinate species. In this case, competition by *Fucus* and *Ascophyllum* and exploitation by *Littorina* are both pushing the community toward reduced species richness.

Lubchenko's research improved our understanding of how trophic interactions can affect community structure. Her work demonstrated that the influence of consumers upon the structure of food webs depends upon their feeding preferences, the density of local consumer populations, and the relative competitive abilities of prey species. While Lubchenko moved the field well beyond the conceptual view held by ecologists when Paine first proposed the keystone species hypothesis, one basic element of the original hypothesis remained: Consumers can exert substantial control over food web structure; they can act as keystones.

Can predators act as keystone species in environments other than the intertidal zone? Two examples of keystone species in riverine and terrestrial environments follow.

# Fish as Keystone Species in River Food Webs

Mary Power (1990) tested the possibility that fish can significantly alter the structure of food webs in rivers. She conducted her research on the Eel River in northern California, where most precipitation falls during October to April, sometimes producing torrential winter flooding. During the summer, however, the flow of the Eel River averages less than 1 m³ per second.

(a)

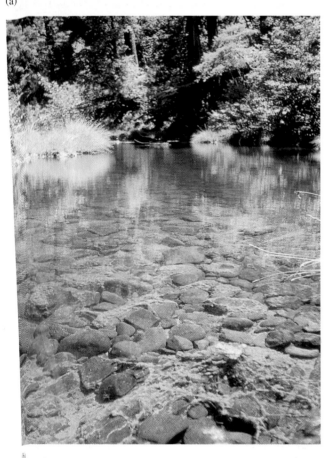

(b)

**Figure 17.10** Seasonal changes in biomass and growth form of benthic alge in the Eel River, California: (*a*) in early summer, June 1989; (*b*) in late summer, August 1989.

In early summer, the boulders and bedrock of the Eel River is covered by a turf of the filamentous alga *Cladophora* (fig. 17.10). However, the biomass of the algae declines by midsummer and what remains has a ropy, prostrate growth form and a "webbed" appearance. These mats of *Cladophora* support dense populations of larval midges in the fly family Chironomidae. The chironomid, *Pseudochironomus richardsoni*, is particularly

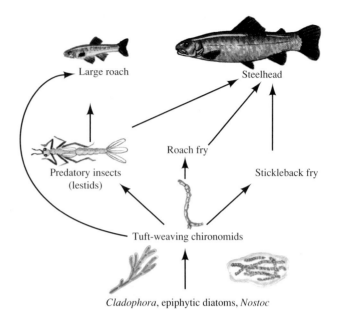

*Cladophora*, epiphytic diatoms, *Nostoc*

**Figure 17.11** Food web associated with algal turf during the summer in the Eel River, California.

abundant. *Pseudochironomus* feeds on *Cladophora* and other algae and weaves the algae into retreats, altering their appearance in the process.

Chironomids are eaten by predatory insects and the young (known as *fry*) of two species of fish: a minnow called the California roach, *Hesperoleucas symmetricus,* and three-spined sticklebacks, *Gasterosteus aculeatus.* These small fish are eaten by young steelhead trout, *Oncorhynchus mykiss.* Steelhead and large roach eat predatory invertebrates, and large roach also feed directly upon benthic algae. These interactions form the Eel River food web pictured in figure 17.11.

Power asked whether or not the two top predators in the Eel River food web, roach and steelhead, significantly influence web structure. She tested the effects of these fish on food web structure by using 3 mm mesh to cage off 12 areas 6 m$^2$ in the riverbed. The mesh size of these cages prevented the passage of large fish but allowed free movement of aquatic insects and stickleback and roach fry. Power excluded fish from six of her cages and placed 20 juvenile steelhead and 40 large roach in each of the other six cages. These fish densities were within the range observed around boulders in the open river.

Significant differences between the exclosures and enclosures soon emerged. Algal densities were initially similar; however, enclosing fish over an area of streambed significantly reduced algal biomass (fig. 17.12). In addition, the *Cladophora* within cages with fish had the same ropy, webbed appearance as *Cladophora* in the open river.

How do predatory fish decrease algal densities? The key to answering this question lies with the Eel River food web (see fig. 17.11). Predatory fish feed heavily on predatory insects, young roach, and sticklebacks. Lower densities of these smaller predators within the enclosures decreased predation on chironomids. Higher chironomid density increased feeding pressure of these herbivores on algal populations. This explanation is supported by Power's estimate that enclosures contained lower densities of predatory insects and fish fry and higher densities of chironomids (fig. 17.13). By enclosing and excluding fish from sections of the Eel River, Power, like Paine and Lubchenko, who worked in the intertidal zone, demonstrated that fish act as keystone species in the Eel River food web.

All of the examples that we have discussed so far have been aquatic. Do terrestrial communities also contain keystone species? An increasing body of evidence indicates that they do.

# The Effects of Predation by Birds on Herbivory

Let's move now from the Mediterranean climate of the Eel River basin to the boreal forests of northern Sweden, where Ola Atlegrim (1989) studied the influence of birds on herbivo-

rous insects and insect-caused plant damage. It appears that insectivorous birds may act as keystones in boreal forests through their effects on populations of herbivorous insects.

Atlegrim studied the food web associated with the bilberry, *Vaccinium myrtillus,* which is a dominant understory shrub in many boreal forests in northern Sweden. The insects that commonly feed on *Vaccinium* include caterpillars of the moth families Geometridae and Tortricidae and the larvae of the Hymenoptera known as sawflies. The geometrid and sawfly larvae feed on the bilberry from exposed positions, while the tortricid larvae bind leaves together with silk to form a shelter within which they feed. Because populations of these larvae can reach high densities, they can do considerable damage to *Vaccinium.* However, Atlegrim observed that larval insect densities peak when many insectivorous birds are feeding insects to their young and posed the following questions: (1) Do birds reduce the density of insect larvae feeding on *Vaccinium?* (2) Do birds have different effects on larvae feeding from exposed versus concealed positions? (3) Does predation by birds reduce larval insect damage to the shoots of *Vaccinium?*

Atlegrim's study sites were located approximately 20 km northwest of Umeå in northern Sweden. He established five study areas in forests ranging from 70 to 120 years old. At each study area, he established 10 study plots 4 m$^2$ and built a bird exclosure over 5 of them. Exclosures consisted of 40 mm$^2$ nylon mesh supported by a wooden frame.

Atlegrim took care to ensure that he could attribute any experimental effects to the exclusion of birds. His exclosures excluded birds but allowed small predaceous mammals, such as shrews, and predaceous invertebrates to move freely into and out of the study plots. He also kept track of the densities of these alternative predators by periodically sampling them with pitfall traps. Why was this an important aspect of Atlegrim's study? In the absence of predation by birds, higher densities of herbivorous insects might have attracted higher numbers of other predators, that is, produced a localized numerical response (see chapter 10). Atlegrim also measured the intensity of sunlight within his exclosures and in adjacent control plots. Why was this aspect of the study necessary? If the exclosures created significant shading, physical effects

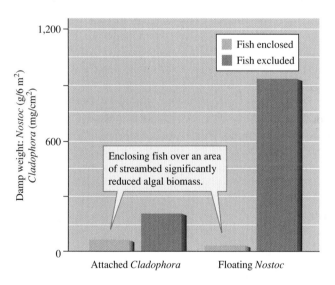

**Figure 17.12** The influence of juvenile steelhead and California roach on benthic algal biomass in the Eel River (data from Power 1990).

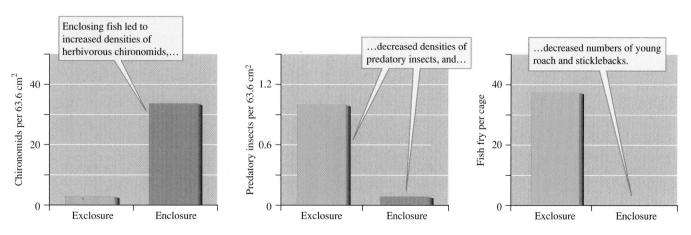

**Figure 17.13** Effect of juvenile steelhead and roach on numbers of insects and young (fry) roach and sticklebacks (data from Power 1990).

*Investigating the Evidence*

# Using Confidence Intervals to Compare Populations

In chapter 15 we reviewed how to calculate confidence intervals for the true population mean as:

$$\text{Confidence interval for } \mu = \bar{X} \pm s_{\bar{X}} \times t$$

Here, we will use the confidence intervals calculated from samples of two populations to create a visual comparison of the populations.

Suppose you are studying the recovery of the food web of a mountain stream from the effects of a flood. Part of your study involves estimating the biomass of each of the consumers in the food web. One of the species you are studying is *Neothremma alicia,* a small species of caddisfly, an insect in the Order Trichoptera, that spends its larval stage grazing on diatoms living on the tops of stones in swift mountain streams. In your study site there are two forks in a stream, one which flooded two months before you took your sample, and one which did not flood. The two forks of the stream are similar in every other way. The following are the dry weights (milligrams) of *N. alicia* that you collected in 0.1 m² quadrats in each study stream:

| Quadrat Number | | | | | | | | |
| 1 | 2 | 3 | 4 | 5 | 6 | 7 | 8 | 9 |
|---|---|---|---|---|---|---|---|---|
| Flooded (mg) | | | | | | | | |
| 4.83 | 3.00 | 3.63 | 1.20 | 2.97 | 1.17 | 1.95 | 0.98 | 1.46 |
| Unflooded (mg) | | | | | | | | |
| 7.08 | 5.18 | 5.97 | 3.64 | 5.14 | 3.05 | 4.23 | 3.14 | 3.71 |

Using these data, we can calculate the means and standard errors for each sample, with the following results:

Flooded stream (*f = flooded*):

$$\bar{X}_f = 2.354 \text{ mg} / 0.1 \text{ m}^2$$

$$S_{\bar{X}_f} = \frac{s_f}{\sqrt{n}} = \frac{1.329 \text{ mg} / 0.1 \text{ m}^2}{3} = 0.443 \text{ mg} / 0.1 \text{ m}^2$$

Unflooded stream (*u = unflooded*):

$$\bar{X}_u = 4.571 \text{ mg} / 0.1 \text{ m}^2$$

$$S_{\bar{X}_u} = \frac{s_u}{\sqrt{n}} = \frac{1.371 \text{ mg} / 0.1 \text{ m}^2}{3} = 0.457 \text{ mg} / 0.1 \text{ m}^2$$

Now using our degrees of freedom ($n - 1$) and a level of confidence of 0.95, we find that our critical value of Student's *t* is 2.31. Using this *t* value, we can calculate a confidence interval for each population:

$$\text{Confidence interval for } \mu = \bar{X} \pm s_{\bar{X}} \times t$$

Flooded stream:

$$\mu_f = 2.354 \text{ mg}/0.1 \text{ m}^2 \pm 0.443 \text{ mg}/0.1 \text{ m}^2 \times 2.31$$

$$\mu_f = 2.354 \text{ mg}/0.1 \text{ m}^2 \pm 1.023 \text{ mg}/0.1 \text{ m}^2$$

Unflooded stream:

$$\mu_u = 4.571 \text{ mg}/0.1 \text{ m}^2 \pm 0.457 \text{ mg}/0.1 \text{ m}^2 \times 2.31$$

$$\mu_u = 4.571 \text{ mg}/0.1 \text{ m}^2 \pm 1.056 \text{ mg}/0.1 \text{ m}^2$$

These sample means and confidence intervals are plotted in figure 1. Recall from chapter 15 (p. 387) that the true population means for each of the study populations has a 95% chance of falling somewhere within the 95% confidence intervals. Now notice that the 95% confidence intervals for the two samples do not overlap. This indicates that there is less than a 5% chance that the two samples were drawn from populations of *N. alicia* with the same mean biomass per unit area. In other words, we have a basis for saying that there is a statistically significant difference in biomass of *N. alicia* in the two study streams. In chapter 18, we will take a somewhat different approach to making a statistical comparison of these two populations (p. 452).

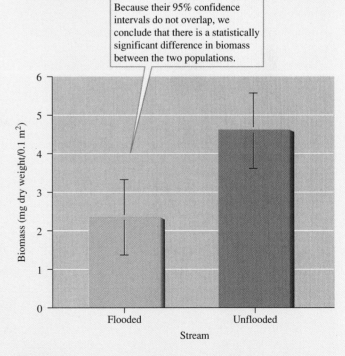

Because their 95% confidence intervals do not overlap, we conclude that there is a statistically significant difference in biomass between the two populations.

**Figure 1** Means and 95% confidence intervals of biomass of *Neothremma alicia* (Class Insecta, Order Trichoptera) in a recently flooded and an unflooded stream in the Rocky Mountains.

alone (see chapters 4–6) could have affected the distributions of herbivorous insects. Finally, Atlegrim measured the density of *Vaccinium* shoots to verify similar densities in exclosure and control plots. His measurements showed that levels of light, densities of nonavian predators, and densities of *Vaccinium* shoots were similar on exclosure and control plots.

Exclosures increased larval insect density an average of 63% across all study sites (fig. 17.14). So, the answer to Atlegrim's first question is yes. Insectivorous birds reduce the densities of herbivorous insect larvae feeding on *Vaccinium*. However, as Atlegrim predicted, some herbivorous larvae are more vulnerable to insectivorous birds than are others. Sawfly and geometrid larvae,

which feed in exposed positions, were significantly higher within exclosures, while the densities of tortricid larvae, which feed in their constructed shelters, showed no effects of bird exclusion. Higher densities of herbivorous insect larvae translated directly into higher levels of damage to *Vaccinium*.

What other piece of information might increase our confidence that the differences between exclosure and control plots were due to bird predation? One of the most significant bits of evidence would be direct observations of birds feeding on the control plots. Atlegrim observed three bird species feeding on control plots: Hazel hen chicks, *Tetrastes bonasia,* great tits, *Parus major,* and pied flycatcher, *Ficedula hypoleuca.*

Insectivorous birds also reduce insect populations and insect damage on plants in midlatitude forests in North America. Robert Marquis and Christopher Whelan (1994) used 3.8 cm white nylon mesh to exclude birds from 30 white oak, *Quercus alba,* saplings at the Tyson Research Center in Eureka, Missouri, during the growing seasons of 1989 and 1990. The bird community at this site includes several dozen species composed of a shifting mix of spring migrants and summer and spring residents.

The researchers sprayed another set of 30 white oak saplings each week with a pyrethroid insecticide. They also handpicked any remaining herbivorous insects from these trees. A third set of white oaks, the control, was not manipulated.

Marquis and Whelan's caged plants were populated by larger numbers of herbivorous insects and experienced significantly greater insect damage (fig. 17.15). These results are consistent with those of Atlegrim's earlier study. Marquis and Whelan also measured the biomass of each of their trees in 1990 and 1991. They found that the biomasses of sprayed and uncaged white oaks were significantly higher than the biomass of caged white oaks. In other words, the higher densities

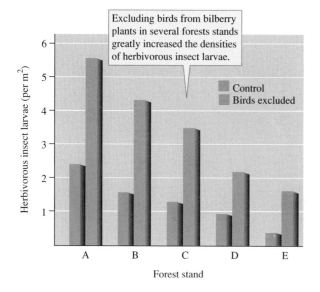

**Figure 17.14** Effect of insectivorous birds on herbivorous insect populations on *Vaccinium myrtillus* (data from Atlegrim 1989).

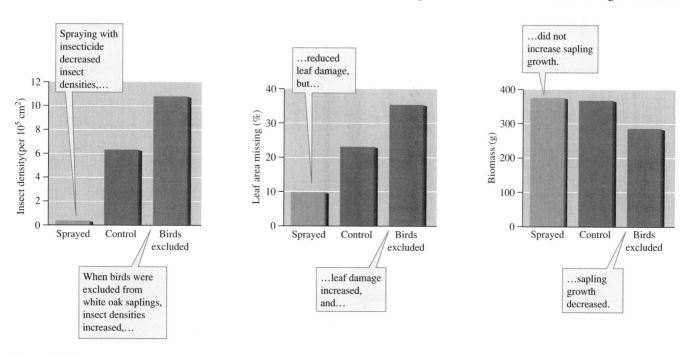

**Figure 17.15** Effect of insectivorous birds on herbivorous insect populations, leaf damage, and sapling growth in white oaks (data from Marquis and Whelan 1994).

of herbivorous insects on the trees from which birds were excluded reduced their growth rates. One implication of these results is that insectivorous birds increase the growth rates of temperate forest trees.

Many studies of food webs and keystone species have been done since Robert Paine's classic study of the intertidal food web at Mukkaw Bay, Washington. The studies have revealed a great deal of biological diversity, which has prompted biologists to ask what characterizes keystone species. This reflection is necessary to avoid the possibility that the term may become so inclusive that it becomes meaningless. The conclusions reached by a conference designed to address this question are summarized in figure 17.16 (Power et al. 1996). Keystone species are those that, despite low biomass, exert strong effects on the structure of the communities they inhabit. As we shall see in the following Concept Discussion, those strong effects are not always positive.

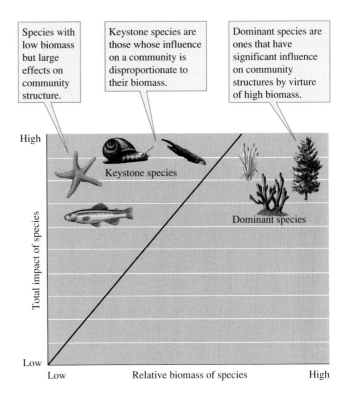

**Figure 17.16** What is a keystone species (data from Power et al. 1996)?

## CONCEPT DISCUSSION

### Exotic Predators

**Exotic predators can collapse and simplify the structure of food webs.**

# Introduced Fish: Predators That Simplify Aquatic Food Webs

People have moved all sorts of species around the planet, but one of the most commonly introduced groups of organisms is fish. Introduced fish often substantially change the food webs of the water bodies where they are introduced. For instance, introduced fishes have devastated the native fishes of Lake Atitlan and Gatun Lake in Central America. In both these cases, introduced predaceous fish completely reshaped the lake food web, producing a less diverse community. The dramatic impact of exotic species on communities may result from their being outside the evolutionary experience of the prey populations in the local community.

Today we are witnessing what may be the greatest devastation ever wrought by an introduced predator. That predator is the Nile perch, *Lates nilotica,* and the aquatic system is Lake Victoria, one of the great lakes of East Africa. Lake Victoria is approximately 69,000 km$^2$ and lies right on the equator. Despite its large surface area, the lake is relatively shallow. The deepest point is approximately 100 m, and most of the lake is less than 60 m deep.

Lake Victoria harbors one of the greatest concentrations of fish species in the world, and the Nile perch may be producing the greatest extinction of vertebrate animals to occur in modern times. The fish fauna of the lake, which included more than 400 species, is being rapidly reduced to a community

dominated by a handful of species. Just three species now dominate the fish catches around Lake Victoria: the introduced Nile perch, the introduced Nile tilapia, *Oreochromis niloticus,* and a single native species, the omena, *Rastrineobola argentea,* a planktivorous minnow of the open waters of the lake.

Hundreds of fish species appear on their way to extinction because humans introduced *one* more fish species into a lake already containing over 400 species. The Nile perch is a predaceous fish native to East Africa, where it attains a length of nearly 2 m and a weight of 60 kg. It may have eliminated species from other East African lakes in the distant past, since lakes with Nile perch have fewer species of fish in the family Cichlidae than do lakes without it. Nile perch were introduced to Lake Victoria around 1954, along with several other fish species from surrounding rivers and lakes. However, Nile perch remained a minor component of the fish fauna for nearly two decades, and then in the early 1980s, its population exploded.

As the population of Nile perch grew, the populations of nearly all other fish species in the lake declined. The only native fish that increased during this period was the planktivorous omena. K. Ligtvoet and F. Witte (1991) used food webs to represent the changes in the Lake Victoria fish community. As you can see in figure 17.17, the original food web consisted of a complex mixture of fishes belonging to several families and orders. Three species of fish dominate the present food web. Notice that Nile perch appear twice in the present food web, as a top predator and as a middle level predator. In other words, there is a substantial amount of cannibalism going on in the community, with small Nile perch serving as a trophic link between insect larvae and large Nile perch. As impressive as

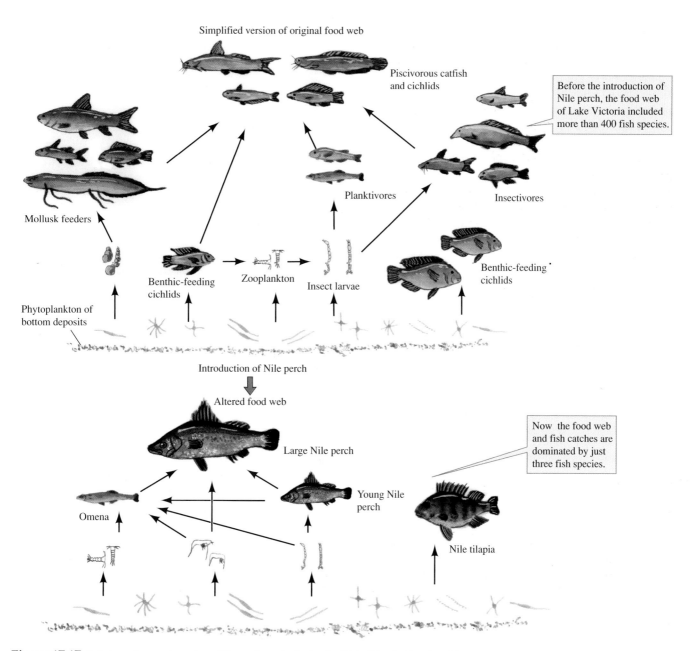

**Figure 17.17**   Influence of an exotic predator, Nile perch, on the food web of Lake Victoria (data from Ligtvoet and Witte 1991).

the contrast shown in figure 17.17 may be, keep in mind that it is a gross simplification and shorthand for the trophic interactions of over 400 species of fish. For a comparison, look back at Winemiller's food webs for Caño Volcán, a food web with only 10 species (see fig. 17.4). The original Lake Victoria food web contained 40 times the number of fish species in the Caño Volcán food web.

The Nile perch has had a major effect on the food web of Lake Victoria and is contributing to a massive extinction. However, can we attribute all these changes in the biota of the lake to this one predator? Les Kaufman (1992) points out that the changes in the Lake Victoria fish community coincide with other changes in the ecosystem. For instance, concentrations of dissolved oxygen have declined significantly.

Lake Victoria water was once oxygenated from top to bottom. Before 1978, organisms requiring oxygen lived in the deepest portions of the lake, which experienced periods of reduced oxygen concentration but were not anoxic. Now, however, the water below about 30 m is often essentially devoid of oxygen.

Depletion of oxygen has produced massive fish kills. Biologists had observed some fish kills at the surface of Lake Victoria but discovered many others only when they began exploring the bottom with a remote-operated vehicle. Images recorded by this vehicle in 1987 showed that there were no fish below about 30 m and that the bottom was littered with dead fish. Massive fish kills, including kills of Nile perch, have become a regular occurrence in Lake Victoria.

Maybe the Nile perch has not caused all of the changes in Lake Victoria. Perhaps the changes are ultimately a consequence of changes in the ecosystem, driven by heavy nutrient additions from the surrounding human population and consequent eutrophication of the lake (see chapter 3). However, as we shall see in chapter 18, predaceous fish like Nile perch can produce changes in ecosystem functioning that may in turn strongly influence populations and communities. Though we look for single causes of complex phenomena, ecological systems are affected by a complex interplay between biotic and abiotic factors. Some of these mutual influences will become more apparent in chapters 18 and 19 as we begin our discussion of the ecology of ecosystems.

## CONCEPT DISCUSSION

### Mutualistic Keystones

**Mutualists can act as keystone species.**

While our earlier discussions of keystone species have emphasized the roles of predators as keystone species, many other kinds of organisms can act as keystone species. Returning to the classification of Power and colleagues shown in figure 17.16, the only requirements for keystone status is that the species in question have relatively low biomass in the community and that it has a high impact on community structure. Increasingly, ecologists are discovering that many mutualistic species meet these requirements. One such group are the cleaner fishes on coral reefs.

## A Cleaner Fish as a Keystone Species

Many species of fish on coral reefs clean other fish of ectoparasites. This relationship, which involves the cleaner fish and its clients, has been shown to be a true mutualism. One of the most widely distributed cleaner fish in the Indo-Pacific region is the cleaner wrasse, *Labroides dimidiatus* (see fig. 15.19a). The feeding activity of cleaner wrasses is intense. Alexandra Grutter of the University of Queensland, Australia, has shown that a single cleaner wrasse can remove and eat 1,200 parasites from client fishes per day. She also performed experiments (Grutter 1999) that documented that fish on reefs without cleaner wrasses harbor approximately four times the number of parasitic isopods as those living on reefs with cleaner wrasses.

What effect might cleaning activity by *L. dimidiatus* have on the diversity of fish on coral reefs? This is the question addressed with a series of field experiments by Redouan Bshary of the University of Cambridge. Bshary studied the effects of cleaner wrasses on reef fish diversity at Ras Mohammed National Park, Egypt (Bshary 2003). The study area consists of a sandy bottom area approximately 400 m from

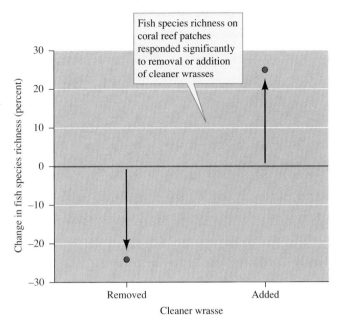

**Figure 17.18** Results of experimental and natural removals or additions of cleaner wrasses. *Labroides dimidiatus,* to reef patches in the Red Sea (data from Bshary 2003).

shore dotted with reef patches in water depths from 2 to 6 m. Bshary chose 46 reef patches separated from other patches by at least 5 m of sandy bottom. He identified and counted the fish species present during dives on these reefs and noted the presence or absence of cleaner wrasses on each reef patch. Bshary recorded 29 natural disappearances or appearances of cleaner wrasses during his study. In addition, he performed experimental removals of cleaner wrasses from reefs and introductions of these cleaners to reef patches where there were none.

Bshary followed the responses of the fish community to natural disappearances and experimental removals and natural colonization and experimental introductions. In doing so, he gained insights into the influence of these tiny mutualists on reef fish diversity. Figure 17.18 summarizes the responses of fish communities on reef patches 4 months following the natural or experimental, addition or removal of cleaner wrasses. Bshary observed a median reduction in fish species richness of approximately 24% where cleaner wrasses disappeared or were removed. Where cleaner wrasses were added, either naturally or experimentally, he observed a median increase in fish species richness of 24%. Bshary's results indicate that the cleaner wrasse acts as a keystone species on the coral reefs of the Red Sea. Mutualists that act as keystone species have also been found on land.

## Seed Dispersal Mutualists as Keystone Species

It appears that ants that disperse seeds have a significant influence on the structure of plant communities in the species-rich fynbos of South Africa. Caroline Christian (2001) observed that native ants disperse 30% of the seeds in the shrublands of

**Figure 17.19** The Argentine ant, *Linepithema humile*, has invaded and disrupted ant communities in many geographic regions. In the fynbos of South Africa, invading Argentine ants are displacing keystone ant species, which threatens the exceptional plant diversity of the fynbos.

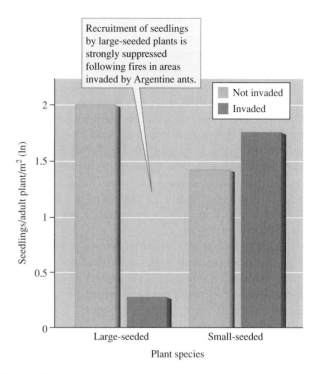

**Figure 17.20** A comparison of recruitment of seedlings following fire in areas invaded by Argentine ants and areas not invaded shows the effects of the displacement of native seed-dispersing ants by Argentine ants (data from Christian 2001).

the fynbos. The plants attract the services of these dispersers with food rewards on the seeds called elaiosomes. However, the Argentine ant, *Linepithema humile* (fig. 17.19), which do not disperse seeds, has invaded these shrublands. Christian documented how the invading Argentine ants have displaced (as they have in other regions) many of the native ant species in the fynbos. In addition, she discovered that the native ant species most impacted by Argentine ant invasion are those species most likely to disperse larger seeds.

Seed-dispersing ants are important to the persistence of fynbos plants because they bury seeds in sites where they are safe from seed-eating rodents and from fire. Fires are characteristic of Mediterranean shrublands such as the fynbos, and seeds are the only life stage of many fynbos plants to survive fires. Consequently, ant dispersal is critical to the survival of many plant species. In a comparison of seedling recruitment following fire, Christian found substantial reductions in seedling recruitment by plants producing large seeds in areas invaded by Argentine ants (fig. 17.20). Meanwhile small-seeded plants, whose dispersers are less affected by Argentine ants, showed no reduction in recruitment following fire. Christian's results, like Bshary's, reveal the influence of mutualists acting as keystone species within the communities they occupy. Other studies are revealing the importance of other mutualists, such as pollinators and mycorrhizal fungi, as keystone species.

## APPLICATIONS & TOOLS

### Humans as Keystone Species

People have long manipulated food webs both as a consequence of their own feeding activities and by introducing or deleting species from existing webs. In addition, many of these manipulations have focused on keystone species. Consequently, either consciously or unwittingly, people have, themselves, acted as keystone species in communities.

## The Empty Forest: Hunters and Tropical Rain Forest Animal Communities

The current plight of the tropical rain forest is well known. However, Kent Redford (1992) points out that with few exceptions, most studies of human impact on the tropical rain forest have concentrated on direct effects of humans on vegetation, mainly on deforestation. Redford expands our view by examining the effects of humans on animals. The picture that emerges from this analysis is that humans have so reduced the population densities of rain forest animals in many areas that they no longer play their keystone roles in the system, a situation Redford calls "ecologically extinct."

Redford estimated that subsistence hunting, a major source of protein for many rural people, results in an annual death toll of approximately 14 million mammals and 5 million birds and reptiles within the Brazilian Amazon. He estimated further that commercial hunters, seeking skins, meat, and feathers, kill an additional 4 million animals annually. Consequently, the total take by hunters within the Brazilian Amazon is approximately 23 million individual animals. However, this figure underestimates the total number of animal deaths, since many wounded animals escape from hunters only to die. Including those fatally wounded animals that escape, Redford places the annual deaths within the Brazilian Amazon at approximately 60 million animals.

Hunters generally concentrate on a small percentage of larger bird and mammal species, however. For instance, Redford estimated that at Cocha Cashu Biological Station in Manu National Park, located in the Amazon River basin in eastern Peru, hunters concentrate on 9% of the 319 bird species and 18% of the 67 mammal species. Because hunters generally concentrate on the larger species, this small portion of the total species pool makes up about 52% of the total bird biomass and approximately 75% of the total mammalian biomass around Manu National Park (fig. 17.21).

As impressive as all these numbers are, there remains a critical question: Do hunters reduce the local densities of the birds and mammals they hunt? The answer is yes. Redford estimated that moderate to heavy hunting pressure in rain forests reduces mammalian biomass by about 80% to 93% and bird biomass by about 70% to 94%.

There may be cause for concern, however, that goes beyond the losses of these immense numbers of animals. As you might expect, many large rain forest mammals and birds may act as keystone species (fig. 17.22). If so, their decimation will have effects that ripple through the entire community. The first to suggest a keystone role for the large animals preferred by rain forest hunters was John Terborgh (1988), who presented his hypothesis in a provocative essay titled, "The Big Things That Run the World."

Terborgh's hypothesis has been supported by a variety of studies. He observed that in the absence of pumas and jaguars on Barro Colorado Island, Panama, medium-sized mammal species are over 10 times more abundant than in areas still supporting populations of these large cats. R. Dirzo and A. Miranda (1990) compared two forests in tropical southern Mexico, one in which hunting had eliminated most of the large mammals and one in which most of the large mammals were still present. The comparison was stark. In the absence of large mammals such as peccaries, jaguars, and deer, the researchers found forests carpeted with undamaged plant seedlings and piled with uneaten and rotting fruits and nuts, signs of a changing forest. Such observations prompted Redford to warn, "We must not let a forest full of trees fool us into believing all is well." Tropical rain forest conservation must also include the large, and potentially keystone, animal species that are vulnerable to hunting by humans.

## Ants and Agriculture: Keystone Predators for Pest Control

In 1982, Stephen Risch and Ronald Carroll published a paper describing how the predaceous fire ant, *Solenopsis geminata*,

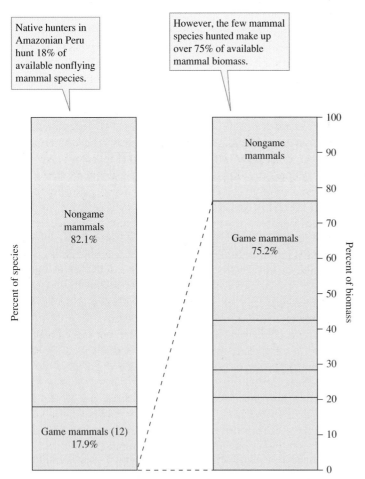

**Figure 17.21** Highly selective hunting by Amazonian natives (data from Redford 1992).

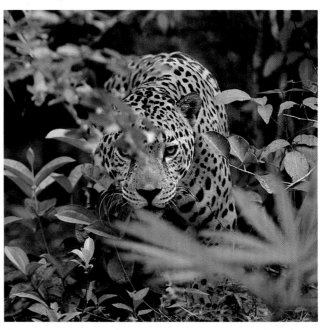

**Figure 17.22** Large predators such as this jaguar may act as keystone species in tropical rain forests.

acts as a keystone predator in the food web of the corn-squash agroecosystem in southern Mexico. While "natural enemies" had been used to control insect pests for some time, Risch and Carroll put these efforts into a community context. They drew conceptual parallels between biological control of insects with natural enemies and studies of the influences of keystone species, citing studies of the influences of herbivores on plant communities and the effects of predators on intertidal communities. In their own experiments, Risch and Carroll demonstrated how predation by *Solenopsis* in the corn-squash agroecosystem reduces the number of arthropods and the arthropod diversity (fig. 17.23). This study showed how *Solenopsis* could act as a keystone species to the benefit of the agriculturist.

The conceptual breakthrough represented by the work of Risch and Carroll is impressive. However, their work had been anticipated, 1,700 years earlier, by farmers in southern China. H. Huang and P. Yang (1987) cite Ji Han, who, in A.D. 304, wrote "Plants and Trees of the Southern Regions" in which he included the following:

> The Gan (mandarin orange) is a kind of orange with an exceptionally sweet and delicious taste . . . In the market, the natives of Jiao-zhi [southeastern China and North Vietnam] sell ants stored in bags of rush mats. The nests are like thin silk. The bags are all attached to twigs and leaves, which, with the ants inside the nests, are for sale. The ants are reddish-yellow in color, bigger than ordinary ants. In the south, if the Gan trees do not have this kind of ant, the fruits will be damaged by many harmful insects and not a single fruit will be perfect.

Now, 17 centuries after the observations of Ji Han, we know this ant as the citrus ant, *Oecophylla smaragdina*. The use of this ant to control herbivorous insects in citrus orchards was unknown outside of China until 1915. In 1915, Walter Swingle, a plant physiologist who worked for the U.S. Department of Agriculture, was sent to China to search for varieties of oranges resistant to citrus canker, a disease that was devastating citrus groves in Florida. While on this trip, Swingle came across a small village where the main occupation of the people was growing ants for sale to orange growers. The ant was the same one described by Ji Han in A.D. 304.

*Oecophylla* is one of the weaver ants, which use silk to construct a nest by binding leaves and twigs together. These ants spend the night in their nest. During the day, the ants spread out over the home tree as they forage for insects. Farmers place a nest in a tree and then run bamboo strips between trees so that the ants can have access to more than one tree. The ants will eventually build nests in adjacent trees and can colonize an entire orchard.

The ants harvest protein and fats when they gather insects from their home tree, but they have other needs as well. They also need a source of liquid and carbohydrates, and they get these materials by cultivating Homoptera, known as soft-scale insects or mealy bugs, which produce nectar. The ants

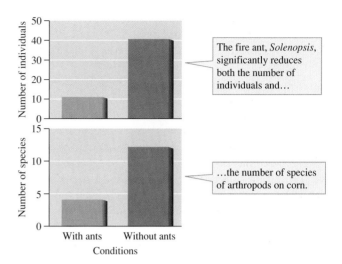

The fire ant, *Solenopsis*, significantly reduces both the number of individuals and...

...the number of species of arthropods on corn.

**Figure 17.23** Effect of *Solenopsis geminata* on the arthropod populations on corn (data from Risch and Carroll 1982).

and soft-scale insects have a mutualistic relationship in which the ants transport the insects from tree to tree and protect them from predators. In return the ants consume the nectar produced by the soft-scale insects. Because of this mutualism with the soft-scale insect, which can itself be a serious pest of citrus, several early agricultural scientists expressed skepticism that *Oecophylla* would be an effective agent for pest control in citrus. They suggested that the use of this ant could produce infestations by soft-scale insects.

Despite these criticisms, all Chinese citrus growers interviewed insisted that *Oecophylla* is effective at pest control and that the damage caused by soft-scale insects is minor. Research done by Yang appears to have solved this apparent contradiction. Comparing orange trees treated with chemical insecticides to those protected by *Oecophylla*, Yang recorded higher numbers of soft-scale insects in the trees tended by ants. However, these higher numbers did not appear to cause serious damage to the orange trees. When Yang inspected the soft-scale insects closely, he found that they were heavily infested with the larvae of parasitic wasps. He also found that the ants did not reduce populations of lacewing larvae and ladybird beetles, predators that feed on soft-scale insects. Huang and Yang concluded that *Oecophylla* is effective at pest control because while it attacks the principal, larger pests of citrus, it does not reduce populations of other predators that attack the smaller pests of citrus, such as soft-scale insects, aphids, and mites (fig. 17.24).

The association between *Oecophylla* and citrus trees seems similar to that between ants and acacias (see chapter 15). There is a difference, however. Humans maintain *Oecophylla* as a substantial component of the food web in citrus orchards. Not only have specialized farmers historically cultivated and distributed the ants, *Oecophylla* must also be protected from the winter cold. The ant cannot survive the winter in southeast China in orange trees. Consequently, farmers must generally provide shelter and food for the ants during winter.

**Figure 17.24** While pests in this North American orange orchard are controlled mainly by chemical insecticides, weaver ants have been used to control insect pests of orange orchards in China for over 17 centuries.

The labor and expense of maintaining these ants through the winter may be reduced by mixed plantings of orchard trees. Farmers in Shajian village in the Huaan district of southeast China have successfully maintained *Oecophylla* over the winter in mixed plantings of orange and pomelo trees. During winter the ants are mostly in pomelo trees, which are larger and have thicker foliage than orange trees, characteristics that reduce cooling rates on winter nights. In this situation, farmers do not have to add new nests of *Oecophylla* each spring. Gradually, the ant has become integrated into the mixed citrus and pomelo orchards and requires little special care from the farmers.

The farmers of southeast China have employed *Oecophylla* as a keystone species in a complex citrus-based food web for a long time. However, the results would not be the same with just any ant species. The citrus growers required a species that acts in a particular way. One wonders how long farmers of the region had experimented with this species before Ji Han wrote his account of their activities in A.D. 304.

# SUMMARY

**A food web summarizes the feeding relations in a community.** The earliest work on food webs concentrated on simplified communities in areas such as the Arctic islands. However, researchers such as Charles Elton (1927) soon found that even these so-called simple communities included very complex feeding relations. The level of food web complexity increased substantially, however, as researchers began to study complex communities. Studies of the food webs of tropical freshwater fish communities revealed highly complex networks of trophic interaction that persisted even in the face of various simplifications. A focus on strong interactions can simplify food web structure and identify those interactions responsible for most of the energy flow in communities.

**The feeding activities of a few keystone species may control the structure of communities.** Robert Paine (1966) proposed that the feeding activities of a few species have inordinate influences on community structure. He predicted that some predators may increase species diversity by reducing the probability of competitive exclusion. Manipulative studies of predaceous species have identified many keystone species, including starfish and snails in the marine intertidal zone and fish in rivers. On land, birds exert substantial influences on communities of their arthropod prey. Jane Lubchenko (1978) demonstrated that the influence of consumers on community structure depends upon their feeding preferences, their local population density, and the relative competitive abilities of prey species. Keystone species are those that, despite low biomass, exert strong effects on the structure of the communities they inhabit.

**Exotic predators can collapse and simplify the structure of food webs.** Introduced fishes have devastated the native fishes of Lake Atitlan and Gatun Lake in Central America. Introduction of the Nile perch is rapidly reducing the species-rich fish fauna of Lake Victoria to a community dominated by a handful of species. The influence of the Nile perch on the fish community of Lake Victoria is enmeshed with massive changes in the lake's ecosystem.

**Mutualists can act as keystone species.** Experimental studies have shown that cleaner fish, species that remove parasites from other fish, act as keystone species on coral reefs. Removing cleaner fish produces a decline in reef fish species richness. Ants that disperse plant seeds in the fynbos of South Africa have been shown to have major influences on plant community structure. Where invading ants have displaced the mutualistic dispersing ants, the plant community suffers a decline in species richness following fires. Other mutualistic organisms that may act as keystone species include pollinators and mycorrhizal fungi.

Humans have acted as keystone species in communities. People have long manipulated food webs both as a consequence of their own feeding activities and by introducing or deleting species from existing food webs. In addition, many of these manipulations have been focused on keystone species. Hunters in tropical rain forests have been responsible for removing keystone animal species from large areas of the rain forests of Central and South America. Chinese farmers have used ants as keystone predators to control pests in citrus orchards for over 1,700 years.

# Review Questions

1. You could argue that the classical food web of Bear Island included several communities, each with its own food web. What were some of the different communities that Summerhayes and Elton (1923) included in their web? On the other hand, because the Bear Island food web includes significant movement of energy (food) and nutrients between what many ecologists might consider to be separate communities, what does their food web say about the distinctness of what we call communities?

2. Winemiller (1990) deleted "weak" trophic links from one set of food webs that he described for fish communities in Venezuela (see fig. 17.4). What was his criterion for designating weak interactions? Earlier, Paine (1980) suggested that ecologists could learn something by focusing on "strong" links in communities. How did Paine's criterion for determining a strong link differ from Winemiller's?

3. What is a keystone species? Paine (1966, 1969) experimented with two starfish that act as keystone species in their intertidal communities along the west coast of North America and in New Zealand. Describe how the intertidal communities in these two areas are similar. Describe the differences between these two communities and the differences in the design of Paine's experiments in these two areas.

4. Explain how the experiments of Lubchenko (1978) showed that feeding preferences, population density, and competitive relations among food species all potentially contribute to the influences of "keystone" consumers on the structure of communities. What refinements did the work of Lubchenko add to the keystone species hypothesis?

5. When Power (1990) excluded predaceous fish from her river sites, the density of herbivorous insect larvae (chironomids) decreased. Use the food web described by Power to explain this response.

6. Using Tscharntke's food web (1992) shown in figure 17.5, predict which species would be most affected if you excluded the bird at the top of the web, *Parus caeruleus*. What species would be affected less? Assume that *P. caeruleus* is a keystone species in this community.

7. Atlegrim (1989) and also Marquis and Whelan (1994) showed that birds in high latitudes and temperate forests reduce insect populations. The results of this research suggest that birds act as keystone species in some communities. What else would we need to know about the birds in these communities before we could conclude that they are keystones in the strict sense? (Hint: Consider figure 17.16.)

8. Notice that in the study by Marquis and Whelan (1994) the biomass of uncaged *Q. alba* was as great as that of sprayed individuals. In other words, spraying protected oak seedlings as well as birds. If spraying can control herbivorous forest insects, why rely on birds to improve tree growth? What advantages does predation by birds have over spraying?

9. Some paleontologists have proposed that overhunting caused the extinction of many large North American mammals at the end of the Pleistocene about 11,000 and 10,000 years ago. The hunters implicated by paleontologists were a newly arrived predatory species, *Homo sapiens*. Offer arguments for and against this hypothesis.

10. All the keystone species work we have discussed in chapter 17 has concerned the influences of animals on the structure of communities. Can other groups of organisms act as keystones? What about parasites and pathogens?

# Suggested Readings

Balirwa, J. S., C. A. Chapman, L. C. Chapman, I. G. Cowx, K. Geheb, L. Kaufman, R. H. Lowe-McConnell, O. Seehausen, J. H. Wanink, R. L. Welcomme, and F. Witte. 2003. Biodiversity and fishery sustainability in the Lake Victoria Basin: an unexpected marriage? *BioScience* 53:703–15.

*A provocative analysis, suggesting that fishing pressure on Nile perch in Lake Victoria may be a key to preserving surviving remnants of the native fish fauna.*

Bever, J. D., P. A. Schultz, A. Pringle, and J. B. Morton. 2001. Arbuscular mycorrhizal fungi: more diverse than meets the eye, and the ecological tale of why. *BioScience* 51:923–31.

*Provides insights into the hidden and dynamic world of arbuscular mycorrhizal fungi.*

Brown, J. H., T. G. Whitham, S. K. M. Ernest, and C. A. Gehring. 2001. Complex species interactions and the dynamics of ecological systems: long-term experiments. *Science* 93:643–50.

*Excellent review of long-term experiments that have revealed some of the complex interactions occurring in biological communities.*

Bshary, R. 2003. The cleaner wrasse, *Labroides dimidiatus,* is a key organism for reef fish diversity at Ras Mohammed National Park, Egypt. *Journal of Animal Ecology* 72:169–72.

*A well-designed experimental study that reveals the strong influence of a key mutualist on community diversity.*

Christian, C. E. 2001. Consequences of a biological invasion reveal the importance of mutualism for plant communities. *Nature* 413:635–39.

*Fascinating study of how seed dispersal mutualisms play a central role in maintaining the diversity within a species-rich shrubland in South Africa.*

Cohen, J. E., T. Jonsson, and S. R. Carpenter. 2003. Ecological community description using the food web, species abundance, and body size. *Proceedings of the National Academy of Sciences of the United States of America* 100:1781–86.

*State of the art use of food webs to characterize a biological community.*

Grutter, A. S. 1999. Cleaner fish really do clean. *Nature* 398:672–73.

*Ingenious experiment to show for the first time that cleaner fish reduce the population density of ectoparasites on the fish that they clean.*

Kaufman, L. 1992. Catastrophic change in species rich freshwater ecosystems. *BioScience* 42:846–58.

Ligtvoet, W. and F. Witte. 1991. Perturbation through predator intro-duction: effects on the food web and fish yields in Lake Victoria (East Africa). In O. Ravera, ed. *Terrestrial and Aquatic Ecosystems: Perturbation and Recovery.* New York: Ellis Horwood.

*These two papers review the recent catastrophic changes in the Lake Victoria fish fauna following the introduction of Nile perch.*

Lubchenko, J. 1978. Plant species diversity in a marine intertidal community: importance of herbivore food preference and algal competitive abilities. *American Naturalist* 112:23–39.

*An elegant model of field research on the role of a keystone con-sumer in the intertidal zone.*

Marquis, R. J. and C. J. Whelan. 1994. Insectivorous birds increase growth of white oak through consumption of leaf-chewing insects. *Ecology* 75:2007–14.

*Careful and creative study design allow Marquis and Whelan to tease out the effects of insectivorous birds on the intensity of herbivory and provide hints of the potential role of birds as key-stone species in temperate forests.*

Power, M. E. 1990. Effects of fish on river food webs. *Science* 250:811–14.

*This paper carries the keystone species concept to a temperate river community. A concise, readable account of research on the workings of a freshwater benthic community.*

Power, M. E., D. Tilman, J. A. Estes, B. A. Menge, W. J. Bond, L. S. Mills, G. Daily, J. C. Castilla, J. Lubchenko, and R. T. Paine. 1996. Challenges in the quest for keystones. *BioScience* 46:609–20.

*This paper provides a review of the keystone species concept and of the vast body of research related to keystone species. It summarizes the thoughts of some of today's leading ecologists on an important ecological concept.*

Redford, K. H. 1992. The empty forest. *BioScience* 42:412–22.

Terborgh, J. 1988. The big things that run the world: a sequel to E. O. Wilson. *Conservation Biology* 2:402–3.

*These two papers review the influence of humans on popula-tions of large animals in rain forests and the possible conse-quences to rain forest ecology.*

van der Heijden, M. G. A., J. N. Klironomos, M. Ursic, P. Moutoglis, R. Streitwolf-Engel, T. Boller, A. Wiemken, and I. R. Sanders. 1998. Mycorrhizal fungal diversity determines plant biodiver-sity; ecosystem variability and productivity. *Nature* 396:69–72.

*Ambitious study on two continents of the effects of mycorrhizal fungal diversity on plant diversity.*

# On the Net

Visit this textbook's accompanying website at www.mhhe.com/ecology (click on the book's title) to take advantage of practice quizzing, study/writing tips, timely news articles, and additional URLs for research on the topics in this chapter.

Community Ecology

Field Methods for Studies of Ecosystems

Food Webs

Species Abundance, Diversity, and Complexity

Invasive and Introduced Species

Endangered Species

Legislation Regarding Endangered Species

*Chapter*

# 18

# Primary Production and Energy Flow

The interactions between organisms and their environments are fueled by complex fluxes and transformations of energy. Sunlight shines down on the canopy of a forest—some is reflected, some is converted to heat energy, and some is absorbed by chlorophyll. Infrared radiation is absorbed by the molecules in organisms, soil, and water, increasing their kinetic state and raising the temperature of the forest. Forest temperature affects the rate of biochemical reactions and transpiration by forest vegetation.

Forest plants use photosynthetically active solar radiation, or PAR (see chapter 6), to synthesize sugars. The plants use some of this fixed energy to meet their own energy needs. Some fixed energy goes directly into plant growth: to produce new leaves, to lengthen the tendrils of vines, to grow new root hairs, and so forth. Some fixed energy is stored as nonstructural carbohydrates, which act as energy stores in roots, seeds, or fruits. Photosynthesis may increase forest biomass.

A portion of the energy fixed by forest vegetation is consumed by herbivores, some is consumed by detritivores, and some ends up as soil organic matter. Energy fixed by forest vegetation powers bird flight through the forest canopy and fuels the muscle contractions of earthworms as they burrow through the forest soil. The forest vegetation is sunlight transformed, as are all the associated bacteria, fungi, and animals and all their activities (fig. 18.1).

We can view a forest as a system that absorbs, transforms, and stores energy. In this view, physical, chemical, and biological structures and processes are inseparable. When we look at a forest (or stream or coral reef) in this way we view it as an ecosystem. An ecosystem is a biological community plus all of the abiotic factors influencing that community. The term *ecosystem* and its definition were first proposed in 1935 by the British ecologist Arthur Tansley. We first encountered Tansley in chapter 13, where we discussed his early work on the soil requirements of *Galium* species. Sometime during his exploration of nature, he realized the importance of considering organisms and their environment as an integrated system. Tansley wrote: "Though the organisms may claim our primary interest, . . . we cannot separate them from their special environment, with which they form one physical system. It is the [eco]systems so formed which, from the point of view of the ecologist, are the basic units of nature on the face of the earth."

Ecosystem ecologists study the flows of energy, water, and nutrients in ecosystems and, as suggested by Tansley, pay as much attention to physical and chemical processes as they do to biological ones. Some fundamental areas of interest for ecosystem ecologists are primary production, energy flow, and nutrient cycling. We will discuss the first two topics in chapter 18 and nutrient cycling in chapter 19.

We saw in chapter 6 how the photosynthetic machinery of plants uses solar energy to synthesize sugars. In chapter 6 we considered photosynthesis from the perspective of the individual grass, tree, or cactus. Here we step back from the biochemical and physiological details of photosynthesis and back even from the individual organism to look at photosynthesis at the level of the whole ecosystem.

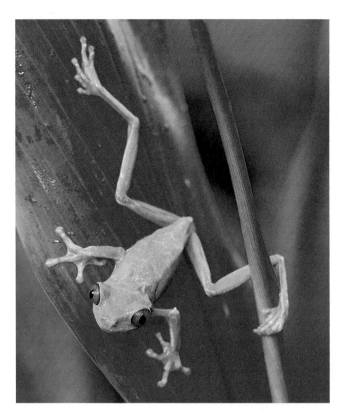

**Figure 18.1** In most ecosystems, sunlight provides the ultimate source of energy to power all biological activity, such as the singing of this tree frog and the growth of the plant on which it sits.

**Primary production** is the fixation of energy by autotrophs in an ecosystem. The **rate of primary production** is the amount of energy fixed over some interval of time. Ecosystem ecologists distinguish between gross and net primary production. **Gross primary production** is the total amount of energy fixed by all the autotrophs in the ecosystem. **Net primary production** is the amount of energy left over after autotrophs have met their own energetic needs. Net primary production is gross primary production minus respiration by primary producers; it is the amount of energy available to the consumers in an ecosystem. Ecologists have measured primary production in a variety of ways but mainly as the rate of carbon uptake by primary producers or by the amount of biomass or oxygen produced.

We discussed feeding biology from a variety of perspectives in previous chapters. In chapter 6, we examined the biology of herbivores, detritivores, and carnivores. In chapter 14, we discussed the ecology of exploitation, and in chapter 17, we used food webs as a means of representing the trophic structure of communities. Ecosystem ecologists are also concerned with trophic structure but have taken a different approach than population and community ecologists.

Ecosystem ecologists have simplified the trophic structure of ecosystems by arranging species into trophic levels based on the predominant source of their nutrition. A **trophic level** is a position in a food web and is determined by the number of transfers of energy from primary producers to that level. Primary producers occupy the first trophic level in ecosystems since they convert inorganic forms of energy, principally light, into biomass. Herbivores and detritivores are often called primary consumers and occupy the second trophic level. Carnivores feeding on herbivores and detritivores are called secondary consumers and occupy the third trophic level. Predators that feed on carnivores occupy a fourth trophic level. Since each trophic level may contain several species, in some cases hundreds, an ecosystem perspective simplifies trophic structure.

Primary production, the conversion of inorganic forms of energy into organic forms, is a key ecosystem process. All consumer organisms, including humans, depend upon primary production for their existence. Because of its importance and because rates of primary production vary substantially from one ecosystem to another, ecosystem ecologists study the factors controlling rates of primary production in ecosystems.

Patterns of natural variation in primary production provide clues to the environmental factors that control this key ecosystem process. Experiments test the importance of those controls. In chapter 18, we discuss the major patterns of variation in primary production in terrestrial and aquatic ecosystems and key experiments designed to determine the mechanisms producing those patterns. In the last sections of chapter 18, we examine patterns of energy flow through ecosystems.

## CONCEPTS

- **Terrestrial primary production is generally limited by temperature and moisture.**
- **Aquatic primary production is generally limited by nutrient availability.**
- **Consumers can influence rates of primary production in aquatic and terrestrial ecosystems.**
- **Energy losses limit the number of trophic levels in ecosystems.**

## CONCEPT DISCUSSION

### Patterns of Terrestrial Primary Production

**Terrestrial primary production is generally limited by temperature and moisture.**

As we surveyed the major terrestrial biomes in chapter 2, you probably got a sense of the geographic variation in rates of primary production. Perhaps you also developed a feeling for the major environmental correlates with that variation. The

variables most highly correlated with variation in terrestrial primary production are *temperature* and *moisture*. Highest rates of terrestrial primary production occur under warm, moist conditions.

## Actual Evapotranspiration and Terrestrial Primary Production

Michael Rosenzweig (1968) estimated the influence of moisture and temperature on rates of primary production by plotting the relationship between annual net primary production and annual actual evapotranspiration. Annual **actual evapotranspiration (AET)** is the total amount of water that evaporates and transpires off a landscape during the course of a year and is measured in millimeters of water per year. The AET process is affected by both temperature and precipitation. The ecosystems showing the highest levels of primary production are those that are warm and receive large amounts of precipitation. Conversely, ecosystems show low levels of AET either because they receive little precipitation, are very cold, or both. For instance, both hot deserts and tundra exhibit low levels of AET.

Figure 18.2 shows Rosenzweig's plot of the positive relationship between net primary production and AET. Tropical forests show the highest levels of net primary production and AET. At the other end of the spectrum, hot, dry deserts and

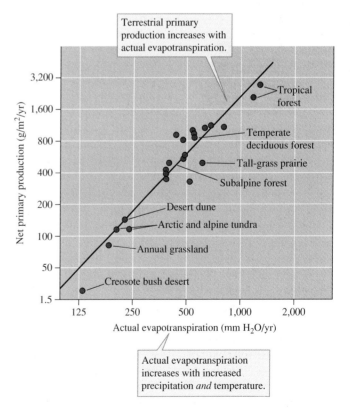

**Figure 18.2** Relationship between actual evapotranspiration and net aboveground primary production in a series of terrestrial ecosystems (data from Rosenzweig 1968).

cold, dry tundra show the lowest levels. Intermediate levels occur in temperate forests, temperate grasslands, woodlands, and high-elevation forests. Figure 18.2 shows that AET accounts for a significant proportion of the variation in annual net primary production among terrestrial ecosystems.

Rosenzweig's analysis attempts to explain variation in primary production across the whole spectrum of terrestrial ecosystems. What controls variation in primary production within similar ecosystems? O. E. Sala and his colleagues (1988) at Colorado State University explored the factors controlling primary production in the central grassland region of the United States. Their study was based on data collected by the U.S. Department of Agriculture Soil Conservation Service at 9,498 sites. To make this large data set more manageable, the researchers grouped the sites into 100 representative study areas.

The study areas extended from Mississippi and Arkansas in the east to New Mexico and Montana in the west and from North Dakota to southern Texas. Primary production was highest in the eastern grassland study areas and lowest in the western study areas. This east–west variation corresponds to the westward changes from tall-grass prairie to short-grass prairie that we reviewed in chapter 2. Sala and his colleagues found that this east–west variation in primary production among grassland ecosystems correlated significantly with the amount of rainfall (fig. 18.3).

Compare the plot by Sala and his colleagues (fig. 18.3) with the one constructed by Rosenzweig (fig. 18.2). How are they similar? How are they different? Both graphs have primary production plotted on the vertical axis as a dependent variable. However, while the Rosenzweig plot includes ecosystems ranging from tundra to tropical rain forest, the plot by Sala and his colleagues includes grasslands only. In addition, different variables are plotted on the horizontal axes of the two graphs. While Rosenzweig plotted actual evapotranspiration, which depends upon temperature and precipitation, Sala and his colleagues plotted precipitation only. They found that including temperature in their analysis did not improve their ability to predict net primary production. Why do you think precipitation alone was sufficient to account for most of the variation in grassland production? A likely reason is that warm temperatures occur during the growing season at all of the study areas included by Sala and his colleagues. In contrast, Rosenzweig's study areas vary widely in growing season temperature.

These researchers found strong correlations between AET or precipitation and rates of terrestrial primary production. However, their models did not completely explain the variation in primary production among the study ecosystems. For instance, in figure 18.2 ecosystems with annual AET levels of 500 to 600 mm of water showed annual rates of primary production ranging from 300 to 1,000 g per square meter. In figure 18.3, grassland ecosystems receiving 400 mm of annual precipitation had annual rates of primary production ranging from about 100 to 250 g per square meter. These differences in primary production challenge ecologists for an explanation.

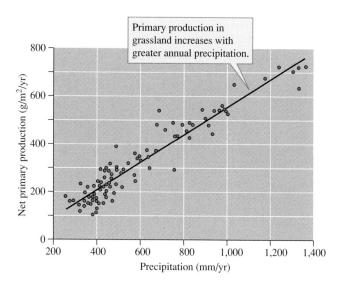

**Figure 18.3** Influence of annual precipitation on net aboveground primary production in grasslands of central North America (data from Sala et al. 1988).

## Soil Fertility and Terrestrial Primary Production

Significant variation in terrestrial primary production can be explained by differences in soil fertility. Farmers have long known that adding fertilizers to soil can increase agricultural production. However, it was not until the nineteenth century that scientists began to quantify the influence of specific nutrients, such as nitrogen (N) or phosphorus (P), on rates of primary production. Justus Liebig (1840) pointed out that nutrient supplies often limit plant growth. He also suggested that nutrient limitation to plant growth could be traced to a single limiting nutrient. This hypothetical control of primary production by a single nutrient was later called "Liebig's Law of the Minimum." We now know that Liebig's perspective was too simplistic. Usually several factors, including a number of nutrients, simultaneously affect levels of terrestrial primary production. However, his work led the way to a concept that remains true today; variation in soil fertility can significantly affect rates of terrestrial primary production.

Liebig's work, and most practical experience prior to Liebig, concerned the productivity of agricultural ecosystems. Do nutrients influence rates of primary production in other ecosystems, such as the tundra or deserts, where human manipulation has been less prominent? Ecologists have demonstrated the significant influence of nutrients on terrestrial primary production through numerous experiments involving addition of nutrients to natural ecosystems.

Ecologists have increased primary production by adding nutrients to a wide variety of terrestrial ecosystems, including arctic tundra, alpine tundra, grasslands, deserts, and forests. For instance, Gaius Shaver and Stuart Chapin (1986) studied the potential for nutrient limitation in arctic tundra. They added commercial fertilizer containing nitrogen, phosphorus, and

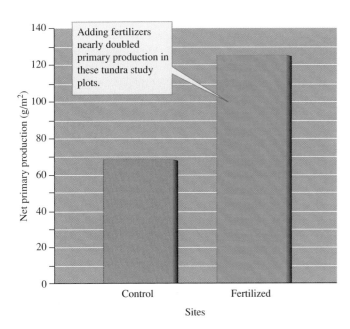

**Figure 18.4** Effect of addition of nitrogen, phosphorus, and potassium on net aboveground primary production in arctic tundra (data from Shaver and Chapin 1986).

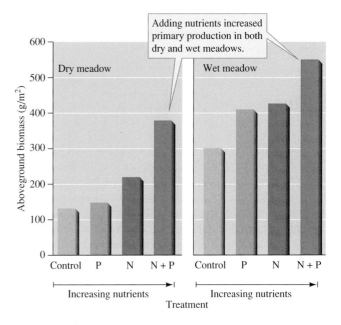

**Figure 18.5** Effect of adding phosphorus (P) and/or nitrogen (N) on aboveground primary production in two environments in alpine tundra (data from Bowman et al. 1993).

potassium to several tundra ecosystems in Alaska. They made a single application of fertilizer to half of their experimental plots and two applications to the remaining experimental plots.

Shaver and Chapin measured net primary production at their control and experimental sites 2 to 4 years after the first nutrient additions. Nutrient additions increased net primary production (by 23%–300%) at all of the study sites. The response to fertilization was substantial and clear at most study sites. Four years after the initial application of fertilizer, net primary production on Kuparuk Ridge was twice as high on the fertilized plots compared to the unfertilized control plots (fig. 18.4).

Nutrient additions to alpine tundra indicate that the response of ecosystems to nutrient addition is affected by prior nutrient availability. William Bowman and his colleagues (1993) added nutrients to the alpine tundra on Niwot Ridge, Colorado. They conducted their experiment in adjacent dry alpine and wet alpine meadows at an elevation of 3,510 m. One of four treatments was applied in both the dry and wet alpine meadows: (1) control (no nutrient additions), (2) nitrogen added, (3) phosphorus added, and (4) nitrogen and phosphorus added. The researchers then measured soil nitrogen and phosphorus concentrations and annual net primary production in each study plot.

Initial concentrations of both nitrogen and phosphorus were higher in the wet meadow ecosystem. And while fertilizing raised the concentrations of both nitrogen and phosphorus in the dry meadow, fertilizing the wet meadow raised the concentration of nitrogen but not phosphorus.

Fertilizing produced greater increases in primary production in the dry meadow than in the wet meadow. Adding nitrogen to the dry meadow increased primary production by 63%. Adding nitrogen *and* phosphorus increased primary produc-

tion by 178%. In contrast, the wet meadow only showed relatively small but statistically significant responses to the additions of both nitrogen and phosphorus (fig. 18.5).

Bowman and his colleagues suggest that these results show that nitrogen is the main nutrient limiting net primary production in the dry meadow and that nitrogen and phosphorus jointly limit net primary production in the wet meadow. They also suggest that light, not nutrients, may limit net primary production in the wet meadow. In other words, the higher biomass in the wet meadow may have produced enough shading to inhibit the growth response of some species to nutrient additions.

Experiments such as these have shown that despite the major influence of temperature and moisture on rates of primary production in terrestrial ecosystems, variation in nutrient availability can also have measurable influence. As we shall see in the next section, nutrient availability is the main factor limiting primary production in aquatic ecosystems.

## CONCEPT DISCUSSION

### Patterns of Aquatic Primary Production

**Aquatic primary production is generally limited by nutrient availability.**

Limnologists and oceanographers have measured rates of primary production and nutrient concentrations in many lakes and at many coastal and oceanic study sites. These studies have produced one of the best documented patterns in the biosphere: the positive relationship between nutrient availability and rate of primary production in aquatic ecosystems.

## Patterns and Models

A quantitative relationship between phosphorus, an essential plant nutrient, and phytoplankton biomass was first described for a series of lakes in Japan (Hogetsu and Ichimura 1954, Ichimura 1956, Sakamoto 1966). The ecologists studying this relationship found a remarkably good correspondence between total phosphorus and phytoplankton biomass.

Later, Dillon and Rigler (1974) described a similar positive relationship between phosphorus and phytoplankton biomass for lake ecosystems throughout the Northern Hemisphere. Remarkably, the slopes of the lines describing the relationship between phosphorus and phytoplankton biomass for the Japanese and Canadian lakes were nearly identical (fig. 18.6).

The data from Japan and North America strongly support the hypothesis that nutrients, particularly phosphorus, control phytoplankton biomass in lake ecosystems. However, what is the relationship between phytoplankton biomass and the rate of primary production? This relationship was explored by Val Smith (1979) for 49 lakes of the north temperate zone. The data from these lakes showed a strong positive correlation between chlorophyll concentrations and photosynthetic rates (fig. 18.7). Smith also examined the relationship between total phosphorus concentration and photosynthetic rate directly. Aquatic ecologists have extended these correlational studies of the relationship between nutrient availability and primary production by manipulating nutrient availability in entire lake ecosystems.

## Whole Lake Experiments on Primary Production

The Experimental Lakes Area was founded in northwestern Ontario, Canada, in 1968, as a place in which aquatic ecologists could manipulate whole lake ecosystems (Mills and Schindler 1987, Findlay and Kasian 1987). For instance, ecologists manipulated nutrient availability in a lake called Lake 226. They used a vinyl curtain to divide Lake 226 into two 8 ha basins each containing about 500,000 m³ of water. Think about these numbers for a second. This was a huge experiment! Each subbasin of Lake 226 was fertilized from 1973 to 1980. The researchers added a mixture of carbon in the form of sucrose and nitrate to one basin and carbon, nitrate, and phosphate to the other basin. They stopped fertilizing the lakes after 1980 and then studied the recovery of the Lake 226 ecosystem from 1981 to 1983.

Both sides of Lake 226 responded significantly to nutrient additions. Prior to the manipulation, Lake 226 supported about the same biomass of phytoplankton as two reference lakes (fig. 18.8). However, when experimenters began adding nutrients,

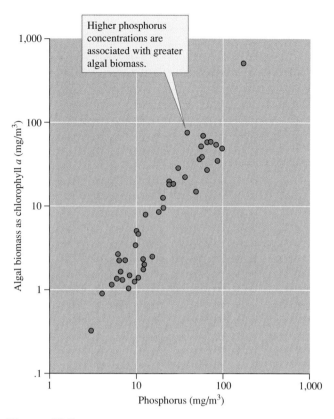

**Figure 18.6** Relationship between phosphorus concentration and algal biomass in north temperate lakes (data from Dillon and Rigler 1974).

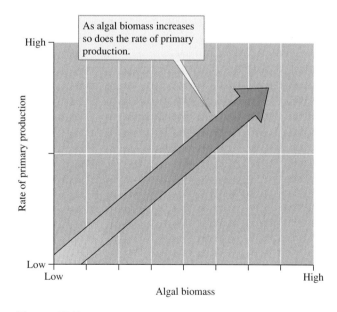

**Figure 18.7** Relationship between algal biomass and rate of primary production in temperate zone lakes (data from Smith 1979).

the phytoplankton biomass in Lake 226 quickly surpassed that in the reference lakes. Phytoplankton biomass remained elevated in Lake 226 until the experimenters stopped adding fertilizer at the end of 1980. Then, from 1981 to 1983 the phytoplankton biomass in Lake 226 declined significantly.

In conclusion, both correlations—between phosphorus concentration and rate of primary production, and whole lake experiments, involving nutrient additions—support the generalization that nutrient availability controls rates of primary production in freshwater ecosystems. Now, let's examine the evidence for this relationship in marine ecosystems.

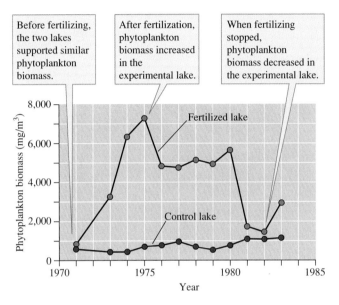

Before fertilizing, the two lakes supported similar phytoplankton biomass.

After fertilization, phytoplankton biomass increased in the experimental lake.

When fertilizing stopped, phytoplankton biomass decreased in the experimental lake.

**Figure 18.8** A whole lake experiment shows the effect of nutrient additions on average phytoplankton biomass (data from Findlay and Kasian 1987).

# Global Patterns of Marine Primary Production

The geographic distribution of net primary production in the sea indicates a positive influence of nutrient availability on rates of primary production. Oceanographers have observed that the highest rates of primary production by marine phytoplankton are generally concentrated in areas with higher levels of nutrient availability (fig. 18.9). The highest rates of primary production are concentrated along the margins of continents over continental shelves and in areas of upwelling. Along continental margins nutrients are renewed by runoff from the land and by biological or physical disturbance of bottom sediments. As we saw in chapter 3, the upwelling that brings nutrient-laden water from the depths to the surface is concentrated along the west coasts of continents and around the continent of Antarctica, areas that appear dark red on figure 18.9, indicating high to very high rates of primary production.

Meanwhile, the central portions of the major oceans show low levels of nutrient availability and low rates of primary production. The main source of nutrient renewal in the surface waters of the open ocean is vertical mixing. Vertical mixing is generally blocked in open tropical oceans by a permanent thermocline. Consequently, the surface waters of open tropical oceans contain very low concentrations of nutrients and show some of the lowest rates of marine primary production.

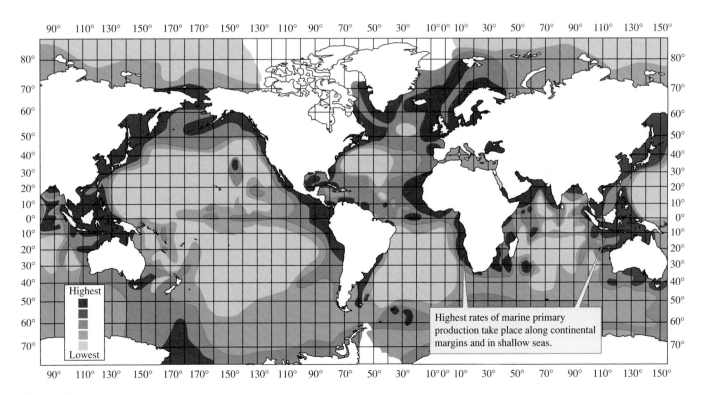

Highest rates of marine primary production take place along continental margins and in shallow seas.

**Figure 18.9** Geographic variation in marine primary production (data from F.A.O. 1972).

What is the experimental evidence for nutrient limitation of marine primary production? Some of the most thorough studies have been conducted in the Baltic Sea. For instance, Edna Granéli and her colleagues (1990) have used nutrient enrichment to test whether nutrient availability limits primary production in the Baltic Sea.

In a test using a single algal species, Granéli added nutrients to filtered seawater from a series of study sites. She added nitrate to one experimental group, phosphates to another, and nothing to a third group of flasks (fig. 18.10). Notice that the flasks with additional nitrate showed increased chlorophyll *a* concentrations at all sites, while the flasks with additional phosphate had chlorophyll *a* concentrations very similar to the control flasks. What do these results indicate? They suggest that the rate of primary production in the Baltic Sea is limited by nutrients. However, in contrast to most freshwater lakes, the limiting nutrient appears to be nitrogen, not phosphorus.

Granéli did similar enrichment studies along a series of stations in the Kattegat, the Belt Sea, and the Skagerrak, where the salinity approaches that of the open ocean. However, in this second series of experiments, she used indigenous phytoplankton rather than a single standardized test species. Once again the concentrations of chlorophyll *a* were higher in the flasks to which nitrate had been added while the control and phosphate treatment groups were virtually indistinguishable. Again, the results indicate nitrogen limitation along virtually the entire study area.

There have been no experiments done in the marine environment that are equivalent to the whole lake manipulations at the Experimental Lakes Area (e.g., Schindler 1990). However, in one experiment, researchers were able to alter the nutrient inputs and concentrations in Himmerfjärden, Sweden, a brackish water coastal inlet of the Baltic Sea with a surface area of 195 km$^2$ (see fig. 18.10). (For comparison, the lake subbasins manipulated in the whole lake experiments were < 0.1 km$^2$.) The results of this manipulative experiment indicate that nitrogen limitation of primary production can shift to phosphorus limitation by altering nitrogen:phosphorus ratios. Increasing additions of phosphorus to Himmerfjärden reinforced nitrogen limitation, while decreasing phosphorus additions and increasing nitrogen additions led to increased phosphorus limitation.

Dillon and Rigler suggested that limnologists pay attention to the scatter of points around lines showing a relationship between nutrient concentrations and phytoplankton biomass (F.A.O. 1972). We call that scatter of points *residual variation*. Residual variation is that proportion of variation not explained by the independent variable, in this case, by nutrient concentration. Dillon and Rigler suggested that environmental factors besides nutrient availability significantly influence phytoplankton biomass. One of those factors is the intensity of predation on the zooplankton that feed on phytoplankton. As we shall see in the next Concept Discussion section, consumers can influence rates of primary production in both terrestrial and aquatic ecosystems.

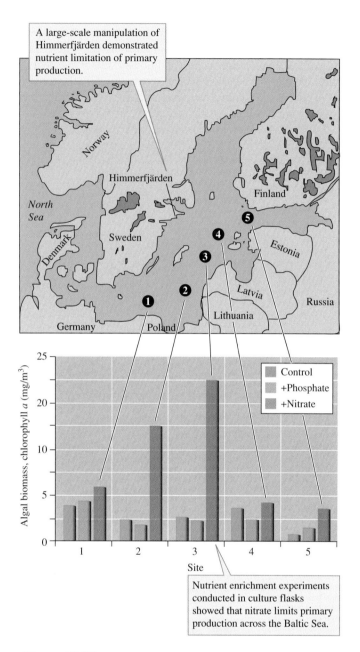

**Figure 18.10** Nitrate control of primary production in the Baltic Sea (data from Granéli et al. 1990).

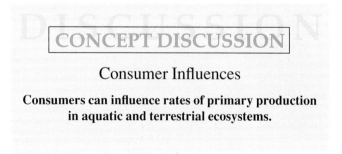

## CONCEPT DISCUSSION

### Consumer Influences

**Consumers can influence rates of primary production in aquatic and terrestrial ecosystems.**

In the first section of chapter 18, we emphasized the effects of physical and chemical factors on rates of primary production. More recently, ecologists have discovered that primary production is also affected by consumers. Ecologists refer to the

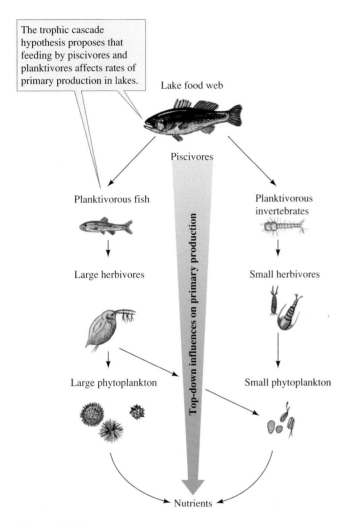

The trophic cascade hypothesis proposes that feeding by piscivores and planktivores affects rates of primary production in lakes.

Lake food web

Piscivores

Planktivorous fish

Planktivorous invertebrates

Large herbivores

Small herbivores

Large phytoplankton

Small phytoplankton

Top-down influences on primary production

Nutrients

**Figure 18.11** The trophic cascade hypothesis.

influences of physical and chemical factors, such as temperature and nutrients, on ecosystems as **bottom-up controls.** The influences of consumers on ecosystems are known as **top-down controls.** In the previous two Concept Discussion sections we discussed bottom-up controls on rates of primary production. Here we discuss top-down control.

## Piscivores, Planktivores, and Lake Primary Production

Stephen Carpenter, James Kitchell, and James Hodgson (1985) proposed that while nutrient inputs determine the potential rate of primary production in a lake, piscivorous and planktivorous fish can cause significant deviations from potential primary production. In support of their hypothesis, Carpenter and his colleagues (1991) cited a negative correlation between zooplankton size, an indication of grazing intensity, and primary production.

Carpenter and Kitchell (1988) proposed that the influences of consumers on lake primary production propagate through food webs. Since they visualized the effects of con-

sumers coming from the top of food webs to the base, they called these effects on ecosystem properties "trophic cascades." The trophic cascade hypothesis (fig. 18.11) is very similar to the keystone species hypothesis (see chapter 17). However, notice that the trophic cascade model is focused on the effects of consumers on ecosystem processes, such as primary production, and not on their effects on species diversity.

Carpenter and Kitchell (1993) interpreted the trophic cascade in their study lakes as follows: Piscivores, such as largemouth bass, feed on planktivorous fish and invertebrates. Because of their influence on planktivorous fish, largemouth bass indirectly affect populations of zooplankton. By reducing populations of planktivorous fish, largemouth bass reduce feeding pressure on zooplankton and zooplankton populations. Large-bodied zooplankton, the preferred prey of size-selective planktivorous fish (see chapter 6), soon dominate the zooplankton community. A dense population of large zooplankton reduces phytoplankton biomass and the rate of primary production. This interpretation of the trophic cascade is consistent with the negative correlation between zooplankton body size and primary production reported by Carpenter and his research team. This hypothesis is summarized in figure 18.12.

Carpenter and Kitchell tested their trophic cascade model by manipulating the fish communities in two lakes and using a third lake as a control. Figure 18.13 shows the overall design of their experiment. Two of the lakes contained substantial populations of largemouth bass. A third lake had no bass, due to occasional winterkill, but contained an abundance of planktivorous minnows. The researchers removed 90% of the largemouth bass from one experimental lake and put them into the other. They simultaneously removed 90% of the planktivorous minnows from the second lake and introduced them to the first. They left a reference lake unmanipulated as a control.

The responses of the study lakes to the experimental manipulations support the trophic cascade hypothesis (fig. 18.13). Reducing the planktivorous fish population led to reduced rates of primary production. In the absence of planktivorous minnows, the predaceous invertebrate *Chaoborus* became more numerous. *Chaoborus* fed heavily upon the smaller herbivorous zooplankton, and the herbivorous zooplankton assemblage shifted in dominance from small to large species. In the presence of abundant, large herbivorous zooplankton, phytoplankton biomass and rate of primary production declined.

Adding planktivorous minnows produced a complex ecological response. Increasing the planktivorous fish population led to increased rates of primary production. However, though the researchers increased the population of planktivorous fish in this experimental lake, they did so in an unintended way. Despite the best efforts of the researchers, a few bass remained. So, by introducing a large number of minnows they basically fed the remaining bass. An increased food supply combined with reduced population density induced a strong numerical response by the bass population (see chapter 10). The manipulation increased the reproductive rate of the remaining largemouth bass 50-fold, producing an abundance of young largemouth bass that feed voraciously on zooplankton.

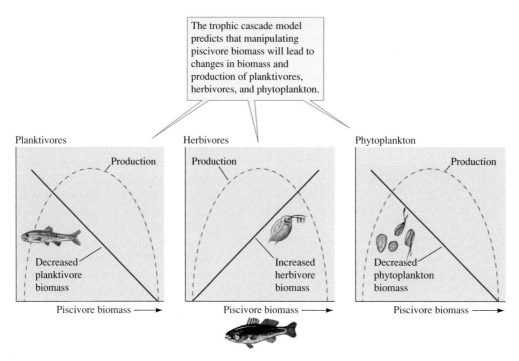

**Figure 18.12** Predicted effects of piscivores on planktivore, herbivore, and phytoplankton biomass and production (data from Carpenter, Kitchell, and Hodgson 1985).

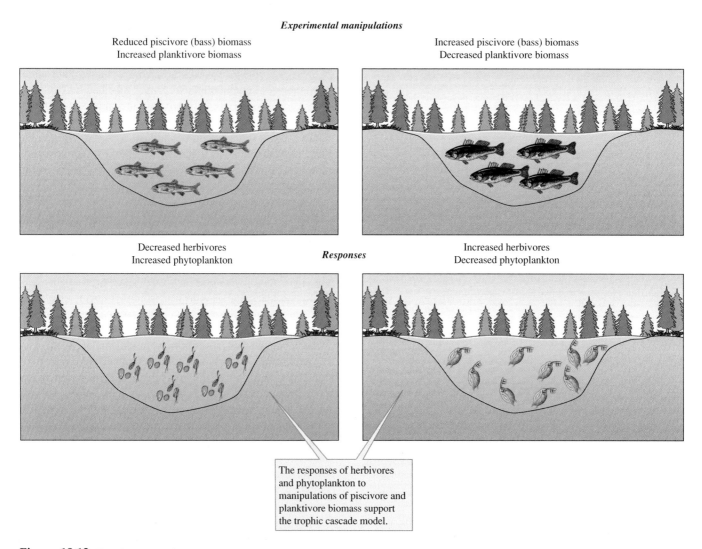

**Figure 18.13** Experimental manipulations of ponds and responses.

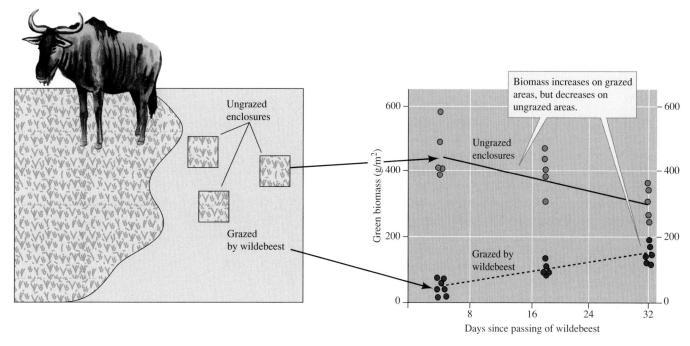

**Figure 18.14**  Growth response by grasses grazed by wildebeest (data from McNaughton 1976).

The lake ecosystem responded to the increased biomass of planktivorous fish (young largemouth bass) as predicted at the outset of the experiment. The biomass of zooplankton decreased sharply, the average size of herbivorous zooplankton decreased, and phytoplankton biomass and primary production increased.

The results of these whole lake experiments show that the trophic activities of a few species can have large effects on ecosystem processes. However, the majority of trophic cascades described by ecologists have been in aquatic ecosystems with algae as primary producers. This pattern prompted Donald Strong (1992) to ask, "Are trophic cascades all wet?" Strong suggested that trophic cascades most likely occur in ecosystems of lower species diversity and reduced spatial and temporal complexity. These are characteristics of many aquatic ecosystems. Despite these restrictions, consumers have significant effects on rates of primary production in some terrestrial ecosystems; one of those is the Serengeti grassland ecosystem.

# Grazing by Large Mammals and Primary Production on the Serengeti

The Serengeti-Mara, a 25,000 km² grassland ecosystem that straddles the border between Tanzania and Kenya, is one of the last ecosystems on earth where great numbers of large mammals still roam freely. Sam McNaughton (1985) reported estimated densities of the major grazers in the Serengeti that included 1.4 million wildebeest, *Connochaetes taurinus albujubatus*, 600,000 Thomson's gazelle, *Gazella thomsonii*, 200,000 zebra,

*Equus burchelli,* 52,000 buffalo, *Syncerus caffer,* 60,000 topi, *Damaliscus korrigum,* and large numbers of 20 additional grazing mammals. McNaughton estimated that these grazers consume an average of 66% of the annual primary production on the Serengeti. In light of this estimate, the potential for consumer influences on primary production seems very high.

Over two decades of research on the Serengeti ecosystem in Tanzania led McNaughton to appreciate the complex interrelations of abiotic and biotic factors there. For instance, both soil fertility and rainfall stimulate plant production and the distributions of grazing mammals. However, grazing mammals also affect water balance, soil fertility, and plant production.

As you might predict, the rate of primary production on the Serengeti is positively correlated with the quantity of rainfall. However, McNaughton (1976) also found that grazing can increase primary production. He fenced in some areas in the western Serengeti to explore the influence of herbivores on production. The migrating wildebeest that flooded into the study site grazed intensively for 4 days, consuming approximately 85% of plant biomass.

During the month after the wildebeest left the study area, biomass within the enclosures decreased, while the biomass of vegetation outside the enclosures increased (fig. 18.14). Grazing increases the growth rate of many grass species, a response to grazing called *compensatory growth*. The mechanisms underlying compensatory growth include lower rates of respiration due to lower plant biomass, reduced self-shading, and improved water balance due to reduced leaf area.

The compensatory growth observed by McNaughton was highest at intermediate grazing intensities (fig. 18.15). Apparently, light grazing is insufficient to produce compensatory growth and very heavy grazing reduces the capacity of the

*Investigating the Evidence*

# Comparing Two Populations with the *t*-test

In chapter 17 we used confidence intervals to compare the biomasses of two populations of the caddisfly, *Neothremma alicia*. That comparison indicated that the population living in a stream that had flooded recently had a lower biomass per unit area. Here, we will use the same samples of the two populations to test for significant differences using a *t*-test, a method for statistical comparison of two samples. The *t*-test involves calculating the statistic *t* and comparing the value with a table of critical values of *t*. The hypothesis is that the populations from which the samples were drawn have the same mean, and the alternative hypothesis is that the population means are different. If the calculated *t* statistic is *less than* the critical value of *t*, the hypothesis that the populations are not different is accepted. If the calculated *t* statistic is *greater than* the critical value of *t*, this hypothesis is rejected.

As an example, let's use the *t*-test to compare our two samples of *N. alicia*:

| Quadrat Number | | | | | | | | |
|---|---|---|---|---|---|---|---|---|
| 1 | 2 | 3 | 4 | 5 | 6 | 7 | 8 | 9 |
| Flooded (mg) | | | | | | | | |
| 4.83 | 3.00 | 3.63 | 1.20 | 2.97 | 1.17 | 1.95 | 0.98 | 1.46 |
| Unflooded (mg) | | | | | | | | |
| 7.08 | 5.18 | 5.97 | 3.64 | 5.14 | 3.05 | 4.23 | 3.14 | 3.71 |

The statistic *t* for this comparison is calculated as:

$$t = \frac{|\bar{X}_f - \bar{X}_u|}{s_{\bar{X}_f - \bar{X}_u}}$$

In this equation:

$\bar{X}_f$ = mean of sample from flooded stream
   = 2.354 mg/0.1 m$^2$

$\bar{X}_u$ = mean of sample from unflooded stream
   = 4.571 mg/0.1 m$^2$

$s_{\bar{X}_f - \bar{X}_u}$ = the standard error of the difference between means,

which is calculated as:

$$s_{\bar{X}_f - \bar{X}_u} = \sqrt{\frac{s_p^2}{n_f} + \frac{s_p^2}{n_u}}$$

Here:

$n_f$ = number of quadrats in sample from the flooded stream
$n_u$ = number of quadrats in sample from the unflooded stream
$s_p^2$ = pooled estimate of the variance

The pooled estimate of the variance for our two samples is calculated as:

$$s_p^2 = \frac{SS_f + SS_u}{DF_f + DF_u}$$

In this equation:

$SS_f$ = sum of squares for the sample from the flooded stream
$SS_u$ = sum of squares for the sample from the unflooded stream
$DF_f$ = degrees of freedom for the sample from the flooded stream
$DF_u$ = degrees of freedom for the sample from the unflooded stream

plant to recover. The large grazing mammals of the Serengeti have substantial influences on its rate of primary production. As McNaughton put it, "African ecosystems cannot be understood without close consideration of the large mammals. These animals interact with their habitats in complex and powerful patterns influencing ecosystems for long periods."

What McNaughton and his colleagues described is essentially a trophic cascade in a terrestrial environment where the feeding activities of consumers have a major influence on ecosystem properties. The Serengeti is now an exceptional terrestrial ecosystem but it was not always so. As we saw in chapter 2, the extensive grasslands of North America and Eurasia were also once populated by vast herds of mammalian grazers. Historians estimate that the population of North American bison in the middle of the nineteenth century numbered up to

60 million. Such a dense concentration of grazers must have had significant influences upon the grassland ecosystems of which they were part. It appears that terrestrial consumers, as well as the aquatic ones studied by Carpenter and Kitchell, can have important influences on primary production.

In the Serengeti, lions are the top predators. Though they are occasionally killed by hyenas, there are no predators that depend principally upon hunting lions as a source of energy. In the ponds studied by Carpenter and Kitchell, largemouth bass were the top carnivores. The number of trophic levels in ecosystems ranges from two to five or six, perhaps seven or eight in exceptional ecosystems. In any case, ecosystems have a limited number of trophic levels. What limits the number of trophic levels? We will consider the factors that limit the number of trophic levels in ecosystems in the next Concept Discussion.

We calculated the sum of squares for a sample in chapter 6 (p. 162) as:

$$\text{Sum of squares} = \Sigma(X - \overline{X})^2$$

Using this equation, we can calculate the sums of squares for the two populations, with the following results:

$$SS_f = 14.138 \ (\text{mg}/0.1 \ \text{m}^2)^2$$
$$SS_u = 15.031 \ (\text{mg}/0.1 \ \text{m}^2)^2$$

and degrees of freedom, which we considered in chapter 11 (p. 287), is $n - 1 = 8$ for both samples. Using $DF = 8$ for both populations, we can calculate the pooled estimate of the variance as follows:

$$s_p^2 = \frac{14.138 \ (\text{mg}/0.1 \ \text{m}^2)^2 + 15.031 \ (\text{mg}/0.1 \ \text{m}^2)^2}{8+8}$$
$$s_p^2 = 1.823 \ (\text{mg}/0.1 \ \text{m}^2)^2$$

Using this value, we can now calculate the standard deviation of the difference between means:

$$s_{\overline{X}_f - \overline{X}_u} = \sqrt{\frac{s_p^2}{n_f} + \frac{s_p^2}{n_u}}$$

$$s_{\overline{X}_f - \overline{X}_u} = \sqrt{\frac{1.823\left(\text{mg}/0.1 \ \text{m}^2\right)^2}{9} + \frac{1.823\left(\text{mg}/0.1 \ \text{m}^2\right)^2}{9}}$$

$$s_{\overline{X}_f - \overline{X}_u} = \sqrt{0.203\left(\text{mg}/0.1 \ \text{m}^2\right)^2 + 0.203\left(\text{mg}/0.1 \ \text{m}^2\right)^2}$$

$$s_{\overline{X}_f - \overline{X}_u} = .637 \ \text{mg}/0.1 \ \text{m}^2$$

Now, we have all the values we need to calculate $t$:

$$t = \frac{\left|\overline{X}_f - \overline{X}_u\right|}{s_{\overline{X}_f - \overline{X}_u}}$$

$$t = \frac{\left|2.354 \ \text{mg}/0.1 \ \text{m}^2 - 4.571 \ \text{mg}/0.1 \ \text{m}^2\right|}{0.637 \ \text{mg}/0.1 \ \text{m}^2}$$

$$t = 3.480$$

At this point in our $t$-test, we need to compare the calculated $t$ with the appropriate critical value. Again, we need to know two factors: the desired level of significance, which will be $P < 0.05$, and the degrees of freedom. In this case we use the pooled degrees of freedom:

$$DF_{pooled} = DF_f + DF_u = 8 + 8 = 16$$

The critical value of Student's $t$ for $P < 0.05$ and $DF = 16$, is 2.12. Since our calculated value of $t$, 3.480, is greater than this critical value, the probability that the population means are the same is *less than* 0.05. Therefore, we reject the hypothesis that the mean biomass of *N. alicia* per unit area is the same in the two streams and accept the alternative hypothesis that the mean biomass of this caddisfly differs in the two streams.

## CONCEPT DISCUSSION

### Trophic Levels

**Energy losses limit the number of trophic levels in ecosystems.**

We began chapter 18 with a partial and highly qualitative energy budget for a forest: Sunlight shines down on the canopy of a forest—some is reflected, some is converted to heat energy, and some is absorbed by chlorophyll. The energy budgets of ecosystems reveal that with each transfer or conversion of energy, some energy is lost. To verify that these losses have the potential to limit the number of trophic levels in ecosystems, we need to quantify the flow of energy through ecosystems. One of the very first ecologists to quantify the flux of energy through ecosystems was Raymond Lindeman.

## A Trophic Dynamic View of Ecosystems

Raymond Lindeman (1942) received his Ph.D. from the University of Minnesota in 1941, where his studies of the ecology of Cedar Bog Lake led him to a view of ecosystems far ahead of its time. Lindeman went from Minnesota to Yale University, where his association with G. E. Hutchinson from

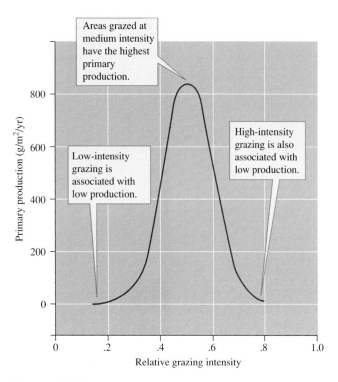

**Figure 18.15** Grazing intensity and primary production of Serengeti grassland (data from McNaughton 1985).

1941 to 1942 led to the publication of a revolutionary paper with the provocative title, "The Trophic-Dynamic Aspect of Ecology." In this paper, Lindeman articulated a view of ecosystems centered on energy fixation, storage, and flows that remains influential to this day. Like Tansley before him, Lindeman pointed out the difficulty and artificiality of separating organisms from their environment and promoted an ecosystem view of nature. Lindeman concluded that the ecosystem concept is fundamental to the study of **trophic dynamics,** which he defined as the transfer of energy from one part of an ecosystem to another.

Lindeman suggested grouping organisms within an ecosystem into trophic levels: primary producers, primary consumers, secondary consumers, tertiary consumers, and so forth. In this scheme, each trophic level feeds on the one immediately below it. Energy enters the ecosystem as primary producers engage in photosynthesis and convert solar energy into biomass. As energy is transferred from one trophic level to another, energy is lost due to limited assimilation, respiration by consumers, and heat production. As a result of these losses, the quantity of energy in an ecosystem decreases with each successive trophic level, forming a pyramid-shaped distribution of energy among trophic levels. Lindeman called these trophic pyramids "Eltonian pyramids," since Charles Elton (1927) was the first to propose that the distribution of energy among trophic levels is shaped like a pyramid.

Figure 18.16 shows the distribution of annual primary production among trophic levels in Cedar Bog Lake and in Lake Mendota, Wisconsin. Energy losses at each trophic level determine the trophic structure of these two ecosystems. As predicted by Elton, the distribution of energy across trophic levels in both lakes is shaped like a pyramid. As suggested at the beginning of this section, the number of trophic levels is limited in both lakes. Lake Mendota includes four trophic levels, while Cedar Bog Lake includes just three.

Following Lindeman's pioneering work, many other ecologists studied energy flow within ecosystems. One of the most comprehensive of these later studies focused on the Hubbard Brook Experimental Forest in New Hampshire.

## Energy Flow in a Temperate Deciduous Forest

James Gosz and his colleagues (1978) studied energy flow in the Hubbard Brook Experimental Forest, which is managed for research by the U.S. Forest Service. They concentrated their efforts on a stream catchment called watershed 6, which was left undisturbed so it could serve as a control for experi-

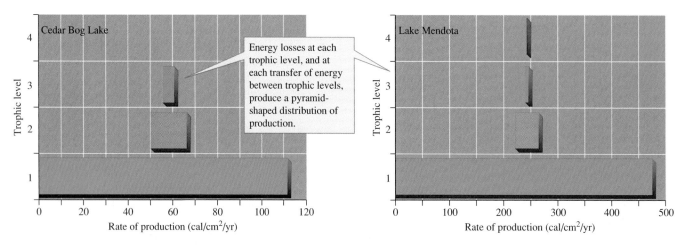

**Figure 18.16** Annual production by trophic level in two lakes (data from Lindeman 1942).

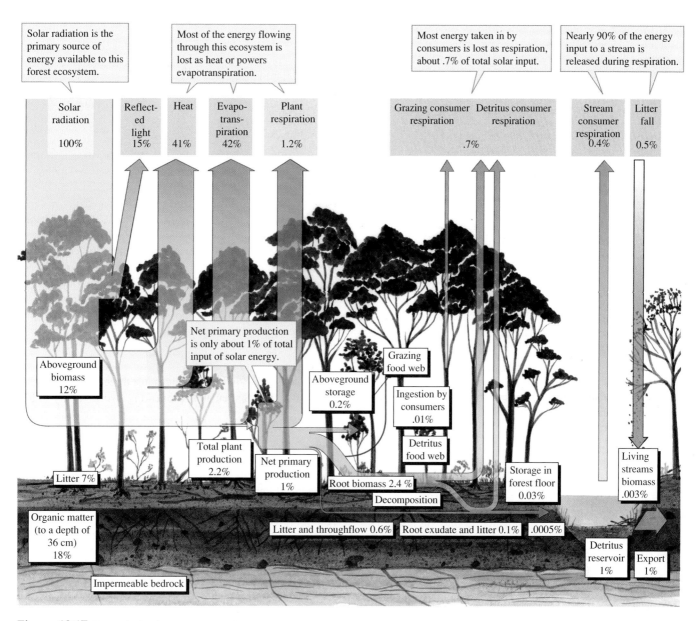

**Figure 18.17** Energy budget for a temperate deciduous forest (data from Gosz et al. 1978).

mental studies on other stream catchments. The energy flow in the Hubbard Brook Experimental Forest was quantified as kilocalories (kcal) per square meter per year. The results of the analysis are shown in figure 18.17.

First let's examine the distribution of organic matter among the major components of the Hubbard Brook ecosystem. The largest single pool of energy in the forest, 122,442 kcal/m², occurred as dead organic matter. Most of the dead organic matter, 88,120 kcal/m², was organic matter in the upper 36 cm of soil. The remainder, 34,322 kcal/m², occurred as plant litter on the forest floor. Total living-plant biomass amounted to 71,420 kcal/m², of which 59,696 kcal/m² was stored in aboveground biomass and 11,724 kcal/m² as belowground biomass.

The total standing stock of energy occurring as dead organic matter and living plant biomass was 193,862 kcal/m². This estimate by Gosz and his colleagues dwarfs the energy

stored in all other portions of the ecosystem. For instance, the energetic content of a caterpillar population during a severe population outbreak amounted to only 160 kcal/m². However, even this amount far exceeds the total energetic content of all vertebrate biomass. The researchers estimated that the total energetic content of the most numerous vertebrates, including chipmunks, mice, shrews, salamanders, and birds, amounted to less than 1 kcal/m². Now that we have inventoried the major standing stocks of energy, let's look at energy flow through the Hubbard Brook Forest.

The main source of energy for the ecosystem is solar radiation. The total input of solar energy to the study area during the growing season was estimated to be 480,000 kcal/m² (expressed as 100% in fig. 18.17). Of this total energy input, 15% was reflected, 41% was converted to heat, and 42% was absorbed during evapotranspiration. About 2.2% of the solar input was fixed by plants as gross primary production. Plant respiration accounted for 1.2%, leaving about 1% as net primary production. In other words, only about 1% of the solar input to the Hubbard Brook ecosystem was available to the herbivores and detritivores that made up the second trophic level.

About 1,199 kcal/m$^2$ of net primary production in the Hubbard Brook Forest went into plant growth. Herbivores consumed only about 41 kcal/m$^2$, approximately 1% of net primary production. Most of the energy available to consumers, approximately 3,037 kcal/m$^2$, occurred as surface litter fall. About 150 kcal/m$^2$ of the litter fall was stored as organic matter on the forest floor. The remainder was used by consumers. An additional source of detritus, amounting to 437 kcal/m$^2$, occurred belowground as root exudate and litter. Most of the energy consumed by grazers and detritivores, approximately 3,353 kcal/m$^2$, was lost as consumer respiration.

Now let's go back to the concept that started this section: Energy losses limit the number of trophic levels in ecosystems. The energy budget carefully constructed by Gosz and his colleagues gives us a basis for understanding this concept. Net primary production in the Hubbard Brook Forest ecosystem was less than 1% of the input of solar energy. In other words, over 99% of the solar energy available to the Hubbard Brook was unavailable for use by a second trophic level. Of the net primary production available to consumers, approximately 96% is lost as consumer respiration. This leaves very little for a third trophic level. It is such losses with each transfer of energy in a food chain that limit the number of trophic levels. As these losses between trophic levels accumulate, eventually there is insufficient energy left over to support a viable population at a higher trophic level.

The top predator on the African savanna is the lion. We might imagine predators fierce enough to prey on lions, but the energetics of energy conversion and transfer within ecosystems would preclude such a predator.

We can see from the studies of Gosz and his colleagues and others that ecosystems depend upon an outside input of energy. Ecosystems store some energy in the form of dead organic matter and biomass, but most energy flows through. As we shall see in chapter 19, however, ecosystems recycle elements such as nitrogen and sulfur. In the Applications & Tools section we review how forms of these and other elements can be used as a tool to determine the trophic structure of ecosystems.

## APPLICATIONS & TOOLS

### Using Stable Isotope Analysis to Trace Energy Flow Through Ecosystems

How do ecologists study the flow of energy through ecosystems? First, they identify the organisms that make up the biological part of the ecosystem. Then, they determine the feeding habits of consumers. They may identify consumers down to species or assign them broader taxonomic categories. Next, they

assign organisms to trophic levels and determine (1) the biomass of each trophic level, (2) the rate of energy or food intake by each trophic level, (3) the rate of energy assimilation, (4) the rate of respiration, and (5) rates of loss of energy to predators, parasites, etc. Finally, ecologists combine their information on individual trophic levels to construct a trophic pyramid such as that constructed by Lindeman (see fig. 18.16) or an energy flow diagram such as that by Gosz and his colleagues (see fig. 18.17).

One of the fundamental steps in constructing a trophic pyramid or energy flow diagram is assigning organisms to trophic levels. While this task may sound easy, for most organisms, it is not. Most assignments are based on studies of feeding habits. If food items are easily identified and feeding habits are well studied and do not change significantly over time or from place to place, you may accurately identify feeding relations and assign organisms to trophic levels. However, if feeding habits are variable or if food items are difficult to identify, it may be difficult to assign organisms accurately to a particular trophic level. One of the most useful tools for making such assignments is stable isotope analysis (see chapter 5, p. 139).

## Trophic Levels of Tropical River Fish

The food webs shown in figure 17.4 (see p. 421), which were described by Kirk Winemiller (1990), indicate the great complexity of feeding relations among the fish of tropical rivers. Historically, ecologists have constructed such food webs using the stomach contents of the fish under study. However, an ecologist examining stomach contents learns what the fish has recently consumed but not what the fish will assimilate. This is a particular problem for detritus-feeding fish.

David Jepsen and Kirk Winemiller (2002) turned to stable isotope analysis for help in their studies of tropical river food webs. Jepsen and Winemiller studied the trophic ecology of fish living in four rivers in Venezuela. They chose rivers that represented a wide range of productivity. Three of the rivers—the Apure, Aguaro, and Cinaruco—drain tropical savannas, called llanos, in Venezuela. Of these, the Apure River, which supports dense stands of emergent and floating aquatic plants, is the most nutrient-rich and most productive. The Aguaro River, which supports submerged aquatic plants and algae, is clearer and somewhat less productive. Meanwhile, the Cinaruco River is a nutrient-poor sandy bottom river with almost no aquatic plants. The least productive river in the study was the Pasimoni River, which drains dense rain forest growing on highly weathered bedrock.

Jepsen and Winemiller found that the same families of fish inhabit the four study rivers and that some fish species live in all four rivers. They used stomach contents, dentition, and literature information to group the fish species in the study rivers into four feeding categories: herbivores, detritivores, omnivores, and piscivores. Herbivores were fish species reported in the literature to feed predominantly on coarse plant material. Detritivores were fish that feed by sucking or scraping particulate organic matter from submerged surfaces. Omnivores were fish that consume a mixture of plant material and invertebrates. The fourth feeding group, piscivores, feed on fish, either by consuming them whole or biting off pieces of flesh, the feeding technique of piranhas.

Once they had assigned each fish species to a feeding category using traditional methods, Jepsen and Winemiller reclassified them using concentrations of stable isotopes. They used $^{15}$N to determine the trophic level of each fish species. Because $^{15}$N is enriched with each transfer from a lower to a higher trophic level, this stable isotope is useful for determining the trophic position of a consumer. Jepsen and Winemiller first determined the concentration of $^{15}$N in the materials at the base of the food web in each study river. They next reassigned each fish species in each river to a trophic level using enrichment of $^{15}$N. In their reassignment, Jepsen and Winemiller assumed that $^{15}$N is enriched 2.8 parts per thousand with each transfer in the food web.

Jepsen and Winemiller's stable isotope analysis revealed a great deal of variation within and among river fish communities that would have remained hidden if they had used traditional approaches only. For instance, $^{15}$N analysis demonstrated that two morphologically similar catfish that are assumed to feed on detritus and algae actually fed at two different trophic levels in two different rivers. One species fed at the base of the food web, which was the expected trophic position for the species, while the other species fed as an omnivore, feeding both on material at the base of the food web and on animal consumers. In another case, the same species of piscivorous fish fed at different tropic levels in different rivers. In two rivers, the $^{15}$N levels in the tissues of this putative piscivore indicated that it fed more as an omnivore than as a specialist on the consumption of fish. These are a few examples of the new information that stable isotope analysis is providing on trophic interactions. The next example shows how stable isotope analysis has been used to verify energy sources.

# Using Stable Isotopes to Identify Sources of Energy in a Salt Marsh

The main energy source in a salt marsh in eastern North America is primary production by the salt marsh grass *Spartina,* most of which is consumed as detritus. The detritus of *Spartina* is carried into tidal creeks at high tide, where it is consumed by a variety of organisms, including crabs, oysters, and mussels. However, *Spartina* is not the only potential source of food for these organisms. The waters of the salt marsh also contain organic matter from upland plants and carry phytoplankton. How much might these other food sources contribute to energy flow through the salt marsh ecosystem?

Bruce Peterson, Robert Howarth, and Robert Garritt (1985) used stable isotopes to determine the relative contributions of *Spartina,* phytoplankton, and upland plants to the nutrition of the ribbed mussel, *Geukinsia demissa,* a dominant filter-feeding species in New England salt marshes. The researchers pointed out that determining the trophic structure of salt marshes is difficult because detritus from different sources is difficult to identify visually, because there are several potential sources of detritus, and because organisms may frequently change their feeding habits. It is difficult to accurately quantify the relative contributions of alternative energy sources to a species like *Geukinsia* using traditional methods. Those methods will also probably miss transient dietary switches entirely.

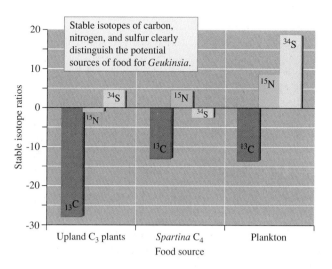

**Figure 18.18** Isotopic content of potential food sources for the ribbed mussel, *Geukinsia demissa,* in a New England salt marsh (data from Peterson, Howarth, and Garritt 1985).

As a solution for these problems, Peterson and his colleagues used the ratios of stable isotopes of carbon, nitrogen, and sulfur to assess the relative contributions of alternative food sources to the nutrition of the mussel. They used the stable isotopes of these three elements because their ratios are different in phytoplankton, upland $C_3$ plants (see chapter 6), and *Spartina,* a $C_4$ grass (fig. 18.18). Upland plants, with a $\delta^{13}C = -28.6$ $^0/_{00}$, are the most depleted of $^{13}C$, while *Spartina,* with a $\delta^{13}C = -13.1$ $^0/_{00}$, is the least depleted. Stable isotopes of sulfur and nitrogen are also distributed differently among these potential energy sources. For instance, *Spartina,* with a $\delta^{34}S = -2.4$ $^0/_{00}$, has the lowest relative concentration of $^{34}S$, while plankton, with a $\delta^{34}S = +18.8$ $^0/_{00}$, has the highest concentration of $^{34}S$.

Because of these differences in isotopic concentrations, the researchers were able to identify the relative contributions of potential food sources to the diet of the mussel (fig. 18.19). Their analyses showed that *Geukinsia* gets most of its energy from plankton and *Spartina* but that the relative contributions of these two food sources depends upon location. In the interior of the marsh, the mussel feeds mainly on *Spartina,* while near the mouth of the marsh it depends mainly on plankton. This is an example of how analyses of stable isotopes can provide us with a window to the otherwise hidden biology of species.

# Food Habits of Prehistoric Human Populations

Stable isotope analysis is also helping archeologists to reconstruct the history of our own species. For instance, stable isotopes have provided insights into the trophic position of

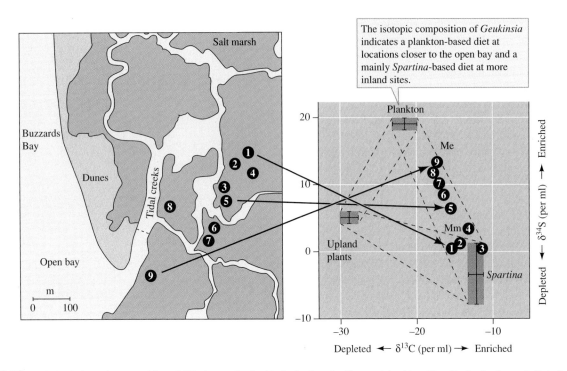

**Figure 18.19** Variation in isotopic composition of ribbed mussels, *Geukinsia demissa,* by distance inland in a New England salt marsh (data from Peterson, Howarth, and Garritt 1985).

humans in prehistoric ecosystems. The people of Central and South America began to cultivate corn, *Zea mays,* about 6,000 to 7,000 years ago, and the farming of corn eventually spread into North America.

Corn appears in the archaeological record of the forested regions of eastern North America about 2,000 years ago (van der Merwe and Vogel 1978). However, the amount remains low for a considerable time, and archeologists have been uncertain of when corn became significant in the diets of the human populations of these regions. Because corn is a C$_4$ grass, its tissues are relatively enriched with $^{13}$C compared to the C$_3$ plants that the Native American population had relied upon prior to the introduction of corn. Therefore, analyses of the carbon isotopes of human remains can give insights into the impact of corn farming on human nutrition. The $^{13}$C content of the collagen of human skeletons suggests that corn made a minor contribution to human diets in these regions for almost 1,000 years. Then, about 1,000 years ago the contribution of corn to human nutrition in the study region reached 24% and began to increase exponentially (fig. 18.20). By A. D. 1300, corn made up 69% to 75% of the diet of some populations in the Mississippi River valley.

Stable isotope analyses have been used to analyze the diets of other prehistoric human populations. For instance, stable isotope analysis has shown that about 6,000 years ago the diets of human populations in what is now Denmark shifted from a predominately marine diet to one dominated by terrestrial foods. Without the tool of stable isotope analysis, it would be much more difficult to accurately estimate timing of these significant shifts in the trophic ecology of prehistoric human populations.

Stable isotope analyses continue to improve our understanding of energy flow through ecosystems. While energy

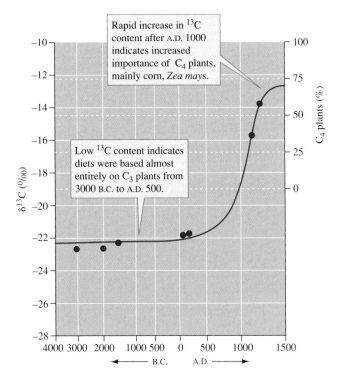

**Figure 18.20** Concentration of $^{13}$C in bone collagen indicates dietary composition of prehistoric Native Americans living in temperate forests in eastern North America (data from van der Merwe 1982).

flows through ecosystems in a one-way path, the elements, or nutrients, upon which organisms depend are recycled and may be used over and over again. The cycling of these nutrients is the subject of chapter 19.

We can view a forest, a stream, or an ocean as a system that absorbs, transforms, and stores energy. In this view, physical, chemical, and biological structures and processes are inseparable. When we look at natural systems in this way we view them as ecosystems. An *ecosystem* is a biological community plus all of the abiotic factors influencing that community.

*Primary production,* the fixation of energy by autotrophs, is one of the most important ecosystem processes. The rate of primary production is the amount of energy fixed over some interval of time. Gross primary production is the total amount of energy fixed by all the autotrophs in the ecosystem. Net primary production is the amount of energy left over after autotrophs have met their own energetic needs.

**Terrestrial primary production is generally limited by temperature and moisture.** The variables most highly correlated with variation in terrestrial primary production are temperature and moisture. Highest rates of terrestrial primary production occur under warm, moist conditions. Temperature and moisture conditions can be combined in a single measure called annual actual evapotranspiration, or AET, which is the total amount of water that evaporates and transpires off a landscape during the course of a year. Annual AET is positively correlated with net primary production in terrestrial ecosystems. However, significant variation in terrestrial primary production results from differences in soil fertility.

**Aquatic primary production is generally limited by nutrient availability.** One of the best documented patterns in the biosphere is the positive relationship between nutrient availability and rate of primary production in aquatic ecosystems. Phosphorus concentration usually limits rates of primary production in freshwater ecosystems, while nitrogen concentration usually limits rates of marine primary production.

**Consumers can influence rates of primary production in aquatic and terrestrial ecosystems.** Piscivorous fish can indirectly reduce rates of primary production in lakes by reducing the density of plankton-feeding fish. Reduced density of planktivorous fish can lead to increased density of herbivorous zooplankton, which can reduce the densities of phytoplankton and rates of primary production. Intense grazing by large mammalian herbivores on the Serengeti increases annual net primary production by inducing compensatory growth in grasses.

**Energy losses limit the number of trophic levels in ecosystems.** Ecosystem ecologists have simplified the trophic structure of ecosystems by arranging species into trophic levels based upon the predominant source of their nutrition. A trophic level is determined by the number of transfers of energy from primary producers to that level. As energy is transferred from one trophic level to another, energy is lost due to limited assimilation, respiration by consumers, and heat production. As a result of these losses, the quantity of energy in an ecosystem decreases with each successive trophic level, forming a pyramid-shaped distribution of energy among trophic levels. As losses between trophic levels accumulate, eventually there is insufficient energy to support a viable population at a higher trophic level.

Stable isotope analysis can be used to trace the flow of energy through ecosystems. The ratios of different stable isotopes of important elements such as nitrogen and carbon are generally different in different parts of ecosystems. As a consequence, ecologists can use isotopic ratios to study the trophic structure and energy flow through ecosystems. Stable isotope analysis has helped quantify dietary composition of wild populations and the major sources of energy used by prehistoric human populations.

# Review Questions

1. Population, community, and ecosystem ecologists study structure and process. However, they focus on different natural characteristics. Contrast the important structures and processes in a forest from the perspectives of population, community, and ecosystem ecologists.

2. M. Huston (1994b) pointed out that the well-documented pattern of increasing annual primary production from the poles to the equator is strongly influenced by the longer growing season at low latitudes. The following data are from table 14.10 in Huston. The data cited by Huston are from Whittaker and Likens (1975).

| Forest Type | Annual NPP (t/ha/yr) | Length of Growing Season (months) | Monthly NPP (t/ha/mo) |
|---|---|---|---|
| Boreal forest | 8 | 3 | 2.7 |
| Temperate forest | 13 | 6 | ? |
| Tropical forest | 20 | 12 | ? |

Complete the missing data to compare the *monthly* production of boreal, temperate, and tropical forests. How does this short-term perspective of primary production in high-, middle-, and low-latitude forests compare to an annual perspective? How does the short-term perspective change our perception of tropical versus high-latitude forests?

3. Many migratory birds spend approximately half the year in temperate forests during the warm breeding season and the other half of the year in tropical forest. Given the analyses you made in question 2, which forest appears to be more productive from the perspective of these migratory birds?

4. Field experiments demonstrate that variation in soil fertility influences terrestrial primary production. However, we cannot say that nutrients exert primary control. That role is still attributed to temperature and moisture. Why do ecologists still attribute the main control of terrestrial primary production to temperature and moisture? Consider the difference in primary production between arctic tundra and tropical forest (see fig. 18.2) and the extent to which nutrient additions (Shaver and Chapin 1988) changed primary production in tundra.

5. Shaver and Chapin (1988) pointed out that though the tundra ecosystems they studied consistently increased primary production in response to fertilization, individual species and growth forms showed more variation in response. Some species and growth forms showed no response, while others decreased production on the fertilized plots. What do these differences in response say about using the responses of individual species to predict responses at the ecosystem level? What about the reverse—can we predict the responses of individual species or growth forms from ecosystem-level responses?

6. Compare the pictures of trophic structure that emerged from our discussions of food webs in chapter 17 with those in chapter 18. What are the strengths of each perspective? What are their limitations?

7. Suppose you are studying a community of small mammals that live on the boundary between a riverside forest and a semi-desert grassland. One of your concerns is to discover the relative contributions of the grassland and the forest to the nutrition of small mammals living between the two ecosystems. Design a research program to find out. (Hint: The grassland is dominated by $C_4$ grasses and the forest by $C_3$ plants.)

8. Most of the energy that flows through a forest ecosystem flows through detritus-based food chains, and the detritus consists mainly of dead plant tissues (e.g., leaves and wood). In contrast, most of the energy flowing through a pelagic marine or freshwater ecosystem flows through grazing food chains with phytoplankton constituting the major primary producers. Ecologists have determined that on average, a calorie or joule of energy takes only several days to pass through the pelagic ecosystem but a quarter of a century to pass through the forest ecosystem. Explain.

9. In chapter 17, we examined the influences of keystone species on the structure of communities. In chapter 18 we reviewed trophic cascades. Discuss the similarities and differences between these two concepts. Compare the measurements and methods of ecologists studying keystone species versus those studying trophic cascades.

10. The studies of nutrient limitation of aquatic primary production that we reviewed focused almost entirely on lakes within the temperate zone. Suppose you are an ecologist interested in determining whether primary production in tropical lakes is subject to similar control by nutrient availability. Design a study to find out what controls rates of primary production in tropical lakes. Use all the sources of information at your disposal, including published research, surveys of natural variation, and large- and small-scale experiments.

# Suggested Readings

Aber, J. D. and J. M. Melillo. 1991. *Terrestrial Ecosystems*. Philadelphia: Saunders College Publishing.

> *This book provides an excellent general introduction to ecosystem ecology.*

Bell, T., W. E. Neill, and D. Schluter. 2003. The effect of temporal scale on the outcome of trophic cascade experiments. *Oecologia* 134:578–86.

> *This research reviews a series of works, examining the influence of time on the persistence of trophic cascade effects.*

Carpenter, S. R. and J. F. Kitchell. 1993. *The Trophic Cascade in Lakes*. Cambridge, England: Cambridge University Press.

> *An engaging synthesis of the authors' work on trophic cascades. The book outlines one of the pioneering large-scale experiments in ecology.*

Codispoti, L. A. 1997. The limits to growth. *Nature* 387:237–38.

Falkowski, P. G. 1997. Evolution of the nitrogen cycle and its influence on the biological sequestration of $CO_2$ in the ocean. *Nature* 387:272–75.

> *This pair of papers reviews the evidence for nitrogen, phosphorus, and iron limitation of marine primary production on short and long timescales. Their discussions demonstrate that there is much to learn even in regard to well-established generalizations, such as nutrient limitation of aquatic primary production.*

Golley, F. B. 1993. *A History of the Ecosystem Concept*. New Haven: Yale University Press.

> *This book is a valuable and readable resource for anyone interested in the development of the discipline of ecosystem ecology.*

Gosz, J. R., R. T. Holmes, G. E. Likens, and F. H. Bormann. 1978. The flow of energy in a forest ecosystem. *Scientific American* 238(3):92–102.

> *A summary of one of the most thorough characterizations of energy flow in an ecosystem that has ever been done.*

Jones, J. I. and C. A. Sayer. 2003. Does the fish-invertebrate-periphyton cascade precipitate plant loss in shallow lakes? *Ecology* 84:2155–67.

> *An intriguing paper that reports on a trophic cascade in the littoral zone of lakes, involving invertebrate-feeding fish, invertebrates, periphyton, and littoral zone plants.*

Lindeman, R. L. 1942. The trophic-dynamic aspect of ecology. *Ecology* 23:399–418.

> *A classic paper that set the foundation for research on production and energy flow in ecosystems.*

McNaughton, S. J., R. W. Ruess, and S. W. Seagle. 1988. Large mammals and process dynamics in African ecosystems. *BioScience* 38:794–800.

> *Summary of the influences of large grazing mammals on the savanna grasslands of Africa—a fascinating parallel to the trophic cascades of lakes.*

Meserve, P. L., D. A. Kelt, W. B. Milstead, and J. R. Gutiérrez. 2003. Thirteen years of shifting top-down and bottom-up control. *BioScience* 53:633–46.

*This long-term study of a terrestrial ecosystem in Chile documents a shifting influence of top-down and bottom-up control of community structure that depends on large-scale climate dynamics.*

Pataki, D. E., D. S. Ellsworth, R. D. Evans, M. Gonzalez-Meler, J. King, S. W. Leavitt, G. Lin, R. Matamala, E. Pendall, R. Siegwolf, C. van Kessel, and J. R. Ehleringer. 2003. Tracing changes in ecosystem function under elevated carbon dioxide conditions. *BioScience* 53:805–18.

*The authors provide an excellent review of the use of isotopes to study ecosystem function.*

Peterson, B. J. and B. Fry. 1987. Stable isotopes in ecosystem studies. *Annual Review of Ecology and Systematics* 18:293–320.

*Solid introduction to the use of stable isotopes in ecosystem studies.*

Polis, G. A., A. L. W. Sears, G. R. Huxel, D. R. Strong, and J. Maron. 2000. When is a trophic cascade a trophic cascade? *Trends in Ecology & Evolution* 15:473–75.

*This is a thought-provoking call for consistency in the way ecologists use the term "trophic cascade."*

Schmitz, O. J., P. A. Hamback, and A. P. Beckerman. 2000. Trophic cascades in terrestrial systems: a review of the effects of carnivore removals on plants. *American Naturalist* 155:141–53.

*A review that supports the applicability of the trophic cascade hypothesis to terrestrial as well as aquatic systems.*

# On the Net

Visit this textbook's accompanying website at www.mhhe.com/ecology (click on the book's title) to take advantage of practice quizzing, study/writing tips, timely news articles, and additional URLs for research on the topics in this chapter.

Community Ecology

Field Methods for Studies of Ecosystems

Food Webs

Primary Productivity

Photosynthesis

Savannah

Temperate Forests

# 19

# Nutrient Cycling and Retention

## OUTLINE

xchanges of nutrients between organisms and their environment is one of the essential aspects of ecosystem function. A diatom living in the surface waters of a lake absorbs an ion of phosphate from the surrounding water. It incorporates the phosphate into some of its DNA as it replicates its chromosomes during cell division. A few hours later, one of the diatom's daughter cells is eaten by a cladoceran, an algae-feeding member of the zooplankton. The cladoceran incorporates the phosphate into a molecule of ATP. The cladoceran lives 2 days more and then is eaten by a planktivorous minnow. Within the minnow, the phosphate is combined with a lipid to form a phospholipid molecule in the cell membrane of one of the minnow's neural cells. A few weeks later, the minnow is eaten by a northern pike and the phosphate is incorporated into part of the pike's skeleton. During the following winter, the pike dies and its tissues are attacked by bacteria and fungi that gradually decompose the pike, including its skeleton. During decomposition, the phosphate is dissolved in the surrounding water. The following spring the very same ion of phosphate is taken up by another diatom, completing its cycle through the lake ecosystem (fig. 19.1).

In chapter 18, we saw that energy makes a one-way trip through ecosystems. In contrast, elements such as phosphorus (P), carbon (C), nitrogen (N), potassium (K), and iron (Fe) are used over and over. Elements that are required for the development, maintenance, and reproduction of organisms are called *nutrients*. Ecologists refer to the use, transformation, movement, and reuse of nutrients in ecosystems as **nutrient cycling.** Because of the physiological importance of nutrients, their relative scarcity, and their influence on rates of primary production, nutrient cycling is one of the most significant ecosystem processes studied by ecologists. Three nutrient cycles play especially prominent roles: the *phosphorus cycle,* the *nitrogen cycle,* and the *carbon cycle.* In the next few pages we review the major features of each of these cycles.

## The Phosphorus Cycle

Phosphorus is essential to the energetics, genetics, and structure of living systems. For instance, phosphorus forms part of the ATP, RNA, DNA, and phospholipid molecules. While of great biological importance, phosphorus is not very abundant in the biosphere. Consequently, phosphorus cycling has received a great deal of attention from ecosystem ecologists.

In contrast to carbon and nitrogen, the global phosphorus cycle does not include a substantial atmospheric pool (fig. 19.2). The largest quantities of phosphorus occur in mineral deposits and marine sediments. Sedimentary rocks that are especially rich in phosphorus are mined for fertilizer and applied to agricultural soils. Soil may contain substantial quantities of phosphorus. However, much of the phosphorus in soils occurs in chemical forms not directly available to plants.

Phosphorus is slowly released to terrestrial and aquatic ecosystems through the weathering of rocks. As phosphorus is released from mineral deposits, it is absorbed by plants and recycled within ecosystems. However, much phosphorus is washed into rivers and eventually finds its way to the oceans, where it will remain in dissolved form until eventually finding its way to the ocean sediments. Ocean sediments will be

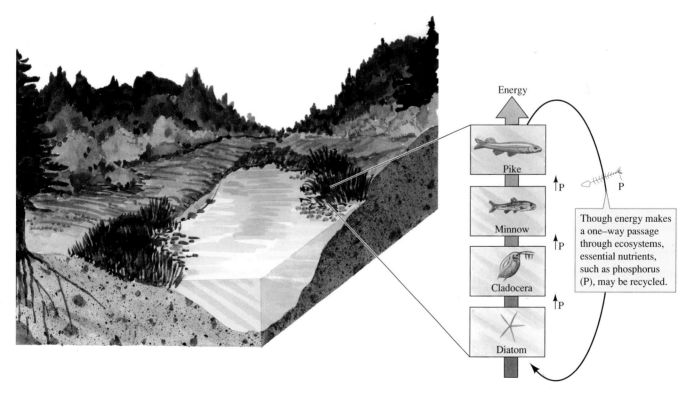

**Figure 19.1** Phosphorus cycle in a lake ecosystem.

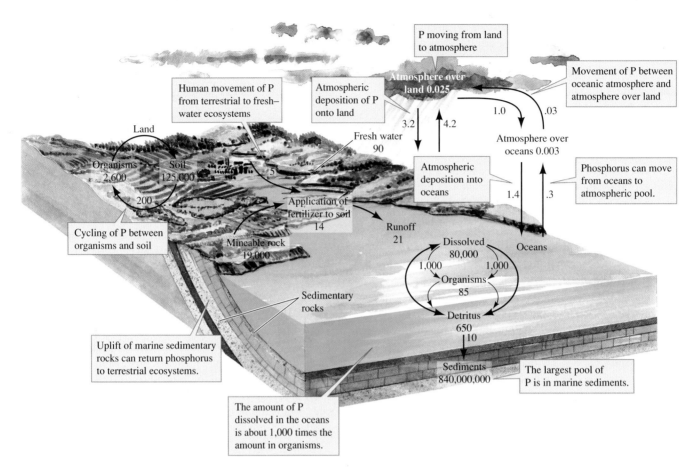

**Figure 19.2** The phosphorus cycle. Numbers are $10^{12}$ g P or fluxes as $10^{12}$ g P per year (data from Schlesinger 1991, after Richey 1983, Meybeck 1982, Graham and Duce 1979).

eventually transformed into phosphate-bearing sedimentary rocks that through geological uplift can form new land. William Schlesinger (1991) points out that the phosphorus released by the weathering of sedimentary rocks has made at least one passage through the global phosphorus cycle.

## The Nitrogen Cycle

Nitrogen is important to the structure and functioning of organisms. It forms part of key biomolecules such as amino acids, nucleic acids, and the porphyrin rings of chlorophyll and hemoglobin. In addition, as we saw in chapter 18, nitrogen supplies may limit rates of primary production in marine and terrestrial environments. Because of its importance and relative scarcity, nitrogen has drawn a great deal of attention from ecosystem ecologists.

Like the carbon cycle, the nitrogen cycle also includes a major atmospheric pool in the form of molecular nitrogen, $N_2$ (fig. 19.3). However, only a few organisms can use this atmospheric supply of molecular nitrogen directly. These organisms, called nitrogen fixers, include (1) the cyanobacteria, or blue-green algae, of freshwater, marine, and soil environments, (2) free-living soil bacteria, (3) bacteria associated with the roots of leguminous plants, and (4) actinomycetes

bacteria, associated with the roots of alders, *Alnus,* and several other species of woody plants.

Because of the strong triple bonds between the two nitrogen atoms in the $N_2$ molecule, nitrogen fixation is an energy-demanding process. During nitrogen fixation, $N_2$ is reduced to ammonia, $NH_3$. Nitrogen fixation takes place under aerobic conditions in terrestrial and aquatic environments, where nitrogen-fixing species oxidize sugars to obtain the required energy. Nitrogen fixation also occurs as a physical process associated with the high pressures and energy generated by lightning. Ecologists propose that all of the nitrogen cycling within ecosystems ultimately entered these cycles through nitrogen fixation by organisms or lightning. There is a relatively large pool of nitrogen cycled in the biosphere but only a small entryway through nitrogen fixation.

Once nitrogen is fixed by nitrogen-fixing organisms, it becomes available to other organisms within an ecosystem. Upon the death of an organism, the nitrogen in its tissues can be released by fungi and bacteria involved in the decomposition process. These fungi and bacteria release nitrogen as ammonium, $NH_4^+$, a process called ammonification. Ammonium may be converted to nitrate, $NO_3^-$, by other bacteria in a process called nitrification. Ammonium and nitrate can be used directly by bacteria, fungi, or plants. The nitrogen in dead organic matter can also be used directly by mycorrhizal fungi, which can be

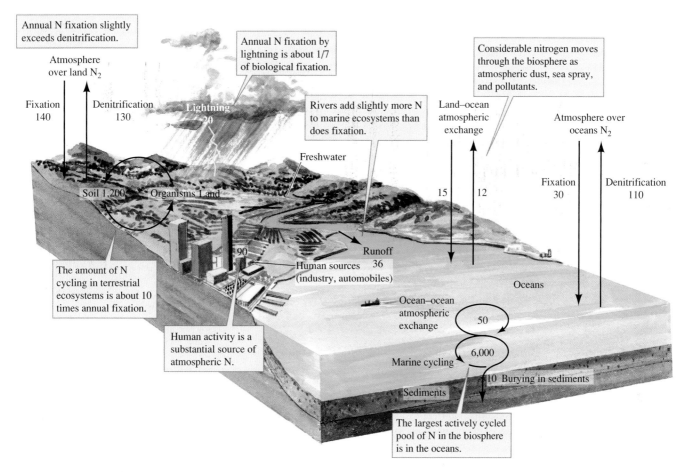

**Figure 19.3** The nitrogen cycle. Numbers represent fluxes as $10^{12}$ g N per year (data from Schlesinger 1991, after Söderlund and Rosswall 1982).

passed on to plants. The nitrogen in bacterial, fungal, and plant biomass may pass on to populations of animal consumers or back to the pool of dead organic matter, where it will be recycled again.

Nitrogen may exit the organic matter pool of an ecosystem through denitrification. Denitrification is an energy-yielding process that occurs under anaerobic conditions and converts nitrate to molecular nitrogen, $N_2$. The molecular nitrogen produced by denitrifying bacteria moves into the atmosphere and can only reenter the organic matter pool through nitrogen fixation. Ecologists estimate that the mean residence time of fixed nitrogen in the biosphere is about 625 years. They estimate that the mean residence time of phosphorus in the biosphere is on the order of thousands of years.

# The Carbon Cycle

Carbon is an essential part of all organic molecules, and, as constituents of the atmosphere, carbon compounds such as carbon dioxide, $CO_2$, and methane, $CH_4$, substantially influence global climate. This connection between atmospheric carbon and climate has drawn all nations of the planet into discussions of the ecology of carbon cycling.

Carbon moves between organisms and the atmosphere as a consequence of two reciprocal biological processes: photosynthesis and respiration (fig. 19.4). Photosynthesis removes $CO_2$ from the atmosphere, while respiration by primary producers and consumers, including decomposers, returns carbon to the atmosphere in the form of $CO_2$. In aquatic ecosystems, $CO_2$ must first dissolve in water before being used by aquatic primary producers. Once dissolved in water, $CO_2$ enters a chemical equilibrium with bicarbonate, $HCO_3^-$, and carbonate, $CO_3^-$. Carbonate may precipitate out of solution as calcium carbonate and may be buried in ocean sediments.

While some carbon cycles rapidly between organisms and the atmosphere, some remains sequestered in relatively unavailable forms for long periods of time. Carbon in soils, peat, fossil fuels, and carbonate rock would generally take a long time to return to the atmosphere. During modern times, however, fossil fuels have become a major source of atmospheric $CO_2$ as humans have tapped into fossil fuel supplies to provide energy for their economic systems.

Ecosystem ecologists study the factors controlling the movement, storage, and conservation of nutrients within ecosystems. You can see broad outlines of these processes in figures 19.2, 19.3, and 19.4. However, much remains to be learned, especially concerning the factors controlling rates of nutrient

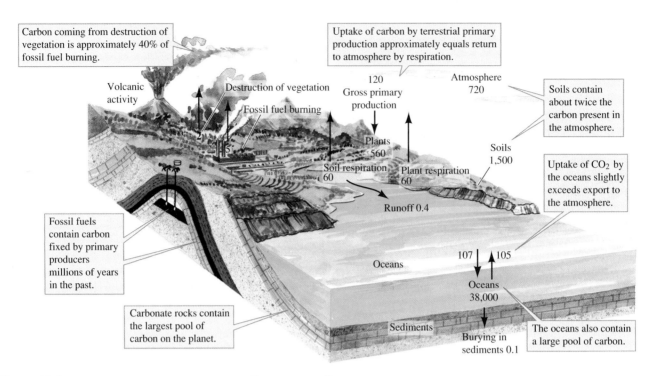

**Figure 19.4** The carbon cycle. Numbers are storage as $10^{15}$ g or fluxes as $10^{15}$ g per year (data from Schlesinger 1991).

exchange within and between ecosystems. Nutrient exchange is substantially affected by the process of decomposition, which is the subject of the next Concept Discussion section.

- **Decomposition rate is influenced by temperature, moisture, and chemical composition of litter and the environment.**
- **Plants and animals can modify the distribution and cycling of nutrients in ecosystems.**
- **Disturbance increases nutrient loss from ecosystems.**

### Rates of Decomposition

**Decomposition rate is influenced by temperature, moisture, and chemical composition of litter and the environment.**

The rate at which nutrients, such as nitrogen and phosphorus, are made available to the primary producers of terrestrial ecosystems is determined largely by the rate at which nutrient

supplies are converted from organic to inorganic forms. This conversion from organic to inorganic form is called **mineralization.** Mineralization takes place principally during **decomposition,** which is the breakdown of organic matter accompanied by the release of carbon dioxide. Consequently, ecosystem ecologists consider decomposition as a key ecosystem process.

The rate of decomposition in terrestrial ecosystems is significantly influenced by temperature, moisture, and the chemical composition of both plant litter and the environment. The chemical characteristics of plant litter that influence decomposition rates include nitrogen concentration, phosphorus concentration, carbon:nitrogen ratio, and lignin content. Ecologists have studied how several of these variables affect rates of leaf decomposition in Mediterranean ecosystems.

## Decomposition in Two Mediterranean Woodland Ecosystems

Antonio Gallardo and José Merino (1993) studied how chemical and physical factors affect rates of decomposition of leaf litter in two Mediterranean woodland ecosystems. Their study sites were located at Doñana Biological Reserve and Monte La Sauceda in southwestern Spain. The mean annual temperature at the two sites differs by only 0.5°C to 16.7°C at Doñana versus 16.2°C at Monte La Sauceda. Both study sites experience Mediterranean climates with wet winters and dry summers (see chapter 2). However, they differ significantly in average annual rainfall. While Doñana Biological Reserve receives about 500 mm of rain annually, Monte La Sauceda receives about 1,600 mm. This

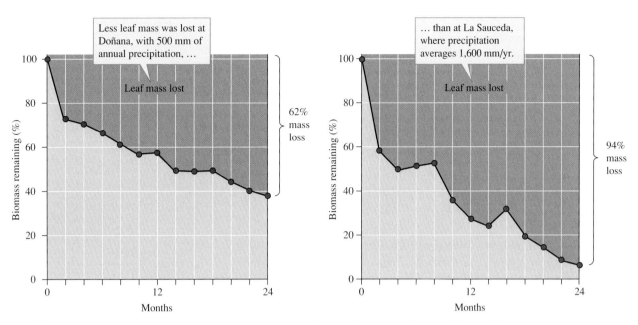

**Figure 19.5** Decomposition of *Fraxinus angustifolia* leaves at wetter and drier sites (data from Gallardo and Merino 1993).

difference in precipitation results from a difference in elevation (see chapter 2). The study site at the Doñana Biological Reserve is located at approximately 20 m elevation, while the elevation at La Sauceda is 432 m. These two sites were ideally suited to study the effects of moisture on rates of decomposition.

Gallardo and Merino also explored the effects of litter chemistry on decomposition by including leaves from nine species of native trees and shrubs that differed significantly in chemical composition. Chemical differences among leaves included differences in concentrations of tannins, lignin, nitrogen, and phosphorus. You may remember from chapter 2 that many of the native plants from areas with a Mediterranean climate produce tough or sclerophylous leaves. Gallardo and Merino also explored the influence of leaf toughness on decomposition rate. They estimated toughness by measuring the amount of force required for a 1.2 mm diameter rod to penetrate the leaves of each species.

Approximately 2 g of air-dried leaves from each of the study species was put into several nylon mesh "litter bags" and placed at the Doñana Biological Reserve and at Monte La Sauceda. The litter bags had a mesh size of 1 mm—small enough to reduce the loss of small leaves, yet large enough to permit aerobic microbial activity and entry of small soil invertebrates. Every 2 months, Gallardo and Merino retrieved litter bags from each study site. They followed this routine for 2 years.

In the laboratory, the researchers measured the mass of leaf tissue remaining in each of 6 to 10 replicate litter bags for each leaf species. Figure 19.5 shows that the amount of leaf mass lost by ash leaves, *Fraxinus angustifolia,* was much higher at Monte La Sauceda. This higher decomposition rate probably reflects the higher precipitation at that site.

Though all types of leaves decomposed faster at Monte La Sauceda, differences in decomposition rates among leaf species were similar at the two sites. For instance, the leaves

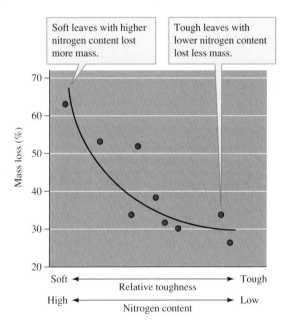

**Figure 19.6** Influence of leaf toughness and nitrogen content on decomposition (data from Gallardo and Merino 1993).

of ash, *Fraxinus,* showed the greatest mass loss at both study sites, while the oak, *Quercus lusitanica,* showed the lowest mass loss at both study sites. Differences in mass loss by the nine plant species reflected differences in the physical and chemical characteristics of their leaves. Gallardo and Merino found that the best predictor of mass loss at the Doñana site was the ratio of toughness to nitrogen content, toughness/%N, and that mass loss was a power function of this ratio:

$$\text{mass} = 545[\text{toughness/N}]^{-0.38}$$

This is the equation for the line shown in figure 19.6. This equation indicates that tougher leaves with lower concentrations of nitrogen decomposed at a lower rate.

The greater mass losses at Monte La Sauceda demonstrate a positive influence of moisture on rates of decomposition, while differences in decomposition rates among leaf species show the influences of chemical composition on decomposition. Even toughness, which is a physical property, is a consequence of chemical composition, especially the concentration of lignin. As we will see in the next example, the ratio of lignin concentration to nitrogen content in leaves is also highly correlated with decomposition rates in temperate forest ecosystems.

# Decomposition in Two Temperate Forest Ecosystems

Jerry Melillo, John Aber, and John Muratore (1982) used litter bags to study leaf decomposition in a temperate forest in New Hampshire. Their study species were beech, *Fagus grandifolia,* sugar maple, *Acer saccharum,* paper birch, *Betula papyrifera,* red maple, *Acer rubrum,* white ash, *Fraxinus americana,* and pin cherry, *Prunus pennsylvanica.* They also compared their results with decomposition of leaves from white pine, chestnut oak, white oak, red maple, and flowering dogwood in a temperate forest in North Carolina.

In both the New Hampshire and North Carolina forests, the researchers found a negative correlation between the leaf mass remaining after 1 year of decomposition and the ratio of lignin to nitrogen concentrations in leaves, % lignin:% N. In other words, leaves with higher lignin:nitrogen ratios lost less mass during the year-long study. As you can see in figure 19.7, the amount of leaf mass remaining was lower at the North Carolina site than at the New Hampshire site. What factors were responsible for these higher rates of decomposition at the North Carolina site? Melillo and his colleagues suggested that higher nitrogen availability in the soils at the North Carolina site may contribute to the higher rates of decomposition observed there. However, higher temperatures at the North Carolina site may also contribute to higher decomposition rates.

Studies in both temperate and Mediterranean regions suggest that rates of decomposition are positively correlated with temperature and moisture. Can we combine these two factors into one? In chapter 18, we reviewed how ecologists studying the effect of climate on terrestrial primary production combined temperature and precipitation into a single measure called actual evapotranspiration, or AET. V. Meentemeyer (1978) analyzed the relationship between AET and decomposition and found a significant positive relationship (fig. 19.8).

If decomposition rates increase with increased evapotranspiration, how would you expect rates of decomposition in tropical and temperate ecosystems to compare? As you probably predicted, rates of decomposition are generally higher in tropical ecosystems. The average annual mass loss in tropical forests shown in figure 19.9 is 120%, or three times the average rate measured in temperate forests. These higher rates probably reflect the effects of higher AET in tropical forests and indicate complete decomposition in less than a year.

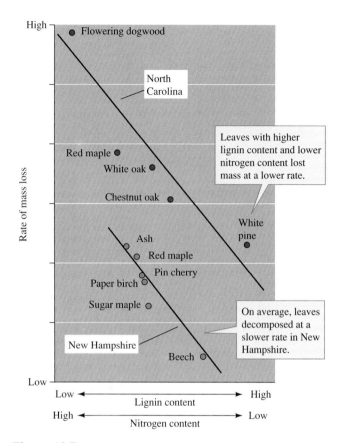

**Figure 19.7**  Influence of lignin and nitrogen content of leaves on decomposition (data from Melillo, Aber, and Muratore 1982).

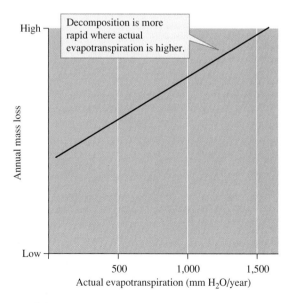

**Figure 19.8**  Relationship between actual evapotranspiration and decomposition (data from Meentemeyer 1978).

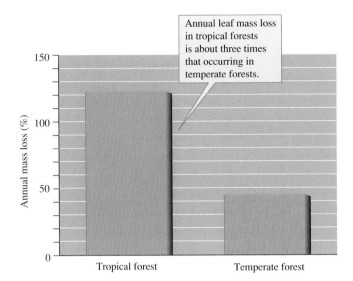

**Figure 19.9** Decomposition in tropical and temperate forests (data from Anderson and Swift 1983).

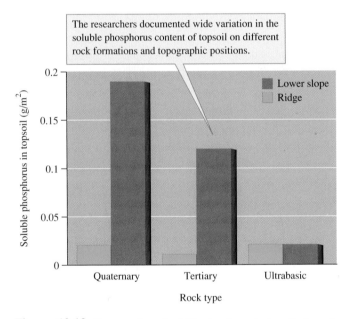

**Figure 19.10** Concentrations of soluble phosphorus in topsoils formed on three rock types and at two topographic positions in Borneo (data from Takyu, Aiba, and Kitayama 2003).

Soil nutrient content has also been shown to have a strong positive effect on rates of nutrient cycling in tropical forests. Three forest ecologists, Masaaki Takyu, Shin-Ichiro Aiba, and Kanehiro Kitayama, took advantage of natural variation in nutrient content on different geologic formations and different topographic situations to explore the factors influencing tropical rain forest functioning in Borneo (Takyu, Aiba, and Kitayama 2003). Takyu, Aiba, and Kitayama established research sites on ridges, which tend to have soils with lower nutrient content, and lower slopes, where nutrient content is higher, on three different rock types. The younger Quaternary sedimentary rock in their study area was approximately 30,000 to 40,000 years old and

soils developing on these rocks tended to have higher nutrient content compared to soils on the other rock types in the study, especially on lower slopes (fig. 19.10). Two older rock types were both approximately 40 million years old. One, a Tertiary sedimentary rock, supported soils that were considerably more fertile than the other, a Tertiary ultrabasic rock.

Takyu, Aiba, and Kitayama's study clearly demonstrated the influence of soil fertility on rates of decomposition and nutrient cycling. Because all study sites were at approximately the same elevation and all were on south-facing aspects, the research team was able to isolate the influences of geologic conditions, especially soil characteristics. Takyu, Aiba, and Kitayama found higher rates of aboveground net primary production, higher rates of litter fall, and higher rates of decomposition on sites with higher concentrations of soluble phosphorus in topsoil, particularly on soils formed on the lower slopes of Quaternary and Tertiary rock formations. These results show that while climate may have a primary influence on decomposition rates, within climatic regions nutrient availability has an ecologically significant effect on decomposition and nutrient cycling rates.

In summary, decomposition in terrestrial ecosystems is influenced by moisture, temperature, soil fertility, and the chemical composition of litter, especially the concentrations of nitrogen and lignin. With the obvious exception of moisture, these factors also influence decomposition rates in aquatic ecosystems, which we examine next.

# Decomposition in Aquatic Ecosystems

Jack Webster and Fred Benfield (1986) reviewed what was known about the decomposition of plant tissues in freshwater ecosystems. Among the most important variables that emerged from their analysis were leaf species, temperature, and nutrient concentrations in the aquatic ecosystem.

Webster and Benfield summarized the rates of leaf breakdown for 596 types of woody and nonwoody plants decaying in aquatic ecosystems and found that the average daily breakdown rate varied more than tenfold. As in terrestrial ecosystems, the chemical composition of litter significantly influences rates of decomposition in aquatic ecosystems.

Mark Gessner and Eric Chauvet (1994) studied the influence of litter chemistry on the rate of leaf decomposition. Their study site was a stream in the French Pyrenees. They included the leaves of several species of trees: alder, *Alnus glutinosa,* ash, *Fraxinus excelsior,* beech, *Fagus sylvatica,* wild cherry, *Prunus avium,* hazel, *Corylus avellana,* sycamore, *Platanus hybrida,* and evergreen oak, *Quercus ilex.* The researchers found that leaves with a higher lignin content decomposed at a slower rate (fig. 19.11). What causes this? Gessner and Chauvet showed that a higher lignin content inhibits colonization of leaves by fungi, the main organisms responsible for decomposition of leaves in streams.

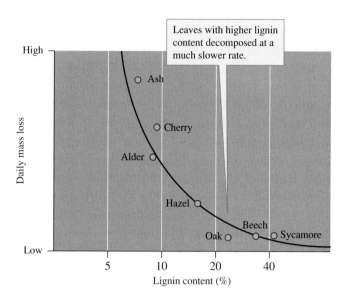

**Figure 19.11** Lignin content of leaves and decomposition in an aquatic ecosystem (data from Gessner and Chauvet 1994).

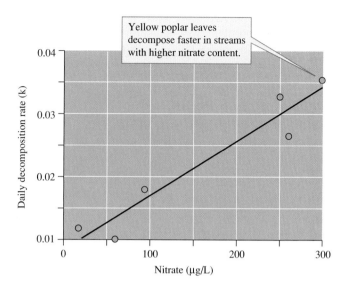

**Figure 19.12** Stream nitrate and decomposition of *Liriodendron* leaves (data from Suberkropp and Chauvet 1995).

The nutrient content of stream water can also influence rates of decomposition. Keller Suberkropp and Eric Chauvet (1995) studied how water chemistry affects rates of leaf decomposition, using the leaves of yellow poplar, *Liriodendron tulipifera*. They placed the leaves in several temperate zone streams differing in water chemistry. The leaves in their litter bags decayed as an exponential function of time, following the relationship:

$$m_t = m_0 e^{-kt}$$

where:

$m_t$ is the mass of leaves at time $t$

$m_0$ is the initial mass of leaves

$e$ is the base of the natural logarithms

$t$ is time in days

$k$ is the daily rate of mass loss

We can use the constant $k$ as an index of decay rate under particular environmental conditions. Suberkropp and Chauvet found that $k$ varied significantly among their study streams. It turned out that leaves decayed faster, that is, had higher $k$, in streams with higher concentrations of nitrates (fig. 19.12). This result is consistent with the suggestion by Melillo's research team that higher rates of decomposition at one of their study sites was due to higher availability of soil nitrogen.

Amy Rosemond and several colleagues of the University of Georgia conducted a similar study in streams draining a tropical forest in Costa Rica (Rosemond et al. 2002). They placed leaves of *Ficus insipida,* a tree that commonly grows along streams in Central America, at 16 stream sites that varied substantially in phosphorus concentration. Leaf decompo-

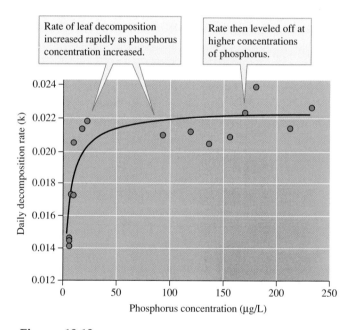

**Figure 19.13** Phosphorus concentration of stream water and rate of decomposition of *Ficus insipida* leaves in tropical streams (courtesy of Amy Rosemond).

sition rate increased markedly as phosphorus concentration increased to about 20 μg per liter, after which decomposition rate leveled off (fig. 19.13).

As in terrestrial ecosystems, litter chemistry and nutrient availability in the environment affect decomposition rates in aquatic ecosystems. The patterns discussed in this section emphasize the role played by the physical and chemical environment in the process of decomposition. As we shall see in the next Concept Discussion, however, animals and plants can also significantly affect the nutrient dynamics of ecosystems.

| Information |
| Questions |
| Hypothesis |
| Predictions |
| Testing ✓ |

*Investigating the Evidence*

## Assumptions for Statistical Tests

In chapter 18 (p. 452) we compared samples from two populations using the *t*-test to judge whether there was a statistically significant difference between the populations. While the *t*-test is one of the most valuable tools for comparisons of pairs of samples, like any tool there are situations where it is appropriate to use a *t*-test and others where it is not. The *t*-test is based on a number of assumptions, as are other statistical tests.

The first assumption of the *t*-test is that each of the samples is drawn from a population with a **normal distribution.** We first considered this assumption in chapter 3 (p. 60), when we discussed the assumptions underlying calculating the sample mean as a way of estimating the average, or typical, in a population. As we saw, the sample mean was appropriate for one population (a sample of seedling heights) but not appropriate for another (a sample of stream invertebrate densities) that we considered. A normal distribution is also assumed for calculating 95% confidence intervals (p. 430) and for regression analysis (p. 213).

Let's consider the assumption of a normal distribution in a bit more detail. A normal distribution has a particular shape. As shown in figure 1, a normal distribution is bell-shaped and proportioned in such a way that predictable percentages of the observations, or measurements, will fall within one, two, or three standard deviations of the mean (see chapter 6, p. 162).

If the characteristic of interest is not normally distributed, then we cannot be certain, for instance, that a 95% confidence interval will be accurate or that two sample means compared using a *t*-test are statistically different. Another requirement of the *t*-test is that the populations being compared have equal variances. Fortunately, many of the kinds of measurements made by ecologists, such as weights of individuals, body lengths or lengths of appendages, running speeds, or rates of photosynthesis, have normal distributions. In addition, the fit of measurements to a normal distribution does not have to be exact. For example, the *t*-test will produce reliable results if the distribution of measurements is fairly symmetrical around the mean. Also, the *t*-test can give reliable results with some differences in variances, as long as the sizes of samples being compared are similar.

However, there are some important attributes of ecological systems that are not distributed normally. These include population densities (numbers per unit area) of plants or animals, proportions of different species in a community, percentage of time that an animal spends in different activities, and exponential rates of litter decay. One way to analyze such data is to use statistical methods that do not assume a normal distribution. We entered this area in chapter 3 (p. 60) where we discussed the sample median. We will explore this area further in chapters 20 to 22.

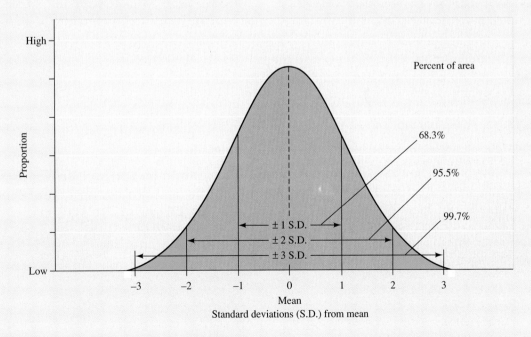

**Figure 1**  Example of a normal distribution, showing the percentages of observations included within one, two , or three standard deviations of the mean.

CONCEPT DISCUSSION

### Organisms and Nutrients

**Plants and animals can modify the distribution and cycling of nutrients in ecosystems.**

How much do particular organisms affect ecosystem processes? With the threats of global change and mass extinction looming ever larger, this is one of the most important questions of our time. We do not yet know enough to offer a satisfying answer to this important question, but the information we do possess indicates that individual plant and animal species can substantially influence the distribution and dynamics of nutrients within ecosystems.

## Nutrient Cycling in Streams

Before we consider how stream animals influence the dynamics of nutrient turnover in streams, we have to consider some special features of this ecosystem. As we saw in chapter 3, the most distinctive feature of stream and river ecosystems is water flow. Jack Webster (1975) was the first to point out that because nutrients in streams are subject to downstream transport, there is little nutrient cycling in one place. Water currents move nutrients downstream. Webster suggested that rather than a stationary cycle, stream nutrient dynamics are better represented by a spiral. He coined the term **nutrient spiraling** to describe stream nutrient dynamics (fig. 19.14).

As an atom of a nutrient completes a cycle within a stream, it may pass through several ecosystem components such as an algal cell, an invertebrate, a fish, or a detrital frag-

ment. Each of these ecosystem components may be displaced downstream by current and therefore contribute to nutrient spiraling. The length of stream required for an atom of a nutrient to complete a cycle is called the **spiraling length.** Spiraling length is related to the rate of nutrient cycling and average velocity of nutrient movement downstream. Denis Newbold and his colleagues (1983) represented spiraling length, $S$, as:

$$S = VT$$

where $V$ is the average velocity at which a nutrient atom moves downstream and $T$ is the average time for a nutrient atom to complete a cycle. If velocity, $V$, is low and the time to complete a nutrient cycle, $T$, is short, nutrient spiraling length is short. Where spiraling lengths are short, a particular nutrient atom may be used many times before it is washed out of a stream system.

The tendency of an ecosystem to retain nutrients is called **nutrient retentiveness.** In stream ecosystems, retentiveness is inversely related to spiraling length. Short spiraling lengths are equated with high retentiveness and long spiraling lengths with low retentiveness. Any factors that influence spiraling length affect nutrient retention by stream ecosystems.

### Stream Invertebrates and Spiraling Length

Nancy Grimm (1988) showed that aquatic macroinvertebrates significantly increase the rate of nitrogen cycling in Sycamore Creek, Arizona. Streams in the arid American Southwest support high levels of macroinvertebrate biomass. Grimm estimated invertebrate population densities as high as 110,000 individuals per square meter and dry biomass as high as 9.62 g per square meter. More than 80% of macroinvertebrate biomass in Sycamore Creek was made up of species that feed on

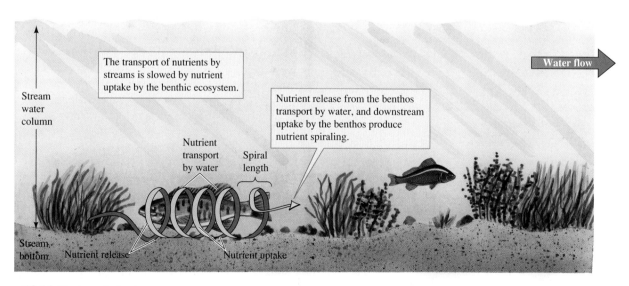

**Figure 19.14** Nutrient spiraling in streams.

small organic particles, a feeding group that stream ecologists call *collector-gatherers.* The collector-gatherers of Sycamore Creek are dominated by two families of mayflies, Baetidae and Tricorythidae, and one family of Diptera, Chironomidae.

Grimm quantified the influence of macroinvertebrates on the nitrogen dynamics in the creek, where primary production is limited by nitrogen availability. She developed nitrogen budgets for stream invertebrates, mainly insect larvae and snails, by quantifying their rates of nitrogen ingestion, egestion (defecation), excretion, and accumulation during growth. By combining these rates with her estimates of macroinvertebrate biomass, Grimm was able to estimate the contribution of macroinvertebrates to the nutrient dynamics of the Sycamore Creek ecosystem.

Her measurements indicated that macroinvertebrates could play an important role in nutrient spiraling. To determine whether they play such a role, what information do we need? We need to know how much of the available nitrogen they ingest. If invertebrates ingest a large proportion of the nitrogen pool, then their influences on nitrogen spiraling may be substantial. Grimm measured the nutrient retention of Sycamore Creek as the daily difference between nitrogen inputs and outputs in her study area. These measurements showed an average rate of retention of nitrogen as 250 mg per square meter per day.

Grimm set this rate of retention as 100% and then expressed her estimates of flux rates as a percentage of this total (fig. 19.15). Nitrogen ingestion rates by macroinvertebrates averaged about 131%. How can ingestion rates be greater than 100%? What this means is that the collector-gatherers in the study stream reingest nitrogen in their feces. This is a well-known habit of detritivores, many of which gain more nutritional value from their food by processing it more than once.

Grimm suggests that rapid recycling of nitrogen by macroinvertebrates may increase primary production in Sycamore Creek. Stream macroinvertebrates excreted and recycled 15% to 70% of the nitrogen pool as ammonia. By their high rates of feeding on the particulate nitrogen pool and their high rates of excretion of ammonia, the macroinvertebrates of the creek reduce the $T$ in the equation for spiral length, $S = VT$. This effect coupled with the 10% of nitrogen tied up in macroinvertebrate biomass, which reduces $V$, reduces the nitrogen spiral length and increases the nutrient retentiveness of Sycamore Creek.

## The Effect of Species Richness on Litter Breakdown

In a laboratory study of leaf breakdown by stonefly leaf shredders (see chapter 3, p. 76), Micael Jonsson and Björn Malmqvist (2003) demonstrated that rates of leaf breakdown, a key process in nutrient cycling, are faster in the presence of more shredder species. Jonsson and Malmqvist collected 6 species of shredder stoneflies from streams near Umeå, in north-central Sweden and maintained them in aquaria with the leaf litter they ate until the experiments began. The leaves used in the experiments were all of alder, *Alnus incana.* Alder leaves contain relatively high concentrations of nitrogen and are highly palatable to detritivores.

To examine how shredder diversity might affect rate of leaf breakdown, Jonsson and Malmqvist created assemblages of 1, 3, 4, or 6 species of stoneflies. In the treatment with a single species, Jonsson and Malmqvist set up six replicated populations of each species for a total of 36 single-species trials. The three-, four-, and six-species treatments were replicated 10 times each. Each replicate for these treatments was created by drawing 3, 4, or 6 stonefly species at random from the pool of 6 species. To keep shredder densities constant across these trials, the three-, four-, and six-species treatments included, respectively, 4, 3, or 2 individuals per species for a total of 12 stoneflies. The stoneflies in each experimental unit were provided with a measured mass of leaves and monitored for 48 days.

The results obtained by Jonsson and Malmqvist (fig. 19.16) clearly indicate a positive effect of increased number of shredder species on rate of leaf mass loss. The increase in rate of mass loss in the three-, four-, and six-species treatments ranged from 21% to 27%. The effect, however, leveled off at 3 species. Still, in

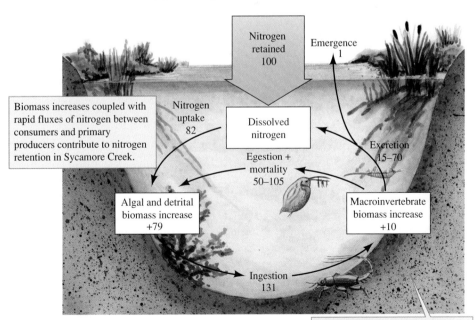

**Figure 19.15** Nitrogen fluxes in Sycamore Creek, Arizona (data from Grimm 1988).

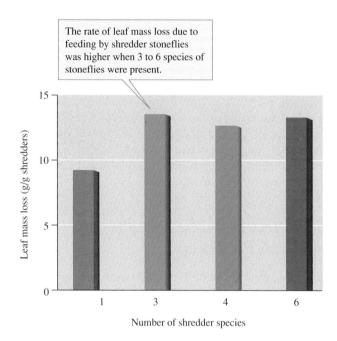

**Figure 19.16** Relationship between number of stonefly shredder species in laboratory aquaria and leaf mass loss during a 48-day experiment (data from Jonsson and Malmqvist 2003).

this high latitude stream system, consumer species richness appears to have a positive effect on rates of organic matter processing and nutrient cycling.

As we shall see in the next section, consumers can also substantially affect nutrient cycling in terrestrial ecosystems.

# Animals and Nutrient Cycling in Terrestrial Ecosystems

As we saw in chapter 16, burrowing animals, such as prairie dogs and pocket gophers, affect local plant diversity. These burrowers also alter the distribution and abundance of nitrogen within their ecosystems.

Pocket gophers can significantly affect their ecosystems because, as we discussed in chapter 16, their mounds may cover as much as 25% to 30% of the ground surface. This deposition represents a massive reorganization of soils and a substantial energy investment, since the cost of burrowing is 360 to 3,400 times that of aboveground movements. Estimates of the amount of soil deposited in mounds by gophers range from 10,000 to 85,000 kg per hectare per year.

Nancy Huntly and Richard Inouye (1988) found that pocket gophers altered the nitrogen cycle at the Cedar Creek Natural History Area in Minnesota by bringing nitrogen-poor subsoil to the surface (fig. 19.17). The result was greater horizontal heterogeneity in nitrogen availability and greater heterogeneity in light penetration. These effects on the nitrogen cycle in prairie ecosystems help explain some of the positive influences that pocket gophers have on plant diversity.

April Whicker and James Detling (1988) found that the feeding activities of prairie dogs also influence the distribution

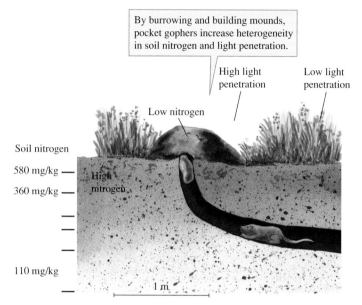

**Figure 19.17** Pocket gophers and ecosystem structure (data from Huntly and Inouye 1988).

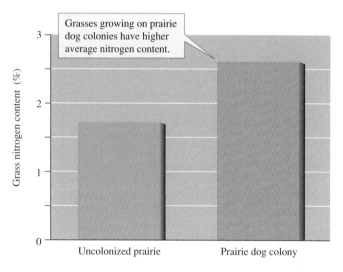

**Figure 19.18** Early season nitrogen content of grasses growing on uncolonized prairie and on a young prairie dog colony (data from Whicker and Detling 1988).

of nutrients within prairie ecosystems. This should not be surprising since these researchers estimate that prairie dogs consume or waste 60% to 80% of the net annual production from the grass-dominated areas around their colonies. One result of this heavy grazing is that aboveground biomass is reduced by 33% to 67% and the young grass tissue that remains is higher in nitrogen content (fig. 19.18). This higher nitrogen content may influence the behavior of bison, which spend a disproportionate amount of their time grazing near prairie dog colonies.

Bison and other large herbivorous mammals, such as moose and African buffalo, may also influence the cycling of nutrients within terrestrial ecosystems. Sam McNaughton and his colleagues (1988) report a positive relationship between grazing intensity and the rate of turnover of plant biomass in the Serengeti Plain of eastern Africa. Figure 19.19 suggests

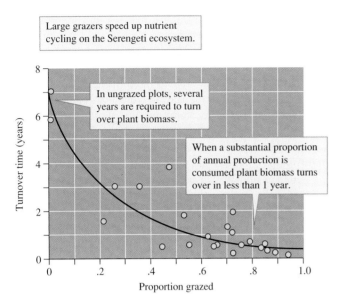

**Figure 19.19** Effect of grazing on time required for turnover of plant biomass on the Serengeti ecosystem (data from McNaughton, Ruess, and Seagle 1988).

that increased grazing increases the rate of nutrient cycling. How will nutrient cycling occur in the absence of grazing by these large herbivores? Without grazing, nutrient cycling occurs more slowly through decomposition and through the feeding of small herbivores.

McNaughton has built a model for nutrient cycling in grasslands that distinguishes between decomposition and grazing. He proposes that having a grazer pathway in an ecosystem speeds up the rate of nutrient cycling. Consequently, the large herbivores of the Serengeti are functionally similar to the collector-gatherer invertebrates of streams (see fig. 19.15). Both groups of organisms speed up the rate of nutrient cycling in their ecosystems. As we shall see, plants can also have substantial influences on the nutrient dynamics of ecosystems.

# Plants and the Nutrient Dynamics of Ecosystems

Plants are not simply the passive recipients of influences from the physical environment or from animals and microbes. Introduced plant species show clearly how plants modify ecosystems. Several species of introduced plants are currently modifying the fynbos of South Africa.

## *The South African Fynbos*

The fynbos is a temperate shrubland or woodland with a Mediterranean climate known for its exceptionally high plant diversity (see chapter 2) and low soil fertility. Two species of *Acacia* were introduced into this environment to stabilize shifting sand dunes in the coastal lowlands. One of these species, *Acacia saligna,* has invaded lowland sand plains within the fynbos, where it has become a dominant plant.

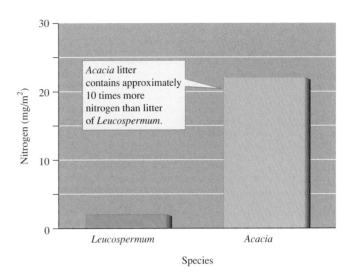

**Figure 19.20** Nitrogen content of plant litter under a native shrub, *Leucospermum purile,* and an introduced shrub, *Acacia saligna,* in the South African fynbos (data from Witkowski 1991).

*A. saligna* grows as a small tree or shrub 3 to 10 m in height, tall enough to overgrow the shorter native fynbos vegetation. In the process of invading the fynbos, *Acacia* has altered the area's nutrient dynamics in several ways, including increasing rates of decomposition, elevating rates of plant litter production, and elevating the nitrogen content of soils.

Edward Witkowski (1991) demonstrated these effects by comparing nutrient dynamics under the canopy of a native shrub, *Leucospermum purile,* in the family Protaceaceae with the nutrient dynamics under the canopy of the introduced *Acacia.* He explored several aspects of nutrient cycling, including soil nutrient content, litter production and decomposition, and nutrient return to soils.

Witkowski found that the amounts of litter produced by *Acacia* and *Leucospermum* were approximately equal. However, the nutrient content of the litter produced by these two species differed significantly (fig. 19.20). In addition, Witkowski found that the concentration of nitrogen in soil was consistently higher under *Acacia.*

Where did the higher levels of nitrogen found under the canopy of *Acacia* come from? It is likely that nitrogen fixation by the plant is at least partly responsible. *Acacia* species are members of the legume family, which form mutualistic relationships with nitrogen-fixing bacteria. Through its bacterial mutualists, *Acacia* can use atmospheric $N_2$, a source of nitrogen unavailable to the native *Leucospermum.* By producing litter rich in nitrogen, *Acacia* is gradually changing the quantity and dynamics of nitrogen in the fynbos ecosystem.

## *An Introduced Tree and Hawaiian Ecosystems*

The Hawaiian Islands have been intensely invaded by both animals and plants. The native flora included approximately 1,200 species, of which over 90% were found nowhere else on earth.

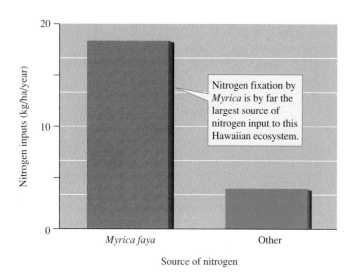

**Figure 19.21**   Nitrogen enrichment of Hawaiian ecosystems by an introduced tree, *Myrica faya* (data from Vitousek and Walker 1989).

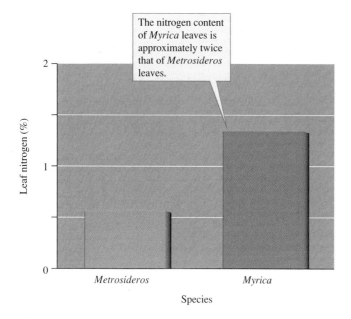

**Figure 19.22**   Leaf nitrogen content of a native tree, *Metrosideros polymorpha,* and an introduced tree, *Myrica faya* (data from Vitousek and Walker 1989).

Humans have added approximately 4,600 exotic species to the Hawaiian flora. As you might expect, many of these exotic species significantly affect native populations, communities, and ecosystems. Peter Vitousek and Lawrence Walker (1989) found that an invading nitrogen-fixing tree, *Myrica faya,* is altering the nitrogen dynamics of ecosystems in Hawaii.

*Myrica* is a native of the Canary, Azores, and Madeira Islands in the North Atlantic, where it grows on lava flows and volcanic soils. The tree was introduced to Hawaii in the late 1800s as an ornamental or medicinal plant and then, in the 1920s and 1930s, was planted widely by the Hawaiian Territorial Department of Forestry for watershed reclamation. The species now occurs on five of the largest of the Hawaiian Islands. *Myrica* in Hawaii is highly invasive, quickly spreading over thousands of hectares. It has the potential to modify the nutrient dynamics of Hawaiian ecosystems because it maintains a mutualistic relationship with a nitrogen-fixing fungus.

Vitousek and Walker studied the influence of *Myrica* on ecosystem properties in the Hawaii Volcanoes National Park. Their study site was located near the summit of Kilauea Volcano at an elevation of 1,100 to 1,250 m. In one of their study areas, *Myrica* is codominant with the native tree *Metrosideros polymorpha.*

Vitousek and Walker estimated the rate of nitrogen fixation by *Myrica* in order to assess the rate at which the species adds nitrogen to the ecosystem. They put the contributions of nitrogen fixation by the tree into context by also estimating the other nitrogen inputs to the ecosystem. They identified several indigenous sources of nitrogen fixation that included lichens, bacteria, and algae. Another major source of nitrogen to their study ecosystem was nitrogen in precipitation.

In the area where *Myrica* is codominant with the native tree *Metrosideros,* nitrogen fixation by *Myrica* is clearly the single largest source of nitrogen input to the ecosystem (fig. 19.21). As you might expect, the leaf tissues of *Myrica* contain a higher concentration of nitrogen than those of *Metrosideros* (fig. 19.22). The higher nitrogen content of the leaves is associated with a higher rate of decomposition and greater nitrogen release during decomposition. The result of these processes is increased nitrogen content in the soils associated with *Myrica.*

In summary, invasive plants can substantially alter the nutrient dynamics of ecosystems. Vitousek and Walker suggest that the effects of invading species on ecosystem properties may offer the opportunity to trace the influences of individual species through the ecosystem and back to the population. Native species undoubtedly have similar effects. However, the effects of invading species are often more apparent, particularly where they are changing fundamental characteristics of entire ecosystems.

The introduction of exotic plants to ecosystems may be considered as a type of disturbance. In the next Concept Discussion section, we review how disturbance increases rates of nutrient loss from ecosystems.

## CONCEPT DISCUSSION

### Disturbance and Nutrients

**Disturbance increases nutrient loss from ecosystems.**

In the previous Concept Discussion we saw how macroinvertebrates may increase nutrient retention by stream ecosystems. In this section, we consider evidence that disturbance reduces nutrient retention.

# Disturbance and Nutrient Loss from the Hubbard Brook Experimental Forest

In chapter 1 (see p. 8), we reviewed the classic study by Gene Likens and Herbert Bormann and their colleagues (1970) that demonstrated biological influences on nutrient loss from forested ecosystems. By clear-cutting the forest on one of their study basins these researchers revealed the role of vegetation in regulating nutrient losses from a temperate forest ecosystem.

The increased rates of nutrient loss following forest cutting were dramatic. The connection between forest cutting and increased nutrient output is shown clearly by plotting nutrient concentrations in the streams draining experimental and control stream basins. Figure 19.23 shows the highly significant increases in nitrate losses following deforestation.

What message can we take away from this pioneering experiment? The important point is that vegetation in the northern hardwoods ecosystem significantly affects rates of nutrient loss by the ecosystem. When Likens and Bormann cut the forest, they removed those biological controls. Similar controls have been demonstrated in other ecosystems.

Peter Vitousek and his colleagues (1979, 1982) studied the effects of disturbance and environmental conditions on rates of nitrogen loss from forest ecosystems. Their study sites included 8 deciduous forests and 11 coniferous forests growing in conditions ranging from acidic to neutral soils, and from wet climates in Washington and Oregon to a dry site in New Mexico. The study also included cold subalpine forests in New Hampshire, New Mexico, and Washington. A major goal of the research team was to identify the factors increasing the risk of nitrogen loss in response to disturbance. Including forests from a wide range of environmental conditions provided the potential to identify the site-specific factors that may promote either loss or retention of nitrogen in the face of disturbance.

Vitousek and his colleagues emphasized nitrogen losses for several reasons. First, as we saw in chapter 18, nitrogen can limit the rate of primary production in terrestrial ecosystems. In addition, as we saw in the experiments of Likens and Bormann, losses of nitrogen are often greater than losses of other nutrients.

Vitousek and his team created "trenched" plots by digging trenches 1 m deep around several 1 m² plots and lining the trenches with plastic to prevent the regrowth of tree roots into the study plots. The researchers also cut any plants within the plots and kept them free of vegetation throughout the study by weeding periodically. They studied the responses of the study plots to trenching by sampling soil water for nitrogen. Any nitrogen occurring in soil water represented potential loss of nitrogen from the forest ecosystem to groundwater, which might eventually find its way to stream water.

What were the similarities and differences between the experiment by Vitousek and his colleagues and that by Likens and Bormann? A major difference was that the trenching experiment was conducted on a small scale and did not disturb the forest canopy. An intact canopy kept the input of solar energy to the forest floor constant. Therefore temperature changes were minimized. Two similarities were that both experiments stopped most uptake of nutrients by plants and both increased the concentrations of nitrate in soil water.

Vitousek and his team found that trenching increased the concentrations of nitrate in soil water up to 1,000 times but that the amount of nitrate in soil water varied greatly from one forest to another. The lowest losses of nitrate occurred in the dry forest site in New Mexico and in the cold alpine environments of New Mexico and Washington. Generally, nitrate losses were greatest at the sites with combinations of temperature and moisture that promoted rapid decomposition.

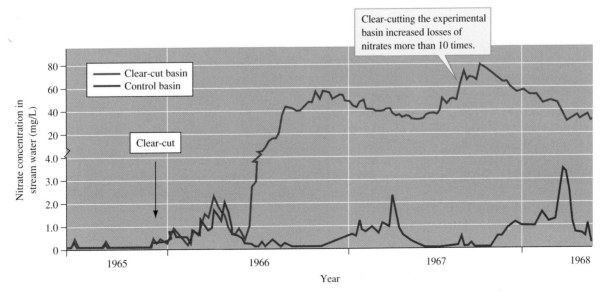

**Figure 19.23** Deforestation and nitrate loss from a deciduous forest ecosystem (data from Likens et al. 1970).

What do these results suggest about the role of vegetation in preventing losses of nitrogen from forest ecosystems? Over the short term, at least, uptake by vegetation is most important in ecosystems with fertile soils and warm, moist conditions during the growing season. Since these are conditions that would promote high rates of plant growth, trees at these sites should be able to rapidly reestablish control of nitrogen loss following disturbance.

Now let's consider how disturbance affects nutrient losses from stream ecosystems, where nutrient loss appears to be highly episodic and associated with disturbance during flooding.

# Flooding and Nutrient Export by Streams

How do the nutrient dynamics of stream ecosystems respond to variations in streamflow? Judy Meyer and Gene Likens (1979) examined the long-term dynamics of phosphorus in Bear Brook, a stream ecosystem in the Hubbard Brook Experimental Forest. They found that during periods of average flow, the ratio of annual phosphorus inputs to exports varied from 0.56 to 1.6 and that the balance depended upon stream discharge. Meyer and Likens found that exports were highly episodic and associated with periods of high flow.

How did Meyer and Likens determine the phosphorus dynamics of Bear Brook ecosystem? They measured the geological and meteorological inputs and the geological exports of phosphorus in the stream, which they divided into three size fractions: (1) dissolved phosphorus, $< 0.45\ \mu m$, (2) phosphorus associated with fine particles, $0.45\ \mu m$ to 1 mm, and (3) phosphorus associated with coarse particles, $> 1$ mm.

Meyer and Likens inventoried the movement and storage of phosphorus in the Bear Brook ecosystem. They measured the inputs of dissolved phosphorus at 12 seeps (areas of groundwater input) along the length of the stream. They also measured meteorological inputs, which included precipitation and forest litter falling or blowing into the stream. The only significant export of phosphorus in the ecosystem was transport with streamflow. The researchers measured the amount of particulate matter transported by the Bear Brook ecosystem by collecting the organic matter deposited behind the weir (a small dam used to measure streamflow) on the stream and by collecting organic matter captured by nets set at seven sites. They estimated the amount of organic matter stored by removing all the organic matter in 42 randomly located 1 m² areas of stream bottom.

During 1974 to 1975, Meyer and Likens estimated an almost exact balance between inputs and exports of phosphorus, with approximately 1,250 mg of phosphorus per square meter of input and approximately 1,300 mg of phosphorus per square meter of output. Despite a balance between input and output, their data indicated significant transformation of phosphorus-size fractions. Phosphorus inputs to Bear Brook were almost evenly divided between dissolved (28%), fine particulate (37%), and coarse particulate (35%) fractions. However, 62% of exports were fine particulates. Clearly phys-

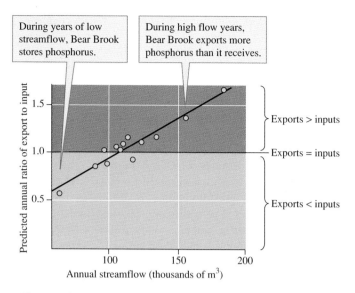

**Figure 19.24**  Annual streamflow and ratio of phosphorus export to input in Bear Brook, New Hampshire (data from Meyer and Likens 1979).

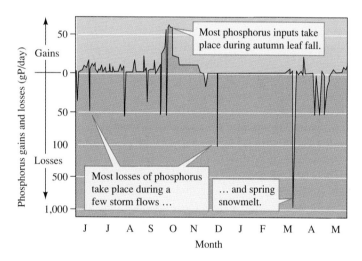

**Figure 19.25**  Daily gains and losses of phosphorus (P) by the Bear Brook ecosystem from 1974 to 1975 (data from Meyer and Likens 1979).

ical and biological processes converted dissolved and coarse particulate forms of phosphorus into fine particulate forms.

Meyer and Likens used their estimates to reconstruct the long-term phosphorus dynamics of Bear Brook. During 1974 to 1975, the ratio of phosphorus export to input was almost exactly one (1.04). However, the Meyer and Likens model of phosphorus dynamics indicated considerable year-to-year variation in this ratio during the period from 1963 to 1975, which included the wettest and driest years in the 20-year precipitation record for the area. The predicted ratio of output to input over this interval was highly correlated with annual streamflow. The ratio ranged from 0.56 in the driest year to 1.6 in the wettest year (fig. 19.24). In other words, during the driest year only 56% of phosphorus inputs were exported, while during the wettest year exports amounted to 160% of inputs. In wet years the stream ecosystem's standing stocks of phosphorus were reduced by high levels of export.

The patterns of inputs and exports of phosphorus from Bear Brook were highly pulsed during Meyer and Likens' study (fig. 19.25). They estimated that from 1974 to 1975, 48% of total annual input of phosphorus to Bear Brook entered during 10 days and that 67% of exports left the ecosystem during 10 days. The annual peak in phosphorus input was associated with autumn leaf fall, and an annual pulse of export was associated with spring snowmelt. Most phosphorus export, however, was irregular because it was driven by flooding caused by intense storms that may occur during any month of the year. If we consider floods as a source of disturbance, the behavior of stream ecosystems is consistent with the generalization that disturbance increases the loss of nutrients from ecosystems.

Aquatic ecologists study the nutrient dynamics of aquatic ecosystems like Bear Brook because, as we saw in chapter 18, nutrient availability is a key regulator of aquatic primary production. As we shall see in the Applications & Tools section, nutrient enrichment of ecosystems by human activity is a worldwide problem.

# APPLICATIONS & TOOLS

## Altering Aquatic and Terrestrial Ecosystems

Human activity increasingly affects the nutrient cycles of ecosystems. Agriculture and forestry may remove nutrients from ecosystems. However, increasingly, human activity enriches ecosystems with nutrients, especially with nitrogen (see chapter 23) and phosphorus. The main source of nitrogen enrichment is air pollution due to burning of fossil fuels and intensive applications of fertilizer by farmers. In the temperate coastal forests of southern Chile, far from urban and industrial centers, inputs of nitrogen amount to about 0.1 to 1.0 kg per hectare per year. In contrast, in the Netherlands, with its high population density and intense agriculture, precipitation adds nitrogen to forest ecosystems at rates up to 60 kg per hectare per year.

Humans are also a major source of nutrient input to aquatic ecosystems. Benjamin Peierls and his colleagues (1991) examined the relationship of human population density within river basins and nitrate concentration and export by 42 major rivers. These rivers, which deliver approximately 37% of the total freshwater flow to the oceans, support human population densities ranging from 1 to 1,000 individuals per square kilometer.

Peierls noted that while the concentration and export of nitrate by rivers is affected by complex biotic, abiotic, and anthropogenic factors, a single variable, human population density, explains most of the variation in nitrate concentration and export (fig. 19.26). The most probable sources of nitrate enrichment of river ecosystems are sewage disposal, atmospheric deposition, agriculture, and deforestation, all of which generally increase with increased human population density.

Human disturbance also increases export of phosphorus from aquatic catchments. There is a significant relationship

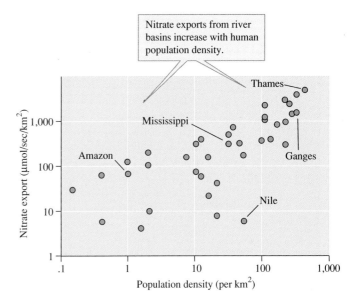

**Figure 19.26** Human population density and nitrate export from river basins (data from Peierls et al. 1991).

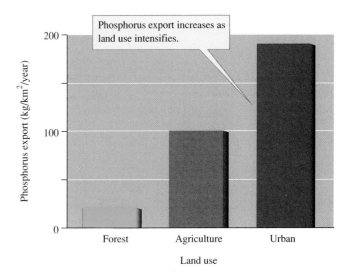

**Figure 19.27** Land use and phosphorus export from stream basins (data from Reckhow and Simpson 1980).

between land use and export of phosphorus from catchments in southern Ontario, Canada. Conversion of forest to agriculture approximately quadruples phosphorus export. Meanwhile, phosphorus exports from urban catchments are nearly two times higher than exports from agricultural catchments and approximately nine times exports from forested catchments (fig. 19.27).

What are the ecological effects of increased nutrient inputs to ecosystems? One of the most obvious, negative consequences is the impoverishment of biological communities. As we saw in chapter 16, terrestrial and aquatic ecosystems enriched with nutrients generally support a lower diversity of primary producers. As nutrients become abundant, light may be left as the single limiting resource. In the absence of disturbance, the nutrient-enriched community is eventually dominated by the best competitors for light. This hypothesis is

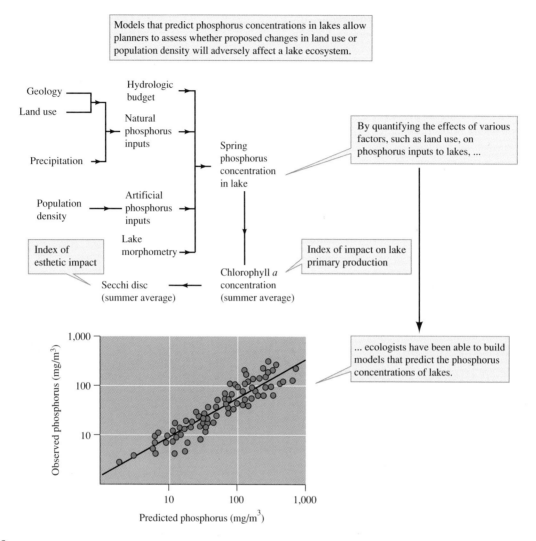

Models that predict phosphorus concentrations in lakes allow planners to assess whether proposed changes in land use or population density will adversely affect a lake ecosystem.

By quantifying the effects of various factors, such as land use, on phosphorus inputs to lakes, ...

Index of impact on lake primary production

Index of esthetic impact

... ecologists have been able to build models that predict the phosphorus concentrations of lakes.

**Figure 19.28** Models that predict the phosphorus concentrations of lakes (data from Dillon and Rigler 1975, Peters 1991).

supported by the long-term fertilization experiments at Rothamsted, England (see fig. 16.16).

Nutrient enrichment of ecosystems, particularly with nitrogen, may also be causing massive local extinctions of fungi. John Jaenike (1991) cites declines in mushroom species diversity of 40% to 80% in Germany, Austria, the Netherlands, and the Czech Republic. Similar declines may also be occurring in North America, but the fungi of North America have not been studied carefully enough to make an assessment. Jaenike suggests that the most significantly affected fungi are the ectomycorrhizal species that form associations with trees. It appears that fertilization is somehow interfering with the mutualistic relationship between trees and their mycorrhizal associates. This hypothesis is consistent with the studies of Nancy Johnson (1993) (see chapter 15), which suggested that fertilization alters the mutualistic relationship between plants and mycorrhizal fungi. The loss of mycorrhizal fungi may threaten the long-term survival of entire forest ecosystems.

The main factors affecting phosphorus concentration in lakes include geology, land use, precipitation, hydrologic budget,

lake morphometry, and human population density. Using these factors, researchers have been able to predict the phosphorus concentrations of lakes with reasonable accuracy (fig. 19.28).

How can a model that predicts phosphorus concentration be useful to land managers? Remember that phosphorus concentration translates directly into lake trophic status, which, in turn, influences the lake's suitability for a variety of human uses. With such a model, the land manager can predict the potential impact of changes in land use or urbanization within a lake basin and determine whether a significant change in lake trophic status is likely. These quantitative predictions by the nutrient loading model can be used to make land management decisions. R. Peters (1991) and D. Schindler (1987) suggested that the worldwide use of these loading models to manage phosphorus inputs into aquatic ecosystems is one of the best examples of how ecological knowledge has been used to solve an environmental problem.

In summary, we know that there is a direct connection between human activity and nutrient enrichment of ecosystems and that nutrient enrichment has a number of negative consequences. How have land managers addressed this significant and complex problem? One of the most successful tools available to land managers has been the nutrient loading model (Peters 1991), especially models that predict the concentration of phosphorus in lakes. As we saw in chapter 18, phosphorus usually limits primary production in freshwater lakes. High concentrations of phosphorus produce eutrophic conditions, including algal blooms, reduced water clarity, and lowered dissolved oxygen. These effects generally reduce the esthetic quality of lakes and restrict their suitability for recreation.

R. A. Vollenweider (1969) pioneered the development of nutrient loading models, which quickly became the basis for management of lake basins around the world. Land managers have used nutrient loading models to regulate the level of human activity in lake basins, particularly the development of summer cottages, in order to forestall the negative impacts of eutrophication.

# SUMMARY

The elements organisms require for development, maintenance, and reproduction are called *nutrients*. Ecologists refer to the use, transformation, movement, and reuse of nutrients in ecosystems as *nutrient cycling*. Nutrient cycling is one of the most ecologically significant processes studied by ecosystem ecologists. The *carbon, nitrogen,* and *phosphorus cycles* have played especially prominent roles in studies of nutrient cycling.

**Decomposition rate is influenced by temperature, moisture, and chemical composition of litter and the environment.** The rate of decomposition affects the rate at which nutrients, such as nitrogen and phosphorus, are made available to primary producers. Rates of decomposition in terrestrial ecosystems are higher under warm, moist conditions. The rate of decomposition in terrestrial ecosystems increases with nitrogen content and decreases with the lignin content of litter. The chemical composition of litter and the availability of nutrients in the surrounding environment also influence rates of decomposition in aquatic ecosystems.

**Plants and animals can modify the distribution and cycling of nutrients in ecosystems.** The dynamics of nutrients in streams are best represented by a spiral rather than a cycle. The length of stream required for an atom of a nutrient to complete a cycle is called the spiraling length. Stream macroinvertebrates can substantially reduce spiraling length of nutrients in stream ecosystems. Animals can also alter the distribution and rate of nutrient cycling in terrestrial ecosystems. Some experiments have shown a positive correlation between shredder species richness and leaf mass loss. Nitrogen-fixing plants increase the quantity and rates of nitrogen cycling in terrestrial ecosystems.

**Disturbance increases nutrient loss from ecosystems.** Vegetation exerts substantial control on nutrient retention by terrestrial ecosystems. Vegetative controls on nutrient loss from forest ecosystems appear to be most important in environments that are warm and moist during the growing season. Vegetative controls appear to be less important in cold and/or dry environments. Nutrient loss by stream ecosystems is highly pulsed and associated with disturbance by flooding.

Nutrient enrichment by humans is altering aquatic and terrestrial ecosystems. Nitrate concentration and export by the earth's major rivers correlate directly with human population density. Human disturbance also increases export of phosphorus from aquatic catchments. Nutrient enrichment appears to be reducing the diversity of plants and fungi in terrestrial ecosystems. Land managers around the world use nutrient loading models to predict and manage the impact of land use on aquatic ecosystems.

# Review Questions

1. Of all the naturally occurring elements in the biosphere, why have the cycles of carbon, nitrogen, and phosphorus been so intensively studied by ecologists? (Hint: Think about the kinds of organic molecules of which these elements are constituents. Also think back to our discussions, in chapter 18, of the influences of nitrogen and phosphorus on rates of primary production.)

2. Parmenter and Lamarra (1991) studied decomposition of fish and waterfowl carrion in a freshwater marsh. During the course of their studies they found that the soft tissues of both fish and waterfowl decomposed faster than the most rapidly decomposing plant tissues. Explain the rapid decomposition of these animal carcasses.

3. Review figure 18.2, in which Rosenzweig (1968) plotted the relationship between actual evapotranspiration and net primary production. How do you think that decomposition rates change

across the same ecosystems? Using what you learned in chapter 19, design an experiment to test your hypothesis.

4. Melillo, Aber, and Muratore (1982) suggested that soil fertility may influence the rate of decomposition in terrestrial ecosystems. Design an experiment to test this hypothesis. If you test for the effects of soil fertility, how will you control for the influences of temperature, moisture, and litter chemistry?

5. Many rivers around the world have been straightened and deepened to improve conditions for navigation. Side effects of these changes include increased average water velocity and decreased movement of water into shallow riverside environments such as eddies and marginal wetlands. What are the probable influences of these changes on nutrient spiraling length? Use the model of Newbold et al. (1983) in your discussion.

6. Likens and Bormann (1995) found that vegetation substantially influences the rate of nutrient loss from small stream catchments in the northern hardwood forest ecosystem. How do vegetative biomass and rates of primary production in these forests affect their capacity to regulate nutrient loss? How much do you think vegetation affects nutrient movements in desert ecosystems?

7. McNaughton, Ruess, and Seagle (1988) proposed that grazing by large mammals increases the rate of nitrogen cycling on the savannas of East Africa. Explain how passing through a large mammal could increase the rate of breakdown of plant biomass. In chapter 18 we also saw how grazing mammals may increase the rate of primary production on the savanna. How might the disappearance of the large mammals of East Africa affect ecosystem processes on the savanna?

8. The fynbos of South Africa is famous for the exceptional diversity of its plant community. Witkowski (1991) showed that invading *Acacia* are enriching the fynbos soil with nitrogen. How might enriching soil nitrogen affect plant diversity in this ecosystem? What mechanisms would likely produce your predicted changes?

9. Kauffman and his colleagues (1993) estimated that burning the tropical forest at their study site resulted in the loss of approximately 21 kg per hectare of phosphorus. This quantity is about 11% to 17% of the total pool of phosphorus. If total annual inputs of phosphorus to the ecosystem, mainly by rain and "dry fall," amount to about 0.2 kg per hectare per year (Murphy and Lugo 1986), how long would it take these inputs to make up for a single agricultural burn such as that created by Kauffman and his colleagues? Assuming a constant rate of loss, how many burns would it take to totally exhaust existing supplies?

10. If rates of decomposition are higher in ecosystems with higher nutrient availability, how should nutrient enrichment affect rates of decomposition? Because of its effects on fungal diversity, could nutrient enrichment of ecosystems affect rates of decomposition differently over the short term versus the long term?

# Suggested Readings

Augustine, D. J. and D. A. Frank. 2001. Effects of migratory grazers on spatial heterogeneity of soil nitrogen properties in a grassland ecosystem. *Ecology* 82: 3149–62.

*This study demonstrates the influence of native grazers on the distribution of soil nitrogen at small to large scales.*

Fisk, M. C., D. R. Zak, and T. R. Crow. 2002. Nitrogen storage and cycling in old- and second-growth hardwood forests. *Ecology* 83: 73–87.

*A very detailed comparison of nitrogen fluxes in second-growth and old-growth temperate deciduous forests which reveals a great deal of complexity and no simple contrast between forests of different ages.*

Gessner, M. O. and E. Chauvet. 1994. Importance of stream microfungi in controlling breakdown rates of leaf litter. *Ecology* 75:1807–17.

Melillo, J. M., J. D. Aber, and J. F. Muratore. 1982. Nitrogen and lignin control of hardwood leaf litter decomposition dynamics. *Ecology* 63:621–26.

*Excellent companion papers that discuss the effects of litter chemistry on rates of decomposition in aquatic and terrestrial environments.*

Grimm, N. B. 1988. Role of macroinvertebrates in nitrogen dynamics of a desert stream. *Ecology* 69:1884–93.

McNaughton, S. J., R. W. Ruess, and S. W. Seagle. 1988. Large mammals and process dynamics in African ecosystems. *BioScience* 38:794–800.

*These two papers offer interesting parallels in the influence of consumers on nutrient cycles in a desert stream and a tropical savanna.*

Jonsson, M. and B. Malmqvist. 2003. Mechanisms behind positive diversity effects on ecosystem functioning: testing the facilitation and interference hypotheses. *Oecologia* 134:554–59.

*This experimental study demonstrates the mechanisms underlying the positive relationship between diversity of detritivores and leaf processing rates.*

Likens, G. E. and F. H. Bormann. 1995. *Biogeochemistry of a Forested Ecosystem.* 2d ed. New York: Springer-Verlag.

*A 32-year record of the structure and dynamics of the Hubbard Brook ecosystem—a worthwhile companion and update to the original papers reporting on the Hubbard Brook experiment.*

Likens, G. E., F. H. Bormann, N. M. Johnson, D. W. Fisher, and R. S. Pierce. 1970. Effects of forest cutting and herbicide treatment on nutrient budgets in the Hubbard Brook watershed-ecosystem. *Ecological Monographs* 40:23–47.

*A detailed report of the classic Hubbard Brook watershed experiment.*

Meyer, J. L. and G. E. Likens. 1979. Transport and transformation of phosphorus in a forest stream ecosystem. *Ecology* 60:1255–69.

*A classic paper on the effects of disturbance by flooding on nutrient budgets of streams—a must for anyone interested in the nutrient dynamics of streams.*

Newbold, J. D., J. W. Elwood, R. V. O'Neill, and A. L. Sheldon. 1983. Phosphorus dynamics in a woodland stream ecosystem: a study of nutrient spiraling. *Ecology* 64:1249–65.

*One of the foundation studies on nutrient spiraling in streams that helped change ecologists' views on the dynamics of nutrients in streams.*

Peierls, B. L., N. F. Caraco, M. L. Pace, and J. J. Cole. 1991. Human influence on river nitrogen. *Nature* 350:386–87.

*Concise study that provides compelling evidence of the influence of humans on nitrogen transport by the world's rivers.*

Perakis, S. S. and L. O. Hedin. 2001. Fluxes and fates of nitrogen in soil of an unpolluted old-growth temperate forest, southern Chile. *Ecology* 82:2245–60.

*Fascinating work on one of the rarest ecosystem types on earth: an unpolluted, old-growth temperate forest. The work documents the exceptional retentiveness of old-growth temperate forests.*

Schlesinger, W. H. 1991. *Biogeochemistry: An Analysis of Global Change.* New York: Academic Press.

*Excellent introduction to the ecology of nutrient cycles, combining a great deal of detail with a well-written presentation.*

Takyu, M., S.-I. Aiba, and K. Kitayama. 2003. Changes in biomass, productivity and decomposition along topographical gradients under different geological conditions in tropical lower montane forests on Mount Kinabalu, Borneo. *Oecologia* 134:397–404.

*A fascinating study in which the scientists were able to isolate the influences of geologic formation and soils from that of climate on rates of primary production and decomposition in a tropical rain forest.*

Tyrell, T. and C. S. Law. 1997. Low nitrate:phosphate ratios in the global ocean. *Nature* 387:793–96.

*The results of this analysis suggest rates of denitrification in the oceans need to be reassessed. Higher than expected rates of denitrification may require adjustments of rates of nitrogen flux from the oceans to the atmosphere.*

Uliassi, D. D. and R. W. Ruess, 2002. Limitations to symbiotic nitrogen fixation in primary succession on the Tanana River floodplain. *Ecology* 83:88–103.

*Uliassi and Ruess demonstrated that phosphorus availability limits rate of nitrogen fixation in alder stands in an Alaskan floodplain ecosystem.*

# On the Net INTERNET

Visit this textbook's accompanying website at www.mhhe.com/ecology (click on the book's title) to take advantage of practice quizzing, study/writing tips, timely news articles, and additional URLs for research on the topics in this chapter.

Community Ecology

Field Methods for Studies of Ecosystems

Species Abundance, Diversity, and Complexity

Food Webs

Nutrient Cycling

Succession and Stability

Water Use and Management

Ecosystem Management

# Succession
# and Stability

The first recorded visit to Glacier Bay gave no hint of its eventual contributions to our understanding of biological communities and ecosystems. In 1794, Captain George Vancouver visited the inlet to what is today called Glacier Bay, Alaska (fig. 20.1). He could not pass beyond the inlet to the bay, however, because his way was blocked by a mountain of ice. Vancouver and Vancouver (1798) described the scene as follows: "The shores of the continent form two large open bays which were terminated by compact solid mountains of ice, rising perpendicularly from the water's edge, and bounded to the north by a continuation of the united lofty frozen mountains that extend eastward from Mount Fairweather."

In 1879, John Muir explored the coast of Alaska, relying heavily on Vancouver's earlier descriptions. Muir (1915) commented in his journal that Vancouver's descriptions were excellent guides except for the area within Glacier Bay. Where Vancouver had met "mountains of ice," Muir found open water. He and his guides from the Hoona tribe paddled their canoe through Glacier Bay in rain and mist, feeling their way through uncharted territory. They eventually found the glaciers, which Muir estimated had retreated 30 to 40 km up the glacial valley since Vancouver's visit 85 years earlier.

Muir found no forests at the upper portions of the bay. He and his party had to build their campfires with the stumps and trunks of long-dead trees exposed by the retreating glaciers. Muir recognized that this "fossil wood" was a remnant of a forest that had been covered by advancing glaciers centuries earlier. He also saw that plants quickly colonized the areas uncovered by glaciers and that the oldest exposed areas, where Vancouver had met his mountains of ice, already supported forests.

Muir's observations in Glacier Bay were published in 1915 and read the same year by the ecologist William S. Cooper. Encouraged by Muir's descriptions, Cooper visited Glacier Bay in 1916 in what was the beginning of a lifetime of study. Cooper saw Glacier Bay as the ideal laboratory for the study of ecological **succession,** the gradual change in plant and animal communities in an area following disturbance or the creation of new substrate. Glacier Bay was ideal for the study of succession because the history of glacial retreat could be accurately traced back to 1794 and perhaps farther.

Cooper ultimately made four expeditions to Glacier Bay. His work and that of later ecologists produced a detailed picture of succession there. Several species of plants colonize an area during the first 20 years after it is

**Figure 20.1** Glacier Bay, Alaska.

exposed by the retreating glacier. These plants, the first in a successional sequence, form a **pioneer community.** The most common members of the pioneer community are horsetail, *Equisetum varietaum,* willow herb, *Epilobium latifolium,* willows, *Salix* sp., cottonwood seedlings, *Populus balsamifera,* mountain avens, *Dryas drummondii,* and Sitka spruce, *Picea sitchensis.*

About 30 years after an area is exposed, the pioneer community gradually grades into a community dominated by mats of *Dryas,* a dwarf shrub. These *Dryas* mats also contain scattered alder, *Alnus crispa, Salix, Populus,* and *Picea.* Then, about 40 years after glacial retreat, the community changes into a shrub-thicket dominated by *Alnus.* Soon after the closure of the *Alnus* thicket, however, *Populus* and *Picea* will grow above it, covering about 50% of the area on sites 50 to 70 years old.

In 75 to 100 years, succession leads to a forest community dominated by *Picea.* Mosses carpet the understory of this spruce forest and here and there grow seedlings of western hemlock, *Tsuga heterophylla,* and mountain hemlock, *Tsuga mertensiana.* Eventually, the population of *Picea* declines and the forests are dominated by *Tsuga.* On landscapes with shallow slopes these hemlock forests eventually give way to muskeg, a landscape of peat bogs and scattered tussock meadows.

Because succession around Glacier Bay occurs on newly exposed geological substrates, not significantly modified by organisms, ecologists refer to this process as **primary succession.** Primary succession also occurs on newly formed volcanic surfaces such as lava flows. In areas where disturbance destroys a community without destroying the soil, the subsequent succession is called **secondary succession.** For instance, secondary succession occurs after agricultural lands are abandoned or after a forest fire.

Succession generally ends with a community whose populations remain stable until disrupted by disturbance. This late successional community is called the **climax community.** The nature of the climax community depends upon environmental circumstances. The communities we discussed in chapter 2—temperate forests, grasslands, etc.—were essentially the climax communities for each of the climatic regimes that we considered. The climax community around Glacier Bay is determined by the prevailing climate and local topography. On well-drained, steep slopes the climax community is hemlock forest. In poorly drained soil on shallow slopes the climax community is muskeg.

Studies of succession show that communities and ecosystems are not static but constantly change in response to disturbance, environmental change, and their own internal dynamics. In many cases, the general direction of change in community structure and ecosystem processes is predictable, at least over the short term. The patterns of change in community and ecosystem properties during succession and the mechanisms responsible for those changes are subjects covered in chapter 20. We also consider a companion topic, community and ecosystem stability.

## CONCEPTS

- **Community changes during succession include increases in species diversity and changes in species composition.**
- **Ecosystem changes during succession include increases in biomass, primary production, respiration, and nutrient retention.**
- **Mechanisms that drive ecological succession include facilitation, tolerance, and inhibition.**
- **Community stability may be due to lack of disturbance or community resistance or resilience in the face of disturbance.**

## CONCEPT DISCUSSION

### Community Changes During Succession

**Community changes during succession include increases in species diversity and changes in species composition.**

Some of the most detailed studies of ecological succession have focused on succession leading to a forest climax. Though primary and secondary forest succession require different amounts of time, the changes in species diversity that occur in each appear remarkably similar.

# Primary Succession at Glacier Bay

We have already reviewed the basic patterns of primary succession around Glacier Bay. Now we return to Glacier Bay to examine successional changes in species diversity and composition. William Reiners, Ian Worley, and Donald Lawrence (1971) studied changes in plant diversity during succession at Glacier Bay. They worked at sites carefully chosen for similarity in physical features but differing substantially in age. Their eight study sites were below 100 m elevation, were on glacial till, an unstratified and unsorted material deposited by a glacier, and all had moderate slopes. The study sites ranged in age, that is, time since glacial retreat, from 10 to 1,500 years.

Their youngest site, which was approximately 10 years old, supported a pioneer community of scattered *Epilobium, Equisetum,* and *Salix.* Site 2 was about 23 years old and supported a mix of pioneer species and clumps of *Populus* and *Dryas.* Site 3, which was approximately 33 years old, sup-

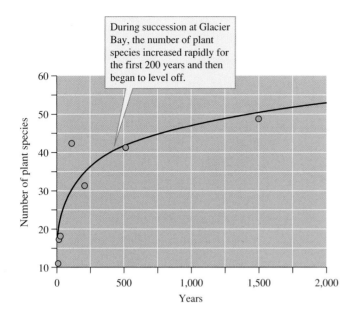

**Figure 20.2** Change in plant species richness during primary succession at Glacier Bay, Alaska (data from Reiners, Worley, and Lawrence 1971).

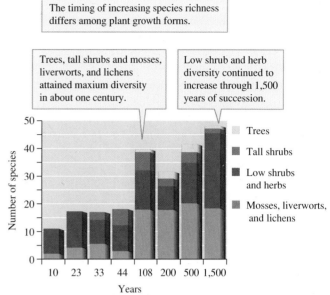

**Figure 20.3** Succession of plant growth forms at Glacier Bay, Alaska (data from Reiners, Worley, and Lawrence 1971).

ported a mat of *Dryas* enclosing clumps of *Salix, Populus,* and *Alnus*. Site 4 was 44 years old and was dominated by a mat of *Dryas* with few open patches. Site 5, which was approximately 108 years old, was dominated by a thicket of *Alnus* and *Salix* with enough emergent *Populus* and *Picea* to form a partial canopy. Site 6 was a 200-year-old forest of *Picea*. Using geological methods, Reiners and his colleagues dated site 7 at 500 years and site 8 at 1,500 years. Both sites were located on Pleasant Island, which because it is located outside the mouth of Glacier Bay had escaped the most recent glaciation, which had destroyed the forests along the bay. Site 7 was an old forest of *Tsuga* that contained a few *Picea*. Site 8 was a muskeg with scattered lodgepole pines, *Pinus contorta*.

The total number of plant species in the eight study sites increased with plot age. As you can see in figure 20.2, species richness increased rapidly in the early years of succession at Glacier Bay and then more slowly during the later stages, approaching a possible plateau in species richness.

Not all groups of plants increased in diversity throughout succession. Figure 20.3 shows that while the species richness of mosses, liverworts, and lichens reached a plateau after about a century of succession, the diversity of low shrubs and herbs continued to increase throughout succession. In contrast, the diversity of tall shrubs and trees increased until the middle stages of succession and then declined in later stages.

The pattern of increased species richness withstand age that Reiners and his colleagues described for the successional sequence around Glacier Bay is one that we will see several times in the examples that follow. However, the tempo of succession is far different. The late successional climax community at Glacier Bay was 1,500 years old. In the following example of secondary succession, the climax forest community was 150 to 200 years old, approximately one-tenth the age of the climax community studied at Glacier Bay.

# Secondary Succession in Temperate Forests

The Piedmont Plateau of eastern North America includes some of the most convenient places to study secondary succession. The deciduous forests of this region were intensively cleared and cultivated beginning approximately three centuries ago. As fields were abandoned and new forest areas cleared, the region was progressively converted to a patchwork of abandoned fields of various ages interspersed with a few areas of undisturbed forest.

This situation provided Henry J. Oosting (1942) with study sites in virtually every stage of secondary succession for his now classic studies of succession in the Piedmont Plateau of North Carolina. David Johnston and Eugene Odum (1956) described the pattern of succession on the Piedmont Plateau as follows. The first species to colonize and dominate abandoned fields are crabgrass, *Digitaria sanguinalis,* and horseweed, *Erigeron canadense*. During the second year of succession, the fields are dominated by either aster, *Aster pilosis,* or ragweed, *Ambrosia artemisiifolia*. A few years later the field is covered by broomsedge, *Andropogon virginicus,* with a scattering of other shrubs and small trees. Pine seedlings may appear as early as the third year and may form a closed canopy in 10 to 15 years. Pine seedlings cannot grow in the shade of larger pines but the seedlings of many deciduous trees can. Consequently, 40- to 50-year-old pine forests generally have a well-developed understory of young deciduous trees. These deciduous trees, especially oak, *Quercus,* and hickory, *Carya,* become the dominant trees by about 150 years, and the pines decline to a few scattered individuals. Because *Quercus* and *Carya* can reproduce in their own shade, the late successional oak-hickory forest is considered the climax stage.

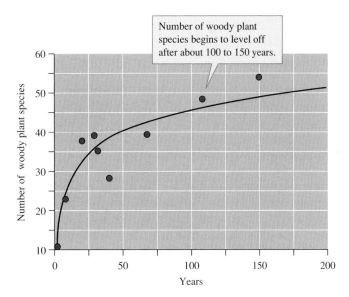

**Figure 20.4** Change in woody plant species richness during secondary forest succession in eastern North America (data from Oosting 1942).

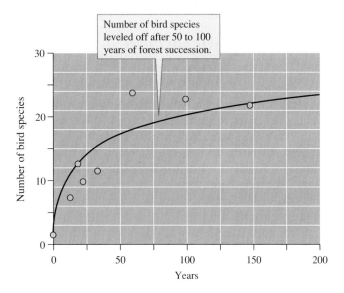

**Figure 20.5** Change in number of breeding bird species during secondary forest succession (data from Johnston and Odum 1956).

Oosting's data show that the number of woody plant species increased during secondary succession on the Piedmont Plateau (fig. 20.4). The successional sequence began with a single species of woody plant invading soon after fields were abandoned and began to level off at 50 to 60 species after approximately 150 years. This increase in species richness follows a logarithmic pattern like the one we saw at Glacier Bay.

How does animal diversity change across the same successional sequence? Johnston and Odum studied the birds living on thirteen 20-acre study sites ranging in successional age from 1 to 150 years and supporting vegetation ranging from grassland to mature oak-hickory forest. The increase in bird diversity across this successional sequence closely paralleled the increase in woody plant diversity observed by Oosting (fig. 20.5). During the grass-forb stage of succession the bird community consisted

**Figure 20.6** Succession in the intertidal zone involves colonization and competition for limited space among species as different as attached marine algae, sea anemones, mussels, and barnacles.

of two species generally associated with grasslands. In the grass-shrub stage, the diversity of birds increased to 8 to 13 species. In 25- to 35-year-old pine forests, the diversity of birds was 10 to 12 species. Pine forests of 60 to 100 years supported 23 to 24 bird species. Johnston and Odum recorded 22 bird species in old-growth oak-hickory forests.

In summary, over a period of approximately one and a half centuries abandoned fields in eastern North America undergo successional changes in plant and bird communities that involve changes in species composition and increases in species diversity. Similar successional changes occur in the marine intertidal zone but on an even shorter timescale: approximately one and a half years instead of one and a half centuries.

# Succession in Rocky Intertidal Communities

When we discussed the influence of disturbance on local species diversity in chapter 16, we saw how an intertidal boulder stripped of its cover of attached organisms was soon colonized by algae and barnacles (fig. 20.6). Looking back on that pattern of com-

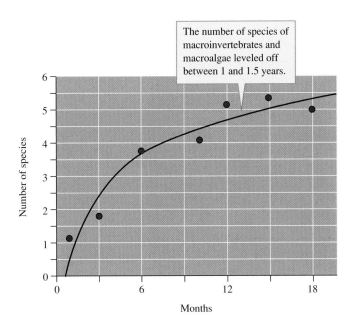

**Figure 20.7** Succession in number of macroinvertebrate and macroalgae species on intertidal boulders (data from Sousa 1979a).

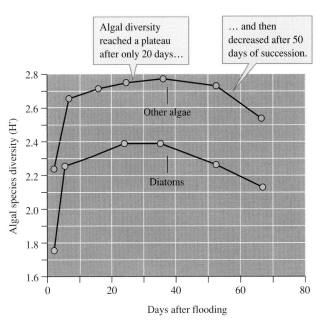

**Figure 20.8** Algal species diversity during succession in Sycamore Creek, Arizona (data from Fisher et al. 1982).

munity change we can now see it as an example of ecological succession. Wayne Sousa (1979a, 1979b) showed that the first species to colonize open space on intertidal boulders were a green alga in the genus *Ulva* and the barnacle *Chthamalus fissus*. The next arrivals were several species of perennial red algae: *Gelidium coulteri, Gigartina leptorhynchos, Rhodoglossum affine,* and *Gigartina canaliculata.* Finally, if there was no disturbance for 2 to 3 years, *G. canaliculata* grew over the other species and dominated 60% to 90% of the space.

Sousa explored succession on intertidal boulders with several experiments. In one of them, he followed succession on small boulders that he had cleaned and stabilized. As in forest succession, the number of species increased with time (fig. 20.7). Notice in the figure that the average number of species increased until about 1 to 1.5 years and then leveled off at about five species.

Primary forest succession around Glacier Bay may require about 1,500 years, and secondary forest succession on the Piedmont Plateau takes about 150 years. Meanwhile the successional changes described by Sousa occurred within about 1.5 years. In the next example, ecological succession within a desert stream occurs in less than 2 months.

## Succession in Stream Communities

Rapid succession has been well documented in Sycamore Creek, Arizona, which has been studied for nearly two decades by Stuart Fisher and his colleagues (1982). Sycamore Creek, a tributary of the Verde River, lies approximately 32 km northeast of Phoenix, Arizona, where it drains approximately 500 km$^2$ of mountainous desert terrain. Evaporation nearly equals precipitation within the Sycamore Creek catchment, so flows

are generally low and often intermittent. However, the creek is subject to frequent flash floods with sufficient power to completely disrupt the community and initiate succession.

Fisher's research team reported on the successional events following one such flood. Intense floods occurred on Sycamore Creek on August 6, 12, and 16, of 1979, with peak flows of 7, 3, and 2 m$^3$ per second. Floods of this intensity mobilize the stones and sand of the stream, scouring some areas and depositing sediments in others. In the process, most stream organisms are destroyed. The three floods of August 1979, eliminated approximately 98% of algal and invertebrate biomass in Sycamore Creek.

In 63 days following these floods, Fisher and his colleagues observed rapid changes in both the diversity and composition of algae and invertebrates. Patterns among primary producers were especially clear. Two days after the floods, the majority of the stream bottom consisted of bare sand with some patches of diatoms. Five days after the flood, diatoms covered about half the streambed. Within 13 to 22 days, diatoms almost completely covered the stream bottom. Other algae, especially blue-green algae and mats consisting of a mixture of the green alga *Cladophora* and blue-green algae, appeared in significant quantities by day 35. By day 63, the bottom of Sycamore Creek consisted of a patchwork of areas dominated by diatoms, blue-green algae, and mats of *Cladophora* and blue-green algae. The diversity of diatoms and other algae, as measured by H' (see chapter 16), leveled off after only 5 days and then began to decline after about 50 days (fig. 20.8).

Invertebrate diversity was strongly influenced by a single dominant species of crane fly, family Tipulidae, *Cryptolabis* sp. (fig. 20.9). Large numbers of *Cryptolabis* larvae in Sycamore Creek depressed H' diversity for all collections except for the collection on day 35, when most of the population emerged as adults. Throughout their collections over the 63-day period of

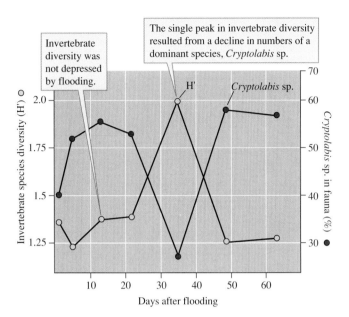

**Figure 20.9** Invertebrate species diversity during succession in Sycamore Creek, Arizona (data from Fisher et al. 1982).

the study, the researchers reported that they collected 38 to 43 species of aquatic invertebrates out of a total of 48 species collected during their studies. In other words, most macroinvertebrate species survived the flood.

Where did these invertebrate species find refuge from the devastating floods of August 1979? The invertebrate community of Sycamore Creek is dominated by insects whose adults are terrestrial. During the floods of August, many adult insects were in the aerial stage and the flood passed under them. These aerial adults were the source of most invertebrate recolonization of the flood-devastated Sycamore Creek.

As we have just seen, ecological succession involves predictable changes in community structure. Succession also leads to predictable changes in ecosystem structure and function. These ecosystem changes with succession are the subject of the next Concept Discussion section.

## CONCEPT DISCUSSION

### Ecosystem Changes During Succession

**Ecosystem changes during succession include increases in biomass, primary production, respiration, and nutrient retention.**

As succession changes the diversity and composition of communities, ecosystem properties change as well. In the last section, we saw how plant and animal community structure changes during primary and secondary succession. In this section, we review evidence that many ecosystem properties

also change during succession. For instance, many properties of soils, such as the nutrient and organic matter content, change during the course of succession.

## Ecosystem Changes at Glacier Bay

Stuart Chapin and his colleagues (1994) documented substantial changes in ecosystem structure during succession at Glacier Bay. They focused on four study areas of approximately 2 km² each. Their first site had been deglaciated about 5 to 10 years and was in the pioneer stage. Their second site had been deglaciated 35 to 45 years and was dominated by a mat of *Dryas*. *Dryas* was just beginning to invade this site when it was studied by Reiners' group more than 20 years earlier. The third site had been deglaciated about 60 to 70 years and was in the alder, *Alnus,* stage. This site was studied by Reiners when it was a young thicket of alder and by Cooper when it was in the pioneer stage. The fourth site studied by Chapin and his colleagues had been deglaciated 200 to 225 years earlier and was a forest of spruce, *Picea,* as it was when studied by Reiners and Cooper.

Chapin and his research team measured changes in several ecosystem characteristics across these study sites. One of the most fundamental characteristics was the quantity of soil. Total soil depth and the depth of all major soil horizons all show significant increases from the pioneer community to the spruce stage (fig. 20.10).

Several other ecologically important soil properties also changed during succession at the Glacier Bay study sites. As figure 20.11 shows, the organic content, moisture, and nitrogen concentrations of the soil all increased substantially. Over the same successional sequence, soil bulk density, pH, and phosphorus concentration all decreased.

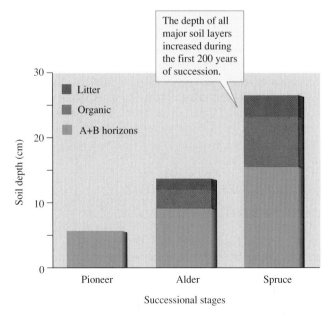

**Figure 20.10** Soil building during primary succession at Glacier Bay, Alaska (data from Chapin et al. 1994).

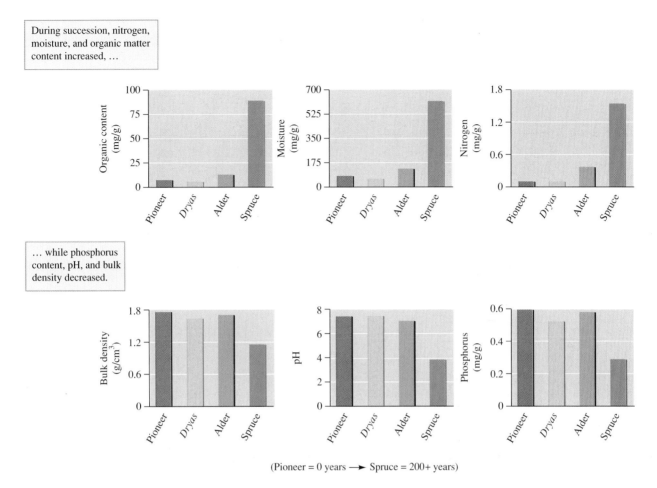

During succession, nitrogen, moisture, and organic matter content increased, …

… while phosphorus content, pH, and bulk density decreased.

(Pioneer = 0 years ⟶ Spruce = 200+ years)

**Figure 20.11**   Changes in soil properties during succession at Glacier Bay, Alaska (data from Chapin et al. 1994).

Why are these changes in soil properties important? They demonstrate that succession involves more than just changes in the composition and diversity of species. Terrestrial succession changes key ecosystem properties. Changes in soil properties are important because soils are the foundation upon which terrestrial ecosystems are built.

We can also see from these ecological studies that the physical and biological properties of ecosystems are inseparable. Organisms acting upon mineral substrates contribute to the building of soils upon which spruce forests eventually grow around Glacier Bay. Soils, in turn, strongly influence the kinds of organisms that grow in a place.

# Four Million Years of Ecosystem Change

The detailed knowledge of ecosystem change that has emerged through studies at Glacier Bay, Alaska, is impressive. However, the sequence of ages represented by the study sites at Glacier Bay, what ecologists call a **chronosequence,** are limited. In 1794, when Captain George Vancouver encountered a wall of ice at the mouth of Glacier Bay, the island of Kauai in the Hawaiian Island chain supported forest ecosystems growing on soils that had developed on lava flows

that were over 4 million years old. The Hawaiian Islands have formed over a hot spot on the Pacific tectonic plate and have been transported on that plate to the northwest, forming a chain of islands that vary greatly in age. The youngest island in the group is the big island of Hawaii, which is currently growing over the hot spot. The big island is made up of volcanic rocks that vary from fresh lava flows to flows that are approximately 150,000 years old. Meanwhile, the islands to the northwest are sequentially older. As in Glacier Bay, teams of ecologists have probed the chronosequence represented by the Hawaiian Island chain for information on ecosystem development. However, in Hawaii the chronosequence spans not hundreds of years but millions.

Lars Hedin, Peter Vitousek, and Pamela Matson (2003) examined nutrient distributions and losses on a chronosequence of forest ecosystems on the islands of Hawaii, Molokai, and Kauai. The youngest ecosystems, which were on Hawaii, had developed on basaltic lava flows that were 300, 2,100, 20,000, and 150,000 years old. The study site on Molokai had developed on rocks that were 1,400,000 years old and the oldest study site, which was on Kauai, was 4,100,000 years old. All sites currently have an average annual temperature of about 16°C and receive approximately 2,500 mm of precipitation annually. They also all support forest communities dominated by the native tree, *Metrosideros polymorpha.*

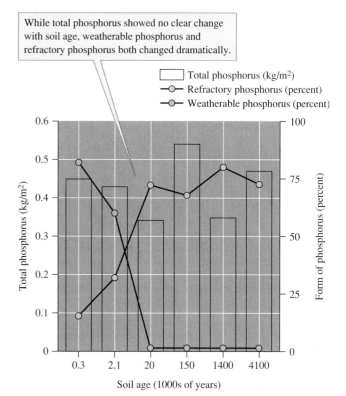

**Figure 20.12** Changes in the organic carbon and total nitrogen content of soils developing on Hawaiian lava flows ranging in age from 300 to 4.1 million years old (data from Hedin, Vitousek, and Matson 2003).

**Figure 20.13** Changes in the total phosphorus and percentages of total phosphorus in weatherable and refractory (low availability) forms in soils developing on Hawaiian lava flows ranging in age from 300 to 4.1 million years old (data from Hedin, Vitousek, and Matson 2003).

Over the chronosequence represented by their six study sites, Hedin, Vitousek, and Matson encountered significant changes in a wide range of soil features. Earlier studies had demonstrated that primary production in the Hawaiian forest ecosystems is limited by nitrogen early in succession and by phosphorus later in succession. Organic matter, which is absent from fresh lava, increased in soils over the first 150,000 years of the chronosequence (fig. 20.12). As we saw, analogous increases in soil organic matter also occur over the course of succession at Glacier Bay (see fig. 20.11). However, in the Hawaiian chronosequence, organic matter was lower at the 1.4 and 4.1 million-year-old sites. Figure 20.12 also shows that changes in soil nitrogen content followed almost precisely the pattern exhibited by soil organic matter.

The pattern of change in the total phosphorus content of soils was remarkably different (fig. 20.13). The total amount of phosphorus in soils showed no obvious pattern of change with site age. However, the forms of phosphorus changed substantially over the chronosequence. Weatherable mineral phosphorus was largely depleted by 20,000 years. Meanwhile the percentage of soil phosphorus in refractory forms, which are not readily available to plants, increased, varying from 68% to 80% of total phosphorus across ecosystems that had developed on lava flows 20,000 years old or older. On these older soils, primary production is limited by phosphorus availability.

Hedin, Vitousek, and Matson found changes in rates of nutrient loss across the chronosequence. Over the course of 4 million years of ecosystem development, these tropical forest ecosystems show progressively higher rates of nitrogen loss but decreased rates of phosphorus loss (fig. 20.14). In

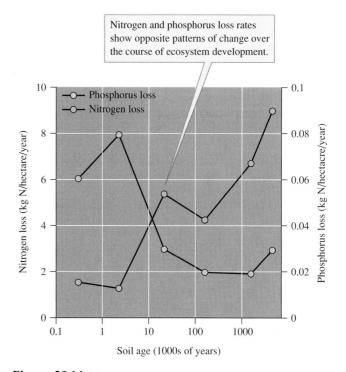

**Figure 20.14** Nitrogen and phosphorus loss rates from soils developing on Hawaiian lava flows ranging in age from 300 to 4.1 million years old (data from Hedin, Vitousek, and Matson 2003).

other words, for approximately 2,000 years these ecosystems are highly retentive of nitrogen but as nitrogen content increases in their soils, they begin to lose nitrogen at a higher rate. Most losses are due to leaching to groundwater. In contrast, as phosphorus becomes progressively less available, and eventually limiting to primary production, in these ecosystems, they become more retentive of phosphorus. As we shall see in the next example, intact vegetative cover may play a key role in nutrient retention in forest ecosystems.

# Recovery of Nutrient Retention Following Disturbance

In chapters 1 and 19, we saw how felling trees in the Hubbard Brook Experimental Forest substantially increased nutrient losses. Let's look again at this experiment by Bormann and Likens (1981) to see what it showed about succession and nutrient retention.

Briefly, Bormann and Likens monitored a control and an experimental stream catchment for 3 years prior to their experimental treatment. They then cut the forest on their experimental catchment and suppressed regrowth of vegetation with herbicides for 3 years (Likens et al. 1978). By suppressing vegetative growth, they delayed succession.

When herbicide applications were stopped, succession proceeded and nutrient losses by the forest ecosystem decreased dramatically. As you can see in figure 20.15, the herbicide suppressed vegetative growth on the experimental catchment for at least 3 consecutive years. It was during this period that the experimental catchment lost large quantities of nutrients, including calcium, potassium, and nitrate.

When herbicide applications stopped in 1969, Likens's group observed simultaneous increases in primary production and decreases in nutrient loss. However, the researchers point out that uptake by vegetation cannot account completely for reduced nutrient loss and that losses of calcium, potassium, and nitrate all peaked during the time when herbicide was still being applied. They suggest that some of the reduced losses during this period can be attributed to reduced amounts of these nutrients in the ecosystem. In other words, nutrient losses were reducing nutrient pools. However, vegetative uptake is clearly implicated since once succession was allowed to occur, nutrient losses from the experimental catch-

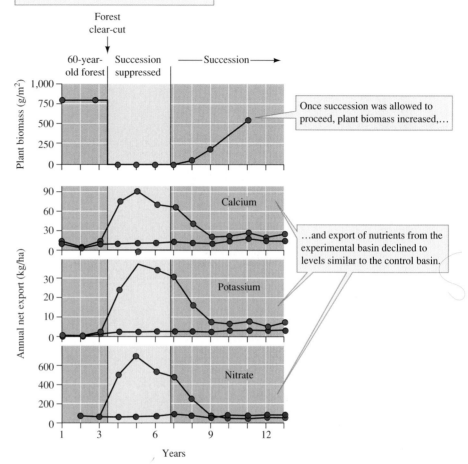

**Figure 20.15** Succession following deforestation and nutrient retention (data from Likens et al. 1978).

ment declined rapidly. Though losses of nitrate returned to predisturbance levels within 4 years, calcium and potassium losses remained elevated above predisturbance levels even after 7 years of forest succession.

## A Model of Ecosystem Recovery

As a result of their observations on the Hubbard Brook Experimental Forest, Bormann and Likens proposed a model for recovery of ecosystems from disturbance (fig. 20.16). Their "biomass accumulation model" divides the recovery of a forest ecosystem from disturbance into four phases: (1) a reorganization phase of 10 to 20 years, during which the forest loses biomass and nutrients, despite accumulation of living biomass; (2) an aggradation phase of more than a century, when the ecosystem accumulates biomass, eventually reaching peak biomass; (3) a transition phase, during which biomass declines somewhat from the peak reached during the aggradation phase; and (4) a steady state phase, when biomass fluctuates around a mean level.

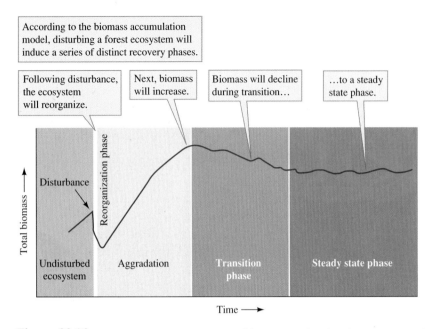

According to the biomass accumulation model, disturbing a forest ecosystem will induce a series of distinct recovery phases.

Following disturbance, the ecosystem will reorganize.

Next, biomass will increase.

Biomass will decline during transition…

…to a steady state phase.

**Figure 20.16** The biomass accumulation model of forest succession (data from Bormann and Likens 1981).

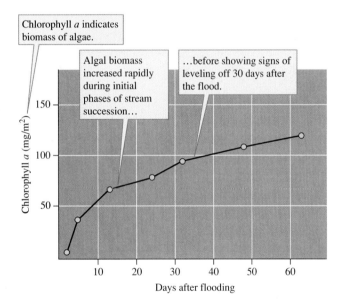

Chlorophyll *a* indicates biomass of algae.

Algal biomass increased rapidly during initial phases of stream succession…

…before showing signs of leveling off 30 days after the flood.

**Figure 20.17** Changes in biomass during stream succession (data from Fisher et al. 1982).

How well does the biomass accumulation model represent the process of forest succession? Does a similar sequence of stages occur during succession in other ecosystems? For instance, do ecosystems eventually reach a steady state? The generality of the biomass accumulation model can be tested on ecosystems, such as Sycamore Creek, Arizona, that undergo rapid succession. Such ecosystems give the ecologist the chance to study multiple successional sequences. As we will see in the following example, the patterns of ecosystem change during succession on Sycamore Creek suggest that several ecosystem features eventually reach a steady state.

# Succession and Stream Ecosystem Properties

Patterns similar to those proposed by the biomass accumulation model were recorded by Fisher's research group during just 63 days of postflood succession in Sycamore Creek, Arizona. Algal biomass increased rapidly for the first 13 days following disturbance and then increased more slowly from day 13 to day 63 (fig. 20.17). Sixty-three days after the flood, algal biomass showed clear signs of leveling off. The biomass of invertebrates, the chief animal group in Sycamore Creek, increased rapidly for 22 days following the flood and then, like the algal portion of the ecosystem, began to level off.

Ecosystem metabolic parameters showed even clearer signs of leveling off before the end of the 63-day study (fig. 20.18). Gross primary production (see chapter 18), measured as grams of $O_2$ produced per square meter per day, increased very rapidly until day 13, increased more slowly between days 13 and 48, and then leveled off between days 48 and 63. Total ecosystem respiration, measured as oxygen consumption per square meter per day, increased quickly for only 5 days after the flood and then began to level off. Respiration by invertebrates, which at its maximum represented about 20% of total ecosystem respiration, leveled off by day 63.

Nancy Grimm (1987) studied nitrogen dynamics in Sycamore Creek following floods that occurred from 1981 to 1983. As in the earlier studies by Fisher and his colleagues (1982), Grimm found that during succession, algal biomass and whole ecosystem metabolism quickly reached a maximum and then leveled off, as did the quantity of nitrogen in the system.

In addition, however, Grimm examined patterns of nitrogen retention during stream succession. She estimated the nitrogen budget in each of her study reaches by comparing the nitrogen inputs at the upstream end to nitrogen outputs at the downstream end. Each 60 to 120 m study reach began where subsurface flows upwelled to the surface and ended downstream, where water disappeared into the sand. Grimm used the ratio of dissolved inorganic nitrogen entering the study reach in the upwelling zone to the amount leaving at the lower end as a measure of nitrogen retention by the stream ecosystem.

Figure 20.19 shows that in the early stages of succession, approximately equal amounts of dissolved inorganic nitrogen entered and left Grimm's study reaches. What do equal levels of input and output indicate regarding nutrient retention? A balance between input and output means that the ecosystem shows no, or zero, retention. The level of retention increased rapidly during succession, leveling off at nearly 200 mg N per square meter per day, about 28 days after a flood. In other

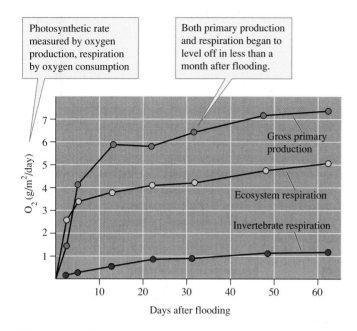

Photosynthetic rate measured by oxygen production, respiration by oxygen consumption

Both primary production and respiration began to level off in less than a month after flooding.

Gross primary production

Ecosystem respiration

Invertebrate respiration

**Figure 20.18** Ecosystem processes during succession in Sycamore Creek, Arizona (data from Fisher et al. 1982).

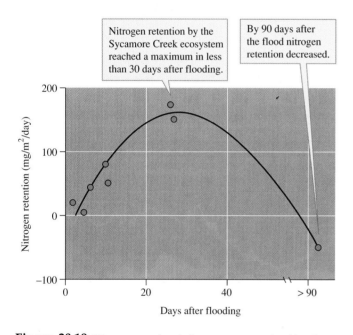

Nitrogen retention by the Sycamore Creek ecosystem reached a maximum in less than 30 days after flooding.

By 90 days after the flood nitrogen retention decreased.

**Figure 20.19** Nitrogen retention during stream succession (data from Grimm 1987).

words, the study reach was accumulating 200 mg N per square meter per day. Then, between 28 days and 90 days after the flood, the study reach showed progressively lower retention until it eventually exported a little more dissolved inorganic nitrogen than came in with groundwater.

The results of Grimm's study raise several questions. First, what mechanisms underlie retention? Grimm attributes most retention by the Sycamore Creek ecosystem to uptake by algae and invertebrates, since levels of nitrogen retention are consistent with the rates at which nitrogen was accumu-

lated by algal and animal populations. What causes the stream reaches to eventually export nitrogen? Grimm suggested that at 90 days postflood her study sites may have stopped accumulating biomass or may have even begun to lose biomass. A loss of biomass in the later stages of succession is consistent with the predictions of the Bormann and Likens biomass accumulation model.

The major point here is that succession, which produces changes in species composition and species diversity, also changes the structure and function of ecosystems ranging from forests to streams. However, we are left with a major question concerning this important ecological process. What mechanisms drive succession? Ecologists have proposed that the mechanisms underlying succession may fall into one of three categories. Those mechanisms are the subject of the next Concept Discussion section.

## CONCEPT DISCUSSION

### Mechanisms of Succession

**Mechanisms that drive ecological succession include facilitation, tolerance, and inhibition.**

An early model for successional change proposed by F. E. Clements (1916) emphasized the role of facilitation as a driver of ecological succession. Later, Joseph Connell and Ralph Slatyer (1977) proposed three alternative models of succession: (1) facilitation, (2) tolerance, and (3) inhibition. This classic paper stimulated ecologists to think beyond facilitation and to also consider tolerance and inhibition as mechanisms underlying successional change (fig. 20.20).

### Facilitation

The **facilitation model** proposes that many species may attempt to colonize newly available space but only certain species, with particular characteristics, are able to establish themselves. These species, capable of colonizing new sites, are called pioneer species. According to the facilitation model, pioneer species modify the environment in such a way that it becomes less suitable for themselves and more suitable for species characteristic of later successional stages. In other words, these early successional species "facilitate" colonization by later successional species. Early successional species disappear as they make the environment less suitable for themselves and more suitable for other species. Replacement of early successional species by later successional species continues in this way until resident species no longer facilitate colonization by other species. This final stage in a chain of facilitations and replacements is the climax community.

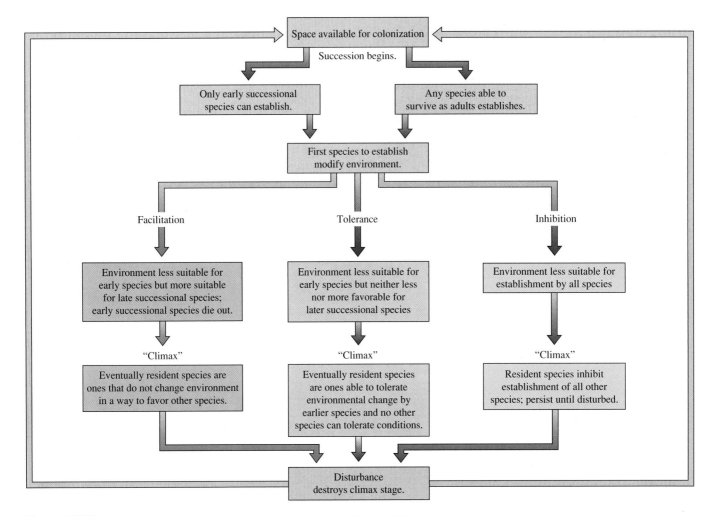

**Figure 20.20**  Alternative successional mechanisms (data from Connell and Slatyer 1977).

## *Tolerance*

How does the **tolerance model** differ from the facilitation model? First, the initial stages of colonization are not limited to a few pioneer species. Juveniles of species dominating at climax can be present from the earliest stages of succession. Second, species colonizing early in succession do not facilitate colonization by species characteristic of later successional stages. They do not modify the environment in a way that makes it more suitable for later successional species. Later successional species are simply those tolerant of environmental conditions created earlier in succession. The climax community is established when the list of tolerant species has been exhausted.

## *Inhibition*

Like the tolerance model, the **inhibition model** assumes that any species that can survive in an area as an adult can colonize the area during the early stages of succession. However, the inhibition model proposes that the early occupants of an area modify the environment in a way that makes the area less suitable for both early and late successional species. Simply, early

arrivals inhibit colonization by later arrivals. Later successional species can only invade an area if space is opened up by disturbance of early colonists. In this case, succession culminates in a community made up of long-lived, resistant species. The inhibition model assumes that late successional species come to dominate an area simply because they live a long time and resist damage by physical and biological factors.

Which of these models does the weight of evidence from nature support? As you will see in the following examples, most studies of succession support the facilitation model, the inhibition model, or some combination of the two.

## Successional Mechanisms in the Rocky Intertidal Zone

What mechanisms drive succession by algae and barnacles in the intertidal boulder fields studied by Sousa (see fig. 20.7)? The alternative mechanisms proposed by Sousa were those of Connell and Slatyer: facilitation, tolerance, and inhibition. Sousa used a series of experiments to test for the occurrence of these alternative mechanisms. He conducted his first exper-

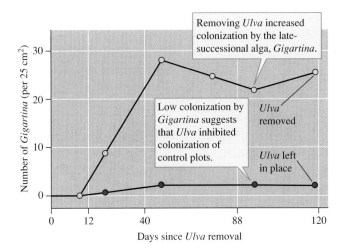

**Figure 20.21** Evidence for inhibition of later successional species (data from Sousa 1979a).

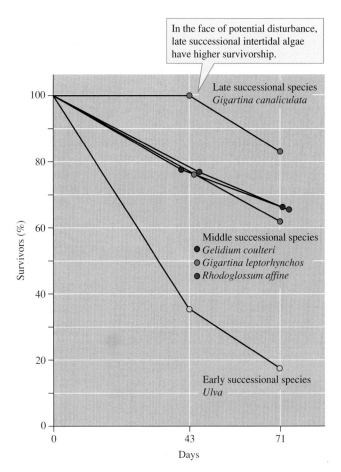

**Figure 20.22** Survivorship of early, middle, and late successional species (data from Sousa 1979b).

iments on 25 cm² plots on concrete blocks placed in the intertidal zone. In this experiment, Sousa explored the influence of *Ulva* on recruitment by later successional red algae by keeping *Ulva* out of four experimental plots and leaving four other control plots undisturbed. This experiment showed that *Ulva* strongly inhibits recruitment by red algae (fig. 20.21).

In a second set of experiments, Sousa studied the effects of the middle successional species *Gigartina leptorhynchos* and *Gelidium* on establishment of the late successional *Gigartina canaliculata*. He selectively removed middle successional species from a set of four experimental plots while simultaneously monitoring another set of four control plots. These experiments were conducted in 100 cm² areas on natural substrate, dominated by either *G. leptorhynchos* or *Gelidium*. When Sousa removed these middle successional species, the experimental plots were quickly reinvaded by *Ulva* and eventually by significantly higher densities of *G. canaliculata,* the late successional species. The effects of these successional algae support the inhibition model for succession.

The inhibition model of succession proposes that early successional species are more vulnerable to a variety of physical and biological factors causing mortality. If algal succession in the intertidal boulder fields studied by Sousa follows the inhibition model, then early successional species should be more vulnerable to various sources of mortality.

Sousa addressed the question of relative vulnerability of algal species with several experiments. In one, he studied the relative vulnerability of intertidal algae to physical stress, especially exposure to air, intense sunlight, and drying wind. He studied the vulnerabilities of the five dominant algal species in his study area by tagging 30 individuals of each species and monitoring their survivorship for 2 months during a period when low tide occurred during the afternoon, when air temperatures are highest. The results of this study show that the early successional species, *Ulva*, had lower survivorship than the middle or late successional species (fig. 20.22).

Sousa also designed several different field and laboratory experiments to explore differential vulnerability to herbi-

vores. The results of all these experiments indicated that the early successional species *Ulva* is more vulnerable to herbivores than later successional species. These results and those of the several other manipulations performed by Sousa support the inhibition model of succession.

Some studies of intertidal succession, however, have demonstrated facilitation. Teresa Turner (1983) pointed out that the bulk of intertidal studies had supported the inhibition model and that the few studies documenting facilitation had shown that facilitation was not obligate. However, she went on to report a case of obligate facilitation during intertidal succession.

Turner described the successional sequence at her Oregon study site as follows. High waves during winter storms create open space in the lower intertidal zone. In May, these open areas are colonized by *Ulva,* the same early colonist of open areas in Sousa's study area, over 1,000 km south of Turner's study site. *Ulva* is eventually replaced by several middle successional species, especially the red algae *Rhodomela larix, Cryptosiphonia woodii,* and *Odonthalia floccosa.* Through this middle stage, the pattern of succession appears much as that in the intertidal boulder field studied by Sousa. However, in the lower intertidal area studied by Turner, the dominant late successional species was not an alga but a flowering plant, the surfgrass *Phyllospadix scouleri.*

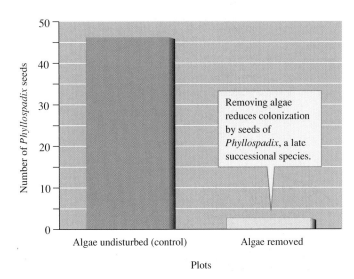

**Figure 20.23** Evidence for facilitation of colonization by an intertidal plant, *Phyllospadix scouleri* (data from Turner 1983).

Turner proposed that recruitment of *Phyllospadix* by seeds depends upon the presence of macroscopic algae. The seeds of *Phyllospadix* are large and bear two parallel, barbed projections. These projections hook and hold the seeds to attached algae. From this attached position, the seed germinates, first producing leaves and then roots by which the plant will anchor itself to the underlying rock. Once established, *Phyllospadix* spreads and consolidates space by vegetative growth.

Turner tested whether recruitment by *Phyllospadix* is facilitated by attached algae by clearing eight 0.25 m² plots of all attached algae. She then compared the number of new *Phyllospadix* seeds in these plots with the number in eight nearby control plots. The control plots remained undisturbed with their algal populations intact except that all *Phyllospadix* seeds were removed at the start of the study. Turner's control areas were dominated by the red alga *Rhodomela larix*, a species prominent in the middle successional stages in her study area and to which *Phyllospadix* seeds attach.

Turner set up and manipulated her study plots in September and then checked them the following March, after the period of seed dispersal. Over the fall and winter a brown alga, *Phaeostrophion irregulare*, colonized the removal plots but the bladelike form of this species apparently does not allow attachment by *Phyllospadix* seeds. When Turner checked the removal and control plots, she found a total of 48 seeds, 46 on the control plots (all attached to *Rhodomela*) and 2 on the removal plots (fig. 20.23). Both seeds on the removal plots were attached to two isolated branches of *Rhodomela* that had sprouted from remnant holdfasts.

During 3 years, Turner systematically searched an area of about 200 m² for *Phyllospadix* seeds and found a total of 298. All were attached to algae. These data support the hypothesis that middle successional algae facilitate recruitment and establishment of *Phyllospadix* and that this facilitation is obligate. As a consequence of Turner's study and others, we can say that facilitation and inhibition occur during intertidal suc-

cession. Other research, which we review in the next example, has shown that facilitation and inhibition also occur during forest succession.

# Successional Mechanisms in Forests

We now turn from succession in the marine intertidal zone, a place where succession occurs in a matter of a few years, to succession in temperate forests. Temperate forest succession takes hundreds of years to complete and so cannot be observed directly within the period of a typical research project. Therefore, most research on the mechanisms driving succession in temperate forests has focused on the earliest stages.

## Mechanisms in Old Field Succession

Catherine Keever (1950) studied succession on old fields of the Piedmont Plateau of North Carolina. She conducted some of the earliest experiments on the mechanisms regulating the early stages of succession in temperate forests. As we saw earlier in chapter 20, the first species to colonize and dominate abandoned fields is generally crabgrass, *Digitaria sanguinalis*. *Digitaria* is usually followed by horseweed, *Erigeron canadense*. During the second year of succession, many fields are dominated by aster, *Aster pilosis*. A few years later the fields are covered by broomsedge, *Andropogon virginicus*. The objective of Keever's study was to determine the causes of this early pattern of species replacements.

Keever's experiments showed that *Erigeron* inhibits the growth of *Aster* and so the replacement of *Erigeron* by *Aster* follows the inhibition model. However, *Aster* stimulates the growth of its successor, *Andropogon,* so this second species replacement follows the facilitation model. Therefore, early succession on the Piedmont Plateau of eastern North America appears to involve a mixture of mechanisms. As the following examples show, a complex mixture of mechanisms is also involved in primary succession of forests on volcanic and glacial substrates.

## Mechanisms in Primary Succession on a Volcanic Substrate

In 1980, Mount St. Helens in the state of Washington erupted with a massive avalanche of rock and snow and a lateral blast of rock and steam superheated to 300°C. The blast spread at speeds of approximately 300 km per hour and carried rocks varying from the size of sand grains to boulders several meters in diameter. The eruption was followed by massive mudflows 50 to 60 m deep and deposition of volcanic ash. Geologists estimated that the volcano ejected up to 50 million m³ of material. The disturbance created by Mount St. Helens set the stage for primary succession that has been carefully followed by a large group of scientists.

The eruption of Mount St. Helens destroyed approximately 600 km² of forests made up of trees up to 60 m in height. In the farthest reaches of the blast zone, vegetation reestablished rapidly from surviving seeds and root fragments. However, immediately north of the volcano there was a 20 km² area of nearly complete devastation known as the pumice plains. A combination of avalanche debris and a blanket of hot volcanic ash and pumice killed all plant life and created an area where ecologists could study primary succession on a barren substrate. Two common species colonizing the pumice plains were pearly everlasting, *Anaphalis margaritacea,* and fireweed, *Epilobium angustifolium.* The seeds of both these species are dispersed by wind and readily colonize over long distances. However, the third common species, a perennial lupine, *Lupinus lepidus,* does not disperse easily. Consequently a few seeds or root fragments of this plant probably survived to found the populations that soon grew up on the pumice plains.

William Morris and David Wood (1989) studied the relative influences of facilitation, tolerance, and inhibition on early succession on the pumice plains. They concentrated on the effects of lupine, a nitrogen-fixing plant that increases the nitrogen content of the soils where it grows on the pumice plains. They examined how this species affects survival, growth, and flowering by *Anaphalis* and *Epilobium.*

In one of their experiments, Morris and Wood tested the hypothesis that live lupine inhibit survival and growth of seedlings but that dead lupine improves seedling growth and survival. This experiment included four treatments: (1) "control" plots on barren ground, (2) "live" treatments where *Anaphalis* and *Epilobium* seedlings were planted within patches of vigorous lupine, (3) "mulch" treatments where lupine was killed by a nonpersistent herbicide 3 weeks before planting *Anaphalis* and *Epilobium* seedlings and the litter of lupine left in place, and (4) "no mulch" treatments where lupine was also sprayed with the herbicide but all litter was removed before planting *Anaphalis* and *Epilobium* seedlings.

What information could the mulch and no mulch treatments provide? Morris and Wood designed these treatments to distinguish between potential belowground and aboveground influences of lupine. Remember that this plant increases the nitrogen content of soils, a potential belowground influence. In addition, it also may have aboveground effects because it adds a surface mulch of dead leaves and stems that reduces the rate of water loss from soils. Any effect of lupine on seedling performance could be the result of aboveground effects, belowground effects, or a combination of the two.

The experiment gave results consistent with both inhibition and facilitation. Seedling survival during the first growing season was highest on the barren control plots, which suggests inhibition. During the second growing season, however, the pattern of survival was reversed and the proportion of surviving seedlings was lowest on the control plots, which suggests facilitation. Contrary to what Morris and Wood expected,

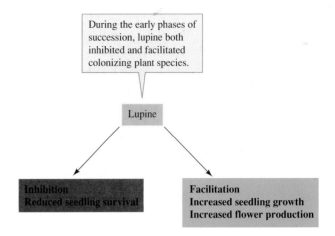

**Figure 20.24**  Effect of the lupine on other plant species colonizing the blast zone of Mount St. Helens.

seedlings survived at about the same rate on the mulch, no mulch, and live treatments.

Lupine had a positive effect on the growth of both *Anaphalis* and *Epilobium,* a result consistent with the facilitation model. The seedlings of both *Anaphalis* and *Epilobium* were larger on the plots with lupine than on control plots. However, during the first growing season, *Anaphalis* grew larger in mulched plots than it did in plots with no mulch or with live lupine.

Patterns of flowering partially support the facilitation hypothesis. *Anaphalis* produced more flowers in the plots with lupine than in the control plots. However, during both the first and second growing seasons *Anaphalis* produced fewer flowers in plots with live lupine than in the no mulch and mulch plots, suggesting that live lupine may partially inhibit flowering by *Anaphalis. Epilobium* did not produce many flowers until the second growing season and then flowered at about the same rates on control and treatment plots.

In summary, we can see that Morris and Wood found a complex blend of influences of lupine on other plant species during the early stages of primary succession on Mount St. Helens (fig. 20.24). A similar blending of mechanisms appears to occur during primary succession at Glacier Bay.

## Mechanisms of Primary Succession Following Deglaciation

The complex mechanisms underlying succession were well demonstrated by the detailed studies of Chapin's research team (1994). They combined field observations, field experiments, and greenhouse experiments to explore the mechanisms underlying primary succession at Glacier Bay, Alaska. Like Morris and Wood, they found that no single factor or mechanism determines the pattern of primary succession at Glacier Bay, Alaska.

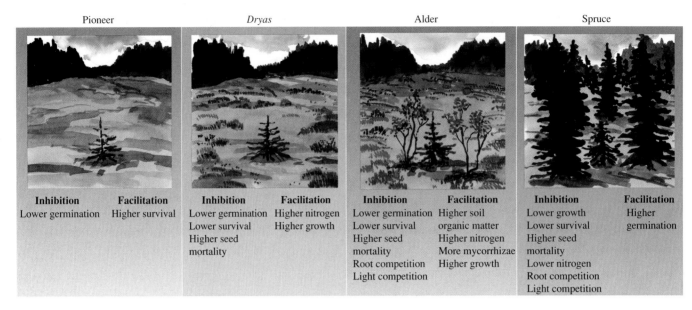

**Figure 20.25** Inhibition and facilitation of spruce during the major successional stages at Glacier Bay, Alaska (data from Chapin et al. 1994).

Figure 20.25 summarizes the complex influences of four successional stages on establishment and growth of spruce seedlings. During the pioneer stage, there is some inhibition of spruce germination. Any spruce seedlings that become established, however, have high survivorship but low growth rates. Spruce seedling growth rates and nitrogen supplies are increased somewhat during the *Dryas* stage. However, this facilitation during the *Dryas* stage is offset by poor germination and survivorship, along with increased seed predation and mortality.

Strong facilitation of spruce seedlings first occurs in the alder stage. During this stage, germination and survivorship remain low and seed mortality, root competition, and light competition are significant. However, these inhibitory effects are offset by increased soil organic matter, nitrogen, mycorrhizal activity, and growth rates. The net effect of alder on spruce seedlings is facilitation.

In the spruce stage, the net influence on spruce seedlings is inhibitory. Germination is high during the spruce stage but this is counterbalanced by several inhibitory effects. Growth rates and survivorship are low and nitrogen availability is reduced. In addition, seed predation and mortality, root competition, and light competition are all high.

These results combined with those of Morris and Wood on Mount St. Helens remind us that nature is far more complex and subtle than models such as that proposed by Connell and Slatyer. However, the Connell and Slatyer model challenged ecologists to think more broadly about succession and to go out and conduct field tests of alternative successional mechanisms. Their response produced today's improved understanding of the process of ecological succession.

In this and the previous two Concept Discussion sections, we have discussed community and ecosystem changes and the mechanisms producing those changes. In the next Concept Discussion section, we consider a companion topic: community and ecosystem stability.

## CONCEPT DISCUSSION

### Community and Ecosystem Stability

**Community stability may be due to lack of disturbance or community resistance or resilience in the face of disturbance.**

## Some Definitions

The simplest definition of **stability** is the absence of change. A community or ecosystem may be stable for a variety of reasons. One reason may be that there has been no disturbance. For instance, the benthic communities of the deep sea may remain stable over long periods of time because of constant physical conditions. The type of stability resulting from an absence of disturbance, if it exists, is not particularly interesting to ecologists.

Ecologists are more interested in how communities and ecosystems may remain stable even when exposed to potential disturbance. Consequently ecologists generally define stability as the persistence of a community or ecosystem in the face of disturbance. Stability may result from two very different characteristics. **Resistance** is the ability of a community or ecosystem to maintain structure and/or function in the face of potential disturbance. However, stability may also result from the ability of a community to return to its original structure after a disturbance. The ability to bounce back after disturbance is called **resilience.** A resilient community or ecosystem may be completely disrupted by disturbance but quickly return to its former state.

What causes communities and ecosystems to be resilient? The phenomenon of resilience takes us back to succession. Remember that we defined succession as the gradual change in plant and animal communities in an area following disturbance or the creation of new substrate. It is succession that restores a community disrupted by disturbance. Succession is the basis for resilience.

Ecologists ask many questions about stability. Are some communities and ecosystems more resistant than others? What factors determine differences in resistance among communities and ecosystems? Are some ecosystems and communities more resilient than others? What factors determine the rate of recovery of community structure and ecosystem processes following disturbance? However, few studies have been conducted at scales appropriate to address these questions. One of the main problems faced by ecologists interested in community and ecosystem stability is the need to conduct detailed studies over a long period of time. There are a few studies that meet this requirement; one of them is the Park Grass Experiment.

# Lessons from the Park Grass Experiment

The Park Grass Experiment is the prototype of all long-term experimental studies in ecology. It was started at the Rothamsted Experimental Station in Hertfordshire, England, between 1856 and 1872. The purpose of the experiment was to study the effects of several fertilizer treatments on the yield and structure of a hay meadow community. Because the Park Grass Experiment has continued without interruption for nearly one and a half centuries, it provides one of the most valuable records of long-term community dynamics. That record provides some unique insights into the nature of community stability.

Jonathan Silvertown (1987) used data from the Park Grass Experiment to respond to the suggestion that existing studies do not conclusively demonstrate that any ecological community is stable. Silvertown pointed out that the Park Grass Experiment is one of the few studies of terrestrial communities that have been carried out in sufficient detail and over sufficient time to provide a test of stability that meets the rigorous requirements suggested by Connell and Sousa.

The composition of the plant community at the Park Grass Experiment has been monitored since 1862. This record reveals at least one level of stability. Over this period, virtually no new species have colonized the meadow. Changes in the community have occurred as a consequence of increases or decreases in species already present in the meadow at the beginning of the experiment.

Silvertown used variation in community composition as a measure of stability. He represented composition as the proportion of the community consisting of grasses, legumes, or other species. The analysis of composition was restricted to the period from 1910 to 1948 to avoid the early period of the experiment when the meadow community was adjusting to the various fertilizer treatments. Figure 20.26 shows the relative proportions of grasses, legumes, and other plants on plots receiving three different treatments: plot 3, no fertilizer; plot 7, P, K, Na, and Mg; and

plot 14, N, P, K, Na, and Mg. The differences in vegetation on the three plots were mostly produced by the different fertilizer treatments and developed early in the Park Grass Experiment.

The proportion of grasses, legumes, and other plants in the study plots varied from year to year, mainly in response to variation in precipitation. Despite this annual variation, figure 20.26 indicates that the proportions of three plant groups remained remarkably similar over the interval of the study. A quantitative analysis of trends in biomass revealed no significant changes in the biomass of the three plant groups in plots 3 and 7 and only a minor, but statistically significant, decrease in the biomass of grasses on plot 14. In other words, the data presented in figure 20.26 show remarkable stability in the proportion of grasses, legumes, and other species.

Does the stability of Silvertown's three major groups of plants in the Park Grass Experiment hold up if we examine community structure at the species level? It turns out that while the proportions of grasses, legumes, and other species remained fairly constant, populations of individual species changed substantially. Mike Dodd and his colleagues (1995) used census data from 1920 to 1979 to examine plant population trends. The result of their analysis showed that some species increased in abundance, some decreased, some showed no trend, while others increased and then decreased (fig. 20.27).

The contrasting results obtained by Silvertown and by Dodd's project suggest that whether a community or ecosystem appears stable may depend upon how we view it. At a very coarse level of resolution, the Park Grass community has remained absolutely stable. It was a meadow community when the Park Grass Experiment began in 1856 and it remains so today. When Silvertown increased the resolution to distinguish between grasses, legumes, and other species, the community again appeared stable. However, when Dodd and his colleagues increased the resolution still further and examined trends in the abundances of individual species, the Park Grass community no longer appeared stable.

Are there stable natural communities? The answer to this question may depend upon how you make your measurements. The ecologist interested in addressing any question concerning community stability is faced with several practical problems. Generally, an adequate study requires a great deal of time, which limits the possibility of replication. One solution to this problem is to study communities and ecosystems, such as Sycamore Creek, Arizona, that undergo more frequent disturbance and show relatively rapid recovery. These systems offer the opportunity to compare recoveries from multiple disturbances.

# Replicate Disturbances and Desert Stream Stability

Numerous studies of disturbance and recovery in Sycamore Creek, Arizona, have produced a highly detailed picture of community, ecosystem, and population responses. This detailed picture suggests that ecologists have just begun to probe the subtleties of ecological stability. For instance, one study shows

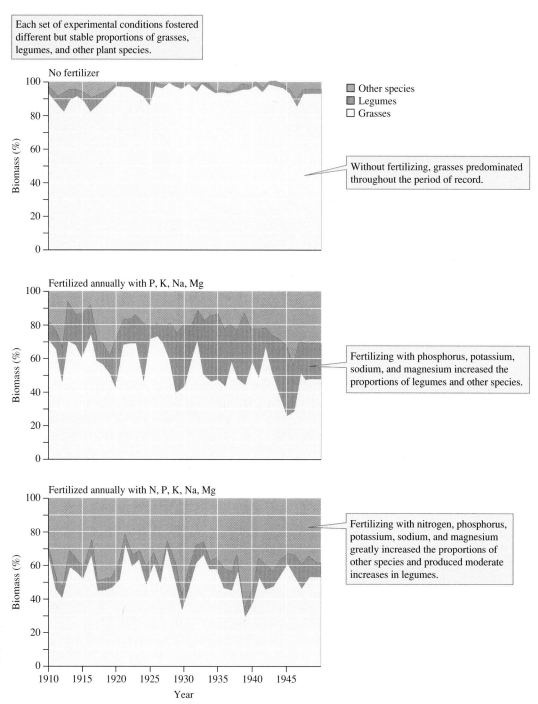

**Figure 20.26** Proportions of grasses, legumes, and other plant species under three experimental conditions (data from Silvertown 1987).

that resistance in the spatial structure of the Sycamore Creek ecosystem underlies spatial variation in ecosystem resilience. Maury Valett and his colleagues (1994) studied the interactions between surface and subsurface waters in Sycamore Creek in order to study the influence of these linkages on ecosystem resilience. They tested the hypothesis that ecosystem resilience is higher where hydrologic linkages between the surface and subsurface water increase the supply of nitrogen. They proposed a controlling role for nitrogen because it is the nutrient that limits primary production in Sycamore Creek.

Valett and his colleagues intensely studied two stream sections at middle elevations in the 500 km² Sycamore Creek catchment. They measured the flow of water between the surface and subsurface along these reaches with devices called *piezometers*. Piezometers can be used to measure the vertical hydraulic gradient, which indicates the direction of flow between surface water and water flowing through the sediments of a streambed. Positive vertical hydraulic gradients indicate flow from the streambed to the surface in areas called upwelling zones. Negative vertical hydraulic gradients indi-

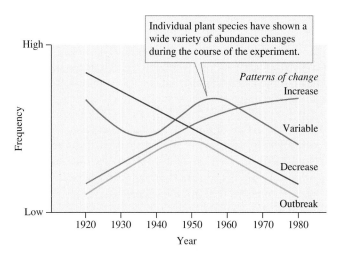

**Figure 20.27** Patterns of species abundance during 60 years of the Park Grass Experiment (data from Dodd et al. 1995).

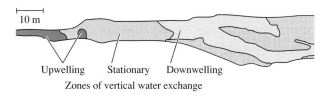

**Figure 20.28** Patterns of upwelling and downwelling in a reach of Sycamore Creek, Arizona (data from Valett et al. 1994).

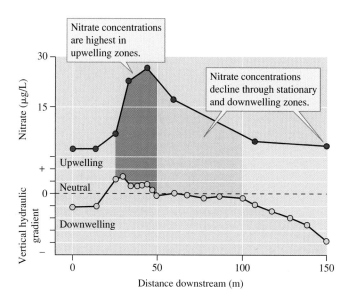

**Figure 20.29** Relationship of nitrate to vertical hydraulic gradient in Sycamore Creek, Arizona (data from Valett et al. 1994).

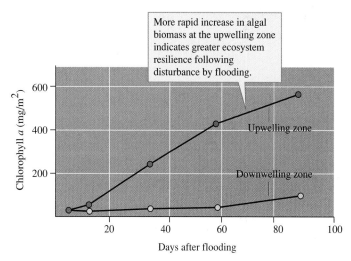

**Figure 20.30** Changes in algal biomass, measured as chlorophyll *a*, following flooding at upwelling and downwelling zones (data from Valett et al. 1994).

cate flow from the surface to the streambed, which occurs in downwelling zones. Zero vertical hydraulic gradients indicate no net exchange between surface waters and water flowing within the sediments. Areas with zero vertical hydraulic gradients are called stationary zones.

Valett and his colleagues measured vertical hydraulic gradient along the lengths of both study sections, producing hydrologic maps for both. The upper end of each study reach was an upwelling zone. The middle reaches were stationary zones and the lower reaches were downwelling zones. Figure 20.28 shows the distributions of these zones across one of the study reaches.

The concentration of nitrate in surface water in the two study reaches varies directly with vertical hydraulic gradient (fig. 20.29). Upwelling zones, which are fed by nitrate-rich waters upwelling from the sediments, have the highest concentrations of nitrate. Nitrate concentrations gradually decline with distance downstream through the stationary and downwelling zones.

The higher concentrations of nitrate in the upper reaches of each study section are associated with higher algal production. Figure 20.30 indicates that algal biomass accumulates at a higher rate in upwelling zones compared to downwelling zones. Valett and his colleagues used rate of algal biomass accumulation as a measure of rate of recovery from disturbance. Because the rate at which algal biomass accumulates in upwelling zones is so much higher than in downwelling zones, they concluded that the rate of ecosystem recovery is higher in

upwelling zones. This pattern supports their hypothesis that algal communities in upwelling zones are more resilient.

The team also found that while flash floods devastated the biotic community, the spatial arrangement of upwelling, stationary, and downwelling zones remained stable. In other words, this aspect of the spatial structure of the Sycamore Creek ecosystem is highly resistant to flash flooding. The location of upwelling, stationary, and downwelling zones remained stable in the face of numerous intense floods.

The spatial stability of the Sycamore Creek ecosystem in the face of potential disturbance is an example of ecosystem resistance. However, what is the source of this stable spatial structure? This spatial stability can be explained by considering geomorphology, especially the distribution of bedrock. Subsurface water is forced to the surface in areas where bedrock lies close to the surface. Upwelling zones in Sycamore

| Information |
| Questions |
| Hypothesis |
| Predictions |
| Testing ✓ |

*Investigating the Evidence*

# Variation Around the Median

The question we consider now is how to represent variation in samples drawn from populations in which measurements or observations do not have normal distributions. When analyzing normally distributed measurements, depending on our purpose, we can estimate and represent variation using the range, variance, standard deviation, standard error, or 95% confidence interval. However, most of these indices of variation are not appropriate for non-normal distributions.

To help us consider how to represent variation when analyzing non-normal distributions, let's return to a sample of mayfly nymphs that we considered in chapter 3 (table 1). Suppose you are studying the recovery of this population following disturbance by a flash flood. The sample was taken from the south fork of Tesuque Creek, New Mexico, a high mountain stream of the southern Rocky Mountains. This fork had flooded one year before the sample was taken.

**Table 1** Number of *Baetis bicaudatus* nymphs in 0.1 m² benthic samples from the disturbed fork of Tesuque Creek, New Mexico

| Quadrats: low to high | | | | | | | | | | | |
|---|---|---|---|---|---|---|---|---|---|---|---|
| 1 | 2 | 3 | 4 | 5 | 6 | 7 | 8 | 9 | 10 | 11 | 12 |

| Number of nymphs | | | | | | | | | | | |
|---|---|---|---|---|---|---|---|---|---|---|---|
| 2 | 2 | 2 | 3 | 3 | 4 | 5 | 6 | 6 | 8 | 10 | 126 |

Now consider the following sample that was taken on the same date, but from an undisturbed fork of the same stream.

**Table 2** Number of *Baetis bicaudatus* nymphs in 0.1 m² benthic samples from the undisturbed fork of Tesuque Creek, New Mexico.

| Quadrats: low to high | | | | | | | | | | | |
|---|---|---|---|---|---|---|---|---|---|---|---|
| 1 | 2 | 3 | 4 | 5 | 6 | 7 | 8 | 9 | 10 | 11 | 12 |

| Number of nymphs | | | | | | | | | | | |
|---|---|---|---|---|---|---|---|---|---|---|---|
| 12 | 30 | 32 | 35 | 37 | 38 | 42 | 48 | 52 | 58 | 71 | 79 |

In chapter 3 we determined the median density of *B. bicaudatus* in the disturbed fork (table 1) as:

Sample median = 4 + 5 = 4.5 *B. bicaudatus* per 0.1 m² quadrat

The median density of *B. bicaudatus* in the undisturbed fork (table 2) is:

Sample median = 38 + 42 = 40 *B. bicaudatus* per 0.1 m² quadrat

The median indicates that the density of *B. bicaudatus* is ten times higher in the undisturbed fork. Now, how can we represent the variation around these medians? One common method to represent variation in cases such as these is to divide the samples into four equal parts, called quartiles, and use the range of measurements between the upper bound of the lowest quartile and the lower bound of the highest quar-

tile. This representation of variation in a sample is called the **interquartile range.** In table 3, the data in tables 1 and 2 have been divided into quartiles with different colors:

**Table 3** Number of *Baetis bicaudatus* nymphs in 0.1 m² benthic samples from the undisturbed and disturbed forks of Tesuque Creek, New Mexico. The first, second, third, and fourth quartiles are shaded orange, yellow, green, and blue, respectively.

| Quadrats: low to high | | | | | | | | | | | |
|---|---|---|---|---|---|---|---|---|---|---|---|
| 1 | 2 | 3 | 4 | 5 | 6 | 7 | 8 | 9 | 10 | 11 | 12 |

| Number of nymphs, disturbed fork | | | | | | | | | | | |
|---|---|---|---|---|---|---|---|---|---|---|---|
| 2 | 2 | 2 | 3 | 3 | 4 | 5 | 6 | 6 | 8 | 10 | 126 |

| Number of nymphs, undisturbed fork | | | | | | | | | | | |
|---|---|---|---|---|---|---|---|---|---|---|---|
| 12 | 30 | 32 | 35 | 37 | 38 | 42 | 48 | 52 | 58 | 71 | 79 |

| Quartiles | **1st** | 2nd | **3rd** | 4th |
|---|---|---|---|---|

Notice that the interquartile range for the undisturbed fork is from 32 to 58; for the disturbed fork, the interquartile range is 2 to 8. Notice that 50% of the quadrat counts in each sample fall within this range. The medians and interquartile ranges for each of the populations are plotted in figure 1, which shows that they do not overlap. However, is there a statistically significant difference in density in the two stream forks? To answer that question, we will need a method for comparing samples that does not assume a normal distribution. We will make that comparison in chapter 21.

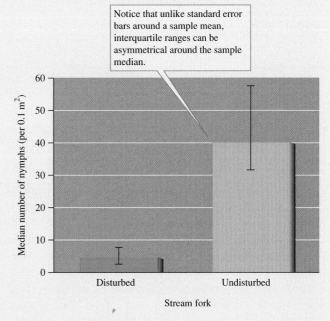

Notice that unlike standard error bars around a sample mean, interquartile ranges can be asymmetrical around the sample median.

**Figure 1** Medians and interquartile ranges of mayfly nymphs. *Baetis bicaudatus,* in 0.1 m² quadrats in disturbed and undisturbed forks of Tesuque Creek, New Mexico.

**Figure 20.31** Detecting change in plant populations using repeat photography: (*a*) MacDougal Crater, Sonora, Mexico, in 1907, (*b*) in 1959, (*c*) in 1972, and (*d*) in 1984.

Creek are located in such areas, and since flooding does not move bedrock, the locations of upwelling zones are stable. Therefore, this aspect of ecosystem stability is controlled by landscape structure. Consequently, the ecologist trying to understand the organization and dynamics of the Sycamore Creek ecosystem must consider the structure of the surrounding landscape. Landscape ecology is the subject of chapter 21.

## APPLICATIONS & TOOLS

### Using Repeat Photography to Detect Long-Term Change

While some graduate students look over their shoulders, Raymond Turner and Julio Betancourt of the U.S. Geological Survey carefully examine a photograph of a desert landscape taken about 100 years earlier. Their goal is to take another photograph of the same scene to document long-term change in the plant community. To do so they must return to the same location and take a photograph from exactly the same spot.

The larger landmarks such as hills and ridges will help them find the general location, but they need finer-scale reference points to locate the exact spot. Turner finally indicates a small boulder about 30 cm in diameter in the foreground, saying, "This should get us close and those small junipers will help orient the cameras." Betancourt agrees. The students are incredulous that someone should think that they can find a small boulder and two small trees after a century. However,

long practice at repeat photography has taught Turner what can be found after a century in the arid lands of the American Southwest.

A field trip later takes the group to the general area of the site. After a careful search, Betancourt finds the remains of the two junipers. They have died sometime during the last half century. Next, Turner finds the small boulder. They use a few more landmarks to orient the camera and then position it within about 1 m of the spot from which the century-old photo was taken.

Using techniques such as these, Ray Turner and his colleagues have produced a very useful photographic record of vegetation changes from throughout the southwestern United States and northwestern Mexico. For instance, a series of repeat photographs beginning in 1907 document substantial vegetation change in MacDougal Crater in northern Sonora, Mexico (Turner 1990). The crater is about 137 m deep and was formed by a volcanic eruption about 200,000 years ago. MacDougal Crater is protected by its steep walls from livestock and other human impacts. This protection removes the possibility that observed changes in vegetation might be the result of human influences.

Figure 20.31 (p. 505) shows a series of photographs taken of MacDougal Crater from 1907 to 1984. While most changes depicted by these photographs are subtle, there is one obvious change in the lower left corner, the location of a population of saguaro cactus, *Carnegiea gigantea*. The saguaro, which appear as small stick figures in the photo, increase in number between the 1907 photograph and the 1959 photograph. Though difficult to see with the naked eye, the saguaros are clearly visible with a magnifying glass. Get a magnifying glass and compare the numbers of *Carnegiea* in the 1907 and 1959 photographs.

(a)

(b)

(c)

**Figure 20.32** Details of plant population biology from repeat photography: (*a*) saguaro cactus in MacDougal Crater, Sonora, Mexico, in 1959, (*b*) same scene in 1984, and (*c*) in 1998. By 1998, the two cactus in the foreground of the 1959 photo had died and fallen.

Close-up photos reveal even more detail. Figure 20.32 shows photographs taken in 1959, 1984, and 1998. The growing conditions were so poor in 1959 that the live saguaro in the photograph formed permanent constrictions on its stems that are still visible in the 1984 photograph. This saguaro died between 1984 and 1998. The dead shrubs in the 1984 photograph are the remains of creosote bushes, *Larrea tridentata,* that apparently died in response to the same drought that formed the constrictions on the saguaro stems.

Using repeat photographs, Turner was able to quantify changes in the plant community of MacDougal Crater. One of the changes he documented was a decrease in the population of *Larrea* and an increase in the population of saguaros (fig. 20.33). From 1907 to 1986 the number of *Larrea* in Turner's study area decreased from 103 to 48. Over the same interval, the number of saguaros increased from 38 to 159 in 1972 and then declined to 140 by 1986.

Some of the most important questions asked by ecologists concern changes in the distribution and abundance of organisms. Repeat photography is an easily overlooked tool that is helping to document changes in plant distribution and abundance during the past century.

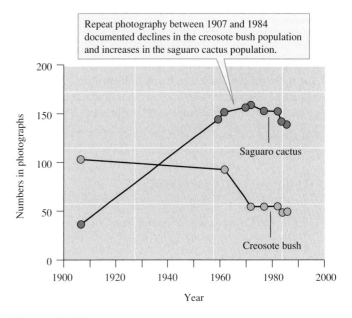

**Figure 20.33** Changes in populations of creosote bushes and saguaro cactus determined by repeat photography (data from Turner 1990).

# SUMMARY

*Succession* is the gradual change in plant and animal communities in an area following disturbance or the creation of new substrate. *Primary succession* occurs on newly exposed geological substrates not significantly modified by organisms. *Secondary succession* occurs in areas where disturbance destroys a community without destroying the soil. Succession generally ends with a climax community whose populations remain stable until disrupted by disturbance.

**Community changes during succession include increases in species diversity and changes in species composition.** Primary forest succession around Glacier Bay may require about 1,500 years, while secondary forest succession on the Piedmont Plateau takes about 150 years. Meanwhile, succession in the intertidal zone requires 1 to 3 years and succession within a desert stream occurs in less than 2 months. Despite the great differences in the time required, all these successional sequences show increased species diversity over time.

**Ecosystem changes during succession include increases in biomass, primary production, respiration, and nutrient retention.** Succession at Glacier Bay produces changes in several ecosystem properties, including increased soil depth, organic content, moisture, and nitrogen. Over the same successional sequence, several soil properties show decreases, including soil bulk density, pH, and phosphorus concentration. During ecosystem development on lava flows in Hawaii,

organic matter and nitrogen content of soils increased over the first 150,000 years and then declined by 1.4 and 4.1 million years. Weatherable mineral phosphorus in soils was largely depleted on lava flows 20,000 years. The percentage of soil phosphorus in refractory form made up the majority of phosphorus on lava flows 20,000 years old or older. Nitrogen losses from these ecosystems increased over time, while phosphorus losses decreased. Succession at the Hubbard Brook Experimental Forest increased nutrient retention by the forest ecosystem. Several ecosystem properties change predictably during succession in Sycamore Creek, Arizona, including biomass, primary production, respiration, and nitrogen retention.

**Mechanisms that drive ecological succession include facilitation, tolerance, and inhibition.** Most studies of succession support the facilitation model, the inhibition model, or some combination of the two. Both facilitation and inhibition occur during intertidal succession. Facilitation and inhibition also occur during secondary and primary forest succession.

**Community stability may be due to lack of disturbance or community resistance or resilience in the face of disturbance.** Ecologists generally define stability as the persistence of a community or ecosystem in the face of disturbance. Resistance is the ability of a community or ecosystem to maintain structure and/or function in the face of potential disturbance. The ability to bounce back after disturbance is called resilience. A resilient community or ecosystem may be completely disrupted by disturbance but quickly return to its former state. Studies of the Park Grass Experiment suggest that our perception of stability is affected by the scale of measurement. Studies in Sycamore Creek indicate that resilience is sometimes influenced by resource availability and that resistance may result from landscape-level phenomena.

Repeat photography can be used to detect long-term ecological change. Most successional sequences and most community and ecosystem responses to climatic change take place over very long periods of time. Repeat photography has become a valuable tool to help ecologists study these long-term changes.

# Review Questions

1. As we saw in figure 20.5, Johnston and Odum (1956) documented substantial change in the richness of bird species in a successional sequence going from the earliest stages in which the plant community was dominated by grasses and forbs to mature oak-hickory forests. Use MacArthur's (see chapter 16) studies (1958, 1961) of foliage height diversity and bird diversity to explain the patterns of diversity increase observed by Johnston and Odum.

2. Would you expect the number of species to remain indefinitely at the level shown in figure 20.7? Space on large stable boulders in Sousa's study site are dominated by the algal *G. canaliculata* and support 2.3 to 3.5 species, not the 5 shown in figure 20.7. Explain. (Hint: How long did Sousa follow his study boulders?)

3. The successional studies in Sycamore Creek produced patterns of variation in diversity that differed significantly from those observed during primary succession at Glacier Bay (see fig. 20.2), old field succession on the Piedmont Plateau (see fig. 20.4), or algal and barnacle succession in the intertidal zone (see fig. 20.7). The main difference was that Fisher and colleagues (1982) observed initial increases in species diversity followed by declines. In contrast, studies of forest and intertidal succession showed increases in diversity but no obvious declines. What may have been responsible for these different results? How might have differences in the longevity of species contributed to the different patterns observed by researchers? (Hint: Think about what we might observe in the other communities if they were studied for a longer period of time.)

4. In most studies of forest succession such as that of Reiners and colleagues (1971) and Oosting (1942), researchers study succession by comparing sites of various ages. This approach is called a "space for time substitution." What are some major assumptions of a space for time substitution? What contribution might the Glacier Bay system make to testing some of those assumptions? Why is this approach often necessary? What advantages for studying succession are offered by systems like Sycamore Creek?

5. The rapid succession shown by the Sycamore Creek ecosystem is impressive. How might natural selection influence the life cycles of the organisms living in Sycamore Creek? Imagine a creek that floods about twice per century. How quickly would you expect the community and ecosystem to recover following one of these rare floods? Explain your answer in terms of natural selection by flooding on the life cycles of organisms.

6. In the studies of mechanisms underlying succession, ecologists have found a great deal of evidence for both facilitation and inhibition. However, they have found little evidence for the tolerance model. Explain this lack of support for the tolerance model.

7. When Mount St. Helens erupted it created a gradient in disturbance. In the pumice plains studied by Morris and Wood (1989), the devastation was almost total. The extent of disturbance was much less in the farthest reaches of the blast zone. How might the rate of forest succession be related to intensity of disturbance around Mount St. Helens? Design a study to test your ideas, including a hypothetical map of the blast zone, the location of study sites, a list of the variables you would measure, a timetable for your study (assume you or your successors study the system for as long as you like), and a list of results that would support or contradict your hypothesis.

8. Ecological succession has been compared to the development of an organism and the climax community to a kind of superorganism. F. E. Clements (1916, 1936) was the best-known proponent of this idea, and H. A. Gleason (1926, 1939) the best-known early opponent of the idea of a community as a kind of superorganism. Gleason proposed that species are distributed independently of each other and that most overlaps in distributions are the result of coincidence, not mutual interdependence. Most modern ecologists hold a view more similar to that of Gleason. Which of the following graphs showing hypothetical distributions of species along an environmental gradient supports the superorganismic view of communities? How does the other graph support the individualistic view of species held by Gleason? (*A, B, C,* and *D* represent the distributions of species along an environmental gradient.)

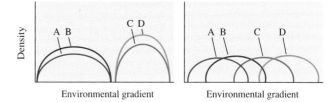

9. Species have come and gone in response to changing global climates during the history of the earth. Some of the mass extinctions of the past have resulted in the deaths of over 90% of existing species. What do these biological changes suggest about the long-term stability of the species composition of climax communities?

10. Succession seems to lead to predictable changes in community and ecosystem structure. Predict the characteristics of a frequently disturbed community/ecosystem versus a largely undisturbed community/ecosystem. What do your predictions suggest about a future biosphere increasingly disturbed by a growing human population? How does the intermediate disturbance hypothesis (see chapter 16) figure into your answer?

# Suggested Readings

Bruno, J. F., J. J. Stachowicz, and M. D. Bertness. 2003. Inclusion of facilitation into ecological theory. *Trends in Ecology & Evolution* 18:119–25.

*In this article, Bruno, Stachowicz, and Bertness present a strong case for the influence of facilitation as a major factor influencing community structure.*

Chadwick, O. A., L. A. Derry, P. M. Vitousek, B. J. Huebert, and L. O. Hedin. 1999. Changes in sources of nutrients during four million years of ecosystem development. *Nature* 397:491–97.

*Fascinating study of the long-term shift in Hawaiian forest ecosystems from nitrogen limitation to phosphorus limitation. The researchers document dependence of the older ecosystems on wind-transported phosphorus in marine aerosols and in dust from central Asia.*

Chapin, F. S., III, L. R. Walker, C. L. Fastie, and L. C. Sharman. 1994. Mechanisms of primary succession following deglaciation at Glacier Bay, Alaska. *Ecological Monographs* 64:149–75.

*This study focused on the mechanisms underlying primary succession at Glacier Bay, Alaska.*

Connell, J. H. and R. O. Slatyer. 1977. Mechanisms of succession in natural communities and their role in community stability and organization. *The American Naturalist* 111:1119–44.

*One of the most influential papers written on the subject of succession. This paper gave ecologists a robust conceptual framework for studying the mechanisms of selection.*

Crews, T. E., L. M. Kurina, and P. M. Vitousek. 2001. Organic matter and nitrogen accumulation and nitrogen fixation during early ecosystem development in Hawaii. *Biogeochemistry* 52:259–79.

*The authors document substantial change in ecosystem properties during the course of succession on 10- 52- and 142-year old lava flows. The paper provides an interesting comparison to succession at Glacier Bay, Alaska.*

Fisher, S. G. 1990. Recovery processes in lotic ecosystems: limits of successional theory. *Environmental Management* 14:725–36.

*A very thought-provoking discussion of successional theory as it applies to stream ecosystems.*

Flory, E. A. and A. M. Milner. 2000. Macroinvertebrate community succession in Wolf Point Creek, Glacier Bay National Park, Alaska. *Freshwater Biology* 44:465–80.

*This study documents nearly 16 years of change in the benthic invertebrate community of a creek created by a receding gla- cier. The pace of succession is in marked contrast to that in Sycamore Creek, Arizona.*

Hedin, L. O., P. M. Vitousek, and P. A. Matson. 2003. Nutrient losses over four million years of tropical forest development. *Ecology* 84:2231–55.

*The authors document major sources of nutrient loss from Hawaiian ecosystems in meticulous detail.*

Kohls, S. J., D. D. Baker, C. van Kessel, and J. O. Dawson. 2003. An assessment of soil enrichment by actinorhizal $N_2$ fixation using delta $^{15}N$ values in a chronosequence of deglaciation at Glacier Bay, Alaska. *Plant and Soil* 254:11–17.

*Stable isotope analysis verifies the importance of nitrogen fixa- tion to the process of nitrogen accumulation during primary succession at Glacier Bay, Alaska.*

Likens, G. E. and F. H. Bormann. 1995. *Biogeochemistry of a Forested Ecosystem*. 2d ed. New York: Springer-Verlag.

*A very readable summary of the long-term changes in nutrient cycling during secondary succession at the Hubbard Brook Experimental Forest.*

Odum, E. P. 1969. The strategy of ecosystem development. *Science* 164:262–70.

*A stimulating synthesis of ideas on the nature of ecological succession—one of the foundations for work on this topic that continues to inspire researchers.*

Reiners, W. A., I. A. Worley, and D. B. Lawrence. 1971. Plant diver- sity in a chronosequence at Glacier Bay, Alaska. *Ecology* 52:55–69.

*Reiners and his colleagues continued and extended the studies of Cooper.*

Stachowicz, J. J. 2001. Mutualism, facilitation, and the structure of ecological communities. *BioScience* 51:235–46.

*The author uses many clear, well-documented examples to sup- port his proposal that mutualism and facilitation play major roles in structuring biological communities.*

Tilman, D. 1996. Biodiversity: population versus ecosystem stabil- ity. *Ecology* 77:350–63.

*With this paper, David Tilman reports on a connection that ecolo- gists have long suspected—the connection between species diver- sity and stability.*

# On the Net

Visit this textbook's accompanying website at www.mhhe.com/ ecology (click on the book's title) to take advantage of practice quizzing, study/writing tips, timely news articles, and additional URLs for research on the topics in this chapter.

Community Ecology

Field Methods for Studies of Ecosystems

Succession and Stability

Temperate Forests

Rocky Shore Communities

Freshwater Habitats

Ecosystem Management

# SECTION VI
## Large-Scale Ecology

*"In the end, we conserve only what we love.
We will love only what we understand."*

Baba Dioum, Senegalese poet

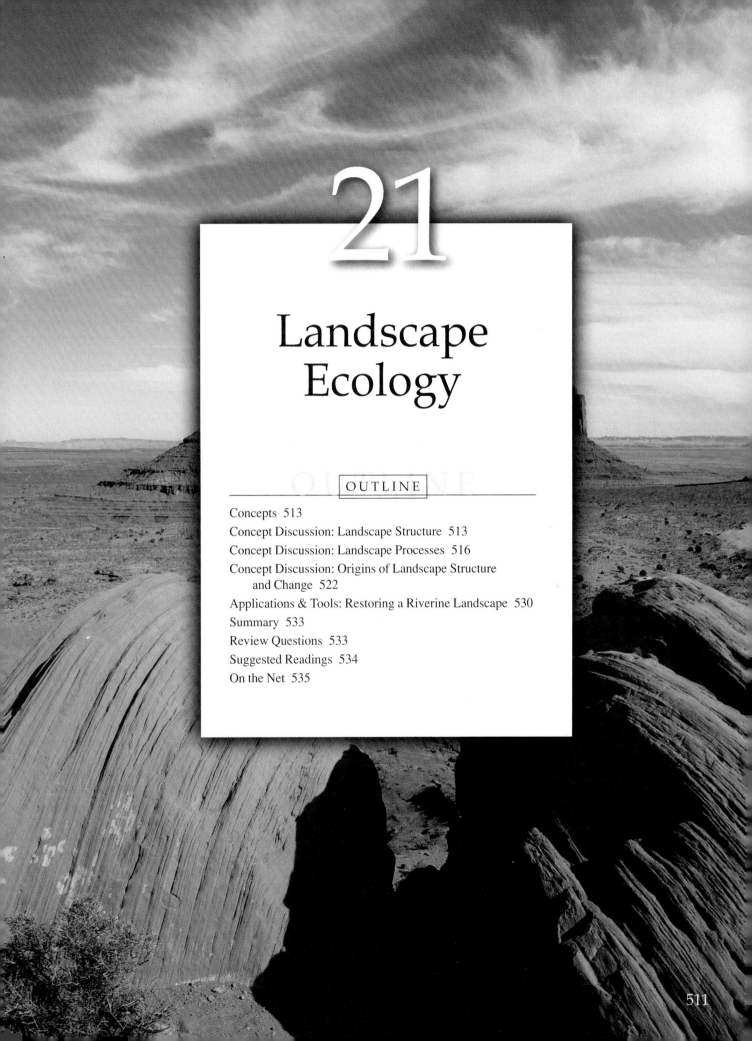

# 21

# Landscape
# Ecology

landscape perspective has informed human activities for centuries. Thirteen centuries ago, an emperor of Japan named Jomei stood on the top of Mount Kagu in Nara Prefecture and looked out upon the surrounding lands. It was Jomei's custom, as it had been for previous emperors, to climb to the top of Mount Kagu each spring and survey the surroundings (fig. 21.1). Because he wrote the following poem, we know some of Jomei's thoughts:

> Countless are the mountains in Yamato,
> But perfect is the heavenly hill of Kagu;
> When I climb it and survey my realm,
> Over the wide lake the gulls are on the wing;
> A beautiful land it is, the Land of Yamato.

Tadahiko Higuchi (1983) cited this poem in his book on the visual and spatial structure of landscapes. He wondered about the significance of the emperor's trip to the top of Mount Kagu and asked, "Was he [the emperor] merely looking at the scenery, or did he have some other purpose?" We cannot answer Higuchi's question with certainty, but history may provide some clues.

Long before the emperor Jomei stood upon Mount Kagu, the people from agricultural communities all over Japan did something similar. Each spring, they climbed a hill near their communities and looked over the surrounding countryside. This custom is called *kunimi,* which means see (*mi*) the domain (*kuni*). The Chinese character for *kuni,* which shows an area surrounded by boundaries, suggests that the purpose of these excursions was to look over a particular property, probably the property owned by the community. Eventually, this ancient custom became more ritualized and was practiced exclusively by the leader of each community and then only by the emperor.

What were these emperors, community leaders, and farmers doing on their promontories each spring? What were they looking at? There are many possible reasons for the custom, including religious, esthetic, economic, and political reasons. Regardless of the precise origin and motivation for the ritual, the custom of "seeing the domain" was a source of information about the landscape upon which these people depended, information of potential value to emperor, community leader, or individual farmer alike.

Imagine yourself standing on the top of a mountain as Japanese emperors once did. What would you see on the surrounding plains? You would see villages, fields, paths, roads, streams, woods, and much more. What would you do with this information? From these heights you could see if fields were well kept, if seedling density was uniform across recently planted fields, if villages were well maintained, if forests were encroaching on cleared properties, and so forth. Ecologists define **landscape ecology** as the study of landscape structure and processes. Given this definition, *kunimi,* or domain viewing, may be the first written record of the practice of landscape ecology.

Though landscape ecology is young as a scientific discipline, people seem to have always placed value on a landscape perspective. The recent emergence of the discipline may be a rediscovery of the practical value of a landscape perspective for understanding and managing nature. The emergence of landscape ecology coincides with the widespread availability of aerial photos (fig. 21.2) and satellite images (fig. 21.3).

**Figure 21.2** Aerial photography has made the perspective sought by early Japanese emperors accessible on all landscapes, such as this agricultural landscape in Japan, not just those near heights such as Mount Kagu.

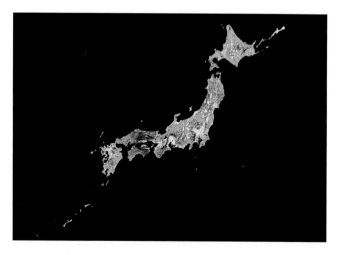

**Figure 21.3** Images of the earth's surface, such as this one of Japan, provide perspectives of landscapes not accessible before the development of satellite-based remote sensing.

**Figure 21.1** The views from hills and mountains, such as this one, were used traditionally in ancient Japan to survey the surrounding lands.

These modern technologies offer overhead views of all land-scapes, not just those, such as Jomei's landscape around Mount Kagu, that are conveniently located near heights.

Landscape ecology focuses on an organizational scale above that addressed by community and ecosystem ecology. To a landscape ecologist, a **landscape** is a heterogeneous area composed of several ecosystems. The ecosystems in a land-scape generally form a mosaic of visually distinctive patches. These patches are called **landscape elements.** The elements in a mountain landscape typically include forests, meadows, bogs, streams, ponds, and rock outcrops. An agricultural land-scape might include fields, fence lines, hedgerows, a patch of woods, a farm yard, and dirt lane. An urban landscape usually includes parks, industrial districts, residential areas, high-ways, and sewage treatment works.

Landscape ecologists study landscape structure, process, and change. In earlier chapters, we discussed structure, process, and change in populations, communities, and ecosys-tems. Structure, process, and change in landscapes form the core of chapter 21.

## CONCEPTS

- **Landscape structure includes the size, shape, com-position, number, and position of different ecosys-tems within a landscape.**
- **Landscape structure influences processes such as the flow of energy, materials, and species between the ecosystems within a landscape.**
- **Landscapes are structured and change in response to geological processes, climate, activities of organ-isms, and fire.**

## CASE DISCUSSION

### Landscape Structure

**Landscape structure includes the size, shape, composition, number, and position of different ecosystems within a landscape.**

Much of ecology focuses on studies of structure and process; landscape ecology is no exception. We are all familiar with the structure, or anatomy, of organisms. In chapter 9 we discussed the structure of populations, and in chapters 16 to 20 we considered the structure of communi-ties and ecosystems. However, what constitutes landscape structure? **Landscape structure** consists mainly of the size, shape, composition, number, and position of ecosys-

tems within a landscape. As you look across a landscape you can usually recognize its constituent ecosystems as distinctive patches, which might consist of woods, fields, ponds, marshes, or towns. The patches within a landscape form the mosaic that we call landscape structure.

Most questions in landscape ecology require that ecolo-gists quantify landscape structure. The following examples show how this has been done on some landscapes and how some aspects of landscape structure are not obvious without quantification.

## The Structure of Six Landscapes in Ohio

In 1981, G. Bowen and R. Burgess published a quantitative analysis of several Ohio landscapes. These landscapes consisted of forest patches surrounded by other types of ecosystems. Six of the 10 km by 10 km areas analyzed are shown in figure 21.4. If you look carefully at this figure you see that the landscapes, which are named after nearby towns, differ considerably in total forest cover, the number of forest patches, the average area of patches, and the shapes of patches. Some of the landscapes are well forested, and others are not. Some contain only small patches of forest, while others include some large patches. In some landscapes, the forest patches are long and narrow, while in others they are much wider. These general differences are clear enough, but we would find it difficult to give more precise descriptions unless we quantified our impressions.

First, let's consider total forest cover. Forest cover varies substantially among the six landscapes. The Concord land-scape, with 2.7% forest cover, is the least forested. At the other extreme, forest patches cover 43.6% of the Washington landscape. Differences between these extremes are clear, but what about some of the less obvious differences. Compare the Monroe and Somerset landscapes (fig. 21.4) and try to esti-mate which is more forested and by how much. Somerset may appear to have greater forest cover, but how much more? You may be surprised to discover that Somerset, with 22.7% forest cover, has twice the forest cover of the Monroe landscape, which includes just 11.8% forest cover (fig. 21.5). This sub-stantial difference could mean the difference between persist-ence and local extinction for some forest species.

Now let's examine the size of forest patches in each of the landscapes. Again, the median area of forest patches differs sig-nificantly across the landscapes. The smallest median areas are in the Monroe landscape, 3.6 ha, and the Concord landscape, 4.1 ha. The Washington landscape has the largest median patch area.

Now, look back at figure 21.4 and try to estimate which of the landscapes contains the greatest number, or highest density, of forest patches. The Somerset landscape, with 244 forest patches, has the highest patch density, and the Monroe landscape, with 180 patches, has the next highest density of forest patches. Obviously, the Concord landscape has the lowest density of for-est patches, with only 46. The Boston landscape, with 86 forest patches, contains the next lowest density of forest patches.

Quantifying landscape structure may reveal relationships not apparent visually. Compare your impression of the landscapes shown here to quantitative representations of some attributes presented in figures 21.5 and 21.6.

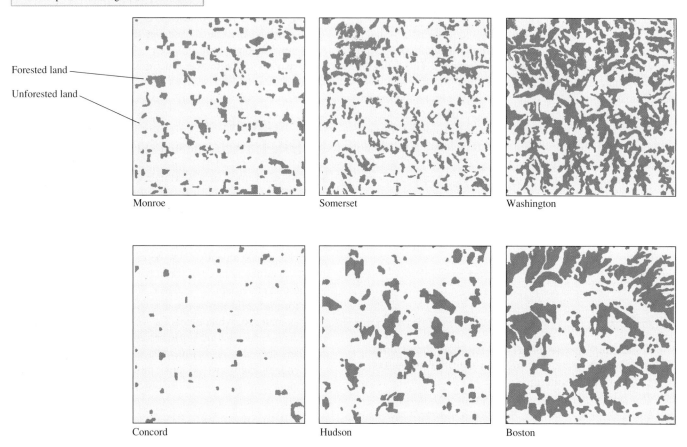

Forested land

Unforested land

Monroe          Somerset          Washington

Concord          Hudson          Boston

**Figure 21.4**    Forest fragments, shown as dark green, in six landscapes in Ohio (data from Bowen and Burgess 1981).

Quantifying landscape structure may reveal relationships not apparent visually. Compare the visual impression of figure 21.4 to the following.

The Somerset landscape has twice the forest cover as the Monroe landscape.

The Boston landscape has nearly 75% of the forest cover on the Washington landscape.

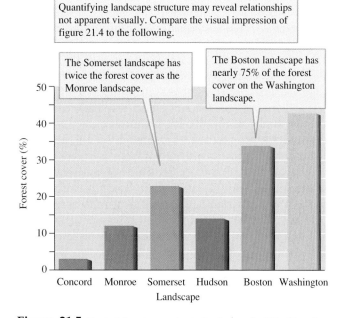

**Figure 21.5**    Percent forest cover in six landscapes in Ohio (data from Bowen and Burgess 1981).

Now let's look at a more subtle feature of landscape structure, patch shape. Bowen and Burgess quantified patch shape by the ratio of patch perimeter to the perimeter (circumference) of a circle with an area equal to that of the patch. Their formula was:

$$S = \frac{P}{2\sqrt{\pi A}}$$

where:

$S$ = patch shape

$P$ = patch perimeter

$A$ = patch area

How do you translate differences in the value of this index into shape? If $S$ is about equal to one, the patch is approximately circular. Increasing values of $S$ indicate less circular patch shapes. High values of $S$ generally indicate elongate patches and a long perimeter relative to area.

Bowen and Burgess calculated the shapes, $S$, for the forest patches in each of their landscapes and then determined the median shape for each (fig. 21.6). The Concord landscape,

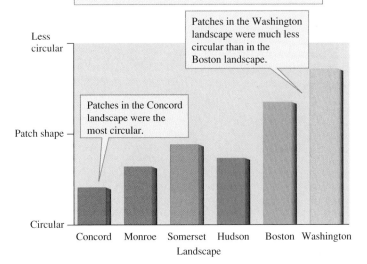

Figure 21.6  Relative shapes of forest patches in six landscapes in Ohio (data from Bowen and Burgess 1981).

with a median $S$ of 1.16, contains the most circular patches of the six landscapes. The Washington landscape, with a median $S$ of 1.6, contains the least circular patches. As we shall see in the next example, landscape ecologists have developed methods for representing landscape structure that go well beyond the classical methods used by Bowen and Burgess.

Until recently, geometry, which means "earth measurement," could only offer rough approximations of complex landscape structure. Today, an area of mathematics called *fractal geometry* can be used to quantify the structure of complex natural shapes. Fractal geometry was developed by Benoit Mandelbrot (1982) to provide a method for describing the dimensions of natural objects as diverse as ferns, snowflakes, and patches in a landscape. Fractal geometry offers unique insights into the structure of nature.

# The Fractal Geometry of Landscapes

During the development of fractal geometry, Mandelbrot asked a deceptively simple question: "How long is the coast of Great Britain?" This is analogous to estimating the perimeter of a patch in a landscape. Think about this question. At first, you might expect there to be only one, exact answer. For simple shapes with smooth outlines such as squares and circles, the assumption of a single answer is approximately correct. However, an estimate of the perimeter of a complex shape often depends upon the size of the measuring device. In other words, if you measure the coastline of Great Britain, you will find that your measurement depends upon the size of the ruler you use. If you were to step off the perimeter of

Great Britain in 1 km lengths, which is like using a ruler 1 km long, you would get a smaller estimate than if you made your measurements with a 100 m ruler. If you measured the coastline with a 10 cm ruler you would get an even larger estimate of the perimeter. The reason that a larger ruler gives a smaller estimate is that the large ruler misses many of the nooks and crannies along the coast. These smaller features show up in estimates made with smaller rulers.

Mandelbrot's answer to his question about the British coastline was, "Coastline length depends on the scale at which it is measured!" We can see the ecological significance of this finding by considering some of its consequences to organisms. Bruce Milne (1993) measured the coastline of Admiralty Island off the coast of southeastern Alaska. He made his measurements from the perspective of two very different animal residents of the island, bald eagles and barnacles.

Milne considered how the measured length of Admiralty Island's coastline depends upon the length of the measuring device. Figure 21.7 plots ruler length on the horizontal axis and estimated length of coastline on the vertical axis. The straight line that joins the dots slopes downward to the right. As Mandelbrot suggested, the estimated coastline length decreases as ruler length increases.

Now, what "ruler" are bald eagles and barnacles using? The distribution of eagle nests around Admiralty Island are about 0.782 km apart. This measurement of internest distance gives us an estimate of the length of coastline required by a bald eagle territory on the island. In contrast, barnacles range

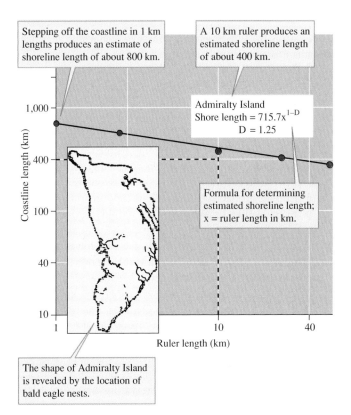

Figure 21.7  Relationship between ruler length and the measured length of the coastline of Admiralty Island, Alaska (data from Milne 1993).

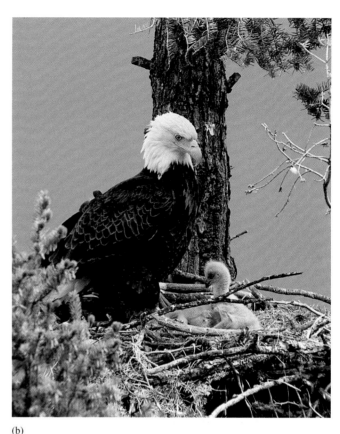

(a)                                        (b)

**Figure 21.8** Perspective on landscapes: fractal geometry tells us that the length of coastline accessible to (*a*) the oil molecules spilling from the hold of an oil tanker, such as the *Exxon Valdez* (the larger ship shown in this photo) is much greater than that used by (*b*) bald eagles.

from 1 to a few centimeters in basal diameter and they are sedentary. Barnacles only need a small area of solid surface to attach themselves and are often packed side by side along a rocky shore. Milne estimated that an individual barnacle requires about 2 cm (0.00002 km) of coastline.

Milne assumed that the eagles are, in effect, using a ruler 0.782 km long to step off the perimeter of the island and that barnacles use a ruler 0.00002 km long. Milne's analysis estimates that from the eagle's perspective, the perimeter of Admiralty Island is just a bit over 760 km. However, to a barnacle stepping off the coastline with its tiny ruler, the perimeter is over 11,000 km! Any of us would probably have assumed that the barnacle population "sees" a lot more of the spatial complexity around Admiralty Island. However, without Mandelbrot's fractal geometry, it would be difficult to predict that the difference in island perimeter for eagles and barnacles would be as great as 760 versus 11,000 km. At the conclusion of his analysis, Milne challenges us to imagine how long the coastline of Admiralty Island must be from the perspective of crude oil molecules. This is the length of coastline that determines the cost of a thorough cleanup after oil spills like that of the *Exxon Valdez* (fig. 21.8).

As in other areas of science, describing aspects of landscape structure, such as the length of the coastline of Admiralty Island or the size, shape, and number of forest patches in Ohio landscapes, is not an end in itself. Landscape ecologists study

landscape structure because it influences landscape processes and change. These are the next topics in the following Concept Discussions.

## CONCEPT DISCUSSION

### Landscape Processes

**Landscape structure influences processes such as the flow of energy, materials, and species between the ecosystems within a landscape.**

Landscape ecologists study how the size, shape, composition, number, and position of ecosystems in the landscape affect **landscape processes.** Though less familiar than physiological and ecosystem processes, landscape processes are responsible for many important ecological phenomena. In chapter 20, we saw how landscape structure, especially the location of shallow bedrock, controls the exchange of nutrients between subsurface and surface waters and local rates of primary production in Sycamore Creek, Arizona. As we will see in the following examples, landscape structure affects other ecologically important processes such as the dispersal of organisms, local population density, extinction of local populations, and the chemical composition of lakes.

# Landscape Structure and the Dispersal of Small Mammals

Landscape ecologists have proposed that landscape structure, especially the size, number, and isolation of habitat patches, can influence the movement of organisms between potentially suitable habitats. For instance, populations of desert bighorn sheep live in the isolated mountain ranges of the southwestern United States and northern Mexico, with individuals moving frequently among the ranges (fig. 21.9). The group of subpopulations of desert bighorn sheep living in an area such as the deserts of southern California constitute a metapopulation (see p. 000). The rate of movement of individuals between such subpopulations can significantly affect the persistence of a species in a landscape.

Human activity often produces habitat fragmentation, which occurs where a road cuts through a forest, a housing development eliminates an area of shrubland, or tracts of tropical rain forest are cut to plant pastures. Because habitat fragmentation is increasing, ecologists study how landscape structure affects the movements of organisms, movements that might mean the difference between population persistence and local extinction.

James Diffendorfer, Michael Gaines, and Robert Holt (1995) studied how patch size affects the movements of three small mammal species: cotton rats, *Sigmodon hispidus,* prairie voles, *Microtus ochrogaster,* and deer mice, *Peromyscus maniculatus.* They divided a 12 ha prairie landscape in Kansas into eight 5,000 m$^2$ areas. The prairie vegetation was mowed to maintain three patterns of fragmentation (fig. 21.10). The least fragmented areas consisted of large, 50 m by 100 m patches. The areas with medium fragmentation each contained 6 medium 12 m by 24 m patches. The most fragmented landscapes contained 10 or 15 small 4 m by 8 m patches.

The researchers predicted that animals would move farther in the more fragmented landscapes consisting of small habitat patches. In fragmented landscapes, individuals must move farther to find mates, food, and cover. They also predicted that animals would stay longer in the more isolated patches within fragmented landscapes. Consequently, the proportion of animals moving would decrease with habitat fragmentation.

The rodent populations were monitored on the study site by trapping them with Sherman live traps twice each month from August 1984 to May 1992. When trapped for the first time, the sex of each individual was determined and the animal was fitted with an ear tag with a unique number. The researchers also weighed, recorded the location of, and checked the reproductive condition of each animal trapped. Over the course of their 8-year study, Diffendorfer, Gaines, and Holt amassed a data set consisting of 23,185 captures. They used these data to construct movement histories for individual animals in order to test their predictions. They expressed movements as *mean square distances,* a measurement that estimates the size of an individual's

(a)

(b)

**Figure 21.9** Fragmented landscapes: (*a*) the small isolated mountain ranges of the southwestern United States and northern Mexico provide habitats for populations of (*b*) desert bighorn sheep, which move frequently between the mountain ranges of the region.

home range. A home range is the area that an animal occupies on a daily basis.

The behavior of two of the three study species supports the hypothesis that small mammals move farther in more fragmented landscapes. As predicted, *Peromyscus* and *Microtus,* living in

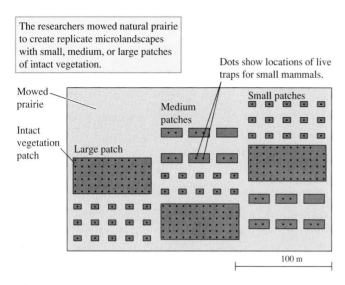

**Figure 21.10** Experimental landscape for the study of small mammal movements (data from Diffendorfer, Gaines, and Holt 1995).

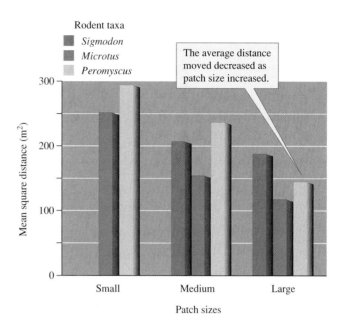

small patches, moved farther than individuals living in medium or large patches (fig. 21.11). However, the movements of *Sigmodon* in medium and large patches did not differ significantly.

The proportion of *Sigmodon, Microtus,* and *Peromyscus* moving within the 5,000 m$^2$ experimental areas supported the hypothesis that animal movements decrease with habitat fragmentation (fig. 21.11). A larger proportion of *Sigmodon* moved within large patch areas than moved within areas with medium patches. Because few *Sigmodon* were captured within small patch areas, their movements within these areas could not be analyzed. A larger proportion of *Microtus* and *Peromyscus* moved within large and medium patches than moved within small patches.

In summary, this experiment shows a predictable relationship between landscape structure and the movement of organisms across landscapes. As the following example shows, those movements may be crucial to maintaining local populations.

# Habitat Patch Size and Isolation and the Density of Butterfly Populations

Ilkka Hanski, Mikko Kuussaari, and Marko Nieminen (1994) found that the local population density of the Glanville fritillary butterfly, *Melitaea cinxia,* is significantly affected by the size and isolation of habitat patches. The researchers studied a metapopulation of these butterflies on Åland Island in southwestern Finland. Their study site consisted of 15.5 km$^2$ of countryside, a landscape consisting of small farms, cultivated fields, pastures, meadows, and woods (fig. 21.12). Within this landscape, habitat suitable for the butterfly consisted of patches of their larval food plant, *Plantago lanceolata,* which generally occurs in pastures and meadows.

There were 50 patches of potential habitat within the study area. Forty-two of these patches were occupied by the

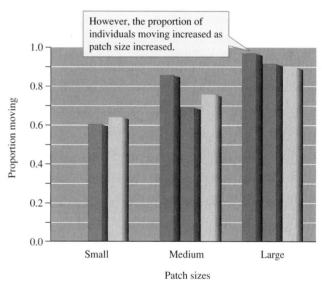

**Figure 21.11** Influence of patch size on small mammal movements within experimental landscapes (data from Diffendorfer, Gaines, and Holt 1995).

butterflies in 1991. The patches ranged in area from 12 to 46,000 m$^2$ and supported populations ranging from 0 to 2,190 individuals. The habitat patches also varied in their degree of isolation from other habitat patches. The distance from habitat patches to the nearest occupied patch varied from 30 m to 1.6 km. However, Hanski and his colleagues found that, from a statistical perspective, the best index of isolation combined distances to neighboring habitat patches and the numbers of butterflies living on those patches.

Habitat patch area influenced both the size and density of the populations. Total population size within a patch increased with patch area. However, population density decreased as patch area increased (fig. 21.13). Thus, though large habitat patches supported larger numbers of individuals than smaller patches, population density was lower on large patches.

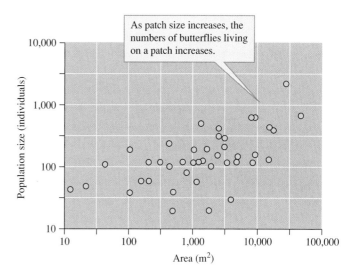

**Figure 21.12** Much of the landscape of southwestern Finland consists of a patchwork of pastures, meadows, and woods.

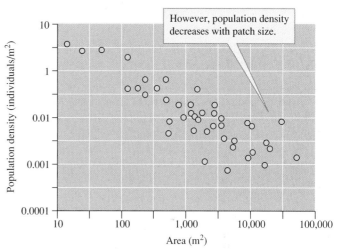

**Figure 21.13** Relationship between habitat patch area and population size and density of the butterfly *Melitaea cinxia* in a landscape on Åland Island, Finland (data from Hanski, Kuussaari, and Nieminen 1994).

The team also found that more isolated patches supported lower densities of butterflies. Isolation influences local population density in these populations because local populations are partly maintained by immigration of *Melitaea* from other patches. For instance, during 1 week of sampling the butterflies in one patch, about 15% of the males and 30% of the females were recaptures from surrounding patches.

This experiment determined that area and isolation of patches strongly influence the size and density of *Melitaea* populations. One conclusion that we can draw from these patterns is that landscape structure is important for understanding the distribution and abundance of the butterflies. It turns out that landscape structure also affects the persistence of local populations. Between 1991 and 1992, Hanski and his colleagues recorded three extinctions of local populations and five colonizations of new habitat patches. All these extinctions and colonizations occurred in small patches with small populations.

The vulnerability of small populations to extinction has been well documented in populations of desert bighorn sheep in the southwestern United States. Joel Berger (1990) explored the relationship of population size to local extinctions in isolated populations of desert bighorn sheep using records from 129 populations in five states: California, Colorado, Nevada, New Mexico, and Texas. Berger found that local populations with fewer than 50 individuals became extinct in less than 50 years, while populations of 51 to 100 individuals became extinct in about 60 years. Populations of more than 100 individuals persisted for at least 70 years.

The studies by Diffendorfer and colleagues, Hanski and colleagues, and Berger show that the movement of organisms and the characteristics of local populations are significantly influenced by landscape structure.

# Habitat Corridors and Movement of Organisms

One long-standing approach to reducing the negative impact of fragmentation and isolation on populations has been to connect

habitat fragments with corridors of similar habitat. Despite the logic of the approach, evidence for the effectiveness of habitat corridors to promote movement of organisms between habitat patches has been scant. Experimental evidence, in particular, has been lacking. Recently, however, a number of experimental studies have filled many of the gaps in our understanding of the effects of habitat corridors on movement by organisms.

Nick Haddad and Kristen Baum (Haddad 1999, Haddad and Baum 1999) studied the influence of corridors on the movements of butterflies associated with early successional habitats. Their study site was the Savannah River Site, South Carolina, a National Environmental Research Park, where they created patches of open habitat in dense 40 to 50-year-old forests of pine. Patches of open habitat were squares, 128 m on a side, with an area of 1.64 ha, the approximate size of forestry clear-cuts in the surrounding region. With the help of the staff of the Savannah River Institute, Haddad and Baum created 27 openings, 8 that were isolated and 19 that were connected by

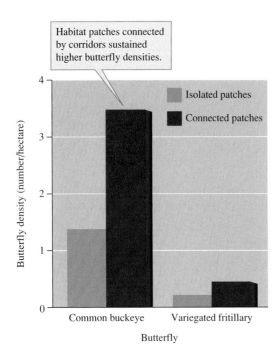

Habitat patches connected by corridors sustained higher butterfly densities.

**Figure 21.14** Densities of common buckeye, *Junonia coenia,* and variegated fritillary, *Euptoieta claudia,* butterflies in early successional patches connected by corridors of open habitat or isolated by surrounding pine forest (data from Haddad and Baum 1999).

corridors of open habitat to other patches. Within each patch all trees were removed and the slash (debris) was burned.

In a study of movements of butterflies between patches, Haddad (1999) focused his attention on two butterfly species: the common buckeye, *Junonia coenia,* and the variegated fritillary, *Euptoieta claudia.* Both these species are specialized for life in early successional habitats and avoid pine forests. Haddad used mark and recapture techniques to study butterfly movements. He marked a total of 1,260 common buckeye butterflies and 189 variegated fritillaries. Haddad subsequently recaptured 239 common buckeye butterflies and 47 variegated fritillaries. His results showed clearly that corridors increased the frequency of movements by both butterfly species between patches. In a companion study Haddad and Baum (1999) also documented higher densities of both butterfly species in open habitats connected by corridors (fig. 21.14).

However, the influences of corridors on movement of organisms between open habitats within the Savannah River Site extend far beyond butterflies. Research conducted by a team of ten investigators (Tewksbury et al. 2002) also showed higher rates of movement by common buckeye and variegated fritillary butterflies between patches connected by corridors. They also discovered higher rates of pollination of plants growing in connected patches and higher rates of seed dispersal, mainly by birds. In summary, these studies and many others have now shown that habitat corridors can facilitate the movement of organisms between otherwise isolated habitat fragments.

As we will see in the next example, landscape structure can also influence the characteristics of ecosystems.

# Landscape Position and Lake Chemistry

Katherine Webster and her colleagues (1996) at the Center for Limnology at the University of Wisconsin and the U.S. Geological Survey explored how the position of a lake in a landscape affects its chemical responses to drought. Drought can affect a wide range of lake ecosystem properties, including nutrient cycling and the concentrations of dissolved ions. However, all lakes do not respond in the same way to drought. For instance, while drought increased the concentration of dissolved substances in Lake 239 at the Experimental Lakes Area in Ontario, it decreased them in Nevins Lake, Michigan.

Webster and her colleagues set out to determine whether the contrasting chemical responses of lakes to drought can be explained by the position of the lake in the landscape. They worked in northern Wisconsin, where they defined the landscape position of a lake as its location within a hydrologic flow system. The team quantified the position of a lake within a hydrologic flow system as the proportion of total water inflow supplied by groundwater.

The sources of water for a lake are precipitation, surface water, and groundwater flow. Different lakes receive different proportions of their water from these sources, and these proportions depend upon a lake's position in the landscape. Figure 21.15 shows a series of lakes along a hydrologic flow system in northern Wisconsin. Morgan Lake, which receives the bulk of its water from precipitation, occupies the upper end of this continuum. Lakes such as this one occupy high points in the hydrologic flow system and are called "hydrologically mounded" lakes. These lakes are sources of water for the rest of the hydrologic flow system. Crystal Lake and Sparkling Lake, which occupy intermediate positions within the hydrologic flow system and receive significant inflows of groundwater, are "groundwater flow through" lakes. Finally, at the lower end of the flow system, are the "drainage" lakes that receive significant surface drainage as well as groundwater drainage.

The important point here is that the positions of these lakes in the landscape determine the proportion of water they receive as groundwater. Webster and her colleagues estimated that Morgan Lake receives no groundwater inflow, while Trout Lake, at the lower end of the hydrologic flow system, receives 35% of its inflow as groundwater. The main source of water for a lake determines its response to drought.

The responses of these seven lakes to a drought were studied from 1986 to 1990. As you might expect, the levels of the lakes dropped during this 4-year drought. However, the amount of drop in lake level was related to a lake's position in the landscape (fig. 21.16). The level of Morgan Lake, at the upper end of the hydrologic flow system, dropped 0.7 m, while the levels of Vandercook, Big Muskellunge, Crystal, and Sparkling Lakes, in the middle of the hydrologic flow system, dropped 0.9 to 1.0 m. Meanwhile, the levels of Trout and Allequash Lakes, the two drainage lakes at the lower end of the hydrologic flow system, dropped very little.

| Information | |
|---|---|
| Questions | *Investigating the Evidence* |
| Hypothesis | |
| Predictions | **Comparison of Two Samples Using a Rank Sum Test** |
| Testing ✓ | |

Suppose you are studying the exchange of organic matter between forests and streams and the landscape you are studying is a mosaic of patches of two forest types: deciduous and coniferous. Part of your study involves determining whether there is a difference in the amount of detritus in streams draining patches of deciduous forest versus those draining coniferous forest. In an initial phase of the study, you take random measurements of the amounts of detritus (g dry weight per m$^2$) in two streams: one draining a deciduous forest patch and one draining a coniferous forest patch:

| Measurements | | | | | | |
|---|---|---|---|---|---|---|
| 1 | 2 | 3 | 4 | 5 | 6 | 7 |
| Deciduous forest | | | | | | |
| 40.6 | 34.2 | 366.5 | 26.9 | 23.1 | 42.8 | 51.1 |
| Coniferous forest | | | | | | |
| 161.1 | 123.5 | 182.3 | 216.6 | 110.9 | 121.2 | 542.4 |

Your hypothesis is that there is no difference in the amounts of detritus that these two streams contain. However, it turns out that the distribution of detritus within the streams is not normal, and so a sample mean will not accurately reflect the typical amount of detritus per square meter. Also, a *t*-test is not appropriate for making a statistical comparison of detritus standing stock in the two ecosystems. The alternative is to use a statistical test that does not assume a normal distribution and compares medians *not* means. One such procedure is the Mann-Whitney test, which uses ranks of measurements or observations made in two populations, rather than the measurements themselves to make a statistical comparison. Here are the same data ordered (ranked) from smallest to largest:

We can now calculate the Mann-Whitney statistic $U$ for the two streams. Let's begin with the stream draining the deciduous forest:

$$U_d = (n_d)(n_c) + \left[\frac{(n_d)(n_d+1)}{2}\right] - T_d$$

$$U_d = (7)(7) + \left[\frac{(7)(7+1)}{2}\right] - 34$$

$$U_d = 49 + 28 - 34$$

$$U_d = 43$$

The Mann-Whitney statistic for the coniferous stream can be calculated in the same way as:

$$U_c = (n_d)(n_c) + \left[\frac{(n_c)(n_c+1)}{2}\right] - T_c$$

Or more simply as:

$$U_c = (n_d)(n_c) - U_d$$

$$U_c = (7)(7) - 43$$

$$U_c = 6$$

At this point in the Mann-Whitney procedure, the larger of the two $U$ values is compared to a table of critical values (appendix C). The applicable critical values are determined by significance level, generally $P < 0.05$, and the sample sizes, $n_1$ and $n_2$, which in this case are $n_d = 7$ and $n_c = 7$. Examining table C.1, we find that the critical value of the Mann-Whitney test statistic for our comparison is 41. Since $U_d = 43$ is greater than 41, we reject the hypothesis that the two streams contain the same standing stock of detritus and accept the alternative hypothesis that the standing stocks of detritus in these two particular streams are different. Can we conclude from this study that streams draining deciduous versus coniferous forests contain different amounts of detritus? At this point, we cannot, and the reasons for that are discussed in chapter 22 (see p. 551).

| Measurements (deciduous patch) | Ranks | Measurements (coniferous patch) | Ranks |
|---|---|---|---|
| 23.1 | 1 | 110.9 | 7 |
| 26.9 | 2 | 121.2 | 8 |
| 34.2 | 3 | 123.5 | 9 |
| 40.6 | 4 | 161.1 | 10 |
| 42.8 | 5 | 182.3 | 11 |
| 51.1 | 6 | 216.6 | 12 |
| 355.6 | 13 | 542.4 | 14 |
| $n_d = 7$ (measurements) | $T_d = \Sigma$ ranks = 34 | $n_c = 7$ (measurements) | $T_c = \Sigma$ ranks = 71 |

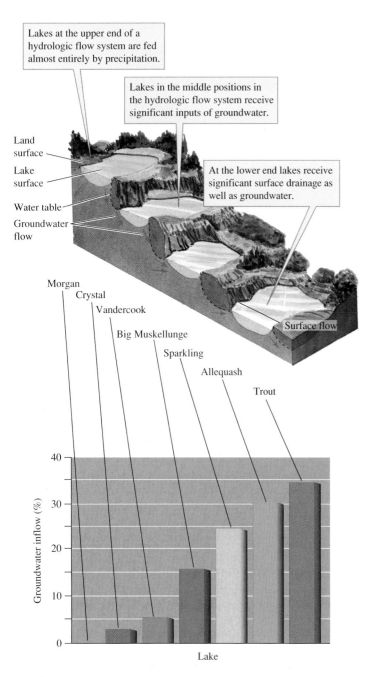

Lakes at the upper end of a hydrologic flow system are fed almost entirely by precipitation.

Lakes in the middle positions in the hydrologic flow system receive significant inputs of groundwater.

At the lower end lakes receive significant surface drainage as well as groundwater.

Land surface

Lake surface

Water table

Groundwater flow

Surface flow

**Figure 21.15** Lake position in the landscape and proportion of water received as groundwater (data from Webster et al. 1996).

Landscape position also significantly influenced a lake's chemical responses to the drought. The *concentrations* of dissolved ions such as calcium ($Ca^{2+}$) and magnesium ($Mg^{2+}$) increased in the majority of the lakes. However, the increase in ion concentration was highest at the upper and lower ends of the hydrologic flow system. Meanwhile, the combined *mass* of $Ca^{2+}$ and $Mg^{2+}$ increased in the three lakes at the lower end but did not change in Morgan Lake, at the upper end of the flow system, and either decreased or did not change in the lakes occupying the middle portions of the hydrologic flow system (fig. 21.16).

The researchers concluded that the increased mass of $Ca^{2+}$ and $Mg^{2+}$ seen at the lower end of the hydrologic flow system was due to an increased proportion of inflows from groundwater and surface water, sources rich in $Ca^{2+}$ and $Mg^{2+}$. The declines in mass of $Ca^{2+}$ and $Mg^{2+}$ in Big Muskellunge Lake are likely due to reduced inflow of ion-rich groundwater. The stability of $Ca^{2+}$ and $Mg^{2+}$ mass in Morgan Lake was attributed to its isolation from the groundwater flow system. Morgan Lake receives almost no groundwater even during wet periods. Regardless of the mechanisms, the chemical responses of these lakes to the drought were related to their positions in the landscape.

In the first section of this chapter, we reviewed the concept of landscape structure. In this section, we explored the connection between landscape structure and landscape processes. But what creates landscape structure? Landscape structure, like the structure of populations, communities, and ecosystems, changes in response to an interplay between dynamic processes. We explore the sources of landscape structure and change in the next Concept Discussion section.

## CONCEPT DISCUSSION

### Origins of Landscape Structure and Change

**Landscapes are structured and change in response to geological processes, climate, activities of organisms, and fire.**

What creates the patchiness we see in landscapes? Many forces combine in numerous ways to produce the patchiness that we call landscape structure. In this section, we review examples of how geological processes, climate, organisms, and fire contribute to landscape structure.

## Geological Processes, Climate, and Landscape Structure

The geological features produced by processes such as volcanism, sedimentation, and erosion provide a primary source of landscape structure. For instance, the alluvial deposits along a river valley provide growing conditions different from those on thin, well-drained soils on nearby hills. A volcanic cinder cone in the middle of a sandy plain offers different environmental conditions than the surrounding plain (fig. 21.17). Distinctive ecosystems may develop on each of these geological surfaces, creating patchiness in the landscape. In the following example, we shall see how distinctive soils contribute to vegetative patchiness in a Sonoran Desert landscape.

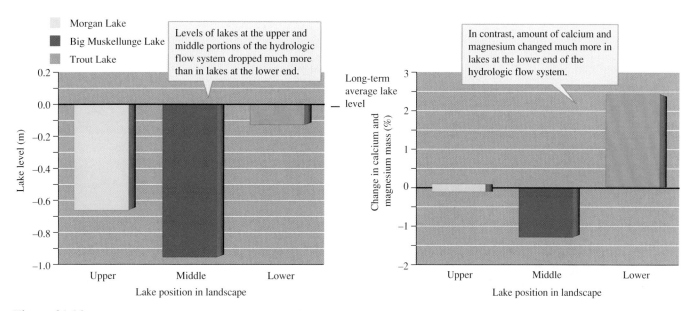

**Figure 21.16** Lake position in a hydrologic flow system and response to a severe drought (data from Webster et al. 1996).

**Figure 21.17** Geological features such as the volcanic cinder cone in the middle of a sandy plain add structure to the landscape by adding a geological surface with distinctive physical and chemical properties.

## Soil and Vegetation Mosaics in the Sonoran Desert

The Sonoran Desert includes many long, narrow mountain ranges separated by basins or valleys. The mountains and basins in this region originated in movements of the earth's crust that began 12 to 15 million years ago. As the mountains were uplifted and the adjacent basins subsided, erosion removed materials from the mountain slopes. This eroded material was deposited in the surrounding basins, forming sloping plains, or *bajadas,* at the bases of the mountains. Sediment deposits in these basins may be over 3 km deep.

From a distance, the bajadas of the Sonoran Desert may appear to be uniform environments, especially against the backdrop of a rugged desert mountain (see fig. 16.2). However, Joseph McAuliffe (1994) has shown that bajadas in the Sonoran Desert near Tucson, Arizona, consist of a complex mosaic of distinctive **landforms.** His studies have shown that intermittent erosion and deposition operating over the past 2 million years have produced a complex landscape.

McAuliffe established study sites on the bajadas of three mountain ranges. At each site he studied soils and plant distributions. In all three study areas, he found a wide range of soil types and plant distributions that corresponded closely to soil age and structure.

Let's look at some of the patterns McAuliffe found on the bajada associated with the northern end of the Tucson Mountains. Going from left to right in figure 21.18, the first soils you see are of early Pleistocene age and are approximately 1.8 to 1.9 million years old. Going northward along the bajada, to the right in figure 21.18, the next soils in the sequence date from the middle to late Pleistocene and are hundreds of thousands of years old. These soils are followed by Holocene deposits that are less than 11,000 years old and are associated with an ephemeral desert water course called Wildhorse Wash. Near the Holocene soils, McAuliffe found soils that dated from the late Pleistocene. These soils were 25,000 to 75,000 years old.

In the space of a few kilometers, McAuliffe found patches of soil that were (1) almost 2 million years old, (2) hundreds of thousands of years old, (3) tens of thousands of years old, and (4) less than 11,000 years old. Because soil-building processes occur over long periods of time, these soils of vastly different ages also differ substantially in structure. Figure 21.19 shows McAuliffe's drawings of typical profiles of Holocene, middle to late Pleistocene, and early Pleistocene soils. The Holocene soils had low amounts of clay and calcium carbonate ($CaCO_3$) and poorly developed soil horizons. They also lacked a *caliche layer,* a hardpan soil horizon formed by precipitation of $CaCO_3$. Middle to late Pleistocene soils had a much higher clay content than Holocene soils, and early Pleistocene soils contained even

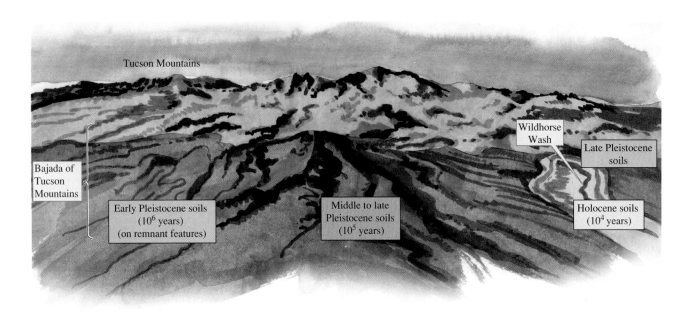

**Figure 21.18** Soil ages on an outwash plain, or bajada, associated with the Tucson Mountains, Arizona; colors used only to show locations of different soils in landscape (data from McAuliffe 1994).

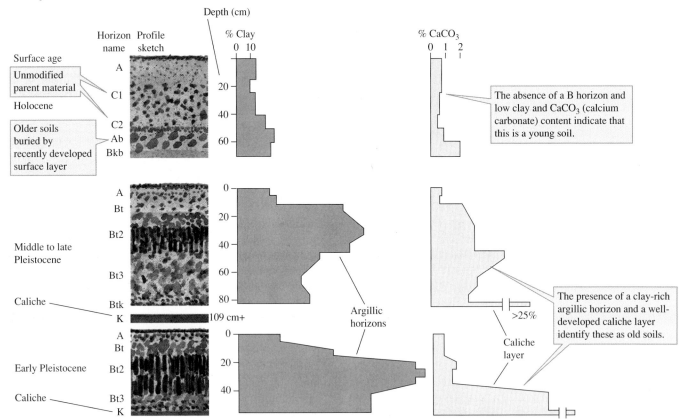

**Figure 21.19** Structural features of young to old desert soils on the Tucson Mountains bajada (data from McAuliffe 1994).

more clay. These clay layers in the older soil profiles are called **argillic horizons.** Middle to late and early Pleistocene soils also contained more $CaCO_3$ and were underlain by a thick layer of caliche.

These differences in soil structure influence the distributions of perennial plants across the Tucson Mountain bajada (fig. 21.20). McAuliffe found that the relative abundances of two shrubs, *Larrea tridentata* and *Ambrosia deltoidea*, accounted for much of the variation in perennial plant distributions. Plant distributions map clearly onto soils of different ages. *Ambrosia* is most abundant on middle to late Pleistocene soils. *Larrea* dominates on Holocene soils and on early Pleistocene soils. Other perennial plant species dominated mainly on the eroding side slopes of early Pleistocene soils.

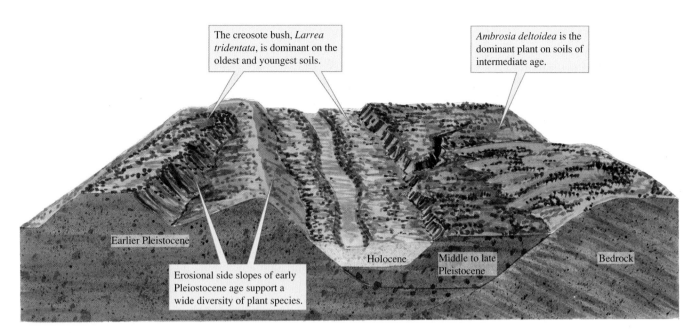

The creosote bush, *Larrea tridentata*, is dominant on the oldest and youngest soils.

*Ambrosia deltoidea* is the dominant plant on soils of intermediate age.

Earlier Pleistocene

Erosional side slopes of early Pleiostocene age support a wide diversity of plant species.

Holocene

Middle to late Pleistocene

Bedrock

**Figure 21.20** Association between vegetation and soils of different ages and structure on the Tucson Mountains bajada; colors used only to show locations of different soils in landscape (data from McAuliffe 1994).

## Climate and Landscape Structure

Climate is a major determinant of landscape structure. Our review of major terrestrial environments in chapter 2 showed a clear connection between landscape structure and climate. Climate determines whether the potential ecosystem in an area will be tundra, temperate forest, or desert. It also sets the baseline for aquatic ecosystems. Climate determines whether rivers flood once a year, twice a year, or at irregular intervals. As climate changes, landscapes change. The advances and retreats of glaciers have shaped whole continents, creating lakes and plains, transporting soils, and carving mountain valleys. Wetting and drying cycles have changed the distribution extent of rain forest and savanna in the Amazon River basin.

We can also see the signature of climate on a very local scale. Let's go back to the soils studied by McAuliffe and review some of the effects of climate. The soil mosaic along the bajada east of the Tucson Mountains consists of patches of material deposited during floods originating in these mountains from nearly 2 million years ago to less than 11,000 years ago. The deposits were laid down during times when the climate produced intense storms that caused flooding and erosion. Materials eroded from mountain slopes were deposited as alluvium on the surrounding bajadas.

These alluvial deposits were gradually changed, and these changes were dependent upon climate. Two of the prominent features of the older soils studied by McAuliffe were the formation of a clay-rich argillic horizon and the formation of a $CaCO_3$-rich caliche layer. Both these soil features are the result of water transport. Clay particles are transported as a colloidal suspension, while the $CaCO_3$ is transported in dissolved form. Consequently, the clays precipitate out of suspension higher in the soil profile than the $CaCO_3$. The result is the layering of an argillic horizon over a caliche layer as shown in figure 21.19.

Water, working on alluvial deposits, is responsible for the soil structure observed by McAuliffe, but it was water delivered to the landscape under particular climatic conditions. We can get a clue about those conditions by observing some soil characteristics. We know that argillic horizons are deposited by water. However, the soils described by McAuliffe also offer clues that the action of water was highly episodic. The argillic horizon in these soils is red, and this red color is the result of a buildup of iron oxides. Oxidation of iron could have only occurred in an oxidizing environment. Because soil saturated with water quickly becomes anoxic, the presence of oxidized iron in the argillic horizon indicates that these soils were formed when conditions were intermittently wet. In other words, the soils along the bajada of the Tucson Mountains formed under particular climatic conditions. Different climatic conditions would have produced different soils and, perhaps, different plant distributions.

While geological processes and climate set the basic template for landscape structure, the activities of organisms can be an additional source of landscape structure and change. In the following example, we consider how the activities of humans and other species can change landscape structure.

## Organisms and Landscape Structure

Organisms of all sorts influence the structure of landscapes. While the following discussion focuses on the influences of animals, plants create much of the distinctive patchwork we call landscape structure. For an example of how plants can change landscape structure, think back to chapter 19, where we discussed Edward Witkowski's studies (1991) of how *Acacia* affects the South African fynbos. We focused on the effects of the

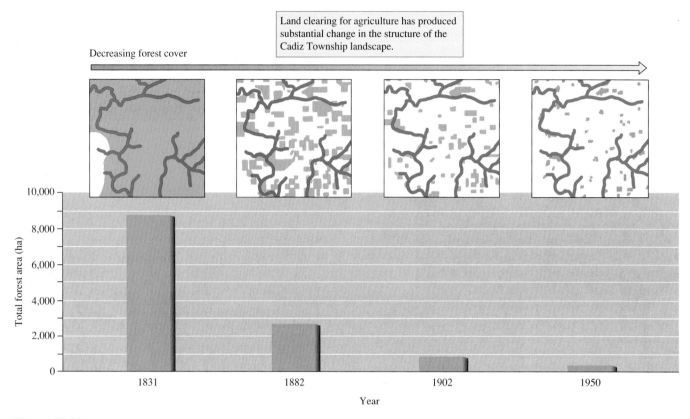

**Figure 21.21** Human-caused change in forest cover in Cadiz Township, Wisconsin (data from Curtis 1956, maps after Curtis 1956).

plant on the quantity of soil nitrogen and rates of decomposition. Now let's take a landscape perspective of the effects of *Acacia*. As this plant invaded the fynbos, it created distinctive patches where nitrogen availability and decomposition rates are higher. By adding these distinctive ecosystem patches, *Acacia* has altered landscape structure.

Many studies of landscape change have focused on the conversion of forest to agricultural landscapes. In North America, an often-cited example of this sort of landscape change is that of Cadiz Township, Green County, Wisconsin (fig. 21.21). In 1831, approximately 93.5% of Cadiz Township was forested. By 1882 the percentage of forested land had decreased to 27% and by 1902 forest cover had fallen to less than 9%. Between 1902 and 1950 the total area of forest decreased again to 3.4%. Similar changes in landscape structure have been observed throughout the midwestern region of the United States. However, in some other forested regions of North America and Europe, the pattern of recent landscape change has been different.

In eastern North America, many abandoned farms have reverted to forest and in these landscapes forest cover has increased. Recent increases in forest cover have also been observed in some parts of northern Europe. One such area is the Veluwe region in the central Netherlands. Maureen Hulshoff (1995) reviewed the landscape changes that have occurred in the Veluwe region during the past 1,200 years. The Veluwe landscape was originally dominated by a mixed deciduous forest. Then, from A.D. 800 to 1100, people gradually occupied the area and cut the forest. Consequently, forests were gradually converted to heath-

lands, which are landscapes dominated by low shrubs and used for livestock foraging. Later, small areas of cropland were interspersed with the extensive heathlands. During the tenth and eleventh centuries some areas were devegetated completely and converted to areas of drifting sand. The problem of drifting sand continued to increase until the end of the nineteenth century, when the Dutch government began planting pine plantations on the Veluwe landscape, a practice that continued into the twentieth century.

Figure 21.22 shows the changes in the composition of the Veluwe landscape from 1845 to 1982. The greatest change over this period was a shift in dominance from heathlands to forests. In 1845, heathlands made up 66% of the landscape, while forests constituted 17%. By 1982, coverage by heathlands had fallen to 12% of the landscape and forest coverage had risen to 64%. The figure also shows modest but ecologically significant changes in the other landscape elements. The area of drift sand reached a peak in 1898 and then dropped and held steady at 3% to 4% from 1957 to 1982. Urban areas established a significant presence beginning in 1957. Finally, coverage by agricultural areas has varied from 9% to 16% over the study interval, the least variation shown by any of the landscape elements.

As total coverage by forest and heathlands changed within the Veluwe landscape, the number and average area of forest and heath patches also changed. These changes indicate increasing fragmentation of heathlands and decreasing fragmentation of forests. For instance, between 1845 and 1982, the number of forest patches declined, while the average area of forest patches increased. During this period, the number of heath patches

The most substantial change in this landscape in the Netherlands was a shift from predominantly heathland to predominantly forest.

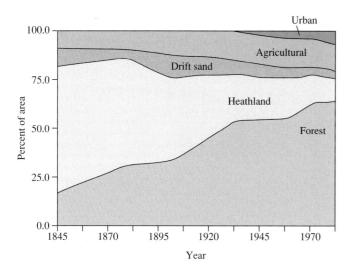

**Figure 21.22** Change in a Dutch landscape (data from Hulshoff 1995).

increased until 1957. Between 1957 and 1982, the number of heath patches decreased as some patches were eliminated. The average area of heath patches decreased rapidly between 1845 and 1931 and then remained approximately stable from 1931 to 1982.

During the period that Cadiz Township in Wisconsin was losing forest cover, this landscape element was increasing in the Veluwe district of the Netherlands. These two examples show how human activity has changed landscape structure. However, what forces drive human influences on landscapes? In both Cadiz Township and the Veluwe landscape, the driving forces were economic. A developing agricultural economy converted Cadiz Township from forest to farmland. The Veluwe landscape was converted from heathland to forest as the local sheep-raising economy collapsed in response to the introduction of synthetic fertilizers and inexpensive wool from Australia.

As we enter the twenty-first century, economically motivated human activity continues to change the structure of landscapes all over the globe. We examine current trends in land cover at the global scale in chapter 23. Before we do that, however, let's examine the effects of some other species on landscape structure.

Many animal species modify landscape structure (fig. 21.23). African elephants feed on trees and often knock them down in the

(a)

(b)

(c)

(d)

**Figure 21.23** Species with significant impacts on landscape structure. (*a*) African elephants control the extent of tree cover in some landscapes. (*b*) Alligators build and maintain ponds in wetland landscapes. (*c*) Feeding and burrowing by kangaroo rats introduce added patchiness into desert landscapes. (*d*) Termite mounds add distinctive landscape features.

process. As a consequence, these elephants can gradually change woodland to grassland. Alligators maintain ponds in the Florida Everglades, a landscape element upon which many species depend to survive droughts. Small species can also change landscapes. Kangaroo rats, *Dipodomys* spp., of the American Southwest dig burrow systems that modify the structure of the soil, the distribution of nutrients, and the distribution of plants to such an extent that the result is recognizable from aerial photos. Similar effects on landscape structure are created by termites and ants.

One of the most adept modifiers of landscapes is the beaver, *Castor canadensis* (fig. 21.24). Beavers alter landscapes by cutting trees, building dams on stream channels, and flooding the surrounding landscape. Beaver dams increase the extent of wetlands in the landscape, alter the hydrologic regime of the catchment, and trap sediments, organic matter, and nutrients. The selective cutting of trees adds patchiness to the plant community and reduces the abundance of tree species preferred as food. These effects add several novel ecosystems to the landscape.

These influences of beavers on landscape structure once shaped the face of entire continents. At one time, beavers modified nearly all the temperate stream valleys in the Northern Hemisphere. The range of beavers in North America extended from arctic tundra to the Chihuahuan and Sonoran Deserts of northern Mexico, a range of approximately 15 million km$^2$. Before European colonization, the North American beaver population numbered 60 to 400 million individuals. However, fur trappers eliminated beavers from much of their historical range and nearly drove them to extinction. With protection, North American beaver populations are recovering and large areas once again show the influence of beavers on landscape structure.

Carol Johnston and Robert Naiman and their colleagues have carefully documented the substantial effects of beavers on landscape structure (e.g., Naiman et al. 1994). Much of their work has focused on the effects of beavers on the 298 km$^2$ Kabetogama Peninsula in Voyageurs National Park, Minnesota. Following their near extermination, beavers reinvaded the Kabetogama Peninsula beginning about 1925. From 1927 to 1988 the number of beaver ponds on the peninsula increased from 64 to 834, a change in pond density from 0.2 to 3.0 per square kilometer. Over this 63-year period, the area of new ecosystems created by beavers, including beaver ponds, wet meadows, and moist meadows, increased from 200 ha, about 1% of the peninsula, to 2,661 ha, about 13% of the peninsula. Foraging by beavers altered another 12% to 15% of upland areas.

Beaver activity has changed the Kabetogama Peninsula from a landscape dominated by boreal forest to a complex mosaic of ecosystems. Figure 21.25 shows how beavers have changed a 45 km$^2$ catchment on the peninsula. These maps show that, between 1940 and 1986, beavers increased landscape complexity within this catchment. Similar changes have occurred over nearly the entire peninsula.

Naiman and his colleagues quantified the effects of beaver over 214 km$^2$, or 72%, of the Kabetogama Peninsula. Within this area, there are about 2,763 ha of low-lying area that can be impounded by beavers. In 1927, the majority of the landscape, 2,563 ha, was dominated by forest. In 1927,

**Figure 21.24** Beavers are among nature's most active landscape engineers.

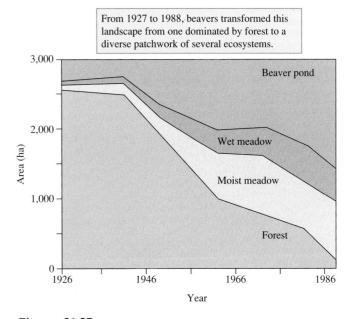

From 1927 to 1988, beavers transformed this landscape from one dominated by forest to a diverse patchwork of several ecosystems.

**Figure 21.25** Beaver-caused landscape changes on the Kabetogama Peninsula, Minnesota (data from Naiman et al. 1994).

moist meadow, wet meadow, and pond ecosystems covered only 200 ha. By 1988, moist meadows, wet meadows, and beaver ponds covered over 2,600 ha and boreal forest was limited to 102 ha. Between 1927 and 1988, beavers transformed most of the landscape.

The changes in landscape structure induced by beavers substantially alter landscape processes such as nutrient retention. Beaver activity between 1927 and 1988 increased the quantity of most major ions and nutrients in the areas affected by impoundments (fig. 21.26). The total quantity of nitrogen increased by 72%, while the amounts of phosphorus and potassium increased by 43% and 20%, respectively. The quantities of calcium, magnesium, iron, and sulfate stored in the landscape were increased by even greater amounts.

Naiman and his colleagues offer three possible explanations for increased ion and nutrient storage in this landscape: (1) beaver ponds and their associated meadows may trap materials eroding

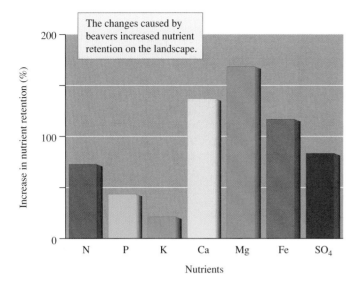

**Figure 21.26** Nutrient retention on the Kabetogama Peninsula after alteration by beavers (data from Naiman et al. 1994).

from the surrounding landscape, (2) the rising waters of the beaver ponds may have captured nutrients formerly held in forest vegetation, and (3) the habitats created by beavers may have altered biogeochemical processes in a way that promotes nutrient retention. Whatever the precise mechanisms, beaver activity has substantially altered landscape structure and processes on the Kabetogama Peninsula.

# Fire and the Structure of a Mediterranean Landscape

Fire contributes to the structure of landscapes ranging from tropical savanna to boreal forest. However, fire plays a particularly prominent role in regions with a Mediterranean climate. As we saw in chapter 2, terrestrial ecosystems in regions with Mediterranean climates, which support Mediterranean woodlands and shrublands, are subject to frequent burning. Hot, dry summers combined with vegetation rich in essential oils create ideal conditions for fires, which can be easily ignited by lightning or humans. In regions with a Mediterranean climate, fire is responsible for a great deal of landscape structure and change.

Richard Minnich (1983) used satellite photos to reconstruct the fire history of southern California and northern Baja California, Mexico, from 1971 to 1980, and found that the landscapes of both areas consist of a patchwork of new and old burns. Though these regions experience similar Mediterranean climates and support similar natural vegetation, their fire histories diverged significantly in the early twentieth century. For centuries, natural lightning-caused fires burned, sometimes for months, until they went out naturally. In addition, Spanish and Anglo-American residents would set fire to the land routinely to improve grazing for cattle and sheep. Then, early in the twentieth century, various government agencies in southern

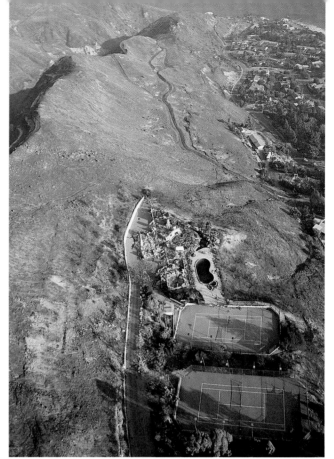

**Figure 21.27** Areas of Mediterranean shrubland in southern California periodically burn over large areas, destroying human habitations in the process.

California began to suppress fires in order to protect property within an increasingly urbanized landscape.

Minnich proposed that the different fire histories of southern California and northern Baja California might produce landscapes of different structure. He suggested that fire suppression allowed more biomass to accumulate and set the stage for large, uncontrollable fires. His specific hypothesis was that the average area burned by wildfires would be greater in southern California.

Minnich tested his hypothesis using satellite images taken from 1972 to 1980 (fig. 21.27). He found that between 1972 and 1980 the total area burned in the two regions was fairly similar (fig. 21.28). However, the size of burns differed significantly between the two regions. The frequency of small burns below 1,000 ha was higher in northern Baja California, while large burns above 3,000 ha were more frequent in southern California. Consequently, median burn size in southern California, 3,500 ha, was over twice that observed in Baja California, 1,600 ha (fig. 21.28).

These results are consistent with Minnich's hypothesis, but do they show conclusively that differences in fire management in southern California and Baja California have produced a difference in burn area? Other factors may contribute to the observed differences in the fire mosaic, including climatic differences, differences in age structure of vegetation, and topographic differences. The exploration of fire's influence on the structure of Mediterranean landscapes continues.

In this section, we have seen how geological processes, climate, the activities of organisms, and fire can contribute to landscape structure and change. Because human activity has

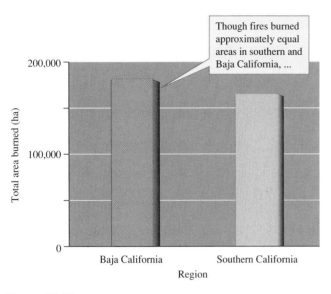
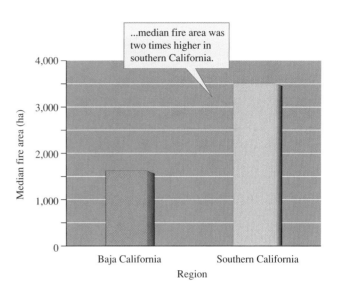

**Figure 21.28** Characteristics of fires in the Mediterranean landscapes of southern and Baja California from 1972 to 1980 (data from Minnich 1983).

often greatly altered landscape structure, there is growing interest in landscape restoration. That is the subject that we take up in the Applications & Tools section.

## APPLICATIONS & TOOLS

### Restoring a Riverine Landscape

Rivers and their floodplains form a complex, highly dynamic landscape that includes river, riparian forest, marsh, oxbow lake, and wet meadow ecosystems. Historically, these ecosystems actively exchanged organisms, inorganic nutrients, and organic energy sources. The key linkage between these landscape elements was periodic flooding.

Floods connect rivers with their associated floodplain ecosystems, and the rates and timing of many ecological processes are determined by the "flood-pulse" (Junk, Bayley, and Sparks 1989, Bayley 1995). Floods determine the form of the riverine landscape by depositing silt on floodplains, isolating oxbow lakes, and creating new river channels. Floods increase rates of decomposition and nutrient cycling in floodplain environments. Many species of river fish use floodplains as spawning and nursery grounds and many riparian plants require flooding for germination and establishment.

Over the past century, water management by building dams, channelizing rivers, constructing flood levees, and diverting water for irrigation has cut the historic connections between most rivers and their floodplains. However, there is growing recognition of the value of these historic connections for maintaining water quality and for supporting biological diversity. Consequently, governments all over the world have begun programs of river restoration. Some of the most ambitious of these projects focus on the Rhine River in Germany

and the Kissimmee River in south Florida, which we discuss in the following section.

## Riverine Restoration: The Kissimmee River

The Kissimmee River flows from its headwaters in Lake Kissimmee southward into Lake Okeechobee. The historical landscape of the Kissimmee River included a highly braided, meandering channel that flowed approximately 166 km from headwaters to mouth. Periodic flooding by the river flooded its 1.5 to 3 km wide floodplain, inundating several different types of ecosystems, including oxbow lakes and four major types of marshes. The river flooded approximately 94% of its floodplain during about half of the year. During some periods the Kissimmee floodplain would remain completely flooded for 2 to 4 years. The Kissimmee flooded more of its floodplain and for longer periods than any other river in North America.

Before flood control, the relationship of the Kissimmee River to its floodplain was similar to that of large tropical rivers such as the Amazon River in South America and Niger River in Africa. This tight linkage between river and floodplain was very important to the birds, fish, and aquatic invertebrates of the Kissimmee. The landscape supported 48 species of fish, 16 species of wading birds, 22 species of ducks and other water birds, and hundreds of species of aquatic invertebrates, the lives of most of which were tied to the Kissimmee's annual flooding cycle. The flood pulse was also critical to nutrient cycling and to maintaining high water quality.

Rapid human population growth in the early 1940s, followed by extensive flooding that lasted from 1947 to 1948, created pressures for flood control on the Kissimmee River. From 1962 to 1970 the river was converted from a braided, meandering river to a series of five reservoirs connected by canals (fig. 21.29). Flow out of Lake Kissimmee and along the

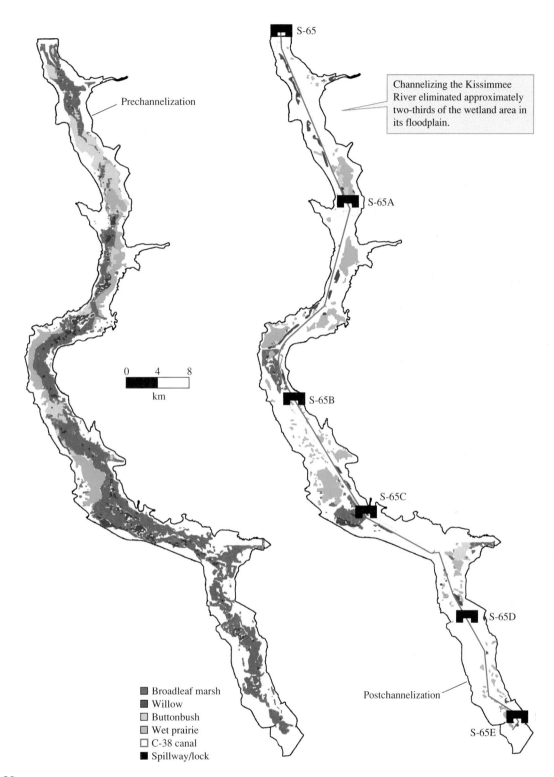

Prechannelization

Channelizing the Kissimmee
River eliminated approximately
two-thirds of the wetland area in
its floodplain.

S-65

S-65A

0    4    8
km

S-65B

S-65C

S-65D

Postchannelization

■ Broadleaf marsh
■ Willow
■ Buttonbush
■ Wet prairie
□ C-38 canal
■ Spillway/lock

S-65E

**Figure 21.29**  Channelization and wetland loss in the Kissimmee River floodplain (data from Toth et al. 1995).

canals is controlled by six separate flow control structures. These flood control measures transformed the 166 km meandering Kissimmee River into a canal 9 m deep, 100 m wide, and 90 km long (fig. 21.30). They also eliminated 14,000 ha of wetlands within the Kissimmee landscape. Most of the former river channels were either filled in by materials dug to form the canal system, dried out, or reduced to such low flows that they were choked by vegetation, especially by introduced species of floating water plants such as water hyacinth, *Eichhornia crassipes.*

**Figure 21.30** Channelizing greatly simplified the structure of the Kissimmee River. Compare the straight artificial channel on the right of this photo with the remnant meandering channels on the left.

**Figure 21.31** Restoration of the Kissimmee River landscape includes refilling approximately 35 km of artificial channel to restore flows to original meandering channels.

These environmental changes had a severe impact on many populations and ecosystem processes. Wintering waterfowl populations declined by 92%. The population of largemouth bass, an important sport fish, declined. Largemouth bass were replaced by nongame species that tolerate low oxygen concentrations. Populations of riverine invertebrates were reduced and replaced by invertebrates of lake and pond ecosystems. Eliminating the flood pulse greatly reduced the exchange of nutrients, organic matter, and organisms between the river and floodplain ecosystems. Stabilized water levels nearly eliminated spawning and foraging habitats for adult fish and refuge and rearing areas for larval and juvenile fish. Ecologists estimated that drying the floodplain wetlands along the Kissimmee resulted in a loss of 6 billion freshwater prawns, *Palaemonetes paludosus.*

The public soon recognized the negative ecological consequences of the Kissimmee flood control project and exerted political pressure to restore the river to premanagement conditions (Cummins and Dahm 1995, Koebel 1995, Toth et al. 1995, Dahm et al. 1995). These pressures eventually produced the Kissimmee restoration project. A limited test restoration was conducted from 1984 to 1990. The project included fluctuating water levels within one of the reservoirs to return flooding to about 1,080 ha of floodplain. Water managers built weirs on the canal to divert water into remnant river channels and created a series of marshes.

The Kissimmee River floodplain landscape showed dramatic responses to these initial restoration efforts. Native vegetation that had historically dominated the system responded positively, while exotic vegetation and vegetation from uplands showed signs of decline. Flow through remnant channels transported accumulated detritus into the canal system. Riverine invertebrates recolonized remnant river channels, and fish quickly moved into the flooded areas. Fish requiring higher levels of dissolved oxygen again dominated restored areas. Large numbers of waterfowl also quickly moved into the new marshes.

In response to these encouraging results, water managers began a far more ambitious plan to restore a large portion of the Kissimmee River system. The restoration project will take about 15 years and restore about 70 km of river channels to a more natural condition and about 11,000 ha of wetlands. The projected costs of repairing the damage to the Kissimmee landscape is $500 million. The Kissimmee restoration will consist of two parts. First, water managers will regulate water in the headwaters of the river to include a more natural flood pulse. Second, much of the Kissimmee canal will be eliminated and water will be returned to the braided river channels. Several kilometers of the Kissimmee canal will be filled (fig. 21.31), and some water control structures will be removed. In addition, 14 km of river channel that were filled during the excavation of the canal will be excavated.

The first phase of the Kissimmee River restoration was completed in 2001. This phase included filling 12 km of the 90 km drain canal, which required 9.2 million m$^3$ of earth fill. Restoration crews also constructed two sections of new river channel totaling 2.4 km in length and reestablished 24 contiguous km of river. They also used explosives to demolish one of the main water control structures on the drain canal.

The keys to this restoration effort involve restoration of landscape structure and landscape processes. First, water managers will restore historical landscape structure by restoring the complex channel network and reestablishing the historic floodplain marshes. Second, they will restore the dominant landscape process, the historic flood pulse. Restoration of flooding will reestablish the hydrologic connections between the river and floodplain ecosystems, connections that promote exchanges of nutrients, energy, and species among ecosystems in the Kissimmee River landscape.

This is the largest landscape restoration project ever undertaken. The restoration of the Kissimmee River landscape is also one of the largest ecological experiments ever conducted and will be a significant test of the predictive ability of ecological theory. The project includes models of expected population, community, ecosystem, and landscape responses as well as careful monitoring of those responses. As it goes forward, ecologists will watch to see how well their predictions compare to the actual responses of the system. The lessons learned on the Kissimmee River will help with future efforts aimed at restoring the structure and processes of damaged landscapes.

## SUMMARY

A landscape is a heterogeneous area composed of several ecosystems. The ecosystems making up a landscape generally form a mosaic of visually distinctive patches. These patches are called *landscape elements*. *Landscape ecology* is the study of landscape structure and processes.

**Landscape structure includes the size, shape, composition, number, and position of different ecosystems within a landscape.** Most questions in landscape ecology require that ecologists quantify landscape structure. Until recently, however, geometry, which means "earth measurement," could offer only rough approximations of complex landscape structure. Today, an area of mathematics called fractal geometry can be used to quantify the structure of complex natural shapes. One of the findings of fractal geometry is that the length of the perimeter of complex shapes depends upon the size of the device used to measure the perimeter. One implication of this result is that organisms of different sizes may use the environment in very different ways.

**Landscape structure influences processes such as the flow of energy, materials, and species between the ecosystems within a landscape.** Landscape ecologists have proposed that landscape structure, especially the size, number, and isolation of habitat patches, can influence the movement of organisms between potentially suitable habitats. The group of subpopulations living on such habitat patches make up a metapopulation. Studies of the movements of small mammals in a prairie landscape show that a smaller proportion of individuals moves in more fragmented landscape but that the individuals that do move will move farther. The local population density of the Glanville fritillary butterfly, *Melitaea cinxia,* is lower on larger and on isolated habitat patches. Small populations of this butterfly and desert bighorn sheep are more vulnerable to local extinction. Habitat corridors have been shown to increase rates of movement among isolated habitat patches. The source of water for lakes in a Wisconsin lake district is determined by their positions in the landscape, which in turn determine their hydrologic and chemical responses to drought.

**Landscapes are structured and change in response to geological processes, climate, activities of organisms, and fire.** Geological features produced by processes such as volcanism, sedimentation, and erosion interact with climate to provide a primary source of landscape structure. In the Sonoran Desert, plant distributions map clearly onto soils of different ages and form a vegetative mosaic that closely matches soil mosaics. This mosaic will gradually shift as geological processes and climate gradually change the soil mosaic. While geological processes and climate set the basic template for landscape structure, the activities of organisms, from plants to elephants, can be an additional source of landscape structure and change. Economically motivated human activity changes the structure of landscapes all over the globe. Beavers can quickly change landscape structure and processes over large regions. Fire contributes to the structure of landscapes ranging from tropical savanna to boreal forest. However, fire plays a particularly prominent role in regions with a Mediterranean climate.

Because human activity has often altered landscape structure and processes in undesirable ways, there is growing pressure and interest in landscape restoration. Some of the most ambitious current restoration efforts focus on the restoration of riverine landscapes. Rivers and their floodplains form a complex, highly dynamic landscape that includes river, riparian forest, marsh, oxbow lake, and wet meadow ecosystems. Over the past century, water management by building dams, channelizing rivers, constructing flood levees, and diverting water for irrigation has cut the historic connections between most rivers and their floodplains. The restoration of the Kissimmee River landscape is also one of the largest ecological experiments ever conducted and will be a significant test of the predictive ability of ecological theory.

## Review Questions

1. How does landscape ecology differ from ecosystem and community ecology? What questions might an ecosystem ecologist ask about a forest? What questions might a community ecologist ask about the same forest? Now, what kinds of questions would a landscape ecologist ask about a forested landscape?

2. How should the *area* of forest patches in an agricultural landscape affect the proportion of bird species in a community that

are associated with forest edge habitats? How should patch area affect the presence of birds associated with forest interiors?

3. The green areas represent forest fragments surrounded by agriculture. Landscapes 1 and 2 contain the same total forest area. Will landscape 1 or 2 contain more forest interior species? Explain.

4. How might the *shapes* of forest patches in a landscape affect the proportion of birds in the community associated with forest

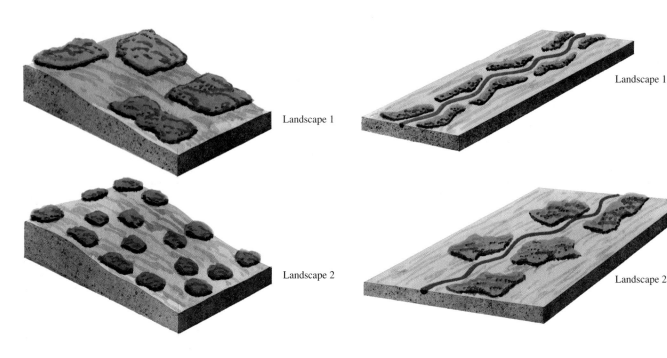

Landscape 1

Landscape 2

Landscape 1

Landscape 2

edge habitat? How might patch shape affect the presence of birds associated with forest interior?

5. Consider the following options for preserving patches of riverside forest. Again, the two landscapes contain the same total area of forest but the patches in the two landscapes differ in shape. Which of the two would be most dominated by forest edge species?

6. How does the concept of metapopulations differ from the perspective of populations that we discussed in section III? (Hint: Think of the spatial contexts of these two views of populations.)

7. How do the positions of patches in a landscape affect the movement of individuals among habitat patches and among portions of a metapopulation? Again, consider the hypothetical landscapes shown in question 5. Which of the two landscapes would promote the highest rate of movement of individuals between forest patches? Can you think of any circumstances in which it might be desirable to reduce the movement of individuals across a landscape? (Hint: Think of the potential threat of pathogens that are spread mainly by direct contact between individuals within a population.)

8. Use fractal geometry and the niche concept (see chapters 13 and 16) to explain why the canopy of a forest should accommodate

more species of predaceous insects than insectivorous birds. Assume that the numbers of bird and predaceous insect species are limited by competition. (Milne's study [1993] of barnacles and bald eagles on Admiralty Island should provide a beginning for your argument.)

9. Analyses such as Milne's comparison (1993) of bald eagles and barnacles demonstrate that organisms of different sizes interact with the environment at very different spatial scales. With this in mind consider the experiments of Diffendorfer and colleagues (1995) on the influence of habitat fragmentation on movement patterns of small mammals. Think about the size of their experimental study area (see fig. 21.10). How might a manipulation of this size have affected the movements of prairie birds? How would their manipulation have affected the movements of ground-dwelling beetles?

10. How do the activities of animals affect landscape heterogeneity? You might use either beaver or human activity as your model. What parallels can you think of between the influence of animal activity on landscape heterogeneity and the intermediate disturbance hypothesis? Which is concerned with the effect of disturbance on species diversity?

# Suggested Readings

Bayley, P. B. 1995. Understanding large river-floodplain ecosystems. *BioScience* 45:153–58.

*Bayley provides an excellent introduction to riverine landscapes, including their structure and dynamics.*

Bennett, A. F., K. Henein, and M. Gray. 1994. Corridor use and the elements of corridor quality: chipmunks and fencerows in a farmland mosaic. *Biological Conservation* 68:155–65.

*This is a fascinating account of how landscape structure can affect the movement of animals. There are many implications for conservation biology here.*

Berggren, Å, B. Birath, and O. Kindvall. 2002. Effect of corridors and habitat edges on dispersal behavior, movement rates, and movement angles in Roesel's bush-cricket (*Metrioptera roeseli*). *Conservation Biology* 16:1562–69.

Berggren, Å, A. Carlson, and O. Kindvall. 2001. The effect of landscape composition success, growth rate and dispersal in introduced bush-crickets *Metrioptera roeseli*. *Journal of Animal Ecology* 70:663–70.

*These two studies give a highly detailed assessment of how landscape structure, particularly corridors, affect behavior and population persistence. The studies span several years and employ a small arthropod as a model, experimental organism.*

Buijse, A. D., H. Coops, M. Staras, L. H. Jans, G. J. Van Geest, R. E. Grift, B. W. Ibelings, W. Oosterberg, and F. C. J. M. Roozen.

2002. Restoration strategies for river floodplains along large lowland rivers in Europe. *Freshwater Biology* 47:889–907.

*An excellent review of the topic of river restoration from a European perspective, focusing mainly on the Rhine and Danube.*

Cummins, K. W. and C. N. Dahm. 1995. Restoring the Kissimmee. *Restoration Ecology* 3:147–48.

Dahm, C. N., K. W. Cummins, H. M. Valett, R. L. Coleman. 1995. An ecosystem view of the restoration of the Kissimmee River. *Restoration Ecology* 3:225–38.

*Two articles from a whole issue of the journal Restoration Ecology dedicated to restoration of the Kissimmee landscape. The entire issue is worth reading by anyone interested in this historic restoration project.*

Forman, R. T. T. 1995. *Land Mosaics.* Cambridge, England: Cambridge University Press.

*Forman presents a comprehensive and up-to-date discussion of landscape ecology—excellent for the student wanting additional depth and breadth in the subject.*

Fuchs, U. and B. Statzner. 1990. Timescales for the recovery potential of river communities after restoration: lessons to be learned from smaller streams. *Regulated Rivers: Research & Management* 5:77–87.

*Fuchs and Statzner present a thought-provoking discussion of points that must be considered when attempting river restoration. Their creative analysis derives useful information from more thorough studies of smaller streams.*

Haddad, N. M. 1999. Corridor and distance effects on interpatch movements: a landscape experiment with butterflies. *Ecological Applications* 9:612–22.

Haddad, N. M. and K. A. Baum. 1999. An experimental test of corridor effects on butterfly densities. *Ecological Applications* 9:623–33.

*Well-designed, landscape experiments demonstrating the positive effects of corridors to butterfly dispersal and population density.*

Hanski, I., M. Kuussaari, and M. Nieminen. 1994. Metapopulation structure and migration in the butterfly *Melitaea cinxia. Ecology* 75:747–62.

*An account of an excellently designed and executed study of a butterfly population in a complex landscape. Hanski and his colleagues show how landscape structure can exert substantial influences on a population.*

Johnston, C. A. and R. J. Naiman. 1990. Aquatic patch creation in relation to beaver population trends. *Ecology* 71:1617–21.

Johnston, C. A. and R. J. Naiman. 1990. The use of a geographic information system to analyze long-term landscape alteration by beaver. *Landscape Ecology* 4:5–19.

*These two papers demonstrate the amazing capacity of beavers—some of the most effective ecosystem engineers—to alter landscape structure.*

Malanson, G. P. 1993. *Riparian Landscapes.* Cambridge, England: Cambridge University Press.

*A concise treatment of riparian landscapes. Malanson's book is ideal for anyone interested in a landscape perspective of riverine systems.*

McAuliffe, J. R. 1994. Landscape evolution, soil formation, and ecological patterns and processes in Sonoran Desert bajadas. *Ecological Monographs* 64:111–48.

*McAuliffe provides a model for studying the interaction of plants with long-term geological processes—an approach pioneered by Henry Cowles a century before but long ignored by most ecologists.*

National Research Council. 1992. *Restoration of Aquatic Ecosystems.* Washington, D.C.: National Academy Press.

*This book by the National Research Council provides a detailed study of restoration of aquatic ecosystems. This is an ideal source for considering the interface between management policy and science.*

Tewksbury, J. J., D. J. Levey, N. M. Haddad, S. Sargent, J. L. Orrock, A. Weldon, B. J. Danielson, J. Brinderhoff, E. I. Damschen, and P. Townsend. 2002. Corridors affect plants, animals and the interaction in fragmented landscapes. *Proceedings of the National Academy of Sciences of the United States of America* 99:12923–12926.

*An experimental study that goes beyond the usual focus to show the positive effects of corridors on critical species interactions.*

Whalen, P. J., L. A. Toth, J. W. Koebel, and P. K. Strayer. 2002. Kissimmee River restoration: a case study. *Water Science and Technology* 45:55–62.

*Concise update on the status, statistics, and projected costs of the Kissimmee River restoration project.*

# On the Net

Visit this textbook's accompanying website at www.mhhe.com/ecology (click on the book's title) to take advantage of practice quizzing, study/writing tips, timely news articles, and additional URLs for research on the topics in this chapter.

Landscape Ecology
Field Methods for Studies of Ecosystems
Water Use and Management

Land Use Planning
Land Use: Forests and Rangelands

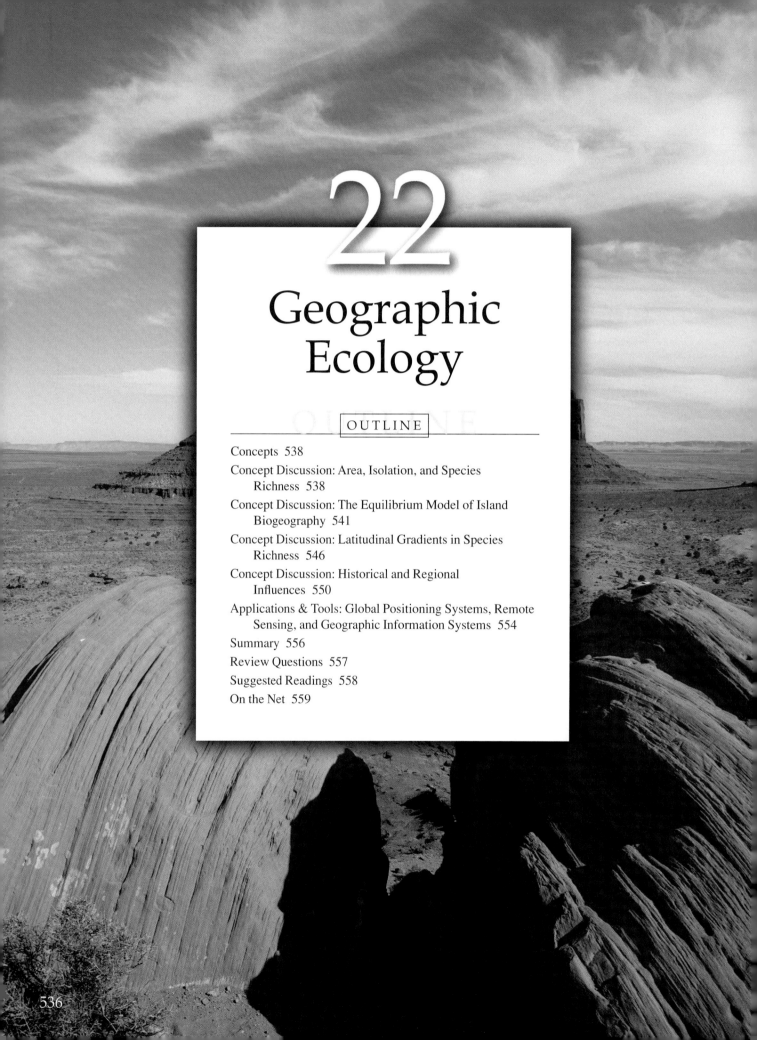

# 22

# Geographic
# Ecology

Geographic ecology began on June 5, 1799, as Alexander von Humboldt and Aimé Bonpland sailed out of the port of Coruña in northwest Spain. Their small Spanish ship managed to slip past a British naval blockade and sail on, first to the Canary Islands and then to South America. Humboldt was a Prussian engineer and scientist and Bonpland was a French botanist. Humboldt came equipped with the finest scientific instruments of the time and was prepared to systematically survey the lands that he and Bonpland would visit. He wrote a letter to a friend a few hours before his ship left port outlining his purpose for the expedition: "I shall try to find out how the forces of nature interreact upon one another and *how the geographic environment influences plant and animal life* [emphasis added]."

Humboldt and Bonpland carried passports issued by the court of King Carlos IV of Spain, giving them permission to conduct scientific studies throughout the Spanish Empire, which then stretched from California to Texas in North America and south to the tip of South America. They had complete access to a vast area of the earth's surface that was essentially unexplored scientifically, and they put that access to productive use. Because their discoveries were so numerous and their explorations so thorough, Simón Bolívar, the liberator of most of Spanish America, referred to Humboldt as "the discoverer of the New World."

Humboldt's expedition was one of the most ambitious scientific explorations of the age. During the course of their expedition Humboldt and Bonpland traveled nearly 10,000 km through South and North America. They traveled on foot, by canoe, or on horseback, visiting latitudes ranging between 12° S and 52° N. They also climbed to nearly 5,900 m on the slopes of Chimborazo; the highest ascent by anyone in history up to that time (fig. 22.1).

The physical feats of their expedition, however, never took precedent over their scientific purpose. For instance, on their climb of Chimborazo, they faced the uncertain dangers of high altitude. Yet, as blood oozed from their lips and gums, Humboldt and Bonpland recorded the altitudinal distributions of plants and animals. Later, Humboldt organized their observations of climate and plant distributions into ingenious visual representations of plant geography. What he did not accomplish, he inspired others to do so. One of those inspired to follow in Humboldt's footsteps was Charles Darwin. Darwin said that his reading of Humboldt's expedition to South America set the course of his whole life.

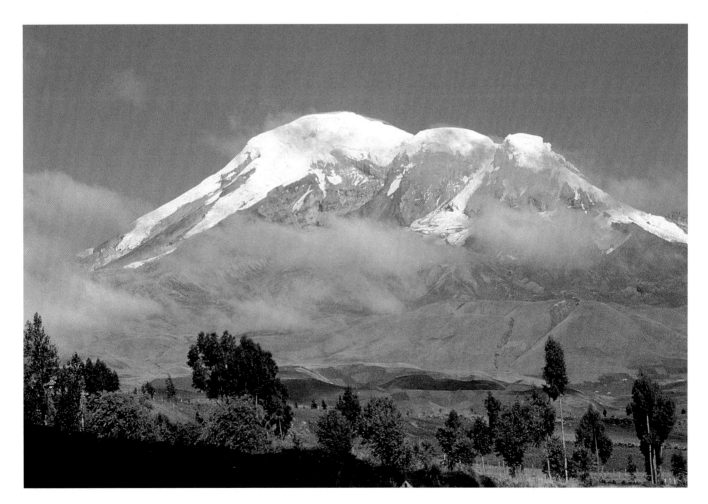

**Figure 22.1** On the slopes of Chimborazo, a 6,310 m high volcanic peak in the Andes Mountains of Ecuador, Alexander von Humboldt and Aimé Bonpland meticulously recorded the altitudinal distributions of plants.

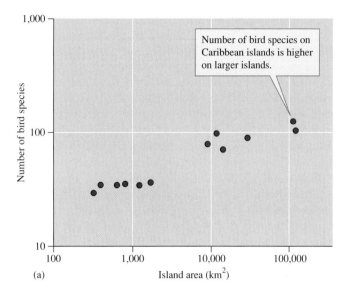

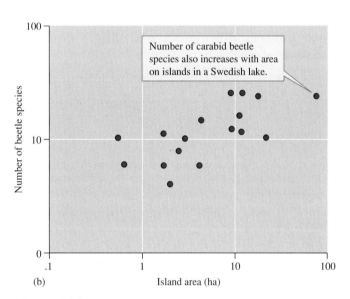

**Figure 22.2** Relationship between island area and number of species (data from Preston 1962a, Nilsson, Bengtsson, and Ås 1988).

Robert H. MacArthur (1972) defined geographic ecology as the "search for patterns of plant and animal life that can be put on a map." MacArthur's map might include an archipelago of islands, a region, or a series of continents. Somewhere above the level of local community and ecosystem ecology and even above the level of landscapes is the realm of **geographic ecology.** Though geographic ecology began long before MacArthur, with explorers such as Humboldt, Darwin, and Wallace, MacArthur put an indelible quantitative stamp on the field when he and E. O. Wilson published their first models of island biogeography.

The development of geographic ecology continues as new generations of scientists equipped with a diversity of tools, both ancient and modern, search for the elusive patterns that can be put on maps. There is a heightened sense of urgency, however, as the prospect of global climate change presses us to understand the forces controlling large-scale patterns of biological diversity and the geographic ranges of species. The breadth of

geographic ecology is as vast as its subject. Consequently, we concentrate our discussions in chapter 22 on just a few aspects of the field: island biogeography, latitudinal patterns of species diversity, and the influences of large-scale regional and historical processes on biological diversity.

## CONCEPTS

- On islands and habitat patches on continents, species richness increases with area and decreases with isolation.
- Species richness on islands can be modeled as a dynamic balance between immigration and extinction of species.
- Species richness generally increases from middle and high latitudes to the equator.
- Long-term historical and regional processes significantly influence the structure of biotas and ecosystems.

## CONCEPT DISCUSSION

### Area, Isolation, and Species Richness

**On islands and habitat patches on continents, species richness increases with area and decreases with isolation.**

## Sampling Area and Number of Species

A quantitative relationship between area and number of species was first developed by Olof Arrhenius (1921), a pioneer in the area of geographic ecology. Arrhenius made his observations on islands near Stockholm, Sweden, where he worked within several plant communities that he identified by names such as *herb–Pinus wood and shore association.* He counted the number of species within areas of various sizes and then developed a mathematical description of the relationship between area sampled and number of plant species. However, Arrhenius worked at scales much smaller than the geographic focus of chapter 22. To see the first quantitative work on geographic patterns, we have to move to a later time.

## Island Area and Species Richness

Frank Preston (1962a) examined the relationship between number of species and the area of islands in the West Indies. As shown in figure 22.2*a,* the fewest bird species live on the smallest islands and the most on the largest islands, which are

Cuba and Hispaniola. The relationship between island area and number of species is not just a property of bird assemblages. Sven Nilsson, Jan Bengtsson, and Stefan Ås (1988) explored patterns of species richness among woody plants, carabid beetles, and land snails on 17 islands in Lake Mälaren, Sweden. The islands ranged in area from 0.6 to 75 ha and were all forested. The researchers were careful to choose islands that showed few or no signs of human disturbance. One of the results of their study was that island area was the best single predictor of species richness in all three groups of organisms. Figure 22.2b shows the relationship found between island area and number of carabid beetles.

When most of us think of islands, the picture that generally comes to mind is a small bit of land in the middle of an ocean. However, many habitats on continents are so isolated that they can be considered as islands.

## Habitat Patches on Continents: Mountain Islands

The many isolated mountain ranges that extend across the Great Basin and southwestern regions of North America are now continental islands. During the late Pleistocene, 11,000 to 15,000 years ago, forest and woodland habitats extended unbroken from the Rocky Mountains to the Sierra Nevada in California. Then as the Pleistocene ended and the climate warmed, forest and alpine habitats contracted to the tops of the high mountains scattered across the American Southwest. As montane habitats retreated to higher elevations, woodland, shrubland, grassland, or desert scrub vegetation invaded the lower elevations. As a consequence of these changes, once-continuous forest and alpine vegetation was converted to a series of islandlike habitat patches associated with mountains and therefore called *montane*.

As montane vegetation contracted to mountaintops, montane animals followed. Mark Lomolino, James H. Brown, and Russell Davis (1989) studied the diversity of montane mammals on isolated mountains in the American Southwest. They focused on the distributions of 26 species of nonflying forest mammals that occur on 27 montane islands. They chose mountain ranges that had been studied thoroughly enough so that their mammal faunas were well known. The list of species, which included shrews, ermine, squirrels, chipmunks, and voles, was limited to species that show a strong association with montane environments.

The team found that montane mammal richness was positively correlated with habitat area. As figure 22.3 shows, the area of the 27 montane islands ranged from less than 7 km² to over 10,000 km², while the number of montane mammals on them ranged from 1 to 16.

## Lakes as Islands

Lakes can also be considered as habitat islands—aquatic environments isolated from other aquatic environments by land. However, lakes differ widely in their degree of isolation.

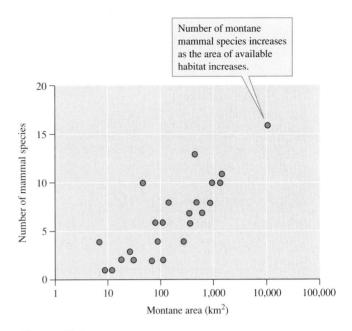

**Figure 22.3** Area of montane habitat and number of montane mammal species on isolated mountain ranges in the American Southwest (data from Lomolino, Brown, and Davis 1989).

Seepage lakes, which receive no surface drainage, are completely isolated, while drainage lakes, which have stream inlets and/or outlets, are less isolated (see chapter 21).

William Tonn and John Magnuson (1982) studied patterns of species composition and richness among fish inhabiting lakes in northern Wisconsin. They focused their research on 18 lakes in the Northern Highlands Lake District of Wisconsin and Michigan. The study was conducted in Vilas County, Wisconsin, which includes over 1,300 lakes (fig. 22.4). With so many lakes at their disposal, Tonn and Magnuson could match lakes carefully for a variety of characteristics. All 18 study lakes had similar bottom substrates and similar maximum depths. However, the lakes spanned a considerable range of surface area (2.4–89.8 ha). Ten of the lakes were drainage lakes or spring fed and eight lakes were seepage lakes. Eight lakes had a history of low oxygen content during winter.

Tonn and Magnuson collected a total of 23 species, 22 in summer and 18 in winter. If we combine their winter and summer collections on each lake and plot total species richness against area, there is a significant positive relationship (fig. 22.4). Once again, we see that the number of species increases with the area of an insular environment. However, these researchers worked with a single lake district. Is there a relationship between lake area and diversity when lakes from several regions are included in the analysis?

Clyde Barbour and James H. Brown (1974) studied patterns of species richness across a worldwide sample of 70 lakes. The lakes in their sample ranged in area from 0.8 to 436,000 km², while the number of fish species ranged from 5 to 245. Barbour and Brown also found a positive relationship between area and fish species richness.

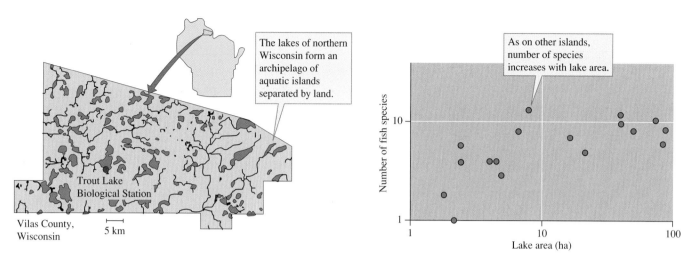

**Figure 22.4**  Lake area and number of fish species in lakes of northern Wisconsin (data from Tonn and Magnuson 1982).

# Island Isolation and Species Richness

There is often a negative relationship between the isolation of an island and the number of species it supports. However, because organisms differ substantially in dispersal rates, an island that is very isolated for one group of organisms may be completely accessible to another group.

## Marine Islands

Robert MacArthur and Edward O. Wilson (1963) found evidence that isolation reduces bird diversity on Pacific islands. In figure 22.5, islands less than 800 km from New Guinea, a potential source of colonists to the other islands, are represented by gold dots; islands greater than 3,200 km from New Guinea are represented by red dots; islands at intermediate distances from New Guinea are represented by blue dots. How does figure 22.5 indicate an effect of isolation on species richness? Compare the relative numbers of species on islands of approximately equal area but that lie at different distances from New Guinea. On average, those closest to New Guinea support a larger number of bird species than those at intermediate or far distances.

Comparative studies of diversity patterns on islands remind us that different organisms have markedly different dispersal abilities. Mark Williamson (1981) summarized the data for the relationship between island area and species richness for various groups of organisms inhabiting the Azore and Channel Islands. The Azore Islands lie approximately 1,600 km west of the Iberian Peninsula, while the Channel Islands are very near the coast of France. While vastly different in distance from mainland areas, both island groups experience moist temperate climates and have biotas that are of European origin. Consequently, a comparison of their biotas should reveal the potential influence of isolation on diversity. Figure 22.6 shows Williamson's summary of species area relationships for ferns and fernlike plants (pteridophytes) and

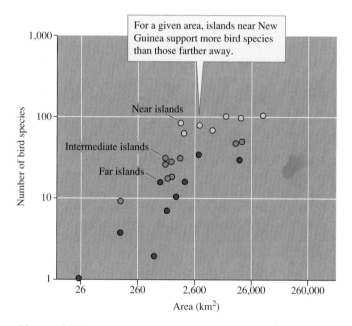

**Figure 22.5**  Distance from New Guinea and birds species richness on Pacific islands (data from MacArthur and Wilson 1963).

land- and water-breeding birds. Both groups of organisms show a positive relationship between island area and diversity on both the Channel and Azore Islands. However, while birds show a clear influence of isolation on diversity, pteridophytes do not. Notice that bird species richness is lower on Azore Islands compared to Channel Islands of similar size. Meanwhile, pteridophyte diversity is similar on islands of comparable size in the two island groups.

These differences in pattern show that the 1,600 km of ocean between the Azore Islands and the European mainland reduces the diversity of birds but not pteridophytes. These differences in the effect of isolation reflect differences in the dispersal rates of these organisms. While land birds must fly across water barriers, pteridophytes produce large quantities of light spores that are easily dispersed by wind. One species

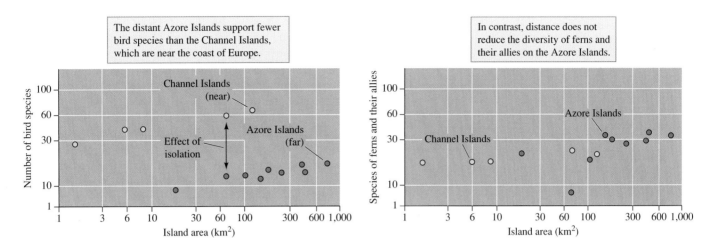

**Figure 22.6** Influence of isolation on diversity of birds and ferns and their allies on the Channel and Azore Islands (data from Williamson 1981).

of pteridophyte, bracken fern, has naturally established populations throughout the globe, including New Guinea, Britain, Hawaii, and New Mexico. When we consider the potential effects of isolation on diversity, we must also consider the dispersal capabilities of the study organisms.

## Isolation and Habitat Islands on Continents

Has an isolation effect been observed across habitat islands on continents? Lomolino, Brown, and Davis, whom we have already discussed, found a strong negative relationship between isolation and the number of montane mammal species living on mountaintops across the American Southwest. They measured isolation as the distance of a mountain range from potential sources of colonists. Colonists can immigrate from the southern Rocky Mountains, with 23 montane mammal species, or from the Mogollon Rim, with 16 montane mammal species. For mountains closer to the southern Rocky Mountains, isolation was measured as the distance from the mountain island to the nearest point of the Rocky Mountains. If the mountain was closer to the Mogollon Rim, the researchers used a composite measure of distance that weighted the distances to the Mogollon Rim and the southern Rocky Mountains by the diversities of their montane mammals. Figure 22.7 shows the results of their analysis, which shows a strong negative relationship between isolation (distance to species source areas) and species richness of montane mammals. The analysis clearly shows that both area and isolation affect the diversity of mammals on habitat patches on continents.

What does the negative effect of isolation on these mountain mammal populations indicate about local assemblages? The higher diversity of mountain islands near the southern Rocky Mountains and/or to the Mogollon Rim suggests that montane mammals continue to disperse across the intervening woodlands and grasslands. Without immigration, near and far mountains of equal area would support similar numbers of montane mammal species. This result suggests that the diversity of organisms on islands is maintained by a dynamic process that involves ongoing

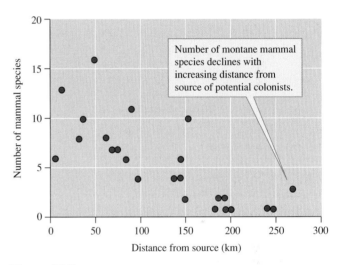

**Figure 22.7** Distance from large montane areas and number of montane mammal species on isolated mountain ranges of the American Southwest (data from Lomolino, Brown, and Davis 1989).

colonization. A dynamic rather than a static view of island diversity is the foundation of an equilibrium model of island diversity that we discuss in the next Concept Discussion section.

## CONCEPT DISCUSSION

### The Equilibrium Model of Island Biogeography

**Species richness on islands can be modeled as a dynamic balance between immigration and extinction of species.**

The examples we just reviewed show clear relationships between species richness and island area and isolation. When confronted with such a pattern, scientists look for explanatory

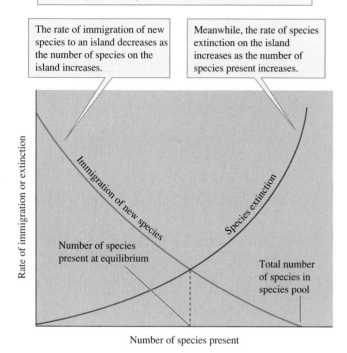

According to the equilibrium model of island biogeography, the number of species on an island is determined by a balance between species immigration and extinction.

The rate of immigration of new species to an island decreases as the number of species on the island increases.

Meanwhile, the rate of species extinction on the island increases as the number of species present increases.

*Immigration of new species*

*Species extinction*

Rate of immigration or extinction

Number of species present at equilibrium

Total number of species in species pool

Number of species present

**Figure 22.8** Equilibrium model of island biogeography (data from MacArthur and Wilson 1963).

mechanisms. What mechanisms might increase species richness on large islands and reduce richness on small and isolated islands? MacArthur and Wilson (1963, 1967) proposed a model that explained patterns of species diversity on islands as the result of a balance between rates of immigration and extinction (fig. 22.8). This model is called the *equilibrium model of island biogeography.*

Figure 22.8 shows that the model presents rates of immigration and extinction as a function of numbers of species on islands. How might rates of immigration and extinction be influenced by the numbers of species on an island? To answer this question, we need to understand what MacArthur and Wilson meant by rates of immigration and extinction. They defined the *rate of immigration* as the rate of arrival of *new* species on an island. *Rate of extinction* was the rate at which species went extinct on the island. MacArthur and Wilson reasoned that rates of immigration would be highest on a new island with no organisms, since every species that arrived at the island would be new. Then as species began to accumulate on an island, the rate of immigration would decline since fewer and fewer arrivals would be new species. They called the point at which the immigration line touches the horizontal axis *P* because it is the point representing the entire "pool" of species that might immigrate to the island.

How might numbers of species on an island affect the rate of extinction? MacArthur and Wilson predicted that the rate of extinction would rise with increasing numbers of

species on an island for three reasons: (1) the presence of more species creates a larger pool of potential extinctions, (2) as the number of species on an island increases, the population size of each must diminish, and (3) as the number of species on an island increases, the potential for competitive interactions between species will increase.

Since the immigration line falls and the extinction line rises as number of species increases, the two lines must cross as shown in figure 22.8. What is the significance of the point where the two lines cross? The point where the two lines cross predicts the number of species that will occur on an island. Thus, the equilibrium model represents the diversity of species on islands as the result of a dynamic balance between immigration and extinction.

MacArthur and Wilson used the equilibrium model to predict how island size and isolation should affect rates of immigration and extinction. They proposed that the rate of immigration is mainly determined by an island's distance from a source of immigrants; for example, the distance of an oceanic island from a mainland. They proposed that rates of extinction on islands would be determined mainly by island size. These predictions are represented in figure 22.9. Notice that the figure predicts that large, near islands will support the greatest number of species, while small, far islands will

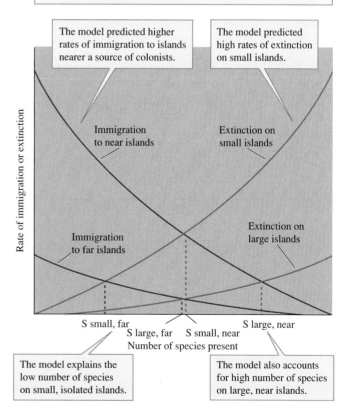

The equilibrium model of island biogeography explained variation in number of species on islands by the influences of isolation and area on rates of immigration and extinction.

The model predicted higher rates of immigration to islands nearer a source of colonists.

The model predicted high rates of extinction on small islands.

Rate of immigration or extinction

Immigration to near islands

Extinction on small islands

Immigration to far islands

Extinction on large islands

S small, far

S large, far    S small, near

S large, near

Number of species present

The model explains the low number of species on small, isolated islands.

The model also accounts for high number of species on large, near islands.

**Figure 22.9** Island distance and area and rates of immigration and extinction (data from MacArthur and Wilson 1963).

support the lowest number of species. The model predicts that small, near islands and large, far islands will support intermediate numbers of species.

The predictions of the equilibrium model of island biogeography are consistent with the patterns of island diversity reviewed in the previous section. Large islands hold more species than small islands, and islands near sources of immigrants hold more species than islands far from sources of immigrants. We should expect the equilibrium model to be consistent with known variation in species richness across islands since MacArthur and Wilson designed their model to explain the known patterns. Did the equilibrium model make any new predictions? The main new predictions were (1) that island diversity is the outcome of a highly dynamic balance between immigration and extinction and (2) that the rates of immigration and extinction are determined mainly by the isolation and area of islands. In other words, the equilibrium model predicts that the species composition on islands is not static but changes over time. Ecologists call this change in species composition **species turnover.**

# Species Turnover on Islands

Turnover of bird species was demonstrated on the California Channel Islands by Jared Diamond (1969). Diamond surveyed the birds of the nine California Channel Islands in 1968, approximately 50 years after an earlier survey by A. B. Howell. The islands range in area from less than 3 to 249 km$^2$ and lie 12 to 61 km from the coast of southern California (fig. 22.10). Howell had thoroughly censused all of the islands except for San Miguel and Santa Rosa Islands, where he had difficulty getting permission to do bird surveys. In his later study, Diamond had full access to all the islands and was able to survey all land and water birds.

The results of Diamond's study support the equilibrium model of island biogeography. The number of bird species inhabiting the California Channel Islands remained almost constant over the 50 years between the two censuses. However, this stability in numbers of species was the result of an approximately equal number of immigrations and extinctions on each

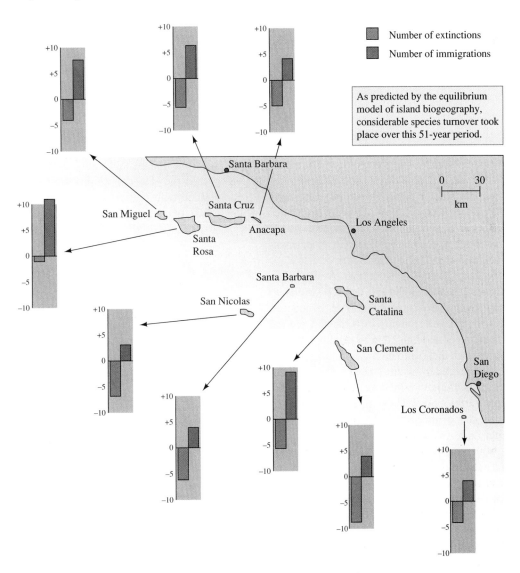

**Figure 22.10** Extinction and immigration of bird species on the California Channel Islands between 1917 and 1968 (data from Diamond 1969).

**Figure 22.11** The mangrove islands in the Florida Keys, which number in the thousands, are convenient places to test the equilibrium model of island biogeography.

of the islands (fig. 22.10). Diamond's study is an excellent example of how theory can guide field ecology. He discovered the dynamics underlying the diversity of birds on the California Channel Islands because he went out to test the MacArthur-Wilson equilibrium model of island biogeography. Additional insights into this model have been provided by experiments.

## Experimental Island Biogeography

As Diamond conducted his surveys of the California Channel Islands, Daniel Simberloff and Edward O. Wilson were engaged in experimental studies of mangrove islands in the Florida Keys (Wilson and Simberloff 1969, Simberloff and Wilson 1969). The Florida Keys support very large stands of mangroves, which are dominated by the red mangrove, *Rhizophora mangle.* Many of these stands occur as small islands that lie hundreds of meters from the nearest large patch of mangroves (fig. 22.11). Simberloff and Wilson chose eight of these small mangrove islands for their experimental study. Their study islands were roughly circular and varied from 11 to 18 m in diameter and 5 to 10 m in height. The distance of islands from large areas of mangroves that could act as a source of colonists varied from 2 to 1,188 m.

The main fauna inhabiting the small mangrove islands of the Florida Keys are arthropods, chiefly insects. Simberloff and Wilson estimated that of the approximately 4,000 species of insects in the Florida Keys, about 500 species inhabit mangroves. Of these 500 species, about 75 commonly live on small mangrove islands. In addition to insects, the mangroves supported 15 species of spiders and other arthropods. The number of insect species on the experimental islands averaged 20 to 40 and the number of spider species ranged from 2 to 10.

Simberloff and Wilson chose two of the islands to act as controls and designated the six others as experimental islands. They carefully surveyed all the islands prior to defaunating the experimental islands. The islands were defaunated by enclosing

them with a tent and then fumigating with methyl bromide. Fumigating was done at night to avoid heat damage to the mangrove trees. Simberloff and Wilson examined the trees immediately after fumigating and found that, with the possible exception of some wood-boring insect larvae, all arthropods had been killed. They followed recolonization by periodically censusing the arthropods on each island for approximately 1 year.

The number of species recorded on the two control islands was virtually identical at the beginning and end of the experiment. Though the number did not change significantly over the period of study, Simberloff and Wilson reported that species composition changed considerably. In other words, there had been species turnover on the control islands, a result consistent with the equilibrium model of island biogeography.

The equilibrium model was also supported by the recolonization studies of the experimental islands. Following defaunation, the number of arthropod species increased on all of the islands. All the islands, except the farthest island, eventually supported about the same number of species as they did prior to defaunation (fig. 22.12). Again, however, the composition of arthropods on the islands was substantially different, indicating species turnover. Species turnover is also indicated by the colonization histories of individual islands, which include many examples of species appearing and then disappearing from the community.

Island colonization can be followed either by removing the organisms from existing islands, as Simberloff and Wilson did when they defaunated their mangrove islands, or by creating new islands. Many new islands formed in a large lake in southern Sweden when the level of the lake was dropped at the end of the nineteenth century. Fortunately, some biologists recognized the rare opportunity offered by the new islands and studied their colonization by plants. These studies have continued for a century.

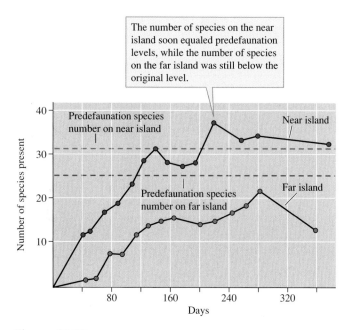

**Figure 22.12** Colonization curves for two mangrove islands that were "near" and "far" from sources of potential colonists (data from Simberloff and Wilson 1969).

# Colonization of New Islands by Plants

The site of this long-term study is Lake Hjälmaren, which covers about 478 km² in Sweden (fig. 22.13). The level of Lake Hjälmaren was lowered 1.3 m between 1882 and 1886 and exposed many new islands. The first plant surveys of the new islands were conducted in 1886, and the islands were surveyed

again in 1892, from 1903 to 1904, from 1927 to 1928, and from 1984 to 1985. Håkan Rydin and Sven-Olov Borgegård (1988) summarized the earlier surveys of these new islands and conducted their own surveys in 1985. The result was a unique long-term record of the colonization of 40 islands.

The study islands vary in area from 65 m² to over 25,000 m² and support a limited diversity of plants. Rydin and Borgegård estimated that approximately 700 species of plants occur around Lake Hjälmaren. Of these 700 plant species, the number recorded on individual islands during the first century of their existence varied from 0 to 127. As expected, this variation in species richness correlated positively with island area over the entire history of the islands and accounted for 44% to 85% of the variation in species richness among islands. Measures of island isolation accounted for 4% to 10% of the variation in plant species richness among islands through the 1903–4 census. Island isolation did not account for significant variation in species richness among islands in subsequent censuses.

Rydin and Borgegård used the censuses of 30 islands to estimate rates of plant immigration and extinction (fig. 22.13). The historical record documents many immigrations and extinctions. There has been a slight excess of immigrations over extinctions on small- and medium-sized islands during the entire 100 years of record. What do these higher rates of immigration indicate? They show that small and medium islands continue to accumulate species. In contrast, large islands attained approximately equal rates of immigration and extinction sometime between 1928 and 1985. Over this period, approximately 30 plants became extinct on each large island and another 30 new species arrived. In other words, it appears that the number of species may have reached equilibrium on large islands.

The observed patterns of colonization were consistent with the predictions of the equilibrium model of island biogeography. Plant species richness on the islands of Lake Hjälmaren, like arthropod richness on the mangrove islands studied by Simberloff and Wilson, appears to be maintained by a dynamic interplay between immigration and local extinction. Many studies support the basic predictions of the equilibrium model of island biogeography. However, many questions remain.

For instance, why do larger islands support more species? Is the greater species richness on large islands due to a direct effect of area or do large islands support higher species richness because they include a greater diversity of habitats? Rydin and Borgegård found that measures of habitat diversity on the study islands accounted for only 1% to 2% of the variation in plant species richness. However, they point out that while some large islands with few habitats support low numbers of plant species, some small islands, with diverse habitats, support higher species richness than would be expected on the basis of area alone. The researchers point out that it is very difficult to separate the effects of habitat diversity from the effects of area. As we shall see in the next example, there is at least one experiment that came close to demonstrating that species richness on islands can be directly affected by area.

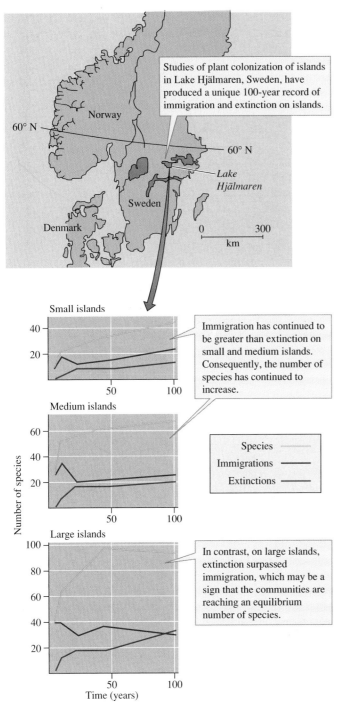

Studies of plant colonization of islands in Lake Hjälmaren, Sweden, have produced a unique 100-year record of immigration and extinction on islands.

Immigration has continued to be greater than extinction on small and medium islands. Consequently, the number of species has continued to increase.

In contrast, on large islands, extinction surpassed immigration, which may be a sign that the communities are reaching an equilibrium number of species.

Species ——— 
Immigrations ——— 
Extinctions ———

**Figure 22.13**  Species number, immigration, and extinction on 30 islands in Lake Hjälmaren, Sweden (data from Rydin and Borgegård 1988).

# Manipulating Island Area

Daniel Simberloff (1976) tested the effect of island area on species richness experimentally. He surveyed the arthropods inhabiting nine mangrove islands that ranged in area from 262 to 1,263 m². The distance of these islands from large areas of mangrove forest ranged from 2 to 432 m. The islands were up to five times the size of the mangrove islands fumigated by Simberloff and Wilson in their earlier study of recolonization and so contained a larger number of arthropod species.

Simberloff kept one island as a control, while reducing the area of the eight other islands by 32% to 76%. Island area was reduced during low tide by removing whole sections of the islands. Workers cut mangroves off below the high tide level and loaded the cut trees and branches on a barge. They then moved the cut material away from the island, where they sank it into deeper water (green mangrove wood sinks). Simberloff reduced the area of four experimental islands twice and the area of the other four experimental islands once only.

The results of Simberloff's experiment show a positive relationship between area and species richness. In all cases where island area was reduced, species richness decreased (fig. 22.14). Meanwhile, species richness on the control island, which was not changed in area, increased slightly. Additional insights are offered by the contrasting histories of islands whose areas were reduced once and those whose areas were reduced twice. For instance, the area of Mud 2 island was reduced from 942 to 327 m² and the richness of its arthropod fauna fell from 79 to 62 species. The area of Mud 2 was not reduced further, and its arthropod richness remained almost constant. Meanwhile, the islands whose area was reduced twice lost species with each reduction in area. Simberloff's results showed that area itself, without increased habitat heterogeneity, has a positive influence on species richness.

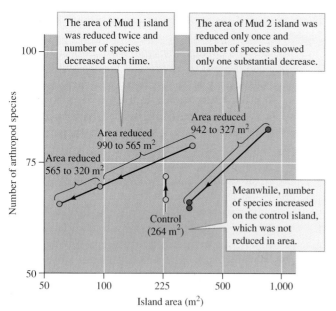

**Figure 22.14** Effect of reducing mangrove island area on number of arthropod species (data from Simberloff 1976).

# Island Biogeography Update

The equilibrium theory of island biogeography has had a major influence on the disciplines of biogeography and ecology. However, much has been discovered in the 40 years since MacArthur and Wilson proposed their theory. For instance, James Brown and Astrid Kodric-Brown (1977) showed how higher rates of immigration to near islands can reduce extinction rates. As a consequence, we now know that, contrary to the original MacArthur-Wilson model, island distance from sources of colonists can influence rates of extinction. Similarly, Mark Lomolino (1990) also extended the original model when he proposed the target hypothesis, demonstrating that island area can have a significant effect on rates of immigration to islands. Brown and Lomolino (2000) pointed out that we have also discovered that species richness is not in equilibrium on many islands. In addition, we now know that species richness on islands is affected by differences among species groups in their speciation, colonization, and extinction rates. And perhaps most significantly, area and isolation are only two of several environmental factors that affect species richness on islands. Brown and Lomolino suggest that we may be on the eve of another revolution in theories that will replace the MacArthur-Wilson model. If so, it will be the result of research largely inspired by their theory as well as by our fascination for the islands themselves.

Experiments on islands such as those of Simberloff and Wilson demonstrate the value of an experimental approach to answering ecological questions. However, there are important ecological patterns that occur over such large scales that experiments are virtually impossible. The ecologists who study these large-scale patterns must rely on other approaches. In the next Concept Discussion section, we discuss one of these important large-scale patterns, latitudinal variation in species richness.

DISCUSSION

## CONCEPT DISCUSSION

### Latitudinal Gradients in Species Richness

**Species richness generally increases from middle and high latitudes to the equator.**

Most groups of organisms are more species-rich in the tropics. This well-known increase in species richness toward the equator became apparent by the middle of the eighteenth century as taxonomists, led by Carolus Linnaeus, described tropical species sent back to Europe by explorers. These explorers and later naturalists, such as Humboldt, Darwin, and Wallace, described overwhelming biological diversity in the tropics. Today, two and a half centuries later, we are still trying to catalog this diversity and do not even know within an order of magnitude its full extent.

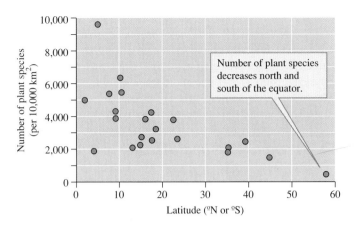

Figure 22.15 Variation in number of vascular plant species with latitude in the Western Hemisphere (data from Reid and Miller 1989).

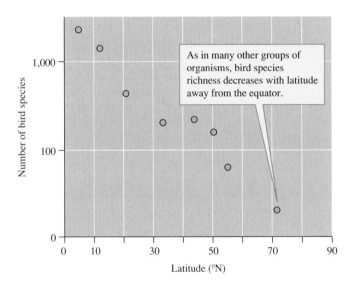

Figure 22.16 Latitudinal variation in number of bird species from Central to North America (data from Dobzhansky 1950).

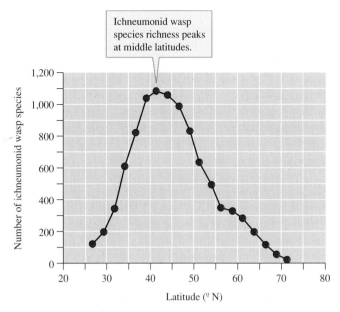

Figure 22.17 An exception to the general decline in species number with latitude: latitudinal variation in ichneumonid wasp species richness (data from Janzen 1981).

Figures 22.15 and 22.16 show examples of how plant species richness (Reid and Miller 1989) and bird species richness (Dobzhansky 1950) decrease toward the poles. Despite some exceptions to the equatorial peak in species diversity (fig. 22.17), the pattern of increased numbers of species in the tropics is pervasive and dramatic. This pattern challenges ecologists for an explanation. Many mechanisms have been proposed to explain latitudinal gradients in species richness. James H. Brown (1988) grouped the hypotheses proposed to explain geographic gradients in species richness into six categories.

## Time Since Perturbation

The *time since perturbation hypothesis* proposes that there are more species in the tropics because the tropics are older and they are disturbed less frequently. That is, more species occur in the tropics because (1) there has been more time for speciation and (2) less frequent perturbation reduces extinction rates. The proponents of this hypothesis assume that the tropics have

remained relatively stable while middle and high latitudes have been repeatedly disrupted by the advance and retreat of glaciers. However, in chapter 16 we saw that intermediate levels of disturbance may increase local diversity. In counterpoint to this hypothesis, Joseph Connell (1978) proposed that the extraordinary diversity of tropical rain forests and coral reefs is maintained by frequent disturbance.

## Productivity

The authors of the *productivity hypothesis* observe that two of the most diverse environments on earth, coral reefs and tropical rain forests, are also extraordinarily productive. This hypothesis proposes that high productivity contributes to high species richness. It assumes that with more energy to divide among organisms, specialized consumers will have larger populations. Since larger populations generally have lower probabilities of extinction than smaller populations, extinction rates should be lower in more productive environments. However, Brown points out that this hypothesis must somehow explain the reduction in species diversity that accompanies nutrient enrichment and increased primary production (see chapter 16).

## Environmental Heterogeneity

The *environmental heterogeneity hypothesis* proposes that the tropics contain more species because they are more heterogeneous than temperate regions. Daniel Janzen (1967) and George Stevens (1989) pointed out that, compared to high-latitude species, most tropical species occur in far fewer environments along altitudinal and latitudinal gradients.

Michael Rosenzweig (1992), however, cautioned that we cannot consider habitat structure independently of the organisms living in a region. Species within more diverse communities tend to subdivide the environment more finely. Consequently, species diversity and habitat heterogeneity are not necessarily independent factors. For instance, when G.W. Cox and Robert Ricklefs (1977) estimated the number of habitats used by birds in Panama versus four Caribbean islands, they found an inverse relationship between numbers of species and the number of habitats used by the birds. In other words, birds appeared to restrict their habitat use in the presence of more species.

## Favorableness

The *favorableness hypothesis* proposes that the tropics provide a more favorable environment than do high latitudes. As we saw in chapter 2, the variation in temperature in high-latitude environments is much greater than in tropical environments. Biologists have proposed that the correspondence between low diversity and the physical variability of high latitudes is no accident. While many species are well adapted to physically harsh conditions, most species on earth are not. Biologists have proposed that physically extreme environments restrict the diversity of organisms. Brown also pointed out that many of the most physically extreme environments are small and isolated. As we saw when we discussed island biogeography, small habitat area and isolation are correlated with reduced species richness.

## Niche Breadths and Interspecific Interactions

Biologists have tried to explain latitudinal gradients in species diversity with several hypotheses concerning relative niche breadths and interspecific interactions. Their hypotheses have included:

1. Tropical species are limited more by biological factors than by physical factors.
2. Tropical species are affected more by interspecific interactions than by intraspecific interactions.
3. The niches of tropical species overlap more than those of higher-latitude species and so tropical species compete more intensively.
4. Tropical species are more specialized, that is, have narrower niches, and so compete less intensively than species at higher latitudes.
5. Tropical species are more subject to controls by predators, parasites, and pathogens.
6. Compared to temperate species, tropical species are involved in more mutualistic interactions.

Brown suggested that these hypotheses present the ecologist with a number of difficulties. Notice that some of these hypotheses are contradictory. In addition, they are difficult to test and do not address the primary differences between the tropics and higher latitudes. For instance, even if the niches of tropical species differ consistently from those of species at higher latitudes, we must determine the causes of those differ-

ences. Brown suggests that biological processes such as competition and predation must play a secondary role in determining species diversity gradients. He proposes that the ultimate causes of geographic gradients in species richness must be physical differences between the tropics and higher latitudes. The following hypothesis uses differences in physical settings to explain latitudinal patterns of diversity.

## Differences in Speciation and Extinction Rates

Ultimately the number of species in a particular area reflects the rate at which new species have been added to the species pool minus the rate at which they have disappeared. Species are added to species pools by either immigration or speciation. However, Rosenzweig (1992) proposed that when we consider the diversity of whole biogeographic provinces, immigration can be largely discounted and speciation will be the primary source of new species. Species are removed from species pools by extinction. So, tropical species richness is greater than at higher latitudes because the tropics have experienced higher rates of speciation and/or lower rates of extinction. However, Brown would remind us here that we need to determine the physical mechanisms that produce differences in speciation and extinction rates in tropical versus higher latitudes.

# Area and Latitudinal Gradients in Species Richness

John Terborgh (1973) and Michael Rosenzweig (1992) proposed that the greater species richness of the tropics can be explained by the greater area covered by tropical regions. It may not be immediately apparent that the tropics, which mainly occupy the area between the tropics of Cancer and Capricorn, include a greater area of both land and water than do higher latitudes. The reason for this is that the typical world map is based upon the *Mercator Projection,* a projection that increases the apparent area at high latitudes. However, if you look at a world globe you will immediately see that the tropical areas of the earth constitute a vast area.

Is there a greater land surface area in the tropics? Rosenzweig quantified the amount of land surface area in various latitudinal zones using a computer map of the earth. He divided the globe into tropical (± 26° of latitude), subtropical (26°–36°), temperate (36°–46°), boreal (46°–56°), and tundra (> 56°). He then measured the area of land within these latitudinal zones and found that the area of land within the tropics far exceeds that of other areas (fig. 22.18).

Not only is there more land (and water) at tropical latitudes, but in addition, temperatures are more uniform across this tropical belt. This pattern was put in the context of geographic ecology by Terborgh, who plotted mean annual temperatures against latitude. As figure 22.19 shows, there is little difference in mean annual temperatures between about 0° and 25° latitude. Because this

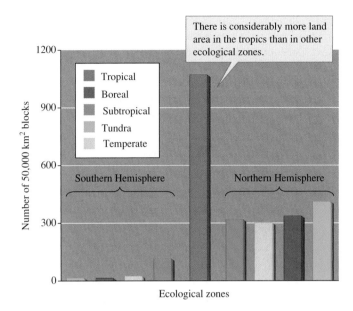

**Figure 22.18**  Land area in five latitudinal biomes (data from Rosenzweig 1992).

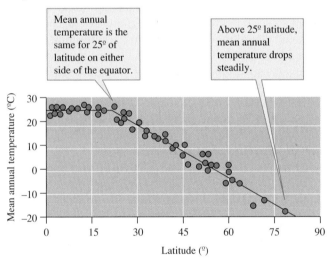

**Figure 22.19**  Mean annual temperature by latitude (data from Rosenzweig 1992, after Terborgh 1973).

temperature pattern occurs both north and south of the equator, mean annual temperature changes little over about 50° of latitude within the tropics. However, above 25° latitude, mean annual temperature declines linearly with latitude. What is the biological significance of this latitudinal pattern of temperature variation? One implication is that tropical organisms can disperse over large areas and not meet with significant changes in temperature.

How do patterns of temperature variation affect rates of speciation and extinction? Rosenzweig proposed that the larger area of tropical regions should reduce extinction rates in two ways. First, large, physically similar areas will allow tropical species to be distributed over a larger area. Within these larger areas, there should be more refuges in which to survive environmental disturbances. Because of their larger range, tropical species should also have greater total population sizes. Larger populations are less likely to become extinct.

Rosenzweig also proposed that larger species ranges should increase rates of allopatric speciation. He reasoned that geographic barriers, such as mountain ranges or deep canyons, are more likely to form within large species ranges than within small species ranges. Therefore, since geographic isolation initiates allopatric speciation, speciation rates are likely to be higher in tropical regions.

Earlier in chapter 22, we saw that species richness increases with island area. However, do larger continents also harbor more species? We explore this question in the following example.

# Continental Area and Species Richness

Karl Flessa (1975, 1981) was the first to examine the relationship between continental area and species richness. He found a strong positive relationship between mammalian richness and the area of continents, large islands, and island groups. Flessa found a significant positive relationship whether his index of mammalian richness was total number of orders, families, or genera. James H. Brown (1986) performed an analysis similar to Flessa's; however, he restricted his analysis to the five major continents plus two large tropical islands, Madagascar and New Guinea. Brown also excluded flying mammals and analyzed patterns of mammalian diversity at the level of genera and species. Like Flessa, Brown found a strong positive relationship between mammalian richness and area (fig. 22.20). Madagascar, with the smallest land area, supports the lowest mammalian richness. Eurasia, with the largest land area, supports the greatest mammalian diversity. The continents of Australia, South America, North America, and Africa, with intermediate areas, contain intermediate levels of mammalian richness.

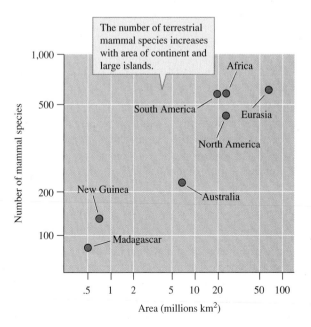

**Figure 22.20**  Relationship between area of continents and large islands and number of nonflying terrestrial mammals (data from Brown 1986).

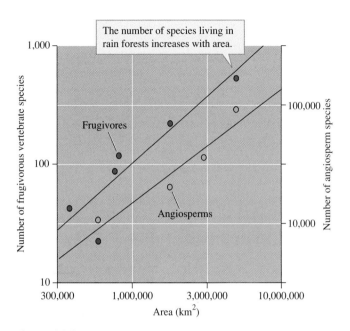

**Figure 22.21** Rain forest area, from Australia to Amazonia, and numbers of flowering plant (angiosperm) species and number of fruit-eating (frugivorous) vertebrate species (data from Rosenzweig 1992).

What do these analyses by Flessa and Brown have to do with higher tropical species richness? Rosenzweig proposed that the greater area of the tropics (see fig. 22.18) is a primary cause of the higher diversity. If differences in area produce differences in species richness, then we should see a positive relationship between continental area and species richness. Flessa and Brown have shown such a relationship. Now, let's go back to the tropics.

Do tropical regions with different areas differ in species richness? If Rosenzweig's explanation for the greater tropical diversity is correct, tropical regions of different areas should support different levels of biological diversity. Rosenzweig examined patterns of diversity among fruit-eating mammals and plants in tropical rain forests ranging from Australia to Amazonia. The result was a strong positive relationship between area and diversity, a result that supports the area hypothesis (fig. 22.21). The smallest area of tropical rain forest, Australia, contains the smallest number of fruit-eating mammal and plant species. Amazonia, with the largest rain forest area, contains the greatest number of fruit-eating mammal and plant species.

In summary, many factors may contribute to higher tropical species richness, including (1) time since perturbation, (2) productivity, (3) environmental heterogeneity, (4) favorableness, (5) niche breadths and interspecific interactions, and (6) differences in speciation and extinction rates. However, several lines of evidence support the hypothesis that differences in surface area play a primary role in determining latitudinal gradients in species richness. Can we conclude that we fully understand the mechanisms underlying latitudinal gradients in diversity? Brown concluded that while we are close to understanding the mechanisms controlling variation in species richness across islands, "The distributions of species

and higher taxa within continents are more complex and for the most part remain to be deciphered." As we shall see in the next Concept Discussion section, some of that unexplained complexity is due to historical and regional differences between the continents.

**CONCEPT DISCUSSION**

### Historical and Regional Influences

**Long-term historical and regional processes significantly influence the structure of biotas and ecosystems.**

Area and isolation explain much of the variation in species diversity and composition across islands. Area appears to account for much of the variation in biological diversity across continents. Additional variation in local diversity appears to be due to differences in habitat heterogeneity, disturbance, predation, and successional age of the local community, factors that we discussed in chapters 16, 17, and 20. However, as the following examples show, these factors are not adequate to explain many geographic differences in biological diversity and community organization. Robert Ricklefs (1987) pointed out that in many cases, unique historical and geographic factors appear to have produced significant regional differences in species richness.

## Exceptional Patterns of Diversity

There are major differences in species richness that cannot be explained by differences in area. For instance, consider the regions with Mediterranean climates that we discussed in chapter 2 (see fig. 2.22), which support Mediterranean woodlands and shrublands. Such regions include the Cape region of South Africa (90,000 $km^2$), southwestern Australia (320,000 $km^2$), and the California Floristic Province (324,000 $km^2$). These regions have similar climates but differ significantly in area. Which of these areas should contain the greatest number of species? The positive relationship between area and species richness that we have seen repeatedly earlier in chapter 22 leads us to predict that southwestern Australia and the California Floristic Province, with more than three times the area of the Cape region of South Africa, will contain the greatest biological diversity. Southwestern Australia and the California Floristic Province have the same area and approximately the same number of species. However, as figure 22.22 shows, the Cape region, the smallest area, contains more than twice the number of plant species as the other two regions.

The failure of area to explain a significant regional diversity pattern is not unique to this example. For instance, Roger Latham and Robert Ricklefs (1993) reported a striking contrast

Information
Questions
Hypothesis
Predictions
Testing ✓

*Investigating the Evidence*

## Sample Size Revisited

In chapter 5 (see p. 129) we considered the number of samples necessary to obtain a reasonably precise estimate of the number of species in two simple communities. In chapter 16 (see p. 403) we reconsidered the same question in relation to very complex communities, concluding that, in some situations, the sampling efforts required for precise estimates of the number of species, must be intense. In general, the sample size necessary to detect statistically significant differences, or effects, increases with the variation of the system under study. Here we consider another question, "What determines sample size?" Another way of putting this question is, "What is a replicate observation or measurement?"

For small-scale studies, the answers to these questions are clear. For instance, in laboratory studies of the running performance of an animal species, or the photosynthetic rate of a plant species, the number of individuals measured would determine sample size. In an experimental field study of the effects of nitrogen availability on plant diversity, the number of field plots in which the investigator manipulated soil nitrogen would determine sample size. However, the answers to these questions may not be as obvious, as ecologists begin to address larger-scale ecological problems. For example, in chapter 21, we compared the standing stocks of detritus in two streams (see p. 521), one that drained a deciduous forest and one that drained a coniferous forest, and found significant differences between the two streams. In that comparison, the number of measurements of detritus standing stock in each stream, which was 7, determined the sample size.

However, based on the comparison of the two study streams, can we conclude that streams draining conifer forests, *in general,* contain higher standing stocks of detritus

compared to deciduous forest streams? We concluded the discussion in chapter 21 by saying that we cannot reach such a general conclusion. Why not? The basic reason is that the study outlined in chapter 21 included only one stream draining each type of forest. In other words, relative to the general relationship between type of forest and amount of detritus in associated streams, our sample size was one. Even if we made 100 measurements of detritus in the two study streams and, as a consequence, obtained very precise estimates of the amount of detritus that each held, the sample size relative to the broader question would still be one stream of each type.

How do we increase sample size for such a study? To do so we would need to locate and study several streams associated with deciduous forests and coniferous forests. Ideally, we would sample beyond a particular landscape and include streams in several landscapes throughout a region. The number of *different* streams of each type that we sampled within the region would determine the sample size. To make statements beyond the regional scale, we would need to sample several regions within a continent. The requirements of ecological research at very large scales soon taxes the limited resources of any single investigator or team of investigators. As a consequence, ecologists studying at large spatial scales increasingly turn to computer-based systems for data gathering and analysis, a topic discussed in the Applications & Tools section in chapter 22. Another way to increase our ability to make inferences is to utilize information gathered and published by other research teams. Doing so puts local and regional studies in broader contexts. In chapter 23, we discuss some approaches to searching this literature (p. 575).

in diversity of temperate zone trees that cannot be explained by an area effect. As we saw in chapter 2, the temperate forest biome covers approximately equal areas in Europe (1.2 million km², eastern Asia (1.2 million km²), and eastern North America (1.8 million km²). The species area relationship would lead us to predict that these three regions would support approximately equal levels of biological diversity. However, figure 22.23 shows that eastern Asia contains nearly three times more tree species than eastern North America and nearly six times more species of trees than Europe.

Let's discuss another example, this time involving birds. In chapter 16 we reviewed several studies that showed a positive relationship between foliage height diversity and bird species diversity. These studies of bird communities showed

that greater vertical heterogeneity in the plant community (foliage height diversity) is associated with greater bird species diversity in areas as widespread as northeastern North America, Puerto Rico, Panama, and Australia (e.g., see fig. 16.10). However, biologists have found that this positive relationship is not universal.

John Ralph (1985) studied the relationship between foliage height diversity and bird species diversity in northern Patagonia, Argentina. He chose study sites distributed along a 50 km moisture gradient in which annual precipitation ranged from 200 to 2,500 mm annually. This precipitation gradient produced a gradient in vegetative structure from grasslands and shrublands through beech forests. Contrary to what had been found in other regions, Ralph found a negative correlation

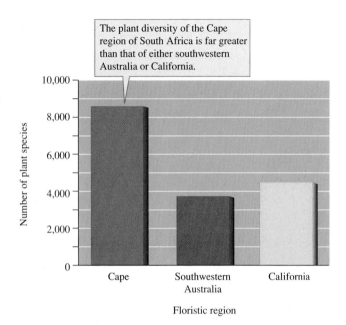

**Figure 22.22** Number of plant species living in three regions with Mediterranean climates (data from Bond and Goldblatt 1984).

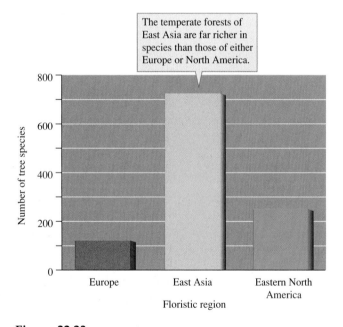

**Figure 22.23** Number of tree species in three temperate forest regions (data from Latham and Ricklefs 1993).

between foliage height diversity and bird species diversity (fig. 22.24). He recorded higher diversity in shrub habitats than in beech forests with greater foliage height diversity.

# Historical and Regional Explanations

How can we explain these exceptional patterns of biological diversity? What mechanisms produced these patterns that are

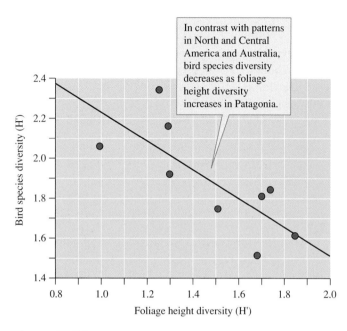

**Figure 22.24** Foliage height diversity and bird species diversity in Patagonia (data from Ralph 1985).

contrary to generalizations discussed in chapter 22 and earlier chapters? In each case, it appears that geography and history offer convincing explanations.

## The Cape Floristic Region of South Africa

Pauline Bond and Peter Goldblatt (1984) attributed the unusual species richness of the Cape floristic region to several historic and geographic factors. Selection for a distinctively Mediterranean flora in southern Africa began during the late Tertiary period, about 26 million years ago. At that time, the climate became progressively cooler and drier, conditions that selected for succulence, fire resistance, and smaller, sclerophyllous leaves. The initial sites for evolution of the Cape flora were likely in south-central Africa, not in the Cape region itself. At that time, Africa lay farther south and the Cape region had a cool, moist climate and supported an evergreen forest.

As Africa drifted northward, the climate of southern Africa became more arid and the ancestors of today's Cape flora gradually migrated toward the Cape region. By the time Africa neared its present latitudinal position during the late Pliocene, about 3 million years ago, southern Africa was very arid and the Cape region had a Mediterranean climate. Bond and Goldblatt suggest that plant speciation within this region was promoted by the highly dissected landscape, the existence of a wide variety of soil types, and repeated expansion, contraction, and isolation of plant populations during the climatic fluctuations of the Pleistocene. They suggest that extinction rates were reduced by the existence of substantial refuge areas, even during times of peak aridity.

## *The Diversity of Temperate Trees*

How did eastern Asia, eastern North America, and Europe, three temperate regions of approximately equal area and climate, end up with such different numbers of tree species? Latham and Ricklefs offer persuasive geographic and historical reasons. They propose that we need to consider what trees in the three regions faced during the last glacial period and how those conditions may have affected extinction rates.

Refer to chapter 2 and study the distributions of temperate forest in eastern Asia, eastern North America, and Europe (see fig. 2.28). Now examine the distributions of mountains in eastern Asia, eastern North America, and Europe shown in figure 2.37. Notice that while there are no mountain barriers to north–south movements of organisms in eastern Asia and eastern North America, the mountains in Europe form barriers that are oriented east to west. Now imagine what happened to a tree species as glaciers began to advance during the last ice age and the climate of Europe became progressively colder. Temperate trees would have had their southward retreat largely cut off by mountain ranges running east to west.

This hypothesis proposes that the lower species richness of European trees has been at least partly a consequence of higher extinction rates during glacial periods. How would you test this hypothesis? Latham and Ricklefs searched the fossil record for extinctions in the three regions. They estimated the number of genera that have become extinct in the three regions during the last 30 to 40 million years. Their analysis showed that most of the plant genera that once lived in Europe have become extinct. A larger proportion of genera has become extinct in Europe than in either eastern Asia or eastern North America (fig. 22.25).

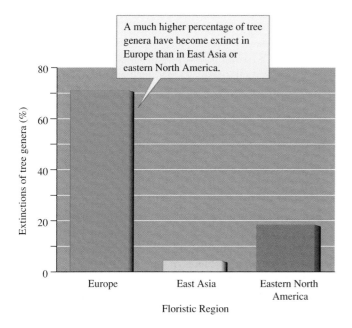

**Figure 22.25** Extinctions of tree genera in Europe, East Asia, and eastern North America since the middle Tertiary period (data from Latham and Ricklefs 1993).

Now consider eastern North America. The only mountain range, the Appalachians, runs north to south. Consequently, in eastern North America, temperate trees had an avenue of retreat in the face of advancing glaciers and cooling climate. The movement of temperate tree populations in the face of climate change has been well documented by paleontologists such as Margaret Davis (see fig. 10.17). There are also no mountain barriers in eastern Asia, where temperate trees can migrate even farther south than in eastern North America.

Higher rates of extinction during glacial periods can explain the lower diversity of trees in Europe. However, why does eastern North America include fewer tree species than eastern Asia? Latham and Ricklefs conclude that the fossil record and present-day distributions of temperate trees indicate that most temperate tree taxa originated in eastern Asia. These Asian taxa subsequently dispersed to Europe and North America. In addition, after the dispersal routes between eastern Asia and eastern North America were closed off, speciation continued in Asia, producing several endemic Asian genera. In other words, there are fewer tree species in eastern North America because most taxa originated in eastern Asia and never dispersed to North America.

## *Bird Diversity in the Beech Forests of South America*

Why do the structurally complex beech forests of South America support lower bird diversity than the simpler shrub habitats? Dolph Schluter and Robert Ricklefs (1993) suggested that the lower bird diversity in beech forests may be due to the restricted geographic distribution of these temperate forests in South America. The shrub habitats where Ralph (1985) recorded the greatest diversity of birds occupy a much larger area within South America.

In addition to their small area, South American beech forests are also isolated from subtropical and tropical forests by vast areas of arid and semiarid vegetation. The biota of these forests includes endemic species of frogs, birds, and mammals, which suggests that South American beech forests have been isolated for a long time. Schluter and Ricklefs suggested that small area and isolation may account for the low diversity of birds in South American beech forests. They cite the relationship found by Ralph as "a convincing demonstration of how effects of local habitat on species diversity may be superseded by regional ones."

In summary, much geographic variation in species richness can be explained by historical and regional processes. The ecologist interested in understanding patterns of diversity at large spatial scales must consider processes occurring over similarly large scales and over long periods of time. As we shall see in chapter 23, a large-scale, long-term perspective is also essential for understanding global ecology.

## APPLICATIONS & TOOLS

Global Positioning Systems, Remote Sensing, and Geographic Information Systems

In 1972, Robert MacArthur defined geographic ecology as the study of patterns you can put on a map. Spatial distributions that can be put on maps are still the center of geographic ecology, but the nature of "maps" has changed tremendously. Modern tools have revolutionized the field. Today, geographic ecologists generally record their data on geographic information systems, which are computer-based systems that store, analyze, and display geographic information. In addition, the geographic ecologists of today also have access to more information of greater accuracy because of remote sensing and global positioning systems.

## Global Positioning Systems

What is the location? This is one of the most basic questions the geographer can ask. Scientists, engineers, navigators, and explorers have spent centuries devising methods to measure elevation, latitude, and longitude. Recent technological advances have improved the accuracy of these measurements.

Alexander von Humboldt, the founder of geographic ecology, would appreciate these recent technological advances. As he explored South and North America, he carefully determined the latitude, longitude, and elevation of important geographic features. For instance, Humboldt was particularly interested in verifying the existence and location of a waterway called the Casiquiare Canal. The Casiquiare reportedly connected the Orinoco River with the Rio Negro, which flows into the Amazon. A connection between two major river systems would make the Casiquiare unique, but its existence was widely doubted.

Humboldt halted his expedition at the junction of the Casiquiare and the Rio Negro so that he could record the latitude and longitude. Biting insects tormented the explorers as they waited for nightfall. Luckily, that night the clouds parted and Humboldt could see the stars well enough to take sightings and determine their position. At other times, he was not so lucky. He once waited for nearly a month for the weather to clear sufficiently to make his sightings on the stars. Today, equipped with a global positioning system, Humboldt could have determined the latitude and longitude of the junction of the Casiquiare and the Rio Negro any time he wished, regardless of weather.

A **global positioning system** determines locations on the earth's surface, including latitude, longitude, and altitude, using satellites as reference points. These satellites, which orbit the earth at a height of about 21,000 km, continuously transmit their position and the time. The satellites keep track of

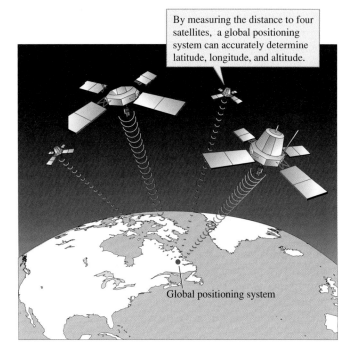

By measuring the distance to four satellites, a global positioning system can accurately determine latitude, longitude, and altitude.

Global positioning system

**Figure 22.26** Global positioning systems determine latitude, longitude, and altitude by measuring the distance from several satellites.

time with an extremely accurate atomic clock that loses or gains 1 second in about 30,000 years. A global positioning system receives the signals broadcast by these satellites. Because the system also includes an extremely accurate clock, the time required for the satellite signal to reach the receiver can be used as a measure of the distance between the two. With measurements of the distance to four satellites, a global positioning system can determine the latitude, longitude, and altitude of any point on earth with great accuracy (fig. 22.26).

While navigation satellites and global positioning systems can accurately locate places on the ground, other satellites provide a wealth of other information about those localities. These "remote sensing" satellites transmit pictures of the earth that are extremely valuable to ecologists.

## Remote Sensing

**Remote sensing** refers to gathering information about an object without direct contact with it, mainly by gathering and processing electromagnetic radiation emitted or reflected by the object. Using this definition, the original remote sensor was the eye. However, we generally associate remote sensing with technology that extends the senses, technology ranging from binoculars and cameras to satellite-mounted sensors.

Remote sensing satellites are generally fitted with electro-optical sensors that scan several bands of the electromagnetic spectrum. These sensors convert electromagnetic radiation into electrical signals that are in turn converted to digital values by a computer. These digital values can be used to construct an image. The earliest of the *Landsat* satellites monitored four bands of electromagnetic radiation, two bands of visible light (0.5–0.6 μm and 0.6–0.7 μm), and two bands in the near infrared (0.7–0.8 μm and 0.8–1.1 μm). From this beginning, satellite imaging systems have gotten progressively more sophisticated both in terms of the number of wavelengths scanned and the spatial resolution.

Satellite-based remote sensing has produced detailed images of essentially every square meter of the earth's surface. These images provide very useful information to ecologists, especially for landscape and geographic ecology. Ecologists have used remote sensing to monitor the biomass of vegetation using indices of "greenness." In the arid American Southwest, vegetative biomass indicates the positions of moist mountain areas.

## Mountain Islands in the American Southwest

Norman Roller and John Colwell (1986) reviewed how satellites gathering information at a coarse level of resolution could be useful for conducting ecological surveys, particularly when those surveys concern large geographic areas. Earlier in chapter 22, we discussed one such survey conducted by Lomolino and his colleagues on the mammals of montane "islands" in the American Southwest. In their study, these researchers relied principally on a 1982 map of the vegetation of the area. They could have also used satellite imagery from the same region to delineate their study areas.

Roller and Colwell included a satellite image of the American Southwest in their paper (fig. 22.27). The red patches within New Mexico and Arizona correspond mainly to the forested montane islands studied by Lomolino and his colleagues. Some other areas also show up as red due to irrigated agriculture near Phoenix, Arizona, and along the Colorado and Gila Rivers in Arizona and the Pecos River and Rio Grande in New Mexico. The yellow, green, and blue areas support progressively less vegetation. Satellite estimates can be made frequently and offer the possibility of detecting changes in plant biomass. This is particularly important in a region like the American Southwest that experiences high year-to-year variability in precipitation and plant production.

## Marine Primary Production from Space

Mary Jane Perry (1986) demonstrated the use of remote sensing to study marine primary production. In chapter 21, we concentrated on patchiness in terrestrial environments. However, marine environments are also highly patchy, especially in regard to primary production (fig. 22.28). Perry pointed out that we know little about interannual variation in marine primary production, particularly in the open ocean.

The main reasons we know little about the dynamics of production in the open ocean are that (1) the open ocean ecosystem is so vast, covering approximately 332 million km$^2$, (2) there are limited numbers of oceanographic ships and other sea-based sampling devices, and (3) open ocean studies are expensive. Perry proposed that satellite remote sensing of ocean color is the best tool at our disposal for studying regional and global marine primary production. Figure 22.28 shows a remote sensing image taken along the west coast of North

**Figure 22.28** Concentrations of chlorophyll *a* and temperature along the west coast of North America determined from satellite imagery. In the image on the left, yellow and red indicate areas of high phytoplankton biomass. These coincide with cooler sea surface temperatures associated with upwelling, which are shown in the right image by violet, purple, and blue colors (Perry 1986).

**Figure 22.27** Islands of high vegetative biomass appear in this satellite image of the American Southwest as dark red. These are mainly found on mountaintops in this region and along river valleys (Roller and Colwell 1986).

America. The colors indicate concentrations of chlorophyll *a*, a measure of phytoplankton biomass. The image indicates a high degree of patchiness; higher concentrations of phytoplankton occur in a narrow coastal band. The figure also shows that the coastal waters supporting high phytoplankton biomass are also colder than offshore areas. What process is suggested by the combination of cold water and high phytoplankton biomass? The combination suggests coastal upwelling, which brings nutrient-rich deep water to the surface.

Perry reports that though oceanographers had long known that phytoplankton populations are patchy, they had no idea of the complexity of that patchiness. Some phytoplankton patchiness occurs far out at sea, and there is also a high degree of temporal variation in patch location. Patches of high production that form along coastlines can move offshore at rates of 2 to 7 km per day. It would be impossible for a ship moving at less than 30 km per hour to capture such spatially and temporally complex patterns of production. However, the satellite that took the image shown in figure 22.28 can scan a path 1,600 km wide with a resolution of about 1 km$^2$ and resample the same area at 5- to 6-day intervals.

This example and the previous one show that satellite-based remote sensing can gather large amounts of data over large areas. This ability solves some of the sampling problems associated with studying large-scale ecological phenomena. However, these large quantities of data create another problem. Ecologists need a system for storing, sorting, analyzing, and displaying these large quantities of geographic information. This is the problem addressed by geographic information systems.

# Geographic Information Systems

In the days of Humboldt, geographers often had too little data. Today, with new tools for gathering great quantities of information, geographers and geographic ecologists can be overwhelmed by data. **Geographic information systems,** computer-based systems for storing, sorting, analyzing, and displaying geographic data, are designed to handle large quantities of data. Sometimes geographic information systems are confused with computerized mapmaking. While these systems can produce maps, they do much more. Much of population ecology is concerned with understanding the factors controlling the distribution and abundance of organisms. However, the geographic context of populations has been often lost. Geographic information systems preserve this geographic information. Because they preserve geographic context, the systems provide ecologists with a valuable tool for exploring large-scale population responses to climate change.

As we shall see in chapter 23, rapid global change challenges the field of ecology to address large-scale environmental questions. As ecologists address these compelling questions, geographic information systems, global positioning systems, and remote sensing will be increasingly valued parts of their tool kit.

*Geographic ecology* focuses on large-scale patterns of distribution and diversity of organisms, such as island biogeography, latitudinal patterns of species diversity, and the influences of large-scale regional and historical processes on biological diversity.

# SUMMARY

**On islands and habitat patches on continents, species richness increases with area and decreases with isolation.** Larger oceanic islands support more species of most groups of organisms than small islands. Isolated oceanic islands generally contain fewer species than islands near mainland areas. Many habitats on continents are so isolated that they can be considered as islands. Species richness on habitat islands, such as mountain islands in the American Southwest, increases with area and decreases with isolation. Lakes can also be considered as habitat islands. They are aquatic environments isolated from other aquatic environments by land. Fish species richness generally increases with lake area. Species richness is usually negatively correlated with island isolation. However, because organisms differ substantially in dispersal rates, an island that is very isolated for one group of organisms may be completely accessible to another group.

**Species richness on islands can be modeled as a dynamic balance between immigration and extinction of species.** The equilibrium model of island biogeography proposes that the difference between rates of immigration and

extinction determines the species richness on islands. The equilibrium model of island biogeography assumes that rates of species immigration to islands are mainly determined by distance from sources of immigrants. The model assumes that rates of extinction on islands are determined mainly by island size. The predictions of the equilibrium model of island biogeography are supported by observations of species turnover on the islands and by colonization studies of mangrove islands in Florida and new islands in Lake Hjälmaren, Sweden.

**Species richness generally increases from middle and high latitudes to the equator.** Most groups of organisms are more species-rich in the tropics. Many factors may contribute to higher tropical species richness, including (1) time since perturbation, (2) productivity, (3) environmental heterogeneity, (4) favorableness, (5) niche breadths and interspecific interactions, and (6) differences in speciation and extinction rates. Several lines of evidence support the hypothesis that differences in surface area play a primary role in determining latitudinal gradients in species richness.

**Long-term historical and regional processes significantly influence the structure of biotas and ecosystems.** Much geographic variation in species richness can be explained by historical and regional processes. Some exceptional situations that seem to have resulted from unique historical and regional processes include the exceptional species richness of the Cape floristic region of South Africa, the high species richness of temperate trees in east Asia, and the low bird diversity in beech forests of South America. The ecologist interested in understanding large-scale patterns of species richness must consider processes occurring over similarly large scales and over long periods of time.

Global positioning systems, remote sensing, and geographic information systems are important tools for effective geographic ecology. A global positioning system determines locations on the earth's surface, including latitude, longitude, and altitude, using satellites as reference points. Remote sensing satellites are generally fitted with electro-optical sensors that scan several bands of the electromagnetic spectrum. These sensors convert electromagnetic radiation into electrical signals that are in turn converted to digital values by a computer. These digital values can be used to construct an image. Geographic information systems are computer-based systems that store, analyze, and display geographic information. Global positioning systems, remote sensing, and geographic information systems are increasingly valuable parts of the ecologist's tool kit. Ecologists are using these new tools to study large-scale, dynamic ecological phenomena such as interannual variation in regional terrestrial primary production, dynamics of marine primary production, and potential population responses to climate change.

# Review Questions

1. The following data (Preston 1962a) give the area and number of bird species on islands in the West Indies:

| Island | Area | Log10 Area | # Species | Log # Species |
|---|---|---|---|---|
| Cuba | 43,000 | 4.633 | 124 | 2.093 |
| Isle of Pines | 11,000 | 4.041 | 89 | 1.949 |
| Hispaniola | 47,000 | 4.672 | 106 | 2.021 |
| Jamaica | 4,470 | 3.650 | 99 | 1.996 |
| Puerto Rico | 3,435 | 3.536 | 79 | 1.898 |
| Bahamas | 5,450 | 3.736 | 74 | 1.869 |
| Virgin Islands | 465 | 2.667 | 35 | 1.544 |
| Guadalupe | 600 | 2.778 | 37 | 1.568 |
| Dominica | 304 | 2.483 | 36 | 1.556 |
| St. Lucia | 233 | 2.367 | 35 | 1.544 |
| St. Vincent | 150 | 2.176 | 35 | 1.544 |
| Grenada | 120 | 2.079 | 29 | 1.462 |

The numbers are expressed in two ways: as simple measurements and counts and as the logarithms of area and numbers of species. Use these data to plot your own species–area relationship. Plot area on the horizontal axis and number of species on the vertical axis. First plot the simple measurements of area and species number on one graph, and then plot the logarithms of area and species number on another graph. Which gives you the tightest relationship between area and species richness?

2. Refer to figure 22.5, which MacArthur and Wilson (1963) used to show how isolation affects species richness on islands. Find a detailed map of the Pacific Ocean and locate New Guinea. Next locate as many of the "near," "intermediate," and "far" islands on the map as you can. This will give you a better sense of the distances represented by the islands. How do the numbers of species on near, intermediate, and far islands support the hypothesis that island isolation tends to reduce species richness?

3. We discussed how Diamond (1969) documented immigrations and extinctions on the California Channel Islands by comparing his censuses of the birds of the islands with the birds recorded over 50 years earlier. Disregarding the numbers for San Miguel and Santa Rosa Islands, which were not well censused in 1917, Diamond showed that an average of approximately six bird species became extinct on California Channel Islands between 1917 and 1968. During the same period, an average of approximately five new bird species immigrated to the islands. Diamond suggested that his estimates of immigration and extinction were likely underestimates of the actual rates. Explain why his comparative study produced underestimates of rates of immigration and extinction.

4. Diamond's estimates (1969) of numbers of species immigrating and numbers that became extinct (six versus five) were virtually identical. Is this near equality in numbers of extinction and immigration consistent with the equilibrium model of island biogeography? Explain.

5. Suppose you are about to study the bird communities on the islands shown on the right, which are identical in area but lie at different distances from the mainland.

According to the equilibrium model of island biogeography, which of the islands should experience higher rates of immigration? What does the equilibrium model of island biogeography predict concerning relative rates of extinction on the two islands?

6. Now, suppose you are going to study the bird communities on the islands shown on the right, which lie equal distances from the mainland but differ in area. According to the equilibrium model of island biogeography, what should be the relative rates of immigration to the two islands? On which islands should rates of extinction be lowest? Explain.

7. Review the major hypotheses proposed to explain the higher species richness of tropical regions compared to temperate and high-latitude regions. How are each of these hypotheses related to relative rates of speciation and extinction in tropical regions and temperate and high-latitude regions?

8. Explain how speciation and extinction rates might be affected by the area of continents. What evidence is there to support your explanation? What does the influence of area on rates of extinction and speciation have to do with higher species richness in tropical regions compared to temperate and high-latitude regions?

9. Ricklefs (1987) pointed out that many large-scale contrasts in species richness and composition cannot be explained by local processes such as competition and predation. Ricklefs proposed that differences in history and geography can leave a unique stamp on regional biotas. The mammals of Australia, including kangaroos, koalas, and duck-billed platypuses, must be one of the best-known examples of a unique biota. How have history and geography, as opposed to local processes, combined to produce this unique assemblage of mammals?

10. Most examples of regional and latitudinal variation in species richness cited in chapter 22 have been terrestrial. Consider regional variation in marine biotas. Like birds on land, fish are one of the best-studied groups of marine organisms. Moyle and Cech (1982) cite the following patterns of fish species richness:

| Atlantic and Gulf Coasts of North America | | Pacific Coast of North America | |
|---|---|---|---|
| **Area** | **Species** | **Area** | **Species** |
| Texas | 400 | Gulf of California | 800 |
| South Carolina | 350 | California | 550 |
| Cape Cod | 250 | Canada | 325 |
| Gulf of Maine | 225 | | |
| Labrador | 61 | | |
| Greenland | 34 | | |

As you can see, fish species richness decreases northward on both coasts. However, the Pacific coast generally supports a larger number of species. This contrast may be another situation requiring historical- and geographic-level explanations. Explore and explain this contrast in species richness using information from the fields of marine biology, oceanography, and ichthyology. Moyle and Cech (1982) and Briggs (1974) are good starting points.

# Suggested Readings

Bellwood, D. R. and T. P. Hughes. 2001. Regional-scale assembly rules and biodiversity of coral reefs. *Science* 292:1532–34.

*An analysis of coral reef diversity that indicates that habitat area within the tropics accounts for more variation in species richness and composition than does latitude.*

Botting, D. 1973. *Humboldt and the Cosmos.* New York: Harper & Row.

*Botting provides a fascinating narrative of Alexander von Humboldt's life and his epic expedition with Aimé Bonpland to the Americas—the first major modern foray into geographic ecology.*

Brown, J. H. 1995. *Macroecology.* Chicago: University of Chicago Press.

*Brown establishes a fresh framework for studies of large-scale ecology. This book looks boldly into the future.*

Brown, J. H. and M. V. Lomolino. 2000. Concluding remarks: historical perspective and the future of island biogeography theory. *Global Ecology and Biogeography* 9:87–92.

*An update and historical perspective on the equilibrium theory of island biogeography from two scientists that have made major contributions to the theory.*

Buzas, M. A., L. S. Collins, and S. J. Culver. 2002. Latitudinal difference in biodiversity caused by higher tropical rate of increase. *Proceedings of the National Academy of Sciences of the United States of America* 99:7841–7843.

*A study that uncovers evidence that rates of speciation are higher in the tropics.*

Fine, P. V. A. 2001. An evaluation of the geographic area hypothesis using the latitudinal gradient in North American tree diversity. *Evolutionary Ecology Research* 3:413–28.

*Detailed analysis showing that tree species richness does not increase smoothly from high latitudes to the tropics but increases gradually outside of the tropics and then increases in step as the tropics are entered.*

Huston, M. 1994. *Biological Diversity.* New York: Cambridge University Press.

*Huston gives an insightful review of patterns of biological diversity at all scales. His discussions of geographic patterns and processes are excellent.*

Lyons, S. K. and M. R. Willig. 2002. Species richness, latitude, and scale-sensitivity. *Ecology* 83:47–58.

*Quantitative analysis that suggests that tropical marsupials and bats have larger ranges than marsupials and bats outside the tropics.*

MacArthur, R. H. 1972. *Geographical Ecology.* New York: Harper & Row.

MacArthur, R. H. and E. O. Wilson. 1967. *The Theory of Island Biogeography.* Princeton, N.J.: Princeton University Press.

*In these books, MacArthur and Wilson established the theoretical foundation for modern studies of geographic ecology.*

Nilsson, S. G., J. Bengtsson, and S. Ås. 1988. Habitat diversity or area *per se*? Species richness of woody plants, carabid beetles and land snails on islands. *Journal of Animal Ecology* 57:685–704.

*An excellent study of patterns of species richness in several groups of organisms. A model study for field investigations of species richness patterns on islands.*

Qian, H., J.-S. Song, P. Krestov, Q. Guo, Z. Wu, X. Shen, and X. Guo. 2003. Large-scale phytogeographical patterns in East Asia in relation to latitudinal and climatic gradients. *Journal of Biogeography* 30:129–141.

*Detailed analysis of large-scale patterns of diversity and composition of 2,808 genera of plants in 45 regions in East Asia from southern China to northern Russia and including Korea and Japan. Results show strong influences of latitude on plant diversity.*

Ricklefs, R. E. and D. Schluter. 1993. *Species Diversity in Ecological Communities*. Chicago: University of Chicago Press.

*This collection of papers provides wide-ranging discussions of species diversity patterns. It is especially valuable for its treatment of geographic and historical influences on patterns of species richness.*

Rosenzweig, M. L. 1992. Species diversity gradients: we know more and less than we thought. *Journal of Mammalogy* 73:715–30.

Rosenzweig, M. L. 1995. *Species Diversity in Space and Time*. New York: Cambridge University Press.

*In both his 1992 paper and his 1995 book, Rosenzweig provides provocative analyses of one of the most stimulating subjects in ecology—spatial variation in species diversity.*

Simberloff, D. S. and E. O. Wilson. 1969. Experimental zoogeography of islands: the colonization of empty islands. *Ecology* 50:278–96.

Wilson, E. O. and D. S. Simberloff. 1969. Experimental zoogeography of islands: defaunation and monitoring techniques. *Ecology* 50:267–78.

*Two classic papers that describe some of the most ambitious experiments done in island biogeography—ideal for anyone interested in large-scale field experiments.*

# On the Net

Visit this textbook's accompanying website at www.mhhe.com/ecology (click on the book's title) to take advantage of practice quizzing, study/writing tips, timely news articles, and additional URLs for research on the topics in this chapter.

Geographic Ecology
Animal Population Ecology
Species Preservation

Species Abundance, Diversity, and Complexity
Biomes and Environmental Habitats

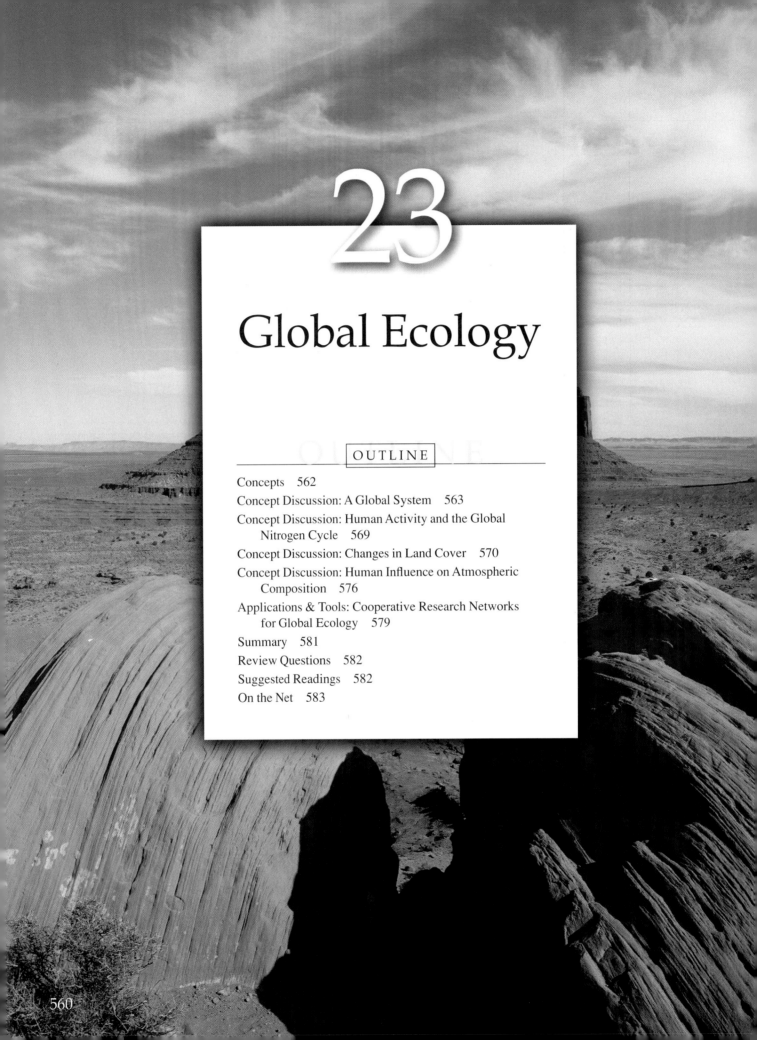

# 23

# Global Ecology

uring the final days of December 1968, the *Apollo 8* mission to the moon transmitted images of the earth rising above the moon's horizon. For the first time, in color, we could see how earth appears from our nearest neighbor in the solar system (fig. 23.1). The human response to the sight of the earth framed by a bleak lunar landscape is captured by the words of the *Apollo 8* astronauts: "The earth from here is a grand oasis in the . . . vastness of space."

That image, of earth as a shining blue ball against the blackness of space, instantaneously changed the perspective that most people held of the planet and made it easier to think of the earth as a single ecological system.

As we enter the twenty-first century, it is important that we keep that perspective alive. The rapid pace of global change challenges ecologists to study ecological phenomena at a global scale. Peter Vitousek (1994) pointed out that this is the first generation in history with the tools to examine how humanity has changed the earth. However, he also reminds us that this may be the last generation that has a chance to affect the course of those changes significantly.

Because many global-scale phenomena are mediated through the atmosphere, let's look briefly at the structure and origins of earth's atmospheric system.

**Figure 23.1**  Oasis in space: earthrise over the moon's horizon.

# The Atmospheric Envelope and the Greenhouse Earth

The earth is wrapped in an atmospheric envelope that makes the biosphere a hospitable place for life as we know it. Clean, dry air at the earth's surface is approximately 78.08% nitrogen, 20.94% oxygen, 0.93% argon, 0.03% carbon dioxide, and less than 0.00005% ozone. Air also contains variable concentrations of water vapor and trace quantities of helium, hydrogen, krypton, methane, and neon. The concentrations of these gases change with altitude. The highest concentrations of atmospheric gases occur in the **troposphere,** a layer extending from the earth's surface to an altitude of 9 to 16 km. However, ozone is most concentrated in the **stratosphere,** which extends from the troposphere outward to an altitude of about 50 km. Above the troposphere are two other layers, the **mesosphere** and the **thermosphere.**

The atmosphere surrounding the earth significantly modifies earth's environment. For instance, the atmosphere reduces the amount of ultraviolet light that reaches the surface of the earth. This shielding by the atmosphere is performed principally by ozone, a trace gas with an extremely important function. The atmosphere also helps keep the surface of the earth warm, a phenomenon called the **greenhouse effect.**

How does the greenhouse effect work? The wavelengths and intensities of energy radiated by the earth into space indicate an object with a temperature of about –18°C. However, the average temperature at the earth's surface is about 15°C. This 33°C difference between predicted and actual temperature results from heat trapped near the earth's surface by the

atmosphere (fig. 23.2). This heat is trapped by the greenhouse gases, which include water vapor, carbon dioxide, methane, ozone, nitrous oxide, and chlorofluorocarbons. Notice that several of these greenhouse gases are products of biological activity. Like the glass of a greenhouse, these gases absorb infrared radiation emitted by a solar-heated earth and reemit most of that energy back to the earth.

Let's look briefly at a budget of solar energy for the earth. About 30% of the solar energy shining on earth is reflected back into space by clouds, by particles in the atmosphere, or by the surface of the earth. Approximately 70% of the solar energy shining on the earth is absorbed either by the atmosphere or by the earth's surface. This energy is reemitted as infrared radiation. Some of the infrared radiation from the atmosphere is radiated into space, and some is radiated toward the surface of the earth. Most of the infrared radiation from the earth's surface is absorbed by greenhouse gases in the atmosphere and radiated back to the earth's surface. By radiating infrared radiation back to the earth's surface, greenhouse gases trap heat energy and raise the earth's surface temperature.

We should remember that the atmosphere is not static. The atmosphere and the oceans are in continuous motion as a consequence of the uneven heating of the earth's surface. In chapters 2 and 3 we reviewed the major patterns of atmospheric and oceanic circulation (see figs. 2.4, 2.5, and 3.5). These circulatory systems link the various regions of the globe into a single physical system by moving heat energy and materials from one part of the biosphere to another. Because of global circulation, pollutants produced in one part of the globe eventually reach all parts of the globe.

Biological activity is largely responsible for the current composition of earth's atmosphere, particularly the concentrations of oxygen, carbon dioxide, and methane. While the effect of life on atmospheric composition is usually associated with

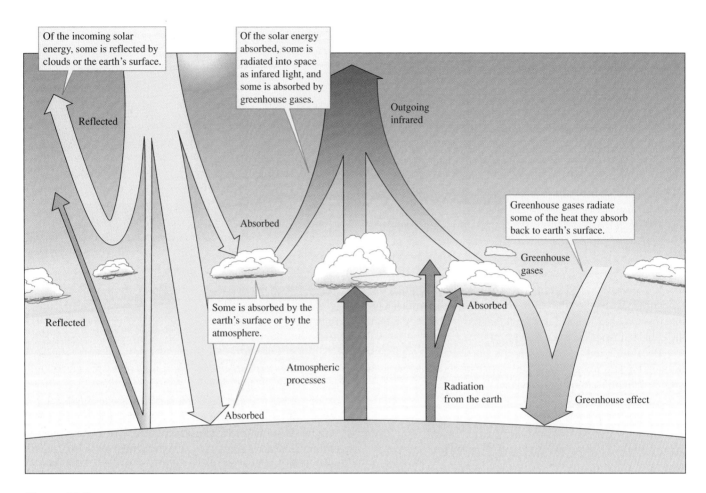

Of the incoming solar energy, some is reflected by clouds or the earth's surface.

Reflected

Of the solar energy absorbed, some is radiated into space as infared light, and some is absorbed by greenhouse gases.

Outgoing infrared

Absorbed

Greenhouse gases radiate some of the heat they absorb back to earth's surface.

Greenhouse gases

Reflected

Some is absorbed by the earth's surface or by the atmosphere.

Absorbed

Atmospheric processes

Radiation from the earth

Greenhouse effect

Absorbed

**Figure 23.2**  The greenhouse effect: heat trapping by earth's atmosphere.

human activity, major biological influences on the atmosphere began about 2 billion years ago. The early atmosphere lacked oxygen and likely contained higher concentrations of carbon dioxide and hydrogen. Atmospheric oxygen began increasing about 2 billion years ago with the appearance of oxygen-producing photosynthesis.

James Walker (1986) referred to the shift in atmospheric composition that eventually produced today's aerobic atmosphere as "the most severe pollution episode in the history of the earth." Why did he refer to oxygen as a pollutant? To answer this question we need to remember what sorts of organisms inhabited the earth of 2 billion years ago. The first life-forms appeared about 3.5 billion years ago. Geological evidence indicates that the atmosphere when these early organisms lived was free of oxygen. Consequently, the earliest organisms were anaerobes, for which oxygen is a deadly poison. As we shall see, the composition of the atmosphere is once again changing, this time in response to human activity.

Our discussion begins with a large-scale atmosphere-ocean system that has global effects on ecological systems. From this general discussion of climatic systems we review some key human influences on the biosphere. Vitousek (1994) discussed three aspects of global change caused by human

activity: (1) changes in the nitrogen cycle, (2) changes in landscapes, and (3) changes in atmospheric $CO_2$. Because these environmental changes may ultimately influence global climate and biological diversity (fig. 23.3), it is important to understand their causes and interactions.

## CONCEPTS

- **The El Niño Southern Oscillation is a large-scale atmospheric and oceanic phenomenon that influences ecological systems on a global scale.**
- **Human activity has greatly increased the quantity of fixed nitrogen cycling through the biosphere.**
- **Rapid changes in global patterns of land use threaten biological diversity.**
- **Human activity is increasing the atmospheric concentration of $CO_2$, which may be increasing global temperatures.**

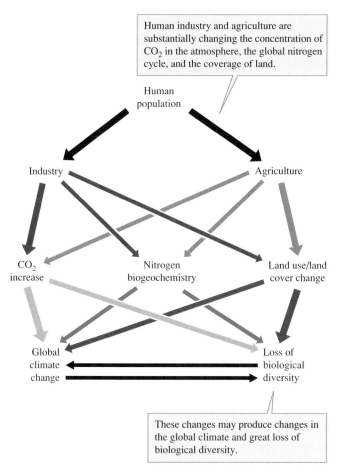

Human industry and agriculture are substantially changing the concentration of $CO_2$ in the atmosphere, the global nitrogen cycle, and the coverage of land.

Human population

Industry

Agriculture

$CO_2$ increase

Nitrogen biogeochemistry

Land use/land cover change

Global climate change

Loss of biological diversity

These changes may produce changes in the global climate and great loss of biological diversity.

**Figure 23.3** Some causes and potential consequences of global environmental change (data from Vitousek 1994).

## CONCEPT DISCUSSION

### A Global System

**The El Niño Southern Oscillation is a large-scale atmospheric and oceanic phenomenon that influences ecological systems on a global scale.**

Large-scale atmospheric and oceanic systems exert global-scale influences on ecological systems. One of the most thoroughly studied of these large-scale systems is the *El Niño Southern Oscillation*. The name **El Niño** originated when this climatic system seemed limited to the west coast of South America. During an El Niño, a warm current appears off the west coast of Peru, generally during the Christmas season (El Niño refers to the Christ child). The term **Southern Oscillation** refers to an oscillation in atmospheric pressure that extends across the Pacific Ocean. Before we discuss the behavior and effects of the El Niño Southern Oscillation, let's

review the historical origins of our present-day knowledge of the system. This knowledge, which we take for granted today, took most of the twentieth century to gather.

## The Historical Thread

In 1904, a British mathematician named Gilbert Walker was appointed Director General of Observatories in India. Walker arrived in India shortly after a disastrous famine from 1899 to 1900 caused by crop failures during a drought. This tragic event led him to search for a way to predict the rainfall associated with the Asian monsoons. Walker (1924) eventually found a correspondence between barometric pressure across the Pacific Ocean and the amount of rain falling during the monsoons. He found that reduced barometric pressure in the eastern Pacific was accompanied by increased barometric pressure in the western Pacific. In a similar fashion, when the barometric pressure fell in the western Pacific, it rose in the eastern Pacific. Walker called this oscillation in barometric pressure the Southern Oscillation.

Today, meteorologists monitor the state of the Southern Oscillation with the Southern Oscillation Index. The value of the index is determined by the difference in barometric pressure between Tahiti and Darwin, Australia (fig. 23.4). Walker noticed that low values of the Southern Oscillation Index were associated with drought in Australia, Indonesia, India, and parts of Africa. Walker also suggested that winter temperatures in Canada were somehow connected to the Southern Oscillation. His studies led him to a global perspective on

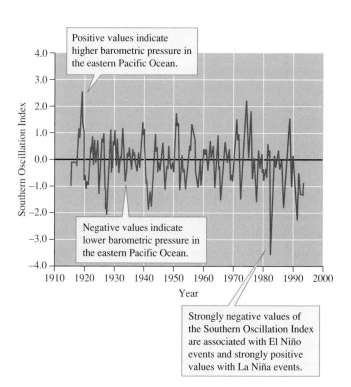

Positive values indicate higher barometric pressure in the eastern Pacific Ocean.

Negative values indicate lower barometric pressure in the eastern Pacific Ocean.

Strongly negative values of the Southern Oscillation Index are associated with El Niño events and strongly positive values with La Niña events.

**Figure 23.4** The Southern Oscillation Index shows the difference in barometric pressures between Tahiti and Darwin, Australia.

climate, a perspective well ahead of his time. Walker was highly criticized for suggesting a climatic link between such widely separated regions. However, he did not waver from his view. He assured his critics that the climatic connection between regions would eventually be explained when the proper measurements could be made.

The connection between Walker's Southern Oscillation and patterns of ocean temperature during El Niños was eventually described by Jacob Bjerknes (1966, 1969), a professor at the University of California at Los Angeles. Bjerknes began developing his capacity for large-scale perspectives as a young scientist in Norway, where he worked on the dynamics of storms in temperate regions. His later work produced a global perspective that connected El Niño with the Southern Oscillation, which Walker had studied 40 years earlier.

Bjerknes was able to connect the Southern Oscillation with El Niño because of a fortuitous coincidence. A strong El Niño from 1957 to 1958 coincided with the International Geophysical Year, during which oceanographic vessels made simultaneous observations across the Pacific and Indian Oceans. For the first time, scientists had extensive oceanographic data during a strong El Niño. Those data showed that the warm waters associated with El Niño were not limited to the west coast of South America but extended far out into the Pacific Ocean.

Bjerknes proposed that the gradient in sea surface temperature across the central Pacific Ocean produces a large-scale atmospheric circulation system that moves in the plane of the equator, as shown in figure 23.5. Air over the warmer western Pacific rises, flows eastward in the upper atmosphere, and then sinks over the eastern Pacific. This air mass then flows westward along with the southeast trade winds, gradually warming and gathering moisture. This westward-flowing air eventually joins the rising air in the western Pacific. As this warm and moist air rises, it forms rain clouds. Bjerknes called this atmospheric system **Walker circulation** after Sir Gilbert Walker.

Bjerknes, like Walker before him, possessed a global perspective before it became commonplace. Eugene Rasmusson (1985) referred to Bjerknes' model, which coupled oceanic and atmospheric circulation, as a "grand hypothesis." This hypothesis stimulated decades of research and has led to a greatly enhanced understanding of the El Niño Southern Oscillation.

## El Niño Today

The El Niño Southern Oscillation is a highly dynamic, large-scale weather system that involves variation in sea surface temperature and barometric pressure across the Pacific and Indian Oceans. We discuss the El Niño Southern Oscillation here because recent discoveries show that this system drives a great deal of climatic variability around the globe. This system affects the climate of North America, South America, Australia, southern Asia, Africa, and parts of southern Europe (fig. 23.6). This climatic variability has substantial influences on the distribution of organisms, structure of communities, and ecosystem processes.

During the mature phase of an El Niño, the sea surface in the eastern tropical Pacific Ocean is much warmer than average and the barometric pressure over the eastern Pacific is lower than average. The combination of warm sea surface temperatures and low barometric pressure promotes the formation of storms over the eastern Pacific Ocean. These storms bring increased precipitation to much of North and South America. During an El Niño, the sea surface in the western Pacific is cooler than average and the barometric pressure is higher than average. These conditions produce drought over much of the western Pacific region, including Australia.

Periods of lower sea surface temperature and higher than average barometric pressure in the eastern tropical Pacific have been named **La Niñas.** La Niña brings drought to much of North and South America. During La Niñas, a pool of warm seawater moves far into the western Pacific. This warm water combined with lower barometric pressures in the western Pacific Ocean generates many storms. Consequently, La Niña brings higher than average precipitation to the western Pacific. It appears that La Niñas and El Niños represent opposite extremes in the El Niño Southern Oscillation cycle.

While often associated with the tropics, the influence of the El Niño Southern Oscillation extends well into temperate regions. For instance, El Niños are associated with higher than average precipitation over much of the western and southeastern United States and adjacent regions of Mexico. La Niñas are consistently associated with drought over most of the same region. The El Niño Southern Oscillation also affects temperatures over large geographic areas. During El Niños, much of the northern United States, Canada, and Alaska are much warmer than average. During La Niñas, these regions are colder than average. As you might expect, this global climate system affects ecological systems around the globe.

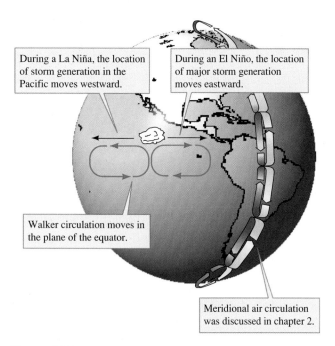

During a La Niña, the location of storm generation in the Pacific moves westward.

During an El Niño, the location of major storm generation moves eastward.

Walker circulation moves in the plane of the equator.

Meridional air circulation was discussed in chapter 2.

**Figure 23.5** Walker circulation, El Niño, and La Niña.

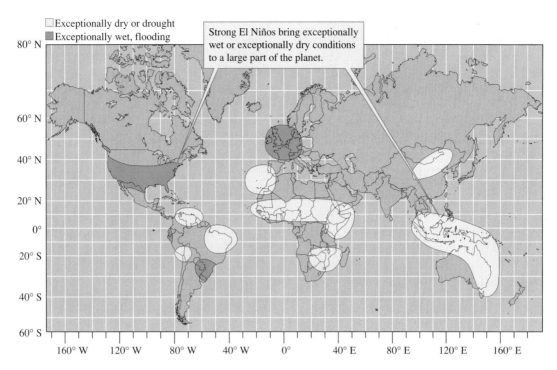

**Figure 23.6**  Effects of the exceptionally strong El Niño of 1982 to 1983 on patterns of global precipitation (data from Diaz and Kiladis 1992).

# El Niño and Marine Populations

Some of the most dramatic ecological responses to El Niño occur in marine populations along the west coast of South America. Long before the recent discovery of the global extent of its effects, El Niño was known to produce declines in coastal populations of anchovies and sardines and the seabirds that feed upon them. How does El Niño induce these population declines? They are produced by changes in the pattern of sea surface temperatures and coastal circulation. Figure 23.7*b* shows sea surface temperatures off the west coast of South America during average conditions. Notice that under average conditions, coastal waters are relatively cool along most of the west coast of South America and that a tongue of cool water extends westward toward the open Pacific Ocean. This cool water is brought to the surface by upwelling. Upwelling along the coast is driven by the southeast trade winds, while the offshore upwelling is driven by the east winds of the Walker circulation.

With the onset of an El Niño, the easterly winds slacken and the pool of warm water in the western Pacific moves eastward. Eventually this pool of warm water reaches the west coast of South America and then moves north and south along the coast (fig. 23.7*a*). During the mature phase of an El Niño, the warm surface water along the west coast of South America shuts off upwelling. Consequently, the supply of nutrients that upwelling usually delivers to surface waters is also shut off. A lower nutrient supply reduces primary production by phytoplankton. This decline in primary production reduces the supply of food available to consumers in the coastal food web and is followed by declines in populations of fish and their predators.

Remote sensing of phytoplankton pigments in surface waters around the Galápagos Islands shows that the 1982–83 El Niño reduced average primary production and dramatically changed the location of production "hot spots." Figure 23.8 shows the concentration of phytoplankton pigments, mainly chlorophyll *a*, around the Galápagos Islands before and during the 1982–83 El Niño. The image made on February 1, 1983, shows the normal patterns when the southeast trade winds produce upwelling on the western sides of the islands. These upwelling areas, which are shown in red and orange, are especially apparent west of the large islands of Fernandina and Isabela. Large areas of high phytoplankton production extended for approximately 150 km west of these islands. On February 1, the mean pigment concentration for surface waters around the Galápagos archipelago was 0.30 mg per cubic meter.

A shift in wind direction reduced average phytoplankton biomass and shifted the location of high production areas. After February 1, the trade winds became progressively weaker and variable in direction. By March 28, the pigment concentration had increased back to 0.28 mg per cubic meter. However, the areas of high phytoplankton biomass had shifted far to the east.

Changes in the rate and distribution of primary production, such as those shown in figure 23.8, induced reproductive failure, migration, and widespread death among seabird populations in the Galápagos Islands and along the west coast of South America during the 1982–83 El Niño. Many seabirds abandoned their nests with the onset of this El Niño and migrated either north or south along the coast of South America. Virtually no birds reproduced and most of the migrating birds starved. Population declines were dramatic. The adult populations of three seabird species on the coast of Peru declined from 6.01 million to 330,000 between March 1982 and May 1983, a population decline of approximately 95%.

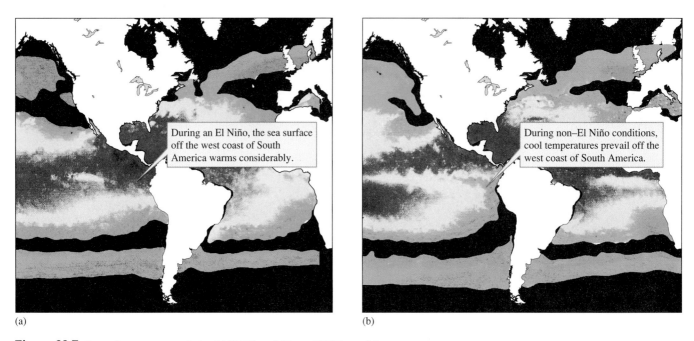

**Figure 23.7** Sea surface temperature during (*a*) El Niño and (*b*) non–El Niño conditions.

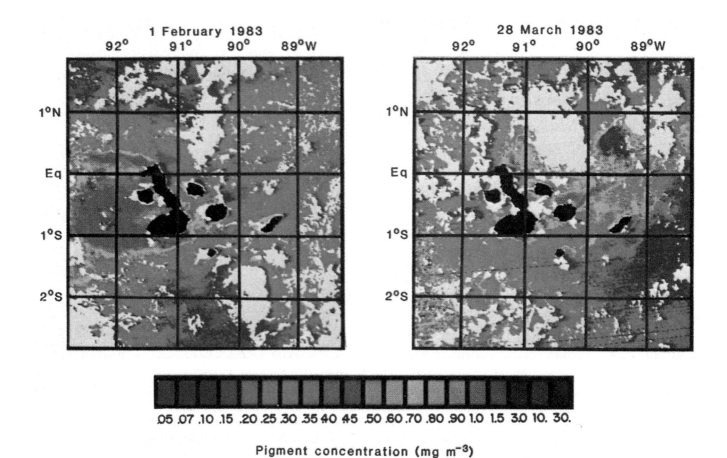

**Figure 23.8** El Niño and areas of high marine primary production around the Galápagos Islands. Notice that the areas of high production were west of the islands on February 1 and east of them on March 28 (data from Feldman, Clark, and Halpern 1984).

**Figure 23.9** Populations of sea lions such as these in the Galápagos Islands are decimated when the coming of El Niño reduces marine primary production.

The 1982–83 El Niño also had a major impact on fur seal and sea lion populations, mainly through reductions in food supply (fig. 23.9). Fur seals and sea lions both feed on fish. The main food fish for the South American fur seal, *Arctocephalus australis,* is the anchoveta, *Engraulis ringens. Engraulis* normally lives at depths of 0 to 40 m. However, during the 1982–83 El Niño, it moved away from fur seal colonies to cooler water at depths of up to 100 m.

In response, fur seals dived deeper and shifted their diets to other fishes. Both on the mainland and on the Galápagos, female fur seals increased their foraging time. Since females are away from their young while foraging, the pups in both populations did not get enough food and all died. On the Galápagos, nearly 100% of mature male fur seals died, while the mortality of adult females and nonterritorial males was approximately 30%. A large fur seal colony at Punta San Juan, Peru, declined from 6,300 to 4,200 individuals.

As the previous examples show, El Niño has well-documented effects on marine populations along the coast of South America. However, as we shall see in the following examples, El Niño can have ecological effects that extend far from the west coast of South America.

# El Niño and the Great Salt Lake

El Niño can influence weather in continental areas far from the central Pacific Ocean (see fig. 23.6). The exceptionally strong El Niño of 1982 to 1983 was the source of many moisture-bearing storms that penetrated deep into the interior of North America. These storms substantially increased precipitation within the basin of the Great Salt Lake. Another wet period, within this lake basin, soon followed with the El Niño of 1986 to 1987. The effects of increased precipitation on the Great Salt Lake ecosystem were dramatic. Between 1983 and 1987, the level of the lake rose 3.7 m and its salinity decreased from 100 to 50 g per liter.

Wayne Wurtsbaugh and Therese Smith Berry of Utah State University documented the ecological responses to these physical changes (Wurtsbaugh and Smith Berry 1990, Wurtsbaugh 1992). Prior to the 1982–83 El Niño, high salinity limited the

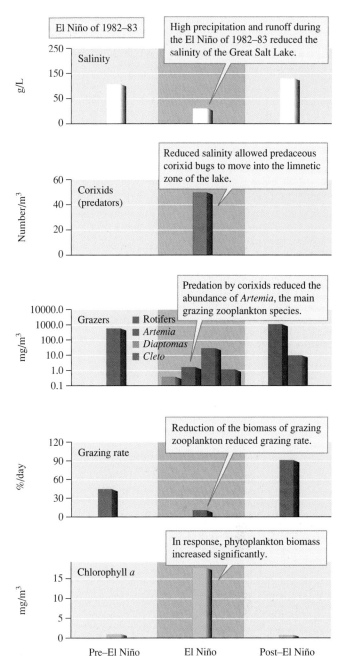

**Figure 23.10** The El Niño of 1982 to 1983 created conditions for a trophic cascade (data from Wurtsbaugh 1992).

zooplankton community of the Great Salt Lake to a few salt-tolerant species, especially the brine shrimp, *Artemia franciscana.* How high was the salinity of the lake? For comparison, the salinity of the open ocean is about 34 to 35 g per liter. At 100 g per liter, the lake's salinity was nearly three times that of seawater. However, by 1985 to 1987 the lake's salinity dropped 50 g per liter, only 50% higher than seawater, and the open lake was invaded by the predaceous insect *Trichocorixa verticalis.*

Wurtsbaugh and Smith Berry found that the invasion of the predatory insects induced a trophic cascade (see chapter 18) in the pelagic zone of the Great Salt Lake (fig. 23.10). Predation reduced the population of brine shrimp from approximately 12,000 to 74 per cubic meter. Though other grazing zooplankton moved into the lake when its salinity dropped, the overall grazing rate was still greatly reduced. As predicted by the trophic cascade model, phytoplankton biomass increased significantly. This increase in phytoplankton biomass was accompanied by reduced water transparency and greatly reduced nutrient concentrations in lake water.

Then from 1987 to 1990, the level of the lake fell 2.8 m and the salinity returned to approximately 100 g per liter. With this increase in salinity nearly all the changes observed in the Great Salt Lake ecosystem were reversed. *Trichocorixa* was eliminated from the open lake, brine shrimp populations increased, and phytoplankton biomass declined. This example shows how a large-scale climate system can control local community and ecosystem structure and processes. This example also suggests that ecological responses to global climate change are complicated by biological interactions and phenomena such as trophic cascades and keystone species effects.

# El Niño and Terrestrial Populations in Australia

The effects of El Niño and La Niña on Australian weather generally mirror their effects on South and North America. El Niño brings drought to Australia while La Niña brings abundant rainfall. Because of the El Niño Southern Oscillation, much of Australia alternates between periods of scarcity and plenty. This environmental fluctuation has profound effects on populations of animals and plants.

## Episodic Establishment by Perennial Plants

Many plants infrequently establish new cohorts of seedlings. Episodic plant establishment is particularly common in arid and semiarid regions. Graham Harrington (1991) studied the effects of soil moisture on survival of the narrow-leaf hopbush, *Dodonaea attenuata,* in a semiarid grass and shrub community in New South Wales, Australia. Over a 97-year period from 1884 to 1981, there were apparently only three periods of widespread establishment of this plant: during the 1890s, in 1952, and in 1974. All three of these periods were associated with the La Niña side of the El Niño Southern Oscillation. This association suggests that plant community structure over much of Australia is significantly influenced by the El Niño Southern Oscillation. However, more quantitative studies of plant establishment in Australia will be required before we can fully understand its relationship to this climate system. A great deal more quantitative information is available on the effects of climate on kangaroo populations.

## El Niño and Kangaroo Populations

The influences of the El Niño Southern Oscillation on Australian populations are clearly reflected in the biology of the red kangaroo, *Macropus rufus.* This animal, which can reach a weight of

**Figure 23.11** Populations of red kangaroos, *Macropus rufus,* are substantially influenced by El Niño and La Niña.

93 kg and a length of about 2.5 m, is the largest of the kangaroos and the largest of the Australian herbivores (fig. 23.11). As we saw in chapter 9, red kangaroos occupy nearly the entire arid and semiarid interior of Australia (see fig. 9.2). The range occupied by red kangaroos is a region where occasional moist periods are interspersed with severe droughts. The reproductive biology of the red kangaroo is not tied to any fixed seasonal cycle but responds rapidly to changing environmental conditions, particularly increased rainfall and vegetative production.

Neville Nicholls (1992) described the reproductive response of red kangaroos to a wet period as follows. During wet years, when food is plentiful, female red kangaroos will simultaneously have a juvenile or "joey" following her, a younger offspring in her pouch nursing, and a quiescent embryo that will enter the pouch as soon as the current occupant leaves. As soon as this replacement occurs, the mother mates again, producing another embryo that will develop to the point where it is ready to enter the pouch. These relay tactics enable female red kangaroos to produce independent offspring at 240-day intervals.

Under marginal conditions, females continue to reproduce. However, most young die soon after they leave the pouch. If food becomes even more scarce, females will stop lactating and the young die in the embryo stage. Red kangaroos stop breeding only in response to severe, prolonged droughts. During droughts, these kangaroos wander widely in search of food. Females range over areas greater than 18 km$^2$ and males over areas of about 36 km$^2$. During severe droughts, females, which may reach sexual maturity as young as 15 to 20 months old, may not breed until they are over 3 years old.

When abundant rains finally come, the female hormonal system responds rapidly. The kangaroos breed quickly and young enter the pouch within 60 days of the onset of significant rainfall. Nicholls suggests that this strategy ensures a short interval between the return of good conditions and recruitment of young into the population. By reproducing large numbers of young when conditions are favorable, the animals increase the size of the adult population that will face each drought induced by El Niño.

S. Cairns and G. Grigg (1993) quantified how rainfall in southern Australia affected populations of red kangaroos (fig. 23.12). The study population reached a peak of 2,175,000 in 1981

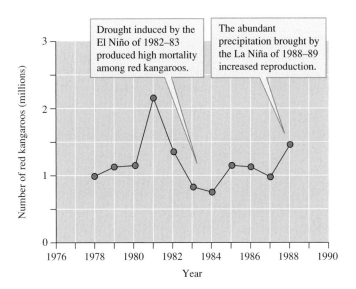

Drought induced by the El Niño of 1982–83 produced high mortality among red kangaroos.

The abundant precipitation brought by the La Niña of 1988–89 increased reproduction.

**Figure 23.12** El Niño, La Niña, and population dynamics of the red kangaroo (data from Cairns and Grigg 1993).

and then declined to 745,000 by 1984. This decline was a response to drought associated with the 1982–83 El Niño. The population rebounded in 1985, held approximately steady through 1986, but then declined somewhat in 1987, during another El Niño. Then, in the last of the records analyzed by Cairns and Grigg, the population began to grow rapidly in response to the abundant rains and plant production induced by the 1988–89 La Niña. This record suggests a tight coupling of *M. rufus* populations to the El Niño Southern Oscillation. This coupling is a mirror image of the influences of El Niño on Galápagos finch populations (see fig. 11.17).

In many situations, biologists studying local populations must consider the influences of large-scale systems such as the El Niño Southern Oscillation. Increasingly, ecologists must consider the effects of human modifications of the global environment. Because of the increasing size of the human population and the intensity of human activity, we are rapidly modifying global nutrient cycles, the face of the land, and even the composition of the atmosphere. The effects of the growing human population on the global environment are the subject of the next three Concept Discussion sections.

## CONCEPT DISCUSSION

### Human Activity and the Global Nitrogen Cycle

**Human activity has greatly increased the quantity of fixed nitrogen cycling through the biosphere.**

When we reviewed the nitrogen cycle in chapter 19, we saw that nitrogen enters the cycle through the process of nitrogen fixation. For millions of years, the only organisms that could fix nitrogen were nitrogen-fixing bacteria and some actinomycete fungi. Then, as humans developed intensive agricul-

tural and industrial processes that fix nitrogen, we began to manipulate the nitrogen cycle on a massive scale.

How has human activity altered the nitrogen cycle? To address this question, we need to review the sources and amounts of nitrogen fixed in the absence of human manipulation. Vitousek (1994) summarized the natural background levels of nitrogen fixation as follows. The nitrogen fixed in terrestrial environments by free-living nitrogen-fixing bacteria and nitrogen-fixing plants totals approximately 100 terragrams (Tg) of nitrogen (N) per year (1 Tg = $10^{12}$ g). Nitrogen fixation in marine environments adds an additional 5 to 20 Tg N per year; fixation by lightning adds about 10 Tg N per year. These estimates of nonhuman sources of fixed nitrogen total approximately 130 Tg N per year.

Human additions to the nitrogen cycle now exceed historical sources of fixed nitrogen. One of the traditional ways that humans have manipulated the nitrogen cycle is by planting agricultural land with nitrogen-fixing crops. At some point, agriculturists learned that rotating legumes such as alfalfa and soybeans with grains such as oats and maize could increase crop yields. We now know that those increased grain yields are due mainly to nitrogen additions to the soil by the legumes. Vitousek estimated that nitrogen-fixing crops fix about 30 Tg N per year. Agriculturists also apply nitrogen fertilizers produced through industrial processes that fix nitrogen. The nitrogen fixed by the fertilizer industry amounts to more than 80 Tg N per year. Finally, Vitousek estimated that the internal combustion engines in cars, trucks, and other conveyances emit about 25 Tg N per year as oxides of nitrogen. V. Smil (1990) estimated the total emission of nitrogen from all combustion of fossil fuels, including coal-fired electrical generation as well as internal combustion engines, at 35 Tg N per year. The main point here is that nitrogen fixation resulting from human activity fixes more nitrogen (135–145 Tg N versus 130 Tg N) than all other sources of fixed nitrogen combined (fig. 23.13).

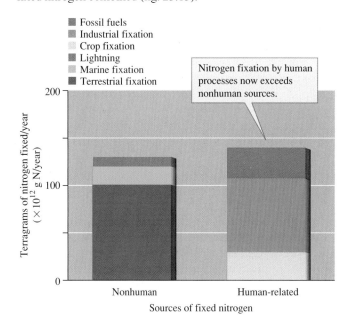

■ Fossil fuels
■ Industrial fixation
□ Crop fixation
■ Lightning
■ Marine fixation
■ Terrestrial fixation

Nitrogen fixation by human processes now exceeds nonhuman sources.

**Figure 23.13** Human and nonhuman sources of fixed nitrogen (data from Vitousek 1994).

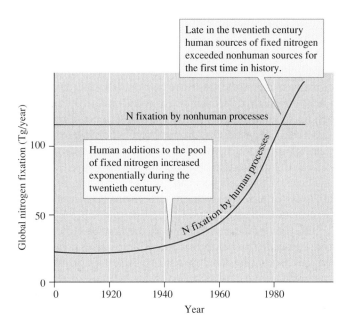

**Figure 23.14** Increase in nitrogen fixation by human processes during the twentieth century (data from Vitousek 1994).

The massive human contribution to the global nitrogen cycle is a recent phenomenon. For instance, the industrial production of fertilizers dates from the early twentieth century, and Vitousek estimated that 50% of all the commercial fertilizer produced prior to 1993 was applied to land between 1982 and 1993. Figure 23.14 shows that human contributions to the global nitrogen cycle have increased exponentially.

What are some of the consequences of these human-induced alterations to the global nitrogen cycle? As we saw in chapter 19, nitrogen additions to ecosystems are associated with local reductions in plant and fungal diversity. Nitrogen enrichment appears to alter the mutualistic relationship between plants and mycorrhizal fungi. Consequently, nitrogen enrichment over large regions threatens the health and survival of entire ecosystems. The health of forests near industrial areas has been in rapid decline. By creating environmental conditions favorable to some species and unfavorable to others, large-scale nitrogen enrichment threatens biological diversity. As we shall see in the next Concept Discussion section, however, changes in land cover pose a more direct threat to biological diversity.

## CONCEPT DISCUSSION

### Changes in Land Cover

**Rapid changes in global patterns of land use threaten biological diversity.**

Humans have changed the face of the earth. Human activities, mainly agriculture and urbanization, have significantly altered one-third to one-half of the ice-free land surface of the earth. Marshes have been drained and filled to build urban areas or airports. Tropical forests have been cut and converted to pasture. The courses of rivers have been changed. The Aral Sea in central Asia has been so starved for water that it is nearly dry. As shown in figure 23.3, Vitousek suggested that changes in land cover may be the greatest single threat to biological diversity. Let's review some of the changes in land cover and the mechanisms that make landscape changes such a powerful threat to biological diversity. A widely cited example of land cover change is the cutting of tropical forests.

## Tropical Deforestation

The cutting and clearing of tropical forests continues at an alarming rate. This deforestation is alarming because tropical forests support half or more of earth's species and appear to influence global climate. Eliminating these forests may lead to losses of thousands of potentially useful species and substantial changes in world climate. In the face of worldwide concern, we need accurate estimates of the state of tropical forests. However, estimates of deforestation rates vary widely.

How much tropical forest is there? How much tropical forest has been cut? What are the current rates of deforestation? David Skole and Compton Tucker (1993) provide us with answers to some of these questions. These researchers reported that tropical forest occurs in 73 countries and once covered 11,610,350 km². However, three-fourths of the world's tropical forests occur in just 10 countries (fig. 23.15). The largest single tract of tropical forest, nearly one-third of the total, occurs in Brazil. Brazil is also the country with the highest rate of deforestation.

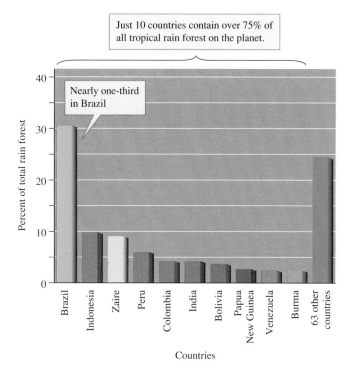

**Figure 23.15** Distribution of rain forest area by country (data from Skole and Tucker 1993).

(a)                                        (b)                                        (c)

**Figure 23.16** Information on tropical deforestation from satellite images: deforestation in Rondônia State, Brazil (light areas), in (*a*) 1975, (*b*) 1986, and (*c*) 1992.

While there has been general agreement that rates of deforestation in Brazil are high, estimates of those rates vary widely. Skole and Tucker set out to provide an accurate estimate of deforestation rates in the Amazon Basin. They based their estimate on photographs taken by *Landsat* satellites in 1978 and 1988. The images they used, *Landsat* Thematic Mapper photos, provide high-resolution information. As you can see in figure 23.16, Thematic Mapper photos clearly show areas of deforestation, regrowth on deforested plots, and areas of isolated forest. Skole and Tucker entered these high-resolution images into a geographic information system (see chapter 22), which they used to create computerized maps of deforestation within the Amazon Basin.

Let's look at the maps in figure 23.17 and see what they tell us. First, notice that large areas, colored light brown in the map, were not forested. These areas, concentrated in the southeastern Amazon Basin, have a semiarid climate and support scrubby vegetation. Some areas, shown in light violet, were covered by clouds and could not be analyzed. Skole and Tucker used the 1978 image to estimate the amount of deforestation that had occurred prior to 1978. They then compared the 1978 and 1988 photos to determine the amount of deforestation during that decade. They divided the Amazon Basin into 16 km by 16 km squares for their analysis. One of those areas is enlarged in an inset on figure 23.17. These insets show the amount of deforestation in 1978 and in 1988. Notice that the deforested area in the inset increased significantly between 1978 and 1988.

Skole and Tucker used their analyses of 16 km by 16 km areas to estimate the percentage of the land surface that had been deforested across the entire Amazon Basin. On their maps, white indicates completely forested areas, while various colors indicate increasing degrees of deforestation. At one end of their spectrum, gray indicates 0.25% to 5% deforested; at the other end, red indicates 90% to 100% deforested. Notice that the color of the area covered by the inset is purple on the 1978 map and green on the 1988 map. Change in color on the map of the entire basin from 1978 to 1988 reflects the increase in deforested area during that period.

How much of the Brazilian Amazon has been deforested? Skole and Tucker estimated that by 1978, 78,000 km$^2$ had been deforested. They also estimated that the annual rate of deforestation between 1978 and 1988 was about 15,000 km$^2$ per year. While this estimate indicates considerable deforestation, it is considerably lower than earlier estimates that ranged from 21,000 to 80,000 km$^2$ per year. Skole and Tucker estimated that the total area deforested by 1988 was 230,000 km$^2$. This estimate, which was slightly lower than the official estimate by the Brazilian government, is probably the most accurate estimate of deforestation within the Amazon Basin made to date.

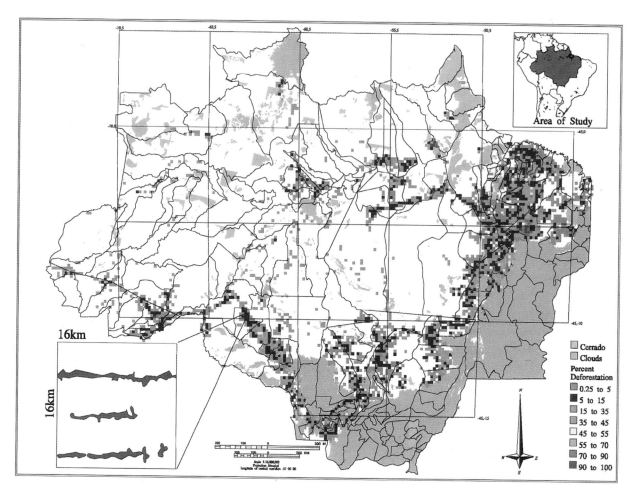

(a)

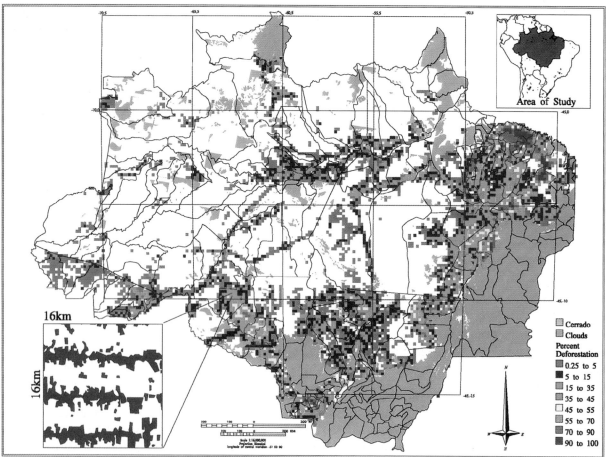

(b)

**Figure 23.17** Deforestation in Amazonia between (*a*) 1978 and (*b*) 1988 (data from Skole and Tucker 1993).

**Figure 23.18** Forest fragments left by clear-cutting forest from the surrounding landscape have very different physical environments than intact forest.

## Edge Effects and Forest Fragmentation

The area of forest removed does not give a complete picture of the ecological effects of deforestation. When a tract of forest is cut, the adjacent forest is affected by changes in the physical environment along its edges, by reduced habitat area, and by isolation. Let's look at the nature of these "edge effects" in Amazonian forest fragments.

In 1979, Brazil's National Institute for Research in Amazonia and the World Wildlife Fund began a long-term study of tropical forest fragmentation. This research project took advantage of a Brazilian law that requires that 50% of land developed in the Amazon Basin remain forested. The researchers worked with ranchers to leave forested tracts in particular areas to facilitate research on the ecological influences of forest fragment size and isolation. The fragments studied were 1, 10, 100, and 200 ha (fig. 23.18). These were compared to areas of 1, 10, 100, and 1,000 ha in undisturbed forest.

When a small fragment of forest is isolated by cutting the surrounding forest, its edge is exposed to greater amounts of solar radiation and wind. Wind and sun combine to change the physical environment within forest fragments. The physical environment along forest edges is hotter and drier and the intensity of solar radiation higher. These physical changes, in turn, affect the structure of the forest community. Tree mortality is higher along the edges of forest fragments and the forest overstory decreases while the thickness of the understory vegetation increases. Fragmentation also decreases the diversity of many animal groups, including monkeys, birds, bees, and carrion and dung beetles. Some of these reductions in animal populations may have significant impacts on key ecological processes such as pollination and decomposition.

Because edge effects, isolation, and reduced habitat area negatively affect biological diversity within tropical forest fragments, Skole and Tucker extended their analysis of deforestation in the Amazon Basin to include theses effects. They assumed that edge effects extend for 1 km from the forest edge. The results of adding

a 1 km edge effect to their deforestation analysis are shown in figure 23.19. This map is the 1988 image shown in figure 23.17, with edge effects added. As you can see, a lot more area of figure 23.19 is red, indicating 90% to 100% impact of deforestation. The inset shows the same area depicted in figure 23.17 but with edge effects added. Only tiny areas of white, indicating no effects of deforestation, remain. Edge effects increased the area of Amazonian forest affected by deforestation from 230,000 to 588,000 km$^2$.

## A Global Perspective

Skole and Tucker provided a detailed picture of land cover changes in the Amazon Basin. However, what is the global rate of tropical deforestation? No one can answer this question precisely, but the best estimates indicate that 52% to 64% of tropical deforestation occurs outside Brazil. Therefore, a conservative estimate of the global rate of deforestation from 1978 to 1988 would be approximately 30,000 km$^2$ per year.

How much land cover change is occurring outside the tropics? Though most people have focused on tropical deforestation, massive deforestation has occurred in temperate and boreal regions. As we saw in chapter 2, the temperate forest regions of Europe, eastern China, Japan, and North America support some of the densest human populations on earth. Large areas of Europe were deforested by the Middle Ages (Williams 1990), and much of the forest of eastern North America was cut by the middle 1800s (see fig. 21.21). The majority of old-growth temperate forests in northwestern North America has been cut (fig. 23.20), and the remaining old-growth forests are threatened by deforestation. In addition, vast areas of boreal forest are being cut in Russia and Canada. As you can see, deforestation is not limited to tropical regions.

R. Kates, B. Turner, and W. Clark (1990) estimated that human activity has transformed approximately half the ice-free land cover of the earth. In the process, many of the major terrestrial biomes of the earth (see chapter 2) have been highly fragmented. Others, such as tropical dry forest, have been nearly eliminated by conversion to agriculture. Because of the negative effect of reduced area on diversity (see chapter 22), these massive land conversions present a major threat to global diversity. That is why Vitousek suggested that human-caused changes to land cover constitute the greatest direct threat to global diversity (see fig. 23.3). However, land cover changes also have the potential to contribute, directly and indirectly, to rapid global climate change. One of the ways that deforestation may affect global climate is by altering the influence of forests on atmospheric concentrations of $CO_2$. The connection between atmospheric composition, human activity, and global climate is the subject of the next Concept Discussion section.

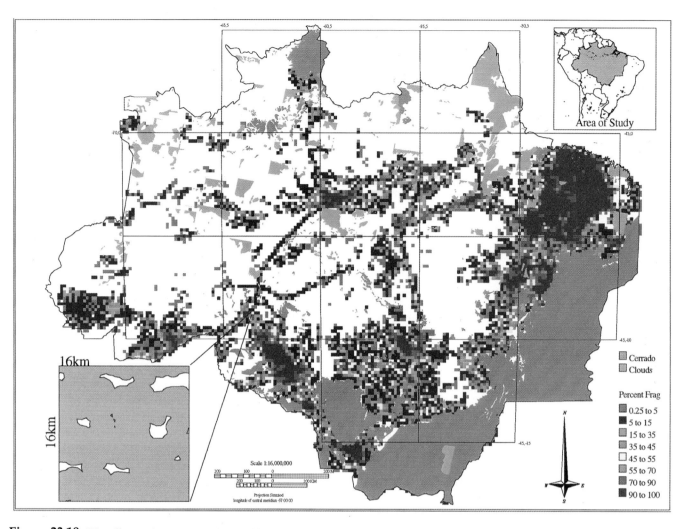

**Figure 23.19** Edge effect produces a greater degree of impact of deforestation than is apparent from the area of clear-cuts alone. Compare to figure 23.17*b* (data from Skole and Tucker 1993).

**Figure 23.20** Deforestation in the temperate forests of Washington.

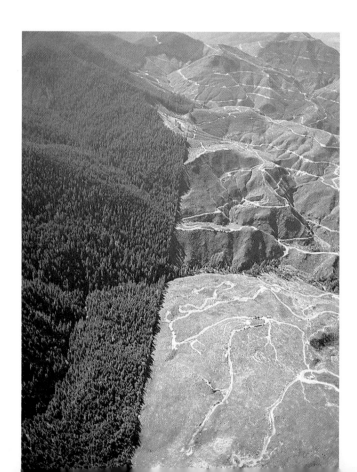

| Information | ✓ |
|---|---|
| Questions | |
| Hypothesis | |
| Predictions | |
| Testing | |

**Investigating the Evidence**

# Discovering What's Been Discovered

Throughout this series of discussions of investigating the evidence, we have emphasized one main source of evidence—original research. While original research is the foundation on which science rests, our emphasis has neglected one of the most valuable sources of information, the published literature. A researcher in any discipline needs to keep up with developments in his or her areas of interest and in related areas. In addition, some researchers may use published literature to weigh the evidence for or against some hypothesis or theory. In the section in chapter 13 titled, "Evidence for Competition in Nature," (see p. 340), we reviewed three studies that took such an approach. Students of ecology may use published literature to learn more about a particular subject, to read additional papers by researchers whose work interests them, or to do literature surveys in support of their own independent research. As pointed out in the preface of this book, however, the explosive pace of scientific discovery makes staying current very difficult. Fortunately, there are now many databases and searching tools that can help.

The databases available for searching ecological literature are far too many to review here. So, we'll focus on three contemporary ones, *Biosis, Cambridge Scientific Abstracts,* and *SciSearch,* which are widely available in university libraries and include many journals of significance to ecology. Some of their characteristics are listed in the accompanying table.

The important point here is that these databases provide access to millions of published papers often covering several decades of research. Of course, few people would want to spend the time laboriously sorting through all those articles. Fortunately, each of these databases includes a powerful search tool that will help you locate articles of interest. Let's consider some basic tips on how to use these tools to conduct an effective search.

You should generally begin your search by summarizing your subject or research interest. Next divide your subject into major concepts or key terms. Be sure to think of alternative terms for the same subject, for example, beetles or Coleoptera, daisy or Asteraceae, competition or interference. Next determine the time period in which you are interested. For instance, do you want only the most current literature on your subject or do you want all literature available in the database.

Once you have your terms listed and have selected a time period, try a search using one or more terms. If you get too many references, too few references, or unwanted references, you can use Boolean Logical Operators to adjust your search. The main Boolean Logical Operators are **and, or,** and **not.** The operator **or** will *broaden* your search and will generally yield more references. For instance, the search specified by "daisy **or** Asteraceae" will retrieve references containing *either* daisy *or* Asteraceae. In contrast the operator **and** will *narrow* your search. The search specified by "daisy **or** Asteraceae **and** desert" will retrieve references containing *either* daisy *or* Asteraceae but restrict the list of references to those concerned with these flowers in *desert areas.* The search specified by "daisy **or** Asteraceae **and** alpine would yield literature on these flowers in alpine zones. If you want to exclude certain types of references from your search, you may choose to use the operator **not.** For example the search "daisy **or** Asteraceae **not** sunflower" will exclude references that include the term sunflower.

Another useful tool for refining searches is the wild card. A wild card is used to locate references including a particular word or term with alternative endings. For example, you may encounter references to the insect order to which beetles belong as Coleoptera, coleopteran, or coleopterans. In the three databases listed below, an asterisk, *, is generally used as a wild card. In all of these databases, the search term *coleoptera** would locate references which included the terms Coleoptera, coleopteran, or coleopterans. Similarly the search term *dais** would retrieve references to both daisy and daisies.

This review is intended to suggest only general guidelines to searching literature. There are many other databases besides the ones listed here, and the creators of all of them work very hard to improve the functioning of their products. As a consequence, the operating details of the various databases are highly dynamic. Therefore, you should periodically review the tips and instructions provided with any database that you might use. The main point of this discussion is to open a door to the rich world of ecological literature, to the world of discovery. Exploring that world can quickly extend your knowledge of the discipline of ecology far beyond the introduction provided by this textbook.

Selected electronic databases that cover the ecological literature

| Database name | Dates of coverage | Types of literature | Relevant subject coverage |
|---|---|---|---|
| Biosis www.biosis.org | Variable | 5,000 scientific journals plus conference proceedings | Biology, ecology, agriculture, botany, environment, microbiology, zoology |
| Cambridge Scientific Abstracts www.csa.com | Variable | 6,000 scientific journals plus conference proceedings | Environmental science, water resources, geology, toxicology |
| SciSearch http://library.dialog.com/bluesheets | Variable | 5,000 scientific journals | Biology, environment (including ecology), earth sciences |

## CONCEPT DISCUSSION

### Human Influence on Atmospheric Composition

**Human activity is increasing the atmospheric concentration of $CO_2$, which may be increasing global temperatures.**

Industrial activity has increased steadily since about the year 1800. Over the same period, atmospheric $CO_2$ has increased steadily. The evidence discussed here shows that most of this atmospheric increase is due to the burning of fossil fuels. Vitousek pointed out that recent increases in atmospheric $CO_2$ concentration are likely to affect global climate and will certainly affect the biota of all terrestrial ecosystems. The effect of human activity on atmospheric $CO_2$ and other gases is one of the most thoroughly studied aspects of global ecology.

The concentration of $CO_2$ in the atmosphere has been dynamic over much of earth's history. Scientists have very carefully reconstructed atmospheric composition by studying air bubbles trapped in ice. As ice built up on glaciers in places such as Greenland and Antarctica, air spaces within the ice preserved a record of the ancient atmosphere. A record of atmospheric composition during the last 160,000 years was extracted and analyzed by a joint team of scientists from France and the former Soviet Union (Lorius et al. 1985, Barnola et al. 1987). This international team studied a 2,083 m core of ice drilled by Soviet scientists and engineers near the Antarctic station of Vostok. Vostok, located in eastern Antarctica at a latitude of over 78° S, has a mean annual temperature of –55°C, ideal conditions for preserving samples of the atmosphere in ice. The Vostok research station sits on the high antarctic plateau, where the ice is about 3,700 m thick. The amazing physical feat of extracting such a long ice core in such difficult physical circumstances is equaled by the dramatic climatic record contained within the Vostok ice core.

To extract air trapped within ice, scientists place sections of an ice core into a chamber and create a vacuum, removing traces of the current atmosphere in the process. The ice, still under vacuum, is then crushed and the air it contains is released into the chamber. Sampling devices then measure the $CO_2$ concentration of the air released from the ice. The Barnola team made 66 measurements of $CO_2$ along the length of the Vostok ice core. The scientists made measurements of $CO_2$ every 25 m along the length of the ice core from about 850 m depth to the bottom of the core. These lower sections of the core correspond to ages from 50,000 to 160,000 years. Because there were many fractures in the core above 850 m depth, the upper portion of the core was generally sampled at intervals greater than 25 m.

Figure 23.21 shows the variation in $CO_2$ concentration revealed by the Vostok ice core. The core indicated two very large fluctuations in atmospheric $CO_2$ concentration. Overall, it shows that $CO_2$ concentrations have oscillated between low concentrations of approximately 190 to 200 parts per million (ppm) and high concentrations of 260 to 280 ppm. About 160,000 years ago, the atmospheric concentration of $CO_2$ was less than 200 ppm. This early period in the Vostok ice core corresponds to an ice age. Then, about 140,000 years ago, the atmospheric concentration of $CO_2$ began to rise abruptly. This rise in $CO_2$ corresponds to a warmer interglacial period. High levels of $CO_2$ persisted until about 120,000 years ago. The concentration of $CO_2$ then declined and remained at relatively low concentrations until about 13,000 years ago, when atmospheric $CO_2$ again increased abruptly.

Notice that the fluctuations in $CO_2$ within the Vostok ice core correspond to variation in temperature (fig. 23.21). The periods of low $CO_2$ correspond to the low temperatures experienced during ice ages, while the periods of high $CO_2$ correspond to warmer, interglacial periods.

The most recent measurements in the Vostok ice core are about 2,000 years old. How has atmospheric $CO_2$ varied during the most recent 2,000 years? W. Post and colleagues (1990) assembled atmospheric $CO_2$ records from a number of sources to estimate atmospheric concentrations during the last 1,000 years (fig. 23.22). The first 1,000 years of the record come from the South Pole ice core, which was analyzed by Ulrich Siegenthaler and colleagues (1988) of the University of Bern, Switzerland. This record shows that the concentration of $CO_2$ remained relatively constant for approximately 800 years. Another study at the University of Bern provided a $CO_2$ record for the most recent 200 years (Friedli et al. 1986).

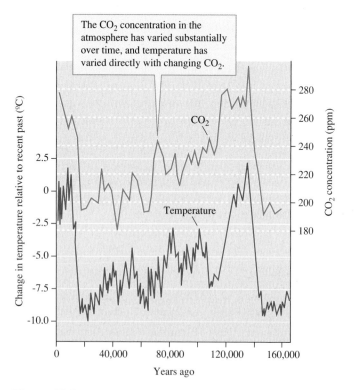

**Figure 23.21** A 160,000-year record of atmospheric $CO_2$ concentrations and temperature change (data from Barnola et al. 1987).

This part of the $CO_2$ record comes from the Siple ice core, from Siple Station at about 75° S latitude. While the Siple ice core does not allow us to look as far back in time as the Vostok record, it provides a very detailed estimate of recent concentrations of atmospheric $CO_2$. H. Friedli and colleagues dated the beginning of the Siple record at about A.D. 1744. At that time, about two and a half centuries ago, the atmospheric concentration of $CO_2$ was about 277 ppm. This estimated concentration is almost identical to those made by the Siegenthaler team for the same time period using the South Pole ice core. Therefore, both the South Pole and Siple ice cores indicate that the $CO_2$ concentration in the middle 1700s was approximately the same as at the end of the Vostok record, about 2,000 years earlier.

The Siple record showed that $CO_2$ increased exponentially from 1744 to 1953. The Friedli team estimated the 1953 concentration of $CO_2$ at 315 ppm. However, the trace in $CO_2$ concentrations shown in figure 23.22 extends beyond 1953 and above 315 ppm. Where do these later measurements come from? These later $CO_2$ concentrations are direct measurements made on Mauna Loa, Hawaii, by Charles Keeling and his associates over a period of about 40 years (Keeling and Whorf 1994).

Keeling's measurements complement the ice core data from the Vostok, South Pole, and Siple stations in two ways. First, they extend the record into the present. Second, they help validate the measurements of $CO_2$ made from the ice cores. How do Keeling's measurements lend credence to the ice core data? Look carefully at the plot of $CO_2$ concentrations shown in figure 23.22. Notice that two of the measurements made from the Siple ice core overlap the period when Keeling and his team made measurements at Mauna Loa. Notice also that the two estimates made independently by Keeling at Mauna Loa and by Friedli and his colleagues from the Siple ice core are almost identical.

The data in figure 23.22 indicate that during the nineteenth and twentieth centuries the concentration of atmospheric $CO_2$ increased dramatically. This period of increase coincides with the Industrial Revolution. However, what evidence is there that human activity caused this observed increase? Vitousek provided evidence by pointing out that the annual increase in atmospheric carbon in the form of $CO_2$ is about 3,500 Tg (1 Tg = $10^{12}$ g), while the annual burning of fossil fuels releases about 5,600 Tg carbon as $CO_2$. So, fossil fuel burning alone produces more than enough $CO_2$ to account for recent increases in atmospheric concentrations.

If we look carefully at the pattern of $CO_2$ increase between 1860 and 1960 we find additional evidence for a human influence. Figure 23.23 shows three interruptions in the otherwise steady increase in the burning of fossil fuels. Those periods correspond to three major disruptions of global economic activity: World War I, the Great Depression, and World War II. At the end of each of these major global upheavals, the increase in atmospheric $CO_2$ resumed. These patterns provide circumstantial evidence that humans are responsible for the modern increase in atmospheric $CO_2$. However, there is also direct evidence.

Additional evidence that human industrial activity is at the heart of recent increases in atmospheric $CO_2$ comes from analyses of atmospheric concentrations of various carbon isotopes (see chapter 18). One of the most useful carbon isotopes for determining the contribution of fossil fuels to atmospheric $CO_2$ is radioactive $^{14}C$. Because $^{14}C$ has a half-life of 5,730 years, fossil fuels, which have been buried for millions of years, contain very little of this carbon isotope. Consequently, burning fossil fuel adds $CO_2$ to an atmosphere that has little $^{14}C$. If fossil fuel additions are a major source of increased atmospheric $CO_2$, then the relative concentration of $^{14}C$ in the atmosphere should be declining.

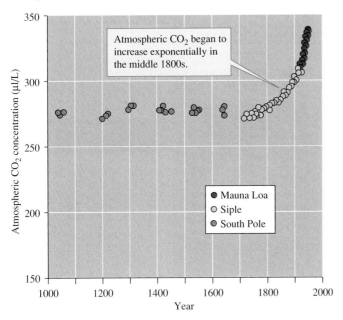

**Figure 23.22** A 1,000-year atmospheric $CO_2$ record (data from Post et al. 1990).

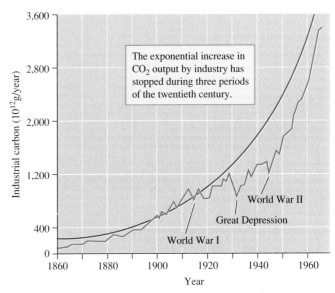

**Figure 23.23** Deviations from recent exponential increases in fossil fuel burning (data from Bacastow and Keeling 1974).

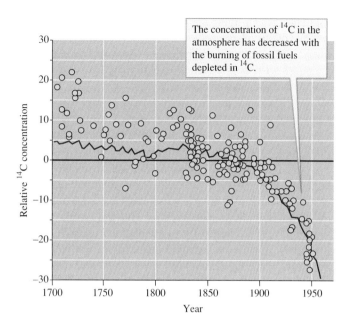

**Figure 23.24** The Suess effect (data from Bacastow and Keeling 1974).

A recent decline in atmospheric $^{14}C$ was first described by Hans Suess (1955), a scientist with the U.S. Geological Survey. Suess made his discovery by analyzing the $^{14}C$ content of wood. He analyzed the $^{14}C$ content of wood laid down by single trees at various times during their growth. He found that annual growth rings laid down in the late 1800s had significantly higher concentrations of $^{14}C$ than those laid down in the 1950s. Suess proposed that the $^{14}C$ content in wood was being progressively reduced because burning of fossil fuels was reducing the atmospheric concentration of $^{14}C$. Because of his pioneering work, reduced atmospheric $^{14}C$ as a consequence of fossil fuel burning is called the **Suess effect.**

Robert Bacastow and Charles Keeling (1974) compiled $^{14}C$ data from several studies of $^{14}C$ in trees and plotted the date when the wood was formed against the relative $^{14}C$ content of the wood. As figure 23.24 shows, the concentration of $^{14}C$ was fairly stable from A.D. 1700 until about 1850. After 1850, $^{14}C$ concentrations in wood declined significantly. The line shows the predictions of $^{14}C$ made by a model built by Bacastow and Keeling. Their model made these predictions based upon global patterns of fossil fuel burning and estimated rates of exchange of carbon between the ocean, the earth's biota, and the atmosphere.

What can we conclude from this evidence? Several things are clear. First, the concentration of $CO_2$ in the atmosphere has varied widely during the last 160,000 years and closely parallels variation in global temperatures. High levels of atmospheric $CO_2$ have corresponded to higher global temperatures. Second, the atmospheric concentration of $CO_2$ has increased substantially in the past two centuries. This modern increase has exceeded all levels reached during the past 160,000 years. Third, there is little doubt that the present levels of $CO_2$ in the atmosphere are strongly influenced by the burning of fossil fuels.

# Depletion and Recovery of the Ozone Layer

In 1985, scientists of the British Antarctic Survey had discovered a major reduction in the amount of ozone, $O_3$, in the stratosphere over the Antarctic. Stratospheric ozone absorbs potentially harmful ultraviolet light, particularly UV-B light, or radiation. Because high-energy, UV-B radiation is capable of destroying biological molecules and damaging living tissue, the ozone layer is critical for the well-being of life on earth. The British scientific team also analyzed historical measurements of ozone, which demonstrated clearly that the total amount of ozone over the Antarctic had been declining since the 1970s. Depletion of earth's ozone layer was not the first sign of human influence on the environment. However, it was a clear and dramatic indication that human impact on the environment had achieved truly global proportions.

Though the ozone hole was centered over the Antarctic far from most human population centers, its discovery generated widespread concern. Perhaps the greatest fear was that the breakdown of the ozone layer over the Antarctic might be a prelude to breakdown of the protective ozone layer over the entire earth, endangering humans as well as crops, wild plants, and animals. Other scientists had warned that the ozone layer was threatened by human activities. However, it was the discovery of the ozone hole that aroused world concern and stimulated international action. Attention was quickly focused on stopping the production of chlorofluorocarbons, or CFCs, organic chemicals containing carbon, chlorine, and fluorine that were widely used as refrigerants. Because CFCs are very stable molecules, their concentrations in the atmosphere gradually increased after their introduction in the 1930s. By the 1970s, the concentrations of CFCs had been increased sufficiently that they could be detected everywhere.

Chlorofluorocarbon molecules circulate in the lower atmosphere long enough to eventually move into the stratosphere, where they are exposed to a great deal more highly energetic ultraviolet light. As CFCs break down, they release chlorine, which can act as a catalyst to destroy ozone molecules. A single chlorine atom released in the stratosphere can continue to destroy ozone molecules until it is removed by some atmospheric process. Therefore, a small amount of chlorine released in the stratosphere can deplete the ozone layer substantially.

World concern over the dangers associated with ozone depletion prompted the 1987 Montreal Protocol on Substances that Deplete the Ozone Layer, which has been signed by 180 countries. The goal of the Montreal Protocol is to reduce and eventually eliminate emissions of human-generated substances that deplete ozone and may represent a model for international cooperation on a complex environmental problem. As a result of the protocol, global production of CFCs has been reduced from over one million tons annually to less than 50,000 tons in 2003.

How has the ozone hole over the Antarctic changed since its discovery in 1985? It has continued to grow larger, reaching its maximum area, so far, in 2000, when it covered nearly 29.2

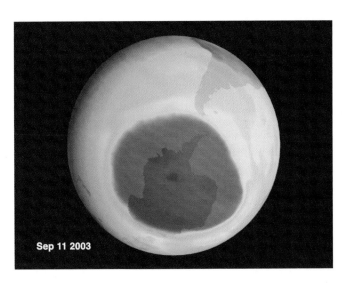

**Figure 23.25** The ozone hole in September, 2003. Even as scientists verified that the rate of ozone depletion was declining, the second largest ozone hole ever formed over Antarctica.

million km$^2$. In 2002, the Antarctic ozone hole closed quickly, suggesting that the ozone layer was recovering. However, the second largest ozone hole was recorded in 2003, when it reached 28.2 million km$^2$ (fig. 23.25). Encouragingly, the year 2003 also saw the first reported evidence that the ozone layer is recovering, when several scientists reported (Newchurch et al. 2003) evidence for a slowdown in stratospheric ozone loss from 1997 to 2003. It appears that cooperation by the international community in the banning of CFCs has begun to reverse the process of ozone depletion. Recovery of the stratospheric ozone layer will likely take at least another half century. However, the news of ozone recovery says clearly that we not only have the capacity to seriously damage the biosphere but where we have the will, we can also act to restore it.

## The Future

How will human-induced changes in atmospheric composition affect ecological systems? Unfortunately, we cannot say for sure and find ourselves in about the same position as Roger Revelle and Hans Suess, who wrote in 1957: "Human beings are now carrying out a large scale geophysical experiment of a kind that could not have happened in the past nor be reproduced in the future. Within a few centuries we are returning to the atmosphere and oceans the concentrated organic carbon stored in sedimentary rocks over hundreds of millions of years. This experiment, if adequately studied and documented, may yield a far-reaching insight into the processes determining weather and climate."

We, just as Revelle and Suess, do not know what the results of the "experiment" will be. They suggested that scientists vigorously study global dynamics of $CO_2$ in order to understand the influence of atmospheric $CO_2$ on weather and climate. Ecologists need to study ecological responses to the

rapid change in global climate that may occur and also how increased atmospheric $CO_2$ directly affects ecological systems from populations through landscapes. In the next section, we discuss one promising approach to studying ecological responses to global change.

## APPLICATIONS & TOOLS

### Cooperative Research Networks for Global Ecology

If global climate changes rapidly, how will biological systems respond? How will such change affect individual organisms, populations, communities, ecosystems, or even whole biomes? While it is easy to ask such questions and to recognize their significance, providing adequate answers is not so easy. Responses to past climatic changes provide some clues but only for a limited number of species and biological processes.

The present possibility of rapid climate change poses a substantial challenge to the scientific community and requires the application of new tools and approaches. Some new tools, such as fractal geometry (see chapter 21), involve new conceptual or analytical methods. Other new tools will consist of high technology, such as remote sensing satellites (see chapter 22) and supercomputers to run complicated global models. New devices, often employing the most recent technological developments, are becoming more and more common in the tool kit of ecologists. However, as impressive as this hardware may be, some of the most important developments required for global-scale ecological research may involve changes in the "culture" of science.

To effectively address the complex problems presented by global change, scientists must appreciate a variety of scientific disciplines and be capable of working effectively within multidisciplinary teams. The large scale of global change also requires that scientists work as cooperative, international teams. Such teams can conduct studies at spatial and temporal scales impossible for an individual researcher.

The U.S. Long-Term Ecological Research (LTER) network fosters large-scale ecological research. One of the central purposes of this research network is to foster cooperative, interdisciplinary research over large geographic areas. These sites include tropical forests, arctic tundra, temperate forests, grasslands, coastal ecosystems, deserts, and two cities. They extend from sea level to elevations over 4,000 m and from the Arctic to the Antarctic (fig. 23.26). This network can be used for many purposes, but, increasingly, it is being used to assess the impact that climate change may have on biological systems. Even this ambitious national effort, however, is inadequate for the task at hand. Therefore, the LTER network is forming long-term partnerships with research programs around the globe.

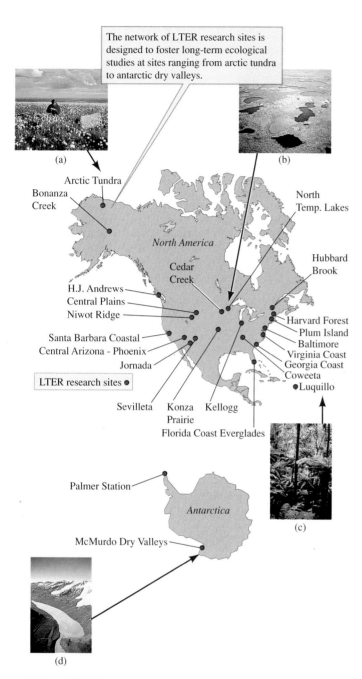

**Figure 23.26** The U.S. Long-Term Ecological Research (LTER) network.

ern, central, and eastern Europe, eastern Asia, Australia, and New Zealand. Several well-developed national research programs already exist in each of these areas, and many other programs are organizing quickly. The concrete goals of the meeting were to initiate exchanges of scientists and data and to foster global-scale comparisons and modeling.

The meeting produced several recommendations:

1. Foster worldwide communication and access to information among scientists engaged in long-term ecological research.

2. Develop a global directory of long-term ecological research sites.

3. Encourage development of additional long-term ecological research sites worldwide.

4. Develop appropriate standardized sampling and study designs, paying particular attention to conducting research at scales appropriate to the question being addressed.

These recommendations capture what this group of scientists regarded as minimal requirements for global ecological research. By 2003, 25 countries were members of the ILTER and 19 more were in the process of developing networks of their own, with the prospect of ILTER membership.

A network of research sites can yield much more information about large-scale and long-term phenomena than can any single site. Arranging research sites into an organized array allows for synchronized measurements and provides the opportunity for observing how the ecological effects of climate change may propagate across regions or even across the entire globe. Such networks would be capable of identifying the geographic extent of a climatic event, when it began, when it ended, and the types of ecological responses that occurred in various regions. Such a network could also identify which areas are most responsive to climatic change and which are least responsive. None of this information could be determined from studies conducted at a single site.

Modern developments such as remote sensing, super-computers, and the global network of computer communications will make an international long-term ecological network more effective. As Peter Vitousek pointed out, this is the first generation to have the tools to study ecology at the scale of the entire globe. These international networks of scientists working cooperatively on global ecological problems signal a change in the culture of science that emphasizes information sharing and an open, multidisciplinary team approach to research. These are the advancements that can make global ecological studies a reality. Ecologists increasingly recognize the urgency of the challenge posed by global change and that addressing ecological problems at a global scale requires an approach to research that is itself global.

The first international long-term ecological research (ILTER) workshop was held in 1993 (Nottrott, Franklin, and Vande Castle 1994). Scientists from around the world met to establish active interactions and collaboration between the LTER network and long-term ecological research programs from around the globe. The geographic areas represented included North America, Central and South America, west-

Chapter 23 focuses on global-scale processes and phenomena, including large-scale weather systems and global change induced by humans. We are the only species that exerts global-scale influences on the environment.

The earth is wrapped in an atmospheric envelope that makes the biosphere a hospitable place for life as we know it. The earth's atmosphere reduces the amount of ultraviolet light reaching the surface. The atmosphere also helps to keep the surface of the earth warm through the *greenhouse effect*. The surface of the earth is kept warmer than it would be by the greenhouse gases, including water vapor, methane, ozone, nitrous oxide, chlorofluorocarbons, and carbon dioxide.

**The El Niño Southern Oscillation is a large-scale atmospheric and oceanic phenomenon that influences ecological systems on a global scale.** The El Niño Southern Oscillation is a highly dynamic, large-scale weather system that involves variation in sea surface temperature and barometric pressure across the Pacific and Indian Oceans. During the mature phase of an El Niño, the sea surface in the eastern tropical Pacific Ocean is much warmer than average and the barometric pressure over the eastern Pacific is lower than average. El Niño brings increased precipitation to much of North and parts of South America and drought to the western Pacific. Periods of lower sea surface temperature and higher than average barometric pressures in the eastern tropical Pacific have been named La Niñas. La Niña brings drought to much of North and South America and higher than average precipitation to the western Pacific. The variation in weather caused by the El Niño Southern Oscillation has dramatic effects on marine and terrestrial populations around the world.

**Human activity has greatly increased the quantity of fixed nitrogen cycling through the biosphere.** For millions of years, the only organisms that could fix nitrogen were nitrogen-fixing bacteria and some actinomycete fungi. The total amount of nitrogen fixed by these historical sources is approximately 130 Tg N per year. The nitrogen now fixed as a consequence of human activity is about 135 to 145 Tg N per year, more than all nonhuman sources of fixed nitrogen combined. Large-scale nitrogen enrichment may threaten biological diversity by creating environmental conditions favorable to some species at the expense of others.

**Rapid changes in global patterns of land use threaten biological diversity.** Human activities, mainly agriculture and urbanization, have significantly altered one-third to one-half of the ice-free land surface of the earth. A widely cited example of land cover change is tropical deforestation. From 1978 to 1988, the rate of deforestation in the Amazon Basin of Brazil averaged about 15,000 $km^2$ per year. By 1988, the total area deforested within the Amazon Basin was 230,000 $km^2$. By adding in edge effects and the effects of isolation, the area of Amazonian forest affected by deforestation increases from 230,000 $km^2$ to 588,000 $km^2$. The global rate of tropical deforestation from 1978 to 1988 was about 30,000 $km^2$ per year. Massive deforestation has also occurred outside of the tropics. Because of the negative effect of reduced habitat area on diversity, these massive land conversions present a major threat to global biological diversity.

**Human activity is increasing the atmospheric concentration of $CO_2$, which may be increasing global temperatures.** Analyses of air trapped in ice shows that the concentration of $CO_2$ in the atmosphere has varied widely during the last 160,000 years and closely parallels variation in global temperatures. High levels of atmospheric $CO_2$ have corresponded to higher global temperatures. The buildup of atmospheric $CO_2$ during the past two centuries has reached levels of atmospheric $CO_2$ not equaled in the past 160,000 years. There is little doubt that the present level of $CO_2$ in the atmosphere is strongly influenced by burning of fossil fuels. Increases in atmospheric $CO_2$ concentration are likely to affect global climate and the structure and processes of ecological systems from populations through landscapes.

Cooperative research networks aid global ecology. The present possibility of rapid climate change poses a substantial challenge to the scientific community. Studying ecology at a global scale requires that scientists develop new tools and approaches. New devices, often employing the most recent technological developments, are becoming more and more common in the tool kit of ecologists. However, some of the most important developments required for global-scale research may involve changes in the "culture" of science. The complexity and large scale of global change requires that scientists work in multidisciplinary, national, and international teams. International networks of scientists now work on global-scale ecological problems in a research environment that emphasizes information sharing and a team approach to research.

# Review Questions

1. Ecologists are now challenged to study global ecology. The apparent role played by humans in changing the global environment makes it imperative that we understand the workings of the earth as a global system. However, this study requires approaches that are significantly different from those that can be applied to traditional areas of ecological study. Historically, much of ecology focused on small areas and short-term studies. What are some of the main differences between global ecology and, for instance, the study of interspecific competition (see chapter 13) or forest succession (see chapter 20)? How will these differences affect the design of studies at the global scale?

2. Geologists, atmospheric scientists, and oceanographers have been conducting global-scale studies for some time. What role will information from these disciplines play in the study of global ecology? Why will global ecological studies generally be pursued by interdisciplinary teams? How can ecologists play a useful role in global studies?

3. What changes in sea surface temperatures and atmospheric pressures over the Pacific Ocean accompany El Niño? What physical changes accompany La Niña? How do El Niño and La Niña affect precipitation in North America, South America, and Australia?

4. Review evidence that the El Niño Southern Oscillation significantly influences populations around the globe. Much of our discussion in chapter 23 focused on the effects of the El Niño Southern Oscillation on populations. Considering our discussions in chapters 18 and 19 of physical controls on rates of terrestrial primary production and decomposition, how does the El Niño Southern Oscillation likely affect these ecosystem processes in Australia or the American Southwest? How would you test your ideas?

5. In chapter 23, we briefly discussed how humans have more than doubled the quantity of fixed nitrogen cycling through the biosphere. In chapter 15 we reviewed studies by Nancy Johnson (1993) on the effects of fertilization on the mutualistic relationship between mycorrhizal fungi and grasses. The increases in fixed nitrogen cycling through the biosphere, particularly that portion deposited by rain, are analogous to a global-scale fertilization experiment. Reasoning from the results of Johnson's study, how should increased fixed nitrogen supplies affect the relationship between mycorrhizal fungi and their plant partners? How would you test your ideas?

6. As we saw in chapters 18 and 19, nitrogen availability seems to control the rates of several ecosystem processes. How should nitrogen enrichment affect rates of primary production and decomposition in terrestrial, freshwater, and marine environments? How could you test your ideas? What role might geographic comparisons play in your studies?

7. Ecologists predict that global diversity is threatened by land use change and by the reductions in habitat area and the fragmentation that accompany land use change. Vitousek (1994) suggested that land use change may be the greatest current threat to biological diversity (see fig. 23.3). What role do studies of diversity on islands and species area relationships on continents (see chapter 22) play in these predictions?

8. Skole and Tucker (1993) documented the rate and extent of recent deforestation in the Amazon Basin in Brazil. This is a prominent example of the land cover changes that likely threaten biological diversity. However, as we saw in chapter 16, Bush, Piperno, and Colinvaux (1989) documented agricultural activity in New World tropical forests beginning at least 6,000 years ago. What makes present-day deforestation in the Amazon Basin different from this historical activity? What does the long history of agriculture in the Amazon Basin suggest about the potential for coexistence of agriculture and biological diversity there?

9. Review the long-term atmospheric $CO_2$ record as revealed by studies of air trapped in ice cores. What is the evidence that burning of fossil fuels is responsible for recent increases in atmospheric $CO_2$ concentrations?

10. What evidence is there that variation in atmospheric $CO_2$ concentration is linked to variation in global temperatures? In recent years the governments of most countries of the world have been working hard to develop international agreements to regulate $CO_2$ emissions. Why are these governments concerned? How might rapid changes in global temperatures lead to the extinction of large numbers of species? How might changes in global temperatures affect agriculture around the world?

# Suggested Readings

Bierregaard, R. O., Jr., T. E. Lovejoy, V. Kapos, A. A. dos Santos, and R. W. Hutchings. 1992. The biological dynamics of tropical rainforest fragments. *BioScience* 42:859–66.

Malcolm, J. R. 1994. Edge effects in central Amazonian forest fragments. *Ecology* 75:2438–45.

*These two papers give a solid introduction to the ecological problem of forest fragmentation in the Amazon Basin.*

Cairns, S. C. and G. C. Grigg. 1993. Population dynamics of red kangaroos (*Macropus rufus*) in relation to rainfall in the south Australian pastoral zone. *Journal of Applied Ecology* 30:444–58.

Glynn, P. W. 1988. El Niño-Southern Oscillation 1982–1983: nearshore population, community, and ecosystem responses. *Annual Review of Ecology and Systematics* 19:309–45.

Wurtsbaugh, W. A. 1992. Food-web modification by an invertebrate predator in the Great Salt Lake (USA). *Oecologia* 89:168–75.

*Dramatic demonstrations of the effects of El Niño on a wide range of aquatic and terrestrial populations and ecosystems. The wide distribution of the affected systems and the magnitude of their responses underscore the ecological significance of this large-scale climate system.*

Clark, D. A., S. C. Piper, C. D. Keeling, and D. B. Clark. 2003. Tropical rain forest tree growth and atmospheric carbon dynamics linked to interannual temperature variation during 1984–2000. *Proceedings of the National Academy of Sciences of the United States of America* 100:5852–57.

*Elevated temperatures over a 16-year period have been correlated with reduced growth by tropical rain forest trees at La*

*Selva, Costa Rica. Thus, global warming could reduce the rate of carbon uptake by tropical forests, which would increase the rate of $CO_2$ accumulation in the atmosphere.*

Diaz, H. F. and V. Markgraf. 1992. *El Niño Historical and Paleoclimatic Aspects of the Southern Oscillation.* Cambridge, England: Cambridge University Press.

Philander, S. G. 1990. *El Niño, La Niña, and the Southern Oscillation.* New York: Academic Press.

*Two very good introductions to the El Niño Southern Oscillation and its effects on global climate.*

Franklin, J. F., C. S. Bledsoe, and J. T. Callahan. 1990. Contributions of the long-term ecological research program. *BioScience* 40:509–23.

Magnuson, J. J. 1990. Long-term ecological research and the invisible present. *BioScience* 40:495–501.

Swanson, F. J. and R. E. Sparks. 1990. Long-term ecological research and the invisible place. *BioScience* 40:502–8.

*These three papers in the same issue of* BioScience *describe the organization and purpose of the U.S. LTER network and the philosophical underpinnings of its founding.*

Johnson, D., C. D. Campbell, J. A. Lee, T. V. Callaghan, and D. Gwynn-Jones. 2002. Arctic microorganisms respond more to elevated UV-B radiation than $CO_2$. *Nature* 416:82–83.

*This study reveals unexpected effects of UV-B radiation on soil microorganisms in the Arctic. The recorded effects, in response to simulated ozone thinning, may reduce the capacity of Arctic ecosystems to function as carbon dioxide sinks.*

Keeling, C. D. and T. P. Whorf. 2003. Atmospheric $CO_2$ records from sites in the SIO air sampling network. In *Trends: A Compendium of Data on Global Change.* Carbon Dioxide Information Analysis Center, Oak Ridge National Laboratory, U.S. Department of Energy, Oak Ridge, TN, U.S.A. (available online).

*A concise synopsis of the record of carbon dioxide rise determined from Mauna Loa, Hawaii. The online site includes the monthly data record from 1958 to 2002.*

Kinzig, A. P. and R. H. Socolow. 1994. Human impacts on the nitrogen cycle. *Physics Today* 47(11):24–31.

*Kinzig and Socolow provide a thorough and readable outline of how human populations affect the global nitrogen cycle.*

Morrison, S. A. and D. T. Bolger. 2002. Variation in a sparrow's reproductive success with rainfall: food and predator-mediated processes. *Oecologia* 133:315–24.

*The authors contrast reproductive output by rufous-crowned sparrows during average, El Niño, and La Niña years. Their work reveals indirect effects mediated through the effect of weather on predator activity.*

Newchurch, M. J., E.-S. Yang, D. M. Cunnold, G. C. Reinsel, J. M. Zawodny, and J. M. Russell III. 2003. Evidence for slowdown in stratospheric ozone loss: first stage of ozone recovery. *Journal of Geophysical Research* 108(D16), 4507, doi:10.1029/2003JD003471, 2003 (published online).

*Researchers provide the first evidence for a slowdown in the rate of depletion of earth's ozone layer.*

Schneider, S. H. 1989. The changing climate. *Scientific American* 261:70–79.

*This paper is a first-rate introduction to the subject of global climate change. It is an excellent preparation for more technical papers in this field of study.*

Vitousek, P. M. 1994. Beyond global warming: ecology and global change. *Ecology* 75:1861–76.

*Vitousek's insightful paper provides a broad perspective from which to consider the phenomenon of global change—a must-read for anyone interested in global-scale environmental problems.*

# On the Net

Visit this textbook's accompanying website at www.mhhe.com/ecology (click on the book's title) to take advantage of practice quizzing, study/writing tips, timely news articles, and additional URLs for research on the topics in this chapter.

Global Ecology

Field Methods for Studies of Ecosystems

Atmosphere, Climate, and Weather

El Niño

La Niña

Global Warming

Nutrient Cycling

Tropical Forests and Extinction

Ecosystem Management

# Appendix A

# Abbreviations Used in This Text

| | |
|---|---|
| °C | degrees Centigrade |
| $^0/_{00}$ | grams of salt per kilogram of water |
| λ | geometric rate of increase |
| kPa | kilopascal |
| MPa | megapascal |
| N | newton |
| Pa | pascal |
| | |
| cm | centimeter |
| $cm^2$ | square centimeter |
| $cm^3$ | cubic centimeter |
| | |
| g | gram |
| kg | kilogram |
| mg | milligram |
| mg/L | milligrams per liter |
| μg/L | micrograms per liter |
| Tg | terragrams |
| | |
| ha | hectare |
| | |
| km | kilometer |
| $km^2$ | square kilometer |
| $km^3$ | cubic kilometer |
| | |
| m | meter |
| $m^2$ | square meter |
| $m^3$ | cubic meter |
| μm | micrometer |
| mm | millimeter |
| nm | nanometer |
| | |
| μmol | micromole |
| | |
| L | liter |
| ml | milliliter |
| μl | microliter |
| | |
| $Ca^{2+}$ | calcium ions |
| $Cl^-$ | chloride ion |
| Fe | iron |
| K | potassium |
| Mg | magnesium |
| $Mg^{2+}$ | magnesium ions |
| N | nitrogen |
| Na | sodium |

| | |
|---|---|
| $Na^+$ | sodium ion |
| P | phosphorus |
| | |
| AET | actual evapotranspiration |
| PAR | photosynthetically active radiation |
| | |
| °S | degrees south |
| °N | degrees north |
| | |
| CPOM | coarse particulate organic matter |
| FPOM | fine particulate organic matter |
| PDSI | Palmer Drought Severity Index |
| | |
| MEI | metabolizable energy intake |
| | |
| $\bar{X}$ | sample mean |
| | |
| n | sample size |
| | |
| $\Sigma X$ | sum of measurements or observations |
| | |
| $\Psi$ | water potential of a solution |
| $\Psi_{solutes}$ | reduction in water potential due to dissolved substances |
| $\Psi_{matric}$ | reduction in water potential due to matric forces within plant cells |
| $\Psi_{pressure}$ | reduction in water potential due to negative pressure created by water evaporating from leaves |
| $\Psi_{soil}$ | water potential of soil |
| $\Psi_{plant}$ | water within plant cells |
| | |
| $\delta X$ | ± the relative concentration of the heavier isotope, for example: D, $^{13}C$, $^{15}N$, or $^{34}S$ in $^0/_{00}$ |
| | |
| $H_s$ | total heat stored in the body of an organism |
| $H_m$ | heat gained from metabolism |
| $H_{cd}$ | heat gained or lost through conduction |
| $H_{cv}$ | heat lost or gained by convection |
| $H_r$ | heat gained or lost through electromagnetic radiation |
| $H_e$ | heat lost through evaporation |
| | |
| $MgH_2O/L$ | milligrams of water per liter |
| $MgH_2O/m^3$ | milligrams of water per cubic meter |
| | |
| $gH_2O/m^3$ | grams of water per cubic meter |

| | |
|---|---|
| $W_d$ | water taken by drinking |
| $W_f$ | water taken in with food |
| $W_a$ | water absorbed from the air |
| $W_e$ | water lost by evaporation |
| $W_s$ | water lost with various secretions and excretions |
| $W_r$ | water taken from soil by roots |
| $W_t$ | water lost by transpiration |
| $W_i$ | internal water |
| | |
| PEP | phosphoenolpyruvate |
| PGA | phosphoglyceric acid |
| RuBP | ribulose bisphosphate |
| | |
| $NH_4^+$ | ammonium |
| $NO_2^-$ | nitrite |
| $NO_3^-$ | nitrate |
| Fe | elemental iron |
| $Fe^{2+}$ | ferrous iron |
| CO | carbon monoxide |
| | |
| $P_{max}$ | maximum rate of photosynthesis |
| $I_{sat}$ | irradiance required to saturate photosynthesis |
| | |
| $N_{e1}$ | number of prey 1 encountered per unit of time |
| $E_1$ | energy gained by feeding on an individual prey 1 minus the costs of handling |
| $C_s$ | cost of searching for the prey |
| $H_1$ | time required for "handling" an individual of prey 1 |
| | |
| $S^2$ | sample variance |
| | |
| RAPD | randomly amplified polymorphic DNA |
| | |
| $h^2$ | heritability of a trait |
| $V_G$ | genetic variance |
| $V_P$ | phenotypic variance |
| $V_E$ | variance in phenotype due to environmental effects on the phenotype |
| $V_T$ | total variation |
| $V_{GE}$ | variation due to gene-by-environment interactions |
| $V_R$ | unexplained residual variation |
| | |
| X | independent variable |
| Y | dependent variable |
| | |
| a | Y intercept |
| b | regression coefficient; slope of the line |
| | |
| PCR | polymerase chain reaction |
| | |
| $R_0$ | net reproductive rate |
| T | generation time |
| r | per capita rate of increase |

| | |
|---|---|
| $I_{max}$ | maximum per capita rate of increase; intrinsic rate of increase |
| N | population size |
| $N_t$ | number of individuals at time $t$ |
| $N_0$ | initial number of individuals |
| e | base of natural logarithms |
| | |
| $\chi^2$ | chi-square |
| $O$ | observed frequency of a particular phenotype |
| $E$ | expected frequency |
| K | carrying capacity |
| | |
| GSI | gonadosomatic index |
| | |
| $N_1$ and $N_2$ | population sizes of species 1 and 2 |
| $K_1$ and $K_2$ | carrying capacities of species 1 and 2 |
| $r_{max1}$ and $r_{max2}$ | intrinsic rates of increase for species 1 and 2 |
| | |
| $s_{\bar{x}}$ | standard error |
| | |
| s | sample standard deviation |
| | |
| $N_h$ | number of hosts |
| $N_p$ | number of parasites or predators |
| | |
| $\alpha$ | significance level |
| $\mu$ | true population mean |
| $t$ | value of the statistic $t$, either calculated or determined from a Student's $t$ table |
| | |
| H' | the value of the Shannon-Wiener diversity index |
| $p_i$ | the proportion of the $i$th species |
| $\log_e$ | the natural logarithm |
| | |
| kcal | kilocalories |
| | |
| $R_{sample}$ | the isotopic ratio in the sample, for example, $^{13}C:^{12}C$ or $^{15}N:^{14}N$ |
| $R_{standard}$ | the isotopic ratio in the standard, for example, $^{13}C:^{12}C$ or $^{15}N:^{14}N$ |
| | |
| $m_t$ | mass of leaves at time $t$ |
| $m_0$ | initial mass of leaves |
| $e$ | base of the natural logarithms |
| k | daily rate of mass loss |
| | |
| U | Mann-Whitney statistic |
| | |
| CFC | chlorofluorocarbons |

# Appendix B

# List of Chapter Concepts

**1  Introduction: What Is Ecology?**

**2  Life on Land**

- Uneven heating of the earth's spherical surface by the sun and the tilt of the earth on its axis combine to produce predictable latitudinal variation in climate.
- Soil structure results from the long-term interaction of climate, organisms, topography, and parent mineral material.
- The geographic distribution of terrestrial biomes corresponds closely to variation in climate, especially prevailing temperature and precipitation.

**3  Life in Water**

- The hydrologic cycle exchanges water among reservoirs.
- The biology of aquatic environments corresponds broadly to variations in physical factors such as light, temperature, and water movements and to chemical factors such as salinity and oxygen.

**4  Temperature Relations**

- Macroclimate interacts with the local landscape to produce microclimatic variation in temperature.
- Most species perform best in a fairly narrow range of temperatures.
- Many organisms have evolved ways to compensate for variations in environmental temperature by regulating body temperature.
- Many organisms survive extreme temperatures by entering a resting stage.

**5  Water Relations**

- The movement of water down concentration gradients in terrestrial and aquatic environments determines the availability of water to organisms.
- Terrestrial plants and animals regulate their internal water by balancing water acquisition against water loss.
- Marine and freshwater organisms use complementary mechanisms for water and salt regulation.

**6  Energy and Nutrient Relations**

- Organisms use one of three main sources of energy: light, organic molecules, or inorganic molecules.
- The rate at which organisms can take in energy is limited.
- Optimal foraging theory attempts to model how organisms feed as an optimizing process.

**7  Social Relations**

- Mate choice by one sex and/or competition for mates among individuals of the same sex can result in selection for particular traits in individuals, a process called sexual selection.

- The evolution of sociality is generally accompanied by cooperative feeding, defense of the social group, and restricted reproductive opportunities.

**8  Population Genetics and Natural Selection**

- Phenotypic variation among individuals in a population results from the combined effects of genes and environment.
- The Hardy-Weinberg equilibrium model helps identify evolutionary forces that can change gene frequencies in populations.
- Natural selection is the result of differences in survival and reproduction among phenotypes.
- The extent to which phenotypic variation is due to genetic variation determines the potential for evolution by natural selection.
- Random processes, such as genetic drift, can change gene frequencies in populations, especially in small populations.

**9  Population Distribution and Abundance**

- The physical environment limits the geographic distribution of species.
- On small scales, individuals within populations are distributed in patterns that may be random, regular, or clumped; on larger scales, individuals within a population are clumped.
- Many populations are subdivided into subpopulations called metapopulations.
- Population density declines with increasing organism size.
- Commonness and rarity of species are influenced by populations size, geographic range, and habitat tolerance.

**10  Population Dynamics**

- A survivorship curve summarizes the pattern of survival in a population.
- The age distribution of a population reflects its history of survival, reproduction, and potential for future growth.
- A life table combined with a fecundity schedule can be used to estimate net reproductive rate ($R_0$), geometric rate of increase, ($\lambda$), generation time (T), and per capita rate of increase ($r$).
- Dispersal can increase or decrease local population densities.

**11  Population Growth**

- In the presence of abundant resources, populations can grow at geometric or exponential rates.
- As resources are depleted, population growth rate slows and eventually stops; this is known as logistic population growth.
- The environment limits population growth by changing birth and death rates.
- On average, small organisms have higher rates of per capita increase, $r_{max}$, and more variable populations, while large organisms have lower rates of per capita increase and less variable populations.

## 12 Life Histories

- Because all organisms have access to limited energy and other resources, there is a trade-off between the number and size of offspring; those that produce larger offspring are constrained to produce fewer, while those that produce smaller offspring may produce larger numbers.
- Where adult survival is lower, organisms begin reproducing at an earlier age and invest a greater proportion of their energy budget into reproduction; where adult survival is higher, organisms defer reproduction to a later age and allocate a smaller proportion of their resources to reproduction.
- The great diversity of life histories may be classified on the basis of a few population characteristics. Examples include fecundity or number of offspring, $m_x$, survival, $l_x$, relative offspring size, and age at reproductive maturity, $\alpha$.

## 13 Competition

- Studies of intraspecific competition provide evidence for resource limitation.
- The niche reflects the environmental requirements of species.
- Mathematical and laboratory models provide a theoretical foundation for studying competitive interactions in nature.
- Competition can have significant ecological and evolutionary influences on the niches of species.

## 14 Exploitation: Predation, Herbivory, Parasitism, and Disease

- Exploitation weaves populations into a web of relationships that defy easy generalization.
- Predators, parasites, and pathogens influence the distribution, abundance, and structure of prey and host populations.
- Predator-prey, host-parasite, and host-pathogen relationships are dynamic.
- To persist in the face of exploitation, hosts and prey need refuges.

## 15 Mutualism

- Plants benefit from mutualistic partnerships with a wide variety of bacteria, animals, and fungi.
- Reef-building corals depend upon mutualistic relationships with algae and animals.
- Theory predicts that mutualism will evolve where the benefits of mutualism exceed the costs.

## 16 Species Abundance and Diversity

- Most species are moderately abundant; few are very abundant or extremely rare.
- A combination of the number of species and their relative abundance defines species diversity.
- Species diversity is higher in complex environments.
- Intermediate levels of disturbance promote higher diversity.

## 17 Species Interactions and Community Structure

- A food web summarizes the feeding relations in a community.
- The feeding activities of a few keystone species may control the structure of communities.
- Exotic predators can collapse and simplify the structure of food webs.
- Mutualists can act as keystone species.

## 18 Primary Production and Energy Flow

- Terrestrial primary production is generally limited by temperature and moisture.
- Aquatic primary production is generally limited by nutrient availability.
- Consumers can influence rates of primary production in aquatic and terrestrial ecosystems.
- Energy losses limit the number of trophic levels in ecosystems.

## 19 Nutrient Cycling and Retention

- Decomposition rate is influenced by temperature, moisture, and chemical composition of litter and the environment.
- Plants and animals can modify the distribution and cycling of nutrients in ecosystems.
- Disturbance increases nutrient loss from ecosystems.

## 20 Succession and Stability

- Community changes during succession include increases in species diversity and changes in species composition.
- Ecosystem changes during succession include increases in biomass, primary production, respiration, and nutrient retention.
- Mechanisms that drive ecological succession include facilitation, tolerance, and inhibition.
- Community stability may be due to lack of disturbance or community resistance or resilience in the face of disturbance.

## 21 Landscape Ecology

- Landscape structure includes the size, shape, composition, number, and position of different ecosystems within a landscape.
- Landscape structure influences processes such as the flow of energy, materials, and species between the ecosystems within a landscape.
- Landscapes are structured and change in response to geological processes, climate, activities of organisms, and fire.

## 22 Geographic Ecology

- On islands and habitat patches on continents, species richness increases with area and decreases with isolation.
- Species richness on islands can be modeled as a dynamic balance between immigration and extinction of species.
- Species richness generally increases from middle and high latitudes to the equator.
- Long-term historical and regional processes significantly influence the structure of biotas and ecosystems.

## 23 Global Ecology

- The El Niño Southern Oscillation is a large-scale atmospheric and oceanic phenomenon that influences ecological systems on a global scale.
- Human activity has greatly increased the quantity of fixed nitrogen cycling through the biosphere.
- Rapid changes in global patterns of land use threaten biological diversity.
- Human activity is increasing the atmospheric concentration of $CO_2$, which may be increasing global temperatures.

# Appendix C

# Statistical Tables

## Table C.1

Critical Values of Student's *t*

| DF | α = 0.10 | α = 0.05 | α = 0.02 | α = 0.01 | DF | α = 0.10 | α = 0.05 | α = 0.02 | α = 0.01 |
|---|---|---|---|---|---|---|---|---|---|
| 1 | 6.31 | 12.71 | 31.82 | 63.66 | 22 | 1.72 | 2.07 | 2.51 | 2.82 |
| 2 | 2.92 | 4.31 | 6.96 | 9.92 | 24 | 1.71 | 2.06 | 2.49 | 2.80 |
| 3 | 2.35 | 3.18 | 4.54 | 5.84 | 26 | 1.71 | 2.06 | 2.48 | 2.78 |
| 4 | 2.13 | 2.78 | 3.75 | 4.60 | 28 | 1.70 | 2.05 | 2.47 | 2.76 |
| 5 | 2.01 | 2.57 | 3.36 | 4.03 | 30 | 1.70 | 2.04 | 2.46 | 2.75 |
| 6 | 1.94 | 2.45 | 3.14 | 3.71 | 35 | 1.69 | 2.03 | 2.44 | 2.72 |
| 7 | 1.89 | 2.36 | 3.00 | 3.50 | 40 | 1.68 | 2.02 | 2.42 | 2.70 |
| 8 | 1.86 | 2.31 | 2.90 | 3.36 | 45 | 1.68 | 2.01 | 2.41 | 2.69 |
| 9 | 1.83 | 2.26 | 2.82 | 3.25 | 50 | 1.68 | 2.01 | 2.40 | 2.68 |
| 10 | 1.81 | 2.23 | 2.76 | 3.17 | 60 | 1.67 | 2.00 | 2.39 | 2.66 |
| 11 | 1.80 | 2.20 | 2.72 | 3.11 | 70 | 1.67 | 1.99 | 2.38 | 2.65 |
| 12 | 1.78 | 2.18 | 2.68 | 3.06 | 80 | 1.66 | 1.99 | 2.37 | 2.64 |
| 13 | 1.77 | 2.16 | 2.65 | 3.01 | 90 | 1.66 | 1.99 | 2.37 | 2.63 |
| 14 | 1.76 | 2.14 | 2.62 | 3.00 | 100 | 1.66 | 1.98 | 2.36 | 2.63 |
| 15 | 1.75 | 2.13 | 2.60 | 2.95 | 120 | 1.66 | 1.98 | 2.36 | 2.62 |
| 16 | 1.75 | 2.12 | 2.58 | 2.92 | 150 | 1.66 | 1.98 | 2.35 | 2.61 |
| 17 | 1.74 | 2.11 | 2.57 | 2.90 | 200 | 1.65 | 1.97 | 2.35 | 2.61 |
| 18 | 1.73 | 2.10 | 2.55 | 2.88 | 300 | 1.65 | 1.97 | 2.34 | 2.59 |
| 19 | 1.73 | 2.09 | 2.54 | 2.86 | 500 | 1.65 | 1.96 | 2.33 | 2.59 |
| 20 | 1.72 | 2.09 | 2.53 | 2.85 | ∞ | 1.65 | 1.96 | 2.33 | 2.58 |

The above values were computed as described by Zar (1996: App 18). More extensive tables of Student's *t* are found in Rohlf and Sokal (1995:7) and Zar (1996: App 18–19).
From: Brower, Zar and von Ende. *Field and Laboratory Methods for General Ecology,* 4/e. Dubuque, Iowa: McGraw-Hill, 1990. 10.

# Table C.2

Critical Values of the Mann-Whitney Test Statistic

$\alpha = 0.10$

| $n_1$ | $n_2=2$ | 3 | 4 | 5 | 6 | 7 | 8 | 9 | 10 | 11 | 12 | 13 | 14 | 15 | 16 | 17 | 18 | 19 | 20 |
|---|---|---|---|---|---|---|---|---|---|---|---|---|---|---|---|---|---|---|---|
| 2 | | | | 10 | 12 | 14 | 15 | 17 | 19 | 21 | 22 | 24 | 25 | 27 | 29 | 31 | 32 | 34 | 36 |
| 3 | | 9 | 12 | 14 | 16 | 19 | 21 | 23 | 26 | 28 | 31 | 33 | 35 | 38 | 40 | 42 | 45 | 47 | 49 |
| 4 | | 12 | 15 | 18 | 21 | 24 | 27 | 30 | 33 | 36 | 39 | 42 | 45 | 48 | 50 | 53 | 56 | 59 | 62 |
| 5 | 10 | 14 | 18 | 21 | 25 | 29 | 32 | 36 | 39 | 43 | 47 | 50 | 54 | 57 | 61 | 65 | 68 | 72 | 75 |
| 6 | 12 | 16 | 21 | 25 | 29 | 34 | 38 | 42 | 46 | 50 | 55 | 59 | 63 | 67 | 71 | 76 | 80 | 84 | 88 |
| 7 | 14 | 19 | 24 | 29 | 34 | 38 | 43 | 48 | 53 | 58 | 63 | 67 | 72 | 77 | 82 | 86 | 91 | 96 | 101 |
| 8 | 15 | 21 | 27 | 32 | 38 | 43 | 49 | 54 | 60 | 65 | 70 | 76 | 81 | 87 | 92 | 97 | 103 | 108 | 113 |
| 9 | 17 | 23 | 30 | 36 | 42 | 48 | 54 | 60 | 66 | 72 | 78 | 84 | 90 | 96 | 102 | 108 | 114 | 120 | 126 |
| 10 | 19 | 26 | 33 | 39 | 46 | 53 | 60 | 66 | 73 | 79 | 86 | 93 | 99 | 106 | 112 | 119 | 125 | 132 | 138 |
| 11 | 21 | 28 | 36 | 43 | 50 | 58 | 65 | 72 | 79 | 87 | 94 | 101 | 108 | 115 | 122 | 130 | 137 | 144 | 151 |
| 12 | 22 | 31 | 39 | 47 | 55 | 63 | 70 | 78 | 86 | 94 | 102 | 109 | 117 | 125 | 132 | 140 | 148 | 156 | 163 |
| 13 | 24 | 33 | 42 | 50 | 59 | 67 | 76 | 84 | 93 | 101 | 109 | 118 | 126 | 134 | 143 | 151 | 159 | 167 | 176 |
| 14 | 25 | 35 | 45 | 54 | 63 | 72 | 81 | 90 | 99 | 108 | 117 | 126 | 135 | 144 | 153 | 161 | 170 | 179 | 188 |
| 15 | 27 | 38 | 48 | 57 | 67 | 77 | 87 | 96 | 106 | 115 | 125 | 134 | 144 | 153 | 163 | 172 | 182 | 191 | 200 |
| 16 | 29 | 40 | 50 | 61 | 71 | 82 | 92 | 102 | 112 | 122 | 132 | 143 | 153 | 163 | 173 | 183 | 193 | 203 | 213 |
| 17 | 31 | 42 | 53 | 65 | 76 | 86 | 97 | 108 | 119 | 130 | 140 | 151 | 161 | 172 | 183 | 193 | 204 | 214 | 225 |
| 18 | 32 | 45 | 56 | 68 | 80 | 91 | 103 | 114 | 125 | 137 | 148 | 159 | 170 | 182 | 193 | 204 | 215 | 226 | 237 |
| 19 | 34 | 47 | 59 | 72 | 84 | 96 | 108 | 120 | 132 | 144 | 156 | 167 | 179 | 191 | 203 | 214 | 226 | 238 | 250 |
| 20 | 36 | 49 | 62 | 75 | 88 | 101 | 113 | 126 | 138 | 151 | 163 | 176 | 188 | 200 | 213 | 225 | 237 | 250 | 262 |

$\alpha = 0.05$

| $n_1$ | $n_2=2$ | 3 | 4 | 5 | 6 | 7 | 8 | 9 | 10 | 11 | 12 | 13 | 14 | 15 | 16 | 17 | 18 | 19 | 20 |
|---|---|---|---|---|---|---|---|---|---|---|---|---|---|---|---|---|---|---|---|
| 2 | | | | | | | 16 | 18 | 20 | 22 | 23 | 25 | 27 | 29 | 31 | 32 | 34 | 36 | 38 |
| 3 | | | | 15 | 17 | 20 | 22 | 25 | 27 | 30 | 32 | 35 | 37 | 40 | 42 | 45 | 47 | 50 | 52 |
| 4 | | | 16 | 19 | 22 | 25 | 28 | 32 | 35 | 38 | 41 | 44 | 47 | 50 | 53 | 57 | 60 | 63 | 66 |
| 5 | | 15 | 19 | 23 | 27 | 30 | 34 | 38 | 42 | 46 | 49 | 53 | 57 | 61 | 65 | 68 | 72 | 76 | 80 |
| 6 | | 17 | 22 | 27 | 31 | 36 | 40 | 44 | 49 | 53 | 58 | 62 | 67 | 71 | 75 | 80 | 84 | 89 | 93 |
| 7 | | 20 | 25 | 30 | 36 | 41 | 46 | 51 | 56 | 61 | 66 | 71 | 76 | 81 | 86 | 91 | 96 | 101 | 106 |
| 8 | 16 | 22 | 28 | 34 | 40 | 46 | 51 | 57 | 63 | 69 | 74 | 80 | 86 | 91 | 97 | 102 | 108 | 113 | 119 |
| 9 | 18 | 25 | 32 | 38 | 44 | 51 | 57 | 64 | 70 | 76 | 82 | 89 | 95 | 101 | 107 | 114 | 120 | 126 | 132 |
| 10 | 20 | 27 | 35 | 42 | 49 | 56 | 63 | 70 | 77 | 84 | 91 | 97 | 104 | 111 | 118 | 125 | 132 | 138 | 145 |
| 11 | 22 | 30 | 38 | 46 | 53 | 61 | 69 | 76 | 84 | 91 | 99 | 106 | 114 | 121 | 129 | 136 | 143 | 151 | 158 |
| 12 | 23 | 32 | 41 | 49 | 58 | 66 | 74 | 82 | 91 | 99 | 107 | 115 | 123 | 131 | 139 | 147 | 155 | 163 | 171 |
| 13 | 25 | 35 | 44 | 53 | 62 | 71 | 80 | 89 | 97 | 106 | 115 | 124 | 132 | 141 | 149 | 158 | 167 | 175 | 184 |
| 14 | 27 | 37 | 47 | 57 | 67 | 76 | 86 | 95 | 104 | 114 | 123 | 132 | 141 | 151 | 160 | 169 | 178 | 188 | 197 |
| 15 | 29 | 40 | 50 | 61 | 71 | 81 | 91 | 101 | 111 | 121 | 131 | 141 | 151 | 161 | 170 | 180 | 190 | 200 | 210 |
| 16 | 31 | 42 | 53 | 65 | 75 | 86 | 97 | 107 | 118 | 129 | 139 | 149 | 169 | 179 | 181 | 191 | 202 | 212 | 222 |
| 17 | 32 | 45 | 57 | 68 | 80 | 91 | 102 | 114 | 125 | 136 | 147 | 158 | 169 | 180 | 191 | 202 | 213 | 224 | 235 |
| 18 | 34 | 47 | 60 | 72 | 84 | 96 | 108 | 120 | 132 | 143 | 155 | 167 | 178 | 190 | 202 | 213 | 225 | 236 | 248 |
| 19 | 36 | 50 | 63 | 76 | 89 | 101 | 114 | 126 | 138 | 151 | 163 | 175 | 188 | 200 | 212 | 224 | 236 | 248 | 261 |
| 20 | 38 | 52 | 66 | 80 | 93 | 106 | 119 | 132 | 145 | 158 | 171 | 184 | 197 | 210 | 222 | 235 | 248 | 261 | 273 |

*continued*

589

## Table C.2

Critical Values of the Mann-Whitney Test Statistic—continued

| | | | | | | | | | $\alpha = 0.01$ | | | | | | | | | | |
|---|---|---|---|---|---|---|---|---|---|---|---|---|---|---|---|---|---|---|---|
| $n_1$ | $n_2 = 2$ | 3 | 4 | 5 | 6 | 7 | 8 | 9 | 10 | 11 | 12 | 13 | 14 | 15 | 16 | 17 | 18 | 19 | 20 |
| 2 | | | | | | | | | | | | | | | | | | | 38 |
| 3 | | | | | | | | 27 | 30 | 33 | 35 | 38 | 41 | 43 | 46 | 49 | 52 | 54 | 57 |
| 4 | | | | | 24 | 28 | 31 | 35 | 38 | 42 | 45 | 49 | 52 | 55 | 59 | 62 | 66 | 69 | 72 |
| 5 | | | | 25 | 29 | 34 | 38 | 42 | 46 | 50 | 54 | 58 | 63 | 67 | 71 | 75 | 79 | 83 | 87 |
| 6 | | | 24 | 29 | 34 | 39 | 44 | 49 | 54 | 59 | 63 | 68 | 73 | 78 | 83 | 87 | 92 | 97 | 102 |
| 7 | | | 28 | 34 | 39 | 45 | 50 | 56 | 61 | 67 | 72 | 78 | 83 | 89 | 94 | 100 | 105 | 111 | 116 |
| 8 | | | 31 | 38 | 44 | 50 | 57 | 63 | 69 | 75 | 81 | 87 | 94 | 100 | 106 | 112 | 118 | 124 | 130 |
| 9 | | 27 | 35 | 42 | 49 | 56 | 63 | 70 | 77 | 83 | 90 | 97 | 104 | 111 | 117 | 124 | 131 | 138 | 144 |
| 10 | | 30 | 38 | 46 | 54 | 61 | 69 | 77 | 84 | 92 | 99 | 106 | 114 | 121 | 129 | 136 | 143 | 151 | 158 |
| 11 | | 33 | 42 | 50 | 59 | 67 | 75 | 83 | 92 | 100 | 108 | 116 | 124 | 132 | 140 | 148 | 156 | 164 | 172 |
| 12 | | 35 | 45 | 54 | 63 | 72 | 81 | 90 | 99 | 108 | 117 | 125 | 134 | 143 | 151 | 160 | 169 | 177 | 186 |
| 13 | | 38 | 49 | 58 | 68 | 78 | 87 | 97 | 106 | 116 | 125 | 135 | 144 | 153 | 163 | 172 | 181 | 190 | 200 |
| 14 | | 41 | 52 | 63 | 73 | 83 | 94 | 104 | 114 | 124 | 134 | 144 | 154 | 164 | 174 | 184 | 194 | 203 | 213 |
| 15 | | 43 | 55 | 67 | 78 | 89 | 100 | 111 | 121 | 132 | 143 | 153 | 164 | 174 | 185 | 195 | 206 | 216 | 227 |
| 16 | | 46 | 59 | 71 | 83 | 94 | 106 | 117 | 129 | 140 | 151 | 163 | 174 | 185 | 196 | 207 | 218 | 230 | 241 |
| 17 | | 49 | 62 | 75 | 87 | 100 | 112 | 124 | 136 | 148 | 160 | 172 | 184 | 195 | 207 | 219 | 231 | 242 | 254 |
| 18 | | 52 | 66 | 79 | 92 | 105 | 118 | 131 | 143 | 156 | 169 | 181 | 194 | 206 | 218 | 231 | 243 | 255 | 268 |
| 19 | | 54 | 69 | 83 | 97 | 111 | 124 | 138 | 151 | 164 | 177 | 190 | 203 | 216 | 230 | 242 | 255 | 268 | 281 |
| 20 | 38 | 57 | 72 | 87 | 102 | 116 | 130 | 144 | 158 | 172 | 186 | 200 | 213 | 227 | 241 | 254 | 268 | 281 | 295 |

The values in the above table are derived, with permission of the publisher, from the extensive tables of Milton (1964, *J. Amer. Statist. Assoc.* 59:925–934). See Zar (1996:App 86–97) for some sample sizes and significance levels not included above.
From: Rpt. in Brower, Zar and von Ende. *Field and Laboratory Methods for General Ecology,* 4/e. Dubuque, Iowa: McGraw-Hill, 1990. 13–14.

## Table C.3

### Critical Values of Chi-Square

| DF | $\alpha = 0.10$ | $\alpha = 0.05$ | $\alpha = 0.025$ | $\alpha = 0.01$ | DF | $\alpha = 0.10$ | $\alpha = 0.05$ | $\alpha = 0.02$ | $\alpha = 0.01$ |
|----|------|------|------|------|----|------|------|------|------|
| 1 | 2.706 | 3.841 | 5.024 | 6.635 | 21 | 29.615 | 32.671 | 35.479 | 38.932 |
| 2 | 4.605 | 5.991 | 7.378 | 9.210 | 22 | 30.813 | 33.924 | 36.781 | 40.289 |
| 3 | 6.251 | 7.815 | 9.348 | 11.345 | 23 | 32.007 | 35.172 | 38.076 | 41.638 |
| 4 | 7.779 | 9.488 | 11.143 | 13.277 | 24 | 33.196 | 36.415 | 39.364 | 42.980 |
| 5 | 9.236 | 11.070 | 12.833 | 15.086 | 25 | 34.382 | 37.652 | 40.646 | 44.314 |
| 6 | 10.645 | 12.592 | 14.449 | 16.812 | 26 | 35.563 | 38.885 | 41.923 | 45.642 |
| 7 | 12.017 | 14.067 | 16.013 | 18.475 | 27 | 36.741 | 40.113 | 43.195 | 46.963 |
| 8 | 13.362 | 15.507 | 17.535 | 20.090 | 28 | 37.916 | 41.337 | 44.461 | 48.278 |
| 9 | 14.684 | 16.919 | 19.023 | 21.666 | 29 | 33.711 | 39.087 | 42.557 | 45.722 |
| 10 | 15.987 | 18.307 | 20.483 | 23.209 | 30 | 40.256 | 43.773 | 46.979 | 50.892 |
| 11 | 17.275 | 19.675 | 21.920 | 24.725 | 31 | 41.422 | 44.985 | 48.232 | 52.191 |
| 12 | 18.549 | 21.026 | 23.337 | 26.217 | 32 | 42.585 | 46.194 | 49.480 | 53.486 |
| 13 | 19.812 | 22.362 | 24.736 | 27.688 | 33 | 43.745 | 47.400 | 50.725 | 54.776 |
| 14 | 21.064 | 23.685 | 26.119 | 29.141 | 34 | 44.903 | 48.602 | 51.966 | 56.061 |
| 15 | 22.307 | 24.996 | 27.488 | 30.578 | 35 | 46.059 | 49.802 | 53.203 | 57.302 |
| 16 | 23.542 | 26.296 | 28.845 | 32.000 | 36 | 47.212 | 50.998 | 54.437 | 58.619 |
| 17 | 24.769 | 27.587 | 30.191 | 33.409 | 37 | 48.363 | 52.192 | 55.668 | 59.893 |
| 18 | 25.989 | 28.869 | 31.526 | 34.805 | 38 | 49.513 | 53.384 | 56.896 | 61.162 |
| 19 | 27.204 | 30.144 | 32.852 | 36.191 | 39 | 50.660 | 54.572 | 58.120 | 62.428 |
| 20 | 28.412 | 31.410 | 34.170 | 37.566 | 40 | 51.805 | 55.758 | 59.342 | 63.691 |

The above values were computed as described by Zar (1996: App 16). More extensive tables of chi-square are found in Rohlf and Sokal (1995:24–25) and Zar (1996:App 13–16). Values of chi-square for degrees of freedom ($v$) greater than 40 may be approximated very accurately (Zar, 1984:482), as follows:

$$\chi^2_{\alpha,v} = v\left(1 - 2/9v + c\sqrt{2/9v}\right)^3,$$

where the appropriate value of $c$ is

| $\alpha =$ | 0.10 | 0.05 | 0.025 | 0.01 |
|----|------|------|------|------|
| $c =$ | 1.28155 | 1.64485 | 1.95996 | 2.32635 |

From: Brower, Zar and von Ende. *Field and Laboratory Methods for General Ecology,* 4/e. Dubuque, Iowa: McGraw-Hill, 1990. 15.

# Glossary

**abundance** the total number of individuals, or biomass, of a species present in a specified area. *(p. 228)*

**abyssal zone** a zone of the ocean depths between 4,000 and 6,000 m. *(p. 56)*

**acclimation** physiological adjustment to change in a particular environmental factor, such as temperature or salinity. *(p. 97)*

**actual evapotranspiration (AET)** the amount of water lost from an ecosystem to the atmosphere due to a combination of evaporation and transpiration by plants. *(p. 443)*

**adaptation** an evolutionary process that changes anatomy, physiology, or behavior, resulting in an increased ability of a population to live in a particular environment. The term is also applied to the anatomical, physiological, or behavioral characteristics produced by this process. *(p. 198)*

**adhesion-adapted** a term applied to seeds with hooks, spines, or barbs that disperse by attaching to passing animals. *(p. 303)*

**age distribution** the distribution of individuals among age groups in a population; often called age structure. *(p. 256)*

**A horizon** a biologically active soil layer consisting of a mixture of mineral materials, such as clay, silt, and sand, as well as organic material, derived from the overlying O horizon; generally characterized by leaching. *(p. 22)*

**allele** one of the alternative forms of the same gene. *(p. 199)*

**allele frequencies** the proportions of alternative forms of a gene (alleles) in a population of organisms. *(p. 207)*

**allopatric** describes the condition in which populations or species have nonoverlapping geographic ranges. *(p. 338)*

**allopolyploidy** a process of speciation initiated by hybridization of two different species. *(p. 330)*

**allozyme** alternative form of a particular enzyme which differs structurally but not functionally from other allozymes coded for by different alleles at the same locus. *(p. 216)*

**arbuscular mycorrhizal fungi (AMF)** mycorrhizae in which the mycorrhizal fungus produces arbuscules (sites of exchange between plant and fungus), hyphae (fungal filaments), and vesicles (fungal energy storage organs within root cortex cells). *(p. 376)*

**arbuscule** a bush-shaped organ on an endomycorrhizal fungus that acts as a site of material exchange between the fungus and its host plant. *(p. 376)*

**archaea** prokaryotes distinguished from bacteria on the basis of structural, physiological, and other biological features. *(p. 146)*

**argillic horizon** a subsoil characterized by an accumulation of clays. *(p. 524)*

**aril** a fleshy covering of some seeds that attracts birds and other vertebrates, which act as dispersers of such seeds. *(p. 303)*

**atoll** a circle of low islands and coral reefs encircling a lagoon, generally formed on a submerged mountain called a seamount. *(p. 61)*

**autotroph** an organism that can synthesize organic molecules using inorganic molecules and energy from either sunlight (photosynthetic autotrophs) or from inorganic molecules, such as hydrogen sulfide (chemosynthetic autotrophs). *(p. 145)*

**balanced growth** cell growth in which all cell constituents, such as nitrogen, carbon, and DNA, increase at approximately the same rate. *(p. 384)*

**barrier reef** a long ridgelike reef that parallels the mainland and is separated from it by a deep lagoon. *(p. 61)*

**Batesian mimicry** evolution of a nonnoxious species to resemble a poisonous or inedible species. *(p. 153)*

**bathypelagic zone** a zone within the deep ocean that extends from about 1,000 to 4,000 m. *(p. 56)*

**behavioral ecology** study of the relationships between organisms and environment that are mediated by behavior. *(p. 171)*

**benthic** an adjective referring to the bottom of bodies of waters such as seas, lakes, or streams. *(p. 56)*

**B horizon** a subsoil in which materials leached from above, generally from the A horizon, accumulate. May be rich in clay, organic matter, iron, and other materials. *(p. 22)*

**biome** biomes are distinguished primarily by their predominant plants and are associated with particular climates. They consist of distinctive plant formations such as the tropical rain forest biome and the desert biome. *(p. 16)*

**biosphere** the portions of earth that support life; also refers to the total global ecosystem. *(p. 3)*

**birthrate** the number of new individuals produced in a population generally expressed as births per individual or per thousand individuals in the population. *(p. 262)*

**boreal forest** northern forests that occupy the area south of arctic tundra. Though dominated by coniferous trees they also contain aspen and birch. Also called taiga. *(p. 39)*

**bottom-up control** control of a community or ecosystem by physical or chemical factors such as temperature or nutrient availability. *(p. 449)*

**bundle sheath** structure, which surrounds the leaf veins of $C_4$ plants, made up of cells, where four-carbon acids produced during carbon fixation are broken down to three-carbon acids and $CO_2$. *(p. 149)*

**caliche** a calcium carbonate-rich hardpan soil horizon; the extent of caliche formation can be used to determine the age of desert soils. *(p. 32)*

**CAM (crassulacean acid metabolism) photosynthesis** a photosynthetic pathway largely limited to succulent plants in arid and semiarid environments, in which carbon fixation takes place at night, when lower temperatures reduce the rate of water loss during $CO_2$ uptake. The resulting four-carbon acids are stored until daylight, when they are broken down into pyruvate and $CO_2$. *(p. 149)*

**carnivore** an organism that consumes flesh; approximately synonymous with predator. *(p. 150)*

**carrying capacity (K)** the maximum population of a species that a particular ecosystem can sustain. *(p. 282)*

**caste** a group of individuals that are physically distinctive and engage in specialized behavior within a social unit, such as a colony. *(p. 190)*

**$C_4$ photosynthesis** in $C_4$ photosynthesis, $CO_2$ is fixed in mesophyll cells by combining it with phosphoenol pyruvate, or PEP, to produce a four-carbon acid. Plants using $C_4$ photosynthesis are generally more drought tolerant than plants employing $C_3$ photosynthesis. *(p. 147)*

**character displacement** changes in the physical characteristics of a species' population as a consequence of natural selection for reduced interspecific competition. *(p. 338)*

**chemosynthetic** refers to autotrophs that use inorganic molecules as a source of carbon and energy. *(p. 145)*

**chi-square ($X^2$)** a statistic used to measure how much a sample distribution differs from a theoretical distribution. *(p. 287)*

**C horizon** a soil layer composed of largely unaltered parent material, little affected by biological activity. *(p. 22)*

**chronosequence** a series of communities or ecosystems representing a range of ages or times since disturbance. *(p. 491)*

**climate diagram** a standardized form of representing average patterns of variation in temperature and precipitation that identifies several ecologically important climatic factors such as relatively moist periods and periods of drought. *(p. 18)*

**climax community** a community that occurs late in succession whose populations remain stable until disrupted by disturbance. *(p. 486)*

**clumped distribution** a pattern of distribution in a population in which individuals have a much higher probability of being found in some areas than in others; in other words, individuals are aggregated rather than dispersed. *(p. 233)*

**clutch size** the number of eggs laid by a bird, reptile, amphibian, or fish. The term is also sometimes applied to the number of seeds produced by a plant. *(p. 264)*

**cohort** a group of individuals of the same age. *(p. 256)*

**cohort life table** a life table based on individuals born (or beginning life in some other way) at the same time. *(p. 256)*

**colonization cycle** the situation in which stream populations are maintained through a dynamic interplay between downstream drift and upstream dispersal. *(p. 271)*

**combined response** the combined effect of functional and numerical responses by consumers on prey populations; determined by multiplying the number of prey eaten per predator times the number of predators per unit area, giving the number of prey eaten per unit area. Combined response is generally expressed as a percentage of the total number of prey. *(p. 365)*

**community** an association of interacting species living in a particular area; also often defined as all of the organisms living in a particular area. *(p. 398)*

**community structure** attributes of a community such as the number of species or the distribution of individuals among species within the community. *(p. 398)*

**comparative method** a method for reconstructing evolutionary processes and mechanisms that involves comparisons of different species or populations in a way that attempts to isolate a particular variable or characteristic of interest, while randomizing the influence of confounding, or confusing, variables on the variable of interest across the species or populations in the study. *(p. 189)*

**competition coefficient** a coefficient expressing the magnitude of the negative effect of individuals of one species on individuals of a second species. *(p. 331)*

**competitive exclusion principle** the principle that two species with identical niches cannot coexist indefinitely. *(p. 328)*

**competitive plants** according to Grime (1977) competitive plants occupy environments where disturbance intensity is low and the intensity of stress is also low. *(p. 313)*

**conduction** the movement of heat between objects in direct physical contact. *(p. 99)*

**confidence interval** a range of values within which the true population mean occurs with a particular probability called the level of confidence. *(p. 387)*

**convection** the process of heat flow or transfer to a moving fluid, such as wind or flowing water. *(p. 99)*

**Coriolis effect** a phenomenon caused by the rotation of the earth, which produces a deflection of winds and water currents to the right of their direction of travel in the Northern Hemisphere and to the left of their direction of travel in the Southern Hemisphere. *(p. 18)*

**$C_3$ photosynthesis** the photosynthetic pathway used by most plants and all algae, in which the product of the initial reaction is phosphoglyceric acid, or PGA, a three-carbon acid. *(p. 147)*

**decomposition** the breakdown of organic matter accompanied by the release of carbon dioxide and other inorganic compounds; a key process in nutrient cycling. *(p. 466)*

**density** the number of individuals in a population per unit area. *(p. 228)*

**density-dependent factor** biotic factors in the environment, such as disease and competition, are often called density-dependent factors because their effects on populations may be related to, or depend upon, local population density. *(p. 285)*

**density-independent factor** abiotic factors in the environment, such as floods and extreme temperature, are often called density-independent factors because their effects on populations may be independent of population density. *(p. 285)*

**dependent variable** the variable traditionally plotted on the vertical or "Y" axis of a scatter plot. *(p. 213)*

**desert** an arid biome occupying approximately 20% of the land surface of the earth in which water loss due to evaporation and transpiration by plants exceeds precipitation during most of the year. *(p. 30)*

**detritivore** organisms that feed on nonliving organic matter, usually on the remains of plants. *(p. 150)*

**diffusion** transport of material due to the random movement of particles; net movement is from areas of high concentration to areas of low concentration. *(p. 121)*

**directional selection** a form of natural selection that favors an extreme phenotype over other phenotypes. *(p. 210)*

**disruptive selection** a form of natural selection that favors two or more extreme phenotypes over the average phenotype in a population. *(p. 210)*

**distribution** the natural geographic range of an organism or the spatial arrangement of individuals in a local population. *(p. 228)*

**disturbance** grime (1977) defined disturbance from the perspective of plants as any process that limits plants by destroying plant biomass. Sousa (1984) also defined disturbance from an organismic perspective as any discrete, punctuated killing, displacement, or damaging of one or more individuals (or colonies) that directly or indirectly creates an opportunity for new individuals (or colonies) to become established. White and Pickett (1985) defined disturbance more broadly as any relatively discrete event that disrupts ecosystem, community, or population structure and changes resources, substrate availability, or the physical environment. *(p. 313)*

**DNA sequencing** methods for determining the sequence of nucleic acids in DNA molecules. *(p. 222)*

**drift** the active or passive downstream movement of stream organisms. *(p. 271)*

**drought** an extended period of dry weather during which precipitation is reduced sufficiently to damage crops, impair the functioning of natural ecosystems, or cause water shortages for human populations. *(p. 47)*

**ecology** study of the relationships between organisms and the environment. *(p. 2)*

**ecosystem** a biological community plus all of the abiotic factors influencing that community. *(p. 6)*

**ecotone** a spatial transition from one type of ecosystem to another; for instance, the transition from a woodland to a grassland. *(p. 9)*

**ecotype** locally adapted and genetically distinctive population within a species. *(p. 201)*

**ectomycorrhizae (ECM)** an association between a fungus and plant roots in which the fungus forms a mantle around roots and a netlike structure around root cells. *(p. 376)*

**ectotherm** an organism that relies mainly on external sources of energy for regulating body temperature. *(p. 100)*

**elaiosome** a structure on the surface of some seeds generally containing oils attractive to ants, which act as dispersers of such seeds. *(p. 303)*

**electrophoresis** an analytical technique involving the separation of molecules in an electrical field. *(p. 221)*

**El Niño** a large-scale coupled oceanic-atmospheric system that has major effects on climate worldwide. During an El Niño, the sea surface temperature in the eastern Pacific Ocean is higher than average and barometric pressure is lower. *(p. 563)*

**endemic** a term applied to populations or species that are found in a particular locality, for instance an island, and nowhere else. *(p. 216)*

**endotherm** an organism that relies mainly on internal sources of energy for regulating body temperature. *(p. 100)*

**energy budget** the rate at which an organism takes in energy relative to the rate at which it expends energy; gives an indication of the amount of energy available for functions such as reproduction. *(p. 5)*

**epilimnion** the warm, well-lighted surface layer of lakes. *(p. 78)*

**epipelagic zone** the warm, well-lighted surface layer of the oceans. *(p. 56)*

**epiphyte** a plant, such as an orchid, that grows on the surface of another plant but is not parasitic. *(p. 7)*

**equilibrium** a state of balance in a system in which opposing factors cancel each other. *(p. 409)*

**estivation** a dormant state that some animals enter during the summer; involves a reduction of metabolic rate. *(p. 111)*

**estuary** the lowermost part of a river, which is under the influence of the tides and is a mixture of seawater and freshwater. *(p. 68)*

**eusociality** highly specialized sociality generally including (1) individuals of more than one generation living together, (2) cooperative care of young, and (3) division of individuals into sterile, or nonreproductive, and reproductive castes. *(p. 183)*

**eutrophic** a term applied to lakes, and sometimes to other ecosystems, with high nutrient content and high biological production. *(p. 80)*

**evaporation** the process by which a liquid changes from liquid phase to a gas, as in the change from liquid water to water vapor. *(p. 100)*

**evolution** a process that changes populations of organisms over time. Since evolution ultimately involves changes in the frequency of heritable traits in a population, we can define evolution more precisely as a change in gene frequencies in a population. *(p. 198)*

**exploitation** an interaction between species that enhances the fitness of the exploiting individual—the predator, the pathogen, etc.—while reducing the fitness of the exploited individual—the prey, host, etc. *(p. 347)*

**exponential population growth** population growth that produces a J-shaped pattern of population increase. In exponential population growth, the change in numbers with time is the product of the per capita rate of increase, $r$, and population size, $N$. *(p. 280)*

**extrafloral nectary** nectar-secreting glands found on structures other than flowers, such as leaves. *(p. 382)*

**facilitation model** according to the facilitation model, pioneer species modify the environment in such a way that it becomes less suitable for themselves and more suitable for species characteristic of later successional stages. *(p. 495)*

**facultative mutualism** a mutualistic relationship between two species that is not required for the survival of the two species. *(p. 375)*

**fecundity** the number of eggs or seeds produced by an organism. *(p. 300)*

**fecundity schedule** a table of birthrates for females of different ages in a population. *(p. 262)*

**female** sex that produces larger more energetically costly gametes (eggs or ova). *(p. 172)*

**fetch** the longest distance over which wind can blow across a body of water; directly related to the maximum size of waves that can be generated by wind. *(p. 330)*

**fitness** the number of offspring contributed by an individual relative to the number of offspring produced by other members of the population. Ultimately defined as the relative genetic contribution of individuals to future generations. *(p. 172)*

**flood pulse concept** a theory of river ecology identifying periodic flooding as an essential organizer of river ecosystem structure and functioning. *(p. 75)*

**food web** a summary of the feeding relationships within an ecological community. *(p. 419)*

**forb** herbaceous plants other than graminoids. *(p. 303)*

**fringing reef** a coral reef that forms near the shore of an island or continent. *(p. 61)*

**functional response** an increase in animal feeding rate, which eventually levels off, that occurs in response to an increase in food availability. *(p. 158)*

**fundamental niche** the physical conditions under which a species might live, in the absence of interactions with other species. *(p. 328)*

**genetic drift** change in gene frequencies in a population due to chance or random events. *(p. 207)*

**geographic ecology** the study of ecological structure and process at large geographic scales; sometimes defined as the study of ecological patterns that can be put on a map. *(p. 538)*

**geographic information system** a computer-based system that stores, analyzes, and displays geographic information, generally in the form of maps. *(p. 556)*

**geometric population growth** population growth in which generations do not overlap and in which successive generations differ in size by a constant ratio. *(p. 279)*

**geometric rate of increase** ($\lambda$) the ratio of the population size at two points in time: $\lambda = N_{t+1}/N_t$, where $N_{t+1}$ is the size of the population at some future time and $N_t$ is the size of the population at some earlier time. *(p. 262)*

**germination** the sprouting of seeds. *(p. 304)*

**global positioning system (GPS)** a device that determines locations on the earth's surface, including latitude, longitude, and altitude, using radio signals from satellites as references. *(p. 554)*

**gonadosomatic index (GSI)** an index of reproductive effort calculated as ovary weight divided by body weight and adjusted for the number of batches of offspring produced per year. *(p. 307)*

**graminoids** grasses and grasslike plants, such as sedges and rushes. *(p. 303)*

**granivore** An animal that feeds chiefly on seeds. *(p. 337)*

**greenhouse effect** warming of the earth's atmosphere and surface as a result of heat trapped near the earth's surface by gases in the atmosphere, especially water vapor, carbon dioxide, methane, ozone, nitrous oxide, and chlorofluorocarbons. *(p. 561)*

**gross primary production** the total amount of energy fixed by all the autotrophs in an ecosystem. *(p. 442)*

**growth form** See *life-form.*

**guild** a group of organisms that make their living in a similar way; for example, the seed-eating animals in a desert, the fruit-eating birds in a tropical rain forest, or the filter-feeding invertebrates in a stream. *(p. 398)*

**gyre** a large-scale, circular oceanic current that moves to the right in the Northern Hemisphere and to the left in the Southern Hemisphere. *(p. 57)*

**hadal zone** the deepest parts of the oceans, below about 6,000 m. *(p. 56)*

**haplodiploidy** sex inheritance in which males are haploid and females are diploid. *(p. 192)*

**Hardy-Weinberg principle** a principle that in a population mating at random in the absence of evolutionary forces, allele frequencies will remain constant. *(p. 207)*

**herbivore** a heterotrophic organism that eats plants. *(p. 150)*

**heritability** the proportion of total phenotypic variation in a trait attributable to genetic variation; determines the potential for evolutionary change in a trait. *(p. 210)*

**hermaphrodite** an individual capable of producing both sperm or pollen and eggs or ova. *(p. 172)*

**heterotroph** an organism that uses organic molecules both as a source of carbon and as a source of energy. *(p. 146)*

**hibernation** a dormant state, involving reduced metabolic rate, that occurs in some animals during the winter. *(p. 111)*

**homeotherm** an organism that uses metabolic energy to maintain a relatively constant body temperature; such organisms are often called warm-blooded. *(p. 100)*

**hydrologic cycle** the sun-driven cycle of water through the biosphere through evaporation, transpiration, condensation, precipitation, and runoff. *(p. 53)*

**hyperosmotic** a term describing organisms with body fluids with a lower concentration of water and higher solute concentration than the external environment. *(p. 121)*

**hyphae** long, thin filaments that form the basic structural unit of fungi. *(p. 376)*

**hypolimnion** the deepest layer of a lake below the epilimnion and thermocline. *(p. 78)*

**hypoosmotic** a term describing organisms with body fluids with a higher concentration of water and lower solute concentration than the external environment. *(p. 121)*

**hyporheic zone** a zone below the benthic zone of a stream; a zone of transition between surface, streamwater flow, and groundwater. *(p. 73)*

**inbreeding** mating between close relatives. Inbreeding tends to increase levels of homozygosity in populations and often results in offspring with lower survival and reproductive rates. *(p. 218)*

**inclusive fitness** overall fitness, which is determined by the survival and reproduction of an individual, plus the survival and reproduction of genetic relatives of the individual. *(p. 184)*

**independent variable** the variable traditionally plotted on the horizontal or "X" axis of a scatter plot. *(p. 213)*

**inhibition model** a model of succession that proposes that early occupants of an area modify the environment in a way that makes the area less suitable for both early and late successional species. *(p. 496)*

**insectivore** a heterotrophic organism that eats insects. *(p. 83)*

**interference competition** form of competition involving direct antagonistic interactions between individuals. *(p. 183)*

**interquartile range** a range of measurements which includes the middle 50% of the measurements or observations in a sample, bounded by the lowest value of the highest 25% of measurements and the highest value of the lowest 25% of measurements. *(p. 504)*

**intersexual selection** sexual selection occurring when members of one sex choose mates from among the members of the opposite sex on the basis of some anatomical or behavioral trait, generally leading to the elaboration of that trait. *(p. 173)*

**interspecific competition** competition between individuals of different species. *(p. 324)*

**intertidal zone** See *littoral zone.*

**intrasexual selection** sexual selection in which individuals of one sex compete among themselves for mates. *(p. 173)*

**intraspecific competition** competition between individuals of the same species. *(p. 324)*

**intrinsic rate of increase** the maximum per capita rate of population increase; may be approached under ideal environmental conditions for a species. *(p. 280)*

**irradiance** the level of light intensity, often measured as photon flux density. *(p. 157)*

**$I_{sat}$** the irradiance required to saturate the photosynthetic capacity of a photosynthetic organism. *(p. 157)*

**isoclines of zero population growth** lines, in the graphical representation of the Lotka-Volterra competition model, where population growth of the species in competition is zero. *(p. 331)*

**isosmotic** a term describing organisms with body fluids containing the same concentration of water and solutes as the external environment. *(p. 121)*

**isozymes** all enzymes with the same biochemical function that are produced by different or the same loci. *(p. 219)*

**iteroparity** reproduction that involves production of an organism's offspring in two or more events, generally spaced out over the life time of the organism. *(p. 312)*

**keystone species** species that, despite low biomass, exert strong effects on the structure of the communities they inhabit. *(p. 423)*

**kin selection** selection in which individuals increase their inclusive fitness by helping increase the survival and reproduction of relatives (kin) that are not offspring. *(p. 184)*

**K selection** a term referring to the carrying capacity of the logistic growth equation; a form of natural selection that favors more efficient utilization of resources such food and nutrients. K selection is predicted to be strongest in those situations where a population lives as densities near carrying capacity much of the time. *(p. 311)*

**La Niña** the opposite of an El Niño. During a La Niña, the sea surface temperature in the eastern Pacific Ocean is lower than average and barometric pressure is higher. *(p. 564)*

**landform** any distinctive feature of the earth's surface. *(p. 523)*

**landscape** an area of land containing a patchwork of ecosystems. *(p. 513)*

**landscape ecology** the study of landscape structure and processes. *(p. 512)*

**landscape elements** the ecosystems in a landscape, which generally form a mosaic of visually distinctive patches. *(p. 513)*

**landscape process** the exchange of materials, energy, or organisms among the ecosystems that make up a landscape. *(p. 516)*

**landscape structure** the size, shape, composition, number, and position of ecosystems within a landscape. *(p. 513)*

**large-scale phenomena** phenomena of a geographic scale rather than a local scale. *(p. 233)*

**level of confidence** one minus the significance level, α, which is generally 0.05, for example, level of confidence = 1–0.05 = 0.95. *(p. 387)*

**life-form** the life-form of a plant is a combination of its structure and its growth dynamics. Plant life-forms include trees, vines, annual plants, sclerophyllous vegetation, grasses, and forbs. *(pp. 302, 399)*

**life history** the adaptations of an organism that influence aspects of its biology such as the number of offspring it produces, its survival, and its size and age at reproductive maturity. *(p. 299)*

**life table** a table of age-specific survival and death, or mortality, rates in a population. *(p. 256)*

**limnetic zone** the open lake beyond the littoral zone. *(p. 78)*

**lithosol** soils very low in organic matter and composed of rock fragments. *(p. 31)*

**littoral zone** the shallowest waters along a lake or ocean shore; where rooted aquatic plants may grow in lakes. *(p. 55)*

**loci** (plural of **locus**) the location along the length of a particular chromosome where a gene is located. *(p. 216)*

**logistic equation** $dN/dt = r_m N(K - N/K)$ *(p. 283)*

**logistic population growth** a pattern of growth that produces a sigmoidal, or S-shaped, population growth curve; population size levels off at carrying capacity (K). *(p. 282)*

**macroclimate** the prevailing climate for a region. *(p. 90)*

**male** sex that produces smaller less costly gametes (sperm or pollen). *(p. 172)*

**mangrove forest** a forest of subtropical and tropical marine shores dominated by salt-tolerant woody plants, such as *Rhizophora* and *Avicennia*. *(p. 68)*

**masting** the synchronous production of large quantities of fruits by trees, such as oaks and beech. *(p. 365)*

**matric force** a force resulting from water's tendency to adhere to the walls of containers such as cell walls or the soil particles lining a soil pore. *(p. 123)*

**Mediterranean woodland and shrubland** a biome associated with mild, moist winter conditions and usually with dry summers between about 30° and 40° latitude. The vegetation of this biome is usually characterized by small, tough (sclerophyllous) leaves and adaptations to periodic fire. This biome is found around the Mediterranean Sea and in western North America, Chile, southern Australia, and southern Africa. This biome is known by many local names such as chaparral, garrigue, maquis, and fynbos. *(p. 33)*

**meristematic tissue** tissue made up of the actively dividing cells responsible for plant growth. *(p. 349)*

**mesopelagic zone** a middle depth zone of the oceans, extending from about 200 to 1,000 m. *(p. 56)*

**mesosphere** a layer in the earth's atmosphere, extending from 64 to 80 km above the earth's surface; temperatures drop steeply with altitude in this atmospheric layer. *(p. 561)*

**metabolic heat** energy released within an organism during the process of cellular respiration. *(p. 99)*

**metabolic water** water released during oxidation of organic molecules. *(p. 125)*

**metalimnion** a depth zone between the epilimnion and hypolimnion characterized by rapid decreases in temperature and increases in water density with depth. Often used synonymously with the term *thermocline*. *(p. 78)*

**metapopulation** a group of subpopulations living in separate locations with active exchange of individuals among subpopulations. *(p. 240)*

**microclimate** a small-scale variation in climate caused by a distinctive substrate, location, or aspect. *(p. 91)*

**microsatellite DNA** sequence of tandemly repetitive DNA, 10 to 100 base pairs long. *(p. 205)*

**mineralization** the breakdown of organic matter from organic to inorganic form during decomposition. *(p. 466)*

**monogamous species** species that mate with only one other individual during their life or at least during one reproductive cycle. *(p. 189)*

**Müllerian mimicry** comimicry among several species of noxious organisms. *(p. 153)*

**mutualism** interactions between individuals of different species that benefit both partners. *(p. 375)*

**mycorrhizae** a mutualistic association between fungi and the roots of plants. *(p. 24)*

**natal territory** territory where an individual was born. *(p. 185)*

**natural history** the study of how organisms in a particular area are influenced by factors such as climate, soils, predators, competitors, and evolutionary history, involving field observations rather than carefully controlled experimentation or statistical analyses of patterns. *(p. 15)*

**natural selection** differential reproduction and survival of individuals in a population due to environmental influences on the population; proposed by Charles Darwin as the primary mechanism driving evolution. *(p. 198)*

**negative phototaxis** movement of an organism away from light. *(p. 348)*

**neritic zone** a coastal zone of the oceans, extending to the margin of a continental shelf, where the ocean is about 200 m deep. *(p. 56)*

**net primary production** the amount of energy left over after autotrophs have met their own energetic needs (gross primary production minus respiration by primary producers); the amount of energy available to the consumers in an ecosystem. *(p. 442)*

**net reproductive rate ($R_0$)** the average number of offspring produced by an individual in a population. *(p. 262)*

**niche** the environmental factors that influence the growth, survival, and reproduction of a species. *(p. 328)*

**nonequilibrial theory** theories of ecological systems that do not assume equilibrial conditions. *(p. 73)*

**normal distrbution** a bell-shaped distribution, proportioned so that predictable proportions of observations or measurements fall within one,

two, or three standard deviations of the mean. *(p. 471)*

**nucleotide** the basic building blocks of nucleic acids, which are made up of a five-carbon sugar (deoxyribose or ribose), a phosphate group, and a nitrogenous base (guanine, cytosine, adenine, or thymine). *(p. 220)*

**numerical response** change in the density of a predator population in response to increased prey density. *(p. 270)*

**nutrient** chemical substance required for the development, maintenance, and reproduction of organisms. *(p. 6)*

**nutrient cycling** the use, transformation, movement, and reuse of nutrients in ecosystems. *(p. 463)*

**nutrient retentiveness** the tendency of an ecosystem to retain nutrients. *(p. 472)*

**nutrient spiraling** a representation of nutrient dynamics in streams, which, because of downstream displacement of organisms and materials, are better represented by a spiral than a cycle. *(p. 472)*

**obligate mutualism** a mutualistic relationship in which species are so dependent upon the relationship that they cannot live in its absence. *(p. 375)*

**oceanic zone** the open ocean beyond the continental shelf with water depths generally greater than 200 m. *(p. 56)*

**O (organic) horizon** the most superficial soil layer containing substantial amounts of organic matter, including whole leaves, twigs, other plant parts, and highly fragmented organic matter. *(p. 22)*

**oligotrophic** a term generally referring to lakes of low nutrient content, abundant oxygen, and low primary production. *(p. 80)*

**omnivore** a heterotrophic organism that eats a wide range of food items, usually including both animal and plant matter. *(p. 83)*

**optimal foraging theory** theory that attempts to model how organisms feed as an optimizing process, a process that maximizes or minimizes some quantity, such as energy intake or predation risk. *(p. 160)*

**optimization** a process that maximizes or minimizes some quantity. *(p. 161)*

**osmosis** diffusion of water down its concentration gradient. *(p. 121)*

**Palmer Drought Severity Index (PDSI)** an index of drought that uses temperature and precipitation to represent moisture conditions in a region relative to long-term average temperature and precipitation within the region. *(p. 47)*

**parasite** an organism that lives in or on another organism, called the host, deriving benefits from it; parasites typically reduce the fitness of the host, but do not generally kill it. *(p. 347)*

**parasitoid** an insect whose larva consumes its host and kills it in the process; parasitoids are functionally equivalent to predators. *(p. 347)*

**pathogen** any organism that induces disease, a debilitating condition, in their hosts; common pathogens include viruses, bacteria, and protozoans. *(p. 347)*

**pelagic** a term referring to marine life zones or organisms above the bottom; for instance, tuna are *pelagic* fish that live in the *epipelagic* zone of the oceans. *(p. 56)*

**per capita rate of increase** usually symbolized as $r$, equals per capita birthrate minus per capita death rate: $r = b - d$. *(p. 265)*

**phase transition** a change in the state of matter such as from a liquid state to a solid state or from a solid to a gas; involves changes in the organization of molecules and their kinetic state. *(p. 9)*

**phenology** the study of the relationship between climate and the timing of ecological events such as the date of arrival of migratory birds on their wintering grounds, the timing of spring plankton blooms, or the onset and ending of leaf fall in a deciduous forest. *(p. 318)*

**pheromone** chemical substance secreted by some animals for communication with other members of their species. *(p. 399)*

**philopatry** a term, which means literally "love of place," used to describe the tendency of some organisms to remain in the same area throughout their lives. *(p. 185)*

**photon flux density** the number of photons of light striking a square meter surface each second. *(p. 147)*

**photosynthesis** process in which the photosynthetic pigments of plants, algae, or bacteria absorb light and transfer their energy to electrons; the energy carried by these electrons is used to synthesize ATP and NADPH, which in turn serve as donors of electrons and energy for the synthesis of sugars. *(p. 96)*

**photosynthetic** a term describing organisms capable of photosynthesis. *(p. 145)*

**photosynthetically active radiation (PAR)** wavelengths of light between 400 and 700 nm that photosynthetic organisms use as a source of energy. *(p. 146)*

**phreatic zone** the region below the hyporheic zone of a stream; contains groundwater. *(p. 73)*

**phytoplankton** microscopic photosynthetic organisms that drift with the currents in the open sea or in lakes. *(p. 58)*

**pioneer community** the first community, in a successional sequence of communities, to be established following a disturbance. *(p. 486)*

**piscivore** a predator that eats fish. *(p. 83)*

**pistil** female organ of a flower. *(p. 181)*

**P$_{max}$** maximum rate of photosynthesis for a particular species of plant growing under ideal physical conditions. *(p. 157)*

**poikilotherm** organisms whose body temperature varies directly with environmental temperatures; commonly called cold-blooded. *(p. 100)*

**polygynous species** species in which a single male mates with more than one female during a breeding season. *(p. 189)*

**polymorphic locus** a locus, or gene, that occurs as more than one allele, each of which synthesizes a different allozyme. *(p. 301)*

**population** a group of individuals of a single species inhabiting a specific area. *(p. 228)*

**population genetics** study of the genetics of populations. *(p. 205)*

**positive phototaxis** movement of an organism toward light. *(p. 348)*

**predator** a heterotrophic organism that kills and eats other organisms for food; usually an animal that hunts and kills other animals for food. *(p. 347)*

**predator satiation** a defensive tactic in which prey reduce their individual probability of being eaten by occurring at very high densities; predators can only capture and eat so many prey and so become satiated when prey are at very high densities. *(p. 365)*

**primary production** the fixation of energy by autotrophs in an ecosystem. *(p. 442)*

**primary succession** succession on newly exposed geological substrates, not significantly modified by organisms; for instance on newly formed volcanic lava or on substrate exposed during the retreat of a glacier. *(p. 486)*

**principle of allocation** the principle that if an organism allocates energy to one function, such as growth or reproduction, it reduces the amount of energy available to other functions, such as defense. *(p. 160)*

**prokaryotes** organisms with cells that have no membrane-bound nucleus or organelles. The prokaryotes include the bacteria and the archaea. *(p. 146)*

**psychrophilic** organisms that live and thrive at temperatures below 20°C. *(p. 98)*

**radiation** the transfer of heat through electromagnetic radiation, mainly infrared light. *(p. 99)*

**random distribution** a distribution in which individuals within a population have an equal chance of living anywhere within an area. *(p. 233)*

**range** the difference between the largest and smallest values in a set of measurements or observations. *(p. 162)*

**rank-abundance curve** a curve that portrays the number of species in a community and their relative abundance; constructed by plotting the relative abundance of species against their rank in abundance. *(p. 401)*

**rate of primary production** the amount of energy fixed by the autotrophs in an ecosystem over some interval of time. *(p. 442)*

**realized niche** the actual niche of a species whose distribution is restricted by biotic interactions such as competition, predation, disease, and parasitism. *(p. 328)*

**regression coefficient** the slope of a regression line. *(p. 213)*

**regression line** the line that best fits the relationship between two variables, X and Y. *(p. 213)*

**regular distribution** a distribution of individuals in a population in which individuals are uniformly spaced. *(p. 233)*

**relative humidity** a measure of the water content of air relative to its content at saturation; relative humidity = water vapor density/saturation water vapor density × 100. *(p. 119)*

**remote sensing** gathering information about an object without direct contact with it, mainly by gathering and processing electromagnetic radiation emitted or reflected by the object; such measurements are typically made from remote sensing satellites. *(p. 554)*

**reproductive effort** the allocation of energy, time, and other resources to the production and care of offspring, generally involving reduced allocation to other needs such as maintenance and growth. *(p. 306)*

**resilience** the capacity to recover structure and function after disturbance; a highly resilient community or ecosystem may be completely disrupted by disturbance but quickly returns to its former state. *(p. 500)*

**resistance** the capacity of a community or ecosystem to maintain structure and/or function in the face of potential disturbance. *(p. 500)*

**resource competition** intraspecific or interspecific competition for limited resources, generally not involving direct antagonistic interactions between individuals. *(p. 325)*

**resource limitation** limitation of population growth by resource availability. *(p. 324)*

**restriction enzymes** the enzymes produced by bacteria to cut up foreign DNA, used in DNA studies to cut DNA molecules at particular places called restriction sites. *(p. 220)*

**restriction fragments** the DNA fragments resulting from the cutting of a DNA molecule by a restriction enzyme. *(p. 221)*

**restriction sites** the particular locations where a restriction enzyme cuts a DNA molecule. *(p. 220)*

**rhodopsin** light-absorbing pigments found in the eyes of animals and in bacteria and archaea. *(p. 146)*

**riparian vegetation** vegetation growth along rivers or streams. *(p. 94)*

**riparian zone** the transition between the aquatic environment of a river or stream and the upland terrestrial environment, generally subject to periodic flooding and elevated groundwater table. *(p. 73)*

**river continuum concept** a model that predicts change in physical structure, dominant organisms, and ecosystem processes along the length of temperate rivers. *(p. 76)*

**r selection** a term referring to the per capita rate of increase; a form of natural selection favoring higher population growth rate. r selection is predicted to be strongest in disturbed habitats. *(p. 311)*

**ruderals** plants or animals that live in highly disturbed habitats and that may depend on disturbance to persist in the face of potential competition from other species. *(p. 313)*

**salinity** the salt content of water. *(p. 58)*

**salt marsh** a marine shore ecosystem dominated by herbaceous vegetation, found mainly along sandy shores from temperate to high latitudes. *(p. 68)*

**sample mean** the average of a sample of measurements or observations; an estimate of the true population mean. *(p. 21)*

**sample median** the middle value in a series of measurements or observations, chosen so that there are equal numbers of measurements in the series that are larger than the median and smaller than the median. *(p. 60)*

**saturation water vapor pressure** the pressure exerted by the water vapor in air that is saturated with water vapor. *(p. 120)*

**scatterhoarded** a term applied to seeds gathered by mammals and stored in scattered caches or hoards. *(p. 303)*

**secondary succession** succession where disturbance has destroyed a community without destroying the soil; for instance, forest succession following a forest fire or logging. *(p. 486)*

**selection coefficient(s)** the relative selection costs or benefits (decreased or increased fitness) associated with a particular biological trait. *(p. 389)*

**self-incompatibility** incapacity of a plant to fertilize itself; such plants must receive pollen from another plant in order to develop seeds. *(p. 181)*

**self-thinning** reduction in population density as a stand of plant increases in biomass, due to intraspecific competition. *(p. 326)*

**self-thinning rule (–3/2)** a rule resulting from the observation that plotting the average weight of individual plants in a stand against density often produces a line with an average slope of approximately –3/2. *(p. 326)*

**semelparity** reproduction that involves production of all of an organism's offspring in one event, generally over a short period of time. *(p. 312)*

**sexual selection** results from differences in reproductive rates among individuals as a result of differences in mating success due to intrasexual selection, intersexual selection, or a mixture of the two forms of sexual selection. *(p. 173)*

**sigmoidal population growth curve** an S-shaped pattern of population growth, with population size leveling off at the carrying capacity of the environment. *(p. 282)*

**size-selective predation** prey selection by predators based on prey size. *(p. 154)*

**small-scale phenomena** phenomena that take place on a local scale. *(p. 233)*

**sociality** group living generally involving some degree of cooperation between individuals. *(p. 183)*

**sociobiology** a branch of biology concerned with the study of social relations. *(p. 171)*

**solifluction** the slow movement of tundra soils down slopes as a result of annual freezing and thawing of surface soil and the actions of water and gravity. *(p. 43)*

**Southern Oscillation** an oscillation in atmospheric pressure that extends across the Pacific Ocean. *(p. 563)*

**spate** sudden flooding in a stream. *(p. 271)*

**species diversity** a measure of diversity that increases with species evenness and species richness. *(p. 400)*

**species evenness** the relative abundance of species in a community or collection. *(p. 400)*

**species richness** the number of species in a community or collection. *(p. 400)*

**species turnover** changes in species composition on islands resulting from some species becoming extinct and others immigrating. *(p. 543)*

**spiraling length** the length of stream required for an atom of a nutrient to complete a cycle from release into the water column to reentry into the benthic ecosystem. *(p. 472)*

**stability** the persistence of a community or ecosystem in the face of disturbance, usually as a consequence of a combination of resistance and resilience. *(p. 500)*

**stabilizing selection** a form of natural selection that acts against extreme phenotypes; can act to impede changes in populations. *(p. 209)*

**stable age distribution** a population in which the proportion of individuals in each age class is constant. *(p. 262)*

**stable isotope analysis** analysis of the relative concentrations of stable isotopes, such as $^{13}C$ and $^{12}C$, in materials; used in ecology to study the flow of energy and materials through ecosystems. *(p. 139)*

**stamen** male organ of a flower. *(p. 181)*

**standard deviation** the square root of the variance. *(p. 162)*

**standard error** an estimate of variation among means of samples drawn from a population. *(p. 356)*

**static life table** a life table constructed by recording the age at death of a large number of individuals; the table is called static because the method involves a snapshot of survival within a population during a short interval of time. *(p. 256)*

**stratosphere** a layer of earth's atmosphere that extends from about 16 km to an altitude of about 50 km. *(p. 561)*

**stream order** a numerical classification of streams whereby they occur in a stream drainage network. In this classification, headwater streams are first-order streams, joining of two first-order streams forms a second-order stream, joining of two second-order streams forms a third-order stream, and so forth. *(p. 73)*

**stress** in general, stress consists of any strong negative environmental conditions that induce physiological responses in an organism or alter the structure of functioning of an ecosystem. Grime (1977) defined stress in relation to plants as external constraints that limit the rate of dry matter production, or growth, of all or part of the vegetation. *(p. 313)*

**stress-tolerant plants** plants that live under conditions of high stress but low disturbance. *(p. 313)*

**strong interactions** feeding activities of a few species that have a dominant influence on community structure. *(p. 422)*

**succession** the gradual change in plant and animal communities in an area following disturbance or the creation of new substrate. *(p. 485)*

**Suess effect** reduced concentration of $^{14}C$ in the atmosphere as a consequence of fossil fuel burning. *(p. 578)*

**survivorship curve** a graphical summary of patterns of survival in a population. *(p. 257)*

**sympatric** describes the condition in which populations or species have overlapping geographic ranges. *(p. 338)*

**taiga** northern forests that occupy the area south of arctic tundra. Though dominated by coniferous trees they also contain aspen and birch. Also called boreal forest. *(p. 39)*

**temperate forest** deciduous or coniferous forests generally found between 40° and 50° of latitude, where annual precipitation averages anywhere from about 650 mm to over 3,000 mm; this biome receives more winter precipitation than temperate grasslands. *(p. 37)*

**temperate grassland** grasslands growing in middle latitudes that receive between 300 and 1,000 mm of annual precipitation, with maximum precipitation usually falling during the summer months. *(p. 35)*

**thermal neutral zone** the range of environmental temperatures over which the metabolic rate of a homeothermic animal does not change. *(p. 105)*

**thermocline** a depth zone in a lake or ocean through which temperature changes rapidly with depth, generally about 1°C per meter of depth. *(p. 56)*

**thermophilic** a term applied to organisms that tolerate or require high-temperature environments. *(p. 98)*

**thermosphere** the outer layer of the earth's atmosphere beginning approximately 80 km above the earth's surface. *(p. 561)*

**tolerance model** a model of succession in which initial stages of colonization are not limited to a few pioneer species, juveniles of species dominating at climax can be present from the earliest stages of succession, and species colonizing early in succession do not facilitate colonization by species characteristic of later successional stages. Later successional species are simply those tolerant of environmental conditions early in succession. *(p. 496)*

**top-down control** the control or influence of consumers on ecosystem processes. *(p. 449)*

**torpor** a state of low metabolic rate and lowered body temperature. *(p. 111)*

**trophic (feeding) biology** the study of the feeding biology of organisms. *(p. 145)*

**trophic dynamics** the transfer of energy from one part of an ecosystem to another. *(p. 454)*

**trophic level** trophic position in an ecosystem; for instance primary producer, primary consumer, secondary consumer, tertiary consumer, and so forth. *(p. 442)*

**tropical dry forest** a broadleaf deciduous forest growing in tropical regions having pronounced wet and dry seasons; trees drop their leaves during the dry season. *(p. 25)*

**tropical rain forest** a broadleaf evergreen forest growing in tropical regions where conditions are warm and wet year-round. *(p. 23)*

**tropical savanna** a tropical grassland dotted with scattered trees; characterized by pronounced wet and dry seasons and periodic fires. *(p. 28)*

**troposphere** a layer of the atmosphere extending from the earth's surface to an altitude of 9 to 16 km. *(p. 561)*

***t*-test** a statistical test used to compare pairs of samples of measurements with approximately, normal distributions and equal variances. *(p. 452)*

**tundra** a northern biome dominated by mosses, lichens, and dwarf willows, receiving low to moderate precipitation and having a very short growing season. *(p. 42)*

**type I survivorship curve** a pattern of survivorship in which there are high rates of survival among young and middle-aged individuals followed by high rates of mortality among the aged. *(p. 259)*

**type II survivorship curve** a pattern of survivorship characterized by constant rates of survival throughout life. *(p. 259)*

**type III survivorship curve** a pattern of survivorship in which a period of extremely high rates of mortality among the young is followed by a relatively high rate of survival. *(p. 259)*

**upwelling** movement of deeper ocean water to the surface; occurs most commonly along the west coasts of continents and around Antarctica. *(p. 57)*

**vapor pressure deficit** the difference between the actual water vapor pressure and the saturation water vapor pressure at a particular temperature. *(p. 120)*

**variance** a measure of variation in a population or a sample from a population. *(p. 162)*

**vesicle** storage organ in vesicular-arbuscular mycorrhizal fungi. *(p. 376)*

**Walker circulation** a large-scale atmospheric circulation system that moves in the plane of the equator. *(p. 564)*

**water potential** the capacity of water to do work, which is determined by its free energy content; water flows from positions of higher to lower free energy. Increasing solute concentration decreases water potential. *(p. 121)*

**water vapor pressure** the atmospheric pressure exerted by the water vapor in air; increases as the water vapor in air increases. *(p. 120)*

**zonation of species** pattern of separation of species into distinctive vertical habitats or zones. *(p. 67)*

**zooplankton** animals that drift in the surface waters of the oceans or lakes; most zooplankton are microscopic. *(p. 58)*

# References

**Adams, P. A. and J. E. Heath.** 1964. Temperature regulation in the sphinx moth, *Celerio lineata. Nature* 201:20–22.

**Allen, E. B. and M. F. Allen.** 1986. Water relations of xeric grasses in the field: interactions of mycorrhizae and competition. *New Phytologist* 104:559–71.

**Anderson, J. M. and M. J. Swift.** 1983. Decomposition in tropical forests. In S. L. Sutton, T. C. Whitmore, and A. C. Chadwick. eds. *Tropical Rain Forest: Ecology and Management.* Oxford: Blackwell Scientific Publications.

**Angilleta, M. J., Jr.** 2001. Thermal and physiological constraints on energy assimilation in a widespread lizard (*Sceloporus undulatus*). *Ecology* 82:3044–56.

**Arrhenius, O.** 1921. Species and area. *Journal of Ecology* 9:95–99.

**Atlegrim, O.** 1989. Exclusion of birds from bilberry stands: impact on insect larval density and damage to the bilberry. *Oecologia* 79:136–39.

**Bacastow, R. and C. D. Keeling.** 1974. Atmospheric carbon dioxide and radiocarbon in the natural carbon cycle: II. changes from A.D. 1700 to 2070 as deduced from a geochemical model. In G. M. Woodwell and E. V. Pecan. eds. *Carbon and the Biosphere.* BHNL/CONF 720510. Springfield, Va.: National Technical Information Service.

**Bailey, N. T. J.** 1951. On estimating the size of mobile populations from recapture data. *Biometrika* 38:293–306.

**Bailey, N. T. J.** 1952. Improvements in the interpretation of recapture data. *Journal of Animal Ecology* 21:120–27.

**Baird, D., I. Barber, and P. Calow.** 1990. Clonal variation in general responses of *Daphnia magna* Straus to toxic stress. I. Chronic life-history effects. *Functional Ecology* 4:399–407.

**Baker, C. S., J. M. Straley, and A. Perry.** 1992. Population characteristics of individually identified humpback whales in southeastern Alaska: summer and fall 1986. *Fishery Bulletin* 90:429–37.

**Baker, M. C., L. R. Mewaldt, and R. M. Stewart.** 1981. Demography of white-crowned sparrows (*Zonotrichia leucophrys nuttalli*). *Ecology* 62:636–44.

**Baldwin, J. and P. W. Hochachka.** 1970. Functional significance of isoenzymes in thermal acclimation: acetylcholinesterase from trout brain. *Biochemical Journal* 116:883–87.

**Barbour, C. D. and J. H. Brown.** 1974. Fish species diversity in lakes. *American Naturalist* 108:473–89.

**Barnes, R. S. K. and R. N. Hughes.** 1988. *An Introduction to Marine Ecology.* Oxford: Blackwell Scientific Publications.

**Barnola, J. M., D. Raynaud, Y. S. Korotkevich, and C. Lorius.** 1987. Vostok ice core provides 160,000-year record of atmospheric CO₂. *Nature* 329:408–14.

**Baur, B. and A. Baur.** 1993. Climatic warming due to thermal radiation from an urban area as possible cause for the local extinction of a land snail. *Journal of Applied Ecology* 30:333–40.

**Bayley, P. B.** 1995. Understanding large river-floodplain ecosystems. *BioScience* 45:153–58.

**Béjà, O., L. Aravind, E. V. Koonin, M. T. Suzuki, A. Hadd, L. P. Nguyen, S. B. Jovanovich, C. M. Gates, R. A. Feldman, J. L. Spudich, E. N.** Spudich, and E. F. Delong. 2000. Bacterial rhodopsin: evidence for a new type of phototrophy in the sea. *Science* 289:1902–06.

**Béjà, O., E. N. Spudich, J. L. Spudich, M. Leclerc, and E. F. Delong.** 2001. Proteorhodopsin phototrophy in the ocean. *Nature* 411:786–89.

**Béjà, O., M. T. Suzuki, J. F. Heidelberg, W. C. Nelson, C. M. Preston, T. Hamada, J. A. Eisen, C. M. Fraser, and E. F. Delong.** 2002. Unsuspected diversity among marine aerobic anoxygenic phototrophs. *Nature* 415:630–33.

**Bennett, K. D.** 1983. Postglacial population expansion of forest trees in Norfolk, UK. *Nature* 303:164–67.

**Berger, J.** 1990. Persistence of different-sized populations: an empirical assessment of rapid extinctions in bighorn sheep. *Conservation Biology* 4:91–98.

**Berry, J. and O. Björkman.** 1980. Photosynthetic response and adaptation to temperature in higher plants. *Annual Review of Plant Physiology* 31:491–543.

**Bertschy, K. A. and M. G. Fox.** 1999. The influence of age-specific survivorship on pumpkinseed sunfish life histories. *Ecology* 80:2299–2313.

**Bethel, W. M. and J. C. Holmes.** 1977. Increased vulnerability of amphipods to predation due to altered behavior induced by larval parasites. *Can J. Zoology* 55:110–115.

**Bjerknes, J.** 1966. A possible response of the atmospheric Hadley circulation to equatorial anomalies of ocean temperature. *Tellus* 18:820–29.

**Bjerknes, J.** 1969. Atmospheric teleconnections from the equatorial Pacific. *Monthly Weather Review* 97:163–72.

**Bloom, A. J., F. S. Chapin III, and H. A. Mooney.** 1985. Resource limitation in plants—an economic analogy. *Annual Review of Ecology and Systematics* 16:363–92.

**Boag, P. T. and P. R. Grant.** 1978. Heritability of external morphology in Darwin's finches. *Nature* 274:793–94.

**Boag, P. T. and P. R. Grant.** 1984a. Darwin's finches on Isla Daphne Major, Galápagos: breeding and feeding ecology in a climatically variable environment. *Ecological Monographs* 54:463–89.

**Boag, P. T. and P. R. Grant.** 1984b. The classical case of character release: Darwin's finches (*Geospiza*) on Isla Daphne Major, Galápagos. *Biological Journal of the Linnean Society* 22:243–87.

**Bobbink, R. and J. H. Willems.** 1987. Increasing dominance of *Brachypodium pinnatum* (L.) Beauv. in chalk grasslands: a threat to a species-rich ecosystem. *Biological Conservation* 40:301–14.

**Bobbink, R. and J. H. Willems.** 1991. Effect of different cutting regimes on the performance of *Brachypodium pinnatum* in Dutch chalk grassland. *Biological Conservation* 56:1–21.

**Bogdanov, L. V. and N. G. Gagal'chii.** 1986. Intraspecific variation in the Asian ladybug *Harmonia axyridis* Pall. near Vladivostok. *Soviet Journal of Ecology* 17:108–113.

**Bollinger, G.** 1909. *Zur Gastropodenfauna von Basel und Umgebung.* Ph.D. dissertation, University of Basel, Switzerland.

**Bolser, R. C. and M. E. Hay.** 1996. Are tropical plants better defended? Palatability and defenses of temperate vs. tropical seaweeds. *Ecology* 77:2269–86.

**Bond, P. and P. Goldblatt.** 1984. Plants of the Cape Flora. *Journal of South African Botany,* Supplementary Volume No. 13.

**Bonner, J. T.** 1965. *Size and Cycle: An Essay on the Structure of Biology.* Princeton, N.J.: Princeton University Press.

**Bormann, F. H. and G. E. Likens.** 1981. *Pattern and Process in a Forested Ecosystem.* New York: Springer-Verlag.

**Bormann, F. H. and G. E. Likens.** 1994. *Pattern and Process in a Forested Ecosystem.* New York: Springer-Verlag.

**Bowen, G. W. and R. L. Burgess.** 1981. A quantitative analysis of forest island pattern in selected Ohio landscapes. ORNL/TM 7759. Oak Ridge National Laboratory, Oak Ridge, Tenn.

**Bowman, W. D., T. A. Theodose, J. C. Schardt, and R. T. Conant.** 1993. Constraints of nutrient availability on primary production in two alpine tundra communities. *Ecology* 74:2085–97.

**Braatne, J. H., S. B. Rood, and P. E. Heilman.** 1996. Life history, ecology and conservation of riparian cottonwoods in North America. In R. F. Stettler, J. H. D. Bradshaw, P. F. Heilman, P. E. and T. M. Hinckley. eds. *Biology of* Populus *and its Implications for Conservation and Management.* Ottawa: NRC Research Press.

**Brenchley, W. E.** 1958. *The Park Grass Plots at Rothamsted.* Harpende: Rothamsted Experimental Station.

**Briggs, J. C.** 1974. *Marine Zoogeography.* New York: McGraw-Hill.

**Brisson, J. and J. F. Reynolds.** 1994. The effects of neighbors on root distribution in a creosote bush (*Larrea tridentata*) population. *Ecology* 75:1693–702.

**Brock, T. D.** 1978. *Thermophilic Microorganisms and Life at High Temperatures.* New York: Springer-Verlag.

**Brower, J., J. Zar, and C. von Ende.** 1990. *Field and Laboratory Methods for General Ecology,* 4 ed. Dubuque, Ia.: Mc-Graw-Hill.

**Brown, J. H.** 1984. On the relationship between abundance and distribution of species. *American Naturalist* 130:255–79.

**Brown, J. H.** 1986. Two decades of interaction between the MacArthur-Wilson model and the complexities of mammalian distributions. *Biological Journal of the Linnean Society* 28:231–51.

**Brown, J. H.** 1988. Species diversity. In A. A. Meyers and P. S. Giller. eds. *Analytical Biogeography.* London: Chapman and Hall.

**Brown, J. H. and A. Kodric-Brown.** 1977. Turnover rates in insular biogeography: effects of immigration on extinction. *Ecology* 58:445–49.

**Brown, J. H. and M. V. Lomolino.** 2000. Concluding remarks: historical perspective and the future of island biogeography theory. *Global Ecology and Biogeography* 9:87–92.

**Brown, J. H. and J. C. Munger.** 1985. Experimental manipulation of a desert rodent community: food addition and species removal. *Ecology* 66:1545–63.

**Brown, J. H., D. W. Mehlman, and G. C. Stevens.** 1995. Spatial variation in abundance. *Ecology* 76:2028–43.

**Bshary, R.** 2003. The cleaner wrasse, *Labroides dimidiatus,* is a key organism for reef fish diversity at Ras

Mohammed National Park, Egypt. *Journal of Animal Ecology* 72:169–72.

Bush, M. B. and P. A. Colinvaux. 1994. Tropical forest disturbance: paleoecological records from Darien, Panama. *Ecology* 75:1761–68.

Bush, M. B., D. R. Piperno, and P. A. Colinvaux. 1989. A 6,000 year history of Amazonian maize cultivation. *Nature* 340:303–5.

Bush, M. B., D. R. Piperno, P. A. Colinvaux, P. E. De Oliveira, L. A. Krissek, M. C. Miller, and W. E. Rowe. 1992. A 14,300-yr paleoecological profile of lowland tropical lake in Panama. *Ecological Monographs* 62:251–75.

Byers, J. E. 2002. Competition between two estuarine snails: implications for invasions of exotic species. *Ecology* 81:1225–39.

Byers, J. E. and L. Goldwasser. 2001. Exposing the mechanism and timing of impact of nonindigenous species on native species. *Ecology* 82:1330–43.

Cairns, S. C. and G. C. Grigg. 1993. Population dynamics of red kangaroos (*Macropus rufus*) in relation to rainfall in the south Australian pastoral zone. *Journal of Applied Ecology* 30:444–58.

Calder, W. A. 1994. When do hummingbirds use torpor in nature? *Physiological Zoology* 67:1051–76.

Calow, P. and G. E. Petts. 1992. *The Rivers Handbook.* London: Blackwell Scientific Publications.

Carey, F. G. 1973. Fishes with warm bodies. *Scientific American.* 228:36–44.

Carpenter, F. L., M. A. Hixon, C. A. Beuchat, R. W. Russell, and D. C. Patton. 1993. Biphasic mass gain in migrant hummingbirds: body composition changes, torpor, and ecological significance. *Ecology* 74:1173–82.

Carpenter, S. R. and J. F. Kitchell. 1988. Consumer control of lake productivity. *BioScience* 38:764–69.

Carpenter, S. R. and J. F. Kitchell. 1993. *The Trophic Cascade in Lakes.* Cambridge, England: Cambridge University Press.

Carpenter, S. R., J. F. Kitchell, and J. R. Hodgson. 1985. Cascading trophic interactions and lake productivity. *BioScience* 35:634–39.

Carpenter, S. R., T. M. Frost, J. F. Kitchell, T. K. Kratz, D. W. Schindler, J. Shearer, W. G. Sprules, M. J. Vanni, and A. P. Zimmerman. 1991. Patterns of primary production and herbivory in 25 North American lake ecosystems. In J. Cole, G. Lovett, and S. F. Findlay. eds. *Comparative Analyses of Ecosystems: Patterns, Mechanisms, and Theories.* New York: Springer-Verlag.

Carroll, S. P. and C. Boyd. 1992. Host race radiation in the soapberry bug: natural history with the history. *Evolution* 46:1052–69.

Carroll, S. P., H. Dingle, and S. P. Klassen. 1997. Genetic differentiation of fitness-associated traits among rapidly evolving populations of the soapberry bug. *Evolution* 51:1182–88.

Carroll, S. P., S. P. Klassen, and H. Dingle. 1998. Rapidly evolving adaptations to host ecology and nutrition in the soapberry bug. *Evolutionary Ecology* 12:955–68.

Carruthers, R. I., T. S. Larkin, H. Firstencel, and Z. Feng. 1992. Influence of thermal ecology on the mycosis of a rangeland grasshopper. *Ecology* 73:190–204.

Case, T. J. 1976. Body size differences between populations of the chuckwalla, *Sauromalus obesus. Ecology* 57:313–23.

Caughley, G. 1977. *Analysis of Vertebrate Populations.* New York: John Wiley & Sons.

Caughley, G., J. Short, G. C. Grigg, and H. Nix. 1987. Kangaroos and climate: an analysis of distribution. *Journal of Animal Ecology* 56:751–61.

Chapin, F. S., III, L. R. Walker, C. L. Fastie, and L. C. Sharman. 1994. Mechanisms of primary succession following deglaciation at Glacier Bay, Alaska. *Ecological Monographs* 64:149–75.

Chapman, R. N. 1928. The quantitative analysis of environmental factors. *Ecology* 9:111–12.

Chapman, V. J. 1977. *Wet Coastal Ecosystems.* Amsterdam: Elsevier Scientific Publishing.

Charnov, E. L. 1973. *Optimal Foraging: Some Theoretical Explorations.* Ph.D. Dissertation, University of Washington, Seattle.

Charnov, E. L. 2002. Reproductive effort, offspring size and benefit-cost ratios in the classification of life histories. *Evolutionary Ecology Research* 4:1–10.

Charnov, E. L., J. Maynard Smith, and J. J. Bull. 1976. Why be a hermaphrodite? *Nature* 263:125–26.

Chiariello, N. R., C. B. Field, and H. A. Mooney. 1987. Midday wilting in a tropical pioneer tree. *Functional Ecology* 1:3–11.

Christian, C. E. 2001. Consequences of a biological invasion reveal the importance of mutualism for plant communities. *Nature* 413:635–39.

Clausen, J., D. D. Keck, and W. M. Hiesey. 1940. *Experimental Studies on the Nature of Species. I. The Effect of Varied Environments on Western North American Plants.* Washington, D.C.: Carnegie Institution of Washington, Publication no. 520.

Clements, F. E. 1916. *Plant Succession: An Analysis of the Development of Vegetation.* Washington, D.C.: Carnegie Institution of Washington, Publication 242.

Clements, F. E. 1936. Nature and structure of the climax. *Journal of Ecology* 24:252–84.

Coghlan, A. 1993. Horse dung may move sewage mountain. *New Scientist* 137:17.

Coley, P. D. and T. M. Aide. 1991. Comparison of herbivory and plant defenses in temperate and tropical broad-leaved forests. In P. W. Price et al. eds. *Plant-Animal Interactions: Evolutionary Ecology in Tropical and Temperate Regions.* New York: John Wiley & Sons.

Connell, J. H. 1961a. The effects of competition, predation by *Thais lapillus* and other factors on natural populations of the barnacle, *Balanus balanoides. Ecological Monographs* 31:61–104.

Connell, J. H. 1961b. The influence of interspecific competition and other factors on the distribution of the barnacle *Chthamalus stellatus. Ecology* 42:710–23.

Connell, J. H. 1974. Ecology: field experiments in marine ecology. In R. N. Mariscal. ed. *Experimental Marine Biology.* New York: Academic Press.

Connell, J. H. 1975. Some mechanisms producing structure in natural communities: a model and evidence from field experiments. In M. L. Cody and J. Diamond. eds. *Ecology and Evolution of Communities.* Cambridge, Mass.: Harvard University Press.

Connell, J. H. 1978. Diversity in tropical rain forests and coral reefs. *Science* 199:1302–10.

Connell, J. H. 1983. On the prevalence and relative importance of interspecific competition: evidence from field experiments. *American Naturalist* 122:661–96.

Connell, J. H. and R. O. Slatyer. 1977. Mechanisms of succession in natural communities and their role in community stability and organization. *The American Naturalist* 111:1119–44.

Cooper, P. D. 1982. Water balance and osmoregulation in a free-ranging tenebrionid beetle, *Onymacris unguicularis,* of the Namib Desert. *Journal of Insect Physiology* 28:737–42.

Cooper, W. S. 1923. The recent ecological history of Glacier Bay, Alaska. *Ecology* 4:93–128, 223–46, 355–65.

Cooper, W. S. 1931. A third expedition to Glacier Bay, Alaska. *Ecology* 12: 61–95.

Cooper, W. S. 1939. A fourth expedition to Glacier Bay, Alaska. *Ecology* 20: 130–55.

Coppock, D. L., J. K. Detling, J. E. Ellis, and M. I. Dyer. 1983. Plant herbivore interactions in a North American mixed-grass prairie. I. effects of black-tailed prairie dogs on intraseasonal aboveground plant biomass and nutrient dynamics and plant species diversity. *Oecologia* 56:1–9.

Coupland, R. T. and Johnson, R. E. 1965. Rooting characteristics of native grassland species in Saskatchewan. *Journal of Ecology* 53:475–507.

Cox, G. W. and R. E. Ricklefs. 1977. Species diversity, ecological release, and community structure in Caribbean landbird faunas. *Oikos* 29:60–66.

Culp, J. M. and G. J. Scrimgeour. 1993. Size dependent diel foraging periodicity of a mayfly grazer in streams with and without fish. *Oikos* 68:242–50.

Cummins, K. W. and C. N. Dahm. 1995. Restoring the Kissimmee. *Restoration Ecology* 3:147–48.

Curtis, J. T. 1956. The modification of mid-latitude grasslands and forests by man. In W. L. Thomas, Jr. ed. *Man's Role in Changing the Face of the Earth.* Chicago: University of Chicago Press.

Dahm, C. N., K. W. Cummins, H. M. Valett, R. L. Coleman. 1995. An ecosystem view of the restoration of the Kissimmee River. *Restoration Ecology* 3:225–38.

Damuth, J. 1981. Population density and body size in mammals. *Nature* 290:699–700.

Darwin, C. 1839. *Journal of Researches into the Geology and Natural History of the Various Countries Visited During the Voyage of H.M.S. 'Beagle' Under the Command of Captain FitzRoy, R.N., From 1832–1836.* London: Henry Colborn. [also available on the Internet—see "On the Net" section.]

Darwin, C. 1842. *The Structure and Distribution of Coral Reefs.* London: Smith, Elder and Company. Reprinted by the University of California Press, Berkeley, 1962.

Darwin, C. 1859. *The Origin of Species by Means of Natural Selection, or the Preservation of Favored Races in the Struggle for Life.* New York: Modern Library [also available on the Internet—see "On the Net" section.]

Darwin, C. 1862. On the two forms, or dimorphic condition, in the species of *Primula,* and on their remarkable sexual relations. In P. H. Barrett. ed. *The Collected Papers of Charles Darwin.* Chicago: University of Chicago Press.

Darwin, C. 1871. *The Descent of Man, and Selection in Relation to Sex.* London: John Murray.

Davis, M. B. 1981. Quaternary history and the stability of forest communities. In D. C. West, H. H. Shugart, and D. B. Botkin. eds. *Forest Succession: Concepts and Application.* New York: Springer-Verlag.

Davis, M. B. 1983. Quaternary history of deciduous forests of eastern North America and Europe. *Annals of the Missouri Botanical Garden* 70:550–63.

Davis, M. B. 1989. Retrospective studies. In G. E. Likens. ed. *Long-Term Studies in Ecology.* New York: Springer-Verlag.

Dawson, T. E., S. Mambelli, A. H. Plamboeck, P. H. Templer, and K. P. Tu. 2002. Stable isotopes in plant ecology. *Annual Review of Ecology and Systematics* 33:507–99.

Deevey, E. S. 1947. Life tables for natural populations of animals. *Quarterly Review of Biology* 22:283–314.

Denno, R. F. and G. K. Roderick. 1992. Density-related dispersal in planthoppers: effects of interspecific crowding. *Ecology* 73:1323–34.

Diamond, J. M. 1969. Avifaunal equilibria and species turnover rates on the Channel Islands of California. *Proceedings of the National Academy of Sciences* 64:57–63.

Diaz, H. F. and G. N. Kiladis. 1992. Atmospheric teleconnections associated with the extreme phases of the Southern Oscillation. In H. F. Diaz and V. Markgraf. eds. *El Niño Historical and Paleoclimatic Aspects of the Southern Oscillation.* Cambridge, England: Cambridge University Press.

Diffendorfer, J. E., M. S. Gaines, and R. D. Holt. 1995. Habitat fragmentation and movements of three small mammals (*Sigmodon, Microtus,* and *Peromyscus*). *Ecology* 76:827–39.

Dillon, P. J. and F. H. Rigler. 1974. The phosphorus-chlorophyll relationship in lakes. *Limnology and Oceanography* 19:767–73.

Dillon, P. J. and F. H. Rigler. 1975. A simple method for predicting the capacity of a lake for development based on lake trophic status. *Journal of the Fisheries Research Board of Canada* 32:1519–31.

Dirzo, R. and A. Miranda. 1990. Contemporary neotropical defaunation and forest structure, function, and diversity—a sequel to John Terborgh. *Conservation Biology* 4:444–47.

Dobzhansky, T. 1937. *Genetics and the Origin of Species.* New York: Columbia University Press.

Dobzhansky, T. 1950. Evolution in the tropics. *American Scientist* 38:209–21.

Dodd, A. P. 1940. *The Biological Campaign Against Prickly Pear. Commonwealth Prickly Pear Board.* Brisbane: Government Printer.

Dodd, A. P. 1959. The biological control of prickly pear in Australia. In A. Keast, R. L. Crocker, and C. S. Christian. eds. *Biogeography and Ecology in Australia.* Monographs in Biology, Volume VIII.

Dodd, M., J. Silvertown, K. McConway, J. Potts, and M. Crawley. 1995. Community stability: a 60-year record of trends and outbreaks in the occurrence of species in the Park Grass Experiment. *Journal of Ecology* 83:277–85.

DOE. 2000. The Human Genome Project Information. http://www.ornl.gov/ TechResources/Human_Genome/home.html

Douglas, M. R. and P. C. Brunner. 2002. Biodiversity of central alpine *Coregonus* (Salmoniformes): impact of one-hundred years of management. *Ecological Applications* 12:154–72.

Edney, E. B. 1953. The temperature of woodlice in the sun. *Journal of Experimental Biology* 30:331–49.

Ehleringer, J. R. 1980. Leaf morphology and reflectance in relation to water and temperature stress. In N. C. Turner and P. J. Kramer. eds. *Adaptations of Plants to Water and High Temperature Stress.* New York: Wiley-Interscience.

Ehleringer, J. R. and C. Clark. 1988. Evolution and adaptation in *Encelia* (Asteraceae). In L. D. Gottlieb and S. K. Jain. eds. *Plant Evolutionary Biology.* London: Chapman and Hall.

Ehleringer, J. R., O. Björkman, and H. A. Mooney. 1976. Leaf pubescence: effects on absorptance and photosynthesis in a desert shrub. *Science* 192:376–77.

Ehleringer, J. R., J. Roden, and T. E. Dawson. 2000. Assessing ecosystem–level water relations through stable isotope ratio analyses. In O. E. Sala, R. B. Jackson, H. A. Mooney, and R. W. Howarth. eds. *Methods in Ecosystem Science.* New York: Springer.

Ehleringer, J. R., S. L. Phillips, W. S. F. Schuster, and D. R. Sandquist. 1991. Differential utilization of summer rains by desert plants. *Oecologia* 88:430–34.

Elton, C. 1924. Periodic fluctuations in the numbers of animals: their causes and effects. *British Journal of Experimental Biology* 2:119–63.

Elton, C. 1927. *Animal Ecology.* London: Sidgewick & Jackson.

Endler, J. A. 1980. Natural selection on color patterns in *Poecilia reticulata. Evolution* 34:76–91.

Endler, J. A. 1986. *Natural Selection in the Wild.* Princeton, N.J.: Princeton University Press.

Endler, J. A. 1995. Multiple-trait coevolution and environmental gradients in guppies. *Trends in Ecology & Evolution* 10:22–29.

F. A. O. 1972. *Atlas of the Living Resources of the Sea.* 3d ed. Rome: F. A. O.

Feldman, G., D. Clark, and D. Halpern. 1984. Satellite color observations of the phytoplankton distribution in the eastern equatorial Pacific during the 1982–1983 El Niño. *Science* 226:1069–71.

Fenchel, T. 1974. Intrinsic rate of natural increase: the relationship with body size. *Oecologia* 14:317–26.

Fietz, J., F. Tataruch, K. H. Dausmann, and J. U. Ganzhorn. 2003. White adipose tissue composition in the free-ranging fat-tailed dwarf lemur (*Cheirogaleus medius;* Primates), a tropical hibernator. *Journal of Comparative Physiology B: Biochemical Systematic and Environmental Physiology* 173:1–10.

Findlay, D. L. and S. E. M. Kasian. 1987. Phytoplankton community responses to nutrient addition in Lake 226, Experimental Lakes Area, northwestern Ontario. *Canadian Journal of Fisheries and Aquatic Sciences* 44(Suppl. 1):35–46.

Fisher, S. G., L. J. Gray, N. B. Grimm, and D. E. Busch. 1982. Temporal succession in a desert stream ecosystem following flash flooding. *Ecological Monographs* 52:93–110.

Fitter, A. and R. K. M. Hay. 1987. *Environmental Physiology of Plants.* London: Academic Press.

Flessa, K. W. 1975. Area, continental drift and mammalian diversity. *Paleobiology* 1:189–94.

Flessa, K. W. 1981. The regulation of mammalian faunal similarity among continents. *Journal of Biogeography* 8:427–38.

Forbes, S. A. 1887. The lake as a microcosm. *Bulletin of the Peoria Scientific Association.* Reprinted in the *Bulletin of the Illinois State Natural History Survey* 15 (1925):537–50.

Frank, P. W., C. D. Boll, and R. W. Kelly. 1957. Vital statistics of laboratory cultures of *Daphnia pulex* De Geer as related to density. *Physiological Zoology* 30:287–305.

Frankham, R. 1997. Do island populations have less genetic variation than mainland populations? *Heredity* 78:311–327.

Frankham, R. and K. Ralls. 1998. Inbreeding leads to extinction. *Nature* 392:441–42.

Frazer, N. B., J. W. Gibbons, and J. L. Greene. 1991. Life history and demography of the common mud turtle *Kinosternon subrubrum* in South Carolina, USA. *Ecology* 72:2218–31.

Friedli, H., H. Lötscher, H. Oeschger, U. Siegenthaler, and B. Stauffer. 1986. Ice core record of the $^{13}C/^{12}C$ ratio of atmospheric $CO_2$ in the past two centuries. *Nature* 324:237–38.

Friedmann, H. 1955. The honey-guides. *Bulletin of the United States National Museum* 208:1–292.

Gallardo, A. and J. Merino. 1993. Leaf decomposition in two Mediterranean ecosystems of southwest Spain: influence of substrate quality. *Ecology* 74:152–61.

Gaston, K. J. 1996. The multiple forms of the interspecific abundance-distribution relationship. *Oikos* 76:211–20.

Gaston, K. J., T. M. Blackburn, J. J. D. Greenwood, R. D. Gregory, R. M. Quinn, and J.H. Lawton. 2000. Abundance-occupancy relationships. Journal of Applied Ecology 37:39–59.

Gause, G. F. 1934. *The Struggle for Existence.* Baltimore: Williams & Wilkins. Reprinted by Hafner Publishing Company, New York, 1969.

Gause, G. F. 1935. Experimental demonstration of Volterra's periodic oscillation in the numbers of animals. *Journal of Experimental Biology* 12:44–48.

Gauslaa, Y. 1984. Heat resistance and energy budget in different Scandinavian plants. *Holarctic Ecology* 7:1–78.

Gessner, M. O. and E. Chauvet. 1994. Importance of stream microfungi in controlling breakdown rates of leaf litter. *Ecology* 75:1807–17.

Gibbs, H. L. and P. R. Grant. 1987. Ecological consequences of an exceptionally strong El Niño event on Darwin's finches. *Ecology* 68:1735–46.

Gibbs, R. J. 1970. Mechanisms controlling world water chemistry. *Science* 170:1088–90.

Gleason, H. A. 1926. The individualistic concept of the plant association. *Torrey Botanical Club Bulletin* 53:7–26.

Gleason, H. A. 1939. The individualistic concept of the plant association. *American Midland Naturalist* 21:92–110.

Glynn, P. W. 1983. Crustacean symbionts and the defense of corals: coevolution of the reef? In M. H. Nitecki. ed. *Coevolution.* Chicago: University of Chicago Press.

Gosz, J. R., R. T. Holmes, G. E. Likens, and F. H. Bormann. 1978. The flow of energy in a forest ecosystem. *Scientific American* 238(3):92–102.

Graça, M. A. S. and F. X. Ferrand de Almeida. 1983. Contribuição para o conhecimento da lontra (*Lutra lutra* L.) num sector da bacia do Rio Mondego. *Ciencia Biológica* (Contribution to the knowledge of the otter (*Lutra lutra* L.) in a sector of the Mondego River basin.) (Portugal) 5:33–42.

Graham, W. F. and R. A. Duce. 1979. Atmospheric pathways of the phosphorus cycle. *Geochimica et Cosmochimica Acta* 43:1195–1208.

Granéli, E., K. Wallström, U. Larsson, W. Granéli, and R. Elmgren. 1990. Nutrient limitation of primary production in the Baltic Sea area. *Ambio* 19:142–51.

Grant, B. R. and P. R. Grant. 1989. *Evolutionary Dynamics of a Natural Population.* Chicago: University of Chicago Press.

Grant, P. R. 1986. *Ecology and Evolution of Darwin's Finches.* Princeton, N.J.: Princeton University Press.

Grassle, J. F. 1973. Variety in coral reef communities. In O. A. Jones and R. Endean. eds. *Biology and Geology of Coral Reefs.* Vol. 2. New York: Academic Press.

Grassle, J. F. 1991. Deep-sea benthic biodiversity. *BioScience* 41:464–69.

Grice, G. D. and A. D. Hart. 1962. The abundance, seasonal occurrence and distribution of the epizooplankton between New York and Bermuda. *Ecological Monographs* 32:287–309.

Grime, J. P. 1977. Evidence for the existence of three primary strategies in plants and its relevance to ecological and evolutionary theory. *American Naturalist* 111:1169–94.

Grime, J. P. 1979. *Plant Strategies and Vegetation Processes.* New York: John Wiley & Sons.

Grimm, N. B. 1987. Nitrogen dynamics during succession in a desert stream. *Ecology* 68:1157–70.

Grimm, N. B. 1988. Role of macroinvertebrates in nitrogen dynamics of a desert stream. *Ecology* 69:1884–93.

Grinnell, J. 1917. The niche-relationships of the California Thrasher. *Auk* 34:427–33.

Grinnell, J. 1924. Geography and evolution. *Ecology* 5:225–29.

Grosholz, E. D. 1992. Interactions of intraspecific, interspecific, and apparent competition with host-pathogen population dynamics. *Ecology* 73:507–14.

Gross, J. E., L. A. Shipley, N. T. Hobbs, D. E. Spalinger, and B. A. Wunder. 1993. Functional response of herbivores in food-concentrated patches: tests of a mechanistic model. *Ecology* 74:778–91.

Grutter, A. S. 1999. Cleaner fish really do clean. *Nature* 398:672–73.

Gunderson, D. R. 1997. Trade-off between reproductive effort and adult survival in oviparous and viviparous fishes. *Canadian Journal of Fisheries and Aquatic Sciences* 54:990–998.

Gurevitch, J., L. L. Morrow, A. Wallace, and J. S. Walsh. 1992. A meta-analysis of competition in field experiments. *American Naturalist* 140:539–72.

Haddad, N. M. 1999. Corridor and distance effects on interpatch movements: a landscape experiment with butterflies. *Ecological Applications* 9:612–22.

Haddad, N. M. and K. A. Baum. 1999. An experimental test of corridor effects on butterfly densities. *Ecological Applicatons* 9:623–33.

Hadley, N. F. and T. D. Schultz. 1987. Water loss in three species of tiger beetles (*Cicindela*): correlations with epicuticular hydrocarbons. *Journal of Insect Physiology* 33:677–82.

Hadley, N. F., A. Savill, and T. D. Schultz. 1992. Coloration and its thermal consequences in the New Zealand tiger beetle *Neocicindela perhispida. Journal of Thermal Biology* 17:55–61.

Hairston, N. G., Sr. 1989. *Ecological Experiments: Purpose, Design, and Execution.* Cambridge, England: Cambridge University Press.

Hall, J. B. and M. D. Swaine. 1976. Classification and ecology of closed-canopy forest in Ghana. *Journal of Ecology* 64:913–51.

Hamilton, W. D. 1964. The genetical evolution of social behavior, I and II. *Journal of Theoretical Biology* 7:1–52.

Hansen, K. T., R. Elven, and C. Brochmann. 2000. Molecules and morphology in concert: tests of some hypotheses in arctic *Potentilla* (Rosaceae). *American Journal of Botany* 87:1466–79.

Hanski, I. 1982. Dynamics of regional distribution: the core and satellite hypothesis. *Oikos* 38:210–21.

**Hanski, I., M. Kuussaari, and M. Nieminen.** 1994. Metapopulation structure and migration in the butterfly *Melitaea cinxia*. *Ecology* 75:747–62.

**Hardie, K.** 1985. The effect of removal of extraradical hyphae on water uptake by vesicular-arbuscular mycorrhizal plants. *New Phytologist* 101:677–84.

**Hardy, G. H.** 1908. Mendelian proportions in a mixed population. *Science* 28:49–50.

**Harper, J. L., P. H. Lovell, and K. G. Moore.** 1970. The shapes and sizes of seeds. *Annual Review of Ecology and Systematics* 1:327–356.

**Harrington, G. N.** 1991. Effects of soil moisture on shrub seedling survival in a semi-arid grassland. *Ecology* 72:1138–49.

**Haukioja, E., K. Kapiainen, P. Niemelä, and J. Tuomi.** 1983. Plant availability hypothesis and other explanations of herbivore cycles: complementary or exclusive alternatives? *Oikos* 40:419–32.

**Havens, K. E.** 1994. A preliminary characterization of the Lake Okeechobee (Florida, USA) food web. *Bulletin of the North American Benthological Society* 11:97.

**Heath, J. E. and P. J. Wilkin.** 1970. Temperature responses of the desert cicada, *Diceroprocta apache* (Homoptera, Cicadidae). *Physiological Zoology* 43:145–54.

**Hedin, L. O., P. M. Vitousek, and P. A. Matson.** 2003. Nutrient losses over four million years of tropical forest development. *Ecology* 84:2231–55.

**Hegazy, A. K.** 1990. Population ecology and implications for conservation of *Cleome droserifolia*: a threatened xerophyte. *Journal of Arid Environments* 19:269–82.

**Heinrich, B.** 1979. *Bumblebee Economics*. Cambridge, Mass.: Harvard University Press.

**Heinrich, B.** 1984. Strategies of thermoregulation and foraging in two vespid wasps, *Dolichovespula maculata* and *Vespula vulgaris. Journal of Comparative Physiology* B154:175–80.

**Heinrich, B.** 1993. *The Hot-Blooded Insects*. Cambridge, Mass.: Harvard University Press.

**Hengeveld, R.** 1988. Mechanisms of biological invasions. *Journal of Biogeography* 15:819–28.

**Hengeveld, R.** 1989. *Dynamics of Biological Invasions*. New York: Chapman and Hall.

**Heron, A. C.** 1972a. Population ecology of a colonizing species: the pelagic tunicate *Thalia democratica* I. individual growth rate and generation time. *Oecologia* 10:269–93.

**Heron, A. C.** 1972b. Population ecology of a colonizing species: the pelagic tunicate *Thalia democratica* II. population growth rate. *Oecologia* 10:293–312.

**Heske, E. J., J. H. Brown, and S. Mistry.** 1994. Long-term experimental study of a Chihuahuan Desert rodent community: 13 years of competition. *Ecology* 75:438–45.

**Higuchi, T.** 1983. *The Visual and Spatial Structure of Landscapes*. Cambridge, Mass.: MIT Press.

**Hillis, D. M., B. K. Mable, A. Larson, S. K. Davis, and E. A. Zimmer.** 1996. Nucleic acids IV: sequencing and cloning. In D. M. Hillis, C. Moritz, and B. K. Mable. eds. *Molecular Systematics*. Sunderland, Mass.: Sinauer Associates, Inc.

**Hillis, D. M., C. Moritz, and B. K. Mable.** 1996. *Molecular Systematics*. Sunderland, Mass.: Sinauer Associates, Inc.

**Hofkin, B. V., D. K. Koech, J. Ouma, and E. S. Loker.** 1991a. The North American crayfish *Procambarus clarkii* and the biological control of schistosome-transmitting snails in Kenya: laboratory and field investigations. *Biological Control* 1:183–87.

**Hofkin, B. V., G. M. Mkoji, D.K. Koech, and E. S. Loker.** 1991b. Control of schistosome-transmitting snails in Kenya by the North American crayfish *Procambarus clarkii. American Journal of Tropical Medicine and Hygiene* 45:339–44.

**Hogetsu, K. and S. Ichimura.** 1954. Studies on the biological production of Lake Suwa. 6. The ecological studies in the production of phytoplankton. *Japanese Journal of Botany* 14:280–303.

**Hölldobler, B. and E. O. Wilson.** 1990. *The Ants*. Cambridge, Mass.: The Belknap Press of Harvard University Press.

**Holling, C. S.** 1959. The components of predation as revealed by a study of small mammal predation of the European pine sawfly. *The Canadian Entomologist* 91:293–320.

**Houde, A. E.** 1997. *Sex, Color, and Mate Choice in Guppies*. Princeton, N.J.: Princeton University Press.

**Howe, W. H. and F. L. Knopf.** 1991. On the imminent decline of the Rio Grande cottonwoods in central New Mexico. *Southwestern Naturalist* 36:218–24.

**Huang, H. T. and P. Yang.** 1987. The ancient cultured citrus ant. *BioScience* 37:665–67.

**Hubbell, S. P. and L. K. Johnson.** 1977. Competition and nest spacing in a tropical stingless bee community. *Ecology* 58:949–63.

**Huber, R., H. Huber, and K. O. Stetter.** 2000. Towards the ecology of hyperthermophiles: biotopes, new isolation strategies and novel metabolic properties. *FEMS Microbiology Reviews* 24:615–23.

**Hudson, P. J., A. P. Dobson, and D. Newborn.** 1992. Do parasites make prey vulnerable to predation? *Journal of Animal Ecology* 61: 681–92.

**Huffaker, C. B.** 1958. Experimental studies on predation: dispersion factors and predator-prey oscillations. *Hilgardia* 27:343–83.

**Hughes, T. P.** 1996. Demographic approaches to community dynamics: a coral reef example. *Ecology* 77:2256–60.

**Hulshoff, R. M.** 1995. Landscape indices describing a Dutch landscape. *Landscape Ecology* 10:101–11.

**Huntly, N. and R. Inouye.** 1988. Pocket gophers in ecosystems: patterns and mechanisms. *BioScience* 38:786–93.

**Huston, M.** 1980. Soil nutrients and tree species richness in Costa Rican forests. *Journal of Biogeography* 7:147–57.

**Huston, M.** 1994a. Biological diversity, soils, and economics. *Science* 262:1676–79.

**Huston, M.** 1994b. *Biological Diversity*. New York: Cambridge University Press.

**Hutchinson, G. E.** 1957. Concluding remarks. *Cold Spring Symposia on Quantitative Biology* 22:415–27.

**Hutchinson, G. E.** 1959. Homage to Santa Rosalia or why are there so many kinds of animals? *American Naturalist* 93:145–59.

**Hutchinson, G. E.** 1961. The paradox of the plankton. *American Naturalist* 95:137–45.

**Hutchinson, G. E.** 1978. *An Introduction to Population Ecology*. New Haven, Conn.: Yale University Press.

**Ichimura, S.** 1956. On the standing crop and productive structure of phytoplankton community in some lakes of central Japan. *Japanese Botany Magazine Tokyo* 69:7–16.

**Inouye, D. W. and O. R. Taylor, Jr.** 1979. A temperate region plant-ant-seed predator system: consequences of extrafloral nectar secretion by *Helianthella quinquenervis. Ecology* 60:1–7.

**Iriarte, J. A., W. L. Franklin, W. E. Johnson, and K. H. Redford.** 1990. Biogeographic variation of food habits and body size of the American puma. *Oecologia* 85:185–90.

**Isack, H. A. and H.-U. Reyer.** 1989. Honeyguides and honey gatherers: interspecific communication in a symbiotic relationship. *Science* 243:1343–46.

**Ishii, K. and K. Marumo.** 2002. Microbial diversity in hydrothermal systems and their influence on geological environments. *Resource Geology* 52:135–46.

**Jackson, J. B. C.** 1991. Adaptation and diversity of reef corals. *BioScience* 41:475–82.

**Jaenike, J.** 1991. Mass extinction of European fungi. *TREE* 6:174–75.

**Jakobsson, A. and O. Eriksson.** 2000. A comparative study of seed number, seed size,seedling size and recruitment in grassland plants. *Oikos* 88:494–502.

**Janzen, D. H.** 1966. Coevolution of mutualism between and acacias in Central America. *Evolution* 20:249–75.

**Janzen, D. H.** 1967. Why mountain passes are higher in the tropics. *American Naturalist* 101:233–49.

**Janzen, D. H.** 1967a. Fire, vegetation structure, and the ant x acacia interaction in Central America. *Ecology* 48:26–35.

**Janzen, D. H.** 1967b. Interaction of the bull's-horn acacia (*Acacia cornigera* L.) with an ant inhabitant (*Pseudomyrmex ferruginea* F. Smith) in eastern Mexico. *The University of Kansas Science Bulletin* 47:315–558.

**Janzen, D. H.** 1978. Seeding patterns of tropical trees. In P. B. Tomlinson and M. H. Zimmermann. eds. *Tropical Trees as Living Systems*. Cambridge, England: Cambridge University Press.

**Janzen, D. H.** 1981. The peak in North American ichneumonid species richness lies between 38° and 42°. *Ecology* 62:532–37.

**Janzen, D. H.** 1981a. Guanacaste tree seed-swallowing by Costa Rican range horses. *Ecology* 62:587–92.

**Janzen, D. H.** 1981b. *Enterolobium cyclocarpum* seed passage rate and survival in horses, Costa Rican Pleistocene seed dispersal agents. *Ecology* 62:593–601.

**Janzen, D. H.** 1985. Natural history of mutualisms. In D. H. Boucher. ed. *The Biology of Mutualism: Ecology and Evolution*. London: Croom Helm.

**Jarvis, J. U. M.** 1981. Eusociality in a mammal: cooperative breeding in naked mole-rat colonies. *Science* 212:571–573.

**Jenny, H.** 1980. *The Soil Resource*. New York: Springer-Verlag.

**Jepsen, D. B. and K. O. Winemiller.** 2002. Structure of tropical river food webs revealed by stable isotope ratios. *Oikos* 96:46–55.

**Johnson, N. C.** 1993. Can fertilization of soil select less mutualistic mycorrhizae. *Ecological Applications* 3:749–57.

**Johnston, D. W. and E. P. Odum.** 1956. Breeding bird populations in relation to plant succession on the Piedmont of Georgia. *Ecology* 37:50–62.

**Jonsson, M. and B. Malmqvist.** 2003. Mechanisms behind positive diversity effects on ecosystem functioning: testing the facilitation and interference hypotheses. *Oecologia* 134:554–59.

**Jordan, C. F.** 1985. Soils of the Amazon rain forest. In G. T. Prance and T. E. Lovejoy. eds. *Amazonia*. Oxford: Pergamon Press.

**Junk, W. J., P. B. Bayley, and R. E. Sparks.** 1989. The flood pulse concept in river-floodplain systems. *Canadian Special Publication in Fisheries and Aquatic Sciences* 106:110–27.

**Kairiukstis, L. A.** 1967. In J. L. Tselniker (ed.) Svetovoi rezhim fotosintez i produktiwnost lesa. (Light regime, photosynthesis and forest productivity) Nauka, Moscow.

**Kallio, P. and L. Kärenlampi.** 1975. Photosynthesis in mosses and lichens. In J. P. Cooper. ed. *Photosynthesis and Productivity in Different Environments*. Cambridge, England: Cambridge University Press.

**Karr, J. R.** 1991. Biological integrity: a long-neglected aspect of water resource management. *Ecological Applications* 1:66–84.

**Karr, J. R. and D. R. Dudley.** 1981. Ecological perspective on water quality goals. *Environmental Management* 5:55–68.

**Kaser, S. A. and J. Hastings.** 1981. Thermal physiology of the cicada, *Tibicen duryi. American Zoologist* 21:1016.

**Kates, R. W., B. L. Turner II, and W. C. Clark.** 1990. The great transformation. In B. L. Turner II, W. C. Clark, R. W. Kates, J. F. Richards, J. T. Mathews, and W. B. Meyer. eds. *The Earth as Transformed by Human Action*. Cambridge, England: Cambridge University Press.

**Katona, S. K.** 1989. Getting to know you. *Oceanus* 32:37–44.

**Kauffman, J. B., R. L. Sanford Jr., D. L., Cummings, I. H. Salcedo, and E. V. S. B. Sampaio.** 1993. Biomass and nutrient dynamics associated with slash fires in neotropical dry forests. *Ecology* 74:140–51.

**Kaufman, L.** 1992. Catastrophic change in species rich freshwater ecosystems. *BioScience* 42:846–58.

**Keeler, K. H.** 1981. A model of selection for facultative nonsymbiotic mutualism. *American Naturalist* 118:488–98.

**Keeler, K. H.** 1985. Benefit models of mutualism. In D. H. Boucher. ed. *The Biology of Mutualism: Ecology and Evolution.* London: Croom Helm.

**Keeling, C. D. and T. P. Whorf.** 1994. Atmospheric $CO_2$ records from sites in the SIO air sampling network. In T. A. Boden, D. P. Kaiser, R. J. Sepan- ski, and F. W. Stoss. eds. *Trends' 93: A Compendium of Data on Global Change.* ORNL/CDIAC-65. Oak Ridge, Tenn.: Carbon Dioxide Information Analysis Center, Oak Ridge National Laboratory.

**Keever, Catherine.** 1950. Causes of succession on old fields of the Piedmont, North Carolina. *Ecological Monographs* 20:230–50.

**Keith, L. B.** 1963. *Wildlife's Ten-year Cycle.* Madison, Wis.: University of Wisconsin Press.

**Keith, L. B.** 1983. Role of food in hare population cycles. *Oikos* 40:385–95.

**Keith, L. B., J. R. Cary, O. J. Rongstad, and M. C. Brittingham.** 1984. Demography and ecology of a declining snowshoe hare population. *Wildlife Monographs* 90:1–43.

**Kempton, R. A.** 1979. The structure of species abundance and measurement of diversity. *Biometrics* 35:307–21.

**Kettlewell, H. B. D.** 1959. Darwin's missing evidence. *Scientific American* 200:48–53.

**Kevan, P. G.** 1975. Sun-tracking solar furnaces in high arctic flowers: significance for pollination and insects. *Science* 189:723–26.

**Kidron, G. J., E. Barzilay, and E. Sachs.** 2000. Microclimate control upon sand microbiotic crusts, western Negev Desert, Israel. *Geomorphology* 36:1–18.

**Killingbeck, K. T. and W. G. Whitford.** 1996. High foliar nitrogen in desert shrubs: an important ecosystem trait or defective desert doctrine? *Ecology* 77:1728–37.

**Klemmedson, J. O.** 1975. Nitrogen and carbon regimes in an ecosystem of young dense ponderosa pine in Arizona. *Forest Science* 21:163–68.

**Knutson, R. M.** 1974. Heat production and temperature regulation in eastern skunk cabbage. *Science* 186:746–47.

**Knutson, R. M.** 1979. Plants in heat. *Natural History* 88:42–47.

**Kodric-Brown, A.** 1993. Female choice of multiple male criteria in guppies: interacting effects of dominance, coloration and courtship. *Behavioral Ecology and Sociobiology* 32:415–420.

**Koebel, J. W., Jr.** 1995. A historical perspective on the Kissimmee River Restoration Project. *Restoration Ecology* 3:149–59.

**Kolber, Z. S., C. L. Van Dover, R. A. Niederman, and P. G. Falkowski.** 2000. Bacterial photosynthesis in surface waters of the open ocean. *Nature* 407:177–79.

**Komai, T. and Y. Hosino.** 1951. Contributions to the evolutionary genetics of the lady-beetle, *Harmonia.* II Microgeographic variations. *Genetics* 36:382–390.

**Korpimäki, E.** 1988. Factors promoting polygyny in European birds of prey—a hypothesis. *Oecologia* 77:278–85.

**Korpimäki, E. and K. Norrdahl.** 1991. Numerical and functional responses of kestrels, short-eared owls, and long-eared owls to vole densities. *Ecology* 72:814–26.

**Krebs, C. J., R. Boonstra, S. Boutin, and A. R. E. Sinclair.** 2001. What drives the 10-year cycle of snowshoe hares? *BioScience* 51:25–36.

**Krebs, C. J., S. Boutin, R. Boonstra, A. R. E. Sinclair, J. N. M. Smith, M. R. T. Dale, K. Martin, and R. Turkington.** 1995. Impact of food and predation on the snowshoe hare cycle. *Science* 269:1112–15.

**Lack, D.** 1947. *Darwin's Finches.* Cambridge, England: Cambridge University Press.

**Lamberti, G. A. and V. H. Resh.** 1983. Stream periphyton and insect herbivores: an experimental study of grazing by a caddisfly population. *Ecology* 64:1124–35.

**Larcher, W.** 1995. *Physiological Plant Ecology.* 3d ed. Berlin: Springer.

**Latham, R. E. and R. E. Ricklefs.** 1993. Continental comparisons of temperate-zone tree species diversity. In R. E. Ricklefs and D. Schluter. eds. *Species Diversity in Ecological Communities.* Chicago: University of Chicago Press.

**Lawton, J. H., D. E. Bignell, B. Bolton, G. F. Bloemers, P. Eggleton, P. M. Hammond, M. Hodda, R. D. Holt, T. B. Larsen, N. A. Mawdsley, N. E. Stork, D. S. Srivastava, and A. D. Watt.** 1998. Biodiversity inventories, indicator taxa and effects of habitat modification in tropical forest. *Nature* 391:72–76.

**Lebo, M. E., J. E. Reuter, C. R. Goldman, C. L. Rhodes, N. Vucinich, and D. Mosely.** 1993. Spatial variations in nutrient and particulate matter concentrations in Pyramid Lake, Nevada, USA, during a dry period. *Canadian Journal of Fisheries and Aquatic Science* 50:1045–54.

**Ledig, F. T., V. Jacob-Cervantes, P. D. Hodgskiss, and T. Eguiluz-Piedra.** 1997. Recent evolution and divergence among populations of a rare Mexican endemic, Chihuahua spruce, following Holocene climatic warming. *Evolution* 51:1815–27.

**Leonard, P. M. and D. J. Orth.** 1986. Application and testing of an index of biotic integrity in small, cool-water streams. *Transactions of the American Fisheries Society* 115:401–14.

**Levang-Brilz, N. and M. E. Biondini.** 2002. Growth rate, root development and nutrient uptake of 55 plant species from the Great Plains Grasslands, USA. *Plant Ecology* 165:117–44.

**Leverich, W. J. and D. A. Levin.** 1979. Age-specific survivorship and reproduction in *Phlox drummondii. American Naturalist* 113:881–903.

**Liebig, J.** 1840. *Chemistry in its Application to Agriculture and Physiology.* London: Taylor and Walton.

**Ligon, D.** 1999. *The Evolution of Avian Mating Systems.* Oxford: Oxford University Press.

**Ligon, J. D. and S. H. Ligon.** 1978. Communal breeding in green woodhoopoes as a case for reciprocity. *Nature* (London) 276: 496–98.

**Ligon, J. D. and S. H. Ligon.** 1982. The cooperative breeding behavior of the green woodhoopoe. *Scientific American* 247: 126–34.

**Ligon, J. D. and S. H. Ligon.** 1989. Green Woodhoopoe. In I. Newton. ed. *Lifetime Reproduction in Birds.* London: Academic Press Ltd.

**Ligon, J. D. and S. H. Ligon.** 1991. Green Woodhoopoe: life history traits and sociality. In P. B. Stacey and W. D. Koenig. ed. *Long-term Studies of Behavior and Ecology.* Cambridge: University Press.

**Ligtvoet, W. and F. Witte.** 1991. Perturbation through predator introduction: effects on the food web and fish yields in Lake Victoria (East Africa). In O. Ravera. ed. *Terrestrial and Aquatic Ecosystems: Perturbation and Recovery.* New York: Ellis Horwood.

**Likens, G. E. and F. H. Bormann.** 1995. *Biogeochemistry of a Forested Ecosystem.* 2d ed. New York: Springer-Verlag.

**Likens, G. E., F. H. Bormann, N. M. Johnson, D. W. Fisher, and R. S. Pierce.** 1970. Effects of forest cutting and herbicide treatment on nutrient budgets in the Hubbard Brook watershed-ecosystem. *Ecological Monographs* 40:23–47.

**Likens, G. E., F. H. Bormann, R. S. Pierce, and W. A. Reiners.** 1978. Recovery of a deforested ecosystem. *Science* 199:492–96.

**Lilleskov, E. A., T. J. Fahey, T. R. Horton, and G. M. Lovett.** 2002. Belowground ectomycorrhizal fungal community change over a nitrogen deposition gradient in Alaska. *Ecology* 83:104–15.

**Lindeman, R. L.** 1942. The trophic-dynamic aspect of ecology. *Ecology* 23:399–418.

**Lindström, E. R., H. Andrén, P. Angelstam, G. Cederlund, B. Hörnfeldt, L. Jäderberg, P.-A. Lemnell, B. Martinsson, K. Sköld, and J. E. Swenson.** 1994. Disease reveals the predator: sarcoptic mange, red fox predation, and prey populations. *Ecology* 75:1042–49.

**Lomolino, M. V.** 1990. The target hypothesis—the influence of island area on immigration rates of non–volant mammals. *Oikos* 57:297–300.

**Lomolino, M. V., J. H. Brown, and R. Davis.** 1989. Island biogeography of montane forest mammals in the American Southwest. *Ecology* 70:180–94.

**Long, S. P. and C. F. Mason.** 1983. *Saltmarsh Ecology.* Glasgow: Blackie.

**Lorius, C., J. Jouzel, C. Ritz, L. Merlivat, N. I. Barkov, Y. S. Korotkevich, and V. M. Kotlyakov.** 1985. A 150,000-year climatic record from antarctic ice. *Nature* 316:591–96.

**Losos, J. B., K. I. Warheit, and T. W. Schoener.** 1997. Adaptive differentiation following experimental island colonization in *Anolis* lizards. *Nature* 387:70–73.

**Lotka, A. J.** 1925. *Elements of Physical Biology.* Baltimore, Md.: Williams and Wilkins.

**Lotka, A. J.** 1932a. Contribution to the mathematical theory of capture. I. Conditions for capture. *Proceedings of the National Academy of Science* 18:172–200.

**Lotka, A. J.** 1932b. The growth of mixed populations: two species competing for a common food supply. *Journal of the Washington Academy of Sciences* 22:461–69.

**Lubchenko, J.** 1978. Plant species diversity in a marine intertidal community: importance of herbivore food preference and algal competitive abilities. *American Naturalist* 112:23–39.

**MacArthur, R. H.** 1958. Population ecology of some warblers of northeastern coniferous forests. *Ecology* 39:599–619.

**MacArthur, R. H.** 1972. *Geographical Ecology.* New York: Harper & Row.

**MacArthur, R. H. and E. R. Pianka.** 1966. On optimal use of a patchy environment. *American Naturalist* 100:603–9.

**MacArthur, R. H. and E. O. Wilson.** 1963. An equilibrium theory of insular zoogeography. *Evolution* 17:373–87.

**MacArthur, R. H. and E. O. Wilson.** 1967. *The Theory of Island Biogeography.* Princeton, N.J.: Princeton University Press.

**MacArthur, R. H. and J. W. MacArthur.** 1961. On bird species diversity. *Ecology* 42:594–98.

**MacLachlan, A.** 1983. Sandy beach ecology—a review. In A. McLachlan and T. Erasmus. eds. *Sandy Beaches as Ecosystems.* The Hague: Dr. W. Junk Publishers.

**MacLulich, D. A.** 1937. Fluctuation in the numbers of the varying hare (*Lepus americanus*). *University of Toronto Studies in Biology Series No. 43.*

**Mahoney, J. M. and S. B. Rood.** 1998. Streamflow requirements for cottonwood seedling recruitment—an integrative model. *Wetlands* 18:634–45.

**Mandelbrot, B.** 1982. *The Fractal Geometry of Nature.* New York: W. H. Freeman.

**Margulis, L. and R. Fester.** 1991. *Symbiosis as a Source of Evolutionary Innovation: Speciation and Morphogenesis* Cambridge, Mass.: MIT Press.

**Marquis, R. J. and C. J. Whelan.** 1994. Insectivorous birds increase growth of white oak through consumption of leaf-chewing insects. *Ecology* 75:2007–14.

**Marshall, D. L. and M. W. Folsom.** 1991. Mate choice in plants: an anatomical to population perspective. *Annual Review of Ecology and Systematics* 22:37–63.

**Marshall, D. L. and M. W. Folsom.** 1992. Mechanisms of nonrandom mating in wild radish. In R. Wyatt. ed. *Ecology and Evolution of Plant Reproduction: New Approaches.* New York: Chapman and Hall.

**Marshall, D. L. and O. S. Fuller.** 1994. Does nonrandom mating among wild radish plants occur in the field as well as in the greenhouse? *American Journal of Botany* 81:439–445.

**Marshall, D. L., M. W. Folsom, C. Hatfield, and T. Bennett.** 1996. Does interference competition among pollen grains occur in wild radish? *Evolution* 50:1842–48.

Marshall, D. L. 1990. Non-random mating in a wild radish, *Raphanus sativus. Plant Species Biology* 5:143–56.

Martikainen, P. and J. Kouki. 2003. Sampling the rarest: threatened beetles in borealforest biodiversity inventories. *Biodiversity and Conservation* 12:1815–31.

May, R. M. 1975. Patterns of species abundance and diversity. In M. L. Cody and J. M. Diamond. eds. *Ecology and Evolution of Communities.* Cambridge, Mass.: Harvard University Press.

May, R. M. 1989. Honeyguides and humans. *Nature* 338:707–8.

McAuliffe, J. R. 1994. Landscape evolution, soil formation, and ecological patterns and processes in Sonoran Desert bajadas. *Ecological Monographs* 64:111–48.

McNaughton, S. J. 1976. Serengeti migratory wildebeest: facilitation of energy flow by grazing. *Science* 191:92–94.

McNaughton, S. J. 1985. Ecology of a grazing ecosystem: the Serengeti. *Ecological Monographs* 55:259–94.

McNaughton, S. J., R. W. Ruess, and S. W. Seagle. 1988. Large mammals and process dynamics in African ecosystems. *BioScience* 38:794–800.

Meentemeyer, V. 1978. An approach to the biometeorology of decomposer organisms. *International Journal of Biometeorology* 22:94–102.

Melillo, J. M., J. D. Aber, and J. F. Muratore. 1982. Nitrogen and lignin control of hardwood leaf litter decomposition dynamics. *Ecology* 63:621–26.

Mendel, G. 1866. Versuche über Pflanzen-Hybriden (Experiments in plant hybridization). *Verhandlungen des Naturforschenden Vereines, Abhandlungen, Brünn* 4:3–47.

Mertz, D. B. 1972. The *Tribolium* model and the mathematics of population growth. *Annual Review of Ecology and Systematics* 3:51–106.

Messier, F. 1994. Ungulate population models with predation: a case study with the North American moose. *Ecology* 75:478–88.

Meybeck, M. 1982. Carbon, nitrogen, and phosphorus transport by world rivers. *American Journal of Science* 282:401–50.

Meyer, J. L. and G. E. Likens. 1979. Transport and transformation of phosphorus in a forest stream ecosystem. *Ecology* 60:1255–69.

Middleton, I. 1984. Are plant toxins aimed at decomposers? *Experientia* 40:299–301.

Miller, R. B. 1923. First report on a forestry survey of Illinois. *Illinois Natural History Bulletin* 14:291–377.

Mills, E. L., J. H. Leach, J. T. Carlton, and C. L. Secor. 1994. Exotic species and the integrity of the Great Lakes. *BioScience* 44:666–76.

Mills, K. H. and D. W. Schindler. 1987. Preface. *Canadian Journal of Fisheries and Aquatic Sciences* 44(Suppl.1):3–5.

Milne, B. T. 1993. Pattern analysis for landscape evaluation and characterization. In M. E. Jensen and P. S. Bourgeron. eds. *Ecosystem Management: Principles and Applications.* Gen. Tech. Report PNW-GTR-318. Portland, Ore.: U.S. Department of Agriculture Forest Service, Pacific Northwest Research Station.

Milne, B. T., A. R. Johnson, T. H. Keitt, C. A. Hatfield, J. David, and P. T. Hraber. 1996. Detection of critical densities associated with piñon-juniper woodland ecotones. *Ecology* 77:805–21.

Milton, R. C. 1964. An extended table of critical values for the Mann-Whitney (Wilcoxon) two-sample statistic. *Journal of the American Statistical Association* 59:925–34.

Minnich, R. A. 1983. Fire mosaics in southern California and northern Baja California. *Science* 219:1287–94.

Molles, M. C., Jr. 1978. Fish species diversity on model and natural reef patches: experimental insular biogeography. *Ecological Monographs* 48:289–305.

Moore, J. 1983. Responses of an avian predator and its isopod prey to an acanthocephalan parasite. *Ecology* 64:1000–1015.

Moore, J. 1984a. Altered behavioral responses in intermediate hosts—an acanthocephalan parasite strategy. *American Naturalist* 123:572–77.

Moore, J. 1984b. Parasites that change the behavior of their host. *Scientific American* 250:108–15.

Moran, P. A. P. 1949. The statistical analysis of the sunspot and lynx cycles. *Journal of Animal Ecology* 18:115–16.

Morita, R. Y. 1975. Psychrophilic bacteria. *Bacteriological Reviews* 39:144–67.

Morris, W. F. and D. M. Wood. 1989. The role of lupine in succession on Mount St. Helens: facilitation or inhibition? *Ecology* 70:697–703.

Morse, D. H. 1980. Foraging and coexistence of spruce-woods warblers. *Living Bird* 18:7–25.

Morse, D. H. 1989. *American Warblers.* Cambridge, Mass.: Harvard University Press.

Mosser, J. L., A. G. Mosser, and T. D. Brock. 1974. Population ecology of *Sulfolobus acidocaldarius.* I. Temperature strains. *Archives for Microbiology* 97:169–79.

Moyle, P. B. and J. J. Cech Jr. 1982. *Fishes and Introduction to Ichthyology.* Englewood Cliffs, N.J.: Prentice Hall.

Muir, J. 1915. *Travels in Alaska.* Boston: Houghton Mifflin.

Müller, K. 1954. Investigations on the organic drift in north Swedish streams. *Reports of the Institute of Freshwater Research of Drottningholm* 35:133–48.

Müller, K. 1974. Stream drift as a chronobiological phenomenon in running water ecosystems. *Annual Review of Ecology and Systematics* 5:309–23.

Munger, J. C. and J. H. Brown. 1981. Competition in desert rodents: and experiment with semipermeable exclosures. *Science* 211:510–12.

Murie, A. 1944. The wolves of Mount McKinley. *Fauna of the National Parks of the U.S., Fauna Series No. 5.* Washington, D.C.: U.S. Department of the Interior, National Park Service.

Murphy, P. G. and A. E. Lugo. 1986. Ecology of tropical dry forest. *Annual Review of Ecology and Systematics* 17:67–88.

Muscatine, L. and C. F. D'Elia. 1978. The uptake, retention, and release of ammonium by reef corals. *Limnology and Oceanography* 23:725–34.

Nadkarni, N. M. 1981. Canopy roots: convergent evolution in rainforest nutrient cycles. *Science* 214:1023–24.

Nadkarni, N. M.1984a. Biomass and mineral capital of epiphytes in an *Acer macrophyllum* community of a temperate moist coniferous forest, Olympic Peninsula, Washington State. *Canadian Journal of Botany* 62:2223–28.

Nadkarni, N. M. 1984b. Epiphyte biomass and nutrient capital of a neotropical elfin forest. *Biotropica* 16:249–56.

Naiman, R. J., G. Pinay, C. A. Johnston, and J. Pastor. 1994. Beaver influences on the long-term biogeochemical characteristics of boreal forest drainage networks. *Ecology* 75:905–21.

Neilson, R. P., G. A. King, and G. Koeper. 1992. Toward a rule-based biome model. *Landscape Ecology* 7:135–47.

Neilson, R. P. 1995. A model for predicting continental-scale vegetation distribution and water balance. *Ecological Applications* 5:362–85.

Newbold, J. D., J. W. Elwood, R. V. O'Neill, and A. L. Sheldon. 1983. Phosphorus dynamics in a woodland stream ecosystem: a study of nutrient spiraling. *Ecology* 64:1249–65.

Newchurch, M. J., E.–S. Yang, D. M. Cunnold, G. C. Reinsel, J. M. Zawodny, and J. M. Russell III. 2003. Evidence for slowdown in stratospheric ozone loss: first stage of ozonerecovery. *Journal of Geophysical Research* 108(D16), 4507,doi:10.1029/2003JD003471, 2003 (published online).

Nicholls, N. 1992. Historical El Niño/Southern Oscillation variability in the Australasian region. In H. F. Diaz and V. Markgraf. eds. *El Niño Historical and Paleoclimatic Aspects of the Southern Oscillation.* Cambridge, England: Cambridge University Press.

Nilsson, S. G., J. Bengtsson, and S. Ås. 1988. Habitat diversity or area *per se*? Species richness of woody plants, carabid beetles and land snails on islands. *Journal of Animal Ecology* 57:685–704.

Nobel, P. S. 1977. Internal leaf area and cellular $CO_2$ resistance: photosynthetic implications of variations with growth conditions and plant species. *Physiologia Plantarum* 40:137–44.

Nottrott, R. W., J. F. Franklin, and J. R. Vande Castle. 1994. *International Networking in Long-Term Ecological Research.* Seattle: U.S. LTER Network Office, University of Washington.

O'Donoghue, M. S. Boutin, C. J. Krebs, and E. J. Hofer. 1997. Numerical responses of coyotes and lynx to the snowshoe hare cycle. *Oikos* 80:150–62.

O'Donoghue, M. S. Boutin, C. J. Krebs, G. Zuleta, D. L. Murray, and E. J. Hofer. 1998. Functional responses of coyotes and lynx to the snowshoe hare cycle. *Ecology* 79:1193–1208.

O'Dowd, D. J. and A. M. Gill. 1984. Predator satiation and site alteration following fire: mass reproduction of alpine ash (*Eucalyptus delegatensis*) in southeastern Australia. *Ecology* 65:1052–66.

Oosting, H. J. 1942. An ecological analysis of the plant communities of Piedmont, North Carolina. *The American Midland Naturalist* 28:1–126.

Orel, V. 1996. *Gregor Mendel: The First Geneticist.* Oxford: Oxford University Press.

Osawa, N. and T. Nishida. 1992. Seasonal variation in elytral colour polymorphism in *Harmonia axyridis* (the ladybird beetle): the role of non-random mating. *Heredity* 69:297–307.

Packer, C. and A. E. Pusey. 1982. Cooperation and competition within coalitions of male lions: kin selection or game-theory? *Nature* 296:740–42.

Packer, C. and A. E. Pusey. 1983. Cooperation and competition in lions: reply. *Nature* 302:356.

Packer, C. and A. E. Pusey. 1997. Divided we fall: cooperation among lions. *Scientific American* 276(5):52–59.

Packer, C., D. A. Gilbert, A. E. Pusey, and S. J. O'brien. 1991. A molecular genetic analysis of kinship and cooperation in African lions. *Nature* 351:562–65.

Paine, R. T. 1966. Food web complexity and species diversity. *American Naturalist* 100:65–75.

Paine, R. T. 1969. A note on trophic complexity and community stability. *American Naturalist* 103:91–93.

Paine, R. T. 1976. Size-limited predation: an observational and experimental approach with the *Mytilus-Pisaster* interaction. *Ecology* 57:858–73.

Paine, R. T. 1980. Food webs: linkage, interaction strength and community infrastructure. *Journal of Animal Ecology* 49:667–85.

Paine, R.T. 1971. A short-term experimental investigation of resource partitioning in a New Zealand rocky intertidal habitat. *Ecology* 52:1096–1106.

Pappers, S. M., G. van der Velde, N. J. Ouborg, and J. M. van Groenendael. 2002. Genetically based polymorphisms in morphology and life history associated with putative host races of the water lily leaf beetle *Galerucella nymphaeae. Evolution* 56:1610–21.

Park, T. 1948. Experimental studies of interspecific competition. I. Competition between populations of flour beetles *Tribolium confusum* Duval and *Tribolium castaneum* Herbst. *Ecological Monographs* 18:267–307.

Park, T. 1954. Experimental studies of interspecific competition. II. Temperature, humidity and competition in two species of *Tribolium. Physiological Zoology* 27:177–238.

Park, T., D. B. Mertz, W. Grodzinski, and T. Prus. 1965. Cannibalistic predation in populations of flour beetles. *Physiological Zoology* 38:289–321.

Park, Y. -M. 1990. Effects of drought on two grass species with different distribution around coastal sand dunes. *Functional Ecology* 4:735–41.

Parmenter, R. R. and V. A. Lamarra. 1991. Nutrient cycling in a freshwater marsh: the decomposition of fish and waterfowl carrion. *Limnology and Oceanography* 36:976–87.

**Parmenter, R. R., C. A. Parmenter, and C. D. Cheney.** 1989. Factors influencing microhabitat partitioning among coexisting species of arid-land darkling beetles (Tenebrionidae): behavioral responses to vegetation architecture. *The Southwestern Naturalist* 34: 319–29.

**Pearcy, R. W.** 1977. Acclimation of photosynthetic and respiratory carbon dioxide exchange to growth temperature in *Atriplex lentiformis* (Torr.) Wats. *Plant Physiology* 59:795–99.

**Pearcy, R. W. and A. T. Harrison.** 1974. Comparative photosynthetic and respiratory gas exchange characteristics of *Atriplex lentiformis* (Torr.) Wats. in coastal and desert habitats. *Ecology* 55:1104–11.

**Pearson, O. P.** 1954. Habits of the lizard *Liolaemus multiformis multiformis* at high altitudes in southern Peru. *Copeia* 1954:111–16.

**Peckarsky, B. L.** 1980. Behavioral interactions between stoneflies and mayflies: behavioral observations. *Ecology* 61:932–43.

**Peckarsky, B. L.** 1982. Aquatic insect predator-prey relations. *BioScience* 32:261–66.

**Peierls, B. L., N. F. Caraco, M. L. Pace, and J. J. Cole.** 1991. Human influence on river nitrogen. *Nature* 350:386–87.

**Perry, M. J.** 1986. Assessing marine primary production from space. *BioScience* 36:461–67.

**Peters, R. H.** 1991. *A Critique for Ecology.* Cambridge, England: Cambridge University Press.

**Peters, R. H. and K. Wassenberg.** 1983. The effect of body size on animal abundance. *Oecologia* 60:89–96.

**Peterson, B. J., R. W. Howarth, and R. H. Garritt.** 1985. Multiple stable isotopes used to trace the flow of organic matter in estuarine food webs. *Science* 227:1361–63.

**Philander, S. G.** 1990. *El Niño, La Niña, and the Southern Oscillation.* New York: Academic Press.

**Phillips, D. L. and J. A. MacMahon.** 1981. Competition and spacing patterns in desert shrubs. *Journal of Ecology* 69:97–115.

**Pianka, E. R.** 1970. On r and K selection. *American Naturalist* 102:592–97.

**Pianka, E. R.** 1972. r and K selection or b and d selection. *American Naturalist* 106:581–88.

**Post, W. M., T.-H. Peng, W. R. Emanuel, A. W. King, V. H. Dale, and D. L. DeAngelis.** 1990. The global carbon cycle. *American Scientist* 78:310–26.

**Power, M. E.** 1990. Effects of fish on river food webs. *Science* 250:811–14.

**Power, M. E., D. Tilman, J. A. Estes, B. A. Menge, W. J. Bond, L. S. Mills, G. Daily, J. C. Castilla, J. Lubchenco, and R. T. Paine.** 1996. Challenges in the quest for keystones. *BioScience* 46:609–20.

**Preston, F. W.** 1948. The commonness, and rarity, of species. *Ecology* 29:254–83.

**Preston, F. W.** 1962a. The canonical distribution of commonness and rarity: part I. *Ecology* 43:185–215.

**Preston, F. W.** 1962b. The canonical distribution of commonness and rarity: part II. *Ecology* 43:410–32.

**Rabinowitz, D.** 1981. Seven forms of rarity. In H. Synge. ed. *The Biological Aspects of Rare Plant Conservation.* New York: John Wiley & Sons.

**Raine, N. E., P. Willmer, and G. N. Stone.** 2002. Spatial structuring and floral avoidance behaviour prevent ant-pollinator conflict in a Mexican ant-acacia. *Ecology* 83:3086–96.

**Ralph, C. J.** 1985. Habitat association patterns of forest and steppe birds of northern Patagonia, Argentina. *The Condor* 87:471–83.

**Rasmusson, E. M.** 1985. El Niño and variations in climate. *American Scientist* 73:168–77.

**Reckhow, K. H. and J. T. Simpson.** 1980. A procedure using modeling and error analysis for the prediction of lake phosphorus concentration from land use information. *Canadian Journal of Fisheries and Aquatic Science* 37:1439–48.

**Redford, K. H.** 1992. The empty forest. *BioScience* 42:412–22.

**Reid, W. V. and K. R. Miller.** 1989. *Keeping Options Alive: The Scientific Basis for Conserving Biodiversity.* Washington, D.C.: World Resources Institute.

**Reilly, S. B., D. W. Rice, and A. A. Wolman.** 1983. Population assessment of the gray whale, *Eschrichtius robustus*, from California shore censuses, 1967–80. *Fishery Bulletin* 81:267–81.

**Reiners, W. A., I. A. Worley, and D. B. Lawrence.** 1971. Plant diversity in a chronosequence at Glacier Bay, Alaska. *Ecology* 52:55–69.

**Revelle, R. and H. E. Suess.** 1957. Carbon dioxide exchange between atmosphere and ocean and the question of an increase of atmospheric $CO_2$ during the past decades. *Tellus* 9:18–27.

**Reysenbach, A.-L., M. Ehringer, and K. Hershberger.** 2000. Microbial diversity at 83°C in Calcite Springs, Yellowstone National Park: another environment where the Aquificales and "Korarchaeota" coexist. *Extremophiles* 4:61–67.

**Rice, D. W. and A. A. Wolman.** 1971. *The Life History and Ecology of the Gray Whale (Eschrichtius robustus).* Special publication no. 3. The American Society of Mammalogists.

**Richey, J. E.** 1983. The phosphorus cycle. In B. Bolin and R. B. Cook. eds. *The Major Biogeochemical Cycles and Their Interaction.* New York: John Wiley & Sons.

**Ricklefs, R. E.** 1987. Community diversity: relative roles of local and regional processes. *Science* 235:167–71.

**Risch, S. J. and C. R. Carroll.** 1982. Effect of a keystone predaceous ant, *Solenopsis geminata*, on arthropods in a tropical agroecosystem. *Ecology* 63:1979–83.

**Roberston, G. P., M. A. Huston, F. C. Evans, and J. M. Tiedje.** 1988. Spatial variability in a successional plant community: patterns of nitrogen availability. *Ecology* 69:1517–24.

**Rohlf, F. J., and R. R. Sokal.** 1995. *Statistical Tables.* 3rd edition. San Francisco: W. H.Freeman and Co.

**Roland, J., N. Keyghobadi, and S. Fownes.** 2000. Alpine Parnassius butterfly dispersal: effects of landscape and population size. *Ecology* 81:1642–53.

**Roller N. E. G. and J. E. Colwell.** 1986. Coarse-resolution satellite data for ecological surveys. *BioScience* 36:468–75.

**Root, T.** 1988. *Atlas of Wintering North American Birds.* Chicago: University of Chicago Press.

**Rosemond, A. D., C. M. Pringle, A. Ramirez, M. J. Paul, and J. L. Meyer.** 2002. Landscape variation in phosphorus concentration and effects on detritus-based streams. *Limnology and Oceanography* 47:278–289.

**Rosenzweig, M. L.** 1968. Net primary productivity of terrestrial environments: predictions from climatological data. *American Naturalist* 102:67–84.

**Rosenzweig, M. L.** 1992. Species diversity gradients: we know more and less than we thought. *Journal of Mammalogy* 73:715–30.

**Roy, B. A.** 1993. Floral mimicry by a plant pathogen. *Nature* 362:56–58.

**Rugh, D. J., R. C. Hobbs, J. A. Lerczak, and J. M. Beiwick.** 2003. Preliminary Estimates of Abundance of the Eastern North Pacific Stock of Gray Whales 1997 to 2002. Paper SC/55/BRG13 submitted to the International Whaling Commission, Scientific Committee.

**Rydin, H. and S-O. Borgegård.** 1988. Plant species richness on islands over a century of primary succession: Lake Hjälmaren. *Ecology* 69:916–27.

**Ryer, C. H., A. Lawton, R. J. Lopez, and B. L. Olla.** 2002. A comparison of the functional ecology of visual vs. nonvisual foraging in two planktivorous marine fishes. *Canadian Journal of Fisheries and Aquatic Sciencs* 59:1305–14.

**Ryther, J. H.** 1969. Photosynthesis and fish production in the sea. *Science* 166:72–76.

**Saccheri, I., M. Kuussaari, M. Kankare, P. Vikman, W. Fortelius, I. Hanski.** 1998. Inbreeding and extinction in a butterfly metapopulation. *Nature* 392:491–94.

**Sakamoto, M.** 1966. Primary production by phytoplankton community in some Japanese lakes and its dependence on lake depth. *Archive für Hydrobiologie* 62:1–28.

**Sala, O. E., W. J. Parton, L. A. Joyce, and W. K. Laurenroth.** 1988. Primary production of the central grassland regions of the United States. *Ecology* 69:40–45.

**Sarukhán, J. and J. L. Harper.** 1973. Studies on plant demography: *Ranunculus repens* L. and *R. acris* L. I. population flux and survivorship. *Journal of Ecology* 61:675–716.

**Schenk, H. J. and R. B. Jackson.** 2002. The global biogeography of roots. *Ecological Monographs* 72:311–28.

**Schindler, D. W.** 1987. Detecting ecosystem responses to anthropogenic stress. *Canadian Journal of Fisheries and Aquatic Sciences* 44:6–25.

**Schindler, D. W.** 1990. Experimental perturbations of whole lakes as tests of hypotheses concerning ecosystem structure and function. *Oikos* 57:25–41.

**Schlesinger, W. H.** 1991. *Biogeochemistry: An Analysis of Global Change.* New York: Academic Press.

**Schluter, D. and R. E. Ricklefs.** 1993. Species diversity: an introduction to the problem. In R. E. Ricklefs and D. Schluter. eds. *Species Diversity in Ecological Communities.* Chicago: University of Chicago Press.

**Schluter, D., T. D. Price, and P. R. Grant.** 1985. Ecological character displacement in Darwin's finches. *Science* 227:1056–59.

**Schmidt-Nielsen, K.** 1964. *Desert Animals: Physiological Problems of Heat and Water.* Oxford: Clarendon Press.

**Schmidt-Nielsen, K.** 1969. The neglected interface. The biology of water as a liquid-gas system. *Quarterly Review of Biophysics* 2:283–304.

**Schmidt-Nielsen, K.** 1983. *Animal Physiology: Adaptation and Environment.* 3d ed. Cambridge, England: Cambridge University Press.

**Schneider, D. W. and J. Lyons.** 1993. Dynamics of upstream migration in two species of tropical freshwater snails. *Journal of the North American Benthological Society* 12:3–16.

**Schoener, T. W.** 1983. Field experiments on interspecific competition. *American Naturalist* 122:240–85.

**Scholander, P. F., R. Hock, V. Walters, F. Johnson, and L. Irving.** 1950. Heat regulation in some arctic and tropical mammals and birds. *Biological Bulletin* 99:237–58.

**Scholten, M. C. T. and J. Rozema.** 1990. The competitive ability of *Spartina anglica* on Dutch salt marshes. In A. J. Gray and P. E. M. Benham. eds. *Spartina anglica-a Research Review.* London: HMSO.

**Scholten, M. C. T., P. Blaaww, M. Stroedenga, and J. Rozema.** 1987. The impact of competitive interactions on the growth and distribution of plant species in saltmarshes. In A. H. L. Huiskes et al. eds. *Vegetation Between Land and Sea.* Dordrecht: W. Junk.

**Schultz, T. D., M. C. Quinlan, and N. F. Hadley.** 1992. Preferred body temperature, metabolic physiology, and water balance of adult *Cicindela longilabris:* a comparison of populations from boreal habitats and climatic refugia. *Physiological Zoology* 65:226–42.

**Schumacher, H.** 1976. *Korallenriff.* Munich: BLV Verlagsgellschaft mbH.

**Seiwa, K. and K. Kikuzawa.** 1991. Phenology of tree seedlings in relation to seed size. *Canadian Journal of Botany* 69:532–38.

**Serrano, D. and J. L. Tella.** 2003. Dispersal within a spatially structured population oflesser kestrels: the role of spatial isolation and conspecific attraction. *Journal of Animal Ecology* 72:400–10.

**Setälä, H. and V. Huhta.** 1991. Soil fauna increase *Betula pendula* growth: laboratory experiments with coniferous forest floor. *Ecology* 72:665–71.

**Shaver, G. R. and F. S. Chapin III.** 1986. Effect of fertilizer on production and biomass of tussock tundra, Alaska, U.S.A. *Arctic and Alpine Research* 18:261–68.

**Sherman, P.W., J. U. M. Jarvis, and S. H. Braude.** 1992. Naked mole rats. *Scientific American* 257(8):72–78.

**Shine, R. and E. L. Charnov.** 1992. Patterns of survival, growth, and maturation in snakes and lizards. *American Naturalist* 139:1257–69.

**Siegenthaler, U., H. Friedli, H. Loetscher, E. Moor, A. Neftel, H. Oeschger, and B. Stauffer.** 1988. Stable-isotope ratios and concentrations of $CO_2$ in air from polar ice cores. *Annals of Glaciology* 10:151–56.

**Silvertown, J.** 1987. Ecological stability: a test case. *American Naturalist* 130:807–10.

**Simberloff, D. S.** 1976. Experimental zoogeography of islands: effects of island size. *Ecology* 57:629–48.

**Simberloff, D. S. and E. O. Wilson.** 1969. Experimental zoogeography of islands: the colonization of empty islands. *Ecology* 50:278–96.

**Sinclair, A. R. E.** 1977. *The African Buffalo.* Chicago: University of Chicago Press.

**Sinervo, B. and C. M. Lively.** 1996. The Rock-paper-scissors game and the evolution of alternative male strategies. *Nature* 380:240–43.

**Skole, D. and C. Tucker.** 1993. Tropical deforestation and habitat fragmentation in the Amazon: satellite data from 1978 to 1988. *Science* 260:1905–10.

**Smil, V.** 1990. Nitrogen and phosphorus. In B. L. Turner II, W. C. Clark, R. W. Kates, J. F. Richards, J. T. Mathews, and W. B. Meyer. eds. *The Earth as Transformed by Human Action.* Cambridge, England: Cambridge University Press.

**Smith, C. L. and J. C. Tyler.** 1972. Space resource sharing in a coral reef fish community. Natural History Museum of Los Angeles County, *Science Bulletin* 14:125–70.

**Smith, V. H.** 1979. Nutrient dependence of primary productivity in lakes. *Limnology and Oceanography* 24:1051–64.

**Soares, A. M. V. M., D. J. Baird, and P. Calow** 1992. Interclonal variation in the performance of *Daphnia magna* Straus in chronic bioassays. *Environmental Toxicology and Chemistry* 11:1477–83.

**Söderlund, R. and T. Rosswall.** 1982. The nitrogen cycles. In O. Hutzinger. ed. *The Handbook of Environmental Chemistry,* vol. 1, part B, *The Natural Environment and the Biogeochemical Cycles.* New York: Springer-Verlag.

**Sousa, W. P.** 1979a. Disturbance in marine intertidal boulder fields: the nonequilibrium maintenance of species diversity. *Ecology* 60:1225–39.

**Sousa, W. P.** 1979b. Experimental investigations of disturbance and ecological succession in a rocky intertidal algal community. *Ecological Monographs* 49:227–54.

**Sousa, W. P.** 1984. The role of disturbance in natural communities. *Annual Review of Ecology and Systematics* 15:353–91.

**Spector, W. S.** 1956. *Handbook of Biological Data.* Philadelphia: W. B. Saunders.

**Stevens E. D., J. W. Kanwisher, and F. G. Carey.** 2000. Muscle temperature in free-swimming giant Atlantic bluefin tuna (*Thunnus thynnus* L.). *Journal of Thermal Biology* 25:419–23.

**Stevens, G. C.** 1989. The latitudinal gradient in geographical range: how so many species coexist in the tropics. *American Naturalist* 133:240–56.

**Stevens, O. A.** 1932. The number and weight of seeds produced by weeds. *American Journal of Botany* 19:784–94.

**Stimson, J.** 1990. Stimulation of fat-body production in the polyps of the coral *Pocillopora damicornis* by the presence of mutualistic crabs of the genus *Trapezia. Marine Biology* 106:211–18.

**Strong, D. R.** 1992. Are trophic cascades all wet? Differentiation and donor-control in speciose ecosystems. *Ecology* 73:747–54.

**Suberkropp, K. and E. Chauvet.** 1995. Regulation of leaf breakdown by fungi in streams: influences of water chemistry. *Ecology* 76:1433–45.

**Suess, H. E.** 1955. Radiocarbon concentration in modern wood. *Science* 122:415–17.

**Sugihara, G.** 1980. Minimal community structure: an explanation of species abundance patterns. *American Naturalist* 116:770–87.

**Summerhayes, V. S. and C. S. Elton.** 1923. Contribution to the ecology of Spitsbergen and Bear Island. *Journal of Ecology* 11:214–86.

**Takyu, M., S.-I. Aiba, and K. Kitayama.** 2003. Changes in biomass, productivity and decomposition along topographical gradients under different geological conditions in tropical lower montane forests on Mount Kinabalu, Borneo. *Oecologia* 134:397–404.

**Tan, C. C.** 1946. Mosaic dominance in the inheritance of color patterns in the lady-bird beetle, *Harmonia axyridis. Genetics* 31:195–210.

**Tan, C. C. and J. C. Li.** 1934. Inheritance of the elytral color patterns of the lady-bird beetle, *Harmonia axyridis* Pallas. *American Naturalist* 68:252–65.

**Tansley, A. G.** 1917. On competition between *Galium saxatile* L. (*G. hercynicum* Weig.) and *Galium sylvestre* Poll. (*G. asperum* Schreb.) on different types of soil. *Journal of Ecology* 5:173–79.

**Tansley, A. G.** 1935. The use and abuse of vegetational concepts and terms. *Ecology* 16:284–307.

**Taper, M. L. and T. J. Case.** 1992. Coevolution among competitors. *Oxford Series in Evolutionary Biology.*

**Teal, J. and M. Teal.** 1969. *Life and Death of the Salt Marsh.* Boston: Little, Brown.

**Terborgh, J.** 1973. On the notion of favorableness in plant ecology. *American Naturalist* 107:481–501.

**Terborgh, J.** 1988. The big things that run the world: a sequel to E. O. Wilson. *Conservation Biology* 2:402–3.

**Terra, L. S. W.** Unpublished Light Trap Data. Vila do Conde, Portugal: Estação Aquícola.

**Tewksbury, J. J., D. J. Levey, N. M. Haddad, S. Sargent, J. L. Orrock, A. Weldon, B. J.Danielson, J. Brinderhoff, E. I. Damschen, and P. Townsend.** 2002. Corridors affectplants, animals and the interactions in fragmented landscapes. *Proceedings of theNational Academy of Sciences of the United States of America.* 99:12923–26.

**Thomson, D. A. and C. E. Lehner.** 1976. Resilience of a rocky intertidal fish community in a physically unstable environment. *Journal of Experimental Marine Biology and Ecology* 22:1–29.

**Thoreau, H. D.** 1854. *Walden.* G. S. Haight ed. Reprinted 1942. New York: W. J. Black.

**Thornhill, R.** 1981. Panorpa (Mecoptera: Panorpidae) scorpionflies: systems for understanding resource-defense polygyny and alternative male reproductive efforts. *Annual Review of Ecology and Systematics* 12:355–386.

**Thornhill, R.** 1984. Scientific methodology in entomology. *Florida Entomologist* 67:74–96.

**Thornhill, R. and J. Alcock.** 1983. *The Evolution of Insect Mating Systems.* Cambridge, Mass.: Harvard University Press.

**Tilman, D.** 1977. Resource competition between planktonic algae: an experimental and theoretical approach. *Ecology* 58:338–48.

**Tilman, D.** 1994. Competition and biodiversity in spatially structured habitats. *Ecology* 75:2–16.

**Tilman, D. and M. L. Cowan.** 1989. Growth of old field herbs on a nitrogen gradient. *Functional Ecology* 3:425–38.

**Todd, A. W. and L. B. Keith.** 1983. Coyote demography during a snowshoe hare decline in Alberta. *Journal of Wildlife Management* 47:394–404.

**Tonn, W. M. and J. J. Magnuson.** 1982. Patterns in the species composition and richness of fish assemblages in northern Wisconsin lakes. *Ecology* 63:1149–66.

**Toolson, E. C.** 1987. Water profligacy as an adaptation to hot deserts: water loss rates and evaporative cooling in the Sonoran Desert cicada, *Diceroprocta apache* (Homoptera, Cicadidae). *Physiological Zoology* 60:379–85.

**Toolson, E. C. and N. F. Hadley.** 1987. Energy-dependent facilitation of transcuticular water flux contributes to evaporative cooling in the Sonoran Desert cicada, *Diceroprocta apache* (Homoptera, Cicadidae). *Journal of Experimental Biology* 131:439–44.

**Tosi, J. and R. F. Voertman.** 1964. Some environmental factors in the economic development of the tropics. *Economic Geography* 40:189–205.

**Toth, L., A. D. Albrey Arrington, M. A. Brady, and D. A. Muszick.** 1995. Conceptual evaluation of potential factors affecting restoration of habitat structure within the channelized Kissimmee River ecosystem. *Restoration Ecology* 3:160–80.

**Toumey, J. W. and R. Kienholz.** 1931. Trenched plots under forest canopies. *Yale University School of Forestry Bulletin* 30:1–31.

**Tracy, C. R.** 1999. Differences in body size among chuckwalla (*Sauromalus obesus*) populations. *Ecology* 80:259–71.

**Tracy, R. L. and G. E. Walsberg.** 2000. Prevalence of cutaneous evaporation in Merriam's kangaroo rat and its adaptive variation at the subspecific level. *Journal of Experimental Biology* 203:773–81.

**Tracy, R. L. and G. E. Walsberg.** 2001. Intraspecific variation in water loss in a desert rodent, *Dipodomys merriami. Ecology* 82:1130–37.

**Tracy, R. L. and G. E. Walsberg.** 2002. Kangaroo rats revisited: re-evaluating a classic case of desert survival. *Oecologia* 133:449–57.

**Tschanrtke, T.** 1992. Cascade effects among four trophic levels: bird predation on galls affects density-dependent parasitism. *Ecology* 73:1689–98.

**Turner, R. M.** 1990. Long-term vegetation change at a fully protected Sonoran Desert site. *Ecology* 71:464–77.

**Turner, T.** 1983. Facilitation as a successional mechanism in a rocky intertidal community. *The American Naturalist* 121:729–38.

**Turner, T. F. and J. C. Trexler.** 1998. Ecological and historical associations of gene flow in darters (Teleostei: Percidae). *Evolution* 52: 1781–1801.

**U.S. Bureau of the Census, International Data Base.** http://www.census.gov/pub/ipc/www/ idbnew.html.

**Ueno, H., Y. Sato, and K. Tsuchida.** 1998. Colour-associated mating success in a polymorphic ladybird beetle, *Harmonia axyridis. Functional Ecology* 12:757–61.

**United Nations Population Information Network.** http://www.un.org/popin/.

**USGS.** 1995. The cranes: status survey and conservation action plan, whooping crane (*Grus americana*). http://www.npsc.nbs.gov.

**USGS.** 2000. Monitoring Grizzly Bear Populations Using DNA. http://www.mesc.nbs.gov/glacier/beardna.htm

**Utida, S.** 1957. Cyclic fluctuations of population density intrinsic to the host-parasite system. *Ecology* 38:442–49.

**Valett, H. M., S. G. Fisher, N. B. Grimm, and P. Camill.** 1994. Vertical hydrologic exchange and ecological stability of a desert stream ecosystem. *Ecology* 75:548–60.

**Vancouver, G. and J. D. Vancouver.** 1798. *A Voyage of Discovery to the North Pacific Ocean.* London: G. G. and J. Robinson.

**van der Merwe, N. J.** 1982. Carbon isotopes, photosynthesis, and archaeology. *American Scientist* 70:596–606.

**van der Merwe, N. J. and J. C. Vogel.** 1978. $^{13}C$ content of human collagen as a measure of prehistoric diet in woodland North America. *Nature* 276:815–16.

**Vannote, R. L., G. W. Minshall, K. W. Cummins, J. R. Sedell, and C. E. Cushing.** 1980. The river continuum. *Canadian Journal of Fisheries and Aquatic Sciences* 37:130–37.

**Verhulst, P. F. and A. Quetelet.** 1838. Notice sur la loi que la population suit dans son accroissement. *Corresponce in Mathematics and Physics* 10:113–21.

**Vitousek, P. M.** 1994. Beyond global warming: ecology and global change. *Ecology* 75:1861–76.

**Vitousek, P. M. and L. R. Walker.** 1989. Biological invasion by *Myrica faya* in Hawaii: plant demography, nitrogen fixation, ecosystem effects. *Ecological Monographs* 59:247–65.

Vitousek, P. M., J. R. Gosz, C. C. Grier, J. M. Melillo, and W. A. Reiners. 1982. A comparative analysis of potential nitrification and nitrate mobility in forest ecosystems. *Ecological Monographs* 52:155–77.

Vitousek, P. M., J. R. Gosz, C. C. Grier, J. M. Melillo, W. A. Reiners, and R. L. Todd. 1979. Nitrate losses from disturbed ecosystems. *Science* 204:469–74.

Vollenweider, R. A. 1969. Möglichkeiten und Grenzen elementarer Modelle der Stoffbilanz von Seen. *Archive für Hydrobiologie* 66:1–36.

Volterra, V. 1926. Variations and fluctuations of the number of individuals in animal species living together. Reprinted 1931. In R. Chapman. *Animal Ecology.* New York: McGraw-Hill.

Walker, G. T. 1924. Correlation in seasonal variations of weather, no. 9: a further study of world weather. *Memoirs of the Indian Meteorology Society* 24:275–332.

Walker, J. C. G. 1986. *Earth History: The Several Ages of the Earth.* Boston: Jones and Bartlett Publishers.

Walter, H. 1985. *Vegetation of the Earth.* 3d ed. New York: Springer-Verlag.

Ward, J. V. 1985. Thermal characteristics of running waters. *Hydrobiologia* 125:31–46.

Watwood, M. E. and C. N. Dahm. 1992. Effects of aquifer environmental factors on biodegradation of organic contaminants. *In Proceedings of the International Topical Meeting on Nuclear and Hazardous Waste Management Spectrum '92.* La Grange Park, Ill.: American Nuclear Society.

Webster, J. R. 1975. Analysis of potassium and calcium dynamics in stream ecosystems on three southern Appalachian watersheds of contrasting vegetation. Ph.D. thesis, University of Georgia, Athens.

Webster, J. R. and E. F. Benfield. 1986. Vascular plant breakdown in freshwater ecosystems. *Annual Review of Ecology and Systematics* 17:567–94.

Webster, K. E., T. K. Kratz, C. J. Bowser, J. J. Magnuson, W. J. Rose. 1996. The influence of landscape position on lake chemical responses to drought in northern Wisconsin. *Limnology and Oceanography* 41:977–84.

Werner, E. E. and G. G. Mittelbach. 1981. Optimal foraging: field tests of diet choice and habitat switching. *American Zoologist* 21:813–29.

Westoby, M. 1984. The self-thinning rule. *Advances in Ecological Research* 14:167–255.

Westoby, M., M. Leishman, and J. Lord. 1996. Comparative ecology of seed size and dispersal. *Philosophical Transactions of the Royal Society of London Series B* 351:1309–18.

Wetzel, R.G. 1975. *Limnology.* Philadelphia: W. B. Saunders.

Whicker, A. D. and J. K. Detling. 1988. Ecological consequences of prairie dog disturbances. *BioScience* 38:778–85.

White, C. S. and J. T. Markwiese. 1994. Assessment of the potential for *in sutu* bioremediation of cyanide and nitrate contamination at a heap leach mine in central New Mexico. *Journal of Soil Contamination* 3:271–83.

White, J. 1985. The thinning rule and its application to mixtures of plant populations. In J. White. ed. *Studies in Plant Demography.* New York: Academic Press.

White, J. and J. L. Harper. 1970. Correlated changes in plant size and number in plant populations. *Journal of Ecology* 58:467–85.

White, P. S. and S. T. A. Pickett. 1985. Natural disturbance and patch dynamics: an introduction. In S. T. A. Pickett and P. S. White. eds. *The Ecology of Natural Disturbance and Patch Dynamics.* New York: Academic Press.

Whittaker, R. H. 1956. Vegetation of the Great Smoky Mountains. *Ecological Monographs* 26:1–80.

Whittaker, R. H. 1965. Dominance and diversity in land plant communities. *Science* 147:250–60.

Whittaker, R. H. and G. E. Likens. 1973. The primary production of the biosphere. *Human Ecology* 1:299–369.

Whittaker, R. H. and G. E. Likens. 1975. The biosphere and man. In *Primary Productivity of the Biosphere.* New York: Springer-Verlag.

Whittaker, R. H. and W. A. Niering. 1965. Vegetation of the Santa Catalina Mountains, Arizona: a gradient analysis of the south slope. *Ecology* 46:429–52.

Wiebe, H. H., R. W. Brown, T. W. Daniel, and E. Campbell. 1970. Water potential measurement in trees. *BioScience* 20:225–26.

Williams, K. S., K. G. Smith, and F. M. Stephen. 1993. Emergence of 13-yr periodical cicadas (Cicadidae, *Magicicada*): phenology, mortality, and predator satiation. *Ecology* 74:1143–52.

Williams, M. 1990. Forests. In B. L. Turner II, W. C. Clark, R. W. Kates, J. F. Richards, J. T. Mathews, and W. B. Meyer. eds. *The Earth as Transformed by Human Action.* Cambridge, England: Cambridge University Press.

Williamson, M. 1981. *Island Populations.* Oxford: Oxford University Press.

Willmer, P. G. and G. N. Stone. 1997. Ant deterrence in *Acacia* flowers: How aggressive ant-guards assist seed-set in *Acacia* flowers. *Nature* 388:165–67.

Wilson, E. O. 1980. Caste and division of labor in leafcutter ants (Hymenoptera: Formicidae: *Atta*), I: The overall pattern in *A. sexdens. Behavioral Ecology and Sociobiology* 7:143–56.

Wilson, E. O. and D. S. Simberloff. 1969. Experimental zoogeography of islands: defaunation and monitoring techniques. *Ecology* 50:267–78.

Winemiller, K. O. 1990. Spatial and temporal variation in tropical fish trophic networks. *Ecological Monographs* 60:331–67.

Winemiller, K. O. 1992. Life history strategies and the effectiveness of sexual selection. *Oikos* 63:318–27.

Winemiller, K. O. 1995. Fish ecology. pp. 49–65 In Vol. 2 *Encyclopedia of Environmental Biology.* New York: Academic Press, Inc.

Winemiller, K. O. and K. A. Rose. 1992. Patterns of life-history diversification in North American fishes: implications for population regulation. *Canadian Journal of Fisheries and Aquatic Sciences* 49:2196–2218.

Winston, M. L. 1992. Biology and management of Africanized bees. *Annual Review of Entomology* 37:173–93.

Witkowski, E. T. F. 1991. Effects of invasive alien acacias on nutrient cycling in the coastal lowlands of the Cape fynbos. *Journal of Applied Ecology* 28:1–15.

Wurtsbaugh, W. A. 1992. Food-web modification by an invertebrate predator in the Great Salt Lake (USA). *Oecologia* 89:168–75.

Wurtsbaugh, W. A. and T. Smith Berry. 1990. Cascading effects of decreased salinity on the plankton, chemistry, and physics of the Great Salt Lake (Utah). *Canadian Journal of Fisheries and Aquatic Science* 47:100–109.

Yoda, K., T. Kira, H. Ogawa, and K. Hozumi. 1963. Intraspecific competition among higher plants. XI. Self-thinning in overcrowded pure stands under cultivated and natural conditions. *Journal of Biology Osaka City University* 14:107–29.

Zar, J. H. 1984. *Biostatistical Analysis.* 2nd Edition. Englewood Cliffs, NJ: PrenticeHall.

Zar, J. H. 1996. *Biostatistical Analysis.* 3rd Edition. Upper Saddle River, NJ: PrenticeHall.

# Credits

## Interior Design

Title Page/Half-Title: Daryl Bensen/Masterfile; Dedication: PhotoDisc; About the Author: Daryl Bensen/Masterfile; BTOC Header: Digital Vision; TOC: Digital Vision, Corbis; Preface: Digital Vision; Tables: PhotoDisc; Appendix A Header: Digital Vision, Corbis; Appendix B Header: Artville, Digital Vision, Corbis; Glossary Header: Corbis; References Header: Digital Vision, Corbis; Credits Header: Digital Vision; Index Header: Digital Vision, Corbis.

## Section Openers

1: © Seasons/Eyewire/Getty; 2: © Vol. 44 Nature/PhotoDisc/Getty; 3: © PhotoDisc Website/Getty; 4: Vol. 44 Nature/PhotoDisc/Getty; 5: © Vol. 112 Sea Life/PhotoDisc/Getty; 6: © Vol. 16 U.S. Landmarks & Travel/PhotoDisc/Getty.

## Chapter 1

Figure 1.1(1): Data: AVHRR, NDVI, Seawifs, MODIS, NCEP, DMSP and Sky2000 star catalog; AVHRR and Seawifs texture: Reto Stockli; Visualization: Marit Jentoft-Nilsen, VAL, NASA GSFC; 1.1(2): Jacques Descloitres, MODIS Rapid Response Team, NASA/GSFC; 1.1(3–4): © Peter Johnson/Corbis Images; 1.1(5): © Tony Wilson-Bligh; Papilio/Corbis Images; 1.1(6): © Fritz Polking; Frank Lane Picture Agency/Corbis Images; 1.1(7): © Corbis/Vol. #6; 1.1(8): © PhotoDisc Website; 1.6: Courtesy, Nalini Nadkarni/Photo by Dennis Paulson; 1.9 (all): Courtesy of Gretchen D. Jones and Ester F. Wilson, USDA, ARS, APMRU; 1.10: © Alexander Lowry/Photo Researchers.

## Chapter 2

Figure 2.1: © Preston J. Garrison/Visuals Unlimited; 2.9: © Frans Lanting/Minden Pictures; 2.12(1): © Frans Lanting/Minden Pictures; 2.12(2): © Michio Hoshino/Minden Pictures; 2.15: © John Shaw/Bruce Coleman; 2.17: © Frans Lanting/Minden Pictures; 2.18: Courtesy, James Sanderson; 2.20a: © Eric Toolson; 2.20b: © C. Haagner/Bruce Coleman; 2.21: © Carr Clifton/Minden Pictures; 2.23: © Wardene Weisser/Bruce Coleman; 2.24: © Jim Brandenburg/Minden Pictures; 2.26: © Larry Simpson/Photo Researchers; 2.27: © Carr Clifton/Minden Pictures; 2.29: © Michio Hoshino/Minden Pictures; 2.30: © L. Veisman/Bruce Coleman; 2.32: © Roger A. Powell/Visuals Unlimited; 2.33: © Michio Hoshino/Minden Pictures; 2.35: © Steve McCutcheon/Visuals Unlimited; 2.36: © John Shaw/Bruce Coleman; 2.39a: © Oldrich Karasek/Peter Arnold; 2.39b: © Francois Gohier/Photo Researchers.

## Chapter 3

Figure 3.1: © Spacescapes/PhotoDisc; 3.3: © Bios/Seitre/Peter Arnold; 3.7a: © Kelvin Aitken/Peter Arnold; 3.7b: © Fred McDonnaughey/Photo Researchers; 3.8: © Emory Kristof/National Geographic Image Collection; 3.10: © Randy Morse/Tom Stack & Associates; 3.11: © Mike Severns/Tom Stack & Associates; 3.15: © Charles V. Angelo/Photo Researchers; 3.16: © Jim Zipp/Photo Researchers; 3.18 (both): © Ted Clutter/Photo Researchers; 3.19: © Scott Blackman/Tom Stack & Associates; 3.20: © Michael P. Gadomski/Photo Researchers; 3.21: © P. Robles Gil/Bruce Coleman; 3.23: © Larry Lefever/Grant Heilman Photography; 3.27: © Robert Perron/Visuals Unlimited; 3.28: © Frans Lanting/Minden Pictures; 3.31a: © Larry Lefever/Grant Heilman Photography; 3.31b: © Walt Anderson/Visuals Unlimited; 3.35: © Peter Arnold/Peter Arnold, Inc.; 3.41 (both): © Runk/Schoenberger/Grant Heilman Photography.

## Chapter 4

Figure 4.1: © Steve McCutcheon/Visuals Unlimited; 4.2: © Walt Anderson/Visuals Unlimited; 4.4a: © Brian Parker/Tom Stack & Associates; 4.4b: © Edward Hodgson/Visuals Unlimited; 4.32: © Frans Lanting /Minden Pictures.

## Chapter 5

Figure 5.19a: © Doug Sokell/Tom Stack & Associates; 5.19b: © John D. Cunningham/Visuals Unlimited; 5.21a: © Doug Sokell/Visuals Unlimited; 5.21b: © Leonard Lee Rue III/Photo Researchers; 5.26 (both): From Toolson and Hadley: "Energy-Dependent facilitation of transcuticular water flux contributes to evaporative cooling in the Sonoran Desert cicada, Diceroprocta apache (Homoptera: Cidadiae)." *The Journal of Experimental Biology,* 1987, vol. 131, pages 439–444. Figs. 2A and 2B. Photo courtesy of Dr. Eric Toolson.

## Chapter 6

Figure 6.1: © D. Holden Bailey/Tom Stack & Associates; 6.8: © Mitsuaki Iwago/Minden Pictures; 6.13a: © Runk/Schoenberger/Grant Heilman Photography; 6.13b: © John Colwell/Grant Heilman Photography.

## Chapter 7

Figure 7.1: © Gregory G. Dimijian/Photo Researchers; 7.2a: © Thomas Kitchin/Tom Stack; 7.2b: © John Anderson/Animals, Animals; 7.3: © Dr. Paul A. Zahl/Photo Researchers; 7.9: © L. West/Photo Researchers; 7.14: © Richard Parker/ Photo Researchers; 7.18: © Kerry T. Givens/Bruce Coleman; 7.21: © Mitsuaki Iwago/Minden Pictures; 7.24: © Mark W. Moffett/Minden Pictures; 7.25: © Raymond A. Mendez/Animals, Animals.

## Chapter 8

Figure 8.1: © Fritz Polking/Peter Arnold; 8.2: © Angelina Lax/Photo Researchers; 8.3: © A. Scott Earle; 8.5: © Bud Lehnhausen/Photo Researchers; 8.7: © Ed Reschke/Peter Arnold; 8.11: © Tom McHugh/Photo Researchers; 8.12: © David M. Schleser/Photo Researchers; 8.16: © Joe McDonald/Animals, Animals; 8.24: © Ken Pilsbury/Natural Visions; 8.25a–c: Clausen, J., D. D. Keck, and W. Hiesey. 1940. Experimental Studies On The Nature of species. Vol. I. Effect of Varied Environments on Western North American Plants. P.8. Carnegie Institution of Washington Publication No. 520. Reproduced by permission of the Carnegie Institution of Washington.

## Chapter 9

Figure 9.1a: © Frans Lanting/Minden Pictures; 9.1b: © W. Perry Conway/Tom Stack & Associates; 9.12: © Richard Weymouth Brooks/Photo Researchers; 9.19: © Jay Cossey; 9.21: © Gregory G. Dimijian/Photo Researchers; 9.26: © W. Perry Conway/Tom Stack & Associates; 9.27: © Michio Hoshino/Minden Pictures; 9.28a–b: © Tom Fernald, College of the Atlantic, Bar Harbor, Maine.

## Chapter 10

Figure 10.1: © Leonard Lee Rue III/Photo Researchers; 10.20a–b: Courtesy, Daniel W. Schneider and John Lyons.

## Chapter 11

Figure 11.1a: © M.I. Walker/Science Source/Photo Researchers; 11.1b: © James Bell/Science Source/Photo Researchers; 11.16a–b, 11.20: © Peter R. Grant, Princeton University.

## Chapter 12

Figure 12.1: © Donald Specker/Animals, Animals; 12.2: © David M. Schleser/Photo Researchers; 12.6: © Frans Lanting/Minden Pictures; 12.14: © Breck P. Kent/Animals, Animals; 12.19(1): © David Schleser/Photo Researchers; 12.19(2): Corbis Digital Stock; 12.25: © Chris McLaughlin/Animals, Animals.

## Chapter 13

Figure 13.1: © Fred McConnaughey/Photo Researchers; 13.2: © George Haling/Photo Researchers; 13.12: Courtesy, Mark McCorry, University College Dublin; 13.18: © Heather Angel; 13.22, 13.23a–b: © Dr. James H. Brown.

## Chapter 14

Figure 14.1: © Mark Newman/Photo Researchers; 14.4a–b, 14.5: Photographs by Dr. B. A. Roy, reprinted by permission from Nature 362: 56–58; 14.10a–b: Courtesy, Dr. Gary A. Lamberti; 14.11 (both): Photographs reproduced with permission of the Department of Natural Resources and Mines, Australia; 14.25: © Gregory G. Dimijian/Photo Researchers.

## Chapter 15

Figure 15.1: © Anthony Mercieca/Photo Johnson; 15.2a: © Dr. Nancy Collins Johnson; 15.2b: Courtesy, Randy Molina/U.S. Forest Service; 15.7: © Robert & Linda Mitchell; 15.16: © Fred Bavendam/Minden Pictures; 15.19a: © Fred Bavendam/Minden Pictures; 15.19b: © Carr Clifton/Minden Pictures; 15.19c: © Thomas Hovland/Grant Heilman Photography; 15.19d: © Leonard Lee Rue III/Photo Researchers; 15.20: © Nigel Dennis/Photo Researcher.

## Chapter 16

Figure 16.1: © Charlie Ott/Photo Researchers; 16.2: © Carr Clifton/Minden Pictures.

## Chapter 17

Figure 17.1: © Francois Gohier/Photo Researchers; 17.10a–b: Courtesy Mary E. Power, UC Berkeley; 17.19: © Mark Moffett/Minden Pictures; 17.22: © Tom & Pat Leeson/Photo Researchers; 17.24: © Robert Villani/Peter Arnold.

## Chapter 18

Figure 18.1: © Michael Sewell/Peter Arnold.

## Chapter 20

Figure 20.1: © Carr Clifton/Minden Pictures; 20.6: © Jim Zipp/Photo Researchers; 20.31a: © D.T. MacDougal; 20.31b–c: © J.R. Hastings; 20.31d: © R.M. Turner; 20.32a: © J.R. Hastings; 20.32b–c: © R.M. Turner.

## Chapter 21

Figure 21.1: © Harvey Lloyd/Corbis Images; 21.2: © Cary Wolinsky/Stock Boston; 21.3: © Restec, Japan/Science Photo Library/Photo Researchers; 21.8a: © AP/Wide World Photos; 21.8b: © Thomas Kitchin/Tom Stack & Associates; 21.9a: © Manfred Gottschalk/Tom Stack & Associates; 21.9b: © Tom & Pat Leeson/Photo Researchers; 21.12: © Rainer K. Lampinen/Panoramic Images; 21.17: © Ken M. Johns/Photo Researchers; 21.23a: © Frans Lanting/Minden Pictures; 21.23b: © Albert Visage/Peter Arnold; 21.23c: © Alford W. Cooper/Alford W. Cooper; 21.23d: © John Shaw/Bruce Coleman; 21.24: © Mark Newman/Tom Stack & Associates; 21.27: © Steve Starr/Stock Boston; 21.30, 21.31: Reprinted, by permission, from the South Florida Water Management District, a public corporation of the State of Florida.

## Chapter 22

Figure 22.1: © Norman Owen Tomalin/Bruce Coleman; 22.11: © H.W. Kitchin/Photo Researchers; 22.27: From Roller, Norman E.G., Colwell, John E. July/August 1986. "Coarse-resolution Satellite Data for Ecological Surveys," *BioScience,* Vol. 36 (7): 473, fig. 2. © 1986 American Institute of Biological Sciences; 22.28 (both): From Perry, Mary Jane. July/August 1986. "Assessing Marine Primary Production from Space." *BioScience,* 36 (7): 462, fig. 1. © 1986 American Institute of Biological Sciences.

## Chapter 23

Figure 23.1: © NASA; 23.8 (both): Gene Feldman, et al. Nov. 30, 1984. "Satellite color observations of the phytoplankton distribution in the eastern equatorial Pacific during the 1982–1983 El Nino," *Science,* 226: 1069–1071. Fig. 2A. ©1984 American Association for the Advancement of Science; 23.9: © Tui De Roy/Minden Pictures; 23.11: © Mitsuaki Iwago/Minden Pictures; 23.16a–c: USGS / EROS; 23.18: © Richard O. Bierregaard, Jr./University of North Carolina, Biology Department; 23.19: From Skole, D., Tucker, C. June 25, 1993. "Tropical deforestation and habitat fragmentation in the Amazon," *Science* 260: 1908. © 1993 American Association for the Advancement of Science. Map courtesy Walter Chomentowski; 23.20: © D. Dancer/Peter Arnold; 23.25: © NASA, Goddard Space Flight Center; 23.26a, c, d: Courtesy of the U.S. Long-Term Ecological Research Network; 23.26b: Courtesy John J. Magnuson.

# Index

*Note:* Page references followed by the letters *f* and *t* indicate figures and tables, respectively.

## A

A horizon, 22, 22*f*
Aber, John, 468
Abundance, species, 399–400. *See also*
  Rarity
    in biological integrity assessment, 84
    classification of, 244–250, 399–400
    estimation of, 247–250
    exploitation and, 351–356
    lognormal distribution of,
      399–400, 400*f*
    in plants, moisture and, 239–240
    of prey, in optimal foraging theory,
      160–161
    range and, 244–246
    relative (evenness), 250, 399, 400–402
Abyssal zone, 56, 57*f*
Acacia, 30*f*
    distribution of, 380
    and landscape structure, 526
    mutualism with ants, 379–382, 380*f*,
      381*f*, 382*f*
*Acacia hindsii,* mutualism with ants,
      381–382, 382*f*
*Acacia saligna,* introduction of, and
      nutrient dynamics,
      475, 475*f*
*Acanthaster planci. See* Crown-of-thorns
      sea star
Acanthocephalans, effects on hosts,
      348–349
Acapulco (Mexico), climate diagram for,
      26*f*, 27
Acclimation, 97
*Acer rubrum. See* Red maple
*Acer saccharum. See* Sugar maple
Acetylcholinesterase, temperature and,
      95, 95*f*
Active channels, 73, 74*f*
Actual evapotranspiration (AET)
    and decomposition rates, 468, 468*f*
    in primary productivity, 443–444, 443*f*
Adams, Phillip, 108
Adaptation
    definition of, 198
    in lizards, 211–212
    in soapberry bugs, 214–215, 214*f*, 215*f*
Adelaide (Australia), climate diagram
      for, 34*f*
*Adelina tribolii,* effects on host
      competition, 351, 351*f*
Adhesion-adapted seeds, 303
*Adiantum decorum. See* Maidenhair fern
Admiralty Island, length of coastline of,
      515–516, 515*f*
Adzuki bean, 362
Adzuki bean weevil, population cycles of,
      parasitism and,
      361–363, 362*f*
*Aedes* spp., osmoregulation by, 137
Aerial photography, 512, 512*f*
AET. *See* Actual evapotranspiration
African buffalo
    grazing by, and primary
      production, 451
    sigmoidal growth by, 282, 283*f*

African elephant
    and landscape structure,
      527–528, 527*f*
    life history of, 312*f*, 315
    predation on, 368*f*
African lion, cooperative breeding in,
      186–188, 187*f*, 188*f*
Africanized bees, expansion of,
      266–268, 266*f*
Age distribution, 256–257, 260–262
    climate and, 261–262
    for human populations, 293, 293*f*
    stable, 262
    in stable *vs.* declining populations,
      260–261
Aggressiveness, and bee colony
      distribution, 233–235
Agriculture
    history of, 413–415, 458
    landscape structure in, 526, 526*f*
    in Mediterranean woodland/
      shrubland, 35
    mutualism in, 390
    and nitrogen cycle, 569–570
    pest control by ants in, 436–438
    in temperate forests, 39
    in temperate grassland, 36–37, 37*f*
    in tropical dry forests, 27
    in tropical rain forests, 25
*Agropyron smithii,* water balance in,
      mycorrhizae and,
      376–377, 377*f*
Aiba, Shin-Ichiro, 469
Air. *See* Atmosphere
*Alces alces. See* Moose
Alder, leaf decomposition in, 469, 473
Alewife, introduction of, 82
Alfalfa, self-thinning in, 326*f*
Algae
    abundance of, relative, 250, 250*f*
    consumer effects on, 352–353, 352*f*,
      353*f*, 354*f*, 425–429,
      426*f*, 427*f*
    in coral reefs, 63
    diversity of, 408
    in intertidal succession, 497–498
    niches of, 405–406
    productivity of, nutrients and,
      446, 446*f*
Alien species. *See* Introduced species
Alkaloids, toxic, in tropical *vs.* temperate
      plants, 151, 151*f*, 152
Allele frequencies, 206–208, 301
Alleles, 199
Allen, Edie, 376
Allen, Michael, 376
Allenspark (Colorado), climate diagram
      for, 46*f*
Alligators, and landscape structure,
      527*f*, 528
Allocation. *See* Energy, allocation of
Allopatric speciation, 549
Allopatric species, 338
Allopolyploidy, 330
Allozymes, 216, 219, 301
*Alpheus lottini,* mutualism with corals, 385
Alpine. *See* Mountain(s)

Alpine ash, predator satiation by, 366, 366*f*
Altitude
    measurement of, 554
    and microclimates, 91
    and temperature, 45, 46*f*, 91
Altricial birds, life histories of, 316, 316*f*
*Ambrosia deltoidea,* soil age and,
      524–525, 525*f*
American crow, distribution of, 238, 238*f*
American robin, survival pattern of,
      258, 258*f*
AMF. *See* Arbuscular mycorrhizal fungi
Ammonification, 464
Ammonium
    in chemosynthesis, 155, 156*f*
    coral excretion of, zooxanthellae and,
      385, 385*f*
    in nitrogen cycle, 464
*Anaphalis margaritacea. See* Pearly
      everlasting
Anchoveta, El Niño and, 567
Andrews Forest, climate diagram for,
      38, 38*f*
*Andropogon gerardii. See* Big
      bluestem grass
Angilletta, Michael, 95–96, 103
Anglerfish, deep-sea, 55, 55*f*
Animals. *See also specific animals*
    diversity of, in aquatic *vs.* terrestrial
      ecosystems, 59, 59*f*
    functional responses of, 158–159, 158*f*
    genetic variation in, 202–205
    hunting of (*See* Hunting)
    and landscape structure, 527–529
    and nutrient cycling, 473–475
    optimal foraging theory on,
      160–161
    performance of, temperature
      and, 94–96
    seeds dispersed by, 303
    size of, and population density,
      243, 243*f*
    thermoregulation by, 102–108
    water balance in, 124, 125*f*
      acquisition in, 124, 125–126
      conservation in, 128–132
      loss in, 124
*Anolis* spp., adaptive change in, 211–212
*Anolis sagrei. See* Brown anole
*Anoplopoma fimbria. See* Sablefish
Ant(s)
    eusociality in, 189–193
    mutualism with
      acacia in, 379–382, 380*f*,
        381*f*, 382*f*
      aspen sunflower in, 382–384,
        383*f*, 384*f*
      introduced species and, 435
      models of, 389–390
    pest control by, 436–438
    seed predation by, 365–366
    seeds dispersed by, 303–304, 434–435
Antarctic pelagic food web, 419, 420*f*
*Apis melifera,* evolution of, 266
*Apis melifera scutellata,* 266–267
Apollo 8 mission, 561
Aposematic colors, as prey defense, 153

Appalachian Mountains, forests in, history
      of, 8–9
*Aprostocetus calamarius,* in food web,
      422, 422*f*
*Aprostocetus gratus,* in food web, 422, 422*f*
Aquatic ecosystems, 52–85. *See also*
      *specific ecosystems*
    complexity of, 406
    decomposition in, 469–470
    diversity of phyla in, 59, 59*f*
    health of, assessment of, 82–85
    hydrologic cycle in, 53–54, 54*f*
    microclimates of, 93–94
    primary production in, 445–448
    salt balance in, 137–138
    thermal stability of, 93–94, 94*f*
    thermoregulation in, 106–107
    water balance in, 119, 136–138
    water movement in, 120–121
*Arabis* spp., pathogen effects on,
      349–350, 350*f*
Arbuscular mycorrhizal fungi (AMF),
      376, 376*f*
Arbuscules, 376
*Arbutus arizonica. See* Arizona madrone
Archaea, 146
*Archanara geminipuncta,* in food web,
      422, 422*f*
Arctic fox, thermal neutral zone of, 105
Arctic islands, food web on, 419, 421*f*
Arctic plants, thermoregulation by,
      101, 102*f*
Arctic species, thermal neutral zone of,
      105–106, 106*f*
Arctic tundra. *See* Tundra
*Arctocephalus australis. See* South
      American fur seal
Argentine ant, introduction of, 435, 435*f*
Argillic horizons, 524, 524*f*, 525
*Arianta arbustorum*
    hatching success of, temperature and,
      113–114, 114*f*
    local extinction of, climate change
      and, 113–114, 114*f*
Aril, 303
Arizona madrone, abundance of, moisture
      and, 239, 239*f*
*Armadillidium vulgare,* parasite effects on,
      348–349, 349*f*
Arrhenius, Olof, 538
*Artemesia frigida,* roots of, 127, 127*f*
*Artemia franciscana,* El Niño and, 567–568
Ås, Stefan, 539
*Ascophyllum* spp., competition by, 427
Asexual reproduction, 172
Ash, leaf decomposition in, 467, 467*f*, 468
Ash Meadows killifish, rarity of, 247
Ash Meadows milk vetch, rarity of, 247
Ash Meadows stick-leaf, rarity of, 247
Asian lady beetles
    nonrandom mating by, 208
    variation in, 206–208, 206*f*
*Asio flammeus. See* Short-eared owl
*Asio otus. See* Long-eared owl
Aspects, and microclimate, 91, 91*f*
Aspen sunflower, mutualism with ants,
      382–384, 383*f*, 384*f*